# Methods in Enzymology

Volume 294

ION CHANNELS

Part C

# METHODS IN ENZYMOLOGY

EDITORS-IN-CHIEF

John N. Abelson Melvin I. Simon

DIVISION OF BIOLOGY
CALIFORNIA INSTITUTE OF TECHNOLOGY
PASADENA, CALIFORNIA

FOUNDING EDITORS

Sidney P. Colowick and Nathan O. Kaplan

*Methods in Enzymology*

*Volume 294*

# *Ion Channels Part C*

EDITED BY


*P. Michael Conn*

OREGON REGIONAL PRIMATE RESEARCH CENTER
BEAVERTON, OREGON







ACADEMIC PRESS
San Diego London Boston New York Sydney Tokyo Toronto

This book is printed on acid-free paper. ∞



Academic Press
*a division of Harcourt Brace & Company*
525 B Street, Suite 1900, San Diego, California 92101-4495, USA
http://www.academicpress.com

Academic Press Limited
24-28 Oval Road, London NW1 7DX, UK
http://www.hbuk.co.uk/ap/

International Standard Book Number: 0-12-182195-1

PRINTED IN THE UNITED STATES OF AMERICA
98 99 00 01 02 03 MM 9 8 7 6 5 4 3 2 1

# Table of Contents


Contributors to Volume 294 . . . . . . . . . . . . . . . . . . . . . ix

Preface . . . . . . . . . . . . . . . . . . . . . . . . . . . . . xv

Volumes in Series . . . . . . . . . . . . . . . . . . . . . . . . xvii


## Section I. Physical Methods


1. Optical Measurements of Calcium Signals in Mammalian Presynaptic Terminals — Peter Saggau, Richard Gray, and John A. Dani — 3

2. Fluorescent Techniques for Measuring Ion Channel Activity — Gönül Veliçelebi, Kenneth A. Stauderman, Mark A. Varney, Michael Akong, Stephen D. Hess, and Edwin C. Johnson — 20

3. Tagging Potassium Ion Channels with Green Fluorescent Protein to Study Mobility and Interactions with Other Proteins — Edwin S. Levitan — 47

4. Spin Label Electron Spin Resonance and Fourier Transform Spectroscopy for Structural/Dynamic Measurements on Ion Channels — Derek Marsh — 59

5. Nuclear Magnetic Resonance Spectroscopy of Peptide Ion Channel Ligands: Cloning and Expression as Aid to Evaluation of Structural and Dynamic Properties — Michael D. Reily, Anne M. Bokman, James Offord, and Patrick McConnell — 92

6. Ligand Binding Methods for Analysis of Ion Channel Structure and Function — Steen E. Pedersen, Monica M. Lurtz, and Rao V. L. Papineni — 117

7. Three-Dimensional Structure of Membrane Proteins Determined by Two-Dimensional Crystallization, Electron Cryomicroscopy, and Image Analysis — Mark Yeager, Vinzenz M. Unger, and Alok K. Mitra — 135

8. Drug-Dependent Ion Channel Gating by Application of Concentration Jumps Using U-Tube Technique — Frank Bretschneider and Fritz Markwardt — 180

9. Voltage Clamp Biosensors for Capillary Electrophoresis — Owe Orwar, Kent Jardemark, Cecilia Farre, Ingemar Jacobson, Alexander Moscho, Jason B. Shear, Harvey A. Fishman, Sheri J. Lillard, and Richard N. Zare — 189

10. Ion Channels as Tools to Monitor Lipid Bilayer–Membrane Protein Interactions: Gramicidin Channels as Molecular Force Transducers — O. S. Andersen, C. Nielsen, A. M. Maer, J. A. Lundbæk, M. Goulian, and R. E. Koeppe II — 208

## Section II. Purification and Reconstitution

11. Purification and Reconstitution of Epithelial Chloride Channel Cystic Fibrosis Transmembrane Conductance Regulator — Mohabir Ramjeesingh, Elizabeth Garami, Ling-Jun Huan, Kevin Galley, Canhui Li, Yanchun Wang, and Christine E. Bear — 227

12. Purification, Characterization, and Reconstitution of Cyclic Nucleotide-Gated Channels — Robert S. Molday and Laurie L. Molday — 246

13. Purification and Heterologous Expression of Inhibitory Glycine Receptors — Bodo Laube and Heinrich Betz — 260

14. Purification and Functional Reconstitution of High-Conductance Calcium-Activated Potassium Channel from Smooth Muscle — Maria L. Garcia, Kathleen M. Giangiacomo, Markus Hanner, Hans-Günther Knaus, Owen B. McManus, William A. Schmalhofer, and Gregory J. Kaczorowski — 274

15. Reconstitution of Native and Cloned Channels into Planar Bilayers — Isabelle Favre, Ye-Ming Sun, and Edward Moczydlowski — 287

16. Iodide Channel of the Thyroid: Reconstitution of Conductance in Proteoliposomes — Philippe E. Golstein, Abdullah Sener, Fernand Colin, and Renaud Beauwens — 304

17. Nystatin/Ergosterol Method for Reconstituting Ion Channels into Planar Lipid Bilayers — Dixon J. Woodbury — 319

18. Isolation of Transport Vesicles That Deliver Ion Channels to the Cell Surface — Saroj Sattsangi and William F. Wonderlin — 339

## Section III. Second Messengers and Biochemical Approaches

19. Protein Phosphorylation of Ligand-Gated Ion Channels — Andrew L. Mammen, Sunjeev Kamboj, and Richard L. Huganir — 353

20. Analysis of Ion Channel Associated Proteins — Michael Wyszynski and Morgan Sheng — 371

21. Signal Transduction through Ion Channels Associated with Excitatory Amino Acid Receptors — Kiyokazu Ogita and Yukio Yoneda — 385

22. Secondary Messenger Regulation of Ion Channels/Plant Patch Clamping — Sarah M. Assmann and Lisa Romano — 410

## Section IV. Special Channels

23. ATP-Sensitive Potassium Channels — M. Schwanstecher, C. Schwanstecher, F. Chudziak, U. Panten, J. P. Clement IV, G. Gonzalez, L. Aguilar-Bryan, and Joseph Bryan — 445

24. Mechanosensitive Channels of Bacteria — Paul Blount, Sergei I. Sukharev, Paul C. Moe, Boris Martinac, and Ching Kung — 458

25. Simplified Fast Pressure-Clamp Technique for Studying Mechanically Gated Channels — Don W. McBride, Jr., and Owen P. Hamill — 482

26. Virus Ion Channels — David C. Ogden, I. V. Chizhmakov, F. M. Geraghty, and A. J. Hay — 490

27. Ion Channels in Microbes — Yoshiro Saimi, Stephen H. Loukin, Xin-Liang Zhou, Boris Martinac, and Ching Kung — 507

28. Design and Characterization of Gramicidin Channels — DENISE V. GREATHOUSE, ROGER E. KOEPPE II, LYNDON L. PROVIDENCE, SUNDARAM SHOBANA, AND OLAF S. ANDERSEN 525

29. Functional Analyses of Aquaporin Water Channel Proteins — PETER AGRE, JOHN C. MATHAI, BARBARA L. SMITH, AND GREGORY M. PRESTON 550

## Section V. Toxins and Other Membrane Active Compounds

30. Pore-Blocking Toxins as Probes of Voltage-Dependent Channels — ROBERT J. FRENCH AND SAMUEL C. DUDLEY, JR. 575

31. *Conus* Peptides as Probes for Ion Channels — J. MICHAEL MCINTOSH, BALDOMERO M. OLIVERA, AND LOURDES J. CRUZ 605

32. Scorpion Toxins as Tools for Studying Potassium Channels — MARIA L. GARCIA, MARKUS HANNER, HANS-GÜNTHER KNAUS, ROBERT SLAUGHTER, AND GREGORY J. KACZOROWSKI 624

33. Potassium Ion Channel Inactivation Peptides — RUTH D. MURRELL-LAGNADO 640

34. Interactions of Snake Dendrotoxins with Potassium Channels — WILLIAM F. HOPKINS, MARGARET ALLEN, AND BRUCE L. TEMPEL 649

35. Use of Planar Lipid Bilayer Membranes for Rapid Screening of Membrane Active Compounds — TAJIB A. MIRZABEKOV, ANATOLY Y. SILBERSTEIN, AND BRUCE L. KAGAN 661

## Section VI. Reagent and Information Sources

36. Antibodies to Ion Channels — ANGELA C. VINCENT, IAN K. HART, ASHWIN PINTO, AND F. ANNE STEPHENSON 677

37. Internet Information on Ion Channels: Issues of Access and Organization — EDWARD C. CONLEY 704

AUTHOR INDEX . . . . . . . . . . 733

SUBJECT INDEX . . . . . . . . . . 769

# Contributors to Volume 294

Article numbers are in parentheses following the names of contributors.
Affiliations listed are current.


PETER AGRE (29), *Departments of Biological Chemistry and Medicine, Johns Hopkins University School of Medicine, Baltimore, Maryland 21205-2185*

L. AGUILAR-BRYAN (23), *Department of Medicine, Baylor College of Medicine, Houston, Texas 77030*

MICHAEL AKONG (2), *SIBIA Neurosciences, Inc., La Jolla, California 92037*

MARGARET L. ALLEN (34), *Departments of Otolaryngology and Pharmacology, University of Washington School of Medicine, Seattle, Washington 98195-7923*

OLAF S. ANDERSEN (10, 28), *Department of Physiology and Biophysics, Cornell University Medical College, New York, New York 10021*

SARAH M. ASSMANN (22), *Department of Biology, Pennsylvania State University, University Park, Pennsylvania 16802*

CHRISTINE E. BEAR (11), *Research Institute, Hospital for Sick Children, and University of Toronto, Toronto, Ontario, Canada M5G 1X8*

RENAUD BEAUWENS (16), *Laboratory of Physiology and Pathophysiology, School of Medicine, Université Libre de Bruxelles, B-1070 Brussels, Belgium*

HEINRICH BETZ (13), *Department of Neurochemistry, Max-Planck-Institute for Brain Research, D-60528 Frankfurt, Germany*

PAUL BLOUNT (24), *Department of Physiology, University of Texas, Southwestern Medical Center, Dallas, Texas 75235*

ANNE M. BOKMAN (5), *Debiopharm S.A., Developpments Biologiques et Pharmaceutiques, CH-1000 Lausanne 9, Switzerland*

FRANK BRETSCHNEIDER (8), *Department of Applied Physiology, University of Ulm, D-89069 Ulm, Germany*

JOSEPH BRYAN (23), *Department of Cell Biology, Baylor College of Medicine, Houston, Texas 77030*

I. V. CHIZHMAKOV (26), *National Institute for Medical Research, London NW7 1AA, United Kingdom*

F. CHUDZIAK (23), *Institüt für Pharmakologie und Toxikologie, Universität Braunschweig, 38106 Braunschweig, Germany*

J. P. CLEMENT IV (23), *Department of Cell Biology, Baylor College of Medicine, Houston, Texas 77030*

FERNAND COLIN (16), *Laboratory of Biomedical Physics, School of Medicine, Université Libre de Bruxelles, B-1070 Brussels, Belgium*

EDWARD C. CONLEY (37), *Molecular Pathology, Ion Channel/Gene Expression Group, University of Leicester Medical Research Council, Leicester LE1 9HN, United Kingdom*

LOURDES J. CRUZ (31), *Marine Science Center, University of the Philippines, Diliman, Quezon City, Philippines*

JOHN A. DANI (1), *Division of Neuroscience, Baylor College of Medicine, Houston, Texas 77030*

SAMUEL C. DUDLEY, JR. (30), *Division of Cardiology, Emory University/VAMC, Decatur, Georgia 30033*

CECILIA FARRE (9), *Department of Chemistry, Göteborg University, S-41296 Göteborg, Sweden*

ISABELLE FAVRE (15), *Department of Pharmacology, Yale University School of Medicine, New Haven, Connecticut 06520*

HARVEY A. FISHMAN (9), *Stanford Medical School, Stanford University, Stanford, California 94305-5080*

Robert J. French (30), *Department of Physiology and Biophysics, University of Calgary, Calgary, Alberta, Canada T2N 4N1*

Kevin Galley (11), *Research Institute, Hospital for Sick Children, Toronto, Ontario, Canada M5G 1X8*

Elizabeth Garami (11), *Research Institute, Hospital for Sick Children, Toronto, Ontario, Canada M5G 1X8*

Maria L. Garcia (14, 32), *Department of Membrane Biochemistry and Biophysics, Merck Research Laboratories, Rahway, New Jersey 07065*

F. M. Geraghty (26), *National Institute for Medical Research, London NW7 1AA, United Kingdom*

Kathleen M. Giangiacomo (14), *Biochemistry Department, Temple University School of Medicine, Philadelphia, Pennsylvania 19140*

Philippe E. Golstein (16), *Laboratory of Physiology and Pathophysiology, School of Medicine, Université Libre de Bruxelles, B-1070 Brussels, Belgium*

G. Gonzalez (23), *Department of Cell Biology, Baylor College of Medicine, Houston, Texas 77030*

M. Goulian (10), *Center for Studies in Physics and Biology, Rockefeller University, New York, New York 10021*

Richard Gray (1), *Division of Neuroscience, Baylor College of Medicine, Houston, Texas 77030*

Denise V. Greathouse (28), *Department of Chemistry and Biochemistry, University of Arkansas, Fayetteville, Arkansas 72701*

Owen P. Hamill (25), *Department of Physiology and Biophysics, University of Texas Medical Branch, Galveston, Texas 77555*

Markus Hanner (14, 32), *Department of Membrane Biochemistry and Biophysics, Merck Research Laboratories, Rahway, New Jersey 07065*

Ian K. Hart (36), *Neurosciences Group, Institute of Molecular Medicine, John Radcliffe Hospital, Headington, Oxford OX3 9DU, United Kingdom*

A. J. Hay (26), *National Institute for Medical Research, London NW7 1AA, United Kingdom*

Stephen D. Hess (2), *SIBIA Neurosciences, Inc., La Jolla, California 92037*

William F. Hopkins (34), *Neurex Corporation, Menlo Park, California 94025*

Ling-Jun Huan (11), *Research Institute, Hospital for Sick Children, Toronto, Ontario, Canada M5G 1X8*

Richard L. Huganir (19), *Department of Neuroscience, Howard Hughes Medical Institute, Johns Hopkins University School of Medicine, Baltimore, Maryland 21205*

Ingemar Jacobson (9), *Astra Hässle AB, S-43183 Mölndal, Sweden*

Kent Jardemark (9), *Department of Anatomy and Cell Biology, Göteborg University, S-41296 Göteborg, Sweden*

Edwin C. Johnson (2), *SIBIA Neurosciences, Inc., La Jolla, California 92037*

Gregory J. Kaczorowski (14, 32), *Department of Membrane Biochemistry and Biophysics, Merck Research Laboratories, Rahway, New Jersey 07065*

Bruce L. Kagan (35), *Department of Psychiatry, UCLA Neuropsychiatric Institute and West Los Angeles Veterans Administration Medical Center, Los Angeles, California 90024*

Sunjeev Kamboj (19), *Department of Neuroscience, Johns Hopkins University School of Medicine, Baltimore, Maryland 21205*

Hans-Günther Knaus (14, 32), *Institute for Biochemical Pharmacology, University of Innsbruck, A-6020 Innsbruck, Austria*

Roger E. Koeppe II (10, 28), *Department of Chemistry and Biochemistry, University of Arkansas, Fayetteville, Arkansas 72701*

Ching Kung (24, 27), *Laboratory of Molecular Biology and Department of Genetics, University of Wisconsin, Madison, Wisconsin 53706*

Bodo Laube (13), *Department of Neurochemistry, Max-Planck-Institute for Brain Research, D-60528 Frankfurt, Germany*

EDWIN S. LEVITAN (3), *Department of Pharmacology, University of Pittsburgh School of Medicine, Pittsburgh, Pennsylvania 15261*

CANHUI LI (11), *Research Institute, Hospital for Sick Children, Toronto, Ontario, Canada M5G 1X8*

SHERI J. LILLARD (9), *Department of Chemistry, Stanford University, Stanford, California 94305-5080*

STEPHEN H. LOUKIN (27), *Laboratory of Molecular Biology, University of Wisconsin, Madison, Wisconsin 53706*

J. A. LUNDBÆK (10), *Department of Neuropharmacology, Novo-Nordisk A/S, DK-2760 Maløv, Denmark*

MONICA M. LURTZ (6), *Department of Veterinary Pathobiology, University of Minnesota, St. Paul, Minnesota 55108*

A. M. MAER (10), *Department of Physiology and Biophysics, Cornell University Medical College, New York, New York 10021*

ANDREW L. MAMMEN (19), *Department of Neuroscience, Johns Hopkins University School of Medicine, Baltimore, Maryland 21205*

FRITZ MARKWARDT (8), *Julius-Bernstein-Institute for Physiology, Martin-Luther-University Halle, D-06097 Halle/Saale, Germany*

DEREK MARSH (4), *Abteilung Spektroskopie, Max-Planck-Institut für biophysikalische Chemie, D-37070 Göttingen, Germany*

BORIS MARTINAC (24, 27), *Department of Pharmacology, University of Western Australia, Nedland, Australia*

JOHN C. MATHAI (29), *Departments of Biological Chemistry and Medicine, John Hopkins University School of Medicine, Baltimore, Maryland 21205-2185*

DON W. MCBRIDE, JR. (25), *Department of Physiology and Biophysics, University of Texas Medical Branch, Galveston, Texas 77555*

PATRICK MCCONNELL (5), *Department of Molecular Biology, Parke-Davis Pharmaceutical Research, Division of Warner Lambert Company, Ann Arbor, Michigan 48105*

J. MICHAEL MCINTOSH (31), *Department of Biology and Psychiatry, University of Utah, Salt Lake City, Utah 84112-0840*

OWEN B. MCMANUS (14), *Department of Membrane Biochemistry and Biophysics, Merck Research Laboratories, Rahway, New Jersey 07065*

TAJIB A. MIRZABEKOV (35), *Department of Pathology, Division of Human Retrovirology, Dana Farber Cancer Institute, Harvard Medical School, Boston, Massachusetts 02115*

ALOK K. MITRA (7), *Department of Cell Biology, The Scripps Research Institute, La Jolla, California 92037*

EDWARD MOCZYDLOWSKI (15), *Departments of Pharmacology and Cellular and Molecular Physiology, Yale University School of Medicine, New Haven, Connecticut 06520*

PAUL C. MOE (24), *Laboratory of Molecular Biology and Department of Bacteriology, University of Wisconsin, Madison, Wisconsin 53706*

LAURIE L. MOLDAY (12), *Department of Biochemistry and Molecular Biology, University of British Columbia, Vancouver, British Columbia V6T 1Z3, Canada*

ROBERT S. MOLDAY (12), *Department of Biochemistry and Molecular Biology, University of British Columbia, Vancouver, British Columbia V6T 1Z3, Canada*

ALEXANDER MOSCHO (9), *McKinsey and Company, 80333 Munich, Germany*

RUTH D. MURRELL-LAGNADO (33), *Department of Pharmacology, University of Cambridge, Cambridge CB2 1QJ, United Kingdom*

C. NIELSEN (10), *Department of Physiology and Biophysics, Cornell University Medical College, New York, New York 10021*

JAMES OFFORD (5), *Department of Molecular Biology, Parke-Davis Pharmaceutical Research, Division of Warner Lambert Company, Ann Arbor, Michigan 48105*

DAVID C. OGDEN (26), *National Institute for Medical Research, London NW7 1AA, United Kingdom*

KIYOKAZU OGITA (21), *Department of Pharmacology, Faculty of Pharmaceutical Sciences, Setsunan University, Osaka 573-01, Japan*

BALDOMERO M. OLIVERA (31), *Department of Biology, University of Utah, Salt Lake City, Utah 84112-0840*

OWE ORWAR (9), *Department of Chemistry, Göteborg University, S-41296 Göteborg, Sweden*

U. PANTEN (23), *Institüt für Pharmakologie und Toxikologie, Universität Braunschweig, 38106 Braunschweig, Germany*

RAO V. L. PAPINENI (6), *Division of Molecular Virology, Baylor College of Medicine, Houston, Texas 77030*

STEEN E. PEDERSEN (6), *Department of Molecular Physiology and Biophysics, Baylor College of Medicine, Houston, Texas 77030*

ASHWIN PINTO (36), *Neurosciences Group, Institute of Molecular Medicine, John Radcliffe Hospital, Headington, Oxford OX3 9DU, United Kingdom*

GREGORY M. PRESTON (29), *Clinical Diagnostics Division, Johnson and Johnson Company, Rochester, New York 14650-2117*

LYNDON L. PROVIDENCE (28), *Department of Physiology and Biophysics, Cornell University Medical College, New York, New York 10021*

MOHABIR RAMJEESINGH (11), *Research Institute, Hospital for Sick Children, Toronto, Ontario, Canada M5G 1X8*

MICHAEL D. REILY (5), *Department of Chemistry, Parke-Davis Pharmaceutical Research, Division of Warner Lambert Company, Ann Arbor, Michigan 48105*

LISA ROMANO (22), *Department of Biology, Pennsylvania State University, University Park, Pennsylvania 16802*

PETER SAGGAU (1), *Division of Neuroscience, Baylor College of Medicine, Houston, Texas 77030*

YOSHIRO SAIMI (27), *Laboratory of Molecular Biology, University of Wisconsin, Madison, Wisconsin 53706*

SAROJ SATTSANGI (18), *Department of Pharmacology and Toxicology, Robert C. Byrd Health Sciences Center, West Virginia University, Morgantown, West Virginia 26506-9223*

WILLIAM A. SCHMALHOFER (14), *Department of Membrane Biochemistry and Biophysics, Merck Research Laboratories, Rahway, New Jersey 07065*

C. SCHWANSTECHER (23), *Institüt für Pharmakologie und Toxikologie, Universität Braunschweig, 38106 Braunschweig, Germany*

M. SCHWANSTECHER (23), *Institüt für Pharmakologie und Toxikologie, Universität Braunschweig, 38106 Braunschweig, Germany*

ABDULLAH SENER (16), *Laboratory of Experimental Medicine, School of Medicine, Université Libre de Bruxelles, B-1070 Brussels, Belgium*

JASON B. SHEAR (9), *Department of Chemistry and Biochemistry, University of Texas, Austin, Texas 78712*

MORGAN SHENG (20), *Howard Hughes Medical Institute and Department of Neurobiology, Massachusetts General Hospital and Harvard Medical School, Boston, Massachusetts 02114*

SUNDARAM SHOBANA (28), *Department of Physiology and Biophysics, Cornell University Medical College, New York, New York 10021*

ANATOLY Y. SILBERSTEIN (35), *Institute of Experimental and Theoretical Biophysics, Russian Academy of Science, Puschino, Moscow Region 142292, Russia*

ROBERT SLAUGHTER (32), *Department of Membrane Biochemistry and Biophysics, Merck Research Laboratories, Rahway, New Jersey 07065*

BARBARA L. SMITH (29), *Departments of Biological Chemistry and Medicine, Johns Hopkins University School of Medicine, Baltimore, Maryland 21205-2185*

KENNETH A. STAUDERMAN (2), *SIBIA Neurosciences, Inc., La Jolla, California 92037*

F. ANNE STEPHENSON (36), *Neurosciences Group, Institute of Molecular Medicine, John Radcliffe Hospital, Headington, Oxford OX3 9DU, United Kingdom*

SERGEI I. SUKHAREV (24), *Department of Zoology, University of Maryland, College Park, Maryland 20742*

YE-MING SUN (15), *Department of Pharmacology, Yale University School of Medicine, New Haven, Connecticut 06520*

BRUCE L. TEMPEL (34), *Departments of Otolaryngology and Pharmacology, University of Washington School of Medicine, Seattle, Washington 98195-7923*

VINZENZ M. UNGER (7), *Department of Cell Biology, The Scripps Research Institute, La Jolla, California 92037*

MARK A. VARNEY (2), *SIBIA Neurosciences, Inc., La Jolla, California 92037*

GÖNÜL VELIÇELEBI (2), *SIBIA Neurosciences, Inc., La Jolla, California 92037*

ANGELA C. VINCENT (36), *Neurosciences Group, Institute of Molecular Medicine, John Radcliffe Hospital, Headington, Oxford OX3 9DU, United Kingdom*

YANCHUN WANG (11), *Research Institute, Hospital for Sick Children, Toronto, Ontario, Canada M5G 1X8*

WILLIAM F. WONDERLIN (18), *Department of Pharmacology and Toxicology, Robert C. Byrd Health Sciences Center, West Virginia University, Morgantown, West Virginia 26506-9223*

DIXON J. WOODBURY (17), *Department of Physiology, Wayne State University School of Medicine, Detroit, Michigan 48201*

MICHAEL WYSZYNSKI (20), *Department of Neurobiology, Harvard Medical School, Boston, Massachusetts 02114*

MARK YEAGER (7), *Department of Cell Biology, The Scripps Research Institute, La Jolla, California 92037*

YUKIO YONEDA (21), *Department of Pharmacology, Faculty of Pharmaceutical Sciences, Setsunan University, Osaka 573-01, Japan*

RICHARD N. ZARE (9), *Department of Chemistry, Stanford University, Stanford, California 94305-5080*

XIN-LIANG ZHOU (27), *Laboratory of Molecular Biology, University of Wisconsin, Madison, Wisconsin 53706*

# Preface

The rapid growth of interest and research activity in ion channels is indicative of their fundamental importance in the maintenance of the living state. A brief examination of the table of contents of this volume of *Methods in Enzymology* and of its companion Volume 293 reveals methods that range from the molecular and biochemical level to the cellular, tissue, and whole cell level. These volumes supplement Volume 207 of this series.

Authors have been selected based on research contributions in the area about which they have written and their ability to describe their methodological contributions in a clear and reproducible way. They have been encouraged to make use of graphics, comparisons to other methods, and to provide tricks and approaches that make it possible to adapt methods to other systems.

I would like to express my appreciation to the contributors for providing their contributions in a timely fashion, to the Editorial Advisory Board for guidance, to the staff of Academic Press for helpful input, and to Mary Rarick and Patti Williams for assisting with the remarkable amount of clerical work needed to coordinate the chapters included.

P. MICHAEL CONN

# METHODS IN ENZYMOLOGY

VOLUME I. Preparation and Assay of Enzymes
*Edited by* SIDNEY P. COLOWICK AND NATHAN O. KAPLAN

VOLUME II. Preparation and Assay of Enzymes
*Edited by* SIDNEY P. COLOWICK AND NATHAN O. KAPLAN

VOLUME III. Preparation and Assay of Substrates
*Edited by* SIDNEY P. COLOWICK AND NATHAN O. KAPLAN

VOLUME IV. Special Techniques for the Enzymologist
*Edited by* SIDNEY P. COLOWICK AND NATHAN O. KAPLAN

VOLUME V. Preparation and Assay of Enzymes
*Edited by* SIDNEY P. COLOWICK AND NATHAN O. KAPLAN

VOLUME VI. Preparation and Assay of Enzymes (*Continued*)
Preparation and Assay of Substrates
Special Techniques
*Edited by* SIDNEY P. COLOWICK AND NATHAN O. KAPLAN

VOLUME VII. Cumulative Subject Index
*Edited by* SIDNEY P. COLOWICK AND NATHAN O. KAPLAN

VOLUME VIII. Complex Carbohydrates
*Edited by* ELIZABETH F. NEUFELD AND VICTOR GINSBURG

VOLUME IX. Carbohydrate Metabolism
*Edited by* WILLIS A. WOOD

VOLUME X. Oxidation and Phosphorylation
*Edited by* RONALD W. ESTABROOK AND MAYNARD E. PULLMAN

VOLUME XI. Enzyme Structure
*Edited by* C. H. W. HIRS

VOLUME XII. Nucleic Acids (Parts A and B)
*Edited by* LAWRENCE GROSSMAN AND KIVIE MOLDAVE

VOLUME XIII. Citric Acid Cycle
*Edited by* J. M. LOWENSTEIN

VOLUME XIV. Lipids
*Edited by* J. M. LOWENSTEIN

VOLUME XV. Steroids and Terpenoids
*Edited by* RAYMOND B. CLAYTON

VOLUME XVI. Fast Reactions
*Edited by* KENNETH KUSTIN

VOLUME XVII. Metabolism of Amino Acids and Amines (Parts A and B)
*Edited by* HERBERT TABOR AND CELIA WHITE TABOR

VOLUME XVIII. Vitamins and Coenzymes (Parts A, B, and C)
*Edited by* DONALD B. MCCORMICK AND LEMUEL D. WRIGHT

VOLUME XIX. Proteolytic Enzymes
*Edited by* GERTRUDE E. PERLMANN AND LASZLO LORAND

VOLUME XX. Nucleic Acids and Protein Synthesis (Part C)
*Edited by* KIVIE MOLDAVE AND LAWRENCE GROSSMAN

VOLUME XXI. Nucleic Acids (Part D)
*Edited by* LAWRENCE GROSSMAN AND KIVIE MOLDAVE

VOLUME XXII. Enzyme Purification and Related Techniques
*Edited by* WILLIAM B. JAKOBY

VOLUME XXIII. Photosynthesis (Part A)
*Edited by* ANTHONY SAN PIETRO

VOLUME XXIV. Photosynthesis and Nitrogen Fixation (Part B)
*Edited by* ANTHONY SAN PIETRO

VOLUME XXV. Enzyme Structure (Part B)
*Edited by* C. H. W. HIRS AND SERGE N. TIMASHEFF

VOLUME XXVI. Enzyme Structure (Part C)
*Edited by* C. H. W. HIRS AND SERGE N. TIMASHEFF

VOLUME XXVII. Enzyme Structure (Part D)
*Edited by* C. H. W. HIRS AND SERGE N. TIMASHEFF

VOLUME XXVIII. Complex Carbohydrates (Part B)
*Edited by* VICTOR GINSBURG

VOLUME XXIX. Nucleic Acids and Protein Synthesis (Part E)
*Edited by* LAWRENCE GROSSMAN AND KIVIE MOLDAVE

VOLUME XXX. Nucleic Acids and Protein Synthesis (Part F)
*Edited by* KIVIE MOLDAVE AND LAWRENCE GROSSMAN

VOLUME XXXI. Biomembranes (Part A)
*Edited by* SIDNEY FLEISCHER AND LESTER PACKER

VOLUME XXXII. Biomembranes (Part B)
*Edited by* SIDNEY FLEISCHER AND LESTER PACKER

VOLUME XXXIII. Cumulative Subject Index Volumes I–XXX
*Edited by* MARTHA G. DENNIS AND EDWARD A. DENNIS

VOLUME XXXIV. Affinity Techniques (Enzyme Purification: Part B)
*Edited by* WILLIAM B. JAKOBY AND MEIR WILCHEK

VOLUME XXXV. Lipids (Part B)
*Edited by* JOHN M. LOWENSTEIN

VOLUME XXXVI. Hormone Action (Part A: Steroid Hormones)
*Edited by* BERT W. O'MALLEY AND JOEL G. HARDMAN

VOLUME XXXVII. Hormone Action (Part B: Peptide Hormones)
*Edited by* BERT W. O'MALLEY AND JOEL G. HARDMAN

VOLUME XXXVIII. Hormone Action (Part C: Cyclic Nucleotides)
*Edited by* JOEL G. HARDMAN AND BERT W. O'MALLEY

VOLUME XXXIX. Hormone Action (Part D: Isolated Cells, Tissues, and Organ Systems)
*Edited by* JOEL G. HARDMAN AND BERT W. O'MALLEY

VOLUME XL. Hormone Action (Part E: Nuclear Structure and Function)
*Edited by* BERT W. O'MALLEY AND JOEL G. HARDMAN

VOLUME XLI. Carbohydrate Metabolism (Part B)
*Edited by* W. A. WOOD

VOLUME XLII. Carbohydrate Metabolism (Part C)
*Edited by* W. A. WOOD

VOLUME XLIII. Antibiotics
*Edited by* JOHN H. HASH

VOLUME XLIV. Immobilized Enzymes
*Edited by* KLAUS MOSBACH

VOLUME XLV. Proteolytic Enzymes (Part B)
*Edited by* LASZLO LORAND

VOLUME XLVI. Affinity Labeling
*Edited by* WILLIAM B. JAKOBY AND MEIR WILCHEK

VOLUME XLVII. Enzyme Structure (Part E)
*Edited by* C. H. W. HIRS AND SERGE N. TIMASHEFF

VOLUME XLVIII. Enzyme Structure (Part F)
*Edited by* C. H. W. HIRS AND SERGE N. TIMASHEFF

VOLUME XLIX. Enzyme Structure (Part G)
*Edited by* C. H. W. HIRS AND SERGE N. TIMASHEFF

VOLUME L. Complex Carbohydrates (Part C)
*Edited by* VICTOR GINSBURG

VOLUME LI. Purine and Pyrimidine Nucleotide Metabolism
*Edited by* PATRICIA A. HOFFEE AND MARY ELLEN JONES

VOLUME LII. Biomembranes (Part C: Biological Oxidations)
*Edited by* SIDNEY FLEISCHER AND LESTER PACKER

VOLUME LIII. Biomembranes (Part D: Biological Oxidations)
*Edited by* SIDNEY FLEISCHER AND LESTER PACKER

VOLUME LIV. Biomembranes (Part E: Biological Oxidations)
*Edited by* SIDNEY FLEISCHER AND LESTER PACKER

Volume LV. Biomembranes (Part F: Bioenergetics)
*Edited by* Sidney Fleischer and Lester Packer

Volume LVI. Biomembranes (Part G: Bioenergetics)
*Edited by* Sidney Fleischer and Lester Packer

Volume LVII. Bioluminescence and Chemiluminescence
*Edited by* Marlene A. DeLuca

Volume LVIII. Cell Culture
*Edited by* William B. Jakoby and Ira Pastan

Volume LIX. Nucleic Acids and Protein Synthesis (Part G)
*Edited by* Kivie Moldave and Lawrence Grossman

Volume LX. Nucleic Acids and Protein Synthesis (Part H)
*Edited by* Kivie Moldave and Lawrence Grossman

Volume 61. Enzyme Structure (Part H)
*Edited by* C. H. W. Hirs and Serge N. Timasheff

Volume 62. Vitamins and Coenzymes (Part D)
*Edited by* Donald B. McCormick and Lemuel D. Wright

Volume 63. Enzyme Kinetics and Mechanism (Part A: Initial Rate and Inhibitor Methods)
*Edited by* Daniel L. Purich

Volume 64. Enzyme Kinetics and Mechanism (Part B: Isotopic Probes and Complex Enzyme Systems)
*Edited by* Daniel L. Purich

Volume 65. Nucleic Acids (Part I)
*Edited by* Lawrence Grossman and Kivie Moldave

Volume 66. Vitamins and Coenzymes (Part E)
*Edited by* Donald B. McCormick and Lemuel D. Wright

Volume 67. Vitamins and Coenzymes (Part F)
*Edited by* Donald B. McCormick and Lemuel D. Wright

Volume 68. Recombinant DNA
*Edited by* Ray Wu

Volume 69. Photosynthesis and Nitrogen Fixation (Part C)
*Edited by* Anthony San Pietro

Volume 70. Immunochemical Techniques (Part A)
*Edited by* Helen Van Vunakis and John J. Langone

Volume 71. Lipids (Part C)
*Edited by* John M. Lowenstein

Volume 72. Lipids (Part D)
*Edited by* John M. Lowenstein

VOLUME 73. Immunochemical Techniques (Part B)
*Edited by* JOHN J. LANGONE AND HELEN VAN VUNAKIS

VOLUME 74. Immunochemical Techniques (Part C)
*Edited by* JOHN J. LANGONE AND HELEN VAN VUNAKIS

VOLUME 75. Cumulative Subject Index Volumes XXXI, XXXII, XXXIV–LX
*Edited by* EDWARD A. DENNIS AND MARTHA G. DENNIS

VOLUME 76. Hemoglobins
*Edited by* ERALDO ANTONINI, LUIGI ROSSI-BERNARDI, AND EMILIA CHIANCONE

VOLUME 77. Detoxication and Drug Metabolism
*Edited by* WILLIAM B. JAKOBY

VOLUME 78. Interferons (Part A)
*Edited by* SIDNEY PESTKA

VOLUME 79. Interferons (Part B)
*Edited by* SIDNEY PESTKA

VOLUME 80. Proteolytic Enzymes (Part C)
*Edited by* LASZLO LORAND

VOLUME 81. Biomembranes (Part H: Visual Pigments and Purple Membranes, I)
*Edited by* LESTER PACKER

VOLUME 82. Structural and Contractile Proteins (Part A: Extracellular Matrix)
*Edited by* LEON W. CUNNINGHAM AND DIXIE W. FREDERIKSEN

VOLUME 83. Complex Carbohydrates (Part D)
*Edited by* VICTOR GINSBURG

VOLUME 84. Immunochemical Techniques (Part D: Selected Immunoassays)
*Edited by* JOHN J. LANGONE AND HELEN VAN VUNAKIS

VOLUME 85. Structural and Contractile Proteins (Part B: The Contractile Apparatus and the Cytoskeleton)
*Edited by* DIXIE W. FREDERIKSEN AND LEON W. CUNNINGHAM

VOLUME 86. Prostaglandins and Arachidonate Metabolites
*Edited by* WILLIAM E. M. LANDS AND WILLIAM L. SMITH

VOLUME 87. Enzyme Kinetics and Mechanism (Part C: Intermediates, Stereochemistry, and Rate Studies)
*Edited by* DANIEL L. PURICH

VOLUME 88. Biomembranes (Part I: Visual Pigments and Purple Membranes, II)
*Edited by* LESTER PACKER

VOLUME 89. Carbohydrate Metabolism (Part D)
*Edited by* WILLIS A. WOOD

VOLUME 90. Carbohydrate Metabolism (Part E)
*Edited by* WILLIS A. WOOD

VOLUME 91. Enzyme Structure (Part I)
*Edited by* C. H. W. HIRS AND SERGE N. TIMASHEFF

VOLUME 92. Immunochemical Techniques (Part E: Monoclonal Antibodies and General Immunoassay Methods)
*Edited by* JOHN J. LANGONE AND HELEN VAN VUNAKIS

VOLUME 93. Immunochemical Techniques (Part F: Conventional Antibodies, Fc Receptors, and Cytotoxicity)
*Edited by* JOHN J. LANGONE AND HELEN VAN VUNAKIS

VOLUME 94. Polyamines
*Edited by* HERBERT TABOR AND CELIA WHITE TABOR

VOLUME 95. Cumulative Subject Index Volumes 61–74, 76–80
*Edited by* EDWARD A. DENNIS AND MARTHA G. DENNIS

VOLUME 96. Biomembranes [Part J: Membrane Biogenesis: Assembly and Targeting (General Methods; Eukaryotes)]
*Edited by* SIDNEY FLEISCHER AND BECCA FLEISCHER

VOLUME 97. Biomembranes [Part K: Membrane Biogenesis: Assembly and Targeting (Prokaryotes, Mitochondria, and Chloroplasts)]
*Edited by* SIDNEY FLEISCHER AND BECCA FLEISCHER

VOLUME 98. Biomembranes (Part L: Membrane Biogenesis: Processing and Recycling)
*Edited by* SIDNEY FLEISCHER AND BECCA FLEISCHER

VOLUME 99. Hormone Action (Part F: Protein Kinases)
*Edited by* JACKIE D. CORBIN AND JOEL G. HARDMAN

VOLUME 100. Recombinant DNA (Part B)
*Edited by* RAY WU, LAWRENCE GROSSMAN, AND KIVIE MOLDAVE

VOLUME 101. Recombinant DNA (Part C)
*Edited by* RAY WU, LAWRENCE GROSSMAN, AND KIVIE MOLDAVE

VOLUME 102. Hormone Action (Part G: Calmodulin and Calcium-Binding Proteins)
*Edited by* ANTHONY R. MEANS AND BERT W. O'MALLEY

VOLUME 103. Hormone Action (Part H: Neuroendocrine Peptides)
*Edited by* P. MICHAEL CONN

VOLUME 104. Enzyme Purification and Related Techniques (Part C)
*Edited by* WILLIAM B. JAKOBY

VOLUME 105. Oxygen Radicals in Biological Systems
*Edited by* LESTER PACKER

VOLUME 106. Posttranslational Modifications (Part A)
*Edited by* FINN WOLD AND KIVIE MOLDAVE

VOLUME 107. Posttranslational Modifications (Part B)
*Edited by* FINN WOLD AND KIVIE MOLDAVE

VOLUME 108. Immunochemical Techniques (Part G: Separation and Characterization of Lymphoid Cells)
*Edited by* GIOVANNI DI SABATO, JOHN J. LANGONE, AND HELEN VAN VUNAKIS

VOLUME 109. Hormone Action (Part I: Peptide Hormones)
*Edited by* LUTZ BIRNBAUMER AND BERT W. O'MALLEY

VOLUME 110. Steroids and Isoprenoids (Part A)
*Edited by* JOHN H. LAW AND HANS C. RILLING

VOLUME 111. Steroids and Isoprenoids (Part B)
*Edited by* JOHN H. LAW AND HANS C. RILLING

VOLUME 112. Drug and Enzyme Targeting (Part A)
*Edited by* KENNETH J. WIDDER AND RALPH GREEN

VOLUME 113. Glutamate, Glutamine, Glutathione, and Related Compounds
*Edited by* ALTON MEISTER

VOLUME 114. Diffraction Methods for Biological Macromolecules (Part A)
*Edited by* HAROLD W. WYCKOFF, C. H. W. HIRS, AND SERGE N. TIMASHEFF

VOLUME 115. Diffraction Methods for Biological Macromolecules (Part B)
*Edited by* HAROLD W. WYCKOFF, C. H. W. HIRS, AND SERGE N. TIMASHEFF

VOLUME 116. Immunochemical Techniques (Part H: Effectors and Mediators of Lymphoid Cell Functions)
*Edited by* GIOVANNI DI SABATO, JOHN J. LANGONE, AND HELEN VAN VUNAKIS

VOLUME 117. Enzyme Structure (Part J)
*Edited by* C. H. W. HIRS AND SERGE N. TIMASHEFF

VOLUME 118. Plant Molecular Biology
*Edited by* ARTHUR WEISSBACH AND HERBERT WEISSBACH

VOLUME 119. Interferons (Part C)
*Edited by* SIDNEY PESTKA

VOLUME 120. Cumulative Subject Index Volumes 81–94, 96–101

VOLUME 121. Immunochemical Techniques (Part I: Hybridoma Technology and Monoclonal Antibodies)
*Edited by* JOHN J. LANGONE AND HELEN VAN VUNAKIS

VOLUME 122. Vitamins and Coenzymes (Part G)
*Edited by* FRANK CHYTIL AND DONALD B. MCCORMICK

VOLUME 123. Vitamins and Coenzymes (Part H)
*Edited by* FRANK CHYTIL AND DONALD B. MCCORMICK

VOLUME 124. Hormone Action (Part J: Neuroendocrine Peptides)
*Edited by* P. MICHAEL CONN

VOLUME 125. Biomembranes (Part M: Transport in Bacteria, Mitochondria, and Chloroplasts: General Approaches and Transport Systems)
*Edited by* SIDNEY FLEISCHER AND BECCA FLEISCHER

VOLUME 126. Biomembranes (Part N: Transport in Bacteria, Mitochondria, and Chloroplasts: Protonmotive Force)
*Edited by* SIDNEY FLEISCHER AND BECCA FLEISCHER

VOLUME 127. Biomembranes (Part O: Protons and Water: Structure and Translocation)
*Edited by* LESTER PACKER

VOLUME 128. Plasma Lipoproteins (Part A: Preparation, Structure, and Molecular Biology)
*Edited by* JERE P. SEGREST AND JOHN J. ALBERS

VOLUME 129. Plasma Lipoproteins (Part B: Characterization, Cell Biology, and Metabolism)
*Edited by* JOHN J. ALBERS AND JERE P. SEGREST

VOLUME 130. Enzyme Structure (Part K)
*Edited by* C. H. W. HIRS AND SERGE N. TIMASHEFF

VOLUME 131. Enzyme Structure (Part L)
*Edited by* C. H. W. HIRS AND SERGE N. TIMASHEFF

VOLUME 132. Immunochemical Techniques (Part J: Phagocytosis and Cell-Mediated Cytotoxicity)
*Edited by* GIOVANNI DI SABATO AND JOHANNES EVERSE

VOLUME 133. Bioluminescence and Chemiluminescence (Part B)
*Edited by* MARLENE DELUCA AND WILLIAM D. MCELROY

VOLUME 134. Structural and Contractile Proteins (Part C: The Contractile Apparatus and the Cytoskeleton)
*Edited by* RICHARD B. VALLEE

VOLUME 135. Immobilized Enzymes and Cells (Part B)
*Edited by* KLAUS MOSBACH

VOLUME 136. Immobilized Enzymes and Cells (Part C)
*Edited by* KLAUS MOSBACH

VOLUME 137. Immobilized Enzymes and Cells (Part D)
*Edited by* KLAUS MOSBACH

VOLUME 138. Complex Carbohydrates (Part E)
*Edited by* VICTOR GINSBURG

VOLUME 139. Cellular Regulators (Part A: Calcium- and Calmodulin-Binding Proteins)
*Edited by* ANTHONY R. MEANS AND P. MICHAEL CONN

VOLUME 140. Cumulative Subject Index Volumes 102–119, 121–134

VOLUME 141. Cellular Regulators (Part B: Calcium and Lipids)
*Edited by* P. MICHAEL CONN AND ANTHONY R. MEANS

VOLUME 142. Metabolism of Aromatic Amino Acids and Amines
*Edited by* SEYMOUR KAUFMAN

VOLUME 143. Sulfur and Sulfur Amino Acids
*Edited by* WILLIAM B. JAKOBY AND OWEN GRIFFITH

VOLUME 144. Structural and Contractile Proteins (Part D: Extracellular Matrix)
*Edited by* LEON W. CUNNINGHAM

VOLUME 145. Structural and Contractile Proteins (Part E: Extracellular Matrix)
*Edited by* LEON W. CUNNINGHAM

VOLUME 146. Peptide Growth Factors (Part A)
*Edited by* DAVID BARNES AND DAVID A. SIRBASKU

VOLUME 147. Peptide Growth Factors (Part B)
*Edited by* DAVID BARNES AND DAVID A. SIRBASKU

VOLUME 148. Plant Cell Membranes
*Edited by* LESTER PACKER AND ROLAND DOUCE

VOLUME 149. Drug and Enzyme Targeting (Part B)
*Edited by* RALPH GREEN AND KENNETH J. WIDDER

VOLUME 150. Immunochemical Techniques (Part K: *In Vitro* Models of B and T Cell Functions and Lymphoid Cell Receptors)
*Edited by* GIOVANNI DI SABATO

VOLUME 151. Molecular Genetics of Mammalian Cells
*Edited by* MICHAEL M. GOTTESMAN

VOLUME 152. Guide to Molecular Cloning Techniques
*Edited by* SHELBY L. BERGER AND ALAN R. KIMMEL

VOLUME 153. Recombinant DNA (Part D)
*Edited by* RAY WU AND LAWRENCE GROSSMAN

VOLUME 154. Recombinant DNA (Part E)
*Edited by* RAY WU AND LAWRENCE GROSSMAN

VOLUME 155. Recombinant DNA (Part F)
*Edited by* RAY WU

VOLUME 156. Biomembranes (Part P: ATP-Driven Pumps and Related Transport: The Na,K-Pump)
*Edited by* SIDNEY FLEISCHER AND BECCA FLEISCHER

VOLUME 157. Biomembranes (Part Q: ATP-Driven Pumps and Related Transport: Calcium, Proton, and Potassium Pumps)
*Edited by* SIDNEY FLEISCHER AND BECCA FLEISCHER

VOLUME 158. Metalloproteins (Part A)
*Edited by* JAMES F. RIORDAN AND BERT L. VALLEE

VOLUME 159. Initiation and Termination of Cyclic Nucleotide Action
*Edited by* JACKIE D. CORBIN AND ROGER A. JOHNSON

VOLUME 160. Biomass (Part A: Cellulose and Hemicellulose)
*Edited by* WILLIS A. WOOD AND SCOTT T. KELLOGG

VOLUME 161. Biomass (Part B: Lignin, Pectin, and Chitin)
*Edited by* WILLIS A. WOOD AND SCOTT T. KELLOGG

VOLUME 162. Immunochemical Techniques (Part L: Chemotaxis and Inflammation)
*Edited by* GIOVANNI DI SABATO

VOLUME 163. Immunochemical Techniques (Part M: Chemotaxis and Inflammation)
*Edited by* GIOVANNI DI SABATO

VOLUME 164. Ribosomes
*Edited by* HARRY F. NOLLER, JR., AND KIVIE MOLDAVE

VOLUME 165. Microbial Toxins: Tools for Enzymology
*Edited by* SIDNEY HARSHMAN

VOLUME 166. Branched-Chain Amino Acids
*Edited by* ROBERT HARRIS AND JOHN R. SOKATCH

VOLUME 167. Cyanobacteria
*Edited by* LESTER PACKER AND ALEXANDER N. GLAZER

VOLUME 168. Hormone Action (Part K: Neuroendocrine Peptides)
*Edited by* P. MICHAEL CONN

VOLUME 169. Platelets: Receptors, Adhesion, Secretion (Part A)
*Edited by* JACEK HAWIGER

VOLUME 170. Nucleosomes
*Edited by* PAUL M. WASSARMAN AND ROGER D. KORNBERG

VOLUME 171. Biomembranes (Part R: Transport Theory: Cells and Model Membranes)
*Edited by* SIDNEY FLEISCHER AND BECCA FLEISCHER

VOLUME 172. Biomembranes (Part S: Transport: Membrane Isolation and Characterization)
*Edited by* SIDNEY FLEISCHER AND BECCA FLEISCHER

VOLUME 173. Biomembranes [Part T: Cellular and Subcellular Transport: Eukaryotic (Nonepithelial) Cells]
*Edited by* SIDNEY FLEISCHER AND BECCA FLEISCHER

VOLUME 174. Biomembranes [Part U: Cellular and Subcellular Transport: Eukaryotic (Nonepithelial) Cells]
*Edited by* SIDNEY FLEISCHER AND BECCA FLEISCHER

VOLUME 175. Cumulative Subject Index Volumes 135–139, 141–167

VOLUME 176. Nuclear Magnetic Resonance (Part A: Spectral Techniques and Dynamics)
*Edited by* NORMAN J. OPPENHEIMER AND THOMAS L. JAMES

VOLUME 177. Nuclear Magnetic Resonance (Part B: Structure and Mechanism)
*Edited by* NORMAN J. OPPENHEIMER AND THOMAS L. JAMES

VOLUME 178. Antibodies, Antigens, and Molecular Mimicry
*Edited by* JOHN J. LANGONE

VOLUME 179. Complex Carbohydrates (Part F)
*Edited by* VICTOR GINSBURG

VOLUME 180. RNA Processing (Part A: General Methods)
*Edited by* JAMES E. DAHLBERG AND JOHN N. ABELSON

VOLUME 181. RNA Processing (Part B: Specific Methods)
*Edited by* JAMES E. DAHLBERG AND JOHN N. ABELSON

VOLUME 182. Guide to Protein Purification
*Edited by* MURRAY P. DEUTSCHER

VOLUME 183. Molecular Evolution: Computer Analysis of Protein and Nucleic Acid Sequences
*Edited by* RUSSELL F. DOOLITTLE

VOLUME 184. Avidin–Biotin Technology
*Edited by* MEIR WILCHEK AND EDWARD A. BAYER

VOLUME 185. Gene Expression Technology
*Edited by* DAVID V. GOEDDEL

VOLUME 186. Oxygen Radicals in Biological Systems (Part B: Oxygen Radicals and Antioxidants)
*Edited by* LESTER PACKER AND ALEXANDER N. GLAZER

VOLUME 187. Arachidonate Related Lipid Mediators
*Edited by* ROBERT C. MURPHY AND FRANK A. FITZPATRICK

VOLUME 188. Hydrocarbons and Methylotrophy
*Edited by* MARY E. LIDSTROM

VOLUME 189. Retinoids (Part A: Molecular and Metabolic Aspects)
*Edited by* LESTER PACKER

VOLUME 190. Retinoids (Part B: Cell Differentiation and Clinical Applications)
*Edited by* LESTER PACKER

VOLUME 191. Biomembranes (Part V: Cellular and Subcellular Transport: Epithelial Cells)
*Edited by* SIDNEY FLEISCHER AND BECCA FLEISCHER

VOLUME 192. Biomembranes (Part W: Cellular and Subcellular Transport: Epithelial Cells)
*Edited by* SIDNEY FLEISCHER AND BECCA FLEISCHER

VOLUME 193. Mass Spectrometry
*Edited by* JAMES A. MCCLOSKEY

VOLUME 194. Guide to Yeast Genetics and Molecular Biology
*Edited by* CHRISTINE GUTHRIE AND GERALD R. FINK

VOLUME 195. Adenylyl Cyclase, G Proteins, and Guanylyl Cyclase
*Edited by* ROGER A. JOHNSON AND JACKIE D. CORBIN

VOLUME 196. Molecular Motors and the Cytoskeleton
*Edited by* RICHARD B. VALLEE

VOLUME 197. Phospholipases
*Edited by* EDWARD A. DENNIS

VOLUME 198. Peptide Growth Factors (Part C)
*Edited by* DAVID BARNES, J. P. MATHER, AND GORDON H. SATO

VOLUME 199. Cumulative Subject Index Volumes 168–174, 176–194

VOLUME 200. Protein Phosphorylation (Part A: Protein Kinases: Assays, Purification, Antibodies, Functional Analysis, Cloning, and Expression)
*Edited by* TONY HUNTER AND BARTHOLOMEW M. SEFTON

VOLUME 201. Protein Phosphorylation (Part B: Analysis of Protein Phosphorylation, Protein Kinase Inhibitors, and Protein Phosphatases)
*Edited by* TONY HUNTER AND BARTHOLOMEW M. SEFTON

VOLUME 202. Molecular Design and Modeling: Concepts and Applications (Part A: Proteins, Peptides, and Enzymes)
*Edited by* JOHN J. LANGONE

VOLUME 203. Molecular Design and Modeling: Concepts and Applications (Part B: Antibodies and Antigens, Nucleic Acids, Polysaccharides, and Drugs)
*Edited by* JOHN J. LANGONE

VOLUME 204. Bacterial Genetic Systems
*Edited by* JEFFREY H. MILLER

VOLUME 205. Metallobiochemistry (Part B: Metallothionein and Related Molecules)
*Edited by* JAMES F. RIORDAN AND BERT L. VALLEE

VOLUME 206. Cytochrome P450
*Edited by* MICHAEL R. WATERMAN AND ERIC F. JOHNSON

VOLUME 207. Ion Channels
*Edited by* BERNARDO RUDY AND LINDA E. IVERSON

VOLUME 208. Protein–DNA Interactions
*Edited by* ROBERT T. SAUER

VOLUME 209. Phospholipid Biosynthesis
*Edited by* EDWARD A. DENNIS AND DENNIS E. VANCE

VOLUME 210. Numerical Computer Methods
*Edited by* LUDWIG BRAND AND MICHAEL L. JOHNSON

VOLUME 211. DNA Structures (Part A: Synthesis and Physical Analysis of DNA)
*Edited by* DAVID M. J. LILLEY AND JAMES E. DAHLBERG

VOLUME 212. DNA Structures (Part B: Chemical and Electrophoretic Analysis of DNA)
*Edited by* DAVID M. J. LILLEY AND JAMES E. DAHLBERG

VOLUME 213. Carotenoids (Part A: Chemistry, Separation, Quantitation, and Antioxidation)
*Edited by* LESTER PACKER

VOLUME 214. Carotenoids (Part B: Metabolism, Genetics, and Biosynthesis)
*Edited by* LESTER PACKER

VOLUME 215. Platelets: Receptors, Adhesion, Secretion (Part B)
*Edited by* JACEK J. HAWIGER

VOLUME 216. Recombinant DNA (Part G)
*Edited by* RAY WU

VOLUME 217. Recombinant DNA (Part H)
*Edited by* RAY WU

VOLUME 218. Recombinant DNA (Part I)
*Edited by* RAY WU

VOLUME 219. Reconstitution of Intracellular Transport
*Edited by* JAMES E. ROTHMAN

VOLUME 220. Membrane Fusion Techniques (Part A)
*Edited by* NEJAT DÜZGÜNEŞ

VOLUME 221. Membrane Fusion Techniques (Part B)
*Edited by* NEJAT DÜZGÜNEŞ

VOLUME 222. Proteolytic Enzymes in Coagulation, Fibrinolysis, and Complement Activation (Part A: Mammalian Blood Coagulation Factors and Inhibitors)
*Edited by* LASZLO LORAND AND KENNETH G. MANN

VOLUME 223. Proteolytic Enzymes in Coagulation, Fibrinolysis, and Complement Activation (Part B: Complement Activation, Fibrinolysis, and Nonmammalian Blood Coagulation Factors)
*Edited by* LASZLO LORAND AND KENNETH G. MANN

VOLUME 224. Molecular Evolution: Producing the Biochemical Data
*Edited by* ELIZABETH ANNE ZIMMER, THOMAS J. WHITE, REBECCA L. CANN, AND ALLAN C. WILSON

VOLUME 225. Guide to Techniques in Mouse Development
*Edited by* PAUL M. WASSARMAN AND MELVIN L. DEPAMPHILIS

VOLUME 226. Metallobiochemistry (Part C: Spectroscopic and Physical Methods for Probing Metal Ion Environments in Metalloenzymes and Metalloproteins)
*Edited by* JAMES F. RIORDAN AND BERT L. VALLEE

VOLUME 227. Metallobiochemistry (Part D: Physical and Spectroscopic Methods for Probing Metal Ion Environments in Metalloproteins)
*Edited by* JAMES F. RIORDAN AND BERT L. VALLEE

VOLUME 228. Aqueous Two-Phase Systems
*Edited by* HARRY WALTER AND GÖTE JOHANSSON

VOLUME 229. Cumulative Subject Index Volumes 195–198, 200–227

VOLUME 230. Guide to Techniques in Glycobiology
*Edited by* WILLIAM J. LENNARZ AND GERALD W. HART

VOLUME 231. Hemoglobins (Part B: Biochemical and Analytical Methods)
*Edited by* JOHANNES EVERSE, KIM D. VANDEGRIFF, AND ROBERT M. WINSLOW

VOLUME 232. Hemoglobins (Part C: Biophysical Methods)
*Edited by* JOHANNES EVERSE, KIM D. VANDEGRIFF, AND ROBERT M. WINSLOW

VOLUME 233. Oxygen Radicals in Biological Systems (Part C)
*Edited by* LESTER PACKER

VOLUME 234. Oxygen Radicals in Biological Systems (Part D)
*Edited by* LESTER PACKER

VOLUME 235. Bacterial Pathogenesis (Part A: Identification and Regulation of Virulence Factors)
*Edited by* VIRGINIA L. CLARK AND PATRIK M. BAVOIL

VOLUME 236. Bacterial Pathogenesis (Part B: Integration of Pathogenic Bacteria with Host Cells)
*Edited by* VIRGINIA L. CLARK AND PATRIK M. BAVOIL

VOLUME 237. Heterotrimeric G Proteins
*Edited by* RAVI IYENGAR

VOLUME 238. Heterotrimeric G-Protein Effectors
*Edited by* RAVI IYENGAR

VOLUME 239. Nuclear Magnetic Resonance (Part C)
*Edited by* THOMAS L. JAMES AND NORMAN J. OPPENHEIMER

VOLUME 240. Numerical Computer Methods (Part B)
*Edited by* MICHAEL L. JOHNSON AND LUDWIG BRAND

VOLUME 241. Retroviral Proteases
*Edited by* LAWRENCE C. KUO AND JULES A. SHAFER

VOLUME 242. Neoglycoconjugates (Part A)
*Edited by* Y. C. LEE AND REIKO T. LEE

VOLUME 243. Inorganic Microbial Sulfur Metabolism
*Edited by* HARRY D. PECK, JR., AND JEAN LEGALL

VOLUME 244. Proteolytic Enzymes: Serine and Cysteine Peptidases
*Edited by* ALAN J. BARRETT

VOLUME 245. Extracellular Matrix Components
*Edited by* E. RUOSLAHTI AND E. ENGVALL

VOLUME 246. Biochemical Spectroscopy
*Edited by* KENNETH SAUER

VOLUME 247. Neoglycoconjugates (Part B: Biomedical Applications)
*Edited by* Y. C. LEE AND REIKO T. LEE

VOLUME 248. Proteolytic Enzymes: Aspartic and Metallo Peptidases
*Edited by* ALAN J. BARRETT

VOLUME 249. Enzyme Kinetics and Mechanism (Part D: Developments in Enzyme Dynamics)
*Edited by* DANIEL L. PURICH

VOLUME 250. Lipid Modifications of Proteins
*Edited by* PATRICK J. CASEY AND JANICE E. BUSS

VOLUME 251. Biothiols (Part A: Monothiols and Dithiols, Protein Thiols, and Thiyl Radicals)
*Edited by* LESTER PACKER

VOLUME 252. Biothiols (Part B: Glutathione and Thioredoxin; Thiols in Signal Transduction and Gene Regulation)
*Edited by* LESTER PACKER

VOLUME 253. Adhesion of Microbial Pathogens
*Edited by* RON J. DOYLE AND ITZHAK OFEK

VOLUME 254. Oncogene Techniques
*Edited by* PETER K. VOGT AND INDER M. VERMA

VOLUME 255. Small GTPases and Their Regulators (Part A: Ras Family)
*Edited by* W. E. BALCH, CHANNING J. DER, AND ALAN HALL

VOLUME 256. Small GTPases and Their Regulators (Part B: Rho Family)
*Edited by* W. E. BALCH, CHANNING J. DER, AND ALAN HALL

VOLUME 257. Small GTPases and Their Regulators (Part C: Proteins Involved in Transport)
*Edited by* W. E. BALCH, CHANNING J. DER, AND ALAN HALL

VOLUME 258. Redox-Active Amino Acids in Biology
*Edited by* JUDITH P. KLINMAN

VOLUME 259. Energetics of Biological Macromolecules
*Edited by* MICHAEL L. JOHNSON AND GARY K. ACKERS

VOLUME 260. Mitochondrial Biogenesis and Genetics (Part A)
*Edited by* GIUSEPPE M. ATTARDI AND ANNE CHOMYN

VOLUME 261. Nuclear Magnetic Resonance and Nucleic Acids
*Edited by* THOMAS L. JAMES

VOLUME 262. DNA Replication
*Edited by* JUDITH L. CAMPBELL

VOLUME 263. Plasma Lipoproteins (Part C: Quantitation)
*Edited by* WILLIAM A. BRADLEY, SANDRA H. GIANTURCO, AND JERE P. SEGREST

VOLUME 264. Mitochondrial Biogenesis and Genetics (Part B)
*Edited by* GIUSEPPE M. ATTARDI AND ANNE CHOMYN

VOLUME 265. Cumulative Subject Index Volumes 228, 230–262

VOLUME 266. Computer Methods for Macromolecular Sequence Analysis
*Edited by* RUSSELL F. DOOLITTLE

VOLUME 267. Combinatorial Chemistry
*Edited by* JOHN N. ABELSON

VOLUME 268. Nitric Oxide (Part A: Sources and Detection of NO; NO Synthase)
*Edited by* LESTER PACKER

VOLUME 269. Nitric Oxide (Part B: Physiological and Pathological Processes)
*Edited by* LESTER PACKER

VOLUME 270. High Resolution Separation and Analysis of Biological Macromolecules (Part A: Fundamentals)
*Edited by* BARRY L. KARGER AND WILLIAM S. HANCOCK

VOLUME 271. High Resolution Separation and Analysis of Biological Macromolecules (Part B: Applications)
*Edited by* BARRY L. KARGER AND WILLIAM S. HANCOCK

VOLUME 272. Cytochrome P450 (Part B)
*Edited by* ERIC F. JOHNSON AND MICHAEL R. WATERMAN

VOLUME 273. RNA Polymerase and Associated Factors (Part A)
*Edited by* SANKAR ADHYA

VOLUME 274. RNA Polymerase and Associated Factors (Part B)
*Edited by* SANKAR ADHYA

VOLUME 275. Viral Polymerases and Related Proteins
*Edited by* LAWRENCE C. KUO, DAVID B. OLSEN, AND STEVEN S. CARROLL

VOLUME 276. Macromolecular Crystallography (Part A)
*Edited by* CHARLES W. CARTER, JR., AND ROBERT M. SWEET

VOLUME 277. Macromolecular Crystallography (Part B)
*Edited by* CHARLES W. CARTER, JR., AND ROBERT M. SWEET

VOLUME 278. Fluorescence Spectroscopy
*Edited by* LUDWIG BRAND AND MICHAEL L. JOHNSON

VOLUME 279. Vitamins and Coenzymes (Part I)
*Edited by* DONALD B. MCCORMICK, JOHN W. SUTTIE, AND CONRAD WAGNER

VOLUME 280. Vitamins and Coenzymes (Part J)
*Edited by* DONALD B. MCCORMICK, JOHN W. SUTTIE, AND CONRAD WAGNER

VOLUME 281. Vitamins and Coenzymes (Part K)
*Edited by* DONALD B. MCCORMICK, JOHN W. SUTTIE, AND CONRAD WAGNER

VOLUME 282. Vitamins and Coenzymes (Part L)
*Edited by* DONALD B. MCCORMICK, JOHN W. SUTTIE, AND CONRAD WAGNER

VOLUME 283. Cell Cycle Control
*Edited by* WILLIAM G. DUNPHY

VOLUME 284. Lipases (Part A: Biotechnology)
*Edited by* BYRON RUBIN AND EDWARD A. DENNIS

VOLUME 285. Cumulative Subject Index Volumes 263, 264, 266–284, 286–289

VOLUME 286. Lipases (Part B: Enzyme Characterization and Utilization)
*Edited by* BYRON RUBIN AND EDWARD A. DENNIS

VOLUME 287. Chemokines
*Edited by* RICHARD HORUK

VOLUME 288. Chemokine Receptors
*Edited by* RICHARD HORUK

VOLUME 289. Solid Phase Peptide Synthesis
*Edited by* GREGG B. FIELDS

VOLUME 290. Molecular Chaperones
*Edited by* GEORGE H. LORIMER AND THOMAS BALDWIN

VOLUME 291. Caged Compounds
*Edited by* GERARD MARRIOTT

VOLUME 292. ABC Transporters: Biochemical, Cellular, and Molecular Aspects
*Edited by* SURESH V. AMBUDKAR AND MICHAEL M. GOTTESMAN

VOLUME 293. Ion Channels (Part B)
*Edited by* P. MICHAEL CONN

VOLUME 294. Ion Channels (Part C)
*Edited by* P. MICHAEL CONN

VOLUME 295. Energetics of Biological Macromolecules (Part B)
*Edited by* GARY K. ACKERS AND MICHAEL L. JOHNSON

VOLUME 296. Neurotransmitter Transporters
*Edited by* SUSAN G. AMARA

VOLUME 297. Photosynthesis: Molecular Biology of Energy Capture
*Edited by* LEE MCINTOSH

VOLUME 298. Molecular Motors and the Cytoskeleton (Part B)
*Edited by* RICHARD B. VALLEE

VOLUME 299. Oxidants and Antioxidants (Part A)
*Edited by* LESTER PACKER

VOLUME 300. Oxidants and Antioxidants (Part B)
*Edited by* LESTER PACKER

VOLUME 301. Nitric Oxide: Biological and Antioxidant Activities (Part C) (in preparation)
*Edited by* LESTER PACKER

VOLUME 302. Green Fluorescent Protein (in preparation)
*Edited by* P. MICHAEL CONN

VOLUME 303. cDNA Preparation and Display (in preparation)
*Edited by* SHERMAN M. WEISSMAN

VOLUME 304. Chromatin (in preparation)
*Edited by* PAUL M. WASSERMAN AND ALAN P. WOLFFE

# Section I

# Physical Methods

# [1] Optical Measurements of Calcium Signals in Mammalian Presynaptic Terminals

*By* PETER SAGGAU, RICHARD GRAY, and JOHN A. DANI

## Introduction

An action potential arriving at the presynaptic terminal opens voltage-dependent $Ca^{2+}$ channels, and the rapid influx of calcium can initiate neurotransmitter release. In addition, the activity-dependent accumulation of calcium participates in mechanisms that adjust the strength and efficacy of the synapse on multiple time scales. Intracellular stores and ligand-gated ion channels can contribute to presynaptic calcium signals. Although the participation of calcium in presynaptic events can often be inferred from indirect measurements, in some cases it is necessary to have direct quantitative data describing the amplitude and time course of the calcium signal. A major problem when trying to obtain those data is that central nervous system (CNS) presynaptic terminals are extremely small. Most CNS presynaptic terminals are about 1 $\mu$m in diameter. In spite of the difficulties, neuroscientists have taken on the technical challenge because presynaptic mechanisms are important for normal synaptic communication and for a wide range of modulatory events. For instance, synaptic modulation and long-term synaptic changes are thought to be the cellular correlates of learning and memory.[1–5]

Techniques have been developed to measure the amount of $Ca^{2+}$ that passes through an ion channel.[6–8] Those approaches required the simultaneous measurement of membrane current and intracellular $Ca^{2+}$ for single cells. Whole-cell patch-clamp techniques were used to measure current, and intracellular $Ca^{2+}$ was monitored with a fluorescent indicator, Fura-2. Those studies yielded the fraction of current carried by $Ca^{2+}$ through various ion channels. Thus, simple measurements of current in known external solutions could be used to calculate the $Ca^{2+}$ influx through specific chan-

[1] D. V. Madison, R. C. Malenka, and R. A. Nicoll, *Annu. Rev. Neurosci.* **14,** 379 (1991).
[2] R. Malinow, *Science* **252,** 722 (1991).
[3] A. Aiba, C. Chen, K. Herrup, C. Rosenmund, C. F. Stevens, and S. Tonegawa, *Cell* **79,** 365 (1994).
[4] S. G. Grant and A. J. Silva, *Trends Neurosci.* **17,** 71 (1994).
[5] Y. Goda and C. F. Stevens, *Current Biol.* **6,** 375 (1996).
[6] R. Schneggenburger, Z. Zhou, A. Konnerth, and E. Neher, *Neuron* **11,** 133 (1993).
[7] S. Vernino, M. Rogers, K. Radcliffe, and J. A. Dani, *J. Neurosci.* **14,** 5514 (1994).
[8] M. Rogers and J. A. Dani, *Biophys. J.* **68,** 501 (1995).



0076-6879/99 $30.00

nels. However, direct current measurements at small presynaptic terminals are extremely difficult. Therefore, less invasive optical methods have been developed to cope with the problems presented by presynaptic terminals. These techniques are especially important for following the amplitude and dynamics of calcium signals. In that way, optical measurements can contribute uniquely to our understanding of processes that are directly or indirectly initiated by calcium entry into the presynaptic terminal.[9–15]

This review describes the general optical methodologies for measuring $Ca^{2+}$ signals from a population of presynaptic terminals and from a single presynaptic terminal. The approach for the single-terminal measurements uses a simplified technique for the unusual case presented by the large mossy fiber presynaptic terminals that contact CA3 pyramidal neurons in the hippocampus.

## General Considerations for Recording Optical Indicators of $Ca^{2+}$ Ions

Ion-sensitive optical indicators chemically act like buffers that form complexes with their target ions. These complexes have optical properties that are different from the free indicators. The difference in spectroscopic properties forms the basis for optical measurements of ionic concentrations. Basic mechanisms and calibration of these indicators are quite similar, independent of the type of target ion. Out of the various ion species involved in neural activity, calcium receives the most attention and has the most varied roles.

The most prominent calcium-sensitive indicators are tetracarboxylic substances. These indicators are based on the calcium buffer BAPTA,[16] which was developed from the buffer ethylene glycol-bis($\beta$-aminoethyl ether)$N,N,N',N'$-tetraacetic acid (EGTA). By adding various chromophores, BAPTA was converted to a large family of fluorescent calcium indicators, with Fura-2 being the most prominent.[17]

Tetracarboxylic calcium-sensitive indicators are strongly negatively charged, which prevents them from crossing cell membranes. To measure intracellular ion concentrations, these indicators have to be either pressure injected or iontophoresed into the cells by a micropipette. Alternatively,

[9] W. G. Regehr and D. W. Tank, *Neuron* **7,** 451 (1991).
[10] W. G. Regehr and D. W. Tank, *J. Neurosci. Methods* **37,** 111 (1991).
[11] M. J. O'Donovan, S. Ho, G. Sholomenko, and W. Yee, *J. Neurosci. Methods* **46**(2), 91 (1993).
[12] R. S. Zucker, *J. Physiol.* **87,** 25 (1993).
[13] M. B. Feller, K. R. Delaney, and D. W. Tank, *J. Neurophysiol.* **76,** 381 (1996).
[14] B. L. Sabatini and W. G. Regehr, *Nature* **384,** 170 (1996).
[15] S. R. Sinha, L. G. Wu, and P. Saggau, *Biophys. J.* **72,** 637 (1997).
[16] R. Y. Tsien, *Biochem.* **19,** 2396 (1980).
[17] G. Grynkiewicz, M. Poenie, and R. Y. Tsien, *J. Biol. Chem.* **260,** 3440 (1985).

membrane-permeant acetoxymethyl ester derivatives of the tetracarboxylic indicators can be loaded into the cell from the bathing solution. These compounds have their calcium-binding carboxylic ends converted to the acetoxymethyl ester (AM) form.[18] These uncharged esters enter the cell by diffusing across the membrane, and then, cytoplasmic esterases cleave the AM revealing the negative charges. The charge tends to trap the indicator inside the cell. The most widely used membrane-permeant $Ca^{2+}$ indicator is Fura-2 AM.

*Instrumentation for Recording Optical Signals*

After loading neurons with an appropriate indicator, calcium-dependent fluorescence can be detected using epifluorescence techniques. The kind of electrical recordings that have to be performed during the optical measurements determines whether an upright or an inverted microscope is preferred.

Upright microscopes can be used to make electrical recordings from individual neurons in brain slices by using infrared differential interference contrast (DIC) optics to improve visual control while applying patch-clamp techniques.[19] With submerged preparations such as brain slices, it is necessary to use water immersion objective lenses; otherwise, movements at the air/solution interface cause optical disturbances that introduce significant noise into the signal from the ion indicator. To keep a relatively long working distance, water immersion lenses are limited in magnification ($\leq 63\times$) and numerical aperture ($\leq 0.9$). The relatively short working distance of available immersion lenses (<2 mm) requires that micropipettes must be positioned at very shallow angles (<25 deg), making electrical recordings more difficult (Fig. 1A).

Inverted microscopes offer some experimental advantages, especially for studying monolayer tissue cultured cells. Optical signals can be improved by using oil-immersion lenses with higher magnification ($\leq 100\times$) and numerical aperture ($\leq 1.4$). Furthermore, access from above for micropipettes is not limited by the objective lens. Unfortunately, inverted microscopes cannot be used for visual control in brain slices because the tissue is too opaque to look through from the bottom to a pipette or cell near the top of the slice (Fig. 1B).

Independent of the type of microscope, appropriate spectroscopic equipment has to be employed (Fig. 1). To excite a high proportion of the total small amount of fluorescent indicator that is present intracellularly, an intense illumination system is needed. Because xenon bulbs (e.g., XBO

[18] R. Y. Tsien, *Nature* **290,** 527 (1981).

[19] G. J. Stuart, H. U. Dodt, and B. Sakmann, *Pflugers. Arch.* **423,** 511 (1993).

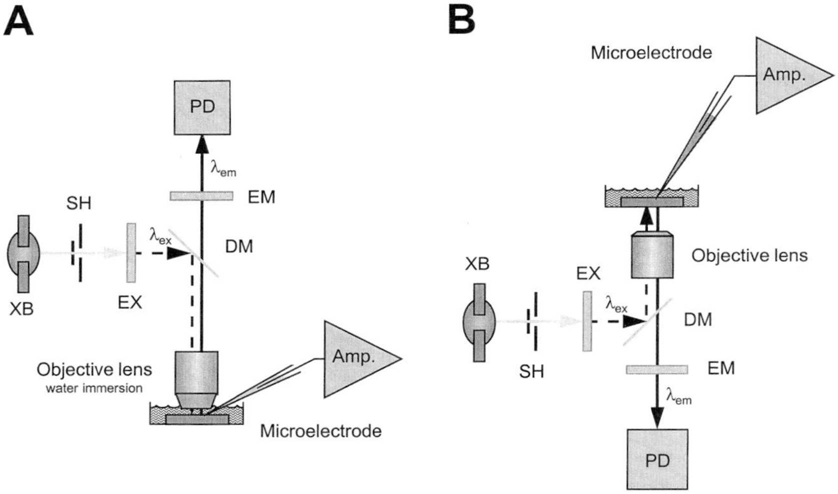


FIG. 1. Instrumentation for epifluorescence measurements. (A) Upright microscope. (B) Inverted microscope. Abbreviations: EM, emission filter; EX, excitation filter; DM, dichroic mirror; PD, photodetector; SH, shutter; XB, xenon bulb; $\lambda_{em}$, emission wavelength; $\lambda_{ex}$, excitation wavelength.

75/W2, Osram) have a more homogeneous emission spectrum and better noise performance, they are commonly preferred over mercury bulbs. Fast electromechanical shutters (e.g., Uniblitz, Vincent, Rochester, NY) are used to precisely control exposure of the preparation to excitation light, which can cause tissue damage or bleaching of the indicator. If calcium indicators require near-UV excitation (e.g., Fura-2, Indo-1, Molecular Probes, Eugene, OR), a quartz illuminator is strongly recommended because the transmission of normal lenses dramatically drops at wavelengths below 350 nm. An epifluorescence filter set that matches the indicator spectra is crucial to optimize the fluorescent signal. These filter sets consist of an excitation filter, a dichroic mirror, and an emission filter. The excitation filter is a bandpass filter that selects the part of the wide spectrum emitted from the xenon bulb that matches the absorbance of the indicator. The dichroic mirror separates excitation and emission light by reflecting light below a certain wavelength while passing light above that value. The emission filter is matched to the fluorescence spectrum of the indicator to prevent any residual excitation light from reaching the detector. Although emission filters with long-pass characteristics are often sufficient, bandpass filters also block long wavelengths outside the fluorescence spectrum. Bandpass filters are preferred because the emission detectors are commonly sensitive to infrared wavelengths.

If no spatial resolution is needed, then the calcium signal can be measured with a simple photodetector such as a silicon photodiode or a photomultiplier tube. In most cases a photodiode is preferred over a photomultiplier for two reasons. First, the quantum efficiency of a photodiode is about 90%, meaning that almost every photon received generates an electron. The quantum efficiency of a photomultiplier is <15%. The low efficiency of photomultipliers should not be confused with their high internal gain (~1000), which is useful for very weak or very fast signals. Second, photomultipliers cost about 10 times more than low-noise photodiodes (e.g., S1336-18BK, Hamamatsu, Brightwater, NJ).

Independent of the type, the photodetector is connected to a current-to-voltage converter and amplifier. In the case of a photodiode, the photocurrents are commonly in the low nanoampere range and fluorescence transients in the percent range can be expected. Therefore, similar low-noise considerations as with patch-clamp amplifiers have to be met. Indeed, a straightforward approach to obtain some preliminary optical recordings is to use a silicon photodiode connected to a patch-clamp head-stage with a 1- to 5-GΩ feedback resistor.[20]

### *Accounting for Changes in Fluorescent Signal*

When measuring fluorescence signals, it is necessary to control for long-term changes in static fluorescence as well as short-term bleaching of the indicator during each exposure to excitation light. Long-term changes in fluorescence are mostly due to accumulated bleaching, which can be minimized to the recording periods by using a fast electromechanical shutter. Residual bleaching and other alterations in indicator concentration influence measures of absolute fluorescence ($F$), but this problem can be circumvented by measuring the fractional change of fluorescence ($\Delta F/F$). This approach normalizes changes in fluorescence to the resting fluorescence. Changes in fluorescence can be obtained either by ac-coupled measurements or by calculating the difference between two dc-coupled measurements. In both cases, tissue autofluorescence has to be corrected separately.

### *Calibrating Optical Signals*

Ion-sensitive indicators bind to free ions and form a complex. Binding of free indicator [X] and free calcium [$Ca^{2+}$] and unbinding of bound calcium [CaX] are dictated by the on-rate ($k_{on}$) and off-rate ($k_{off}$), which

[20] F. J. Sigworth, *in* "Single-Channel Recording" (B. Sakmann and E. Neher, eds), pp. 3–35. Plenum Press, New York, 1983.

are $k_{on} \approx 10^8/M{\cdot}sec$ and $k_{off} \approx 20/sec$ for Fura-2. The rates of binding and unbinding are equal at equilibrium:

$$k_{on}[X][Ca^{2+}] = k_{off}[CaX]$$

This equation can be solved for the concentration of free calcium, which is what we want to measure:

$$[Ca^{2+}] = K_d \frac{[CaX]}{[X]} \tag{1}$$

where the dissociation constant ($K_d$) is approximately equal to 200 n$M$ for Fura-2.

Both the free and the bound indicator are fluorescent. Therefore, the total fluorescence ($F$) equals the sum of the fluorescence of the free indicator ($F_X = F'_X\,[X]$) and the bound indicator ($F_{CaX} = F'_{CaX}\,[CaX]$). Each component is the product of the molar fluorescence ($F'_X$, $F'_{CaX}$) times the concentration of the species ([X], [CaX]). Assume the fluorescence of the indicator dramatically increases on binding calcium. Then, the total fluorescence will be at an extreme low under calcium-free conditions ($F_{min} = F'_X[X_T]$) and will be at an extreme high when all the indicator has bound calcium (i.e., no free indicator is present, $F_{max} = F'_{CaX}[CaX] = F'_{CaX}[X_T]$), where $[X_T]$ is the total concentration of the indicator ($[X_T] = [X] + [CaX]$). The actual fluorescence ($F$) has to be corrected for these extreme values ($F_{min}$, $F_{max}$):

$$F - F_{min} = (F'_{CaX} - F'_X)[CaX]$$
$$F_{max} - F = (F'_{CaX} - F'_X)[X]$$

By rearranging and substituting in Eq. (1), the calibration equation can be written in terms of the measured values:

$$[Ca^{2+}] = K_d \frac{F - F_{min}}{F_{max} - F} \tag{2}$$

When indicators are employed that shift their fluorescence spectra on binding to calcium (e.g., the excitation spectrum of Fura-2 or the emission spectrum of Indo-1), then measurements can be made at two distinct wavelengths ($\lambda_1$, $\lambda_2$) to obtain a ratio ($R = F_{\lambda_1}/F_{\lambda_2}$).[17] This protocol corrects for variations in the concentration of indicator, the optical path length, and the instrumental parameters. Ratio measurements are also more sensitive because the calcium-dependent changes in fluorescence at both wavelengths are usually of opposite sign. For example, with Fura-2, a $[Ca^{2+}]$ increase at the lower excitation wavelength ($\lambda_1 = 340$ nm) increases the fluorescence, and a $[Ca^{2+}]$ increase at the higher excitation wavelength ($\lambda_2 = 380$ nm) decreases the fluorescence. Expanding the calibration equation [Eq. (2)] for dual-wavelength indicators gives

$$[Ca^{2+}] = K_d^* \frac{R - R_{min}}{R_{max} - R} \quad (3)$$

where $K_d^* = K_d(F_{max}/F_{min})$, $R_{min}$ is the ratio in 0 calcium, and $R_{max}$ is the ratio in a saturating concentration of calcium. $R_{min}$ and $R_{max}$ are often most accurate when they are obtained under conditions that approximate the experimental milieu.

Under some experimental conditions it is impractical to perform true ratio measurements. An alternative is to perform hybrid measurements. First, the resting calcium concentration is determined by a ratio measurement; then, the change in the calcium concentration, $\Delta[Ca^{2+}]$, is optically followed at only a single wavelength.[21] This protocol can be used to measure fast calcium signals; however, it requires invariant indicator concentration during recording (e.g., insignificant bleaching of the indicator). By differentiating the single wavelength calibration equation [Eq. (2)] and rearranging, we obtain the following result in terms of measurable quantities:

$$\Delta[Ca^{2+}] = (K_d + [Ca^{2+}]) \frac{\Delta F/F}{(F_{max} - F)/F} \quad (4)$$

where $\Delta F/F$ is the fractional change in fluorescence and $(F_{max} - F)/F = \Delta F_{max}/F$ is the maximal fractional change from resting to saturating calcium concentrations.

### *Selecting Appropriate Indicator*

A good overview of available indicators and their properties can be found in the Molecular Probes catalog.[22] Selection of an indicator should be based on at least three considerations: (1) expected range of $[Ca^{2+}]$ to be measured, (2) calibration of the absolute calcium concentration, and (3) possible range of excitation wavelengths based on available instrumentation.

[21] D. B. Jaffe, D. Johnston, N. Lasser-Ross, J. E. Lisman, H. Miyakawa, and W. N. Ross, *Nature* **357,** 244 (1992).

[22] R. P. Haugland, "Handbook of Fluorescent Probes and Research Chemicals." Molecular Probes, Eugene, Oregon, 1996.

1. The indicator must be selected so that its dissociation constant ($K_d$) matches the range of calcium concentrations to be measured. Figure 2 illustrates that near $K_d$ the change in fluorescence is linearly dependent on the calcium concentration. In presynaptic terminals the $Ca^{2+}$ concentration never falls below a resting level, but the concentration undergoes large transient rises. In that case, it is useful to choose an indicator with a $K_d$ that is about 10-fold higher than the resting concentration. If we assume a presynaptic resting calcium level of about 100 n$M$, then an indicator with a $K_d$ of about 1 $\mu M$ would allow us to monitor concentrations up to 10 $\mu M$ (Fig. 2). When selecting an indicator by its dissociation constant, one has to keep in mind that the buffer kinetics of the indicator also depend on the on-rate ($k_{on}$) and off-rate ($k_{off}$). If a 10-$\mu M$ concentration of an indicator with an on-rate of $k_{on} = 10^8/M \cdot \text{sec}$ experiences an instantaneous 5-$\mu M$ calcium concentration increase, then the time constant to reach equilibrium $Ca^{2+}$ binding is $1/\tau = k_{on}([Ca^{2+}] + [X]) + k_{off}$, yielding $\tau = 625$ $\mu s$. The corresponding time constant for decay after instantaneously removing $Ca^{2+}$ is $\tau_{off} = 50$ ms (Fig. 3).

2. An absolute calibration is best performed with indicators that allow for ratio measurements at two excitation or emission wavelengths. Otherwise, calibration curves have to be generated under identical conditions as during the experiment, meaning that instrumental parameters such as indicator concentration, optical path length, and illumination intensity have to be constant. For intracellular indicators, identical calibration and test conditions are not easily achieved and might require additional tests such

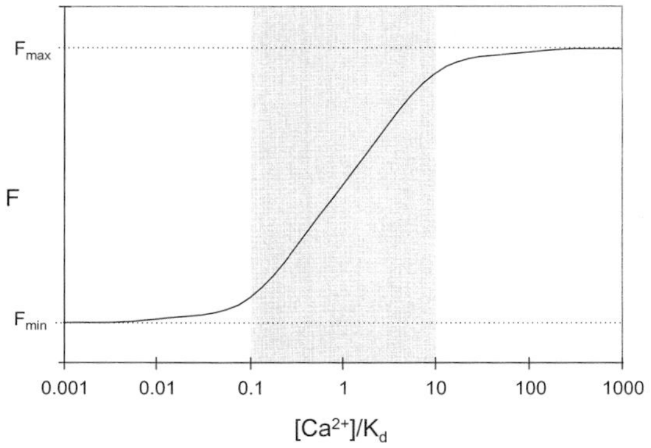


FIG. 2. Calibration curve of a fluorescent calcium buffer. The calcium concentration, $[Ca^{2+}]$, is normalized by the dissociation constant, $K_d$, in relation to the total fluorescence, $F$. Within a concentration range of 0.1–10 $K_d$, the fluorescence will change rather linearly with $[Ca^{2+}]$. Even at extreme calcium concentrations, the measured fluorescence falls within the limits set by $F_{min}$ to $F_{max}$. Note that the ordinate is a linear scale.

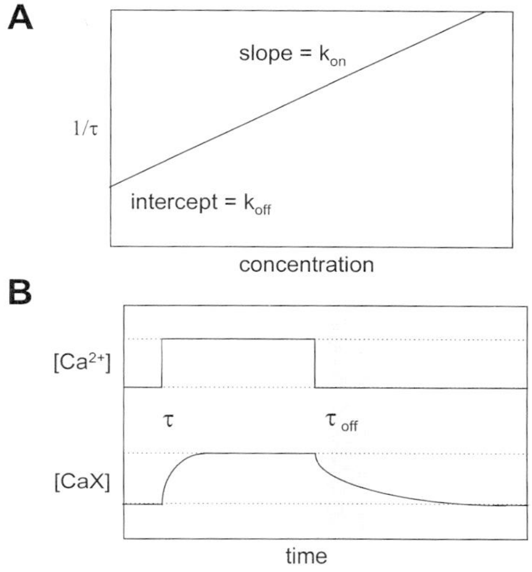


FIG. 3. Kinetic parameters of an optical indicator. (A) The on-rate, $k_{on}$, and off-rate, $k_{off}$, for $Ca^{2+}$ binding to the indicator can be estimated graphically, where the abscissa is the concentration of the rate-limiting species. (B) Time constants for formation ($\tau$) and decay ($\tau_{off}$) of [CaX] in response to a step increase and a step decrease in $[Ca^{2+}]$.

as absorbance measurements. If an absolute calibration is not necessary, then single wavelength indicators are useful and measurements of fractional fluorescence ($\Delta F/F$) are sufficient, given that an autofluorescence correction is performed.

3. The type of indicator also can be limited by the available instrumentation. For example, ratio measurements at two emission wavelengths can be made with the calcium indicator Indo-1. The excitation wavelength, however, is lower for Indo-1 ($\approx$335 nm) than for Fura-2 (340 and 380 nm). As mentioned earlier, UV excitation below 350 nm cannot be efficiently applied without quartz optics.

## Preparing Hippocampal Brain Slices for Optical Measurements

The hippocampal glutamatergic synapses between the dentate granule neurons and the CA3 pyramidal neurons and between the CA3 and the CA1 pyramidal neurons are frequently studied in brain slices prepared from young adult guinea pigs or rats.[19] Animals are anesthetized with a mixture of ketamine, acepromazine, and xylazine and, then, are quickly decapitated. Brains are immediately removed and dropped into ice-cold artificial cerebrospinal fluid (ACSF) of the following composition (in m$M$):

124 NaCl, 5 KCl, 2.5 $CaCl_2$, 1.2 $MgCl_2$, 22 $NaHCO_3$, 1.25 $NaH_2PO_4$, and 10 D-glucose. The pH of all solutions is maintained at 7.4 by constant equilibration with 95% (v/v) $O_2$ and 5% (v/v) $CO_2$. The quality of neurons from adult animals (rats > 40 days) can be improved by perfusing with an ice-cold saline just before decapitation.[23,24] The chest is opened and gravity-fed saline is perfused through the left ventricle: saline (in m$M$), 110 choline chloride, 2.5 KCl, 1.25 $NaH_2PO_4$, 25 $NaHCO_3$, 0.5 $CaCl_2$, 7 $MgCl_2$, 7 dextrose, 1.30 ascorbate, and 2.4 pyruvic acid. After dissecting the hippocampus, 200–400 $\mu$m transverse hippocampal brain slices are prepared in ice-cold ACSF on a vibrating tissue slicer and stored in ACSF. After a resting period of about 1 hr, the slices are transferred to a continuously perfused recording chamber that is temperature controlled.

## Measurement of $Ca^{2+}$ in Populations of Presynaptic Terminals

To measure an optical signal from presynaptic terminals requires that the contribution to the signal from presynaptic structures be isolated from other contributions. The most straightforward approach is to block the postsynaptic contribution to the optical signal with specific receptor antagonists. With this approach the loading procedure is simplified because a membrane-permeant indicator can be added to the bath to load the cells indiscriminately.[25,26] When required, more sophisticated techniques can be used to load a particular population of presynaptic terminals in order to perform studies with intact synaptic transmission.

### *Loading Technique for Populations of Presynaptic Terminals*

Loading all structures by applying a membrane-permeant indicator to the bath can be sufficient for some studies. For greater specificity, however, it is necessary to load only a portion of the presynaptic neurons in one area of the hippocampus. In that case, local perfusion[10] or extracellular injection[27,28] of a membrane-permeant calcium indicator (e.g., Fura-2 AM) onto a presynaptic pathway results in local uptake into axons. After the AM ester is cleaved intracellularly, the indicator ideally is trapped and will eventually reach the presynaptic terminals of interest.

The following procedure describes the Fura-2 loading of presynaptic

[23] G. C. Newman, F. E. Hospod, H. Qi, and H. J. Patel, *Neurosci. Methods* **61,** 33 (1995).
[24] J. C. Magee and D. Johnston, *Science* **275,** 209 (1997).
[25] P. Saggau, R. D. Sheridan, and A. Ogura, *Eur. J. Neurosci.* **Suppl. 1,** 14.5 (1988).
[26] G. Hess and U. Kuhnt, *Neuroreport* **3,** 361 (1992).
[27] P. Saggau, L. G. Wu, and I. Yehezkely, *Biophys. J.* **61,** 2941 (1992).
[28] L. G. Wu and P. Saggau, *J. Neurosci.* **14,** 645 (1994).

terminals from CA3 neurons that form synapses onto CA1 neurons. With minor modifications, this procedure can be employed with other indicators or different preparations. We routinely use submerged brain slices, but this is not a prerequisite for this approach. Fura-2 AM (50 μg, Molecular Probes, Eugene, OR) is dissolved in 5 μl dimethyl sulfoxide (DMSO) containing 10% Pluronic acid (Molecular Probes); 50 μl ACSF is added, and the mixture is sonicated for about 1 min. The final solution contains about 1 m*M* Fura-2 AM, 10% DMSO, and 1% Pluronic acid. This solution is left in the dark at room temperature for about 1 hr before use. A fire-polished (patch) pipette with a tip diameter of about 2 μm is filled with this solution and connected to a pressure injection device (e.g., Picospritzer II, General Valve, Fairfield, NJ). Under visual control, with a low magnification lens or a dissection microscope, the tip of the pipette is positioned in the *stratum radiatum* (SR) of area CA1 and lowered to 200–250 μm below the surface of a 400-μm-thick hippocampal slice. A small amount (<1 μl) of Fura-2 AM solution is pressure injected into the extracellular space; then, the pipette is withdrawn. The success of the injection can be monitored with low-intensity illumination. A small fluorescent spot should be visible at the injection site throughout the experiment. Caution is required because it appears that fluorescent calcium indicators are particularly susceptible to bleaching during the loading phase. Thus, visual inspection during the loading phase with strong excitation light must be kept at a minimum. Fura-2 AM will locally fill axons, and the trapped indicator will diffuse to the nerve terminals within 1–2 hr (Fig. 4).

### *Recording Optical Signals from Populations of Presynaptic Terminals*

After loading a brain slice preparation following the above protocol, fluorescence emerging from presynaptic terminals filled with calcium indicator can be detected using epifluorescence techniques. We have used an inverted microscope with an oil-immersion objective lens (e.g., 50×, NA 0.9, Zeiss, Thornwood, NY), but an upright microscope with a water-immersion lens (e.g., 63×, NA 0.9, Zeiss) also can be used. For simultaneous patch-clamp recordings under visual control,[19] an upright microscope with DIC optics and a longer working-distance lens (e.g., 40×, NA 0.75) is needed. When recording from CA3 presynaptic terminals in the CA1 region of the hippocampus, fluorescence is detected from a limited area with a 150-μm diameter in SR. This area is about 500 μm away from the injection site, and it contains many synapses (Fig. 4). When calcium signals from a presynaptic terminal population are to be recorded without distinguishing individual terminals, a simple photodetector without spatial resolution is sufficient. The presynaptic origin of calcium signals can be verified

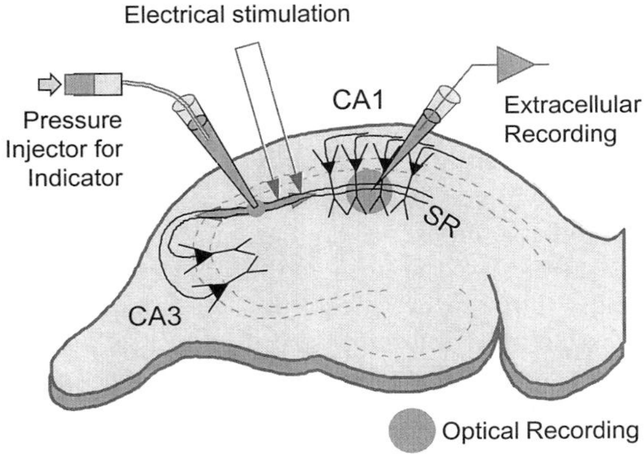


FIG. 4. Loading of a population of presynaptic terminals. Pressure is used to inject a small amount of calcium indicator in its AM form into *stratum radiatum* of the CA3, where the axons are located on their way to the CA1 region. The indicator that is taken up into the axons diffuses both anterogradely and retrogradely, eventually filling CA3 presynaptic terminals onto CA1 pyramidal cells. The positions of the optical recording, the stimulation electrode, and the electrical recording electrode are shown.

by comparing recordings under test conditions with recordings obtained under control conditions with postsynaptic transmission blocked (Fig. 5).

*Calibration Technique for Populations of Presynaptic Terminals*

In most cases, it is sufficient to record transient changes in presynaptic calcium that are related to synaptic transmission. As can be seen in Fig. 2, the change in fluorescence is quite linear within 100-fold change of the calcium concentration. To compare measurements, fluorescence transients have to be normalized for the static fluorescence after being corrected for the autofluorescence of the preparation. These fluorescence transients are expressed as $\Delta F/F$. The amplitude of the Ca transient is measured as the difference between the maximal concentration and the resting concentration. Corrections for bleaching during the recording are as shown in Fig. 6.

If the absolute $[Ca^{2+}]$ must be determined, then the optical signals can be best calibrated by forming the ratio ($R$) between two consecutive fluorescence recordings taken at two different wavelengths (e.g., 340/380 nm excitation with Fura-2). Care must be taken, however, that fluorescence intensities do not change significantly during these measurements, which could happen due to bleaching of the indicator under the intense illumination. It is best to bracket the measurements: measure at 380 nm, next at 340 nm and, then, back at 380 nm. The extreme values, $R_{min}$ and $R_{max}$, can

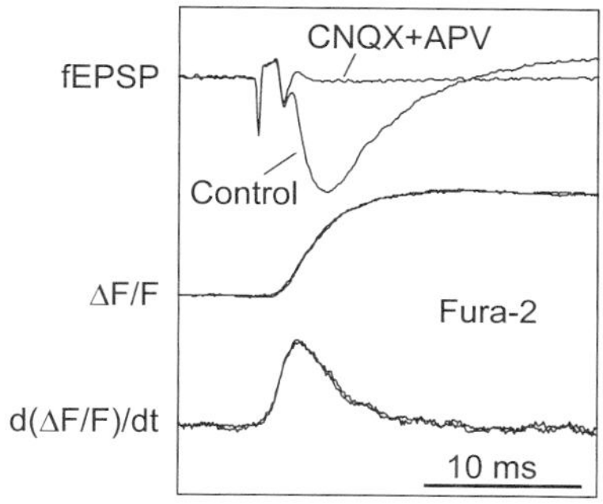


FIG. 5. Optical calcium signals from a population of presynaptic terminals. Single action potentials were elicited with a bipolar electrode in the CA3 axons. The calcium signals ($\Delta F/F \propto [Ca^{2+}]$) were optically recorded in the area of CA1 pyramidal cell dendrites together with field EPSPs (Control). During blockade of synaptic transmission (with CNQX + APV), only the presynaptic action potential can be detected while the calcium signal remains unchanged (the two signals overlay). The calcium influx into presynaptic terminals is illustrated by taking the derivative of the calcium signal ($d(\Delta F/F)dt \propto I_{Ca}$). Note that the calcium signals during intact and blocked presynaptic transmission are superimposed.

be determined using a set of solutions with known calcium concentrations (e.g., calcium calibration buffer kit II, Molecular Probes) to which constant amounts of free calcium indicator are added. The background fluorescence of the brain slice has to be measured before the injection of Fura-2 AM and corrected before computing ratios. Calibrated signals can be computed by using Eq. (3).

A more practical way to obtain calibrated signals is to employ hybrid measurements. To do this, the resting calcium level is obtained first, by means of a ratio measurement as described above. The experimental $Ca^{2+}$ transient is then simply and rapidly followed with a single excitation wavelength (usually 380 nm for Fura-2). The maximal change in fluorescence ($F_{max}$) is obtained later at the same excitation wavelength (e.g., 380 nm) by applying a saturating calcium concentration. Then fractional changes in fluorescence ($\Delta F/F$) can be obtained, and a calibration of the signal can be computed by using Eq. (4).

## Measurement of $Ca^{2+}$ in a Single Presynaptic Terminal

To describe measurements from a single presynaptic terminal, we will focus on the special case presented by mossy fiber terminals in the CA3 region of the hippocampus. Axonal projections from granule cells of the dentate gyrus end in mossy fiber terminals (MFTs) that form synapses onto CA3 pyramidal neurons in the hippocampus. These presynaptic terminals

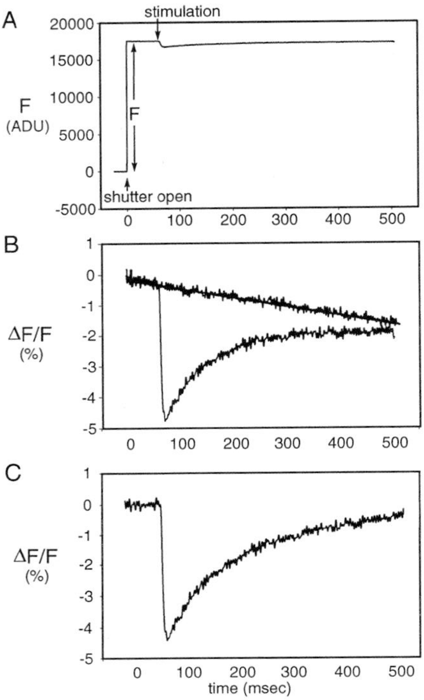


FIG. 6. (A) Uncorrected photodiode electrical output from a mossy fiber terminal (MFT) filled with Fura-2. A large increase in fluorescence (F) is measured when the shutter opens and the MFT is illuminated at 380 nm. Stimulation of the granule cell axons (downward arrow) causes a decrease in the Fura-2 signal, indicating an increase in intraterminal [$Ca^{2+}$]. The baseline autofluorescence level (before opening the shutter) is set to zero while focusing on an area of the slice adjacent to the filled MFT. (B) $\Delta F/F$ is calculated from the trace after shutter opening. A bleaching trace is taken without stimulation, fit with a straight line, and subtracted from the traces during stimulation. (C) The $\Delta F/F$ trace is shown again after correcting for bleaching and autofluorescence. In this case the fluorescence change is plotted as the data were obtained, but some researchers invert the signal at 380 nm, showing an upward deflection to indicate an increase in [$Ca^{2+}$].

have several unique features that make them especially amenable to synaptic studies. The features that are most important for the methods described here are that they are large in diameter (3–8 $\mu$m), and they are localized on the proximal apical dendrites of the CA3 neuron cell bodies. Because of their large size and location, they are easily identified with a light microscope, and their large volume can hold sufficient dye to provide measurable optical signals from individual presynaptic terminals.

*Loading Technique Applicable for Mossy Fiber Presynaptic Terminals*

Three methods were tested to load mossy fiber terminals with Fura-2: the local-perfusion method as described by Regehr and Tank[10]; the local injection method described by Wu and Saggau[28]; and a third simple method requiring only brief incubation of the whole slice in Fura-2 AM described by Gray *et al.*[29] The third method has some advantages: it is faster and much easier than the other methods, and it loads most MFTs well. The first two methods also loaded MFTs, but by their nature fewer MFTs were loaded, greatly reducing the number of potential MFTs for recording.

A stock solution of Fura-2 AM (Molecular Probes) is made by dissolving 50 $\mu$g of Fura-2 AM in 40 $\mu$l of DMSO with 20% Pluronic F-127 (Molecular Probes). This stock solution is added to 3 ml of oxygenated ACSF, which is then sonicated for 10 sec at low power (Kontes 9110001 ultrasonic cell disruptor, Vineland, NY) and spun for 15 sec in a low-speed benchtop centrifuge (Beckman Microfuge, Palo Alto, CA).

Individual slices are then removed from the holding chamber and placed in a 35-mm petri dish containing about 3 ml of ACSF with 8.3 $\mu M$ of Fura-2 AM. The slice is incubated in the solution at 30°C for 10–15 min. The slice should then be rinsed gently with (at least) 10 volume changes of ACSF before being transferred to the experimental chamber on the microscope.

The loading procedure preferentially loads 3- to 8-$\mu$m-diameter structures in *stratum lucidum,* consistent with the size and location of MFTs. Longer incubation times will also begin to load granule cell bodies in the dentate gyrus and diffuse structures in the hilus. Incubation times of up to 2 hr do not strongly label CA3 pyramidal neurons because they do not seem to load well with Fura-2 AM.

Using fluorescent illumination, MFTs can be located in the narrow band along the proximal apical dendrites of the CA3 neurons. Alternating between viewing the brightly stained MFTs with a silicon intensified target (SIT) camera (Dage-MTI SIT 6 GLX, Michigan City, IN) and the DIC image with a Newvicon camera (Hamamatsu 2400) allows candidate MFTs to be chosen for optical recording.

*Recording Optical Signals from a Single Mossy Fiber Terminal*

After a brain slice has been loaded with indicator, calcium signals from single MFTs can be recorded by means of epifluorescence techniques. We have used an upright microscope with a water-immersion objective lens (e.g., 40×, NA 0.75, Zeiss). The fluorescence signal from the $Ca^{2+}$ indicator is directed through a filter (e.g., 480-nm long-pass filter) to the camera

[29] R. Gray, A. S. Rajan, K. A. Radcliffe, M. Yakehiro, and J. A. Dani, *Nature* **383,** 713 (1996).

port on the microscope to a photodiode detector (e.g., Hamamatsu S1336-18BK). A modified field diaphragm also is valuable to limit excitation illumination only to a circle 10–15 $\mu$m in diameter near the single MFT. In the simplest case, the fluorescence signal can be monitored only with 380-nm excitation. The waveform is determined as $\Delta F/F$, calculated from the raw fluorescence signal ($F$) after correcting for the background autofluorescence.[10,29]

Changes in presynaptic $[Ca^{2+}]$ can be measured during electrical stimulation of the mossy fiber pathway. Presynaptic stimulation can be induced with a bipolar tungsten stimulating electrode connected to a constant-current stimulus isolator placed in the upper blade of the granule cell body layer. Single stimuli (100-$\mu$s duration) are applied and the optical signal is recorded simultaneously. To be sure that only a presynaptic signal is being monitored, postsynaptic glutamate receptors can be inhibited with APV (100 $\mu M$) and CNQX (20 $\mu M$).

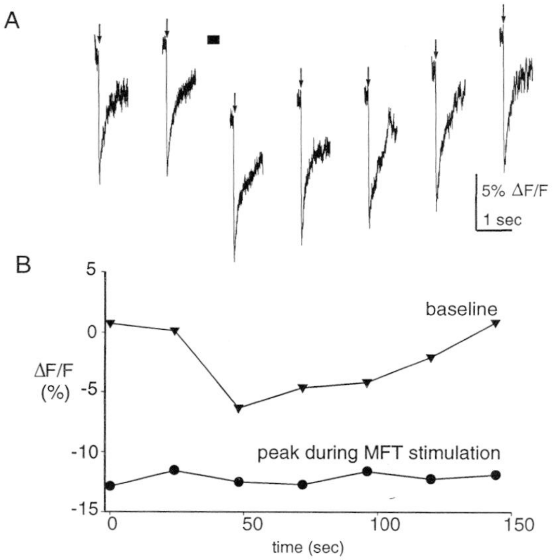


FIG. 7. (A) Time-compressed optical signals before, during, and after a 5-sec puffer application of 20 $\mu M$ nicotine (solid bar) onto a Fura-2 loaded MFT. The actual time between each trace was 24 sec. Each rapid downward deflection is the optical response to electrical stimulation of the mossy fiber pathway. Traces were corrected for background fluorescence and dye bleaching. All electrolyte solutions contained APV (100 $\mu M$) and CNQX (20 $\mu M$) to block postsynaptic glutamate receptors. (B) Summary of the changes in baseline fluorescence (solid triangles) that are due to synaptic stimulation (solid circles). The results indicate that nicotine caused a transient increase in the baseline intraterinal $[Ca^{2+}]$, but had little effect on the average $Ca^{2+}$ entry during synaptic stimulation.

### *Calibration Technique for Single Presynaptic Terminal*

All optical traces collected with the photodiode should be corrected for autofluorescence and bleaching. The autofluorescence (in this case the intrinsic optical signal given off by the hippocampus) can be measured in an area of the slice away from stained MFTs and subtracted from each optical signal trace. Bleaching is a time-dependent decrease in the optical signal during illumination. Traces should be corrected for bleaching by fitting a straight line to traces collected without stimulation of the mossy fiber pathway and, then, subtracting those values from each trace (Fig. 6). If the bleaching trace cannot be described well by a straight line, multiple bleaching traces (16–64) should be collected, averaged, low-pass filtered, and subtracted from each data trace.

Two measurements should be made from each trace of optical data: the baseline fluorescence, corresponding to the "resting" $Ca^{2+}$ level, and the transient due to $Ca^{2+}$ entry through voltage-gated Ca channels as the action potential invades the terminal. An example of an optical recording from a single MFT that made use of the approach described is shown in Fig. 7. Calcium influx into the MFT was evoked by electrical stimulation and by application of $<20\ \mu M$ nicotine.[29] Control experiments that blocked the action potential, using tetrodotoxin ($\approx 0.5\ \mu M$) and $Cd^{2+}$ ($\approx 200\ \mu M$), eliminated the electrically evoked $Ca^{2+}$ signal.

## Comments

Calcium plays many roles in neuronal mechanisms, from directly activating neurotransmitter release to initiating many forms of modulation and plasticity. In many cases, optical approaches offer the best or only alternative for monitoring $Ca^{2+}$ signals without significantly disrupting local biological events. A further experimental strength is gained when optical measurements can be combined with patch-clamp electrophysiology to control and measure electrical properties. Because of the experimental advantages and the increased availability of indicators, optical approaches have come to provide valuable methods for dissecting the details of neuronal signaling.

## Acknowledgments

This work was supported by NIH grants from NINDS (NS21229, NS33147 and NS29871) and from the NIDA (DA09411).

# [2] Fluorescence Techniques for Measuring Ion Channel Activity

*By* GÖNÜL VELIÇELEBI, KENNETH A. STAUDERMAN, MARK A. VARNEY, MICHAEL AKONG, STEPHEN D. HESS, and EDWIN C. JOHNSON

## Introduction

Changes in intracellular free calcium concentration ($[Ca^{2+}]_i$) play a crucial role in cellular physiology. A number of cell surface receptors and channels are known to regulate $[Ca^{2+}]_i$ through different molecular mechanisms. Therefore, the functional and pharmacologic properties of many of these cell surface receptors and ion channels can be studied effectively by measuring changes in $[Ca^{2+}]_i$ in intact cells. For our drug discovery efforts, we have targeted several ion channel and receptor systems that play different roles in neuronal physiology and pathophysiology. These molecular targets include voltage- and ligand-gated ion channels: the human neuronal voltage-gated calcium channels (VGCCs),[1–4] ligand-gated nicotinic acetylcholine receptor channels (NAChRs),[5,6] ionotropic *N*-methyl-D-aspartic acid (NMDA),[7,8] $\alpha$-amino-3-hydroxy-5-methyl-4-isoxazolepropionic acid (AMPA)- and kainate-type excitatory amino acid receptor (EAA) channels. All of these channels mediate elevation of $[Ca^{2+}]_i$ via $Ca^{2+}$ influx from the extracellular medium upon depolarization or activation by agonist.

[1] M. E. Williams, D. H. Feldman, A. F. McCue, R. Brenner, G. Veliçelebi, S. B. Ellis, and M. M. Harpold, *Neuron* **8,** 71 (1992a).

[2] M. E. Williams, P. F. Brust, D. H. Feldman, S. Patthi, S. Simerson, A. Maroufi, A. F. McCue, G. Veliçelebi, S. B. Ellis, and M. M. Harpold, *Science* **257,** 389 (1992b).

[3] P. F. Brust, S. Simerson, A. F. McCue, C. R. Deal, S. Schoonmaker, M. E. Williams, G. Veliçelebi, E. C. Johnson, M. M. Harpold, and S. B. Ellis, *Neuropharmacology* **32,** 1189 (1993).

[4] M. E. Williams, L. M. Marubio, C. R. Deal, M. Hans, P. F. Brust, L. H. Philipson, R. J. Miller, E. C. Johnson, M. M. Harpold, and S. B. Ellis, *J. Biol. Chem.* **269,** 22347 (1994).

[5] K. J. Elliott, S. B. Ellis, K. J. Berckhan, A. Urrutia, L. E. Chavez-Noriega, E. C. Johnson, G. Veliçelebi, and M. M. Harpold, *J. Mol. Neurosci.* **7,** 217 (1996).

[6] L. E. Chavez-Noriega, J. H. Crona, M. S. Washburn, A. Urrutia, K. J. Elliott, and E. C. Johnson, *J. Pharmacol. Exp. Ther.* **280,** 346 (1997).

[7] S. D. Hess, L. P. Daggett, J. Crona, C. Deal, C.-C. Lu, A. Urrutia, L. Chavez-Noriega, S. B. Ellis, E. C. Johnson, and G. Veliçelebi, *J. Pharmacol. Exp. Ther.* **278,** 808 (1996).

[8] M. A. Varney, C. Jachec, C. Deal, S. D. Hess, L. P. Daggett, R. Skvoretz, M. Urcan, J. H. Morrison, T. Moran, E. C. Johnson, and G. Veliçelebi, *J. Pharmacol. Exp. Ther.* **279,** 367 (1996).




0076-6879/99 $30.00

Our strategy for drug discovery starts with the cloning of cDNAs encoding the specific subtypes of the targeted human ion channels and proceeds to stable expression of functional channels in mammalian host cells, thereby generating subtype-specific cellular targets for drug screening. To facilitate the identification of novel subtype-selective ligands, we sought to develop functional cell-based assays that can detect both competitive agonists and antagonists as well as allosteric activators and inhibitors. First, it was necessary to ascertain that the activity of the targeted receptors and channels in the functional assay displayed the expected pharmacologic profile observed in native systems. Therefore, the assay had to be sensitive and rapid enough to detect the activation or inhibition of ligand- and voltage-gated channels as well as G-protein-coupled receptors. Since the experimental protocol was to be performed with intact cells, it was also desirable that the assay be compatible with testing cells plated on 96-well microtiter dishes. Finally, in order to be a valuable screening tool, the functional assay had to be adaptable to automation so that the assay could be carried out with a higher throughput.

One of the functional assays we have established that meets these criteria measures changes in $[Ca^{2+}]_i$. Calcium-sensing fluorescent dyes offer the most facile means of monitoring changes in $[Ca^{2+}]_i$ both in cell populations and in individual cells. The ability to measure $[Ca^{2+}]_i$ in real time without disrupting cells represents a significant technical advance in cell biology pioneered and led by Roger Tsien.[9] The design, synthesis, and characterization of these fluorescent reagents have been described and reviewed extensively.[10,11] The most commonly used calcium-sensitive dyes, Fluo-3, Fura-2, and Indo-1, are structural analogs of the highly selective calcium chelators ethylene glycol-bis($\beta$-aminoethyl ether) N,N,N′,N′-tetraacetic acid (EGTA) and 1,2-bis(2-aminophenoxy) ethane-N,N,N′,N′-tetraacetic acid (BAPTA). Typically, the acetoxymethyl ester form of the dye readily permeates into the cell, wherein the ester group is hydrolyzed by intracellular esterases, generating the calcium-sensitive form of the free dye that is trapped inside the cell. Fura-2 and Fluo-3 have different affinities for calcium as well as different absorption and fluorescence properties. Fluo-3 is a single excitation wavelength dye, and on binding $Ca^{2+}$ ($K_d$ = 390 n$M$ at pH 7.2, 22°)[12] undergoes more than a 100-fold increase in fluorescence without a change in excitation/emission spectra.[12,13] The unbound dye is

[9] R. Y. Tsien, *Nature* **290,** 527 (1981).

[10] R. Y. Tsien, *TINS* **11,** 419 (1988).

[11] R. Y. Tsien, *Methods Cell Biol.* **30,** 27 (1989).

[12] R. P. Haugland, *in* "Handbook of Fluorescent Probes and Research Chemicals" (M. T. Z. Spence, ed.). Molecular Probes, Inc., Eugene, Oregon, 1996.

[13] J. P. Y. Kao, A. T. Harootunian, and R. Y. Tsien, *J. Biol. Chem.* **264,** 8179 (1989).

almost nonfluorescent whereas $Ca^{2+}$-bound Fluo-3 absorbs and fluoresces in the visible range ($\lambda_{Ex}$ = 506 nm and $\lambda_{Em}$ = 526 nm).[12,14] Fura-2 is a ratiometric dye that undergoes a shift in its excitation spectrum on binding $Ca^{2+}$ ($K_d$ = 145 n$M$ at pH 7.2, 22°).[12] As determined in cuvette-based measurements, the excitation maximum ($\lambda_{Ex}$) shifts from 362 nm for the unbound dye to 335 nm for the calcium–dye complex with minimal change in its emission maximum ($\lambda_{Em}$) at 510 nm.[12,15] Because the excitation peak at a saturating concentration of $Ca^{2+}$ exhibits an apparent red shift in microscope-based measurements in Fura-2-loaded cells,[16] the shift in excitation maximum is typically monitored by excitation at 350 and 385 nm, and the ratio of fluorescence resulting from these two excitations is used to correct for variations in dye concentration, cell thickness, and cell number.

We have used both Fluo-3 and Fura-2 to develop sensitive and rapid assays to measure changes in $[Ca^{2+}]_i$ in cells expressing recombinant ion channels and receptors. We describe here the experimental methods with particular emphasis on the validation of the assay for human VGCCs, NAChRs, and NMDA receptors.

## Experimental Procedures

### *Reagents and Instrumentation*

Fluo-3-acetoxymethyl ester (Fluo-3 AM) and fura-2-acetoxymethyl ester (Fura-2 AM) are purchased from Molecular Probes, Inc. (Eugene, OR). The stock solutions of Fluo-3 AM and Fura-2 AM are first prepared in dimethyl sulfoxide (DMSO) and Pluronic F127, then diluted in HEPES-buffered saline (HBS: 125 m$M$ NaCl, 5 m$M$ KCl, 0.62 m$M$ $MgSO_4$, 1.8 m$M$ $CaCl_2$, 20 m$M$ HEPES, 6 m$M$ glucose, pH 7.4) to give final concentrations of 1.0% DMSO and 0.07% Pluronic F127.

The Fluo-3 fluorescence measurements are performed at 0.33-sec intervals using a fluorimeter capable of reading a 96-well microtiter plate one well at a time (Cambridge Technology, Inc., Watertown, MA). The fluorimeter is equipped with an excitation bandpass filter of 485 ± 20 nm, and an emission bandpass filter at 530 ± 20 nm. Solutions are added to each well either manually with a pipette or automatically with a Digiflex (Titertek, Huntsville, AL) dispenser. For manual additions, the lid of the fluorimeter is briefly opened to allow access to the microtiter plate; hence, approximately

[14] A. Minta, J. P. Y. Kao, and R. Y. Tsien, *J. Biol. Chem.* **264,** 8171 (1989).
[15] G. Grynkiewicz, M. Poenie, and R. Y. Tsien, *J. Biol. Chem.* **260,** 3440 (1985).
[16] P. A. Negulescu and T. E. Machen, *Methods Enzymol.* **192,** 38 (1990).

a 5-sec gap is introduced into the corresponding fluorescence recording after agonist addition (e.g., Fig. 8; see later section). Automatic additions are performed by a computer-controlled Digiflex apparatus equipped with a dispensing line of polyethylene tubing that is positioned directly above the microtiter well. Additions by the Digiflex facilitate automation of the assay and eliminate the gaps in fluorescence recordings, which can be important for extremely rapid responses.

The Fura-2 fluorescence measurements are performed at 0.25-sec intervals per dual excitation (ratio pair) using a customized plate-imaging fluorimeter capable of recording the fluorescence output from all 96 wells simultaneously [SIBIA Neurosciences/Science Applications International Corporation (SIBIA/SAIC), San Diego, CA]. The fluorimeter is equipped with two excitation bandpass filters of 350 ± 15 nm and 385 ± 15 nm and an emission filter of 535 ± 50 nm. The solutions are added simultaneously to all wells of the 96-well microplate by an integrated 96-channel pipettor (Carl Creative Systems, Inc., Harbor City, CA). All washing and aspiration procedures are carried out using a 96-well automatic plate washer (Bio-Tek Instruments, Inc., Winooski, VT). Liquid and plate handling are performed in automated fashion using a Zymark robot (Model ZB0021, Zymark Corporation, Hopkinton, MA).

### *Measurement of $[Ca^{2+}]_i$ by Plate-Based Fluo-3/Fura-2 Assay*

The following assay protocol was optimized for use with the transfected human embryonic kidney (HEK293 cells) described here. Adaptation to other cell lines may require optimization of several obvious experimental parameters, such as cell number, dye concentration, dye loading time and temperature, cell washing conditions, and instrument settings such as gain and integration time.

1. The cells are plated on poly(D-lysine)-coated 96-well microtiter plates at a density of 1–2 × $10^5$ cells/well. Twenty-four hours after plating, the culture medium is aspirated and the cells washed twice with 250 $\mu$l HBS. Subsequently, 200 $\mu$l HBS is added to each well and background fluorescence ($F_{bkg}$) recorded.

2. The buffer is removed and the cells are incubated with 30–100 $\mu$l of 20 $\mu M$ Fluo-3 AM for 1–2 hr at 20°, or 1 $\mu M$ Fura-2 AM for 1 hr at room temperature. For Fluo-3, we have calculated that these loading conditions typically result in a maximal fluorescence value ($F_{max}$) of approximately 15,000 arbitrary units compared to a background of 1000 units. It has previously been shown that if cells are loaded with a $Ca^{2+}$-sensitive dye (e.g., Fura-2 or Fluo-3) to achieve a final intracellular concentration of

approximately 100 $\mu M$, a good fluorescence signal can be obtained without excessively buffering cytosolic $Ca^{2+}$.[17]

3. Unloaded dye is removed by aspiration and each well is washed with 250 $\mu$l HBS. Next, HBS (180 $\mu$l) is added to each well and following a 2-min recovery period, the basal fluorescence ($F_b$) is recorded for 3 sec. The recovery period should be less than 10 min to minimize time-dependent extrusion of the dye from the cells. Next, 20 $\mu$l test drug solution is added to each well, and the fluorescence ($F_{res}$) recorded for 40–60 sec (e.g., Figs. 1, 5, and 8) at 0.33- or 0.25-sec intervals for Fluo-3 or Fura-2, respectively. This test is carried out to determine if the drug has intrinsic agonist properties.

4. Two minutes after the agonist test, $F_b$ is recorded for another 3 or 5 sec before adding a reference agonist to each well and recording $F_{res}$ for another 40–60 sec. This test is performed to determine whether the test compound has antagonist properties, because an antagonist will reduce the response to the reference agonist in this test.

5. For Fluo-3 measurements, at the end of the agonist and antagonist tests, $F_{max}$ is determined by adding Triton X-100 to each well to a final concentration of 0.1%. The contents of the wells are mixed by drawing the contents of the well up and down in a pipette tip five times. After an incubation of approximately 2–5 min, $F_{max}$ is recorded. To obtain the minimum fluorescence ($F_{min}$), $MnCl_2$ is added to a final concentration of 10 m$M$. Again, the contents of the wells are mixed, and the $F_{Mn}$ value measured after 2 min. For Fura-2 measurements, empirically determined calibration values are used (see below).

6. In those instances where activation of a cell line elicits a very large $[Ca^{2+}]_i$ signal, the peak of fluorescence ($F_{res}$) can be higher than the $F_{max}$ determined after lysis of the cells with Triton X-100. In part, because the fluorimeter detects the fluorescence emission most efficiently from the bottom of the cell plate, the signal from a layer of cells at the bottom of the plate can be larger than that from a column of cell lysate following solubilization with Triton X-100. We have found that a cocktail of ionomycin/carbonyl cyanide p-(trifluoromethoxy)phenyl hydrazone (FCCP)/carbachol can be used with HEK293 cells to obtain a more precise estimate of $F_{max}$ and $F_{Mn}$. In this protocol, $F_{max}$ was recorded after the cells were incubated for 2–4 min in HBS, pH 9.0, containing 21.8 m$M$ $CaCl_2$, 5 $\mu M$ ionomycin, 20 $\mu M$ FCCP, and 100 $\mu M$ carbachol. The higher pH and $CaCl_2$ are used to optimize the activity of ionomycin, whereas FCCP and carbachol are added to release $Ca^{2+}$ from intracellular stores and to reduce energy-dependent intracellular buffering. The $F_{Mn}$ values are

[17] E. Neher, *Neuropharmacology* **34,** 1423 (1995).

determined using the same $MnCl_2$ quench procedure described in step 5, except that another 2- to 4-min incubation period is included after addition of the $MnCl_2$.

*Data Analysis*

1. The peak and basal $[Ca^{2+}]_i$ concentrations for the Fluo-3 results were calculated as described by Kao *et al.*[13] First, $F_{bkg}$ was subtracted from $F_{res}$, $F_{Mn}$, and $F_{max}$. Next, $F_{min}$ was calculated according to Eq. (1):

$$F_{min} = F_{Mn}/8 \tag{1}$$

Finally, the corrected fluorescence values were used in Eq. (2):

$$[Ca^{2+}] \text{ (in n}M) = K_d[(F_{res} - F_{min})/(F_{max} - F_{res})] \tag{2}$$

where $K_d$ is 390 n$M$ for Fluo-3.[12]

2. For Fura-2, the peak and basal $[Ca^{2+}]_i$ concentrations were calculated as described by Grynkiewicz *et al.*[15] using Eq. (3):

$$[Ca^{2+}] \text{ (in n}M) = K_d\,[(R_{res} - R_{min})/(R_{max} - R_{res})](S_{385}) \tag{3}$$

where $K_d$ is 145 n$M$ for Fura-2[12]; $R_{min}$ and $R_{max}$ represent the fluorescence ratios ($F_{350}/F_{385}$) determined as described above for $F_{min}$ and $F_{max}$, and $S_{385}$ represents the ratio of $F_{385}$ for unbound dye to $F_{385}$ for $Ca^{2+}$-saturated dye.

For our instrument, the $R_{min}$ and $R_{max}$ values were empirically determined from calcium standards as 0.11 and 2.90, respectively. More specifically, two plates were prepared containing, in each well, 180 $\mu$l of either 10 m$M$ EGTA or 100 $\mu M$ $CaCl_2$ in 10 m$M$ $Na_2HPO_4$, pH 7.5. Fura-2 (free acid) was added to all wells to a final concentration of 0.1 $\mu M$, and the $F_{350}$ and $F_{385}$ values were recorded for all 96 wells in both plates. The $R_{min}$ and $R_{max}$ values were determined from the ratio of average $F_{350}$ to average $F_{385}$ for the EGTA and $CaCl_2$ plates, respectively. The $S_{385}$ value was determined from the ratio of average $F_{385}$ on the EGTA plate to average $F_{385}$ on the $CaCl_2$ plate.

3. For agonist-evoked responses, all fluorescence determinations were performed in four replicate wells. The receptor-mediated $[Ca^{2+}]_i$ responses were quantitated by calculating either the ratio of peak $[Ca^{2+}]_i$ after drug addition to the basal $[Ca^{2+}]_i$ prior to drug addition (P/B), or by calculating the difference between peak $[Ca^{2+}]_i$ and basal $[Ca^{2+}]_i$ (P − B).

*Development of Stable Cell Lines Expressing Recombinant Human Ion Channels*

The stable cell lines expressing VGCCs, NAChRs, and NMDA receptors were established using protocols similar to those described for human metabotropic receptors.[18,19] Briefly, using the calcium phosphate coprecipitation method,[20] host cells, typically HEK293 cells, were transfected with cDNA expression plasmids encoding one or more individual subunits of human VGCCs, nicotinic, NMDA, AMPA, or kainate receptors. A plasmid encoding the neomycin gene (e.g., pSV2*neo*) was also introduced into the cells to serve as a selectable marker. Two days later, the transfected cells were selected in Dulbecco's modified Eagle's medium containing 10% dialyzed fetal bovine serum, 2 m*M* L-glutamine, 100 U/ml penicillin G, 100 $\mu$g/ml streptomycin, and 500 $\mu$g/ml G418. All tissue culture reagents were obtained from GIBCO-BRL (Grand Island, NY). The G418-resistant cells were cloned by two rounds of limiting dilution, and the clones were identified using the $[Ca^{2+}]_i$ assay. The clones exhibiting the most robust $[Ca^{2+}]_i$ responses were characterized further in more detailed pharmacologic studies. The stability of the expression of the recombinant receptor or ion channel in the selected clonal cell line was ascertained by monitoring the magnitude and the pharmacology of the $[Ca^{2+}]_i$ response for 30–40 passages in culture. In addition, we have also expressed most of these channels transiently in HEK293 cells and detected robust $[Ca^{2+}]_i$ signals on activation. An example with the NAChRs is shown later in Fig. 12.

### Pharmacologic Validation of Assay

The fluorescent dye-based $[Ca^{2+}]_i$ assay was validated in each target receptor system before it was utilized in drug screening. To this end, cell lines stably expressing specific subtypes of the different ion channels were characterized pharmacologically using reference compounds in the Fluo-3- or the Fura-2-based $[Ca^{2+}]_i$ assay. The results obtained in each case were compared with those obtained by measuring inward currents using electrophysiologic techniques. We present selected results here that demonstrate several salient features of the assay. For ease of organization, we have grouped the different voltage- and ligand-gated channels according to the fraction of the inward current carried by $[Ca^{2+}]_i$, based on reported

[18] L. Daggett, A. I. Sacaan, M. Akong, S. Rao, S. D. Hess, C. Liaw, A. Urruita, C. Jachec, S. B. Ellis, J. Dreessen, T. Knopfel, G. B. Landwehrmeyer, C. M. Testa, A. B. Young, M. Varney, E. C. Johnson, and G. Veliçelebi, *Neuropharmacology* **34,** 871 (1995).

[19] F. F. Lin, M. Varney, A. I. Sacaan, C. Jachec, L. P. Daggett, S. Rao, T. Whisenant, P. Flor, R. Kuhn, J. A. Kerner, D. Standaert, A. B. Young, and G. Veliçelebi, *Neuropharmacology* **36,** 917 (1997).

[20] R. E. Kingston, *in* "Current Protocols in Molecular Biology" (F. M. Ausubel, R. Brent, R. E. Kingston, D. D. Moore, J. G. Seidman, J. A. Smith, and K. Struhl, eds.), p. 9.1.4. John Wiley & Sons, New York, 1996.

results.[21,22] However, additional factors such as number of recombinant receptors expressed per cell, kinetics of the $[Ca^{2+}]_i$ response, desensitization properties, and assay conditions may also affect the absolute magnitude of $[Ca^{2+}]_i$ signals in these cell lines.

## Ion Channels with High Fractional $Ca^{2+}$ Response

In this group, we included ion channels in which $Ca^{2+}$ influx accounts for greater than 10% of the inward current.

### *Voltage-Gated Calcium Channels*

Neuronal voltage-gated calcium channels (VGCCs) are multimeric channel complexes composed of an $\alpha_1$-, an $\alpha_2\delta$-, and a $\beta$-type subunit that are activated by depolarization of the membrane. Several isoforms of the $\alpha_1$ and the $\beta$ subunit and several splice variants of genes representing all three subunits have been cloned. Specific subtypes of VGCCs are composed of different combinations of the three types of subunits and exhibit unique biophysical and pharmacologic properties (for review, see Ref. 23) We have generated several cell lines stably expressing different combinations of the three human VGCC subunits. Depolarization of the VGCC-expressing cell lines by addition of KCl causes influx of extracellular $Ca^{2+}$, and the ensuing increase in $[Ca^{2+}]_i$ can be readily detected in real time by an increase in Fluo-3 or Fura-2 fluorescence. Representative kinetic traces for the $\alpha_{1A\text{-}2}\alpha_{2b}\delta\beta_{4a}$ (P/Q-type VGCC), $\alpha_{1B\text{-}1}\alpha_{2b}\delta\beta_{3a}$ (N-type VGCC), and $\alpha_{1E\text{-}3}\alpha_{2b}\delta\beta_{1b}$ (possibly R-type VGCC) subtypes activated by KCl are shown in Fig. 1. The $[Ca^{2+}]_i$ response to KCl addition reached a peak within 10 sec in each cell line. The magnitude of the $[Ca^{2+}]_i$ response was comparable in all three cell lines (approximately 500 n*M* above basal levels), but there were apparent differences in the decay kinetics of the $[Ca^{2+}]_i$ response, with the slowest decay observed in the cell line expressing the $\alpha_{1B\text{-}1}\alpha_{2b}\delta\beta_{3a}$ subtype.

We used the fluorescent dye-based $[Ca^{2+}]_i$ assay to characterize the pharmacologic properties of the three VGCC subtypes. Various reagents and several peptide toxins have been identified that display differential interaction with the $\alpha_{1A}$, $\alpha_{1B}$, and $\alpha_{1E}$ subtypes (for review, see Ref. 24). The results obtained in the fluorescent dye-based $[Ca^{2+}]_i$ assay (Fig. 2) agree

[21] M. Rogers and J. Dani, *Biophys. J.* **68,** 501 (1995).

[22] N. Burnashev, Z. Zhou, E. Neher, and B. Sakmann, *J. Physiol.* **485.2,** 403 (1995).

[23] W. A. Catterall, *Ann. Rev. Biochem.* **64,** 493 (1995).

[24] B. M. Olivera, G. P. Miljanich, J. Ramachandran, and M. E. Adams, *Annu. Rev. Biochem.* **63,** 823 (1994).

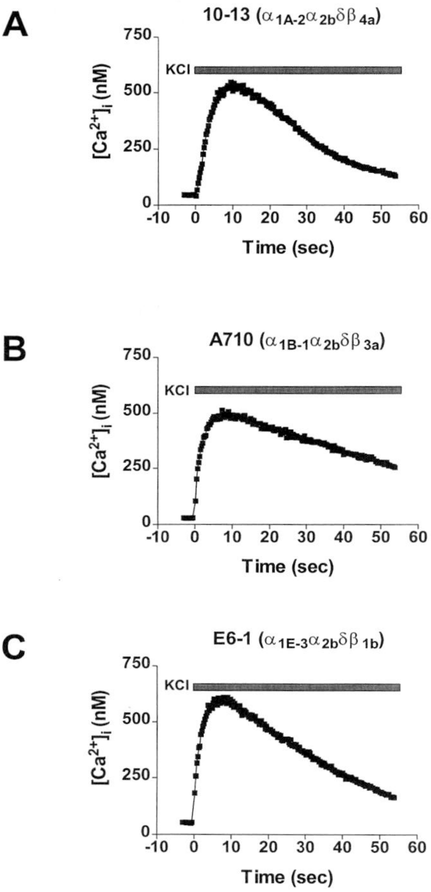


FIG. 1. Time course of the KCl-induced $[Ca^{2+}]_i$ signal in HEK293 cells stably expressing human VGCCs. HEK293 cells were stably transfected with cDNAs encoding the indicated subunits of human VGCCs, and clonal cell lines were established as described in the text. (A) 10-13 cell line expressing $\alpha_{1A\text{-}2}\alpha_{2b}\delta\beta_{4a}$. (B) A710 cell line expressing $\alpha_{1B\text{-}1}\alpha_{2b}\delta\beta_{3a}$. (C) E6-1 cell line expressing $\alpha_{1E\text{-}3}\alpha_{2b}\delta\beta_{1b}$. The cells were loaded with Fluo-3 as described in the text, and the fluorescence was monitored with a plate-reading fluorimeter. Data points were measured at approximately 0.33-sec intervals. The first few seconds of data were recorded with cells bathed in HBS. At time zero, depolarization buffer containing 70 m$M$ KCl and 4 m$M$ $CaCl_2$ (final concentrations) was added automatically using the Digiflex dispenser, and the response was monitored for the next 55 sec. At the end of the experiment, $[Ca^{2+}]_i$ was calculated from $F_{Max}$ and $F_{Mn}$ values determined with the Triton X-100 method. The points represent mean values (error bars omitted for clarity) of quadruplicate determinations from one experiment.

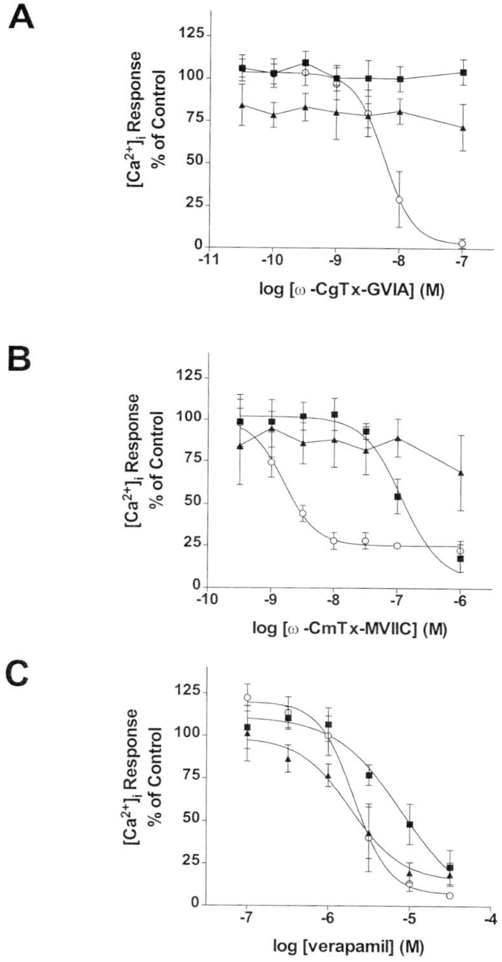


FIG. 2. Pharmacologic characterization of cell lines stably expressing human VGCCs. The cells expressing specific VGCC subtypes were loaded with Fura-2 as described in the text, and the fluorescence was monitored with the SIBIA/SAIC imaging system. The cells were preincubated with $\omega$-CgTx-GVIA (A), $\omega$-CmTx-MVIIC (B), or verapamil (C) for 5–10 min prior to activation with 70 m$M$ KCl. The $y$ axis shows the response amplitude as a percent of the positive control $[Ca^{2+}]_i$ response to 70 m$M$ KCl with no antagonist added. Points are the mean ± SD of quadruplicate determinations from a representative experiment. Curves were fit to points by Prism software (GraphPad, Inc., San Diego, CA). In those instances where curves could not be fit to the points, the points were connected with lines. –■– 10-13 cell line expressing $\alpha_{1A\text{-}2}\alpha_{2b}\delta\beta_{4a}$; –○– A710 cell line expressing $\alpha_{1B\text{-}1}\alpha_{2b}\delta\beta_{3a}$; –▲– E6-1 cell line expressing $\alpha_{1E\text{-}3}\alpha_{2b}\delta\beta_{1b}$.

with the established pharmacology of the three subtypes. For example, $\omega$-conotoxin GVIA ($\omega$-CgTx-GVIA) specifically inhibited the KCl-induced $[Ca^{2+}]_i$ response in the $\alpha_{1B-1}\alpha_{2b}\delta\beta_{3a}$-expressing cell line with an $IC_{50}$ of $8.9 \pm 1.4$ n*M* (mean $\pm$ SEM) without any detectable effect on cell lines expressing $\alpha_{1A-2}\alpha_{2b}\delta\beta_{4a}$ or $\alpha_{1E-3}\alpha_{2b}\delta\beta_{1b}$ (Fig. 2A). Another peptide toxin, $\omega$-CmTx-MVIIC, blocked the KCl-induced response in cells expressing $\alpha_{1A-2}\alpha_{2b}\delta\beta_{4a}$ and $\alpha_{1B-1}\alpha_{2b}\delta\beta_{3a}$ channels with $IC_{50}$ values of $98 \pm 12$ n*M* and $5.8 \pm 1.7$ n*M*, respectively, without significantly affecting the $[Ca^{2+}]_i$ response in cells expressing $\alpha_{1E-3}\alpha_{2b}\delta\beta_{1b}$ channels (Fig. 2B). Finally, verapamil, a relatively nonselective blocker, displayed a slight preference for cells expressing $\alpha_{1B-1}\alpha_{2b}\delta\beta_{3a}$ ($IC_{50} = 3.5 \pm 0.8$ $\mu M$) or $\alpha_{1E-3}\alpha_{2b}\delta\beta_{1b}$ ($IC_{50} = 5.7 \pm 1.4$ $\mu M$) channels compared to cells expressing $\alpha_{1A-2}\alpha_{2b}\delta\beta_{4a}$ channels ($IC_{50} = 32 \pm 4$ $\mu M$) (Fig. 2C). These results indicated that the fluorescent dye-based $[Ca^{2+}]_i$ assay was sensitive and rapid enough to detect the changes in $[Ca^{2+}]_i$ resulting from the activation of VGCCs. Furthermore, the $[Ca^{2+}]_i$ response displayed the expected subtype-specific pharmacologic profile observed by whole-cell recording, thus, validating the assay as a means to screen unknown compounds for their activity on human VGCC subtypes.

The fluorescent dye-based $[Ca^{2+}]_i$ assay can be more sensitive than electrophysiologic methods for measuring small ionic currents that decay very slowly or incompletely. For example, in HEK293 cells stably expressing $\alpha_{1A-2}\alpha_{2b}\delta\beta_{3a}$ VGCCs, we were able to detect voltage-gated $Ba^{2+}$ or $Ca^{2+}$ currents in only 5 out of 78 cells, and none of the currents were greater than 120 pA (Fig. 3A). On the other hand, with the fluorescent dye-based $[Ca^{2+}]_i$ assay, we could reliably detect KCl-induced $[Ca^{2+}]_i$ responses of 150–200 n*M* (Fig. 3B). The higher sensitivity of the fluorescence-based $[Ca^{2+}]_i$ measurements can most easily be explained by the fact that these particular VGCCs inactivate relatively slowly ($\tau_{inact} \approx 690$ ms) and incompletely, with a noninactivating component of approximately 80%. In the fluorescent dye-based $[Ca^{2+}]_i$ assay, the depolarization stimulus, i.e., 70 m*M* KCl, is maintained throughout the measurement, and $Ca^{2+}$ continually enters the cell and accumulates, resulting in a detectable $[Ca^{2+}]_i$ change. In contrast, electrophysiologic measurements monitor only the flux of current across the membrane over a much shorter time scale. Consequently, a very small but incompletely inactivating $Ca^{2+}$ current can be seen as a robust signal with fluorescent dye-based $[Ca^{2+}]_i$ measurements. Neher and Augustine[25] have described in detail the effects of $[Ca^{2+}]_i$ indicators in prolonging and amplifying the $Ca^{2+}$ current signal by means of $Ca^{2+}$ accumulation and buffering $[Ca^{2+}]_i$ in bovine adrenal chromaffin cells.

[25] E. Neher and G. J. Augustine, *J. Physiol.* **450,** 273 (1992).

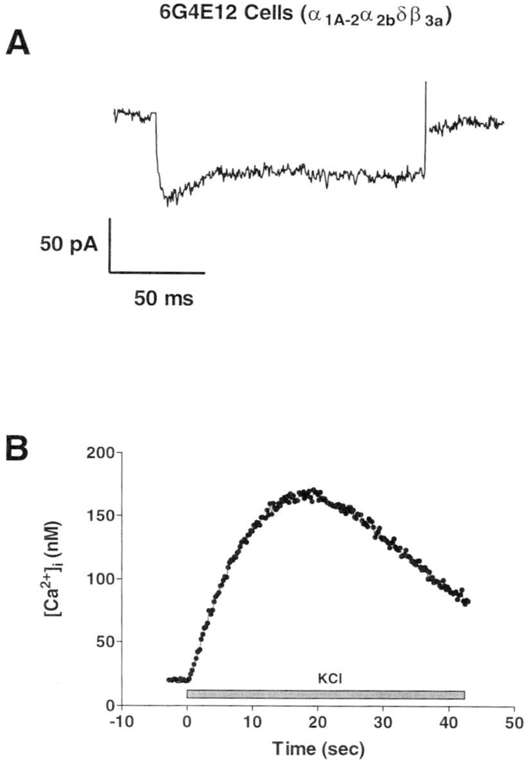


FIG. 3. Comparison of the electrophysiologic current and fluorescent dye-based $[Ca^{2+}]_i$ signal in HEK293 cells stably expressing $\alpha_{1A-2}\alpha_{2b}\delta\beta_{3a}$ VGCCs. In (A), the $Ba^{2+}$ (15 m*M*) current was recorded from a cell depolarized to +10 mV from a holding potential ($V_h$) of −90 mV. In (B), cells loaded with Fluo-3 were depolarized with 70 m*M* KCl.

One limitation of the fluorescent dye-based $[Ca^{2+}]_i$ assay in the VGCC system may arise from the difficulty in controlling the membrane potential and, therefore, the state of voltage-dependent inactivation of the channels. For example, the $\alpha_{1E-3}\alpha_{2b}\delta\beta_{1-3}$-containing neuronal VGCCs undergo steady-state inactivation with membrane depolarization, with approximately one-half of the channels inactivated at −60 mV.[4] Therefore, if the resting membrane potential of the stable cell line expressing the $\alpha_{1E-3}\alpha_{2b}\delta\beta_{1-3}$ channels is more depolarized than −60 mV, the responses to KCl depolarization may be relatively small. Electrophysiologic measurements have revealed that treatment with valinomycin, principally a $K^+$ ionophore, can hyperpolarize the HEK293 cell membrane by 45–70 mV (data not shown). Therefore, this limitation may be overcome by preincubation of the cells with

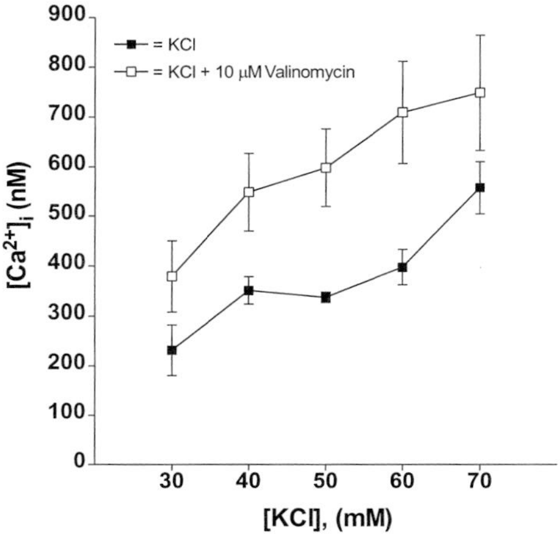


FIG. 4. Effect of valinomycin treatment on the KCl-induced $[Ca^{2+}]_i$ response. The KCl-induced $[Ca^{2+}]_i$ signal was measured in Fluo-3-loaded HEK293 cells expressing $\alpha_{1E-3}\alpha_{2b}\delta\beta_{1b}$-VGCCs. The cells were stimulated with increasing concentrations of KCl either alone or after a 10-min preincubation with 10 $\mu M$ valinomycin. The points represent mean values ($\pm$ SD) of quadruplicate determinations from a representative experiment.

valinomycin (1–10 $\mu M$) before KCl is added. As shown in Fig. 4, in $\alpha_{1E-3}\alpha_{2b}\delta\beta_{1-3}$-expressing HEK293 cells, the $[Ca^{2+}]_i$ response to KCl is higher in valinomycin-treated cells. Because valinomycin acts to hyperpolarize the membrane, it is also possible that some of the increased $[Ca^{2+}]_i$ response is due to the increase in the driving force for $Ca^{2+}$ resulting from the more negative membrane potential.

### *NMDA Receptors*

NMDA receptors are ligand-gated cation channels with up to 11% of the inward current carried by $Ca^{2+}$. These channels are heteromeric complexes composed of at least an R1- and an R2-type subunit and are activated by glutamate and glycine.[22,26] NMDA receptors represent a pharmacologically versatile system containing multiple sites of drug interaction (Fig. 5A). HEK293 cells stably expressing the hNMDAR1A/2A and hNMDAR1A/2B subtypes of human NMDA receptors respond to the application of 100 $\mu M$ glutamate and 30 $\mu M$ glycine with a rise in $[Ca^{2+}]_i$ of approximately 400 n$M$ that reaches maximal levels 10–20 sec after agonist addition and remains elevated for more than 60 sec (Fig. 5B).

Using suitable reference compounds that recognize each of these sites, we have demonstrated that the fluorescent dye-based $[Ca^{2+}]_i$ assay can be

[26] M. Hollmann and S. F. Heinemann, *Annu. Rev. Neurosci.* **17,** 31 (1994).

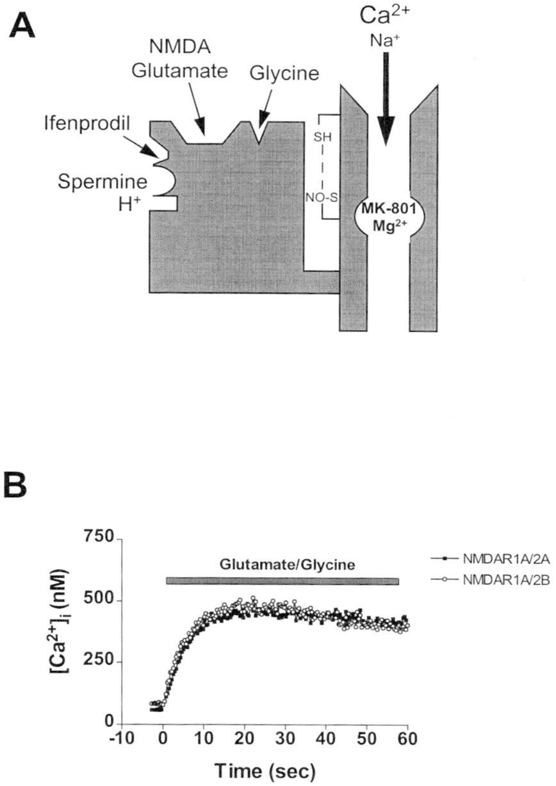


FIG. 5. Agonist-induced $[Ca^{2+}]_i$ responses in HEK293 cells stably expressing human NMDA receptors. (A). Schematic representation of an NMDA receptor. Activation of both the glutamate and glycine site are required for functional activation of the NMDA receptor and subsequent cation entry. In addition, there are several modulatory sites on the receptor. The NMDA receptor channel is inhibited by $Mg^{2+}$ and channel blockers such as MK-801, PCP, memantine, and ketamine. The polyamine site is regulated by spermine, which may overlap with the proton inhibitory site and the ifenprodil binding site (at receptors containing the NMDAR2B subunit). The thiol groups (SH) may react with various forms of nitric oxide to modulate the receptor activity. [This figure is modified from a version published by S. A. Lipton and P. A. Rosenberg,[27] *N. Engl. J. Med.* **330,** 613 (1994).] (B). Representative kinetic traces of $[Ca^{2+}]_i$ in response to 100 $\mu M$ glutamate and 30 $\mu M$ glycine, measured in Fluo-3-loaded HEK293 cells stably expressing hNMDAR1A/2A or hNMDAR1A/2B. Measurements were performed in nominally $Mg^{2+}$-free HBS, and data points show mean values from four replicate wells.

TABLE I
PHARMACOLOGY OF RECOMBINANT HUMAN NMDA RECEPTORS STABLY EXPRESSED IN HEK293 CELLS[a]

| | $EC_{50}$ or $IC_{50}$ value ($\mu M$) | | | | | |
|---|---|---|---|---|---|---|
| | hNMDAR1A/2A | | | hNMDAR1A/2B | | |
| Compound | Mean | SD | *N* | Mean | SD | *N* |
| *Agonists* | | | | | | |
| Glutamate | 0.736 | 0.371 | 8 | 0.578 | 0.222 | 8 |
| NMDA | 8.75 | 3.78 | 8 | 9.43 | 2.51 | 8 |
| *Antagonists* | | | | | | |
| 5,7-DCKA | 0.165 | 0.06 | 6 | 0.943 | 0.586 | 8 |
| (±)CPP | 1.58 | 0.94 | 6 | 6.43 | 3.26 | 8 |
| Ketamine | 18.9 | 15.1 | 5 | 10.1 | 2.96 | 5 |
| Ifenprodil | >30 | | 8 | 0.303 | 0.214 | 8 |

[a] Values are expressed as mean ± SD from *N* experiments.

used to detect effectively the activity of compounds that interact with the receptor through both competitive and noncompetitive mechanisms (Table I). We studied two agonists (glutamate and NMDA) and a competitive antagonist (±)-3-(2-carboxypiperazin-4-yl) propyl-1-phosphonate ((±)CPP) at the glutamate site (Figs. 6A, 6B, and 6C; Table I). At the glycine site, we were not able to determine the potency of glycine as a coagonist of the NMDA receptor, probably because the contaminating levels of glycine present in the assay are sufficiently high to fully saturate the glycine site.[8] The cells, rather than the buffer, appear to be the main source of the contaminating glycine. However, the activity of competitive antagonists at the glycine site can be detected. This is demonstrated by the results obtained for 5,7-dichlorokynurenic acid (5,7-DCKA), a competitive glycine-site antagonist, that inhibits the glutamate/glycine-induced $[Ca^{2+}]_i$ signal in hNMDAR1A/2A- and hNMDAR1A/2B-expressing cells with an $IC_{50}$ of approximately 0.2 and 0.9 $\mu M$, respectively (Fig. 6D; Table I). These values are comparable to $IC_{50}$ values of 0.2 $\mu M$ and 0.4 $\mu M$ determined using two-electrode voltage-clamp recording techniques in *Xenopus* oocytes injected with mRNAs encoding hNMDAR1A/2A or hNMDAR1A/2B, respectively.[7]

In addition, we examined two noncompetitive antagonists that interact at different sites on the NMDA receptor. Ketamine, a channel blocker, inhibited the glutamate/glycine-induced activation of both subtypes of human NMDA receptors in a concentration-dependent manner, with no apparent selectivity between the two subtypes (Fig. 6E; Table I). Another noncompetitive blocker, ifenprodil, displayed marked subtype selectivity,

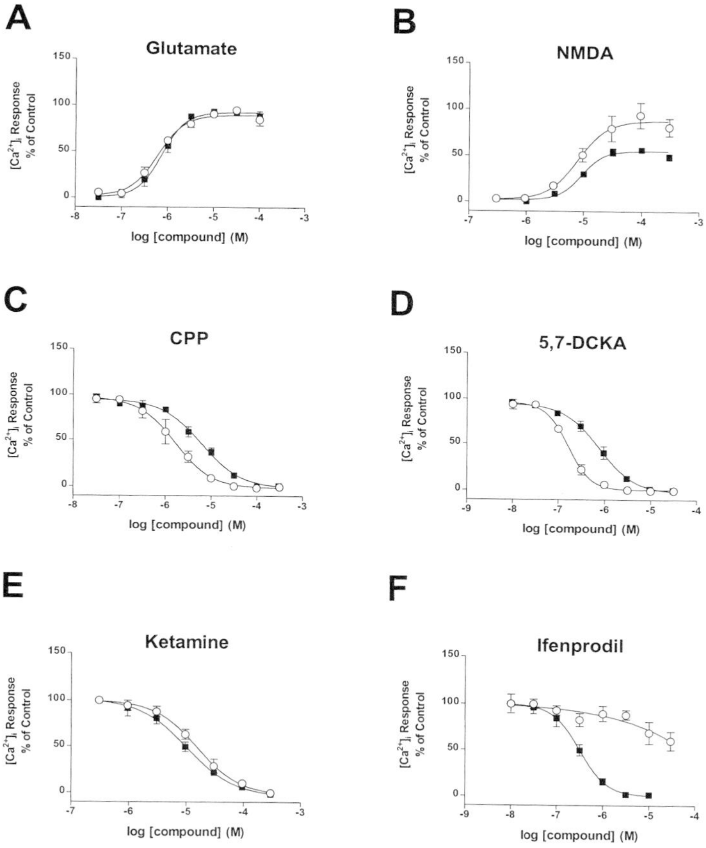


FIG. 6. Pharmacologic profile of HEK293 cells stably expressing hNMDAR1A/2A and hNMDAR1A/2B. Concentration-response curves to (A) glutamate, (B) NMDA, and inhibition curves to (C) (±)CPP, (D) 5,7-DCKA, (E) ketamine, and (F) ifenprodil were constructed from $[Ca^{2+}]_i$ measurements in Fura-2-loaded HEK293 cells stably expressing hNMDAR1A/2A (○) or hNMDAR1A/2B (■). Data represent the mean ± SEM from five to eight separate experiments, each performed in duplicates. For agonists, the data are normalized to the response elicited by 100 $\mu M$ glutamate/30 $\mu M$ glycine. In the antagonist studies, 3 $\mu M$ glutamate and 3 $\mu M$ glycine were used.

with more than 100-fold greater potency for the NMDAR1A/2B subtype compared to the hNMDAR1A/2A (Fig. 6F; Table I). The ifenprodil results underscore the principal advantage of using a functional assay for screening compounds. In contrast to binding studies that require prior knowledge of a recognition site, the functional assay enables identification of subtype-selective compounds that interact with the receptor at novel sites.

## Ion Channels with Moderate Functional $Ca^{2+}$ Response

In this group, we included ion channels in which $Ca^{2+}$ influx accounts for 1–10% of the inward current.

### *AMPA/Kainate Receptors*

AMPA/kainate receptors, also referred to as non-NMDA ionotropic EAA receptors, are ligand-gated cation channels that respond to glutamate. Both AMPA and kainate receptors are multimeric complexes composed of one or more types of subunits. Burnashev *et al.*[22] have reported that recombinant homomeric AMPA receptors composed of GluR1, GluR2(Q) or GluR4 or kainate receptors composed of GluR6(V,C,Q) can flux $Ca^{2+}$ with a fractional $Ca^{2+}$ response of 2–4%. Consistent with this, HEK293 cells stably expressing the hGluR3 subtype of human AMPA receptors and the hGluR6(I,Y,Q) subtype of human kainate receptors respond to agonist application with inward currents that desensitize rapidly (Figs. 7A and 7D). The desensitization can be attenuated by treating the hGluR3- and hGluR6-expressing cells with cyclothiazide (CTZ) or concanavalin A (con A), respectively (Figs. 7B and 7E). Due to rapid desensitization, glutamate-induced $[Ca^{2+}]_i$ response typically cannot be detected in these cells in the absence of CTZ or con A (Figs. 7C and 7F). In contrast, in the presence of CTZ or con A, both hGluR3- and hGluR6-expressing cells, respectively, respond to glutamate with a robust increase in $[Ca^{2+}]_i$ of approximately 500 n*M* above basal levels (Figs. 7C and 7F).

### *Nicotinic Acetylcholine Receptors*

Neuronal nicotinic acetylcholine receptors are also ligand-gated multimeric channel complexes, presumed to be composed of five subunits representing one or more of an $\alpha$-type and a $\beta$-type subunit.[28] Multiple $\alpha$ and $\beta$ subunits have been cloned and coexpressed stably in different combinations in HEK293 cells.[5,29,30] Cells stably expressing the binary combinations of $\alpha_2\beta_4$, $\alpha_3\beta_4$, and $\alpha_4\beta_4$ subunits respond to nicotinic agonists with a relatively rapid rise in $[Ca^{2+}]_i$ (Fig. 8). The kinetics and magnitudes of the $[Ca^{2+}]_i$ responses to a maximally effective agonist concentration were compared in the three cell lines. The largest $[Ca^{2+}]_i$ elevation was measured

[27] S. A. Lipton and P. A. Rosenberg, *N. Engl. J. Med.* **330,** 613 (1994).

[28] D. S. McGehee and L. W. Role, *Ann. Rev. Physiol.* **57,** 521 (1995).

[29] P. Whiting, R. Schoepfer, J. Lindstrom, and T. Priestley, *Mol. Pharmacol.* **40,** 463 (1991).

[30] M. Gopalakrishnan, L. M. Monteggia, D. J. Anderson, E. J. Molinari, M. Piattoni-Kaplan, D. Donnelly-Roberts, S. P. Arneric, and J. P. Sullivan, *J. Pharmacol. Exp. Ther.* **276,** 289 (1996).

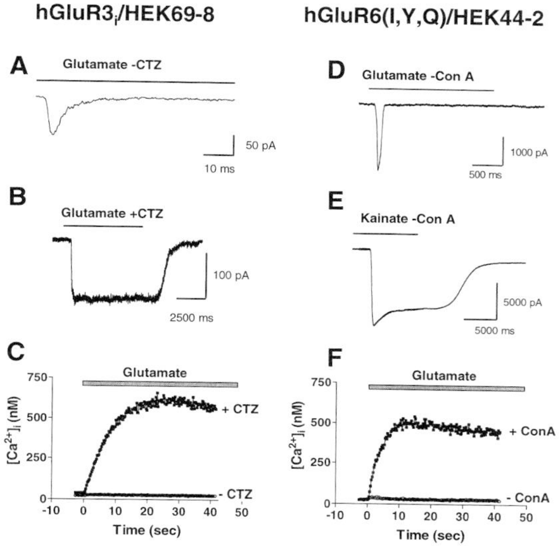


FIG. 7. Comparison of electrophysiologic and fluorescent dye-based $[Ca^{2+}]_i$ measurements of the activation of human AMPA ($hGluR3_i$) or kainate (hGluR6(I,Y,Q)) receptors stably expressed in HEK293 cells. Whole-cell patch-clamp recordings of HEK293 cells stably expressing the human AMPA receptor $hGluR3_i$ (A and B) in the absence (A) and presence (B) of 100 $\mu M$ CTZ in response to a rapid application of 1 m$M$ glutamate. Recordings from the human kainate receptor hGluR6(I,Y,Q) (D and E) in the absence (D) or following a 5-min pretreatment (E) with 0.3 mg/ml con A in response to a rapid application of 1 m$M$ glutamate (D) or 1 m$M$ kainate (E). Cells were held at a resting membrane potential of $-60$ mV. Representative kinetic traces of $[Ca^{2+}]_i$ in response to 1 m$M$ glutamate in HEK293 cells stably expressing $hGluR3_i$ (C) and hGluR6(I,Y,Q) (F). Measurements were performed in Fluo-3-loaded cells. AMPA receptor desensitization was inhibited by 100 $\mu M$ CTZ (C) and kainate receptor desensitization was inhibited by 0.3 mg/ml con A (F) by pretreatment for 5 min with either drug. Data points shown mean values from four replicate wells.

in the $\alpha_2\beta_4$-expressing cell line with a stimulation of more than 1300 n$M$ above basal levels, likely resulting from a higher level of receptor expression in this cell line.

Pharmacologic characterization of the $[Ca^{2+}]_i$ response was performed by determining the potencies of several reference agonists and antagonists in all three cell lines. The concentration–response curves for four nicotinic agonists in the $\alpha_2\beta_4$-expressing cell line are shown in Fig. 9A. In a representative experiment, epibatidine was the most potent agonist on the $\alpha_2\beta_4$ subtype ($EC_{50}$ 1.8 n$M$), followed by cytisine ($EC_{50}$ 483 n$M$, nicotine ($EC_{50}$ 50.5 n$M$), and DMPP ($EC_{50}$ 59.5 n$M$). In the antagonist studies, we tested mecamylamine, *d*-tubocurarine, and dihydro-$\beta$-erythroidine (DH$\beta$E) for

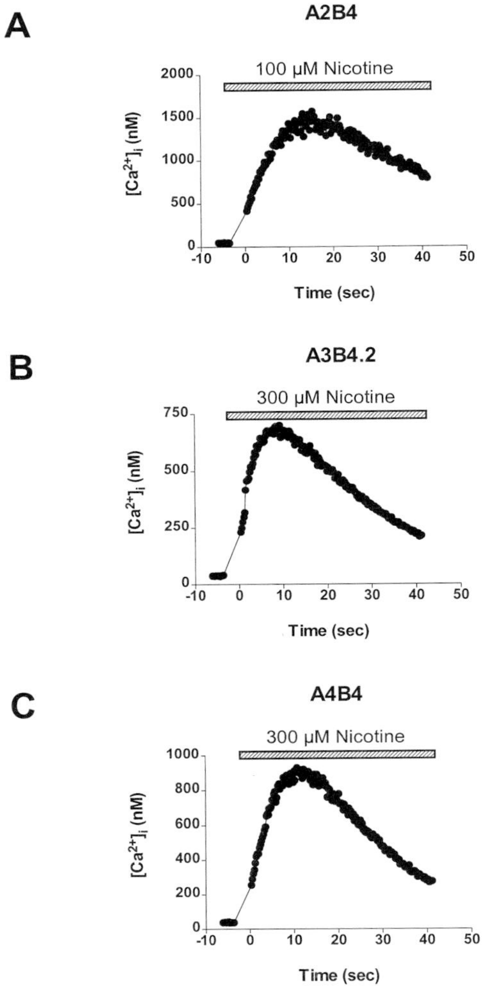


FIG. 8. Nicotine-induced changes of $[Ca^{2+}]_i$ in HEK293 cells stably expressing human NAChR subunits. The $[Ca^{2+}]_i$ measurements were performed in A2B4 (expressing $\alpha_2\beta_4$), A3B4.2 (expressing $\alpha_3\beta_4$), and A4B4 (expressing $\alpha_4\beta_4$) cell lines loaded with Fluo-3. After 10 measurements of basal fluorescence, the lid of the fluorimeter was opened briefly and 40 $\mu$l of nicotine was added manually to 160 $\mu$l of HBS already in the wells to the indicated final concentration. The 5-sec gap at $t = 0$ in each fluorescence record is due to opening and closing of the lid during nicotine addition. Each point represents the mean $[Ca^{2+}]_i$ value (in n$M$) of four individual wells from a 96-well plate (error bars were omitted for clarity). Representative $[Ca^{2+}]_i$ responses are shown for each cell line.

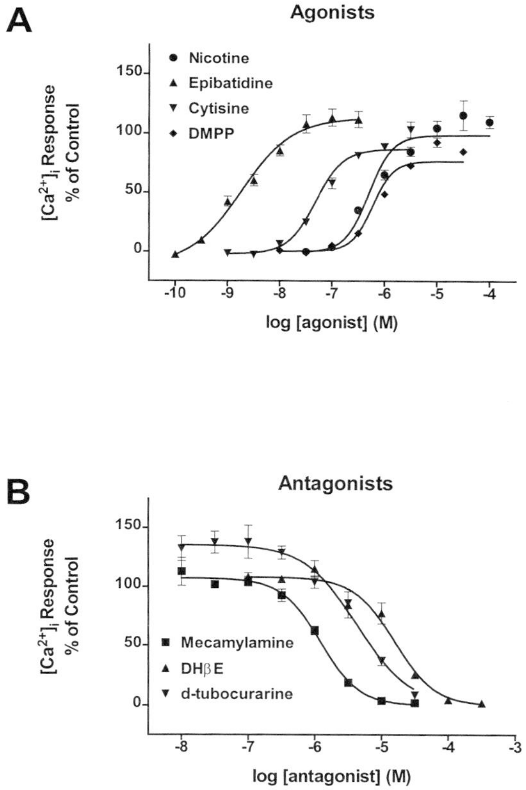


FIG. 9. Pharmacologic profile of A2B4 cells stably expressing human $\alpha_2\beta_4$ neuronal NAChRs. A2B4 cells were loaded with Fura-2 as described in the text, and fluorescence was monitored by the SIBIA/SAIC plate imaging system. The agonist data (A) are shown as a percent of the response to a maximal concentration of nicotine, whereas the antagonist data (B) are shown as a percent of the response to an $EC_{80}$ concentration of nicotine (10 $\mu M$). The points are means ($\pm$ SD) from quadruplicate determinations. The cells were incubated with the antagonists for 5–10 min prior to stimulation with nicotine, and the antagonists did not produce a response by themselves. Note that the compounds were added in a 10× solution containing DMSO, and because the density of this solution is greater than the buffer, the mixture may settle to the bottom of each well, thereby raising the effective concentration of compound. This can alter the apparent potency of the compound by shifting the concentration–response curves to the left.

their ability to inhibit nicotine-induced $[Ca^{2+}]_i$ responses. Again, in a representative experiment, all three antagonists fully inhibited the response to 10 $\mu M$ nicotine, with a rank order of mecamylamine ($IC_{50}$ 1.2 $\mu M$) > *d*-tubocurarine ($IC_{50}$ 4.3 $\mu M$) > DH$\beta$E ($IC_{50}$ 15.5 $\mu M$) (Fig. 9B).

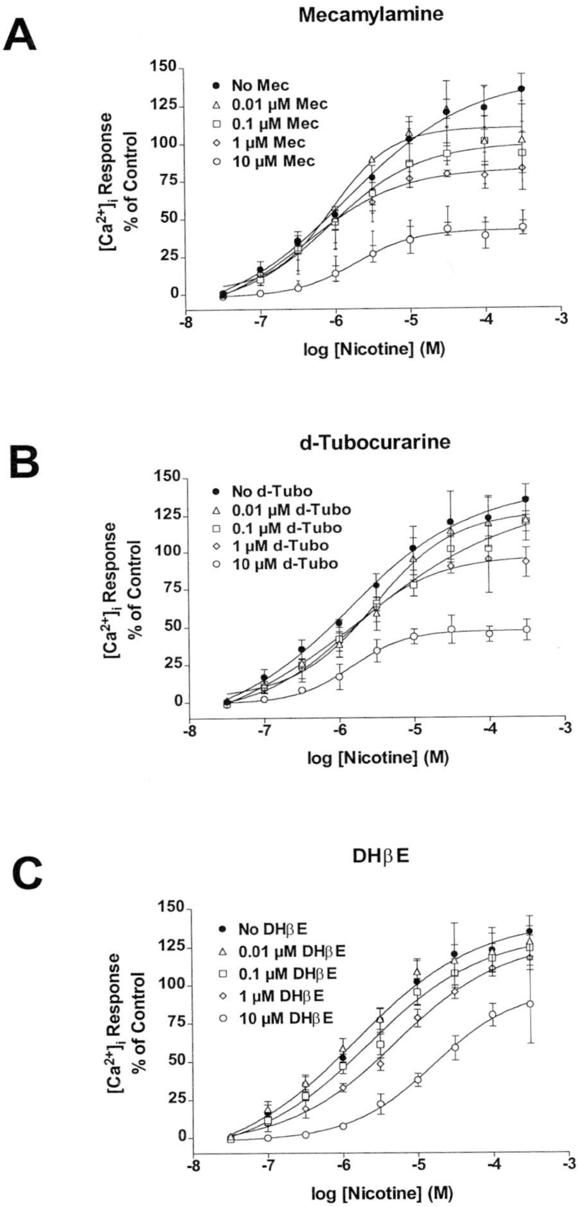


FIG. 10. Schild analysis of the inhibition of $\alpha_2\beta_4$-NAChRs by mecamylamine, *d*-tubocurarine, or DH$\beta$E. A2B4 cells expressing the $\alpha_2\beta_4$-NAChRs were loaded with Fura-2 as described in the text, and fluorescence was monitored by the SIBIA/SAIC plate imaging system. The

The fluorescent dye-based $[Ca^{2+}]_i$ assay can be used to examine the mechanism of action of antagonists using Schild analysis, as shown in Fig. 10 for the three antagonists on the $\alpha_2\beta_4$-expressing cells. Increasing concentrations of mecamylamine (Fig. 10A) and *d*-tubocurarine (Fig. 10B) reduced the efficacy of nicotine without affecting its potency, and thus these two antagonists appeared to act through a noncompetitive mechanism. By contrast, DH$\beta$E behaved as a competitive antagonist since increasing its concentration reduced the potency of nicotine without significantly reducing its efficacy (Fig. 10C).

### *Cyclic Nucleotide-Gated Channels*

Cyclic nucleotide-gated channels (CNGChs) are functionally ligand-gated cation channels that have some structural features of voltage-gated ion channels.[31] These channels are permeable to both mono- and divalent cations, with 2–8% of the current being carried by $Ca^{2+}$.[32] The activation of the CNGChs can be measured using the fluorescent dye-based $[Ca^{2+}]_i$ assay. HEK293 cells stably expressing the rat CNGCh respond to forskolin with a rise in $[Ca^{2+}]_i$ that reaches a peak at approximately 60 sec and decays over the next 60 sec to an elevated, long-lasting plateau phase (Fig. 11A). The effect of forskolin is concentration dependent, with an $EC_{50}$ of approximately 10 $\mu M$ (Fig. 11B), similar to the $EC_{50}$ of forskolin to activate adenylyl cyclase.

## Ion Channels with Low Fractional $Ca^{2+}$ Response

In this group, we included ion channels in which $Ca^{2+}$ influx accounts for less than 1% of the inward current.

Heteromeric and homomeric AMPA receptors containing the unedited GluR2(R) subunit have relatively low permeability to calcium.[33] In these cases, the activation of the channel can be coupled to a $[Ca^{2+}]_i$ signal through coexpression with a recombinant VGCC subtype. The activation

[31] U. B. Kaupp, *TINS* **14,** 150 (1991).
[32] S. Frings, R. Seifert, M. Godde, and U. B. Kaupp, *Neuron* **15,** 169 (1995).
[33] N. Burnashev, H. Monyer, P. Seeburg, and B. Sakmann, *Neuron* **8,** 189 (1992).

---

*y* axis represents the response as a percent of the control response to a maximal concentration of nicotine. Nicotine concentration curves were performed in the presence of increasing concentrations of antagonist (5–10 min preincubation) as indicated on the graphs. Points are means (± SD) of quadruplicate determinations.

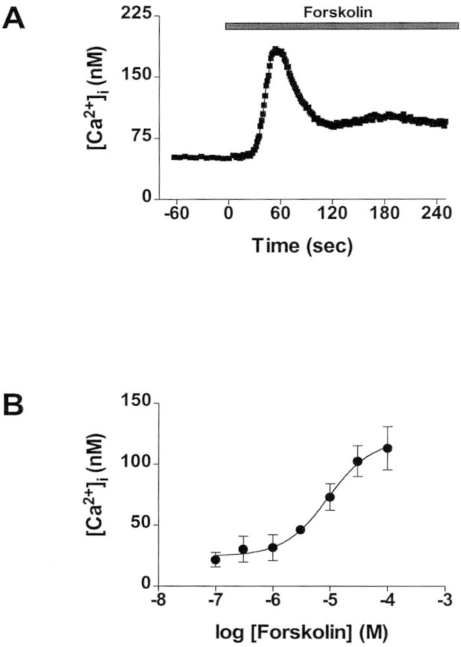


FIG. 11. Time course and pharmacologic characterization of forskolin-induced $[Ca^{2+}]_i$ signal in HEK293 cells stably expressing the rat CNG channel. (A) Representative kinetic trace of $[Ca^{2+}]_i$ in response to 100 $\mu M$ forskolin in the presence of 1 m$M$ IBMX in Fluo-3-loaded HEK293 cells stably expressing the rat CNG channel. Data points represent mean values from four replicate wells. (B) A concentration–response curve to forskolin in the presence of 1 m$M$ IBMX in Fluo-3-loaded HEK293 cells stably expressing the rat CNG channel. Data points represent the mean ± SD from quadruplicate wells from a single experiment representative of two experiments.

of the ligand-gated channel depolarizes the cell membrane, which in turn activates the voltage-gated calcium channel. We tested this approach using one of the channels with moderate relative fractional $Ca^{2+}$ response (the NAChR-$\alpha_3\beta_4$). To this end, we transiently expressed the NAChR-$\alpha_3\beta_4$ in HEK293 cells that stably express the $\alpha_{1B\text{-}2}\alpha_{2b}\delta\beta_{1c}$ subtype of human VGCCs (C1–C4 cell line). The nicotine-evoked $[Ca^{2+}]_i$ responses were compared to transient expression of the NAChR-$\alpha_3\beta_4$ in the host HEK293 cell line. The results revealed that NAChR-$\alpha_3\beta_4$-expressing HEK293 cells responded to the agonist (100 $\mu M$ DMPP) with approximately a 3-fold increase in $[Ca^{2+}]_i$ that reached maximal levels after 20 sec (Fig. 12A). In comparison, in cells expressing both NAChR-$\alpha_3\beta_4$ and N-type VGCCs, the magnitude of

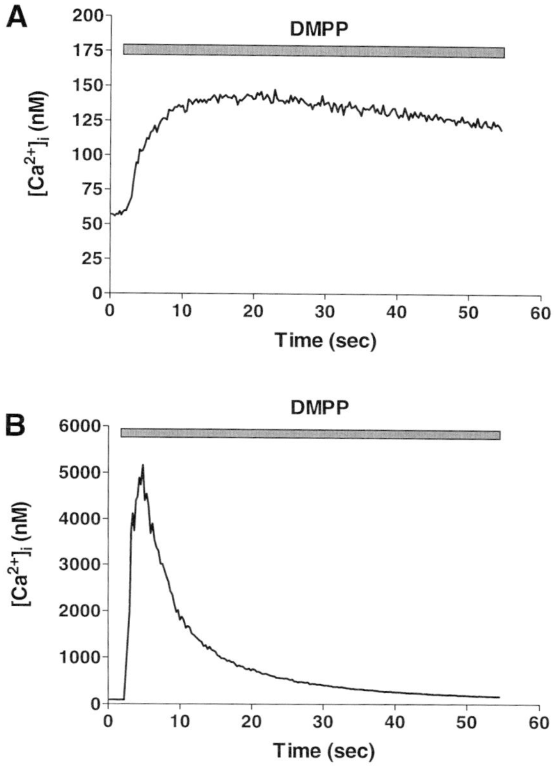


FIG. 12. Coupling the activation of NAChRs to activation of VGCCs by coexpression in HEK293 cells. HEK293 and C1–C4 cells stably expressing the N-type VGCCs ($\alpha_{1B\text{-}2}\alpha_{2b}\delta\beta_{1b}$) were transiently transfected with cDNAs encoding the $\alpha_3$ and $\beta_4$ subunits of human NAChRs. Forty-eight hours after the transfection, each group of cells was loaded with Fluo-3, and the fluorescence was monitored by a plate-reading fluorimeter. As indicated on the graphs, the cells were stimulated with the nicotinic agonist DMPP (100 $\mu M$). Note the larger amplitude and faster kinetics of the response to DMPP in C1–C4 cells expressing the recombinant VGCCs (B) compared to HEK293 cells (A). The plots show the mean response of quadruplicate wells from a representative experiment.

the agonist-induced $[Ca^{2+}]_i$ response was greatly enhanced to approximately 100-fold above basal levels with faster kinetics, reaching maximum levels within 10 sec (Fig. 12B). The pharmacology of this response was also compared in both systems. As shown in Fig. 13, the DMPP-induced $[Ca^{2+}]_i$ response in NAChR-$\alpha_3\beta_4$-expressing HEK293 cells was blocked by a nicotinic antagonist (*d*-tubocurarine) but was not significantly altered by $\omega$-CgTx-GVIA, a specific blocker of the $\alpha_{1B}$-containing VGCCs. By con-

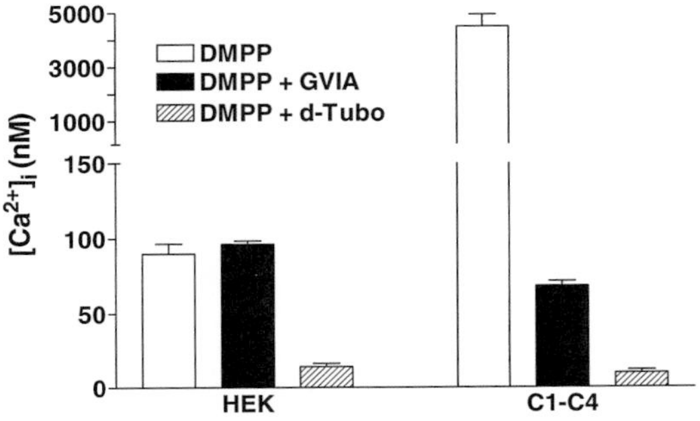


FIG. 13. Comparison of the pharmacology of DMPP-induced $[Ca^{2+}]_i$ signals in HEK293 or in C1–C4 cells transiently expressing $\alpha_3\beta_4$ NAChRs. The data are from the same experiment shown in Fig. 12. The cells were stimulated with 100 $\mu M$ DMPP alone or after a 10-min preincubation with either 1 $\mu M$ $\omega$-CgTx-GVIA (GVIA), a selective antagonist of N-type VGCCs, or 10 $\mu M$ *d*-tubocurarine (d-Tubo), a NAChR antagonist. Note that $\omega$-CgTx-GVIA had no effect on the DMPP-induced $[Ca^{2+}]_i$ signal in the HEK293 cells and that *d*-tubocurarine had no effect on the VGCCs expressed in C1–C4 cells (data not shown).

trast, in $\alpha_3\beta_4$-expessing C1–C4 cells, the DMPP-induced elevation of $[Ca^{2+}]_i$ was sensitive to both *d*-tubocurarine and $\omega$-CgTx-GVIA. To compare the pharmacologic properties of the two systems in greater detail, the potencies of nicotine, cytisine, and DMPP were determined using the $[Ca^{2+}]_i$ assay. There were no significant differences in either the potencies of the three reference agonists or their relative efficacies (Table II). Thus, these results indicated that the activity of NAChR channels and other channels with low relative $Ca^{2+}$ permeability can also be quantitated by measurements of the $[Ca^{2+}]_i$ response resulting from the activation of voltage-gated calcium channels when the two channels are coexpressed in the same host cells. A

TABLE II

COMPARISON OF AGONIST ACTIVITIES IN CELLS EXPRESSING $\alpha_3\beta_4$-NAChR WITH AND WITHOUT COEXPRESSION OF VOLTAGE-GATED $Ca^{+2}$ CHANNELS[a]

| Characteristic | Cell line | Nicotine | Cytisine | DMPP |
|---|---|---|---|---|
| Potency | $\alpha_3\beta_4$ | 51 ± 5 (3) | 46 ± 18 (3) | 16 ± 4 (3) |
| $EC_{50}$ ($\mu M$) | VGCC + $\alpha_3\beta_4$ | 117 ± 74 (3) | 38 (2) | 30 ± 9 (3) |
| Efficacy | $\alpha_3\beta_4$ | 100 (3) | 69 ± 5 (3) | 58 ± 11 (3) |
| % of nicotine response | VGCC + $\alpha_3\beta_4$ | 100 (3) | 41 ± 43 (3) | 44 ± 39 (3) |

[a] Values are expressed as mean ± SD from (*N*) experiments.

similar approach has been used to develop a stable cell line coexpressing recombinant rat NAChRs and L-type VGCCs.[34]

Another way to enhance the $[Ca^{2+}]_i$ signal with receptors or ion chanels that have low permeability to $Ca^{2+}$ is to increase the extracellular concentration of $CaCl_2$. For example, RD (TE671) cells express human neuromuscular-type NAChRs that have a lower relative permeability to $Ca^{2+}$ than neuronal NAChRs.[35,36] When the RD cells were assayed in normal HBS buffer containing 1.8 m*M* $CaCl_2$, suberyldicholine stimulated only a small increase in $[Ca^{2+}]_i$ of approximately 25 n*M* above basal levels compared to 250 n*M* when assayed in HBS containing 21.8 m*M* $CaCl_2$ (data not shown). This effect of $Ca^{2+}$ is likely mediated by the increased driving force for $Ca^{2+}$ entry.

## Limitations of Fluorescent Dye-Based $[Ca^{2+}]_i$ Assay for Measurement of Ion Channel Activity

The fluorescent dye-based $[Ca^{2+}]_i$ assay is most readily adaptable to those ion channels that flux detectable amounts of $Ca^{2+}$ by the techniques described in this report. However, other fluorescent dyes can potentially be utilized in a similar manner to measure the activity of ion channels that flux other ions. For example, sodium-sensitive dyes, such as SBFI and sodium green[12] can be used to measure $Na^+$ flux, and $Cl^-$ flux can be monitored using the fluorescent dye 6-methoxy-*N*-(3-sulfopropyl)quinolinium (SPQ)[37] while changes in membrane potential can be measured using oxonol dyes[38] and aminonaphthylethenylpyridinium (ANEP) dyes.[39] As described here for $Ca^{2+}$-sensing dyes, assays involving other fluorescent dyes would also be optimized and validated to ensure that the particular channel or receptor system of interest displayed the expected pharmacologic characteristics in the assay.

Changes in $[Ca^{2+}]_i$ are not always linear with changes in whole-cell recording currents. Therefore, although the rank order of potency should be the same in both assays, absolute agonist potencies determined from $[Ca^{2+}]_i$ measurements may differ from those determined by electrophysiology. In addition, agonist efficacies may not be the same between the two

[34] E. Stetzer, U. Ebbinghaus, A. Storch, L. Poteur, A. Schrattenholz, G. Kramer, C. Methfessel, and A. Maelicke, *FEBS Lett.* **397,** 39 (1996).

[35] M. Bencherif and R. J. Lukas, *Mol. Cell. Neurosci.* **2,** 52 (1991).

[36] S. Vernino, M. Amador, C. W. Luetje, J. Patrick, and J. Dani, *Neuron* **8,** 127 (1992).

[37] G. R. Ehring, Y. V. Osipchuk, and M. D. Cahalan, *J. Gen. Physiol.* **104,** 1129 (1994).

[38] J. E. González and R. Y. Tsien, *Biophys. J.* **69,** 1272 (1995).

[39] L. M. Loew, L. B. Cohen, J. Dix, E. N. Fluhler, V. Montana, G. Salama, and J.-Y. Wu, *J. Membr. Biol.* **130,** 1 (1992).

functional assays, since this will depend on the receptor number and $Ca^{2+}$ permeability. Thus, a compound that is a partial agonist at a receptor when measured by electrophysiology may appear as a full agonist in the $Ca^{2+}$ assay, provided it activates a sufficient number of receptors to elicit a saturating $[Ca^{2+}]_i$ response.

A $[Ca^{2+}]_i$ signal may not be detected if the ion channels desensitize rapidly, e.g., AMPA receptors (Fig. 7). However, agents that block or slow receptor desensitization may allow the detection of $[Ca^{2+}]_i$ signals (e.g., CTZ for AMPA receptors). In addition, some agonists elicit different levels of receptor desensitization for the same receptor. For example, glutamate induces a rapid desensitization of AMPA receptors, whereas kainate can evoke nondesensitizing or slowly desensitizing responses.[40] Consequently, the fluorescent dye-based $[Ca^{2+}]_i$ assay may indicate differences in efficacy between two agonists that may not be detected in whole-cell recordings when desensitization is not blocked.

In $[Ca^{2+}]_i$ assays performed with a population of cells, the resting membrane potential of the cells is not easily controlled, although some manipulation is possible with ionophores such as valinomycin. Therefore, the detection of compounds that interact in a voltage-dependent manner may be compromised.[41]

## Advantages of Fluorescent Dye-Based $[Ca^{2+}]_i$ Assays

The studies summarized in this report demonstrate the utility and validity of $[Ca^{2+}]_i$ measurements as a means for evaluating the activity of recombinant ion channels stably expressed in mammalian cells. Although not discussed here, these protocols for $[Ca^{2+}]_i$ measurements are readily adaptable to G-protein-coupled receptors, such as the class I metabotropic glutamate receptors, the activation of which results in increased $[Ca^{2+}]_i$ through the stimulation of the phosphoinositide hydrolysis pathway.[18,19] Membrane-permeable $Ca^{2+}$-sensitive fluorescent dyes allow rapid measurement of $[Ca^{2+}]_i$ in whole cells, and detection of the fluorescence output in a plate-reading fluorimeter further facilitates detection of changes in $[Ca^{2+}]_i$ at a resolution of 0.25–0.33 sec. The combination of these features with automation has significantly enhanced the capability to screen drugs directly for functional effects on specific receptor subtypes.

The ability to test compounds in a functional assay offers clear advantages over testing compounds in a binding assay. The latter necessitates the availability of a ligand of sufficient potency and selectivity for a particu-

[40] K. M. Partin, D. K. Patneau, and M. L. Mayer, *Mol. Pharmacol.* **46,** 129 (1994).
[41] J. W. Stocker, L. Nadasdi, R. W. Aldrich, and R. W. Tsien, *J. Neurosci.* **17,** 3002 (1997).

lar site on the receptor. In most cases, the binding assay will only detect those compounds that competitively displace the bound reference ligand. By contrast, a functional assay can identify compounds that interact with the receptor either competitively at the same site as the reference agonist or noncompetitively at another site. This, in turn, markedly increases the potential for discovering ligands with novel structures. Additionally, agonists and antagonists can be readily distinguished in a functional assay and their relative efficacies can be determined without the need for secondary assays. Compounds that are detected in a displacement binding assay must subsequently be evaluated in functional assays to discern agonists from antagonists. For the antagonists, $[Ca^{2+}]_i$ measurements can also be used effectively to determine the mechanism of action. Finally, the use of cell lines expressing specific recombinant receptors as the targets for compound screening facilitates the discovery of subtype-selective compounds. In addition to their potential as effective therapeutic agents, subtype-selective compounds also represent valuable pharmacologic tools to study the role of the specific receptor subtypes in normal and pathophysiologic states.

## Acknowledgments

We especially thank Dr. Michael Harpold for valuable input and continued support and encouragement throughout this work. In addition, we acknowledge Janis Corey-Naeve, Paul Brust, Alison Gillespie, Fen-Fen Lin, Christine Jachec, Susan Simerson, James Crona, Rhonda Skvoretz, Charlie Deal, Robert Siegel, and Carla Suto for contributions to the establishment and validation of the cell lines; Sandy Madigan for critical review of the manuscript; and Karen Payne for assistance in document preparation.

# [3] Tagging Potassium Ion Channels with Green Fluorescent Protein to Study Mobility and Interactions with Other Proteins

*By* EDWIN S. LEVITAN

## Introduction

Interest in understanding cell biological aspects of potassium ion ($K^+$) channel function has grown. Relevant issues include how channels are assembled, targeted to the plasma membrane, localized in different regions of polarized cells (e.g., axons versus dendrites in neurons), and clustered at postsynaptic regions. To date, such studies have employed biochemical and immunohistochemical approaches. For example, tetramerization of



0076-6879/99 $30.00

channel subunits has been assayed with sucrose density centrifugation and gel filtration, cell surface expression by biotinylation of extracellular domains, and clustering by immunofluorescence. These approaches share the property that they can be applied only to disrupted or fixed cells. Thus, it has not been possible to approach cell biological issues by studying functional $K^+$ channels in live cells.

Mobility of channels in the plasma membrane can be studied in live cells once channels are fluorescently tagged. In the past this has been accomplished by analyzing fluorescence recovery after photobleaching (FRAP) (also termed fluorescence photobleaching recovery or FPR) of fluorophores attached to channel-specific drugs. For example, it was demonstrated that $Na^+$ channel mobility, deduced from FRAP of labeled scorpion toxins, differs dramatically in the cell body and the axon hillock of neurons.[1] Similarly, innervation has been shown to alter mobility of these channels in skeletal muscle.[2] Now that channels and channel-binding proteins have been cloned, FRAP could be used to study the structural basis of channel localization and clustering. Furthermore, with an appropriate label, intracellular assembly could be investigated.

Two technical issues have hindered routine application of FRAP to the study of $K^+$ channels. First, there are few subunit-specific drugs. Furthermore, in addition to specificity, such drugs must have long residence times compared to the kinetics of channel movement. Thus, without appropriate labels for specific channels, meaningful FRAP experiments cannot be performed. This limitation has now been breached with the ability to tag $K^+$ channels with bright variants of the green fluorescent protein (GFP). Second, performing conventional spot FRAP experiments required a specially designed setup that is not typically available to channel researchers. This problem can be overcome by performing FRAP studies on scanning laser confocal microscopes that are now found in many departmental or university core facilities. The coupling of GFP-tagging and confocal FRAP approaches has already led to the demonstration that the structural requirements for clustering and immobilizing voltage-gated $K^+$ channels by PSD95 are distinct.[3] It is likely that this approach will also be valuable in studying association of channels to regulatory proteins (e.g., G proteins) and the cytoskeleton. This chapter is aimed at describing (1) how to pick appropriate available GFP variants for generating GFP–$K^+$ channel fusion proteins and

[1] K. J. Angelides, L. W. Elmer, D. Loftus, and E. Elson, *J. Cell Biol.* **106,** 1911 (1988).
[2] K. J. Angelides, *Nature* **321,** 63 (1986).
[3] N. Veyna-Burke, D. Li, K. Takimoto, K. Hoyt, S. Watkins, and E. S. Levitan, *Biophys. J.* **72,** A351 (1997).

(2) how to acquire and analyze FRAP data obtained with a confocal microscope.

## GFP Variants and Commercial Vectors

Green fluorescent proteins are a family of endogenous fluorescent proteins found in a variety of jellyfish. The *Aequoria victoria* GFP has been cloned and its crystal structure solved.[4,5] It is a compact $\beta$ barrel-shaped 27-kDa protein. Interestingly, the fluorophore is encoded by the peptide: cyclization of amino acids 65–67 and dehydrogenation of Y66 produce a fluorophore that emits in the visible spectrum.[6] Two excitation peaks at 395 and 475 nm produce fluorescence at ~510 nm. This fluorescence is quite stable (i.e., GFP is resistant to photobleaching) apparently due to the fact that the fluorophore is buried within the protein and is not accessible to solvent or oxygen.

The utility of GFP has been dramatically enhanced by mutating the native jellyfish sequence. First, it was found that a S65T mutation increased absorbance at the long wavelength 6-fold, attenuated excitation at the short wavelength, and shifted long-wavelength excitation to 489 nm.[7] These changes in spectral properties facilitated efficient detection of S65T GFP with standard wideband fluorescein isothiocyanate (FITC) fluorescence optics already present in a large number of laboratories. This mutation also resulted in more rapid folding of the GFP. A screening mutagenesis strategy then led to the discovery that addition of a F64L mutation increased brightness another 6-fold.[8] Finally, altering codon usage to be more consistent with mammalian genes (i.e., humanizing the gene) led to a 4-fold increase in expression in mammalian cells.[9] Together these advances increased the brightness of GFP by ~140-fold and enabled visualization of relatively rare proteins.

Working with GFPs has been facilitated by commercialization of many vectors. We have used Clontech EGFP fusion protein vectors to label either the N terminus or C terminus of channel proteins. These vectors can be purchased in any frame and are driven by a standard cytomegalovirus promoter. They include a multiple cloning site, a simian virus 40 (SV40) poly(A) site, a consensus Kozak ribosome binding site, and a kanamycin resistance

[4] M. Ormo, A. B. Cubitt, K. Kallio, L. A. Gross, R. Y. Tsien, and S. J. Remington, *Science* **273,** 1392 (1996).
[5] F. Yang, L. G. Moss, and G. N. Phillips, *Nature Biotechnol.* **14,** 1246 (1996).
[6] O. Shimomura, *FEBS Lett.* **104,** 220 (1987).
[7] R. Heim, A. B. Cubitt, and R. Y. Tsien, *Nature* **373,** 663 (1995).
[8] B. P. Cormack, R. Valdivia, and S. Falkow, *Gene* **173,** 33 (1996).
[9] T. Yang, L. Cheng, and S. R. Kain, *Nucleic Acids Res.* **24,** 4592 (1996).

TABLE I
PROPERTIES OF GFP VARIANTS INCORPORATED INTO COMMERCIALLY AVAILABLE FUSION PROTEIN VECTORS[a]

| Variant | Excitation peak λ (nm) | Emission peak λ (nm) | Relative brightness |
|---|---|---|---|
| Wild Type | 395,*475* | ~510 | 1 |
| F99S, M153T, V163A (Clontech GFPuv) | 395,*475* | ~510 | 18 |
| F64L, S65C, I168T (humanized-Quantum rsGFP) | 473 | ~510 | 35 |
| S65T | 489 | ~510 | 6 |
| F64L, S65T (*mut1*, humanized-Clontech EGFP) | 489 | ~510 | 35 |
| F64L, Y66H, V163A (Quantum BFP) | 387 | 450 | — |
| F64L, S65T, Y66H, Y145F (Clontech EBFP) | 380 | 440 | — |

[a] Brightness does not take codon usage into account. Italic font indicates less effective absorption.

gene for bacterial selection. Similar vectors have been developed by Quantum Technologies as well. GFP vectors are also sold by GIBCO-BRL (Gaithersburg, MD), Pharmingen (San Diego, CA), and Packard (Meriden, CT). However, these constructs do not include a multiple cloning site for generation of fusion proteins.

Labeling of multiple distinct proteins in a single cell is now possible with the advent of blue-shifted GFPs (BFPs). Initially, the blue shift was introduced with a Y66H mutation.[10] Subsequent amino acid changes have increased solubility and photostability of the blue-emitting variants (Table I). To date, no side-by-side comparison has been performed with the structurally distinct commercial variants. Vendors have introduced multiple cloning sites into the constructs so that fusion protein vectors are now available. Another strategy for performing colocalization experiments is to use two GFP variants that differ only in their excitation. This may be possible with GFPuv and EGFP clones (Clontech). Unfortunately, this approach is compromised by the fact that codon usage for the GFPuv vector is optimal for bacteria. Furthermore, the EGFP and GFPuv spectra overlap to some degree at all relevant wavelengths. Thus, careful controls will be required to ensure no cross-talk in detection of these two variants. It is likely that the continuous generation of new vectors will yield better GFPuv and

[10] R. Heim, D. C. Prasher, and R. Y. Tsien, *Proc. Natl. Acad. Sci. U.S.A.* **91,** 12501 (1994).

BFP vectors. However, it should be noted that the short wavelength light required for exciting these constructs could be potentially damaging to cells. Therefore, colocalization will be optimal when new variants with further increased excitation wavelengths are developed to yield truly red GFPs. Such mutants would nearly eliminate the problems of photodamage and endogenous background fluorescence in localization of tagged $K^+$ channels.

## Expression of GFP-Tagged $K^+$ Channels

To study GFP-tagged channels in live cells, it is necessary to generate fusion protein constructs, to introduce the plasmids into cells, and to verify expression and function. We have had great success transiently transfecting EGFP-Kv1.4 and EGFP-Kv1.5 into HEK293, Chinese hamster ovary (CHO), and COS cells (ATCC, Manassas, VA). Both calcium phosphate and lipofection reagents (lipofectamine, Tfx-50) have been used routinely. We have employed patch clamping to show that expression of the tagged channels is robust: currents >20 nA per cell are typical. Total cellular GFP–channel protein levels may also be detected by immunoblot with monoclonal and polyclonal anti-GFP antibodies sold by Clontech. Alternatively, channel protein can be quantitated in a fluorimeter with purified GFP proteins used as a standard. This latter approach has been successfully applied with GFP-GIRK1 channels expressed in a few *Xenopus* oocytes.[11]

For the voltage-gated channels, the GFP moiety was fused to the N terminus so that the C-terminal end that interacts with PSD95 was not affected. However, adding GFP blocked N-type inactivation of the Kv1.4 construct presumably by sterically preventing the ball peptide from entering the pore.[3] With GIRK1 channels, placing the GFP on the N terminus blocked G-protein-activated function (E. Levitan, unpublished results, 1996) while a C-terminal GFP construct retained normal gating properties.[11] Thus, deciding which end of the channel to tag must be governed by the subject of study (e.g., clustering by PSD95) and channel-specific idiosyncracies.

The fluorescent signals from our EGFP-based constructs are very strong. Even through standard low numerical aperture objectives, neutral density filters are used to attenuate illumination. This feature has obviated potential problems of background autofluorescence and light scattering from the cells and lipofection reagents. Interestingly, the pattern of labeling is not identical among the cell lines. Most notably, CHO cells seem to retain a significant amount of the protein in cytoplasmic vesicles. The difference

[11] D. E. Logothetis, K. W. Chan, M. Yoshida, and M. Sassaroli, *Biophys. J.* **72,** A11 (1997).

between these cell lines may be due to the presence of storage vesicles in CHO cells that are not abundant in the 293 and COS cells.[12] Operationally, the potential confounding effect of cytoplasmic vesicles makes using 293 cells attractive for measuring mobility in the plasma membrane.

Transient expression of GFP-tagged $K^+$ channels in more differentiated excitable cell lines such as PC12 and AtT-20 cells is also tenable. However, we have found that the signal is lower. This likely reflects a lower copy number of plasmids per cell. This may also explain why we have had difficulty detecting channel fluorescence in primary cultured neurons and cardiac cells. Likewise, we have also found that generating stably transfected cell lines with clearly visible channel expression is problematic. Again, such cells would probably have only one or two copies of the GFP-channel gene in each cell. This limitation does not hold for free EGFP perhaps because translation of cytoplasmic proteins is more efficient than channels and because soluble GFP is very stable. In contrast, the half-life of fusion proteins may be governed by the channel domains. Thus, the simplest approach to working with GFP-tagged $K^+$ channels is to use transient transfection of clonal cell lines.

## Using FRET to Study Protein–Protein Interactions

The BFPs are also useful for conducting FRET (fluorescence resonance energy transfer) experiments. When BFP and GFP are in proximity, they can transfer energy via dipole–dipole interactions. Thus, excitation of a BFP can nonradiatively excite GFP to produce green emission photons. FRET is particulary useful because energy transfer is inversely proportional to the distance between the donor (BFP) and the acceptor (GFP) fluorophores raised to the sixth power. In practice, the fluorophores must be very close (i.e., <60 Å) for FRET to be detectable. Furthermore, small changes in the distance between the two fluorophores will produce a major change in FRET. With the use of GFP variants and FRET, it has been possible to detect proteolysis of a BFP–linker–GFP fusion protein.[13,14] A similar strategy has also been employed to develop a calcium-sensitive dye based on the fact that GFP–calmodulin–BFP fusion protein changes conformation on binding of calcium.[15] For channel researchers, FRET has the promise of allowing detection of changes in association of pore-forming subunits with regulators (e.g., G proteins) and cytoskeletal proteins.

[12] M. A. Bittner, M. K. Bennett, and R. W. Holz, *J. Biol. Chem.* **271,** 11214 (1996).
[13] R. Heim and R. Y. Tsien, *Curr. Biol.* **6,** 178 (1996).
[14] R. D. Mitra, C. M. Silva, and D. C. Youvan, *Gene* **173,** 13 (1996).
[15] V. A. Romoser, P. M. Hinkle, and A. Perschini, *J. Biol. Chem.* **272,** 13270 (1997).

To follow FRET, it is optimal to measure the ratio between the emission GFP and the BFP (F510/F440) after excitation of BFP at 380 nm. Following this ratio is important because the excitation light will directly excite some of the GFP. However, as FRET occurs emission by BFP should drop while emission by GFP should increase. Hence, an increase in the ratio should be seen when BFP- and GFP-tagged proteins directly interact. A failure to detect FRET does not exclude interaction, however. If the proteins are large and the fluorophores are oriented away from each other, distance constraints may come into play. The fact that $K^+$ channels are tetrameric may reduce this problem. On the other hand, the use of large fluorophores could provide steric hindrance and inhibit binding.

How can a channel physiologist perform such experiments with commonly used equipment? The easiest answer is to modify a dual-emission ratio setup used for measuring calcium with Indo-1 or pH with Snarf-1. Such setups can also be purchased from commercial vendors and the University of Pennsylvania Biomedical Instrumentation Group. Imaging could also be performed on a setup with a filter wheel on the emission path so that a camera could alternate between sampling at the two emission wavelengths. Numerous vendors sell software to drive filter wheels and to calculate ratio images.

## Measuring Channel Mobility with Fluorescent Recovery after Photobleaching

FRAP allows one to assay mobility of functional fluorescent channels in live cells. The strategy underlying the technique is that after photobleaching the GFP moiety in a small area of the cell, surrounding unbleached channels can diffuse into the area to produce a recovery of fluorescence. If all channels were freely diffusible, one would expect that recovery would proceed until the concentration of channels was uniform across bleached and unbleached areas. The diffusion coefficient ($D$) is proportional to the square of the spot radius and inversely proportional to the half-time of recovery ($t_{1/2}$).

In the case of channels, typically a very small area and hence a small percentage of channels is bleached. Thus, the level following recovery would be expected to be essentially identical to the level before bleaching. In fact, for real membrane proteins, recovery is never complete (Fig. 1). This implies that two populations of channels are present: one is mobile and accounts for the recovery, whereas the other cannot move on the time scale of the experiment. The latter population is usually referred to as the immobile fraction. However, such proteins do not have to be stationary. They could be moving very slowly or be restricted to moving in small subdomains of

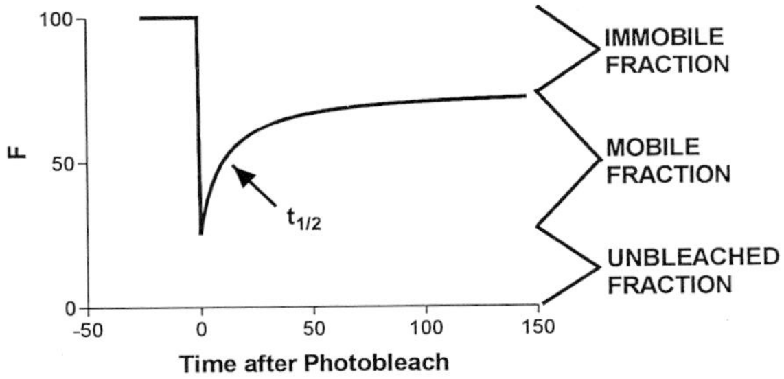


FIG. 1. Analysis of FRAP data. Photobleaching produces a drop in fluorescence in the bleached area. The extent of recovery gives the mobile fraction. In this case 67% of the channels were mobile and 33% were immobile. The kinetics of recovery is characterized $t_{1/2}$. The time course of recovery depends on the shape of the bleached region. The hyperbolic curve shown here is expected after bleaching with a gaussian spot.

membrane. Thus, FRAP experiments reveal the percentage of channels that are freely mobile and their diffusion coefficient.

## Performing FRAP Experiments with Scanning Laser Confocal Microscope

Traditionally, photobleaching has been accomplished by focusing a bright round spot of laser onto the cell with a modified epifluorescence microscope and recovery has been recorded with a video camera or a photomultiplier tube. This experimental design is best suited for flat cells (or for large round cells that approximate flat surfaces) and requires that labeled molecules be limited to the plasma membrane. This is not a problem when a water-soluble fluorophore is introduced from the bathing medium (e.g., labeled antibodies or water-soluble drugs). However, membrane permeant drugs or GFP tagging will label channels inside the cell. If intracellular compartments abut the plasma membrane, both compartments will be sampled. Even more important, performing spot photobleaching experiments is difficult for most channel researchers because the required setup is not commonly available.

These problems can be circumvented by using standard laser confocal microscopes to photobleach channels fused to GFP. Many universities and departments have core imaging facilities equipped with such instruments. Confocal microscopy offers advantages when studying round cells and when labeled channels are distributed among different organelles. By sampling a single plane of the cell (a $\sim$0.6-$\mu$m-deep slice using a small pinhole and a 60× NA 1.4 objective), one can clearly identify regions where different

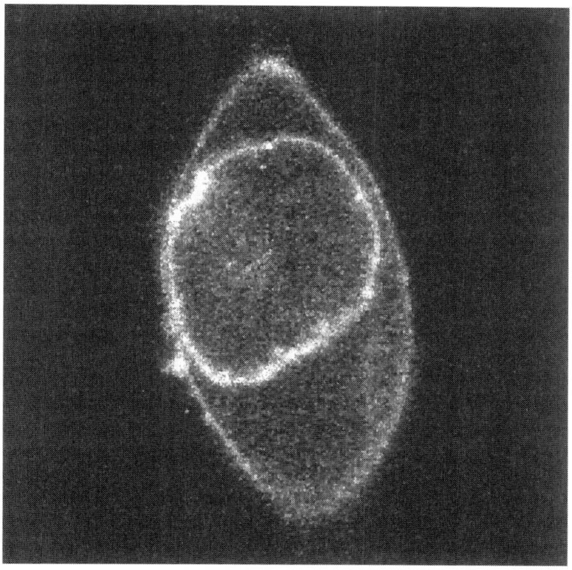

FIG. 2. Confocal image of a GFP-tagged $K^+$ channel expressed in 293 cells. Note that an intracellular compartment, likely the endoplasmic reticulum, abuts the plasma membrane. Regions in which these two compartments cannot be distinguished should be avoided when conducting FRAP experiments.

compartments encounter each other and avoid sampling those areas in FRAP experiments (Fig. 2). The slice through a spherical cell will yield a circle representing plasma membrane channels and, in the case of GFP–Kv1 channels, sometimes an inner circle likely representing subunits in the endoplasmic reticulum. Thus, GFP-tagged channels in different organelles can be studied.

Photobleaching can be produced by increasing illumination by either increasing the power to the laser or, more dramatically, by removing neutral density filters normally used to image GFP. On the Molecular Dynamics SCLM 2000 (Sunnyvale, CA), we usually set the laser power at 10 mW and image with the neutral density filter set at 3% light throughput. Under these conditions, GFP fluorescence is very stable. For photobleaching, rapid repetitive line scans at 100% illumination are used. This illumination should be brief compared to the kinetics of recovery. Because $K^+$ channel diffusion is slow, the change in neutral density can be produced manually. Photobleaching can be produced by a spot of light. However, using a line scan (i.e., a rapidly scanning spot) gives the ability to study FRAP simultaneously

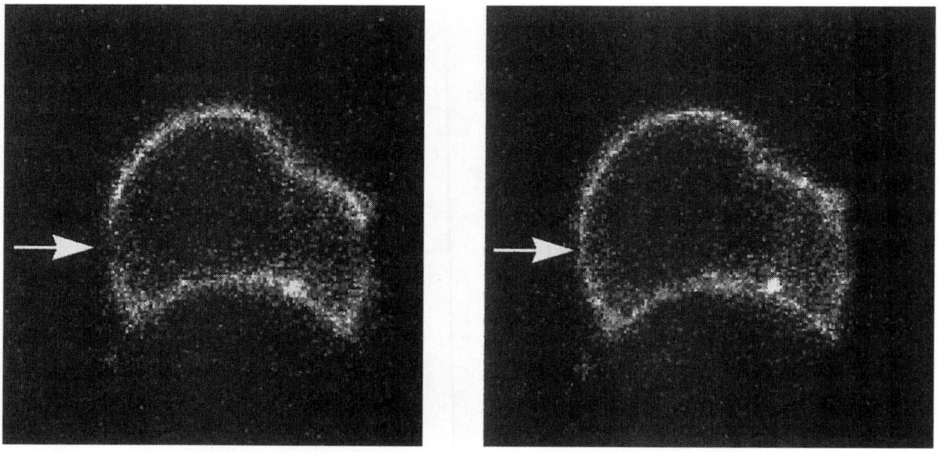

FIG. 3. Confocal images immediately after photobleaching (left) and 10 min after photobleaching (right). Note that there is recovery in the dark photobleached regions. Position of bleaching line indicated by the arrow.

in two regions of the plasma membrane (see Fig. 3). For very rapid diffusion, line scan mode can also be used to follow recovery. However, for the slow movement of $K^+$ channels, we photobleach in line scan mode, but collect full images to measure recovery. This approach allows one to examine signals throughout a confocal section.

## Artifacts, Corrections, and Calibrations

The quantitative relationship between the diffusion coefficient $D$ and $t_{1/2}$ is dependent on the pattern of illumination. Important factors include whether photobleaching is produced by a line or spot and whether the profile of illumination is Gaussian or even. Furthermore, the degree of photobleaching also affects recovery kinetics. Thus, obtaining exact values for $D$ requires application of equations deduced for the experimental design employed.[16,17] Two alternative approaches can be used for interpreting experimentally deduced values of $t_{1/2}$. First, one can take advantage of the fact that $D$ is always inversely proportional to $t_{1/2}$. Therefore, if an experimental manipulation (e.g., mutation of the channels, coexpression of a binding protein, or addition of a second messenger) alters $t_{1/2}$, then

[16] T. M. Jovin and W. L. C. Vaz, *Methods Enzymol.* **172,** 471 (1989).

[17] M. Edidin, *in* "Mobility and Proximity in Biological Membranes" (S. Damjanovich, M. Edidin, J. Szollosi, and L. Tron, eds.), p. 109. CRC Press, Boca Raton, Florida, 1994.

the fold change in $D$ can be deduced without determining its exact value. Alternatively, the diffusion of a known standard can be used to calibrate the setup. Fluorescent spherical beads can be used for this purpose because their diffusion is described by the Stokes–Einstein relation: $D = kT/6\pi\eta r$. This calibration also corrects for the imperfect geometry of real FRAP experiments. Ideally, one would like to photobleach a plane perpendicular to the focal plane. However, although confocal microscopes sample a thin plane, illumination is not confocal. Rather, a double cone of light is produced. By deducing an empirical scaling factor from a known standard, it is possible to determine the real value of $D$.

Another concern is that repetitive scanning and the intense illumination required for photobleaching could cause photodamage. We have performed two sets of controls that suggest that this is not the case. First, we have measured the effect of photobleaching on channel function. Because the GFP being bleached is on the same polypeptide as the channel, we reasoned that channel function should be as proximal as possible an indicator of nonspecific damage. We could not assay channel function on the confocal microscope. Nor could we assay channels in a small photobleached area. Therefore, whole cells were photobleached by exposure to an unattenuated mercury arc lamp light source through a high numerical aperture lens. Because the whole cell was illuminated for a long period of time, damage should be much greater than that produced by a local line scan or spot bleach. Nevertheless, in such experiments, loss of voltage-gated $K^+$ current was far less extensive than loss of fluorescence.[3] Another indication that photodamage is not significant has come from studies of GFP-tagged secretory granules. In those studies, the diffusion coefficients deduced from FRAP experiments and particle tracking (which was performed under low illumination) were similar.[18] Hence, photodamage was not significant with our protocols.

Artifacts can be produced by misuse of the confocal microscope. First, the confocal microscope must be properly aligned. The easiest way to test for proper function of the instrument is to image subresolution beads (e.g., $\leq$100 nm in diameter). XY and Z scans should be obtained. The resultant point spread function should be ellipsoid with long axis in the $Z$ direction. The dimensions of the point spread function (usually measured as the distance between the 50% intensity points encountering and leaving the image) should be compared to the sizes given by the manufacturer for a given pinhole and objective. Furthermore, in the $Z$ direction the point spread function should not be tilted. A nonsymmetric, tilted, or oversized

[18] N. V. Burke, W. Han, D. Li, K. Takimoto, S. C. Watkins, and E. S. Levitan, *Neuron* **19,** 1095 (1997).

point spread function is indicative of a problem in the microscope that must be addressed.

Another potential problem can arise from the photomultipliers used to count photons at each pixel while the laser is scanning the sample. The gain of the photomultiplier can be controlled by altering its voltage setting. It is critical that the photomultiplier be used in a linear response range. Because many instruments use 8-bit digitizers, this means that turning down the sensitivity to avoid saturation compromises the ability to resolve dim objects moving during recovery after photobleaching. Avoiding saturation is also important for the long-term performance of the photomultiplier. Bright laser light can cause permanent damage. Even the bright fluorescence produced during FRAP can lead to a period of reduced sensitivity. This should be avoided if possible by shuttering the photomultiplier or by decreasing its sensitivity during photobleaching. If this cannot be done and there is a global dimunition of fluorescence throughout the sample (i.e., far from the photobleached area), it may indicate that the sensitivity of the photomultiplier is compromised. Resultant data can be corrected for such an effect. However, protecting the photomultiplier from exposure to bright illumination while the sensitivity is turned up is a far better approach.

## Significance of $K^+$ Channel Mobility Measurements

For $K^+$ channels and other intrinsic membrane proteins, free diffusion is dominated by the viscosity of the membrane and is only weakly dependent on protein mass, size, or shape.[19] Therefore, mobility measurements reveal and quantitate cellular mechanisms that limit channel diffusion and localization. By combining GFP-tagging, site-directed mutagenesis, and FRAP approaches, it is possible to study the structural basis of channel movement and distribution in the plasma membrane and subcellular organelles. Furthermore, the influence of interactions with clustering proteins and chaperones can be assayed. Direct interactions between channels and binding partners can also be explored with FRET. Thus, optical approaches will allow channel physiologists to attack biochemical and cell biological questions in live cells instead of detergent extracts.

## Acknowledgments

I thank Drs. Daniel Axelrod (University of Michigan) and Frederick Lanni (Carnegie Mellon University) for insights into FRAP studies. ESL is supported by grants from the NIH (NS32385 and HL55312) and an Established Investigator Award from the American Heart Association.

[19] P. G. Saffman and M. Delbruck, *Proc. Natl. Acad. Sci. U.S.A.* **72,** 3111 (1975).

# [4] Spin-Label Electron Spin Resonance and Fourier Transform Infrared Spectroscopy for Structural/Dynamic Measurements on Ion Channels

*By* DEREK MARSH

## Introduction

This contribution deals with two forms of spectroscopy, spin-label electron spin resonance (ESR) spectroscopy and Fourier transform infrared (FTIR) spectroscopy, that are particularly well adapted to membrane and ion channel studies.[1] FTIR is a branch of vibrational spectroscopy, whereas spin label ESR is a form of magnetic resonance. These two techniques are capable of giving complementary structural and dynamic information and have the advantage that, unlike visible, fluorescence, or UV spectroscopy, they are insensitive to opaque or highly scattering particulate samples. The requirements for the quantity of sample are similar in both cases. Neither has the high sensitivity of, for example, fluorescence spectroscopy, but both have specific advantages in terms of the detailed information that may be obtained. In general, quantities in excess of 100 $\mu$g will be required, although higher sensitivity can be achieved by using attenuated total reflection techniques in infrared spectroscopy and loop–gap resonators, coupled with microwave preamplifiers, in the case of spin-label ESR spectroscopy.

Spin-label ESR is a probe technique that owes much of its success on the one hand to the versatile chemistry of stable nitroxyl free radicals, and on the other hand to the unique environmental, orientational, and motional sensitivity of magnetic resonance.[2] Spin-label spectral line shapes are sensitive to rotational motions on the time scale of those of membrane lipids, whereas the power saturation properties of the spin-label spectra are sensitive to slower motions, including the rotational motion of integral membrane proteins. The dynamic studies, both on lipid–protein interactions and on protein rotational diffusion, can give direct information also on membrane protein structure and assembly.[3] Additionally, there is a range of saturation-based relaxation measurements that gives information on

[1] J. B. C. Findlay and D. Marsh, *in* "Ion Channels—A Practical Approach" (R. Ashley, ed.), p. 241. IRL Press, Oxford, 1995.

[2] D. Marsh and L. I. Horváth, *in* "Advanced EPR. Applications in Biology and Biochemistry" (A. J. Hoff, ed.), p. 707. Elsevier, Amsterdam, 1989.

[3] D. Marsh, *in* "New Comprehensive Biochemistry, Vol. 25, Protein–Lipid Interactions" (A. Watts, ed.), p. 41. Elsevier, Amsterdam, 1993.





0076-6879/99 $30.00

accessibility to polar and apolar relaxants that is directly related to membrane protein structure and assembly.[4–6]

The amide vibrational bands in FTIR have frequencies that are characteristic of the particular hydrogen-bonding pattern of the peptide backbone, i.e., of the secondary structural elements of the protein.[7,8] Dichroic ratios of the intensities of the amide bands with polarized radiation, for aligned membrane samples, give the orientation of the secondary structural elements within the membrane.[9–11]

To date, the ion-channel applications of both spectroscopies are principally to a ligand-gated ion channel, the nicotinic acetylcholine receptor, or to synthetic peptides of sequence identical with those of the putative transmembrane segments of ion channels for which the sequence is known. These, together with the sodium pump and porin channels, will be used as illustrations. Applications to native channels, e.g., the $Na^+$ channel, are to be expected in the near future.

A previous publication related to experimental aspects of the determination of ion channel structure by spectroscopic techniques[1] can be consulted for further background. In the absence of a high-resolution three-dimensional structure of any ion channel, other than gramicidin, which consists of alternating D- and L-amino acids, and the outer membrane porins, spectroscopic studies of ion channel structure and studies on synthetic peptide model systems are particularly valuable.[12]

## Spin-Label Electron Spin Resonance

Usually, either the lipid component of the membrane or the channel protein itself is spin-labeled. Alternatively, both may be spin-labeled in double-labeling experiments to detect mutual spin–spin interactions. The spin-labeled lipid is present at probe concentrations (≤1 mol% relative to total lipid) and is intercalated spontaneously in the membrane on addition from a small volume of concentrated (≥ 1 mg/ml) ethanol solution. In the case of spin-labeled phospholipids, which may form their own vesicles,

[4] W. L. Hubbell and C. Altenbach, *in* "Membrane Protein Structure: Experimental Approaches" (S. H. White, ed.), p. 224. Oxford University Press, New York, 1994.

[5] D. Marsh, *Appl. Magn. Reson.* **3,** 53 (1992).

[6] D. Marsh, *Chem. Soc. Rev.* **22,** 329 (1993).

[7] D. M. Byler and H. Susi, *Biopolymers* **25,** 469 (1986).

[8] S. Krimm and J. Bandekar, *Adv. Protein Chem.* **38,** 181 (1986).

[9] K. J. Rothschild and N. A. Clark, *Biophys. J.* **25,** 473 (1979).

[10] D. Marsh, *Biophys. J.* **72,** 2710 (1997).

[11] L. K. Tamm and S. A. Tatulian, *Q. Rev. Biophys.* **30,** 365 (1997).

[12] D. Marsh, *Biochem. J.* **315,** 345 (1996).

unincorporated spin label is resolved from the spin-labeled membranes by centrifugation and washing. Membrane proteins are spin-labeled on reactive side chains by using nitroxide derivatives of the standard covalent modification reagents of protein chemistry. Again, unbound spin label is removed by centrifugation and washing. Specificity of covalent labeling can be achieved by exploiting differential reactivity or by the introduction of specific residues by site-directed mutagenesis (or directly by synthesis, in the case of peptides), e.g., in cysteine scanning experiments. In certain cases, different spin-labeled side chains can be distinguished by their spectral characteristics, i.e., their mobility.

*Practical Considerations*

Most commercially available 9-GHz continuous-wave ESR spectrometers are suitable for spin-label studies. This normal type of spectrometer uses field modulation (normally at 100 kHz) and phase-sensitive detection, which results in the ESR spectra being displayed as the first derivative with respect to the magnetic field scan. The field modulation amplitude should be less than the line width ($\leq 1$ gauss), except in the case of the nonlinear saturation transfer ESR applications. The spectra are centered in the familiar $g = 2$ region (at ca. 3250–3400 gauss, depending on the exact microwave frequency) and do not require particularly high resolution or a large sweep width of the magnetic field (usually 100 gauss). The standard microwave cavity may be used, but a loop–gap resonator combined with a microwave preamplifier,[13] or a dielectric resonator, may be advantageous when quantities of sample are strictly limited. In standard applications, the microwave power should be set to avoid appreciable saturation ($\leq$10 mW). For saturation transfer ESR studies, moderately saturating microwave powers are used, and in progressive saturation experiments the microwave power is stepped systematically. Sample temperature is controlled by thermostatted gas flow, through a double-wall quartz Dewar that inserts into the microwave cavity. Temperature can be measured with a fine-wire thermocouple positioned close to the sample. Care must be taken to avoid microwave heating in power saturation studies.

For saturation transfer ESR studies, it must be possible to set the phase of the lock-in detector accurately (0.1–1° precision). For progressive saturation studies, accurate calibration of the microwave attenuator is necessary. In both cases, critical coupling of the microwave power into the cavity is essential, and corrections for changes in cavity $Q$-factor with differ-

[13] J. S. Hyde and W. Fronciscz, *in* "Advanced EPR. Applications in Biology and Biochemistry" (A. J. Hoff, ed.), p. 277. Elsevier, Amsterdam, 1989.

ent samples must be made to achieve a given microwave $H_1$ field at the sample.[14]

Further details of the instrumental aspects of ESR can be found in Ref. 15 and of specific applications to spin-labeling in Ref. 16.

The most generally useful disposable sample cells for membrane studies are ≤1-mm-i.d. thin glass capillaries, such as are used for positive-delivery pipettes. These can be readily flame sealed. Quartz sample capillaries are preferred for sensitive applications to reduce dielectric loss (and any background signals), but are seldom used at 9-GHz frequencies. Quartz flat cells with 0.3-mm spacing are the ideally preferred geometry for aqueous samples in a rectangular microwave cavity, but are inconvenient for use with particulate membrane samples, except in their adaptation for oriented planar membrane systems. They are also less easy to thermostat. Sample capillaries can be filled with a drawn-out Pasteur pipette after homogenizing the membrane suspension. The membrane sample is then packed by low-speed centrifugation in a bench-top, or hematocrit, centrifuge. Excess supernatant is removed to minimize microwave losses associated with liquid water, leaving only sufficient buffer to prevent the sample from drying out. A final membrane pellet of 5 mm height is ideal. The sealed sample capillary is accommodated within a standard 4-mm-diameter quartz ESR tube, which contains dry light silicone oil as a thermal buffer. The membrane sample is centered in the microwave cavity so as to experience the maximum microwave and modulation fields. For saturation transfer ESR and progressive saturation experiments, the alignment of the sample capillary along the vertical axis of the microwave cavity is also important.

### *Lipid–Protein Interactions*

A spin label attached to the 14C-atom of stearic acid [14-(4,4-dimethyloxazolidine-*N*-oxyl)stearic acid] is usually optimal for detecting direct interactions of the lipid chains with the intramembranous segments of integral proteins. This gives rise to two-component ESR spectra, in which the lipid spin-label population interacting with the integral protein may be distinguished readily from that in the fluid bilayer regions of the membrane, and quantitated by spectral subtraction.[17] The synthesis of suitable spin-

[14] P. Fajer and D. Marsh, *J. Magn. Reson.* **49,** 212 (1982).

[15] P. F. Knowles, D. Marsh, and H. W. E. Rattle, "Magnetic Resonance of Biomolecules." Wiley-Interscience, London, 1976.

[16] D. Marsh, *in* "Techniques in Lipid and Membrane Biochemistry" (J. C. Metcalfe and T. R. Hesketh, eds.), Vol. B4/III, B426, p. 1. Elsevier, Amsterdam, 1982.

[17] D. Marsh, *in* "Progress in Protein–Lipid Interactions" (A. Watts and J. J. H. H. M. de Pont, eds.), p. 143. Elsevier, Amsterdam, 1985.

labeled fatty acids and the corresponding spin-labeled phospholipids is described in Refs. 18 and 19. Spin-labeled fatty acids are commercially available from Sigma (St. Louis, MO), and spin-labeled phosphatidylcholine phospholipids from Avanti Polar Lipids (Alabaster, AL). Of the commercially available spin-label positional isomers, those labeled on C-12 or C-16 of the acyl chain are the most suitable for the study of lipid–protein interactions.

Quantitation of the relative populations of the two-component ESR spectra is performed by interactive spectral subtraction, using digital libraries of single-component reference spectra. The latter are recorded at closely spaced temperature intervals in order to obtain optimum matching of the spectral line shapes. Spin-labeled samples of the extracted membrane lipid dispersed in buffer can be used to match the fluid bilayer component in the composite two-component spectra. Spin-labeled bilayers of a synthetic phospholipid (e.g., dimyristoylphosphatidylcholine) in the gel phase, i.e., below the chain-melting transition temperature, can be used to match the motionally restricted, protein-interacting component in the composite spectrum. It is advisable to sonicate the samples, in the latter case, in order to obtain a wider range of spectral line shapes and splittings at the different temperatures. Sonicated synthetic lipid bilayer dispersions in the fluid phase, above the chain-melting temperature, may be used to obtain reference spectra for the fluid lipid component in membrane systems reconstituted with the corresponding phospholipid. Spectral subtraction is performed by an interactive protocol.[16,19] A trial difference spectrum is compared with the complementary single-component reference spectra, in order to identify that which best matches the difference spectrum. This reference spectrum is then subtracted from the original composite spectrum, and the resulting difference spectrum compared with the complementary library of reference spectra, in order to choose the optimum match for the latter, and so on. For optimizing the endpoint in subtracting the motionally restricted component, the difference spectrum should be displayed with vertical expansion for sensitive comparison with the complementary reference in the outer wings of the fluid component.

The quantity obtained from the spectral subtractions is the fraction, $f$, of spin-labeled phospholipid that is motionally restricted, i.e., is associated directly with the intramembranous surface of the protein. This is obtained from the subtraction factor, together with double integration of the composite and reference first-derivative spectra that is needed to obtain normalized

[18] D. Marsh and A. Watts, *in* "Lipid–Protein Interactions" (P. C. Jost and O. H. Griffith, eds.), p. 53. Wiley-Interscience, New York, 1982.

[19] D. Marsh and A. Watts, *Methods Enzymol.* **88,** 762 (1982).

intensities. Ideally, the complementary subtractions should yield consistent values of $f$ and $1 - f$, respectively. If not, the two resulting independent values of $f$ may be averaged to compensate for slight errors in the complementary subtractions.

The fractional population, $f$, of motionally restricted lipids yields information on both the stoichiometry and the selectivity of the lipid–protein interaction. The number of lipid association sites on the protein is obtained from the equation for equilibrium lipid exchange association with the protein by[17,20]:

$$N_b = n_t/[1 + K_r(1 - f)/f] \quad (1)$$

where $n_t$ is the lipid–protein ratio in the membrane and $K_r$ is the association constant of the spin-labeled lipid probe with the protein, relative to that of the unlabeled background host lipid. Several studies have shown that $K_r = 1$ for a spin-labeled analog of the background lipid, i.e., that spin-labeling of the lipid does not disturb the thermodynamics of the lipid–protein association. The number, $N_b$, of lipid association sites at the intramembranous surface of the protein may therefore be determined directly by using a spin-labeled phospholipid (usually phosphatidylcholine) for which $K_r = 1$. Chemical analysis is used to obtain the lipid–protein ratio, $n_t$, and then the stoichiometry is simply $N_b = f_o n_t$, where $f_o$ is the fractional motionally restricted population of the nonselective spin label. Specificities of interaction can be determined by measuring $f$ for different spin-labeled lipid species in membranes of fixed lipid/protein ratio, then $K_r = f(1 - f_o)/[f_o(1 - f)]$, using the nonselective spin label as reference. If the lipid–protein ratio can be varied, either by reconstitution or by progressive delipidation of the membrane, the dependence on $n_t$ can be fitted to Eq. (1) to yield values of both $N_b$ and $K_r$ simultaneously. Such lipid–protein titrations require that $n_t > N_b$, in order to avoid nonspecific protein–protein aggregation.

This type of lipid–protein interaction can be directly relevant to channel function. It has been shown that the lipid–protein interface is the site of noncompetitive blocking action of the nicotinic acetylcholine receptor channel by aminated local anaesthetics. This was established both by using spin-labeled local anaesthetic analogs and by competition of unlabeled local anaesthetics with spin-labeled lipids.[21] More generally, the stoichiometry and selectivity of lipid–protein interaction can be used to obtain information

[20] J. R. Brotherus, O. H. Griffith, M. O. Brotherus, P. C. Jost, J. R. Silvius, and L. E. Hokin, *Biochemistry* **20,** 5261 (1981).

[21] L. I. Horváth, H. R. Arias, H. O. Hankovsky, K. Hideg, F. J. Barrantes, and D. Marsh, *Biochemistry* **29,** 8707 (1990).

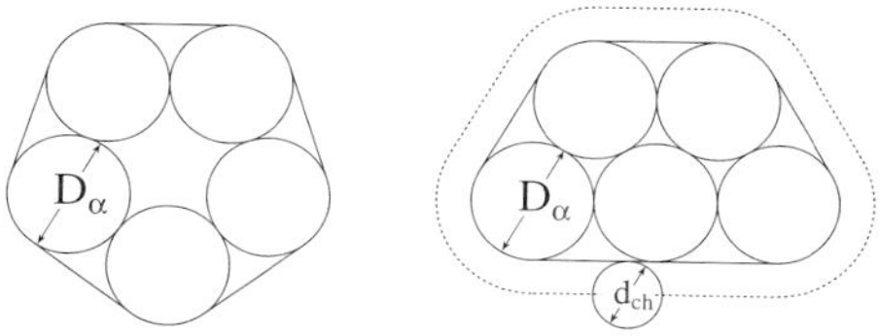


FIG. 1. Assembly of transmembrane $\alpha$ helices in a regular polygonal arrangement (left) and in a helical sandwich arrangement (right). The number of diacyl phospholipids, $N_b$, that can be accommodated as a first shell surrounding the protein in a lipid bilayer membrane is given by Eq. (2), where $D_\alpha$ is the helix diameter, and $d_{ch}$ is the diameter of a lipid chain. For a regular polygonal arrangement, the radius of the central pore is given by: $R_p = D_\alpha[1/\sin(\pi/n_\alpha) - 1]/2$.

on the intramembranous assembly of integral channel proteins. A selectivity for negatively charged lipids locates some of the basic residues of the protein close to the phospholipid headgroups at the membrane surface.[3] This already can constrain the transmembrane disposition of putative channel peptides, given that the secondary structure is established by FTIR,[22] and can also be a very sensitive structural indication of the response to site-specific mutagenesis.[23] These latter examples refer to the single putative transmembrane segment of the IsK (or mink) protein, which on expression induces slowly activating voltage-gated $K^+$ channels.[24]

The lipid–protein stoichiometry is a direct measure of the size of the intramembranous perimeter of the channel protein. This, in turn, is determined by the assembly of the transmembrane segments and also by the degree of protein oligomerization. For a regular polygonal assembly of transmembrane helices, or for a helical sandwich arrangement, the lipid–protein stoichiometry is given by[3,25]:

$$N_b = \pi(D_\alpha/d_{ch} + 1) + n_\alpha D_\alpha/d_{ch} \qquad (2)$$

where $n_\alpha$ is the number of transmembrane helices in the assembly, $D_\alpha =$ 1.0 nm is the diameter of an $\alpha$ helix, and $d_{ch} = 0.48$ nm is the diameter of a lipid chain (see Fig. 1). [For multilayer sandwiches, Eq. (2) is valid for $1 < n_\alpha < 7$, but for $n_\alpha \geq 7$ centered hexagonal arrangements are possible, in which one of the helices blocks the pore. In this case, $n_\alpha$ is replaced by $(n_\alpha - 1)$, etc.] For the $Na^+,K^+$-ATPase ion pump, the number of motionally

[22] L. I. Horváth, T. Heimburg, P. Kovatchev, J. B. C. Findlay, K. Hideg, and D. Marsh, *Biochemistry* **34,** 3893 (1995).

[23] L. I. Horváth, P. F. Knowles, P. Kovatchev, J. B. C. Findlay, and D. Marsh, *Biophys. J.* **73,** 2588 (1997).

[24] T. Takumi, H. Ohkubo, and S. Nakanishi, *Science* **242,** 1042 (1988).

[25] D. Marsh, *Eur. Biophys. J.* **26,** 203 (1997).

restricted lipids per $\alpha\beta$ protomer is found to be $N_b \approx 32$–33, which corresponds reasonably well with the predicted number of transmembrane helices of $n_\alpha = 11$.[20,26,27] For the nicotinic acetylcholine receptor, the number of motionally restricted lipids per $\alpha_2\beta\gamma\delta$ monomer is $N_b \approx 40$,[28] which is less than would be predicted for an all-helical transmembrane domain (i.e., $n_\alpha = 20$). However, electron crystallographic studies suggest that the intramembranous secondary structure of the nicotinic acetylcholine receptor may be of a mixed $\alpha$-$\beta$ type.[29] Equation (2) also applies approximately to oligomerization of protein subunits, where $D$ then becomes the diameter of the subunit and $n$ is the aggregation number.

Alternatively, for a $\beta$-barrel transmembrane protein, the number of lipids that can be accommodated at the perimeter of the $\beta$ barrel is given approximately by[25]:

$$N_b = n_\beta D_\beta/(d_{ch} \cos \gamma_\beta) \qquad (3)$$

where $D_\beta = 0.47$ nm is the interstrand separation, $\gamma_\beta$ is the tilt angle of the $\beta$ strand relative to the membrane normal, and $n_\beta$ is the number of $\beta$ strands in the barrel. Because $\beta$ strands are much more extended peptide structures than are $\alpha$ helices, a tilt of $\gamma_\beta \approx 60°$ is required for the same number of nonpolar residues to be accommodated within the membrane as for an $\alpha$ helix. For such a highly tilted $\beta$-sheet structure of a transmembrane peptide from the IsK channel-associated protein, the value of the lipid stoichiometry found experimentally is $N_b/n_\beta \approx 2$ per strand, in agreement with predictions from Eq. (3).[22,23,30]

## *Saturation Transfer ESR*

The first requirement for a saturation transfer ESR measurement of protein rotational diffusion is the lack of independent segmental motion of the spin label relative to the protein. For optimization, the covalent attachment of the spin label should be as short and as rigid as possible. A series of $\beta$-vinyl ketone reagents has been explored for this purpose,[31,32]

[26] M. Esmann, A. Watts, and D. Marsh, *Biochemistry* **24,** 1386 (1985).

[27] M. Esmann, K. Hideg, and D. Marsh, *in* "The Na$^+$, K$^+$-Pump, Part A: Molecular Aspects. Progress in Clinical and Biological Research" (J. C. Skou, J. G. Norby, A. B. Maunsbach, and E. Esmann, eds.), p. 189. Alan Liss, New York, 1988.

[28] J. F. Ellena, M. A. Blazing, and M. G. McNamee, *Biochemistry* **22,** 5523 (1983).

[29] N. Unwin, *J. Mol. Biol.* **229,** 1101 (1993).

[30] A. Aggeli, N. Boden, Y.-L. Cheng, J. B. C. Findlay, P. F. Knowles, P. Kovatchev, P. J. H. Turnbull, L. I. Horváth, and D. Marsh, *Biochemistry* **35,** 16213 (1996).

[31] M. Esmann, H. O. Hankovszky, K. Hideg, J. A. Pedersen, and D. Marsh, *Anal. Biochem.* **189,** 274 (1990).

[32] M. Esmann, P. C. Sar, K. Hideg, and D. Marsh, *Anal. Biochem.* **213,** 336 (1993).

TABLE I
RECOMMENDED INSTRUMENTAL SETTINGS FOR SATURATION TRANSFER ESR IN SECOND HARMONIC, OUT-OF-PHASE ABSORPTION MODE[a]

| [Spin label] ($\mu M$) | Sample length (mm) | $\langle H_1^2 \rangle^{1/2}$ (gauss) | $H_m$ (gauss) |
|---|---|---|---|
| ≥30 | 5 | 0.25[b] | 5.0 |
| ≤30 | >20 | 0.25 | 10.0 |

[a] Adapted from M. A. Hemminga, P. A. de Jager, D. Marsh, and P. Fajer, *J. Magn. Reson.* **59,** 160 (1984).

[b] These values correspond to microwave powers in the region of 30–40 mW, depending on the sample configuration. The power must be adjusted for differences in the $Q$ factor of the microwave cavity with different samples [P. Fajer and D. Marsh, *J. Magn. Reson.* **49,** 212 (1982)].

of which the indanedione derivative is one of the most suitable.[22] Of the commercially available spin-labeled sulfhydryl reagents, the maleimide with the shortest link and a five-membered pyrrolidine ring, 3-maleimido-2,2,5,5-tetramethyl-1-pyrrolidinyloxy, is recommended for saturation transfer ESR studies (Sigma Chemical Co.).

Standard saturation transfer ESR spectra are recorded as the second harmonic absorption signal detected 90° out of phase with respect to the field modulation (50-kHz modulation with 100-kHz detection), under conditions of partial microwave saturation. Saturation transfer spectra are considerably less intense than normal in-phase spectra, but their intensity increases with slower molecular rotational motion. Moderately high microwave fields, $H_1$, and relatively high modulation fields, $H_m$, are required. Recommendations for instrumental settings are given in Table I; those for a sample of 5-mm length have become standard. Small samples (of 5-mm length) are much preferred to avoid the effects of inhomogeneities in the $H_1$ and $H_m$ fields.[14] Adjustments of the microwave power, $P$, are necessary to maintain a fixed standard $H_1$ field at the sample, depending on the cavity $Q$ factor, such that the product $PQ$ is maintained constant. The $Q$ factor depends on the sample properties and is measured from the frequency separation, $\Delta\nu$, between the points of half-power absorption of the cavity dip in the klystron mode: $Q = \nu_o/\Delta\nu$, where $\nu_o$ is the resonant frequency of the cavity. The $H_1$ field of the sample is calibrated with an easily saturable sample that has narrow Lorentzian lines. For a standard peroxylamine disulfonate solution, the saturation broadening of the first derivative spectrum as a function of $H_1$ is $\Delta H_{pp}^2 = \Delta H_{pp}^{o^2} + (\frac{4}{3})H_1^2$, where $\Delta H_{pp}$ is the peak-to-peak line width that is readily measurable.[33]

[33] R. G. Kooser, W. V. Volland, and J. H. Freed, *J. Chem. Phys.* **50,** 5243 (1969).

A crucial feature of saturation transfer ESR experiments is the accurate setting of the receiver phase. The 90° out-of-phase position is established by minimizing the signal from the sample at subsaturating microwave powers (normally <1 mW). Typically, out-of-phase nulls should be less than 1% of the intensity of the saturation transfer spectrum recorded at high power. Alternatively, a nonsaturable reference sample, which has a spectrum outside the region of interest for spin labels, can be used. A single crystal of a concentrated paramagnetic ion salt, which has a spectrum at room temperature is a suitable reference. Further details of the instrumental and practical aspects of saturation transfer ESR can be found in Ref. 2.

The rotation correlation times or diffusion coefficients are obtained from the line shapes of the spin-label saturation transfer ESR spectra. The line heights, $L''$ and $H''$, at intermediate positions in the low- and high-field regions of the spectrum are sensitive to motion. The ratios of $L''$ and $H''$ to the line heights, $L$ and $H$, of the corresponding peaks at the extrema of the spectrum are used for calibration in terms of the rotational correlation time.[34] Calibrations are also possible for a diagnostic line height ratio, $C'/C$, defined in the center of the spectrum, and in terms of the normalized integrated intensity, $I_{ST}$, of the saturation transfer ESR spectrum.[35] The latter is defined as the ratio of the first integral of the saturation transfer ESR spectrum to the second integral of the conventional, in-phase first-derivative spectrum, and has the advantage of being additive in multicomponent systems. The calibrations of rotational correlation time, $\tau_R$ [$=1/(6D_R)$], are established with spin-labeled hemoglobin in solutions of known viscosity, and may be expressed in the following form[36]:

$$\tau_R^{eff} = k/(P_o - P) + b \quad (4)$$

where $P$ is the diagnostic spectral parameter ($L''/L$, $H''/H$, $C'/C$ or $I_{ST}$), $P_o$ is the limiting value of $P$ for very slow motion, and $k$, $b$ are the calibration constants. The values of these calibrations are given in Table II for each of the diagnostic parameters. The calibrations are established with isotropic solutions and some correction is necessary for the anisotropic, uniaxial rotation of integral proteins in membranes. In the latter case, the correlation time, $\tau_{R\|}$, for rotation around the membrane normal is given by[2,37]:

$$\tau_{R\|} = \frac{1}{2}\tau_R^{eff}\sin^2\theta \quad (5)$$

[34] D. D. Thomas, L. R. Dalton, and J. S. Hyde, *J. Chem. Phys.* **65,** 3006 (1976).
[35] L. I. Horváth and D. Marsh, *J. Magn. Reson.* **54,** 363 (1983).
[36] D. Marsh and L. I. Horváth, *J. Magn. Reson.* **99,** 332 (1992).
[37] B. H. Robinson and L. R. Dalton, *J. Chem. Phys.* **72,** 1312 (1980).

TABLE II
PARAMETERS FOR ROTATIONAL CORRELATION TIME CALIBRATIONS OF DIAGNOSTIC LINE HEIGHT RATIOS[a] AND NORMALIZED INTEGRAL INTENSITIES[b] OF SECOND-HARMONIC SATURATION TRANSFER ESR SPECTRA[c]

| $P$ | Range fitted | | $P_0$ | $k(\mu s)$ | $b(\mu s)$ |
|---|---|---|---|---|---|
| $L''/L$ | 0.2 | 2.0 | 1.825 | 105.6 | 63.8 |
| $H''/H$ | 0.2 | 2.0 | 2.17 | 407 | 210 |
| $C'/C$ | 0.2 | 1.0 | 1.01 | 21.3 | 21.1 |
| | −0.4 | 1.0 | 0.976 | 11.9 | 7.82 |
| $I_{ST}$ | $0.15 \times 10^{-2}$ | $1.0 \times 10^{-2}$ | $1.07 \times 10^{-2}$ | 0.400 | 43.6 |

[a] $P \equiv L''/L$, $H''/H$, or $C''/C$.
[b] $P \equiv I_{ST}$.
[c] According to Eq. (4). Adapted from D. Marsh, *Appl. Magn. Reson.* **3,** 53 (1992).

where $\theta$ is the angle that the $z$-principal axis of the spin-label nitroxide group makes with the membrane normal. For $\theta = 90°$, which is the most probable value (especially for a multiply labeled protein), the rotational correlation time has its maximum value: $\tau_{R\parallel} = \tau_R^{eff}/2$. The latter is therefore an upper estimate for $\tau_{R\parallel}$.

The rotational correlation time is related directly to the volume, $V_m$, of the portion of the protein that is within the membrane[2,38]:

$$\tau_{R\parallel} = 2\eta_m V_m/(3F_{R\parallel}k_B T) \tag{6}$$

where $\eta_m \approx 2.5$–$5$ P is the effective viscosity within the membrane, $k_B$ is Boltzmann's constant, and $T$ is the absolute temperature. For proteins of noncircular cross section, the shape factor $F_{R\parallel}$ in Eq. (6) is given by:

$$F_{R\parallel} = \frac{2(b/a)}{1 + (b/a)^2} \tag{7}$$

where $a$ and $b$ are the semiaxes of the assumed elliptical cross section. For a circular cross section, $F_{R\parallel} = 1$. Clearly from Eq. (6), the rotational correlation time is related directly to the degree of aggregation, $n$, of the channel protein: $\tau_{R\parallel}$ ($n$-mer) $= n\ \tau_{R\parallel}$ (monomer), which makes it possible to study aggregation processes.

In addition, if values are assumed from the membrane viscosity and spin-label orientation, the cross-sectional area, $A_m = V_m/h_m$, where $h_m \approx 4.5$ nm is the membrane thickness, can be estimated. For a regular polygonal arrangement of $n_\alpha$ $\alpha$ helices, the outer radius is given by $R_m = D_\alpha(1 + 1/\sin \pi/n_\alpha)/2$, for example (cf. Fig. 1).

[38] F. Jähnig, *Eur. Biophys. J.* **14,** 63 (1986).

The rotational correlation time in Eq. (6) is determined solely by the volume of the transmembrane section of the protein, because the aqueous viscosity is negligible in comparison with that of the membrane. If, however, the viscosity of the aqueous phase is increased by addition of, e.g., sucrose or glycerol, the volume, $V_o$, of the extramembranous portion of the protein can be determined. The dependence of the rotational correlation on the external viscosity, $\eta_o$, is given by[39]:

$$\tau_{R\|}(\eta_o) = \tau_{R\|}(0) + 2\eta_o V_o/(3F_{R\|}k_B T) \tag{8}$$

where the correlation time for zero external viscosity, $\tau_{R\|}(0)$, is given by Eq. (6). From the ratio of the gradient to the intercept of the viscosity dependence, the ratio $V_o/V_m$ of the volumes of the extra- and intramembranous parts of the protein can be obtained.

The rotational correlation times of the $Na^+,K^+$-ATPase ion pump in native membranes are found to be consistent with a diprotomer or small oligomer.[40] From the dependence on aqueous viscosity, it is estimated that approximately 50–70%, of the protein is external to the membrane, in reasonable agreement with results from low-resolution electron crystallography.[39] For the nicotinic acetylcholine receptor, however, the effective rotational correlation times in native membranes from *Torpedo* electroplax indicate strong immobilization, and the receptor is only free to rotate after removal of the 43-kDa extrinsic protein by alkaline extraction.[41,42]

### *Saturation and Relaxation Enhancements*

There is a range of structurally sensitive spin-label experiments that relies on the spin-lattice relaxation enhancement of spin-labeled proteins by a second paramagnetic species. These measurements give the accessibility of the relaxant to the spin label in the case of the Heisenberg spin exchange interactions, or the distance of closest approach of the relaxant to the spin label in the case of magnetic dipole–dipole interactions with paramagnetic ions. Suitable relaxants are molecular oxygen, which is preferentially concentrated in the hydrophobic interior of membranes, paramagnetic ions, or their complexes, which are confined to the aqueous phase, or a spin-labeled lipid. Microwave saturation experiments are used because these are more sensitive to weak spin–spin interactions than are line broadening experiments (i.e., $T_2$ relaxation).

[39] M. Esmann, K. Hideg, and D. Marsh, *Biochemistry* **33,** 3693 (1994).
[40] M. Esmann, L. I. Horváth, and D. Marsh, *Biochemistry* **26,** 8765 (1987).
[41] A. Rousselet, J. Cartaud, and P. F. Devaux, *Biochim. Biophys. Acta* **648,** 169 (1981).
[42] A. Rousselet, J. Cartaud, P. F. Devaux, and J.-P. Changeux, *EMBO J.* **1,** 439 (1982).

Suitable spin-labeled reagents for labeling membrane proteins are derivatives of maleimide, iodoacetamide, isothiocyanate, or chloromercury compounds, all of which are available from Sigma. The methanethiosulfonate spin label is available from Reanal (Budapest, Hungary).

The most straightforward and generally applicable method of detecting spin-lattice relaxation enhancements is the progressive saturation experiment. The microwave power is steadily increased and departures from the linear increase in signal strength with the $H_1$ field (i.e., square root of the microwave power) are an indication for saturation, which is a measure of the spin-lattice relaxation rate. The dependence of the ESR signal strength on $H_1$ (or microwave power) is given by:

$$S(H_1) = \frac{S_o H_1}{(1 + \gamma_e^2 H_1^2 T_1 T_2)^{\varepsilon}} \tag{9}$$

where $T_1$ is the spin-lattice relaxation time, $T_2$ is the transverse relaxation time (which determines the line widths), and $\gamma_e = 1.761 \times 10^7\ s^{-1}\ gauss^{-1}$ is the gyromagnetic ratio of the electron. For saturation of the integrated intensity of the ESR absorption (double integral of the first-derivative spectrum), the exponent in the denominator of Eq. (9) is $\varepsilon = 1/2$, in all cases. For a Lorentzian line, $\varepsilon = 1$ for the height of the absorption spectra and $\varepsilon = 3/2$ for the height of the first-derivative spectrum. However, inhomogeneous broadening of the Lorentzian lines, which is nearly always present in spin-label spectra, leads to different values of $\varepsilon$, which have to be determined empirically in the two latter cases. It is therefore preferable to use the integrated intensity of the absorption (which is insensitive to inhomogeneous broadening) in fitting the power saturation curves with a fixed value of $\varepsilon = 1/2$.[43,44] The quantity that is obtained from the saturation curves is then the $T_1T_2$ relaxation time product. Mostly, the relaxation enhancement will have little influence on $T_2$ and the $T_1T_2$ product yields the enhancement in spin-lattice relaxation directly. To determine $T_1$ from the $T_1T_2$ product, $T_2$ may be estimated from the (Lorentzian) line width, $\Delta H$ $(=1/\gamma_e T_2)$ (see, e.g., Ref. 45). The latter is important when comparing progressive saturation results from spin labels with different line widths.

Similar instrumental and technical considerations apply to ESR saturation studies in general, as those already enumerated for saturation transfer ESR. Small (5-mm-long) samples, carefully centered in the microwave

[43] T. Páli, L. I. Horváth, and D. Marsh, *J. Magn. Reson.* **A101,** 215 (1993).

[44] D. Marsh, T. Páli, and L. I. Horváth, *in* "Biological Magnetic Resonance, Vol. 14: Spin Labeling: The Next Millenium" (L. J. Berliner, ed.) p. 23. Plenum Publishing Corp., New York, 1998.

[45] D. Stopar, K. A. J. Jansen, T. Páli, D. Marsh, and M. A. Hemminga, *Biochemistry* **36,** 8261 (1997).

cavity, are required, and the microwave $H_1$ field at the sample should be calibrated. Accurate setting of the detector phase is not important for progressive saturation experiments, but is equally important as in saturation transfer ESR for other nonlinear ESR experiments that are sensitive to $T_1$. A typical example of the protocol for analyzing power saturation experiments on site-specific spin-labeled mutants of an integral membrane protein is given in Ref. 45.

In addition, the normalized intensities, $I_{ST}$, of the saturation transfer ESR spectra are sensitive to spin-lattice relaxation. As a first approximation, the value of $I_{ST}$ is linearly proportional to $T_1$.[46] For spin labels that have appreciable saturation transfer ESR intensities, this is therefore a more sensitive means of detecting spin-lattice relaxation enhancements than are progressive saturation experiments [cf. Eq. (9) with $\varepsilon = 1/2$]. Applications of this new nonlinear ESR method have been reviewed recently.[44] A promising alternative nonlinear ESR method for measuring spin-lattice relaxation enhancements is the 90° out-of-phase first harmonic absorption signal.[47] This has stronger signals than the second-harmonic saturation transfer ESR spectra and is less sensitive to $T_2$-relaxation effects. Further, the sensitivity to different ranges of $T_1$ may be tuned by varying the modulation frequency, while still maintaining adequate signal strength of the out-of-phase first harmonic.

The enhancements in spin-lattice relaxation rate, $1/T_1$, depend linearly on the concentration [R] of relaxant[44]:

$$1/T_1 = 1/T_1^{\circ} + k_{RL}[R] \tag{10}$$

where $T_1^{\circ}$ is the value of $T_1$ in the absence of relaxant, and $k_{RL}$ depends on the translational diffusion coefficient and cross section for collision of the relaxant, in the case of Heisenberg spin exchange, or on the distance of closest approach in the case of magnetic dipole–dipole interactions with paramagnetic ions. For a progressive saturation experiment, in which the spin–label line widths are unchanged by the paramagnetic relaxant, Eq. (10) may be written in terms of the $T_1T_2$ product in the presence and absence of relaxant:

$$1/(T_1T_2) = 1/(T_1T_2)^{\circ} + (k_{RL}/T_2^{\circ})[R] \tag{11}$$

where $T_2^{\circ}$ is the constant value of $T_2$. The experimentally determined quantity, i.e., $k_{RL}/T_2^{\circ}$, must therefore be corrected by a factor $1/\Delta H$, when comparing data from spin labels with different line widths.[48] Standard exper-

[46] T. Páli, V. A. Livshits, and D. Marsh, *J. Magn. Reson.* **B113,** 151 (1996).
[47] D. Marsh, V. A. Livshits, and T. Páli, *J. Chem. Soc. Perkin 2,* 2545 (1997).
[48] M. M. E. Snel and D. Marsh, *Biochim. Biophys. Acta* **1150,** 155 (1993).

iments for determining $k_{RL}$ involve comparing the relaxation rate for a sample saturated with argon or nitrogen with that for samples saturated with oxygen or in solutions of paramagnetic ions or their complexes in concentrations up to 30 m*M*.

In the case of Heisenberg spin exchange interactions, i.e., of contact interactions, the values of $k_{RL}$ may be used as a measure of the accessibility of the relaxant to the spin label.[4,48] Molecular oxygen, a paramagnetic relaxant that is preferentially partitioned into the membrane, displays a depth-dependent concentration profile that increases monotonically toward the center of the membrane. Paramagnetic ions, e.g., hexaaquo $Ni^{2+}$, and their charged complexes, e.g., chromium oxalate, are confined to the aqueous phase. Electroneutral paramagnetic ion complexes, e.g., nickel acetyl acetonate or ethylenediaminediacetic acid (NiEDDA), partition into the membrane with a strongly decreasing concentration profile toward the center. The ratio of the relaxation enhancements by oxygen and neutral paramagnetic ion complexes therefore provides a sensitive method to determine the vertical location of a spin-label group within the membrane.[49] In the case of site-directed spin-labeling with cysteine-scanning mutagenesis, more detailed information can be obtained from the relaxation enhancements by oxygen in the membrane. The interaction with oxygen identifies which spin-labeled residues are exposed to the lipid at the hydrophobic surface of the protein and hence the orientation of the transmembrane segments relative to the interior of the protein.[50] Further, in these cases, the relaxation enhancement by oxygen displays a periodicity that is characteristic of the secondary structure of the transmembrane segment, i.e., a 3.6-residue periodicity for an $\alpha$ helix and a 2-residue periodicity for a $\beta$ sheet.

The relaxation enhancement by paramagnetic ions (e.g., $Ni^{2+}$) confined to the aqueous phase of spin labels buried in the membrane takes place by the distance-dependent through-space magnetic dipole–dipole interaction. In this case, the spin label is not accessible to the relaxant and the value of $k_{RL}$ in Eqs. (10) and (11) is given by[51]:

$$k_{RL} = f_m / R_o^m \tag{12}$$

where $R_o$ is the distance of closest approach of the paramagnetic ions to the spin label and the factor $f_m$ contains all other (separation-independent) factors determining the dipolar interaction, including the *g* values of the spin label and paramagnetic ion, and the spin and $T_1$-relaxation time of

[49] C. Altenbach, D. A. Greenhalgh, H. Gobind Khorana, and W. L. Hubbell, *Proc. Natl. Acad. Sci. U.S.A.* **91,** 1667 (1994).

[50] C. Altenbach, T. Marti, H. Gobind Khorana, and W. L. Hubbell, *Science* **248,** 1088 (1990).

[51] T. Páli, R. Bartucci, L. I. Horváth, and D. Marsh, *Biophys. J.* **61,** 1595 (1992).

the paramagnetic ion. For an isolated spin label–paramagnetic ion pair, the exponent in Eq. (12) is $m = 6$, and $R_o$ is then the distance of the spin label from the paramagnetic ion binding site. In the absence of a single unique paramagnetic ion binding site, the interaction with the spin label must be summed over all paramagnetic ions, yielding $m = 4$ for a two-dimensional surface distribution of ions, or $m = 3$ for a three-dimensional bulk distribution of ions. For the latter cases, $R_o$ is the depth of the spin label from the membrane surface. Calibrations of the paramagnetic relaxation enhancement by aqueous $Ni^{2+}$ ions have been made by using phospholipids spin labeled in their acyl chains in gel-phase phospholipid membranes. The decrease in reciprocal intensity $\Delta(1/I_{ST})$ of the saturation transfer ESR spectrum is given by[51]:

$$d\{\Delta(1/I_{ST})\}/d[\mathrm{Ni}^{2+}] = 14.8(\mathrm{m}M^{-1})\left[\frac{1}{R_o^3} - \frac{1}{(d_l - R_o)^3}\right] \tag{13}$$

where $R_o$ (nm) is the distance of the spin label from the membrane surface and $d_l = 4.0$–$4.5$ nm is the membrane thickness. This may be used to give an estimate of $R_o$ from a single saturation transfer ESR measurement on a membrane protein. For greater sensitivity, or when using progressive saturation ESR measurements, it is advisable to calibrate the paramagnetic enhancements by using spin-labeled lipids in the membrane of interest. An illustrative example for the latter is given in Ref. 45.

The exposure to the lipid and the depth in the membrane of a spin label attached to a channel protein may be determined from the interactions with lipid probes bearing the spin label at different positions in the lipid molecule. In these double-labeling experiments, the spin-labeled protein and the spin-labeled lipid should be in roughly comparable amounts to maximize their mutual interactions. The saturation properties of the double-labeled system are then compared with those of the corresponding single-labeled systems. It is essential that the complementary single-labeled systems be identical in composition and level of spin labeling to the double-labeled system. Any reduction in saturation (i.e., enhancement in relaxation) of the double-labeled systems relative to that which would be predicted from the saturation of the same proportions of the single-labeled systems separately, corresponds to spin–spin interactions and therefore to mutual accessibility of the two spin-labeled species.

Prediction of the saturation properties of the double-labeled sample in the absence of mutual spin–spin exchange interactions requires knowledge of the fractional populations, $f_i$, of the two spin-labeled components. This is best obtained from spectral subtractions. The saturation behavior without spin–spin interaction can then be calculated from the linear additivity of the double-integrated spectral intensity $S(H_1)$, in the case of a progressive saturation experiment[52]:

$$S(H_1) = S_o H_1 \left[ \frac{f_1}{(1 + \sigma_1)^{1/2}} + \frac{f_2}{(1 + \sigma_2)^{1/2}} \right] \tag{14}$$

where $\sigma_i = \gamma_e^2 H_1^2 T_{1,i} T_{2,i}$, with $i = 1,2$ [cf. Eq. (9)] and $f_1 + f_2 = 1$. For a saturation transfer ESR experiment with double labeling the situation is similar; the normalized spectral intensity, $I_{ST}$, is simply the weighted sum of the component normalized intensities.[53] For more precise quantitative comparison between double-labeled systems with different positions of lipid spin-labeling, the mutual spin exchange frequency can be calculated as described in Ref. 52. An example of the determination of the depth of penetration of a spin-labeled protein into a membrane containing lipids spin labeled at various depths is given in the same reference.

## Fourier Transform Infrared Spectroscopy

The most relevant applications of FTIR spectroscopy in the field of ion channels are for determination of the secondary structure of the channel and of the orientation of the secondary structural elements composing the channel relative to the membrane normal. Interest therefore centers on the amide vibrational bands of the channel protein. There are several of these characteristic bands from the peptide backbone, but the amide I band centered at 1650 $cm^{-1}$, which is contributed predominantly by a carbonyl stretching vibration (76%), and the amide II band centered at 1540 $cm^{-1}$, which is contributed principally by the NH in-plane bending vibration (43%) and the C–N stretching vibration (29%), are of principal interest. An important feature of the amide II band is that it is shifted by ca. 100 $cm^{-1}$ to a lower wave number, on deuteration of the amide group, because of its large contribution from the N–D in-plane bending vibration. From the point of view of dichroism measurements, these two amide modes are important because the polarizations of their principal vibrational modes are very different. The importance of the amide I and amide II bands lies in their relatively high intensity and the fact that they lie in a region of the spectrum that is relatively free from overlap with other infrared bands. The various secondary structural elements of the channel protein are identified from their characteristic frequencies by using band-narrowing techniques. The relative proportions of these secondary structural elements are then

[52] M. M. E. Snel and D. Marsh, *Biophys. J.* **67,** 737 (1994).

[53] L. I. Horváth, P. J. Brophy, and D. Marsh, *Biophys. J.* **64,** 622 (1993).

quantitated by band-fitting techniques. The orientations of the secondary structural elements are determined by infrared dichroism experiments, by using linearly polarized infrared radiation with oriented membranes. The relative polarizations of the various bands are also an aid to assignment of particular bands to the different secondary structures. For instance, in the amide I region, the band polarization is principally perpendicular to the $\beta$-strand axis in a $\beta$ sheet, and is preferentially oriented parallel to the helix axis for an $\alpha$ helix. Finally, the rates of exchange of the amide protons for deuterons, when the channel assembly is exposed to $D_2O$, are a sensitive indicator of the accessibility of the peptide groups to $D_2O$ and hence of both the tertiary structure of the channel fold and the depth of embedding of the secondary structural elements in the membrane.

### *Practical Considerations*

To a certain extent, the practical aspects of FTIR spectroscopy applied to membranes ion channels have been considered previously.[1] Because the amide infrared bands, as for any condensed-state sample, are intrinsically broad, there are no great demands on the FTIR spectrometers with regard to resolution. Practically any commercial Fourier transform instrument is suitable. Where sensitivity is an issue, cooled detectors are advisable, and for fast kinetics both response time of the detector and the scan speed of the interferometer are important. However, in many cases, mixing times or the speed of flow will be the limiting factor in kinetic studies. The most important feature for ion channel applications, and for membranes in general, is the ability to subtract the background water signal reliably, for which a digital Fourier transform instrument is essential. Water vapor is a perennial problem. This gives rise to very sharp lines over a broad region of spectral interest. These interfering lines can partially be subtracted, or suppressed by mild smoothing, but exclusion of water vapor is the real solution. Care must be taken to purge the spectrometer with dry nitrogen, to minimize the length of time that the sample chamber is exposed to the atmosphere, and to ensure that the sample itself is tightly sealed. An adequate time for purging should be allowed after changing samples; an automatic sample changer with which the sample chamber remains closed can be helpful in this respect. Ideally, vacuum instruments have improved performance with respect to water vapor contamination, but are considerably more costly.

To improve signal-to-noise ratio, many interferograms (typically $\geq 100$) are accumulated automatically for Fourier transformation. Prior to Fourier transformation an apodization function is applied to the accumulated interferogram in order to cut off extraneous high-frequency noise. A sincsquared function is often considered optimum, but triangular apodization has the advantage that it produces only positive side lobes in the transformed spectrum, and produces a near-Gaussian line shape. The latter has considerable advantages when it comes to band narrowing by Fourier deconvolution.

A standard transmission cell for aqueous (or membranous) samples has $CaF_2$ windows separated by a Teflon spacer gasket that is 50 $\mu$m in thickness for samples in $D_2O$ or 10 $\mu$m or less for samples in $H_2O$ (which has a much higher infrared absorbance). Particulate membrane samples in $H_2O$ do pose considerable problems for transmission experiments, and an attenuated total reflection (ATR) setup, which has a short penetration depth, is preferred. Not infrequently, recourse to dried samples is necessary for recording spectra of protonated amide groups. The sample cells are clamped with Viton gaskets and thermostatted by circulating temperature-controlled water through the metal housing. High concentrations ($\geq$1 mg/ml) are required because of the small sample volume (50–100 $\mu$l). Air bubbles must be avoided when filling standard transmission cells.

ATR experiments are most often undertaken with oriented samples, and are most conveniently performed with a horizontal ATR plate (germanium or zinc selenide) that is sealed in a thermostatted housing. Membrane vesicles will absorb and orient spontaneously on the ATR plate, but checks must be made (e.g., by transmission experiments) that proteins are not denatured on the surface. Alternatively, a very thin layer of phospholipid (e.g., dipalmitoylphosphatidylcholine) can be deposited onto the ATR plate, either by spreading and drying from chloroform–methanol (2 : 1 v/v) solution or by Langmuir–Blodgett techniques. The membrane vesicles, or reconstituted membranes, can then be oriented on this thin phospholipid substratum.[54] From the spectroscopic point of view, a germanium ATR plate is preferred for samples in $H_2O$ because of the high refractive index, which gives a shorter depth of penetration of the evanescent infrared wave into the sample. A zinc selenide ATR plate has advantages in dichroism measurements, however, because it maintains the polarization of the incident radiation over several reflections.

For transmission experiments on an oriented sample, the membrane vesicles may be oriented by a combination of centrifugation and/or partial drying. A suitable method "isopotential spin-drying" has been described previously for polarized infrared spectroscopy.[55] AgCl is a suitable orienting substrate for infrared transmission studies.

[54] U. P. Fringeli, H. J. Apell, M. Fringeli, and P. Läuger, *Biochim. Biophys. Acta* **984,** 301 (1989).
[55] N. A. Clark and K. J. Rothschild, *Methods Enzymol.* **88,** 326 (1982).

*Fourier Deconvolution, Band Narrowing, and Band Fitting*

Because the frequency dispersion between the modes from different protein secondary structures is not particularly great for a given amide band, spectral narrowing techniques are required to resolve the overlapping bands from the different modes. Evaluation of the second derivative of the spectrum is one means of band narrowing that is used. This creates sharp peaks at the position of the component bands, at the expense of signal-to-noise ratio, but side bands arising from the second derivative of a strong peak may overlap with or be mistaken for a weaker peak. The main disadvantage of second derivative methods is that they are determined by the shape and not by the intensity of the original spectral envelope, and therefore do not reflect at all the true component band intensities.

Fourier self-deconvolution[56] consists of multiplying the interferogram from the original spectrum by an increasing function that compensates for the decay in the interferogram that is associated with the line broadening of the intrinsic spectral line shape. Fourier transformation then yields a spectrum in which the component lines are narrower and therefore may be better distinguished than in the original spectral envelope. The resultant spectral line shapes are given by the Fourier transform of the apodization function. The parameters that must be specified for the self-deconvolution are the intrinsic line shape, the intrinsic line width, and the factor $K$ by which the lines are narrowed. The intrinsic line shape is usually Lorentzian, unless this has been changed by the original apodization. The intrinsic line width is normally in the range of 17–19 $cm^{-1}$. Suitable values of the band narrowing factor are $K = 2$–3. Conservative degrees of line narrowing are advisable because deconvolution (and other "resolution enhancement" techniques) is accompanied by a degradation in signal-to-noise ratio. Additionally, overdeconvolution can give rise to the appearance of spurious peaks.

An advantage of Fourier deconvolution is that this band narrowing procedure preserves the relative intensities of the component bands. This can help when quantitating the relative populations of different components. In principle, the component line positions that are identified by band narrowing can be used in fitting the band shape of the original spectrum. The fitting parameters are the relative intensities, line shapes, and line widths of the component bands. In practice, the envelope of the amide I band can be rather broad and structureless for large proteins, which can make band fitting of the original spectrum underdetermined even with fixed

[56] J. K. Kauppinen, D. J. Moffatt, H. H. Mantsch, and D. G. Cameron, *Appl. Spectrosc.* **35,** 271 (1981).

band positions. A useful approach, therefore, is first to fit the deconvoluted spectrum and then gradually to decrease the degree of band narrowing down to $K = 1$, with progressive band fitting at each stage.[57] In this way, the final fitting of the original spectrum can be reasonably constrained.

*Amide Band Assignments*

Because deuterium exchange in $D_2O$ shifts the amide II band to an inaccessible spectral region, most secondary structural evaluations are performed on the amide I band. In principle, the amide II band in protonated samples can be used as a check and confirmation of the amide I band assignments, but the amide II region is less well characterized with respect to the contributions from the different secondary structural elements. If the amide II band can be recorded satisfactorily only for dry samples (because of overlapping baseline, etc.), the amide I band can be used to check that no appreciable changes in secondary structure are occurring on drying.

The region of the amide I and amide II bands that are characteristic of particular protein secondary structures in the protonated form are given in Table III. Where appropriate, the different infrared-active modes for each secondary structure are designated by $\nu(\delta, \delta')$, where $\delta$ and $\delta'$ are the phase differences between the vibrations of adjacent intrachain and interchain next-neighbor amide groups, respectively, in a given cooperative vibrational mode. In general, one particular mode will dominate in intensity, and whichever this is alternates between the amide I and amide II bands. This dominant infrared-active mode is indicated by VS or S in Table III, and represents the characteristic frequency of the amide band that is associated with a particular secondary structure. A notable exception is the amide I band for an $\alpha$ helix. In this case, the $\nu_{\|}(0)$ and $\nu_{\perp}(\pm 2\pi/p)$ modes are of comparable intensity and are very closely spaced in frequency. For this reason only a single range of amide I frequencies is given for the $\alpha$ helix in Table III; these correspond to all overlapping $\alpha$-helical modes. The identity of the $\nu(\delta, \delta')$ modes in Table III has been established from their characteristic polarizations, either parallel ($\|$)) or perpendicular ($\perp$) to the chain axis, and from theoretical calculations. For the $\alpha$ helix and antiparallel $\beta$-pleated sheet, the identity of the modes is reasonably well established by both theory and experiment, but for the parallel $\beta$-pleated sheet definitive experimental examples are lacking. Parallel $\beta$-sheets have so far not been identified in integral membrane proteins, but they occur with high abundance in soluble proteins. The amide frequencies of the other secondary

[57] E. Goormaghtigh, V. Cabiaux, and J.-M. Ruysschaert, *Eur. J. Biochem.* **193,** 409 (1990).

TABLE III

INFRARED ABSORPTION BAND POSITIONS IN THE AMIDE I AND AMIDE II REGIONS[a]

| | | Frequency[b] ($cm^{-1}$) | | $\Theta^c$ (°) | |
|---|---|---|---|---|---|
| Secondary structure | Mode | Amide I | Amide II | Amide I | Amide II |
| α Helix | $\nu_\parallel(0)$ | 1650–1660 VS[d] | 1545–1550 VS | 29–40[e] | 75–77[e] |
| | $\nu_\perp(\pm 2\pi/p)^f$ | | 1510–1520 MW | | |
| Antiparallel | $\nu_\parallel(0, \pi)$ | 1680–1695 W | 1524–1530 S | 0 | 0 |
| β sheet | $\nu_\perp(\pi, 0)$ | 1625–1640 VS | 1555 MW | 90 | 90 |
| Parallel | $\nu_\parallel(0, 0)$ | | 1540–1550 S | 0 | 0 |
| β sheet | $\nu_\perp(\pi, 0)$ | 1640–1645 VS | | 90 | 90 |
| Unordered | | 1650–1660 S | 1535–1540 M | — | — |
| β Turns | | 1638–1646 and 1661 | | | |
| β Edge/aggregated strands | | 1680–1700 S<br>1610–1625 S | | | |

[a] Of proteins in $H_2O$. Data assembled from S. Krimm and J. Bandekar, *Adv. Protein Chem.* **38,** 181 (1986); J. Bandekar and S. Krimm, *Biopolymers* **27,** 909 (1988); N. A. Nevskaya and Yu. N. Chirgadze, *Biopolymers* **15,** 637 (1976); Yu. N. Chirgadze and N. A. Nevskaya, *Biopolymers* **15,** 607 and 627 (1976); T. Miyazawa and E. R. Blout, *J. Am. Chem. Soc.* **83,** 712 (1961) for α helix and β sheet; from H. H. Mantsch, A. Perczel, M. Hollosi, and G. D. Fasman, *Biopolymers* **33,** 201 (1993) for β turns; and from J. L. R. Arrondo, A. Muga, J. Castresana, and F. M. Goñi, *Prog. Biophys. Molec. Biol.* **59,** 23 (1993), and M. Jackson and H. H. Mantsch, *Crit. Rev. Biochem. Molec. Biol.* **30,** 95 (1995) for the remainder.

[b] In $D_2O$ the amide I bands are shifted to ca. 5–10 $cm^{-1}$ lower wave number. For the amide II region, the bands are shifted much further downward in $D_2O$ by ca. 100 $cm^{-1}$.

[c] Angle of the transition moment to the long molecular axis, required in interpreting infrared dichroic ratios. For well-separated bands this corresponds directly to the polarizations of the corresponding collective mode, $\nu_\parallel$ or $\nu_\perp$. For strong overlapping bands (notably the α-helix amide I), this corresponds to the orientation of an individual amide group transition moment (see text).

[d] Abbreviations following the frequencies indicate relative band intensities: S, strong; M, medium; W, weak; V, very. Very weak bands are ignored.

[e] T. Miyazawa and E. R. Blout, *J. Am. Chem. Soc.* **83,** 712 (1961); E. M. Bradbury, L. Brown, A. R. Downie, A. Elliot, R. D. B. Fraser, and W. E. Handby, *J. Mol. Biol.* **5,** 230 (1962); M. Tsuboi, *J. Polym. Sci.* **59,** 139 (1962).

[f] $p = 3.6$ is the number of residues per turn of an α helix.

structures that are listed in Table III (random, turns, etc.) are established mainly from experimental studies.

Because the frequency of the amide bands depends on the strength (or length) of the peptide hydrogen bonds, it is not entirely surprising that a range of frequencies can be observed, depending on the degree of perfection of the particular secondary structure. An extreme example is afforded by the "free" peptide hydrogen bonds of proteins or peptides in dimethyl sulfoxide (DMSO), which have an amide I frequency of ca. 1662–1666 $cm^{-1}$.[58,59] A given secondary structure in a protein therefore may not necessarily have the same amide frequency as a homopolypeptide with the corresponding secondary structure. Wherever possible, some indication of this variability is given by the ranges of frequencies quoted in Table III. These ranges therefore include bent helices, imperfect sheets or sheets of limited size, etc. Not surprisingly, unordered peptide structures may exhibit a wide range of amide frequencies, and random coil proteins are characterized by extremely broad infrared bands.

A feature that also should be noted with regard to samples in $D_2O$ is that the amide I frequencies for proteins with fully deuterated peptide groups are shifted downward by ca. 5–10 $cm^{-1}$. The amount of the shift actually observed depends on the degree of H–D exchange, i.e., on the accessibility of the amide groups, and can vary from zero for membrane-embedded structures to complete exchange for unordered structures. For the latter, the amide I frequency is shifted down from 1652–1660 $cm^{-1}$ in $H_2O$ to 1640–1648 $cm^{-1}$ in $D_2O$. This can be an aid to assignment because unordered structures thus become resolved from $\alpha$-helical structures that are not readily exchangeable on (brief) exposure to $D_2O$. In general, secondary structures, other than random, are incompletely exchanged in $D_2O$, especially if they are embedded in the membrane.

In addition to the data given in Table III, an extensive compilation for soluble proteins in $D_2O$ has been given in Ref. 7. This is based on comparison with X-ray crystal structures, and a notable feature is the appearance of bands in the amide I regions characteristic of $\beta$ sheets for proteins that contain little $\beta$ sheet in the crystal. These bands have therefore been associated with extended peptide chains that are not incorporated in any identifiable sheet structure.

Certain other remarks are necessary with respect to Table III. Whereas, the positions of the bands corresponding to $\alpha$ helices and $\beta$ sheets are reasonably well established, those for turns are less certain. A systematic study of the amide I region for the $\beta$ turns in cyclic peptides has attributed bands at 1638–1643 $cm^{-1}$ to amides involved in hydrogen bands that stabilize the turn, and higher frequency bands in the region of 1661 $cm^{-1}$ to amides that are not internally hydrogen bonded and are sterically constrained.[60] Currently, there is also little consensus with regard to $3_{10}$ helices. The most recent work on model peptides in $D_2O$ has identified bands at 1634–1640

[58] M. Jackson and H. H. Mantsch, *Biopolymers* **31,** 1205 (1991).

[59] M. Jackson and H. H. Mantsch, *Biochim. Biophys. Acta* **1078,** 231 (1991).

[60] H. H. Mantsch, A. Perczel, M. Hollosi, and G. D. Fasman, *Biopolymers* **33,** 201 (1993).

$cm^{-1}$ in the amide I region.[61,62] This differs from previous suggestions for $3_{10}$ helices. The bands ascribed to $\beta$ edges or aggregated strands are reasonably well established. These appear very characteristically at the outer extremes of the amide I region, on the thermal denaturation of proteins, where they have been ascribed to aggregation processes. The so-called $\beta$ edge was first identified in concanavalin A[63] and proposed to arise from carbonyl groups at the edge of a $\beta$ sheet that are hydrogen bonded to a different intra- or intermolecular structure. An intramolecular candidate for hydrogen bonding was suggested to be the $\alpha$-helix side chains in $\alpha/\beta$ barrels.[64] The latter structure has been suggested for the transmembrane domain of the ligand-gated nicotinic acetylcholine receptor ion channel.[29]

A significant feature for some integral proteins is that the amide I bands arising from transmembrane helices can occur at higher frequencies (ca. 1660 $cm^{-1}$), outside the normal range that is given for $\alpha$ helices in Table III. The classic example is bacteriorhodopsin,[9] but a similar result is found also for the 16-kDa proton channel that is associated with vacuolar ATPases.[65] It is likely, in analogy with the effects of DMSO (see above; Refs. 58 and 59), that the anomalously high amide I frequencies are associated with more flexible or stretched $\alpha$ helices[11] in which hydrogen bond strengths can be strongly perturbed. Such an effect of hydrogen bond stretching has also been observed to occur on the change in hydrophobic matching with a model transmembrane peptide that takes place at the lipid phase transition.[66]

Further cautions and caveats with regard both to amide I band assignments and to secondary structure quantitation can be found in Refs. 67 and 68.

Analysis of the amide I band from proteolyzed membrane preparations of the nicotinic acetylcholine receptor suggests that the intramembranous structure of this ligand-gated ion channel is composed of both transmembrane $\alpha$ helices and $\beta$ sheets.[69] Similarly, trypsinized $Na^+,K^+$-ATPase, for

[61] S. M. Miick, A. P. Todd, and G. Millhauser, *Biochemistry* **30,** 9498 (1992).

[62] G. Martinez and G. Millhauser, *J. Struct. Biol.* **114,** 23 (1995).

[63] J. L. Arrondo, N. M. Young, and H. H. Mantsch, *Biochim. Biophys. Acta* **952,** 261 (1988).

[64] J. L. R. Arrondo, A. Muga, J. Castresana, and F. M. Goñi, *Prog. Biophys. Mol. Biol.* **59,** 23 (1993).

[65] A. Holzenburg, P. C. Jones, T. Franklin, T. Páli, T. Heimburg, D. Marsh, J. B. C. Findlay, and M. E. Finbow, *Eur. J. Biochem.* **213,** 21 (1993).

[66] Y.-P. Zhang, R. N. A. H. Lewis, R. S. Hodges, and R. N. McElhaney, *Biochemistry* **34,** 2362 (1995).

[67] W. K. Surewicz, H. H. Mantsch, and D. Chapman, *Biochemistry* **32,** 389 (1993).

[68] M. Jackson and H. H. Mantsch, *Crit. Rev. Biochem. Molec. Biol.* **30,** 95 (1995).

[69] U. Görne-Tschelnokow, A. Strecker, C. Kaduk, D. Naumann, and F. Hucho, *EMBO J.* **13,** 338 (1994).

which the transmembrane segments have been identified, also gives an amide I band that has components characteristic of both $\alpha$-helix and $\beta$-sheet structures.[70] A peptide corresponding to the single putative transmembrane segment of the IsK $K^+$-channel protein has a strong tendency to form $\beta$-sheet structures when reconstituted in lipid membranes.[22,30] The amide I band then consists of a single component at 1627 $cm^{-1}$, with a minor component at $\sim$1695 $cm^{-1}$, which is characteristic of antiparallel $\beta$ sheets. Such short peptides may, however, display a plasticity in their secondary structure.[12]

*Amide Hydrogen–Deuterium Exchange*

The kinetics of amide H–D exchange for samples incubated in $D_2O$ can give valuable information on the tertiary structure of a membrane protein, in addition to the qualitative conclusions mentioned above with regard to identification of unordered structures and membrane-embedded segments. Accessibility of the amide groups for exchange may increase with tertiary structural changes, even though the secondary structure, as recorded by the amide band shapes, may be conserved. Most frequently, the exchange kinetics are determined from the decrease in intensity of the amide II band, in the 1550-$cm^{-1}$ region, on the shift of this band to much lower wave numbers. Alternatively, the smaller shifts that occur within the amide I band on deuteration may be quantitated by using difference spectroscopy. The positive area under the difference spectrum is a direct measure of the extent of amide H–D exchange.[71] The second method has the advantage that it is less susceptible to baseline shifts from HDO that may accumulate from leakage of water vapor into the sample during long kinetic runs. Additionally, the difference spectra may be compared to check whether different secondary structural domains are exchanging at different rates. A promising higher resolution approach is to use two-dimensional cross-correlation spectroscopy,[72] which has recently been introduced to establish correlations between different bands of the one-dimensional FTIR spectrum in response to a perturbation such as amide group deuteration.

In general, the time resolution of mixing is a limiting factor for fast exchange kinetics, and the information that can be obtained is the fraction of amide groups that undergoes fast exchange (i.e., within the dead time) and the rate constants (and amplitudes) for the slowly exchanging population. Except for the simplest cases, there will normally be a distribution of rate constants for amide H–D exchange, which in certain cases may be semicontinuous. In this case, it is convenient to express the kinetics of

[70] T. Heimburg, M. Esmann, and D. Marsh, *J. Biol. Chem.* **272,** 25685 (1997).
[71] T. Heimburg and D. Marsh, *Biophys. J.* **65,** 2408 (1993).
[72] A. Gericke, S. J. Gadaleta, J. W. Brauner, and R. Mendelsohn, *Biospectroscopy* **2,** 341 (1996).

amide deuteration in terms of the area under the decay curve of the protonated amide, which for a multiexponential process yields the mean rate constant, $\langle k \rangle$, for exchange[71]:

$$\langle k \rangle^{-1} = \int_0^\infty \left[ \frac{\sum_i A_i \exp(-k_i t)}{\sum_i A_i} \right] dt = \frac{\sum_i A_i / k_i}{\sum_i A_i} \tag{15}$$

The mean rate constant can then be evaluated in terms of various quasi-continuous models, e.g., reaction coupled with (slow) diffusion. The temperature dependence of the mean rate constant may be used to determine further details of the contributing rate processes, but it must be remembered that the relative proportions of amides contributing to the fast and slow regimes may also change with temperature. An instructive example of an application to a membrane system is given in Ref. 71. Within the context of ion channel proteins, relevant issues are the H–D exchange rates of the amides of the pore-lining residues, and whether the exchange of these amides that are situated within the pore can be modulated, e.g., by gating or selective blockers.

Trypsinized membrane preparations of the $Na^+$,$K^+$-ATPase ion pump contain a larger proportion of bands in the amide I region that are attributed to protonated $\alpha$ helices than do native membranes, even after extensive exchange in $D_2O$.[70] This corresponds to the transmembrane section of the protein. In addition, the mean rate constant for H–D exchange of the slowly exchanging population of amide groups in the trypsinized preparations is approximately half that for native membranes.

*Infrared Dichroism*

The infrared dichroic ratios, $R$, are obtained from the ratio of the absorbances, $A_\parallel$ and $A_\perp$, of oriented membrane samples with irradiation polarized parallel and perpendicular, respectively, to the plane of incidence of the infrared beam:

$$R = A_\parallel / A_\perp \tag{16}$$

If necessary, corrections must be made for the imperfect degree of polarization introduced by the infrared polarizer. Inevitably, the dichroic ratio depends on the angle, $i$, which the incident beam makes with the membrane normal. For an ATR setup this angle is fixed and determined by the cut-angle of the ATR plate, which is usually 45° (see Fig. 2, left). For a transmission experiment, the sample must be tilted with respect to the infrared beam, and by using a goniometer various angles of incidence, $i$, can be set (see Fig. 2, right).

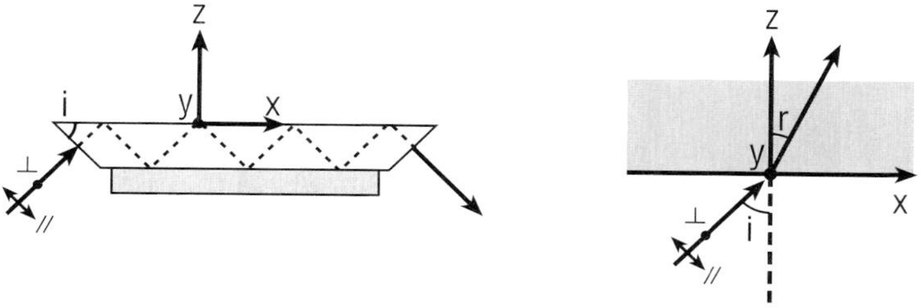


FIG. 2. Geometric relations of the incident radiation to the plane of the substrate on which the sample is oriented in polarized ATR experiments (left) and in a polarized transmission experiment (right). The $z$ axis lies along the normal to the orienting substrate, the $x$ axis is the orthogonal axis in the plane of incidence, and the $y$ axis is perpendicular to the plane of incidence. The angle of incidence, $i$, is determined by the cut-angle of the ATR plate, or by the tilt of the tranmission cell with respect to the incident beam. Polarizations of the infrared beam parallel (‖) and perpendicular (⊥) to the plane of incidence are indicated. [Reprinted from D. Marsh, *Biophys. J.* **72,** 2710 (1997).]

For interpreting the measured dichroic ratios, the integral protein molecules can be assumed, in all cases, to be rotationally disordered in the plane of the membrane. This means that the absorbances, $A_{\parallel}$ and $A_{\perp}$, for the parallel and perpendicularly polarized incident beams are summed over the azimuthal angle in the membrane plane.

With the conventional geometries indicated in Fig. 2, for both ATR and transmission experiments, the resultant dichroic ratio is then given by[10]:

$$R = \frac{E_x^2}{E_y^2} + \frac{E_z^2 \langle M_z^2 \rangle}{E_y^2 \langle M_y^2 \rangle} \tag{17}$$

where $\langle M_z^2 \rangle$ and $\langle M_y^2 \rangle$ are the squares of the components of the transition moment (proportional to the absorption coefficients) in the $z$ and $y$ directions that are indicated in Fig. 2. The angular brackets indicate that these values are summed over the remaining angles specifying orientation of the transmission moment relative to the bilayer normal. It is these latter angles that specify the orientational distribution of the protein in the membrane (see Fig. 3). The components of the electric field vector of the infrared radiation in the membrane, $\mathbf{E} = (E_x, E_y, E_z)$, that appear in Eq. (17) depend both on the angle of incidence, $i$, and on the refractive indices of the various media constituting the sample and its support. These differ for a transmission experiment and an ATR experiment. For the transmission experiment, the components of the electric field amplitudes are given by (from Snell's law of refraction):

$$E_x^2/E_y^2 = 1 - \sin^2 i/n_2^2 \tag{18}$$

$$E_z^2/E_y^2 = \sin^2 i/n_2^2 \tag{19}$$

where $n_2 \sim 1.4$–$1.5$ is the refractive index of the membrane sample at the infrared frequency, and is taken to be close to that for hydrocarbon. For an ATR experiment the electric field amplitudes of the evanescent wave are given by[73,74]:

$$\frac{E_x^2}{E_y^2} = \frac{n_1^2 - n_3^2/\sin^2 i}{n_1^2 - n_3^2/\tan^2 i} = \frac{n_1^2 - 2n_3^2}{n_1^2 - n_3^2} \tag{20}$$

$$\frac{E_z^2}{E_y^2} = \frac{(n_3/n_2)^4\, n_1^2}{n_1^2 - n_3^2/\tan^2 i} = \frac{n_1^2\,(n_3/n_2)^4}{n_1^2 - n_3^2} \tag{21}$$

where $n_1$, $n_2$, $n_3$ are the infrared refractive indices of the ATR plate, the sample, and the bulk medium, respectively. The second equalities in Eqs. (20) and (21) apply to the standard condition $i = 45°$, i.e., to a 45°-cut ATR plate. Typical values of the refractive indices are $n_1 = 4.0$ (germanium) or $n_1 = 2.4$ (zinc selenide) for the ATR plate, and $n_3 = 1.32$–$1.33$ for $H_2O$ and $D_2O$ except at the position of the water absorption band at 1650 cm$^{-1}$, where anomalous dispersion yields a value of $n_3 = 1.29$ for $H_2O$.

A point currently of some controversy is the appropriate formulation for the electric field amplitudes in an ATR sample.[74] For a thin film, the thickness of the sample film is less than the penetration depth (~400 nm) from the ATR plate into the sample of the evanescent infrared wave. The upper phase consisting of bulk medium must then be taken into account, and Eqs. (20) and (21) as given above are applicable. This is the approach normally taken in most ATR membrane studies. If the sample is thicker than the penetration depth of the evanescent infrared wave, then the bulk medium is substituted by the sample, and $n_3 = n_2$ and $n_3/n_2 = 1$ in Eqs. (20) and (21). The latter is known as the "thick-film" approximation and conditions under which this may be more appropriate have been discussed.[11,75] It is certainly possible that the orientation of protein segments that are largely exposed to water may be better evaluated with the latter model.

In the thin-film approximation, the resulting values for the ratios of the electric field amplitudes are $E_x^2/E_y^2 = 0.877$ (0.562) and with $n_2 = 1.43$ are $E_z^2/E_y^2 = 0.828$ (1.060), for a germanium (zinc selenide) 45°-cut ATR plate. In the thick-film approximation, the corresponding values with $n_3 = 1.43$

[73] N. J. Harrick, "Internal Reflection Spectroscopy." Harrick Scientific Corp. Ossining, New York, 1967.

[74] P. H. Axelsen and M. J. Citra, *Prog. Biophys. Molec. Biol.* **66,** 227 (1996).

[75] M. J. Citra and P. H. Axelsen, *Biophys. J.* **71,** 1796 (1996).

become $E_x^2/E_y^2 = 0.853$ (0.450) and $E_z^2/E_y^2 = 1.147$ (1.550) for a germanium (zinc selenide) ATR plate.

*Angular Orientations*

Determination of the orientation of the secondary structural elements relative to the membrane from the infrared dichroic ratios differs in two important respects between $\alpha$ helices and $\beta$ sheets. For the $\alpha$ helix, the assumption of axial symmetry about the helix axis holds to a reasonable degree of approximation. However, the resultant vibrational modes that are polarized parallel and perpendicular to the helix axis have comparable intensities and occur at practically the same frequency so that they cannot be resolved in the amide I band (see Table III). In a $\beta$ sheet, the resultant amide I modes that are polarized parallel and perpendicular to the $\beta$-strand axis are widely separated and a single mode dominates in intensity (see Table III). However, because the $\beta$ strands are fixed within the $\beta$ sheet, there is no axial symmetry of the perpendicularly polarized transition moment about the $\beta$-strand axis. The two cases of $\alpha$ helices and $\beta$ sheets are considered separately.

Assuming axial symmetry for an $\alpha$ helix, the angular distribution of the helix axis about the membrane normal is given by[9,10]:

$$\langle\cos^2\gamma\rangle = \frac{[R_\alpha(\Theta) - E_x^2/E_y^2](1+\cos^2\Theta) - 2(E_z^2/E_y^2)(1-\cos^2\Theta)}{[R_\alpha(\Theta) - E_x^2/E_y^2 + 2E_z^2/E_y^2]\,(3\cos^2\Theta - 1)} \tag{22}$$

where $\gamma$ is the angle between the helix axis and the membrane normal (see Fig. 3, left), and $\Theta$ is the angle that an individual amide transition moment makes with the helix axis. Values of the latter, determined experimentally for $\alpha$-helical polypeptides, are given in Table III. The orientation of an individual amide transition moment is used because the parallel and perpendicularly polarized amide I modes for an $\alpha$ helix are superimposed and therefore the sum of their intensities is recorded.[10] The orientation of an $\alpha$ helix is conventionally expressed in terms of an order parameter, $S_\alpha = \frac{1}{2}(3\langle\cos^2\gamma\rangle - 1)$. Using a value of $\Theta = 39°$ for the amide I band (see Table III) and electric field intensities given above for a 45°-cut germanium ATR plate, in the thin-film approximation, the helix order parameter is given numerically by:

$$S_\alpha = 2.46(R_{\alpha,\mathrm{I}}^{\mathrm{ATR}} - 1.71)/(R_{\alpha,\mathrm{I}}^{\mathrm{ATR}} + 0.78) \tag{23}$$

where $R_{\alpha,\mathrm{I}}^{\mathrm{ATR}}$ is the dichroic ratio measured for the amide I band by ATR. Correspondingly for a transmission experiment, the helix order parameter is given by:

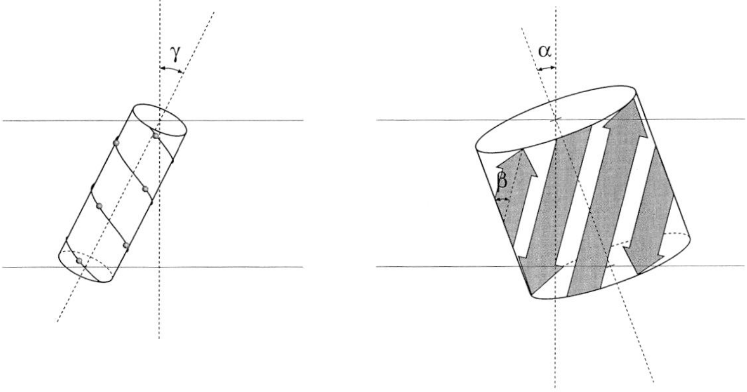


FIG. 3. Orientation of an $\alpha$ helix (left) and a $\beta$ barrel (right) in the membrane. The axis of the $\alpha$ helix is tilted at an angle $\gamma$ to the membrane normal. The axis of the $\beta$ barrel is tilted at an angle $\alpha$ to the membrane normal and the $\beta$ strands are tilted at an angle $\beta$ to the barrel axis. The $\beta$ strands are tilted at an angle $\gamma$ (not indicated) to the membrane normal, where $\cos\gamma = \cos\alpha\cos\beta$. The $\nu_{\parallel}(0,0)$ amide mode is polarized parallel to the helix axis and the $\nu_{\perp}(\pm 2\pi/p)$ modes are polarized perpendicular to the helix axis. The $\nu_{\parallel}(0,\pi)$ amide mode of the antiparallel $\beta$ sheet is polarized parallel to the $\beta$-strand axis and the $\nu_{\perp}(\pi,0)$ mode is polarized perpendicular to the $\beta$-strand axis.

$$S_\alpha = \frac{2.46(R^{\mathrm{T}}_{\alpha,\mathrm{I}} - 1)}{1.47\sin^2 i + R^{\mathrm{T}}_{\alpha,\mathrm{I}} - 1} \tag{24}$$

for $\Theta = 39°$ and $n_2 = 1.43$, and where $i$ is the angle of incidence.

The orientation of the transmembrane $\alpha$-helical segments in the nicotinic acetylcholine receptor channel has been found by ATR dichroism measurements to lie approximately along the membrane normal, $S_\alpha \approx 1$.[69] This result is essentially in agreement with the orientation of the pore-lining helices that are resolved by electron crystallography.[29]

The approximation of axial symmetry about the helix axis holds rigorously for the single transmembrane stretch of a bitopic $\alpha$-helical protein. For a polytopic $\alpha$-helical protein, in which the transmembrane helices are tilted with respect to the longitudinal axis of the whole protein molecule, the approximation only holds good insofar as summation over the helical periodicity gives rise to approximate axial symmetry. The residual degree of nonaxiality may be significant for short helical segments or for comparison of dichroic data with detailed molecular models (see Ref. 76).

In general, because of the nonaxiality of the transition moment, measurement of a single dichroic ratio is insufficient to specify fully the angular

[76] D. Marsh, *Biophys. J.* **75,** 354 (1998).

distribution of a $\beta$-sheet structure.[77] For $\beta$ sheets, it is necessary to measure two dichroic ratios from bands with different orientations of the transition moment.[10] The amide I and amide II bands are suitable for this purpose. The dominant mode of the amide I band is polarized perpendicular to the $\beta$-strand axis, i.e., $\Theta = 90°$, and that of the amide II band is polarized parallel to the strand axis, i.e., $\Theta = 0°$. These dominant modes are reasonably well separated from the complementary modes of opposite polarization, particularly in the case of the amide I band (see Table III).

The principal amide II mode of the $\beta$ sheet is less used than the principal amide I band because it must be recorded for the protonated form in $H_2O$, which may necessitate the use of dried samples (see section on Amide Band Assignments). However, it has the advantage that the transition moment is directed along the $\beta$-strand axis (i.e., $\Theta = 0°$), which automatically ensures axial symmetry in this unique case. The orientational distribution of the $\beta$-strand axes may therefore be determined directly from the dichroic ratio, $R_{\beta,\mathrm{II}}$, of the amide II band alone[10]:

$$\langle \cos^2 \gamma \rangle = \frac{R_{\beta,\mathrm{II}} - E_x^2/E_y^2}{R_{\beta,\mathrm{II}} - E_x^2/E_y^2 + 2E_z^2/E_y^2} \tag{25}$$

where $\gamma$ is the angle the $\beta$-strand axis makes with the membrane normal. In terms of the thin-film approximation, with a germanium ATR plate, the angular distribution of the $\beta$-strand axes is given by (see above):

$$\langle \cos^2 \gamma \rangle = (R_{\beta,\mathrm{II}}^{\mathrm{ATR}} - 0.877)/(R_{\beta,\mathrm{II}}^{\mathrm{ATR}} + 0.779) \tag{26}$$

where $R_{\beta,\mathrm{II}}^{\mathrm{ATR}}$ is the ATR dichroic ratio of the amide II band in a $\beta$ sheet. Correspondingly, for a transmission infrared experiment, the measured dichroic ratio, $R_{\beta,\mathrm{II}}^{\mathrm{T}}$, can be used to give:

$$\langle \cos^2 \gamma \rangle = (R_{\beta,\mathrm{II}}^{\mathrm{T}} - 1 + 0.489 \sin^2 i)/(R_{\beta,\mathrm{II}}^{\mathrm{T}} - 1 + 1.467 \sin^2 i) \tag{27}$$

where $i$ is the angle of incidence and $n_2 = 1.43$ has been assumed.

For tilted $\beta$ sheets, however, the orientational distribution of the $\beta$ strands is not identical with that of the $\beta$ sheets themselves, unless the $\beta$ strands are not tilted within the $\beta$ sheets. In the general case, the dichroic ratio of the amide II band (i.e., $\Theta = 0°$) is given by[10]:

$$R_{\beta,\mathrm{II}} = \frac{E_x^2}{E_y^2} + \frac{2\langle \cos^2 \alpha \rangle \langle \cos^2 \beta \rangle}{1 - \langle \cos^2 \alpha \rangle \langle \cos^2 \beta \rangle} \frac{E_z^2}{E_y^2} \tag{28}$$

and correspondingly for the amide I band (i.e., $\Theta = 90°$) the dichroic ratio is given by[10]:

[77] A. Rodionova, S. A. Tatulian, T. Surrey, F. Jähnig, and L. K. Tamm, *Biochemistry* **24,** 1921 (1995).

$$R_{\beta,\mathrm{I}} = \frac{E_x^2}{E_y^2} + \frac{2\langle\cos^2\alpha\rangle(1 - \langle\cos^2\beta\rangle)}{1 - \langle\cos^2\alpha\rangle(1 - \langle\cos^2\beta\rangle)}\frac{E_z^2}{E_y^2} \tag{29}$$

where $\alpha$ is the angle that the $\beta$ sheet makes with the membrane normal and $\beta$ is the angle of tilt of the $\beta$ strand within the $\beta$ sheet (see Fig. 3, right). Combination of measurements of the amide I and amide II dichroic ratios, $R_{\beta,\mathrm{II}}$ and $R_{\beta,\mathrm{I}}$, respectively, therefore specifies both the angular distribution of the $\beta$ sheets, i.e., $\langle\cos^2\alpha\rangle$, and that of the strand tilt within the sheets, i.e., $\langle\cos^2\beta\rangle$.

The order parameter of the $\beta$ sheets (cf. Fig. 3, right), which is defined by $S_\beta = \frac{1}{2}(3\langle\cos^2\alpha\rangle - 1)$, is then given by[10]:

$$S_\beta = \frac{3}{2}\frac{R_{\beta,\mathrm{I}}^{\mathrm{ATR}} - 0.877}{R_{\beta,\mathrm{I}}^{\mathrm{ATR}} + 0.779} + \frac{R_{\beta,\mathrm{II}}^{\mathrm{ATR}} - 1.705}{R_{\beta,\mathrm{II}}^{\mathrm{ATR}} + 0.779} \tag{30}$$

where $R_{\beta,\mathrm{I}}^{\mathrm{ATR}}$ and $R_{\beta,\mathrm{II}}^{\mathrm{ATR}}$ are the dichroic ratios of the amide I and amide II bands determined by ATR, and the thin-film approximation for the electric field amplitudes with a germanium ATR plate is used. Correspondingly, for a transmission experiment, the order parameter of the $\beta$ sheets is given by:

$$S_\beta = \frac{3}{2}\frac{R_{\beta,\mathrm{I}}^{\mathrm{T}} - 1 + 0.49\sin^2 i}{R_{\beta,\mathrm{I}}^{\mathrm{T}} - 1 + 1.47\sin^2 i} + \frac{R_{\beta,\mathrm{II}}^{\mathrm{T}} - 1}{R_{\beta,\mathrm{II}}^{\mathrm{T}} - 1 + 1.47\sin^2 i} \tag{31}$$

where $i$ is the angle of incidence ($n_2 = 1.43$). Further, the tilt distribution of the strands in the $\beta$ sheets is given by:

$$\langle\cos^2\beta\rangle = \left[1 + \frac{(R_{\beta,\mathrm{I}}^{\mathrm{ATR}} - 0.877)(R_{\beta,\mathrm{II}}^{\mathrm{ATR}} + 0.779)}{(R_{\beta,\mathrm{II}}^{\mathrm{ATR}} - 0.877)(R_{\beta,\mathrm{I}}^{\mathrm{ATR}} + 0.779)}\right]^{-1} \tag{32}$$

or

$$\langle\cos^2\beta\rangle = \left[1 + \frac{(R_{\beta,\mathrm{I}}^{\mathrm{T}} - 1 + 0.49\sin^2 i)(R_{\beta,\mathrm{II}}^{\mathrm{T}} - 1 + 1.47\sin^2 i)}{(R_{\beta,\mathrm{II}}^{\mathrm{T}} - 1 + 0.49\sin^2 i)(R_{\beta,\mathrm{I}}^{\mathrm{T}} - 1 + 1.47\sin^2 i)}\right]^{-1} \tag{33}$$

for an ATR experiment and a transmission experiment, respectively.

The situation is somewhat different, however, for the opposite extreme from a planar $\beta$ sheet, namely, a $\beta$ barrel of *circular* cross section such as is indicated in Fig. 3, right.[11,76] There is then axial symmetry about the $\beta$-barrel axis and Eq. (22) should apply, where $\gamma$ is the angle that the barrel axis makes with the membrane normal, and $\Theta$ is the angle that the resultant transition moment in a $\beta$ strand makes with the barrel axis. For the amide I band of the $\beta$ barrel: $\Theta = 90° - \beta$, and for the amide II band: $\Theta = \beta$ (see Fig. 3, right). The order parameter of the barrel axis, $S_{\mathrm{B}} = \frac{1}{2}(3\langle\cos^2\gamma\rangle - 1)$, is then given by[11,76]:

$$S_B = \frac{2R_{\beta,I}^{ATR} - 3.41}{R_{\beta,I}^{ATR} + 0.78} + \frac{2R_{\beta,II}^{ATR} - 3.41}{R_{\beta,II}^{ATR} + 0.78} \quad (34)$$

instead of Eq. (30), for an ATR experiment. Correspondingly, for a transmission experiment, the order parameter of the $\beta$ barrel is given by:

$$S_B = \frac{2(R_{\beta,I}^{T} - 1)}{R_{\beta,I}^{T} - 1 + 1.47 \sin^2 i} + \frac{2(R_{\beta,II}^{T} - 1)}{R_{\beta,II}^{T} - 1 + 1.47 \sin^2 i} \quad (35)$$

instead of Eq. (31). Further, the tilt distribution of the strands, relative to the barrel axis is given by:

$$\langle \cos^2 \beta \rangle = \frac{1}{3} + \frac{1}{3}\left[1 + \frac{(R_{\beta,I}^{ATR} - 1.71)(R_{\beta,II}^{ATR} + 0.78)}{(R_{\beta,II}^{ATR} - 1.71)(R_{\beta,I}^{ATR} + 0.78)}\right]^{-1} \quad (36)$$

or

$$\langle \cos^2 \beta \rangle = \frac{1}{3} + \frac{1}{3}\left[1 + \frac{(R_{\beta,I}^{T} - 1)(R_{\beta,II}^{T} - 1 + 1.47 \sin^2 i)}{(R_{\beta,II}^{T} - 1)(R_{\beta,I}^{T} - 1 + 1.47 \sin^2 i)}\right]^{-1} \quad (37)$$

instead of Eqs. (32) and (33), respectively. The effects of nonaxiality in $\beta$-barrel oligomers has also been considered.[76]

For the nonspecific OmpF porin, which forms weakly cation-sensitive voltage-gated channels,[78,79] the axis of the $\beta$-barrel structure is inclined to the membrane normal, $S_B \approx 0.69$.[76,80] The $\beta$ strands, also, have a pronounced tilt relative to the axis of the barrel, $\langle \cos^2 \beta \rangle = 0.52$. The latter is in essential agreement with the X-ray crystal structure for this class of porin. The $\beta$-sheet structures of peptides corresponding to the single putative transmembrane segment of the IsK $K^+$-channel protein are more strongly tilted relative to the surface normal in reconstituted membranes: $S_\beta \approx 0.19$.[10,30] The $\beta$ strands are also strongly tilted within the sheets, $\langle \cos^2 \beta \rangle \approx 0.65$, corresponding to an effective orientation of $\beta = 36°$, which perhaps suggests a homogeneous structure with a stagger by one residue between adjacent strands. For the nicotinic acetylcholine receptor, the intramembranous $\beta$ strands display a considerable larger degree of tilt or disorder.[69]

## Note Added in Proof

The first high-resolution structure of a gated ion channel, the SKC1 $K^+$-channel encoded by the KcsA gene from *Streptomyces lividans,* has been determined by X-ray crystallography.[81] The channel is a cone-shaped tetramer of 2-helix bundles, with the residues forming the selectivity filter in an extended conformation. Site-directed spin-labelling ESR has been performed by cysteine scanning of the two helices of this channel, in combination with paramag-

[78] H. Schindler and J. P. Rosenbusch, *Proc. Natl. Acad. Sci. U.S.A.* **75,** 3751 (1978).
[79] J. H. Lakey and F. Pattus, *Eur. J. Biochem.* **186,** 303 (1989).
[80] E. Nabedryk, R. M. Garavito, and J. Breton, *Biophys. J.* **53,** 671 (1988).
[81] D. A. Doyle, J. Morais Cabral, R. A. Pfuetzner, A. Kuo, J. M. Gulbis, S. L. Cohen, B. T. Chait, and R. MacKinnon, *Science* **280,** 69 (1998).

netic relaxation agents.[82] The geometrical constraints obtained from ESR spectroscopy are consistent with the X-ray structure, and spin-spin interactions reveal a large conformational change at the C-terminal end of the pore-lining helix on pH-gating of the channel.

[82] E. Perozo, D. Marien Cortes, and L. G. Cuello, *Nature Struct. Biol.* **5,** 459 (1998).

# [5] Nuclear Magnetic Resonance Spectroscopy of Peptide Ion Channel Ligands: Cloning and Expression as Aid to Evaluation of Structural and Dynamic Properties

*By* MICHAEL D. REILY, ANNE M. BOKMAN, JAMES OFFORD, and PATRICK MCCONNELL

## Introduction

As candidates for structural measurements, ion channels are daunting targets. Their large size and the fact that they are anchored in biological membranes make atomic resolution information unattainable by available techniques. Fortunately, there is a vast array of peptide ligands whose structures are more accessible that bind tightly to the outer vestibules of different ion channels. Peptides from this class are found in a variety of animal venoms and are characterized by a high cysteine content and a conserved backbone fold[1,2] (Fig. 1). Structural information about these ligands is important for understanding their function as pharmacologic tools[3] and as therapeutic entities.[4] The high binding affinity and exquisite selectivity of these ligands suggests that structural inferences can be drawn about the ion channel from an adequate knowledge of the structure of the ligands themselves.[5] Additionally, more direct information may be obtained about the ion channel by studying the ligand bound to a relevant fragment of the channel.[6] These peptides are particularly well suited to nuclear magnetic resonance (NMR) measurements, partly because of their small

[1] L. Narasimhan, J. Singh, C. Humblet, K. Guruprasad, and T. Blundell, *Nature Struct. Biol.* **1,** 850 (1994).

[2] P. K. Pallaghy, K. J. Nielsen, D. J. Craik, and R. S. Norton, *Prot. Sci.* **3,** 1833 (1994).

[3] B. M. Olivera, G. Miljanich, J. Ramachandran, and M. E. Adams, *Annu. Rev. Biochem.* **63,** 823 (1994).

[4] K. Valentino, R. Newcomb, T. Gadbois, T. Singh, S. Bowersox, S. Bitner, A. Justice, D. Yamashiro, B. B. Hoffman, R. Ciaranello, G. Miljanich, and J. Ramachandran, *Proc. Natl. Acad. Sci. U.S.A.* **90,** 7894 (1993).

[5] P. Stampe, L. Kolmakova-Partensky, and C. Miller, *Biochemistry* **33,** 443 (1994).

[6] V. Basus, G. Song, and E. Hawrot, *Biochemistry* **32,** 12290 (1993).



0076-6879/99 $30.00

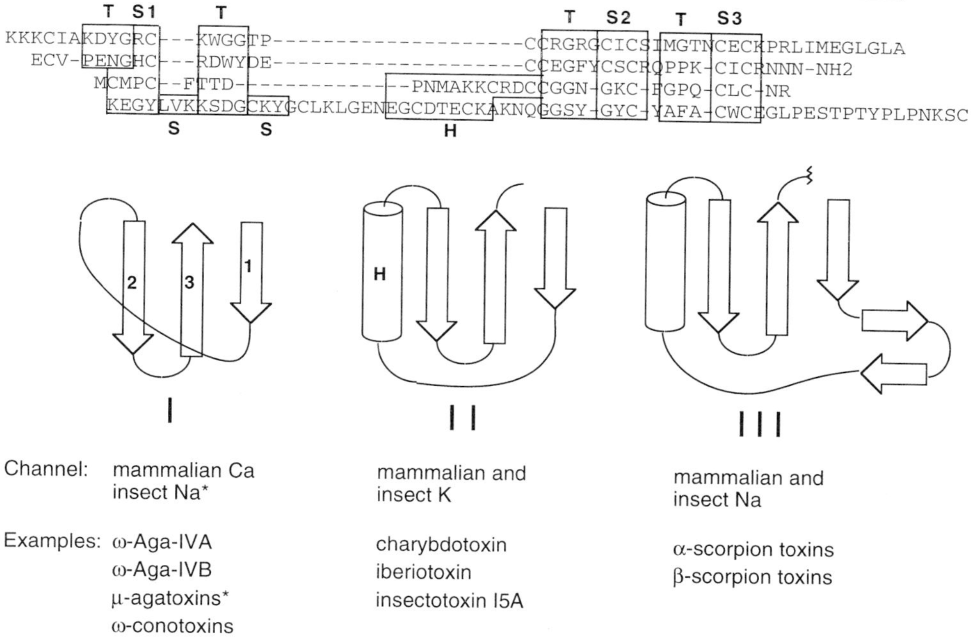


FIG. 1. The primary and secondary structures of several ion channel-active toxins from animal venoms and their specificities. [Reprinted with permission from D. O. Omecinsky, K. E. Holub, M. E. Adams, and M. D. Reily, *Biochemistry* **35,** 2836 (1996). Copyright 1996, American Chemical Society.]

size, typically 20–70 amino acids. Furthermore, their often high net charge confers good aqueous solubility and their many disulfide bridges provide extraordinary structural rigidity for their size.[7] To apply the structural information about these peptides to our understanding of their function and to the structure of the channels they interact with, we must have (1) a large database of peptide ligands whose structures have been determined to high precision and (2) an understanding of the molecular motions that occur within the peptide. In this article, we briefly discuss the aspects of NMR spectroscopy pertinent to the study of peptide toxins that bind to ion channels, and detail an expression and purification protocol that will both enable NMR-active isotope incorporation into these peptides and provide a facile system for producing mutant analogs for structure–function

[7] Y. Nishiuchi, K. Kumagaye, T. Noda, T. X. Watanabe, and S. Sakakibara, *Biopolymers* **25,** 561 (1986).

studies. Finally, we discuss experimental considerations for carrying out NMR studies on this class of peptides.

## NMR Spectroscopy of Peptide Toxins

NMR has been used extensively to measure solution structures of peptide toxins that inhibit ion channels. The many reports of NMR structures of toxins from gastropods (see Mitchell *et al.*[8] for a recent review of conotoxin structures), spiders,[9–14] and scorpions[15–19] alone in solution and complexed with peptides corresponding to their binding sites[6] attest to the applicability of NMR to the study of these peptides. Virtually all of these structures have been determined using well-established methods[20] based on interproton nuclear Overhauser effects (NOEs) and vicinal proton–proton coupling constants. These tightly folded peptides have a large surface area relative to their size and, as a result, conventional NOE-based structure determination yields solution structures that are underdetermined. While the literature contains many structures of this class of molecules, additional experimental constraints could improve the precision and utility of this information.

The dynamics of these molecules is less well understood. An ideal albeit somewhat unrealistic situation would be one in which the ligand is perfectly rigid, thus providing a "plaster cast" of the portion of the ion channel that interfaces with the ligand and an ideal template from which to design

[8] S. S. Mitchell, K. J. Shon, B. Olivera, and C. M. Ireland, *J. Nat. Toxins* **5,** 191 (1996).

[9] D. O. Omecinsky, K. E. Holub, M. E. Adams, and M. D. Reily, *Biochemistry* **35,** 2836 (1996).

[10] M. E. Adams, I. M. Mintz, M. D. Reily, V. Thanabal, and B. P. Bean, *Mol. Pharmacol.* **44,** 681 (1993).

[11] M. D. Reily, K. E. Holub, W. R. Gray, T. M. Norris, and M. E. Adams, *Nature Struct. Biol.* **1,** 853 (1994).

[12] H. Yu, M. K. Rosen, N. A. Saccomano, D. Phillips, R. A. Volkmann, and S. L. Schreiber, *Biochemistry* **32,** 13123 (1993).

[13] J. I. Kim, S. Konishi, H. Iwai, T. Kohno, H. Gouda, I. Shimada, K. Sato, and Y. Arata, *J. Mol. Biol.* **250,** 659 (1995).

[14] M. D. Reily, V. Thanabal, and M. E. Adams, *J. Biomol. NMR* **5,** 122 (1995).

[15] E. Adjadj, V. Naudat, E. Quiniou, D. Wouters, P. Sautiere, and C. T. Craescu, *Eur. J. Biochem.* **246,** 218 (1997).

[16] A. S. Arsen'ev, V. I. Kondakov, V. N. Maiorov, and V. F. Bystrov, *FEBS Lett.* **165,** 57 (1984).

[17] F. Bontems, C. Roumestand, P. Boyot, B. Gilquin, Y. Doljansky, A. Menez, and F. Toma, *Eur. J. Biochem.* **196,** 19 (1991).

[18] E. Blanc, V. Fremont, P. Sizun, S. Meunier, J. Van Rietschoten, A. Thevand, J.-M. Bernassau, and H. Darbon, *Proteins: Struct. Funct. Genet.* **24,** 359 (1996).

[19] M. Dauplais, A. Lecoq, J. Song, J. Cotton, N. Jamin, B. Gilquin, C. Roumestand, C. Vita, C. L. C. de Medeiros, E. G. Rowan, A. L. Harvey, and A. Menez, *J. Biol. Chem.* **272,** 4302 (1997).

[20] K. Wüthrich, "NMR of Proteins and Nucleic Acids." Wiley-Interscience, New York, 1986.

functional epitopes. Unfortunately, nature is not so accommodating and, clearly, despite the highly rigid structure of many venom-derived ion channel binding peptides, there are conformation degrees of freedom within the ligand in solution. This suggests that there are residues that are capable of undergoing relatively low-cost rearrangements to adopt specific geometries imposed by the ligand–channel complex. Residues that are not rigidly anchored in the solution structure are candidates for participation in such an "induced fit" model of the interaction. Thus, a detailed understanding of the motions that occur within biomolecules is critical to our ability to fully utilize structural information gleaned from solution or solid-state studies and to account for dynamic processes that are crucial to function. An affirmation of this is the large number of theoretical and experimental methods that have been developed to elucidate molecular motion. In particular, NMR spectroscopic methods to measure heteronuclear relaxation provide a perturbation-free means of investigating dynamic processes at the atomic level having time scales that range from picoseconds to microseconds.[21]

### *Traditional NMR Approaches*

The pioneering work of Wüthruch and Ernst established two-dimensional (2-D) NMR spectroscopy as a tool for measuring the solution structures of peptides and small proteins. Early methods still in routine use today rely primarily on constraints derived from one- and two-dimensional NMR experiments designed to measure interproton through-space dipolar couplings (NOEs) and through-bond scalar interactions (coupling constants). After the experimental constraints are imposed, a combination of distance geometry[22] and constrained dynamics algorithms[23] is used to generate many structures that satisfy these within the bounds of reasonable protein structure. These structures are then statistically analyzed[24] to access their precision and quality. Such methods are very robust, but require a large number of constraints per residue to provide high-quality structures [$\leq$1.2-Å rmsd (root mean square deviation) for backbone atoms]. Typically, this criterion is easily met with core residues that are tightly constrained by neighboring amino acids on all sides, whereas surface and flexible resi-

[21] R. Brüschweiler *in* "Understanding Chemical Reactivity 8" (R. Tycko, ed.), p. 301. Kluwer Academic Publishers, Dordrecht, 1994.

[22] T. Havel and K. Wüthrich, *Bull. Math. Biol.* **46,** 673 (1984).

[23] A. T. Brünger, G. M. Clore, A. M. Gronenborn, and M. Karplus, *Proc. Natl. Acad. Sci. U.S.A.* **83,** (1986); G. M. Clore and A. M. Gronenborn, *Crit. Rev. Biochem. Mol. Biol.* **24,** 479 (1989).

[24] S. G. Hyberts, M. S. Goldberg, T. F. Havel, and G. Wagner, *Prot. Sci.* **1,** 736 (1992).

dues often provide only a few NOEs and therefore are not well defined from these data alone. The result from a typical application of these homonuclear proton-based methods is shown in Fig. 2 for the agatoxin ω-Aga-IVB, isolated from the venom of the funnel web spider.[14] Regions for which there are many constraints are well defined within the ensemble of experimentally correct structures, whereas residues in loops and most notably in the C-terminal tail are less well defined.

The lack of precision across an ensemble of static structures determined by NMR is a result of insufficient constraints and cannot be assigned *a priori* to molecular motion. For a given atom, the lack of measurable constraints means one of three things. First, NOEs and/or coupling constants cannot be assigned due to spectral overlap. In this case, accidental coincidence of key resonances can prevent critical NOEs from being observed in a 2-D spectrum. This accidental overlap can be overcome by utilizing additional frequency dimensions, as discussed later. Second, the nucleus may not be close enough to any neighbors to give rise to NOEs. Such residues will likely be in contact with bulk solvent and dipole–dipole interactions will not be measurable. Third, rapid internal motion creates multiple environments for the nucleus, which therefore interacts with nuclear dipoles at different sites. Depending on the rate of internal motion and the exact chemical environment, nuclei may show no NOE or attenuated NOEs to sites that are inconsistent with a single static structure. In the first two situations, the nucleus in question may or may not be undergoing rapid internal motion; this must be determined independently. Obviously, structures such as those shown in Fig. 2 would be of much greater benefit if the interstructure precision was higher and the internal motions

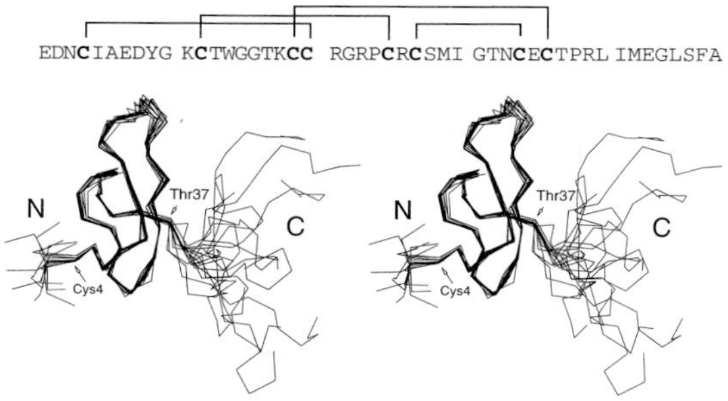


FIG. 2. Carbon-α trace of an ensemble of NMR structures of ω-Aga-IVB. Locations of the C and N termini, Cys-4 and Thr-37, are indicated. Residues for which many long-range NOEs were identified appear as tight groupings in the ensemble, whereas regions that appear disordered gave few or no long-range NOEs.

were better characterized to rationalize ill-defined regions. Greater precision can only be obtained through the use of methods such as those described next to identify more constraints, i.e., more NOEs and coupling constants.

*Heteronuclear-Based NMR Approaches*

Protons are by far the most sensitive NMR active nucleus found in biomolecules, and interproton NOEs will always provide a foundation for NMR structural determinations of small- to medium-sized proteins. However, much additional information can be gleaned from a system by exploiting the NMR properties of attached heavy nuclei, $^{13}C$ and $^{15}N$ in the case of proteins. For example, NOE data can be obtained by resolving the proton–proton NOE measurement based on the Larmor frequency, $\omega$, of an attached carbon and/or nitrogen. These 3-D or 4-D experiments[25] accomplish this by spreading the correlation data out in frequency dimensions corresponding to attached heavy nuclei, thereby minimizing spectral overlap and assignment ambiguities and allowing a larger number of unambiguous NOEs to be extracted from a given system. These heteronuclear-resolved and triple resonance methods require enriched levels of the stable isotopes $^{13}C$ and/or $^{15}N$. Also, developments in the measurement of heteronuclear $^{1}H–^{13}C$ and $^{1}H–^{15}N$ coupling constants[26] have provided a means for increasing the number of constraints per residue and thereby increasing the precision of the resulting structures. At natural abundance of the heavy nuclei, sample concentrations of $\geq 20$ m*M*[27] and $\geq 100$ m*M*[28] are required to measure 3 bond $^{1}H–^{13}C$ and $^{1}H–^{15}N$ coupling constants, respectively. For a peptide the size of $\omega$-Aga-IVA, this corresponds to a requirement of 60 mg (for $^{13}C$) and 250 mg (for $^{15}N$). The low natural abundance of $^{13}C$ (ca. 1%) and $^{15}N$ (ca. 0.3%) dictates that these heteronuclear resolution and filtering techniques are of very limited utility unless enrichment of the protein in the appropriate isotope is feasible.

Within the past decade, NMR has been used increasingly to study the internal motions in peptides and proteins. Methods have been developed that rely on the measurement of relaxation parameters for specific heavy atoms within the protein and relating these to motions that are different

[25] G. M. Clore and A. M. Gronenborn, *Prog. Nucl. Magn. Reson. Spectrosc.* **23,** 43 (1991).

[26] G. Wagner, *Prog. Nucl. Magn. Reson. Spectrosc.* **22,** 101 (1990).

[27] M. D. Reily, V. Thanabal, D. O. Omecinsky, J. B. Dunbar, Jr., A. M. Doherty, and P. L. DePue, *FEBS Lett.* **300,** 136 (1992).

[28] M. D. Reily, V. Thanabal, E. A. Lunney, J. T. Repine, C. C. Humblet, and G. Wagner, *FEBS Lett.* **302,** 97 (1992).

from the overall correlation time[29–32] (typically 1–5 ns for molecules of this size). Reviews by Dayie *et al.*[33] and Kay[34] cover the theoretical and experimental aspects of measuring heteronuclear relaxation in proteins and model-based and model-free methods used to interpret these data in terms of internal motion. The most extensively studied reporter of protein backbone motion is by far relaxation of the backbone $^{15}$N atom, and to a lesser extent $^{13}$C$\alpha$, since the relaxation of these nuclei is dependent primarily on dipolar interactions with their attached protons. In turn, the spectral density functions that determine how efficiently this dipolar relaxation occurs depend on the effective correlation times associated with the motions of the N–H or C–H bond vectors. Commonly measured parameters for elucidating molecular motion include the longitudinal ($T_1$) and transverse ($T_2$ or $T_1\rho$) relaxation times and heteronuclear NOEs, among others. The time constants $T_1$ and $T_2$ characterize the rate of decay back to equilibrium of perturbed magnetization components that are parallel and orthogonal to the applied magnetic field, respectively. The NOE results from cross-relaxation between nuclei (for example, $^{13}$C and $^1$H) in which the $T_1$ of $^{13}$C is dominated by dipolar coupling to the proton. All of these parameters can be sensitive to internal motions that are faster than the overall tumbling rate of the molecule and $T_2$ is also influenced by slower conformational exchange processes. Some of these parameters as they relate to correlation time are shown in Fig. 3. Thus, internal motion that reduces the effective correlation time for a nucleus can be manifested in one or more of these measurable parameters. Reports of carbonyl[33] and $^2$H relaxation[35] as measures of backbone and side-chain motions, respectively, promise to add to the arsenal of perturbation-free probes of protein dynamics using NMR. For peptides and proteins with good solubility (5–10 m*M*), it is possible to measure $^{13}$C$\alpha$ relaxation times at natural abundance[36–38] with a high degree of accuracy. However, because of the low relative abundance of NMR active isotopes for the heavy elements, enrichment with $^{13}$C can be beneficial and is, with rare exceptions, required for $^{15}$N studies.

[29] N. A. Farrow, O. Zhang, A. Szabo, D. A. Torchia, and L. E. Kay, *J. Biomol. NMR* **6,** 153 (1995).
[30] T. Yamazaki, R. Muhandiram, and L. E. Kay, *J. Am. Chem. Soc.* **116,** 8266 (1994).
[31] J. W. Peng and G. Wagner, *Biochemistry* **34,** 16733 (1995).
[32] M. Akke and A. G. Palmer III, *J. Am. Chem. Soc.* **118,** 911 (1996).
[33] K. T. Dayie, G. Wagner, and J.-F. Lefevre, *Annu. Rev. Phys. Chem.* **47,** 243 (1996).
[34] L. E. Kay, *Biochem. Cell Biol.* **75,** 1 (1997).
[35] D. R. Muhandiram, T. Yamazaki, B. D. Sykes, and L. E. Kay, *J. Am. Chem. Soc.* **117,** 11536 (1995).
[36] A. G. Palmer III, M. Rance, and P. E. Wright, *J. Am. Chem. Soc.* **113,** 4371 (1991).
[37] K. Arvidsson, J. Jarvet, P. Allard, and A. Ehrenberg, *J. Biomol. NMR* **4,** 653 (1994).
[38] N. R. Nirmala and G. Wagner, *J. Am. Chem. Soc.* **110,** 7557 (1988).

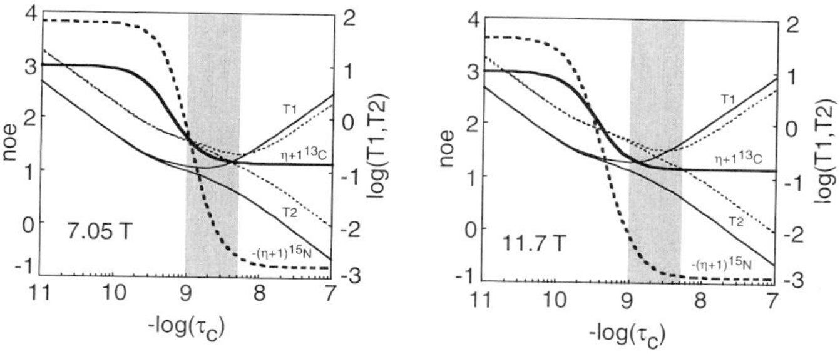


FIG. 3. Heteronuclear NOE, $T_1$ and $T_2$ curves for $^{13}C$ and $^{15}N$ nuclei in proteins plotted as a function of correlation time. The effect of the applied magnetic field on these parameters can be seen by comparing the left and right panels, which depict parameter dependence on $\tau_c$ at 7.05 T (300 MHz $^1H$, 75.4 MHz $^{13}C$, 30.4 MHz $^{15}N$) and 11.75 T (500 MHz $^1H$, 125 MHz $^{13}C$, 50.7 MHz $^{15}N$), respectively. The shaded areas indicate the range of overall correlation times expected for toxin-sized peptides. These curves were generated using equations that included terms for chemical shift anisotropy (CSA) relaxation as described in A. G. Palmer III, M. Rance, and P. E. Wright, *J. Am. Chem. Soc.* **113,** 4371 (1991). The value used for chemical shift anisotropy, $(\sigma_{\parallel} - \sigma_{\perp})$ for $^{15}N$ was $-160$ ppm and for $^{13}C\alpha$, 25 ppm.

## Isotope Labeling Strategies

Taking the above points into consideration, it is desirable to incorporate high levels of $^{13}C$ and $^{15}N$ into these peptides and thereby both improve the precision of solution structures and provide spectroscopic handles for detailed dynamics studies. Molecular biology has advanced contemporaneously with protein NMR techniques and efficient expression of exogenous proteins from microorganisms provides a means of incorporating the isotopes either uniformly or at selected amino acids. By and large, peptide ion channel ligands are small enough to prepare synthetically; however, the cost of uniform incorporation of NMR active isotopes makes this prohibitive for most applications. As a comparison of the relative materials cost of synthetic versus biosynthetic approaches for uniform $^{15}N$ labeling, 1 $\mu$mol of a 50-amino-acid peptide is estimated to cost approximately \$1400 to produce synthetically and between \$30 and \$300 via *Escherichia coli* expression, depending on whether minimal or enriched media is used.[39] For uniform $^{13}C$ and $^{15}N$ labeling, the cost skyrockets to \$15,000 for the

[39] Based on the following assumptions: The yield of crude unfolded, deprotected peptide is 10% (molar basis) for solid-phase synthesis and 100 or 50 mg/liter for bacterial expression in enriched and minimal media, respectively. The refolding efficiency starting with purified reduced and unfolded peptide is 10% for both approaches. Materials costs are based on current prices for commercially available $^{15}N$ uniformly labeled Boc-protected amino acids, $^{15}N$-labeled enriched media and $^{15}NH_4Cl$ for minimal media.

synthetic sample and $750 to $1250 for the recombinant protein.[40] It is also important to note that many fully protected, isotope-labeled amino acids are not available from commercial sources at the present time. Synthetic strategies preclude the labeling of sites with these amino acids without resorting to custom syntheses.

## Bacterial Expression of ω-Aga-IVA Analogs

In addition to the necessity for stable isotope incorporation, two of the many benefits of having a cloned and expressed toxin is the ability (1) to conduct structure–activity relationship (SAR) studies by the creation of point mutants and chimeric peptides using site-directed mutagenesis and (2) to generate milligram to gram quantities of venom components that are only available from natural sources in much smaller amounts. Thorough peptide toxin SARs have been investigated using a solid-phase synthesis approach on the ω-contoxins MVIIA[41] and GIVA,[42] both of which are small (≤27 residue) peptides with three disulfide bonds. Several reports of *E. coli* expression of peptide toxins that bind to ion channels have also appeared with the goal of studying the molecular mechanism of channel activity via site-directed mutagenesis. The scorpion toxins, charybdotoxin[43,44] (37 amino acids) and LqhαIT (62 amino acids),[45] have been expressed either as a soluble fusion protein or as the mature, insoluble protein, respectively. Bacterial expression of a component of sea snake venom, erabutoxin a (66 amino acids)[46] and an extensive SAR has also been reported. Adams[47] has recently reported an independent expression of wild-type ω-Aga-IVA as a secreted fusion protein in *E. coli*. The prospect of a yield high enough to make NMR-active isotope incorporation feasible (2.5–30 mg/liter in the above literature) provided the impetus for the work

[40] Assuming the current cost of uniformly $^{13}C$,$^{15}N$-labeled enriched media and [$^{13}C$]glucose for minimal media.

[41] L. Nadasdi, D. Yamashiro, D. Chung, K. Tarczy-Hornoch, P. Adriaensssens, and J. Ramachandran, *Biochemistry* **34,** 8076 (1995).

[42] M. J. Lew, J. P. Flinn, P. K. Pallaghy, R. Murphy, S. L. Whorlow, C. E. Wright, R. S. Norton, and J. A. Angus, *J. Biol. Chem.* **272,** 12014 (1997).

[43] C.-S. Park, S. F. Hausdorf, and C. Miller, *Proc. Natl. Acad. Sci. U.S.A.* **88,** 2046 (1991).

[44] S. A. Goldstein, D. J. Pheasant, and C. Miller, *Neuron* **12,** 1377 (1994).

[45] N. Zilberberg, D. Gordon, M. Pelhate, M. E. Adams, T. M. Norris, E. Zlotkin, and M. Gurevitz, *Biochemistry* **35,** 10215 (1996).

[46] F. Ducancel, J.-C. Boulain, O. Trémeau, and A. Ménez, *Prot. Eng.* **3,** 139 (1989); M. Hervé, L. Pillet, P. Humbert, O. Trémeau, F. Ducancel, C. Hirth, and A. Ménez, *Eur. J. Biochem.* **208,** 125 (1992); L. Pillet, O. Trémeau, F. Ducancel, P. Drevet, S. Zinn-Justin, S. Pinkasfeld, J.-C. Boulain, and A. Ménez, *J. Biol. Chem.* **268,** 909 (1993).

[47] M. E. Adams, personal communication (1997).

described here toward developing expression, purification, and refolding protocols for a mutant of the P-type calcium channel blocker ω-Aga-IVA. Considering previous structural studies of the ω-agatoxins,[10–14] we were interested in the functional, dynamic, and structural role of the C-terminal tail, which happens to be a somewhat unusual moiety in this peptide class.[3] As a result, we designed a sequence that was CNBr-labile at the N and C termini with Met at positions −1 and 42, yet was protected from additional cleavages by alteration of the wild-type Met at position 30 to Ile. It is to be hoped that these methods will prove to be generally applicable and will provide an avenue for obtaining other isotope-labeled peptide toxins.

*Cloning and Insertion in Expression Vector*

The ω-Aga-IVA[M30I] mutant construct is cloned from overlapping oligonucleotides and amplified via polymerase chain reaction (PCR). A methionine residue is placed in front of the ω-Aga-IVA[M30I] sequence to enable the use of cyanogen bromide to cleave the toxin from the leader peptide. Details of the construct are shown in Fig. 4. Multiple single clones are sequenced on an ABI 373 sequencer to ensure that no errors are introduced in the clone both after the PCR and after ligations into each new expression vector.

Two expression systems were tested for the expression of the ω-Aga-IVA[M30I] mutant. The first was a T7 polymerase-based system. This system, pET30a (Novagen, Madison, WI), utilizes a T7 polymerase for transcription in the *E. coli* strain BL21(DE3), which expresses the polymerase gene. The vector also contains a sequence encoding a series of six histidine residues to allow purification using nickel-chelate columns. The second system, pQE31 (Qiagen, Valencia, CA), also contains a hexahistidine ($His_6$) sequence for purification on nickel columns, but transcription is driven by a T5 polymerase control region. The cells used for expression are SG13009(Prep4) and encode a repressor that tightly regulates the expression of the T5 polymerase.

Initial expression of the ω-Aga-IVA[M30I] construct in the pET30a vector yielded low levels of protein, detectable only by Western blot analysis. A portion of the cloned sequence was recloned into the pQE31 vector. This resulted in a construct containing a relatively long ladder sequence, but still containing a Met residue just in front of the first lysine residue in the ω-Aga-IVA[M30I] mutant sequence. Levels of expression of this construct were much higher, yielding approximately 375 mg of purified unfolded ω-Aga-IVA[M30I] construct per liter of fermentation broth. In both systems the expressed protein was found in the insoluble mass, presumably sequestered by *E. coli* as inclusion bodies, greatly facilitating purification and minimizing degradation.

**A**

T7 Promoter

GCGTAGAGGA TCGAGATCGA TCTCGATCCC GCGAAATTAA TACGACTCAC

Lac Operator — Xba I

TATGGGGAA TTGTGAGCGG ATAAACAATTC CCCTCTAGAA ATAATTTTGT

TTAACTTTAA GAAGGAGATA TACATATGCA CCATCATCAT CATCATTCTT

CTGGTCTGGT GCCACGCGGT TCTGGTATGA AAGAAACCGC TGCTGCTAAA

TTCGAACGCC AGCACATGGA CAGCCCAGAT CTGGGTACCG ACGACGACGA

CAAGGCCATG GCTGATATCG GATCCGAATT CGAGCTCCGT CGACAAGCTT

XhoI

GCGGCCGCAC TCGAGCACCA CCACCACCAC CACTGAGATC CGGCTGCTAA

CAAGCCCGA

**B**

RBS ATG AGA GGA TCG CAT CAT CAT CAT CAT CAT GGA TCC GCA

Kpn I — Hind III

TGC GAG CTC GGT ACC CCG GGT CGA CCT GCA GCC AAG CTT

**C**

Xba I

TCTAGAAAGGAGATATACAT ATG AAG TGC TAC CCG CGC CTG GGT GGT GGT
M K C Y P R L G G G

GGC GGC CGC CTG GTT CCT CGT AAG AAG AAG
G G R L V P R K K K

TGC ATC GCT AAG GAC TAC GGT AGA TGC AAG
C I A K D Y G R C K

TGG GGT GGT ACG CCT TGC TGC AGA GGT AGA
W G G T P C C R G R

GGT TGC ATC TGC TCT ATC ATG GGT ACG AAT
G C I C S I M G T N

TGC GAG TGC AAG CCT AGA CTT ATC ATG GAG
C E C K P R L I M E

XhoI

GGT CTT GGT CTT GCT TAA TAA CTCGAG
G L G L A X X

**D**

10 20 30 40

KKKCIAKDYGRCKWGGTPCCRGRGCICSIMGTNCECKPRLIMEGLGLA

M → I

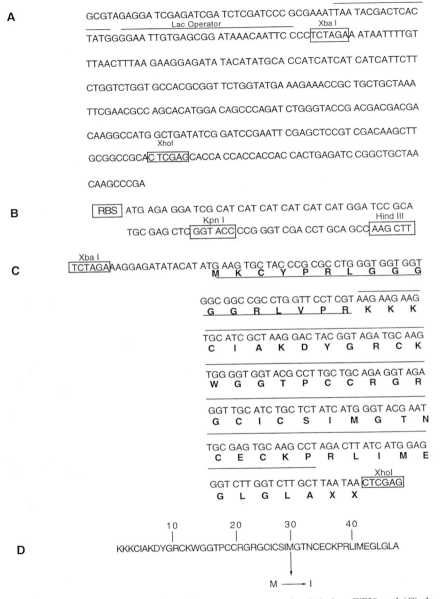

FIG. 4. The total ω-Aga-IVA[M30I] construct sequence for (A) the pET30 and (C) the pQE30 vectors. Cloning sites are marked. (B) The sequence of the synthetic ω-Aga-IVA clone. Underlined sequence corresponds to the sequence of the wild-type toxin, with the [M30I] mutation at position 30 indicated by the box. (D) The protein sequence of ω-Aga-IVA and identification of the chosen mutation.

*Small-Scale Expression in pQE31*

The pQE31 ω-Aga-IVA[M30I] plasmid (Fig. 4) is introduced into competent SG13009(pREP4) cells and transformants are selected on plates containing 100 μg/ml ampicillin and 30 μg/ml kanamycin. A single colony is transferred into 5 ml of LB medium containing 100 μg/ml ampicillin and 30 μg/ml kanamycin and grown to saturation overnight at 37°. Fresh LB medium (1 liter) containing 100 μg/ml ampicillin and 30 μg/ml kanamycin is inoculated with 500 μl of the culture and incubated at 37° for 4–5 hr, to an $A_{600}$ of 0.6 to 0.9. After addition of isopropylthio-β-D-galactoside (IPTG) to a final concentration of 1 m*M*, the culture is further incubated at 37° for another 4–6 hr. Cells are harvested by centrifugation at 5000 rpm at 4°. The cell pellets are either used immediately to isolate the inclusion bodies[48] or stored at −80°.

*Inclusion Body Isolation*

The cell pellets are resuspended in 100 ml of ice-cold lysis buffer (50 m*M* $NaH_2PO_4$, 300 m*M* NaCl, 10 m*M* imidazole, pH 8.0) and lysed with a French press. The cell lysate is then centrifuged at 10,000 *g*, resuspended in 75 ml of buffer (50 m*M* potassium phosphate, 10 m*M* $MgCl_2$) to which 9 units/ml benzonase (American International Chemicals, Natick, MA) and 2500 units/ml lysozyme has been added, and incubated at 37° for 2 hr. Triton X-100 is added to a final concentration of 1% and the solution is stirred overnight at 4°. The suspension is then diluted 2-fold with water and centrifuged at 12,000 *g*. The pellet is resuspended in 150 ml of water and the pH adjusted to pH 7.85. The suspension is stirred at room temperature for 30 min then centrifuged as above. The pellet is once again resuspended in 150 ml of water and the pH of the suspension adjusted to pH 8.4, agitated for 30 min at room temperature, and finally centrifuged again as above. The resulting pellet is finally resuspended in 75 ml of 95% (v/v) ethanol, stirred at room temperature for 30 min, and centrifuged as above. At this point, the inclusion bodies are either solubilized immediately or dried and stored at −80°.

*Large-Scale Expression of ω-Aga-IVA[M30I] Construct*

To scale up the protein production, the expression system is transferred to a 10-liter fermentor. The culture is started by inoculating a 200-ml shake flask [2% (w/v) Lennox L Broth (Becton Dickinson, Cockeysville, MD) containing 100 μg/ml ampicillin and 30 μg/ml kanamycin] with 1 ml of a

[48] Yitzhak Stabinsky, personal communication (1992); see also F. A. O. Marston, P. A. Lowe, M. T. Doel, J. M. Schoemaker, S. White, and S. Angal, *Bio/Technology* **2,** 800 (1984).

frozen 20% (v/v) glycerol stock in the SG13009(pREP4) cells described above and incubating at 30°. After the shake flask becomes cloudy (8 hr), use 1 ml of the shake flask culture to inoculate the fermentor. The media for the fermentor consists of 2% (w/v) of the following: yeast extract (Difco, Detroit, MI), acidicase peptone (Becton Dickinson, Cockeysville, MD), casitone (Difco), gelatone (Difco); 0.2% (w/v) of the following: $KH_2PO_4$ anhydrous, $K_2HPO_4$ anhydrous, and $Na_2HPO_4 \cdot 7H_2O$; 100 $\mu$g/ml ampicillin and 30 $\mu$g/ml kanamycin. The culture is maintained at pH 6.8 with a combination of lactic acid and 0.25 *M* NaOH. Air is sparged into the fermentor at 8 liters/min and the temperature was maintained at 30°. At an $OD_{600}$ of 10, the temperature of the fermentor is adjusted to 37° and recombinant protein expression is induced by addition of IPTG to a concentration of 3.5 m*M*. Three hours after induction the fermentor is harvested using both hollow-fiber microfiltration and centrifugation. The cell paste is stored at −70°.

The cell paste (490 g) is resuspended to 1.2 liters with 50 m*M* sodium phosphate (pH 7.2) containing 10 m*M* $MgCl_2$ and 1% Triton X-100. Benzonase is added to 9.4 units/ml. The cells are lysed by passing the suspension twice through a Dyno-Mill KDL at 100 ml/min. The Dyno-Mill is equipped with a 600-ml chamber containing 500 ml of 0.25- to 0.5-mm lead-free glass beads. The jacket temperature is maintained at −13° and the agitator speed is 4200 rpm.

The lysate suspension is centrifuged at 13,680 *g* at 4° for 90 min. The pellet is resuspended with 750 ml of 50 m*M* potassium phosphate, 50 m*M* EDTA, 10 m*M* $MgCl_2$ (pH 8.0) containing 15 units/ml Benzonase and 2500 units/ml chicken egg white lysozyme (Sigma, St. Louis, MO). The suspension is adjusted to pH 7.5 with NaOH and stirred at 37° for 2 hr. Triton X-100 is added to a concentration of 1% and the suspension is stirred at 4° overnight. The volume of the suspension is adjusted to 1.6 liters with water and insoluble material is pelleted by centrifugation as above. The pellet is washed twice by resuspending with 1 liter of water and pH adjustment with NaOH (7.85 and 8.4 for the first and second wash, respectively) followed by stirring at room temperature for 30 min and centrifugation as above. The pellet is resuspended with 500 ml 95% ethanol, stirred at room temperature overnight and centrifuged as above. The resulting pellet is lyophilized and stored at −70°. The yield of dried inclusion bodies is 12.5 g.

### *Purification of Recombinant ω-Aga-IVA[M30I] Construct*

Initial purification through His-tag affinity chromatography[49] provides a very simple and rapid means for removal of most of the contaminants

[49] M. C. Smith, T. C. Furman, T. D. Ingolia, and C. Pidgeon, *J. Biol. Chem.* **263,** 7211 (1988).

from the inclusion bodies. Acid elution is preferred to the use of imidazole because the construct is quite stable at low pH and subsequent steps required lyophilizations and reversed-phase high-performance liquid chromatography (HPLC) under acidic conditions. Based on HPLC retention times and the appearance of multiple bands in nonreducing polyacrylamide gel electrophoresis (PAGE), it is clear that the Ni-NTA purified protein consists of many different folded conformers and oligomeric species. These multiple species can be converted to a single component by addition of reducing agent. Tris-(2-carboxyethyl)phosphine hydrochloride (TCEP,[50] Molecular Probes, Eugene, OR) was found to be a more convenient reducing agent than either dithiothreitol (DTT) or 2-mercaptoethanol for this application. The primary advantage of TCEP is its ability to reduce rapidly the protein at low pH (pH ~2.5)[51] where protein reoxidization is slow and the reduced protein is stable and highly soluble. The reduction is done such that the solution is ready for reversed-phase HPLC, which removes reducing agents and residual contaminants not removed by the Ni-NTA column.

*Ni-NTA Chromatography.* Inclusion bodies (1.0 g) are added to 500 ml of solubilizing buffer (6 *M* guanidine hydrochloride, 0.1 *M* sodium phosphate, 10 m*M* Tris buffer, pH 8.0) and stirred at room temperature overnight. The solution is centrifuged at 10,000 *g* and the clarified supernatant stirred with 50 ml of Ni-NTA resin (Qiagen) for 1–2 hr. The slurry is then poured into a 30- × 2.5-cm column attached to an FPLC (fast protein liquid chromatography, Pharmacia, Piscataway, NJ) system, where both resistivity and $A_{280}$ are monitored. Once the column is packed, it is washed with two column volumes of solubilizing buffer, followed by two column volumes of wash buffer (0.1 *M* KCl, 0.1 *M* $NaPO_4$ 10 m*M* Tris, pH 8.0), and finally with deionized water until the eluant resistivity becomes <0.01 mS/cm. At this point, the His-tagged protein is eluted with 0.2 *M* acetic acid, until the monitored $A_{280}$ becomes <0.01. The acetic acid fraction was diluted 2-fold with water and lyophilized. Using an estimated extinction coefficient of 7200 $M^{-1}$ at 280 nm, a UV spectrum of the acetic acid fraction indicated that ca. 300 mg of crude expression product is obtained from 1.0 g of inclusion bodies. The resulting powdery residue is dissolved into 5–10 ml of 0.1% trifluoroacetic acid (TFA), and stirred for 12 hr with 5–10 molar excess of TCEP to fully reduce the protein.

*Reversed-Phase HPLC.* HPLC separations are performed on a Varian 5500 using 0.1% TFA in water as buffer A and 0.1% TFA in acetronitrile as buffer B. The reduced ω-Aga-IVA[M30I] construct solubilized in 0.1%

[50] U. T. Ruegg and J. Rudinger, *Methods Enzymol.* **47,** 111 (1977).

[51] J. A. Burns, J. C. Butler, J. Moran, and G. M. Whitesides, *J. Org. Chem.* **56,** 2648 (1991).

TFA is filtered and 75–100 mg is loaded onto a preparative scale (22- × 250-mm) alltima phenyl column (Alltech Associates, Deerfield, IL), which is equilibrated at 25% buffer B. A gradient was run from 25 to 30% buffer B in 30 min at a flow rate of 10 ml/min. This method resolved the reduced ω-Aga-IVA[M30I] construct, which eluted as the major peak, from residual contaminants, trace nickel, and reducing agents. The fractions of interest are lyophilized after evaporation of most of the acetonitrile.

*Refolding and Purification*

Peptide toxins of this class are rich in cysteines, and their tertiary structure is largely determined by the intramolecular disulfide bonds. Thus, successful refolding strategies rely on controlled oxidation under partially denaturing conditions. There are numerous variables that one might experiment with including the type and concentration of chaeotropic agent, temperature, pH, salt concentration, and the amount, identity and oxidation/reduction (ox/red) proportions of redox mediators. Each different protein will undoubtedly have its own set of preferred conditions and several weeks should be set aside to investigate these parameters on a small scale. The protocol described below works well for the ω-Aga-IVA[M30I] construct and is an adaptation of one described by Hillyard *et al.*[52] The refolding buffer (0.5 *M* guanidine hydrochloride, 10 m*M* Tris, 0.2 m*M* glutathione, pH 7.7) is prechilled to 4° and the reduced, HPLC-purified ω-Aga-IVA[M30I] construct is then added (slowly, with rapid stirring) to a final concentration of 0.02 m*M*, and the reaction allowed to proceed at 4° over 5–7 days. The progress of refolding is monitored via reversed-phase HPLC on an analytical phenyl column. Once no additional changes are observed, typically after 5 days, the refolding solution is dialyzed (Spectrapor 3,000 molecular weight cutoff) twice against a 4-fold volume excess of 5 m*M* acetic acid at 4° over 2 days, then against water overnight, and finally lyophilized. The lyophilized powder is then resuspended into 2–5 ml of 0.1% (TFA) and the differently folded conformers are separated on the preparative scale phenyl column previously mentioned, with the same gradient. Fractions are individually collected, lyophilized, and structurally compared to wild-type ω-Aga-IVA using NMR spectroscopy (see below). Figure 5a shows a typical HPLC chromatogram with the fractions containing correctly folded peptide indicated. Typically, 18–20% of the protein is converted into the properly folded material.

[52] D. R. Hillyard, V. D. Monje, I. M. Mintz, B. P. Bean, L. Nadaski, J. Ramachandran, G. Miljanich, A. Azimi-Zoonooz, J. M. McIntosh, L. J. Cruz, J. S. Imperial, and B. M. Olivera, *Neuron* **9,** 69 (1992).

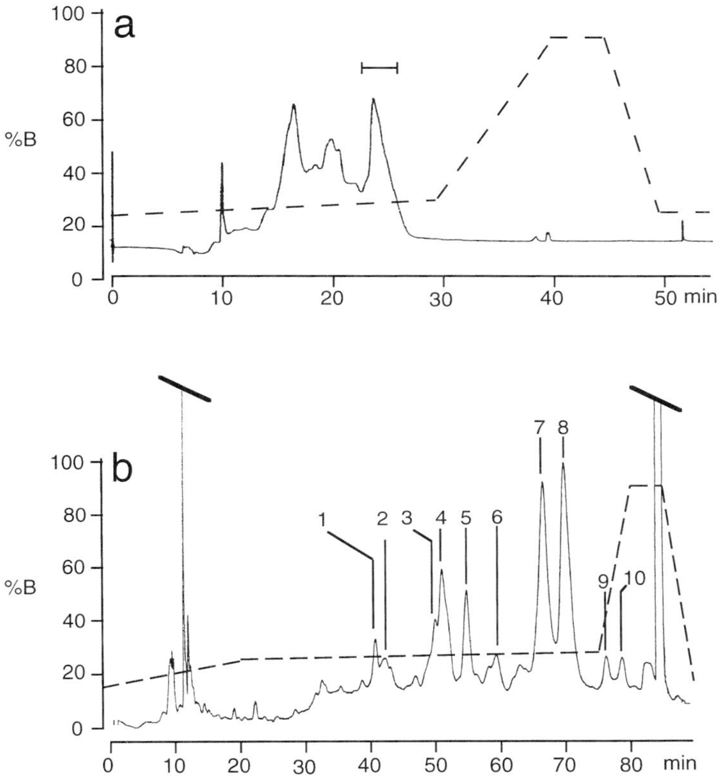


FIG. 5. Purification steps using reversed-phase HPLC. (a) Chromatogram of soluble products from refolding of the full-length ω-Aga-IVA[M30I] construct. The fraction indicated by the horizontal bar had the highest biological activity and gave NMR spectra consistent with properly folded peptide, see text. (b) CNBr cleavage products of the single isolated peak (indicated by the bar) from the top trace. Dotted lines indicate the approximate profile of the gradient used in these chromatograms.

Temperature and pH have a pronounced effect on the refolding of ω-Aga-IVA[M30I] construct. At 25°, and pH values ranging from 6 to 8.2, most of the protein precipitates out of the refolding solution, whereas at 4° the refolding reaction occurs with minimal loss of protein. At pH 8.2, the oxidation is completed after 12 hr at 4°, but much of the protein aggregates and precipitates out of solution. At pH 6.2, oxidation is still not completed after 10 days at 4°. It was found that pH 7.7 is an excellent compromise; the yield is only slightly lower than at pH 6.2 but the oxidation/refolding was completed within 7 days.

*Cleavage from the Leader Peptide*

Removal of a leader peptide can be accomplished by specific proteolysis with thrombin,[53] or factor X,[43] requiring insertion of the appropriate cleavage site into the clone. Problems with this method include artifact residues in the cleaved protein and the possibility of nicking or cleaving of the product at other than the cleavage site.[54] Another concern with the ω-Aga[IVA]M30I] construct is the proximity of the N-terminal Lys and the globular region of the peptide; it seems likely that enzymatic cleavage would be inhibited by the bulky folded domain. Cyanogen bromide cleaves specifically at Met residues and represents an attractive alternative to proteolytic cleavage for these types of systems. Wild-type ω-Aga-IVA contains two Met residues at positions 30 and 42. With the knowledge that both of these Met residues can be replaced by norleucine with full retention of activity,[11] and that the folded protein is stable in 70% TFA, CNBr cleavage was the method of choice. This same rationale would apply to many peptide toxins of this class.

The lyophilized HPLC fractions of the oxidized full-length construct that showed the highest activity and positive NMR analysis (see below and Fig. 5) are dissolved in 70% TFA at a concentration of 2.0 mg/ml with a 1000-fold molar excess of CNBr (~2 g CNBr/100 mg protein). The reaction is then allowed to proceed in the dark, at room temperature, for 6 to 8 hr, diluted 5-fold with water, and lyophilized. The dilution and lyophilization process are repeated three times. For the large-scale CNB solution lyophilizations, a 1.5-liter cold finger is placed in line with one of the vacuum ports and placed in a dry ice–acetone bath as a primary trap for the noxious volatiles, facilitating proper treatment of the reaction by-products and providing some protection for our general-purpose lyophilizer. The material resulting from the final lyophilization is then resuspended into 2–5 ml of 0.1% TFA and subjected to preparative scale reversed-phase HPLC. The digested mixture is separated on the 22-mm phenyl column using a shallow gradient of 0.1% TFA in acetonitrile (15 to 23% in 20 min, 23 to 25% in 55 min at 10 ml/min, Fig. 5b). The various fractions are analyzed by mass spectrometry (MS) and NMR, as described below.

## Characterization of the Peptide

*Biological Assays*

Whenever possible, a functional or binding assay should be used to identify active components in the refolding medium. For the ω-Aga-

[53] D. J. Hakes and J. E. Dixon, *Anal. Biochem.* **202,** 293 (1992).

[54] J. Y. Chang, S. S. Alkan, N. Hilschmann, and D. G. Braun, *Eur. J. Biochem.* **151,** 225 (1985).

IVA[M30I] construct, it was not possible to measure near-wild-type activity in HPLC fractions of the full-length refolded peptide in a rat brain $^{45}Ca$ flux assay. However, fractionation of the solution of a typical refolding experiment resulted in peaks that contained material with a range of $IC_{50}$ values from 5 to 0.3 $\mu M$.[55] Note that the fractions that were most active were still 100-fold less potent than wild-type ω-Aga-IVA, possibly due to the fact that the leader peptide (His tag) was present in all of these samples. CNBr-treated samples lacked the C-terminal tail and were therefore not expected to be active.[56]

*Physical Characteristics*

It is advantageous to characterize all of the HPLC fractions by matrix-assisted laser desorption mass spectrometry (MALDI).[57] This is particularly true after CNBr treatment, where incomplete cleavage products that may or may not be properly folded can easily be identified. Using MALDI, tiny aliquots taken directly from the preparative HPLC fractions yielded accurate ($\pm 0.1\%$) molecular weight information. Fractions of the proper molecular weight can then be analyzed to assess the protein fold using NMR spectroscopy. NMR chemical shifts are sensitive to protein secondary structure[58] and therefore provide a rapid means to determine if a particular fraction is properly folded. By comparing the $^{1}H$ chemical shifts of HPLC isolates against those from an authentic sample of ω-Aga-IVA (Peptides International, Louisville, Kentucky) it was clear that the latest eluting major peak (peak 8 in Figure 5b) was properly folded. Figure 6 compares identical regions of a 2-D total correlation spectroscopy (TOCSY) spectrum for authentic ω-Aga-IVA with spectra from two predominant CNBr cleavage products isolated by HPLC. This region of the 2-D spectrum contains most of the scalar correlations within Cys residues and identifies the chemical shifts of the H$\alpha$ (F1) and H$\beta$ (F2) for key residues. Large deviations from the spectrum of an authentic sample (compare lower and center panels in Fig. 6) indicate conformational changes, probably associated with an incorrect disulfide bonding pattern, whereas the similarity seen in the center and upper panels strongly suggests an identical fold. This "chemical shift mapping" represents a powerful technique to identify the appropriate species after fractionation of a refolding mixture even when biologic activity

[55] J. Geer, personal communication (1997).

[56] M. Kuwada, T. Teramoto, K. Y. Kumagaye, K. Nakajima, T. Watanabe, T. Kawai, Y. Kawakami, T. Niidome, K. Sawada, Y. Nishizawa, and K. Katayama, *Mol. Pharmacol.* **46,** 587 (1994).

[57] F. Hillenkamp, M. Karas, R. C. Beavis, and B. T. Chait, *Anal. Chem.* **63,** 1193A (1991); F. Hillenkamp and M. Karas, *Methods Enzymol.* **193,** 280 (1990).

[58] L. Szilagyi, *Proc. Nucl. Magn. Reson. Spectros.* **27,** 325 (1995).

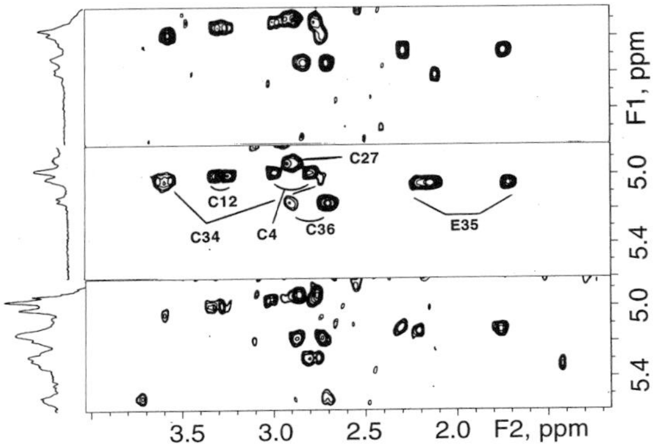


FIG. 6. The Hα–Hβ region of the 500-MHz $^{1}H$–$^{1}H$ 2-D TOCSY spectra for authentic ω-Aga-IVA (center panel), and peaks 5 and 8 from the HPLC separation of the CNBr digest (bottom and top panels, respectively). Assignments for selected cross-peaks are indicated.

assays are inappropriate. Spectra such as those shown in Fig. 6 can be obtained in about 1 hr with 1 mg of sample.

## Experimental Considerations

### *Structure Determination*

The favorable solution characteristics of most of these toxins provide much latitude in choosing conditions under which to carry out NOE-based structure determinations. The solubility is generally quite good and aggregation is rarely a problem. The latter can be identified by observing an increase in the $^{1}H$ NMR line widths as peptide concentration is increased. If ample peptide is available and it is determined that aggregation is not a problem, 5–10 m*M* or higher sample concentrations will provide excellent 2-D NMR results on modern high-field spectrometers operating at 500- or 600-MHz proton frequency. For toxins below 30–40 amino acids, there is probably no benefit to utilizing higher dimensional methods to obtain NOEs. For larger peptides, particularly those with highly flexible regions, or for toxins complexed with channel fragments isotopic labeling would enable the use of 3-D methods to resolve overlapping peaks. The ω-agatoxins are good examples of peptides that would benefit from higher dimensional experiments as the amide protons in the flexible C terminus congregate in a very narrow chemical shift range,[10–14] making extraction of structural information for these amino acids difficult.

These peptides often have some self-buffering capability, so the use of a buffer can usually be avoided. If necessary, one of the buffers generally used for NMR (e.g., acetate-$d_3$, Tris-$d_{11}$, phosphate) can be employed. Freund and Kalbitzer[59] have reported physiologic buffers specifically for use with NMR samples. An ideal pH to work at is one at which the peptide carries a physiologic charge, but is low enough ($\leq 7$) that amide exchange rates for exposed residues do not become too fast.[60] If histidines are not present in the molecule and the other titratable residues have normal p$K_a$ values, a physiologic ionization state should be achieved at about pH 5.5–6.0. Observation of the change in shifts in a 1-D or 2-D TOCSY $^1$H NMR spectrum as a function of pH should allow estimation of p$K_a$ values and the selection of an appropriate pH.

The standard suite of 2-D NMR experiments for resonance assignment and structural determination includes correlated spectroscopy (COSY[61]) and TOCSY[62] for intraresidue connectivities, and nuclear Overhauser effect spectroscopy (NOESY[63]) to measure intra- and interresidue NOEs. The advent of pulsed field gradients has spawned analogs of these experiments with greatly improved water suppression.[64] Indirect NOEs resulting from spin diffusion still present a problem, although it is less severe than in larger proteins. One can treat the homonuclear NOE data using a full relaxation matrix approach,[65] or apply one of the available pulse sequences to suppress spin diffusion during data acquisition. Variants of the NOESY experiment that use spin network editing to eliminate spin diffusion pathways block-decoupled NOESY (BD-NOESY[66]) or which allow NOEs to build up in the presence of a strong spin locking field rotating frame NOE spectroscopy (ROESY[67]) have been used in toxin studies.[9] In addition to computational and experimental methods for ensuring that only direct NOEs are used as constraints, short mixing times ($\leq 100$ ms) should be used to collect the NOESY data from which the principal interproton distances will be derived. Once the proton resonances are assigned, 1-D and 2-D TOCSY and/or NOESY spectra should be recorded at at least three

[59] J. Freund and H. R. Kalbitzer, *J. Biomol. NMR* **5,** 321 (1995).

[60] K. Wüthrich and G. Wagner, *J. Mol. Biol.* **130,** 1 (1979).

[61] M. Rance, O. W. Sørensen, G. Bodenhausen, G. Wagner, R. R. Ernst, and K. Wüthrich, *Biochem. Biophys. Res. Commun.* **117,** 479 (1983).

[62] L. Braunschweiler and R. R. Ernst, *J. Magn. Reson.* **53,** 521 (1983).

[63] A. Kumar, R. R. Ernst, and K. Wüthrich, *Biochem. Biophys. Res. Commun.* **95,** 1 (1980).

[64] V. Sklenar, M. Piotto, R. Leppik, and V. Saudek, *J. Magn. Reson.* **102,** 241 (1993).

[65] V. J. Basus, L. Nadasdi, J. Ramachandran, and G. P. Miljanich, *FEBS Lett.* **370,** 163 (1995).

[66] C. G. Hoogstraten, W. M. Westler, S. Macura, and J. L. Markley, *J. Magn. Reson. Ser. B* **102,** 232 (1993).

[67] A. A. Bothner-By, R. L. Stephens, J. Lee, C. D. Warren, and R. W. Jeanloz, *J. Am. Chem. Soc.* **106,** 811 (1984).

different temperatures to determine proton chemical shift temperature coefficients. These, in turn, can be related to amide proton solvent exposure and their possible involvement in hydrogen bonding,[68] as well as reveal any unusual temperature-dependent conformational features. If 20 m*M* peptide concentrations can be had, then triple-bond $^1$H–$^{13}$C coupling constants can be obtained at natural abundance using $^{13}$C-filtered TOCSY and NOESY experiment.[26–28] For samples where that concentration cannot be achieved, or to obtain 3 bond $^1$H–$^{15}$N coupling constants, uniform isotope labeling with either $^{15}$N or $^{13}$C (or both) will allow extraction of 3 bond heteronuclear coupling constants using normal $^1$H–$^1$H 2D experiments.[69]

*Dynamics Studies*

In addition to the points mentioned above for structural studies, a number of factors must be taken into consideration for measurement of heteronuclear relaxation rates. Relaxation measurements that are undertaken with a primary objective of correlating possible internal motions with the disorder observed in the solution structures should attempt to reproduce the solution structure conditions as closely as possible. Heteronuclear NOEs and both longitudinal ($T_1$) and transverse ($T_2$) relaxation times can easily be measured for most residues. For natural abundance $^{13}$C relaxation experiments, a good working concentration of peptide is 10 m*M* or higher, although data can be obtained on samples as low as 4 m*M* if several weeks of instrument time are available at high fields ($\geq$600 Mhz) *vide infra.* For $^{15}$N relaxation experiments or for $^{13}$C studies where sample amount or solubility is limited, uniform or site-directed labeling will usually be necessary. An experimental approach to increase the overall correlation time of a small peptide and thereby make different relaxation parameters more sensitive to internal motion would be the addition of a viscous cosolvent to the solution. Perdeuterated glycerol is commercially available and its use in small molecule relaxation studies has been reported.[30]

Relaxation studies should usually be carried out at a field strength where maximum sensitivity to picosecond internal motions can be anticipated (i.e., where the product $\omega\tau_c$ is $>>1$, where $\omega$ is the Larmor frequency of the observed nucleus). For this class of peptides the expected correlation times of 1–5 ns would dictate using the highest field available, probably no less than 500 MHz for most experiments (see Fig. 3). Ideally, identical experiments can be done at different field strengths to better sample the spectral density functions. Interest in this field has led to a myriad of pulse sequences that can be used to measure heteronuclear relaxation parameters. Several

[68] T. Higashijima, Y. Jobayashi, U. Nagai, and T. Miyazawa, *Eur. J. Biochem.* **97,** 43 (1979).
[69] A. C. Wang and A. Bax, *J. Am. Chem. Soc.* **118,** 2483 (1996).

groups have led the field in this area, including those of Kay,[30,34,35,70,71] Palmer,[32,36] and Wagner.[31,33,72] Rather than attempt to detail specific sequences generated by these and other groups in this rapidly evolving field, the ensuing discussion deals briefly with, in a general way, the type of parameters one may encounter in implementing them. The reader is referred to the original publications for specific details. In general, similar pulse sequences can be used for $^{15}N$-labeled samples and natural abundance $^{13}C$ measurements. More sophisticated pulse programs have been reported for measuring carbon and nitrogen in uniformly double-labeled samples to eliminate effects due to cross-relaxation and scalar coupling among the heteronuclei[30] and to take advantage of pulsed field gradients for coherence selection.[72–74]

For $T_1$ and $T_2$ measurements, most pulse sequences are designed in such a way that the signal decays to zero for long incremental delays. An advantage of this approach is that a two-parameter fit can be used to analyze the data and the signal decay is independent of the recycle delay. A minimum of six 2-D experiments should be acquired for each determination with incremental delays equally spaced between an initial short delay (5–20 ms) and a long delay of approximately $2T$ where $T$ is the longest expected $T_1$ or $T_2$ for the heavy nucleus under investigation. For delays longer than this, most of the signals will be too weak to quantitate with a given number of scans (i.e., relaxation is nearly complete) and will not contribute to the measurement. For molecules of this size, the longest delays will typically be 0.6–1.2 sec. For $T_2$ measurements, special consideration should be given to construction of the spin-echo sequence. The delay used between 180° carbon pulses in the CPMG sequence should be much less than the 1 bond CH or NH coupling constant, $1/2J_{XH}$, or measured values for $T_2$ will be significantly lower than actual values.[71] As an example, for $^{13}C\alpha$–$^{1}H\alpha$ ($1/2J_{CH} \sim 3.5$ ms) $T_2$ values measured using a 1-ms delay between CPMG components, it is expected that the actual $T_2$ values could be as much as 50% higher than the measured values. Also, Palmer *et al.*[32,36] provide an excellent discussion of the impact of slow (microsecond and millisecond time scale) conformational exchange processes on transverse relaxation. Since $T_1$ and $T_2$ are measured such that the signal decays with increasing incremental delay, the signal-to-noise requirements should be

[70] L. E. Kay, D. A. Torchia, and A. Bax, *Biochemistry* **28,** 8972 (1989).

[71] L. E. Kay, L. K. Nicholson, F. Delaglio, A. Bax, and D. A. Torchia, *J. Magn. Reson.* **97,** 359 (1992).

[72] K. T. Dayie and G. Wagner, *J. Magn. Reson. Ser. A* **111,** 121 (1994).

[73] N. A. Farrow, R. Muhandiram, A. U. Singer, S. M. Pascal, C. M. Kay, G. Gish, S. E. Shoelson, T. Pawson, J. D. Forman-Kay, and L. E. Kay, *Biochemistry* **33,** 5984 (1994).

[74] Y.-C. Li and G. T. Montelione, *J. Magn. Reson. Ser. B* **105,** 45 (1994).

estimated using a long incremental delay to determine an appropriate number of transients needed to produce good data. In most cases 64–512 transients per $T_1$ block should be sufficient. Recycle delays are only critical if inversion-recovery-type pulse sequences are used and then should be at least twice the longest relevant ($^{12}C$ isotopomer) proton $T_1$, about 5 sec.[36] Data sets should all be processed identically; linear prediction or similar methods can be used to extend the second dimension (heteronucleus) time domain to improve digital resolution.[75] Peaks are selected and integrated in the data set with the shortest incremental delay and the same peak footprint then used to integrate all other time points in the same data set. These integrals and incremental delays are then subjected to a two-parameter nonlinear curve fitting program to extract $1/T_1$ or $1/T_2$.

For $^{13}C$–H and $^{15}N$–H NOE experiments two data sets are acquired; one with presaturation of the protons during the recycle delay to allow the NOE to build up and one without presaturation. To ensure that maximum NOE enhancements are obtained, a recycle delay of >5 times the $T_1$ of the heteronucleus to be measured should be used, typically 3–5 sec. Data processing and reduction is similar to that described for $T_1$ and $T_2$ measurements. The enhancement, $\eta + 1$, is calculated as the ratio of the peak intensity with presaturation to the peak intensity without presaturation, $I_s/I_0$. A minimum of two replicates for any of the above mentioned techniques to measure relaxation times or heteronuclear NOEs will provide a means for error estimation and identification of aberrant results.

We have applied some of the methods discussed above to measure $^{13}C\alpha$ relaxation parameters for a concentrated (4 m$M$) sample of wild-type $\omega$-Aga-IVB that was obtained from the venom of *Agelenopsis aperta.*[10] Measurements were carried out at 11.7 and 14.1 T (125- and 150-MHz $^{13}C$ frequency, respectively) using the pulse sequences of Kay *et al.*[70] NOE data were derived from two separate experiments as described above. Data for each determination was signal averaged for 7 days. Figure 7 (bottom) shows a summary of the $^1H\alpha$–$^{13}C\alpha$ NOEs at specific sites in the toxin measured at a magnetic field strength of 14.1 T. It is clear from these data that both termini relax differently from the central portion of the peptide. We can estimate the correlation time of the internal motion, $t_i$, for a given residue using the relationship: $1/\tau_i = 1/\tau_c + 1/\tau_e$, where $\tau_c$ and $\tau_e$ are the overall and effective correlation times, respectively. The disulfide-rich region, residues 4–36, has relatively small NOEs that average $1.2 \pm 0.2$ suggesting an overall correlation time of $\geq$1 ns. This, combined with $T_1$ measurements made at different field strengths (see below), and assuming that this portion of the molecule can be treated as a rigid, isotropically tumbling molecule in which

[75] J. J. Led and H. Gesmar, *Methods Enzymol.* **239,** 318 (1994).

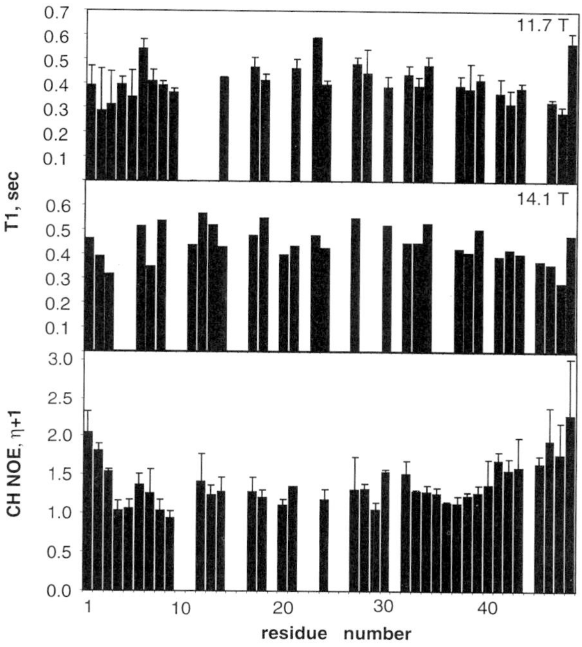


FIG. 7. $^{13}C\alpha$ relaxation parameters for non-Gly residues in wild-type ω-Aga-IVB. (Bottom) $H\alpha$–$C\alpha$ NOEs measured at 11.7 T. (Center and top) Summary of the longitudinal $^{13}C\alpha$ relaxation times at 11.7 and 14.1 T. Error bars indicate the standard deviation between two experiments (only one $T_1$ measurement was made at 14.1 T). In certain cases, parameters could not be measured due to poor signal-to-noise ratio, overlap, or interfering modulation from the HDO resonance; these appear as gaps in the data.

relaxation of the $^{13}C\alpha$ nucleus is dominated by dipolar coupling with the attached $^{1}H\alpha$, an estimated overall correlation time, $\tau_c$, of 4.5 ± 0.5 ns is obtained. The NOEs in the C and N termini range from ~1.2 for residues 37–39 to a maximum of 2.3 for A48, corresponding to effective correlation times ranging from $\tau_e \gg \tau_c$ down to about 500 ps. The average NOE for residues 1–3 and 37–48 of 1.6 ± 0.3 corresponds to an effective correlation time of 1.2 ± 0.5 ns, yielding an average internal correlation time for these highly mobile regions of 800–900 ps.

Longitudinal relaxation times ($T_1$) were also measured for ω-Aga-IVB, and a summary of the results for data collected at carbon frequencies of 150 and 125 MHz is shown in the top two boxes in Fig. 7. The average $T_1$ measured at 125 MHz was 400 ± 80 ms and at 150 MHz was 450 ± 80 ms. The $\tau_c$ of ~4.5 ns is slightly longer than values reported for the similarly sized proteins BTPI[76] and EGF,[77] suggesting that some aggregation may be occurring at this high concentration of peptide. The variability of the $T_1$ measurements from experiment to experiment was greater than for the $^1H\alpha$–$^{13}C\alpha$ NOEs. The data for many of the residues often gave a poor fit to the exponential function due to a poor signal-to-noise ratio in the spectra, despite extensive signal averaging (each set of $T_1$ data took 7 days of spectrometer time). Although the large errors in the data make analysis difficult, some trends can be noted. For example, NOEs increase dramatically in a predictable way as one gets farther away from the Cys-rich regions, whereas differences are present but not so dramatic for the $T_1$ values (see Fig. 7). With the exception of the ultimate residues, the $T_1$ values are shorter than average for the C- and N-terminal residues. If Fig. 3 is used to interpret the $T_1$ data, a change in effective correlation time from 5 to 2–3 ns is consistent with such a decrease in $T_1$ and in keeping with the above analysis of the NOE data. The ultimate residues E1 and A48 both have large $T_1$ values and C–H NOEs, only explained by subnanosecond internal correlation times. Clearly, neither the C–H NOE data nor the $T_1$ data are complete or precise enough to make detailed conclusions about specific dynamic events. However, it is apparent that both measurements are consistent with each other and results in the conclusion that moderately fast internal motions in the terminal amino acids contribute to the lack of $^1H$–$^1H$ NOEs for residues in these regions.[14] Thus we can implicate internal motion faster than the overall correlation time of the peptide to rationalize the variation in the static structures calculated using conventional NMR restraints (Fig. 2).

The natural abundance relaxation studies related above combined with solution structure determinations provide previously unavailable insights into the structure and function of peptide toxins. Although this new information about molecular motions is critical to our ultimate understanding of the importance of flexibility to the peptide's ability to bind to its ion channel, the information presented here is qualitative and more detailed studies are needed to fully characterized molecular motions in $\omega$-Aga-IVB. Natural abundance studies such as these suffer from low sensitivity. In the case of $\omega$-Aga-IVB, we were fortunate because this material is present in unusually copious amounts in the venom of *A. aperta.* Even so, approximately 5000 milkings of this thumbnail-sized spider were required to produce the 11 mg of peptide needed to prepare the sample used in these

[76] P. E. Smith, R. C. van Schaik, T. Szyperski, K. Wüthrich, and W. F. van Gunsteren, *J. Mol. Biol.* **246,** 356 (1995).

[77] B. Celda, C. Biamonti, M. J. Arnau, R. Tejero, and G. T. Montelione, *J. Biomol. NMR* **5,** 161 (1995).

studies. Such experiments would be nicely supplemented by isotope-enriched material. Indeed, access to an expression system such as the one described may be the only means by which structural and dynamics studies can be done at all for less abundant toxins. The 100- to 330-fold improvement in effective sample concentration obtained by $^{13}C$ or $^{15}N$ labeling allows measurements to be conducted at less extravagant concentrations where aggregation is minimized and eliminates the signal-to-noise problems that plagued the present study. Furthermore, natural abundance $^{15}N$ relaxation studies are not feasible and so nitrogen labeling will improve the quality of relaxation data by expanding the spectral frequencies that can be monitored.

NMR spectroscopy is a powerful tool that has been well utilized in the study of the structure of peptide toxins in solution. The availability of recombinant toxins opens the door to an entirely new repertoire of NMR experiments that can be brought to bear on refining this structural knowledge base and expanding it into the realm of molecular dynamics, adding new dimensionality to our understanding of the function of this important class of peptides.

# [6] Ligand Binding Methods for Analysis of Ion Channel Structure and Function

*By* STEEN E. PEDERSEN, MONICA M. LURTZ, and RAO V. L. PAPINENI

## Introduction

The nicotinic acetylcholine receptor (AChR)[1] is an ion channel that is opened by the binding of two molecules of acetylcholine on its extracellular surface.[2] On prolonged exposure to acetylcholine, the channel desensitizes and acquires high affinity for acetylcholine. The equilibrium binding to the two sites appears weakly cooperative, but the two sites are distinct, as shown by the binding of various antagonists that preferentially bind one site versus the other.[3] Two subunits constitute each site: $\alpha\gamma$ and $\alpha\delta$. The heterogeneity of the sites is primarily due to the distinct contributions of

[1] AChR, Nicotinic acetylcholine receptor; $\alpha$-BgTx, $\alpha$-bungarotoxin; DFP, diisopropyl fluorophosphonate.

[2] A. Devillers-Thiery, J. L. Galzi, J. L. Eisele, S. Bertrand, and J.-P. Changeux, *J. Membr. Biol.* **136,** 97 (1993).

[3] R. R. Neubig and J. B. Cohen, *Biochemistry* **18,** 5464 (1979).



0076-6879/99 $30.00

the $\gamma$ and $\delta$ subunits at each site.[4] The sites are allosterically linked to a binding site located within the channel pore itself, the noncompetitive antagonist site. Binding to this site can alter the conformation in favor of either the resting conformation or the desensitized conformation, depending on the ligand. Desensitization, as induced by noncompetitive antagonists, is also marked by increased affinity for agonist binding to the acetylcholine sites. The ability of many cholinergic ligands to bind all three sites further complicates analyses of the linkage.

An increased awareness of these phenomena has permitted binding to the AChR to be understood in more detail, and binding assays are finding greater use in elucidating the structure of the binding sites. In this article we describe several ligand binding techniques: radioligand binding by centrifugation assay, radioligand binding by DE-81 filter binding, and fluorescent ligand binding. In addition, we discuss how to analyze direct binding measurements, indirect binding by competition, and noncompetitive allosteric effects.

## Comparison of Methods

### *Radioligand Binding*

Measuring ligand binding by radioactively labeled ligands has the advantages of high sensitivity, fewer artifacts than fluorescence, and direct determination of the amount of ligand by scintillation or gamma counting. Counting is substantially more sensitive than fluorescence: femtomoles of ligand can be directly detected whereas the limit of detection for fluorescence is near picomoles. There are some drawbacks to using radioactivity: the intrinsic health hazards, detection of $\beta$-emission usually involves handling scintillation cocktails, and all types of radionuclei require significant effort for proper storage and disposal.

### *Fluorescence Binding Measurements*

The advantages of fluorescence measurements echo the limitations of using radioactivity. Changes in fluorescence can be followed at any time resolution that still yields a detectable signal, and the data are obtained immediately. There is usually little need for the special handling and disposal of fluorophores. A primary disadvantage of fluorescence is the lower sensitivity, which substantially limits the usable concentration range of fluorescent ligands. A second disadvantage is that the fluorescence signal

[4] S. E. Pedersen and J. B. Cohen, *Proc. Natl. Acad. Sci. U.S.A.* **87,** 2785 (1990).

must be calibrated routinely for quantitative measurements. Fluorescence yield depends on several factors: instrument optics, the fluorophore, the solution, sample geometry, and the detector. Fluorescence measurements also require a sensitive fluorescence spectrophotometer. Some instruments can be obtained for roughly the same price as a scintillation counter, though research-grade instruments may be substantially more expensive.

### *Selection of Ligand*

A substantial number of radioactive ligands are commercially available that bind the acetylcholine binding sites. The most common is $^{125}$I-labeled $\alpha$-BgTx. Tritiated or $^{14}$C-labeled ligands available include acetylcholine, epibatidine, and nicotine. For the noncompetitive site, however, there are no radioligands currently available. [$^3$H]Phencyclidine was recently discontinued by DuPont/New England Nuclear (Boston, MA). Other ligands that have been used, such a [$^3$H]ethidium or [$^3$H]histrionicotoxin, must be radiolabeled through a radiolabeling service. The fluorescent ligands ethidium, quinacrine, and crystal violet are available through standard sources. Except for derivatives of $\alpha$-BgTx, fluorescent ligands for the acetylcholine binding sites must be synthesized or obtained from a donor.

## Methods

### *Preparation of Membranes*

The preparation of AChR-rich membranes from *Torpedo* electric organ has been described in this series.[5] The procedure we follow is essentially that of Sobel *et al.* with some modifications.[6,7] This preparation can be conveniently scaled up to processing 2 kg of electric organ per batch with yields of several hundred milligrams of membrane protein. The membrane vesicles are near 20% purity in AChR, as assessed by the [$^3$H]acetylcholine binding assay described later (1.5–2 nmol, acetylcholine binding sites per milligram of protein). For many of the assays, particularly the microcentrifugation assay, it is adequate to use lower specific activity membranes, and often it is desirable to do so. Therefore, we usually save a side fraction from the discontinuous sucrose fractionation which contains membranes with specific activity that vary from 0.1–1 nmol acetylcholine binding sites per milligram protein.

[5] A. Chak and A. Karlin, *Methods Enzymol.* **207,** 546 (1992).
[6] A. Sobel, M. Weber, and J.-P. Changeux, *Eur. J. Biochem.* **80,** 215 (1977).
[7] S. E. Pedersen, E. B. Dreyer, and J. B. Cohen, *J. Biol. Chem.* **261,** 13735 (1986).

*Radioligand Binding Assays*

The following procedure describes the microcentrifuge ligand binding assay used for routine binding measurements. It is based on procedures developed and used in Jonathan Cohen's laboratory.[3,8,9] *Torpedo* AChR-rich membranes are diluted into HTPS (250 m$M$ NaCl, 5 m$M$ KCl, 3 m$M$ $CaCl_2$, 2 m$M$ $MgCl_2$, 0.02% $NaN_3$, 20 m$M$ HEPES, pH 7.0) and centrifuged for 30 min at 15 krpm (~19,000 $g$) in a TOMY MTX-150 centrifuge (Peninsula Laboratories, Belmont, CA). This initial centrifugation is to remove light membranes that might not sediment during the later centrifugation step. Protein assays show that typically no more than 10–20% of the membranes are lost in this step. The pellet is resuspended by passing the membranes through a 25-gauge syringe needle several times. The membranes are then treated with 1 m$M$ diisopropyl fluorophosphonate (DFP) for 1 hr at ambient temperature to inactivate acetylcholinesterase. DFP hydrolyzes rapidly in aqueous solution and, therefore, must be diluted from the neat liquid immediately prior to mixing with the membranes. A second 1-hr treatment with 0.1 m$M$ DFP is then carried out and the membranes transferred to ice. Subsequent dilutions and additions are then made in the 0.1 m$M$ DFP solution.

Samples are assembled in Eppendorf-type microcentrifuge tubes and incubated for 30 min at either 4° or ambient temperature. They are then centrifuged for 30 min at 15 krpm (19,000 $g$) in a TOMY microcentrifuge. A sample of the supernatant is retained for counting to determine free ligand and the remainder is removed with a gel-loading pipette tip attached to an aspirator with a trap for collecting the radioactive ligand. Traces of supernatant clinging to the sides of the tubes are adsorbed by a cotton swab. The tubes are then left upside-down on the cotton swab for 15 min to drain any remaining supernatant left on the pellet. The pelleted membranes are dissolved in 100 $\mu$l 10% (w/v) sodium dodecyl sulfate (SDS) by shaking on a vortexer for at least 15 min, and then transferred to a scintillation counting vial; the tube is rinsed with an additional 10 $\mu$l 10% SDS and the rinse added to the scintillation vial. The nonspecific binding is determined by parallel samples including excess carbamylcholine or $\alpha$-bungarotoxin.

The choice of the ligand's specific radioactivity, its concentration in the binding assays, and the concentration of binding sites must be chosen to optimize the signal with regard to the type of information desired. The concentration of binding sites is limited by the need to form a discrete pellet on centrifugation. The quality of data deteriorates if less than 50 $\mu$g

[8] E. K. Krodel, R. A. Beckman, and J. B. Cohen, *Mol. Pharmacol.* **15,** 294 (1979).
[9] S. E. Pedersen, *Mol. Pharmacol.* **47,** 1 (1995).

of membranes are used; using 100 $\mu$g often improves the data. Thus, the lower limit of binding site concentration is determined by the volume desired and the specific binding activity of the membranes. To minimize the concentration of binding sites, we often use a low specific activity membrane fraction in the assay (0.1–1 nmol [$^3$H]acetylcholine binding sites per milligram protein). A typical concentration of binding sites is 20–40 n$M$ in final volume of 1 ml.

*[$^3$H]Acetylcholine Binding.* To monitor the conformational changes of the acetylcholine receptors by noncompetitive antagonists, a high specific radioactivity [$^3$H]acetylcholine (~30 Ci/mmol; available from American Radiolabeled Chemicals Inc., St. Louis, MO) can be used at a concentration substantially lower than both the dissociation constant for acetylcholine and the number of binding sites, as long as only a moderate percentage of the ligand is bound. In this way, the extent of binding is highly sensitive to the conformational equilibrium of the receptor and can be used to determine the conformational preference of the nonradioactive ligand. An example of this is shown in Fig. 1A. The increase in binding is due to the conformational effects of crystal violet that desensitizes the AChR, which is seen as higher affinity binding. At higher concentrations, crystal violet competes directly at the acetylcholine binding sites to reduce binding.

Competitive binding is used to determine the affinity of a ligand that is not radioactive. Nonradioactive ligands are incubated with membranes and an excess of low specific activity acetylcholine (~100 mCi/mmol). In this case it is convenient to keep the [$^3$H]acetylcholine concentration higher than both its $K_D$ and the binding site concentration such that the sites are saturated; 100 n$M$ [$^3$H]acetylcholine is typical with 20- to 40-n$M$ binding sites. The inhibition data will yield a $K_{app}$, the concentration at which the signal is decreased 50%. This value is most easily obtained by fitting the data to an appropriate equation using a nonlinear regression algorithm. It can then be used to calculate the $K_D$ for the inhibitor (see below).

*Binding to Noncompetitive Site.* For [$^3$H]phencyclidine or [$^3$H]ethidium binding to the noncompetitive antagonist binding site, the assay is carried out as described above except that the DFP incubation steps can be omitted. A low concentration of [$^3$H]phencyclidine is convenient for measuring either competition by other noncompetitive antagonists or measuring conformationally driven effects of agonsits or noncompetitive antagonists. Both effects can be seen in Fig. 1B where [$^3$H]phencyclidine binding is altered by the addition of tubocurine: tubocurine binds the acetylcholine binding sites at lower concentrations, resulting in partial conversion to the desensitized conformation of the AChR, which has higher affinity for [$^3$H]phencyclidine as seen by the increase in binding. At higher concentrations, tubocurine competes directly for binding at the noncompetitive antagonist site, producing a loss of [$^3$H]phencyclidine binding. The complete desensiti-

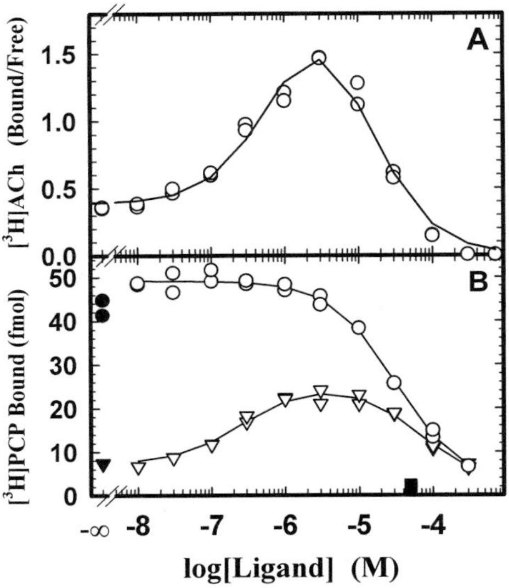


FIG. 1. Allosteric interactions detected by binding assays at low radioligand concentrations. (A) The effect of increasing crystal violet concentrations on [$^3$H]acetylcholine binding was determined using the centrifugation assay described in the text. The assay was carried out in a 1-ml volume containing 100 $\mu$g AChR-rich membranes (18 n*M* acetylcholine binding sites) and 1 n*M* [$^3$H]acetylcholine. The data are plotted as the ratio of the bound ligand to free ligand. Under these conditions, this value is inversely proportional to the $K_D$. The rising phase shows that crystal violet increases the affinity of the acetylcholine binding sites for acetylcholine. The falling phase at higher concentrations reflects direct competition for binding at the acetylcholine binding sites. (B) The effect of varying concentration of tubocurine, a *d*-tubocurarine analog, was measured by [$^3$H]phencyclidine binding using the centrifugation assay. AChR-rich membranes (50 $\mu$g; 200 $\mu$l; 62.5 n*M* AChR) were incubated with 1 n*M* [$^3$H]phencyclidine (43 Ci/mmol) and the indicated concentrations of ligand. The rising phase of binding (▽) shows increased affinity for [$^3$H]phencyclidine as a result of desensitization from tubocurine binding to the $\alpha\gamma$ acetylcholine binding site. Tubocurine competitively inhibits the AChR at higher concentrations resulting in the bell-shaped curve. The direct competition by tubocurine at the noncompetitive site is further illustrated by its inhibition in the presence of 100 $\mu M$ carbamylcholine (○) to block effects at the acetylcholine binding sites. Controls in the absence (▼) and presence (●) of carbamylcholine illustrate the enhanced binding due to the strong desensitizing. The binding in the presence of 50 $\mu M$ proadifen, a noncompetitive antagonist, is also shown (■).

zation can be seen by addition of an agonist, carbamylcholine, which increases binding about 5-fold over the baseline.

*Assays Using $^{125}$I-Labeled α-Bungarotoxin.* $^{125}$I-Labeled-α-bungarotoxin is used extensively for measuring binding to the acetylcholine receptor. A variety of assays have been described; the most common one relies on the separation of free from bound ligand by adsorption of the AChR to DE-81 anion-exchange filter paper.[10] We describe a variation of this assay to measure binding in physiologic buffer. It avoids the use of the detergent Triton X-100, which may contribute to desensitization of the AChR and thereby cause artifacts when assessing the affinity of competing ligands. Because of the intrinsic high affinity of $^{125}$I-labeled $\alpha$-BgTx for the AChR (the $K_D$ is near picomolar) and its long dissociation time (hours), competitive binding assays are not carried out at equilibrium, but rather use an initial rate assay. The assay is carried out for a limited period of time during the linear portion of the association time course. The amount of binding within this time is proportional to the free binding site concentration. Therefore a competing ligand will reduce the rate of [$^{125}$I]$\alpha$-BgTx binding to the extent that is occupies the binding sites. The rate of binding is sensitive to ionic strength,[11] such that an initial rate of binding in low ionic strength buffer can be done in 1–2 min, whereas 45 min is necessary for binding in physiologic saline.

The initial rate assay relies on maintaining a linear rate of binding of [$^{125}$I]$\alpha$-BgTx during the course assay. This requires pseudo first-order kinetic conditions, such that only a small percentage of the added [$^{125}$I]$\alpha$-BgTx is bound and only a minor proportion of the AChR binding sites are bound. A rule of thumb is that the maximum binding should be less than one-half of each site; that is, a quarter of the total binding acetylcholine binding sites.

AChR-rich membranes are suspended to a concentration of 1–2 n*M* in HTPS supplemented with 0.1% bovine serum albumin (BSA) and preincubated with the competing ligand for 30 min or longer. $^{125}$I-Labeled $\alpha$-BgTx (~200 Ci/mmol) is then added to a final concentration of 2 n*M* and the reaction incubated for 45 min. The binding is stopped by dilution with four volumes of 300 n*M* $\alpha$-BgTx in 10 m*M* Tris, 0.1% (w/v) BSA, 0.1% (w/v) Triton X-100 (pH 7.4). This serves to isotopically dilute the [$^{125}$I]$\alpha$-BgTx and prevent further binding and to lower the ionic strength. The latter is necessary for binding to the DE-81 filters. A 60-$\mu$l aliquot of each reaction is spotted on a DE-81 filter pinned to a Styrofoam rack. The filters are allowed to sit no more than a few minutes before transfer to a tray with wash buffer: 10 m*M* Tris, 50 m*M* NaCl, 0.1% Triton X-100 (pH 7.4). The filters are washed twice, in batches, for 15 min each. The length of washing is not critical. More important is the spotting of the sample onto the filters; the exact time of incubation on the filter before transfer to wash buffer is

[10] J. Schmidt and M. A. Raftery, *Anal. Biochem.* **52,** 349 (1973).
[11] J. Schmidt and M. A. Raftery, *J. Neurochem.* **23,** 617 (1974).

relatively unimportant as long as the filters remain damp. If the filters begin to dry before washing, the nonspecific binding will increase dramatically.

The background binding in this assay can easily become too high; it appears to be dependent on the batch of DE-81 filters. A simple solution is to increase the amount of binding by using a higher receptor concentration, as long as the conditions for the assay listed above are still met.

*[$^{125}$I]α-BgTx Binding to BC$_3$H-1 Cells.* BC$_3$H-1 cells express the mouse muscle AChR and provide a convenient system for studying the properties of a mammalian AChR.[12,13] The binding assay can be carried out in 24-well tissue culture plates. Despite the lower receptor number in these cells than are obtained from *Torpedo* membranes, the assay is simpler because the free [$^{125}$I]α-BgTx can be washed off the cells with quite low background retention.

BC$_3$H-1 cells are maintained in Dulbecco's modified Eagle's medium (DMEM) with 20% fetal bovine serum, 100 units/ml penicillin, and 0.1 mg/ml streptomycin. They are seeded (8000–12,000 cells per well) into 24-well tissue culture plates that have been precoated with gelatin (0.2% porcine gelatin, 500 μl per well, for 3 hr before plating). The cells are grown until they reach ~70% confluence and then the medium is changed to Dulbecco's modified Eagle's medium with 8% fetal bovine serum, 2% horse serum, 100 units/ml penicillin, and 0.1 mg/ml streptomycin. This initiates differentiation of the cells and expression of the AChR. The cells are ready for assay after 3 days of incubation with horse serum.

Like the *Torpedo* membrane assay, this assay relies on the initial rate of binding of [$^{125}$I]α-BgTx and its slow dissociation rate. The cells are removed from the incubator and equilibrated to room temperature in a hood for 30 min. The remainder of the assay can be carried out on the bench at ambient temperature. Competing ligands are diluted in the media and then applied to the cells (350 μl) after removing the old media. The cells are incubated for 30 min with the competing ligand on a slowly rotating platform. $^{125}$I-Labeled α-BgTx is then added in a 50-μl aliquot to a final concentration of ~2 n*M* and allowed to incubate for another 45 min. The solution is then aspirated off the cells and the cells washed twice with 0.7 ml Dulbecco's modified Eagle's medium. The bound [$^{125}$I]α-BgTx retained in the well is then transferred to a counting vial after dissolving the cells with 150 μl 1% Triton X-100 for 2 hr. The gelatin coating on the wells is necessary for sticking of the cells to the plates during the incubation and washing procedures, but it also slows the dissolution by Triton X-100. It is

[12] S. Sine and P. Taylor, *J. Biol. Chem.* **254,** 3315 (1979).
[13] S. Sine and P. Taylor, *J. Biol. Chem.* **256,** 6692 (1981).

important to ensure that all the cells are dissolved before transferring to the counting vial.

It is important to establish independently that the incubation time [$^{125}$I]$\alpha$-BgTx is in the linear portion of the association rate curve. The rate of binding can be adjusted by varying the [$^{125}$I]$\alpha$-BgTx concentration. As mentioned earlier, a rule of thumb is that no more than 25% of the total binding sites should be bound and only a minor portion of the [$^{125}$I]$\alpha$-BgTx should be bound. The total number of receptors can be determined by using longer binding time (3 hr) and higher [$^{125}$I]$\alpha$-BgTx concentrations. We routinely use [$^{125}$I]$\alpha$-BgTx with a specific radioactivity of 200 Ci/mmol, but higher activity [$^{125}$I]$\alpha$-BgTx can also be obtained and used in this assay for higher sensitivity.

### *Effects of Noncompetitive Antagonists on Binding to Agonist Sites*

Agonists and competitive antagonists, which bind the acetylcholine binding sites, stabilize the desensitized conformation of the AChR to varying degrees. The extent to which this occurs can be measured indirectly by examining the effect of noncompetitive antagonists on the binding of the acetylcholine site ligands. A variety of noncompetitive antagonists stabilize the desensitized conformation. We have used phencyclidine and proadifen; the former for its high solubility, low partitioning with membranes[14] and well-characterized binding, and the latter for its ability to desensitize to a greater extent.[8] Fewer ligands are available that preferentially stabilize the resting conformation of the AChR. The best is tetracaine.[15]

Noncompetitive ligands are usually added in competitive radioligand binding assays to a concentration near 30 $\mu M$. It is necessary to add sufficient ligand to be well above the $K_D$ of the noncompetitive ligand for the resting conformation of the AChR (near 5 $\mu M$ for proadifen and phencyclidine). However, high concentrations (>100 $\mu M$) can sometimes perturb the lipid bilayer and induce nonspecific effects. This can cause problems in obtaining good pellets in the microcentrifuge assay and can cause dissociation of the $BC_3H$-1 cells from the tissue culture dish. High concentrations of some noncompetitive antagonists also inhibit the rate of [$^{125}$I]$\alpha$-BgTx binding to *Torpedo* AChR-rich membranes.[8] The exact cause for this effect is not known but may be due to desensitization of the AChR.

[14] M. M. Lurtz, M. L. Hareland, and S. E. Pedersen, *Biochemistry* **36,** 2068 (1997).

[15] J. B. Cohen, L. A. Correll, E. B. Dreyer, I. R. Kuisk, D. C. Medynski, and N. P. Strnad, *in* "Molecular and Cellular Mechanisms of Anesthetics" (S. H. Roth and K. W. Miller, eds.), pp. 111–124. Plenum Publishing, New York, NY, 1986.

## Fluorescent Ligand Binding Assays

### *Fluorescent Ligands for Acetylcholine Binding Sites*

Only few fluorescent ligands are available that bind the acetylcholine binding sites. A variety of fluorescent derivatives of $\alpha$-BgTx are commercially available,[16] but are of limited use for binding assays because they do not have substantial changes in fluorescence on binding. Two agonists, dansyl-C6-choline[17] and NBD-5-acylcholine,[18] have been characterized extensively. They share the high affinity and rapid binding characteristics of acetylcholine and are excellent real-time monitors of the conformational transitions of the AChR because their quantum yield changes substantially on binding. The change in signal is enhanced by using energy transfer from tryptophans on the AChR to the bound ligand. Dansyl-choline[19] is an antagonist that binds with somewhat lower affinity but is useful for quantitation of receptor specific activity[3,20]; it is also commercially available.[21] These ligands are useful for competitive binding assays for assessing the affinity of unlabeled ligands. Several caveats render competitive inhibition by fluorescence less useful than the radioligand assays. The concentration of receptor needed for an adequate signal is near 100 n*M*. For competitive inhibition it is desirable to have the ligand in excess, thus requiring several hundred nanomolar concentrations. This will increase the concentration of competing ligand required for complete inhibition. A second caveat is that many competing ligands will absorb at a light of 280–290 nm, the wavelength required for excitation of these ligands. This adds the complication of having to correct for inner filter effects at higher competitor concentrations.

### *Fluorescent Ligands for Noncompetitive Site*

A variety of fluorophores bind the noncompetitive antagonist site. The best characterized ligand is ethidium bromide[22]; others include quinacrine,[23] decidium,[24] and crystal violet. The latter compound has a 200-fold increase

[16] R. P. Haugland, "Handbook of Fluorescent Probes and Research Chemicals." Molecular Probes, Eugene, Oregon, 1996.

[17] T. Heidmann and J.-P. Changeux, *Eur. J. Biochem.* **94,** 255 (1979).

[18] H. Prinz and A. Maelicke, *J. Biol. Chem.* **258,** 10263 (1983).

[19] J. B. Cohen and J.-P. Changeux, *Biochemistry* **12,** 4855 (1973).

[20] C. F. Valenzuela, J. A. Kerr, and D. A. Johnson, *J. Biol. Chem.* **267,** 8238 (1992).

[21] Sigma Chemicals, St. Louis, MO.

[22] J. M. Herz, D. A. Johnson, and P. Taylor, *J. Biol. Chem.* **262,** 7238 (1987).

[23] H.-H. Grünhagen and J.-P. Changeux, *J. Mol. Biol.* **106,** 517 (1976).

[24] D. A. Johnson, R. D. Brown, J. M. Herz, H. A. Berman, G. L. Andreasen, and P. Taylor, *J. Biol. Chem.* **262,** 14022 (1987).

in fluorescence yield on binding the ion channel.[25] Ethidium fluorescence enhancement can be used to measure binding to the noncompetitive antagonist site. Titrations with competing ligands can be conveniently performed in a single cuvette to generate inhibition curves.

### *Fluorescent Assay for Noncompetitive Binding*

Fluorescent measurements on the AChR require a good-quality fluorometer because of the scattering due to membranes. We use research-grade instruments (SLM 8000, Rochester, NY and ISS PC1, Urbana, IL); however less expensive fluorometers are likely to be adequate. For measurements, AChR-rich membranes are sedimented in the ultracentrifuge and resuspended to 200 n*M* in acetylcholine binding sites in HTPS (for fluorescence measurements the sodium azide is omitted from the HTPS). Ethidium is added to a final concentration 250 n*M* and the agonist carbamylcholine is added to 100 $\mu M$. Carbamylcholine serves to desensitize the AChr, which improves the affinity for ethidium, and prevents binding of ethidium to the agonist sites. A 2.5- to 3-ml volume of the suspension is stirred in a 10- × 10-mm cuvette.

Fluorescence is excited with 340 nm light through a visible-absorbing filter (Oriel 59152, Stratford, CT) and is measured at 595 nm through a 540-nm cuton filter (Oriel 59502). The filters improve the signal-to-noise ratio by removing stray light. To measure the affinity of a competing ligand, concentrated solutions can be titrated into the cuvette using a syringe. It is desirable to keep the additions to small volumes to avoid compensating for volume changes and to avoid affecting the equilibrium of the binding of ethidium. It is necessary to wait about 15 min between additions in order to let the slow dissociation of ethidium come to completion. This greatly limits the speed of the assay, but many sets of data can be measured simultaneously in this way.

## Analysis of Ligand Binding

The object of this section is to provide some practical guides for the analysis of direct binding data and competitive binding data. The last section also discusses the use of thermodynamic cycles for interpreting conformational effects and allosteric ligand interactions. The theoretical basis and the derivation for the various equations will not be presented; they can be

[25] M. M. Lurtz and S. E. Pedersen, 1997 unpublished observations.

found elsewhere. The interested reader can find more detail in a number of books and articles.[26,27] A good basic primer has been published by Klotz.[28]

*Analysis of Binding Isotherms*

*Subtraction of Nonspecific Binding.* To analyze binding isotherms, nonspecific binding is subtracted, and the data are then fit by a nonlinear regression algorithm to the equation for single site binding [Eq. (1)]. For the centrifugation assays described earlier, the free ligand concentration is determined directly by counting an aliquot of the supernatant. A common problem in subtracting nonspecific binding in these assays is that the free ligand concentration is not the same in parallel samples for *total binding* and for *nonspecific binding.* In this case, it is incorrect simply to subtract the nonspecific binding from the corresponding samples without inhibitor. If the nonspecific binding is linear with concentration, it can be fit to a line, and the fitted parameters used to calculate the nonspecific for each sample, using the corresponding free concentration of ligand. If the *nonspecific binding* is nonlinear, then the values can be fit to other functions; a hyperbolic equation often works well [$B = A \cdot L/(L + K)$]. Alternatively, the values can be manually interpolated from the *nonspecific binding* data. This problem is illustrated in Fig. 2.

In cases where the concentrations of receptor are the same or higher than the ligand concentrations, substantial overestimates of $K_D$ values can arise from assuming the free ligand concentration to be equal to the concentration of ligand added. Further error results from directly subtracting the parallel samples that define nonspecific binding. This is a particular problem for fluorescence titration assays where it is inconvenient to determine directly the free ligand for each data point. A correction can improve the estimate: if the amount of receptor is known, then the amount of bound ligand can be estimated and subtracted from the total ligand concentration to provide a corrected free ligand concentration. Replotting the data in terms of this new concentration then gives an improved estimate for the $K_D$. This ignores any ligand depletion from partitioning of ligand into the membrane, which may also be significant and result in severalfold errors in $K_D$ estimates. When precise determinations of the $K_D$ by fluorescence are desired, the free ligand concentration can be determined independently

[26] J. Wyman and S. J. Gill, "Binding and Linkage: Function Chemistry of Biological Macromolecules." University Science Books, Mill Valley, California, 1990.

[27] C. R. Cantor and P. R. Schimmel, "Biophysical Chemistry: Part III, The Behavior of Biological Macromolecules." W. H. Freeman Co., San Francisco, 1980.

[28] I. M. Klotz, "Ligand-Receptor Energetics: A Guide for the Perplexed." John Wiley & Sons, New York, 1997.

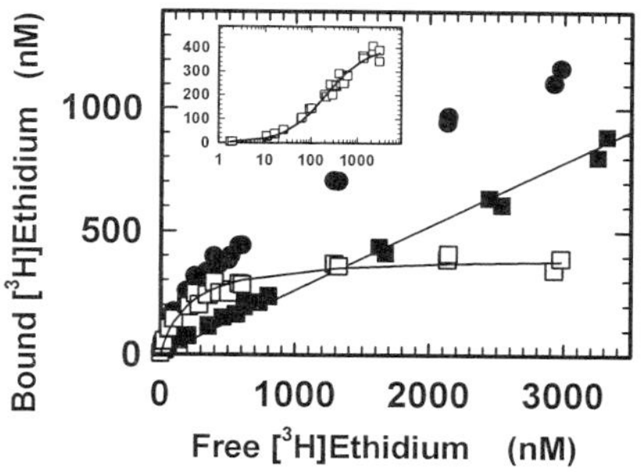


FIG. 2. Subtraction of nonspecific binding by direct determination of free ligand concentrations. AChR-rich membranes (88 $\mu$g; 200 $\mu$l; 100 n$M$ AChR) were incubated with varying concentrations of [$^3$H]ethidium (0.2 Ci/mmol) in the absence (●) and presence (■) of 100 $\mu M$ phencyclidine. After centrifugation of the membranes, the supernatants were counted to determine the free [$^3$H]ethidium concentration. The nonspecific binding (■), with phencyclidine, was fit to a line. The linear parameters were then used to calculate the nonspecific component of binding using the values of free [$^3$H]ethidium for the samples without phencyclidine. This nonspecific component was then subtracted from the binding value to give the specific binding (□). The specific component of binding was then fit to Eq. (1) (solid curve). *Inset*: Semilog plot of the specific binding illustrating the difficulty of adequately demonstrating saturability of binding data. Data that appear to level off in the linear plot do not appear to reach a well-defined plateau value when viewed in a semilog plot. However, obtaining data at higher concentrations is limited by the substantial nonspecific binding.

by removing the bound ligand by centrifugation and measuring the supernatant concentration by fluorescence or by HPLC.[14]

*Curve Fitting.* Prior to the ubiquitous use of computers, binding data were often linearized and plotted in accordance with the Scatchard equation.[29] This method is still useful for qualitative evaluation of the data, but nonlinear regression of the data is now easy to perform with commonly available programs. We routinely use Sigmaplot (SPSS, Jandel Scientific, Chicago, IL). Binding is analyzed using the equation for the binding of a ligand to a single binding site [Eq. (1)], where $RL$ is the measured binding concentration, $R_0$ is the total binding site concentration, $L$ is the *free* ligand concentration. This equation is simply derived from the definition of the dissociation constant, $K_D = R \cdot L/RL$, and the mass balance equation, $R_0 = R + RL$.

$$RL = \frac{R_0 \cdot L}{L + K_D} \tag{1}$$

The maximum binding, $R_0$, and the $K_D$ for binding are extracted from the fit. Most fitting routines will also supply statistical data that indicate

[29] G. Scatchard, *Ann. N.Y. Acad. Sci.* **51,** 660 (1949).

the quality of the fit. It is important that the data demonstrate saturability, otherwise both parameters can have substantial error. As discussed by Klotz,[28] the best plot for visualizing whether the binding has reached saturation is a semilog plot of bound ligand versus the log of the ligand concentration (see Fig. 2, inset). The quality of the fit can usually be determined by comparing the best fit with the data and looking for systematic deviations. This can also be done by plotting the residuals and looking for a pattern. If the residuals are not scattered evenly about the origin, then the fit is poor. Poor fits may arise from a number of problems. If the data do not saturate, the maximum value will be poorly determined and may deviate significantly from the expected value, resulting in error in the $K_D$ as well. The data may reflect components from multiple sites, cooperativity, or heterogeneous samples that lead to a poor approximation by the single site model.

If the data are presumed to bind more than one site, it must be analyzed by other equations. In the case of cooperativity between sites, the data can often be fit to the Hill equation[30] [Eq. (2)]. The value $n$ reflects the degree of apparent cooperativity between two or more sites. In the case of positive cooperativity, the slope of the semilog plot will be steeper than for a single binding site and $n > 1$. For negative cooperativity or for multiple independent sites, $n < 1$. It is difficult to distinguish multiple independent sites from negative cooperativity among the sites by equilibrium binding. If multiple independent sites are suspected, they may also be fit to a sum of terms, each term with the same form as Eq. (1).

$$RL = \frac{R_0}{1 + (K_D/L)^n} \tag{2}$$

*Analysis of Competitive Inhibition Data*

Competitive inhibition to a single site can generally be fit to Eq. (3), which describes the inhibition of binding of $L$ by increasing concentrations of $I$. In this equation, $RL$ is the amount bound observed, $A$ is the amplitude of the change in binding, $Bcg$ is the background or nonspecific binding, and $K_{app}$ is the concentration of $I$ that produces 50% inhibition. To calculate the $K_I$ for the inhibiting ligand (i.e., dissociation constant for the inhibitor $I$) it is necessary to correct for the concentration of the observed ligand $L$ using Eq. (4), where $K_D$ is the dissociation constant for $L$. $K_D$ must be determined independently under similar conditions.

$$RL = \frac{A}{1 + I/K_{app}} + Bcg \tag{3}$$

[30] A. V. Hill, *J. Physiol.* (*Lond.*) **40,** iv (1910).

$$K_I = K_{app} \frac{K_D}{L + K_D} \quad (4)$$

Competitive inhibition data must have well-defined maximum and minimum values in order to obtain reliable values for $K_{app}$ by fitting to Eq. (3): data should extend 2 orders of magnitude on either side of the $K_{app}$. For analyzing compounds of unknown affinity, controls should be included using a known ligand at a concentration that gives complete inhibition. Then, if the data do not extend to complete inhibition, the fit can be forced to the *Bcg* value defined by the control inhibitor.

For fitting data to the acetylcholine binding sites of the AChR, it is often necessary, as well as desirable, to fit the data to inhibition at two distinct sites and obtain a $K_{app}$ for each site. This is accomplished by fitting the data to Eq. (5), which represents inhibition to two independent sites present in equal amounts. At times, it is necessary to fit the data with no prior assumption about the relative amounts of the two sites assumed; this is described by Eq. (6). This situation occurs, for instance, with inhibition of $[^{125}I]\alpha$-BgTx binding by 13′-iodo-*d*-tubocurarine: although the two binding sites are present in equal amounts, the binding of 13′-iodo-*d*-tubocurarine influences the rate of $[^{125}I]\alpha$-BgTx to the second site, and it does not appear represented in equal amplitude[31]:

$$RL = A\left[\frac{1}{1 + I/K_{1app}} + \frac{1}{1 + I/K_{2app}}\right] + Bcg \quad (5)$$

$$RL = \frac{A_1}{1 + I/K_{1app}} + \frac{A_2}{1 + I/K_{2app}} + Bcg \quad (6)$$

When attempting to distinguish binding to two sites, it is important that $K_{1app}$ and $K_{2app}$ differ sufficiently. The binding to two sites of equal intrinsic affinity can be nearly as well described by binding to two sites that differ 4-fold in $K_D$. Therefore, a separation of $K_{1app}$ and $K_{2app}$ of tenfold is often required for reliable determination of the individual constants. A good test is to run both single-site and two-site fits to the data. While there will generally be improvement in the residuals because of the increase in the number of parameters, there should be clear evidence from inspection of the graph that the two-site model provides a better fit to the data before it is interpreted as such. As a rule, the best determination of the error is to repeat the experiment.

[31] S. E. Pedersen and R. V. L. Papineni, *J. Biol. Chem.* **270,** 31141 (1995).

### Thermodynamic Cycle Analysis

Thermodynamic cycles are useful tools for analyzing ligand binding when it is necessary to interpret the effects of ligand structure, receptor mutations, heterotropic allosteric effects, or conformational changes. Thermodynamic cycles simply reflect the first law of thermodynamics, the conservation of energy, and the corollary of path independence: because energy is a state function, the difference in energy between two states is independent of the path. Scheme I shows the version for analyzing the allosteric effects of one ligand ($I$) on a second ligand ($L$). If the two states are considered to be the free receptor with unbound ligand ($R + I + L$) and the ternary complex ($RIL$), then it matters not whether $L$ binds first and then $I$ or vice versa. The free energy of each step is proportional to the log of the equilibrium constant ($\Delta G = -RT \ln K$). Therefore, taking each path and setting the energies equal and removing the logarithms yields $K_{\mathrm{L}} \cdot K'_{\mathrm{I}} = K_{\mathrm{I}} \cdot K'_{\mathrm{L}}$. This can be rewritten $K_{\mathrm{L}}/K'_{\mathrm{L}} = K_{\mathrm{I}}/K'_{\mathrm{I}}$, which says that the ratio of the binding constants for $L$ in the absence and presence of $I$ is the same as the ratio of the binding constants for $I$ in the presence and absence of $L$. For example, if $I$ inhibits the binding of $L$, then $L$ will inhibit the binding $I$ to the same extent. The equation also shows that if three of the constants are known, then the fourth is also determined.

The scheme does not illustrate the underlying conformational changes, but they are implicit in the changes in the equilibrium constants: each constant reflects the binding to the equilibrium distribution of conformations for the receptor, which may be influenced by the presence of the second ligand. It is clear that long-range interactions between the ligands must take place, often mediated by receptor conformational changes, or else $K_{\mathrm{L}}$ would simply equal $K'_{\mathrm{L}}$.

This cycle can be used to analyze several scenarios: the effect of varying concentrations of an allosteric ligand $I$ on the binding of the measured ligand $L$, or the effect of a isotonic concentration of $I$ on the binding affinity of $L$. The latter are described by Eq. (7), where the binding of $L$ is first analyzed by Eq. (1) or by Eqs. (3) and (4), depending on whether the data are obtained by direct binding or by competitive inhibition. The $K$ value

$$
\begin{array}{ccc}
R + L + I & \xrightleftharpoons{K_{\mathrm{L}}} & RL + I \\
K_{\mathrm{I}} \Big\updownarrow & & \Big\updownarrow K'_{\mathrm{I}} \\
RI + L & \xrightleftharpoons{K'_{\mathrm{L}}} & RLI
\end{array}
$$

SCHEME I

obtained equals the right-hand term in the denominator of Eq. (7). If a saturating concentration of $I$ is used, the $K$ value reduces to $K_L K'_I/K_I$, which is simply equal to $K'_L$.

$$RL = R_0 \frac{L}{L + K_L \dfrac{1 + I/K_I}{1 + I/K'_I}} \tag{7}$$

Conversely, if the radioactive (or fluorescent) ligand $L$ is held constant and the effect of varying concentrations of $I$ is measured, then Eq. (7) can be rearranged to give Eq. (8). The data can be fit according to Eq. (3) to yield the maximum, minimum, and $K_{app}$. Those values can then be interpreted according to Eqs. (9)–(11). In this case, the plateau value at high concentrations of $I$ does not reflect nonspecific binding, but rather reflects the binding of $L$ as affected by $I$ [Eq. (11)]. For this reason, it is important to have an independent determination of nonspecific binding using an excess of a ligand known to compete with $L$. This value should be substracted before the analysis.

$$\frac{RL}{R_0} = (B_\infty - B_0)\frac{I}{I + K_{app}} + B_0 \tag{8}$$

$$K_{app} = \frac{K_I K'_I(L + K_L)}{LK_I + K_L K'_I} \tag{9}$$

$$B_0 = \frac{L}{L + K_L} \tag{10}$$

$$B_\infty = \frac{L}{L + K_L K'_I/K_I} \tag{11}$$

*Double-Mutant Thermodynamic Cycle Analysis*

The principle behind the double-mutant thermodynamic cycle analysis has been articulated by Carter *et al.*[32] and was outlined by Ackers and Smith[33] for the study of pairwise interactions of residues in proteins. The pairwise analysis overcomes some of the caveats of single mutation analysis on binding energetics and conformational changes. It can be used to determine whether changes in homologous ligands are independent and can be used to evaluate interactions among specific loci on ligands and receptors.

[32] P. J. Carter, G. Winter, A. J. Wilkinson, and A. R. Fersht, *Cell* **38,** 835 (1984).
[33] G. K. Ackers and F. R. Smith, *Ann. Rev. Biochem.* **54,** 597 (1985).

To illustrate the method, consider ligands $L_1$ and a close analog $L_2$, which differ by a small, discrete structural change, for example, a single methylation in *d*-tubocurarine. $L_1$ and $L_2$ bind to a receptor site, $R_a$. A single mutation in $R_a$ will yield the receptor homolog $R_b$. If the ligand functional group does not interact with the modified receptor amino acid, then the *change* in affinity of the ligand will be *independent* of the receptor modification. If the ligand functional group does interact with the receptor amino acid residue, then the *change* in affinity upon modifying $L_1$ to $L_2$ will be *dependent* on whether the receptor residue is also modified.

This explanation can be formalized using the thermodynamic cycle shown in Scheme II. A binding analysis is performed to measure the equilibrium dissociation constants for both $L_1$ and $L_2$ to both $R_a$ and $R_b$. This will generate four binding constants and the corresponding free energies: $\Delta G_{a1}$, $\Delta G_{a2}$, $\Delta G_{b1}$, and $\Delta G_{b2}$; the subscripts indicate the receptor and the ligand respectively. Each binding energy is represented at a corner of the cycle. The principle of path independence leads to the same constraints discussed above: all paths from $R_aL_1$ to $R_bL_2$ must have the same energy. Therefore $\Delta\Delta G_a + \Delta\Delta G_2 = \Delta\Delta G_1 + \Delta\Delta G_b$, where $\Delta\Delta G_x$ represents the *difference* between two binding energies.

If the two ligand and receptor sites are *independent* of each other then the *change* in ligand binding affinities ($\Delta\Delta G_x$) will be the same in each direction; i.e., $\Delta\Delta G_a = \Delta\Delta G_b$ and $\Delta\Delta G_1 = \Delta\Delta G_2$. In this case, the change in binding energy caused by modifying the ligand ($\Delta\Delta G_a$ and $\Delta\Delta G_b$) is the same for both receptor types. Conversely, the *change* inbinding energy caused by mutating the receptor ($\Delta\Delta G_1$ and $\Delta\Delta G_2$) is the same for either ligand.

If the mutated receptor residue interacts with the altered ligand site, then $\Delta\Delta G_a \neq \Delta\Delta G_b$ and $\Delta\Delta G_1 \neq \Delta\Delta G_2$. However, the cyclic relationship must hold, which can be rewritten as follows: $\Delta\Delta G_a - \Delta\Delta G_b = \Delta\Delta G_1 - \Delta\Delta G_2$. This difference is the interaction energy and reflects the strength of the interaction. The identity of the two differences provides an internal control on the experimental measurements.

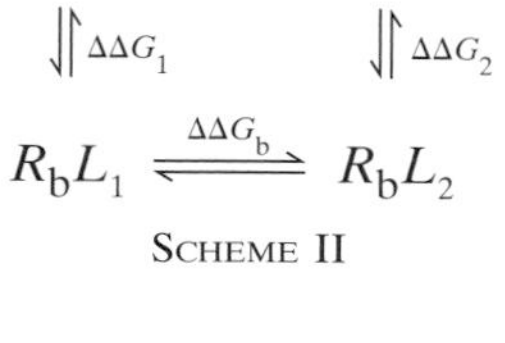


SCHEME II

This type of analysis has been used to study the role of individual hemoglobin residues in conformational transitions on ligand binding.[33] It has also been applied to study the interaction of scorpion agatoxin 2 with the *Shaker* potassium channel and successfully revealed close electrostatic interactions.[34] This method is likely to find further use in analyzing the structures of binding sites that are inaccessible to direct structural determination.

[34] P. Hidalgo and R. MacKinnon, *Science* **268,** 307 (1995).

# [7] Three-Dimensional Structure of Membrane Proteins Determined by Two-Dimensional Crystallization, Electron Cryomicroscopy, and Image Analysis

*By* MARK YEAGER, VINZENZ M. UNGER, and ALOK K. MITRA

## Introduction

The Brookhaven Protein Data Bank now has about 5000 atomic structures available for soluble proteins. This compares with about 20 membrane protein structures, many of which are of the same class. Strategies continue to be developed for growing three-dimensional (3D) crystals of membrane proteins,[1–5] and recent progress has been encouraging.[6,7] Nevertheless, the high-resolution structure analysis of membrane proteins is still a formidable task. In addition, no recombinant eucaryotic membrane protein has as yet been amenable to 3D crystallization. Soluble fragments of membrane proteins have been overexpressed, purified, and examined by conventional X-ray crystallography.[8–10] However, this approach does not allow examina-

[1] W. Kühlbrandt, *Quart. Rev. Biophysics* **21,** 429 (1988).
[2] H. Michel, ed., "Crystallization of Membrane Proteins." CRC Press, Boca Raton, 1991.
[3] F. Reiss-Husson, *in* "Crystallization of Nucleic Acids and Proteins: A Practical Approach" (A. Ducruix and R. Giegé, eds.), p. 175. IRL Press, Oxford, 1992.
[4] R. M. Garavito, D. Picot, and P. J. Loll, *J. Bioenerg. Biomembr.* **28,** 13 (1996).
[5] E. Pebay-Peyroula, G. Rummel, J. P. Rosenbusch, and E. M. Landau, *Science* **277,** 1676 (1997).
[6] R. M. Garavito and S. H. White, *Curr. Opin. Struct. Biol.* **7,** 533 (1997).
[7] C. Ostermeier and H. Michel, *Curr. Opin. Struct. Biol.* **7,** 697 (1997).
[8] J. Wang, Y. Yan, T. P. J. Garrett, J. Liu, D. W. Rodgers, R. L. Garlick, G. E. Tarr, Y. Husain, E. L. Reinherz, and S. C. Harrison, *Nature* **348,** 411 (1990).
[9] A. M. De Vos, M. Ultsch, and A. A. Kossiakoff, *Science* **255,** 306 (1992).



0076-6879/99 $30.00

tion of transmembrane domains, which are involved in signal transduction and transport across membranes. Solid state nuclear magnetic resonance (NMR) spectroscopy has successfully been used to examine the transmembrane domains of membrane proteins.[11] However, the protein must be examined in micelles and the molecular weight limit precludes examination of complex polytopic proteins.

An alternative approach is to grow 2D crystals and use electron cryomicroscopy and image analysis to solve the structure.[12–21] Currently, the structural characterization of the majority of membrane proteins by electron cryocrystallography is indeed a challenging endeavor because the proteins are often present in low abundance in their native membranes, and only a few hundred micrograms of purified recombinant protein may be available. Furthermore, membrane proteins tend to be labile when solubilized in detergent. However, in several cases it has now been possible to grow two-dimensional (2D) crystals in lipid bilayer membranes that can be analyzed by electron cryocrystallography. Moreover, the success in obtaining atomic resolution structures for bacteriorhodopsin[22,23] and the light-harvesting complex II[24] attests that 2D crystallization, electron cryomicroscopy, and image analysis are emerging as powerful approaches for the structural characterization of membrane proteins.

There are several advantages of analyzing 2D crystals by electron cryomicroscopy and image analysis compared with X-ray analysis of 3D crystals.

---

[10] L. Shapiro, A. M. Fannon, P. D. Kwong, A. Thompson, M. S. Lehmann, G. Grübel, J.-F. Legrand, J. Als-Nielsen, D. R. Colman, and W. A. Hendrickson, *Nature* **374,** 327 (1995).

[11] S. J. Opella, *Nat. Struct. Biol.* **4 Suppl.,** 845 (1977).

[12] L. A. Amos, R. Henderson, and P. N. T. Unwin, *Prog. Biophys. Mol. Biol.* **39,** 183 (1982).

[13] M. Stewart, *J. Electron Microsc. Tech.* **9,** 301 (1988).

[14] E. J. Boekema, *Electron Microsc. Rev.* **3,** 87 (1990).

[15] P. A. Bullough, *Electron Microsc. Rev.* **3,** 249 (1990).

[16] W. Kühlbrandt, *Quart. Rev. Biophysics* **25,** 1 (1992).

[17] A. Engel, A. Hoenger, A. Hefti, C. Henn, R. C. Ford, J. Kistler, and M. Zulauf, *J. Struct. Biol.* **109,** 219 (1992).

[18] B. K. Jap, M. Zulauf, T. Scheybani, A. Hefti, W. Baumeister, U. Aebi, and A. Engel, *Ultramicrosc.* **46,** 45 (1992).

[19] W. Chiu and M. F. Schmid, *Curr. Opin. Biotech.* **4,** 397 (1993).

[20] W. Chiu, M. F. Schmid, and B. V. Prasad, *Biophys. J.* **64,** 1610 (1993).

[21] D. L. Dorset, *Acta Cryst.* **B52,** 753 (1996).

[22] R. Henderson, J. M. Baldwin, T. A. Ceska, F. Zemlin, E. Beckmann, and K. H. Downing, *J. Mol. Biol.* **213,** 899 (1990).

[23] Y. Kimura, D. G. Vassylyev, A. Miyazawa, A. Kidera, M. Matsushima, K. Mitsuoka, K. Murata, T. Hirai, and Y. Fujiyoshi, *Nature* **389,** 206 (1997).

[24] W. Kühlbrandt, D. N. Wang, and Y. Fujiyoshi, *Nature* **367,** 614 (1994).

First, a great advantage of this technique is the absence of the phase problem. Because images of the crystals are recorded, phases can be directly calculated by Fourier transformation. Second, the amount of material required for electron crystallography is substantially less than that needed for X-ray crystallography. For instance, our current analysis of gap junction channels is based on membrane specimens that are enriched for the recombinant connexin but still do not show a detectable band on Coomassie-stained sodium dodecyl sulfate (SDS) gels. This compares with the need for several milligrams of protein required for typical 3D crystallization. Furthermore, purification of such large quantities of protein is quite expensive because of the need for substantial amounts of detergents, protease inhibitors, and tissue culture media. For electron cryomicroscopy of membrane proteins, specimens can be preserved in an unstained, frozen-hydrated state within the lipid bilayer so that the native structure is revealed.[25–29] This compares with the analysis of 3D crystals in the presence of detergents and nonphysiologic precipitants. One would assume that the detergent belt around the hydrophobic perimeter of the membrane protein in the 3D crystal would mimic the lipid bilayer environment, but at least one case has been identified in which a dramatic difference occurred when the protein was examined in bilayers versus 3D crystals. The pore-forming toxin $\alpha$-hemolysin assembled as a heptamer in 3D crystals[30] but was hexameric in lipid bilayer membranes.[31,32] With the membrane protein embedded in a lipid bilayer, it is also possible to examine functionally important states. For example, Berriman and Unwin[33] have developed a rapid freezing method that has been successfully used to examine the open conformation of the acetylcholine receptor by exposing the crystals to acetylcholine within milliseconds of plunge-freezing for electron cryomicroscopy.[34]

Similar to X-ray crystallography, the quality of the crystals is a major limitation. To date, most studies using electron cryocrystallography of 2D

[25] K. A. Taylor and R. M. Glaeser, *J. Ultrastruct. Res.* **55,** 448 (1976).

[26] R. A. Milligan, A. Brisson, and P. N. T. Unwin, *Ultramicrosc.* **13,** 1 (1984).

[27] N. Unwin, *Ann. N.Y. Acad. Sci.* **483,** 1 (1986).

[28] W. Chiu, *Ann. Rev. Biophys. Biophys. Chem.* **15,** 237 (1986).

[29] J. Dubochet, M. Adrian, J.-J. Chang, J.-C. Homo, J. Lepault, A. W. McDowall, and P. Schultz, *Quart. Rev. Biophysics* **21,** 129 (1988).

[30] L. Song, M. R. Hobaugh, C. Shustak, S. Chelay, H. Bayley, and J. E. Gouaux, *Science* **274,** 1859 (1996).

[31] A. Olofsson, U. Kavéus, M. Thelestam, and H. Hebert, *J. Ultrastruct. Mol. Struct. Res.* **100,** 194 (1988).

[32] D. M. Czajkowsky and Z. Shao, *Biophys. J.* **72,** A139 (1997).

[33] J. Berriman J, and N. Unwin, *Ultramicros.* **56,** 241 (1994).

[34] N. Unwin, *Nature* **373,** 37 (1995).

crystals only allowed analysis at modest ~6-Å resolution. This is sufficient for revealing the packing of transmembrane $\alpha$-helices within the lipid bilayer, but the connecting loops and $\beta$ structure are often not delineated at this resolution. To date, only a few proteins have been crystallized in 2D with sufficient order that they diffract to atomic resolution. This problem is in part due to disorder in the crystals. However, besides crystal disorder, instrumental effects are also important. Because the phases are computed directly from images, the microscope must be extremely stable so that movement during imaging is on the order of 1 Å per second or less. Such microscopes are indeed expensive, and setting up a laboratory for electron cryocrystallography is substantially more expensive than setting up one for X-ray crystallography. The best microscopes available use a field emission gun as an electron source to generate extremely bright and coherent beams, operate at higher voltages (200–400 kV) so that the depth of field is increased to allow examination of thicker specimens, employ stages that are exquisitely stable to vibrations, and operate at liquid nitrogen or even liquid helium temperatures.

In this review we provide a general outline for the steps involved in the structure analysis of membrane proteins by electron cryocrystallography. We present our results on the analysis of two-dimensional crystals of gap junction channels, aquaporin channels, and histidine-tagged Fab molecules, to exemplify respectively the general methods of *in situ* 2D crystallization, *in vitro* 2D crystallization, and lipid monolayer crystallization. Also discussed are the preparation of frozen-hydrated specimens, performance of low-dose transmission electron cryomicroscopy, and image analysis of 2D crystals.

## Steps in Structural Analysis of Membrane Proteins by Electron Cryocrystallography

Table 1 outlines the strategy for the structural characterization of membrane proteins by electron cryocrystallography, which involves isolation of membranes containing the desired protein, 2D crystallization, electron cryomicroscopy, and computer image processing.[35]

### *Isolation of Membranes*

To our knowledge, our analysis of cardiac gap junction channels is the first example where a polytopic membrane protein has been expressed in a heterologous system and examined by structural methods.[36] The overex-

[35] M. Yeager, *Microsc. Res. Tech.* **31,** 452 (1995).

[36] V. M. Unger, N. M. Kumar, N. B. Gilula, and M. Yeager, *Nature Struct. Biol.* **4,** 39 (1997).

pression of membrane proteins is not at all routine. The reader is referred to Grisshammer and Tate[37] for a state-of-the-art review. For our analysis of gap junctions, a C-terminal truncation mutant (after lysine 263) of rat heart $\alpha_1$Cx43 connexin (designated $\alpha_1$Cx263T) was expressed in a stably transfected baby hamster kidney (BHK) cell line under control of the inducible mouse metallothionein promoter.[38] Freeze-fracture, thin-section, and negative stain electron microscopy of the BHK cell membranes demonstrated that the recombinant protein assembles with the characteristic septalaminar morphology of gap junctions. In addition, dye transfer experiments[39] demonstrated that the gap junctions are functional. Of note, however, is that the levels of expression are quite low. Nevertheless, this amount of protein is sufficient to pursue 2D crystallization experiments because the junctions naturally assemble in specialized membrane domains that can be selectively enriched.

### *Procedure for Isolation of Membranes Enriched for Recombinant Gap Junctions*

#### Induction of Connexin Expression

1. Examine the cells by light microscopy and assess the density. When the cells appear almost confluent, they are ready for induction.
2. Aspirate the medium and replace with fresh medium containing 100 m*M* zinc acetate.

#### Enrichment of Gap Junctions

1. Place a 500-ml JA10 centrifuge bottle into an ice bucket and insert a funnel.
2. Use a cell scraper to dislodge the cells from the bottom of the plates, and decant the media with the suspended cells.
3. *Spin 1* (2 krpm, 5 min, 4°, JA10 rotor, Beckman J-30I centrifuge).
4. Decant the supernatant.
5. Carefully pipette 10 ml of ice-cold HEPES buffer (10 m*M* HEPES, pH 7.5, 0.8% NaCl) into the centrifuge tube so as to not disturb the pellet.
6. Detach the pellet by swirling the centrifuge bottle.
7. Pour the contents with the intact pellet into a 50-ml conical plastic tube (Falcon, Corning).
8. Vortex the Falcon tube to break up the pellet.
9. Increase the volume to 50 ml with ice-cold HEPES buffer.

[37] R. Grisshammer and C. G. Tate, *Quart. Rev. Biophys.* **28,** 315 (1995).

[38] N. M. Kumar, D. S. Friend, and N. B. Gilula, *J. Cell Sci.* **108,** 3725 (1995).

[39] V. M. Unger, D. W. Entrikin, X. Guan, B. Cravatt, N. M. Kumar, R. A. Lerner, N. B. Gilula, and M. Yeager, *Biophys. J.* **72,** A291 (1997).

TABLE I

GENERAL OUTLINE FOR STRUCTURE ANALYSIS OF 2-D CRYSTALS

- Isolation of membranes
  - Protein source
    - High abundance: obtain tissue
    - Low abundance: overexpression in insect cells, BHK cells, yeast cells, etc.
  - Homogenization and low-speed centrifugations to remove soluble components
  - Consider extraction with chaotropic agents (e.g., $KI$ and $Na_2S_2O_3$) to remove extrinsic proteins
  - Sucrose gradient ultracentrifugation to enrich the membrane fraction
- Two-dimensional crystallization
  - *In situ* method
    - Incubate with low concentrations of nonionic detergents, phospholipases, etc. to remove lipids and concentrate protein in bilayer
    - Remove detergent (e.g., by dialysis or Bio-Beads)
  - *In vitro* reconstitution
    - Solubilize protein using high concentrations of nonionic detergents
    - Purify protein using size-exclusion, ion-exchange, hydrophobic interaction, or affinity chromatography
    - Mix purified protein–detergent complexes with synthetic or native lipids that have been solubilized in detergents
    - Remove detergent by dialysis or Bio-Beads
  - Lipid monolayer crystallization
    - Isolate and purify water-soluble protein domain
      - Water-soluble conformation of membrane protein (e.g., annexins)
      - Ectodomain generated by protease cleavage
      - Ectodomain expressed using recombinant DNA methods
    - Incubate protein solution in droplet having lipid monolayer surface
      - Use positively and negatively charged lipids to exploit electrostatic interaction between protein and lipid headgroups
      - Exploit specific affinity between particular headgroup (e.g., nickel lipid) and protein (e.g., hexahistidine tag)
- Transmission Electron Microscopy
  - Deposit membranes on grid by centrifugation or drop adherence
  - Prepare specimen for cryomicroscopy by plunging the grid into ethane slush or embedding specimen in mordants such as glucose, trehalose, or tannin
  - Store grids in liquid nitrogen
  - Transfer grid to stage using cryotransfer system

10. *Spin 2* (2 krpm, 5 min, 4°, GH-3.8 rotor, Beckman GS-6KR centrifuge).
11. Decant the supernatant from the Falcon tube, and add 5 ml of ice-cold HEPES buffer containing 140 $\mu$g/ml phenylmethylsulfonyl fluoride (PMSF) to the pellet.
12. Suspend the pellet by vortex mixing, and increase the total volume to 15 ml with HEPES buffer containing 140 $\mu$g/ml PMSF.
13. Incubate the Falcon tube for 5 min on ice.
14. Sonicate the same three times for 5 sec (Branson sonicator, output setting 6–7, 60% duty cycle) waiting 15 sec between each burst. Keep

TABLE I (*continued*)

| |
|---|
| Perform low dose cryomicroscopy with specimen at temperature of liquid nitrogen or helium |
| Identify large, unbroken membrane sheets, preferably with polygonal edges and with proper ice thickness |
| Record image using minimal dose techniques |
| Search mode: scan grid in overfocused diffraction mode |
| Focus mode: at high magnification (~200,000×) in area just adjacent to region to be photographed located Gaussian focus, select defocus level (e.g., 4,000-Å underfocus), correct for astigmatism, and ensure that there is no drift of stage |
| Exposure mode: record image or diffraction pattern of crystal using dose of $<10e^-/Å^2$ |
| Image appraisal |
| Examine EM negatives |
| Perform optical diffraction and select images that display sharp, bright reflections to high resolution; optimal defocus; minimal drift |
| Digitize crystalline patches identified by optical diffraction |
| Image processing (see Table II) |
| Taper edges of digitized array |
| Compute Fourier transform |
| Index reflections in computed diffraction pattern |
| Refine crystal lattice parameters |
| Correct for crystal lattice distortions |
| Mask diffraction pattern and calculate filtered image |
| Select reference image |
| Calculate cross-correlation map |
| Correct for lattice displacements |
| Extract corrected amplitudes and phases |
| Correct for effects due to CTF and astigmatism |
| Evaluate plane group symmetry from images of untilted crystals |
| Refine phase origin, tilt axis, and tilt angle |
| Merge data from multiple images |
| If amplitudes are derived by |
| Electron diffraction: subtract background due to inelastic scattering |
| Fourier transformation of images: apply inverse temperature factor to correct for resolution dependent fall-off of image amplitudes |
| Interpolate phase and amplitude data in 3D lattice lines |
| Compute 3D map by Fourier inversion with symmetry constraints defined by lattice |
| Use graphics software to examine map |

the sonicator tip at the bottom of the tube to minimize foaming. The final suspension should appear milky and homogeneous.

15. Examine a drop of the suspension by light microscopy to ensure that cell lysis is at least 90%. If there are still numerous intact cells, repeat the sonication for 5 sec.
16. Increase the volume to 35 ml for ultracentrifugation.
17. *Spin 3* (25 krpm, 45 min, 4°, SW28, rotor, Beckman LE-80K ultracentrifuge). Decant the supernatant, and resuspend the pellet in 5 ml of ice-cold HEPES buffer.
18. Add 30 ml of 49% sucrose in HEPES buffer to an SW28 ultracentri-

fuge tube, and pipette the 5 ml of crude membrane homogenate onto the top of the sucrose cushion.

19. *Spin 4* (25 krpm, 45 min, 4°, SW28 rotor, Beckman LE-80K ultracentrifuge).
20. Use a plastic disposable pipette to carefully remove material (e.g., solid PMSF) floating on the top of the tube. Then retrieve the 0/49% interface and transfer to another SW28 tube. The homogenate should appear milky white.
21. Increase the volume to 35 ml with ice-cold sucrose-free HEPES buffer. Use a plastic pipette to mix the solution to dilute the sucrose in the aspirated interface.
22. *Spin 5* (25 krpm, 45 min, 4°, SW28 rotor, Beckman LE-80K ultracentrifuge).
23. Decant the supernatant, add 1-ml ice-cold HEPES buffer containing 140 $\mu$g/ml PMSF, disrupt the pellet by repeated passage through a plastic Eppendorf pipette and homogenize the sample by sonication for a few seconds.
24. The specimen is stable at 4° for ~2 weeks.

*Two-Dimensional Crystallization*

As in any crystallization trial, one must optimize and refine the conditions for crystallization and specimen handling to maximize the degree of order.[40] In general, the variables to be tested to achieve 2D crystallization are similar to those used for 3D crystallization and include evaluation of divalent cations, pH, detergents, lipids, buffers, ionic strength, temperature, precipitants, ligands, and inhibitors. Purity and yield of the protein are maximized, and proteolysis is minimized. We currently use three methods (*in situ* crystallization, *in vitro* reconstitution, and lipid monolayer crystallization) to grow 2D crystals (Fig. 1). Another method is to grow 2D crystals of a detergent-solubilized and purified membrane protein directly on the electron microscope (EM) grid. This approach has been successfully used to grow 2D crystals of the $H^+$-ATPase from *Neurospora crassa*.[41] Crystallization is induced by using buffer conditions that are similar to those used for 3D crystallization (e.g., high concentrations of PEG or ammonium sulfate). In this way crystallization is induced by interactions of the hydrophilic domains of the protein. Some soluble proteins can also be grown as

[40] A. McPherson, "Preparation and Analysis of Protein Crystals," John Wiley and Sons, New York, 1982.

[41] M. Auer, G. A. Scarborough, and W. Kühlbrandt, *Nature* **392,** 840 (1998).

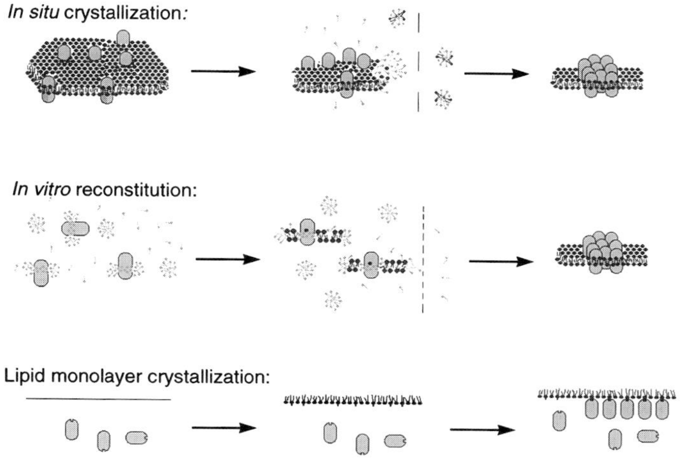


FIG. 1. Schematic diagram for three strategies that have been used to grow 2D crystals. In the method of *in situ crystallization*, the protein is never removed from its native membrane. A requirement for the success of this technique is that the protein must already be fairly concentrated in the membrane. 2D crystallization is induced using gentle conditions that extract membrane lipids without solubilizing the protein. In the method of *in vitro reconstitution*, the protein is solubilized from its native membrane and purified as a complex with detergent micelles. The purified protein–detergent complexes are mixed with lipids that have also been solubilized in detergent. Lipid bilayers are reconstituted by removing the detergent via dialysis. Crystallization is induced by using a lipid : protein ratio that is high enough to promote protein/protein interactions. In the *lipid monolayer crystallization* technique, lipids are spread at an air/water interface so that the aliphatic chains extend into the air, and the lipid headgroups are exposed at the aqueous interface. A soluble protein is present in the aqueous phase. Binding of the protein to the lipid headgroups occurs by electrostatic interactions or by a specific affinity tag. For instance, the lipid monolayer can be doped with a lipid containing a nickel headgroup that will chelate a polyhistidine tag on the protein.

2D crystals, such as catalase,[42,43] tubulin,[44,45] and VP6, the rotavirus inner capsid protein.[46] The progress of 2D crystallization is usually assessed by negative stain electron microscopy.

### *In situ Crystallization*

A special advantage of this approach is that the protein is never removed from its native membrane.[16,17,18,47] However, a requirement for the success

[42] P. N. Unwin, *J. Mol. Biol.* **98,** 235 (1975).
[43] C. W. Akey, M. Szalay, and S. J. Edelstein, *Ultramicrosc.* **13,** 103 (1984).
[44] L. A. Amos and T. S. Baker, *Nature* **279,** 607 (1979).
[45] E. Nogales, S. G. Wolf, and K. H. Downing, *J. Struct. Biol.* **118,** 119 (1997).
[46] G. G. Hsu, A. R. Bellamy, and M. Yeager, *J. Mol. Biol.* **272,** 362 (1997).
[47] M. Yeager, *Acta Cryst.* **D50,** 632 (1994).

of this technique is that the protein must already be fairly concentrated in the membrane or have a tendency to self-associate in patches that can be isolated. 2D crystallization is induced using gentle conditions that extract membrane lipids without solubilizing the protein. Detergents, enzymes, and chaotropes[48] that have been used include deoxycholate,[47,49,50] dodecylmaltoside,[47,49,51] Lubrol,[52] $D_2O$,[53] Tween,[36,54] 1,2-diheptanoyl-*sn*-phosphocholine (DHPC),[36] and phospholipases.[55,56] For the expressed gap junction channels, the order of naturally occurring 2D arrays was improved by sequential exposure to Tween 20 and DHPC (Fig. 2). In addition, exposure to the detergents enriched the preparation by solubilizing nonjunctional protein.

*Procedure for in Situ 2-D Crystallization of Recombinant Gap Junction Channels*

1. The protein concentration of the membrane sample enriched for gap junctions is estimated using the DotMetric assay (Geno Technology Inc.).
2. The volume of the enriched gap junction suspension is adjusted so that the protein concentration is 1 mg/ml, after addition of the following agents to the final concentrations that are indicated: 2.8% Tween 20, 200 m*M* KI, 2 m*M* sodium thiosulfate, 140 $\mu$g/ml PMSF, 50 $\mu$g/ml gentamicin in 10 m*M* HEPES buffer (pH 7.5) containing 0.8% NaCl. Because Tween 20 is contaminated with aldehydes, peroxides, and free acids, it must first be purified by ion-exchange chromatography [resin AG501-X8(D); BioRad, Richmond, CA] immediately before use.
3. Add a magnetic mini-stirring bar to the tube and stir at 27° for 12 hr.
4. After a 12-hr incubation, add solid DHPC (Avanti Polar Lipids, Birmingham, AL) to 13 mg/ml, and stir for an additional 1 hr at 27°.
5. Add 30 ml 25% sucrose (prepared in 10 m*M* HEPES, 0.8% NaCl, pH 7.5) to an SW28 centrifuge tube, and carefully add the entire contents on top of the sucrose cushion.

[48] Y. Hatefi and W. G. Hanstein, *Methods Enzymol.* **31,** 770 (1974).
[49] M. Yeager and N. B. Gilula, *J. Mol. Biol.* **223,** 929 (1992).
[50] R. M. Glaeser, J. S. Jubb, and R. Henderson, *Biophys. J.* **48,** 775 (1985).
[51] E. Gogol and N. Unwin, *Biophys. J.* **54,** 105 (1988).
[52] G. Zampighi and P. N. Unwin, *J. Mol. Biol.* **135,** 451 (1979).
[53] S. S. Sikerwar and N. Unwin, *Biophys. J.* **54,** 113 (1988).
[54] G. F. Schertler, C. Villa, and R. Henderson, *Nature* **362,** 770 (1993).
[55] C. A. Mannella, *Science* **224,** 165 (1984).
[56] C. A. Mannella, *Biochim. Biophys. Acta* **981,** 15 (1989).

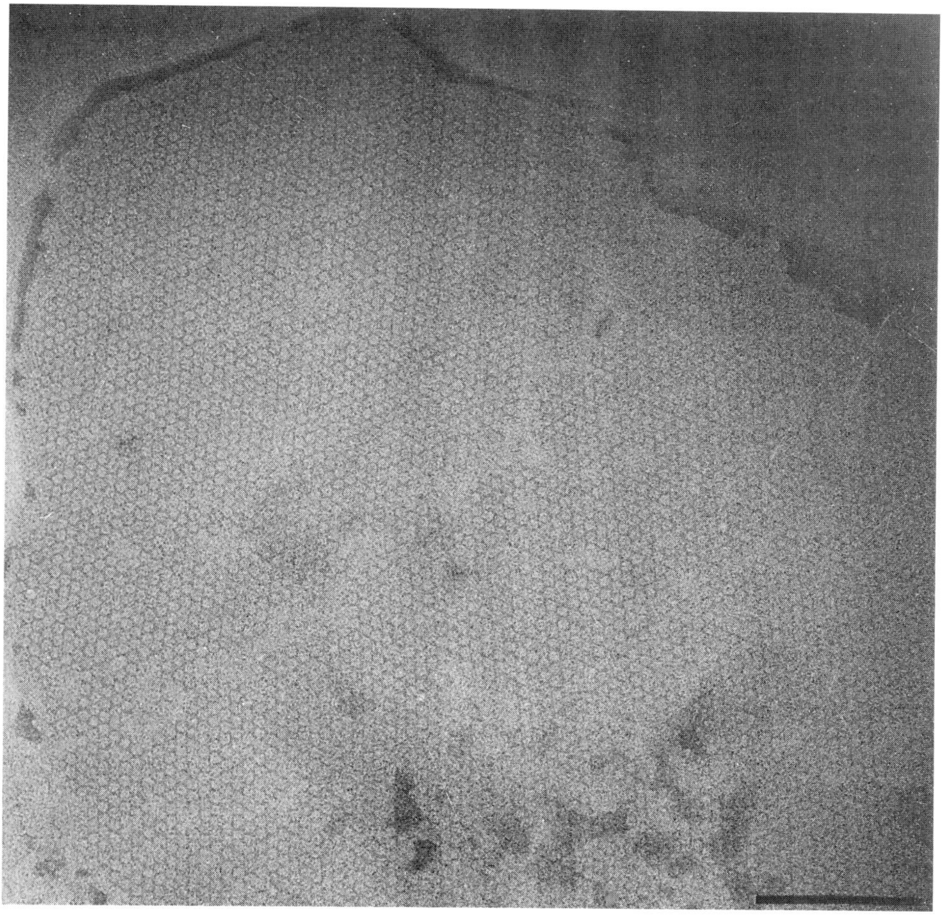

FIG. 2. Electron micrograph of a negatively stained 2D crystal of recombinant gap junction channels formed by a C-terminal truncation mutant of rat heart $\alpha_1$-connexin ($\alpha_1$Cx263T). The hexagonal packing of the channels is clearly seen. Such crystals are grown by *in situ* crystallization in which the membranes are incubated in buffer containing Tween 20 and DHPC to extract lipids. The mottled appearance at the bottom of the crystal may represent regions that are partially solubilized by the detergents. Note that the crystal is made of several mosaic domains, and there is a fracture in the crystal that may have occurred during transfer to the supporting carbon substrate. For membrane proteins with hydrophilic ectodomains, we have found that adherence of the specimen to the carbon support is optimal if the grids are rendered hydrophilic by glow discharge.[99] However, strong interactions between the carbon and the crystals may also deform the crystals. We have found that pretreatment of the grid with a detergent such as Tween or octylglucoside can be used to modulate the adherence of the crystals to the carbon. Bar: 1000 Å.

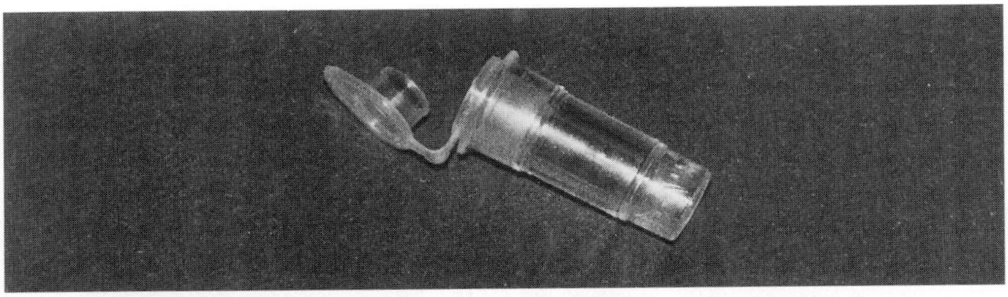

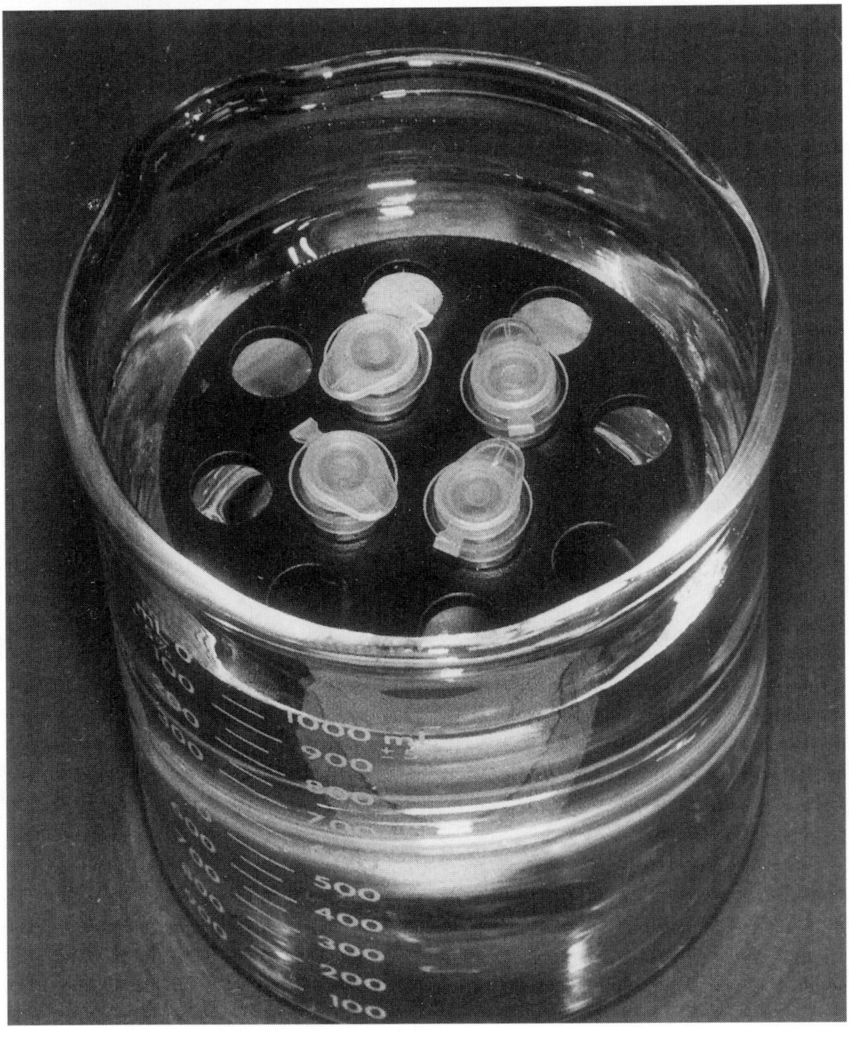
1000 ml
900
500
400
300
200
100

6. *Spin 6* (25 krpm, 1 hr, 4°, SW28 rotor, Beckman LE-80K ultracentrifuge).
7. Carefully pipette off the sucrose so that the detergent meniscus does not touch the pellet, and wipe the tube clean with a Kimwipe to remove any residual sucrose or detergent.
8. Resuspend the pellet in 600–800 $\mu$l of HEPES buffer, and sonicate for a few seconds as above so that the suspension is homogeneous.
9. Dialyze (exclusion limit for the membrane of 100,000) for at least 24 hr against 500 ml of the buffer with one change of the buffer (HEPES, plus 5 ml gentamicin/liter of buffer) at ~12 hr using the Bio-Tech International tubes (Fig. 3).
10. Store the suspension at 4° in a 1.5-ml Eppendorf tube.

In the method of *in vitro reconstitution*, the protein is solubilized from its native membrane and purified as a complex with detergent micelles.[16–18] The purified protein–detergent complexes are mixed with synthetic lipids or extracted native lipids that have also been solubilized in detergent. Lipid bilayers are then reconstituted by removing the detergent by dialysis or by the use of Bio-Beads.[56a] Crystallization is induced by using a lipid : protein ratio that is low enough to promote protein/protein interactions. An important aspect of this approach is that the solubilized protein must be stable for a period that is sufficient to remove detergent by dialysis. A critical variable is the lipid : protein ratio. If the lipid concentration is too high, then the protein will not be sufficiently concentrated in the plane of the lipid bilayer to induce 2D crystallization. Alternatively, if the protein concentration is too high, the protein may aggregate and denature on removal of the detergent. Dedicated glassware should be used for handling all solvents so that they are not contaminated by detergents used for routine glass washing. Lipids to be considered for reconstitution include dimyristoylphosphatidylcholine (DMPC), dioleylphosphatidylcholine (DOPC), DPOPC, where PC stands for a mixture of phosphatidylcholines from soybean, dilaurylphosphatidylcholine (DLPC), OPPC, POPS, DOPE, DPOPE, DOPG, POPG, as well as the lipids from the native membrane. Lipid : cholesterol mixtures are also examined.

[56a] J. J. Lacapere, D. L. Stokes, G. Mosser, J. L. Ranck, G. Leblanc, and J. L. Rigaud, *Ann. N.Y. Acad. Sci.* **834,** 9 (1997).

FIG. 3. Microdialysis chambers available from Bio-Tech International, Inc (Seattle, WA). The bottom of the tubes are fitted with dialysis membranes available in a molecular weight cutoff from 8000 to 500,000 Da. The advantage of this system is that the tube caps can be opened to retrieve small aliquots of the sample during dialysis in order to assess the progression of crystallization. We have cut holes in inert 8.5-cm plastic disks to hold several tubes during dialysis.

For aquaporin I (AQP1, formerly CHIP28, channel-forming integral membrane protein of 28 kDa), the 2D crystals were better ordered if the protein was deglycosylated.[57,58] For the lipids tested, crystals were obtained with only DMPC and DOPC at 27° (Fig. 4). In general, the crystals grown with DOPC were larger and better ordered than for DMPC. In addition, cholesterol had no effect on 2D crystallization. Other groups have had success growing high-resolution 2D crystals of AQP1 using phospholipase treatment[59] as well as reconstitution with *Escherichia coli* lipids.[60,61]

*Procedure for 2-D Crystallization of AQP1 by in Vitro Reconstitution*

1. Lipids are obtained from Avanti Polar Lipids in glass ampules and are prepared as 1% stock solutions in chloroform (Aldrich, Milwaukee, WI, HPLC Grade) and are stored at −30 in Reacti-Vials (Pierce, Rockford, IL) that have Teflon caps with a rubber septum that allows removal of the lipid solution without opening the vial. The lipid solution in the vial is flushed with a gentle stream of dry nitrogen gas as the cap is screwed on.
2. To solubilize the lipids in detergent, an aliquot is transferred to a 5-ml glass test tube and dried under a gentle stream of nitrogen. A buffer containing 4% octyl-$\beta$-D-glucopyranoside (OG, Anatrace), 20 m*M* sodium phosphate (pH 7.0), 0.1 m*M* EDTA, 100 m*M* NaCl, and 0.025% $NaN_3$ is added to the dried lipids to give a final detergent concentration of 1% when solubilized.
3. A micromagnetic stirring barr is added, the tube is flushed with nitrogen and then capped with a cork or sealed with Parafilm. The lipids are solubilized by magnetic stirring overnight at room temperature.
4. Purified human AQP1 is deglycosylated with PNGase F (NEB), and the final purification step involves Q-Sepharose chromatography in a buffer containing 35–80 m*M* OG, 20–50 m*M* sodium phosphate (pH 7.2–7.5), 0.1 m*M* EDTA, and 100 m*M* NaCl.
5. The protein concentration is determined by the Lowry method and adjusted to 1–2 mg/ml.
6. For any crystallization experiment, the total sample volume before dialysis is 50 $\mu$l, and optimal crystallization occurs at lipid : protein ratios between 1 : 1 to 1 : 3 (w/w). Cholesterol at a lipid : cholesterol ratio of 1 : 0 to 1 : 10 has no effect.
7. The solutions are pipetted into microdialysis chambers prepared from

[57] A. K. Mitra, M. Yeager, A. N. van Hoek, M. C. Wiener, and A. S. Verkman, *Biochemistry* **33,** 12735 (1994).

[58] A. K. Mitra, A. N. van Hoek, M. C. Wiener, A. S. Verkman, and M. Yeager, *Nature Struct. Biol.* **2,** 726 (1995).

[59] B. K. Jap and H. Li, *J. Mol. Biol.* **251,** 413 (1995).

[60] T. Walz, B. L. Smith, P. Agre, and A. Engel, *EMBO J.* **13,** 2985 (1994).

[61] T. Walz, D. Typke, B. L. Smith, P. Agre, and A. Engel, *Nature Struct. Biol.* **2,** 730 (1995).

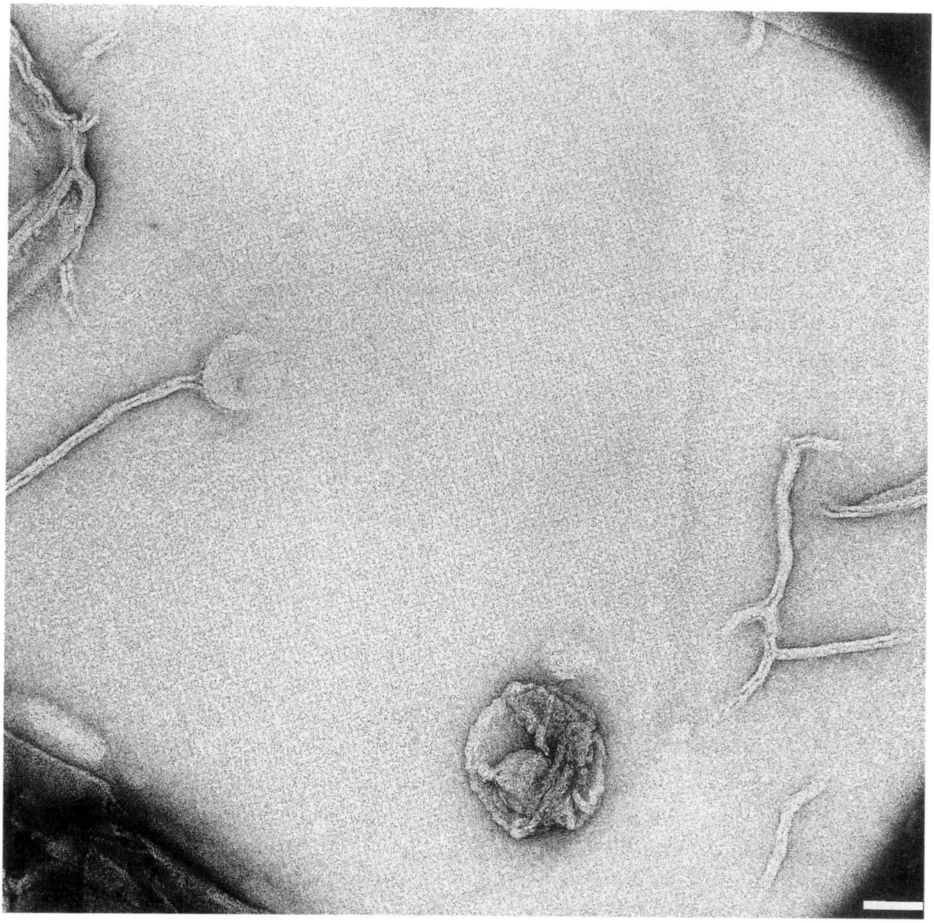

FIG. 4. Electron micrograph of a negatively stained 2D crystal of aquaporin 1 (formerly, CHIP28). Such crystals are grown by *in vitro* reconstitution and are typically 1.5–2.5 $\mu$m in diameter. The square lattice is clearly visible, as well as folds in the crystal and a separate vesicle that has collapsed onto the 2D crystal. [Reproduced from A. K. Mitra, M. Yeager, A. N. van Hoek, M. C. Wiener, and A. S. Verkman, *Biochemistry* **33,** 12735 (1994) by permission of the American Chemical Society.] Bar: 1000 Å.

the caps of Eppendorf tubes (Fig. 5), and dialysis is performed using SpectraPor membranes with a molecular weight cutoff of 12–14 kDa. The microdialysis caps float on the surface of the buffer.

8. Slow detergent dialysis is performed at 27° for 6–8 days against 1-liter volumes of buffer, with four to five buffer changes. For any single dialysis experiment, a single buffer and lipid combination can be tested over a range of lipid : protein ratios.

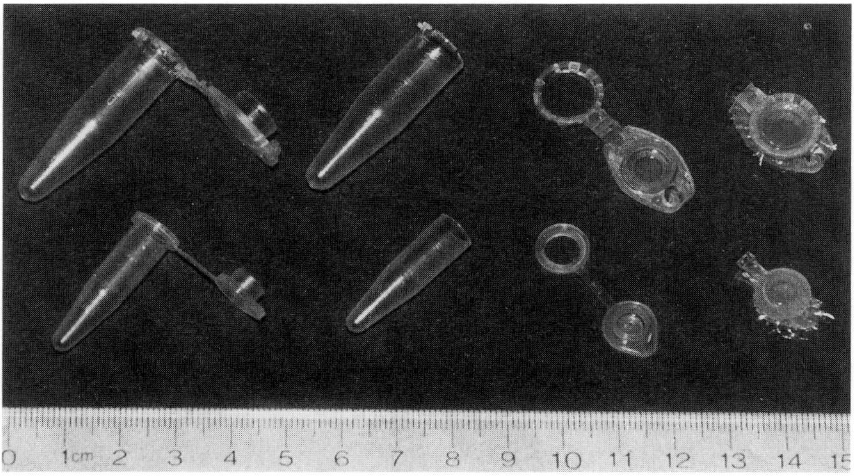


FIG. 5. Microdialysis apparatus constructed from the caps of Eppendorf tubes.[57] A razor blade is used to cut the caps off of the tubes. The cap is inverted and the specimen is added to the well (~10 and ~50 μl for 0.5- and 1.5-ml tube caps, respectively). A piece of dialysis membrane is placed over the well and the ring is snapped closed. The caps will float on the surface of the dialysis buffer. The sample is retrieved by piercing the membrane with a needle and aspirating the sample with a micropipette.

9. Samples are retrieved from the dialysis caps and stored at 4° in 1.5-ml Eppendorf tubes.

In the *lipid monolayer crystallization* technique,[62–64] lipids are spread at an air/water interface so that the aliphatic chains will extend into the air, and the lipid headgroups will be exposed at the aqueous interface. Soluble protein molecules present in the aqueous phase bind to the lipid headgroups via electrostatic interactions or by a specific affinity tag. Negatively charged lipids include stearic and oleic acid (Supelco). Examples of positively charged lipids include octadecylamine (Sigma) and 1,2-diacyl-3-dimethylammonium-propane (Avanti Polar Lipids). Affinity tags can exploit a receptor/ligand or enzyme/substrate interaction (e.g., cholera toxin binding to its receptor[65]; avidin binding to a biotinylated lipid[66]). A method

[62] S. A. Hemming, A. Bochkarev, S. A. Darst, R. D. Kornberg, P. Ala, D. S. Yang, and A. M. Edwards, *J. Mol. Biol.* **246,** 308 (1995).

[63] A. Brisson, A. Olofsson, P. Ringler, M. Schmutz, and S. Stoylova, *Biol. Cell* **80,** 221 (1994).

[64] W. Chiu, A. J. Avila-Sakar, and M. F. Schmid, *Adv. Biophys.* **34,** 161 (1997).

[65] H. O. Ribi, D. S. Ludwig, K. L. Mercer, G. K. Schoolnik, and R. D. Kornberg, *Science* **239,** 1272 (1988).

[66] S. A. Darst, M. Ahlers, P. H. Meller, E. W. Kubalek, R. Blankenburg, H. O. Ribi, H. Ringsdorf, and R. D. Kornberg, *Biophys. J.* **59,** 387 (1991).

of potential general utility exploits the strategy of using a histidine-tagged protein.[67,68] The lipid monolayer can be doped with a synthetic lipid containing a chelating nickel headgroup that will bind to a protein with a hexahistidine tag (Fig. 6). This method would certainly be applicable for the 2D crystallization of soluble ectodomains of membrane proteins generated by recombinant DNA technology or by release from the membrane by proteolysis. A recent adaptation of the monolayer crystallization technique has exploited the propensity of some lipids such as galactosylceramide to form unilamellar tubes that provide a substrate for the crystallization of helical arrays of proteins.[68a,68b]

*Procedure for Lipid Monolayer 2-D Crystallization*

1. Stock solutions of the special lipid (i.e., charged or liganded) in 10% egg phosphatidylcholine are prepared in chloroform and hexane. For 2D crystallization trials the ratio of the special lipid to the bulk lipid is varied from 0 (as a control) to 1. Egg phosphatidylcholine is typically used because it remains fluid over a wide range of temperatures and compressions.
2. A Teflon block that has wells 3 mm in diameter (the diameter of an EM grid) is thoroughly cleaned by storage in chromic and sulfuric acid (Fisher Cleaning Solution). The block is washed by sonication for 30 min in detergent solution (7*X*, Linbro or Versa-Clean, Fisher) and then rinsed for ~30 min in doubly distilled, deionized $H_2O$.
3. To maintain a hydrated atmosphere, filter paper is placed on the bottom of a 5-cm petri dish and thoroughly wetted with buffer. The Teflon block is then inserted.
4. High concentrations of glycerol or salt in the protein solution may impede crystallization. Hence, the sample may need to be diluted just before addition to the well in the Telfon block.
5. Use an Eppendorf pipette to fill the 0.5-mm-deep well with 10 $\mu$l protein solution at a concentration of 50–1000 $\mu$g/ml.
6. Use a glass micropipette to add ~0.5–1 $\mu$l of the lipid solution at a concentration of $\leq$0.5 mg/ml.
7. Cover the petri dish and incubate at the desired temperature (e.g., room temperature or 4°).
8. 2D crystallization on the lipid monolayer can occur quickly (30–60 min) or take days.

[67] E. W. Kubalek, S. F. Le Grice, and P. O. Brown, *J. Struct. Biol.* **113,** 117 (1994).

[68] E. Barklis, J. McDermott, S. Wilkens, E. Schabtach, M. F. Schmid, S. Fuller, S. Karanjia, Z. Love, R. Jones, Y. Rui, X. Zhao, and D. Thompson, *EMBO J.* **17,** 1199 (1997).

[68a] P. Ringler, W. Muller, H. Ringsdorf, and A. Brisson, *Chem. Eur. J.* **3,** 620 (1997).

[68b] E. M. Wilson-Kubalek, R. E. Brown, H. Celia, and R. A. Milligan, *Proc. Natl. Acad. Sci. USA* **95,** 8040 (1998).

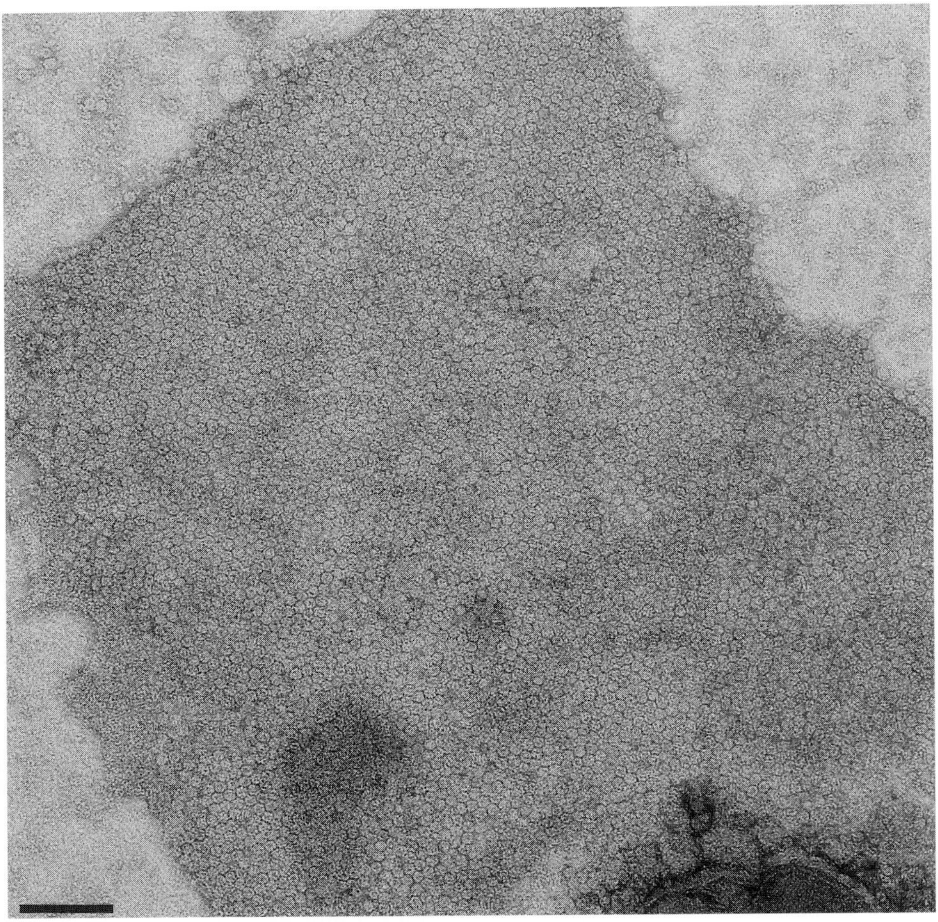

FIG. 6. Electron micrograph of a negatively stained 2D crystal of a histidine-tagged Fab fragment.[100] Such crystals are grown by lipid monolayer crystallization. The hexagonal lattice is clearly visible, and separate, circular Fab oligomers can be seen at the edges of the crystal. The growth of such crystals exploits the specific attachment of the histidine-tagged Fab molecules to lipids containing a nickel headgroup. Bar: 1000 Å.

9. EM grids with a hydrophobic carbon substrate are used since the attachment occurs via the lipid aliphatic chains.
10. EM forceps are used to carefully place the grid on top of the droplet with the carbon side facing the droplet surface.
11. For some systems, a clue whether crystals have formed is that the grid will not rotate or spin when placed on the droplet.

12. The grid is immediately removed from the droplet, and prepared for microscopy.
13. Alternatively, the surface of the droplet can be picked up with a platinum/paladium wire loop via surface tension.[69] The loop has a diameter slightly larger than the EM grid, so that transfer to the grid is accomplished by carefully passing the grid through the loop. This process presumably reduces stress on the crystal during transfer from the air/water interface, thereby reducing deformation and breakage.

*Transmission Electron Cryomicroscopy*

1. Transfer the 2D crystals to the EM grid:

a. By centrifugation of an aliquot of the membrane suspension at 1000 g for 5 min (used for gap junction crystals, Fig. 7).

b. By pipetting a 5-$\mu$l droplet of the sample onto the grid (used for AQP1).

c. As noted above for lipid monolayer crystallization.

2. Prepare frozen-hydrated specimens. (We use an ethane slush as the cryogen because its freezing rate of about $10^6$ deg/sec ensures that the buffer will be vitrified.[29]) To prevent evaporation and concentration of salts during blotting and plunging, high humidity is maintained by working in a cold room or placing the cryoplunger in a controlled humidity chamber.

a. Attach EM forceps to the drop rod, and adjust the position of the Dewar that holds the ethane cup so that the grid will fall into the center of the cup (Fig. 8). If the cryoplunger is not in an enclosed chamber, the grid should be positioned fairly close to the ethane cup so that evaporation does not occur during free fall of the plunger. Now fill the Dewar with liquid nitrogen.

b. Gently blow a stream of ethane gas via a Pasteur pipette into the cup. The ethane will first liquify and then solidify (Fig. 8b, item 1). (**Warning:** Wear safety glasses and be aware that ethane/air mixtures are explosive! Grid freezing should be conducted away from open flames in a well-ventilated, spark-free environment. Avoid any contact with liquid ethane because it will immediately cause severe burns.)

c. A transfer cup with an attached wire loop (Fig. 8b, item 4) is placed in the first Dewar and filled with liquid nitrogen. This transfer well is used to store the grids temporarily until you are ready to fill the grid boxes.

d. A second Dewar (Fig. 9a, item 5) is filled with liquid nitrogen and is used to hold the 50-ml conical plastic tube in which the grid boxes are stored.

[69] F. J. Asturias and R. D. Kornberg, *J. Struct. Biol.* **114,** 60 (1995).

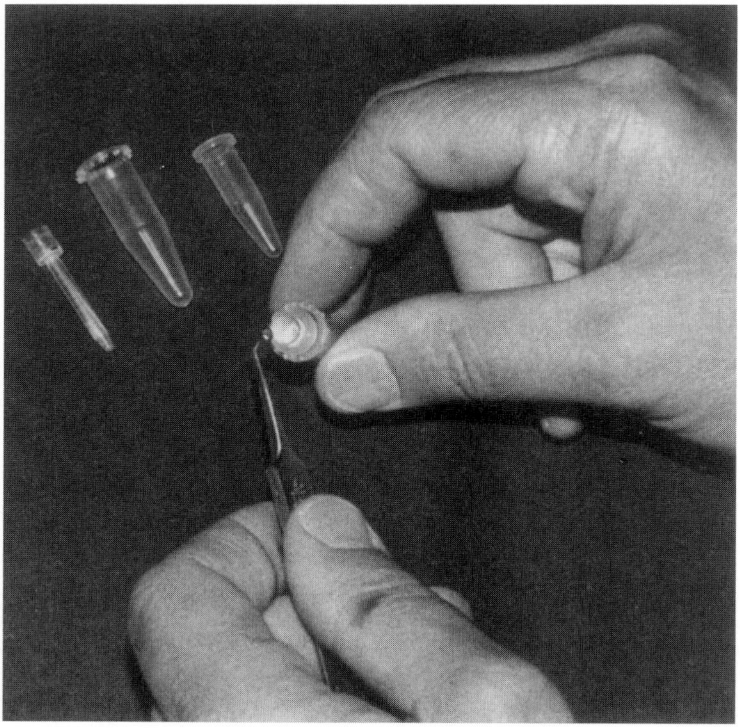

FIG. 7. Plastic buckets for centrifugation of samples onto EM grids. The snap lids of 0.5- and 1.5-ml Eppendorf tubes are removed with a razor blade. The bucket is inserted into the 0.5-ml tube, which is inserted into a 1.5-ml tube, and the system then fits into the rotor of a benchtop microfuge. The buckets accommodate 50–75 $\mu$l of sample, and the grid is placed in the bucket as shown.

e. Pick up the EM grid that has the sample applied, and attach the forceps to the drop rod.

f. Melt the solidified ethane by gently inserting a scalpel blade into the ethane cup until the rod or blade touches the bottom of the cup. Alternatively, the ethane can be warmed by blowing a gentle stream of gaseous ethane into the cup using a Pasteur pipette.

g. If the sample is applied to the EM grid as a droplet, the sample can be pipetted onto the grid with the forceps already attached to the drop rod, and the sample is allowed to settle for ~90 sec. For specimens in which the crystals are pelleted onto the grid, add a ~4-$\mu$l droplet of buffer to the grid so that it does not dry during positioning of the forceps on the drop rod.

h. Blot the grid almost to dryness by pressing a strip of Whatman (Clifton, NJ) #2 filter paper onto the grid. The blotting time will have to be optimized for your specimen in order to have a layer of ice that is thin

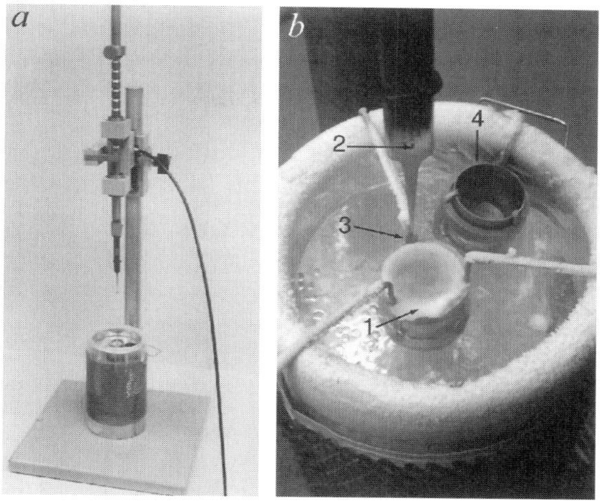


FIG. 8. (a) The original cryoplunger used by Nigel Unwin in his analysis of frozen-hydrated 2D crystals of gap junctions[101] and the acetylcholine receptor.[102] (b) The metal cap from a culture tube (1) is inverted and supported by a wire tripod frame in a Dewar filled with liquid nitrogen. Ethane gas is blown into the inverted cap, which liquefies and then solidifies. A metal rod or ethane gas is used to just melt the solidified ethane so that it is slushy. The EM forceps (2) and grid (3) are attached to a weighted rod, which is centered above the ethane cup. The grid is blotted face-on almost to dryness with filter paper. This can be assessed by the capillary spread of the liquid onto the paper. As the wet circle stops growing in diameter, the paper is removed from the grid, and the grid is immediately plunged into the ethane slush via a foot peddle, which releases a lever holding the weighted rod. The frozen grids are placed in a transfer cup (4) before placement in grid boxes for storage. Elaborations of this basic design allow application of ligands or a change in buffer conditions just before the grid enters the ethane slush.[103] In this way, dynamic states of macromolecular complexes can be trapped. For instance, this approach has been used to capture the intermediates in the photocycle of bacteriorhodopsin,[104] the open state of the acetylcholine receptor,[34] conformational changes that occur with pH activation of viral surface fushion glycoproteins,[105] and conformational states of myosin.[106]

enough for the electron beam to penetrate but thick enough so that the entire specimen is preserved in vitrified buffer. Note that filter paper may contain trace amounts of chemicals such as divalent cations that may affect the structure of your specimen during blotting. If freezing is done in a cold room, blotting is sometimes optimized by preheating the filter paper using a coffee cup warmer to ensure that the filter paper is dry.

i. Just as you remove the filter paper from the grid, immediately press the foot pedal of the cryoplunger to release the drop rod.

j. Detach the forceps from the rod and quickly insert the grid into the transfer cup filled with liquid nitrogen.

k. After freezing four grids, the transfer cup is emptied into a Styrofoam boat that is filled with liquid nitrogen.

l. The lid of a gridbox (Fig. 9b, item 15) is unscrewed with a screwdriver (Fig. 9a, item 7), and the grid box is placed in the boat.

m. In the Styrofoam boat under liquid nitrogen, EM forceps (Fig. 9a, item 8) are used to transfer the grids into the slots of the grid box.

n. When the four grids have been loaded in the grid box, hold the grid with dissecting forceps (Fig. 9a, item 9), and use a screwdriver to tighten the lid of the grid box lid. Do not overtighten the plastic screw because it is fragile at liquid nitrogen temperatures and will break off.

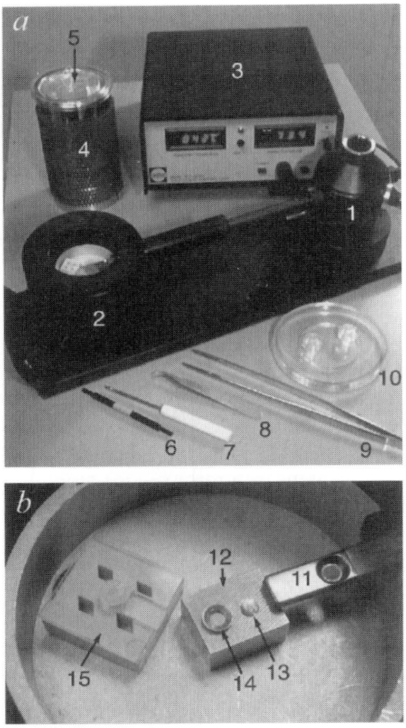


FIG. 9. (a) A Gatan cryostage (1), cryotransfer system (2) and temperature control unit (3). (b) Closeup view of the cryotransfer stage. The numbered items are (4) Dewar filled with liquid nitrogen; (5) 50-ml conical plastic tube for long-term storage of grid boxes; (6) clip ring tool that is used to secure the grid in place at the tip of the cryostage; (7) screwdriver used to loosen and tighten the screw in the lid of the grid box; (8) EM forceps for transferring the grid from the grid box to the cryotransfer system; (9) large forceps for transferring the grid box from the 50-ml conical tube to the cryotransfer system; (1) Plexiglass lid that covers the grid mounting chamber; (11) tip of the cryostage showing the circular well that holds the grid; (12) aluminum block that is positioned under the tip of the cryostage to prevent bubbling of nitrogen into the open well, which makes the grid "dance," thereby impeding rapid deployment of the clip ring; (13) EM grid; (14) clip ring that locks the grid into the well at the tip of the cryostage; (15) boxes that accommodate 4 grids that are prepared by cutting standard 24-grid boxes from Pellco (Redding, CA). A threaded hole is drilled in the center of the grid box to attach a plastic lid.

o. The grid box is then placed in a 50-ml Falcon conical plastic tube (Fig. 9a, item 5) for long-term storage in large ~15-liter Dewars. To facilitate retrieval of the tube from the cane of the storage Dewar, two holes are punched in the cap of the tube through which a length of fishing line is tied. Attach a label to the end of the line identifying the specimen.

3. A cryotransfer system is used to transfer the grid to the cold stage. In our laboratory we use a modified Gatan cryotransfer system (Fig. 9).

a. The Dewar of the Gatan cryostage (Fig. 9a, item 1) should be evacuated by pumping at least overnight on a Gatan dry pumping station or equivalent turbomolecular pump.

b. The electron microscope should be aligned and ready for use with the anticontaminator precooled with liquid nitrogen.

c. Use the attached fishing line to transfer the 50-ml conical storage tube with the grid boxes (Fig. 9a, item 5) from the 15-liter Dewar flask to a Dewar (Fig. 9a, item 4) filled with liquid nitrogen.

d. Fill a Styrofoam boat with liquid nitrogen.

e. Hold the storage tube with your gloved hand, unscrew the cap, and pour the liquid nitrogen and the grid boxes into the Styofoam boat.

f. Insert the cryoholder into the cryotransfer system (Fig. 9a, item 2), and carefully position a 1-cm aluminum block under the tip of the holder (Fig. 9b, item 12) to support the tip when the grid is secured in position (Fig. 9b, item 11).

g. Fill the well in the cryotransfer system and the Dewar on the cryoholder with liquid nitrogen. Cover the well with the Lucite disk (Fig. 9a, item 10) while the holder cools down.

h. Attach the heater (Fig. 9a, item 3) to the Dewar on the specimen holder and turn on the main switch to monitor the temperature. Be sure to keep the heater switch off.

i. When the temperature has reached −190°, unscrew the lid of the grid box and then very quickly transfer the box from the Styrofoam boat to the well in the cryotransfer unit. This transfer should be fast to minimize condensation, which may contaminate the grids.

j. The grid tends to "dance" less in the well at the tip of the cryostage if the liquid nitrogen level is kept just slightly above the tip of the specimen holder. Be careful not to knock the tip of the holder since the metal is soft and the tip is quite brittle when cooled down.

k. Screw the aluminum clip ring (Fig. 9b, item 14) that holds the grid in place onto the threaded end of the clip ring tool (Fig. 9a, item 6), make sure the aluminum cube (Fig. 9b, item 12) is under the tip for support, and press fit the ring over the grid. The tip of the handle of the attachment tool is cooled in liquid nitrogen and is then used to press the ring into place to ensure a snug fit against the grid. (If not snug, the grid may drift during examination.)

l. While waiting for the temperature of the cold stage to stabilize, the tip of the cryoholder is covered with a sliding brass flange to prevent contamination.

m. Cycle the rotary pump on the electron microscope so that it does not turn on during insertion of the cryostage.

4. Insert the cold stage into the microscope (Fig. 10).

a. This step is critical because the cryostage is a very expensive, delicate piece of equipment, and the cryotransfer needs to be done quickly to minimize condensation on the tip of the specimen holder during insertion into the microscope, which will contaminate the grid with cubic and hexagonal ice.

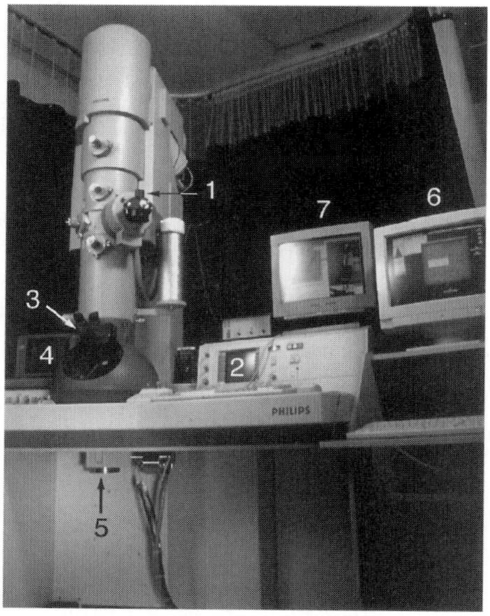


FIG. 10. Philips CM200FEG transmission electron microscope equipped with a Gatan cold stage (1) for electron cryomicroscopy at about $-180°$. The field emission gun provides an extremely bright and coherent source for high-resolution imaging, nominally at 1.2-Å resolution. The control screen (2), the keys, the push buttons and the knobs on the front panel of the microscope give access to the different functions of a computer system which keeps track of all operator commands and directs them to the appropriate device associated with the selected mode. The microcontroller screen (2) serves to provide information about the status of the microscope and also accepts operator commands. XY translation and Z height are controlled with a joystick. The image of the specimen is viewed via binoculars (3) on a fluorescent viewing screen. The settings for beam intensity, magnification, astigmatism, lens currents, etc., are under computer control and can be downloaded from a laptop computer (4) and provide ease of use for low-dose microscopy. The Gatan multiscan $1024 \times 1024$ charge-coupled device (CCD) camera (5) is controlled on-line via computer (6) and is particularly useful for recording electron diffraction patterns.[107,108] With the current technology the optimal resolution for recording of CCD images from beam sensitive biologic specimens is about 10 Å. Therefore, high-resolution images are still recorded on photographic film. Because the microscope is under computer control, automated data collection is feasible via a remote microscopy server (7).

b. For cryostages that are inserted through a side port of the microscope column (e.g., Gatan and Oxford Instruments cryostages), the stage has to be rotated during insertion. During this process liquid nitrogen can spill out of the Dewar of the cold stage. The window on the viewing chamber and the surface of the microscope should therefore be covered with a towel to prevent liquid nitrogen from coming into direct contact with surfaces on the microscope.

c. During insertion of a stage into the microscope, the airlock is evacuated for a defined period of time. We tend to use as short a time as possible to prevent condensation of water which will contaminate the grid. The pumping time must still be sufficient so that the vacuum in the column of the microscope does not deteriorate when the stage is inserted into the column.

d. Liquid nitrogen in the Dewar of the cold stage often boils. To eliminate nitrogen bubbling, a hollow tube equipped with a stopper can be inserted into the nitrogen for a few seconds to bleed off the nitrogen gas.

5. The microscope is set for low-dose cryomicroscopy in which the field being photographed only receives a small dose of electrons (5–10 $e^-/\text{Å}^2$) during the photographic exposure. This is accomplished by performing microscopy under three conditions[70] (Fig. 11):

a. *SEARCH mode:* In search mode the grid is scanned with a beam about one-quarter the size of one grid square to locate regions to photograph. Overfocused diffraction mode is preferred over simply viewing the grid at low magnification with a dim beam because the acceptable brightness is much higher without increasing the electron dose. Hence, it is easier to identify the specimen and evaluate the ice thickness and regions contaminated with hexagonal or cubic ice. Hexagonal ice can be identified by its characteristic hexagonal crystal habit. In diffraction mode, hexagonal ice will generate sharp diffraction spots. Cubic ice is mosaic in appearance and generates a powder pattern when examined in diffraction mode.

b. *FOCUS* mode: In focus mode, a region of the carbon substrate is examined at high magnification (~220,000×) in order to find Gaussian focus, correct for astigmatism, and to check that there is no specimen drift. In focus mode, the beam is deflected about one beam diameter (~3 $\mu$m for an exposure mag of ~45,000×) from the area to be photographed. For examining tilted crystals, it is critical that the beam translation be set parallel with the tilt axis of the grid. On Philips electron microscopes the focus mode allows translation (designated S1 and S2) to either side of the area to be photographed so that either translation can be selected if you happen to translate the beam over a grid bar. In addition, assessing the level of focus on either side of the area to be photographed allows more precise determination of true focus.

[70] R. C. Williams and H. W. Fisher, *J. Mol. Biol.* **52,** 121 (1970).

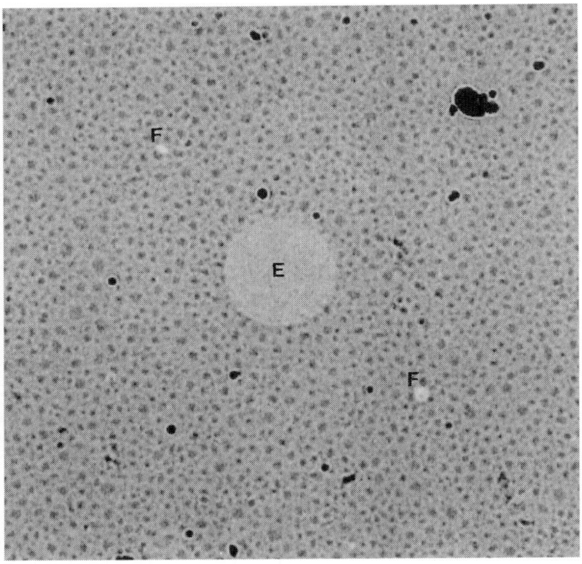


FIG. 11. A low-magnification image of a frozen-hydrated grid showing the technique for low-dose electron microscopy. In focus mode (F) the beam is deflected to a region of carbon on either side of the area to be photographed so that the carbon grain can be examined with a bright beam at high magnification (220,000×) in order to adjust the level of defocus, correct for astigmatism, and check for specimen drift. The translations in focus mode are parallel with the tilt axis of the grid. The exposure (E) is taken at a magnification of ×30,000 to ×60,000 so that the electron dose is 5–10 $e^-/Å^2$. The cobblestone appearance in the background is due to boiling of the ice during this high-dose exposure. The black deposits in the background are contaminating ice crystals.

c. *EXPOSURE mode:* Compared to alloys examined in materials science, biologic specimens are exquisitely sensitive to electron radiation damage. This limits the maximum magnification that can be used to ~60,000×, and images are typically recorded at 30,000× to 60,000×. During low-dose microscopy the specimen is only illuminated during the ~1-sec exposure so that the dose is ~5–10 $e^-/Å^2$. For these conditions the optical density on Kodak film SO163 will be about 1 when the film is developed in full-strength D19 developer for about 10 min. Electron diffraction patterns are typically recorded at a camera length of ~1.7 m using ~20 sec exposures, with a much lower cumulative dose of ~0.4 $e^-/Å^2$.

6. Scan the grid and record images. The Philips electron microscopes have three touch keys that allow shifting from search mode to focus mode to exposure mode.

a. Press the SEARCH key and scan the grid by viewing the image on the fluorescent screen with binoculars or via an attached television camera.

b. Carefully center an area to be photographed. The pointer in the microscope can be inserted for precise centering.

c. Press the FOCUS key to examine the carbon grain in both S1 and S2 positions (Fig. 11). To locate Gaussian focus, view the grain both under- and overfocussed with finer and finer increments of focus. Because contrast is minimized at true focus, the texture of the carbon grain will appear smoothest at true focus (Fig. 12). True focus is located in both the S1 and S2 positions, which may differ slightly. The best estimate for true focus is therefore set at half of the difference between the values for true focus in S1 and S2. Adjustments are also made to correct for astigmatism (Figs. 12 and 13). Having found true focus, the beam is underfocused a desired amount (e.g., −5000 Å using 200-kV electrons). If the microscope is not aligned to be parfocal, an additional defocus difference will occur when shifting from 220,000× in focus mode to the exposure magnification. This shift can sometimes be several thousand ångstoms and must be taken into account when setting the defocus value for the exposure.

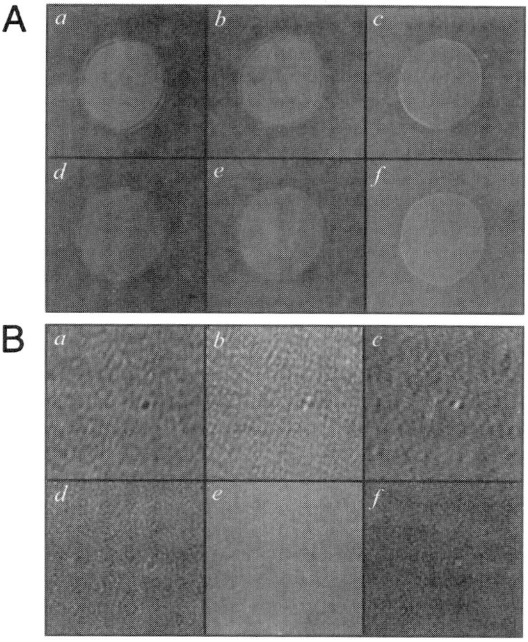


FIG. 12. True focus and astigmatism are determined by evaluation of the carbon grain. Astigmatism is not corrected in *a*, *b*, and *c*, and is corrected in *d*, *e*, and *f*. The images in *a* and *d*, *b* and *e*, and *c* and *f* were recorded at underfocus, true focus, and overfocus, respectively. (A) Astigmatism is easily seen by the asymmetric Fresnel fringes at the edges of a hole in the carbon. Note that the asymmetric arcs are in orthogonal directions when the image is underfocused versus overfocused (compare *a* and *c*). (B) Astigmatism causes the carbon grain to be stretched into "line foci" that are in orthogonal directions when the image is underfocused versus overfocused (compare *a* and *c*). Reproduced with permission of Philips Electron Instruments.

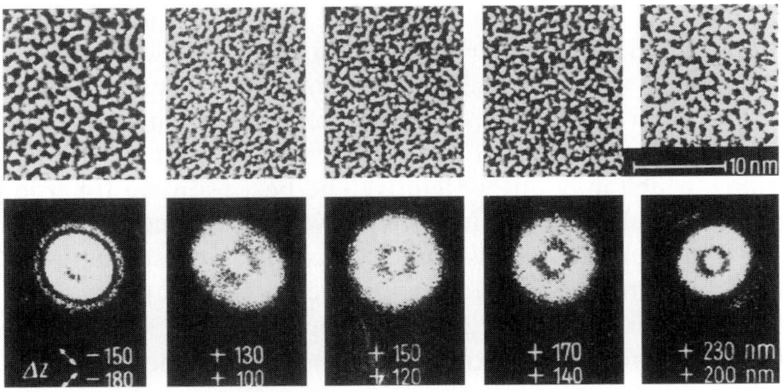


FIG. 13. Electron micrographs and optical transforms of an untilted carbon support film, showing various image defects: (a) specimen drift, (b) misaligned objective aperture generating astigmatism, (c) and (d) charging of the objective aperture. (Reproduced from Ref. 77 with permission of Springer-Verlag.) Note the corresponding distortions in the Thon rings of the optical diffraction patterns that can be used as a diagnostic aid to assess astigmatism and drift. The Thon rings are a consequence of the contrast transfer function (CTF) of the electron microscope, which results from objective lense aberrations such as astigmatism and spherical aberrations, as well as defocus.[109] To achieve contrast in unstained, frozen-hydrated images, micrographs are recorded under conditions where the electron beam is underfocused. The location of the nodes in the Thon rings can be used to determine the level of defocus. Precise determination of the level of defocus is important because phases of the reflections alternate by 180° between successive peaks in the CTF. The CTF is a periodic sin function, and the defocus value corresponding to the first node in the CTF will be given by $\Delta F = [C_s \nu^4 \lambda^3 + 2]/2\ \nu^2 \lambda$, where $C_s$ is the spherical aberration of the objective lens, $\lambda$ is the electron wavelength, and $\nu$ is the spatial frequency. Because the values of $C_s$ and $\lambda$ are very small (usually 2 mm and 0.037 Å for 100-kV electrons), a simple estimation for the defocus level of a micrograph is given by $1/\nu^2\lambda$, where the location of the first CTF node ($\nu$) in the diffraction pattern can be calculated from the display of the Fourier transform.

d. While still in focus mode, check that the specimen is not drifting (Fig. 13), and only then should the photograph be taken. Because the microscope is sensitive to vibration, keep your hands off the microscope during the exposure, do not talk, and sit still.

e. Illumination of the specimen during the exposure may induce movement, which degrades the resolution of the image. This may be overcome by recording the exposure using a focused beam of ~1000 Å (compared with a flood beam of 2–3 μm), which is scanned over the specimen to record the image.[71] Such "spot-scan imaging" can be combined with "dynamic focusing" in which the level of defocus is adjusted during the exposure to correct automatically for the changes in defocus that occur when recording images from tilted specimens.[72]

[71] K. H. Downing, *Science* **251,** 53 (1991).
[72] K. H. Downing, *Ultramicrosc.* **46,** 199 (1992).

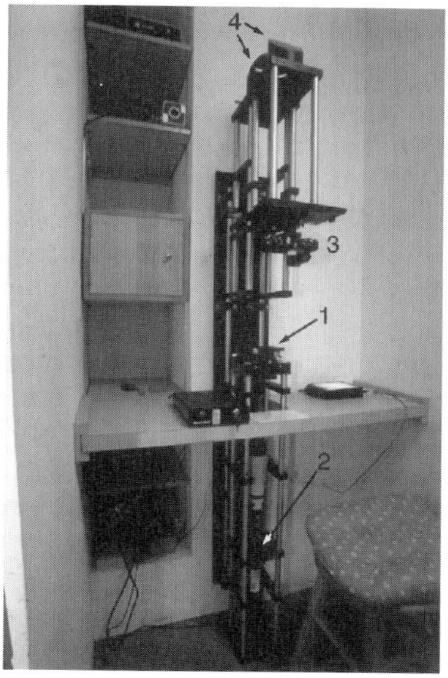


FIG. 14. An optical diffractometer is used to evaluate the quality of the 2D crystals. The photographic negative is placed on the stage (1), which is illuminated with coherent laser light (2). A 35-mm camera (3) can be used to record images or to view the negative via an eyepiece. By removing the objective lens, the diffraction pattern from the negative can be visualized. The mirrors (4) allow a folded design for upright placement in a compact space.[110]

f. The area that you just photographed can now be viewed by pressing the EXPOSURE key.

g. Press the SEARCH key and repeat steps b to e for the next area to be photographed.

*Image Appraisal*

1. Examine EM negatives on a lightbox to assess whether the ice is too thick, whether there is contamination on the grid, or whether the crystal is broken.
2. Evaluate the negatives using an optical diffractometer (Fig. 14) to check whether the crystal was drifting during the exposure. Select the crystals that display sharp, bright reflections to the highest resolution.
3. Digitize crystalline patches identified by optical diffraction using a step size that is appropriate for the resolution expected in the image (Fig. 15). According to principles of signal processing, the Nyquist limit states that if a continuous signal is to be reproduced with fidelity, then the signal has to be sampled at twice the highest frequency in

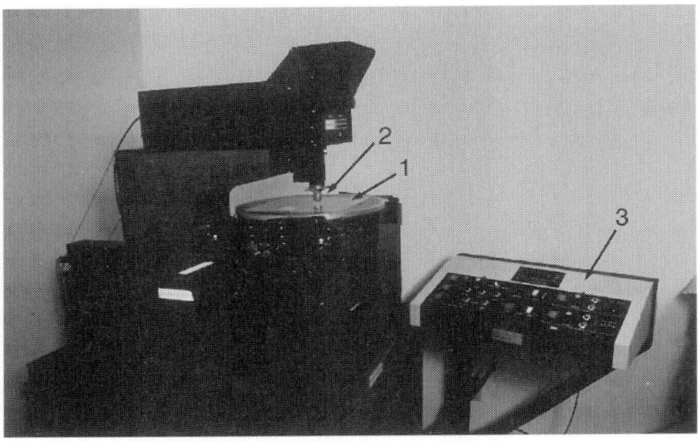


FIG. 15. Images from the best crystals are digitized using a Perkin-Elmer microdensitometer, which generates an XY array of optical densities. The negative is held in place on the glass plate (1) using a suction device. Lenses (2) are located above and below the negative. The instrument has the capability for scanning in 2-$\mu$m increments and has a very high dynamic range. The control panel (3) is interfaced to a computer for automated data collection.

the signal.[73,74] This concept has also been referred to as the Whittaker–Shannon sampling theorem.[75,76] If we anticipate a resolution in the image of ~5 Å, then the image should be sampled with a step size of at least 2.5 Å. For example, micrographs recorded at 45,000 magnification are digitized using a step size of 10 $\mu$m, corresponding to 2.17 Å on the specimen [(10 $\mu$m/46,000) $\times$ ($10^4$ Å/$\mu$m)].

### *Image Processing*

Image processing is an approach to derive an enhanced 2D or 3D reconstruction of a structure based on the analysis of images recorded from 2D projections.[12,13,77–80] For the analysis of 2D crystals recorded by electron

[73] A. Watt and M. Watt, "Advanced Animation and Rendering Techniques, Theory and Practice," p. 112. Addison-Wesley, Wokingham, 1992.

[74] P. A. Lynn and W. Fuerst, "Introductory Digital Signal Processing with Computer Applications," p. 8. John Wiley & Sons, Chichester, 1994.

[75] D. Sayre, *Acta Cryst.* **5,** 843 (1952).

[76] J. E. Mellema, *in* "Computer Processing of Electron Microscope Images" (P. W. Hawkes, ed.), p. 89. Springer-Verlag, Berlin, 1980.

[77] D. L. Misell, *in* "Image Analysis, Enhancement and Interpretation—Practical Methods in Electron Microscopy" (A. M. Glauert, ed.), p. 40. North-Holland Publishing, Amsterdam, 1978.

[78] M. Moody, *in* "Biophysical Electron Microscopy, Basic Concepts and Modern Techniques" (P. W. Hawkes and U. Valdrè, eds.), p. 145. Academic Press, London, 1990.

[79] J. C. Russ, "The Image Processing Handbook," p. 165. CRC Press, Boca Raton, 1992.

[80] B. E. P. Beeston, R. W. Horne, and R. Markham, *in* "Electron Diffraction and Optical Diffraction Techniques—Practical Methods in Electron Microscopy" (A. M. Glauert, ed.), p. 318. North-Holland Publishing, Amsterdam, 1972.

microscopy, there are four aspects to the enhancement: (1) the images are recorded from 2D crystals that represent several thousand copies of the molecular structure; (2) the spots in the diffraction pattern represent the signal, whereas the noise in between the spots can be removed by filtering; (3) the resolution of the data can be extended by methods to straighten or "unbend" the crystal lattice; and (4) the packing of the molecules in the crystal is based on a certain symmetry, which can be enforced in the density map.

We use the MRC suite of programs for image processing of 2D crystals[81–88] which can be operated on DEC Alpha, SGI, or other workstations. A discussion of each program is beyond the scope of this review, but they are described in general in Table II, and their implementation will be described in detail.[89] A general flowchart of the steps with the corresponding programs is shown in Fig. 16, and a summary of the steps is as follows.

1. Use HISTO to generate a histogram of the optical densities in the image.

2. TAPEREDGE is used to wrap around the optical densities at the edges of the image to prevent central spikes in the Fourier transform that arise from the discontinuity in density at the opposite edges of the image.

3. For large images LABEL is used to average adjacent pixels to speed up processing in the refinement of the crystal lattice.

4. FFTRANS is now used to compute the Fourier transform, which can be displayed using XIMDISP (Fig. 17).

5. XIMDISP can also be used to index the reflections in the computed diffraction pattern and iteratively refine the crystal lattice parameters.

6. Correct for crystal lattice distortions (Fig. 18).

a. Select high signal-to-noise reflections that will be used to generate a filtered image.

b. MASKTRANA masks the selected diffraction spots from the background, and a filtered image is generated by inverse Fourier transformation.

c. The image is first filtered tightly using a small box of pixels that includes each spot. The filtered image is then visualized using XIMDISP

[81] R. A. Crowther, R. Henderson, and J. M. Smith, *J. Struct. Biol.* **116,** 9 (1996).

[82] R. A. Crowther and U. B. Sleytr, *J. Ultrastruct. Res.* **58,** 41 (1977).

[83] D. A. Agard, *J. Mol. Biol.* **167,** 849 (1983).

[84] J. Baldwin and R. Henderson, *Ultramicrosc.* **14,** 319 (1984).

[85] R. Henderson, J. M. Baldwin, K. H. Downing, J. Lepault, and F. Zemlin, *Ultramicrosc.* **19,** 147 (1986).

[86] W. A. Havelka, R. Henderson, J. A. W. Heymann, and D. Oesterhelt, *J. Mol. Biol.* **234,** 837 (1993).

[87] J. M. Valpuesta, J. L. Carrascosa, and R. Henderson, *J. Mol. Biol.* **240,** 281 (1994).

[88] W. A. Havelka, R. Henderson, and D. Oesterhelt, *J. Mol. Biol.* **247,** 726 (1995).

[89] V. M. Unger, A. Cheng, and M. Yeager, in preparation.

TABLE II
PROGRAMS USED FOR PROCESSING IMAGES OF 2-D CRYSTALS

| | |
|---|---|
| Image/Map Display and General Processing | |
| XIMDISP | General program to display images, Fourier transforms, cross-correlation maps, etc. |
| HEADER | Prints out information in header record |
| LABEL | Performs various image manipulations such as pixel averaging to reduce image size |
| HISTO | Generates histogram of densities in image |
| FFTRANS | Computes fast Fourier transform |
| BOXIMAGE | Masks selected area within a digitized image |
| TAPEREDGE | Tapers density at edge of image to remove central spikes in transform arising from image boundary |
| TWOFILE | Performs linear combination, or multiplies/divides data in two files |
| Processing of images of two-dimensional crystals | |
| EMTILT | Calculates tilt angles from lattice parameters of 2D crystal |
| MASKTRANA | Masks the Fourier transform in preparation for filtering |
| AUTOCORRL | Performs an autocorrelation calculation and expansion |
| QUADSERCHB | Searches cross-correlation map for the position and height of cross-correlation peaks by profile fitting |
| CCUNBENDD | Corrects image for distortions in the 2D crystal lattice |
| MMBOX | Provides amplitudes and phases from Fourier transform |
| TTBOX | Corrects for tilt transfer function and writes out amplitudes and phases |
| TTMASK | Combines MASKTRAN and TTBOX for processing images of highly tilted crystals |
| CTFAPPLY | Applies contrast transfer function to phase data from MMBOX |
| ORIGTILTD | Determines phase origins, refines tilt geometries, and merges data |
| LATLINED | Performs a least-squares fit to the amplitude and phase data from tilted crystals to derive lattice line |
| LATLINPRESCALE | Corrects image amplitudes for contrast transfer function |
| PREPMKMTZ | Removes unreliable lattice line points from the 3D data set and readjusts figures-of-merit as desired |

to select a region that is visually judged to be the most crystalline. This region is later boxed as the reference area for unbending the lattice.

d. AUTOCORRL is then used to generate an autocorrelation map from a part of the tightly filtered image that is centered around the center of the reference.

e. XIMDISP is used to view the autocorrelation map in order to determine the shape and extent of the profile that will be used for searching the cross-correlation map.

f. MASKTRANA is then used a second time to loosely filter the original image, which will contain the information for correcting imperfections in the lattice.

g. The cross-correlation transform is then generated by TWOFILE, which multiplies the loosely masked image Fourier transform by the com-

TABLE II (*continued*)

| | |
|---|---|
| ALLSPACE | Determines the plane group, origin and beam tilt on a single image of an untilted crystal |
| AVRGAMPHS | Averages projection amplitudes and phases from multiple images |
| SCALIMAMP3D | Calculates inverse B factors for individual images and rescales image-derived amplitudes |
| General utilities programs | |
| FOMSTATS | Adjusts the figure-of-merit values from AVRGAMPHS for derivations from the theoretical phase (projection data only), calculates statistics in resolution bins and overall phase error, adjusts the phase for two-fold constraints, and rejects forbidden reflections |
| PLOTALL | Generates a plot of the phase error for individual Fourier terms of projection data |
| CHINDEXING | Reindexes structure factor lists obtained by MMBOX (useful for images of more highly tilted images where indexing of the original transform can be ambiguous) |
| IDEAL | Generates ideal p1 reference data set for calculating the point spread function |
| PNTSPREAD | Reads the 3D data file and resets the amplitudes to unity and the phases to 0 in order to estimate the resolution |
| Programs to calculate and contour maps | |
| F2MTZ | Converts input data of diverse formats to standard CCP4 *mtz* format |
| FFT | Crystallographic fast Fourier transform program |
| EXTEND | Extends maps/images to multiples of the unit cell |
| N-PLUTO | General contouring and atomic model plotting program |
| O | Software package designed for visualization of density maps and docking atomic models |
| AVS | Software package from Adanced Visualization Systems for graphics rendering of maps: isosurfaces, transparencies, texture mapping, etc. |

plex conjugate of the reference Fourier transform. The inverse Fourier transform generates the cross-correlation map.

h. QUADSERCHB is then used to search the cross-correlation map for the location and height of peaks.

i. CCUNBENDD uses the position of the cross-correlation peaks to generate a map of vectors that provides the direction and magnitude for shifting the image density into best agreement with the density in the reference image. A corrected or "unbent" image is generated by applying these translations to the original image.

j. XIMDISP is now used to display the cross-correlation map to identify the region in best agreement with the reference image. The original image is then reboxed by including only those regions that have a certain correlation coefficient (as a general guideline, e.g., ≥75% for thick specimens or ~30% for thin specimens) with the reference image.

7. The process from steps 3 to 6 can be repeated in several variations to correct even finer lattice deviations in the crystal.

8. After boxing the area defined in step 6j, FFTRANS is then used to generate a transform of the corrected and boxed image, and MMBOX

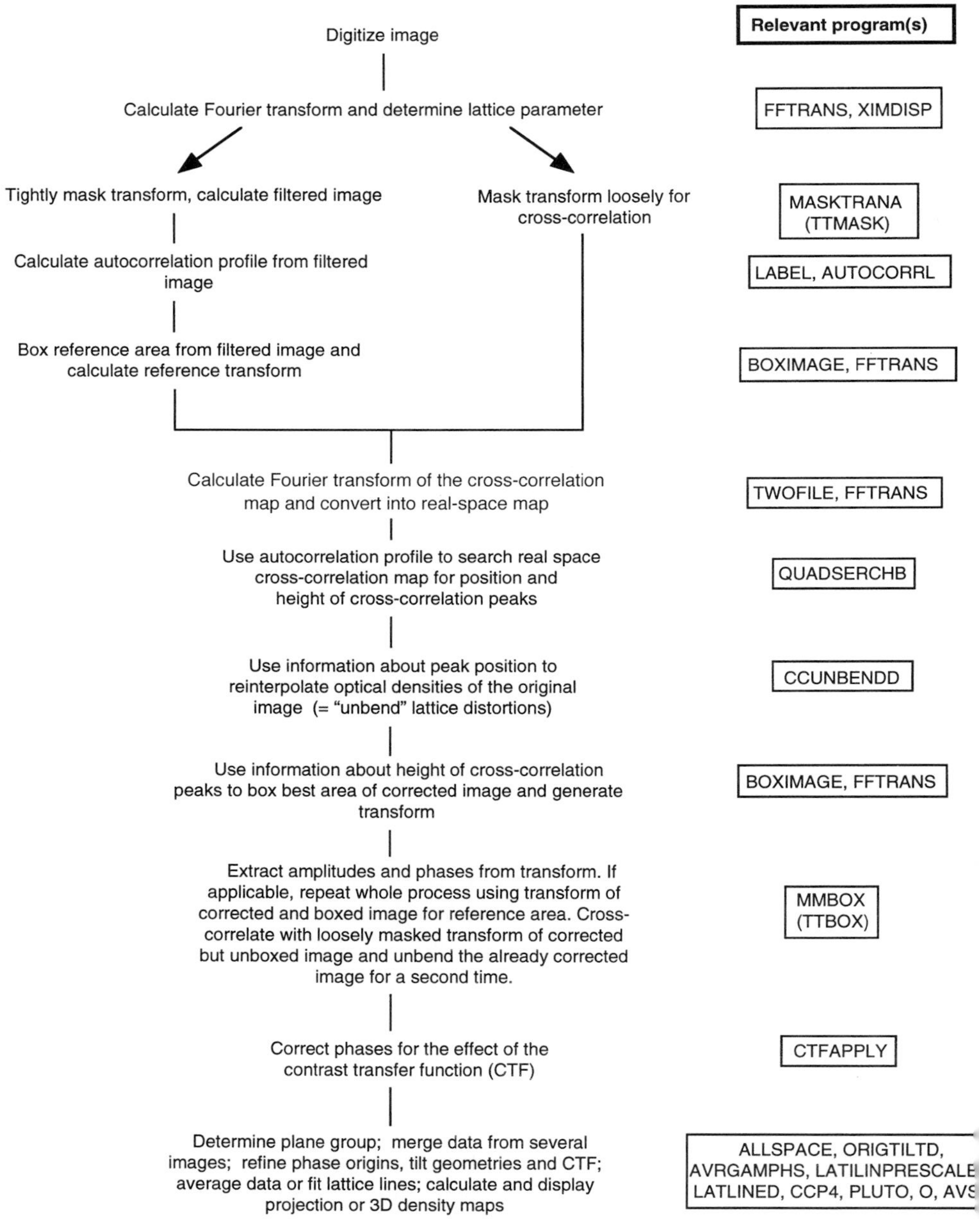


FIG. 16. An outline of the steps used for image processing of 2D crystals. The program titles in the MCR image processing suite are indicated.

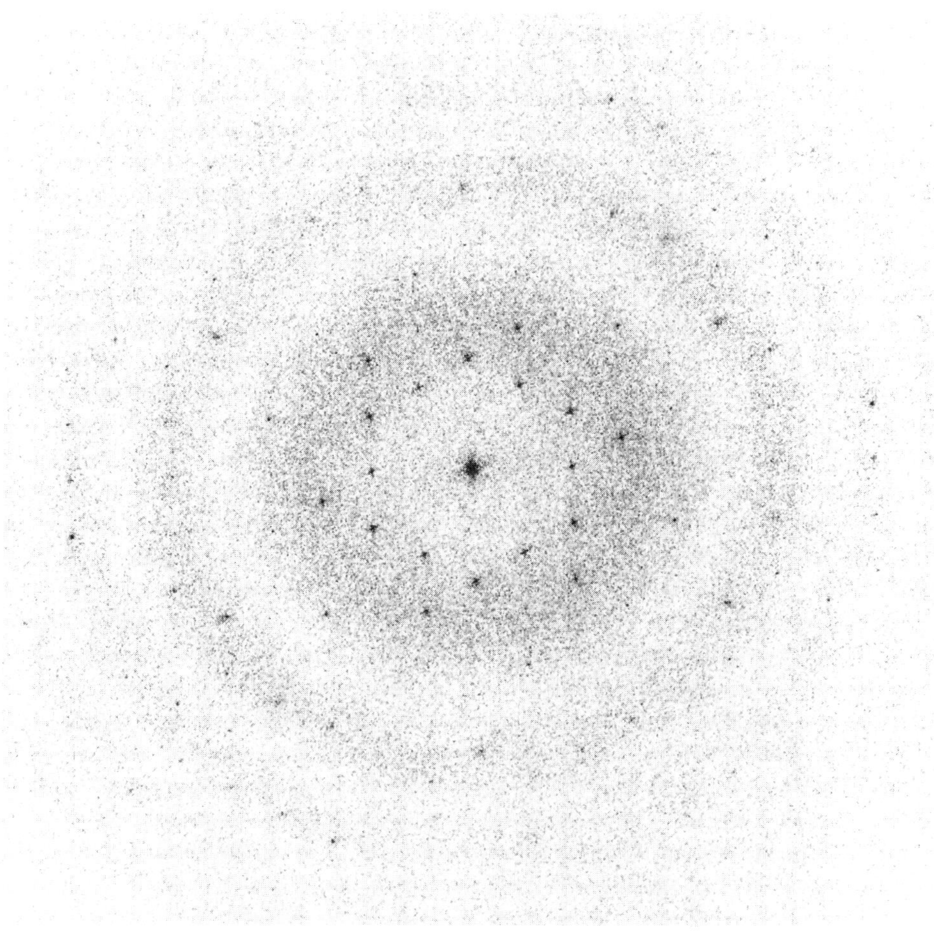

FIG. 17. Computed diffraction pattern from a frozen-hydrated 2D crystal of a recombinant, truncated form of rat heart $\alpha_1$-connexin ($\alpha_1$Cx263T). Spots at the outer edge of the pattern correspond to ~9.5-Å resolution. Note the diffuse rings of density in the background, referred to as Thon rings, that are due to the CTF.

provides a digital readout of the amplitudes and phases that can be used to assess the significance of the final data list and the success of unbending (Fig. 19).

9. Effects on the phases due to the contrast transfer function of the microscope are now corrected using CTFAPPLY (Fig. 20). The parameters needed in the correction are the electron wavelength (which will be dependent on the kilovolts of the microscope), the spherical aberration coefficient of the objective lens, the defocus level of the image, the dimensions of the Fourier transform, the crystal lattice parameters, the magnification, and

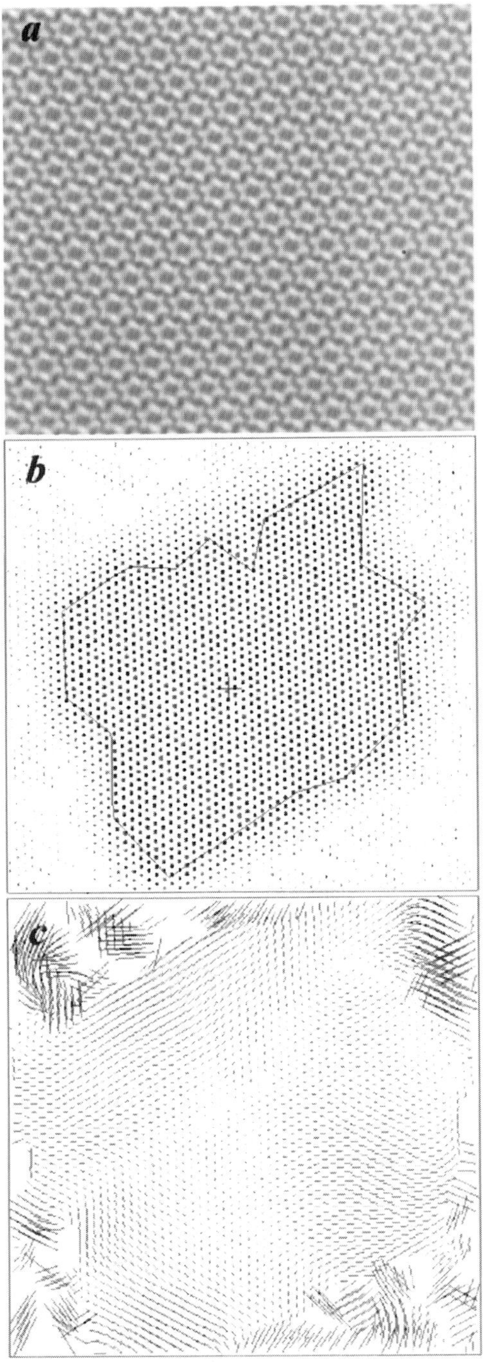
a
b
c

the densitometer scanning step size. Astigmatism will distort the Thon rings into ellipses. Examination of the Fourier transform using XIMDISP can be used to determine the approximate pixel location of the nodes in the Fourier transform along the major and minor axes of the ellipse with respect to the *x* axis of the transform.

10. In determining a 2D projection or a 3D map, all of the corrected Fourier transforms have to be refined to a common phase origin. Furthermore, for an image of a tilted crystal, the tilt geometry needs to be known. An estimate for the location of the tilt axis can be determined by optical diffraction (Fig. 14). The negative is examined using a large aperture to identify the axis along which the Thon rings do not vary in diameter. For images with a high degree of tilt, the tilt axis and tilt angle can be determined using EMTILT, which computes the tilt geometry based on the distortion in the lattice. However, the accuracy of this approach decreases as the tilt angle decreases below ~30°, and for small tilt angles also depends on the symmetry of the specimen. For images with small tilt angles, the tilt angle can be estimated by noting the shift in the nodes of the CTF on opposite sides of the tilt axis.

11. The final merging of the amplitude and phase data from multiple images of tilted crystals is performed using ORIGTILTD.

12. If the amplitudes are collected from electron diffraction patterns, then no corrections need to be made for CTF effects on the amplitude (Fig. 21), and only a background due to inelastic scattering needs to be subtracted.

13. If the amplitudes are computed by Fourier transformation of the images, then the amplitudes are modulated by the CTF, which is corrected using LATLINPRESCALE. In addition, image-derived amplitudes exhibit a resolution dependent fall-off due to drift and charging during the recording of images. This fall-off can be compensated by multiplying the higher resolution amplitudes with an inverse temperature factor.[46,89a,94d]

---

FIG. 18. Correction of crystal lattice distortions in a recombinant gap junction 2D crystal by the method of Henderson *et al.*[85] Strong reflections in the Fourier transform that are in good agreement with the lattice parameters are masked and used to calculate a filtered image at ~15-Å resolution (a). A region in the filtered image that appears to have the least distortion is selected as a reference area (6 × 6 channels in this case), and the cross-correlation function is calculated. The darker symbols in the cross-correlation map (b) indicate regions of high correlation between the reference and the region being compared in the crystal. The error map in (c) shows the translational offsets (×20) with respect to the reference area that are applied to "unbend" the image. Based on the cross-correlation map, the original image is reboxed to include only those areas that have a high cross-correlation coefficient with the reference area of ~75% or greater for generating the final corrected Fourier transform.

| Resolution Range | Before Unbending | | | | | | | | | After Unbending | | | | | | | | | IQ | Before | After |
|---|---|---|---|---|---|---|---|---|---|---|---|---|---|---|---|---|---|---|---|---|---|
| **80-15Å** | 10 | 12 | 13 | 8 | 8 | 7 | 7 | 7 | 8 | 10 | 10 | 16 | 8 | 8 | 6 | 7 | 7 | 8 | 1 | 11 | 14 |
| | 6 | 10 | 20 | 17 | 13 | 8 | 7 | 7 | 6 | 8 | 8 | 13 | 13 | 9 | 8 | 6 | 6 | 8 | 2 | 4 | 6 |
| | 11 | 12 | 22 | 22 | 29 | 16 | 8 | 21 | 9 | 9 | 11 | 22 | 27 | 10 | 6 | 8 | 13 | 7 | 3 | 4 | 3 |
| | 10 | 9 | 19 | 271 | 359 | 73 | 30 | 17 | 9 | 7 | 10 | 11 | 168 | 312 | 45 | 32 | 16 | 12 | 4 | 5 | 0 |
| | 10 | 19 | 37 | 319 | 692 | 284 | 37 | 14 | 10 | 9 | 18 | 32 | 319 | 978 | 242 | 28 | 13 | 8 | 5 | 2 | 2 |
| | 9 | 15 | 18 | 42 | 96 | 115 | 34 | 12 | 11 | 14 | 11 | 33 | 39 | 260 | 144 | 12 | 11 | 8 | 6 | 3 | 1 |
| | 6 | 7 | 8 | 6 | 24 | 25 | 20 | 15 | 10 | 7 | 7 | 7 | 5 | 12 | 34 | 17 | 14 | 6 | 7 | 2 | 2 |
| | 7 | 8 | 8 | 9 | 9 | 9 | 13 | 6 | 11 | 6 | 7 | 9 | 8 | 10 | 12 | 13 | 7 | 8 | 8 | 2 | 6 |
| | 5 | 6 | 7 | 8 | 8 | 11 | 9 | 9 | 8 | 6 | 6 | 7 | 7 | 8 | 12 | 11 | 8 | 8 | 9 | 3 | 2 |
| **15-10Å** | 8 | 8 | 9 | 11 | 12 | 9 | 8 | 8 | 9 | 6 | 7 | 11 | 8 | 7 | 8 | 6 | 7 | 10 | 1 | 0 | 6 |
| | 8 | 10 | 11 | 12 | 17 | 13 | 12 | 10 | 10 | 10 | 6 | 11 | 14 | 15 | 11 | 9 | 10 | 10 | 2 | 3 | 8 |
| | 10 | 8 | 9 | 11 | 15 | 15 | 13 | 9 | 9 | 10 | 7 | 10 | 13 | 11 | 14 | 17 | 12 | 9 | 3 | 9 | 7 |
| | 8 | 11 | 15 | 20 | 41 | 19 | 21 | 12 | 6 | 10 | 9 | 13 | 19 | 27 | 17 | 10 | 10 | 7 | 4 | 3 | 7 |
| | 7 | 7 | 17 | 15 | 29 | 26 | 13 | 10 | 8 | 9 | 9 | 12 | 35 | 68 | 33 | 12 | 8 | 6 | 5 | 3 | 4 |
| | 7 | 7 | 11 | 14 | 15 | 17 | 14 | 12 | 7 | 8 | 8 | 11 | 16 | 20 | 18 | 13 | 8 | 12 | 6 | 3 | 0 |
| | 8 | 7 | 8 | 11 | 13 | 15 | 16 | 14 | 6 | 7 | 7 | 11 | 15 | 11 | 16 | 13 | 10 | 11 | 7 | 2 | 1 |
| | 8 | 8 | 7 | 8 | 9 | 12 | 12 | 11 | 7 | 6 | 6 | 8 | 9 | 11 | 17 | 13 | 10 | 10 | 8 | 12 | 5 |
| | 8 | 4 | 8 | 6 | 7 | 7 | 8 | 9 | 6 | 8 | 6 | 6 | 6 | 8 | 10 | 11 | 7 | 7 | 9 | 10 | 7 |
| **9.9-7.0Å** | 7 | 9 | 9 | 9 | 9 | 8 | 9 | 8 | 7 | 7 | 7 | 9 | 8 | 7 | 8 | 8 | 9 | 7 | 1 | 0 | 3 |
| | 8 | 10 | 10 | 9 | 11 | 9 | 10 | 7 | 7 | 7 | 8 | 9 | 10 | 9 | 10 | 9 | 8 | 7 | 2 | 3 | 12 |
| | 8 | 8 | 8 | 11 | 22 | 18 | 14 | 10 | 8 | 9 | 8 | 9 | 7 | 9 | 8 | 8 | 8 | 8 | 3 | 2 | 9 |
| | 8 | 10 | 11 | 11 | 24 | 32 | 16 | 14 | 11 | 8 | 10 | 10 | 20 | 41 | 17 | 8 | 9 | 8 | 4 | 8 | 5 |
| | 7 | 8 | 9 | 13 | 18 | 25 | 17 | 13 | 9 | 7 | 9 | 10 | 25 | 68 | 36 | 10 | 8 | 8 | 5 | 11 | 6 |
| | 8 | 9 | 7 | 9 | 11 | 14 | 14 | 11 | 10 | 8 | 8 | 8 | 10 | 21 | 15 | 9 | 8 | 8 | 6 | 12 | 2 |
| | 6 | 8 | 8 | 9 | 10 | 9 | 10 | 10 | 9 | 7 | 8 | 10 | 7 | 9 | 10 | 9 | 8 | 8 | 7 | 5 | 7 |
| | 8 | 8 | 8 | 10 | 7 | 6 | 6 | 8 | 9 | 7 | 7 | 8 | 8 | 9 | 9 | 9 | 7 | 8 | 8 | 23 | 23 |
| | 7 | 6 | 7 | 8 | 7 | 8 | 9 | 10 | 8 | 8 | 7 | 7 | 9 | 8 | 7 | 8 | 8 | 10 | 9 | 11 | 8 |
| **6.9-5.8Å** | 8 | 8 | 7 | 7 | 8 | 6 | 6 | 7 | 6 | 8 | 8 | 8 | 8 | 6 | 6 | 7 | 7 | 6 | 1 | 0 | 0 |
| | 8 | 8 | 7 | 7 | 8 | 6 | 7 | 6 | 7 | 7 | 7 | 8 | 8 | 8 | 9 | 8 | 8 | 6 | 2 | 0 | 1 |
| | 8 | 8 | 8 | 6 | 7 | 7 | 8 | 7 | 7 | 7 | 7 | 8 | 8 | 6 | 8 | 8 | 8 | 6 | 3 | 2 | 5 |
| | 7 | 7 | 7 | 7 | 7 | 8 | 7 | 8 | 7 | 7 | 6 | 8 | 10 | 8 | 7 | 6 | 5 | 6 | 4 | 4 | 3 |
| | 7 | 5 | 7 | 8 | 9 | 8 | 7 | 6 | 5 | 9 | 6 | 7 | 10 | 10 | 7 | 8 | 8 | 6 | 5 | 8 | 14 |
| | 7 | 6 | 7 | 6 | 8 | 9 | 6 | 9 | 6 | 5 | 7 | 8 | 7 | 9 | 8 | 8 | 8 | 7 | 6 | 5 | 2 |
| | 6 | 8 | 9 | 6 | 6 | 8 | 7 | 9 | 7 | 7 | 7 | 7 | 8 | 8 | 8 | 7 | 7 | 6 | 7 | 5 | 3 |
| | 6 | 7 | 8 | 5 | 7 | 7 | 5 | 8 | 8 | 6 | 7 | 6 | 7 | 7 | 7 | 7 | 7 | 6 | 8 | 21 | 21 |
| | 7 | 6 | 7 | 7 | 7 | 8 | 7 | 6 | 7 | 6 | 6 | 6 | 7 | 7 | 6 | 7 | 7 | 10 | 9 | 24 | 20 |

FIG. 19. Comparison of the average reflection intensities before and after correction for crystal lattice distortions in specific resolution bands (80–15, 15–10, 9.9–7.0, and 6.9–5.8 Å). Note that the unbending procedure increases the signal intensity and improves the sharpness of the reflections. After unbending, significant data are even detectable in the band form 6.9- to 5.8-Å resolution. The *IQ* value is a measure of signal-to-noise. An *IQ* value of 1 corresponds to a signal-to-noise ratio ≥8, and an *IQ* of 8 corresponds to a signal-to-noise of ~1. Note the increase in the number of reflections with better *IQ* values after correction for crystal lattice distortions.

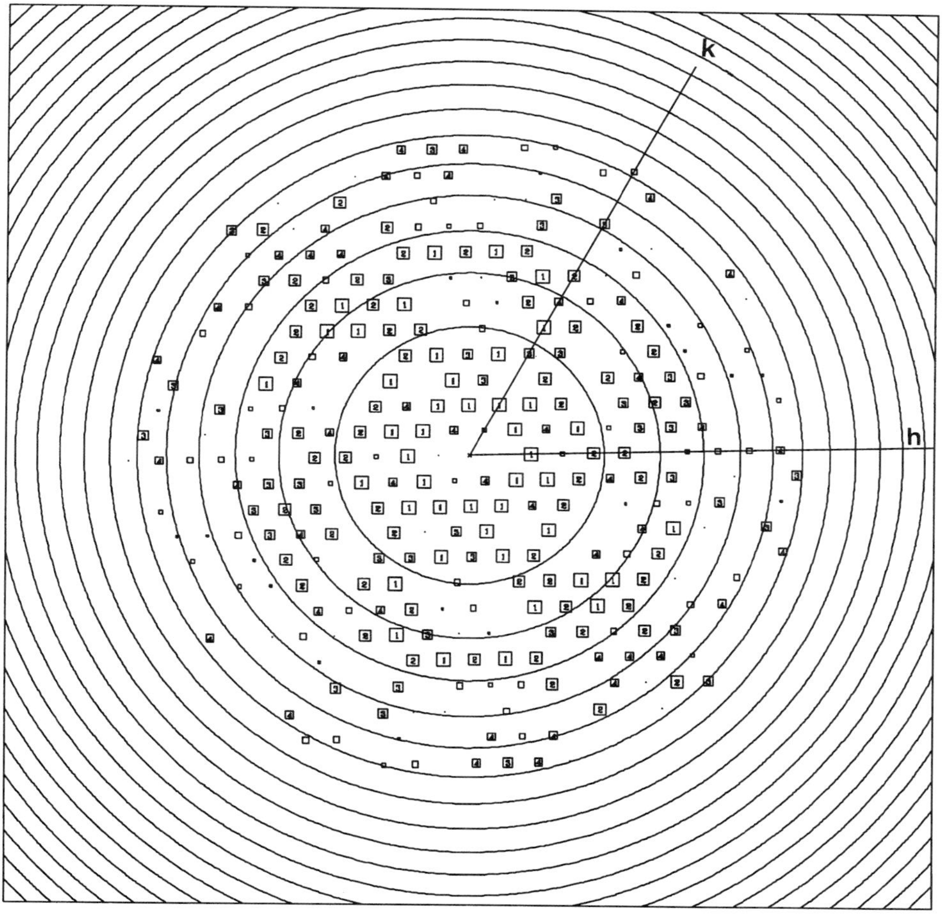


FIG. 20. Signal-to-noise ratios for all individual reflections to ~6-Å resolution derived by Fourier transformation of a corrected image of a frozen-hydrated 2D recombinant gap junction crystal. The rings represent the nodes in the CTF, determined by the level of defocus of the micrograph. The numbers are the IQ values and are inversely related to the signal-to-noise ratio (see Fig. 19).

14. Smooth curves are fitted by a least-squares method to the merged amplitude and phase data using LATLINED (Fig. 22).

15. The lattice line curves are sampled at a spacing of at least $1/T$ (where $T$ is the estimated thickness of the specimen), and the 3D map is computed by inverse Fourier transform using symmetry constraints defined by the lattice.

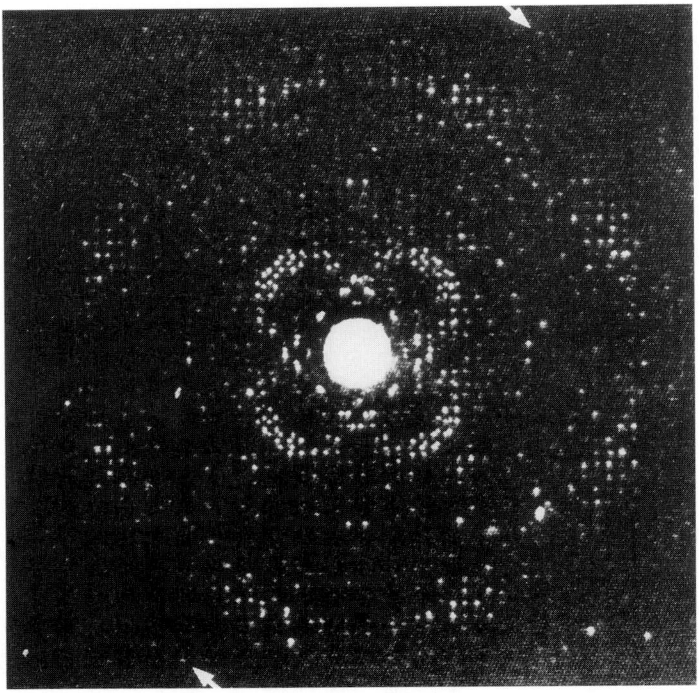

FIG. 21. The AQP1 2D crystals are sufficiently large so that electron diffraction patterns can be directly recorded. The amplitudes in the diffraction pattern are not affected by the CTF since the patterns are recorded at true focus. Hence, the derived amplitudes are much more reliable than those determined by Fourier transformation of images, as in Fig. 17. The arrows identify high-resolution reflections (3.5 Å). [From A. K. Mitra, A. N. van Hoek, M. C. Wiener, A. S. Verkman, and M. Yeager, *Nature Struct. Biol.* **2,** 726 (1995) and reproduced by permission of Nature Publishing.]

16. The map can be visualized using a variety of software packages (CCP4[90]; O[91]; and AVS[92]).

## Interpretation of Maps

The first milestone in the structure analysis is to derive a projection map from images of untilted, unstained crystals. The next step is to derive a 3D map by recording and processing images of tilted crystals. This is

[89a] G. F. Schertler, C. Villa, and R. Henderson, *Nature* **362,** 770 (1993).

[90] CCP4 (Collaborative Computational Project, No. 4) 1994.

[91] T. A. Jones and M. Kjeldgaard, "O Version 5.9, The Manual," Uppsala University, Sweden.

[92] B. Sheehan, M. E. Pique, and M. Yeager, *J. Struct. Biol.* **116,** 99 (1996).

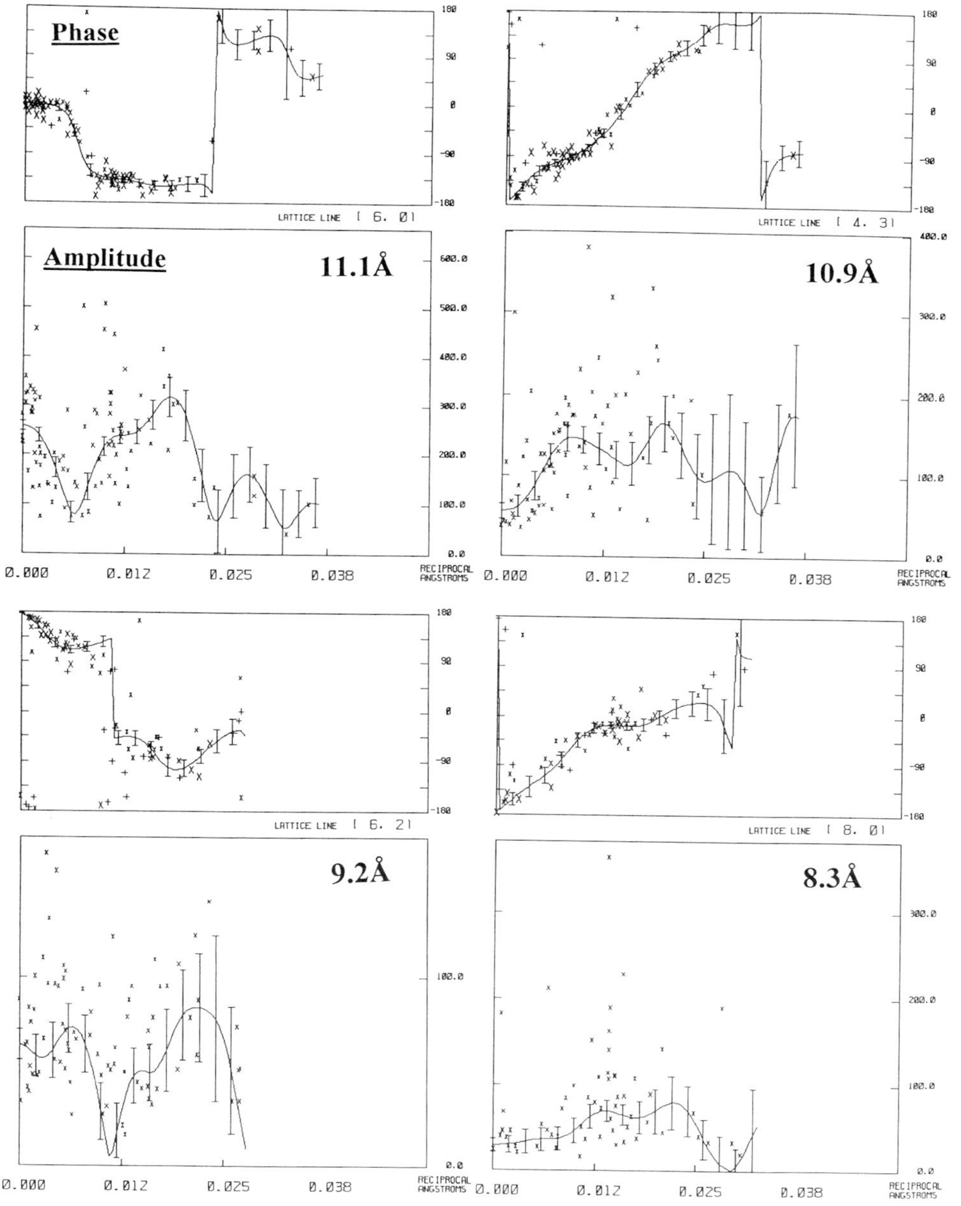


FIG. 22. Examples for phase (top) and amplitude (bottom) modulation along lattice lines of tilted 2D crystals. A least-squares method is used to generate smooth curves through the data points.[83] Inverse Fourier transformation is used to generate a 3D map by sampling the smooth curves at an increment of at least $1/D$, where $D$ is the thickness of the specimen.

particularly challenging because drift and charging will degrade the resolution of the images when the crystals are tilted. The ultimate goal is to derive a map at near-atomic resolution. However, as noted in the introduction, crystal quality may preclude determination of a structure at atomic resolution.

*15–30 Å Resolution*

Even at such a low resolution, important molecular details can be revealed: crystal symmetry, quaternary structure, general molecular boundary, protein/protein interactions, and the location of the aqueous pore in channels. For instance, in our analysis of 2D crystals of rat heart gap junctions,[47,49] the projection maps showed that the channels are formed by a hexameric cluster of subunits with a central channel. A projection map of AQP1 in this resolution range demonstrated that the protein subunits assemble as tetramers.[57] Functionally important sites can be identified by labeling with ligands such as undecagold clusters and antibodies. Furthermore, difference map analysis can also be used to locate specific proteins in a complex.[93]

*10–15 Å Resolution*

In this range, there is not much information to be gleaned. The molecular boundary will be somewhat better defined, but secondary structure is typically not visualized.

*5–10 Å Resolution*

A resolution of ~10 Å is just at the boundary for visualizing $\alpha$ helices, which tend to pack with a spacing of about 10 Å. Since packed $\alpha$ helices roughly perpendicular to the bilayer are a common motif for the transmembrane domains of membrane proteins, even a projection map at 5- to 7-Å resolution may reveal important structural details. For instance, a projection map of a recombinant gap junction channel at 7-Å resolution showed that each subunit in the hexameric connexon contains two $\alpha$ helices that appear roughly perpendicular to the membrane (Fig. 23). In addition, the projection map demonstrated that the docking of the two connexons in forming the complete intercellular channel involves a 30° rotational stagger between the apposed connexons.[36] A 3D map of tetrameric AQP1 channels revealed that each monomer is formed by a barrel of tilted $\alpha$ helices that enclose an aqueous vestibule leading to the selective water channel (Fig. 24, color plate). However, at 5- to 10-Å resolution, it will be difficult to define $\beta$ sheets, and $\alpha$ helices may be difficult to delineate if they are highly tilted.

[93] M. Yeager, J. A. Berriman, T. S. Baker, and A. R. Bellamy, *EMBO J.* **13,** 1011 (1994).

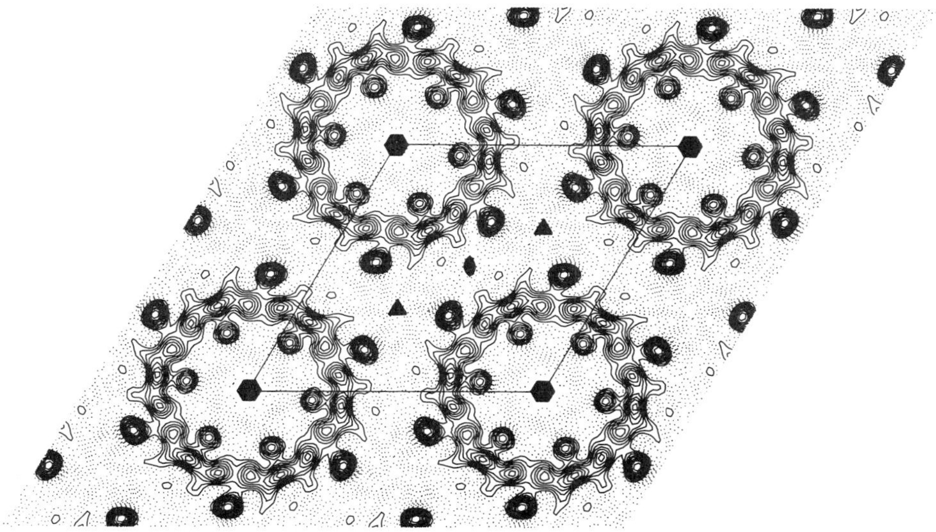

FIG. 23. Contour map of the projected density of recombinant gap junction channels at 7-Å resolution computed with p6 symmetry. The channel is formed by a hexameric cluster of subunits, each of which has three major features: (i) a ring of circular densities centered at 17-Å radius interpreted as $\alpha$ helices that line the channel, (ii) a ring of densities centered at 33-Å radius interpreted as $\alpha$ helices that are most exposed to the lipid, and (iii) a continuous band of density at 25-Å radius separating the two groups of helices. The hexagonal lattice had parameters $a = b = 79.4 \pm 0.3$ Å and $\gamma = 120 \pm 0.3°$. The map was scaled to a maximum density of 250 and contoured in steps of 25.0 (r.m.s. density = 87.4). The solid lines indicate density above the mean. [From V. M. Unger, N. M. Kumar, N. B. Gilula, and M. Yeager, *Nature Struct. Biol.* **4,** 39 (1997) and reproduced by permission of Nature Publishing.]

### *Resolution of 3.5 Å or Better*

Such a resolution will permit tracing of the C$\alpha$ backbone and definition of the amino acid side chains.[22–24] For bacteriorhodopsin, the level of structural detail was comparable to high-resolution maps derived by X-ray crystallography.[5]

In this volume focused on ion channels, let us suppose, for example, that we have successfully derived a density map of a voltage-regulated $K^+$ channel. At 15–20 Å of resolution, a 3D map would confirm whether the channels are tetrameric in the crystal and reveal the external shape of the oligomer. The vestibule to the ion conduction pathway will be detected if it is more than ~15 Å in diameter. At 5- to 7-Å in-plane resolution, the secondary structure of the transmembrane domains could be assessed. If the vertical resolution is comparable, the folding of the cytoplasmic and extracellular loops might be visible if they are ordered to this resolution. In addition, analysis of complexes could reveal the binding sites for toxins and antibodies. To visualize the binding sites for low molecular weight drugs, the reagents would have to be labeled with a heavy atom cluster in order to increase the density of the drug. In addition, time-resolved cryomicroscopy might allow trapping of the open, closed, and inactivated conformations. Insight into the following specific aspects of $K^+$ channel structure would potentially be revealed in a 6-Å resolution 3D map: (1) What is the molecular boundary of the tetrameric channel? (2) Does each $K^+$ channel monomer contain six transmembrane $\alpha$ helices and a $\beta$ hairpin between S5 and S6? (3) What is the molecular topography of the vestibule of the pore? (4) Where is the selectivity filter located? (5) Where is the S4 voltage sensor located? (6) What is the packing arrangement between helices of adjacent subunits that confers stability in the tetrameric assembly? (7) What is the degree of tilt of the putative $\alpha$-helices and their interaction with the S5-S6 loop? At 3.5-Å resolution these points would be answered to the level of individual side chains. Indeed, a recent X-ray structure at 3.2 Å resolution of a bacterial homolog of an inward rectifying $K^+$ channel has provided answers to some of these questions.[93a]

## Conclusions

The use of electron microscopy for the structural characterization of 2D crystals of membrane proteins began in 1975 with the pioneering work of Henderson and Unwin in the analysis of glucose-embedded 2D crystals of bacteriorhodospin.[94] The 6-Å resolution map revealed seven transmembrane $\alpha$ helices and serves as a paradigm for membrane protein structure in the same way that the X-ray structure of myoglobin served as a paradigm for the structure of soluble proteins. Other notable events in the development of electron cryocrystallography were a projection map of bacteriorhodopsin at 3.5-Å resolution,[85] a 3D map at 3.5-Å resolution,[22] and most recently a map at 3.0-Å resolution.[23] To date, LHCII[24] and tubulin[94a] are the only other proteins determined at atomic resolution by electron cryocrystallography. For several membrane proteins, density maps with an in-plane resolution ranging from 6 to 9 Å have resolved protein secondary structure: bacterial porin,[94b] LHCI,[94c] halorhodopsin,[94d] aquaporin I,[95,96,97]

[93a] D. A. Doyle, J. M. Cabral, R. A. Pfuetzner, A. Kuo, J. M. Gulbis, S. L. Cohen, B. T. Chait, and R. MacKinnon, *Science* **280,** 69 (1998).

[94] R. Henderson and P. N. T. Unwin, *Nature* **257,** 28 (1975).

[94a] E. Nogales, S. G. Wolf, and K. H. Downing, *Nature* **391,** 199 (1998).

[94b] B. K. Jap, P. J. Walian, and K. Gehring, *Nature* **350,** 167 (1991).

[94c] S. Karrasch, P. A. Bullough, and R. Ghosh, *EMBO J.* **14,** 631 (1995).

[94d] W. A. Havelka, R. Henderson, and D. Oesterhelt, *J. Mol. Biol.* **247,** 726 (1995).

[95] A. Cheng, A. N. van Hoek, M. Yeager, A. S. Verkman, and A. K. Mitra, *Nature* **387,** 627 (1997).

[96] T. Walz, T. Hirai, K. Murata, J. B. Heymann, K. Mitsuoka, Y. Fujiyoshi, B. L. Smith, P. Agre, and A. Engel, *Nature* **387,** 624 (1997).

[97] H. Li, S. Lee, and B. K. Jap, *Nature Struct. Biol.* **4,** 263 (1997).

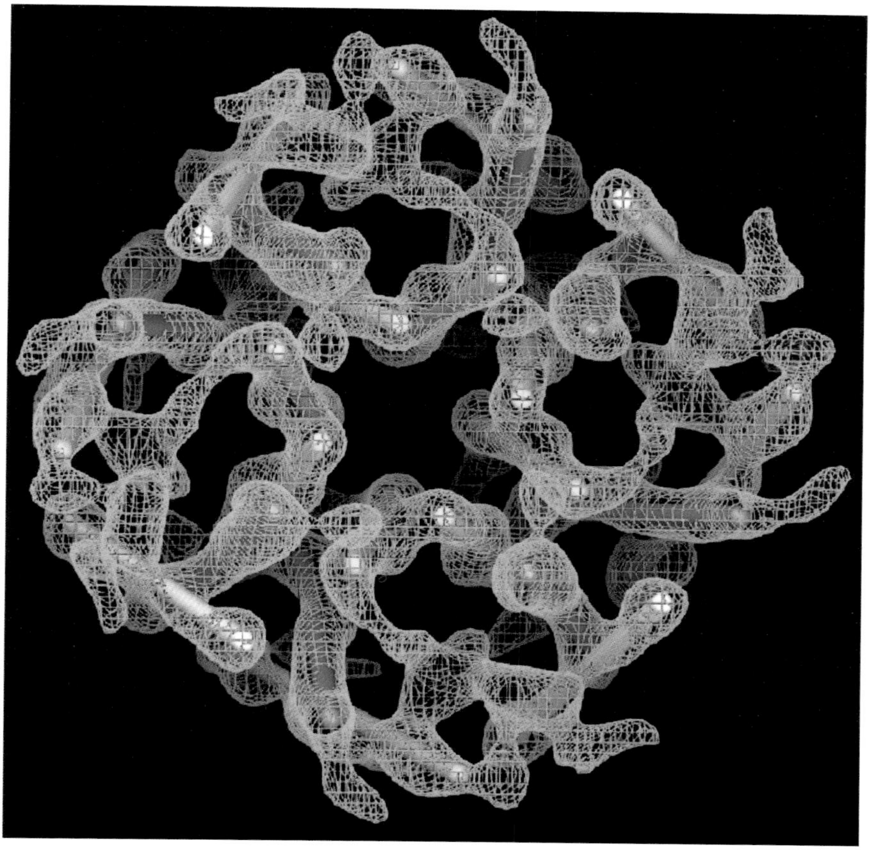

FIG. 24. 3D density map of human erythrocyte AQP1 (previously named CHIP28) at 7-Å in-plane resolution, viewed approximately perpendicular to the plane of the bilayer. The molecules assemble as a tetramer, and each monomer contains a water-selective channel. The rods positioned in the density map trace the approximate paths of six tilted $\alpha$ helices, which form a right-handed barrel that surrounds the aqueous pathway. [From A. Cheng, A. N. van Hoek, M. Yeager, A. S. Verkman, and A. K. Mitra, *Nature* **387,** 627 (1997) and reproduced by permission of Nature Publishing.]

gap junctions,[36] glutathione transferase,[98] vertebrate rhodopsin,[98a] and a bacterial $H^+$-ATPase.[41] In addition, tubular crystals of membrane proteins that manifest helical symmetry offer the advantage of providing all views of the molecule, thereby obviating the need to tilt the specimen. Hence, there is no missing cone of data as is inherent in datasets obtained from tilted 2D crystals. Tubular assemblies of the acetylcholine receptor[34,98b] and the $Ca^{2+}$ATPase of sarcoplasmic reticulum[98c] have yielded complete 3D maps at 9 and 8 Å resolution, respectively. Given all of these achievements, electron cryomicroscopy and image analysis have together emerged as a powerful strategy for the structural characterization of membrane proteins. The electron cryomicroscopes and computational methods that are currently available have the routine potential to achieve a resolution that is near atomic. Specific for membrane protein structure analysis, the critical barriers are obvious—obtaining sufficient protein and growing high-resolution 2D crystals. Given the substantial effort that is being devoted to the technology of membrane protein overexpression and the continuing list of proteins for which high-resolution 2D crystals are available, the future for this field is indeed bright.

[98] H. Hebert, I. Schmidt-Krey, R. Morgenstern, K. Murata, T. Hirai, K. Mitsuoka, and Y. Fujiyoshi, *J. Mol. Biol.* **271,** 751 (1997).

[98a] V. M. Unger, P. A. Hargrave, J. M. Baldwin, and G. F. Schertler, *Nature* **389,** 203 (1997).

[98b] N. Unwin, *J. Mol. Biol.* **229,** 1101 (1993).

[98c] P. Zhang, C. Toyoshima, K. Yonekura, N. M. Green, and D. L. Stokes, *Nature* **392,** 835 (1998).

[99] J. Dubochet, M. Groom, and S. Mueller-Neuteboom, *Adv. Optical and Electron Microscopy* **8,** 107 (1982).

[100] B. Adair, E. Wilson, S. Weiner, T. Kunicki, and M. Yeager, unpublished observations.

[101] P. N. T. Unwin and P. D. Ennis, *Nature* **307,** 609 (1984).

[102] A. Brisson and P. N. T. Unwin, *Nature* **315,** 474 (1985).

[103] J. R. Bellare, H. T. Davis, L. E. Scriven, and Y. Talmon, *J. Electron Microsc. Tech.* **10,** 87 (1988).

[104] S. Subramaniam, M. Gerstein, D. Oesterhelt, and R. Henderson, *EMBO J.* **12,** 1 (1993).

[105] S. D. Fuller, J. A. Berriman, S. J. Butcher, and B. E. Gowen, *Cell* **81,** 715 (1995).

[106] M. Walker, J. Trinick, and H. White, *Biophys J.* **68,** 87S (1995).

[107] J. Brink and W. Chiu, *J. Struct. Biol.* **113,** 23 (1994).

[108] M. B. Sherman, J. Brink, and W. Chiu, *Micron* **27,** 129 (1996).

[109] F. Thon, *Z. Naturforschg.* **21a,** 476 (1966).

[110] E. D. Salmon and D. DeRosier, *J. Microsc.* **123,** 239 (1981).

## Acknowledgments

We thank Michael Whittaker and Alan McPhee for photography and Anchi Cheng and Brian Adair for help with figure preparation. The writing of this review has been supported by grants from the National Institutes of Health (M.Y. and A.K.M.), a Grant-in-Aid from the American Heart Association (A.K.M.), the Donald E. and Delia B. Baxter Foundation (M.Y.), and a postdoctoral fellowship from the American Heart Association (V.M.U.). M.Y. is an established investigator of the American Heart Association and Bristol-Myers Squibb.

# [8] Drug-Dependent Ion Channel Gating by Application of Concentration Jumps using U-Tube Technique

*By* F. BRETSCHNEIDER and F. MARKWARDT

## Introduction: Advantages of Technique

To study the time course of the reaction of ion channels or transporters to the binding of agonists, drugs, or ions *in vitro* rapid application systems are needed with time courses of concentration jumps faster than the kinetics of the reaction of the protein. Furthermore, it is often necessary to apply several different substances while investigating one cell or membrane patch which has a limited life span under the recording conditions.

In this article, an application system is described with solution exchange times in the 100-ms range. The method was introduced by Krishtal and Pidoplichko.[1]

The U-tube application system is inexpensive, easy to install, and easy to clean. The exchange of U-tube solution(s) (UTS) takes only a few seconds. The application of substances can be triggered by the computer, which also records the ionic currents. Therefore, the time relation between concentration jump and ionic currents is known. Only cells (or patches) in the vicinity of the hole of the U-tube and directly downstream of the U-tube in laminar perfused bathing systems are in contact with the UTS. The application of irreversible effectors to cells upstream or far from the hole of the U-tube is avoided. So, the effect of drug applications on several different cells can be measured within one set of cells placed in the bath chamber.

The volume of a new UTS needed for replacement (<0.2 ml) and the flow during application to the cell under investigation (<0.05 ml/min) are so small that only minimal amounts of possibly expensive drug solutions

[1] O. A. Krishtal and V. I. Pidoplichko, *Neurosc.* **5,** 2325 (1980).



0076-6879/99 $30.00

are needed. During times without drug application the U-tube may be perfused simply with distilled water.

## Principle of Function

A U-shaped tube with a hole on its convex side is located at a distance of 50–200 $\mu$m away from the cell under investigation (see Fig. 1). The tube is constantly perfused with an external solution (UTS). The hydrostatic pressure of the UTS is adjusted in such a manner that the pore sucks the bathing solution into the tube if the valve located downstream from the hole is open. Closing the valve stops the suction and applies hydrostatic pressure to the pore. The flow through the hole is reversed and the UTS passes from the tube to bathe the cell (see inset in Fig. 1). The flow out of the hole in the tube is nearly laminar.[1] A complete solution exchange in the vicinity of a cultured cell within 100–200 ms can be achieved using

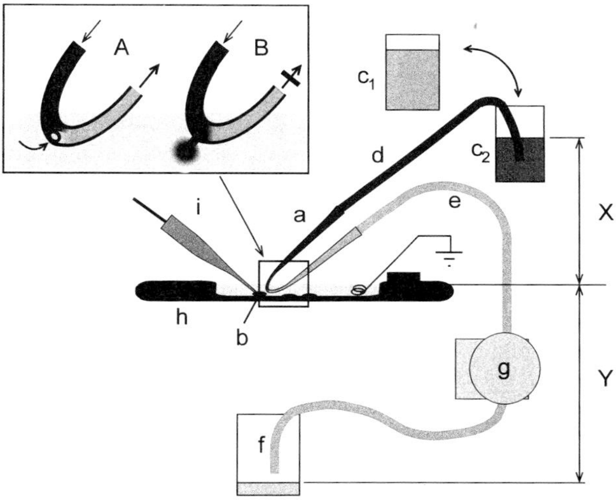


FIG. 1. Scheme of the U-tube system. The U-tube (a) is positioned in the vicinity of the cell under investigation (b). UTS from one of several storing beakers ($c_1$, $c_2$,) flows through the influx arm (d), the U-tube, and together with additional bath solution out of the efflux arm (e) into a collecting beaker (f) if the valve (g) is open. h, Bath chamber with adherent cells; i, patch electrode. The height differences X and Y are explained in the text. *Inset:* Top of the U-tube with its hole. (A) If the valve downstream is open, the flow on the right side ("efflux arm") is greater than on the left ("influx arm"). Therefore, additional bathing solution is sucked into the U-tube (arrow). (B) Closing the valve results in flow of UTS out of the hole into the bath.

this system.[1–5] Reopening of the valve causes the bathing solution to be resuctioned thereby efficiently removing the drug from the vicinity of the cell.

We found that the force for flow and suction in the U-tube can be generated by gravity alone without using a pump, thus avoiding any problems due to discontinuous pumping. The difference in height between the surface of the UTS reservoirs and the hole (surface of the bathing solution) (X in Fig. 1) must be smaller than that between the hole and the lower end of the polyethylene tubing downstream (Y). A net flow through the pore into the U-tube results, thus diluting the UTS with bathing solution in the efflux arm. The exchange rate of the U-tube solution as well as the flow rate during drug application can be carefully controlled by changing A and B. (Fig. 1). The adjusted flow rates may depend on the cost of the applied drug, the speed needed for the solution exchange, and the mechanical stress that can be tolerated by the cells or patches under investigation.

The cells are usually attached to the bottom of the bathing chamber. Some cells may become detached from the bottom and are therefore fixed only by the patch pipette. Such cells may undergo vigorous movements during application of UTS. Fortunately, in many cases this has no effect on the current recording. The solution exchange is essentially complete within 150–300 ms depending mostly on the distance of the cell from the hole of the U-tube and the flow rates. Fenwick *et al.*[2] estimated 20–30 ms for membrane patches positioned within the center of the hole.

The speed of the concentration jump may be increased when performing the solution exchange by displacing the laminar outflow of the U-tube cross to the flow direction. This can be achieved by connecting the holder of the U-tube to a piezo translator.[6,7] The shift of the laminar flow must be large enough to bring the cell or patch fully into and out of the laminar outflow of the U-tube. The power supply of the piezo translator can also be controlled by the recording computer.

The removal of the UTS after reopening of the magnetic valve depends on the flow rate of the bathing solution (about 0.5–2 sec in our system).

A small air bubble sucked into the influx pathway while changing the polyethylene tubing from one UTS source to another indicates the arrival of new UTS at the hole.

[2] E. M. Fenwick, A. Marty, and E. Neher, *J. Physiol.* (*Lond.*) **331,** 577 (1982).
[3] F. Bretschneider, M. Klapperstück, M. Löhn, and F. Markwardt, *Pflügers Arch.* **429,** 691 (1995).
[4] S. J. Robertson, M. G. Rae, E. G. Rowan, and C. Kennedy, *Br. J. Pharmacol.* **118,** 951 (1996).
[5] S. Dryselius, P. E. Lund, and B. Hellman, *Cell. Struct. Funct.* **19,** 385 (1994).
[6] F. Markwardt and G. Isenberg, *J. Gen. Physiol.* **99,** 841 (1992).
[7] C. Franke, H. Hatt, and J. Dudel, *Neurosci. Lett.* **77,** 199 (1987).

To position the U-tube near the cell a holder and a manipulator for making adjustments are necessary (see Fig. 2). We recommend mounting the manipulator for the U-tube holder on the table of the microscope and to position the valve far from this table and the patch pipette.

*Equipment*

Glass capillaries 80 mm long with an inner diameter of 500 $\mu$m are used for U-tubes. Polyethylene tubing that fits on the ends of the capillary (approximately 200- to 600-$\mu$m inner diameter), a holder for the U-tube on a three-dimensional movable micromanipulator, a two-tube magnetic pinch valve (NResearch, Maplewood, NJ), and beakers for storing different UTS are also needed for a U-tube system. A board or holder should be available to position the storing beakers above the bath chamber of the setup. Because we used a cage around the patch-clamp setup, no problems caused by additional noise occurred. It is advisable, however, to ground

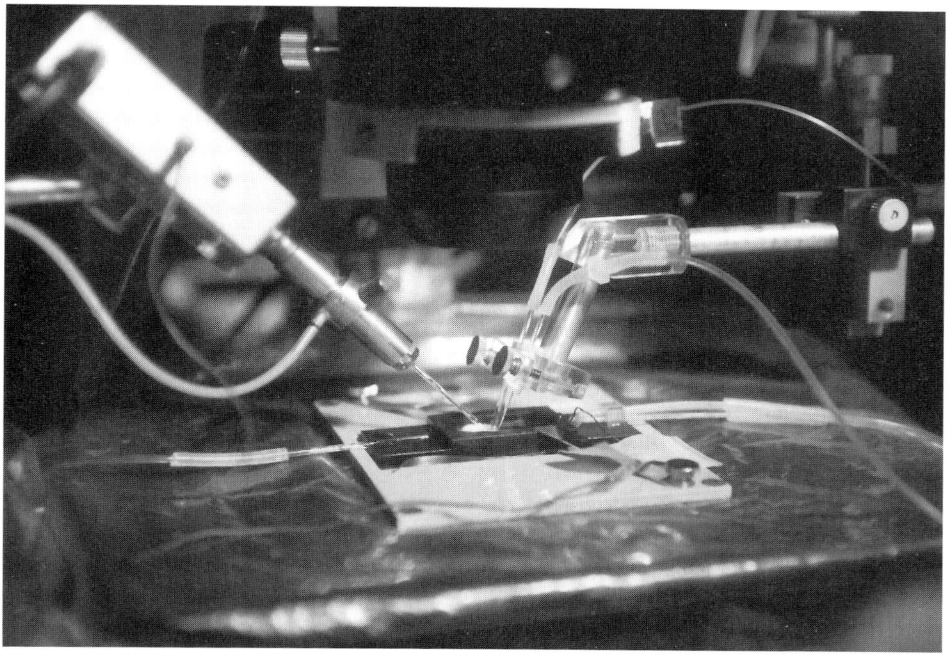

FIG. 2. Central part of a patch-clamp setup. The top of the pipette (left side) and of the U-tube (right side) are positioned in the bathing chamber. The U-tube with polyethylene tubing on its branches (for influx and efflux) is mounted on a holder and connected to a micromanipulator (partially shown).

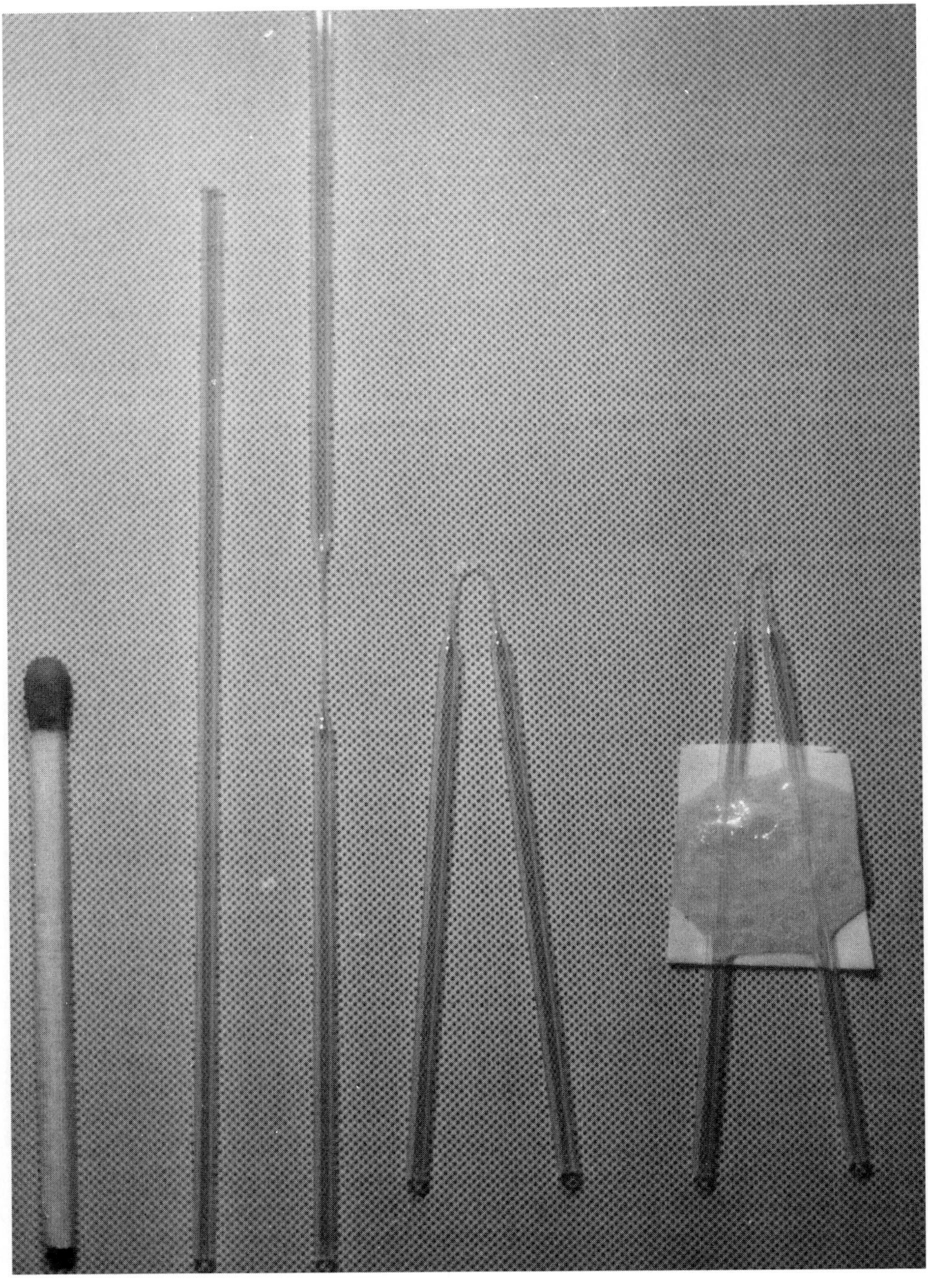

the U-tube system. For switching the valve a power supply triggered by the computer via digital/analog output is recommended.

To construct the U-tube the following supplies are needed: a micropipette puller, a Bunsen burner, and a system for heating a platinum wire including a holder for the U-tube and a microscope (magnification: 20×) for observing the procedure (a patch pipette polisher can be used) as well as a syringe.

### *Preparation of Glass U-Tube*

The glass capillaries are pulled in one step to an inner diameter of about 200 $\mu$m. The narrow part is about 5 mm long. To bend the capillary to form the U-shape hold the narrow part of the capillary near the flame of a Bunsen burner until the distal end moves downward. Be careful that the branches forming the U come into one plane. Fix the branches on a piece of paper with epoxy resin (see Fig. 3).

To make the hole on the top of the U-tube a heating procedure is used (see Fig. 4). The U-tube is inverted and fixed in a holder. Using microscopic controls the tube is positioned above a glowing platinum wire (Fig. 4A). One of the two arms of the U-tube is temporarily closed (for instance, with a closed piece of polyethylene tubing), the other arm is connected with tubing to a syringe filled with air. While generating positive pressure by means of the syringe move the U-tube close to the hot wire. A "nose" or protrusion with a thin wall is formed at the top of the U-tube (Fig. 4B). Then the top of this protrusion is broken using a cool wire (Fig. 4C). Finally the edges of the hole must be polished by further heating (Fig. 4D).

The U-tube is now ready to be attached to the holder. A conventional microelectrode holder is suitable.

### *Testing U-Tube System*

Two polyethylene tubes, one for influx and the other for efflux, are attached to the U-tube and the valve is positioned. With negative pressure at the end of the "efflux arm" (valve open) produced by a syringe, the application system is filled with solution.

Using a microscope to observe the flow out of the hole, the U-tube is perfused with distilled water. The top of the U-tube is placed in the bathing chamber perfused with a physiologic salt solution.

---

FIG. 3. Stages of manufacturing a U-tube. From left to right: a glass pipette in relation to a match, a pipette after one-step pulling, the U-shaped tube with a hole on its top, and the U-tube fixed with epoxy resin on a piece of paper.

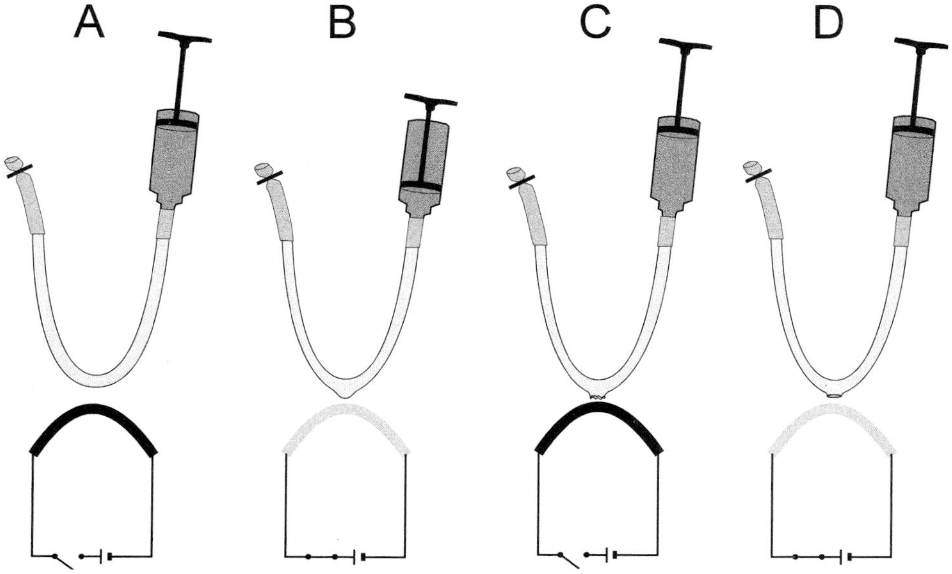


FIG. 4. Producing the hole on the top of the U-tube. (A) The U-tube is positioned above a platinum wire at a distance of about 200 $\mu$m. The branches are closed as described in text. (B) Heating (wire gray) and positive pressure (via the syringe) result in a growing "nose" or protrusion. (C) The top of the bulge is broken with the help of the cool wire (black). (D) Additional heating polishes the edges of the hole to avoid clogging of substances.

If the U-tube system is filled with a diluted salt solution the outflow of the U-tube can be localized by measuring electrical conductivity. This is done by means of an open patch pipette to which a square wave test pulse is delivered by a patch-clamp amplifier.[2]

The operation of the system can be evaluated by recording currents from ion channels selective for a specific ion species ($K^+$ channels are very suitable for this purpose). The bathing and the U-tube solutions should have different ion concentrations. Then the application of UTS and the reapplication of bathing solution shifts the reversal potential of the ionic current through these channels. The time course and the final value of the changing reversal potential are measures of the kinetics and the completeness of the solution exchange, respectively.

*Tips and Applications*

By compressing the tubing using the switching pin of a magnetic pinch valve the "efflux arm" is closed and suction is stopped. A part of the diluted downstream solution is pushed back and flows transiently out of the pore at the top of the U-tube. To avoid this, a second tube (of equal dimension) of the pinch valve is filled with solution, is closed at one end, and is connected as a "blind sack" to the efflux arm of the U-tube (see Fig. 5). When the efflux pathway is shut, the pin of the valve compresses the efflux arm and depresses the blind sack. A shunt volume flows from the efflux arm into the blind sack and not back to the pore. Another possibility for diminishing recoil pressure is to store an air bubble in a blind sack.[5]

The U-tube system may not work properly for several reasons: (1) If the diameter of the narrow part of the U-tube is less than 100 $\mu$m the flow rate in the "influx arm" is very slow. (2) If the hole is too small ($<40$ $\mu$m) cell debris or crystals may clog the tube during the experiments. If this is the case, before starting new experiments each day, clean the application system with $H_2O_2$ (30%) and distilled water to remove crystals and debris. It is also possible to reopen the hole by mechanical cleaning using the flexible top of a tube for filling patch pipettes. In some cases it is useful to alternate pressing and suctioning with a syringe at the end of the efflux arm. (3) If the hole is larger than about 200 $\mu$m in diameter, the exchange of UTS may be very slow. Mostly bathing solution and not UTS is sucked into the efflux pathway during the time the valve is open. (4) Air bubbles in the system tend to reduce the flow rates and/or close the hole. Here also suction via a distal syringe is recommended. Avoid sucking the cell attached to the patch pipette into the hole when manipulating the syringe. On the other hand, patches from cells which are not attached to the bottom of the chamber can be removed by sucking these cells into the U-tube.

The outflow of UTS should reach the cell under investigation and not only the surface of the bathing solution. This can be detected visually by testing the system with distilled water.

The lifetime of a U-tube system is not limited. Sometimes we have used the same system for several weeks.

## Summary

The rapid application system described has been used to study a variety of ionic channels in several different types of single cells.[1–5,8–12]

[8] J. Bormann, *Eur. J. Pharmacol.* **166,** 591 (1989).

[9] E. X. Albuquerque, A. C. S. Costa, M. Alkondon, K. P. Shaw, A. S. Ramona, and Y. Aracava, *J. Recept. Res.* **11,** 603 (1991).

[10] C. M. Gammon, G. S. Oxford, A. C. Allen, K. D. McCarthy, and P. Morell, *Brain Res.* **479,** 217 (1989).

[11] J. Kapur and R. L. Macdonald, *Mol. Pharmacol.* **50,** 458 (1996).

[12] R. J. Evans and C. Kennedy, *Br. J. Pharmacol.* **113,** 853 (1994).

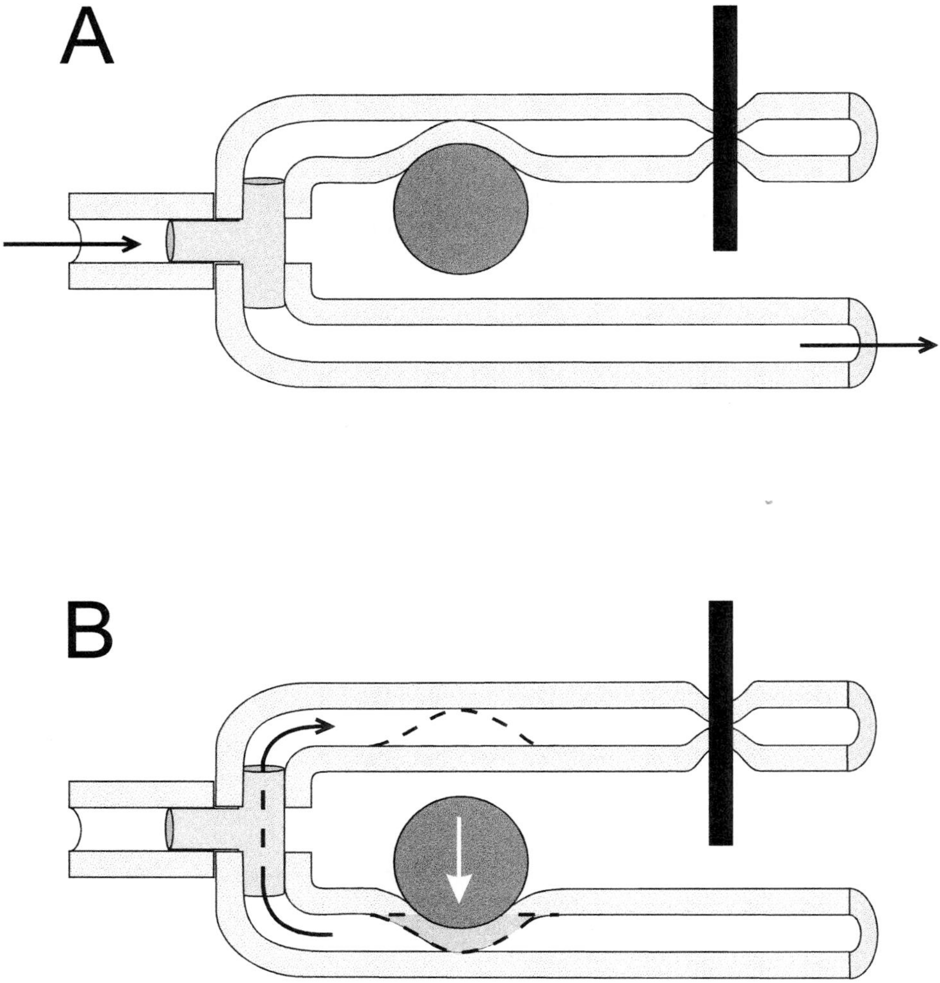


FIG. 5. Principle of avoiding recoil pressure ("blind sack"). (A) When the valve is open (pinch, dark gray, above), solution flows from the left to the right through the efflux arm of the U-tube system. (B) Closing the efflux arm is accompanied by opening the blind sack reservoir. This increases and decreases the volume (marked) in the blind sack and in the efflux arm, respectively, by the same amount. Therefore, this volume flows from the efflux arm into the blind sack thereby avoiding recoil pressure (arrow).

The system is inexpensive, easy to install, and can be used repeatedly. The consumption of UTS, i.e., drugs or agonists, is low. The time interval between switching the valve and the expected effect is often shorter than 150 ms for cells about 8–15 $\mu$m in diameter and is about 20–25 ms for patches positioned in the hole of the U-tube. Time and duration of substance application can be controlled by a computer connected to a digital-analog (D/A) output.

# [9] Voltage-Clamp Biosensors for Capillary Electrophoresis

*By* OWE ORWAR, KENT JARDEMARK, CECILIA FARRE, INGEMAR JACOBSON, ALEXANDER MOSCHO, JASON B. SHEAR, HARVEY A. FISHMAN, SHERI J. LILLARD, and RICHARD N. ZARE

## Introduction

During the past two decades, there have been significant developments in microanalytical separation and sampling techniques for low molecular weight bioactive compounds. However, improvements in highly selective and sensitive detection strategies for these compounds are still needed. The heart of the problem is that many biologically active species do not contain molecular features to allow sensitive detection. Exceptions to this behavior do exist, and such compounds as catecholamines and indoleamines, which are easily oxidized, have been detected using ultramicroelectrodes in single synaptic vesicles following exocytosis.[1] The exocytotic release of serotonin and tryptophan-containing proteins has also been detected using native UV fluorescence following separation.[2] On the other hand, amino acid neurotransmitters such as glutamate, aspartate, glycine, and $\gamma$-aminobutyric acid (GABA) are not easily oxidized, do not have significant absorption cross sections above the water cutoff wavelength, and are virtually nonfluorescent. For their sensitive quantitation, incorporation of chemical reagents such as fluorogenic or fluorescent dyes, electroactive or highly absorbing moieties, or radioactive isotopes are required. Generally, drawbacks with such labeling schemes include lack of selectivity (similar functional groups

[1] R. M. Wightman, J. A. Jankowski, R. T. Kennedy, K. T. Kawagoe, T. J. Schroeder, D. J. Leszczyszyn, J. A. Near, E. J. Diliberto, and O. H. Viveros, *Proc. Natl. Acad. Sci. U.S.A.* **88,** 10754 (1991).

[2] S. J. Lillard, E. S. Yeung, and M. A. McCloskey, *Anal. Chem.* **68,** 2897 (1996).





0076-6879/99 $30.00

will be altered in the same way) and uncertainties in reaction efficiencies. Furthermore, with certain labeling schemes, the biological system under study can be perturbed, possibly to the extent that the data collected are biased. Most importantly, however, any detection technique that measures singularly a physical or chemical property of the analyte, separated from its biological context, will not reveal its biological function.

Low molecular weight neuroactive compounds bind to membrane receptors. Ligand binding can lead to many different events, such as opening of ion channels or activation of G-protein-coupled intracellular cascades. Many times these events can be detected using optical or electrical recording techniques, and the receptor and its coupled reactions form the basis of a biological detector or biosensor.[3,4] The advantages of using biosensors to detect biologically active species include (1) high specificity owing to molecular recognition, (2) high sensitivity because ligand binding triggers biological amplification cascades, (3) no chemical derivatization step, and (4) biological meaning of the detector response.

Fast-acting neurotransmitters activate ligand-gated ion channels. These receptors confer a high degree of selectivity because their binding sites have been genetically engineered to suit a single or a few structurally related ligands. They also offer high sensitivity because, on ligand binding, approximately $10^4$ ions will either enter or leave the cell per millisecond. This flow of ions can be measured with voltage- and patch-clamp techniques, sometimes down to the level of a single receptor. In contrast to traditional detection techniques for microseparations, this biosensor detector measures an electrical current resulting from the interaction between a receptor and a ligand binding to it, leading to meaningful biological information.

Capillary electrophoresis–patch-clamp detection (CE-PC) is a novel technique that can be used for the fractionation and detection of receptor agonists and receptor antagonists in complex and extremely small ($10^{-9}$–$10^{-18}$ liter) sample solutions. CE-PC is a combination of two well-established techniques. A particularly interesting aspect of CE is that it can be useful in solution nanochemistry applications because the inlet of the electrophoresis capillary can be pulled to narrow (submicron) diameters, and can be used to inject extremely small volumes ($10^{-15}$–$10^{-18}$ liters) or even single biopolymers, such as DNA oligomers.[5] Because of this high sampling resolution it is our hope that CE-PC technology can be used to

[3] D. C. Wijesuriya and G. A. Rechnitz, *Biosens. Bioelectr.* **8,** 155 (1993).

[4] H. M. McConnell, J. C. Owicki, J. W. Parce, D. L. Miller, G. T. Baxter, H. G. Wada, and S. Pitchford, *Science,* **257,** 1906 (1992).

[5] D. T. Chiu, A. H. Hsiao, A. Gaggar, R. A. Garza-López, O. Orwar, and R. N. Zare, *Anal. Chem.* **69,** 1801 (1997).

probe neurotransmitter release and dynamics in biological microdomains. Here we present information about how to optimize the union between CE and PC, especially to study ligands that act on receptors belonging to the glutamate–receptor family. Also, the use of *Xenopus laevis* oocytes that have been injected with mRNA and total rat brain RNA, as two-electrode voltage-clamp (VC) detectors in CE, is discussed.

Recently, patch-clamped neuronal cell membranes have been used as *in situ* sensors, so-called "sniffer-patch detectors," to demonstrate release of acetylcholine and glutamate-like compounds from neuronal growth cones and turtle photoreceptors.[6] The sniffer-patch technique offers extremely good spatial resolution, down to $\sim 1\ \mu m^3$, which allows detection of quantal release of neurotransmitters. A major limitation with this technique, however, is its inability to discriminate between multiple ligands that activate the same type of receptor. To achieve this goal without sacrificing spatial resolution, we have coupled both patch- and two-electrode voltage-clamp biosensors on-line to capillary electrophoresis.[7–9] In these systems, ligands are separated electrophoretically, delivered onto the surface of a cell, and detected at a characteristic migration time through the capillary. Such techniques can be optimized for a specific analyte or class of analytes. This optimization is accomplished by using cell types and cell lines that express native or recombinant receptors specific for the analyte of interest. Additionally, it is feasible to identify bioactive components from their electrophoretic migration times in such a manner that the migration time and response of the detector provide a unique signature of the bioactive component.

## Capillary Electrophoresis

Free-solution capillary electrophoresis in its present configuration was introduced by Jorgenson and Lukacs.[10] They showed that highly efficient electrophoretic separations could be performed in narrow-bore (5- to 75-$\mu m$ inner diameter, 100- to 400-$\mu m$ outer diameter) fused silica capillaries (usually 20–100 cm in length). Because thin-walled capillaries efficiently dissipate Joule heat, minimal distortion is imposed on the separated analyte

[6] T. G. J. Allen, *Trends Neurosci.* **20,** 192 (1997).

[7] J. B. Shear, H. A. Fishman, N. L. Allbritton, D. Garigan, R. N. Zare, and R.H. Scheller, *Science* **267,** 74 (1995).

[8] O. Orwar, K. Jardemark, I. Jacobson, A. Moscho, H. A. Fishman, R. H. Scheller, and R. N. Zare, *Science* **272,** 1779 (1996).

[9] K. Jardemark, O. Orwar, I. Jacobson, A. Moscho, and R. N. Zare, *Anal. Chem.* **69,** 3427 (1997).

[10] J. W. Jorgenson and K. D. Lukacs, *Anal. Chem.* **53,** 1298 (1981).

bands, and under ideal circumstances, theoretical plate numbers in excess of $10^6$ can be accomplished.

In CE, the capillary is filled with buffered electrolyte solution (i.e., running buffer) and both ends are immersed in reservoirs that typically contain the same solution. Most commonly, the anode and cathode are placed in the inlet and outlet reservoirs, respectively, and are connected to a high-voltage source.[10,11] Above about pH 2, the bare fused-silica capillary wall is negatively charged. Cations in the electrolyte are attracted to the charged fused silica surface and an electrical double layer is created. When an electric field is applied across the capillary, a sheath of positive ions drags the bulk solution in the cathodic direction, which results in electro-osmotic flow. This electro-osmotic flow acts as a noise-free pump.

The separation mechanism for free-solution CE (termed capillary zone electrophoresis, CZE) is based on the differential electro-osmotic and electrophoretic migration rates of charged species. The electrophoretic mobility $\mu_{ep}(m^2/V \cdot sec)$, can be expressed as:

$$\mu_{ep} = q_i(6\pi\eta r_i)^{-1} \tag{1}$$

where $\eta(N \cdot sec/m^2)$ is the kinematic viscosity of the separation medium, and $r_i(m)$ and $q_i(A \cdot sec)$ are the effective radius and charge, respectively, of the migrating species. The electrophoretic migration velocity in m/sec, is given by

$$\nu_{ep} = \mu_{ep} E \tag{2}$$

where $E$ is the electric field strength (V/m). The electro-osmotic mobility, $\mu_{eo}$, can be expressed as

$$\mu_{eo} = \varepsilon\zeta(4\pi\eta)^{-1} \tag{3}$$

where $\varepsilon(C/V \cdot m)$ is the dielectric constant of the separation medium, and $\zeta(V)$ is the zeta potential, which is the potential difference between the Stern plane and the buffer solution. The electro-osmotic migration velocity is given by

$$\nu_{eo} = \mu_{eo} E \tag{4}$$

and the net migration velocity for a species subjected to both electro-osmosis and electrophoresis is

$$\begin{aligned} \nu_{net} &= (\mu_{eo} + \mu_{ep})E \\ &= \nu_{eo} + \nu_{ep} \end{aligned} \tag{5}$$

[11] S. F. Y. Li, "Capillary Electrophoresis: Principles, Practice and Applications," Elsevier, Amsterdam, 1992.

Under normal operating conditions $|\nu_{eo}| > |\nu_{ep}|$, so that all species—whether possessing positive, neutral, or negative charges—migrate in the same direction past a single detector.

CE is an excellent match to cell-based biosensors because physiologic buffers can be employed as electrolytes and the dimensions of the separation capillary correspond well with those of the sensor. CE can also be used as an injection technique for administering organic and inorganic ionic species into single cells.[12]

In CE-PC and CE-VC, detection is based on the activation of ligand-gated ion channels. On ligand binding to such receptors, ions flow either into or out of the cell, resulting in a measurable current. The transmembrane flux of ions is created by concentration and electrical driving forces, and can be controlled by changing the transmembrane potential and ionic composition of the intracellular and extracellular solutions. When the CE effluent is directed onto the surface of a cell, the electrophoresis electrolyte effectively becomes the extracellular solution. Therefore, proper ionic composition of the CE running buffer is critical because of the requirements set forth by the Nernst equation and the permeability of the ion channel. According to the Nernst equation, the equilibrium potential of an ionic species, $V_{eq}$, can be written as:

$$V_{eq} = \frac{RT}{zF} \ln([X]_o/[X]_i) \tag{6}$$

where $R$ is the gas constant (8.314 J/mol · K), $T$ is the temperature (K), $z$ is the valence of the ion, $F$ is Faraday's constant ($9.6487 \times 10^4$ A · sec/mol), and $[X]_o$ and $[X]_i$ are the extracellular and intracellular concentrations, respectively, of the ionic species. At 20°, $RT/zF$ is about 25 mV for a monovalent cation. For simplicity, external and internal ion activities are assumed to be equal and thus cancel. By knowing the membrane (holding) potential, $V_m$ (which can be set by a command voltage in the patch clamp amplifier), and the reversal potential (i.e., $V_{eq}$) given by the Nernst equation, the ionic current, $I$, through an open channel can be calculated from

$$I = \gamma(V_m - V_{eq}) \tag{7}$$

where $\gamma(S)$ is the conductance of the ion channel. Thus, if the reversal potential is determined for a specific ligand-gated ion-channel system (by selection of internal and external ionic compositions) and a proper holding potential is chosen, then it is possible to optimize the direction and amplitude of the transmembrane current.

[12] S. Nussberger, F. Foret, S. C. Hebert, B. L. Karger, and M. A. Hediger, *Biophys. J.* **70,** 998 (1996).

In our experiments, HEPES–saline buffers are used as electrolytes in the CE capillaries, vials, and as cell bath media. For the CE-PC experiments, the buffer consists of 140 m*M* NaCl, 5 m*M* KCl, 1 m*M* $CaCl_2$, 1 m*M* $MgCl_2$, 10 m*M* HEPES, and 10 m*M* D-glucose (pH 7.4, NaOH). In the CE-VC experiments, the buffer consists of 95 m*M* NaCl, 2 m*M* KCl, 2 m*M* $CaCl_2$, and 10 m*M* HEPES (pH 7.5, NaOH). Other buffers, depending on application or preference, can be used. There are some receptor ligands that require certain molecules or ions as cofactors to activate the ion channel. In these cases, the electrolyte composition must be modified to include such species. For example, in the detection of *N*-methyl-D-aspartate(NMDA)-evoked receptor responses, a $Mg^{2+}$-free glycine-supplemented HEPES–saline buffer is used.

## Fabrication of Fractured Electrophoresis Capillaries

The voltage applied to a CE capillary is typically 10–30 kV, which creates electric field strengths of several hundred volts per centimeter. As mentioned earlier, typical CE setups have the outlet reservoir as the cathodic end, which is usually ground. When cell-based biosensors are used, the CE outlet reservoir is a chamber, which contains the cells and the bath solution, positioned on a microscope. To avoid high-offset potentials operating on the cell membrane during electrophoresis, the CE capillary is fractured and connected to ground 3–5 cm above the outlet. Effectively, this connection to ground functions as a voltage divider and the residual potential is compensated by an offset potential to the patch-clamp amplifier system.[9] Fractured CE capillaries were originally developed for ultrasensitive electrochemical detection.[13] Alternatives to cracking the capillaries might include laser-drilling holes in the capillary walls, as has been used for on-line chemical derivatization in CE[14] or the use of a floating (virtual) ground that keeps the cell bath close to zero potential.[15]

We have developed a simple technique to fracture CE capillaries that efficiently decreases the CE outlet potential. These assemblies are robust and can be used for several months. Fused silica capillaries 10–50 cm long with 20- to 50-$\mu$m inner diameters and 370-$\mu$m outer diameters (Polymicro Technologies Inc., Phoenix, AZ) are used. The outlet end of the capillary is first fixed to a microscope coverslip with a small piece of adhesive tape. A 0.5-in section of glass tubing (e.g., disposable microliter pipette glass capillaries; World Precision Instruments, Saratoga, FL) with an inner diame-

[13] R. A. Wallingford and A. G. Ewing, *Anal. Chem.* **59,** 1762 (1987).
[14] S. L. Pentoney, X. Huang, D. S. Burgi, and R. N. Zare, *Anal. Chem.* **60,** 2625 (1988).
[15] P. Vaughan and M. Trotter, *Can. J. Physiol. Pharmacol.* **60,** 604 (1982).

ter that matches the outer diameter of the CE capillary is threaded coaxially onto the CE capillary from the outlet. This tubing is immobilized on both sides with a small droplet of cyanoacrylate glue (which is also called Super Glue or Crazy Glue) so that the outlet end of the glass tubing is located exactly at the position where the fracture is to be (Fig. 1A). The glue is allowed to cure, and the capillary is then filled with distilled water. A 1-ml syringe, in which the needle has been inserted into Teflon or medical tubing, serves this purpose. The inlet of the capillary is inserted into the tubing (whose inner diameter matches the outer diameter of the capillary) and the capillary is easily filled. The coverslip mount is then placed on the stage of a stereomicroscope.

A ceramic capillary cutter (Polymicro) is used to make a vertical cut through both the outer polyimide coating and the fused silica wall. The size of the incision is difficult to control exactly, but should allow just a

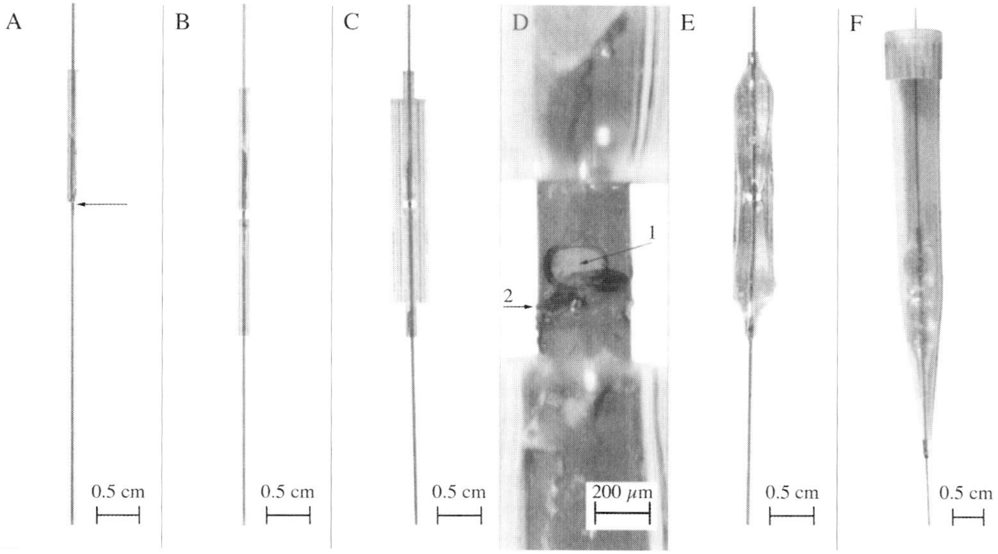


FIG. 1. Construction of a fractured CE capillary. The outlet of the capillary is at the bottom of each frame. (A) Glass tubing is threaded onto the CE capillary and immobilized on both sides. A small incision is made at the position of the arrowhead using a capillary cutter. (B) Another piece of 0.5-in-long glass tubing is threaded onto the CE capillary and immobilized 1–2 mm from the first piece. (C) Two sections of glass tubing are aligned, centered around the incision, and fixed to the pieces of coaxially mounted glass tubing. The area near the fracture is tapped and the fracture is completed. (D) Photomicrograph of the incision produced by the capillary cutter (1) and the crack (2). (E) The assembly is embedded with epoxy. (F) The mount is inserted and fixed with epoxy into a 1-ml plastic pipette tip. (Photographs courtesy of Susanne Orwar.)

small droplet of water to emerge from the capillary interior. By pushing water through the capillary with the syringe, it is possible to check the split ratio between the incision and the outlet. We have found, empirically, that if approximately 50% of the injected volume elutes from the CE outlet, then cracks with good grounding characteristics are produced. Following this, another 0.5-in section of glass tubing is threaded onto the CE capillary. This piece is placed within 1–2 mm from the other glass tubing, centered around the incision, and immobilized at both ends using cyanoacrylate glue (Fig. 1B). When the glue has cured, two sections of 0.5-in glass tubing are aligned in parallel, centered around the incision, and fixed with cyanoacrylate glue to the coaxially mounted pieces of glass tubing (Fig. 1C).

When the glue is cured, tapping around the incision using tweezers or a small metallic rod completes the fracture (Fig. 1D). The assembly is then lifted from the coverslip and embedded with a two-component epoxy glue (Fig. 1E). Application of glue into the fractured area should be avoided. When the epoxy has cured, the mount is inserted and fixed with epoxy resin into a 1-ml plastic pipette tip (Fig. 1F). The apical end of the pipette tip is glued with epoxy onto the capillary. The pipette reservoir is then filled with the same electrolyte used as CE running buffer and connected to ground with a platinum wire. Offset potentials with our system (+12-kV, 40-cm-long, 50-$\mu$m inner diameter capillaries) are about 10–20 mV, which can easily be compensated with the voltage-clamp amplifier.[9]

## Cell Preparation

### *Patch-Clamp Detection*

Because virtually any cell type can be used in CE-PC, we have used acutely isolated neurons, primary cultures, or cultured cell lines as detectors. The cells are cultured and maintained according to standard tissue culture protocols. Suspended cells are preferred for whole-cell recordings and immobilized cells are preferred for outside-out patch recordings. Glass substrates coated with poly(L-lysine) or poly(D-ornithine), for example, can be used to immobilize acutely isolated or nonadherent cells.

Although the use of transfected cells expressing a single recombinant receptor type would allow for novel analyte selection, acutely isolated interneurons from the rat olfactory bulb have been used typically in our studies. These cells can be used to detect inhibitory and excitatory amino acid neurotransmitters and transmitter mimetics. Among other receptors, these cells express inhibitory $GABA_A$ and Gly receptors and excitatory receptors belonging to the glutamate receptor family, such as the (*R,S*)-$\alpha$-amino-3-hydroxy-5-methyl-4-isoxazole propionate (AMPA), kainate, and

NMDA types. The procedure for isolating olfactory bulb interneurons is as follows: newborn or adult rats (10–200 g) are anesthetized in halothane (ISC Chemicals Ltd., Avonmouth, England) and decapitated. The olfactory bulbs are then dissected from the brains, sliced into four pieces, and placed in an incubation chamber. The incubation chamber contains proteases from *Aspergillus oryzae* (2.5 mg/ml), which are dissolved in prewarmed (32°) HEPES–saline buffer containing 140 m*M* NaCl, 5 m*M* KCl, 10 m*M* HEPES, 10 m*M* glucose, 2 m*M* $MgCl_2$ (pH 7.4, NaOH). After 25–30 min, the slices are washed with protease-free buffer solution for 20 min. During both the enzymatic treatment and washing, the solutions are continuously perfused with a gas mixture containing 95% (v/v) $O_2$ and 5% (v/v) $CO_2$.

Following enzymatic treatment and washing, the slices are kept in a HEPES–saline buffer held at 20°, containing 1 m*M* $CaCl_2$, that is continuously perfused with 95% $O_2$ and 5% $CO_2$. The slices are then disintegrated by shear forces with gentle suction through the tip of a fire-polished Pasteur pipette. The cell suspension is then placed in a petri dish and diluted by $Ca^{2+}$-containing (1 m*M*) HEPES–saline buffer. To detect NMDA receptor-mediated responses, a $Mg^{2+}$-free and $Ca^{2+}$-containing HEPES–saline buffer supplemented with 10 $\mu M$ glycine is used. The petri dish is transferred to the microscope stage. Viable interneurons can be harvested up to 6 hr after the interruption of the enzymatic treatment. Chemicals and enzymes are obtained from Sigma Chemical (St. Louis, MO).

*Two-Electrode Voltage-Clamp Detection*

Oocytes from the South African clawed frog (*Xenopus laevis*) are widely used as expression systems for a broad spectrum of proteins.[16] Oocyte detectors with tailormade selectivities and response characteristics can be produced[7] by injection of *in vitro* transcribed mRNA that encode for ligand-gated ion channels.[17,18] The sparse expression of self-made ion channels in these oocytes makes them well suited for functional studies of ligand-gated and voltage-dependent ion channels. Because the literature on the expression of ion channels in *Xenopus* oocytes is comprehensive, this paper is limited to the procedures used for CE-VC detection in our laboratories.

*Xenopus laevis* frogs are available from several companies including Xenopus Ltd. (Nutfield, UK), Nasco Biologicals, Inc. (Fort Atkinson, WI) and Dipl.Biol.-Dipl.Ing., Horst Kähler Institut für Entwicklungsbiologie (Hamburg, Germany). The procedure to prepare *Xenopus* oocytes for two-

[16] J. B. Gurdon, C. D. Lane, H. R. Woodland, and G. Marbaix, *Nature* **233,** 177 (1971).

[17] D. A. Melton, P. A. Kreig, M. R. Rebagliati, T. Maniatis, K. Zinn, and M. R. Green, *Nucleic Acids Res.* **12,** 7035 (1984).

[18] R. Swanson and K. Folander, *Methods Enzymol.* **207,** 310 (1992).

electrode voltage clamp experiments includes isolation of oocytes from the frogs, (preparation and) injection of RNA, and defolliculation. The oocytes have to rest for at least 24 hr between each procedure. In our protocol, *Xenopus* frogs are anesthetized on ice, and the oocytes are surgically removed from the ovarium with forceps. The incision is closed with a suture.

Many different methods exist for isolation of RNA from brain tissue. The method described below, which uses a strong denaturant, was originally devised to isolate RNA from cells that cannot be separated easily into cytoplasm and nuclei (e.g., frozen fragments of tissue), or from cells that are particularly rich in RNases (e.g., pancreatic cells). The method is modified from Glisin *et al.*[19] and Ullrich *et al.*[20] Pieces (0.3 g) of rat (Harlan UK United, Blackthorn, UK) brains are placed in prechilled Falcon tubes containing 4 *M* guanidine isothiocyanate, 25 m*M* sodium acetate (pH 6.0), and 0.14 *M* 2-mercaptoethanol. The cell lysates are homogenized with a Polytron homogenizer (Brinkman Instruments Inc., Westbury, NY) and centrifuged (400 rpm, 2 min) at room temperature. Homogenization shears the nuclear DNA and prevents the formation of an impenetrable mat, which can hinder sedimentation of the RNA. The supernatant is then layered onto a cushion of 5.7 *M* CsCl and 10 m*M* ethylenediaminetetraacetic acid (EDTA) (pH 7.5), prepared in RNase-free water, in a clear RNase-free ultracentrifuge tube (Beckman Instruments Inc., Fullerton, CA). The position of the top of the cushion is marked on the outside of the tube, and the contents are centrifuged at 21° for 21 hr at a speed of 40,000 rpm (SW 55-rotor, Beckman Instruments Inc., Fullerton, CA). A swinging bucket rotor is preferred to a fixed-angle rotor so that the RNA is deposited at the bottom of the tube, rather than along the walls (where it can come into contact with the cell lysate). The centrifuge tube should be handled carefully to avoid any mixing of its contents.

A line is drawn on the outside of each tube 0.5 cm from the bottom. The fluid above the level of the cushion (upper mark) is then removed and discarded. With a fresh, RNase-free pipette, the fluid above the lower mark is removed and discarded, and the bottom of the tube is cut off, just above the level of the remaining fluid, with a heated razor blade or a scalpel. The tube is inverted to allow the fluid to drain onto a pad of Kimwipes, then returned to an upright position (and the RNA pellet verified). The pellet is washed with 70% (v/v) ethanol to remove remaining CsCl, dissolved in 300 $\mu$l of 0.3 *M* sodium acetate (pH 6.0), and the RNA-containing solution is transferred to an Eppendorf tube. To precipitate the RNA, 750 $\mu$l of

[19] V. Glisin, R. Crkvenjakov, and C. Byus, *Biochemistry* **13,** 2366 (1974).

[20] A. Ullrich, J. Shine, J. Chirgwin, R. Pictet, E. Tischer, W. J. Rutter, and H. M. Goodman, *Science* **196,** 1313 (1974).

ice-cold ethanol (99.7%) is added to the tube and stored overnight at $-20°$. To collect the RNA precipitate, the solution is centrifuged (13,000 rpm, 10 min, 4°). The pellet is then washed with 70% ethanol, centrifuged (13,000 rpm, 5 min, 4°), and the ethanol is removed. Because all ethanol has to be removed to avoid poisoning of the oocytes, the RNA pellet is allowed to dry in a vacuum. The RNA is then dissolved in a small volume of RNase-free water and stored at $-80°$ until used.

Stage V or VI oocytes are microinjected with 50 nl total rat brain RNA[21] or *in vitro* transcribed mRNA (~0.1 $\mu$g/$\mu$l) expressing cloned rat serotonin (5-hydroxytryptamine, 5-HT) 5HT1c receptors.[7] Stage II or III oocytes can also be used if a smaller membrane capacitance is desired, which speeds up the clamping process of the oocyte's membrane potential.[22] A microdispenser (Drummond Scientific Company, Brookmall, PA), which is mounted on a micromanipulator, is used for injection of the RNA. The oocyte is penetrated by the microdispenser tip and injected at the vegetal (yellowish) hemisphere. Asymmetrical ("tongue-formed") tips are preferred and the microdispenser needle should be heated in an oven (200°, 16 hr) to be free from RNases. Following injection of RNA, small clusters of oocytes are defolliculated by gentle agitation in 2 mg/ml collagenase (type IA; Sigma) for 2 hr. Alternatively, oocytes can be gently treated with collagenase (1 mg/ml) for 1 hr, then the next day placed in a solution of high osmolarity. Following this step, (after 30 seconds) the follicular cell layer becomes visible through a microscope and can be removed manually with forceps. For proper translation of the injected RNA, the oocytes are then incubated 3–6 days in modified Barth's solution: 88 m*M* NaCl, 1 m*M* KCl, 2.7 m*M* $NaHCO_3$, 10 m*M* HEPES, 0.82 m*M* $MgSO_4$, 0.33 m*M* $Ca(NO_3)_2$, 0.41 m*M* $CaCl_2$, 100 $\mu$g/ml gentamicin (pH 7.5, NaOH) at 19°.

## Capillary Electrophoresis–Patch-Clamp Recording

### *Patch Clamping*

In a patch-clamp experiment a cell is attached to the tip of a glass microelectrode by controlled suction to yield a high-resistance seal.[23,24] It is then possible to choose one of several modes such as outside-out, inside-out, or whole-cell recording. For detailed explanations on the theory and practice of patch-clamping, the reader is referred to Refs. 23 and 24.

[21] A. L. Goldin, *Methods Enzymol.* **207,** 266 (1992).

[22] D. S. Kraffe and H. A. Lester, *Methods Enzymol.* **207,** 340 (1992).

[23] E. Neher and B. Sakmann, *Nature* **260,** 799 (1976).

[24] B. Sakmann and E. Neher, "Single-Channel Recording," 2nd Ed., Plenum Press, New York, 1995.

*Pipette Solutions*

Patch-clamp pipette internal solutions serve to mimic the cytosol. By including or excluding certain ionic species in the pipette electrode solution, and by choosing a proper holding potential, the direction and magnitude of the transmembrane ionic flow can be controlled for a given ion channel. In our experiments we use two kinds of pipette solutions. For detection of ligands (e.g., glutamate, kainate, and NMDA) that activate receptor or ion-channel systems permeable to $Na^+$, $K^+$, and $Ca^{2+}$, a solution consisting of 100 m*M* KF, 2 m*M* $MgCl_2$, 1 m*M* $CaCl_2$, 11 m*M* ethylene glycol bis($\beta$-aminoethyl ether)-*N*,*N*,*N*′,*N*′-tetraacetic acid (EGTA), and 10 m*M* HEPES (pH 7.2, KOH) is used. When preparing this solution, KF must be added gradually to avoid precipitation. For ligands that activate $Cl^-$-mediating receptor–ion-channel complexes (e.g., $GABA_A$ and strychnine-sensitive Gly receptors) a buffer containing 140 m*M* CsCl, 1 m*M* $MgCl_2$, 1 m*M* $CaCl_2$, 11 m*M* EGTA, and 10 m*M* HEPES (pH 7.4, KOH) is used. To maintain cellular energy and redox status, 2 m*M* Mg-ATP and 5 m*M* reduced glutathione can be added to the patch-clamp pipette internal solution.[25] Chemicals can be obtained from Sigma.

*Regeneration of Capillary and Buffer Vial Maintenance*

To obtain efficient electro-osmotic flow and high precision in migration times, it is necessary to keep the silanol groups ionized. The capillary can be regenerated by flushing 1 *M* NaOH followed by 0.1 *M* NaOH, distilled water, and running buffer. It is sometimes advantagenous to include an organic solvent such as methanol or ethanol in the washing step, especially if hydrophobic substances have been injected into the CE capillary.

It is important to change the solution of the reservoir covering the crack in the CE capillary periodically—preferably after each run. Otherwise, receptor ligands might diffuse into the CE capillary and depending on the concentration, either increase the background noise or cause distinctive response features. Likewise, the solution of the inlet buffer vial should be changed periodically, especially if the sample contains high concentrations of ligands. The risk for analyte crossover and ubiquitous injection[26] can cause a continuous supply of receptor ligands into the capillary, again causing activation of receptors in an uncontrollable way.

[25] A. Marty and E. Neher, "Single-Channel Recording," 2nd Ed., p. 45. Plenum Press, New York, 1995.

[26] H. A. Fishman, N. M. Amudi, T. T. Lee, R. H. Scheller and R. N. Zare, *Anal. Chem.* **66,** 2318 (1994).

*Instrumentation*

In our experiments, signals are recorded with a patch-clamp amplifier (Model List L/M EPC, List-Electronics, Darmstadt, Germany or Axopatch 200B, Axon Instruments, Foster City, CA), digitized (20 kHz, PCM 2 A/D VCR adapter, Medical Systems Corp., NY) and then stored on videotape until data analysis. For the production of complete electropherograms, the signals from the videotape are digitized at 2 Hz.

The experimental setup is shown in Fig. 2. An inverted microscope is used for viewing the cells and controlling the experiment. The microscope is placed inside a Faraday cage, and all power supplies or voltage sources are placed outside the cage. The Faraday cage is constructued from 2-mm-thick aluminum plates, and may include sections of aluminum screen to allow easier viewing of the microscope region. The Faraday cage serves to shield the experiment from electronic and magnetic fields that may interfere with the patch-clamp recordings. Shielding becomes especially critical if these experiments are performed in atypical electrophysiology laboratories. Specifically, if the patch-clamp setup is in the vicinity of mechanical pumps and high-powered pulsed lasers, then additional precautions are essential. These include shielding (and grounding) the headstage and pipette holder with aluminum foil, and having an isolated circuit breaker for the patch-clamp amplifier. Also, the microscope lamp should be powered by a voltage source (preferably dc to avoid interference from the line frequency) that

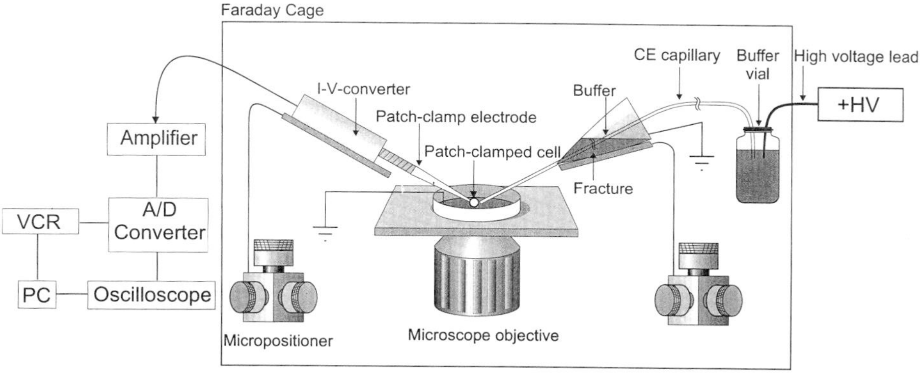


FIG. 2. Patch-clamp detection system for capillary electrophoresis (CE). The inlet of the separation capillary is inserted into a buffer vial, which is connected to a high-voltage power supply. The CE capillary is fractured and connected to ground 3 cm above the outlet. The tip of the patch-clamp electrode is positioned ~5–25 $\mu$m from the capillary outlet with a micropositioner.

is a separate unit from the microscope (thus placed outside the Faraday cage). When these issues are addressed, the rms noise decreases by nearly 10-fold (with lasers on). Further details of shielding patch-clamp experiments from electronic and magnetic noise also can be found elsewhere.[24,27]

Electrophoresis is performed by applying a positive potential (up to 30 kV) to the inlet of the CE capillary with a high-voltage supply (LKB, Bromma, Sweden). The CE capillary and the high-voltage lead (i.e., inlet side) are inserted into a buffer vial conained in a polycarbonate holder with a 1-inch wall thickness. The polycarbonate vial holder is equipped with an interlock for safety precautions against electrical shock. The outlet of the CE capillary is mounted on a micromanipulator, and positioned into the cell-containing buffer at an angle of ~30° with respect to the microscope stage. Injection of sample solution into the CE capillary is performed hydrodynamically by raising the inlet of the capillary 10 cm above the level of the capillary outlet for 5–10 sec. The volume, $v_s$, of a sample hydrodynamically injected into a CE capillary can be calculated using the Hagen–Poiseuille equation:

$$v_s = \Delta P r^4 \pi t (8 \eta L)^{-1} \tag{8}$$

where $\Delta P$ is the pressure drop over the capillary, $r$ the internal radius of the capillary, $t$ the injection time, $\eta$ the viscosity of the injected sample, and $L$ the total length of the capillary. Typically, our injection volumes are in the low-nanoliter range.[26]

The patch-clamp pipette with the cell attached to the tip is placed 5–25 $\mu$m from the center of the outlet of the CE capillary. The pipette is fixed by a headstage that is connected to the patch-clamp amplifier.[24,27] A chlorided silver wire, functioning as a reference electrode, is placed in the cell bath proximal to the CE capillary outlet. It is important to keep this wire well chlorided in order to effectively keep the cell bath at a low potential. This, in turn, will reduce the effect of the capillary-induced potential on the cells. The headstage is firmly mounted on a micromanipulator. It is essential that at least the micromanipulator holding the headstage and patch-clamp pipette be of a high-graduation type to allow exact positioning of the patch-clamped cell at the CE outlet. Other components included in our patch-clamp setup are an oscilloscope, a videotape recorder, an A/D converter, Butterworth and Bessel filters, and a computer system for sampling of data. In addition, a pipette puller for production of patch pipettes and a microforge for fire-polishing the electrode tips are required.

[27] R. A. Levis and J. L. Rae, *Methods Enzymol.* **207,** 14 (1992).

*Example of CE-PC Data*

Figure 3 shows the CE separations of GABA, Glu, and NMDA detected with outside-out patch-clamp detection. In addition to migration times, the characteristics of the current responses elicited by the separated agonists further confirm their identities. For example, analysis (see below) of single-channel conductances, distribution of conductance states, and ion-channel open and shut times yields information that can identify the ligand–receptor interaction.[9] In essence, it is feasible to detect indirectly a single analyte molecule with this technique.[8] When whole-cell patch-clamp detection is employed, current traces of separated components give mean single-channel conductance levels, $\gamma$ (in pS) and corner frequencies, $f_{ci}$ (in Hz).

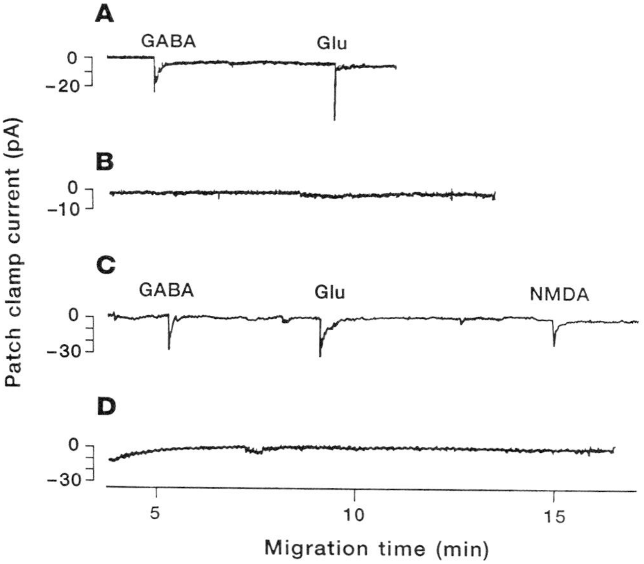


FIG. 3. Inward currents recorded from an outside-out patch following separation of GABA, Glu, and NMDA (250 $\mu M$ of each) by capillary electrophoresis. (A) Separation of GABA and Glu. (B) Separation of HEPES–saline onto same patch used in (A). (C) Separation of GABA, Glu, and NMDA. (D) Separation of HEPES-saline onto same patch used in (C). Current traces were sampled at 2 Hz. The patch-clamp electrode contained CsCl solution. The holding potential was −70 mV and the applied high voltage was +12 kV. Different capillaries were used for the separations and their respective controls. [Reprinted with permission from O. Orwar, K. Jardemark, I. Jacobson, A. Moscho, H. A. Fishman, R. H. Scheller, and R. N. Zare, *Science* **272,** 1779 (1996). Copyright 1996 American Association for the Advancement of Science.]

*Data Analysis*

A strength of the CE-PC detection technique is that it generates a high amount of information that can be used to identify a certain type of receptor interaction and the ligand that caused it. Electrophoretic migration rates together with several parameters pertaining to the electrical properties of the receptor in its activated and quiescent state are used for this purpose. In what follows, a brief account of patch-clamp data analysis that is pertinent to CE-PC is given. It is important to note that rather than characterize receptor physiology, we exploit receptors as detectors in CE and use a few key parameters for coupled receptor–ligand identification. More in-depth aspects on analysis of patch-clamp recorded current traces can be found in Ref. 24.

Electrophoretic migration rates are obtained by analyzing the time it takes from sample injection to detector response. Because peak responses obtained for desensitizing receptors do not track the physical presence (Gaussian concentration profile) of the analyte, the onset of the response is used as an index of a ligand's migration time. For nondesensitizing ligand–receptor interactions, the migration time is noted as the center of the normal-distributed detector response.

Various analyses can be performed on patch-clamp recorded currents. Spectral analysis of whole-cell currents yields information about ion-channel kinetics and the mean single-channel conductance. These parameters correlate to a specific ligand–receptor interaction. For spectral analysis of whole-cell currents, the signal from the videoadapter is filtered with an eight-pole Butterworth filter (bandwidth 1 kHz, −3 dB) and is digitized at 2 kHz. Records are divided into 0.5-sec blocks prior to calculation of the spectral density. The mean power spectrum is calculated by averaging all power spectra obtained (at least 20) from these blocks. Agonist-induced power spectra are subtracted from power spectra obtained during membrane resting conditions. The resulting power spectrum is fitted by a single or double Lorentzian function using a least-squares Levenberg–Marquardt algorithm with proportional weighting.[28] Apparent single-channel conductances can be estimated according to the following equation:

$$\gamma = \sigma^2/[(E - E_r)I_m] \tag{9}$$

where $\sigma^2$ is the current variance, $E$ the holding potential, $E_r$ the reversal potential, and $I_m$ the mean current.

[28] D. Colquhoun and A. Hawkes, *Proc. R. Soc. Lond. B. Biol. Sci.* **199,** 231 (1997).

Unitary ion-channel events can be resolved using outside-out patch-clamp recording.[23,24] Because of their small sizes, these patches have gigaohm (GΩ)-resistance glass-to-membrane seals and extremely low membrane capacitances. In contrast to mean single-channel conductances that are calculated from a population of ion channels, intrinsic properties of the receptor complex such as distributions of conductance levels, transition between different conductance states, and channel open and closed times can be analyzed. The distribution of conductance states in a single receptor type can be displayed in amplitude histograms representing a "fingerprint" of the activated channel.[24]

Current-to-voltage (*I–V*) relationships can be obtained from current responses evoked by continuous superfusion of agonists, and on-the-fly from responses caused by separated analytes. Together with knowledge about the ionic composition of the intracellular (i.e., pipette) and extracellular solutions, reversal potentials and rectification of the *I–V* curve yield information about ion-channel permeability and identity of the activated receptor.[29] For elimination of responses evoked by voltage-dependent ion channels, the *I–V* curve obtained between the responses is subtracted from the ramp obtained during the agonist-activated responses.

## Two-Electrode Voltage-Clamp Recording

Figure 4 shows a two-electrode voltage-clamped *Xenopus* oocyte during a CE experiment. The configuration includes voltage-recording (Fig. 4A) and current-injecting (Fig. 4B) electrodes (both of which penetrate the oocyte), a CE capillary positioned at a short distance from the oocyte membrane (Fig. 4C), and a reference electrode (Fig. 4D).[30] The current-passing and voltage-measuring electrodes are made from borosilicate capillaries and are back-filled with 0.3–3 *M* KCl. The electrodes are mounted on headstages, which, in turn, are connected to a voltage-clamp amplifier. The headstages and electrodes are shielded by a Faraday cage. The oocyte for recording is placed in a small polycarbonate chamber filled with a HEPES buffer solution. The capillary outlet is placed between the electrodes, proximal to the animal hemisphere (dark area) of the oocyte. Procedures for capillary and buffer vial maintenance are as described in the previous section.

[29] B. Hille, "Ionic Channels of Excitable Membranes," Sinauer, Sunderland, Massachusetts, 1992.

[30] W. Stühmer, *Methods Enzymol.* **207,** 319 (1992).

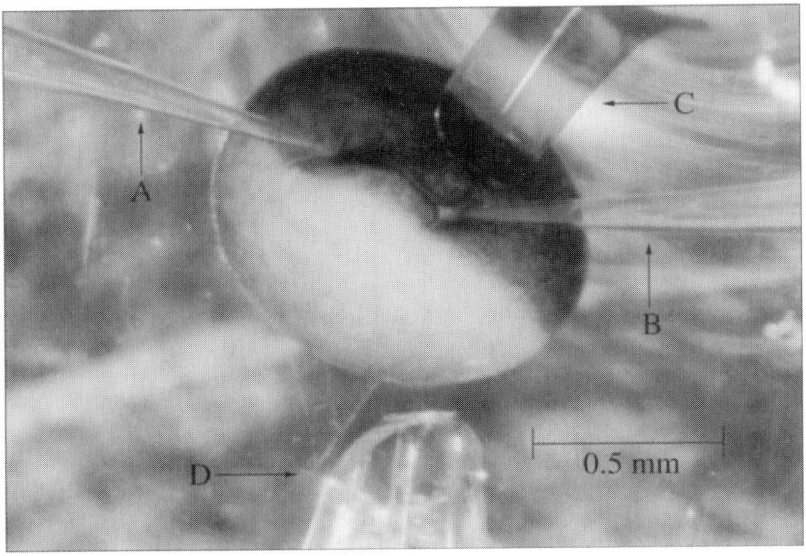


FIG. 4. A photomicrograph showing a xenopus during a capillary electrophoresis experiment. The voltage-recording electrode (A) and a current-injecting electrode (B) penetrate the oocyte membrane. The capillary outlet (C) is placed ~20 $\mu$m from the surface of the oocyte's animal hemisphere. (D) Reference electrode. (Photograph courtesy of Susanne Orwar.)

### *Example of CE-VC Data*

Electropherograms of 5-HT and a buffer blank are shown in Fig. 5. Detection was performed using two-electrode voltage-clamped *Xenopus* oocytes expressing the cloned rat 5HT1c receptor.[7]

The *Xenopus* oocyte detection system is more mechanically stable than the CE-PC system, the latter of which is limited by the relatively short lifetime of the patch pipettes (no longer than 30 min). In addition, an oocyte in the two-electrode voltage-clamp system can be used for several hours and lends itself for quantitative measurements, since multiple doses can be injected onto the same oocyte and reliable dose–response curves can be created. Also, when using preferably stage II and III oocytes with an appropriate sampling rate, power spectra can be obtained and analyzed to yield information about the mean single-channel conductance for a specific ion channel.

## Summary

By coupling CE to voltage-clamp biosensors, a highly selective and sensitive means for analyzing biologically active components in complex

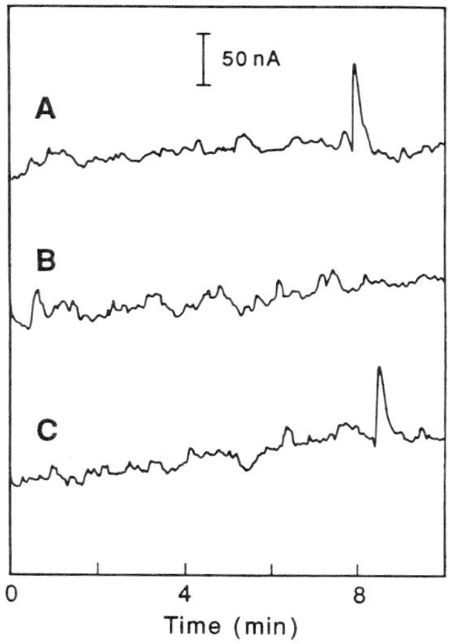


FIG. 5. Electropherograms demonstrating detection of serotonin (5-HT) using CE-VC. The membrane potential of a *Xenopus* oocyte, expressing the cloned rat 5HT1c receptor, was clamped at −70 mV, and the signal current was measured. The traces have been inverted to show typical positive-going CE peaks. (A) 5-HT (100 $\mu M$) was injected and a peak is detected at ~8 min. (B) Blank buffer was injected, and gives no response. (C) 5-HT was injected again and the electropherogram demonstrated a reproducible detected response at approximately the same migration time as in (A). The solutions were separated at ~250 V/cm in a 40-$\mu$m-inside-diameter capillary. [Reprinted with permission from J. B. Shear, H. A. Fishman, N. L. Allbritton, D. Garigan, R. N. Zare, and R. H. Scheller, *Science* **267,** 74 (1995). Copyright 1995 American Association for the Advancement of Science.]

mixtures with minimum sample handling is obtained. In contrast to traditional techniques for chemical analyses, this system can be engineered for sensitivity to specific biological species, thereby improving the response selectivity and greatly simplifying the challenge of chemical analysis. CE-PC and CE-VC are complementary techniques. CE–PC detection is a highly sensitive and information-rich technique that enables the study of electrophoretically separated ligands activating a single ion channel. The technique suffers from short lifetimes of the patch pipettes and is therefore mainly qualitative. CE-VC detection, on the other hand, might become a quantitative technique because the detector cells are viable for several hours. Even if mean single-channel conductances and *I–V* relationships can be obtained with CE-VC, the trade-off with this format is that much less information can be extracted from the current traces compared to patch-clamp recorded data.

## Acknowledgment

This work is supported by the Swedish Natural Science Research Council (NFR), the Swedish Foundation for Strategic Research (SSF), and the U.S. National Institute of Drug Abuse (NIDA). S.J.L. thanks the U.S. National Institutes of Health (NIH) for a postdoctoral fellowship.

# [10] Ion Channels as Tools to Monitor Lipid Bilayer–Membrane Protein Interactions: Gramicidin Channels as Molecular Force Transducers

*By* O. S. Andersen, C. Nielsen, A. M. Maer, J. A. Lundbæk, M. Goulian, and R. E. Koeppe II

## Introduction

Numerous studies show that membrane protein function depends on the bilayer lipid composition.[1,2] Specifically, the function of integral membrane proteins varies with lipid bilayer thickness[3–7] and monolayer equilibrium curvature.[8–11] In most cases there is only modest chemical specificity in these membrane lipid–protein interactions.[1] This lack of chemical specificity, together with the large number of lipid types that are found in the membranes of any given cell, has caused difficulties for attempts to understand how the function of integral membrane proteins is affected by the bilayer lipid composition. These difficulties arose in part because the results were interpreted within the framework of the Singer–Nicolson *fluid mosaic membrane* model.[12]

[1] P. F. Devaux and M. Seigneuret, *Biochim. Biophys. Acta* **822,** 63 (1985).
[2] A. Bienvenüe and J. S. Marie, *Curr. Top. Membr.* **40,** 319 (1994).
[3] M. Caffrey and G. W. Feigenson, *Biochemistry* **20,** 1949 (1981).
[4] A. Johannsson, G. A. Smith, and J. C. Metcalfe, *Biochem. Biophys. Acta* **641,** 416 (1981).
[5] M. Criado, H. Eibl, and F. J. Barrantes, *J. Biol. Chem.* **259,** 9188 (1984).
[6] P. A. Baldwin and W. L. Hubbell, *Biochemistry* **24,** 2633 (1985).
[7] M. F. Brown, *Chem. Phys. Lipids* **73,** 159 (1994).
[8] J. Navarro, M. Toivio-Kinnucan, and E. Racker, *Biochemistry* **23,** 130 (1984).
[9] J. W. Jensen and J. S. Schutzbach, *Biochemistry* **23,** 1115 (1984).
[10] S.-W. Hui and A. Sen, *Proc. Natl. Acad. Sci. U.S.A.* **86,** 5825 (1989).
[11] C. D. McCallum and R. M. Epand, *Biochemistry* **34,** 1815 (1995).
[12] S. J. Singer and G. L. Nicolson, *Science* **175,** 720 (1972).



0076-6879/99 $30.00

The fluid mosaic membrane model evolved from thermodynamic considerations relating to the organization of the main membrane components: phospholipids, cholesterol, and proteins.[12] The guiding principles of the model were, first, that the lipids are organized in a liquid–crystalline bilayer in which integral membrane proteins are embedded and, second, the need to maximize hydrophobic and hydrophilic interactions. These principles have influenced all subsequent work—and led to the notion of hydrophobic coupling between the hydrophobic exterior surface of the membrane-spanning part of integral membrane proteins and the bilayer hydrophobic core. If the hydrophobic coupling were sufficiently strong, the thickness of the bilayer core adjacent to the protein would equal the length of the exterior, hydrophobic surface of the protein. A weakness of the fluid mosaic membrane model was that the lipid bilayer component was assumed to be a passive entity only—a permeability barrier that separated the extracellular and intracellular aqueous phases. This point of view was strengthened by numerous studies on the permeability properties of lipid bilayers to small polar solutes (e.g., Ref. 13), which showed that the lipid bilayer could be approximated as being a ~5 nm thin sheet of liquid hydrocarbon.

The view of the lipid bilayer as a sheet of liquid hydrocarbon led to the notion of bilayer fluidity as an important determinant of protein function. The limitations of this notion were exposed by Lee,[14] who pointed out that the changes in protein function were often associated with changes in the ligand affinity of the protein, which suggests that the membrane lipids somehow affect the conformational preference of bilayer-embedded proteins. Such lipid-dependent effects on the conformation of integral membrane proteins are difficult to impossible to explain within the scope of models that invoke changes in membrane fluidity as being a primary determinant of protein function. If not fluidity, however, what then?

An important, but neglected consequence of the liquid–crystalline organization of lipid bilayers is that one needs to incorporate the bilayer material properties (thickness and compression modulus, curvature and bending modulus) into a description of membrane protein organization and function.[15–17] Similarly, one needs to consider specifically the importance of geometric packing criteria[18] for lipid–protein interactions and protein function.

[13] A. Walter and J. Gutknecht, *J. Membr. Biol.* **77,** 255 (1986).

[14] A. G. Lee, *Prog. Lipid Res.* **30,** 323 (1991).

[15] W. Helfrich, *Z. Naturforsch.* **28C,** 693 (1973).

[16] E. A. Evans and R. M. Hochmuth, *Curr. Top. Membr. Transp.* **10,** 1 (1978).

[17] O. G. Mouritsen and M. Bloom, *Biophys. J.* **46,** 141 (1984).

[18] J. N. Israelachvili, *Biochim. Biophys. Acta* **469,** 221 (1977).

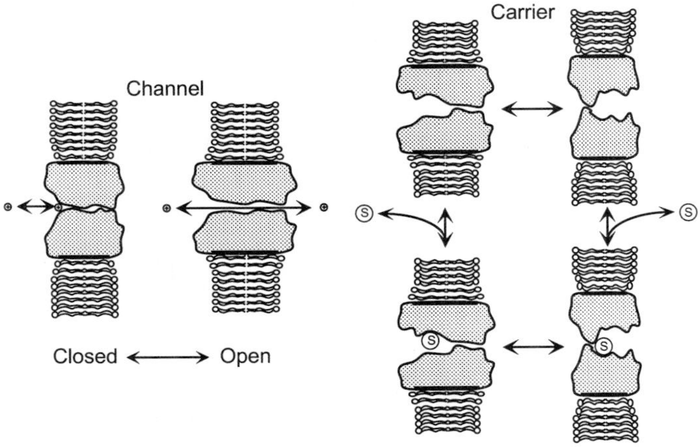


FIG. 1. Solute transfer by membrane-spanning channels and conformational carrier. Solute (ion) transfer through membrane-spanning channels does not involve major changes in channel structure. Solute transfer by conformational carriers is intimately coupled to protein conformational changes, which in turn may perturb the structure of the surrounding bilayer.

## Protein Conformational Changes and Bilayer Perturbations

The bilayer material properties can affect membrane protein function because the sequence of protein (conformational) state changes that are associated with normal protein function may involve changes in protein structure that affect the protein–lipid interface.[19,20] The thermodynamic need to maximize hydrophobic interactions between the membrane-spanning domain of integral membrane proteins and the bilayer core will couple these protein conformational changes to changes in the structure of the immediately surrounding bilayer (Fig. 1). The liquid–crystalline organization of the lipid bilayer means that this perturbation of the bilayer structure has an energetic cost—the membrane deformation energy ($\Delta G^0_{\mathrm{def}}$).[21,22] This becomes important for protein function because the free-energy difference ($\Delta G^0_{\mathrm{tot}}$) between two protein conformations is the sum of contributions from the protein per se ($\Delta G^0_{\mathrm{prot}}$) and terms that arise from the protein's interactions with the environment, which include $\Delta G^0_{\mathrm{def}}$. The material properties of the bilayer, and thus $\Delta G^0_{\mathrm{def}}$, will vary as a function of the membrane

[19] P. N. T. Unwin and P. D. Ennis, *Nature* **307,** 609 (1984).

[20] N. Unwin, C. Toyoshima, and E. Kubalek, *J. Cell. Biol.* **107,** 1123 (1988).

[21] H. W. Huang, *Biophys. J.* **50,** 1061 (1986).

[22] P. Helfrich and E. Jakobsson, *Biophys. J.* **57,** 1075 (1990).

lipid composition, which provides a mechanism for the control of the protein conformational preference and function by the bilayer.

A protein conformational change that involves a change in the hydrophobic length of the membrane-spanning segment of the protein will cause both compression and bending of the two monolayers. The relative contribution of these two, independent modes of membrane deformation to $\Delta G^0_{def}$ depends on the bilayer material properties (thickness and compression modulus; monolayer equilibrium curvature and bending modulus). The equilibrium curvature of a lipid monolayer is determined by the variation of the intermolecular lateral interactions along the molecular axis, which usually is expressed in terms of the effective "shape" of the lipids in the monolayer[23] (Fig. 2). Lipids that have a cylindrical "shape" form planar monolayers with no equilibrium curvature. If the effective cross-sectional area of the polar headgroup region is larger than that of the acyl chains, the monolayer will have a positive equilibrium curvature. If the effective cross-sectional area is less than that of the acyl chains, the monolayer will have a negative curvature.

In bilayers, the two monolayers must have complementary curvatures. Thus, the formation of a (planar) bilayer by lipids that by themselves would tend to form curved monolayers will cause a change in the effective shape of the lipid molecules because of the requirement for a uniform cross-sectional area/molecule across a planar bilayer. The energy that is required to change the lipid "shape" causes a stress in the bilayer. The attractive interactions between the two monolayers in a bilayer mean that one must distinguish between the equilibrium curvature of the monolayer, which is determined by the average lipid shape, and the curvature of the bilayer, which is determined by the coupled monolayers. Whenever the equilibrium monolayer curvature differs from the bilayer curvature, the bilayer will be under a curvature-induced stress.

## Membrane Perturbations and Channel Function

Pharmacologic modification of lipid bilayers that affect their material properties, as monitored by changes in the energetics of gramicidin channel–lipid bilayer interactions, also affect the function of integral membrane proteins.[24] This is shown in Fig. 3, which shows the reversible effect of Triton X-100 on voltage-dependent sodium channels ($\mu_1$ subtype) expressed in human embryonic kidney (HEK293) cells. Triton X-100 promotes a

[23] P. R. Cullis and B. de Kruijff, *Biochim. Biophys. Acta* **559,** 399 (1979).

[24] J. A. Lundbæk, P. Birn, J. Girshman, A. J. Hansen, and O. S. Andersen, *Biochemistry* **35,** 3825 (1996).

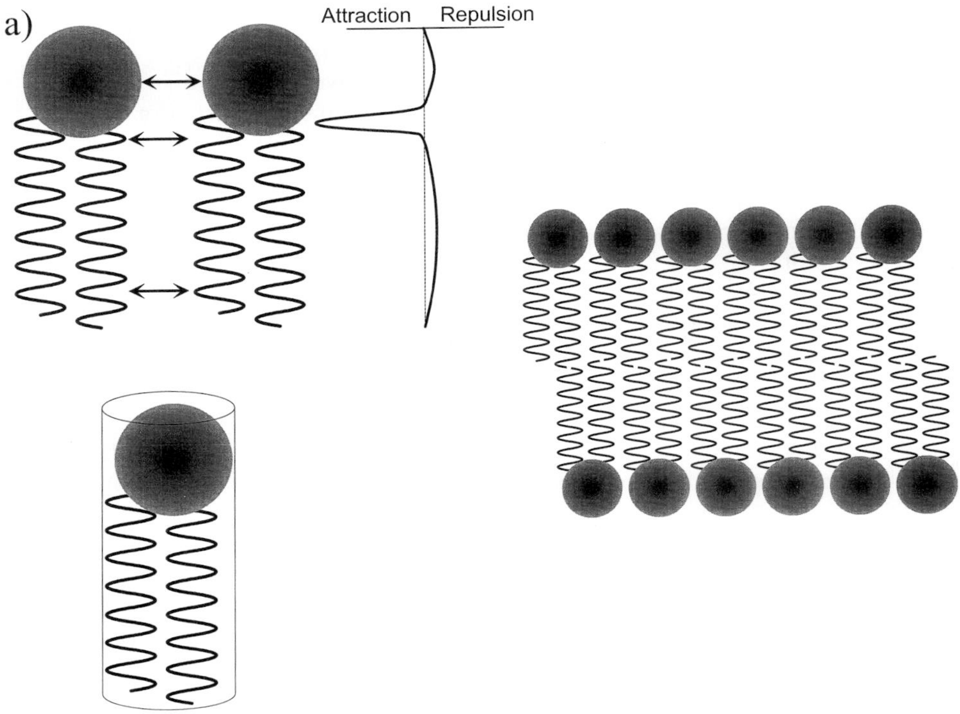


FIG. 2. Shape of lipid molecules and bilayer curvature stress. (a) Formation of "relaxed" bilayers from molecules that in isolation have a "cylindrical shape" as described by the profile of intermolecular interactions along the molecule length (*top* and *bottom*, left-hand side). The isolated monolayer will have zero curvature, and two monolayers form a relaxed bilayer. (b) Formation of bilayers that are under stress from molecules that in isolation have a "cone shape" as described by the profile of intermolecular interactions along the molecule length (*top* and *bottom* left-hand side). The isolated monolayer will have positive curvature, and two monolayers will form a frustrated bilayer in which the individual molecules are forced into an approximately cylindrical shape.

positive monolayer curvature, which on *a priori* grounds would be expected to stabilize gramicidin channels—as is the case.[24,25] Triton also stabilizes reversibly one or more inactivated states in voltage-dependent calcium[24] and sodium channels—and the densensitized state of nicotinic acetylcholine receptors.[26] This commonality suggests that the Triton X-100 alters channel

[25] D. B. Sawyer, R. E. I. Koeppe, and O. S. Andersen, *Biochemistry* **28,** 6571 (1989).
[26] M. Kasai, T. R. Podleski, and J.-P. Changeux, *FEBS Lett.* **7,** 13 (1970).

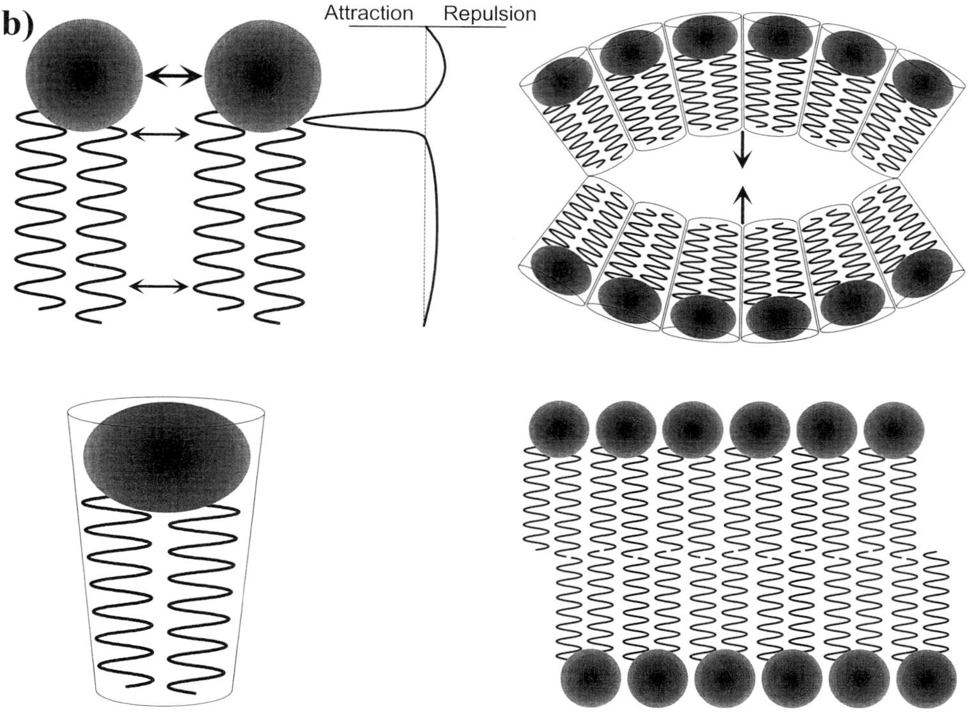


FIG. 2. (*continued*)

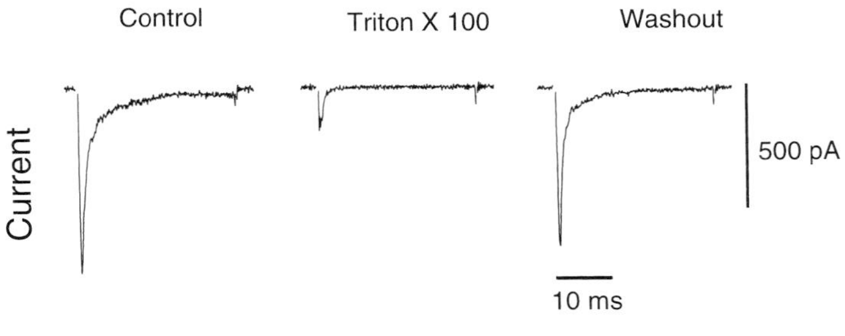


FIG. 3. Effect of Triton X-100 on sodium channel currents. Rat muscle $\mu_1$ sodium channel $\alpha$ subunits were expressed in HEK293 cells. Membrane currents were measured using the whole-cell configuration following a depolarization from $-80$ mV to $-10$ mV. The current signal was corrected for leak conductance and filtered at 3 kHz. The three traces show sodium channel currents before, during, and 8 min after superperfusion with 100 $\mu M$ Triton X-100. The plasmid carrying the $\mu_1$ message was a generous gift from Dr. Gail Mandel (SUNY at Stony Brook).

function by altering some general membrane property, such as the membrane deformation energy associated with the transition from closed or open channel states to inactivated or desensitized states.

## Membrane Deformation Energy

In the simplest description, the magnitude of $\Delta G^0_{\text{def}}$ depends on the bilayer compression–expansion and splay–distortion moduli, the bilayer tension, and the spontaneous monolayer curvature. The different energy contributions can be unified using the theory of liquid crystal elastic deformations,[21,22,27] which allows the interdependent (compression, bending, etc.) contributions to $\Delta G^0_{\text{def}}$ to be evaluated. The interdependence among the components arises because the bilayer responds to an imposed distortion by minimizing the overall deformation free energy by varying both the compression–expansion, spray–distortion, etc., contributions to $\Delta G^0_{\text{def}}$. Each of these contributions varies as a function of the bilayer deformation profile—and a change in the profile that minimizes, say, the compression–expansion contribution, will also affect the magnitude of the splay–distortion contribution.

Subject to the choice of boundary conditions, the $\Delta G^0_{\text{def}}$ associated with a local change in bilayer thickness can be approximated as a quadratic function of the extent of the membrane deformation ($u$), the difference between the membrane hydrophobic thickness ($d$) and the protein's hydrophobic exterior length ($l$)[21,24,27]:

$$\Delta G^0_{\text{def}} = Au^2 = A(d - l)^2 \tag{1}$$

where $A$ is a phenomenologic spring constant associated with the membrane deformation. [Equation (1) is exact for a restricted set of boundary conditions only.[27]]

$\Delta G^0_{\text{def}}$ is comprised of several contributions. Figure 4 shows the variation in $\Delta G^0_{\text{def}}$ and its two major components, $\Delta G^0_{\text{CE}}$ (the compression–expansion component) and $\Delta G_{\text{SD}}$ (the splay–distortion component), as a function of distance from a membrane-spanning gramicidin dimer. The splay–distortion contribution dominates close to the channel–bilayer interface, whereas the compression–expansion contribution is dominant further away from the channel. The relative contributions of the $\Delta G_{\text{CE}}$ and $\Delta G_{\text{SD}}$ components to $\Delta G^0_{\text{def}}$ are comparable. Consequently, a pharmacologically induced change in the magnitude of $\Delta G^0_{\text{def}}$ cannot be attributed solely to a change in $\Delta G_{\text{CE}}$ or $\Delta G_{\text{SD}}$, which has implications for understanding how the lipid bilayer composition and material properties affect protein function. The

[27] C. Nielsen, M. Goulian, and O. S. Andersen, *Biophys. J.* **74,** 1966 (1998).

membrane perturbation extends ~30 Å from the dimer, but most of the deformation energy (about 75%) arises in the region between $r = 10$ Å and $r = 20$ Å, which corresponds to the annulus of lipid molecules that are in immediate contact with the dimer. This implies that one may be able to effect large changes in the membrane deformation energy by rather modest changes in the composition of this boundary layer, which could have implications for understanding how lipid-soluble, or amphipathic, substances affect the conformational preference of (and function) of integral membrane proteins.

The question thus becomes: Is the bilayer deformation energy of sufficient magnitude to affect the protein conformational preference and function? The total deformation free energy for the situation described in Fig. 4 is 11.9 $kT$. The quaternary conformational changes in gap junction channels (with radius $r_0 = 30$ Å) involve changes in the channels' hydrophobic length of 0.3 Å.[19] If the channels (with $l = 30$ Å) are imbedded in a bilayer with $d = 30$ Å (perfect hydrophobic match), the $\Delta G^0_{\text{def}}$ contribution to $\Delta G^0_{\text{tot}}$ for the open ↔ close transition will be ~0.8 $kT$. For the same transition in a membrane with $d = 32$ Å, the $\Delta G^0_{\text{def}}$ contribution will be ~6 $kT$. Even larger deformation energies can result if the spontaneous monolayer

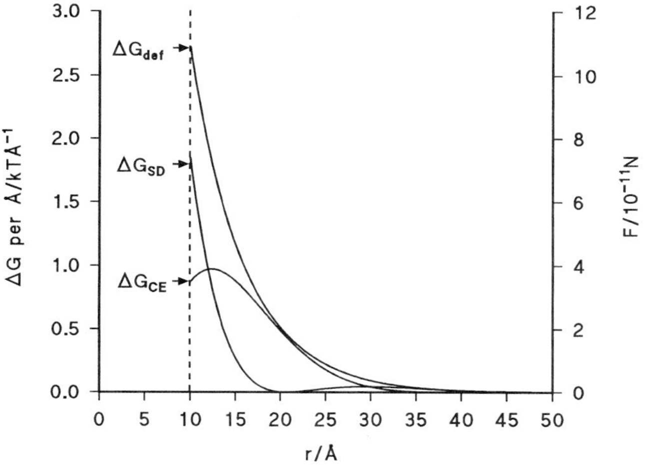


FIG. 4. Membrane deformation energy components per unit lenght as a function of radial distance from the bilayer/channel boundary. The membrane deformation $u = d - l = 4$ Å. $\Delta G_{\text{def}}$ denotes the total deformation energy, $\Delta G_{\text{SD}}$ the splay–distortion component, and $\Delta G_{\text{CE}}$ the compression–expansion component. The splay–distortion modulus $K_c$, the compression–expansion modulus $K_a$ and the membrane thickness $d$ are connected by scaling laws (e.g., Ref. 27). The example was calculated with $K_c = 2.85 \cdot 10^{-10}$ N · Å, $K_a = 2.85 \cdot 10^{-11}$ N/Å, and $d = 28.5$ Å.

curvature is different from zero. These estimates depend, however, on the choice of boundary conditions—and the assumption that the material constants that are determined in "macroscopic" experiments on isolated bilayers are valid to describe events close to the protein–bilayer boundary. It therefore is necessary to have quantitative measurements that probe how the lipid bilayer affects structurally well-defined conformational transitions in membrane proteins.[28]

## Carrier versus Channel: Choice of Reporter Protein

The energetic coupling between proteins and bilayers will, in principle, affect the function of all imbedded proteins. Membrane-spanning channels and conformational carriers, however, catalyze the transmembrane transfer of selected solutes by fundamentally different mechanisms (Fig. 1). In channels, the control of function arises from conformational changes between nonconducting (closed) and conducting (open) states. The individual catalytic events (the transfer of a solute/ion across the membrane) are uncoupled from the protein conformational changes associated with channel gating. In carriers, or ATP-driven pumps, the catalytic event(s) are inextricably coupled to protein conformation changes. Moreover, the continued operation of such membrane proteins reflects the continued cycling through the different kinetic states.

This difference in catalytic mechanism has implications for how the function of channels and carriers is affected by the lipid bilayer. In either case, a change in bilayer material properties (in $\Delta G^0_{\mathrm{def}}$) will alter the equilibrium distribution between different protein conformers. In the case of membrane-spanning channels, the distribution between conducting and nonconducting states (as well as the kinetics of the transitions between these states) will be altered. One can monitor directly the distribution between nonconducting (closed) and conducting (open) channel states (conformations) by measuring the changes in the channel-mediated ionic current, which provides for a direct readout of the number of conducting channels. In the case of conformational carriers, or pumps, a change in the equilibrium constant between the major conformers (the binding site exposed toward the left or the right in Fig. 1) will affect the rate constants for both the left → right and right → left transitions (usually in opposite directions). Systematic studies on how the protein function (turnover rate) is affected by the membrane lipid composition thus may show that the turnover rate is a nonmonotonic function of lipid composition (of

[28] S. M. Gruner, *in* "Biologically Inspired Physics" (L. Peliti ed.), pp. 127–135. Plenum Press, New York, 1991.

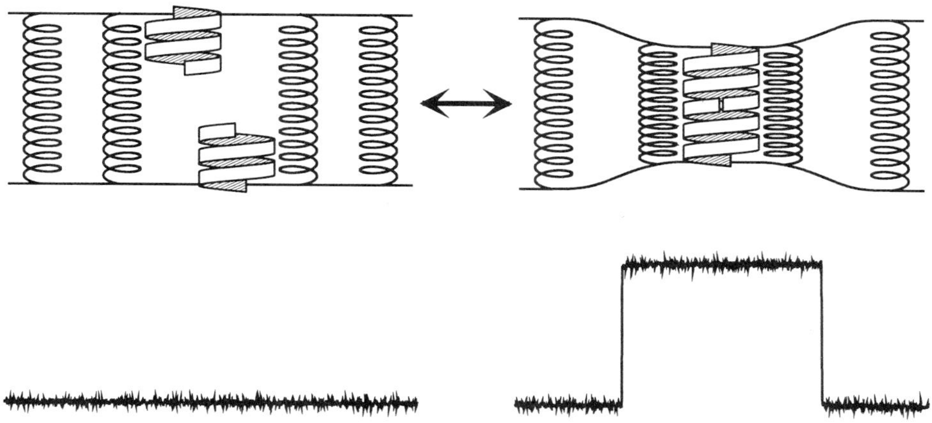

FIG. 5. Schematic representation of gramicidin channel formation by two membrane-inserted $\beta^{6.3}$-helical monomers. The average membrane thickness is larger than the length of the membrane-spanning dimer, and channel formation is associated with a membrane "dimpling." Channel formation can be monitored electrophysiologically by the appearance of the single-channel current events (channel formation is upward, channel dissociation is downward).

$\Delta G^0_{\rm def}$) because of the opposing effects on the left $\rightarrow$ right and the right $\rightarrow$ left rate constants. This complicates attempts to understand how a change in bilayer material properties will affect the carrier function. Ion conducting channels therefore offer advantages not enjoyed by the carriers for attempts to elucidate the basis for bilayer control of protein function.

## Molecular Force Transducers

Among ion permeable channels, the gramicidin monomer $\rightleftharpoons$ dimer equilibrium that is associated with the formation of membrane-spanning gramicidin channels constitutes a reasonably well-defined structural transition in a membrane inclusion (Fig. 5). Standard gramicidin channels are miniproteins formed by the transmembrane assembly[29] of two $\beta^{6.3}$-helical monomers[30] that join at their formyl-NH termini to form the conducting channels (see Refs. 31, 32, and 33 for reviews). Most, if not all, membrane-

[29] A. M. O'Connell, R. E. Koeppe II, and O. S. Andersen, *Science* **250,** 1256 (1990).

[30] K. He, S. J. Ludtke, Y. Wu, H. W. Huang, O. S. Andersen, D. Greathouse, and R. E. I. Koeppe, *Biophys. Chem.* **49,** 83 (1994).

[31] O. S. Andersen and R. E. I. Koeppe II, *Physiol. Rev.* **72,** S89 (1992).

[32] J. A. Killian, *Biochim. Biophys. Acta* **1113,** 391 (1992).

[33] R. E. I. Koeppe and O. S. Andersen, *Annu. Rev. Biophys. Biomol. Struct.* **25,** 231 (1996).

spanning gramicidin dimers are conducting channels,[34] and there is no evidence for chemical specificity in the interactions between gramicidin channels and their host bilayer.[35,36] These properties make the gramicidins suitable for use as molecular force transducers for investigating the mechanical properties of lipid bilayers.

Gramicidin channels can be used as force transducers because channel formation in lipid bilayers with a hydrophobic thickness, $d$, that is different than the hydrophobic length, $l$, of the gramicidin dimer forces the bilayer to "dimple" or "pucker" as it adapts its hydrophobic thickness to the channel length. Any alterations in the ability of the bilayer to adjust to the channel will alter the equilibrium constant for channel formation. A membrane perturbant, such as the odorant limonene, 1-methyl-4-(1-ethylethenyl)cyclohexene, does not interact specifically with gramicidin channels. Nevertheless, limonene has significant effects on gramicidin channels, which can be observed as changes in the average channel lifetime (Fig. 6). This result can be rationalized by considering the effective "shape" of this strongly hydrophobic molecule (cf. Fig. 2). Limonene has no polar moiety so, when incorporated in bilayers, these molecules will induce a lateral pressure in the bilayer interior, and the limonene molecules can be considered to be cone shaped with the base of the cone at the hydrophobic core of the bilayer. A planar membrane formed from phospholipids, doped with limonene, will be in a state of (curvature) stress, because the presence of the limonene will tend to drive the overall membrane shape toward concave surfaces. This curvature stress causes the twofold change in average lifetime that is observed.

## Measuring $\Delta\Delta G^0_{def}$ and Phenomenological Spring Constant

The principle underlying the use of gramicidin channels as molecular force transducers is simple: to monitor how the gramicidin monomer $\rightleftharpoons$ dimer equilibrium is affected by maneuvers that alter the bilayer properties. The practical implementation can be done in several different ways: by measuring the equilibrium constant for channel formation as a change in membrane conductance; by measuring the disjoining force the bilayer imposes on the channels as a change in channel lifetime; and by measuring the equilibrium distribution between channels of different length.

[34] W. R. Veatch, R. Mathies, M. Eisenberg, and L. Stryer, *J. Mol. Biol.* **99,** 75 (1975).

[35] L. L. Providence, O. S. Andersen, D. V. Greathouse, R. E. I. Koeppe II, and R. Bittman, *Biochemistry* **34,** 16404 (1995).

[36] J. Girshman, J. V. Greathouse, R. E. I. Koeppe II, and O. S. Andersen, *Biophys. J.* **73,** 1310 (1997).

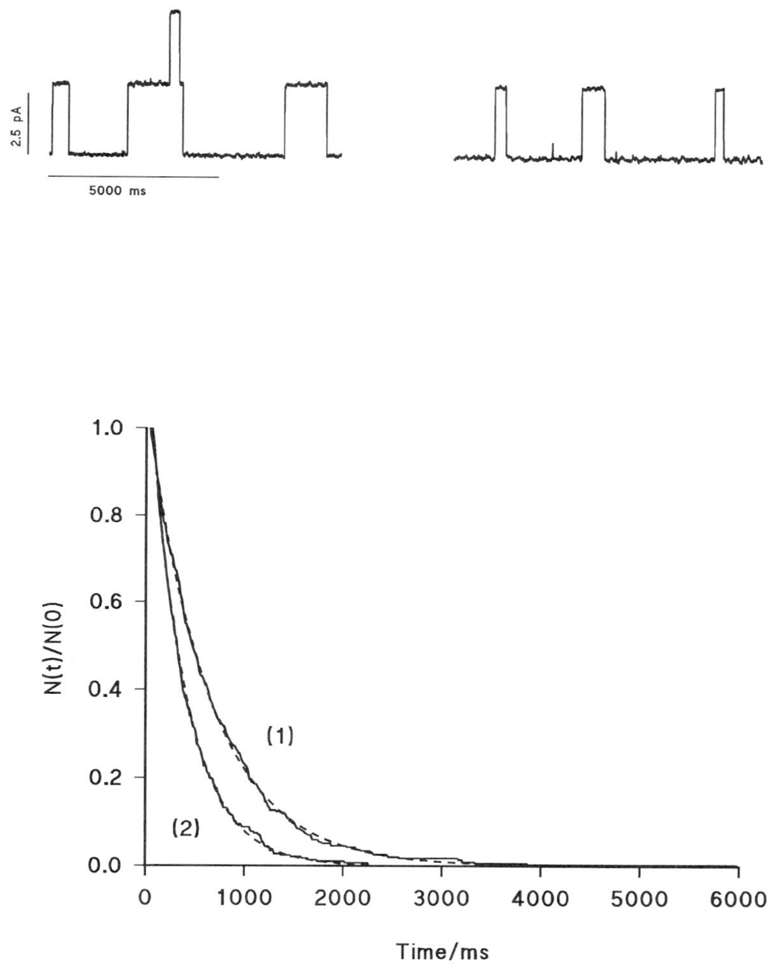


FIG. 6. The effect of limonene on the behavior of gramicidin A channels in dioleoylphosphatidylcholine/*n*-decand bilayers. *Top:* Single-channel current traces before (left) and a few minutes after (right) the addition of 1 m*M* limonene to the aqueous solution on both sides of the bilayer. *Bottom:* Normalized survivor histograms. Curve (1) denotes results in the absence of limonene; curve (2) denotes results obtained in the presence of 1 m*M* limonene. The interrupted curves denote the best fit of a single exponential distribution to the results: $N(t)/N(0) = \exp(-t/\tau)$, where $N(0)$ and $N(t)$ denote the number of channels with lifetimes longer than time zero and time $t$. In the absence of limonene, $\tau = 630$ ms, $N(0) = 365$; in the presence of limonene, $\tau = 370$ ms, $N(0) = 328$. The reduction in $N(0)$ reflects a reduction in the channel appearance rate. Experimental conditions as in Ref. 42; applied potential 200 mV, current signal filtered at 200 Hz, 1 *M* NaCl, and 22°.

*Equilibrium Constant Approach*

The gramicidin dimerization constant $K_D$ is given by

$$K_D = [D]/[M]^2 \tag{2}$$

where [D] and [M] denote the surface densities of gramicidin dimers and monomers. Assuming that $\Delta G^0_{def}$ is the only extrinsic contribution to $\Delta G^0_{tot}$,

$$\begin{aligned} K_D &= \frac{[D]}{[M]^2} = \exp[-\Delta G^0_{tot}/kT] = \exp[-(\Delta G^0_{prot} + \Delta G^0_{def})/kT] \\ &= K^{prot}_D \exp(-\Delta G^0_{def}/kT) \end{aligned} \tag{3}$$

where $k$ is Boltzmann's constant, $T$ the temperature in Kelvin, and

$$K^{prot}_D = \exp(-\Delta G^0_{prot}/kT) \tag{4}$$

For practical use, Eq. (3) is rewritten as

$$\Delta G^0_{def} = -kT \ln\left(\frac{[D]}{K^{prot}_D [M]^2}\right) \tag{5}$$

The gA channel-associated membrane conductance $G$ is proportional to the number of gramicidin channels in the membrane:

$$G = [D] \cdot g \tag{6}$$

where $g$ is the single-channel conductance. Combining Eqs. (5) and (6),

$$\Delta G^0_{def} = -kT \ln\left(\frac{G/g}{K^{prot}_D [M]^2}\right) \tag{7}$$

which provides the desired link between bilayer energetics and electrophysiologic measurements.

In practice, it is most convenient to use Eq. (7) to measure *changes* in $\Delta G^0_{def}$ ($\Delta\Delta G^0_{def}$) in the limit when $[D] \ll [M]$[37,38]:

$$\Delta\Delta G^0_{def} = -kT \ln\left\{\frac{G_{II}/g_{II}}{G_I/g_I}\right\} \tag{8}$$

where the subscripts (I and II) denote the two different experimental situations that are being compared. $\Delta G^0_{def}$ can be modified by pharmacologic means,[37,38] e.g., by the addition of compounds that alter the equilibrium

[37] J. A. Lundbæk and O. S. Andersen, *J. Gen. Physiol.* **104,** 645 (1994).

[38] J. A. Lundbæk, A. M. Maer, and O. S. Andersen, *Biochemistry* **36,** 5695 (1997).

monolayer curvature. Relatively modest modifications of the bilayer properties can change $\Delta G^0_{def}$ by 10–15 kJ/mol[37,38]—indicating that the bilayer deformation energy may be of sufficient magnitude to affect protein function.

The phenomenologic spring constant $A$ [cf. Eq. (1)] scales with the protein radius.[27] Thus, once $A$ is determined for any protein one can estimate its value for other proteins, assuming that the boundary conditions at the protein/lipid contact are similar for the two proteins. The value of $A$ can be estimated in experiments where the extent of the membrane deformation, $u = d - l$, is varied systematically. One can vary the channel length, $l$, at a constant membrane thickness, $d$, by changing the length of the amino acid sequence of gramicidin or by varying the membrane thickness at a constant channel length by changing the length of the phospholipid acyl chains. Using Eq. (1):

$$\Delta G^0 = \Delta G^0_{prot} + \Delta G^0_{def} = \Delta G^0_{prot} + Au^2 = \Delta G^0_{prot} + A(d - l)^2 \quad (9)$$

or

$$\Delta\Delta G^0 = A[(d + \mu - l)^2 - (d - l)^2] = A[(u + \mu)^2 - u^2] = A\mu(2u + \mu) \quad (10)$$

where $\mu$ denotes the change in $u$. Assuming that the single-channel conductance does not vary as a function of $u$, $A$ can be determined from the following relation:

$$\Delta\Delta G^0_{def} = -kT \ln\left\{\frac{G_{u+\mu}}{G_u}\right\} = A\mu(2u + \mu) \quad (11)$$

It is important to note that, as $u + \mu$ increases, the notion of strong hydrophobic coupling eventually will fail (meaning that $u + \mu \neq d - l$) because $\Delta G^0_{def}$, $= A(u + \mu)^2$, will become so large that it becomes advantageous to allow hydrophobic residues to be in direct contact with water. When $u + \mu \neq d - l$, i.e., when there is slippage, Eq. (11) may give rise to an erroneous determination of $A$.

This problem applies to integral membrane proteins as well as gramicidin channels. For example, the effective spring constant for membrane deformations adjacent to an integral membrane protein of radius 30 Å is ~4 kJ/(mol · Å$^2$).[27] For the same protein, the hydrophobic penalty associated with a hydrophobic mismatch is ~20 kJ/(mol · Å). The membrane deformation energy increases as a quadratic function of $u$; the hydrophobic energy increases only as a linear function of the hydrophobic mismatch. This means the incremental deformation energy eventually will exceed the incremental hydrophobic energy, which in this example will occur when $u > 2.5$ Å. Strong hydrophobic coupling thus will fail for larger membrane deformations. In fact, there could be slippage even for rather modest deformations.

Another limitation is the assumption that a change in $u$ will affect only $\Delta G^0_{def}$. Even though the nonconducting $\beta^{6.3}$-helical gramicidin monomers are inserted into the bilayer, there may be an energetic penalty associated with monomer insertion. That may affect the equilibrium distribution between folded (membrane-inserted) and unfolded (membrane-adsorbed) monomer conformations.[39] To the extent this energy penalty varies as a function of the thickness–length mismatch $(d - l)$, it will affect the determination of $\Delta\Delta G^0_{def}$.

*Channel Lifetime Approach*

Rather than the membrane conductance, one can measure the disjoining force the bilayer imposes on the membrane-spanning gramicidin dimers, which affect both the association ($k_1$) and dissociation ($k_{-1}$) rate constants. $k_{-1}$ is of primary interest because $k_{-1} = 1/\tau$, where $\tau$ is the average dimer (channel) lifetime, which is directly measurable.

$$k_{-1} = \frac{1}{\tau_0}\exp(-\Delta G^{\ddagger}/kT) \tag{12}$$

where $\Delta G^{\ddagger}$ is the activation energy for dimer dissociation and $1/\tau_0$ is a frequency factor (in Eyring's transition state theory $1/\tau_0 = kT/h$). The transition state for dimer dissociation occurs when the monomers move a distance $\delta$ apart, and $\Delta G^{\ddagger}$ is the sum of the intrinsic activation energy $\Delta G^{\ddagger}_{prot}$ and the difference in bilayer deformation energy ($\Delta G^{\ddagger}_{def}$) for a deformation of $u - \delta$ and $u$, respectively. Using Eq. (1),

$$\begin{aligned}\Delta G^{\ddagger} &= \Delta G^{\ddagger}_{prot} + \Delta G^{\ddagger}_{def} = \Delta G^{\ddagger}_{prot} + A([u - \delta]^2 - u^2) \\ &= \Delta G^{\ddagger}_{prot} - A(2u - \delta)\delta\end{aligned} \tag{13}$$

cf. Ref. 24, and

$$\tau = \tau_{prot}\exp\{\Delta G^{\ddagger}_{def}/kT\} = \tau_{prot}\exp\{-A(2u - \delta)\delta/kT\} \tag{14}$$

where $\tau_{prot} = \tau_0 \cdot \exp\{\Delta G^{\ddagger}_{prot}/kT\}$. When strong hydrophobic coupling pertains (when $u = d - l$), $A$ can be determined from the variation of $\tau$ as a function of $d$ (J. A. Lundbæk and O. S. Andersen, manuscript in preparation). $A$ is large, again indicating that the bilayer deformation energy associated with a hydrophobic mismatch may be of sufficient magnitude to affect protein function.

[39] N. Mobashery, C. Nielsen, O. S. Andersen, *FEBS Lett.* **412,** 15 (1997).

The advantage of the lifetime approach is that it is more convenient to measure channel lifetimes than the large membrane conductance. In addition, one does not need to consider changes in the energetic cost of monomer insertion into the monolayers. The disadvantage is that the precise value of $\delta$ is not known. The transition state most likely is when the dimer is stabilized by only four intermolecular hydrogen bonds,[24] in which case $\delta =$ 1.6 Å. But that estimate could be off by a factor of 2.

*Heterodimer Formation*

A third approach to measure $\Delta\Delta G^0_{\text{def}}$, or $A$, is to determine the relative stabilizaton of heterodimers formed between two gramicidin analogs that differ in length by, say, two amino acid residues. (The gramicidin sequences must differ in length by an even number of residues, in order to ensure that the heterodimers are stabilized by six hydrogen bonds.[40] The formation of the heterodimeric channels can be described by the reaction (cf. Greathouse *et al.*,[40a] this volume):

$$A_2 + B_2 \rightleftharpoons AB + BA \tag{15}$$

Let the channels formed by two gramicidin analogs, e.g., a 14- and a 16-residue gramicidin (A and B, respectively), differ in length by $2\lambda$. The membrane deformations associated with the formation of $A_2$ and $B_2$ will be $u$ and $u - 2\lambda$, and the deformation associated with heterodimer formation will be $u - \lambda$. Assuming there are no monomer-specific interactions, i.e., that $\Delta G^0_{\text{prot}}$ is the same for all three channel types,

$$\Delta\Delta G^0_{\text{def}} = A\{2(u - \lambda)^2 - [(u - 2\lambda)^2 + u^2]\}/2 = -A\lambda^2 \tag{16}$$

The advantages of this approach are that it is a one-parameter equation, one does not need to know $u(d$ or $l)$, and $\lambda$ is quite well determined (1.6 Å per L-D-dipeptide). The disadvantage is that the assumption that $\Delta G^0_{\text{prot}}$ is invariant among the different channel types is difficult to verify.

## Conclusions

The bilayer and its embedded proteins exert reciprocal effects on each other: protein conformational change $\rightleftharpoons$ bilayer deformation energy. This reciprocity emphasizes the dynamic implications of the hydrophobic coupling between bilayer and proteins. That is, in addition to serving as an

[40] J. T. Durkin, L. L. Providence, R. E. I. Koeppe II, and O. S. Andersen, *J. Mol. Biol.* **231,** 1102 (1993).

[40a] D. V. Greathouse, R. E. I. Koeppe II, L. L. Providence, S. Shobana, and O. S. Andersen, *Methods Enzymol.* **294** [28], 1998 (this volume).

organizing principle for the folding and insertion of membrane proteins, the need to minimize the exposure of hydrophobic groups to water[12] also provides a means for the regulation of protein function by the bilayer. The control of protein function by the membrane lipids is, *to a first approximation,* a "simple" energetic question, which can be addressed using the continuum theory of liquid–crystal deformations with minimal chemical specificity and measured by monitoring changes in channel open probability. Chemical specificity will be important in some cases; but that is likely to be the exception rather than the rule. This simplifies the interpretational problems considerably, because one can disregard specific chemical identity of the numerous lipid types that are present in biological membranes. The situation becomes similar to that for electrified interfaces, where the Gouy–Chapman theory of the diffuse double layer serves as a major organizing principle (e.g., Ref. 41) and where chemical specific interactions are introduced only when absolutely needed.

## Acknowledgments

This work was supported by NIH grants GM21342 (O.S.A.) and GM34968 (R.E.K.), the Danish Research Council (C.N.), NSF (A.M.M.), the William Keck Foundation (M.G.), and the Norman & Rosita Winston Foundation (C.N.).

[41] S. McLaughlin, *Ann. Rev. Biophys. Chem.* **18,** 113 (1989).
[42] O. S. Andersen, *Biophys. J.* **41,** 119 (1983).

# Section II

# Purification and Reconstitution

# [11] Purification and Reconstitution of Epithelial Chloride Channel Cystic Fibrosis Transmembrane Conductance Regulator

*By* MOHABIR RAMJEESINGH, ELIZABETH GARAMI, KEVIN GALLEY, CANHUI LI, YANCHUN WANG, and CHRISTINE E. BEAR

## Introduction

When the cystic fibrosis (CF) gene was first discovered, its protein product, the cystic fibrosis transmembrane conductance regulator (CFTR) was thought to act either as a chloride channel or as a chloride channel regulator.[1] Eventually, the chloride channel activity of CFTR was confirmed using a variety of experimental approaches. First, expression of recombinant CFTR in heterologous cell systems confers the appearance of cAMP-regulated chloride channels.[2–4] Second, mutagenesis of amino acid residues thought to reside in putative membrane-spanning domains causes alterations in single-channel conductance and/or anion selectivity of the conductance conferred with CFTR expression.[5,6] Finally, reconstitution of purified CFTR in planar phospholipid bilayers causes the appearance of cAMP-activated chloride channels, exhibiting biophysical properties identical to those observed in patch-clamp studies of epithelial cell membranes.[7] The chloride channel function of CFTR is currently thought to be critical for the elaboration of salt and water secretion across the epithelial cell lining of the airways, pancreatic ductules, gastrointestinal tract, and reproductive tract.[8]

[1] J. Riordan, J. Rommens, B.-S. Kerem, N. Alon, R. Rozmahel, Z. Grzelczak, J. Zielenski, S. Lok, N. Plavsic, C. Jia-ling, M. Drumm, M. Iannuzzi, F. Collins, and L.-C. Tsui, *Science* **245,** 1066 (1989).

[2] M. Anderson, R. Gregory, S. Thompson, D. Souza, S. Paul, R. Mulligan, A. Smith, and M. Welsh, *Science* **253,** 202 (1991).

[3] N. Kartner, J. Hanrahan, T. Jensen, L. Naismith, S. Sun, C. Ackerley, E. Reyes, L.-C. Tsui, J. M. Rommens, C. E. Bear, and J. R. Riordan, *Cell* **64,** 681 (1991).

[4] J. A. Tabcharani, X.-B. Chang, J. R. Riordan, and J. W. Hanrahan, *Nature* **352,** 628 (1991).

[5] J. A. Tabcharani, X.-B. Chang, J. R. Riordan, and J. W. Hanrahan, *Biophys. J.* **62,** (1992).

[6] M. Anderson and M. Welsh, *Science* **257,** 1701 (1992).

[7] C. Bear, C. Li, N. Kartner, R. Bridges, T. Jensen, M. Ramjeesingh, and J. Riordan, *Cell* **68,** 809 (1992).

[8] M. Welsh, L.-C. Tsui, T. F. Boat, and A. L. Beaudet, *in* "The Metabolic and Molecular Basis of Inherited Disease" (C. R. Scriver, A. L. Beaudet, W. S. Sly, and D. Valle, eds.), pp. 3799–3876. McGraw-Hill, New York, 1995.



0076-6879/99 $30.00

In this chapter, we describe the method we employ to purify and functionally reconstitute CFTR in model membranes.[7,9–11] As previously mentioned, this experimental system has allowed us to define some of the functional properties of the protein which are intrinsic to CFTR by direct biophysical assays. Our most recent studies of the coupling of the catalytic and channel functions of CFTR best illustrate the utility of our reconstitution system. Using purified CFTR, we find that this molecule is not only a chloride channel, but it is also an ATPase.[10] Furthermore, we find that CFTR utilizes the energy released by ATP hydrolysis to fuel the opening and closing of the channel gate. To date, studies of the enzymatic activity of CFTR in cellular membranes have not been possible because of the difficulty in eliminating the background activity of other membrane ATPases. Consequently, future studies of the structural basis for CFTR ATPase activity and the coupling of this catalytic activity with channel gating must be performed using purified protein.

The body of this chapter addresses the methods we use to express, purify, and reconstitute CFTR. Furthermore, we describe the procedures we use to study the function of the reconstituted molecule. We compare two different strategies for CFTR purification from Sf9 cells; our original method, which employs conventional chromatographic techniques,[7] and a novel procedure, which applies metal affinity chromatography to purify a CFTR molecule engineered to possess a polyhistidine tag at its carboxy terminus (CFTR-His)[11] (Fig. 1). In the original method, recombinant CFTR is extracted from Sf9 cells using sodium dodecyl sulfate (SDS). This strong anionic detergent is used because of the difficulty in solubilizing membrane incorporated CFTR using milder detergents. The use of SDS obligates the application of multistep purification and reconstitution protocols.[7] Our novel method capitalizes on the development of a novel family of fluorinated surfactants which can be used in conjunction with metal affinity chromatography. This one-step purification procedure is rapid and leads to the effective purification (>95%) of CFTR-His. Further, following reconstitution, the functional properties of CFTR-His are identical to those described for CFTR protein purified using the original protocol. We predict that this new method for CFTR purification may be applicable to other ion channels and will expedite studies of the structure–function relationships of these membrane proteins.

[9] C. Li, M. Ramjeesingh, and C. E. Bear, *J. Biol. Chem.* **271,** 11623 (1996).

[10] C. Li, M. Ramjeesingh, W. Wang, E. Garami, M. Hewryk, D. Lee, J. M. Rommens, K. Galley, and C. E. Bear, *J. Biol. Chem.* **271,** 28463 (1996).

[11] M. Ramjeesingh, C. Li, E. Garami, L.-J. Huan, M. Hewryk, Y. Wang, K. Galley, and C. Bear, *Biochem. J.* **327,** 17 (1997).

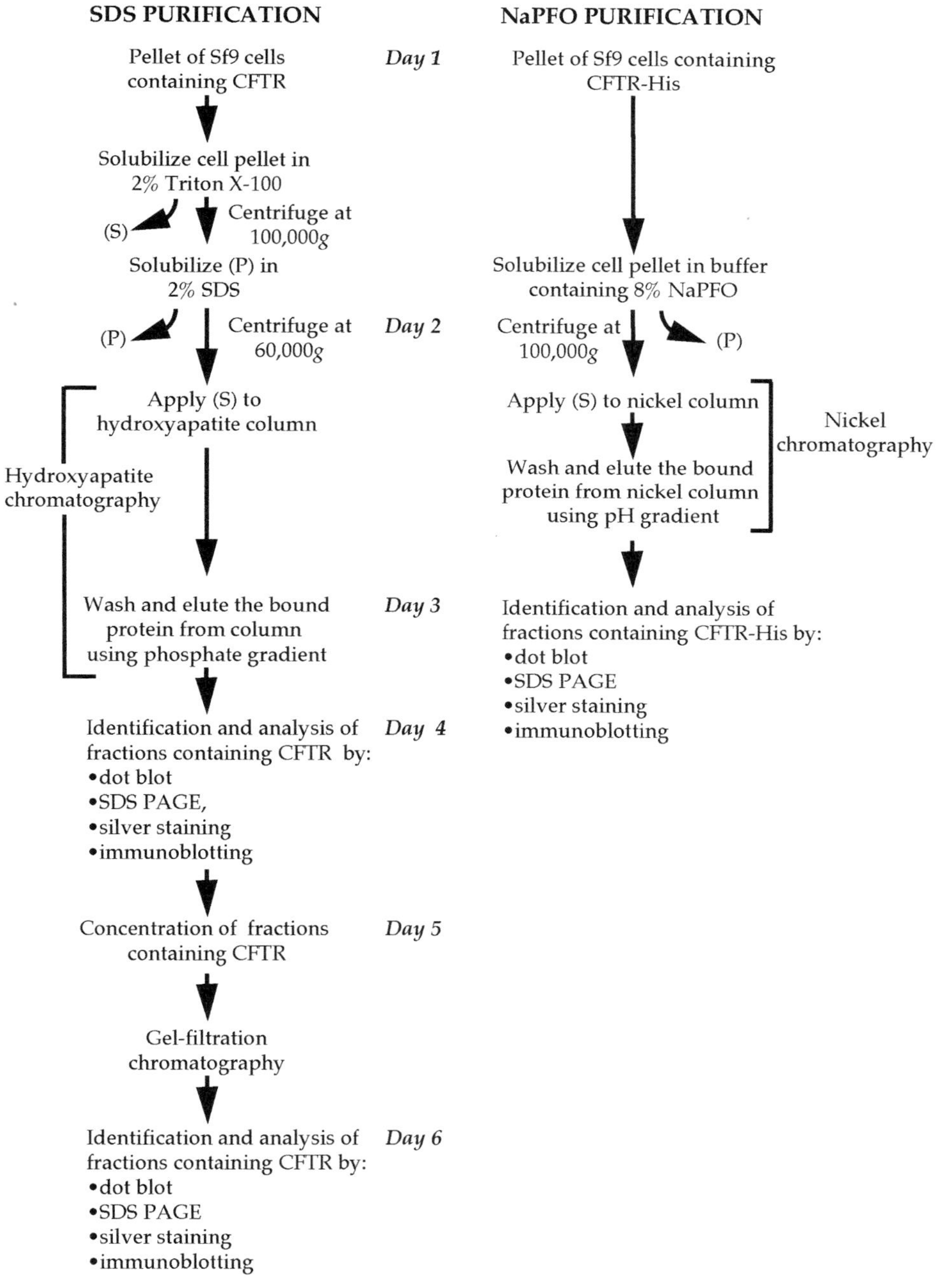


FIG. 1. Flow diagram of two different purification procedures for CFTR. *Left:* Purification scheme for SDS-solubilized CFTR protein. *Right:* Purification scheme for NaPFO-solubilized CFTR-His protein.

## Expression of CFTR in Sf9–Baculovirus System

### *Rationale*

We use the *Spodoptera frugiperda* fall armyworm ovary (Sf9)–baculovirus expression system for production of CFTR and CFTR variants for several reasons. First, functional expression of CFTR can be rapidly confirmed in patch-clamp studies of transfected cells.[3] Second, the yield of recombinant protein is very high: close to 1% of total cellular protein of infected cells is CFTR.[7] Further, since Sf9 insect cells appear to possess relatively permissive quality controls with respect to protein processing, CFTR variants that are misprocessed and fail to reach the cell surface of mammalian cells, such as CFTRΔF508, can reach high levels of surface expression in Sf9 cells. Differences in glycosylation between Sf9–CFTR and CFTR produced in mammalian cells may be a harbinger of the quality control mechanisms that regulate membrane protein expression. Like other proteins produced in Sf9 cells,[3] Sf9–CFTR possesses only core glycosylation, not the complex glycosylation that has been observed for the mature CFTR protein produced in mammalian cells. Because there is no evidence that differences in the degree of glycosylation affect the structure or function of the protein, we use the Sf9-baculovirus system to optimize our chances to obtain large quantities of membrane incorporated CFTR and CFTR mutant proteins.

### *Procedures*

*Construction of Transfer Vectors.* The CFTR open reading frame (ORF) has been subcloned into the baculoviral transfer vector pBlueBac4 (Invitrogen, San Diego, CA) for the purpose of expression in the Sf9 insect cell system. This construct, pBlueBac4-CFTR ORF, and derivatives thereof, are used for all of our recent expressions of CFTR protein. To generate CFTR protein with a polyhistidine tag, a PCR (polymerase chain reaction) product corresponding to the C terminus of CFTR plus the polyhistidine tag ($H_{10}$) was ligated into the pBlueBac4-CFTR ORF construct so as to replace the wild-type 3′ end of the ORF with the polyhistidine tagged 3′ end. The reverse oligonucleotide 5′CTGACGGTACCACTAGTGATG ATGATGATGATGATGATGATGATGATGAAGCCTTGTATCTTG CACC3′ was used in conjunction with the forward oligonucleotide 5′ATG GTGTGTCTTGGGATTCA3′ to create this PCR product, which was then subcloned back into pBlueBac4-CFTR ORF. The entire amplified region was then confirmed by sequencing. Whereas the C terminus of CFTR is QDTRL, the C terminus of CFTR-His is $QDTRLH_{10}$.

*Gene Transfer into Sf9 Cells and Production of Protein.* Recombinant baculovirus is produced in Sf9 cells as previously described,[3] incorporating recent modifications. Sf9 cells are cotransfected with the baculoviral transfer construct (pBlueBac4 encoding CFTR or CFTR-His) and linear baculoviral DNA (Bac-N-Blue DNA, Invitrogen) to produce recombinant CFTR or CFTR-His containing viruses. The supernatant generated from this transfection is used to infect cells and recombinant events are detected as blue plaques. Recombination is confirmed and clones selected for further study on the basis of purity, i.e., lack of contamination by wild-type virus as assessed by PCR analysis of viral supernatant and by detection of CFTR protein as assessed by Western blot analysis of the infected cells. Working stocks of recombinant virus are produced and titred as previously described.[3]

Sf9 cells are grown from frozen stocks purchased from Invitrogen. Initially, cells are grown as a monolayer in TMN-FH complete media [Grace's insect media with L-glutamine hydrolyzate (GIBCO-BRL, Gaithersburg, MD) and yeastolate (GIBCO)], at 27° in a $CO_2$-free incubator. When 90% confluency is reached, cells from $3 \times 75\text{-cm}^2$ flasks are transferred to 100 ml of the suspension culture media consisting of 50% (v/v) TMN-FH, 50% Excell-401 (JRH Biosciences, Lenexa, KS) in 250-ml shaker flasks (Bellco, NJ). The cells are then incubated at 27° with constant shaking at 150 rpm. To maintain cultures continuously, the cells are split when their density reaches $2\text{–}3 \times 10^6$ cells/ml, and reseeded into fresh media (1:1 TMN-FH:Excell 401) at approximately $1 \times 10^6$ cells/ml. Typically, CFTR and CFTR-His protein are produced in 1 liter of suspension culture in 2-liter shaker flasks. For optimal Sf9 cell infection, cell density is approximately $2 \times 10^6$ cell/ml with a viability of 98%. These cells are infected by high-titer viral stocks at a multiplicity of infection of 5 for 48 hr. After this period, approximately 5% of the Sf9 cells are lysed due to infection. Infected cells are harvested from 1 liter of media by centrifugation at 2000*g* at 4° for 20 min and the cell pellet washed once with PBS. The cell pellets are then stored at −80°. Protein has been purified from cell pellets that have been stored in this manner for more than 1 yr with no significant reduction in activity.

## Solubilization and Purification of CFTR

The detergent solubilization of hydrophobic integral membrane proteins is an absolute requirement for their purification. This initially involves the incorporation of the protein into water-soluble micelles by displacing the lipids and other associated molecules bound to the protein with detergent molecules. The protein-detergent micelles can subsequently be manipulated by many of the conventional techniques of protein purification. The choice of detergent is dictated by the degree of hydrophobicity of the protein.

Recombinant CFTR expressed in Sf9 cells is poorly soluble in all detergents tested except for sodium dodecyl sulfate, and the monovalent salts of pentadecafluorooctanoic acid.[11] The purification of CFTR solubilized in these two detergents is described in the following paragraphs.

## *Solubilization and Purification of CFTR in Sodium Dodecyl Sulfate*

*Rationale.* SDS is a strong anionic detergent which is very effective at extracting membrane proteins. Unfortunately, SDS is also a very powerful dissociating membrane detergent that usually results in loss of biological activity of solubilized proteins. It has been shown, however, that integral membrane proteins are resistant to complete denaturation in sodium dodecyl sulfate (SDS). For example, bacteriorhodopsin maintains 50% $\alpha$-helical content in SDS[12] and the detergent seems to promote $\alpha$-helical structure in less complex hydrophobic molecules.[13–15] Because many integral membrane proteins contain a large amount of $\alpha$-helical content and no disulfide bonds, the use of SDS for purification may offer distinct advantages because of its ability to disaggregate hydrophobic proteins during chromatography.[16] Hydroxyapatite, cation-exchange, and gel-filtration chromatographic methods of purification are amenable to the presence of SDS. Successful reactivation of SDS-solubilized proteins has been reported for a number of membrane proteins including bacteriorhodopsin,[12,17] P-glycoprotein,[18] nicotinic acetylcholine receptor,[19] 5′-nucleotidase, and neuraminidase.[16]

### *Procedures*

#### *Sample Preparation in Sodium Dodecyl Sulfate*

1. Frozen Sf9 cell pellets expressing recombinant CFTR are thawed and then resuspended in 150 ml of PBS containing 2% Triton X-100 and a cocktail of protease inhibitors [leupeptin 10 $\mu$g/ml, aprotinin 10 $\mu$g/ml, E-64 [*trans*-epoxysuccinyl-L-leucylamido(4-guanidino)butane] 10 $\mu M$, benzamidine 1 m$M$, dithiothreitol (DTT) 2 m$M$, and $MgCl_2$ 5 m$M$] and DNase I 20 U/ml.

[12] K. S. Huang, H. Bayley, M. J. Liao, E. London, and H. G. Khorana, *J. Biol. Chem.* **256,** 3802 (1981).

[13] C. R. Dawson, A. F. Drake, J. Helliwell, and R. C. Hider, *Biochim. Biophys. Acta* **510,** 75 (1978).

[14] J. C. Steele, Jr., and J. A. Reynolds, *J. Biol. Chem.* **254,** 1633 (1979).

[15] C. S. Wu and J. T. Yang, *Biochemistry* **19,** 2117 (1980).

[16] S. Hjerten, M. Sparrman, and J. Liao, *Biochim. Biophys. Acta* **939,** 476 (1988).

[17] E. London and H. G. Khorana, *J. Biol. Chem.* **257,** 7003 (1982).

[18] M. Dong, F. Penin, and L. G. Baggetto, *J. Biol. Chem.* **271,** 28875 (1996).

[19] W. Hanke, J. Andree, J. Strotmann, and C. Kahle, *Eur. Biophys. J.* **18,** 129 (1990).

2. The suspension is nutated at room temperature for 1 hr, and then centrifuged at 100,000*g* for 2 hr at 4°. The supernatant is discarded and the pellets are transferred into 200 ml of 10 m*M* sodium phosphate, 2% SDS (w/v), and 3% mercaptoethanol (v/v), pH 7.4, and stirred overnight with a magnetic stirrer at room temperature.
3. Undissolved material is pelleted at 60,000*g* for 1 hr and the supernatant is filtered using a 500-ml filtering unit (0.2 or 0.45 $\mu$m) with prefilters to prevent clogging. The filtered sample is then applied to the ceramic hydroxyapatite column.

*Hydroxyapatite Chromatography*

1. SDS-solubilized CFTR sample is applied to the column at a flow rate of 1 ml/min at room temperature.
2. The column is washed with 100 ml of buffer A, followed by 100 ml of buffer B. The column is then washed with 200 ml of a buffer containing 40% of buffer B and 60% of buffer C (400 m*M* final phosphate concentration). This is followed by 100 ml of a 400–600 m*M* phosphate gradient (100% buffer C) to elute CFTR. Washing of the column is continued with an additional 100 ml of buffer C.
3. Three-milliliter fractions from the gradient and subsequent washes are collected and analyzed by dot blot. Immunopositive fractions are further analyzed by Western blot (10 $\mu$l of each fraction) and by silver-stained protein gel (40 $\mu$l of each fraction). CFTR usually elutes as a broad band (about 10 fractions) between 540 and 600 m*M* phosphate. The silver-stained gel validates the quality of separation. A good separation yields a single band corresponding to CFTR and some lower molecular weight bands (see Fig. 2). A less than optimal separation may necessitate the sacrifice of a few of the initial CFTR-containing fractions to ensure the purity of the preparation. The CFTR-containing fractions are pooled and concentrated in a Centriprep 50 concentrator (Amicon, Danvers, MA) to a final volume of 500 $\mu$l.

*Gel-Filtration Chromatography*

1. The concentrated CFTR sample (500 $\mu$l) is applied to a Superose 12 column that has previously been equilibrated with 100 ml of buffer D.
2. The column is eluted with the buffer at a flow rate of 0.5 ml/min and 1-ml fractions are collected (Fig. 3). A Western blot and a silver-stained protein gel of fractions eluting at $V_e/V_o$ of 1.4 to 1.6 are done to verify the protein and its purity. Pure CFTR-containing fractions

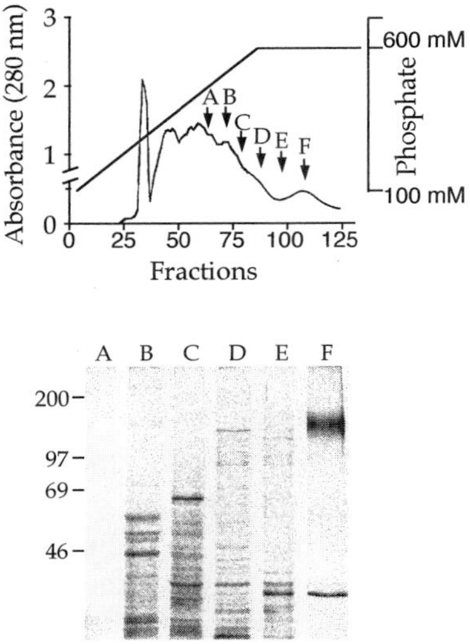


FIG. 2. Hydroxyapatite column chromatography of SDS-solubilized Sf9-CFTR. *Top:* Elution profile with a phosphate gradient generated from two buffers of 100 and 600 m*M* phosphate, each containing 0.15% SDS and 5 m*M* DTT, pH 6.8. *Bottom:* SDS–polyacrylamide gel (6%) of the peaks as indicated in the legend. Protein bands are visualized by silver staining. CFTR protein is eluted in peak F at 600 m*M* phosphate.

are pooled for reconstitution and concentrated on a Centricon 100 concentrator (Amicon) at 15° to a final volume of 500 $\mu$l. The yield of CFTR is quantitated by amino acid analysis. The yield can vary from 15 to 500 $\mu$g of protein depending on the expression level of the protein in the infected Sf9 cells.

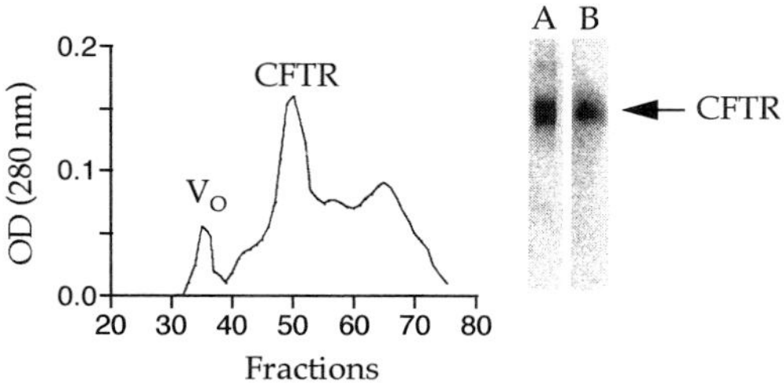


FIG. 3. Superose 12 gel filtration chromatography of CFTR-containing fractions from a hydroxyapatite column. CFTR elutes at $V_e/V_o$ (eluted volume relative to void volume) of 1.4. Purified CFTR runs as a single band on a silver-stained gel (A) and gives a single immunopositive band on a Western blot (B) with CFTR monoclonal antibody (MAb), M3A7. (Generously provided by Dr. N. Kartner, Pharmacology, University of Toronto.)

### *Reagents and Equipment*

*Hydroxyapatite Column.* The column is composed of four Econo-Pac CHT-II cartridges (5 ml each) attached in series (Bio-Rad, Richmond, CA). Initially, the column is washed with 100 ml of buffer C and equilibrated with 100 ml of buffer A before use. Between runs, an additional washing step using 100 ml of 1 *N* NaOH is employed.

*Gel Filtration Column.* Preparative Superose 12 column (Pharmacia, Uppsala, Sweden).

*Equipment.* FPLC (fast protein liquid chromatography, Pharmacia, Piscataway, NJ) or any liquid chromatography system with a pump capable of generating a linear gradient.

### *Buffers*

Buffer A: 10 m*M* sodium phosphate, 0.15% LiDS (lithium dodecyl sulfate), 5 m*M* DTT, 0.025% $NaN_3$ (sodium azide), pH 6.8. Note that this buffer is degassed under vacuum prior to addition of LiDS. The buffer is filtered through a 0.22-$\mu$m Steritop-GP filter (Millipore, Bedford, MA).

Buffer B: 100 m*M* sodium phosphate, 0.15% SDS, 5 m*M* DTT, 0.025% $NaN_3$, pH 6.8. Prepared as above.

Buffer C: 600 m*M* sodium phosphate, 0.15% SDS, 5 m*M* DTT, 0.025% $NaN_3$, pH 6.8, prepared as above. Some heating may be required to solubilize all of the reagents.

Buffer D: 10 m*M* Tris, 5 m*M* DTT, 100 m*M* NaCl, 0.5 m*M* EGTA and 0.25% LiDS, pH 7.8.

### *Solubilization and Purification of CFTR in Sodium Pentadecafluorooctanoic Acid*

*Rationale.* The monovalent salts of pentadecafluorooctanoic acid (PFO) are derived from a novel family of fluorinated surfactants that are much more active than ordinary surfactants due to the hydrophobicity of the fluorocarbon chain. Furthermore, the compatibility of these detergents with biological assays and their ease of removal make them good candidates for purification and functional reconstitution of membrane proteins.[20] More importantly, sodium pentadecafluorooctanoate, unlike SDS does not abolish protein affinity interactions, thereby allowing the purification of a recombinant polyhistidine-tagged CFTR by single-step nickel affinity chromatography.[11]

[20] F. H. Shepherd and A. Holzenburg, *Anal. Biochem.* **224,** 21 (1995).

*Procedures*

*Sample Preparation in NaPFO*

1. Frozen Sf9 cell pellets expressing recombinant CFTR-His are thawed and then resuspended in 150 ml of phosphate-buffered saline (PBS) containing 2% Triton X-100 and a cocktail of protease inhibitors (leupeptin 10 $\mu$g/ml, aprotinin 10 $\mu$g/ml, E64 10 $\mu M$, benzamidine 1 m$M$, DTT 2 m$M$, and $MgCl_2$ 5 m$M$) and DNase I 20 U/ml.
2. The suspension is nutated for 1 hr at room temperature, and then centrifuged at 100,000$g$ for 2 hr at 4°. The supernatant is discarded and the pellets are transferred into 100 ml of 8% NaPFO in 20 m$M$ phosphate, pH 8.0, and stirred overnight with a magnetic stirrer.
3. Undissolved material is pelleted at 60,000$g$ for 1 hr and the supernatant filtered using a 500-ml filtering unit (0.2 or 0.45 $\mu$m) with prefilters to prevent clogging. The filtered sample is then applied to the nickel column.

*Nickel Affinity Chromatography*

1. The freshly regenerated 25-ml nickel column is attached to the FPLC and the CFTR-His containing sample is applied at 1 ml/min.
2. The column is then washed with 60 ml of buffer A. A pH gradient is applied to the column titrating buffer A with buffer B, from 0% buffer B to 100% buffer B in 100 ml.
3. Three-milliliter fractions are collected and analyzed by dot blot. Immunopositive fractions are analyzed by Western blot (10 $\mu$l of each fraction) and by silver-stained protein gel (40 $\mu$l of each fraction) (Fig. 4). CFTR protein that elutes below pH 6.8 is pooled and concentrated in a Centriprep 50 concentrator (Amicon, Danvers, MA) to a final volume of 500 $\mu$l. The yield of CFTR is quantitated by amino acid analysis.

*Reagents and Buffers*

Pentadecafluorooctanoic acid is from Fluorochem (Old Glossop, UK).

*Column.* Twenty-five milliliters of packed nickel chelating resins (Qiagen, CA) are poured and washed with 100 ml of buffer A before use.

*Equipment.* FPLC (Pharmacia) or any liquid chromatography system capable of generating a linear gradient.

*Buffers.* Buffers containing sodium pentadecafluorooctanoate (NaPFO) are prepared by adding the free acid, pentadecafluorooctanoic acid, to the buffer and titrating with NaOH to the required pH.

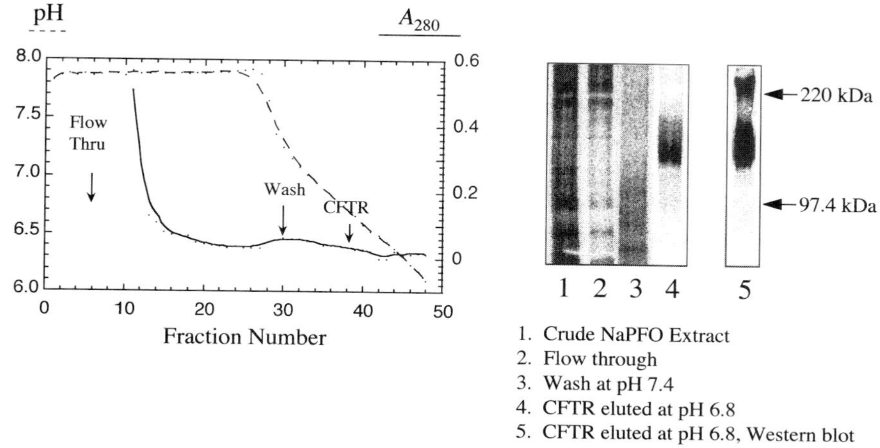


FIG. 4. Immobilized metal ion affinity chromatography of Sf9–CFTR in NaPFO. *Left:* Elution profile with a pH gradient indicated, generated from two buffers of pH 7.4 and 4.0, each containing 25 m*M* phosphate, 100 m*M* NaCl, and 4% NaPFO. *Right:* Silver-stained protein bands after SDS–PAGE (6%) of crude NaPFO membrane extract and fractions as described in the legend. Lane 5 shows immunoreactivity of protein from fraction labeled as lane 4 with CFTR MAb, M3A7.

Buffer A: 20 m*M* phosphate, 4% NaPFO pH 7.8. Note that the buffer is degassed under vacuum prior to NaPFO addition. The buffer is filtered through a Steritop-GP 0.22 $\mu$m filter (Millipore).

Buffer B: 20 m*M* phosphate, 4% NaPFO, pH 4.0, prepared as in buffer A.

## Reconstitution of CFTR

### *Rationale*

Reconstitution of hydrophobic membrane proteins involves the reinsertion of the protein into a phospholipid bilayer or vesicles of known composition. Phospholipid added to the detergent-solubilized protein initially forms mixed lipid–detergent micelles. The critical step in reconstruction is detergent removal from the mixed micelles. As detergent concentration falls below a critical minimum, phospholipid micelles spontaneously change conformation to form bilayer vesicles. Simultaneously, phospholipid displaces detergent at the exposed hydrophobic surfaces of the protein, which subsequently becomes incorporated across the vesicle bilayer. Several methods are available for detergent removal depending on the type of detergent used, such as hydrophobic adsorption chromatography,[21] gel-exclusion chromatography,[22] dilution,[23] or dialysis. Detergent dialysis is the simplest of the methods and relies on the detergent monomer concentration being substantially high compared to the very low concentration of lipid monomers, allowing the detergent to be removed considerably faster than the lipid.[24,25] Detergents with high critical micellar concentration (CMC) and low aggregation number are the easiest to remove by this method.[22] This dialysis method has been adopted for reconstituting SDS-solubilized proteins despite the low CMC and moderately high aggregation number by utilizing a mild detergent exchange step before vesicle reconstitution. Exchange can occur as a separate step prior to the addition of the lipid provided that the protein is soluble in its new detergent.[18] Alternatively, the detergent exchange can occur in the presence of lipids during dialysis.[12,16] We have adopted the latter technique because of the insolubility of CFTR in other detergents. In addition, to further facilitate detergent removal, the SDS concentration is diluted below its CMC prior to dialysis, and a dialysis membrane cutoff of 50 kDa, large enough for SDS micelles to permeate, is used to ensure complete exchange of SDS for cholate.

The choice of lipids for reconstitution of integral membrane proteins is loosely based on (1) ability of the lipids to form liposomes, (2) biochemical requirements of the protein for specific lipids, and (3) capacity to form fusogenic vesicles, suitable for bilayer assays for single-channel activity. The effect of lipids on function of membrane proteins is well documented. P-Glycoprotein, a member of the ABC superfamily of proteins to which CFTR belongs, shows significantly higher ATPase activity in the presence of phosphatidylethanolamine (PE) and phosphatidylserine (PS) than in the presence of phosphatidylcholine (PC).[26] PE vesicles have been reported to be more fusogenic, possibly because the PE head groups are less hydrated than PC head groups.[27–29] However, pure PE does not form vesicles at neutral pH and physiologic ionic strength.[30] On the other hand, vesicle formation can be induced, when PE lipids are doped with other naturally

[21] R. Moriyama, H. Nakashima, S. Makino, and S. Koga, *Anal. Biochem.* **139,** 292 (1984).
[22] A. J. Furth, *Anal. Biochem.* **109,** 207 (1980).
[23] E. Racker, T.-F. Chien, and A. Kandrach, *FEBS Lett.* **57,** 14 (1975).
[24] S. Razin, *Biochim. Biophys. Acta* **265,** 241 (1972).
[25] V. Rhoden and S. M. Goldin, *Biochemistry* **18,** 4173 (1979).
[26] C. A. Doige, X. Yu, and F. J. Sharom, *Biochim. Biophys. Acta* **1146,** 65 (1993).
[27] S. H. White and G. I. King, *Proc. Natl. Acad. Sci. U.S.A.* **82,** 6532 (1985).
[28] L. J. Lis, M. McAlister, N. Fuller, R. P. Rand, and V. A. Parsegian, *Biophys. J.* **37,** 667 (1982).
[29] R. Sundler, N. Duzgunes, and D. Papahadjopoulos, *Biochim. Biophys. Acta* **649,** 751 (1981).
[30] D. Papahadjopoulos and J. C. Watkins, *Biochim. Biophys. Acta* **135,** 639 (1967).

occurring, bilayer-forming lipids such as PS and PC[31–33] and when the reconstitution is performed at low ionic strength.[30] PS, a negatively charged phospholipid, helps to maintain liposome integrity by decreasing the tendency of liposomes to aggregate and has been shown to promote fusion with cultured cells.[34,35]

### *Procedures*

#### *Reconstitution of CFTR from LiDS*

1. Fifty $\mu$g of CFTR protein (after Superose 12 column elution) in LiDS is concentrated with a Centricon 100 concentrator (Amicon) to a final volume of 100 $\mu$l. A tenfold dilution with buffer B and reconcentration to 100 $\mu$l yields CFTR protein with a final LiDS concentration of 0.92 m*M*.
2. The protein is then added to 200 $\mu$l (200 $\mu$g) of the liposome preparation (PE : PS : PC : ergosterol, 5 : 2 : 1 : 1 by weight) and allowed to incubate at room temperature for 1 hr.
3. The lipid–protein mixture is transferred to a dialysis bag (Spectra/Por membrane, molecular weight cutoff 50,000) and dialyzed against 2 liters of buffer A for 17 hr, followed by a 24-hr dialysis against 4 liters of buffer B and another 24 hr of dialysis against buffer C.
4. The reconstituted proteoliposomes are aliquoted and stored under argon at −80°.

#### *Reconstitution of CFTR from NaPFO*

1. CFTR sample (100 $\mu$g) in 25 m*M* phosphate, 4% NaPFO, pH 6.2, is diluted 1 : 10 with buffer B and concentrated in a Centriprep 50 concentrator (Amicon) to a final volume of 500 $\mu$l (molecular weight cutoff 50,000) to 600 $\mu$l. The final NaPFO concentration is 0.4%.
2. Four hundred microliters of liposome preparation containing 4 mg of lipid (PE : PS : PC : ergosterol, 5 : 2 : 1 : 1 by weight) is added to 100 $\mu$g (500 $\mu$l) of CFTR. The mixture is dialyzed (Spectra/Por membrane, molecular weight cutoff 50,000) for 18 hr against 4 liters of buffer B containing 0.025% $NaN_3$ followed by a final dialysis against

[31] P. R. Cullis and B. de Kruijff, *Biochim. Biophys. Acta* **559,** 399 (1987).
[32] R. J. Ho and L. Huang, *J. Immunol.* **134,** 4035 (1985).
[33] T. F. Taraschi, A. T. van der Steen, B. de Kruijff, C. Tellier, and A. J. Verkleij, *Biochemistry* **21,** 5756 (1982).
[34] R. Fraley, R. M. Straubinger, G. Rule, E. Springer, and D. Papahadjopoulos, *Biochemistry* **20,** 6978 (1981).
[35] J. Damen, J. Regts, and G. Scherphof, *Biochim. Biophys. Acta* **712,** 44 (1982).

4 liters of buffer C. The reconstituted liposomes are aliquoted and stored under argon at −80°.

*Liposome Preparation*

1. Three milligrams (300 μl) of the lipid mixture (PE : PS : PC : ergosterol, 5 : 2 : 1 : 1 by weight) in chloroform is dried in a 20-ml Pyrex test tube by argon gas. The tube is rotated during the drying process to allow the bottom of the test tube to be evenly coated with the lipid. Residual moisture is removed by a gentle stream of argon for 1 hr.
2. Buffer B (300 μl) is added to the lipid-coated tube, which is then incubated on ice for 10 min to rehydrate the lipid mixture. The latter is then sonicated in a bath sonicator until the solution is translucent. This usually requires 2–5 min of intermittent sonication. The test tube is kept on ice between sonications to prevent overheating of the lipid.

### *Reagents*

*Lipids.* PE (from egg, Avanti Polar Lipids, Birmingham, AL), PS (from brain, Avanti), and PC (from egg yolk, Avanti), ergosterol (Sigma, St. Louis, MO). Ergosterol is recrystallized from ethanol, dried under vacuum, and stored at −20°. A stock solution of a lipid mixture containing PE : PS : PC : ergosterol (5 : 2 : 1 : 1 by weight) is prepared at a concentration of 10 mg/ml of lipid in chloroform, aliquoted in 1-ml vials with Teflon stoppers (Reacti-Vial, Pierce, Rockford, IL), and stored under argon at −80°. Dialysis membranes (Spectra/Por) are purchased from Spectrum, Houston, TX.

*Buffers*

Buffer A: 8 m*M* HEPES, 0.5 m*M* EGTA, 0.025% $NaN_3$, and 1.5% sodium cholate, pH 7.2.
Buffer B: 8 m*M* HEPES, 0.5 m*M* EGTA, 0.025% $NaN_3$, pH 7.2.
Buffer C: 8 m*M* HEPES, 0.5 m*M* EGTA, pH 7.2.

*Equipment.* A bath sonicator (model G112SP1G, Laboratory Supplies Co., Hicksville, NY).

## Assessment of Functional Properties of Reconstituted CFTR Channels

As mentioned previously, we have shown that purified, reconstituted CFTR functions as a chloride channel with biophysical and regulatory features identical to those reported for a chloride channel located on the apical membrane of epithelial cells.[7] Like the chloride channel in epithelial

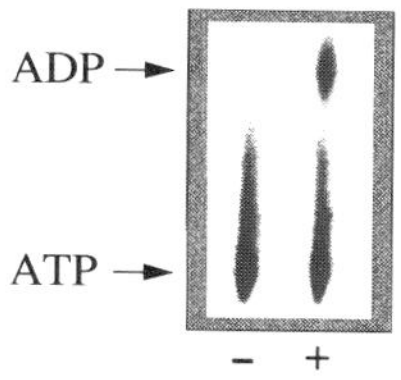


FIG. 5. ATPase activity of purified, reconstituted CFTR measured as the production of ADP from ATP. TLC separation of [α-$^{32}$P]ATP and [α-$^{32}$P]ADP, the upper arrow indicates [α-$^{32}$P]ADP; and the lower arrow [α-$^{32}$P]ATP. The left lane indicates lack of [α-$^{32}$P]ADP production (−) by liposomes alone, incubated with [α-$^{32}$P]ATP for 4 hr at 32°, and the right lane indicates [α-$^{32}$P]ADP production by liposomes containing CFTR–His (100 ng) (+).

cells, purified, reconstituted CFTR requires phosphorylation by protein kinase A (PKA) and ATP hydrolysis for activity.[2,36,37] Once activated, the CFTR chloride channel exhibits a low unitary conductance and slow gating kinetics.[3,4] More recently, we have been using purified, reconstituted CFTR protein in our studies of the structural basis for its regulation by nucleotides.[9] CFTR is a unique ion channel in that it hydrolyzes ATP and utilizes this energy to open and close the gate through which the flux of chloride ion occurs. Currently, we are assessing the effect of mutations within the Walker consensus sequences for nucleotide binding in order to determine the structural basis for CFTR ATPase activity and nucleotide-dependent gating. Each purified, reconstituted CFTR variant is currently being studied with respect to its ATPase activity and channel function.

Measurements of the ATPase activity of a suspension of proteoliposomes provide a rough estimate of the number of CFTR molecules capable of ATPase activity and of the catalytic activity of each CFTR molecule (Fig. 5). The ion-channel activity of each CFTR variant is examined using two different assays: radioisotopic flux assay[9] and single-channel studies following fusion with planar lipid bilayers.[7,9,10] Similar to the ATPase assay, the electrogenic $^{36}Cl^-$ flux assay permits a macroscopic view of the number of channel competent CFTR molecules and their regulation by phosphorylation (Fig. 6). On the other hand, planar lipid bilayer studies permit a detailed examination of the conductance properties and ATP-dependent gating of individual CFTR molecules on phosphorylation. A potential limitation in the study of mutant forms of CFTR using this assay relates to the difficulty in discerning between failure of proteoliposomes to fuse to the

[36] T. Baukrowitz, T.-C. Hwang, A. Nairn, and D. Gadsby, *Neuron* **12,** 473 (1994).
[37] K. L. Gunderson and R. R. Kopito, *Cell* **82,** 231 (1995).

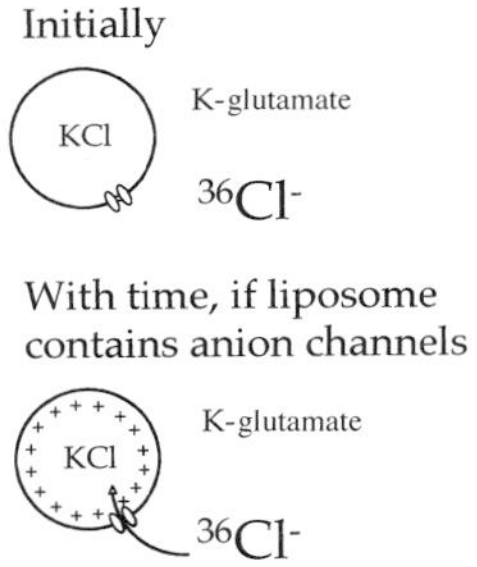


FIG. 6. Schematic representation of proteoliposome flux assay of electrogenic chloride uptake by purified CFTR. In this diagram a single purified CFTR molecule provides the influx path for accumulation of $^{36}Cl^-$. The driving force for $^{36}Cl^-$ uptake is the positive intraliposome potential difference created by the initial creation of a chemical gradient for chloride ion.

bilayer and protein dysfunction. A technique described by Woodbury *et al.*[37a] helps to alleviate this problem.

Addition of the channel forming peptide, nystatin, renders all proteoliposomes fusogenic with the planar lipid bilayer and also permits the detection and confirmation of these fusion events.[10] Conveniently, nystatin only forms an active channel in the presence of the lipids cholesterol or ergosterol. On fusion of the nystatin-and-ergosterol-containing proteoliposomes with planar lipid bilayers, a transient conductance spike is observed, providing a marker for each fusion event. On the other hand, because the bilayer phospholipid mixture contains only PS and PE, ergosterol will diffuse away from nystatin following fusion of the proteoliposomes; hence, the nystatin-channel forming complex will dissipate and will not further contaminate the activity generated by CFTR, the ion channel of interest. This technique has proven to be particularly useful in our study of CFTR variants which we suspect to have low channel open probability. In these cases, the detection of repeated proteoliposome fusion spikes in the absence of channel opening events provides convincing evidence that the variant CFTR protein is not functioning normally. We have provided an example of the utility of the nystatin-mediated fusion technique in our studies of the disease-causing CFTR mutation, CFTRG551D. In Fig. 7, it is possible to see that unlike wild-type CFTR, frequent spikes are observed, indicating the fusion of CFTRG551D proteoliposomes with no evidence of accompanying single-channel openings and closings. Experiments of this type convinced us that the CFTRG551D channel protein had difficulty in opening.

[37a] D. J. Woodbury, *Methods Enzymol.* **294,** [17], 1998 (this volume).

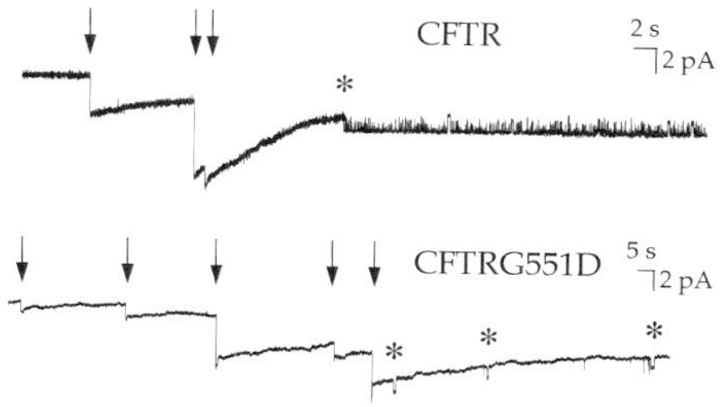


FIG. 7. Examples of the utility of nystatin-mediated liposome fusion for comparison of the channel activity of wild-type CFTR with variant CFTR proteins, i.e., CFTRG551D. Arrows indicate conductance spikes associated with nystatin-mediated fusion events; asterisks indicate appearance of CFTR or CFTRG551D channel openings. For wild-type CFTR protein, three fusion spikes precede the appearance of a channel opening. For CFTRG551D, fusion of several liposomes is verified by the appearance of fusion spikes. The fifth fusion is associated with channel activity which is clearly altered from that observed for the wild-type CFTR protein.

### *Procedures*

*Phosphorylation of CFTR.* Both CFTR channel and catalytic activities are dependent on phosphorylation by PKA. Hence, in our studies of CFTR, the catalytic subunit of PKA is added to the proteoliposomes to phosphorylate the protein. The enzyme must then be removed so that it will not contaminate our subsequent assays of ATPase and channel activity.

1. CFTR (10–30 $\mu$g) reconstituted in phospholipid liposomes is phosphorylated by incubation for 1 hr at room temperature in a reaction mixture comprised of 200 n*M* catalytic subunit of PKA (Promega, Madison, WI), 0.5 m*M* ATP in 50 m*M* Tris, 50 m*M* NaCl, and 5 m*M* $MgCl_2$ at pH 7.5.
2. To remove PKA after the phosphorylation reaction, CFTR proteoliposomes are airfuged twice at 100,000*g* for 30 min. Between spins, the proteoliposomes are washed with buffer containing 50 m*M* Tris and 50 m*M* NaCl at pH 7.2. Alternatively, catalytic subunit of PKA was separated from the proteoliposomes by spin-column chromatography using Sephadex G-50 (Pharmacia).

### *Equipment*

Airfuge ultracentrifuge (Beckman, Palo Alto, CA).

### *Assay of CFTR Activity as an ATPase*

ATPase activity is measured as the production of [$\alpha$-$^{32}$P]ADP from [$\alpha$-$^{32}$P]ATP by purified, reconstituted CFTR (Fig. 5). This ratio is corrected for spontaneous hydrolysis by subtracting the [$\alpha$-$^{32}$P]ADP/[$\alpha$-$^{32}$P]ATP ratio of control liposomes (no CFTR) from the experimental ratio. Radiolabeled ADP and ATP are separated by polyethyleneimine (PEI) chromatography.

1. Proteoliposomes containing CFTR are dispersed by sonicating twice briefly for approximately 10 sec.
2. The ATPase assay is carried out in a 15-$\mu$l reaction mixture containing 100 ng CFTR in phospholipid liposomes, 20 m$M$ Tris, 40 m$M$ NaCl, 5 m$M$ $MgCl_2$, and 10 $\mu$Ci of [$\alpha$-$^{32}$P]ATP (3 Ci/$\mu$mol) and 1 m$M$ nonradioactive ATP at pH 7.5. Reaction mixture is sonicated for approximately 6 sec and then incubated at 32° for 4 hr.
3. The reaction is stopped by addition of 5 $\mu$l of 10% SDS.
4. Prior to application of samples from ATPase reaction, PEI plates are prespotted with 1 $\mu$l of 5 m$M$ nonradioactive ADP. The nonradioactive ADP is allowed to dry at room temperature. Nonradioactive ADP acts as a carrier, enhancing the migration of the [$\alpha$-$^{32}$P]ADP. One-microliter samples from ATPase reaction vials are spotted on a PEI-cellulose plate and developed in 1 $M$ formic acid–0.5 $M$ LiCl.
5. The position and quantity of the radioactive ADP and ATP are ascertained using a Molecular Dynamics PhosphorImager. The data are analyzed using the ImageQuant software package (Molecular Dynamics, CA).

*Reagents*

Radiolabeled ATP: [$\alpha$-$^{32}$P]ATP 10 mCi/ml (Amersham, Oakville, Canada).

PEI cellulose plates: 20- × 20-cm polyethyleneimine cellulose TLC plates with UV indicator (Aldrich, Milwaukee, WI).

## *Assays of CFTR Activity as Chloride Channel*

*Electrogenic Tracer Uptake Assay for Study of CFTR Function.* A concentrative tracer uptake assay is used to characterize the chloride conductance properties of reconstituted CFTR, as developed by Garty *et al.*[38] and modified by Goldberg and Miller[39] (Fig. 6).

*Procedures*

1. Proteoliposomes (100 $\mu$l) are preloaded with 150 m$M$ KCl and centrifuged through Sephadex G-50 columns equilibrated with glutamate-containing salts; potassium glutamate (125 m$M$), sodium glutamate

[38] H. Garty, B. Rudy, and S. J. Karlish, *J. Biol. Chem.* **258,** 13094 (1983).

[39] A. F. Goldberg and C. Miller, *J. Membr. Biol.* **124,** 199 (1991).

(25 m$M$), glutamic acid (10 m$M$), Tris–glutamate (20 m$M$) at pH 7.6, to replace external chloride. The eluted liposomes are diluted to 600 $\mu$l with the above glutamate buffer.

2. Uptake is initiated and quantified by addition of 1.0 $\mu$Ci/mol of $^{36}Cl^{-1}$.
3. Intravesicular $^{36}Cl^-$ is assayed at various time points following separation of 100-$\mu$l liposomes from the external media using a Dowex 1 anion-exchange minicolumn (Sigma).

*Reagents.* Sephadex G-50 (Sigma, $^{36}Cl^-$ (ICN, CA, USA), Dowex 1 ion exchanger, glutamate form (Sigma).

*Buffers*

Buffer A: Potassium glutamate (125 m$M$), sodium glutamate (25 m$M$), glutamic acid (10 m$M$), Tris–glutamate (20 m$M$), pH 7.6.
Buffer B: KCl (50 m$M$), MOPS [3-(N-morpholino)propanesulfonic acid] (10 m$M$), pH 7.2.

### *Planar Bilayer Studies of Liposomes Containing Purified CFTR*

*Procedures*

NYSTATIN INCORPORATION INTO PROTEOLIPOSOMES. As in our previous studies, proteoliposome fusion with planar lipid bilayers is facilitated and detected by the introduction of nystatin (120 $\mu$g/ml), a technique originally described by Woodbury and Miller.[40]

1. Ten microliters of nystatin stock solution (1 mg/ml in methanol) is added to 1 mg (100 $\mu$l) of the lipid mixture in chloroform.
2. The mixture is dried in a 5-ml Pyrex test tube by argon gas. The tube is rotated during the drying process to allow the bottom of the test tube to be evenly coated with the lipid. Residual moisture is removed by spraying with a gentle stream of argon for 1 hr.
3. Buffer B (100 $\mu$l) is added to the lipid-coated tube with incubation on ice for 10 min to rehydrate the lipid mixture.
4. This lipid mixture is then sonicated intermittently in the bath sonicator until the solution is translucent to light. The tube is kept on ice between sonications.
5. An aliquot of the nystatin-containing liposomes is added to an equal volume of the CFTR containing proteoliposomes (from either preparation) in a small Pyrex test tube. The mixture is frozen for 5 min in a dry ice–ethanol slurry, thawed at room temperature and sonicated for 15–20 sec. This freeze–thaw cycle is repeated twice before

[40] D. J. Woodbury and C. Miller, *Biophys. J.* **58,** 833 (1990).

the nystatin-containing proteoliposomes are studied in the planar bilayer chamber.

*Bilayer Formation and Proteoliposome Fusion.* Planar lipid bilayers are formed by painting a 10 mg/ml solution of phospholipid (PE : PS at a ratio of 1 : 1) in *n*-decane over a 200-$\mu$m aperture in a bilayer chamber. Bilayer formation is monitored electrically by observation of the increase in membrane capacitance. In all experiments, bilayer capacitance is greater than 200 pF. Fusion of liposomes is potentiated with the establishment of an osmotic gradient across the lipid bilayer; the *cis* compartment of the bilayer chamber, defined as that compartment to which liposomes are added, contains 300 m*M* KCl, and the *trans* compartment, connected to ground, contains 50 m*M* KCl. Fusion events of nystatin-containing liposomes are indicated by the appearance of transient "nystatin spikes" in bilayer conductance[40] (Fig. 7).

### *CFTR Channel Detection and Analysis*

CFTR channel activity is detected using a bilayer amplifier. Data are recorded and analyzed using pCLAMP 6.0.2 software (Axon Instruments, Burlingame, CA). Prior to analysis of open probability and dwell times, single-channel data are digitally filtered at 100 Hz. Ideal records are created by use of a half-height transition protocol.

*Reagents*

Nystatin (Sigma), *n*-decane (Sigma).

*Equipment.* Bilayer amplifier (custom made by M. Shen, Physics Laboratory, University of Alabama) and bilayer chamber (Warner Instruments, Hamden, CT).

# [12] Purification, Characterization, and Reconstitution of Cyclic Nucleotide-Gated Channels

*By* Robert S. Molday and Laurie L. Molday

## Introduction

Cyclic nucleotide-gated (CNG) channels comprise a family of cation-selective channels that are cooperatively activated or inhibited by the binding of cGMP or cAMP. In vertebrate rod and cone photoreceptor cells and olfactory receptor neurons, these channels play a central role in sensory transduction pathways by controlling the flow of $Na^+$ and $Ca^{2+}$ into the



0076-6879/99 $30.00

cell in response to signal-mediated changes in intracellular cyclic nucleotide levels. CNG channels have also been detected in a wide variety of other vertebrate and invertebrate cells and tissues such as retinal bipolar and ganglion cells, kidney, pineal gland, testis, heart, muscle, *Drosophila* eye and antennas, and carrot cells.[1,2] In some instances, these channels have been cloned, and analysis of their sequences reveals that many are identical or highly similar to the rod, cone, or olfactory CNG channel. The physiologic role of CNG channels in nonsensory cells, however, has yet to be determined.

The CNG channels of rod photoreceptors and olfactory neurons have been most extensively studied at a molecular level.[1–4] The native channels consist of two homologous subunits termed $\alpha$ and $\beta$ (or subunit 1 and 2) that assemble into an oligomeric complex, most likely an $\alpha_2\beta_2$ tetramer. Like voltage-gated channels, the subunits of the photoreceptor and olfactory channel contain six putative transmembrane segments (S1–S6), a voltage sensor-like motif comprising the S4 segment, and a pore region that forms part of the central cavity through which ions move (Fig. 1). In addition, the carboxyl-terminal region of the CNG channels has a cyclic nucleotide binding domain of approximately 130 amino acids that is structurally related to the nucleotide fold of the bacterial catabolite activator protein CAP.

The CNG subunits also exhibit some distinctive structural properties. The $\alpha$ subunit contains a consensus sequence for N-linked glycosylation within a hydrophilic loop linking the S5 membrane-spanning segment to the pore region. This site (Asn-327) has been shown to be glycosylated in the bovine rod subunit.[5] In addition, the $\alpha$ subunits contain a glutamic acid residue (Glu-363 in the bovine rod subunit) within the pore region that plays a crucial role in external divalent cation blockage of the channel.[6] The $\beta$ subunits, on the other hand, are not glycosylated and the rod subunit does not contain a negatively charged residue in the pore region.[7,8]

The $\beta$ subunit of mammalian rod cells is unusual in that it is composed of two distinct parts.[8] The C-terminal part contains the structure features

[1] U. B. Kaupp, *Curr. Opin. Neurobiol.* **5,** 434 (1995).

[2] J. T. Finn, M. E. Grunwald, and K.-W. Yau, *Annu. Rev. Physiol.* **58,** 395 (1996).

[3] R. S. Molday, *Curr. Opin. Neurobiol.* **6,** 445 (1996).

[4] W. N. Zagotta and S. A. Siegelbaum, *Annu. Rev. Neurosci.* **19,** 235 (1996).

[5] P. Wohlfart, W. Haase, R. S. Molday, and N. J. Cook, *J. Biol. Chem.* **267,** 644 (1992).

[6] M. J. Root and R. MacKinnon, *Neuron* **11,** 459 (1993).

[7] T.-Y. Chen, Y.-W. Peng, R. S. Dhallan, B. Ahamed, R. R. Reed, and K.-W. Yau, *Nature* **362,** 764 (1993).

[8] H. G. Körschen, M. Illing, R. Seifert, F. Sesti, A. Williams, S. Gotzes, C. Colville, F. Müller, A. Dosè, M. Godde, L. Molday, U. B. Kaupp, and R. S. Molday, *Neuron* **15,** 627 (1995).

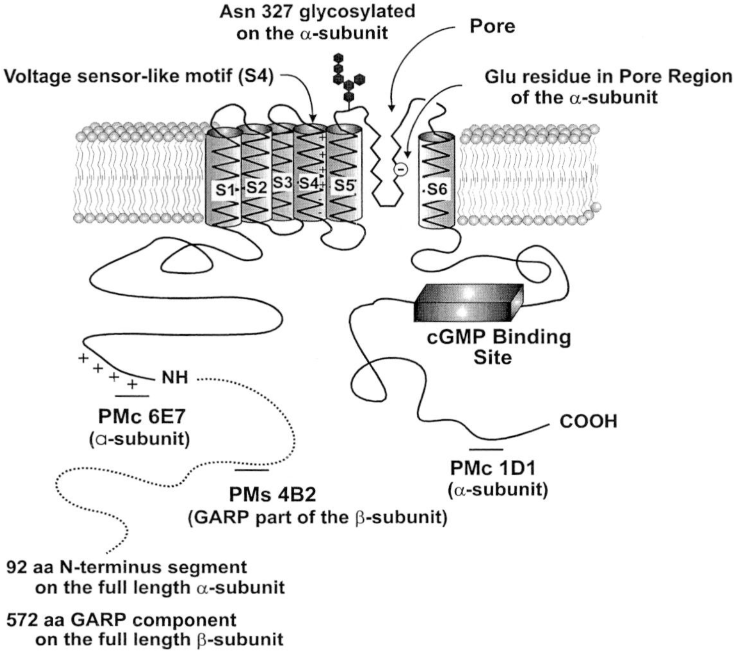


FIG. 1. Current topological model for rod CNG channel subunits. Structural features include six transmembrane segments (S1–S6), a voltage sensor-like motif comprising the S4 transmembrane segment, a pore region between S5 and S6, and a cGMP-binding site. In addition, the $\alpha$ subunit is glycosylated at Asn-327 and the full-length form contains an additional 92 amino acids not found in the $\alpha$ subunit in the channel from rod outer segment (ROS) membranes; the $\beta$ subunit contains a GARP component at its N terminus consisting of 572 amino acids. The binding site for calmodulin has been localized to a segment near the N terminus of the olfactory channel. The calmodulin binding site for the rod channel is on the $\beta$ subunit, but its exact location remains to be determined. The location of the binding sites for monoclonal antibodies PMc 1D1 and PMc 6E7 against the $\alpha$ subunit and PMs 4B2 against the GARP component of the $\beta$ subunits is shown.

characteristic of other CNG channel subunits, whereas the N-terminal part is identical to a glutamic acid rich protein called GARP. The GARP part of the $\beta$ subunit does not appear to be required for channel activity since removal of this segment does not alter the channel activity. Only a small C-terminal part of GARP is present in the $\beta$ subunit recently cloned from testis.[9]

[9] M. Biel, X. Zong, A. Ludwig, A. Sautter, and F. Hofmann, *J. Biol. Chem.* **271,** 6349 (1996).

The $\alpha$ subunits of the rod and olfactory channel assemble into functionally active homotetrameric channels when expressed in HEK293 cells or *Xenopus* oocytes.[1,2,7,10] In contrast, the $\beta$ subunit expressed by itself fails to form functional channels. Coexpression of the $\alpha$ and $\beta$ subunit, however, gives rise to a channel with electrophysiological properties characteristics of native channel. For example, coexpression of the rod $\alpha$ and $\beta$ subunits produces a channel with a rapid opening and closing or flickering behavior, micromolar sensitivity to the pharmacological blocker, L-*cis*-diltiazem, and ion selectivity and modulatory properties characteristic of the CNG channels in native rod photoreceptor membranes.[7,8]

Calmodulin binds to and modulates the activity of the rod and olfactory channels in a calcium-dependent manner.[3,11,12] The sensitivity of the rod channel for cGMP is decreased by about 2-fold, whereas the sensitivity of the olfactory channel for cAMP is reduced by 10 to 20-fold in the presence of $Ca^{2+}$-calmodulin. Binding studies indicate that $Ca^{2+}$-calmodulin binds to the $\beta$ subunit of the rod channel and the $\alpha$ subunit of the olfactory channel.

The cGMP-gated channel of rod photoreceptors, unlike other CNG channels, is expressed in quantities sufficient for biochemical analysis. It has been estimated that the cGMP-gated channel comprises as much as 6% of the protein in rod outer segment (ROS) plasma membranes. The abundance of this channel coupled with the availability of well-established procedures for isolation of ROS membranes from bovine retinas has enabled several laboratories to obtain highly pure channel preparations suitable for structure–function analysis.[13,14] In this article, we describe several procedures that have been used to purify the cGMP-gated channel from ROS membranes and reconstitute the channel into lipid vesicles for functional analysis. The application of these methods to the isolation of CNG channels from other cells systems is discussed.

## Materials and Methods

### *Solutions*

Homogenizing buffer: 20% (w/v) sucrose, 20 m*M* Tris–acetate, pH 7.4, 10 m*M* glucose, 1 m*M* $MgCl_2$.

[10] E. R. Liman and L. B. Buck, *Neuron* **13,** 622 (1994).
[11] Y.-T. Hsu and R. S. Molday, *Nature* **361,** 76–79 (1993).
[12] M. Liu, T.-Y. Chen, B. Ahamed, J. Li, and K.-W. Yau, *Science* **266,** 1348 (1994).
[13] N. J. Cook, W. Hanke, and U. B. Kaupp, *Proc. Natl. Acad. Sci. U.S.A.* **84,** 585 (1987).
[14] Y.-T. Hsu and R. S. Molday, *J. Biol. Chem.* **269,** 29765 (1994).

Hypotonic lysis buffer: 10 m*M* HEPES–KOH, pH 7.4, 1 m*M* ethylenediaminetetraacetic acid (EDTA) and 1 m*M* dithiothreitol (DTT).

3 - [3 - (Cholamidopropyl)dimethylammonio] - 1 - propane sulfonate (CHAPS) solubilization buffer: 10 m*M* HEPES–KOH, pH 7.4, 1 m*M* DTT, 10 m*M* $CaCl_2$, 0.15 *M* KCl, 18 m*M* CHAPS, 2 mg/ml asolectin (soybean phosphatidylcholine, type IV-S; Sigma, St. Louis, MO) and protease inhibitors [0.1 m*M* diisopropylfluorophosphate, 5 $\mu$g/ml aproteinin, 1 $\mu$g/ml leupeptin, and 2 $\mu$g/ml pepstatin or 20 $\mu M$ Pefabloc SC (Boehringer Mannheim, Germany)].

CHAPS column buffer: 10 m*M* HEPES–KOH, pH 7.4, 1 m*M* DTT, 1 m*M* $CaCl_2$, 0.15 *M* KCl, 12 m*M* CHAPS, and 2 mg/ml asolectin.

DEAE elution buffer: 10 m*M* HEPES–KOH, pH 7.4, 1 m*M* DTT, 10 m*M* $CaCl_2$, 0.7 *M* KCl, 12 m*M* CHAPS, and 2 mg/ml asolectin.

Red dye elution buffer: 10 m*M* HEPES–KOH, pH 7.4, 1 m*M* DTT, 10 m*M* $CaCl_2$, 1.8 *M* KCl, 12 m*M* CHAPS, and 2 mg/ml asolectin.

Reconstitution buffer: 10 m*M* HEPES–KOH, pH 7.4, 0.1 *M* KCl, 1 m*M* DTT, 2 m*M* $CaCl_2$, and 10 m*M* CHAPS and 18 mg/ml asolectin.

Dialysis buffer: 10 m*M* HEPES–KOH, pH 7.4, 0.1 *M* KCl, and 2 m*M* $CaCl_2$.

### *Purification of Rod CNG Channel*

A flowchart depicting the steps involved in the purification and characterization of the rod cGMP-gated channel is shown in Fig. 2. Generally, this involves (1) the isolation of ROS membranes from bovine retina, (2) the solubilization of the membranes in CHAPS detergent, and (3) the purification of the channel by calmodulin-Sepharose affinity chromatography, antichannel monoclonal antibody (mAb)-Sepharose affinity chromatography, or a combination of a DEAE ion-exchange and AF red dye adsorption chromatography. The identity and purity of the isolated channel is confirmed by sodium dodecyl sulfate (SDS) gel electrophoresis and Western blotting, and the activity of the channel is assessed by functional reconstitution into lipid vesicles for ion flux assays. Detailed procedures are described below.

*Preparation of Bovine Rod Outer Segment Membranes.* Rod outer segments are isolated from bovine retina by a continuous sucrose gradient centrifugation procedure.[15] Dark-adapted retinas from 100 freshly dissected bovine eyes are immersed in 40 ml of homogenizing buffer and gently shaken for 1 min under dim red light to break off the outer segments. The suspension is then filtered through a Teflon screen or cheesecloth and

[15] R. S. Molday and L. L. Molday, *J. Cell Biol.* **105,** 2589 (1987).

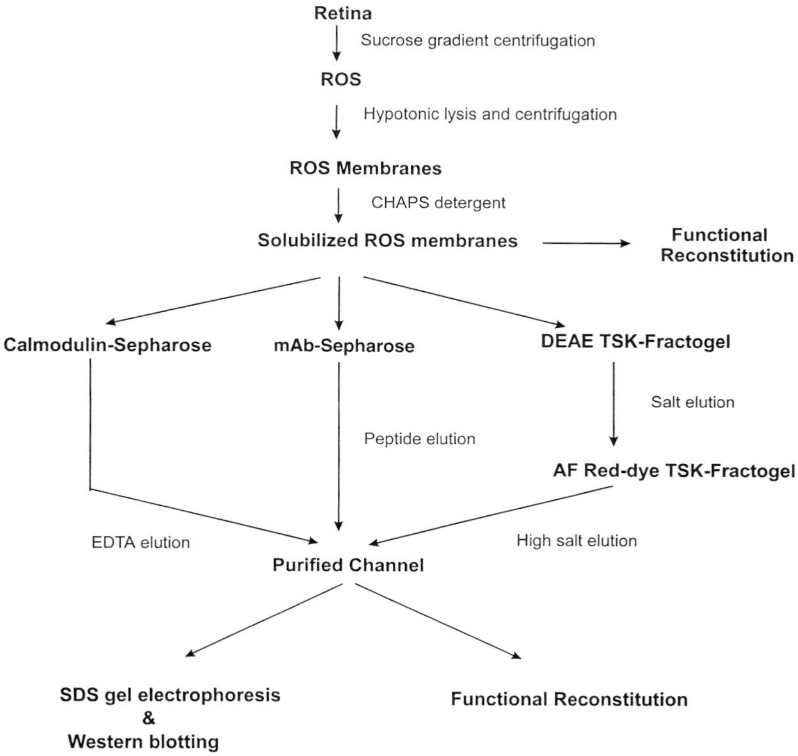


FIG. 2. Diagram showing the essential steps used to isolate the cGMP-gated channel from retina tissue. ROS are isolated from bovine retina by sucrose gradient centrifugation. ROS membranes are obtained by hypotonic lysis of ROS followed by sedimentation. The membranes are then solubilized in CHAPS detergent and the channel is isolated by affinity chromatography on a calmodulin-Sepharose column or a monoclonal antibody (mAb)-Sepharose column or by sequential chromatography on a DEAE ion-exchange column and a red dye adsorption column. The identity and purity of the channel is analyzed by sodium dodecyl sulfate (SDS) gel electrophoresis and Western blotting, and the activity is assessed by functional reconstitution into lipid vesicles for analysis of cGMP-dependent $Ca^{2+}$ efflux.

loaded onto six 20-ml 30–50% (w/v) continuous sucrose gradients prepared in 20 m*M* Tris–acetate buffer, pH 7.4, containing 10 m*M* glucose, and 1 m*M* $MgCl_2$. Centrifugation is carried out in a Beckman SW28 rotor at 25,000 rpm (82,500*g*) for 45 min at 4°. The pink band of ROS from each tube is carefully collected with a syringe, diluted with five volumes of homogenizing solution, and centrifuged at 13,000 rpm (20,000*g*) in a Sorvall SS-34 rotor 5 or 20 min at 4°. The ROS pellet is then resuspended in about 8 ml of homogenizing buffer to obtain a final ROS concentration of 8–10 mg/ml protein. The ROS are either used immediately or stored in light-tight vials at −70°. A yield of 70–80 mg of ROS protein is typically obtained.

Unbleached ROS membranes are stripped of soluble proteins under dim red light by a hypotonic lysis procedure.[15] ROS are suspended in 10 volumes of hypotonic lysis buffer at 4° and centrifuged at 13,000 rpm for 10 min in a Sorvall SS-34 rotor. The membrane pellet is resuspended in the same buffer, and the washing procedure is repeated two more times. The final pellet is resuspended in 10 m*M* HEPES–KOH, pH 7.4 (original volume), to obtain a ROS protein concentration of 7–9 mg/ml as determined by the bicinchoninic acid (BCA) assay (Pierce, Rockford, IL). The lysis and washing procedures are carried out under dim red light to facilitate the removal of transduction and other soluble proteins of the visual cascade system.

*Solubilization of ROS Membranes.* For functional reconstitution studies, ROS membranes are solubilized in CHAPS detergent in the presence of phospholipid as first described by Cook *et al.*[13,16,17] Typically, unbleached ROS membranes are slowly added with stirring to CHAPS solubilization buffer at 4° to obtain a final ROS protein concentration of 1–1.5 mg/ml. For immunoaffinity chromatography, DTT is omitted from the solubilization buffer and subsequent isolation procedures to prevent the loss of antibody from the immunoaffinity matrix by reduction of interchain disulfide bonds. After stirring for 30 min at 4°, residual aggregated material is removed by centrifugation at 15,000 rpm (27,000*g*) for 30 min at 4° in a Sorvall SS34 rotor. Only a small pellet should be observed after this step. The supernatant containing the solubilized channel is exposed to normal light and used either directly for channel reconstitution or for channel purification as described later.

Triton X-100 can be used in place of CHAPS to solubilize ROS membranes for purification of the CNG channel by column chromatography. However, since the critical micellar concentration (CMC) of Triton X-100 is low, this detergent is difficult to remove and, therefore, not generally applicable for functional reconstitution of the channel. When Triton X-100 is used for the isolation of the channel, a final Triton X-100 concentration of 1% is typically used for solubilization and 0.1% for column chromatography.

*Calmodulin Affinity Chromatography.* Calmodulin-Sepharose can be used to obtain highly pure channel preparations from detergent-solubilized ROS membranes in a single chromatographic step.[14] Typically, 20 ml of CHAPS-solubilized ROS membrane (~20–30 mg protein) is added to 2.5 ml of calmodulin-Sepharose (Pharmacia) preequilibrated at 4° with the

[16] N. J. Cook, C. Zeilinger, K.-W. Koch, and U. B. Kaupp, *J. Biol. Chem.* **261,** 17033 (1986).
[17] N. J. Cook and U. B. Kaupp, *Photobiochem. Photobiophys.* **13,** 331 (1986).

CHAPS column buffer. The solution is passed through the calmodulin-Sepharose column at a flow rate of about 0.5 ml/min. After the unbound fraction is collected, the column is washed with 10 volumes of the CHAPS column buffer or until the absorbance at 280 nm returns to a baseline level. The bound fraction containing the rod CNG channel is then eluted with CHAPS column buffer in which the $CaCl_2$ is replaced with 1.5 m*M* EDTA. Elution of protein from the column can be monitored by absorbance at 280 nm and/or SDS gel electrophoresis. Using this procedure, approximately 100 $\mu$g of highly pure channel can be obtained from 25 mg of ROS membrane protein.

*Immunoaffinity Chromatography.* A number of monoclonal antibodies (mAb) have been produced that recognize well-defined epitopes on the $\alpha$ or $\beta$ subunits of the bovine rod CN channels.[18-20] Two of these antibodies, PMc 6E7 against the N-terminal region of the native $\alpha$ subunit (63-kDa polypeptide) and PMs 4B2 against a repeat region of the GARP part of the $\beta$ subunit (240-kDa), have been used to isolate the channel from detergent-solubilized ROS membranes by immunoaffinity chromatography.[14,19] A typical procedure utilizing the PMc 6E7 monoclonal antibody is described later. The methods are generally applicable for other well-characterized antichannel monoclonal antibodies.

PREPARATION OF IMMUNOAFFINITY MATRIX. Three milliliters of packed CNBr-activated Sepharose 2B suspended in 3 ml of 10 m*M* sodium borate buffer, pH 8.4, 0.1 *M* NaCl is added to 6 mg of the purified PMc E67 monoclonal antibody in 3 ml of the same buffer. After gentle mixing for 4 hr at 4°, the Sepharose matrix is washed by low-speed centrifugation, once with 10 volumes of sodium borate/NaCl buffer and three times with 10 volumes each of 10 m*M* Tris-HCl buffer, pH 7.4, containing 50 m*M* glycine and 0.15 *M* NaCl. Since the efficiency of coupling is generally between 80 and 95%, approximately 1.6–1.9 mg of antibody can be coupled per milliliter of packed Sepharose.

COLUMN CHROMATOGRAPHY. A 20-ml solution of the CHAPS-solubilized ROS membranes is applied to a column containing 3 ml of the PMc 6E7-Sepharose equilibrated at 4° in CHAPS column buffer without DTT. The unbound protein is eluted from the column at a flow rate of about 0.5 ml/min and the matrix is then washed with 10 volumes of the CHAPS column buffer or until the 280-nm absorbance returns to baseline. The

[18] N. J. Cook, L. L. Molday, D. Reid, U. B. Kaupp, and R. S. Molday, *J. Biol. Chem.* **265,** 18690 (1989).

[19] L. L. Molday, N. J. Cook, U. B. Kaupp, and R. S. Molday, *J. Biol. Chem.* **265,** 18690 (1990).

[20] R. S. Molday, L. L. Molday, A. Dosè, I. Clark-Lewis, M. Illing, N. J. Cook, E. Eismann, and U. B. Kaupp, *J. Biol. Chem.* **266,** 21917 (1991).

column matrix is then incubated for 30 min with 3 ml of the CHAPS column buffer containing 0.1 mg/ml of the 6E7 competing peptide corresponding to the N terminus of the 63-kDa $\alpha$ subunit (Ser-Asn-Lys-Glu-Gln-Glu-Pro-Lys-Glu-Lys-Lys-Lys-Lys-Lys).[20] Finally, the channel is eluted from the column and fractions are collected and analyzed by SDS gel electrophoresis. The immunoaffinity column can be regenerated by washing in 0.1 *M* acetic acid followed by CHAPS column buffer.

*DEAE Ion-Exchange and Red Dye Affinity Chromatography.* This method, employing two chromatographic steps, was initially developed by Cook *et al.*[13] to isolate the CNG channel from bovine ROS membranes. In the first step, a DEAE column is used to separate the rod CNG channel, the Na/Ca-K exchanger, and several other negatively charged proteins from rhodopsin and other ROS proteins. In the second step, the channel is selectively adsorbed to a red dye matrix, facilitating the removal of the Na/Ca-K exchanger and other containing proteins. The procedure follows.

DEAE-Fractogel TSK chromatography. Approximately, 20 ml of the CHAPS-solubilized ROS membranes is applied to a column containing 2 ml of DEAE-Fractogel TSK (E. Merck, Darmstadt, Germany) preequilibrated at 4° in CHAPS column buffer containing 10 m*M* $CaCl_2$. The column is then slowly eluted at flow rate of 0.5 ml/min, and subsequently washed with the CHAPS column buffer until the absorbance at 280 nm returns to a baseline level. The crude channel complex is then eluted from the column with DEAE elution buffer. It has been suggested that increased yields of channel can be obtained by decreasing the KCl concentration to 0.1 *M* in the CHAPS solubilization buffer and CHAPS column buffer to facilitate the binding of the channel to the DEAE matrix.[21]

AF Red Dye Fractogel TSK Chromatography. The fractions eluted from the DEAE-Fractogel column are pooled and applied directly to a column containing about 2 ml of AF Red-Fractogel TSK (E. Merck) equilibrated with the DEAE elution buffer at 4°. After slowly passing the solution through the column at a flow rate of about 0.5 ml/min, the column is washed with DEAE elution buffer until the absorbance returns to baseline levels. The channel is subsequently eluted with red dye elution buffer. A 110-fold purification of the rod channel with a 32% recovery has been reported using this two-column procedure.[13]

## *Biochemical Characterization of Rod CNG Channel*

The purity of the channel preparation can be assessed by SDS gel electrophoresis using the Laemmli buffer system.[13,14,19] Highly purified rod

[21] N. J. Cook, *Methods Neurosci.* **15,** 271 (1993).

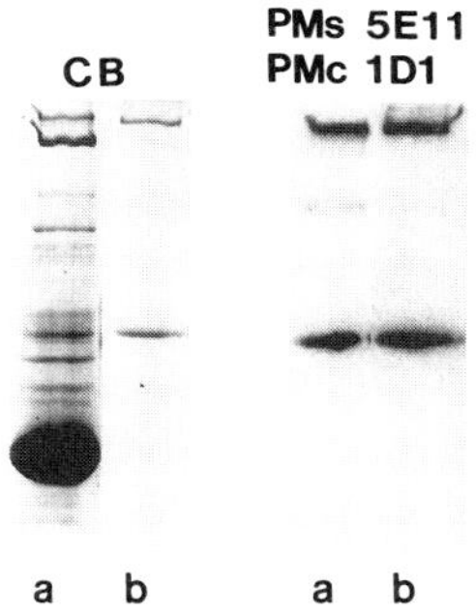


FIG. 3. SDS gel electrophoresis and Western blotting of an immunoaffinity purified preparation of the rod CNG channel. Bovine ROS membranes (a) were solubilized in CHAPS buffer and passed through a PMc 6E7-Sepharose column. After the column was washed to remove unbound protein, the channel was eluted with 0.1 mg of the 6E7 competing peptide. The proteins on the SDS gels were either stained with Coomassie blue (CB) or transferred onto Immobilon membranes and labeled with a mixture of the PMs 5E11 antibody specific for the 240-kDa $\beta$ subunit and the PMc 1D1 antibody specific for the 63-kDa $\alpha$ subunit. [Reproduced from R. S. Molday and Y.-T. Hsu, *Behav. Brain Sci.* **18,** 441 (1995).]

channel preparations consist of a 63- and 240-KDa polypeptide that is intensely stained by either Coomassie blue or silver staining (Fig. 3). Western blots labeled with channel subunit specific antibodies indicate that the 63-kDa polypeptide is the $\alpha$ subunit of the rod channel and the 240-kDa polypeptide is the $\beta$ subunit. The molecular mass of the ROS channel $\alpha$ subunit is significantly lower than the 79.6-kDa molecular mass predicted from its amino acid sequence and the apparent molecular mass of 78 kDa observed for the heterologously expressed subunit. The size difference between the native and full-length $\alpha$ subunit is due to the absence of a 92-amino-acid N-terminal segment in the former. The significance of this photoreceptor cell specific truncation of the $\alpha$ subunit N terminus is not yet understood.[20] In contrast the 240-kDa molecular mass of the $\beta$ subunit determined by SDS–polyacrylamide gel electrophoresis (SDS–PAGE) is significantly higher than its 155-kDa molecular mass calculated from its amino acid sequence. The relatively high content of glutamic acid residues in the GARP part of the $\beta$ subunit has been suggested to be responsible for the anomalous migration of this subunit on SDS polyacrylamide gels.[8]

In addition to the two prominent channel bands, several faintly stained bands are sometimes observed in channel preparations. These bands may represent proteins that either nonspecifically associate with the channel during the solubilization and purification procedure or specifically interact and copurify with the channel. In the case of calmodulin-Sepharose chromatography, several calmodulin binding proteins are isolated along with the channel by this procedure as observed on Western blots labeled with calmodulin.[11,14]

*Functional Reconstitution of Rod CNG Channel*

A detergent dialysis method has been developed to reconstitute effectively the CHAPS-solubilized channel into lipid vesicles for functional analysis.[13,16] This procedure involves the slow removal of the CHAPS detergent from the channel in the presence of excess phospholipid and calcium. The channel is incorporated into the lipid bilayer of the vesicles that form as the detergent is removed, and calcium is trapped inside the vesicles. Calcium present on the external side of the closed vesicles is subsequently removed either by dialysis against calcium-free buffer or by ion-exchange or gel-filtration chromatography in order to create a transmembrane $Ca^{2+}$ gradient. The cGMP-dependent release of calcium from lipid vesicles containing the channel can be readily monitored spectrophotometrically using the metallochromic dye Arsenazo III. Calcium trapped in lipid vesicles that do not contain the channel can be subsequently released using the ionophore A23187.

*Reconstitution Procedure.* Typically, 1 ml of the purified channel (or solubilized ROS membranes) is added to 1 ml of reconstitution buffer and the solution is dialyzed at 4° against 1 liter of dialysis buffer. The dialysis buffer is changed three times over a period of 2 days. Finally, the solution is dialyzed for 6–12 hr at 4° against dialysis buffer that does not contain $CaCl_2$.

*Channel Activity Measurements.* After dialysis, the vesicles (0.2–0.5 ml) are mixed with dialysis buffer containing 75 $\mu M$ Arsenazo III in a cuvette to give a final volume of 2 ml. After recording a baseline in a dual-wavelength spectrophotometer set at 650 and 730 nm, 4 $\mu$l of cGMP is added with stirring to give the final desired concentration (generally between 1 and 150 $\mu M$), and the rate of change in absorbance is recorded for kinetic analysis. The release of total calcium (from lipid vesicles that do not contain a functional channel) is then determined by the addition of 1 $\mu M$ A23187 and monitored spectrophotometrically. The channel activity is determined from the ratio of calcium released by cGMP to total calcium released (calcium released by cGMP + calcium released by A23187) as described by Cook *et al.*[13]

## Discussion

The CNG channel can be effectively purified from CHAPS-solubilized rod outer segment membranes either by affinity chromatography on a calmodulin or antichannel monoclonal antibody matrix or by a combination

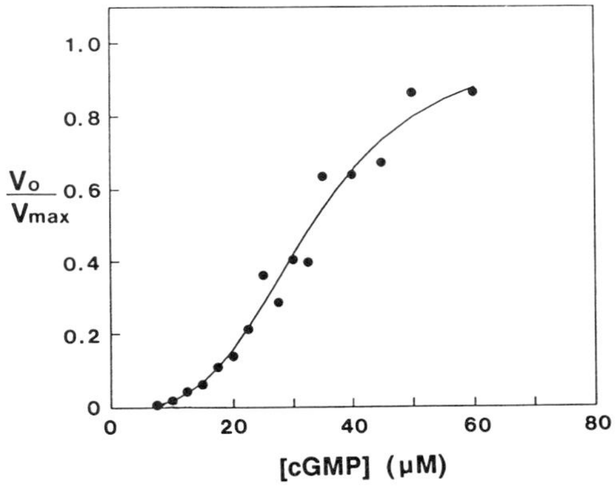


FIG. 4. The dependence of the reconstituted rod CNG channel activity on cGMP concentration. The bovine rod channel purified on a PMc 6E7-Sepharose column was reconstituted into lipid vesicles, and the efflux of $Ca^{2+}$ from vesicles as a function of cGMP concentration was monitored spectrophotometrically using the optical dye Arsenazo III. $V_o/V_{max}$ is the normalized initial velocity for cGMP-dependent $Ca^{2+}$ efflux. The sigmoidal curve was calculated using a $K_m$ of 33 $\mu M$ cGMP and a Hill coefficient of 3.3. [Reproduced from R. S. Molday and Y.-T. Hsu, *Behav. Brain Sci.* **18,** 441–451 (1995).]

of ion-exchange and red dye absorption chromatography. The purified channel obtained by these methods consists of the $\alpha$ and the $\beta$ subunit having an apparent molecular mass of 63 and 240 kDa, respectively, as determined by SDS–PAGE.[8,11,13] In each instance, the purified channel has been reconstituted into lipid vesicles for analysis of cGMP-dependent channel activity.

Kinetic analysis of cGMP-dependent efflux of calcium from reconstituted liposomes indicates that, like the native channel, the reconstituted channel is cooperatively activated by cGMP with a $K_{1/2}$ of 10–30 $\mu M$ and a Hill coefficient of 2.5–3.5 (Fig. 4). It is also modulated by calmodulin in a calcium-dependent manner.[11,14,22] The apparent $K_{1/2}$ for cGMP is increased by about twofold in the presence of $Ca^{2+}$-calmodulin without affecting the cooperativity or $V_{max}$ of the channel.[14]

The reconstituted rod CNG channel, however, differs from the channel present in native ROS membranes or the channel expressed as a heteromeric complex in HEK 293 cells in its sensitivity to the pharmacological inhibitor, L-*cis*-diltiazem. Whereas micromolar concentrations of this agent effectively block the native and heterologously expressed heteromeric rod

[22] R. S. Molday and Y.-T. Hsu, *Behav. Brain Sci.* **18,** 441 (1995).

CNG channel,[7,23,24] little, if any, inhibition is observed for the reconstituted channel.[13] The reason for this difference in sensitivity to L-*cis*-diltiazem is not understood at the present time.

The three methods described here result in a 36- to 110-fold purification of the CNG channel from ROS membranes and yields ranging from 30 to 50% as measured by cGMP-dependent channel activity.[13,14] Some variation in the specific activity of the channel has also been reported. This is likely due to the efficacy of channel reconstitution, rather than the presence of significant contaminants. For example, in two independent studies, the specific activity of the channel reconstituted from ROS membranes and DEAE columns differed by 2-fold, despite the fact that the protein compositions appear identical by SDS gel electrophoresis.[13,14] Differences in the reported channel activity may result from differences in purity of the lipids used in these reconstitution procedures, the freshness of the dissected retina used to isolate ROS, or other experimental parameters.

Each purification method has certain advantages and disadvantages. Calmodulin affinity chromatography involves only a single column step and utilizes commercially available reagents. The channel is eluted from the calmodulin-Sepharose column under relatively mild conditions, i.e., EDTA. However, since calcium has been reported to stabilize the channel,[13] it is important to add back calcium immediately after the channel is eluted from the column to stabilize the detergent-solubilized channel. Although, in principle, calmodulin-Sepharose columns can be reused many times, it has been found that the yield of channel diminishes with use, presumably due to some irreversible binding of the channel and other ROS membranes to the column matrix. A disadvantage of the calmodulin affinity chromatography method is that several other calmodulin binding proteins copurify with the channel on calmodulin-Sepharose columns as visualized in Western blots labeled with radioiodinated calmodulin.[14]

The immunoaffinity procedure is a highly effective and rapid, single-step purification procedure.[11,22] The immunoaffinity purified channel is free of other calmodulin binding proteins; however, as in the case of the other purification procedures, additional proteins sometimes can be seen as faintly stained bands on SDS polyacrylamide gels. One such protein is the Na/Ca-K exchanger that migrates with an apparent molecular mass of 230 kDa and has been suggested to be associated with the channel under some conditions. The channel yields vary considerably with the different antibody-Sepharose matrices. Antibodies with relatively low binding affinities do not effectively bind all the channel, resulting in low yields; antibodies

[23] K.-W. Koch and U. B. Kaupp, *J. Biol. Chem.* **260,** 6788 (1985).

[24] J. H. Stern, U. B. Kaupp, and P. MacLeish, *Proc. Natl. Acad. Sci. U.S.A.* **83,** 1163 (1986).

with exceptionally high binding affinities efficiently bind the channel, but elution of the channel from the column with the competing peptide is usually inefficient, thereby also resulting in low recovery. In our hands, the yields of channel from different monoclonal antibody-Sepharose columns vary from 20–60%.

An important advantage of the antibody-Sepharose reagent is that it can be effectively used for immunoprecipitation studies where the end analysis is simply detection of the channel by SDS gel electrophoresis and Western blotting.[19] Using this approach antibody-Sepharose matrices have been used to immunoprecipitate small quantities of CNG channels from detergent-solubilized ROS membranes and from extracts of HEK293 cells expressing the channel subunits.[8,19] A disadvantage of the immunoaffinity procedure is that antichannel monoclonal antibodies or immunoaffinity matrices are not commercially available at the present time.

The ion-exchange–red dye chromatographic procedure has been successfully used in the initial purification and characterization of the channel from ROS membranes.[13] The ion-exchange step is simple and straightforward, yielding a highly enriched preparation of the channel suitable for functional reconstitution. The negatively charged rod channel and Na/Ca-K exchanger bind to the DEAE column, thereby separating these proteins from the bulk rhodopsin. It is likely that the highly negatively charged GARP region of the $\beta$ subunit is responsible for the binding of the channel to the DEAE ion-exchange matrix. Accordingly, this ion-exchange chromatographic procedure may not be suitable for isolation of the heterologously expressed homeric $\alpha$ subunit or CNG channels from other cells and tissues that do not contain the GARP region of the $\beta$ subunit.[9] The red dye column efficiently removes the Na/Ca-K exchanger and most other contaminating proteins. However, the yield and ability to functionally reconstitute the channel after the red dye column is quite variable, possibly due to differences in batches of the column matrix.[21]

Although the three methods described here have been most widely used, a preliminary report has described the use of an 8-bromo-cGMP-agarose matrix to isolate the rod channel from ROS membranes.[25] Using this procedure, the channel subunit was shown to consist of an 80-kDa polypeptide, instead of the 63- to 68-kDa polypeptide observed by the other three procedures. Further studies are needed to resolve this difference and to further evaluate the usefulness of a cyclic nucleotide affinity column for channel purification.

In principle, the methods described here to purify the cGMP-gated channel from ROS membranes can be used to isolate rod CNG channels

[25] R. Hurwitz and V. Holcombe, *J. Biol. Chem.* **266,** 7975 (1991).

from other cells and tissues. The immunoaffinity method is best suited for this purpose because of the high degree of specificity of monoclonal antibodies and the simplicity of the purification procedures. Immunoaffinity methods can also be used to isolate CNG channels from other cells and tissues once well-characterized monoclonal antibodies are developed. The calmodulin-Sepharose method may also be useful as a first step for the isolation of other CNG channels known to bind calmodulin, such as the olfactory channel.[12] However, since most cells contain many calmodulin binding proteins, such preparations may require additional purification steps in order to obtain highly purified channel preparations.

The future challenge is to obtain relatively large quantities of highly purified CNG channels for high-resolution structural analysis. The methods described here serve as a useful starting point for such studies.

# [13] Purification and Heterologous Expression of Inhibitory Glycine Receptors

*By* BODO LAUBE and HEINRICH BETZ

## Introduction

Glycine is a major inhibitory neurotransmitter in the vertebrate central nervous system, most prominently in brain stem and spinal cord. Its postsynaptic actions are mediated via a pentameric chloride channel protein, the inhibitory glycine receptor (GlyR). Strychnine, a convulsive poison in man and animals, antagonizes glycine activation of the GlyR and has proven to be a highly useful tool to characterize this membrane protein.[1] In adult spinal cord, the GlyR is composed of three copies of ligand binding $\alpha_1$ and two copies of structural $\beta$ subunits. During development and in peripheral tissues, additional $\alpha$-subunit genes ($\alpha_2$–$\alpha_4$) are expressed that display distinct pharmacologic and functional properties.[1] Heterologous expression has allowed their detailed characterization. Here, we describe methods for the purification and expression in heterologous cell systems of the GlyR.

[1] J. Kuhse, H. Betz, and J. Kirsch, *Curr. Opin. Neurobiol.* **5,** 318 (1995).

METHODS IN ENZYMOLOGY, VOL. 294

0076-6879/99 $30.00

## Purification of Glycine Receptor from Mammalian Spinal Cord

Affinity purification of the GlyR has been successfully achieved from rat,[2] mouse,[3] and porcine[4] spinal cord tissue as well as from cells expressing recombinant receptors.[5] All procedures involve adsorption of the detergent-solubilized receptor on aminostrychnine-agarose followed by biospecific elution with the agonist glycine. Routinely this procedure results in highly enriched (>80%) preparations containing the postsynaptic GlyR complex, i.e., its $\alpha$ and $\beta$ subunits together with the anchoring protein gephyrin.[6,7] All steps have to be performed at 4° in the presence of a protease inhibitor mix.[2] After solubilization, phospholipids (phosphatidylcholine) have to be included in all detergent buffers to stabilize the native conformation of the GlyR.[2,8]

### *Preparation of Membranes*

Spinal cord, medulla oblongata, and pons are removed from adult (100–200 g) Wistar rats (or mice or pigs), frozen in liquid nitrogen, and stored at −70°. Crude synaptic membranes are prepared from the frozen tissue after homogenization in >20 volumes of buffer A consisting of 25 m*M* potassium phosphate buffer, pH 7.4, containing 5 m*M* ethylenediaminetetraacetic acid (EDTA), 1 m*M* dithiothreitol (DTT), and protease inhibitors [100 m*M* benzethionium chloride, 1 m*M* benzamidinehydrochloride, 100 m*M* phenylmethylsulfonyl fluoride (PMSF), and 16 mU/ml of aprotinin] using a motor-driven Teflon or Polytron homogenizer. After optional removal of debris and nuclei by centrifugation at 1000*g* for 30 min, membranes are collected at 30,000*g* for 30 min and washed once with buffer A following the protocol described earlier. The resulting membrane pellet is rehomogenized in buffer A containing 0.6 *M* sucrose and centrifuged at 30,000*g* for 60 min. After resuspending the membrane fraction in 10 volumes of buffer A and centrifugation at 48,000*g* for 20 min, the membranes are suspended in three volumes of buffer A, frozen in liquid nitrogen, and stored at −70°. Preparation of membranes from transfected tissue culture

[2] F. Pfeiffer, D. Graham, and H. Betz, *J. Biol. Chem.* **257,** 9389 (1982).
[3] C. M. Becker, I. Hermanns-Borgmeyer, B. Schmitt, and H. Betz, *J. Neurosci.* **6,** 1358 (1986).
[4] D. Graham, F. Pfeiffer, R. Simler, and H. Betz, *Biochemistry* **24,** 990 (1985).
[5] W. Hoch, H. Betz, and C. M. Becker, *Neuron* **3,** 339 (1989).
[6] B. Schmitt, P. Knaus, C. M. Becker, and H. Betz, *Biochemistry* **26,** 805 (1987).
[7] P. Prior, B. Schmitt, G. Grenningloh, I. Pribilla, G. Multhaup, K. Beyreuther, Y. Maulet, P. Werner, D. Langosch, A. Kirsch, and H. Betz, *Neuron* **8,** 1161 (1992).
[8] F. Pfeiffer and H. Betz, *Brain Res.* **226,** 273 (1981).

cells[5,9] basically follows the same protocol. However, the sucrose buffer step is not required.

*Solubilization of Glycine Receptor*

The GlyR has been solubilized and purified from spinal cord membranes using either the nonionic detergent Triton X-100[2,3] or the ionic detergent cholate.[4] Solubilization efficiencies with cholate are usually better; however, the GlyR associated protein gephyrin is largely lost.[4]

For routine purification,[2] membranes are incubated in 25 m*M* $KP_i$, pH 7.4, containing final concentrations of 3% (w/v) Triton X-100, 1 *M* KCl, 5 m*M* neutralized EDTA, 5 m*M* neutralized ethylene glycol bis($\beta$-aminoethyl ether)-*N,N,N′,N′*-tetraacetic acid (EGTA), 5 m*M* DTT, and protease inhibitors. Protein concentrations may range from 1 to 10 mg/ml. The mixture is held on ice and repeatedly homogenized during 1 hr. After centrifugation at 100,000*g* for 1 hr at 4°, the supernatant is carefully decanted, filtered through a porous glass sieve to retain myelin lipids, and used as source of the solubilized GlyR.

*Preparation of Aminostrychnine-Agarose*

2-Aminostrychnine had been commercially available previously; however, no supplier is presently known. The compound therefore has to be synthesized from strychnine by nitration followed by reduction with Sn/HCl as described.[10] For coupling to Affi-Gel 10 (Bio-Rad, Richmond, CA), 20 ml of the gel is washed three times with one volume of dimethylformamide. To the washed resin in 20 ml of dimethylformamide, 500 mg of solid 2-aminostrychnine are added, and the mixture is gently shaken for 2 days at 4° in a brown bottle. After the addition of 200 ml of ethanolamine and incubation for another 2 hr, the gel is washed with 20 ml of dimethylformamide followed by 20 ml of 50 m*M* potassium phosphate buffer, pH 7.4, containing 0.02% sodium azide. After transfer into a column the gel is washed with an additional 400 ml of this buffer and stored in a brown bottle at 4° for up to 2 years.[2] Aliquots are washed extensively with 25 m*M* potassium phosphate buffer ($KP_i$), pH 7.4, containing 1% (w/v) Triton X-100 and 1 *M* KCl (buffer B), before being used for affinity chromatography.

[9] H. Sontheimer, C. M. Becker, P. R. Prittchett, P. R. Schofield, G. Grenningloh, H. Kettenmann, H. Betz, and P. H. Seeburg, *Neuron* **2,** 1491 (1989).

[10] C. R. Mackerer, R. L. Kochman, T. F. Shen, and F. M. Hershenson, *J. Pharmacol. Exp. Ther.* **201,** 326 (1977).

*Affinity Chromatography*

An aminostrychnine-agarose column (1–5 ml) is washed with 10–20 column volumes of buffer B, and then solubilized GlyR (up to ca. 300 ml) is applied to the column with a maximal flow rate of about 2.5 ml/hr of affinity resin. The column is washed extensively with about 20 column volumes of buffer B containing 5 m*M* EDTA, 5 m*M* EGTA, 5 m*M* DTT, 2.5 m*M* phosphatidylcholine and protease inhibitors (buffer C). For elution, one column volume of buffer C containing 200 m*M* glycine (buffer D) is continously pumped through the column for several hours.[2] After elution, a second addition of 1 column volume of buffer D results in a secondary eluate that also contains native GlyR. A final elution with 1 column volume buffer D containing 5 *M* urea yields denatured GlyR polypeptides and regenerates the column for future use in additional purifications. We have been able to repeatedly use the affinity matrix for purification over periods greater than 1 year.

## Heterologous Expression of Glycine Receptor

Expression in heterologous cell systems has been widely used to investigate functional properties of the GlyR by both electrophysiologic recording and biochemical methods.[1] Injection into *Xenopus* oocytes of poly(A)$^+$ RNA prepared from spinal cord generates glycine gated chloride channels in the oocyte membrane whose properties closely correspond to those detected in neurons. Functional GlyRs have also been generated by injection of *in vitro* transcribed rat and human $\alpha$- and $\beta$-subunit mRNAs. Expression of the $\alpha_1$, $\alpha_2$, or $\alpha_3$ subunits alone and in combination with the $\beta$ subunit has shown that both homo- and heterooligomeric GlyRs exist.[1] Transient expression of the recombinant GlyRs has also been achieved by transfection of cultured mammalian cells or using the baculovirus system.

*Oocyte Expression System*

Since Miledi and Sumikawa[11] demonstrated that various types of ion channels and receptors can be expressed in *Xenopus laevis* oocytes injected with poly(A)$^+$ RNA isolated from appropriate tissue, the oocyte expression system has proven highly useful to investigate various aspects of the pharmacology and function of GlyR proteins. These include (1) identification of structure–function relationships by mutational analysis,[12] (2) posttranslational processing and assembly of this multisubunit receptor,[13] (3) compari-

[11] R. E. Miledi and K. Sumikawa, *Biomed. Res.* **3,** 390 (1982).
[12] V. Schmieden, J. Kuhse, and H. Betz, *Science* **262,** 256 (1993).
[13] J. Kuhse, B. Laube, D. Magalei, and H. Betz, *Neuron* **11,** 1049 (1993).

son of the pharmacologic properties of native and recombinant GlyRs,[14] (4) receptor modulation by effectors,[15] and (5) analysis of mutations causing hereditary animal and human motor disorders.[16] The *Xenopus* oocyte expression system is ideally suited for electrophysiology, since the large cells are easy to handle and current amplitudes are high (routinely up to 5 $\mu$A). Biochemical studies in contrast are tedious since large numbers of cells have to be injected individually.

Heterologous expression of the GlyR in *Xenopus* oocytes has been achieved by both cytoplasmic injection of poly(A)$^+$ RNA or *in vitro* synthesized cRNA encoding different subunits and nuclear injection of cDNAs.

## *Isolation of Poly(A)$^+$ RNA from Neural Tissue*

Poly(A)$^+$ RNA suitable for injection can be prepared from spinal cord and brain according to standard methods.[17] However, pitfalls have been encountered with both isolated and *in vitro* transcribed RNA samples due to contamination with RNases. The following general comments on the handling of RNA should help to eliminate this problem: (1) Wear gloves throughout all manipulations (RNases are everywhere). (2) Prepare all buffers with filtrated, autoclaved $H_2O$. All tubes and pipette tips should also be autoclaved. (3) Clean all gel tanks and glass plates thoroughly. A common source of RNase on gel electrophoresis equipment stems from DNA preparations that have been treated with RNase A. (4) Templates used for transcription must be RNase free. DNA minipreparations are suitable if care is taken to remove contaminating RNases. Generally the template is linearized with a restriction endonuclease that cleaves downstream of the RNA polymerase promoter and the DNA inserted into the multiple cloning site. The digested DNA should be purified by proteinase K followed by two phenol–chloroform (1 : 1, v/v) steps and one chloroform extraction followed by ethanol precipitation. (5) It is recommended that RNase inhibitors be included in the transcription reaction.

## *In Vitro Transcription*

RNA suitable for expression in *Xenopus* oocytes can be readily prepared by *in vitro* transcription of GlyR subunit cDNAs with bacteriophage T3 or T7 RNA polymerase in the presence of a 5′cap [P-1,5′-(methyl)guanosine

[14] V. Schmieden, G. Grenningloh, P. Schofield, and H. Betz, *EMBO J.* **8,** 695 (1989).

[15] B. Laube, J. Kuhse, N. Runstroem, A. Kirsch, V. Schmieden, and H. Betz, *J. Physiol. (Lond.)* **483,** 613 (1995).

[16] D. Langosch, B. Laube, N. Rundstroem, V. Schmieden, J. Bormann, and H. Betz, *EMBO J.* **13,** 4223 (1994).

[17] A. L. Goldin and K. Sumikawa, *Methods Enzymol.* **207,** 279 (1992).

P-3,5-guanosine triphosphate] analog.[13] Addition of the 5′cap structure appears to be essential for RNA stability and efficient expression of transcribed RNA in oocytes.[18] Receptor cDNAs should be inserted into suitable cloning vectors, i.e., pBluescript II M13$^{+/-}$, KS/SK (Stratagene, La Jolla, CA), pRc/CMV or pcDNAI/AMP (Invitrogen) plasmids, which provide T3/T7 promoters. Before transcription, the plasmid has to be linearized at a convenient restriction site as follows.

## *Specific Procedures*

### *Linearization of the Vector*

1. Mix 4 pmol of vector DNA, required enzyme, 10 μl buffer (10×), and $H_2O$ to 100 μl.
2. Incubate for 1 hr at 37° (or at the temperature required by the enzyme). Check digest by running 2 μl on a 1% agarose gel together with undigested and linearized plasmid.
3. Add 2 μl of proteinase K (50 μg/ml) and incubate for another 30 min at 37°.
4. Extract twice with 1 volume phenol/chloroform (1 : 1), once with 1 volume chloroform/isoamyl alcohol (24 : 1), and precipitate with 0.1 volume 3 *M* sodium-acetate, pH 5.2, and 2 volumes of ethanol (100%) for 0.5 hr at −20° (or at −70° for 5 min).
5. Wash once with 1 ml of 70% (v/v) ethanol and resuspend in 10 μl TE buffer (10 m*M* Tris, pH 7.5, 0.1 m*M* EDTA).
6. Dot 1 μl on an ethidium bromide agarose plate or measure the OD at 260 nm for quantitation.

### *Transcription*

1. Mix

| | At final concentration |
|---|---|
| 1 μg linearized plasmid (1 μg/μl) | 50 ng/μl |
| 2 μl rATP-, rCTP-, rUTP-, and rGTP-mix (5 m*M* each) | 500 μ*M* |
| 2 μl CAP (5 m*M*) | 500 μ*M* |
| 4 μl transcription buffer 5× (200 m*M* Tris-HCl, pH 7.5, 250 m*M* NaCl, 40 m*M* $MgCl_2$, 10 m*M* spermidine) | |
| 1 μl RNase inhibitor II | 1 U/μl |
| 2 μl SP6, T3, or T7 polymerase (Boehringer, Mannheim, Germany) | 2 U/μl |
| $H_2O$ | to 20 μl |

[18] R. Ben-Azis and H. Soreq, *Nucleic Acids Res.* **18,** 3418 (1990).

2. Incubate for 1 hr at 37°.
3. Add 0.5 μl DNase 1 (10 U/μl) and incubate for 15 min at 37°.
4. Add 1 μl of 0.5 *M* EDTA, pH 8.0, and 80 μl of TE buffer (10 m*M* Tris, pH 7.5, 0.1 m*M* EDTA) or diethyl pyrocarbonate (DEPC)-treated water.
5. Extract once with 1 volume phenol/chloroform (1:1) and 1 volume chloroform/isoamyl alcohol (24:1).
6. Add 1 volume 4 *M* ammonium acetate and 2 volumes of ethanol. Incubate and precipitate the RNA for 15 min at −70° in a dry ice–ethanol mixture.
7. Spin for 30 min at 13,000 rpm and 4° in an Eppendorf centrifuge.
8. Drain the pellet and rinse with 80% ethanol cooled to −20°.
9. Spin for 10 min at 13,000 rpm, dry in a Rotovap (Fischer, Frankfurt, Germany), and resuspend in 10 μl $H_2O$.
10. Determine the concentration of RNA in the sample and store at −70° in 1-μl aliquots.

*Preparation of Oocytes*

*Xenopus laevis* oocytes are egg precursors that on hormonal stimulation become fertilizable eggs. They are stored as paired gonades in the abdominal cavity and can be surgically removed and maintained in Barth's medium [88 m*M* NaCl, 1 m*M* KCl, 0.4 m*M* $CaCl_2$, 0.3 m*M* $Ca(NO_3)_2$, 8.4 m*M* $MgSO_4$, 2.4 m*M* $NaHCO_3$, 10 m*M* Tris, pH 7.2]. Only the large stage V and VI oocytes[19] with a diameter of approximately 1–1.2 mm should be used for electrophysiological experiments. Several ovarian lobes can be removed from female frogs anesthetized by immersion in 1% (w/v) urethane (Sigma, St. Louis, MO) by making a 1- to 2-cm incision in the abdomen. This incision can be rapidly sutured under urethane narcosis with sterile surgical thread and needles; the animals recover rapidly and may be used again as oocyte donors in a later experiment. The ovarial lobes are then mechanically separated in pieces of about 10–50 oocytes by using forceps. For the isolation of single oocytes, connective tissue and blood vessels are removed by using forceps and a pair of microscissors (Dösch, Heidelberg). Subsequently, the layer of follicle cells surrounding the oocyte must be removed carefully (see below) because it constitutes a potential source of problems due to electrical coupling to the oocyte via gap junctions. The surrounding vitelline membrane, a glycoprotein matrix that gives the oocyte structural rigidity, does not impair binding of compounds to the oocyte

[19] J. N. Dumont, *J. Morphol.* **136,** 153 (1972).

surface up to a molecular mass of about 8000 Da and can therefore be left on the oocyte for routine electrophysiology. All dissection procedures should be done in Barth's medium.

There are two ways to remove the follicle cell layers:

1. Extensive collagenase treatment that completely strips off the layers is achieved by immersing the oocyte for 1–3 hr in calcium-free saline containing 2 mg/ml of collagenase Type IIA (Sigma).
2. A less extensive collagenase treatment followed by manual removal of the follicular layer with fine forceps (Inox 5) under binocular control (M3Z; Zeiss, Oberkochen, Germany) may be the better method, since oocytes undergoing the extensive collagenase treatment often show reduced survival times.

A striking feature of *Xenopus* oocytes is their highly characteristic polarity: the animal pole region is of dark green-brownish color whereas the vegetal pole has a bright yellow-white appearance. Polarity is maintained inside the oocyte: the nucleus consistently is found at the animal pole, whereas receptors expressed from exogenous mRNAs accumulate preferentially in the vegetal pole region of the plasmamembrane. This nonhomogeneity has no consequences for whole-cell two-electrode voltage clamping, but great importance when recording GlyRs in different patch-clamp configurations.[20] For achieving high-resistance seals, a clean oocyte surface is essential. Because devitellinized oocytes are extremely fragile and tend to disintegrate on contact with air–water interfaces, the vitelline membrane should only be removed for single-channel recordings requiring a gigaseal. This can be achieved by placing the oocyte in a hypertonic stripping solution (200 m*M* potassium aspartate, 20 m*M* KCl, 1 m*M* $MgCl_2$, 10 m*M* EGTA, 10 m*M* HEPES, pH 7.4) and allowing it to shrink. During shrinking, the vitelline membrane detaches from the plasma membrane and forms a transparent sphere around the oocyte that can be removed with fine forceps. Devitellinized oocytes should be stored in small culture dishes containing a bottom layer of 2% agarose to avoid sticking to the plastic.

### *Injection of Oocytes*

Oocytes of healthy appearance with clearly visible vegetal and animal poles are transferred into a culture dish and allowed to recover for >2 hr.

[20] O. P. Hamill, A. Marty, E. Neher, B. Sakmann, and F. J. Sigworth, *Pflügers Arch.* **391,** 85 (1981).

Thereafter damaged oocytes can be easily identified by visual inspection and discarded. Oocyte injections are performed under a fixed-stage binocular (Zeiss, Leitz) with objective lenses of 6.5–40× and eye pieces of 10× magnification. The binocular is mounted to an injection table with a movable injection chamber and a micromanipulator (Narishige, East Meadow, NY, or Bachofer, Reutlingen, Germany) for positioning the injection capillary, which is connected to a pressure generator (microinjector).

*Pulling and Calibration of Injection Capillaries.* Pull glass capillaries (borosilicate, 2 mm outer diameter; Hilgenberg) with a total volume of 500 nl for multiple injections by using a List electronic two-step puller L/M-3P-A or a similar device. When using the List puller, the capillary is fixed in the upper and lower holders for maximal distance. The real distance of the first pull is given by a disk with a height of 7 mm and made with a current of 20 A. Changing the disk to the 9-mm one allows pulling of the fine tip in the second step with a current of 16.5 A resulting. Pressure injection is useful for injecting volumes >1 nl in a reproducible manner. For RNA injection, volumes of 10–50 nl per oocyte have to be applied. Two different positive pressure modes should be available: a high one for ejection and a low pressure to prevent backfilling of the pipette by capillary action or diffusion (microinjector PV820; Bachofer). One of the most critical factors for injecting reproducible volumes is experience in manufacturing suitable micropipettes. To prevent backfilling of the pipette by capillary action and to avoid difficulties during injection, the pipette tip must be inspected visually and corrected to a size of about 4–8 $\mu$m. With tip sizes of less than 1 $\mu$m, pressure ejection becomes increasingly difficult; this can be partially overcome by cleaning the micropipette with chromic acid solutions or silanization to decrease surface tension. The volume of fluid ejected is strongly dependent on tip size because a reduction or increase in tip size results in remarkable changes in the flow rate. To calculate the volume ejected per pressure application, deposit a drop of fluid on the tip of the micropipette by a single pressure increase. The volume of this drop may be calculated by measuring the radius and assuming the drop to be spherical (100 $\mu$m diameter correspond to ≈500 pl). For injection, we use a manipulator (MD4, Bachofer) where 20 oocytes fixed on an injection table can be injected consecutively. A desirable injection volume is about 50 nl corresponding to 10 ng of RNA per oocyte when using a solution containing 200 ng RNA per microliter. The injection capillary holds a total volume of 500 nl; therefore up to 10 oocytes can be injected with one filled capillary. For transferring the RNA from the tube into the injection needle, 10-$\mu$l siliconized (dichloromethylsilane; Merck, Darmstadt, Germany) and sterilized 10-$\mu$l micropipettes (Brand, Wortheim, Germany) are required.

*Cytoplasmic Injection*

1. Fix the injection capillary in the holder of the micromanipulator and divide its tip into 10 equal sections using a small overhead marker.
2. Fill a 10-$\mu$l micropipette (Brand) with 1 $\mu$l of RNA solution by using a micrometer syringe.
3. Connect a 5-ml injection syringe to the freshly pulled injection capillary and position it close to the RNA containing micropipette.
4. Form a drop of RNA solution at the tip of the micropipette by increasing pressure using the microsyringe.
5. When the pipette is inserted into the fluid, the meniscus can be seen to raise in the tip of the capillary. Create negative pressure to fill the injection needle.
6. Position the oocytes on the injection table.
7. Create a slight positive pressure in the injection capillary and penetrate the oocyte membrane by using the micromanipulator.
8. Increase the pressure until the RNA solution starts to flow.
9. Remove the needle by using the micromanipulator after 50 nl of solution have been injected and proceed to the next oocyte.
10. After injection, incubate all oocytes at 19° (Heraeus, Hanau, Germany) in 35-mm petri dishes (Greiner, Frickenhausen, Germany) containing Barth's medium and change the medium daily.

GlyR subunits express efficiently; thus electrophysiological recording is possible already after 24 hr of injection. Maximal expression levels are reached after 2 days and maintained for up to 3 days. After incubation at 19°, oocytes may be stored at 4° for up to 2 weeks without significant changes in channel behavior and membrane properties. Before recording, any remaining follicular cells should be removed mechanically under the microscope (M3Z, Zeiss) by using forceps.

*Nuclear Injection*

Expression of GlyRs via nuclear injection of cDNAs has the advantage that no preparation and handling of RNA is required. In addition, higher expression rates may be achieved.[21] However, the survival of oocytes after nuclear injection is generally much lower than that obtained on cytoplasmic injection of RNA.

Nuclear injection can be easily performed provided the nucleus of the oocyte is visible. To this end, the nucleus is moved close to the cell surface

[21] O. Taleb and H. Betz, *EMBO J.* **13,** 1318 (1994).

by gentle centrifugation. This can be achieved by putting the oocytes with the dark side up into the wells of a 96-well microtiter plate, adding Barth's medium, and centrifuging the plates at 1000*g* for 15 min at room temperature. Nuclei then will appear as white spots in the dark animal pole region. The plasmid pCis,[22] pRc/CMV, or pcDNAI/Amp (Invitrogen) containing the appropriate GlyR cDNA insert is diluted in water to concentrations of 25–50 ng/$\mu$l; when using an injection volume of ~10 nl, this corresponds to a total amount of 250 pg of plasmid. Injection capillaries are made from Drummond microcaps in two steps using a vertical puller (LM-3P-A; List Electronics) as described before. Then the capillary tip is broken using optical control (microforge; Bachofer) until the desired tip diameter is obtained ($\approx$5 $\mu$m). The capillary is inserted into a micromanipulator (MD4; Bachofer) and connected to a pressure generator (microinjector PV 820; Bachofer) that allows the generation of different injection parameters by varying the time and pressure of injection. The protocols for the calibration of the needle and the injection procedure are similar to those used for cytoplasmic injection. Only the inside volume of the capillary for the nuclear injection is lowered to 100 nl, thus allowing 10 injections of 10 nl. Sometimes the oocytes have to be turned under binocular control (40×) to orient the visible nucleus perpendicularly to the tip of the capillary. Then the tip is introduced into the nucleus, and pressure is activated until the above volume is injected. Subsequently the cells are incubated in Barth's medium at 19° as above.

## Transient Expression in HEK293 Cells

Coprecipitates composed of "calcium phosphate" (hydroxyapatite) and purified DNA have been used for >20 years for the transfer to and expression of genetic information in mammalian cells in culture. This technique has become one of the preferred methods to express recombinant glycine receptors in mammalian cells for analyzing their functional properties.[16] We use a popular immortalized cell line, human embryonic kidney 293 cells (HEK 293; CRL 1573, ATCC, Rockville, MD), and a transfection procedure adapted from Chen and Okayama.[23] Routinely, transfection efficacies of ca. 50% can be achieved. The expression vectors contain a promoter that interacts with factors in the cell line, i.e., a cytomegalovirus (CMV) promoter or other viral promoters (SV40) to provide high transcript levels. Poly(A) addition sites, which are contained in many eukaryotic expression vectors, enhance in general the production of receptors. The

[22] C. M. Gorman, R. D. Gies, and G. McCray, *DNA Prot. Eng. Tech.* **2,** 3 (1990).
[23] C. Chen and H. Okayama, *Mol. Cell. Biol.* **8,** 2745 (1987).

quality of the plasmid DNA is also important; phenol/chloroform extraction and ethanol precipitation should be used for plasmid preparation. DNA purity can be checked by measuring optical densities at 260 vs. 280 nm; the ratio should be ≈2.

Expression by transfecting cells provides some advantages over the *Xenopus* expression system: (1) Biochemical studies are possible, since large numbers of cells can be rapidly obtained. (2) No *in vitro* translation of RNA is required. (3) Control over the intracellular solution is possible during electrophysiologic recording. (4) Fast kinetics can be recorded easily in the whole-cell mode because cell sizes are much smaller than for oocytes.

### *Cultivation of HEK293 Cells*

HEK293 cells can be kept for 20–30 passages when replated before confluence. For electrophysiologic recording, the cells can be seeded onto fibronectin-coated (20 μg/ml fibronectin) plastic or glass coverslips (Thermanox, 13 mm; Nunc, Rochester, NY) in standard 24-well culture dishes with minimum essential medium (MEM). Cultures for biochemical experiments are grown in 30- or 100-mm uncoated tissue culture dishes. The cells are transfected by the calcium phosphate method 48 hr after plating and can be analyzed 48 hr after the transfection. For efficient transfection, the cells must be preconfluent because they have to divide in order to take up and express DNA. One inherent problem is the presence of nontransfected cells; even in good experiments, only about 50% of the cells are transfected. Thus, identification of transfected cells becomes important for electrophysiology; cotransfection with green fluorescence protein[24] (GFP) cDNA can be used for visualization of transfected cells. In addition, under differential interference contrast optics transfected cells often show a granular or vacuolated appearance. To eliminate electrical coupling of the HEK293 cells, an inhibitor of mitosis (14.3 m*M* uridine) can be added to the medium 24 hr after transfection; in addition, isolated cells should be selected for whole-cell recording. For electrophysiology the coverslips with the transfected cells are transferred to the recording chamber where the culture medium can be exchanged by the appropriate extracellular solution for recording.

### *Preparing HEK293 Cells for Transfection*

Routinely, HEK293 cells are cultivated in minimal essential medium (MEM) (GIBCO-BRL, Gaithersburg, MD), containing 10% (v/v) heat-inactivated fetal calf serum (FCS), 10 m*M* L-glutamine, and 100 U/ml

[24] J. Marshall, R. Molloy, G. Moss, J. R. Howe, and T. E. Hughes, *Neuron* **14,** 211 (1993).

penicillin/100 μg/ml streptomycin in an incubator (Heraeus) maintaining a 5% $CO_2$ atmosphere at 37°. Cells are replated every 3 days as follows.

1. Harvest 293 cells from one or several 10-cm plates (Becton Dickinson) containing $8 \times 10^6$ cells by removing the medium and adding 15 drops of trypsine (GIBCO-BRL) per plate for 2 min.
2. Collect the cells with medium and distribute them either into 24-well plates (Becton Dickinson) to obtain $5 \times 10^4$ cells/well, or into 10-cm dishes to obtain $2 \times 10^6$ cells/dish.
3. Let the cells grow for 48 hr.
4. Check one plate to see that cells cover ≈50% of the plate surface.

*Transfection*

1. For the transfection of 4 wells in a 24-well plate (or one 10-cm dish), put into a clear polycarbonate-type tube 1 μg (10 μg) of DNA from a 1 μg/μl stock and add 150 μl (375 μl) water.
2. Then add first 50 μl (125 μl) of 1 *M* $CaCl_2$ and then 200 μl (500 μl) of 2× BBS (see below).
3. Wait 1–3 min and add 1.6 ml (4 ml) $MEM^{+++}$ (precipitation will form if you wait long enough and will be visible with a very faint haze in the solution).
4. Gently add 0.5 ml (5 ml) to each well (dish); avoid dislodging of the cells.
5. Put the plates in 3% $CO_2$ incubator for 24 hr.
6. Remove and discard media, wash with Puck's saline A [PSA/$Ca^{2+}$; EGTA 1 m*M*, $CaCl_2 \cdot 2H_2O$ 0.1 m*M*, glucose 5 m*M*, KCl 5 m*M*, NaCl 150 m*M*, $NaHCO_3$ 3 m*M*, sterilize using 0.2 μ*M* bottle top filter (Becton Dickinson)] and add 0.5 ml (5 ml) fresh $MEM^{+++}$ per well (dish).
7. Place plates in a 5% (v/v) $CO_2$ incubator at 37°.
8. Forty-eight hours after transfection cells should be optimal for recording or harvesting

To increase transfection efficiency, a glycerol shock may be applied 5 hr after the transfection:

1. Remove transfection medium.
2. Incubate the cells for 3 min with a sterile 10% glycerol solution in water and wash 3× with phosphate-buffered saline (PBS, 80 m*M*; $Na_2HPO_4 \cdot 2H_2O$; 20 m*M* $NaH_2PO_4 \cdot H_2O$; 1.4 *M* NaCl; 10×).
3. Add new medium and continue as described before.

2× BBS

1. Weight into beaker using fine balance; 1.07 g BES (*N,N*-bis[2-hydroxyethyl-2-aminoethansulfonic acid) (Sigma), 50 m*M* final; 1.63 g NaCl, 280 m*M* final; 0.0267 g $Na_2HPO_4 \cdot 2H_2O$, 1.5 m*M* final.
2. Add $H_2O$ to 100 ml, stir for 15 min.
3. Add NaOH (from 2 *N* stock) to pH 6.95 at room temperature and wait for 15 min.
4. Check if pH is at 6.95 (add HCl if necessary), sterilize solution using 0.2-$\mu$m filter, and store at $-20°$.

## Baculovirus System

Baculoviruses have proven to be powerful and versatile eukaryotic expression systems, which produce recombinant proteins at high levels of the total insect cell protein. The $\alpha_1$ subunit of the GlyR has been successfully expressed in the baculovirus/Sf9 (*Spodoptera frugiperda* fall armyworm ovary) insect cell system (PharMingen, San Diego, CA) and shown to be functional by both biochemical and electrophysiologic criteria.[25,26] The protocols used are very similar to those described in the supplier's manual and are therefore not dealt with in this article. However, much of the protein produced seems to be localized in intracellular compartments,[26] suggesting that posttranslational processing and/or sorting may constitute a limiting factor of GlyR production in this versatile expression system.

[25] M. Cascio, N. E. Schoppa, R. L. Grodzicki, F. J. Sigworth, and R. O. Fox, *J. Biol. Chem.* **268,** 22135 (1993).

[26] J. Morr, N. Rundstroem, H. Betz, D. Langosch, and B. Schmitt, *FEBS Lett.* **368,** 495 (1995).

# [14] Purification and Functional Reconstitution of High-Conductance Calcium-Activated Potassium Channel from Smooth Muscle

*By* MARIA L. GARCIA, KATHLEEN M. GIANGIACOMO, MARKUS HANNER, HANS-GÜNTHER KNAUS, OWEN B. MCMANUS, WILLIAM A. SCHMALHOFER, and GREGORY J. KACZOROWSKI

## Introduction

Potassium channels constitute the largest and most diverse family of ion channels.[1] These channels can be categorized according to their biophysical and pharmacologic properties, and all share in common a high selectivity for $K^+$ as the permeant ion. As a first approximation, $K^+$ channels can be divided into either voltage-gated or ligand-gated channels, depending on the stimulus that triggers conformational changes leading to channel opening. High-conductance $Ca^{2+}$-activated $K^+$ (maxi-$K^+$) channels are activated by both membrane depolarization and binding of $Ca^{2+}$ to sites at the intracellular face of the channel. Maxi-$K^+$ channels are present in both electrically excitable and nonexcitable cells, and display high conductance and selectivity for $K^+$. These channels are involved in regulation of the excitation–contraction coupling process in smooth muscle, as well as in control of transmitter release from neuroendocrine tissues. The pharmacology of maxi-$K^+$ channels has been developed during the last few years and efforts are continuing to identify novel and selective modulators of this channel family.[2,3]

Development of the molecular pharmacology of maxi-$K^+$ channels was greatly facilitated by discovery of charybdotoxin (ChTX), a peptidyl channel inhibitor that is a minor component of *Leiurus quinquestriatus* var. *hebraeus* venom.[4] ChTX binds in the outer vestibule of maxi-$K^+$ channels with a 1 : 1 stoichiometry and blocks ion conduction by physically occluding the channel pore. Purification and subsequent radiolabeling of ChTX with $Na^{125}I$ to high-specific activity provided a tool with which to identify maxi-$K^+$ chan-

[1] A. Wei, T. Jegla, and L. Salkoff, *Neuropharmacology* **35,** 805 (1996).

[2] M. L. Garcia, M. Hanner, H.-G. Knaus, R. Koch, W. Schmalhofer, R. S. Slaughter, and G. J. Kaczorowski, *Adv. Pharmacol.* **39,** 425 (1997).

[3] G. J. Kaczorowski, H.-G. Knaus, R. J. Leonard, O. B. McManus, and M. L. Garcia, *J. Biomembr. Bioenerg.* **28,** 255 (1996).

[4] M. L. Garcia, H.-G. Knaus, P. Munujos, R. S. Slaughter, and G. J. Kaczorowski, *Am. J. Physiol.* **269,** C1 (1995).





0076-6879/99 $30.00

nels in native tissues and to develop their molecular pharmacology. Another important question concerns the exact molecular composition of native maxi-$K^+$ channels. We know that some properties of the channel, such as $Ca^{2+}$ sensitivity, vary greatly from tissue to tissue. This could be explained by either the presence of distinct pore-forming subunits that would possess different sensitivities to $Ca^{2+}$, or by modulation of the pore-forming subunit by other auxiliary subunits that are closely associated with the pore. This, in fact, is the situation found for $Na^+$ and $Ca^{2+}$ channels; these channels exist *in vivo* as multiple-subunit complexes, where auxiliary proteins have major effects on properties of the subunit that constitutes the pore. To date many of the auxiliary subunits of ion channels have been identified after purification of the channel preparation to homogeneity from native tissues. In fact, it would be difficult to achieve the same results using molecular biological approaches (e.g., expression-cloning techniques) because there is no way to predict which parameter of the channel would be altered by coupling to various auxiliary subunits. Thus, purification of a native channel complex remains a viable approach by which to determine subunit composition of a channel.

In this article, we discuss procedures that have been developed to purify maxi-$K^+$ channels from smooth muscle tissues, as well as how to reconstitute the purified preparation into liposomes for determination of channel activity. In addition, we also discuss ways of obtaining protein sequence information from components of the maxi-$K^+$ channel preparation that are useful for obtaining full-length cDNA clones of these proteins.

## Solubilization of Maxi-$K^+$ Channels

To achieve a functionally active, homogeneous maxi-$K^+$ channel preparation, three parameters must be established. First, a marker for the channel must be identified that can be used to track the protein during purification. For this purpose, we used $^{125}I$-labeled ChTX binding to maxi-$K^+$ channels.[5] Monitoring the effect of other channel modulators on the binding reaction provides a way of assessing the integrity of the channel preparation. Second, it is necessary to identify a source of tissue that possesses significant amounts of the target. Out of the sources that we screened for maxi-$K^+$ channels (i.e., neuronal, skeletal muscle, and smooth muscle tissues), we found that bovine tracheal and aortic smooth muscle displayed the highest density of $^{125}I$-labeled ChTX binding sites. Finally, the detergent used for extraction of the channel from its lipid environment must maintain the structural

[5] J. Vazquez, P. Feigenbaum, V. F. King, G. J. Kaczorowski, and M. L. Garcia, *J. Biol. Chem.* **265,** 15564 (1990).

integrity of the protein in its solubilized form, and the channel complex must remain stable for the extended period of time that is required to carry out all purification steps. Out of all detergents that we tested, only CHAPS and digitonin yielded both channel solubilization and initial toxin binding activity.[6] However, activity was found to be more stable in the presence of digitonin; therefore, this was the detergent of choice for attempting maxi-$K^+$ channel purification.

Purified sarcolemmal membrane vesicles derived from either bovine tracheal or aortic smooth muscle are prepared by methods that are well described in the literature.[7,8] These membranes can be stored at $-70°$ until use without loss of [$^{125}$I]ChTX binding activity. Before solubilization, membranes are thawed and the following protease inhibitors added: 1 m*M* iodoacetamide, 0.1 m*M* phenylmethylsulfonyl fluoride, and 0.1 m*M* benzamidine (these agents are present throughout the entire purification procedure). Digitonin is then added from a 10% (w/v) stock solution to a final concentration of 0.5%, and the mixture is incubated at 4° for 10 min with continuous shaking. It is worth noting that digitonin, being a natural product, can vary significantly in its physical properties between different lots and manufacturers. We have found that digitonin special grade (water-soluble), purchased from Biosynth AG (Skokie, IL), gives highly reproducible results. The easiest way to prepare a 10% (w/v) solution of digitonin is to sonicate the suspension for a few minutes. Once in solution, the material is filtered through 0.2-$\mu$m cellulose acetate filters to eliminate any particulate. At the end of the 10-min exposure to digitonin, the membrane suspension is subjected to centrifugation for 50 min at 180,000*g* at 4°. The supernatant ($S_1$) contains large amounts of contaminant proteins, but no significant [$^{125}$I]ChTX binding activity, and can therefore be discarded. The pellet is homogenized using a glass–Teflon homogenizer in 20 m*M* NaCl, 20 m*M* Tris-HCl, pH 7.4; digitonin is added to a final concentration of 1%; and the suspension is incubated at 4° as indicated earlier. After centrifugation to separate soluble from particulate material, the supernatant ($S_1$) is retained and the pellet is subjected to another extraction with detergent as previously indicated. We have found that for optimal recovery of solubilized [$^{125}$I]ChTX binding sites, the extraction procedure needs to be repeated up to a total of six times. The supernatants from the second to the sixth extraction ($S_{2-6}$) are combined for further processing. Solubilized material

[6] M. Garcia-Calvo, J. Vazquez, M. Smith, G. J. Kaczorowski, and M. L. Garcia, *Biochemistry* **30,** 11157 (1991).

[7] R. S. Slaughter, J. L. Shevell, J. P. Felix, M. L. Garcia, and G. J. Kaczorowski, *Biochemistry* **28,** 3995 (1989).

[8] R. S. Slaughter, A. F. Welton, and D. W. Morgan, *Biochim. Biophys. Acta* **904,** 92 (1987).

TABLE I
PURIFICATION OF ChTX RECEPTOR FROM BOVINE TRACHEAL SMOOTH MUSCLE[a]

| Step | $^{125}$I-ChTX Binding | | Protein | | Specific activity (pmol/mg protein) | Purification (x-fold) |
|---|---|---|---|---|---|---|
| | pmol | % | mg | % | | |
| Membranes | 8014 | | 10067 | | 0.79 | |
| $S_{2-6}$ | 4268 | 100 | 6327 | 100 | 0.67 | 1.0 |
| DEAE-Sepharose | 2712 | 63.5 | 1762 | 27.8 | 1.54 | 2.3 |
| WGA-Sepharose | 2292 | 53.7 | 453 | 7.1 | 5.05 | 7.5 |
| Mono Q | 1627 | 38.1 | 81 | 1.3 | 20.08 | 30.0 |
| Hydroxylapatite | 1138 | 26.6 | 8.5 | 0.13 | 133.88 | 199.8 |
| First sucrose gradient | 801 | 18.7 | 2.6 | 0.04 | 308.07 | 459.8 |
| Mono S | 248 | 5.8 | 0.38 | 0.006 | 652.63 | 974 |
| Second sucrose gradient | 142 | 3.3 | 0.13 | 0.002 | 1092.30 | 1630 |
| | 42[b] | 1.0[b] | 0.032[b] | 0.0005[b] | 1312.50[b] | 1959[b] |

[a] Amounts referred to starting material derived from ~250 cow tracheas. [Reprinted with permission from the American Society of Biochemistry and Molecular Biology, Inc., from M. Garcia-Calvo *et al. J. Biol. Chem.* **269,** 676 (1994).]
[b] Highest specific activity fraction from the gradient.

contains as much as 50–60% of the binding sites present originally in membranes, with a similar specific activity as defined by pmol [$^{125}$I]ChTX binding sites/mg protein (Table I).[9,10]

Determination of [$^{125}$I]ChTX binding to solubilized receptors can be easily accomplished by filtration techniques.[6] For this purpose, solubilized material in 0.05% (w/v) digitonin is incubated with [$^{125}$I]ChTX in a medium consisting of 10 m*M* NaCl, 20 m*M* Tris-HCl, pH 7.4, 0.1% (w/v) bovine serum albumin (BSA), at room temperature until equilibrium conditions are achieved (ca. 1 hr). At the end of the incubation period, 10 $\mu$l of a 50 mg/ml (w/v) $\gamma$-globulin solution is added, followed by addition of 4 ml of 10% (w/v) polyethylene glycol ($M_r \sim 8000$) in 100 m*M* NaCl, 20 m*M* Tris-HCl, pH 7.4. The precipitate is immediately collected onto GF/C glass fiber filters that have been presoaked in 0.5% polyethyleneimine, and filters are rinsed twice with 4 ml of polyethylene glycol quench buffer. For determination of nonspecific binding, parallel incubations are carried out in the presence of 10 n*M* ChTX.

[9] M. Garcia-Calvo, H.-G. Knaus, O. B. McManus, K. M. Giangiacomo, G. J. Kaczorowski, and M. L. Garcia, *J. Biol. Chem.* **269,** 676 (1994).
[10] K. M. Giangiacomo, M. Garcia-Calvo, H.-G. Knaus, T. J. Mullmann, M. L. Garcia, and O. McManus, *Biochemistry* **34,** 15849 (1995).

## DEAE-Sepharose Chromatography

To facilitate the following purification steps, solubilized ChTX receptor is adjusted to 50 m*M* NaCl, and loaded onto a DEAE-Sepharose CL-6B column equilibrated with 50 m*M* NaCl, 20 m*M* Tris-HCl, pH 7.4, 0.1% digitonin. After washing the resin thoroughly with equilibration buffer, bound receptor is eluted batchwise with 170 m*M* NaCl, 20 m*M* Tris-HCl, pH 7.4, 0.1% digitonin. This step allows for >60% recovery of solubilized receptors with more than 2-fold enrichment in [$^{125}$I]ChTX binding activity (Table I). Importantly, a large amount of protein is removed at this step which otherwise would cause a significant decrease in the yield at the next purification step.

## Wheat Germ Agglutinin—Sepharose Chromatography

The DEAE-Sepharose eluted receptor is incubated overnight at 4° with wheat germ agglutinin (WGA)-Sepharose in 200 m*M* NaCl, 20 m*M* Tris-HCl, pH 7.4, 0.1% digitonin. The suspension is then placed in an empty column, and the fluid phase is collected until the WGA-Sepharose resin is packed. The column is washed with 10 bed volumes of equilibration buffer to remove unbound material. Glycoproteins are then eluted with 200 m*M* *N*-acetyl-D-glucosamine present in the equilibration buffer. The eluate is dialyzed against 20 m*M* Tris-HCl, pH 7.4, 0.05% digitonin, then concentrated 20-fold by use of an Amicon (Danvers, MA) ultrafiltration cell, and finally adjusted with NaCl to a final concentration of 200 m*M*. This step allows for recovery of >50% of solubilized ChTX receptors, with >7-fold enrichment in specific activity (Table I). It is important to take care that sufficient amounts of WGA-Sepharose resin be employed to provide for full retention of glycoproteins. Otherwise significant quantities of material will appear in the eluate and this will have to be processed again using more resin. For this reason, the DEAE-Sepharose step is critical in order to remove contaminating material that will decrease the loading capacity of the WGA-Sepharose resin.

## Mono Q Ion-Exchange Chromatography

Mono Q ion-exchange chromatography, although yielding modest enrichment of the ChTX receptor, is crucial in that it allows the subsequent step in purification to proceed optimally. We employ a Mono Q HR 10/10 (Pharmacia, Piscataway, NJ) ion-exchange column equilibrated with 100 m*M* NaCl, 20 m*M* Tris-HCl, pH 7.4, 0.05% digitonin, which is operated at room temperature. Because loading capacity of the column is limited to ca. 100 mg of protein, caution should be used so as not to overload it. Thus, the WGA-Sepharose eluted material, after being filtered through 0.2-$\mu$m filters, is usually applied in several consecutive runs. Elution of bound material is achieved with a linear gradient of NaCl (0.1–0.5 *M*) over 70 min at a flow rate of 2 ml/min; 4-ml fractions are usually collected. Fractions containing higher specific activity of ChTX binding elute between 0.21 and 0.31 *M* NaCl. This ion-exchange chromatography yields ca. 40% recovery of solubilized ChTX receptors, with a 30-fold enrichment in specific activity (Table I).

### *Hydroxylapatite Chromatography*

Fractions from the Mono Q ion-exchange column are adjusted to 80 m*M* sodium phosphate, pH 7.0, filtered through 0.2-$\mu$m filters, and loaded onto a Bio-Gel HPHT (Bio-Rad, Richmond, CA) 100 × 7.8-mm hydroxylapatite column equilibrated with 80 m*M* sodium phosphate, pH 7.0, 10 m*M* NaCl, 0.05% digitonin, at room temperature. Because the loading capacity of this column is limited to ca. 25 mg, it is necessary to be cautious with the amount of material applied. If necessary, the Mono Q eluate should be loaded in several consecutive runs. Bound material is eluted with a linear gradient of 80–160 m*M* sodium phosphate in 10 m*M* NaCl applied within 12 min, followed by a linear gradient from 160 m*M* sodium phosphate in 10 m*M* NaCl to 560 m*M* sodium phosphate, 70 m*M* NaCl applied within 10 min, at a flow rate of 0.5 ml/min. Fractions of 1 ml are collected, and those containing the highest specific activity of ChTX binding, elute between 200 and 440 m*M* sodium phosphate. These fractions contain ≥25% of the solubilized receptors and are enriched ca. 200-fold (Table I).

### *Sucrose Density Gradient Centrifugation*

Fractions from the hydroxylapatite column are dialyzed against 20 m*M* Tris-HCl, pH 7.4, 0.05% digitonin, and concentrated using an Amicon ultrafiltration cell. The material is then applied to a continuous 7–25% (w/v) sucrose gradient in 20 m*M* Tris-HCl, pH 7.4, 0.05% digitonin, separated by centrifugation for 12 hr at 34,000 rpm in a Beckman SW 40 Ti rotor (Palo Alto, CA), and fractionated into 0.6-ml fractions. ChTX binding activity migrates as a large particle with an apparent sedimentation coefficient of 23S. The recovery of solubilized ChTX receptors is ca. 20% with >450-fold enrichment in specific activity (Table I).

## Mono S Ion-Exchange Chromatography

Active fractions from the sucrose density gradient centrifugation step are dialyzed against 20 m*M* 2[N-morpholino]ethanesulfonic acid (MES–

NaOH), pH 6.2, 0.05% digitonin, and loaded onto a Mono S HR 5/5 (Pharmacia, Piscataway, NJ) ion-exchange column that had been equilibrated with the same buffer at room temperature. Bound material is eluted at a flow rate of 0.5 ml/min with a linear gradient of NaCl (0–700 m*M* in 20 min). ChTX binding activity elutes between 120 and 260 m*M* NaCl. Recovery of solubilized receptors is ca. 6% with almost 1000-fold enrichment in specific activity (Table I).

## Sucrose Density Gradient Centrifugation

A density gradient approach was again chosen for the last purification step because it takes advantage of the large mass of the ChTX receptor, thus allowing easy separation from other remaining small molecular weight contaminating protein components. Before applying to a continuous 7–25% (w/v) sucrose gradient, fractions from the Mono S ion-exchange column are dialyzed against 20 m*M* Tris-HCl, pH 7.4, 0.01% digitonin, and the sample is concentrated. Protein components are separated as indicated before. The average recovery of receptors is ca. >3% with >1600-fold enrichment (Table I). The fraction with the highest specific activity is enriched ca. 2000-fold with respect to starting material and represents 1% of initial binding activity. The specific activity of the final preparations, 1.3 nmol [$^{125}$I]ChTX binding sites/mg protein, is very close to the value estimated for a pure preparation consisting of a tetrameric structure of four pore forming subunits and four auxiliary subunits (see below). The recovery of activity at each step of purification approaches 100%. This implies that no substantial loss of activity occurs during the time involved in the purification procedure. This is remarkable since, depending on the amount of starting material used, the whole procedure can take several days. For instance, purification that involves material derived from 250 cow tracheas (Table I) can take up to 10–14 days to be completed. Proteolytic degradation of the pore-forming subunit, however, occurs during the hydroxylapatite chromatography step. This has been noted in immunoblot experiments employing site-directed antibodies raised against different peptide sequences of the protein. Although initially all antibodies recognize a single polypeptide in membrane preparation with the expected $M_r$ (125,000), and continue to do so throughout the first steps in purification, after the hydroxylapatite column, the pattern of recognition changes due to specific cleavages in the C-terminal region of the protein.[11] These cleavages, however, do not cause any loss in [$^{125}$I]ChTX binding activity and, since the

[11] H.-G. Knaus, A. Eberhart, R. O. A. Koch, P. Munujos, W. A. Schmalhofer, J. W. Warmke, G. J. Kaczorowski, and M. L. Garcia, *J. Biol. Chem.* **270,** 22434 (1995).

mass of the receptor remains unchanged, we believe that the fragments remain associated through disulfide bridges. Evidence supporting this hypothesis comes from the fact that if the final preparation is subjected to SDS–PAGE in the absence of reducing agents, all antibodies appear to recognize a single polypeptide corresponding to the full-length protein. We do not know the exact nature of the proteolytic cleavage that takes place at that specific stage in purification, but we speculate that it may be due to some conformational change that occurs in the protein on binding to the hydroxylapatite matrix followed by the action of a protease that copurifies with the receptor. One can also speculate that the proteolytic cleavage is self-inflicted. The carboxy terminus of the $\alpha$ subunit of the bovine smooth muscle maxi-$K^+$ channel contains a domain homologous with serine proteases, although sequence analysis suggests that this domain is probably inactive.[12]

The purified ChTX receptor consists of two subunits, the pore-forming subunit ($\alpha$) and an auxiliary subunit ($\beta$). Detection of the $\beta$ subunit by silver staining after SDS–PAGE is virtually impossible. However, the $\beta$ subunit can be visualized by either of two independent methods. The first involves covalent incorporation of $[^{125}I]$ChTX into the $\beta$ subunit in the presence of a bifunctional cross-linking reagent. As previously demonstrated with intact membranes, incorporation of radioactivity takes place exclusively into a polypeptide with a $M_r$ of 35,000.[6,9,10] On deglycosylation, a final product of 25,600 is obtained, which corresponds to the size of the core protein, 21,200, given that 4400 of this mass is contributed by the radiolabeled toxin. It is also possible to detect the $\beta$ subunit after labeling the purified preparation with $^{125}I$-labeled Bolton–Hunter reagent. The $\beta$ subunit displays an identical time course of deglycosylation as that of the $[^{125}I]$ChTX cross-linked protein, and the apparent molecular weight of the deglycosylated $^{125}I$-labeled Bolton–Hunter labeled core protein is 21,400.[9] Interestingly, the electrophoretic mobility of the $\alpha$ subunit is not altered after *N*-glycanase treatment, indicating that this protein is not glycosylated by N-linked sugars. These findings are in agreement with the postulated transmembrane topology of this protein, in which putative N-linked glycosylation sites are placed on the inner face of the membrane.[13]

The pharmacologic properties of the purified ChTX receptor are identical to those characteristic of the receptor in intact membranes.[9,10] Thus, not only does the affinity of $[^{125}I]$ChTX remain unchanged, but the ability of other agents to modulate the binding reaction is also unchanged. These agents include other related peptidyl inhibitors of maxi-$K^+$ channels, such

[12] G. W. J. Moss, J. Marshall, and E. Moczydlowski, *J. Gen. Physiol.* **108,** 473 (1996).
[13] M. Wallner, P. Meera, and L. Toro, *Proc. Nat. Acad. Sci. U.S.A.* **93,** 14922 (1996).

as iberiotoxin and limbatustoxin, as well as small organic molecules such as tetraethylammonium ion, all of which bind to the outer vestibule of the channel. In addition, ions such as potassium, barium, and cesium, which bind with high affinity to sites located along the ion conduction pathway, display identical $K_i$ values in the purified preparation as those found with intact membranes. Finally, members of the indole diterpene family of maxi-$K^+$ channel inhibitors modulate [$^{125}$I]ChTX binding to the purified receptor in the same fashion as they do in native preparations. For instance, paxilline causes a concentration-dependent stimulation of toxin binding, whereas aflatrem produces full inhibition of the binding reaction. In summary, the procedure described above yields a homogeneous preparation consisting of two subunits that displays all the pharmacologic properties of the native ChTX receptor in smooth muscle.

## Reconstitution of ChTX Receptor into Liposomes

To demonstrate that the purified ChTX receptor is indeed the maxi-$K^+$ channel, it is necessary to reconstitute the preparation in a system appropriate for making electrical recordings of single-channel activity. The biophysical and pharmacologic properties of native maxi-$K^+$ channels have been well characterized after incorporation of channels from membrane vesicle preparations into artificial phospholipid bilayers. However, the presence of digitonin in the purified preparation precludes such recordings due to the instability of the bilayers formed. Therefore, we elected to reconstitute the purified preparation into liposomes first in order to eliminate detergent, and then fuse the liposomes with an artificial lipid bilayer. Because the critical micellar concentration (CMC) of digitonin is very low, it is necessary to break up the micelles with a second detergent, such as 3-[(3-cholamidopropyl)dimethylammonio]-1-propanesulfonic acid (CHAPS). For this purpose, aliquots of purified receptor in 0.05% (w/v) digitonin are incubated on ice with 0.3% BSA, 0.34% L-$\alpha$-phosphatidylcholine, and 0.85% CHAPS for 30 min. We have found that the amount of extract in 0.05% digitonin to achieve optimal reconstitution is critical, and that 300 $\mu$l of material yields good results. The mixture is then applied to a 1-ml Extracti-Gel D column (Pierce, Rockford, IL) equilibrated with 100 m$M$ NaCl, 20 m$M$ HEPES–NaOH, pH 7.4, 0.2% BSA, and eluted with 1.5 ml of 100 m$M$ NaCl, 10 m$M$ $MgCl_2$, 20 m$M$ HEPES–NaOH, pH 7.4. Proteoliposomes are then precipitated by addition of polyethyleneglycol ($M_r \sim 8000$) to give a final concentration of 25%, and collected by centrifugation at 100,000 rpm (Beckman TLA 100.3 rotor) for 20 min. Proteoliposomes are washed once in 100 m$M$ NaCl, 20 m$M$ HEPES–NaOH, pH 7.4, collected by centrifugation as indicated above, resuspended in washing medium, frozen in liquid $N_2$, and stored at $-70°$. There is no loss of biological activity after storage under these conditions for up to several months. The efficiency of reconstitution varies from 25 to 60% as determined by [$^{125}$I]ChTX binding to the proteoliposome preparation. In the absence of digitonin, toxin binding is about 30% of the amount observed in the presence of this detergent, suggesting that the receptor preferentially reconstitutes in the inside-out orientation.[9]

## Fusion of Reconstituted ChTX Receptors into Lipid Bilayers

Proteoliposomes were fused with lipid bilayers composed of either neutral zwitterionic lipids [1-palmitoyl-2-oleoylphosphatidylethanolamine (POPE) and 1-palmitoyl-2-oleoylphosphatidylcholine (POPC) in a 7/3 molar ratio] or a combination of neutral and charged lipids [POPE and 1-palmitoyl-2-oleoylphosphatidylserine (POPS) (Avanti Polar Lipids, Birmingham, AL) in a 1/1 molar ratio]. The lipids were dissolved in decane (50 mg/ml) and applied to a 250-$\mu$m hole in a polycarbonate chamber. The use of charged lipids facilitated fusion of proteoliposomes with the bilayer, but neutral lipids were used in experiments where it was desirable to minimize the effects of lipid surface charge on channel function. The hole in the dry polycarbonate cup was pretreated by application of a small amount of lipid solution that was allowed to dry before adding the aqueous solutions. This step enhanced the stability of the bilayers, which would typically last for many hours. Bilayers were formed by painting a very small volume of lipid solution over the hole and monitoring membrane capacitance. Proteliposome fusion with the bilayer was enhanced by establishing an osmotic gradient across the bilayer. Typically, the proteoliposomes were added to the cis side containing 150 m*M* KCl, 10 m*M* HEPES and 10 $\mu M$ $CaCl_2$, pH 7.20, with KOH. The opposite, trans, side contained 10–25 m*M* KCl, 10 m*M* HEPES, and 10 $\mu M$ $CaCl_2$, pH 7.20, with KOH. Using higher KCl concentrations of up to 1 *M* on the cis side enhanced the rate of vesicle fusion. Direct application of small amounts of the proteoliposome preparation to the bilayer by a glass rod or a microliter pipette is an efficient use of the purified preparation. After channel insertion into the bilayer, the osmotic gradient was collapsed by addition of a small volume of a concentrated KCl solution (2 *M* KCl, 10 m*M* HEPES, pH 7.20) to the trans side, which reduced the probability of further channel insertion. Channels inserted with either polarity, which was determined from the voltage and calcium dependence of channel open probability.

Membrane currents were recorded using commercial voltage-clamp amplifiers (Dagan 3900 and List EPC7, Minneapolis, MN). Patch-clamp amplifiers with relatively large input capacitances (especially those with integrating headstages) worked best because they are stable when connected with a large bilayer source capacitance and provide relatively constant frequency response in the face of experiment-to-experiment variations in bilayer capacitance. The amplifier was connected to small wells filled with 0.2 *M* KCl by silver electrodes coated with silver chloride. These small wells were connected to the bilayer chambers by agar brides containing 0.2 *M* KCl.

## Properties of Reconstituted Channels

After fusing proteoliposomes containing ChTX receptors purified from bovine trachea and aorta with the bilayer, we routinely observed high-conductance channels with the biophysical and pharmacologic properties of smooth muscle maxi-$K^+$ channels. Single-channel conductance was about 250 pS in neutral lipids and 320 pS in charged lipids. The single-channel conductance of the channel purified from aorta measured in neutral lipids was identical with the conductance of native channels from aorta. The conductance of the purified channels was highly selective for potassium over sodium or chloride and increased as the potassium concentration was raised.

Perhaps the most defining characteristic of maxi-$K^+$ channels is their dual regulation by calcium and membrane potential. Open probability of channels purified from either aorta or trachea increased with membrane depolarization with an *e*-fold increase in open probability per 10–12 mV depolarization. Internal calcium shifted this voltage activation curve to the left as expected for native maxi-$K^+$ channels. At a constant membrane potential, calcium increased channel open probability with a Hill coefficient of 2.9 for aortic channels suggesting that three to four or more calcium ions can bind during maximal activation. The distributions of channel open and closed times were described by three and five to six exponential components, respectively, suggesting three open and five to six shut channel states. These kinetic properties are similar to what has been reported for maxi-$K^+$ channels in other tissues[14] and are precisely what is expected for a channel activated by binding of four or more calcium ions. The purified channel showed transitions between discrete gating modes, suggesting that moding behavior is intrinsic to the channel complex.

The pharmacologic properties of the reconstituted channels were as expected for maxi-$K^+$ channels from smooth muscle. Externally applied tetraethylammonium (TEA) caused a rapid block of the channels that resulted in a time-averaged reduction in single-channel conductance. The $K_i$ for block at 0 mV was 193 $\mu M$ and the block increased at negative membrane potentials. The voltage dependence of block was described by a simple model where TEA bound to a site located in the pore and the

[14] O. B. McManus, *J. Bioenerg. Biomembr.* **23,** 537 (1991).

bound particle sensed 19% of the membrane field. ChTX blocked the channels by a bimolecular mechanism with a $K_d$ value of 4.6 n*M*. The mean toxin-blocked time of 30 sec was briefer than the value of 64 sec previously observed for native channels from bovine aorta.[15] The channel behavior of the reconstituted ChTX closely resembles the properties of maxi-$K^+$ channels in native smooth muscle and suggests that the purified complex is sufficient to reconstitute most aspects of channel function.

## Isolation of Subunits and Amino Acid Sequence Analysis

One of the major goals of purifying a native channel is to determine its subunit composition, and obtain partial amino acid sequence from the subunits so that these data can be used to isolate full-length cDNAs. Many methods have been described in the literature for obtaining amino acid sequences from purified proteins. We elected to separate the maxi-$K^+$ channel subunits by SDS–PAGE, electroelute the proteins from the gel, subject them to proteolytic digestion, and purify the fragments by microbore high-performance liquid chromatography (HPLC).

For the $\beta$ subunit, fractions from the final sucrose density gradient centrifugation containing ~31 pmol of [$^{125}$I]ChTX binding sites were dialyzed against 10 m*M* sodium borate, pH 8.8, 0.05% Triton X-100 and then reacted with 50 $\mu$Ci of $^{125}$I-labeled Bolton–Hunter reagent for 15 min on ice. The iodinated sample was separated by SDS–PAGE on 12% gels, and the wet gel was exposed for 30 min to Kodak (Rochester, NY) XAR film to localize the position of the proteins. The area of radioactivity corresponding to the $\beta$ subunit was cut from the gel and electroeluted for 12 hr in 0.1 *M* ammonium acetate, 0.1% SDS. The sample was then dialyzed against 40 m*M* sodium phosphate, pH 7.8, 0.02% SDS for 24 hr, concentrated to 50 $\mu$l, and incubated with 5 $\mu$g of $V_8$ endoproteinase Glu-C for 14 hr at room temperature. The digestion mixture was then loaded onto a Vydac column (Hesperia, CA) (RP-300, 5 $\mu$m, 150 × 2.1 mm) that had been equilibrated with 2% acetonitrite, 10 m*M* trifluoroacetic acid, using an ABI 130A separation system (Foster City, CA). Elution of bound material was achieved in the presence of a linear gradient of 2–99% acetonitrile at a flow rate of 50 $\mu$l/min, and peaks were collected manually. The separated peptides were loaded onto Porton peptide filter supports and subjected to automated Edman degradation employing an integrated microsequencing system (Porton Instruments PI 2090E, Fullerton, CA) with an on-line detection system. Using this approach, we were able to obtain a 28-amino-acid

[15] K. M. Giangiacomo, E. E. Sugg, M. Garcia-Calvo, R. J. Leonard, O. B. McManus, G. J. Kaczorowski, and M. L. Garcia, *Biochemistry* **32,** 2363 (1993).

sequence from one of the peptides that enabled us to construct oligonucleotide probes to screen cDNA libraries and isolate the full-length cDNA coding for the $\beta$ subunit of the maxi-$K^+$ channel.[16]

To obtain internal amino acid sequence from the $\alpha$ subunit, the purified maxi-$K^+$ channel was subjected to SDS–PAGE and the area of the gel corresponding to the $\alpha$ subunit was cut out, and electroeluted as previously described. After electroelution, samples were reduced in the presence of 4% SDS with 1% 2-mercaptoethanol, alkylated with iodoacetic acid, and dialyzed for 18 hr against 6 *M* urea, 10 m*M* sodium phosphate, pH 7.2, containing 20g/liter Dowex AG 1 $\times$ 2 resin. Dialysis continued for 24 hr against 5 m*M* Tris-HCl, pH 8.5, 0.05% CHAPS. Samples were then concentrated 20-fold and incubated with 4 $\mu$g of trypsin (final concentration of 80 $\mu$g/ml) for 15 hr at 37°. The digested $\alpha$ subunit was then loaded onto an Applied Biosystems Aquapore $C_{18}$ column (Foster City, CA) (RP-300, 7 $\mu$m, 100 $\times$ 1 mm), equilibrated with 2% acetonitrile, 7 m*M* trifluoroacetic acid, at a flow rate of 30 $\mu$l/min. Elution was achieved in the presence of a linear gradient from 2 to 98% acetonitrile (0.66%/min), and peaks were collected manually. The separated peptides were loaded onto Porton peptide filter supports and subjected to automated Edman degradation. Most of the peaks did not yield any sequence, but we were able to determine unambiguous amino acid sequence from seven of them.[17] All of the sequences can be aligned in very high homology or even identity with the deduced amino acid sequence of Slo, a maxi-$K^+$ channel protein cloned from different sources and tissues. These data indicate that the $\alpha$ subunit of the purified maxi-K channel from smooth muscle is a member of the Slo family of $K^+$ channels.

## Conclusion

The ion channel superfamily, and in particular the $K^+$ channel family, continues to grow. From cloning of the *Caenorhabditis elegans* genome, it is estimated that perhaps 50–80 potassium channel genes exist in that organism.[1] From the sequence data accumulated so far, eight families of potassium channel genes are conserved between *C. elegans* and vertebrate species. However, the existence of auxiliary subunits of ion channels has usually been demonstrated after biochemical purification of these proteins from native tissues. Maxi-$K^+$ channels are an example where the presence of a $\beta$ subunit was not predictable. Given that maxi-$K^+$ channels found in

[16] H.-G. Knaus, K. Folander, M. Garcia-Calvo, M. L. Garcia, G. J. Kaczorowski, M. Smith, and R. Swanson, *J. Biol. Chem.* **269,** 17274 (1994).

[17] H.-G. Knaus, M. Garcia-Calvo, G. J. Kaczorowski, and M. L. Garcia, *J. Biol. Chem.* **269,** 3921 (1994).

different tissues display distinct sensitivities to $Ca^{2+}$, this could be due to the presence of tissue-specific splice variants of the protein. However, it is now clear that coexpression of the $\beta$ subunit with the pore-forming component markedly alters the $Ca^{2+}$ sensitivity of this channel.[18,19] As well, the $\beta$ subunit has profound effects on some of the pharmacologic properties of maxi-$K^+$ channels. Thus, the maxi-$K^+$ $\beta$ subunit is an integral component of channel function, and by itself, will alter the $Ca^{2+}$ sensitivity of the channel. Similar effects of $\beta$ subunits have been found with other types of ion channels (e.g., voltage-gated Na, $Ca^{2+}$, and $K^+$ channels), and this phenomenon will undoubtedly be rediscovered as other ion channel families are explored. Biochemical purification of the maxi-$K^+$ channel has helped us gain a better understanding of this protein's structure and function. Clearly, some of the ideas and techniques described in this article are directly applicable to other types of ion channel proteins, too.

[18] M. Wallner, P. Meera, M. Ottolia, G. J. Kaczorowski, R. Latorre, M. L. Garcia, E. Stefani, and L. Toro, *Recept. Channels* **3,** 185 (1995).

[19] O. B. McManus, L. M. H. Helms, L. Pallanck, B. Ganetzky, R. Swanson, and R. J. Leonard, *Neuron* **14,** 1 (1995).

# [15] Reconstitution of Native and Cloned Channels into Planar Bilayers

*By* ISABELLE FAVRE, YE-MING SUN, and EDWARD MOCZYDLOWSKI

## Introduction and General Overview

A planar bilayer is an artificial membrane formed across a small hole, ~50 $\mu$m or larger in diameter. The hole on which the membrane is formed is usually placed in a thin plastic partition separating two aqueous compartments, but artificial bilayers may also be formed on a glass micropipette tip. Insertion or incorporation of a channel-forming molecule into such a membrane provides a simple experimental system for electrical recording of channel-mediated currents. Planar bilayer recording of ion channels is practiced for a number of reasons. Frankly, it is a technique that yields incredibly rich mechanistic information on a relatively low budget while offering kaleidoscopic displays of single-channel fluctuations that some workers find delightful, even soothing.

More formally, the following applications and research stratagems have emerged as the principal rationales and justifications for forming planar



0076-6879/99 $30.00

membranes: (1) assessment of the functional channel activity of peptides, purified membrane proteins, or other classes of membrane-active molecules; (2) as a biochemically defined system to investigate single-channel mechanisms and pharmacology at a basic biophysical level; (3) investigation of the influence of lipids and other membrane constituents on channel behavior; (4) as a model system for elementary studies of membrane fusion, capacitance changes, and other membrane-associated phenomena; (5) as a system for routine study of various channels in cellular membranes that are technically difficult to access (e.g., endoplasmic reticulum, bacterial and mitochondrial membranes); (6) use as a membrane dilution approach for obtaining long (>1 hr) single-channel recordings of some channels that are often clustered in high density on cell membranes; (7) for experiments where simultaneous open access to solutions of both sides of the membrane is required; (8) as an assay system for screening complex biological mixtures such as scorpion venoms for the presence of channel-specific toxins, inhibitors, and activators that can be subsequently isolated by purification; and (9) use as a companion system to cellular expression for analyzing the unitary behavior of cloned channels and channel mutants produced by recombinant DNA methodology.

There are a number of excellent books[1,2] and chapters[3-5] that review the historical development of the field of artificial membranes and provide detailed protocols for diverse bilayer techniques that form the basis of contemporary methodology. Our intent in this article is to highlight technical aspects of selected examples where the bilayer approach has been widely applied to functional analysis of major classes of ion channel proteins. Identification of techniques and preparations that have been found to be most generally applicable may facilitate extension of the bilayer approach to other classes of channel proteins that have not yet been thoroughly domesticated.

Following this preview, we first give a detailed protocol for the preparation of plasma membranes from rat skeletal muscle.[6,7] For unknown reasons, this preparation is noted for particularly robust incorporation of maxi $Ca^{2+}$-

[1] C. Miller, ed., "Ion Channel Reconstitution." Plenum Press, New York, 1986.

[2] W. Hanke and W.-R. Schlue, "Planar Lipid Bilayers: Methods and Applications." Academic Press, New York, 1993.

[3] P. Labarca and R. Latorre, *Methods Enzymol.* **207,** 447 (1992).

[4] B. E. Ehrlich, *Methods Enzymol.* **207,** 463 (1992).

[5] A. J. Williams, *in* "Ion Channels: A Practical Approach" (R. H. Ashley, ed.), p. 43. IRL Press, Oxford, 1995.

[6] E. G. Moczydlowski and R. Latorre, *Biochim. Biophys. Acta* **732,** 412 (1983).

[7] X. Guo, A. Uehara, A. Ravindran, S. H. Bryant, S. H. Hall, and E. Moczydlowski, *Biochemistry* **26,** 7546 (1987).

activated $K^+$ channels ($K^+_{Ca}$ channels) and voltage-sensitive $Na^+$ channels ($Na^+_V$ channels). The latter channels may be recorded in planar bilayers in a mode lacking inactivation in the presence of batrachotoxin, a method pioneered by Krueger, French, and colleagues.[8,9] For workers who may be just beginning to delve into bilayer techniques, this preparation is recommended as a demonstration experiment or positive control to "make sure that things are working properly" before venturing into unknown territory. Next, we provide a representative compilation of methods for bilayer incorporation of channel proteins from native tissues. Finally, we summarize new approaches to the study of cloned channels in planar bilayers. We view this last topic as a promising avenue for future research, because it offers the possibility of extending the functional analysis of specific mutations of many types of channel proteins to the bilayer assay system. This would be analogous to single-channel studies of gramicidin derivatives that have provided important molecular insights to the gating and conductance behavior of this simple channel molecule.[10,11]

Before proceeding further, it is worthwhile to briefly survey the landscape of bilayer work as applied to channels. An experimenter faces three major hurdles that must be surmounted in order to routinely collect usable data from a bilayer apparatus. These impediments are (1) electrical noise and high membrane capacitance, (2) membrane formation, and (3) channel incorporation.

Seal resistance, a critical noise factor in patch-clamp recording,[12] is generally not a problem in bilayer work since the electrical resistance of artificial membranes properly formed on plastic partitions usually exceeds 100 GΩ. The major limitation of planar bilayers is "voltage noise," which arises from the high capacitance that is proportional to the membrane area. Low-pass filtering of high-frequency noise is absolutely necessary to resolve single-channel currents from a planar bilayer. Such filtering also attenuates the shortest fluctuations of channels under study and generally means that fast processes cannot be resolved as well in planar bilayer recordings vs. patch-clamp recordings. The only available solution to this noise problem is to reduce the bilayer area by using as small a hole as practical. The tradeoff in using small bilayers is that they can be more difficult to form and

[8] B. K. Krueger, J. F. Worley III, and R. J. French, *Nature* **303,** 172 (1983).

[9] R. J. French, J. F. Worley III, and B. K. Krueger, *Biophys. J.* **45,** 301 (1984).

[10] D. V. Greathouse, R. E. Koeppe, L. L. Providence, S. Shobana, and O. S. Andersen, *Methods Enzymol.* **294,** [28], 1998 (this volume).

[11] O. S. Andersen, C. Nielsen, A. M. Maer, J. A. Lundbaek, M. Goulian, and R. E. Koeppe, *Methods Enzymol.* **294,** [10], 1998 (this volume).

[12] F. J. Sigworth, *in* "Single-Channel Recording" (B. Sakmann and E. Neher, eds.), p. 3. Plenum Press, New York, 1983.

much less amenable to channel incorporation. Relatively low-noise bilayer recordings have been achieved by using specially designed holes (25–80 $\mu$m in diameter) on plastic partitions[13] and also by forming bilayers directly on small-diameter glass micropipettes using painting,[14] "tip-dip,"[15–18] or "bilayer punch"[19] methods. For routine work in our laboratory we use a bilayer diameter of ~200 $\mu$m. This permits recording of well-resolved unitary currents of maxi $K^+_{Ca}$ channels with conductance of ~250 pS at ~1-kHz filtering or batrachotoxin-activated $Na^+_V$ channels with a conductance of ~20 pS at ~200-Hz filtering.

Planar bilayers have enjoyed their widest application in steady-state recording of channel currents at constant voltage or by using linear voltage ramps[20] to monitor the current-voltage behavior of open channels. The study of fast, transient, voltage-activated currents using pulsed, voltage-step protocols presents special problems for bilayer work due to large capacitative transients that are difficult to compensate. This problem can be largely overcome by the use of a specially designed headstage and amplifiers such as those marketed by Axon Instruments (Foster City, CA) and Warner Instrument Corp. (Hamden, CT). At least two groups of researchers have successfully recorded transient currents of voltage-activated $Ca^{2+}$ channels and $K^+$ channels in planar bilayers by using such equipment.[13,21,22]

Contrary to its undeserved reputation as a sorcerer's ritual, the art of forming planar bilayers is not difficult to master. However, it must be said that some students experience frustration when working with bilayers despite noble efforts. For those with the right combination of patience and dexterity, there are two types of basic techniques for forming bilayers[23]

[13] W. F. Wonderlin, A. Finkel, and R. J. French, *Biophys. J.* **58,** 289 (1990).

[14] R. Sitsapesan, R. A. P. Montgomery, and A. J. Williams, *Pflügers Arch. Eur. J. Physiol.* **430,** 584 (1995).

[15] R. Coronado and R. Latorre, *Biophys. J.* **43,** 231 (1983).

[16] B. A. Suarez-Isla, K. Wan, J. Lindstrom, and M. Montal, *Biochemistry* **22,** 2319 (1983).

[17] T. Schuerholz and H. Schindler, *FEBS Lett.* **152,** 187 (1983).

[18] W. Hanke, C. Methfessel, U. Wilmsen, and G. Boheim, *Biochem. Bioeng. J.* **12,** 329 (1984).

[19] O. Andersen, *Biophys. J.* **41,** 119 (1983).

[20] G. Eisenman, R. Latorre, and C. Miller, *Biophys. J.* **50,** 1025 (1986).

[21] R. L. Rosenberg, P. Hess, J. P. Reeves, H. Smilowitz, and R. W. Tsien, *Science* **231,** 1564 (1986).

[22] R. Sherman-Gold, ed., "The Axon Guide for Electrophysiology and Biophysics Laboratory Techniques." Axon Instruments Inc., Foster City, California, 1993.

[23] S. H. White, *in* "Ion Channel Reconstitution" (C. Miller, ed.), p. 3. Plenum Press, New York, 1986.

known as "painting" or "folding." In the painting method,[24] the partition containing the hole initially separates two chambers filled with an appropriately buffered electrolyte solution (e.g., 150 m*M* KCl, 10 m*M*, HEPES–KOH, pH 7.2). A solution of phospholipids dissolved in an alkane solvent, commonly decane, is spread over the hole with a brush, glass capillary rod, or other implement, and the resulting film is allowed to thin. Depending on the size of the hole and particular lipids, such films spontaneously thin over a time course as short as 1 min to produce a lipid bilayer that is sealed to the hole by an annulus of lipid and solvent. The term *black lipid membrane* or BLM is often used to describe such membranes, because by monitoring the thinning process optically, one first observes an iridescent film that transforms into a black hole as the membrane thins to bilayer thickness and no longer reflects light. Bilayer formation is also commonly monitored by measuring the membrane capacitance, which increases to a stable value as the membrane thins. In the folding technique,[25] a bilayer is assembled from a lipid monolayer formed on the surface of the solution by depositing a drop of lipid in a volatile solvent such as pentane. The bilayer is formed by raising the solution above the hole in the partition several times[25] or by dipping the tip of a micropipette through the air–monolayer–water interface several times.[15–18] Folded membranes generally contain much less solvent than painted membranes; however, there is much evidence to suggest that decane in the bilayer phase of painted membranes does not interfere with the function of many types of ion channels. As judged by the number of laboratories routinely using painted vs. folded membranes, it appears that painted membrane technology is relatively easier to apply to incorporation of most types of channel proteins.

The final hurdle, incorporation of channel proteins into planar bilayers, is undoubtedly the most problematic and technically the most difficult to overcome. Although bilayer insertion of small lipophilic molecules and peptides, such as gramicidin, alamethicin, and some detergent-solubilized membrane proteins, can occur directly from the aqueous phase, most large channel proteins must be transferred from a native cellular membrane or a liposome membrane to the planar bilayer.

One approach is to first deposit the protein of interest at the air–water interface by spreading reconstituted proteoliposomes and then to use the folded membrane technique to form the bilayer and thus insert channels. Examples of the successful application of this method have included the

[24] P. Mueller, D. Rudin, H. T. Tien, and W. C. Wescott, *Circulation* **26,** 1167 (1962).
[25] M. Montal and P. Mueller, *Proc. Natl. Acad. Sci. U.S.A.* **69,** 3561 (1972).

nicotinic acetylcholine receptor channel as purified from the *Torpedo* ray electric organ and insect muscle.[26,27]

The second technique makes use of the well-known but incompletely understood process of membrane fusion (not to be confused with cold fusion). This method, which seems to be more generally applicable, involves the fusion of native membrane vesicles or reconstituted liposomes containing the channel of interest to the planar membrane.[28] While bilayer fusion clearly works,[29] a person using this method somehow always has the Goldilockian perception that there is too little fusion or too much fusion and it is never "just right." Several parameters that can be exploited to control the fusion process include $Ca^{2+}$ concentration, particular lipid mixtures (PE and acidic phospholipids favor fusion, PC inhibits), and osmotic gradients (hypertonic on the cis, vesicle-containing side favors fusion). However, incorporation conditions that work best for each type of preparation must be determined empirically. Woodbury and Miller[30,31] have also introduced a general fusion technique involving the use of liposomes containing nystatin and ergosterol. In principle, this method can be applied to any membrane preparation or channel protein and potentially offers a systematic approach for controlling and monitoring the fusion process.

## Rat Muscle T-Tubule Membranes: A Reliable Source of $K^+_{Ca}$ Channels and $Na^+_v$ Channels

Following Miller and Racker's breakthrough[28] in 1976 of achieving reproducible incorporation of $K^+$-selective channels from sarcoplasmic reticulum vesicles into painted lipid bilayers, similar fusogenic membrane preparations were eagerly sought for biophysical studies of channels that mediate the signalling currents of electrically excitable cells. This goal was largely met in 1982 when Latorre *et al.*[32] used a preparation of membrane vesicles derived from the transverse tubule system of rabbit skeletal muscle to incorporate and record unitary behavior of large conductance $K^+_{Ca}$ channels. Similarly, in 1983, Krueger *et al.*[8] described a bilayer preparation for studying single $Na^+_v$ channels in the presence of batrachotoxin, a natural toxin that prevents $Na^+_v$ channel inactivation. These accomplishments led

[26] M. Montal, R. Anholt, and P. Labarca, *in* "Ion Channel Reconstitution" (C. Miller, ed.), p. 157. Plenum Press, New York, 1986.
[27] W. Hanke and H. Breer, *J. Gen. Physiol.* **90,** 855 (1987).
[28] C. Miller and E. Racker, *J. Membr. Biol.* **30,** 283 (1976).
[29] L. Chernomordik, A. Chanturiya, F. Green, and J. Zimmerberg, *Biophys. J.* **69,** 922 (1995).
[30] D. J. Woodbury and C. Miller, *Biophys. J.* **58,** 833 (1990).
[31] D. J. Woodbury, *Methods Enzymol.* **294,** [17], 1998.
[32] R. Latorre, C. Vergara, and C. Hidalgo, *Proc. Natl. Acad. Sci. U.S.A.* **79,** 805 (1982).

to the recognition that, given the right kind of preparation, the bilayer system could be applied to the *in vitro* study of native ion channels resident in all kinds of cellular membranes. Although numerous applications of planar bilayer reconstitution have been described, such preparations are known to vary considerably in their reproducibility and ease of handling. The key factors for the most successful applications seem to be the purity and homogeneity of the membrane preparation, the inherent "fusability" of the preparation as measured by channel incorporation frequency, and the experimenter's ability to identify and select for the channel of interest by controlling the ionic conditions and exploiting pharmacology.

As mentioned earlier, plasma membrane vesicles from rat skeletal membrane[6,7] work very well for the routine study of $K^+_{Ca}$ channels and batrachotoxin-activated $Na^+_v$ channels. The time-tested procedure for purifying such vesicles essentially involves differential centrifugation of homogenized skeletal muscle followed by separation of plasma membranes (probably a mixture of T tubules and surface sarcolemma) and SR membranes on a sucrose density gradient. The relatively rapid method given below is derived from the more sophisticated procedures and studies[33,34] of membrane biochemists, who developed methodology for the separation of muscle membranes by means of painstaking biochemical and electron microscopic analyses. The protocol for this preparation currently used in our laboratory is given next. The excruciating details are given for the convenience of new students, course instructors, or researchers who may just be entering the world of the planar bilayer reconstitution.

### *T-Tubule Membrane Preparation*

Perform the following steps at 4° and keep all solutions ice-cold, except the phenylmethylsulfonyl fluoride (PMSF) solution. Prechill the solutions overnight, especially the large volume of initial homogenization buffer. For a large membrane preparation, use 8–10 adult male Sprague-Dawley rats, each weighing ~300 g, which will provide ~500 g of starting skeletal muscle tissue.

1. Euthanize each rat by exposure to $CO_2$ in a plastic desiccator chamber containing dry ice before decapitation and exsanguination. Lay the animal on its abdomen and start the dissection with a dorsal incision of the skin, from tail to neck along the spine. Continue with a pair of scissors and scalpel to remove the skin from the back and hind limbs. Dissect skeletal muscle from the rear legs, thighs, and back while removing as much fat

[33] M. Rosemblatt, C. Hidalgo, C. Vergara, and N. Ikemoto, *J. Biol. Chem.* **256,** 8140 (1981).
[34] Y. H. Lau, A. H. Caswell, and J.-P. Brunschwig, *J. Biol. Chem.* **252,** 5565 (1977).

and connective tissue as possible. Store the freshly dissected muscle in a beaker containing ice-cold homogenization buffer. Rinse and blot the tissue on paper towels, weigh, and divide it into batches of 100 g. Mince each batch of muscle by cutting it repeatedly with scissors. If an electric chopper is available, process the 100 g batches to ground/chopped meat.

2. In a standard, high-speed blender, add 100 g ground muscle to 300 ml of homogenization buffer. Just before turning on the blender, add 0.2 m*M* PMSF. Homogenize the tissue at the highest speed for 30 sec. Stop the blender and repeat for another 30 sec at high speed.

3. Centrifuge the homogenate in 250-ml bottles for 10 min at 5000 rpm in a Beckman JA14 rotor (2500*g*). Pour the supernatant through several layers of cheesecloth and save it on ice.

4. Combine each pellet in the blender with an original volume of homogenization buffer and 0.2 m*M* PMSF. Rehomogenize at the highest speed twice for 30 sec and centrifuge the homogenate as in step 3. Pour the supernatant through cheesecloth and combine it with the supernatant from step 3 in a large beaker. Measure the total volume of supernatant and add solid KCl to a final concentration of 0.6 *M*. Dissolve the KCl completely and solubilize contractile proteins by continuous stirring for 30 min at 4°. With a starting material of ~500 g muscle the supernatant volume may be close to 1.5 liters at this stage.

5. Pellet the membranes by ultracentrifuging the preparation for 30 min at 40,000 rpm in a Beckman 45Ti or 50.2Ti rotor (~100,000*g*). After each centrifuge run, discard the clear supernatant and save the pellets. Considerable time can be saved by using one or two sets of tubes and adding homogenate on top of membrane pellets from the previous run.

6. After all the supernatant has been processed by ultracentrifugation, pool and resuspend the pellets in homogenization buffer, but do not exceed a final volume of 50 ml. Homogenize the membrane preparation thoroughly in a glass Dounce homogenizer and centrifuge it for 10 min at 8000 rpm in a Beckman JA20 rotor (5000*g*). Save the white supernatant, but exclude the brown mitochondrial pellet. This centrifugation step may be repeated to further reduce mitochondrial contamination.

7. Prepare six tubes containing 30 ml of 32% (w/v) sucrose buffer for centrifugation in a Beckman SW28 swinging bucket rotor. Carefully layer ~8 ml of the whitish supernatant from step 6 on top of the 32% sucrose layer, filling nearly to the top of the tube. Centrifuge the step gradients overnight (12–18 hr) at 25,000 rpm (85,000*g*).

8. Collect and pool the whitish band at the 32% sucrose interface. Dilute this pool with one volume of dilution buffer and pellet the plasma membranes for 60 min in an ultracentrifuge at 100,000*g*. Discard the supernatant and resuspend the final pellet in a small volume (~2 ml) of resuspen-

sion buffer. Transfer the preparation into a small glass Dounce homogenizer and homogenize thoroughly. The final preparation should be homogeneously thick and white. Aliquot the membrane preparation into 50-$\mu$l portions in small microcentrifuge tubes and quick-freeze them by immersion in liquid nitrogen. When stored at $-70°$, this preparation remains active for incorporation of $K^+_{Ca}$ channels and $Na^+_V$ channels for as long as 1 year.

*Solutions*

Homogenization buffer: 300 m*M* sucrose, 20 m*M* MOPS–KOH, pH 7.4, 0.02% $NaN_3$.

32% sucrose gradient: 32% (w/v) sucrose in 20 m*M* MOPS–KOH, pH 7.4.

Dilution buffer: 20 m*M* MOPS–KOH, pH 7.4.

Resuspension buffer: 300 m*M* sucrose, 20 m*M* MOPS–KOH, pH 7.4.

PMSF solution: 200 m*M* PMSF in acetone treated with molecular sieves to remove water. It is best to prepare this stock solution just before use. It must be kept at room temperature to prevent precipitation.

*Note:* In addition to PMSF, other standard protease inhibitors may be added to the homogenization buffer. However, we have found that addition of the calcium chelator, ethylenediaminetetraacetic acid (EDTA), to the solutions dramatically reduces the incorporation activity of $K^+_{Ca}$ channels. The use of calcium chelators in the preparation should be avoided for studies of this channel.

## Preparations for Reconstituting Diverse Types of Channels from Native Tissues

Many different types of ion channels have been studied in planar bilayers with the degree of rigorous characterization that is necessary to ensure that the bilayer data reflect native function. Table I is a partial list of channel preparations that have been studied extensively. Typically, such projects start with a well-characterized membrane preparation from a favorable tissue source and progress toward the routine incorporation of a particular type of channel that can be readily identified. There are also many examples of traditional purification and reconstitution of channel proteins into liposomes followed by fusion with planar bilayers. The examples of Table I are a good source of methods that may be applied to development of new bilayer preparations.

Membrane reconstitution is indispensible for investigating and demonstrating the transport function of membrane proteins. However, there is

TABLE I
MEMBRANE PREPARATIONS FOR INCORPORATING VARIOUS CLASSES OF ION CHANNELS INTO PLANAR BILAYERS FROM NATIVE TISSUES

| Channel type | Membrane preparations | References |
|---|---|---|
| Sarcoplasmic reticulum (SR) $K^+$ channel | SR vesicles from mammalian skeletal muscle | 28, 42 |
| $ClC_0$ $Cl^-$ channel | Plasma membrane vesicles from noninnervated face of *Torpedo* electroplax | 43, 44 |
| $K^+_{Ca}$ channels (BK or maxitype) | Plasma membrane vesicles from mammalian skeletal muscle, brain; smooth muscle from intestine, uterus, and aorta | 32, 45–48, this work |
| $Na^+_v$ channels (activated by batrachotoxin) | Plasma membrane vesicles from mammalian brain, skeletal muscle, and heart | 7, 8, this work |
| $Ca^{2+}_v$ channels (dihydropyridine-activated) | Plasma membrane vesicles from mammalian skeletal and cardiac muscle | 49, 50 |
| Ryanodine-receptor/$Ca^{2+}$-release channel | Junctional SR from skeletal muscle, cardiac SR | 51–54 |
| $InsP_3$-gated $Ca^{2+}$-release channel | Endoplasmic reticulum vesicles from mammalian cerebellum | 55 |
| ATP-sensitive $K^+$ channels | Plasma membranes from mammalian skeletal muscle and vascular smooth muscle | 56–58 |
| Amiloride-sensitive epithelial $Na^+$ channel | A6 kidney cell line from *Xenopus laevis,* mammalian kidney membranes | 59, 60 |

always the hazard that native functional properties have been altered in the course of the reconstitution process. For single-channel assays, there is also the danger that current fluctuations might represent the activity of contaminants in the preparation rather than the protein of interest. Thus, in bilayer research it is imperative that the activity of reconstituted channels be thoroughly characterized with respect to established functional characteristics of the native preparation. For channels, this means that the properties of the channel in a planar bilayer must be carefully evaluated in comparison to those measured by cellular electrophysiology or biochemical assays of function. This responsibility may be difficult to carry out for channels such as $Ca^{2+}$-release channels that reside in intracellular membranes and cannot easily be recorded by patch clamping of the native membrane. Nevertheless, in well-documented systems, investigators have found ways of determining whether bilayer results genuinely reflect physiological behavior. One of the most powerful approaches involves the use of specific pharmacologic agents. For example, the unique blocking effect of charybdotoxin[35] on maxi-$K^+_{Ca}$ channels assayed in planar bilayers or in cellular preparations[36] provides a good tool for cross-validation. In the case of $Ca^{2+}$-release channels, which are under intensive study in planar bilayers,[37–41] sensitivity to the plant alkaloid, ryanodine, or specific activation by inositol 1,4,5-trisphosphate ($InsP_3$) provides an unmistakable identification. Pharmacology is just one of the tools for the molecular fingerprinting of reconstituted channels. Other signature properties that must be cross-verified between cells and bilayers are unitary conductance, ionic selectivity, gating kinetics, and biochemical modulation.

## Methods for Reconstituting Cloned and Heterologously Expressed Channels into Planar Bilayers

In view of the unique advantages of the planar bilayer recording system, one may consider its applications in connection with the heterologous

[35] C. Miller, E. Moczydlowski, R. Latorre, and M. Phillips, *Nature* **313,** 316 (1985).
[36] M. Wallner, P. Meera, M. Ottolia, G. J. Kaczorowski, R. Latorre, M. L. Garcia, E. Stefani, and L. Toro, *Recept. Channels* **3,** 185 (1995).
[37] F. A. Lai, H. P. Erikson, E. Rousseau, Q.-Y. Liu, and G. Meissner, *Nature* **331,** 315 (1988).
[38] R. Coronado, S. Kawano, C. J. Lee, C. Valdivia and H. H. Valdivia, *Methods Enzymol.* **207,** 699 (1992).
[39] A. R. G. Lindsay and A. J. Williams, *Biochim. Biophys. Acta* **1064,** 89 (1991).
[40] J. A. Copello, S. Berg, H. Onoue, and S. Fleischer, *Biophys. J.* **73,** 141 (1997).
[41] I. Bezprozvanny and B. E. Ehrlich, *J. Gen. Physiol.* **104,** 821 (1994).
[42] P. Labarca, R. Coronado, and C. Miller, *J. Gen. Physiol.* **76,** 397 (1980).
[43] M. M. White and C. Miller, *J. Biol. Chem.* **254,** 10161 (1979).
[44] R. E. Middleton, D. J. Pheasant, and C. Miller, *Biochemistry* **33,** 13189 (1994).
[45] P. H. Reinhart, S. Chung, and I. B. Levitan, *Neuron* **2,** 1031 (1989).
[46] X. Cecchi, O. Alvarez, and D. Wolff, *J. Membr. Biol.* **91,** 11 (1986).
[47] L. Toro, J. Ramos-Franco, and E. Stefani, *J. Gen. Physiol.* **96,** 373 (1990).
[48] L. Toro, L. Vaca, and E. Stefani, *Am. J. Physiol.* **260,** H1779 (1991).
[49] H. Affolter and R. Coronado, *Biophys. J.* **48,** 341 (1985).
[50] R. L. Rosenberg, P. Hess, J. P. Reeves, H. Smilowitz, and R. W. Tsien, *Science* **231,** 1564 (1986).
[51] F. A. Lai, H. P. Erikson, E. Rousseau, Q.-Y Liu, and G. Meissner, *Nature* **331,** 315 (1988).
[52] J. Ma, M. Fill, M. Knudson, K. P. Campbell, and R. Coronado, *Science* **242,** 99 (1988).
[53] A. R. G. Lindsay and A. J. Williams, *Biochim. Biophys. Acta* **1064,** 89 (1991).
[54] J. A. Copello, S. Berg, H. Onoue, and S. Fleischer, *Biophys. J.* **73,** 141 (1997).
[55] I. Bezprozvanny and B. E. Ehrlich, *J. Gen. Physiol.* **104,** 821 (1994).
[56] L. Parent and R. J. Coronado, *J. Gen. Physiol.* **94,** 445 (1989).
[57] R. J. Kovacs and M. T. Nelson, *Am. J. Physiol.* **261,** H604 (1991).
[58] M. Ottolia and L. Toro, *J. Membr. Biol.* **153,** 203 (1996).
[59] S. Sariban-Sohraby, R. Latorre, M. Burg, L. Olans, and D. Benos, *Nature* **308,** 80 (1984).
[60] I. I. Ismailov, B. K. Berdiev, and D. J. Benos, *J. Gen. Physiol.* **106,** 445 (1995).

expression of cloned channel proteins. This particular combination of techniques could be very useful for the functional analysis of channels that are naturally expressed at very low levels or in cases where native tissues do not provide an adequate source for protein purification. Also, the incorporation of purified mutant channels into planar bilayers offers the possibility of conducting structure–function investigations in a pristine *in vitro* environment devoid of endogenous cytoplasmic and membrane-associated constituents inherent to cellular expression systems such as *Xenopus* oocytes or eukaryotic cell lines. The particular combination of a controlled membrane environment and site-specific channel mutations may be especially applicable to questions related to channel regulation.

The purpose of this section is to show that such considerations have not escaped the attention of astute channelologists, since a number of examples can be cited where substantial progress has been made in this direction. A rather unique demonstration of the principle is given by the reconstitution of the mitochondrial voltage-dependent anion channel (VDAC) into planar lipid bilayers.[61] The cloned yeast VDAC gene can be subjected to site-directed mutagenesis and such mutants can be expressed in a VDAC-deleted strain of yeast. The recombinant mutant VDAC protein can then be purified from mitochondria of transformed yeast and reconstituted into planar bilayers. This approach has been used to identify amino acid residues that affect the ionic selectivity of the VDAC channel.[61] To our knowledge, the first application to involve a nonorganellar channel protein, was the work of Rosenberg and East[62] reporting planar bilayer incorporation of a recombinant *Shaker* $K_V^+$ channel that was originally cloned from *Drosophila.* Their unique experimental approach involved the use of a cell-free expression system. Shaker $K_V^+$ channel mRNA was translated *in vitro* using a rabbit reticulocyte lysate in the presence of avian microsomal membranes. In this method a source of rough endoplasmic reticulum vesicles is required for proper cotranslational synthesis of membrane proteins. Fusion of the resulting microsomal vesicles containing *in vitro* synthesized *Shaker* protein with planar bilayers resulted in the observation of $K_V^+$ channel activity with functional behavior expected for *Drosophila Shaker* channels.[62]

Clearly, the reconstitution of recombinant channel proteins requires the identification of a good expression system where functional channels are made in high yield and a good method for isolating the channels and efficiently inserting them into bilayers. Table II summarizes various examples of progress toward this goal as gleaned from the recent literature.

[61] M. Colombini, S. Peng, E. Blachly-Dyson, and M. Forte, *Methods Enzymol.* **207,** 432 (1992).
[62] R. L. Rosenberg and J. E. East, *Nature* **360,** 166 (1992).

TABLE II
RECOMBINANT ION CHANNELS RECONSTITUTED INTO PLANAR LIPID BILAYERS

| Recombinant channel | Heterologous expression system | Sucrose gradient fractionation of membrane vesicles | Protein purification and incorporation into liposomes | References |
|---|---|---|---|---|
| *Shaker* $K_v^+$ channel | *In vitro* translation | – | – | 62 |
| | COS cells | + | – | 73 |
| CFTR[a] $Cl^-$ channel | Sf9 cells | – | + | 78 |
| | HEK293 cells | + | – | 74, 75 |
| maxi-$K_{Ca}^+$ channel | *Xenopus* oocytes | + | – | 64 |
| | COS cells | + | – | 76 |
| | HEK293 cells | + | – | 77 |
| $K_{IR}^+$ channel[b] (IRK1) | *Xenopus* oocytes | + | – | 65 |
| $ClC_0$ *Torpedo* $Cl^-$ channel | HEK293 cells | – | + | 80 |
| Ryanodine receptor/ $Ca^{2+}$-release channel | Sf9 cells | – | + | 79 |

[a] Cystic fibrosis transmembrane regulator.
[b] Inwardly rectifying $K^+$ channel.

Aside from the *in vitro* translation method,[62] these examples make use of several different systems for heterologous cellular expression of cloned channels and use two different approaches for membrane reconstitution that mirror those discussed earlier for native channels. One method is to directly isolate microsomal vesicles containing recombinant channels on a sucrose gradient after homogenization of transfected cells. The second method involves solubilization and purification of the recombinant channel protein followed by reconstitution into liposomes in a functionally active form. In the remainder of this chapter, we discuss relevant details of these procedures that may ultimately provide a basis for the systematic study of many different types of recombinant channels in planar bilayers.

The widespread popularity and success of *Xenopus laevis* oocytes as an expression system for channels, transporters, and receptors[63] has led to the consideration of the oocyte membrane as a source of membrane vesicles for planar bilayer incorporation. Thus far two types of cloned channels have been successfully reconstituted after expression in *Xenopus* oocytes: the *Drosophila* $K_{Ca}^+$ channel, Dslo,[64] and the murine inwardly rectifying $K^+$

[63] T. Shih, R. Smith, L. Toro, and A. Goldin, *Methods Enzymol.* **293,** 529 (1998).
[64] G. Pérez, A. Lagrutta, J. P. Adelman, and L. Toro, *Biophys. J.* **66,** 1022 (1994).

channel, IRK1.[65] In the latter example, Aleksandrov *et al.*[65] employed preexisting methods[66,67] for the oocyte membrane preparation, whereas Pérez *et al.*[64] used the following simple procedure. In brief, oocytes expressing recombinant channels were manually disrupted in a high-$K^+$ potassium buffer (400 m*M* KCl, 5 m*M* PIPES, pH 6.8) in the presence of a mixture of proteinase inhibitors. The homogenate was then layered onto a discontinuous sucrose gradient (0.75 ml 20% sucrose and 0.75 ml 50% sucrose in the same high-$K^+$ buffer) and centrifuged at 30,000*g* for 30 min to allow the separation of the membrane fraction from the yolk. A membrane band at the 20–50% sucrose interface was collected, diluted 3-fold with the high-$K^+$ buffer and membrane vesicles were recovered in a final centrifugation step. Clean separation of a membrane fraction from fatty acid contamination appeared to be critical for stability of the planar lipid bilayer after fusion with oocyte microsomal vesicles.[64]

A major drawback with the use of *Xenopus* oocyte membranes is the abundant presence of endogenous $Ca^{2+}$-activated $Cl^-$ channels.[68,69] This latter background conductance is well known to electrophysiologists who use frog oocytes for two-electrode voltage clamping. It has been overcome in two ways, either by intracellular injection of $Ca^{2+}$ chelators or by the elimination of $Cl^-$ from the bath solution whenever possible.[70–72] Substitution of $Cl^-$ salts by salts of the impermeant anion, methansulfonate, was used to mask insertion of the contaminating $Cl^-$ conductance when Dslo $K^+_{Ca}$ channels[64] or IRK1 $K^+$ channels[65] were studied. However, this endogenous conductance does handicap the oocyte expression system and may render it inappropriate for certain applications, e.g., cloned $Cl^-$ channels. Other factors, such as the time-consuming injection and handling of oocytes and the small yield of microsomal vesicles, may also mitigate against the routine use of oocytes for planar bilayer work.

As an alternative approach, a number of researchers have expressed various cloned ion channels in mammalian cell lines such as COS cells and HEK293 cells as a starting source for planar bilayer reconstitution. Patches

[65] A. Aleksandrov, B. Velimirovic, and D. E. Clapham, *Biophys. J.* **70,** 2680 (1996).

[66] W. H. Kinsey, G. L. Decker, and W. J. Lennarz, *J. Cell Biol.* **87,** 248 (1980).

[67] G. J. Bretzel, J. Janeczek, M. Born, J. H. Tiedemann, and H. Tiedemann, *Roux's Arch. Dev. Biol.* **195,** 117 (1986).

[68] T. Takahashi, E. Neher, and B. Sakmann, *Proc. Natl. Acad. Sci. U.S.A.* **84,** 5063 (1987).

[69] M. E. Barish, *J. Physiol.* (*Lond.*) **342,** 309 (1983).

[70] W. Stühmer and A. B. Parekh, *in* "Single-Channel Recording" (B. Sakmann and E. Neher, eds.), 2nd Ed., p 341. Plenum Press, New York, 1995.

[71] P. T. Ellinor, J. Yang, W. A. Sather, J. P. Zhang, and R. W. Tsien, *Neuron* **15,** 1121 (1995).

[72] T. Schlief, R. Schonerr, K. Imoto, and S. H. Heinemann, *Eur. Biophys. J.* **25,** 75 (1996).

are easily excised from the plasma membrane of mammalian cells and, in contrast to oocytes, these latter cell lines do not have a high background of endogenous currents that complicate electrophysiological measurements. The first example of this approach was functional reconstitution of an inactivation-deficient mutant of the *Shaker* $K_v^+$ channel transiently expressed in COS cells.[73] This study was followed by reconstitution of several other kinds of cloned channels: human cystic fibrosis transmembrane conductance regulator (CFTR),[74,75] maxi-$K_{Ca}^+$ channels from mouse[76] and *Drosophila*,[77] and more recently, in our laboratory, the $\mu$1 $Na_v^+$ channel from rat skeletal muscle (see Fig. 1). The basic experimental procedure used by Sun *et al.*[73] to recover membranes from COS cells transiently transfected with the *Shaker* $K_v^+$ channel first involved harvesting the cells in an alkaline buffer containing 150 m*M* KCl, 2 m*M* $MgCl_2$, 5 m*M* ethyleneglycol-bis($\beta$-aminoethylether)-N,N,N′,N′-tetraacetic acid (EGTA), adjusted to pH 10.6 with $NH_4OH$. The harvested cells are then lysed by forcing the cell suspension several times through a small-gauge syringe needle. This homogenate is next layered on a prechilled sucrose step gradient [10-ml sample, 14 ml 20% (w/v) sucrose, 14 ml 38% (w/v) sucrose in 20 m*M* MOPS–KOH, pH 7.1] and centrifuged in a Beckman SW28.1 rotor for 45 min at 25,000 rpm. The turbid band at the 20/38% sucrose interface was collected, diluted four times with cold water, and pelleted. The final membrane pellet fraction was resuspended in a buffer containing 250 m*M* sucrose, 10 m*M* HEPES–KOH, pH 7.3, aliquoted, quickly frozen, and stored at −70°. Similar methods have also been successfully used to recover microsomal vesicles from HEK293 cells expressing $K_{Ca}^+$ channels[76,77] and rat skeletal muscle $Na_v^+$ channels (Y. Sun and E. Moczydlowski, unpublished results 1998, Fig. 1).

In another application involving the reconstitution of CFTR mutants, a more detailed approach to the membrane fractionation of transiently transfected HEK293 cells was undertaken by Xie *et al.*[74,75] The question of membrane origin is especially relevant in the case of CFTR, since the biosynthetic processing of some CFTR mutants is defective. Such mutations result in a failure of newly synthesized CFTR to be properly transported and expressed in the surface plasma membrane leading to an accumulation of the protein in intracellular membranes. By using discontinuous sucrose gradient fractionation, Western blot analysis, and immunoprecipitation,

[73] T. Sun, A. A. Naini, and C. Miller, *Biochemistry* **33,** 9992 (1994).
[74] J. Xie, M. L. Drumm, J. Ma, and P. B. Davis, *J. Biol. Chem.* **270,** 28084 (1995).
[75] J. Xie, M. L. Drumm, J. Zhao, J. Ma, and P. B. Davis, *Biophys. J.* **71,** 3148 (1996).
[76] M. Müller, D. Madan, and I. B. Levitan, *Neuropharmacology* **35,** 877 (1996).
[77] G. W. Moss, J. Marshall, M. Morabito, J. R. Howe, and E. Moczydlowski, *Biochemistry* **35,** 16024 (1996).

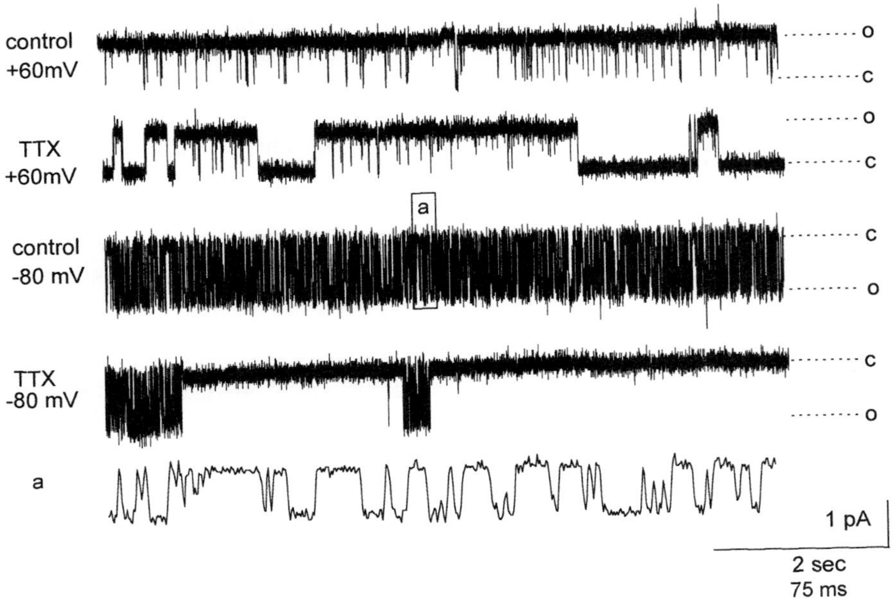


FIG. 1. Planar bilayer recording of a recombinant $\mu$1 $Na_v^+$ channel. cDNA coding for the rat skeletal muscle $\mu$1 $Na_v^+$ channel was subcloned into the mammalian cell expression vector pcDNA3 (Invitrogen, Carlsbad, CA). Human fibroblast HEK293 cells were transfected with the $\mu$1/pcDNA3 vector using the calcium phosphate precipitation method.[82] A cell line of stably transfected cells expressing the $\mu$1 $Na_v$-channel (~5-nA typical peak whole-cell $Na^+$ current) was selected by growth in the presence of the antibiotic G418 (700 $\mu$g/ml). Ten flasks of cells (50% confluent, 150 $cm^2$) were used to prepare membrane vesicles by modifying the procedure of Sun *et al.*[73] previously described for COS cells. The major modifications of this latter method were that harvested cells were fragmented using a glass Dounce homogenizer followed by mild sonication and that KCl in the solutions was substituted by NaCl. Single $Na_v^+$ channels from this preparation were incorporated into planar bilayers in the presence of symmetrical 200 m*M* NaCl, 10 m*M* MOPS–NaOH, pH 7.4, and 0.5 $\mu$*M* batrachotoxin using methods similar to those described previously.[7] The figure shows selected traces from a long (~1-hr) recording of a single recombinant $Na_v^+$ channel (filter frequency 100 Hz). The first and third traces from the top, labeled "control," were respectively taken at holding voltages of +60 mV and −80 mV (physiological convention, extracellular ground) before the addition of tetrodotoxin (TTX). The rapid flickering at −80 mV corresponds to voltage-dependent closing of the channel associated with voltage-activation (c, closed level; o, open level). The second and fourth traces from the top were taken after addition of 20 n*M* TTX to the external side. Long-lived closed states in these traces represent individual TTX-blocking events[7–9] characteristic of batrachotoxin-modified $Na_v^+$ channels. The bottom trace corresponds to a boxed segment of the third trace labeled "a" and is shown at an expanded time scale.

Xie *et al.*[74,75] showed that wild-type CFTR is found in plasma membrane vesicles that mainly band at interfaces of 28/33% sucrose and 33/36% sucrose, whereas mutant CFTR protein is found in intracellular membrane vesicles that band at 36/38.7% and 38.7/43.7% sucrose interfaces. This study demonstrated the feasibility of separating plasma membranes from intracellular membranes and should be useful to researchers attempting to express and reconstitute channel proteins that are located in either of these membrane fractions. In summary, the use of transient or stable transfection in COS or HEK293 cells combined with the isolation of semipurified membrane fractions[73–77] offers an attractive approach for expression, membrane vesicle isolation, and incorporation of cloned and mutant channels into planar lipid bilayers.

While the preceding methods may satisfy ordinary demands of most applications, a true bilayer reductionist insists on the ultimate degree of compositional control. This purist philosophy requires that the cloned channel be heterologously expressed at high levels, purified to biochemical homogeneity, incorporated into liposomes, and then fused into the bilayer. Examples of this connoisseur approach may be cited for CFTR,[78] the ryanodine receptor $Ca^{2+}$-release channel,[79] and the $ClCl_0$ $Cl^-$ channel.[80] The last example[80] is especially instructive because its methods seem to be broadly applicable. In this case, the recombinant $ClCl_0$ $Cl^-$ channel was first expressed by large-scale transient transfection in HEK293 cells.[80] Membrane protein from these cells was solubilized with CHAPS detergent and the $ClCl_0$ $Cl^-$ channel protein was purified by immunoaffinity column chromatography. The detergent was removed by a gel filtration procedure resulting in the formation of reconstituted liposomes, which could then be fused to planar bilayers using the nystatin-mediated fusion technique.[30,31] The relatively straightforward procedures used in this example were developed as an extension of work on the purification and reconstitution of the native voltage-gated $Cl^-$ channel from *Torpedo* electric organ[44,81] and later applied to cloned wild-type and mutant $ClCl_0$ channel expressed in HEK293 cells.[80] A cautiously optimistic outlook envisions that similar strategies may lead the way to the functional reconstitution of other purified channel proteins and also contribute toward two- and three-dimensional crystallization and protein structural determination.

[78] C. E. Bear, C. H. Li, N. Kartner, R. J. Bridges, T. J. Jensen, M. Ramjeesingh, and J. R. Riordan, *Cell* **68,** 809 (1992).

[79] K. Ondrias, A. M. Brillantes, A. Scott, B. E. Ehrlich, and A. R. Marks, *Soc. Gen. Physiol. Ser.* **51,** 29 (1996).

[80] R. E. Middleton, D. J. Pheasant, and C. Miller, *Nature* **383,** 337 (1996).

[81] A. F. Goldberg and C. Miller, *J. Membr. Biol.* **124,** 199 (1991).

[82] C. Chen and H. Okayama, *Mol. Cell. Biol.* **7,** 2745 (1987).

## Acknowledgments

This work was supported by grants to E.M. from the National Institutes of Health (GM51172) and the American Heart Association (95008820). Isabelle Favre was supported by a James Hudson Brown–Alexander Brown Coxe postdoctoral fellowship and an award from the CIBA-GEIGY-Jubiläums-Stiftung.

# [16] Iodide Channel of the Thyroid: Reconstitution of Iodide Conductance in Proteoliposomes

*By* PHILIPPE E. GOLSTEIN, ABDULLAH SENER, FERNAND COLIN, and RENAUD BEAUWENS

## Introduction

The functional unit of the thyroid is the follicle, a cystlike structure lined by a single-layer epithelium, the follicular epithelium, enclosing a central cavity: the colloid. The latter one represents a special extracellular compartment which serves as a reservoir for storage of iodinated thyroglobulin, i.e., a prohormone from which thyroid hormones are released on appropriate stimulus. Thyroglobulin is synthesized within the endoplasmic reticulum of the follicular cells (or thyrocytes) and is secreted into the colloid. Iodide reaches this compartment separately where it is eventually coupled to thyroglobulin. Follicular cells are known to concentrate plasma iodide into their cytosol by uptake via the basolateral $Na^+$–$I^-$ symporter cloned by the group of Dai *et al.*[1] Iodide then diffuses through the apical membrane and reaches the colloid. We hypothesize that this step is mediated by a specific protein, an anion channel, by analogy to intestinal chloride secretion, which exhibits cotransport-mediated uptake at the basolateral border, followed by a hormonally regulated chloride channel at the apical border. However iodide concentration within the thyrocyte is about 1000-fold lower than that of chloride and a unique selectivity toward iodide was therefore suspected. This article reviews the evidence for the existence of an iodide channel in the apical membrane of the thyrocyte.

[1] G. Dai, O. Levy, and N. Carrasco, *Nature* (*Lond.*) **379,** 458 (1996).



0076-6879/99 $30.00

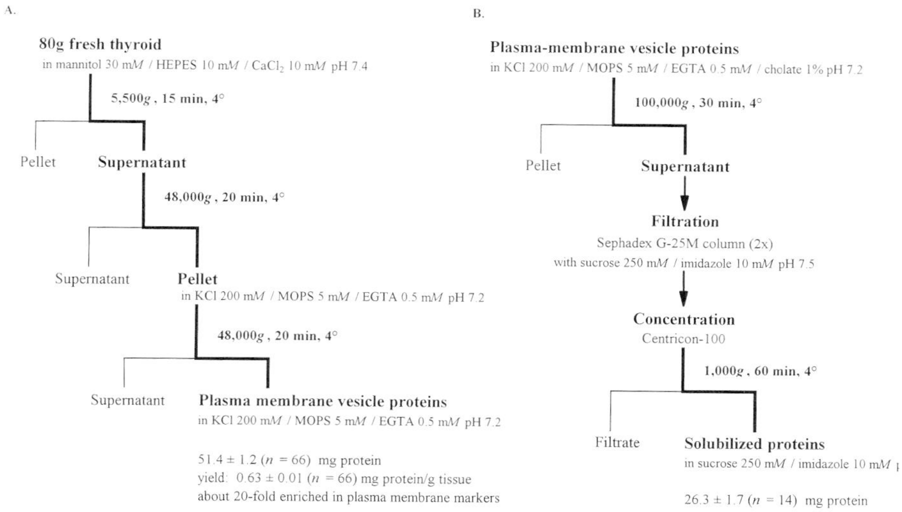


FIG. 1. Flowchart for plasma membrane vesicle (A) preparation and (B) solubilization.

## Experimental Procedures

### *Plasma Membrane Vesicles*

*Preparation.* Thyroid plasma membrane vesicles are prepared from fresh bovine thyroid by differential centrifugation,[2] as illustrated in Fig. 1A. Fresh thyroid (80 g) was homogenized in five volumes of a hypotonic HEPES/Tris buffer (10 m$M$, pH 7.4) containing mannitol (30 m$M$) and $CaCl_2$ (10 m$M$). The homogenate is centrifuged at low speed (5500$g$ for 15 min at 4°) to discard the pellet containing unbroken cells, nuclei, and most of the mitochondria. The supernatant was then centrifuged at high speed (48,000$g$ for 20 min at 4°) and the pellet (about 1 ml) was suspended in 40 ml of either (1) an imidazole/gluconic acid buffer 10 m$M$, pH 7.5, containing sucrose 250 m$M$ for transport assay using the inward potassium gradient; or in (2) potassium gluconate 110 m$M$, KI 15 m$M$, imidazole 10 m$M$ buffered to pH 7.5 with gluconic acid, for transport assay in the presence of the outward iodide gradient; or in (3) a MOPS (morpholinopropanesulfonic acid)/KOH buffer 5 m$M$, pH 7.2, containing KCl 200 m$M$ and ethylene glycol-bis($\beta$-aminoethyl ether)-N,N,N',N'-tetraacetic acid (EGTA) 0.5 m$M$ for solubilization of plasma-membrane proteins. The second high-speed

[2] P. E. Golstein, M. Abramow, J. E. Dumont, and R. Beauwens, *Am. J. Physiol.* **263,** C590 (1992).

centrifugation (48,000*g* for 20 min at 4°) allowed optimal washing of the pellet and yielded plasma membrane vesicles enriched[2] about 20-fold in apical plasma membrane markers ($\gamma$-glutamyltransferase, 5′-nucleotidase, thyroperoxidase), about 15-fold in basolateral plasma membrane marker ($Na^+$,$K^+$-ATPase) and in Golgi membrane marker (glycoprotein $\beta$-D-galactosyltransferase). The vesicles were stored at −80° until use for transport assay or were solubilized for further reconstitution into liposomes.

*Transport Assay.* The presence of the iodide channel into thyroid plasma membrane vesicles was demonstrated by measuring the potential dependent uptake of radiolabeled iodide ($^{125}I^-$) at 4° in the absence of sodium.[2] These selected conditions allowed differentiation of the activity of the channel from other transport processes, such as carrier-mediated (co)transport possibly also present in the membrane vesicles preparation.[3] Indeed, neutral cotransporters or exchangers that are insensitive to the membrane potential are still sensitive to a decrease in temperature because the lipid bilayer can undergo a phase transition from the liquid to the solid state, usually around 15–18°. Electrogenic cotransporter also critically depends on the temperature, and the $Na^+$-dependent cotransporters further require the presence of substantial sodium gradient. So, the electrogenic $Na^+$–$I^-$ cotransporter, located within the basolateral membrane of the thyrocyte is not functional and therefore does not interfere with iodide channel measurement.

Two different procedures are used to generate a potential across the vesicular membrane. The conductive iodide uptake is here defined as the difference in uptake in the presence and absence of a membrane potential. The initial conductive iodide uptake is the value obtained at 30 sec, because this early time is in the linear phase of the uptake.

INWARD POTASSIUM GRADIENT

*Principle.* The inward potassium gradient ($[K^+]_{in}$ = 0 m*M* and $[K^+]_{out}$ = 77 m*M*) is generated by mixing vesicles prepared in the imidazole/gluconic acid buffer, i.e., a potassium- and iodide-free medium, and incubated in a potassium gluconate medium (100 m*M*) containing $^{125}I^-$ (10 $\mu M$, 5 $\mu$Ci/ml). The extravesicle medium is hypertonic relative to the intravesicular medium to prevent swelling and eventual lysis of vesicles. In the absence of a potassium permeability (no valinomycin), the vesicles incorporated a relatively low amount of $^{125}I^-$ even if they have an iodide conductance. Indeed, the infinitesimal entry of iodide into the intravesicular space generates a potential difference that rapidly prevents further $^{125}I^-$ accumulation. On the contrary, addition of valinomycin, a potassium ionophore, increases the potassium permeability about a 1000-fold, generating an inside positive membrane potential that drives a large $^{125}I^-$ uptake.

[3] R. Kinne and K. H. Kinne, *in* "Renal Biochemistry: Cells, Membranes, Molecules" (R. Kinne, ed.), p. 99. Elsevier, Amsterdam, 1985.

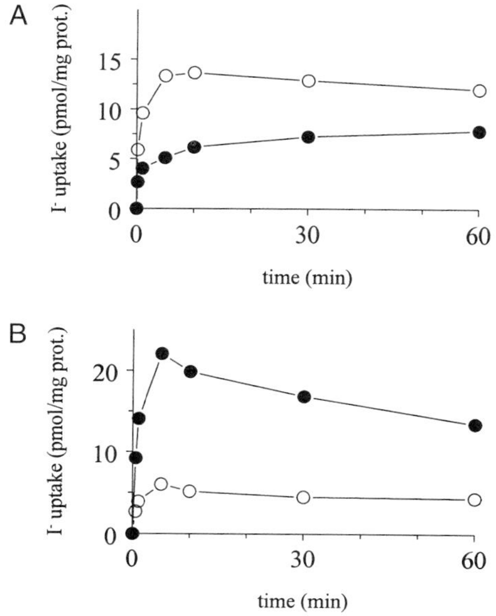


FIG. 2. Conductive iodide uptake as a function of time by the (A) inward $K^+$ gradient method and (B) the outward $I^-$ gradient method. Vesicles were incubated in the absence (filled circles) or in the presence of valinomycin (unfilled circles). $[I^-]_{out}$ = 10 $\mu M$ (A) or 1 $\mu M$ (B).

*$^{125}I^-$ uptake.* Thirty microliters of vesicles (240 $\mu$g of protein) are mixed with 100 $\mu$l of potassium gluconate 100 m*M*, sucrose 250 m*M*, imidazole 10 m*M*, $^{125}I^-$ (10 $\mu M$, 5 $\mu$Ci/ml), pH 7.5 (Fig. 2A). Valinomycin is dissolved in ethanol and added to the vesicles at the final concentration of 20 $\mu$g/mg protein, 2 min before the uptake is started. At appropriate times, the radioactivity incorporated into the vesicles is determined by the rapid filtration technique.[4] The mixture is filtered on a 0.45-$\mu$m nitrate cellulose filter (Millipore, Bedford, MA), and washed twice with 5 ml of an ice-cold Tris-HCl buffer, 1 m*M*, pH 7.5, containing KCl, 250 m*M*. The filter that retained the vesicles is digested in 10 ml of liquid scintillation medium and the radioactivity counted.

OUTWARD IODIDE GRADIENT

*Principle.* The outward iodide gradient ($[I^-]_{in}$ = 15 m*M* and $[I^-]_{out}$ = 1 $\mu M$) is generated by anion-exchange chromatography according to an adaptation of the method developed by Garty *et al.*[5] for sodium transport. The vesicles prepared in potassium gluconate 110 m*M*, KI 15 m*M*, imidazole

[4] U. Hopfer, K. Nelson, J. Perrotto, and K. J. Isselbacher, *J. Biol. Chem.* **248,** 25 (1973).
[5] H. Garty and S. J. D. Karlish, *Methods Enzymol.* **172,** 155 (1989).

10 m$M$, pH 7.5, are eluted on a chilled Dowex (1 × 10) (50–100 mesh, Fluka, Ronkonkoma, NY) column, equilibrated in gluconic acid allowing exchange of extravesicular iodide for gluconate.[6] The eluate is mixed with an equal volume of isotonic incubation medium containing potassium gluconate 125 m$M$ and $^{125}I^-$ (2 $\mu M$, ± 6 $\mu$Ci/ml). As the concentration gradient is set up, iodide can leave the vesicles, which have an iodide conductance. This induces a positive inside membrane potential driving the uptake of $^{125}I^-$ added to the extravesicular medium. Addition of valinomycin in these conditions (i.e., in the presence of potassium) short-circuits the membrane potential and, hence, abolishes the $^{125}I^-$ uptake.[7]

*$^{125}I^-$ uptake.* Two hundred fifty microliters of vesicles (1540 $\mu$g of protein) are eluted with 1 ml of imidazole/gluconic acid buffer on a chilled 4.5-cm Dowex column, equilibrated in gluconic acid (Fig. 2B). The eluate is divided in two volumes of 0.5 ml, for incubation with or without valinomycin, added at the final concentration of 9 $\mu M$, 2 min before the uptake is started. Control vesicles receive the same concentration of ethanol. Both parts of the eluate are mixed with one volume of incubation medium containing potassium gluconate 125 m$M$, imidazole 10 m$M$ and $^{125}I^-$ (2 $\mu M$, ± 6 $\mu$Ci/ml), pH 7.5, yielding a final iodide concentration of 1$\mu M$. Aliquots of 130 $\mu$l (containing 100 $\mu$g of protein) are removed from the radioactive suspension at various times and eluted through a 3-cm ice-cold Dowex column using 1.5 ml of the ice-cold imidazole/gluconic acid buffer. Radioactivity entrapped within the vesicles is counted.

Compared to the rapid filtration technique, this elution method offers these advantages: (1) the background radioactivity is considerably reduced and (2) it obviates the loss of vesicles smaller than the filter pore size. This latter point is especially important for measurement of iodide conductance in proteoliposomes whose estimated size varies from 0.1–0.3 $\mu$m. The initial conductive iodide uptake was 5.7 ± 0.3 pmol/mg protein ($n$ = 8), higher than with the inward gradient method, probably in relationship with the generation of a higher membrane potential difference.

*Proteoliposomes*

Purification an unknown membrane protein requires the development of a simple and accurate test of its activity that can be used at each step of purification. The first challenge is the solubilization of the protein of interest. The second step consists of estimating its activity. We elected to reconstitute the protein into liposomes and to measure their $^{125}I^-$ uptake in various conditions. The parameters of solubilization, reconstitution into

[6] W. Breuer, *Biochim. Biophys. Acta* **1022,** 229 (1990).

[7] P. E. Golstein, A. Sener, and R. Beauwens, *Am. J. Physiol.* **268,** C111 (1995).

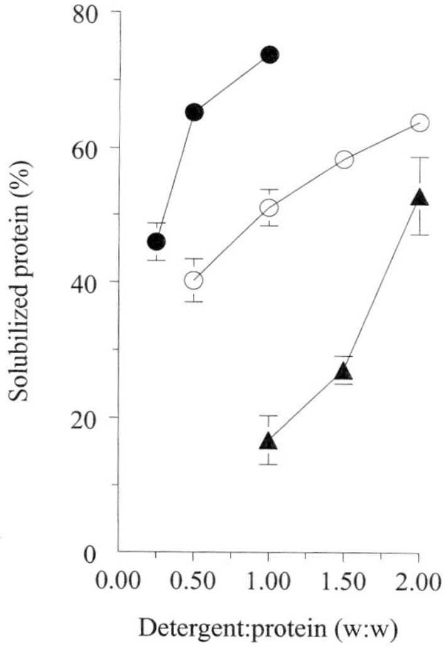


FIG. 3. Effect of different detergents on thyroid plasma membrane protein solubilization. Thyroid plasma membrane proteins were solubilized with the different detergent : protein ratios of sodium cholate (filled circles), CHAPS (unfilled circles), and *n*-octylglucoside (filled triangles). [Reprinted with permission from P. E. Golstein, A. Sener, and R. Beauwens, *Biochem. J.* **312,** 543 (1995). Copyright 1995 Biochemical Society.]

proteoliposomes, and activity measurement of the iodide channel are optimized.

*Preparation*

SOLUBILIZATION OF PLASMA MEMBRANE PROTEINS. Solubilization of a membrane protein is the first step toward its purification. The detergent should ideally solubilize the membrane protein without altering its biological activity and should also have a high critical micellar concentration because it must be easily removed from the medium.[8] Unfortunately, for an unknown membrane protein, there is no absolute rule to predict which detergent would be ideal. Therefore, for the iodide channel, three different detergents are used at different detergent : protein ratios[9]: (1) sodium cholate, an anionic detergent; (2) CHAPS, a zwitterionic detergent; and (3) *n*-octylglucoside, a neutral detergent. Figure 3 shows that sodium cholate achieved the highest solubilization at the lowest detergent : protein ratio. The 0.5 : 1 ratio solubilized 65.9 ± 1.2% of plasma membrane proteins ($n$ = 19).

[8] H. Helenius, D. R. McCaslin, E. Fries, and C. Tanford, *Methods Enzymol.* **56,** 734 (1979).
[9] P. E. Golstein, A. Sener, and R. Beauwens, *Biochem. J.* **312,** 543 (1995).

Membrane proteins are solubilized in high ionic strength medium (200 m*M* KCl), which ensures a most effective solubilization, yet precludes the measurement of the iodide conductance which requires a chloride-free medium. It is therefore imperative to replace the ionic medium with a sucrose medium (imidazole/gluconic acid buffer, 10 m*M*, pH 7.5, containing sucrose, 250 m*M*). Detergent removal and buffer exchange are simultaneously performed by gel filtration on a prepacked Sephadex G-25M column (Sephadex PD-10 column, Pharmacia, Piscataway, NJ). About 30–35 mg of solubilized proteins in 2.5 ml is applied on the column preequilibrated with the same buffer and eluted with imidazole/gluconic acid buffer. The void volume of the column (2.5 ml) is discarded and the protein material in the next 2.5 ml of eluate collected. This operation is repeated to ensure maximal removal of unbound detergent and chloride. Figure 4 illustrates the separation by the Sephadex G-25M column of a 2.5-ml mixture containing bovine serum albumin (40 mg) and $^{125}I^-$ (25,000 cpm) in a MOPS/KOH buffer 5 m*M*, pH 7.2, containing KCl 200 m*M*, EGTA 0.5 m*M*, and cholate 1%. The recovery of both albumin and $^{125}I^-$ is 100%.

Solubilized proteins are then concentrated by ultrafiltration on a Centricon 100-kDa membrane (Amicon, Danvers, MA) to reach a concentration of about 30 mg/ml. This material is stored at −80° until used for proteoliposomes reconstitution.

Liposome Preparation. Liposomes are prepared from phospholipid films conserved at 4° for up to 6 weeks (Fig. 5). The films are a combination of 10 mg of egg phosphatidylcholine (egg PC) and cholesterol in a 4:3

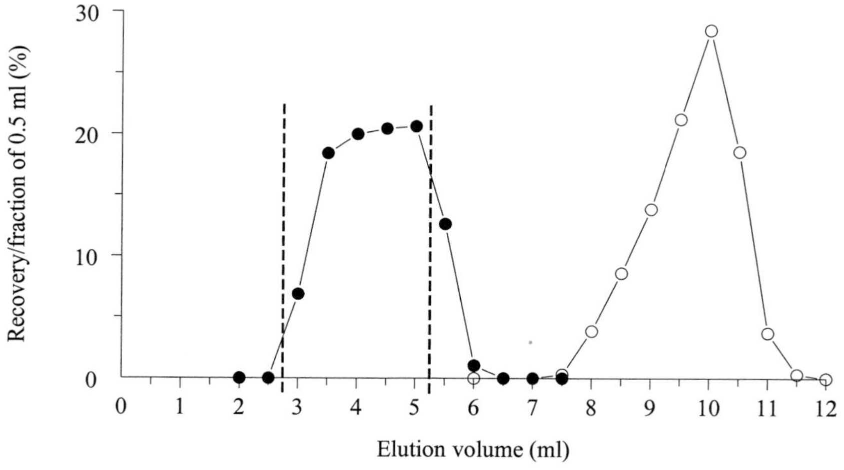


Fig. 4. Gel filtration of a mixture of albumin (filled circles) and $^{125}I^-$ (unfilled circles).

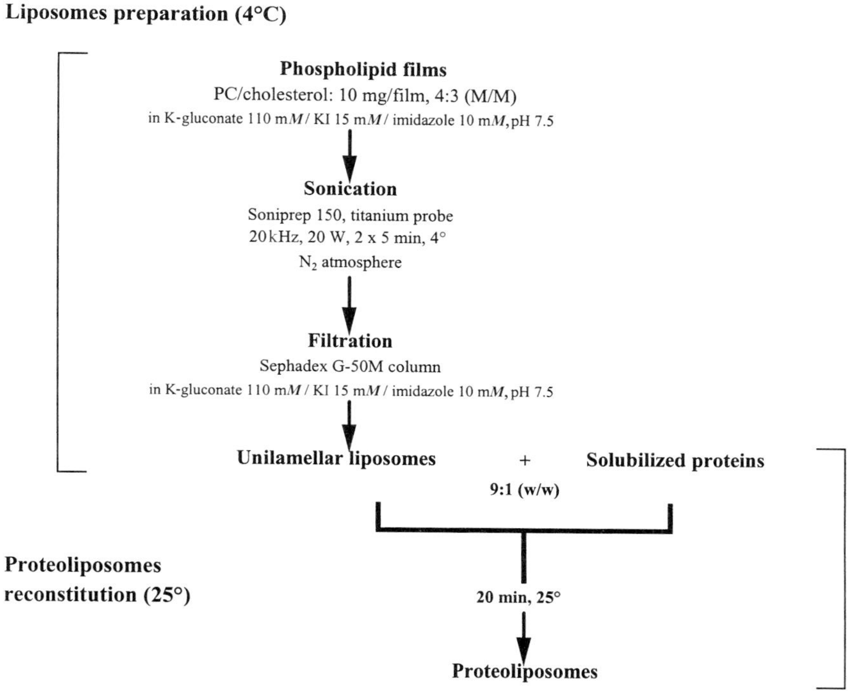


FIG. 5. Flowchart for preparation of liposomes and proteoliposomes.

molar/molar (M : M) ratio. They are prepared in glass tubes at a concentration of 10 mg of lipid/ml chloroform, which was evaporated under nitrogen atmosphere for 30 min and then under vacuum for at least 18 hr at room temperature (20–22°).

For each reconstitution, phospholipid films are freshly suspended in potassium gluconate 110 m*M*, KI 15 m*M*, imidazole 10 m*M*, pH 7.5 (10 mg/ml), and sonicated at 20 kHz, 20 W on ice for two periods of 5 min, under nitrogen (Soniprep 150; titanium probe 9.5 mm in diameter). To eliminate possible titanium fragments, phospholipid aggregates, and large size multilamellar liposomes, the suspension is centrifuged (400*g*, 8 min, 4°) through a Sephadex G-50M column preequilibrated overnight with the same buffer and prepared in a 5-ml syringe plugged with a Whatman GF/C filter. The size of the unilamellar liposomes obtained is estimated by electron microscopy using negative staining; their diameter ranged from 100 to 300 nm.

PROTEOLIPOSOME RECONSTITUTION. Freshly prepared liposomes are incubated with solubilized proteins at a 9 : 1 (w/w) ratio, in a water bath at

25° and vortexed vigorously every 5 min (Fig. 5). Practically, for a volume of 250 $\mu$l, 1400 $\mu$g of lipid and 155 $\mu$g of protein are combined and the volume adjusted to 250 $\mu$l with potassium gluconate 110 m$M$, KI 15 m$M$, imidazole 10 m$M$, pH 7.5.

This method of proteoliposome reconstitution is simple, rapid, and allows optimal control of the media. Gel-filtration chromatography is used to remove the detergent and to adjust the buffer composition simultaneously, before protein insertion into liposomes which are prepared separately. In most current procedures, the phospholipids and the solubilized proteins are mixed in high ionic strength medium containing the detergent. A second step, using Bio-Beads (Bio-Rad, Richmond, CA) and/or dialysis, is needed to remove the detergent and to replace the medium.[10] These procedures are time consuming and do not allow good control of the media.

### *Transport Assay*

OUTWARD IODIDE GRADIENT. The activity of the iodide channel is estimated by the outward iodide gradient method. Two hundred fifty microliters of proteoliposomes (1400 $\mu$g of lipid and 155 $\mu$g of protein) are eluted with 1 ml of imidazole/gluconic acid buffer on a chilled 4.5-cm Dowex column, equilibrated with gluconic acid. The eluate is divided in two volumes of 0.5 ml, for incubation with or without valinomycin, added at the final concentration of 9 $\mu M$, 2 min before the uptake is started. Both parts of the eluate are mixed with one volume of incubation medium containing potassium gluconate 125 m$M$, imidazole 10 m$M$, and $^{125}I^-$ (2 $\mu M$, $\pm$ 6 $\mu$Ci/ml), pH 7.5, yielding a final iodide concentration of 1 $\mu M$. Aliquots of 130 $\mu$l (10 $\mu$g of protein) are removed from the radioactive suspension at various times and eluted through a 3-cm ice-cold Dowex column using 1.5 ml of the ice-cold imidazole/gluconic acid buffer. Radioactivity entrapped within the vesicles is counted.

To determine the optimal cholate : protein ratio, proteoliposomes (90 $\mu$g of lipid) are reconstituted with 10 $\mu$g of protein solubilized with three different cholate : protein ratios of 0.25 : 1, 0.5 : 1, and 1 : 1. The cholate : protein ratio of 0.5 : 1 was selected because it yielded the highest initial conductive iodide uptake, although it solubilized fewer plasma membrane proteins than higher cholate : protein ratios, suggesting more selective solubilization of the iodide channel and/or lower loss of its biological activity (data not shown).

[10] S. Ran and D. J. Benos, *J. Biol. Chem.* **266,** 4782 (1991).

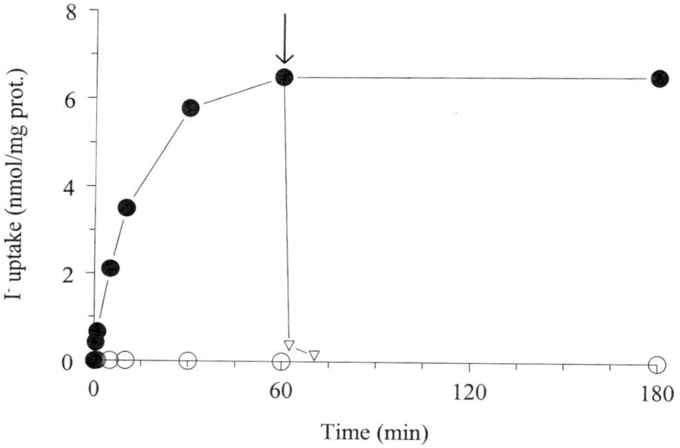


FIG. 6. Time course of the iodide uptake into proteoliposomes. Proteoliposomes were incubated in the absence (filled circles) or presence (unfilled circles) of valinomycin. The arrow indicates addition of valinomycin at 60 min (unfilled triangles).

The time course of the $^{125}I^-$ uptake is presented in Fig. 6. In the absence of valinomycin, the uptake is linear up to 1 min, reaching a plateau after 30 min, which is maintained at least for 180 min. This further validates the use of the value measured at 30 sec to compute the initial conductive iodide uptake. Addition of valinomycin completely abolishes this uptake, demonstrating its potential dependency.

OPTIMAL CONDITIONS FOR TESTING IODIDE CONDUCTANCE

*Lipid dependency.* Proteoliposomes are reconstituted with a low constant amount of protein (10 $\mu$g) and variable amounts of lipid (30–120 $\mu$g) (Fig. 7A). The initial conductive iodide uptake increases linearly between 30 and 90 $\mu$g lipids. In contrast, in liposomes prepared with the same amounts of lipids (30–120 $\mu$g), the initial iodide uptake increases much more slowly, in a direct proportion to the amount of lipid.[9] Therefore, 90 $\mu$g of lipid is selected for reconstitution. The uptakes are quite similar at 30 $\mu$g of lipid, suggesting that the amount of lipids becomes rate limiting for protein insertion.

*Protein dependency.* Proteoliposomes are reconstituted with a constant amount of lipid (90 $\mu$g) and variable amounts of protein, ranging from 5 to 60 $\mu$g (Fig. 7B). The initial conductive iodide uptake increased linearly from 5 to 20 $\mu$g and appeared to saturate above 40 $\mu$g of protein. In proteoliposomes reconstituted with 90 $\mu$g of lipid and 10 $\mu$g of protein, the initial conductive iodide uptake was 304.7 ± 64.1 pmol/mg protein ($n$ = 23 experiments performed on six different preparations).

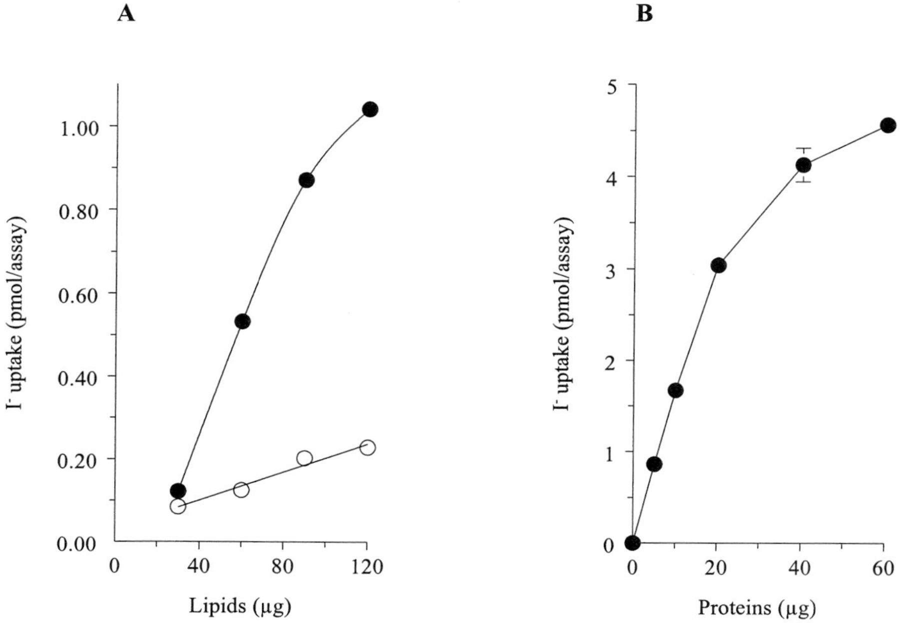


FIG. 7. Conductive iodide uptake (A) into proteoliposomes (filled circles) and liposomes (unfilled circles) as a function of the amount of lipid and (B) into proteoliposomes as a function of the amount of protein inserted.

*Dependence on magnitude of imposed outward iodide gradient.* The initial conductive iodide uptake increases as a function of the magnitude of the imposed iodide gradient, which sets up a membrane potential (positive inside) across the proteoliposomal membrane (Fig. 8). Proteoliposomes (10 $\mu$g of protein and 90 $\mu$g of lipid) are loaded with buffers containing various iodide concentrations (KI 1.5, 15 and 50 m$M$) and potassium gluconate is adjusted to maintain a constant osmolarity. The outward gradient generated by 15 m$M$ intraproteoliposomal iodide appeared sufficient to generate a sizeable signal.

SELECTIVITY OF IODIDE CONDUCTANCE. Proteoliposomes reconstituted with solubilized thyroid membrane proteins (10 $\mu$g of protein and 90 $\mu$g of lipid) exhibit a highly selective conductance for iodide, demonstrated by two different approaches.[7]

*Outward anionic gradient.* Intraproteoliposomal iodide is substituted by different anions (either $Cl^-$, $NO_3^-$, or $SO_4^{2-}$ 15 m$M$), and an outward anionic gradient is set up on a Dowex resin, as for iodide ($[\text{anion}]_{in}$ = 15 m$M$ and $[I^-]_{out}$ = 1 $\mu M$). In each of these conditions, the

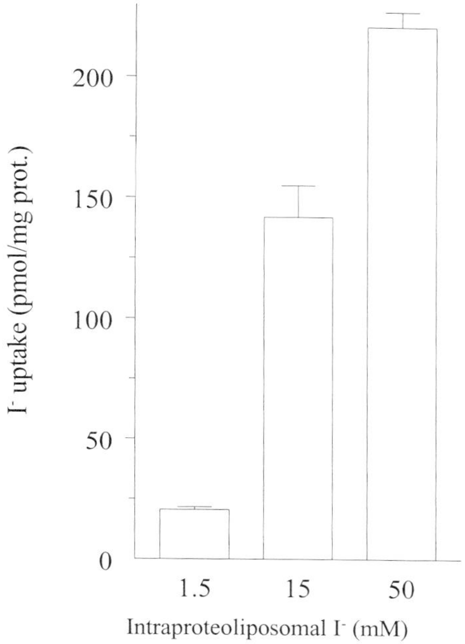


FIG. 8. Effect of the magnitude of the outward iodide gradient imposed on the conductive iodide uptake into proteoliposomes. Intraproteoliposomal iodide concentration was varied while extraproteoliposomal iodide was kept contant at 1 $\mu M$.

initial iodide uptake observed was not significantly different from background, indicating that none of these anions is able to generate a membrane potential and that they do not permeate the iodide channel.

*Extraproteoliposomal anions.* The initial conductive iodide uptake is measured in the presence of an outwardly directed iodide gradient, and different anions ($Cl^-$, $F^-$, $Br^-$, or $NO_3^-$) are added to the extraproteoliposomal medium at 1 m$M$, i.e., in a 1000-fold excess over the iodide concentration. The anions failed to inhibit the iodide uptake, further indicating that they do not permeate the iodide channel.

SPECIFICITY OF IODIDE CONDUCTANCE. The specific nature of an uptake means that it is mediated by a given protein.[9] To demonstrate that the iodide conductance is mediated by some plasma membrane thyroid protein, the initial iodide uptake is measured in different conditions of reconstitution.

The initial conductive iodide uptake is (1) correlated to the amount of solubilized proteins inserted into liposomes (Fig. 7B); (2) mediated by a "native" thyroid membrane protein component, and its denaturation (heat-

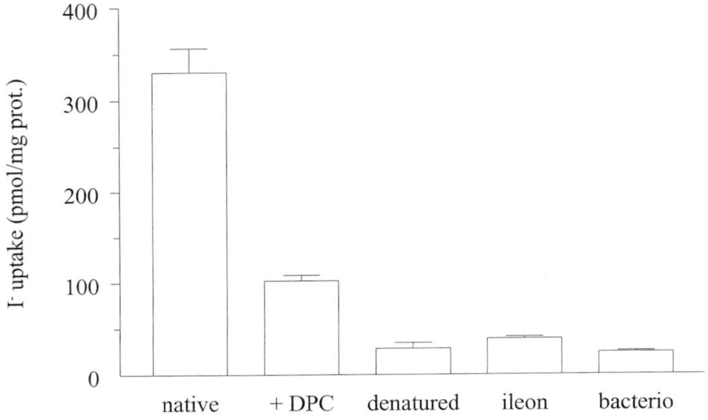


FIG. 9. Conductive iodide uptake into proteoliposomes reconstituted with the different membranes proteins. Thyroid plasma membrane proteins: native ± DPC 100 $\mu M$, or denatured; ileal membrane proteins (ileon) or bacteriorhodopsin (bacterio).

ing for 5 min at 95°), before reconstitution, abolishes the iodide uptake (Fig. 9); (3) not observed when the proteoliposomes are reconstituted with other membrane proteins, like bacteriorhodopsin or bovine ileal membrane proteins (prepared by the $Mg^{2+}$/EDTA precipitation method[11]) (Fig. 9); (4) only observed in proteoliposomes reconstituted with one well-defined fraction of thyroid membrane proteins (molecular mass estimated between 100 and 150 kDa) when the latter ones are separated by size-exclusion chromatography (Fig. 10). The maximal activity of the iodide channel was found in two fractions (10 and 11), corresponding to a molecular mass between 100 and 150 kDa and to about 8% of the membrane proteins. No activity was detected in the other fractions, demonstrating that the iodide conductance is probably mediated by one defined membrane protein; and (5) considerably reduced by the anionic channel inhibitor *N*-phenylanthranilic acid (DPC, 100 $\mu$M) suggesting that the iodide channel shares some properties with other anionic channels (Fig. 9).

*Biophysics of Iodide Channel*

KINETIC PROPERTIES OF IODIDE UPTAKE. The outwardly directed iodide gradient set up by fixing the extraproteoliposomal iodide concentration at 1 $\mu M$ (anion exchange resin) imposes a potential difference across the proteoliposomal membrane permeable to iodide. If the proteoliposomes are relatively impermeable to other ions present (potassium and gluconate),

[11] B. Stieger and H. Murer, *Eur. J. Biochem.* **135,** 95 (1983).

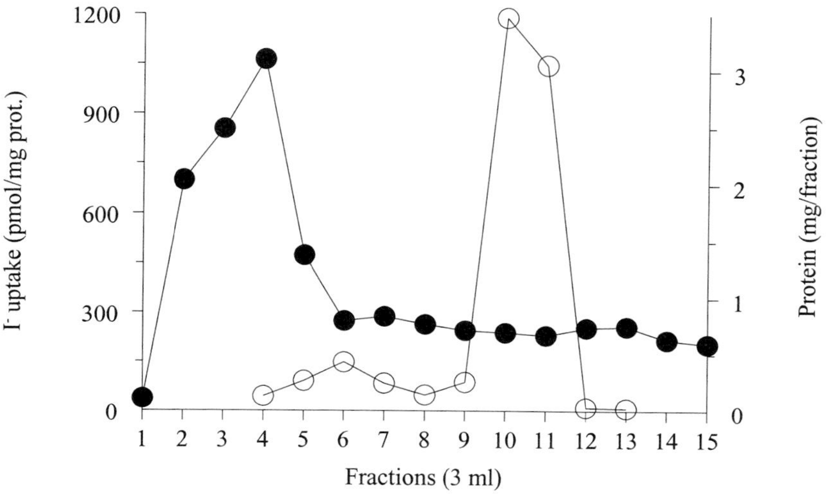


FIG. 10. Conductive iodide uptake in proteoliposomes reconstituted from various fractions obtained by size-exclusion chromatography of solubilized membrane proteins. Solubilized plasma membrane proteins (15–20 mg in a MOPS/KOH buffer 5 m*M*, pH 7.2, containing KCl 200 m*M*, EGTA 0.5 m*M*, and cholate 1%) were loaded on a Superdex 200 Hiload column (2.6 × 60 cm, Pharmacia) equilibrated with a $K_2HPO_4/KH_2PO_4$ buffer 50 m*M*, pH 7.2, containing KCl 200 m*M* and sodium cholate 1%. Elution was performed with the same buffer at a flow rate of 0.75 ml/min and 15 fractions of 3 ml were collected every 4 min, from 51 to 111 min. Each fraction was concentrated, desalted, and buffer was exchanged for an imidazole/gluconic acid buffer. The protein content (filled circles) and the activity of the conductive iodide channel (unfilled circles) were determined in each fraction.

this inside positive membrane potential will drive the uptake of $^{125}I^-$ present in the outside medium. In the absence of valinomycin, the rapid increase of $^{125}I^-$ indicates that the permeability for iodide is dominant over that of any other ion present. The $^{125}I^-$ uptake into proteoliposomes reaches a maximum after 30 min and is maintained for several hours (Fig. 6). This kinetics is in sharp contrast with that observed in vesicles in the same conditions (Fig. 2B). Garty's model predicts that the initial and final slopes of uptake are closely linked to the permeabilities of respectively the "charging" and the "discharging" ions. The stable plateau observed in Fig. 6 indicates that there is no leakage at all of other ions dissipating the electric field. Therefore, the Nernst relationship does apply and the electric potential is determined by the intra- and extraproteoliposomal iodide concentrations. The practical favorable consequences of the situation cannot be overestimated: concentrations and potential are known ($\Delta\Psi$ = 230 mV) and stay constant during the overall course of the experiment as well as the distribution volume, particularly during the initial step when the flux of the labeled ion is measured. The plateau indicates that the specific activity of iodide inside and outside the proteoliposomes is identical.

The dissipation of the electric field can however be triggered by valinomycin. If valinomycin is added before starting the uptake, no $^{125}I^-$ is incorporated into the proteoliposomes (Fig. 6). If valinomycin is added when the plateau is reached, an abrupt discharge is observed, which does not allow a precise study of its kinetics. The latter one should depend on the mean number of molecules of valinomycin in a single proteoliposome, which itself should be a function of its bulk concentration. At 9 $\mu M$ valinomycin, all the proteoliposomes were discharged, indicating that each had captured at least one molecule of valinomycin (Fig. 6).

CHARACTERISTICS OF IODIDE CONDUCTANCE. One might wonder if we are dealing with a channel or an electrogenic transporter. A channel is an integral membrane protein forming a hydrophilic central cavity, more or less selective for a given ion. Ions permeate the channel passively according to their electrochemical potential and dehydrate partially and/or bind reversibly to a specific binding site at the level of the filter of selectivity located inside the channel. These processes are not very sensitive to the temperature and require a low activation energy, around 5 kcal/mol, at 20°.[12] To estimate the activation energy of the iodide conductance observed, the initial conductive iodide uptake was measured on proteoliposomes at different temperatures. The activation energy, between 14 and 25°, is 2 to 3.7 kcal/mol, typical of an ionic channel.

## Conclusion

In summary, a specific membrane protein solubilized by cholate and substantially enriched by size-exclusion chromatography, is able to transport iodide across proteoliposomal membrane in response to an imposed electrochemical gradient. This conductive flux is at least 1000-fold more selective for iodide over other anions tested, in particular over the most abundant chloride anion. Estimation of the activation energy of this transport yielded a value of 2 to 3.7 kcal/mol, quite typical of ionic channel. Hence the protein was called the "iodide channel."

Quite interestingly, the cholate solubilization and proteoliposomal reconstitution procedures did not induce substantial potassium or gluconate permeability. Therefore, in media containing only potassium, iodide, and gluconate, a Nernst equilibrium potential for iodide was actually set up across the proteoliposomal membrane.

[12] B. Hille, *J. Gen. Physiol.* **66,** 535 (1975).

It is expected that future purification of the iodide channel, which is currently ongoing, will unravel its tissue distribution (perhaps not limited to the thyroid), its possible hormonal regulation, and its eventual role in disease states.

Acknowledgment

P. E. Golstein was "Chargé de Recherches" from the "Fonds National de la Recherche Scientifique" (Brussels, Belgium), which supported this work (grant 3.4613.95).

# [17] Nystatin/Ergosterol Method for Reconstituting Ion Channels into Planar Lipid Bilayers

*By* DIXON J. WOODBURY

## Introduction

Reconstitution of membrane channels into planar lipid bilayers is a powerful technique to determine the ion selectivity, transport rate, voltage dependence, and kinetics of membrane proteins. (See also chapter by Ramjeesingh *et al.*,[1] this volume.) The major challenge with using planar bilayers to study ion channels is getting the channel into the bilayer. Some ion channels insert spontaneously into the bilayer when they are added to the aqueous solution next to the bilayer. Channels that insert spontaneously, for example, gramicidin and alamethicin, are easier to study and have been well characterized. However, many ion channels are not soluble in solution and do not insert spontaneously into bilayers.

One method to place these channels into bilayers is to isolate them in vesicles and induce the vesicles to fuse with a bilayer. Usually, this means fracturing the plasma membrane into small pieces, which spontaneously form vesicles, and then delivering the vesicles near the bilayer with the right fusion conditions. The right fusion conditions include an osmotic gradient and a specific type of channel in the vesicle membrane.[1a,2–4]

[1] M. Ramjeesingh, E. Garami, K. Galley, C. Li, Y. Wang, and C. E. Bear, *Methods Enzymol.* **294,** [11], 1998 (this volume).
[1a] C. Miller, P. Arvan, J. N. Telford, and E. Racker, *J. Membr. Biol.* **30,** 271 (1976).
[2] F. S. Cohen, J. Zimmerberg, and A. Finkelstein, *J. Gen. Physiol.* **75,** 251 (1980).
[3] D. J. Woodbury and J. E. Hall, *Biophys. J.* **54,** 345 (1988).
[4] F. S. Cohen, W. D. Niles, and M. H. Akabas, *J. Gen. Physiol.* **93,** 201 (1989).



0076-6879/99 $30.00

Because not all channels induce fusion, it is not generally possible to know if a particular channel-containing vesicle is also capable of fusing. In practice, it is concluded that a vesicle was fusigenic if an ion conductance is observed in the bilayer after vesicle addition. This is achieved by (1) isolating vesicles containing native membrane proteins, (2) establishing an osmotic gradient across a bilayer, (3) adding vesicles to the hyperosmotic side, and (4) observing the bilayer for an increase in conductance. If a new conductance is seen, it is assumed to be due to channels originally in the isolated vesicles. This is a fair conclusion, but it can also be misleading. It may be that the majority of isolated vesicles (containing the major proteins to be studied) are not fusigenic and that the observed conductance is due to a small number of vesicles that contain a contaminating channel that makes these vesicles fusigenic.[5] Consider a hypothetical case of a subcellular vesicle that is not fusigenic (not able to fuse with a bilayer). A preparation of these organelles that is 99% pure would still contain 1% contaminating vesicles, which could be fusigenic. Following addition of $10^9$ of these vesicles only some of the $10^7$ contaminating vesicles fuse with the bilayer and hence only their channels are observed. In other words, even though there are $99 \times 10^7$ copies of every protein in the organelle membrane, none of these ever appear in the bilayer membrane.

One way to avoid this dilemma is to treat the vesicles in such a way as to make them equally fusigenic. With all vesicles equally fusigenic, the frequency of observation of a certain kind of channel should reflect the actual relative abundance of that protein in the vesicle population. All vesicles can be made fusigenic by adding a special component to each vesicle that will guarantee its fusion. With such an addition, the possible variation in the ability of different vesicles to fuse becomes less important. This is how the nystatin/ergosterol fusion method works.[6,7]

For nystatin/ergosterol-induced fusion, nystatin (an antibiotic) and ergosterol (a sterol) are added to the vesicle membrane. These components combine to form ion channels that induce fusion in the presence of a salt gradient. Channels are formed of ~10 nystatin monomers in a barrel-stave arrangement with ergosterol apparently required as a glue to hold the monomers together.[8,9] When the glue is removed, the channels fall apart. Thus, if ergosterol-rich vesicles fuse into an ergosterol-free bilayer, nystatin channels will turn off after ergosterol dissociates from the channel complex

[5] D. J. Woodbury, *Biophys. J.* **65,** 973 (1993).
[6] W. D. Niles and F. S. Cohen, *J. Gen. Physiol.* **90,** 703 (1987).
[7] D. J. Woodbury and C. Miller, *Biophys. J.* **58,** 833 (1990).
[8] A. Cass, A. Finkelstein, and V. Krespi, *J. Gen. Physiol.* **56,** 100 (1970).
[9] M. Moreno-Bello, M. Bonilla-Marin, and C. Gonzalez-Beltran, *Biochim. Biophys. Acta* **944,** 97 (1988).

and diffuses away into the huge excess of bilayer lipid.[7] The ability to turn off the nystatin conductance is important for most applications, where the conductance of nystatin channels would interfere with the determination of conductance properties of the newly reconstituted ion channel protein.

For success in using the nystatin/ergosterol (N/E) method, care must be taken to ensure (1) that fusion-inducing nystatin channels are properly formed in each vesicle, (2) that nystatin channels turn off after fusion, and (3) that conductance spikes signaling vesicle fusion can be observed. This last point is especially important to calculate the abundance of a protein in its original vesicle population. Several additional points are important for vesicle–bilayer fusion using any method, including delivery of vesicles to the bilayer and the biophysical characterization of reconstituted channels. These issues and details for successful formation of N/E vesicles are presented in the following sections.

## Preparation of Fusigenic Artificial (N/E) Vesicles

The first step in the reconstitution of proteins into bilayers by the N/E method is to prepare artificial vesicles or liposomes that can fuse (i.e., are fusigenic). These vesicles are then joined with native vesicles containing the membrane proteins of choice. Alternatively, if the protein is isolated and purified, it can be reconstituted directly into N/E vesicle as described later.

### *Vesicle Lipids*

In my laboratory, we have not tested a wide range of lipid mixtures for their compatibility in forming fusigenic N/E vesicles but have found that 12–25 mol% ergosterol is essential. Typically, N/E vesicles are made of 20 mol% ergosterol (Erg), 20 mol% PS (phosphatidylserine), 10–20 mol% PC (phosphatidylcholine), and the remainder PE (phosphatidylethanolamine). All of our phospholipids are obtained from Avanti Polar Lipids, Inc. (Alabaster, AL). Although commercially available ergosterol is not very pure (~90%) and can be recrystallized from ethanol before use,[10] we routinely use ergosterol from Sigma (St Louis, MO) without recrystallization. Stock ergosterol solutions in chloroform (10 mg/ml) can be stored in the freezer for at least 6 months. Because the mole percent of ergosterol is important, it is judicious (and convenient) to make stock solutions of lipids in chloroform (10 mg/ml) with the desired ratio of lipids. If brain lipids are used at a mole ratio of 2 : 1 : 1 : 1 (PE : PC : PS : Erg) then the corresponding weight or volume percent is 43 : 22 : 23 : 12. When the lipids are thus premixed, the

[10] F. S. Cohen and W. D. Niles, *Methods Enzymol.* **220,** 50 (1993).

ratio of lipids will not change even when some chloroform is lost due to evaporation. These solutions are stored at −20° or −80° and should be used within 3 months. Stock lipid solutions are warmed to room temperature before opening, to reduce water condensation on the cold fluid surface. Nystatin (Sigma) stock solutions are prepared by dissolving 2.5 mg nystatin per milliliter methanol. The methanol must be kept dry (free of water) during storage or nystatin will precipitate out of solution. Only freshly opened methanol should be used to mix nystatin solutions and the nystatin stock solution should be warmed to room temperature before opening—especially on humid days. Mild bath sonication can be used to aid dispersing nystatin in methanol.

To prepare a batch of vesicles, 125–250 $\mu$l of stock lipids and 5 $\mu$l of stock nystatin are dried under a stream of nitrogen or argon. After evaporation 250 $\mu$l of 150 m*M* NaCl, 8 m*M* HEPES (pH 7.2) are added to the test tube so that final nystatin concentration is 50 $\mu$g/ml. The mixture is vortexed for ~5 min and sonicated in a water bath sonicator (at 22°) for ~2 min. The mixture is cloudy after vortexing and should become translucent after the first minute of sonication. The vesicles are frozen in a dry ice–ethanol bath: just before use, the mixture is thawed at room temperature and sized as discussed next.

### *Importance of Vesicle Size for Fusion*

Theoretically, size can affect the fusigenic nature of vesicles according to the law of Laplace and the permeability of the vesicle membrane as previously described.[11,12] Experimentally, we have found in our laboratory that reasonable fusion rates are not observed unless vesicle size is controlled. There are two simple ways to control the size of a vesicle. One is the freeze–thaw–sonicate (FTS) method of Pick[13] and the second is the use of sizing filters as described by MacDonald *et al.*[14]

The FTS method relies on a freeze–thaw cycle to enlarge the vesicles and sonication to decrease vesicle size. During freeze–thaw, individual vesicles rupture and fuse together to form larger vesicles, multilamellar vesicles, and perhaps even extended bilayer sheets. Sonication creates vibrational oscillations within the solution that breaks up larger structures. In this way, extended sheets are broken up into smaller sheets, which spontaneously close to form vesicles, and large vesicles bud into smaller vesicles.

[11] D. J. Woodbury and J. E. Hall, *Biophys. J.* **54,** 1053 (1988).

[12] W. D. Niles, F. S. Cohen, and A. Finkelstein, *J. Gen. Physiol.* **93,** 211 (1989).

[13] U. Pick, *Arch. Biochem. Biophys.* **212,** 186 (1981).

[14] R. C. MacDonald, R. I. MacDonald, B. P. Menco, K. Takeshita, N. K. Subbarao, and L. R. Hu, *Biochim. Biophys. Acta* **1061,** 297 (1991).

The fact that the extent of sonication is critical for successful vesicle–bilayer fusion[14a] supports the notion that vesicle size must be controlled for successful N/E fusion. For vesicles prepared as described above, sonication in a bath-type sonicator for 5–15 sec yields mostly unilamellar vesicles ~250 nm in diameter. Unfortunately, it is not enough to specify just the time of bath sonication. This is because each sonicator is different and how efficiently a bath sonicator transduces electrical energy into usable vibrational energy is dependent on the height of the water in the bath. Before sonication, the water height of the bath must be adjusted (tuned) for optimum power transfer. Because there is no simple way to quantify how much vibrational energy a sonicator transmits to a sample, a qualitative description of the procedure used in my laboratory follows.

The bath sonicator (G112SP1T, Laboratory Supplies Co., Inc., Hicksville, NY) is powered through a variable transformer set to ~85 V. The level of water in the bath is adjusted so that small droplets of water splash from the center of the bath into the air. Aqueous solutions of vesicles (100–250 $\mu$l) are sonicated in a 13- × 100-mm glass test tube (plastic does not work as well). The tube is held several millimeters deep into the center of the bath. Sonication intensity and tube placement are adjusted so that the test tube vibrates just enough to cause a tingling sensation in the fingers. Typically, vesicles are sonicated for ~10 sec. If longer times are used (1–2 min), the test tube becomes warm. Should this occur, sonication must be stopped until the test tube cools down. Sonication can then be continued for the desired time. It is possible to estimate vesicle size following sonication by using a sizing column.[15] These columns are formed from S-1000 gel (Pharmacia LKB Biotechnology, Piscataway, NJ) and can separate particles up to ~100,000,000 Da, which corresponds to a vesicle size of ~300 nm. Vesicles that were oversonicated do not have high fusion rates. These vesicles have an average size of less than 200 nm as estimated on a sizing column. Vesicles that were undersonicated also do not have high fusion rates. These vesicles came out of the column with the void volume and thus have an average size ≥300 nm. N/E vesicles that have a higher rate of fusion usually came off the column near the void volume, but with a broader peak, indicating that their average size is near 300 nm (unpublished results).

The second method to prepare vesicles that are the right size for N/E fusion is to use sizing filters. A polycarbonate membrane filter that has holes of a fixed size is clamped in a holder (Avestin, Inc., Ottawa, Ontario, Canada) to which two 1-ml gas-tight syringes are attached, one on each

[14a] M. L. Kelly and D. J. Woodbury, *Biophys. J.* **70,** 2593 (1996).

[15] Y. Nozaki, D. D. Lasic, C. Tanford, and J. A. Reynolds, *Science* **217,** 366 (1982).

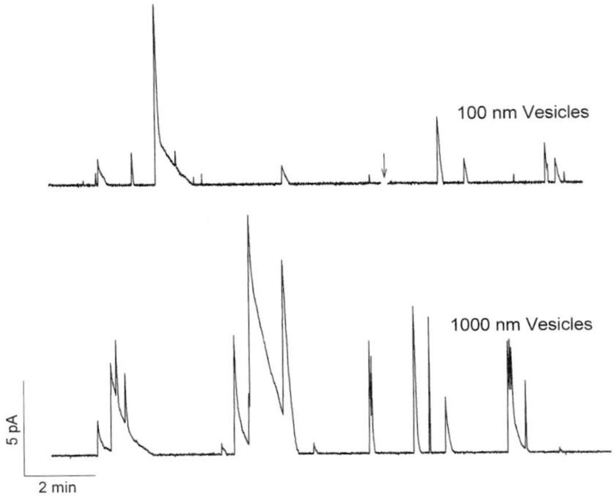


FIG. 1. Current spikes due to the fusion of N/E vesicles with a bilayer. The top trace shows the fusion of 17 vesicles that are 100 nm or smaller (the arrow indicates where more vesicles were added to the *cis* chamber). The bottom trace shows the fusion of 20 vesicles that are 1000 nm or smaller. N/E vesicles were prepared as described in the text but were extruded through different size filters as noted. For both experiments, there was a 150/645 m*M* gradient of NaCl, the *cis* chamber was stirred at 5 Hz, membrane holding potential was −60 mV, and membrane capacitance was ~200 pF.

side. At least 250 μl of vesicles are loaded in one syringe and forced through the filter to the other syringe. The vesicles are then passed back through the filter and into the original syringe. In this manner the vesicles are repeatedly extruded through the filter an odd number of times (usually 21), giving vesicles that are approximately the size of the holes in the filter (or smaller). When vesicles are sized by polycarbonate filters, satisfactory fusion rates are obtained for vesicles that are 100–1000 nm in diameter (Fig. 1). Visible fusion spikes have also been reported for vesicles that are 15–60 nm in diameter.[16] However, some of these fusion spikes are small and very fast.

Figure 2 shows the frequency distribution of spike height due to vesicles extruded through a 100- or 1000-nm filter. The spike height represents the peak current flowing through the bilayer at the moment of fusion. Presumably, the current is due to preformed nystatin channels in the fusing vesicles. If nystatin is uniformly distributed between the membrane of all vesicles, then a vesicle with twice as much membrane should cause a

[16] C. E. Bear, C. H. Li, N. Kartner, R. J. Bridges, T. J. Jensen, M. Ramjeesingh, and J. R. Riordan, *Cell* **68,** 809 (1992).

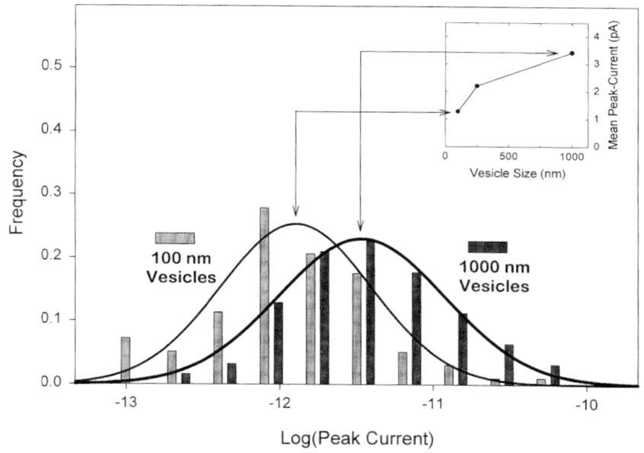


FIG. 2. Bar graph showing the distribution of spike heights for vesicles extruded through 100- or 1000-nm filters. The 100-nm vesicles (gray) tend to cause smaller current spikes than 1000-nm (black) vesicles. These graphs include the data shown in Fig. 1. The smooth lines are best fit Gaussian curves to the bar data and the $x$ axis is logarithmic in peak current. *Inset:* Plot of mean peak current versus vesicle size.

conductance spike twice as large. However, this prediction is not confirmed by the data. Figure 2 (inset) shows that although smaller vesicles tend to induce smaller conductance spikes, the spike height is not a strict function of vesicle surface area (i.e., the square of the radius). The reason for this is unknown, but if N/E vesicles are smaller than 1000 nm even before filtering, this could be a contributing factor.

## Planar Lipid Bilayers

In this section I do not attempt to describe different methods of forming planar lipid bilayers, but will focus on those points that relate to successful N/E fusion. However, note that many of the points covered here are relevant to other methods of vesicle membrane fusion.

### *Lipids*

Planar bilayers (also called black lipid membranes, or BLMs, because they appear black in reflected light) are formed from organic solutions of lipid. Typically, a decane solution of lipid is used since vesicles fuse more readily to decane-containing than to decane-free planar membranes.[17] In our laboratory, we have had good success with a 7:3 mixture of PE and

[17] F. S. Cohen, M. H. Akabas, J. Zimmerberg, and A. Finkelstein, *J. Cell Biol.* **98,** 1054 (1984).

PC at 20–30 mg/ml decane. Other mixtures of lipids have been used with equal success. It is not necessary to use the negatively charged lipid, PS, in the bilayer, or to add $Ca^{2+}$ in the solution. In fact, if PS is used in forming vesicles, as suggested above, it is better not to use $Ca^{2+}$ because this can cause the vesicles to clump or fuse together before they reach the bilayer. $Ca^{2+}$ can be added to the bilayer solution after vesicle–bilayer fusion.

*Preparing Cups with Holes for Forming Bilayers*

More reliable N/E fusion is obtained from bilayers formed on conical holes melted in polystyrene cups as described by Wonderlin *et al.*[18] Their procedure produces holes that keep the bilayer in close contact with the front (*cis*) solution, which is critical to decrease the diffusional barrier of the poorly stirred solution within the hole. Two simplifying modifications are made to their procedure. First, holes are melted in plastic cups with a tapered wire (0.047 inch piano wire) that is heated to 250° by mounting it in a temperature-controlled soldering iron. Second, just before use, holes are treated with ~1 μl of lipid in decane that was applied to the inside of the hole and wicked through the hole by touching the outside of the hole with a corner of a tissue.

N/E fusion works well with cups that have hole sizes between 100 and 250 μm in diameter, although fusion rates are higher with larger membranes. A current-to-voltage transducer circuit is used to monitor bilayer electrical properties. Bilayers formed in such cups typically have capacitances of 30–200 pF and conductances of <0.1 pS/pF.

For unknown reasons, some new cups (5–20%) do not form stable membranes or do not form membranes that allow N/E fusion. This may be due to unobservable irregularities in the hole. Therefore, it is wise to test each cup for successful fusion with N/E vesicles before using it in experiments with membrane proteins. A cup should be discarded if the first two or three separate attempts are unsuccessful (cups are washed between attempts).

Cups can be reused by cleaning them according to the following simplified procedure. After use, cups are soaked overnight in a 4% solution of liquid dishwashing soap. The cups are rinsed for at least 1 hr in warm running tap water and then left to soak in distilled water for several hours or overnight. This rinse and soak cycle is repeated a second time in clean glassware. In this manner, cups can be used multiple times (10–40×).

[18] W. F. Wonderlin, A. Finkel, and R. J. French, *Biophys. J.* **58,** 289 (1990).

However, each exposure to the lipid–decane solution forms small cracks in the plastic surrounding the hole. Eventually, these cracks form conductance pathways for ions and the cup must be discarded.

### *Painting Bilayer Membranes*

A membrane is formed by blowing a small air bubble with the tip of a pipette that has been wetted with the decane–lipid solution, and brushing the bubble over the hole. The pipette is wetted by dipping the tip into the solution and then blowing out (or wicking away) excess fluid in the tip. When a membrane forms, it is two lipid molecules thick and supported on all edges by a thicker torus or annulus. As shown in Fig. 3, the torus is the boundary between the membrane and the edge of the hole in the cup. It is mostly decane with a monolayer of lipid at each solution interface. The width of the torus can vary significantly, but it should not be too thick for fusion experiments (or stirring will sweep vesicles off the membrane and onto the torus). As a rule, the area of a bilayer should be at least 50% of the area of the hole. This means that for a 200-$\mu$m-diameter hole, the torus should be <30 $\mu$m wide. It is easy to determine if the bilayer is 50% of the area of the hole by measuring membrane capacitance, which is proportional to membrane area. The bilayer is large enough if its measured capacitance is at least 50% of the capacitance expected for a membrane the size of the hole. Because the hole size is fixed, it need only be measured once (with

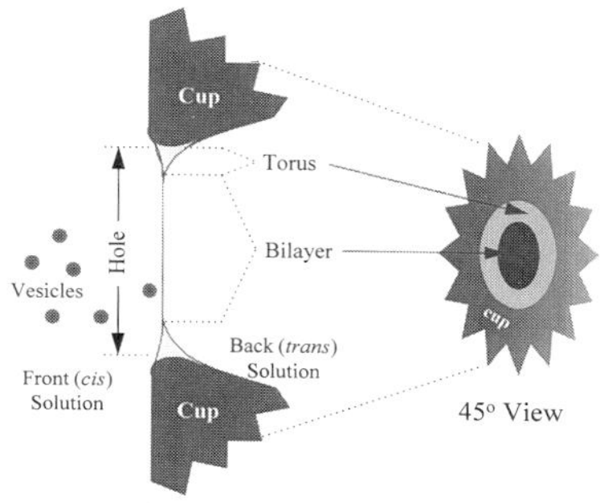


FIG. 3. Cut-out drawing of a plastic cup containing a tapered hole that is filled with a torus and a black lipid membrane. A side view (left) and an off-angle front view are shown. The area of the bilayer (black) and torus (gray) are indicated. The torus should not be too wide or it can interfere with effective delivery of vesicles to the bilayer. The *cis* solution is outside the cup and is the side to which vesicles are added.

a light microscope) and the corresponding capacitance calculated according to the following equation:

$$C_m = C_s A_m = 0.5\ \mu F/cm^2 \times \pi(d/2)^2 = 0.004\ pF/\mu m^2 \times d^2 \qquad (1)$$

where $C_m$ is the capacitance of a membrane, $C_s$ is the specific capacitance of a decane containing membrane, $A_m$ is the surface area of the membrane, and $d$ is the diameter of the hole. For example, a cup with a hole diameter of 250 $\mu$m has a maximum bilayer capacitance of 250 pF, but a cup with a hole diameter of 150 $\mu$m has a maximum bilayer capacitance of only 90 pF.

## Fusion Conditions: Osmotic Gradient and Stirring

Fusion of N/E vesicles is induced by establishing a transbilayer osmotic gradient.[2,11] After a bilayer is formed in symmetrical 150 m*M* (or lower) NaCl solution, an osmotic gradient of at least 400 m*M* is established across the bilayer by adding NaCl on the side to which vesicles will be added (*cis*). This is readily done by adding 100 $\mu$l of a stock solution of 3 *M* NaCl and then withdrawing 100 $\mu$l with the same pipette while stirring. This is repeated one or two more times to form a final concentration of 645–860 m*M* NaCl on the *cis* side. Other salts that are permeable through nystatin channel (e.g., KCl and potassium acetate) can also be used. Note that fusion is dependent strictly on the osmotic gradient, i.e., the difference in salt concentration, not on the ratio of salt concentrations (which sets the Nernst potential). The Nernst potential affects the size of the currents that flow through nystatin and other channels but not fusion directly. Also, we have not observed fusion spikes in solutions that have a pH less than 5.

After the final gradient is set, the baseline conductance and capacitance of the membrane are recorded and the membrane potential is set to $\pm$60 mV. This potential is necessary in order to observe the conductance changes due to fusion of N/E vesicles. A larger potential can be used if desired and either polarity can be used since N/E channels are almost equally permeable to $Na^+$ and $Cl^-$. N/E vesicles (2–25 $\mu$l) are then added on the *cis* side, but away from the membrane. During addition, or soon afterward, the *cis* chamber must be stirred.

As shown in Fig. 1, fusion of N/E vesicles to the planar bilayer is signaled by sudden increases in planar membrane conductance resulting from incorporation of vesicular nystatin channels. The increase in conductance is transient, decaying back to the baseline, because ergosterol, required for the integrity of the nystatin pore,[8,9,19] diffuses throughout the ergosterol-free planar bilayer after vesicle fusion. To obtain reproducible fusion rates,

[19] J. Bolard, *Biochim. Biophys. Acta* **864,** 257 (1986).

it is important to stir the bilayer chamber continually; conversely, fusion can be suspended simply by stopping the stirrer.

Stirring always causes bilayer current noise and therefore several steps must be taken to reduce this noise if fusion events are to be measured while stirring. Two steps to reduce stirring noise are to add a 4-Hz lowpass filter to the current output line and to make sure that the stir flee does not bump the walls while spinning. The most significant step in reducing stirring noise is to stir the *cis* chamber at a slow rotational speed. Speeds of 2–6 Hz adequately stir the chamber, delivering vesicles to the bilayer, while causing minimal electrical noise. There are two techniques for such slow-speed stirrers that are small enough to fit the space that is usually available under a bilayer chamber. One is a computer-controlled induction motor distributed by Biotech Products (Greenwood, IN). This device is sold together with a total bilayer control and data acquisition system and is rather expensive, but allows the experimenter to adjust the stirring speed between 3 and 20 Hz. A much cheaper but inflexible solution is to buy a small 115-rpm motor (Edmund Scientific Co., Barrington, NJ) and attach a small horseshoe magnet to its shaft. Either way, these steps reduce stirring noise to $<0.1$ pA, sufficient to easily observe N/E fusion spikes which are usually $>0.5$ pA at $-60$ mV.

### *Enhancing Vesicle Delivery to the Bilayer*

As described above, stirring is an essential step in driving vesicle fusion. This is most likely due to long diffusion times for large particles such as vesicles. Two tricks can dramatically increase delivery of vesicles to the membrane: ejecting near the membrane and brushing onto the membrane. Both delivery methods also have significant disadvantages.

Ejecting vesicles near the bilayer is useful when the amount of vesicles is limited or when massive amounts of fusion are desired. Pipettes for ejecting vesicles near the bilayer are formed by pulling 1.8-mm-o.d. glass pipettes to $\sim 7$ $\mu$m and using a microforge to smooth and shrink the tip to 4–6 $\mu$m. A right-angle bend is made 4 mm from the tip with a small Bunsen burner. The pipette tip is filled with vesicle solution (5 mg lipid/ml) and loaded into a holder mounted on a three-axis micromanipulator. For ejection, the tip of the pipette is moved within $\sim 40$ $\mu$m of the membrane. This delicate task is simplified by using the pipette as a light pipe and illuminating the tip by shining bright light directly at the other end. Typically, vesicles are ejected for 5–20 sec by applying pressure to a 10-ml syringe connected through tubing to the pipette.

A simpler alternative to the above method can also be used. Instead of a glass pipette, vesicles are added to the bilayer chamber by ejecting

1 μl of vesicle solution (1 mg lipid/ml) about 0.5 mm above the hole. Vesicles are ejected from a "10-μl ultramicro" plastic pipette tip inclined 20–30° away from the normal to the bilayer. The pipette is positioned by sliding it down, along the hard surface of the cup, until the tip is just above the hole. This method also delivers a large amount of vesicles almost directly onto the membrane.

However, with both these methods, especially the last one, there is the risk of liposome-induced conductance steps (LICS). LICS are membrane conductance changes during the first 5 min after vesicle ejection that closely mimic ion channel behavior. LICS are caused by the interactions of lipid structures with a bilayer, but are rare, only occurring when massive amounts of vesicles are ejected near the membrane. Experimenters tempted to use either ejection method are encouraged to review the properties of LICS and the conditions under which they occur.[20]

A second method to deliver vesicles to the membrane easily is to brush them on. After a stable membrane is formed by brushing the hole with a lipid-coated pipette tip, a clean pipette is dipped in the vesicle solution and then used to repaint the membrane by forming a bubble and brushing it across the bilayer. Although extremely easy, brushing has some significant disadvantages. Because the bilayer must reform, its native conductance and capacitance are not known (although it likely will be similar to the values measured before rebrushing). Also, it is not possible to know how many vesicles have fused with the membrane during the brushing, although this information is not always needed.

## Making Native Vesicles Fusigenic

So far, methods have been described to prepare fusigenic (N/E) vesicles, form a bilayer, and fuse N/E vesicles into the bilayer. In this section I describe how native membrane proteins can be prepared for fusion into a bilayer via N/E vesicles. Solubilized and purified proteins are easy to reconstitute and are discussed at the end of this section. Most native proteins are isolated in membrane fractions that contain other "contaminating" proteins. These preparations require a little more work to reconstitute. As a specific example of this latter type, I describe the procedure that my laboratory has used successfully for synaptic vesicles isolated from *Torpedo californica.* The same method works for other native vesicle preparation, such as those made from plasma membrane fragments by homogenization.

[20] D. J. Woodbury, *J. Membr. Biol.* **109,** 145 (1989).

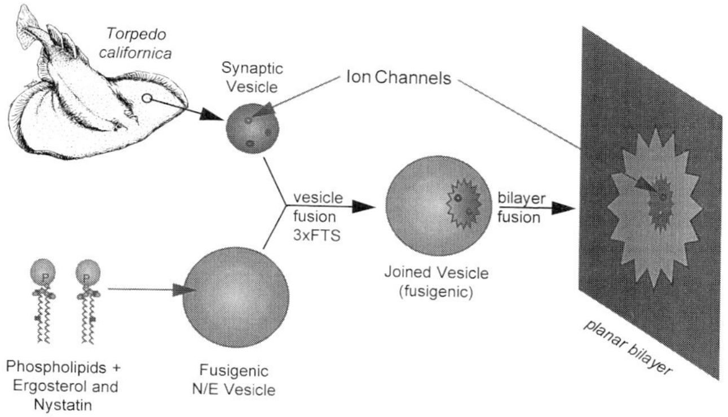


FIG. 4. Outline of the procedure for forming joined vesicles (JVs). JVs are made from fusigenic N/E vesicles and synaptic vesicles (SV) obtained from *Torpedo* electric organ. The two populations of vesicles are combined by three cycles of freezing, thawing, and sonicating (FTS).

Isolated synaptic vesicles (SVs) can be made fusigenic by first incorporating them into N/E vesicles to make a new composite vesicle termed the *joined vesicle,* or JV (Fig. 4). N/E vesicles act as carrier vesicles, combining with synaptic vesicle membrane while retaining their permeability properties that make them fusigenic. A description of the technique is given below; additional details are given elsewhere.[7,14a] The JVs are formed by mixing the two vesicle populations using the FTS reconstitution procedure developed by Kasahara and Hinkle.[21] N/E vesicles are made as above but without nystatin. Specifically, lipid (1.25 mg) is evaporated and sonicated for 1 min in 125 $\mu$l of 300 m$M$ NaCl solution. Synaptic vesicles containing ~60 $\mu$g of lipid (~15 $\mu$g protein) are added to the N/E vesicles. Nystatin is added to the mixture at a final concentration of 50–60 $\mu$g/ml and the mixture is frozen, thawed, and sonicated for 5–15 sec. The freeze–thaw and sonication steps are repeated two more times to disperse the integral membrane protein from the SV membrane into the excess of N/E vesicle lipid. As shown in Fig. 4, the end product is a JV that contains both nystatin channels and protein and membrane fragments from SVs. Note that most of the lipid in JVs comes from N/E vesicles, thus ensuring that their fusion properties are governed by N/E and not the SV. The large dilution also helps to decrease toward 1 the number of SV channels reconstituted into each JV, thus simplifying single-channel measurements.

This method will only work if the two vesicle populations fuse during the FTS step. Fusion will happen only if both populations of vesicles burst

[21] M. Kasahara and P. C. Hinkle, *J. Biol. Chem.* **252,** 7384 (1977).

open during FT. Although small vesicles (such as SVs) are less susceptible to rupture by FT, a large fraction (25–40%) of SVs do rupture following FT.[22] Therefore, as long as three FTS cycles are used, most (~58%) of the small synaptic vesicles will rupture. Since N/E vesicles are added in excess to SV, ruptured SV will most likely fuse with N/E vesicles.

For purified proteins the procedure can be simplified. If the native protein happens to be soluble (few membrane proteins are), then the protein can be added directly to the N/E vesicle preparation just before sonication. However, most purified membrane proteins are solubilized in detergent. For such proteins the following procedure can be used (compare to N/E vesicle preparation above): Dry 250 $\mu$l of stock lipids (with ergosterol but without nystatin) under a stream of nitrogen or argon. Dissolve the lipids in 250 $\mu$l of an aqueous solution such as 150 m*M* NaCl, 8 m*M* HEPES (pH 7.2) containing 2% sodium deoxycholate (we have not tried other detergents). Vortex ~5 min or until the sample becomes clear. Add detergent-solubilized protein to the lipid mix and remove the detergent. Detergent can be removed by dialysis against 1000× volume with three solution changes over 3 days or by using Bio-Beads (SM-2, Bio-Rad, Hercules, CA). Detergent removal with Bio-Beads is simpler and can be done in 1 day. The procedure is to add the detergent and protein-containing mixture to 0.1 g of 20–50 mesh Bio-Beads in a test tube and gently shake at 4° for 3–4 hr. A gel-loading pipette tip is used to remove the solution from the beads. (The beads are too large to enter the tip of the pipette.) This solution is added to 0.1 g of fresh beads and again shaken for 3–4 hr. The detergent-free solution is removed from the second batch of Bio-Beads again with a gel-loading pipette tip; typically, about 200 $\mu$l are recovered. At this point nystatin is added to the test tube so that the final nystatin concentration is 50–60 $\mu$g/ml. The mixture is vortexed for ~5 min and sonicated in a water bath sonicator for ~20 sec. The newly formed protein-containing vesicles are frozen in a dry ice–ethanol bath. Just before use, the sample is thawed and sized as discussed above.

One important parameter that is not given above is the amount of protein to use for a typical vesicle preparation. Usually, this information is given as a lipid-to-protein weight ratio and ratios between 10 and 10,000 to 1 have been used. But, the important ratio is the number of functional protein units to the number of vesicles, i.e., the average number of channels per vesicle. This is especially important when single-channel data are required. For single-channel data, it is desirable to have an average of

[22] D. J. Woodbury and M. Kelly, *Cryobiology* **31,** 279 (1994).

~0.3 active channels per vesicle. With this channel density, a few fusions will likely incorporate one, but not two, channels in the bilayer membrane.

How do you calculate the channel density? The three properties that must be known to calculate channel density are (1) the molecular weight of the functional channel unit, (2) the size of the vesicle, and (3) the molecular weight and surface area of a lipid molecule. Unfortunately, these values can seldom be known with great certainty. With some proteins, an apparent molecular weight may be known from gel electrophoresis. However, the number of subunits that form the functional unit and the number of denatured (nonfunctional) units may not be known. With respect to vesicle size, vesicles made by sonication often have a large diversity of sizes and may be multilamellar. If the vesicles are sized by passing them through a polycarbonate filter, as described above, then their size will be more uniform but still distributed about a mean. Also multilamellar vesicles may still be present if a large-diameter hole is used to size the vesicles (>200 nm). Because of the tighter curvature required, it is less likely that vesicles smaller than 200 nm will be multilamellar (contain another vesicle inside). The third property, molecular weight and surface area of a phospholipid molecule, can be approximated by 765 g/mol and 63 Å$^2$, respectively, for the lipid mix given above. The surface area of phospholipids varies from 48 to 96 depending on the exact type and mix of lipids.[23] Ergosterol is not included in these average numbers since it mostly fills up space between lipid tails.

The first step is to calculate the number of vesicles in a typical preparation. Assuming that 250 $\mu$l of stock lipids are used (12% of which are ergosterol), the number of phospholipid molecules is given as:

$$\frac{0.25\text{ ml} \times 88\%\text{ phospholipid} \times 10\text{ mg/ml} \times 6.02 \times 10^{23}\text{ molecule/mol}}{765\text{ g/mol}}$$
$$= 1.7 \times 10^{18}\text{ molecules of phospholipid} \quad (2)$$

Next, the number of phospholipid molecules per vesicle can be calculated for a vesicle with a diameter of 200 nm as:

$$\frac{2 \times \text{SA}_{\text{vesicle}}}{\text{SA}_{\text{lipid}}} = \frac{2 \times 4\pi(100\text{ nm})^2}{63\text{ A}^2} = 4.0 \times 10^5\text{ lipids/vesicle} \quad (3)$$

[23] R. P. Rand and V. A. Parsegian, *in* "The Structure of Biological Membranes" (P. Yeagle, ed.), pp. 264–265. CRC Press, Boca Raton, Florida, 1992.

Note that the vesicle surface area must be multiplied by 2 since the membrane is formed from a bilayer of lipid. Thus, a standard preparation of vesicles contains:

$$\frac{1.7 \times 10^{18} \text{ lipids}}{4.0 \times 10^{5} \text{ lipids/vesicle}} = 4.3 \times 10^{12} \text{ vesicles} \qquad (4)$$

or $1.7 \times 10^{10}$ vesicles per microliter. So, for a channel protein that has a molecular weight of 100,000 g/mol and is 85% active, the amount of protein that must be added to a batch of vesicles to get 0.3 functional channels per vesicle is given by:

$$\frac{0.3(4.3 \times 10^{12} \text{ molecules})}{0.85(6.02 \times 10^{23} \text{ molecules/mol})}(100{,}000 \text{ g/mol}) = 0.25\ \mu\text{g} \qquad (5)$$

Because 2.5 mg of lipid were used for this example, the lipid-to-protein weight ratio is 10,000 to 1. However, this is a minimum protein estimate. Typically 10–100× more protein is needed since most protein samples are not pure and/or totally functional.

## Contamination Problem: Calculating Natural Abundance of a Channel

One strength of using the planar lipid bilayer method for measuring the properties of ion channels is that it is extremely sensitive and can detect *single* channels. However, this is also a problem because a protein preparation is never 100% pure and contaminating proteins can also be easily reconstituted into the bilayer. For example, even if a protein preparation is 99.99% pure, there would still be ~3,000,000 vesicles with contaminating protein among the $10^{11}$ vesicles added to the bilayer chamber. This is especially a problem if the purified protein is inactive or does not have a measurable ion conductance. In this case, if the N/E fusion method is not used, then the vast majority of vesicle fusions will go undetected and only the rare fusion of a vesicle containing the contaminating active channel is observed. This activity could be mistaken for the purified inactive protein. This detection of "noise" is amplified by the fact that channel-containing vesicles are more fusigenic than vesicles without channels (or vesicles with closed or inactive channels). Thus, it is possible, to observe the fusion of one vesicle containing a contaminating ion channel while millions of vesicles with inactive channels or channels that do not make the vesicle fusigenic are unobserved.

TABLE I
PHYSICAL PROPERTIES OF VESICLES[a]

| Type of vesicle | Diameter[b] (nm) | Surface area ($\mu m^2$) | Total lipid used ($\mu g$) |
|---|---|---|---|
| Synaptic (SV) | 85 | 0.023 | 60 |
| N/E | ~250 | ~0.2 | 1250 |
| Joined (JV) | ~250 | ~0.2 | |

[a] See Fig. 4.
[b] Diameter is defined as the distance across the vesicle from the middle of one membrane to the middle of the other. Diameters of N/E vesicles (sized by sonication) and JV were estimated from an S-1000 sizing column.

One way around the problem of fusing contaminating proteins into bilayers is to find a way to measure the fraction of vesicles that contain each type of channel. Traditionally, this was impossible with the bilayer method since not all vesicles are fusigenic and not all fusions are detectable. However, this is not the case with N/E vesicles. N/E fusion makes it possible to detect every fusion and to count both the number of vesicles that contain channels and those that do not.

My laboratory has published the identification of several different types of ion conductances which are due to proteins in the SV membrane. Because we used N/E fusion, it was possible to estimate the number of copies of each type of channel in an individual SV. Interestingly, it turns out that there are, on average, one to two copies of each channel per synaptic vesicle. As an example of the power of this technique, the data are presented here, with the simplification that all channel types are presented as if they are multiple copies of just one channel type. The total number of channels per synaptic vesicle can be calculated from the average number of channels reconstituted into the bilayer per fusion event. From this average and a calculation of the density of SVs per joined vesicle, JV, the number of channels per SV can be determined. The information necessary to calculate the density of SVs per JV is summarized in Table I. From these numbers, the predicted surface density of one channel in a single SV is 1 per 0.023 $\mu m^2$. As illustrated in Fig. 4 and quantified in Table I, SVs are joined with N/E vesicles to form JVs. The three FTS cycles fuse ~58% of the SVs with N/E vesicles, the remaining 42% apparently remain intact.[22] Therefore, 35 $\mu g$ of the added 60 $\mu g$ of SV lipid combines with the 1250 $\mu g$ of lipid from the N/E vesicles. Assuming that there is just one copy of this channel per SV, its surface density in a JV can be calculated as:

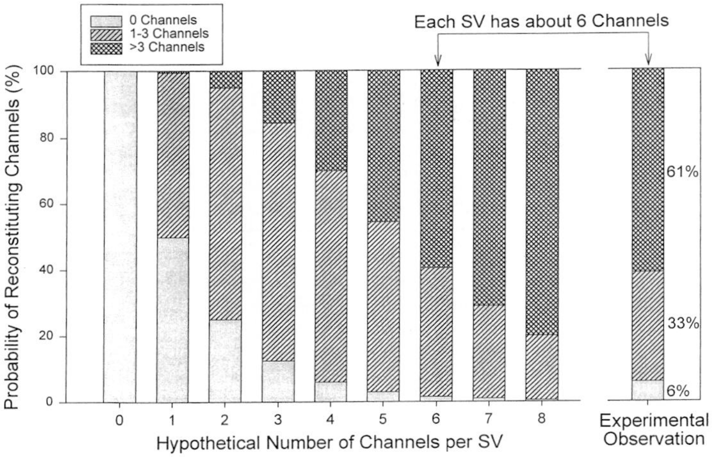


FIG. 5. Hypothetical (left) and experimental (right) probability of reconstituting channels into a bilayer membrane. Note that in 33% of the experiments, 1–3 channels were reconstituted. This most closely matches the prediction of 39% obtained by assuming that a single SV contains six ion channels in its membrane.

$$\frac{35\ \mu\text{g}}{1250\ \mu\text{g} + 35\ \mu\text{g}} \times \frac{0.2\ \mu\text{m}^2/\text{JV}}{0.023\ \mu\text{m}^2/\text{SV}} \times \frac{1\ \text{channel}}{\text{SV}} = 0.23\ \text{channels/JV} \quad (6)$$

If the channel distributes according to a Poisson distribution, this means that 19% of the JVs have one copy of this channel (and 2% have more than one copy). After fusion of three vesicles (the average number of fusions in these experiments) there is a 50–50 chance that there is at least one channel in the membrane.

Similar calculations allow us to compare the experimental frequency of observed channels with predicted values. After about three fusions we observed that 6% of the bilayer membranes contained no channels and 33% contained one to three channels (right column, Fig. 5). The left part of Fig. 5 shows the predicted probabilities of observing channels in the bilayer, for values of one to eight channels per synaptic vesicle. These probabilities are based on the Poisson distribution which assumes that all channels are distributed evenly. Notice that there is a very close match between our data and the assumption that there are six channels in the membrane of each SV. The assumption that there are less than five or more than seven channels does not fit the data.

Taking six as the total number of channels per vesicle, and the fact that the four different channel types were observed at an approximate ratio of 2 : 2 : 1 : 1, we conclude that each synaptic vesicle contains one to two copies of each channel type. This result demonstrates a novel approach to the determination of channel density in isolated vesicles.[14a]

## Determination of Biophysical Properties of Channels Fused into Bilayers

Once an ion channel is successfully reconstituted into the lipid bilayer, many biophysical properties can be determined, in addition to channel abundance as described above. By measuring the membrane current as a function of applied voltage, the conductance of the channel can be found. Also, because fusion conditions include an asymmetric salt gradient, the reversal potential in the fusion solutions indicates the selectivity of the channel for the salts used in the gradient. For example, if a 150/750 m*M* NaCl gradient is established, then a reversal potential of 41 mV on the 150 m*M* side is expected for a channel that passes $Na^+$ but excludes $Cl^-$. Additionally, channel properties as a function of voltage and pharmacologic agents can also be easily determined. These topics are covered elsewhere.[24–27]

## Limitations of N/E Vesicle Fusion

Having extolled the virtues of N/E fusion with the planar lipid bilayer system, it is prudent to note that there are limitations and difficulties with this system. First, it may not be easy to find the appropriate amount to dilute the protein with lipid such that single-channel data can reliably be obtained. There are several reasons for this. If the protein is too dilute, many fusions are required before the protein is observed. If the protein is too concentrated, then too many copies will fuse with the first vesicle. Additionally, it is not always easy to get uniform mixing such that the protein is randomly distributed. For example, if two channels are linked in their native membrane, then they may not randomly distribute when mixed with N/E vesicles (by FTS). This will result in a larger than expected number of vesicles with either no channels or two channels, and very few vesicles with just one channel. FTS cycles may correct this problem, but some proteins may be damaged by more rigorous cycling.

[24] C. Miller, "Ion Channel Reconstitution," pp. 573. Plenum Press, New York, 1986.

[25] B. Hille, "Ionic Channels of Excitable Membranes." Sinauer Associates, Inc., Sunderland, Massachusetts, 1992.

[26] P. Labarca and R. Latorre, *Methods Enzymol.* **207,** 447 (1992).

[27] W. Hanke and W.-R. Schlue, "Planar Lipid Bilayers: Methods and Applications." Academic Press, San Diego, California, 1993.

A second potential problem is that N/E fusion requires adding the nystatin channel to vesicles, and therefore to the bilayer. The procedures presented above are designed to minimize the possibility that nystatin channels will remain active in the membrane more than a minute after fusion. However, this is not always the case. After a large number of fusions, enough ergosterol could be transferred into the bilayer to allow continued formation of nystatin channels. Also, at higher concentrations of nystatin, it can form channels without ergosterol. Generally, a long-lasting nystatin conductance is only seen after 50–100 fusion events and with bigger vesicles. In such cases, the experiment should be aborted.

A third problem is that bigger membranes increase the probability of fusion but also increase membrane noise and decrease time resolution. Thus, there is a balance between acceptable noise levels and acceptable fusion rates. When a small-conductance channel is studied, a smaller membrane (<50 pF) is usually desired to enhance the signal-to-noise ratio. However, this will often require adding a larger number of vesicles to the bilayer in order to get a decent fusion rate. Conversely, if a protein channel is scarce or expensive, it is best to use a large membrane (>150 pF) so that less material will be required.

Additional problems include the use of a solvent (e.g., decane) to spread the bilayer. Although N/E fusion may work with solventless membranes, it has not been tested.

## Summary

The nystatin–ergosterol (N/E) method is described and reviewed. Using this procedure, an experimenter can promote and detect fusion of vesicles with planar lipid bilayers. N/E fusion provides a straightforward mechanism to reconstitute any membrane protein into planar lipid bilayers. Once reconstituted, it is easy to determine the ion selectivity, transport rate, voltage dependence, and kinetics of any conductance caused by the membrane protein. Fusigenic N/E vesicles are made with a mixture of phospholipids, ergosterol, and nystatin. Vesicle size can be adjusted either with sonication or with polycarbonate filters. The best vesicles contain ~20 mol% ergosterol, are ~200 nm in diameter, and are in a solution containing ~50 $\mu$g/ml nystatin. Vesicle fusion requires an osmotic gradient and delivery of vesicles to the bilayer. Vesicle delivery is increased by (1) stirring of the chamber that contains vesicles, (2) larger bilayers, and (3) bilayers that are face-flush with the vesicle-containing solution. Because constant stirring is critical for delivery of vesicles to the bilayer, a system that allows simultaneous stirring and sensitive electrical measurements is desirable.

The main strength of the bilayer technique has always been that the experimenter has control over the milieu of the membrane system. The N/E fusion technique adds to this strength by controlling fusion of vesicles to the bilayer, thus allowing the quantitative transfer of isolated proteins from vesicle to bilayer. The techniques and calculations necessary for successful quantitative reconstitution are given in detail.

Acknowledgments

The author would like to thank Kathie Rognlien and Dr. Marie L. Kelly for help in preparing this work and Dr. Per Stampe at the University of Rochester, NY, for his "mistake" in discovering the brush technique with N/E vesicles. Also, the author wishes to thank Stephen N. Alix and Dr. Carl S. Helrich, Jr., at Goshen College, for their critical reading of the manuscript. This work was supported by NIH grant R29 MH50003.

# [18] Isolation of Transport Vesicles that Deliver Ion Channels to the Cell Surface

*By* SAROJ SATTSANGI and WILLIAM F. WONDERLIN

## Introduction

Ion channels, like other integral membrane proteins, must be transported among intracellular organelles and the plasma membrane within the membranes of transport vesicles. The synthesis and cell-surface expression of new channel proteins probably requires the sequential movement of channel proteins, via transport vesicles, among several compartments, including the endoplasmic reticulum, Golgi apparatus, *trans*-Golgi network (TGN), and the plasma membrane. Following expression in the cell surface, ion channel proteins can be endocytosed in endocytic vesicles and then, perhaps, returned to the cell surface from recycling endosomes. By isolating subpopulations of vesicles that transport channel proteins between specific compartments, we can investigate vesicle-related factors that might regulate the trafficking of ion channels, such as the selective packaging of specific sets of ion channels or the cotransport of channels with proteins that provide targeting information. Furthermore, by characterizing the physiologic properties of channels isolated from transport vesicles at intermediate stages in their synthesis, we might better understand the relationship between the steps in the biosynthesis of ion channels and the acquisition of their mature physiologic properties. We have used two systems to study the properties of voltage-gated ion channels in transport vesicles: transport vesicles iso-



0076-6879/99 $30.00

lated from the squid giant axon and transport vesicles isolated from N1E-115 neuroblastoma cells grown in culture.

## Transport Vesicles Isolated from Squid Giant Axon

Axoplasm extruded from the giant axon of the squid provides a rich source of organelles, among which are transport vesicles that deliver proteins to the axolemma. In the method described next, axoplasm is collected by roller extrusion, and then it is treated with potassium carbonate to disrupt the cytoskeleton and strip proteins from the surface of vesicles. The salt stripping is absolutely necessary for the fusion of vesicles with planar bilayers, and potassium carbonate is much more effective than potassium iodide, which we used in an earlier study.[1] This method provides a crude mixture of organelles from which several types of ion channels, including delayed-rectifier $K^+$ channels, can be efficiently incorporated into planar bilayers. The most likely source of these channels is the pool of transport vesicles that delivers channel proteins to the cell surface, but further investigation will be necessary to confirm this point. This method has been successfully used with axoplasm from *Loligo pealei*[1] and *Loligo opalescens*.[2]

### *Solutions*

Buffer A: 500 m*M* potassium acetate, 10 m*M* HEPES, 100 $\mu M$ ethylenediaminetetraacetic acid (EDTA), pH 7.0

Buffer B: 1 *M* sucrose, 10 m*M* HEPES (free acid), 1 m*M* EDTA, 1 m*M* ethylene glycol-bis($\beta$-aminoethyl ether)-N,N,N′,N′-tetraacetic acid (EGTA), 2 m*M* $MgCl_2$, adjust pH to 7.0 with KOH

### *Method*

1. Remove 5- to 7-cm lengths of giant axon from the squid mantle and store in $Ca^{2+}$-free artificial seawater on ice. One or two axons easily provide enough axoplasm for a day of recording; up to 10–12 axons can be prepared at a single time.

2. Add 100 m*M* potassium carbonate and 1 m*M* phenylmethylsulfonyl fluoride (PMSF) to 10 ml buffer B. This should increase the pH to slightly greater than pH 11.0. This is the chaotropic buffer. Add 2 mg/ml L-ascorbate to 10 ml of buffer B. This is the neutralizing buffer. Acidify the neutralizing buffer by adding about 60 $\mu$l stock glacial acetic acid, so that the solution

[1] W. F. Wonderlin and R. J. French, *Proc. Natl. Acad. Sci. U.S.A.* **88,** 4391 (1991).

[2] W. F. Wonderlin and R. J. French, unpublished observations (1993).

produced by mixing equal volumes of the chaotropic buffer and the neutralizing buffer has a pH of 7.0.

3. Immediately before extrusion, excise 5 mm from the large (proximal) end of the axon and lay the axon onto a Whatman (Clifton, NJ) filter paper with the large end extending 3–5 mm beyond the filter and onto a piece of Parafilm. Be sure that the extruded axoplasm does not come into contact with the filter paper or it will be wicked away. Blot away excess buffer from the cut end, place 20 $\mu$l of buffer A over the cut end, and slowly extrude the axoplasm into buffer A using a roller (a glass pipette covered with plastic shrink tubing works nicely). When the extrusion is complete, gently pull away the axon while using fine-tipped forceps to stabilize the puddle of axoplasm on the Parafilm.

4. After 1 min, carefully remove buffer A and replace it with the chaotropic buffer, 5 $\mu$l/cm axon length. Triturate gently 100 times, avoiding the formation of air bubbles. The axoplasm should liquefy quickly.

5. Dilute the dissociated axoplasm with an equal volume of the neutralizing buffer (i.e., 5 $\mu$l/cm).

6. Transfer the diluted axoplasm to a 1.5-ml Eppendorf tube and centrifuge 10 min at 10,000*g* at 4° to separate large contaminating material. Remove and save the supernatant (you probably will not see a pellet).

7. Use the dissociated axoplasm fresh for incorporation into bilayers or store aliquots under argon or nitrogen at −80°. We have incorporated channels from axoplasm stored with ascorbate and under an inert gas for at least 1 year.

8. We recommend incorporating channels into horizontal planar bilayers, because this method provides the most efficient utilization of small aliqots of axoplasm. A 1-$\mu$l aliquot of the disrupted axoplasm is deposited over a bilayer formed within a small (45- to 70-$\mu$m) aperture in a horizontal partition bathed in symmetric 500 m*M* potassium acetate, allowing it to sink onto the surface of the bilayer. After 1–2 min, a small aliquot (10 $\mu$l) of hypotonic solution (10 m*M* HEPES, pH 7.0) is perfused below the bilayer, which transiently establishes an osmotic gradient. Channel incorporation should be rapid, frequently first indicated by the appearance of a large conductance channel with very brief, spikelike openings.[1]

9. Delayed-rectifier and "spike" $K^+$ channels are the most frequently observed channels, and we have also identified voltage-gated $Na^+$ channels modified by batrachotoxin.[1] We have occasionally observed a large-conductance channel with multiple conductance levels; we believe this channel is probably not from the same population of transport vesicles from which the $K^+$ and $Na^+$ channels are incorporated, based on the absence of any evidence that the large-conductance channel is expressed in the plasma membrane.

## Transport Vesicles Isolated from N1E-115 Neuroblastoma Cells

N1E-115 neuroblastoma cells differentiated by dimethyl sulfoxide (DMSO) display morphologic, neurochemical, and physiologic properties characteristic of neurons.[3] We have developed a method for isolating transport vesicles that deliver proteins to the cell surface of differentiated N1E-115 cells. An important consideration in designing this protocol was that we wanted to isolate selectively the repertoire of carrier vesicles that are actually en route to the cell surface. This precluded homogenization/fractionation techniques that cannot distinguish between vesicles vectorially transported to the cell surface versus stationary organelles, and which might artifactually vesicularize small tubular structures, such as early endosomes.[4] Instead, we modified previously developed techniques for collecting carrier vesicles released in an adenosine triphosphate (ATP)-dependent manner from mechanically perforated cells.[5–7] Modification of these techniques was necessary because highly differentiated neurons, with their long filamentous processes and irregular morphology, are not very amenable to mechanical perforation, especially if the transport of vesicles to the surface of the entire neuron is to be studied. As an alternative, we chemically perforated the plasma membrane with saponin, a detergent that binds cholesterol and forms pores in cholesterol-rich membranes, such as the plasma membrane.[8] These pores are 100 to 1000 nm in diameter, which should be much larger than TGN-derived, cell-surface recycling and synaptic vesicles.[9–13]

By using the method described below, we can effectively separate, by velocity sedimentation on Ficoll gradients, two populations of vesicles that are released from perforated cells, including large vesicles that preferentially transport newly synthesized proteins and small vesicles that preferentially transport recycled proteins to the cell surface. Separation by velocity sedimentation on Ficoll gradients offers an important advantage compared to velocity sedimentation on traditional glycerol or sucrose gradients be-

[3] Y. Kimhi, I. Palfrey, I. Spector, Y. Barak, and U. Z. Littauer, *Proc. Natl. Acad. Sci. U.S.A.* **73,** 462 (1976).

[4] J. Tooze and M. Hollinshead, *J. Cell Biol.* **115,** 635 (1991).

[5] M. K. Bennett, A. Wandinger-Ness, and K. Simons, *EMBO J.* **7,** 4075 (1988).

[6] L. Clift-O'Grady, A. D. Linstedt, A. W. Lowe, E. Grote, and R. B. Kelly, *J. Cell Biol.* **110,** 1693 (1990).

[7] M. Grimes and R. B. Kelly, *J. Cell Biol.* **117,** 539 (1992).

[8] A. Lepers, R. Cacan, and A. Verbert, *Biochimie* **72,** 1 (1990).

[9] P. J. Morin, N. Liu, R. J. Johnson, S. E. Leeman, and R. E. Fine, *J. Neurochem.* **56,** 415 (1991).

[10] I. de Curtis and K. Simons, *Cell* **58,** 719 (1989).

[11] M. K. Bennett and R. H. Scheller, *Annu. Rev. Biochem.* **63,** 63 (1994).

[12] W. Stoorvogel, V. Oorschot, and H. J. Geuze, *J. Cell Biol.* **132,** 21 (1996).

[13] T. C. Südhof, *Nature* **375,** 645 (1995).

cause the higher viscosity of a Ficoll gradient permits the loading of substantially larger volumes of dilute samples.[14] Vesicles released from perforated cells are present in the incubation medium at a very low concentration. By loading larger sample volumes onto Ficoll gradients, it is not necessary to concentrate these vesicles by pelleting, which can introduce serious artifacts, prior to separation by velocity sedimentation. Although isopycnic centrifugation is useful for determining the density of the transport vesicles, it is of little use in separating the two populations of vesicles described below because their densities in isoosmotic solutions are very similar. Note also that the 2- to 3-fold difference in size among the largest and smallest transport vesicles is large enough to permit unambiguous separation by velocity sedimentation, but it is not large enough to permit the same degree of separation by differential centrifugation.

*Solutions*

GGA buffer: 38 m*M* potassium gluconate, 38 m*M* potassium glutamate, 38 m*M* potassium aspartate, 2 m*M* EGTA, 2.5 m*M* $MgCl_2$, 1 m*M* dithiothreitol (DTT), 25 m*M* HEPES, pH 7.2. GGA buffer is the same pseudo-intracellular buffer previously used by Bennet *et al.*[5] to release vesicles from mechanically perforated MDCK cells.

*Method*

1. We grow N1E-115 neuroblastoma cells (passages 18–55) in Dulbecco's modified Eagle's medium [DMEM, 37° and 5% (v/v) $CO_2$] supplemented with 5% (v/v) fetal bovine serum (FBS), penicillin (100 units/ml), streptomycin (100 $\mu$g/ml), Fungizone (2.5 mg/ml), and 20 m*M* HEPES (added from a pH 7.6 stock). Cells to be used for vesicle collection are grown to near confluence in 25-cm$^2$ flasks (60-mm dishes also work well). The cells are seeded at $2 \times 10^5$ cells per 25-cm$^2$ flask, and they are differentiated by changing the medium to 1% FBS/2% DMSO or adding nerve growth factor (NGF) (50 ng/ml) on the third day after seeding. The cells are perforated after 7–9 days in the differentiating medium, at which time they have extended long processes.

2. Immediately before perforation, wash the cells twice with 5 ml cold GGA buffer to remove the growth medium. Incubate the cells for 20 min at 37° in 2 ml GGA buffer containing saponin (75 $\mu$g/ml) and either an ATP-regenerating system (2 m*M* ATP, 8 m*M* creatine phosphate, 10 units/ml creatine kinase) or an ATP-depleting system (apyrase, 30 units/ml). A cocktail of protease inhibitors, including 100 $\mu M$ PMSF and 0.1 $\mu$g/ml each of aprotinin, leupeptin, pepstatin A, and chymostatin, can also be added

[14] W. F. Wonderlin, *Analytical Biochem.* **258,** 74 (1998).

to the incubation buffer without affecting the release of vesicles. Saponin (75 μg/ml) should irreversibly perforate 100% of the cells, measured by loss of trypan blue exclusion, compared to a 1–2% staining of control cells. When we examined the perforated cells by scanning electron microscopy, a large number of variably shaped pores (50–500 nm in diameter) were observed in the plasma membrane. Other than the marked increase in the permeability of the plasma membrane, saponin-perforated cells appeared to be structurally intact when viewed by light microscopy (Hoffman Modulation Optics, Greenvale, NY, 400×). The cytosol was clarified, organelles were visible within the cells, long processes remained intact, and the cells remained attached to the culture dishes.

3. Gently remove the incubation medium; the cells are very fragile after perforation. Centrifuge the incubation medium for 20 min at 7000*g* at 4° to clear any cells or debris that have loosened from the flask during the perforation/incubation period.

4. Load 2 ml of the 7000*g* supernatant onto a 9-ml 8–14% linear Ficoll gradient (w/v, prepared in GGA buffer) formed over a 1 ml 40% sucrose pad in an SW40 tube. Centrifuge at 40,000 rpm for 4 hr at 4° in an SW40 rotor. The 8–14% Ficoll gradient is optimal for separation of the vesicles and should be formed very carefully. We used a Beckman (Fullerton, CA) syringe-type density gradient former with very reproducible results.

5. It is very useful to determine the distribution of membranes within each gradient for comparison with the distribution of specific proteins. The concentration of vesicle membrane and associated protein is too low to permit easy measurement by UV absorbance or protein assay. We have found that TMA-DPH, a cationic, amphiphilic dye, (1-[4-(trimethylammonio)phenyl]-6-phenylhexa-1,3,5-triene *p*-toluene sulfonate, from Molecular Probes, Eugene, OR) provides an extremely sensitive and nondestructive probe for monitoring the distribution of vesicles in the gradient while the gradient is being pumped from the tube. TMA-DPH is virtually nonfluorescent in aqueous medium, but fluoresces intensely when it enters a hydrophobic environment.[15] The predominant hydrophobic environment in the gradients is the lipid core of vesicle membranes that have sedimented into the gradient, and the TMA-DPH fluorescence at a particular location in the gradient should be proportional to the amount (i.e., area) of membrane at that location. Prepare TMA-DPH as a 500 μ*M* stock in dimethylformamide, store at −20°, and add it at 500 n*M* to all gradient components (pad, gradient and sample) prior to centrifugation. After centrifugation, puncture the bottom of the tube and pump the gradient solution through a small flow cell in a fluorimeter (we use a 40-μl glass capillary flow cell

[15] C. Bronner, Y. Landry, P. Fonteneau, and J. Kuhry, *Biochemistry* **25,** 2149 (1986).

in a Turner Model 450 filter fluorimeter, VWR, South Plainfield, NJ) while collecting fractions. Excite at 360 nm and measure emission at 430 nm.

The typical fluorescence profile for vesicles separated by velocity sedimentation on a Ficoll gradient is shown in Fig. 1A. Three peaks are always evident when the cells are perforated in the presence of ATP. The first (left) peak includes rapidly sedimenting (RS) vesicles that accumulate at the interface between the bottom of the Ficoll gradient and the 40% sucrose pad. This peak sometimes consists of two components, a denser component

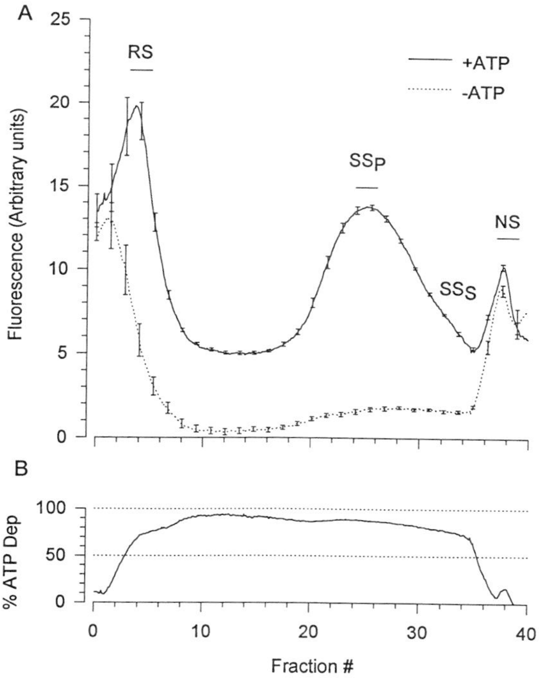


FIG. 1. Separation of populations of vesicles by velocity sedimentation on Ficoll gradients. (A) The average fluorescence profiles (±SEM, $n = 3$) are shown following velocity sedimentation of vesicles released from cells perforated and incubated either in the presence of an ATP-regenerating system (+ATP, 3 gradients) or an ATP-depleting system (−ATP, apyrase, 3 gradients). The fluorescence profile is oriented with the top of the gradients at the right. For each gradient, the fluorescence profile was divided by the cellular protein remaining in the flask after removal of the incubation medium. This provided an approximate correction of each fluorescence profile for the amount of cells in the flask, and this reduced the variability of the averaged profiles. (B) The percentage ATP dependence of the release of vesicles was calculated by dividing the difference between the two fluorescence profiles shown in Part (A) by the amplitude of the +ATP profile.

penetrating the 40% sucrose pad and a lighter component beginning at the sucrose pad–Ficoll interface and extending into the Ficoll gradient. The RS vesicles are always completely separated from a second, slowly sedimenting (SS) population of vesicles that appears as a broad peak beginning near the middle of the gradient and extending to near the top of the gradient. There is often a slight inflection or shoulder approximately midway along the trailing edge of the SS peak, demarcating a subpopulation of smaller-sized vesicles from the rest of the SS population. We designate this region of smaller SS vesicles the $SS_S$ region, and the vesicles near the SS peak are designated $SS_P$. A third peak of fluorescence is located at the top of the gradient (right peak in Fig. 1A), and this peak consists of nonsedimenting (NS) material accumulating at the interface between the sample region and the top of the gradient. The NS peak contains detergent micelles, because a similar peak was observed in control gradients in which GGA buffer containing only saponin was centrifuged (data not shown). We also observed that nonspecific biotinylation of surface proteins at 0° resulted in preferential labeling of proteins collected in the NS fractions (data not shown), indicating that the NS peak might contain fragments of plasma membrane or proteins solubilized from the plasma membrane.

The release of RS and SS vesicles is ATP dependent, as evident in the 75–90% reduction in the lighter half of the RS vesicle peak and the entire SS vesicle peak when the cells were perforated while incubated in an ATP-depleting buffer (GGA, apyrase 30 units/ml, no added ATP) (Fig. 1). A similar reduction in the RS and SS vesicle peaks can be produced by depleting ATP with a hexokinase/glucose system (data not shown). The ATP-dependent release of RS and SS vesicles demonstrates that the vesicles in these peaks are not formed simply by solubilization of membranes by saponin. In contrast, the accumulation of nonsedimenting material at the top of the gradient is not dependent on the presence of ATP. Note that the NS peak observed in the absence of ATP overlaps very little with the SS vesicles observed in the presence of ATP, and, therefore, the SS region can be sampled over much of its range without contamination by the nonsedimenting material at the top of the gradient. Although the breadth of the RS and SS peaks might be expected to result, at least in part, from vesicles beginning their sedimentation at different starting positions within the large sample region, the shape of the fluorescence profile is identical when 1, 2, or 3 ml of sample is loaded onto the gradient (data not shown). Therefore, the shape of the RS and SS peaks most likely reflects the variability of the intrinsic physical properties of the vesicles rather than variability in their starting positions. This low sensitivity of the effectiveness of separation to the volume of sample loaded onto the gradient is a very useful characteristic of the high-viscosity Ficoll gradients. Finally, the fluorescence

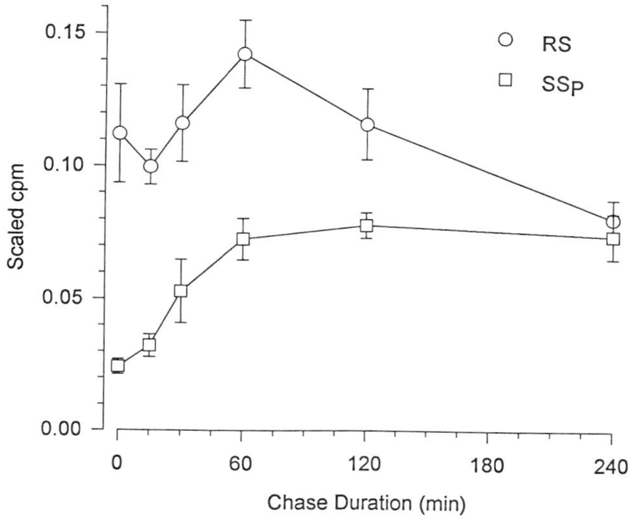


FIG. 2. [$^{35}$S]Methionine pulse-chase labeling. Each flask was washed twice in 5 ml of methionine-free DMEM and then pulsed with 2 ml of methionine-free DMEM containing [$^{35}$S]methionine (160 $\mu$C/ml) for 10 min at 37°. The cells were washed three times in cold chase medium (DMEM supplemented with 2 m*M* cold methionine) and chased at 37° for 0, 15, 30, 60, 120, or 240 min. The average ($\pm$SEM) scaled cpms for four pulse-chase experiments are shown for RS and $SS_P$ fractions. For each fraction, the cpm was divided by the fluorescence to express the radioactivity relative to the amount of membrane present in the fraction. These scaled values were then normalized for each experiment (i.e., each set of time points) to a sum of one; this greatly reduced the variability when the four experiments were averaged. At the end of the 10-min labeling period, a small peak of radiolabeled proteins was immediately apparent in the RS vesicles. This was followed by a larger peak at the 60-min chase and then a decline of radiolabeled proteins in the RS vesicles. In contrast, the $SS_P$ vesicles exhibited a delayed and persistent incorporation of radiolabeled proteins.

signal can be completely eliminated by not including TMA-DPH in the gradient (data not shown), demonstrating that light scattering by vesicles did not contribute to the signal, and this supports our use of TMA-DPH fluorescence for quantifying the relative amount of membranes at different positions in the gradient.

The three peaks separated by velocity sedimentation provide an opportunity to examine the distribution of ion channel proteins among three cellular compartments. The NS peak contains proteins solubilized from the plasma membrane, and it can be probed to determine the cell surface expression of proteins. The RS and SS vesicle populations appear to transport preferentially newly synthesized and recycled proteins, respectively. This conclusion is strongly supported by pulse-chase experiments in which

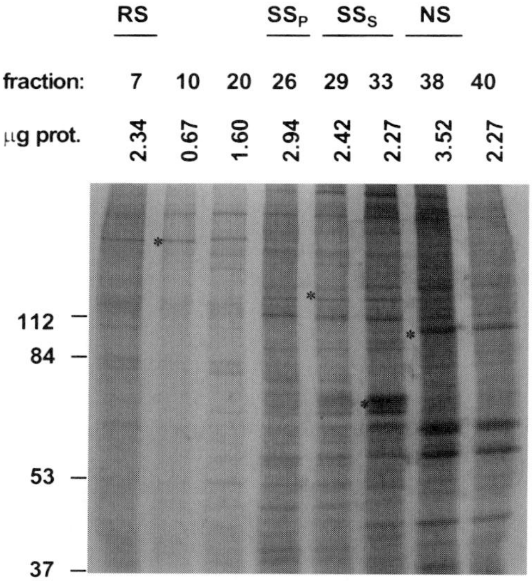


FIG. 3. Accumulation of radiolabeled proteins in carrier vesicles. The cells were incubated in DMEM containing [$^{35}$S]methionine (25 $\mu$C/ml) for 18 hr at 37° prior to perforation and isolation of vesicles by velocity sedimentation. The vesicles were pelleted from the indicated fractions and separated by SDS–PAGE. The amount of protein loaded in each lane is indicated above the lane. The positions of molecular weight standards are indicated at the left of the figure. Although comparable amounts of protein where loaded from most of the fractions, very few labeled proteins were evident in the RS vesicles, but many labeled proteins accumulated in the $SS_S$ vesicles and, to a lesser extent, the $SS_P$ vesicles. The localization of some proteins (bands marked by asterisks) within specific regions of the velocity gradient indicated that there was little cross contamination among different regions of the velocity gradient.

we measured the time course of release of radiolabeled proteins following pulse labeling with [$^{35}$S]methionine (Fig. 2). Proteins transported by RS vesicles were rapidly and transiently labeled, whereas the labeling of proteins carried by SS vesicles was delayed and persistent. SS vesicles, but not RS vesicles, also accumulated radiolabeled proteins over an 18-hr labeling period (Fig. 3), which supports our conclusion that SS vesicles carry recycled proteins. Analysis by electron microscopy and modeling of sedimentation (data not shown) has indicated that the size of RS vesicles (80–90 nm in diameter) is similar to the size of TGN-derived vesicles,[10] and the size of $SS_P$ vesicles (50–65 nm in diameter) is similar to the size of putative general cell surface recycling vesicles.[12] Further study will be required to determine with greater precision the source(s) of RS and SS vesicles. Although our present investigation supports our current conclusion that the RS vesicles preferentially transport newly synthesized proteins and the SS vesicles

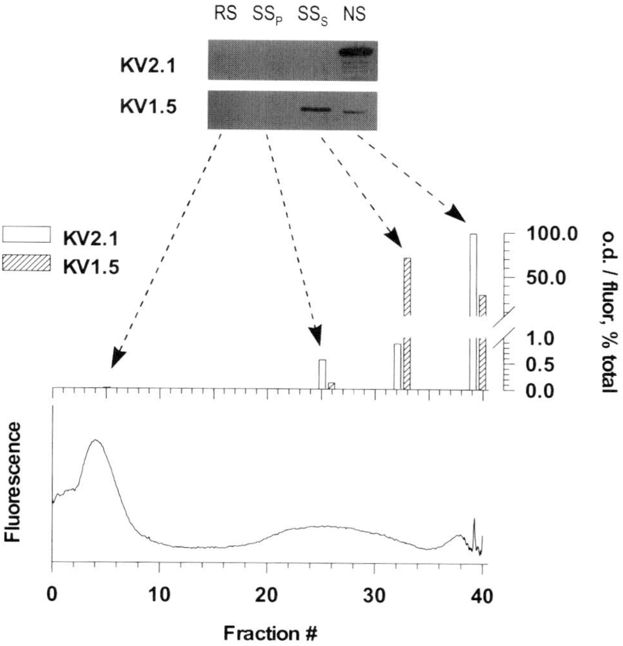


FIG. 4. Distribution of Kv1.5 and Kv2.1 channels among gradient fractions collected from a Ficoll gradient whose fluorescence profile is shown in the lower panel. Channel proteins from four gradient fractions were separated on 8.5% sodium dodecyl sulfate (SDS)–polyacrylamide gels, blotted and blocked. The blots were probed for 90 min with primary antibodies, and the immunoreactivity was detected using an HRPO (horseradish peroxidase) conjugated secondary antibody and the Amersham ECL (enhanced chemiluminescence) detection system. Antibodies were generously provided by Dr. Jim Trimmer, and they included pGEX-drk1 (anti-Kv2.1) and pGEX-K41C (anti-Kv1.5). The integrated optical density for each band was scaled by the fluorescence of the fraction and then displayed as a percentage of the total scaled signal for each protein. The distribution of Kv1.5 and Kv2.1 proteins was markedly different, with nearly all of the Kv2.1 immunoreactivity present in the NS region, whereas the Kv1.5 protein was more prominent in the SS$_S$ vesicles. Note the split in the ordinate scaling for the scaled optical density.

transport recycled proteins, it is still necessary to perform additional experiments (e.g., cell surface labeling) to test whether a specific protein being investigated is recycled or not.

An example of probing the vesicle populations separated on Ficoll velocity gradients for Kv1.5 and Kv2.1 channels is shown in Fig. 4. Samples from four regions of a gradient were separated by SDS–PAGE and probed with antibodies against Kv1.5 and Kv2.1 channel proteins. The Kv1.5 and Kv2.1 proteins were differentially distributed among the peak fractions, with the Kv2.1 proteins present only in the NS (cell surface) fraction, whereas the Kv1.5 proteins were present in both the $SS_S$ and, to a lesser extent, the NS regions, suggesting that they might be recycled to the cell surface. Pulse-chase experiments will be required to test the recycling of Kv1.5 proteins. Both channel proteins were nearly undetectable in the RS fractions. The method we have described above provides ample quantities of vesicle proteins to permit investigations that use highly sensitive techniques, such as immunoblotting, but we have not yet demonstrated that experimental techniques requiring larger quantities of vesicles, such as fusion with planar bilayers, are practical with this method for isolating transport vesicles.

## Summary

The squid giant axon provides a very simple preparation for the collection of bulk quantities of transport vesicles, and this greatly facilitates the physiologic study of ion channels incorporated into planar bilayers from these vesicles. However, this preparation is also limited in the repertoire of transport vesicles that can be studied, and it is not very convenient for some biochemical techniques, such as pulse-chase labeling experiments. Cultured N1E-115 cells, on the other hand, provide a preparation from which a larger repertoire of types of transport vesicles can be isolated, and many biochemical techniques can be applied in conjunction with physiologic studies. Further refinement of the techniques for isolating specific populations of vesicles from cultured cells will provide even greater insight into the role of vesicles in mediating ion channel trafficking.

## Acknowledgments

The authors would like to thank Drs. Robert French and Steve Graber for helpful comments on this paper. This work was supported by a Faculty Development Award for the Pharmaceutical Manufacturers Research Association (W.F.W.) and NSF grant IBN-9319496 (W.F.W.).

# Section III

# Second Messengers and Biochemical Approaches

# [19] Protein Phosphorylation of Ligand-Gated Ion Channels

*By* ANDREW L. MAMMEN, SUNJEEV KAMBOJ, and RICHARD L. HUGANIR

## Introduction

Neurotransmitter receptors mediate signal transduction at the postsynaptic membrane of chemical synapses in the nervous system. The major excitatory and inhibitory neurotransmitter receptors in the brain are ligand-gated ion channels. These receptors directly bind neurotransmitters, resulting in the opening of an intrinsic ion channel. The predominant ligand-gated ion channels in the nervous system are the nicotinic acetylcholine receptor, the glutamate receptors, the $\gamma$-aminobutyric acid ($GABA_A$) receptors and the glycine receptors. Biochemical and electrophysiologic studies of these receptors have demonstrated that they are multiply phosphorylated by a variety of protein kinases.[1] Phosphorylation of these receptors regulates many functional properties, including desensitization, open channel probability, open time, and subcellular targeting.[1] Because of the central role of ligand-gated ion channels in synaptic transmission, protein phosphorylation of these receptors is a major mechanism in the regulation of synaptic transmission and may underlie many forms of synaptic plasticity.[1–3] In this article we review a variety of techniques to examine the role of protein phosphorylation in the regulation of ligand-gated ion channel function. We review general strategies and methods for characterizing the phosphorylation state of ligand-gated ion channels, identifying phosphorylation sites on these channels, and analyzing the physiologic consequences of channel phosphorylation. To facilitate this discussion, we use the glutamate receptor subunit GluR1 as an example throughout this review.

## Biochemical Characterization of Phosphorylation of Ligand-Gated Ion Channels

Ideally, the phosphorylation state of a ligand-gated ion channel should be investigated in the tissue(s) where it is endogenously expressed. For example, the phosphorylation state of glutamate receptors is often analyzed

[1] K. W. Roche, W. G. Tingley, and R. L. Huganir, *Curr. Opin. Neurosci.* **4,** 383 (1994).
[2] L. A. Raymond, C. D. Blackstone, and R. L. Huganir, *Trends Neurosci.* **16,** 147 (1993).
[3] R. A. Nicoll and R. C. Malenka, *Nature* **377,** 115 (1995).



0076-6879/99 $30.00

in brain slice preparations or in primary cultures of central neurons.[4] To identify and characterize specific phosphorylation sites at the biochemical and functional level, however, it is often necessary to study phosphorylation of wild-type and mutant recombinant channels. Because most primary cultures are difficult to transfect and, more importantly, have a background of wild-type channels, wild-type and mutant channels are typically expressed separately in heterologous systems such as the HEK293 or COS cell lines or in *Xenopus* oocytes. These heterologous expression systems simplify the study of channel phosphorylation by allowing biochemical and electrophysiologic comparisons between wild-type and mutant channels. However, to confirm the physiologic relevance of channel phosphorylation and its regulation in heterologous systems, it is always important to compare these findings with results obtained in cells where the channels are natively expressed.

The first step in characterizing the phosphorylation state of a given channel is to determine whether the protein of interest is indeed a substrate for protein kinases. To accomplish this, primary cultures or transfected cells expressing the channel are incubated with ortho[$^{32}$P]phosphate. The cells will incorporate the labeled phosphate into adenosine triphosphate (ATP) at the $\gamma$-phosphate position where it can be transferred to proteins by protein kinases. Care must be taken to allow enough time for the ortho[$^{32}$P]phosphate to reach equilibrium with the intracellular pools of ATP and the protein kinase substrates. For most cell types this takes around 4 hr and can be monitored by examining the $^{32}$P incorporation into the substrate of interest. Following incubation of the cells with ortho[$^{32}$P]phosphate, the cells can be treated with activators of specific protein kinases for short periods of time or left untreated to examine the basal phosphorylation state of the substrate protein. After the cells are isolated, the ion channels are solubilized from the membrane with detergents and immunoprecipitated from the labeled cell extracts. The isolated channel is then analyzed by sodium dodecyl sulfate–polyacrylamide gel electrophoresis (SDS–PAGE), and the dried gels exposed to film to detect $^{32}$P incorporation into the channel.

Below is a protocol for the immunoprecipitation of $^{32}$P-labeled glutamate receptors from transiently transfected HEK293 cells. The same protocol is used for labeling and immunoprecipitating glutamate receptors from primary cultures by eliminating the transfection step. This protocol may also be easily modified for the labeling and immunoprecipitation of other ligand-gated ion channels from a variety of cell types. Because of the relatively large amounts of radioactivity used in these experiments, special

[4] C. Blackstone, T. H. Murphy, S. J. Moss, J. M. Baraban, and R. L. Huganir, *J. Neurosci.* **14,** 7585 (1994).

care must be taken to avoid $^{32}P$ contamination and exposure. We recommend practicing the immunoprecipitations with unlabeled cultures prior to working with the labeled material.

### *Labeling and Cell Extract Preparation*

*Reagents*

Immunoprecipitation buffer (IPB): 10 m*M* sodium phosphate (pH 7.0), 100 m*M* NaCl, 10 m*M* sodium pyrophosphate, 50 m*M* NaF, 1 m*M* sodium orthovanadate, 5 m*M* ethylenediaminetetraacetic acid (EDTA), 5 m*M* EGTA, 1 $\mu M$ okadaic acid, 1 $\mu M$ microcystin-LR, and appropriate protease inhibitors

DAY 1

1. Split HEK293 cells onto 100-mm tissue culture plates.

DAY 2

1. Transfect HEK293 cells with the appropriate cDNA. We typically use the calcium phosphate precipitation method to transfect HEK293 cells with 20 $\mu$g cDNA/plate.[5]

DAY 4

1. Remove the media from the plates and wash the cells twice with 5 ml phosphate-free MEM (Sigma St. Louis, MO).
2. Add 2.5 ml phosphate-free MEM to each plate.
3. Prelabel each plate with 2 mCi/ml of ortho[$^{32}P$]phosphate. We order ortho[$^{32}P$]phosphate as a 50 mCi/ml solution from NEN Dupont and add 100 $\mu$l to each plate. Return the plates to the incubator for 4 hr to allow the label to reach steady-state levels.
4. Activate endogenous protein kinases by treating cells with phorbol esters, forskolin, calcium ionophores, etc. Leave one plate untreated to examine basal phosphorylation.
5. Remove "hot" media and rinse each plate twice with 3–4 ml room temperature phosphate-buffered saline (PBS). Carefully dispose of the liquid radioactive waste from these washes.
6. Add 150 $\mu$l of room temperature IPB with 1% SDS to each plate and scrape with a cell scraper. Leave the extracts in the plates at this stage.
7. Dilute extracts by adding 750 $\mu$l of ice-cold IPB with 1% Triton X-100 to each plate and mix with a cell scraper.

[5] C. D. Blackstone, S. J. Moss, L. J. Martin, A. I. Levey, D. L. Price, and R. L. Huganir, *J. Neurochem.* **58,** 1118 (1992).

8. Transfer the labeled extract to a 15-ml conical tube on ice.
9. Carefully sonicate each sample with a probe sonicator at setting 6 for 20 sec, making sure not to contaminate area with $^{32}P$, and return to ice. Sonication breaks up the DNA, making the solution less viscous and easier to work with. Samples can be frozen at this stage, if desired.

*Immunoprecipitation of Protein from Labeled Extracts*

1. Transfer labeled extract to a 1.5-ml screwtop Eppendorf tube containing 200 $\mu$l of a 1 : 1 slurry of protein A-Sepharose beads (Sigma) suspended in IPB with 1% bovine serum albumin (BSA), 50 $\mu$l preimmune serum, 15 units DNase, and 150 $\mu$g RNase. The addition of DNase and RNase minimizes the contamination of the sample with $^{32}P$-labeled DNA or RNA. Rotate tubes for 1 hr at 4°. This step preabsorbs any protein that may stick nonspecifically to the protein A-Sepharose beads.
2. Centrifuge at 2000 rpm for 1 min and add supernatant to a 1.5-ml screwtop Eppendorf tube containing 200 $\mu$l of the 1 : 1 slurry of protein A-Sepharose beads and the precipitating antibody. Rotate tubes for 2 hr at 4°.
3. Wash the beads with 1 ml each of IPB with 1% Triton X-100 (×2), IPB with 1% Triton X-100 and 500 m*M* additional NaCl (×3), and IPB (×2). Carefully dispose of these radioactive washes.
4. Elute immunoprecipitated material from beads with 150 $\mu$l SDS–PAGE sample buffer. Analyze samples by SDS–PAGE and stain the gel with Coomassie blue. Then, dry the gel between two sheets of cellophane (Bio-Rad, Richmond, CA) and expose to film.

In the experiment shown in Fig. 1, we transfected HEK293 cells with wild-type GluR1, a mutant form of GluR1 in which the serine at residue 845 has been converted to an alanine (GluR1 S845A), and used mock transfected cells as controls. These cells were labeled with ortho[$^{32}P$]phosphate, treated with phorbol ester (a PKC activator), forskolin (which indirectly stimulates PKA activity), and IBMX, and immunoprecipitate with anti-GluR1 antibodies as described above. Figure 1A shows that both the wild-type and mutant receptors are phosphorylated under these conditions. Although we have previously shown that serine-845 is a substrate for phosphorylation PKA[6] (and see below), in this experiment, the mutant receptor

[6] K. W. Roche, R. J. O'Brien, A. L. Mammen, J. Bernhardt, and R. L. Huganir, *Neuron* **16,** 1179 (1996).

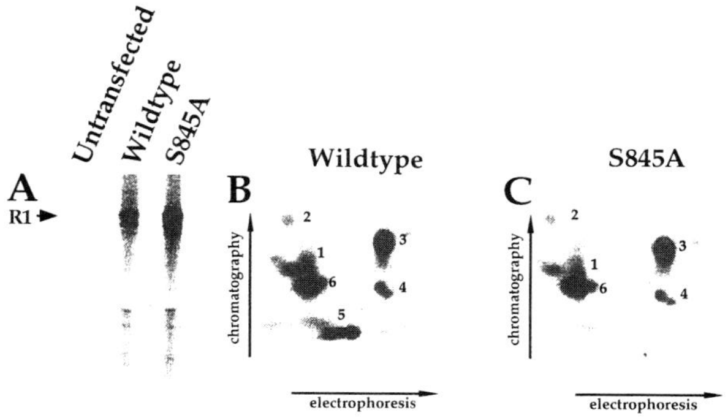


FIG. 1. Immunoprecipitation and mapping of [$^{32}$P]-labeled channels from transfected cells. QT6 cells were transfected with wild-type GluR1, mutant GluR1 (S845A), or mock transfected. These plates were incubated with ortho[$^{32}$P]phosphate for 4 hr, treated with 100 n$M$ phorbol 12-myristate 13-acetate (PMA), 10 $\mu M$ forskolin (FSK), and 100 $\mu M$ IBMX (IBMX) for 15 min. GluR1 was immunoprecipitated from detergent extracts of each plate (A). Two-dimensional phosphopeptide maps were generated from immunoprecipitated GluR1 (B) and GluR1 S845A (C).

appears to be phosphorylated even more robustly than the wild type. This is because GluR1 contains phosphorylation sites for other protein kinases (including PKC[6]) and the mutant receptor was expressed at higher levels than the wild-type receptor in this experiment. This result exemplifies that careful controls to examine the expression level of your protein of interest need to be performed in order to quantitate phosphorylation. No labeled receptor was immunoprecipitated from mock transfected cells.

### *Phosphopeptide Mapping of Ligand-Gated Ion Channels*

The next step in analyzing the phosphorylation of a ligand-gated ion channel involves subjecting the $^{32}$P-labeled phosphoprotein to two-dimensional phosphopeptide map and phosphoamino acid analysis. In the first procedure, the labeled material is excised from the gel and digested with protease. The resulting peptides are spotted onto a cellulose thin-layer chromatography (TLC) plate and subjected to electrophoresis. This separates the proteolytic fragments according to charge. Next, the plate is subjected to ascending chromatography where the peptides are separated according to their solubility in the chromatography buffer. The TLC plates are then exposed to film in order to visualize the phosphopeptides. Each ligand-gated ion channel subunit will yield a unique map depending on the location of its proteolytic cleavage sites and phosphorylated residues.

Below is a protocol for the two-dimensional phosphopeptide map analysis of glutamate receptors. To achieve optimal separation of phosphopeptides generated from other ligand-gated ion channels, it may be necessary to employ different proteases, electrophoresis times, and ascending chromatography buffers. These issues—and phosphopeptide mapping in general—are discussed thoroughly elsewhere.[7]

### *Proteolytic Digestion*

#### *Reagents*

Destain solution: 25% methanol/10% acetic acid in $H_2O$

1× Trypsin solution: 0.3 mg/ml trypsin TPCK (Sigma) dissolved in 50 m*M* $NH_4HCO_3$

10× Trypsin solution: 3 mg/ml trypsin TPCK dissolved in 50 m*M* $NH_4HCO_3$

#### *Procedure*

1. Cut the $^{32}P$-labeled ligand-gated ion channel out of the dried acrylamide gel and place in a glass scintillation vial.
2. Wash each protein-containing gel fragment with 20 ml destain solution for 30 min (×3).
3. Wash with 20 ml 50% methanol for 30 min (×2).
4. Transfer the gel fragment to an Eppendorf tube and dry down in a Speed-Vac, approximately 2 hr.
5. To each tube, add 1 ml of 1× trypsin solution and incubate overnight at 37°. The next morning add 100 $\mu$l of a 10× trypsin solution and incubate an additional 3–4 hr at 37°.
6. Remove the supernatant from the tube and save. Add 1 ml $H_2O$ to the gel fragment-containing tube and incubate 1 hr at 37°.
7. Combine the supernatants and dry down in a Speed-Vac. Resuspend the dried material in 0.5 ml $H_2O$ and dry down again; repeat three times to completely sublimate the $NH_4HCO_3$.
8. Resuspend the dried material in 10 $\mu$l $H_2O$ and microcentrifuge at 14,000 rpm for 10 min.
9. Remove the supernatant to a fresh Eppendorf tube and count in a scintillation counter (Cerenkov method).

[7] W. J. Boyle, P. V. D. Geer, and T. Hunter, *Methods Enzymol.* **201,** 110 (1991).

*Two-Dimensional Thin Layer Chromatography*

*Reagents*

Electrophoresis buffer (pH 3.4): acetic acid : pyridine : $H_2O$ at 19 : 1 : 89
Chromatography buffer: pyridine : butanol : acetic acid : $H_2O$ at 15 : 10 : 3 : 12
Basic fuchsin solution: 1 mg/ml solution
Phenol red solution: 1 mg/ml solution

*Procedure*

1. On a 20- × 20-cm Kodak (Rochester, NY) 13255 cellulose TLC sheet, make a single pencil mark at the sample origin, 10 cm from each side and 4 cm from the bottom of the sheet. Make two additional pencil marks 5 cm from each side and 4 cm from the bottom of the sheet.
2. Spot at least 250 cpm of the sample at the sample origin. However, do not spot more than about 2000 cpm of the sample per plate. Use only a fraction of the 10-$\mu$l sample if possible because overloading of the sample can cause streaking of the phosphopeptides. The sample should be spotted 1 $\mu$l at a time with drying in between (a hair dryer works much faster than air drying). Avoid gouging the TLC plate when spotting the sample.
3. Spot 1 $\mu$l each of basic fuchsin and phenol red solutions at the origin.
4. Prepare two sheets of 25- × 25-cm Whatman paper. Place a pencil mark 6.5 cm from the bottom and 12.5 cm from each side of one piece of Whatman paper; using the mark as its center, cut a 3-cm-diameter hole in this piece of Whatman paper. Place the spotted TLC plate on top of the intact piece of Whatman paper. Prewet the Whatman paper with the hole with pH 3.4 electrophoresis buffer. Lay the prewetted Whatman paper on the TLC plate so that the spotted sample appears in the middle of the hole. With a Pasteur pipette, dribble pH 3.4 electrophoresis buffer on the top piece of Whatman paper until the entire TLC plate underneath is wet. To avoid movement of the sample away from the origin, one should dribble electrophoresis buffer around the perimeter of the hole such that the diffusing buffer reaches the spotted sample from all sides simultaneously.
5. While still wet, place the TLC plate in an electrophoresis tank containing the electrophoresis buffer with the spotted sample in the middle between the two electrodes. Run at 500 V until the dyes reach the marks 5 cm from the edges of the plate (about 1.5 hr).

6. Remove the plate and allow to dry completely.
7. Place the plate in an ascending chromatography chamber with the bottom of the plate submerged in the chromatography buffer. When the buffer is 1 cm from the top of the plate (4–6 hr) remove it and allow to dry completely.
8. Wrap the TLC plate in Saran wrap and expose to film. For faster results place the wrapped plate in a PhosphoImager cassette.

The two-dimensional phosphopeptide map generated for each ligand-gated ion channel represents a highly reproducible "phosphopeptide fingerprint" of that channel. In the case of wild-type GluR1, phosphopeptide maps of the immunoprecipitated receptor from phorbol ester and forskolin treated HEK293 cells include seven distinct phosphopeptides (Figure 1B). However, this does not necessarily imply that there are seven distinct phosphorylation sites on GluR1; incomplete protease digestion is common and often generates multiple phosphopeptides, which include the same phosphorylation site. In fact, we have shown that GluR1 phosphopeptides 3, 4, and 6 all include the same PKC phosphorylation site.[6]

*Phosphoamino Acid Analysis of Ligand-Gated Ion Channels*

Serine, threonine, and tyrosine are each substrates for protein phosphorylation, and determining which of these amino acids are phosphorylated is an important step in characterizing the phosphorylation of a ligand-gated ion channel. To perform phosphoamino acid analysis, follow the protocol for proteolytic digestion of the sample described above and then use the isolated sample for acid hydrolysis to individual amino acids.

*Phosphoamino Acid Analysis Protocol*

*Reagents*

Electrophoresis buffer (pH 1.9): formic acid : acetic acid : $H_2O$ at 1 : 10 : 89

Electrophoresis buffer (pH 3.4): acetic acid : pyridine : $H_2O$ at 19 : 1 : 89

*Procedure*

1. Add at least 500 cpm of the sample to a 16- × 75-mm black-topped Kimax tube with a Teflon screw cap containing 0.5 ml 6 *N* HCl. Blow $N_2$ gently over the liquid before screwing the lid on tightly.
2. Place in a 105° oven for 1–2 hr and then transfer to a microcentrifuge tube and Speed-Vac until dry.
3. Resuspend the sample in 0.5 ml $H_2O$ and redry.
4. Resuspend the sample in 10 μl $H_2O$ and vortex 30 sec. Spin the

tube at 14,000 rpm for 10 min and transfer the supernatant to a second microcentrifuge tube.

5. Prepare a Kodak TLC 13255 cellulose sheet as follows: Mark the two side edges of the sheet with a pencil 4 cm from the bottom. Lay a ruler across the sheet between the two marks and, starting 2 cm from the side, make pencil marks every 4 cm. Up to five samples may be spotted at these five origins. Make additional marks 5 and 14 cm above where the samples will be spotted.
6. Spot the samples 1 $\mu$l at a time with drying in between.
7. Prepare fresh phosphoserine, phosphothreonine, and phosphotyrosine standards at 10 mg/ml in $H_2O$. Spot 1 $\mu$l of each on top of each spotted sample. Also spot 1 $\mu$l of phenol red on each sample.
8. Cut three pieces of Whatman paper: 25 × 25 cm, 25 × 17 cm, and 25 × 5 cm. Lay the TLC plate on the large piece of Whatman paper and prewet the others with the pH 1.9 electrophoresis buffer. Lay the 25- × 17-cm Whatman paper over the TLC plate such that its lower edge is 1.5 cm above the spotted samples. Lay the 25- × 5-cm Whatman paper such that its upper edge is 1.5 cm below the spotted samples. Drip pH 1.9 solution onto the prewetted Whatman papers until the TLC plate underneath is wet. The buffer should gradually diffuse from the upper and lower pieces of Whatman paper to wet the samples without moving them.
9. While wet, place the TLC plate in an electrophoresis tank containing pH 1.9 electrophoresis buffer. The samples should be nearest the cathode. Electrophorese the samples at 500 V until the phenol red dye reaches the first pencil mark, 5 cm above where the samples were spotted.
10. Without allowing it to dry, transfer the plate to an electrophoresis tank containing pH 3.5 electrophoresis buffer and electrophorese at 500 V until the phenol red dye reaches the next pencil mark, 14 cm above where the samples were spotted.
11. Dry the TLC plate, then dip in ninhydrin (1% in acetone) and allow to dry. The phosphoamino acid standards will turn purple in about 15 min. Phosphoserine migrates at the front, followed by phosphothreonine and, finally, phosphotyrosine.
12. Cover the plate with Saran wrap and expose to film.

*Fusion Protein Phosphorylation Studies*

Ligand-gated ion channels are composed of relatively large polypeptides which may contain many serines, threonines, and tyrosines. Thus, determining which residues are phosphorylated may seem like a daunting task.

However, if the protein kinase that phosphorylates the channel has been identified, one can begin by searching the protein for the appropriate protein kinase consensus sites.[8] Furthermore, if the topology of the channel in the membrane has been determined, candidate sites can be narrowed to those that are present on intracellular regions of the channel. Once regions likely to include phosphorylation sites have been identified, it is often useful to make fusion proteins corresponding to these regions and phosphorylate them *in vitro* with purified protein kinases. If the channel of interest has large continuous domains which might contain phosphorylation sites, it is helpful to generate several smaller fusion proteins (50–100 residues long) which span the length of the domain. These fusion proteins can then be analyzed by two-dimensional phosphopeptide mapping and the resulting maps compared to those generated from native channels or full-length recombinant channels expressed *in vivo*.

## *In Vitro Fusion Protein Phosphorylation*

### *Reagents*

5× PKA reaction buffer: 50 m*M* HEPES, pH 7.0, 100 m*M* $MgCl_2$

5× PKC and CaMKII reaction buffer: 50 m*M* HEPES, pH 7.0, 50 m*M* $MgCl_2$, 5 m*M* $CaCl_2$

ATP mixture: 250 μ*M* ATP (2000 cpm/pMol)

### *Procedure*

1. Add 10 μl (1 μg) of a 0.1 mg/ml solution of fusion protein substrate (dissolved in PBS) to a 1-ml microcentrifuge tube on ice.
2. Add 20 μl of the appropriate 5× reaction buffer to each tube.
3. For PKC reactions add 10 μl of a 500 μg/ml phosphatidylserine, 50 μg/ml diolein solution in $H_2O$. For CaMKII reactions add 10 μl of a 0.3 mg/ml calmodulin solution in $H_2O$.
4. Add $H_2O$ to bring volume to 90 μl.
5. Add 9 μl ATP mixture.
6. Begin reaction by adding 1 μl of a 100 μg/ml solution of the desired protein kinase and incubate at 30° for 30 min. Terminate the reactions by adding 50 μl of 3× sample buffer and analyze by SDS–PAGE. Stain the gel with Coomassie blue, dry the gel between two sheets of cellophane, and expose to film. Cut out the appropriate band from the gel to perform phosphopeptide map and phosphoamino acid analysis as described above.

[8] R. B. Pearson and B. E. Kemp, *Methods Enzymol.* **200,** 62 (1991).

When comparing fusion protein maps with maps from full-length channels, one should look for phosphopeptides that are of similar shape and comigrate on TLC plates. Figure 2B shows the map of a fusion protein corresponding to the C terminus of GluR1 phosphorylated *in vitro* with purified PKA. Note the presence of a phosphopeptide cluster that closely resembles and seems to comigrate with phosphopeptide 5 from full-length receptor (Fig. 1B). This finding suggests that phosphopeptide 5 may be contained within this fusion protein and that it may be a substrate for PKA phosphorylation. (To confirm that phosphopeptides from the two preparations actually comigrate, it is often useful to spot both on the same TLC plate and look to see whether they overlap when processed together.) Also note the presence of an additional phosphopeptide not observed in maps of the full-length receptor (indicated by an arrow). Such spurious phosphopeptides are often seen when proteins are phosphorylated *in vitro* and may represent phosphorylation at sites that are normally extracellular or otherwise inaccessible to protein kinases.

### *Site-Specific Mutagenesis of Phosphorylation Sites*

Once phosphorylation sites have been narrowed to small domains within a channel and the identity of the phosphorylated amino acid has been determined by phosphoamino acid analysis, candidate serines, threonines, and tyrosines can be targeted for point mutation. Typically, alanines are substituted for serines and threonines whereas phenylalanines are substituted for tyrosines. The resulting mutant recombinant receptors are then expressed in transfected cells and phosphopeptide maps generated as pre-

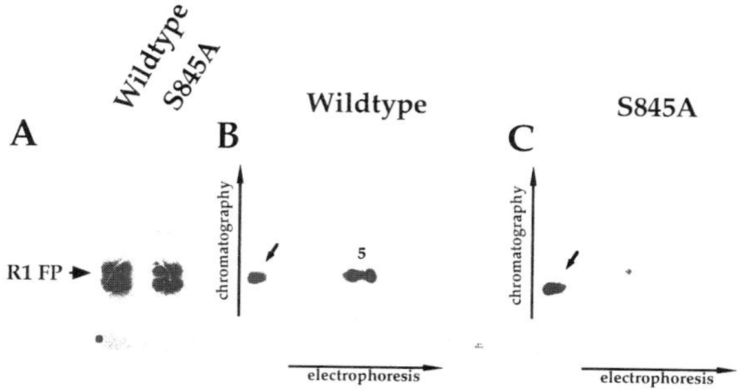


FIG. 2. *In vitro* phosphorylation and mapping of fusion proteins. Fusion proteins corresponding to the C termini of GluR1 and GluR1 S845A were phosphorylated *in vitro* with purified PKA and analyzed by SDS–PAGE (A). Two-dimensional phosphopeptide maps were generated from the PKA phosphorylated GluR1 (B) and GluR1 S845A (C) C-terminal fusion proteins.

viously described. In the case of GluR1, for example, mutating Ser-845 to Ala abolished phosphopeptide 5 (compare Figs. 1B and 1C). Mutant fusion proteins may also be created, phosphorylated *in vitro*, and screened by two-dimensional map analysis. Figure 2C shows that a phosphopeptide map of the S845A C-terminal GluR1 fusion protein phosphorylated *in vitro* with purified PKA does not include phosphopeptide 5. Taken together with the finding that phosphopeptide maps of GluR1 from cells which are not treated with forskolin do not include phosphopeptide 5 (not shown), these results suggest that PKA phosphorylates Ser-845 on GluR1. In general, the absence of wild-type phosphopeptides from maps of mutant channels is taken as strong evidence that the mutated residues are sites of protein phosphorylation.

*Phosphorylation Site-Specific Antibodies*

After a specific channel residue has been identified as a phosphorylation site, its phosphorylation state in a number of different preparations and under a wide variety of conditions is often of interest. To facilitate such studies, phosphorylation site-specific antibodies which recognize the channel only when the residue of interest is phosphorylated can be generated. These antibodies reduce the difficulty and expense associated with phosphopeptide mapping and allow the rapid screening of many samples by Western blot analysis. To generate such antibodies, phosphopeptides with chemically phosphorylated serine, threonine, or tyrosine residues must be synthesized and injected into rabbits. We have found that 12-mers with the phosphorylated residue at the sixth position work well. Including a lysine at the end of the phosphopeptide to facilitate coupling to the carrier thyroglobulin is also helpful (see Ref. 9 for a detailed discussion of phosphopeptide synthesis methods).

Phosphorylation site-specific antibodies can be separated from antibodies that recognize the nonphosphorylated peptide by loading the serum on an affinity column containing the nonphosphorylated equivalent of the phosphopeptide antigen. The phosphorylation site-specific antibodies should be contained in the flow-through along with other nonspecific antibodies which fail to bind the nonphosphorylated peptide. Often, the serum contains very few antibodies that recognize the nonphosphorylated peptide, and this first purification step may not be necessary. To purify the antibody of interest away from nonspecific antibodies, the flow-through from the first column (or crude serum) may be loaded onto an affinity column containing the phosphopeptide antigen. After washing the column, the

[9] W. G. Tingley, M. D. Ehlers, K. Kameyama, C. Doherty, J. B. Ptak, C. T. Riley, and R. L. Huganir, *J. Biol. Chem.* **272,** 5157 (1997).

phosphorylation site-specific antibodies can be eluted with standard methods and used for Western blot analysis.

Figure 3 demonstrates the utility of a phosphorylation site-dependent antibody that recognizes GluR1 only when Ser-845 is phosphorylated (GluR1 845-P). QT6 cells were transfected with wild-type or mutant GluR1 and treated with phorbol ester, forskolin and IBMX, or vehicle. Membrane extracts were prepared from each sample, separated by SDS–PAGE, blotted onto PVDF membrane, and analyzed by Western blot. Forskolin treatment caused an increase in GluR1 845-P labeling of wild-type channel, confirming that PKA phosphorylates GluR1 on Ser-845. However, when the blot was treated with lambda phosphatase prior to application of the antibody, GluR1 845-P did not label GluR1 from forskolin-treated cells. Furthermore, when QT6 cells were transfected with the GluR1 S845A channel and treated with forskolin, the GluR1 845-P antibody did not recognize the mutant channel. These two results demonstrate the phosphorylation state dependence of the GluR1 845-P antibody. A different antibody

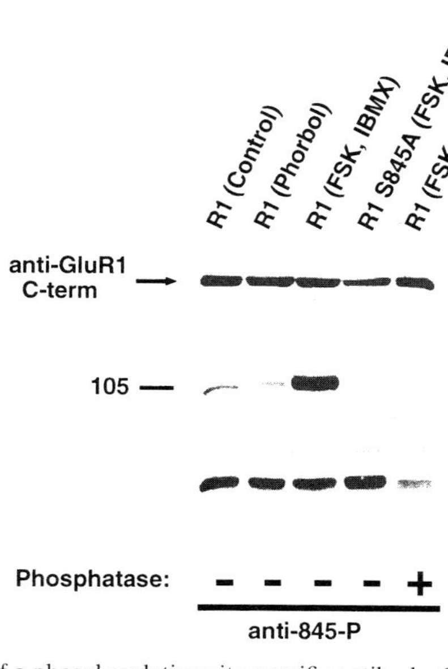


FIG. 3. Specificity of a phosphorylation site-specific antibody. QT6 cells expressing GluR1 or GluR1 S845A were treated with control solution (Control), 100 n$M$ phorbol 12-myristate 13-acetate (PMA), or 10 $\mu M$ forskolin and 100 $\mu M$ IBMX (FSK, IBMX) for 15 min as indicated. Membranes were prepared and run on SDS–PAGE, transferred to PVDF membranes, and immunoblotted with the phosphorylation site-specific antibody GluR1 845-P or an antibody to the C terminus of GluR1.

raised against the carboxy terminus of GluR1 shows that equal amounts of the channel were loaded in each lane.

Phosphorylation site-specific antibodies are especially useful for examining the phosphorylation state of ligand-gated ion channels in preparations that are not amenable to labeling with ortho[$^{32}$P]phosphate. For example, the GluR1 845-P antibody may be used to compare the degree of GluR1 Ser-845 phosphorylation in the brains of rats that have undergone different treatments. These antibodies are also potentially useful for examining the subcellular distribution of phosphorylated channels by immunocytochemistry.

## Functional Effects of Ligand-Gated Ion Channel Phosphorylation

A variety of electrophysiologic studies have provided evidence that ligand-gated ion channels are modulated by protein phosphorylation. However, it is not clear from these studies whether this modulation occurs via the direct phosphorylation of the channel itself. To address this question, patch-clamp studies using wild-type and mutant receptors expressed in heterologous systems are necessary. The physiologic properties of recombinant wild-type receptors are studied before and after phosphorylation and compared to various phosphorylation site mutant receptors.

### *Stimulation of Phosphorylation in Whole-Cell Recordings*

The functional effects of phosphorylation have been investigated by applying pharmacologic agents that activate kinases (e.g., phorbol esters or forskolin) to primary cultures or transfected cells. This approach has been particularly useful in establishing roles for PKA- and PKC-mediated phosphorylation in modulating ligand-gated ion channel function (for example, see Refs. 10–12). An alternative approach is to apply the kinase itself to the intracellular side of the channel.[6,13–17] This requires the isolation of

[10] P. Greengard, J. Jen, A. C. Nairn, and C. F. Stevens, *Science* **253,** 1135 (1991).

[11] M. J. Courtney and D. G. Nicholls, *J. Neurochem.* **59,** 983 (1992).

[12] H. Markram and M. Segal, *J. Physiol.* (*Lond.*) **457,** 491 (1992).

[13] S. J. Moss, T. G. Smart, C. D. Blackstone, and R. L. Huganir, *Science* **257,** 661 (1992).

[14] L. A. Raymond, C. D. Blackstone, and R. L. Huganir, *Nature* **261,** 637 (1993).

[15] L.-Y. Wang, F. A. Taverna, X. D. Huang, J. F. MacDonald, and D. R. Hampson, *Science* **259,** 1173 (1993).

[16] L. Chen and L. Y. Huang, *Nature* **356,** 521 (1992).

[17] J. L. Yakel, P. Vissavajjhala, V. A. Derkach, D. A. Brickey, and T. R. Soderling, *Proc. Natl. Acad. Sci. U.S.A.* **92,** 1376 (1995).

a constitutively active form of the kinase or coapplication of the native kinase with appropriate activators.

We have found that simply supplementing the intracellular solution with the kinase can be problematic since it may diffuse via the tip into the cell before a control response can be established. Furthermore, it is often difficult to obtain a gigaseal when a protein is present in the intracellular solution. This problem can be circumvented by tip-filling the electrode with a control intracellular solution (lacking the kinase) and back-filling the pipette with the intracellular solution containing kinase. Consistent volumes of tip-filling and back-filling solutions should be used to preventing intercell variability in the measured effect of the kinase and the latency of its onset.

Intracellular patch perfusion is a preferred method for introducing kinases to the inside of cells. It involves injection of intracellular solution containing the kinase (and a suitable carrier protein) directly to the inside of cells via tubing placed close to the tip of the patch electrode. We have found this method to be particularly successful: It avoids problems in gigaohm seal formation, since intracellular injection always occurs after "breaking through" into the whole-cell configuration. Furthermore, it allows an unambiguous measurement of a baseline response in the absence of kinase. Placing the intracellular perfusion tubing near the tip of the electrode ($<2$ mm) should allow relatively rapid entry of solution into the cell, due to diffusion and the fluid turbulence created by pressure injection. For example, dye (trypan blue) can be seen near the tip of the electrode within seconds of injection into a patch electrode. Measurable changes in GluR1 whole-cell currents after intracellular application of PKA[6] or a constitutively active form of CaMKII (unpublished data) are seen in as little as 2 min.

The internal patch-perfusion tubing is easily fabricated from polypropylene tubing, which can be melted and gently drawn out to the desired diameter and length. For these experiments we use special electrode holders with an additional inlet for the patch-perfusion tubing purchased from Adams and List Associates, Ltd. Several other manufacturers produce suitable holders. These electrode holders are similar in most respects to standard holders, however, due to the noise created by the proximity of the perfusion tubing and the electrode wire, it may be difficult to obtain low-noise single-channel recordings using this method.

The solution containing the kinase can be expelled from the intracellular perfusion tubing by applying pressure via a syringe. The final 50–100 $\mu$l of this tubing (leading into the electrode holder) should be detachable. This part of the tubing is filled with the kinase containing solution. The rest of the tubing that connects this 50–100 $\mu$l section to the syringe should be filled with mineral oil. The latter is noncompressible and prevents volume changes within the intracellular perfusion tubing from occurring when the patch electrode experiences changes in pressure. Thus, intracellular perfusion occurs smoothly and without formation of bubbles.

### *"Intracellular" Solution Composition*

Because various intracellular components are required for phosphorylation/dephosphorylation reactions, it is important that they be preserved or at least "mimicked" during extended recordings. For this reason, some experimenters use an "ATP-regenerating solution,"[18] which, as the name implies, mimics the cellular ATP-generating machinery. In addition to ATP (2–4 m*M*), this solution is supplemented with creatine phosphokinase (50 U/ml) and phosphocreatine (20 m*M*). Use of this solution has allowed NMDA and AMPA-receptor mediated responses to be recorded over extended durations without appreciable "rundown," which appears to be mediated by dephosphorylation.[15]

The intracellular solution should be kept on ice to prevent degradation of labile contents. As far as is practical, the kinase containing intracellular solution should also be kept on ice. It is difficult to keep the kinase in the patch-perfusion tubing cool, therefore only small volumes should be used to fill this tubing. The tube should frequently be flushed and replenished with newly thawed and diluted kinase. How often the kinase containing solution is replaced depends on its stability at room temperature. For our experiments using constitutively active CaMKII or PKA, we limit the use of a batch of kinase to 2 hr.

An additional consideration for these experiments is that proteins bind to glass by way of multiple nonspecific interactions with the charged glass surface of the patch electrode. Thus, in addition to adding the kinase of interest to the intracellular solution, it is necessary to saturate most of the nonspecific glass–protein interactions by adding a "carrier protein" such as bovine serum albumin or creatine phosphokinase.

### *Basal Phosphorylation of Ligand-Gated Ion Channels*

Sites of basal phosphorylation in glutamate receptors have been detected in neurons and HEK293 cells.[4,6] The level of basal phosphorylation may depend on the cell type in which the receptor is expressed and factors such as the composition of the tissue culture serum (for example, serum often contains thrombin, which can activate PKC in many cell types). Basal phosphorylation may make the detection of any functional effect of an

[18] P. Forscher and G. S. Oxford, *J. Gen. Physiol.* **85,** 743 (1985).

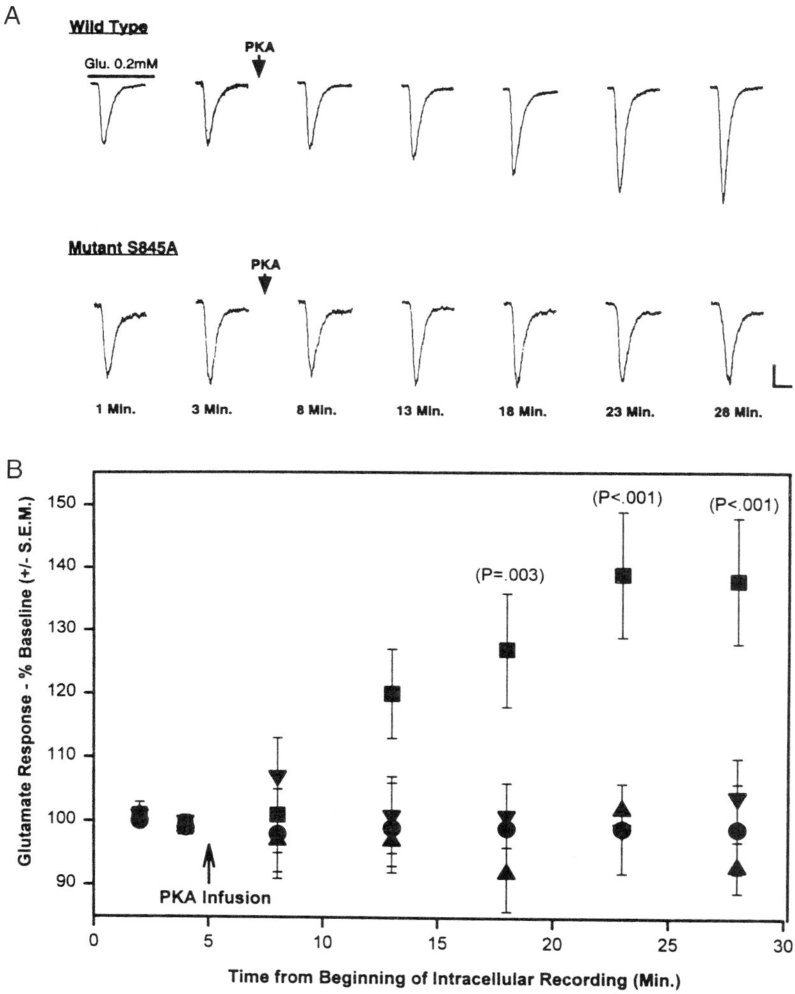


FIG. 4. Functional effects of GluR1 phosphorylation by PKA. Wild-type and mutant GluR1 responses activated by glutamate in HEK293 cells (duration of glutamate application is indicated by solid line). The calibrations bars indicate 100 pA and 10 ms. PKA (20 $\mu$g/ml) was infused after holding the cells for ~5 min. In wild-type GluR1-expressing cells, the responses become significantly larger after ~15 min (A and B, ■). This effect appeared to be specifically mediated by PKA: the PKA buffer caused no potentiation (B, ●) and PKA inhibitor peptide ($PKI_{5-24}$ amide; 200 $\mu$g/ml) blocked potentiation (B, ▲). In addition, the GluR1 S845A mutant was not potentiated by PKA (A and B, ▼). Symbols and vertical lines indicate data gathered from 4–12 cells ± SEM.

exogenous kinase difficult or impossible. If, on the other hand, sites are already stoichiometrically phosphorylated by basal kinase activity, functional changes may be detected in the presence of phosphatases.

In the case of GluR1 for example, the C-terminal S831 site was found to be basally phosphorylated in HEK293 cells, making it difficult to investigate the functional effect of phosphorylation at this site. However, S831 was not phosphorylated in the QT6 quail fibroblast cell line. Detailed biochemical experiments showed that this site was phosphorylated by CaMKII and PKC. Furthermore, electrophysiologic studies revealed that receptors expressed in these cells can be strongly potentiated when exposed to a constituitively active form of CaMKII (unpublished observations).

### *Determining Functional Effect of Phosphorylation at Specific Sites*

Biochemical analysis using wild-type and mutant receptors has allowed phosphorylation sites to be specifically identified. To establish a physiologic role for phosphorylation at particular sites, responses of wild-type and mutant receptors are measured. For example, the effect of PKA phosphorylation of GluR1 at S845 has been determined using internal kinase perfusion. This site is robustly phosphorylated when the receptor is exposed to forskolin and IBMX. Furthermore, infusion of PKA into the cell caused a significant potentiation of the peak response of the channel, while the S845A mutant was neither phosphorylated by PKA, nor underwent potentiation in the presence of PKA (see Fig. 4). That the observed increased in the GluR1-mediated response was due to the specific action of PKA was confirmed using a peptide inhibitor of PKA which blocked the potentiation (Fig. 4B). These results strongly suggest that PKA phosphorylation in Ser-845 directly potentiates GluR1 function.

## Conclusion

In this article we have attempted to describe a biochemical and electrophysiologic approach to investigating ligand-gated ion channel phosphorylation. As a model we have used phosphorylation of the GluR1 receptor subunit, but the principles and concepts should be applicable to other classes of ligand-gated ion channels. These studies provide us with a better understanding of the regulation of ligand-gated ion channels by protein phosphorylation, which may be a major mechanism for the modulation of synaptic function.

# [20] Analysis of Ion Channel Associated Proteins

*By* MICHAEL WYSZYNSKI and MORGAN SHENG

## Introduction

Ion channels are typically heterooligomeric protein complexes composed of several distinct subunits. Ion channels in turn interact with a variety of intracellular proteins including clustering proteins, cytoskeletal proteins, protein kinases, and G proteins. The distinction between a true channel "subunit" and an ion channel "associated protein" can be somewhat arbitrary, especially if the primary function of the protein is not known. For the purposes of this article, however, we define ion channel associated proteins as those that are not directly involved in channel function (e.g., they do not contribute to the formation of the channel pore or alter significantly the biophysical properties of the channel). Rather ion channel associated proteins include those that are involved in localization, cytoskeletal anchoring, or regulation of their associated ion channels.

Ion channel associated proteins can be initially identified by many means, including biochemical copurifcation (e.g., acetylcholine receptor (AChR) and rapsyn, glycine receptors and gephyrin, *Shaker*-type $K^+$ channels and their $\beta$-subunits),[1] genetic interactions (e.g., degenerin channels and their interacting proteins in *Caenorhabditis elegans*; reviewed in Ref. 2), and the yeast two-hybrid system.[2a] In this chapter we consider the approaches for confirming and analyzing an interaction between an ion channel and its putative interacting protein, using as a primary example the association between PSD-95 and *Shaker* $K^+$ channels and *N*-methyl-D-aspartate (NMDA) receptors (see Fig 1). This interaction was first identified by the yeast two-hybrid system.[3-5]

## General Considerations

After a protein is identified as a putative channel associated protein via yeast two-hybrid screening or other kinds of affinity-based screening, the

[1] M. Sheng and E. Kim. *Curr. Opin. Neurobiol.* **6,** 602 (1996).
[2] J. Garcia-Anoveros and D. Corey, *Annu. Rev. Neurosci.* **20,** 567 (1997).
[2a] M. Niethammer and M. Sheng, *Methods Enzymol.* **293,** [7] (1998).
[3] E. Kim, M. Niethammer, A. Rothschild, Y. N. Jan, and M. Sheng, *Nature* **378,** 85 (1995).
[4] H.-C. Kornau, L. T. Schenker, M. B. Kennedy, and P. H. Seeburg, *Science* **269,** 1737 (1995).
[5] M. Niethammer, E. Kim, and M. Sheng, *J. Neurosci.* **16,** 2157 (1996).



0076-6879/99 $30.00

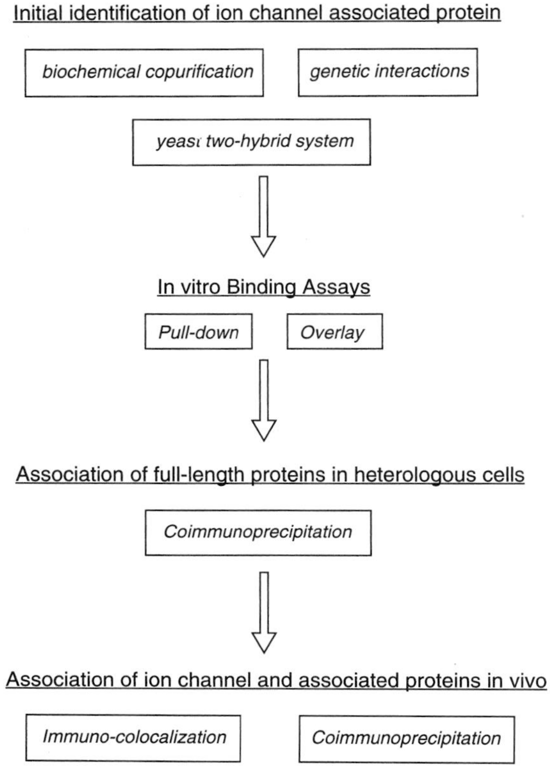


FIG. 1. A scheme that summarizes a general approach for analyzing putative ion channel associated proteins.

most important task is to confirm the authenticity of the interaction *in vivo*. The most significant tests of *in vivo* association of two proteins in common use are coimmunoprecipitation (a convenient and stringent way to show biochemical copurification) and double-label immunofluorescent colocalization (a direct way to show that they are in the same subcellular location, a prerequisite if they are to interact directly). Ultimately, genetic experiments provide critical evidence for an *in vivo* interaction and offer important clues to its functional significance. For instance, genetic deletion of the channel-associated protein could result in mislocalization of the ion channel or in abnormal regulation of channel properties. Of course, genetic approaches are essential to determine the function of the interaction at the organismal level. Before undertaking the more laborious *in vivo* studies, however, it is reasonable to conduct *in vitro* binding experiments to confirm an interaction between an ion channel and its suspected partner.

## *In Vitro* Binding of Recombinant Proteins

*In vitro* binding is useful to confirm an interaction initially identified by the yeast two-hybrid system. Although interactions detected by the yeast two-hybrid system almost invariably involve direct binding between the interacting proteins, *in vitro* binding assays using purified proteins can confirm that the interaction requires no other ancillary protein. More importantly, an *in vitro* binding assay offers the opportunity to measure directly the affinity of binding.

Typically, *in vitro* binding assays are performed with the relevant parts of the proteins produced in bacteria as recominant fusion proteins. We describe here two commonly used *in vitro* binding assays: a filter overlay assay (sometimes known as a Far-Western) and a "pull-down" or coprecipitation assay. In filter overlay assays, one protein (protein X) is subjected to SDS–PAGE and transferred to nitrocellulose filters. The other protein (protein Y) is applied in solution to the filter and its binding to protein X is visualized (through radioactive or indirect chemiluminescent labeling and autoradiography) as a "band on a gel." Thus, in filter overlays, protein Y is used as a probe in analogous fashion to antibodies in a Western blot. In the "pull-down" assay, protein X is attached to a solid support (e.g., agarose bead) and incubated with protein Y in solution. If proteins X and Y associate, pelleting of the beads by centrifugation will result in the precipitation of both X and Y. The presence of Y can then be assayed by immunoblotting the precipitated proteins.

### *Filter Overlay Assay*

The example of a filter overlay assay provided here is that of PSD-95 binding to the C-terminus of NMDA receptor NR2 and *Shaker* $K^+$ channel subunit proteins.[5] The first two PDZ domains of PSD-95 were produced as a hexahistidine–fusion protein (using pRSETB, Invitrogen) in bacteria and purified by nickel-nitrilotriacetic acid (Ni-NTA) chromatography (using Probond resin, Invitrogen, Carlsbad, CA). A hexahistidine ($H_6$)-labeled fusion protein of PSD-95 ($H_6$–PSD-95) is here used as the probe. The NR2 C-terminal tails were fused to glutathione *S*-transferase (vector pGEX, Pharmacia, Piscataway, NJ) and overexpressed in bacteria. It is not necessary to purify the GST fusions since they will be separated by SDS–PAGE and transferred to a nitrocellulose filter to act as targets for binding by $H_6$-PSD95. Specificity of binding is then determined by the position on the filter where the hexahistidine–fusion protein binds.

1. Electrophorese total bacterial lysates containing GST fusion proteins of the NR2 C-terminus (~40 kDa) and GST fusion proteins of the

Kv1.4 C-terminus (~37 kDa) in a 10% SDS–PAGE gel. Usually we electrophorese 1–5 $\mu$g of lysate protein and aim to have approximately 100 ng of target protein available for binding. To confirm the specificity of the interaction, GST alone or GST fused to unrelated protein sequences should also be used as targets.

2. Transfer the resolved proteins from the acrylamide gel to a nitrocellulose filter at 35 V/cm for 2 hr in transfer buffer [50 m*M* Tris–base, 40 m*M* glycine, 1 m*M* ethylenediaminetetraacetic acid (EDTA), 20% (v/v) methanol, pH 8.3]. To facilitate the transfer of proteins larger than 100 kDa, 0.1% (w/v) SDS may also be included in the transfer buffer.

3. To check that the transfer was successful, visualize the proteins transferred to the nitrocellulose filter by incubating the filter in Ponceau S protein stain. The molecular weight markers should also be identified and marked at this stage.

4. The denatured filter-bound proteins may require partial or complete renaturation of their conformations for successful binding to the recombinant fusion proteins serving as probe. To renature the filter bound proteins, incubate the nitrocellulose filter in buffer A (10 m*M* HEPES, 60 m*M* KCl, 1 m*M* EDTA, and 1 m*M* 2-mercaptoethanol) containing 6 *M* guanidine hydrochloride at 4° for 10 min. Repeat the incubations sequentially at 4° for 10 min using decreasing concentrations of guanidine hydrochloride (3, 1.5, 0.75, 0.38, 0.19, 0.1, and 0 *M*). The buffer A solutions containing guanidine hydrochloride are reusable and should be stored at 4°.

5. Block the filter in buffer B (25 m*M* HEPES, 120 m*M* KCl, 1 m*M* EDTA, and 0.2% Triton X-100) containing 5% nonfat dry milk (NFDM) at 4° for 1 hr with shaking.

6. Repeat step 5 using buffer B containing 1% NFDM.

7. For the protein binding reaction, overlay blocked filter with buffer B containing 10 mg/ml bovine serum albumin (BSA) (to suppress nonspecific binding by the protein probe) and 1 $\mu$g/ml of the purified hexahistidine–fusion protein of the first two PDZ domains from PSD-95. Typically, 3–5 ml of the overlay solution is sufficient for covering a nitrocellulose filter with a surface area of 50 $cm^2$. Incubate at 4° overnight.

8. Next day, bring the overlaid filter to room temperature by incubating at room temperature for 30 min.

9. Wash the filter with buffer B for 10 min at room temperature. Repeat wash twice.

10. Rinse the filter briefly with TBST [10 m*M* Tris (pH 7.4), 150 m*M* NaCl, 0.1% Tween 20].

11. To visualize the bound hexahistidine–fusion protein by Western blotting, incubate the filter in antibody dilution solution (TBST containing 2% BSA and 2% horse or goat serum) containing antibodies directed against

an epitope tag in the hexahistidine leader sequence (1:10,000 dilution of mouse monoclonal anti-T7.Tag antibodies, Novagen, Milwaukee, WI) at 4° overnight. Next morning, wash the filter with TBST containing 5% NFDM at room temperature for 10 min. Repeat the wash with TBST alone. Incubate the washed filter in antibody dilution solution containing a 1:1000 dilution of anti-mouse HRP-conjugated antibodies (Amersham, Arlington Heights, IL) at room temperature for 30 min, wash twice with TBST at room temperature for 10 min, and perform enhanced chemiluminescence (ECL, Amersham).

12. To show the position and relative abundance of the GST–fusion proteins in each lane, strip the filter of antibodies by incubating the filter in stripping buffer [62 m*M* Tris (pH 6.7), 2% (w/v) SDS, and 0.1% mercaptoethanol] at 60° for 30 min. Block the filter with TBST containing 5% NFDM at room temperature for 30 min, and perform Western blotting as described above using 0.5 $\mu$g/ml rabbit anti-GST antibodies (Santa Cruz Biotechnology, Santa Cruz, CA).

### *"Pull-Down" Assay*

The example of a pull-down assay provided here is that of $\alpha$-actinin-2 binding to NMDA receptors.[6] The pull-down assay can be applied to extracts from native tissues (as illustrated here) or to extracts from transfected heterologous cells or even semipurified preparations of bacterial fusion proteins. Obviously, "pull-down" of protein Y by protein X from a complex mixture (such as a brain extract) does not indicate that there is a direct interaction between X and Y. On the contrary, it is typical to obtain a heterogeneous mixture of proteins by this method, and it is important to check the specificity of the precipitation and to confirm it in as stringent conditions as possible.

The coprecipitation or pull-down of ion channel associated proteins from rat brain extracts begins with the preparation of a brain membrane fraction from whole brain homogenates.

1. Dissect the whole brain from 1 adult rate (age ~ 6 weeks, weight ~ 150–200 g).
2. Mince the rat brain on ice with a razor.
3. Homogenize the minced brain tissue in 3 ml of ice-cold homogenization buffer [0.32 *M* sucrose, 10 m*M* Tris (pH 7.4), 1 m*M* EDTA, and various protease inhibitors] using a Dounce homogenizer. For protease inhibitors, we typically use a cocktail consisting of 1 m*M*

[6] M. Wyszynski, J. Lin, A. Rao, E. Nigh, A. H. Beggs, A. M. Craig, and M. Sheng, *Nature* **385,** 439 (1997).

phenylmethylsulfonyl fluoride (PMSF) and 10 μg/ml each of pepstatin, aprotinin, and leupeptin.

4. Transfer the homogenate to a 15-ml conical centrifuge tube, centrifuge at 1000*g* for 10 min at 4°, and separate the supernatant from the insoluble pellet.
5. Resuspend by homogenization the insoluble pellet in 1 ml of ice-cold homogenization buffer, repeat centrifugation, and separate the supernatant from the remaining insoluble material.
6. Pool supernatants from steps 4 and 5 and perform a high-speed centrifugation at 60,000 *g* at 4° for 1 hr in an ultracentrifuge.
7. The supernatant obtained from this centrifugation may be considered the crude "soluble" brain fraction.
8. Resuspend the insoluble pellet in 8 ml of ice-cold resuspension buffer [20 m*M* Tris (pH 7.4), 1 m*M* EDTA] with various protease inhibitors.
9. Incubate on ice for 10 min and pellet by centrifugation at 60,000*g* at 4° for 30 min.
10. Resuspend the pellet in 3 ml of ice-cold resuspension buffer. This resuspension represents the crude "membrane" fraction.

Following preparation of the rat brain membrane fraction, ion channel associated proteins in the membrane fractions are made soluble by detergent extraction. Alternatively, the soluble brain fraction can be used for the pull-down if the putative ion channel associated protein is present in this fraction.

11. Dilute 200 μl of membrane fraction in 200 μl of ice-cold extraction buffer [80 m*M* Tris (pH 8.0), 300 m*M* NaCl, 1 m*M* EDTA, and 2% Triton X-100].
12. Allow extraction to proceed at 4° for 1 hr with mixing by inversion.
13. Centrifuge at 60,000*g* at 4° for 30 min.
14. Save the supernatant containing the extracted protein frozen at −80°C.

The "pull-down" assay also requires the affinity-purification of a recombinant fusion protein containing the ion channel protein of interest. This fusion protein will be used to affinity-purify (i.e., pull-down) proteins from the detergent extract prepared in step 14.

15. Prepare GST–fusion protein of the NR1 C-terminus using glutathione-Sepharose beads (Pharmacia) by standard purification protocols. Do not release the GST–fusion protein from the glutathione-Sepharose beads.

16. Incubate GST–fusion protein (~100 μg) attached to the glutathione-Sepharose beads with Triton X-100 extracts of rat brain (~200 μg protein from step 14) at 4° for 1 hr.
17. Wash the beads three times by resuspending protein bound glutathione-Sepharose beads with phosphate-buffered saline (PBS) containing 0.1% Triton X-100.
18. After washing, elute the bound proteins from the beads with SDS sample buffer, resolve by SDS–PAGE, and visualize the α-actinin "pulled-down" by immunoblotting with α-actinin-2 specific antibodies.[6]

As in the overlay assay, the specificity of the precipitation needs to be determined by using glutathione-Sepharose beads coupled to GST alone or GST fused to unrelated protein sequences as control affinity matrices. Further, the relative amount of GST fusion proteins used should be equivalent, and can be assayed by anti-GST immunoblotting as described earlier for the filter overlay assay.

## Association of Full-Length Proteins in Heterologous Cells

The assays described above, the filer overlay or "pull-down" assays, are essentially tests for *in vitro* association between proteins, one or both of which is typically produced as a recombinant bacterial fusion protein containing only one fragment of the protein. These are useful assays when beginning to confirm a putative interaction between two proteins because they are relatively easy and do not require specific antibodies against either protein. It is obviously important, however, to confirm that a *full-length* ion channel protein can interact with a *full-length* channel associated protein in a *cellular* environment. This can be conveniently done by coexpression in a heterologous cell by transfection, and coimmunoprecipitation of the two expressed proteins. Such an approach requires the complete coding region of the cDNAs for the two putatively interacting proteins, and also specific antibodies to both proteins. In the absence of specific antibodies, one or both proteins can be tagged with an epitope tag [such as *myc* or hemagglutinin (HA) tags] which can be recognized by a commercially available monoclonal antibody. In the following protocol, the example given is that of Kv1.4 (a *Shaker*-type $K^+$ channel) binding to PSD-95 in heterologous (COS-7) cells.[7]

[7] E. Kim, and M. Sheng, *Neuropharmacology* **35,** 993 (1996).

*Coimmunoprecipitation of Full-Length Proteins Expressed in Heterologous Cells*

Coimmunoprecipitation of proteins transiently expressed in heterologous cells requires transfection of cultured cells, detergent solubilization of the expressed proteins (since ion channels are intrinsic membrane proteins), and immunoprecipitation employing antibodies that are specific for the expressed proteins.

1. Plate COS-7 cells in tissue dishes (35 mm) for a confluence of 50–70%.
2. Next day, the cells should be 70–80% confluent. Prepare a DNA transfection mix of the eukaryotic expression constructs for full-length PSD-95 and Kv1.4 by adding equivalent amounts of each plasmid (no more than 1 $\mu$g total DNA) to 200 $\mu$l OPTI-MEM (GIBCO-BRL, Gaithersburg, MD) and 5 $\mu$l of lipofectamine (GIBCO-BRL). Allow the DNA/lipofectamine complexes to form for 30 min at room temperature. Following this incubation, add 800 $\mu$l of OPTI-MEM to each transfection mix.
3. Wash the cells thoroughly with 2 ml of OPTI-MEM.
4. Remove the OPTI-MEM wash and quickly add the DNA/lipofectamine OPTI-MEM mix (~1 ml) to the washed cells.
5. Incubate the cells in the DNA/lipofectamine mix in 5% $CO_2$ at 37° for 5 hr.
6. Exchange the DNA/lipofectamine mix for normal medium and incubate the cells in 10% $CO_2$ at 37°. Depending on the efficiency of the eukaryotic expression vector employed and on the protein expressed, efficient expression (up to 1% of total cellular protein) of the protein takes 24–48 hr.

Prior to immunoprecipitation, the transfected cells are harvested from the cell culture dish, and the expressed proteins extracted with nondenaturing detergents.

7. Wash the cells twice with 2 ml of ice-cold phosphate-buffered saline.
8. Following the last wash, quickly add to the dish 350 $\mu$l of ice-cold extraction buffer [50 m*M* Tris (pH 7.4), 150 m*M* NaCl, 1 m*M* EDTA, plus the extraction detergent of choice]. For the solubilization of PSD-95 and Kv1.4 from heterologous cells we use RIPA [1% (v/v) Nonidet P-40 (NP-40), 0.5% (w/v) deoxychlorate, and 0.1% (w/v) SDS].
9. On ice, flush cells off the bottom of the dish by triturating the extract buffer over the surface of the dish.
10. Transfer the cell suspension to a 1.5-ml microcentrifuge tube. Allow extraction to proceed with shaking for 1 hr at 4°.
11. Pellet insoluble material by centrifugation at 60,000*g* at 4° for 30 min. The supernatant contains the detergent-extracted proteins (soluble fraction). The pellet contains the detergent-insoluble proteins (insoluble

fraction). Prior to immunoprecipitation, perform Western blotting analysis to determine the efficiency of solubilization by comparing the amount of expressed protein in the insoluble fraction to the amount of protein in the soluble fraction. Efficient immunoprecipitation requires that the protein be efficiently solubilized by detergent (at least 10–30%).

The immunoprecipitation of protein complexes from cell extract requires either the expression of epitope-tagged proteins or antibodies which specifically recognize the proteins of interest.

12. To 100-$\mu$l aliquots of the detergent extract (input), add the immunoprecipitating antibodies to a final concentration of 5 $\mu$g/ml.

In this example, guinea pig polyclonal antibodies specific for either PSD-95[3] or rabbit polyclonal antibodies to Kv1.4[8] are used. Reciprocal immunoprecipitation using antibodies specific for either of the binding partners is useful in confirming a positive result; e.g., immunoprecipitation of PSD-95 leads to the coimmunoprecipitation of Kv1.4 and immunoprecipitation of Kv1.4 leads to the coimmunoprecipitation of PSD-95.

Perform a negative control for the immunoprecipitation with the final 100-$\mu$l aliquot of detergent extract by immunoprecipitating with a nonspecific antibody such as rabbit IgG or the same quantity of an unrelated antibody. The nonspecific antibody should not immunoprecipitate either PSD-95 or Kv1.4 from the extracted fraction.

13. Incubate primary antibody and detergent extract at 4° for 1–2 hr with mixing by inversion.

14. Add 20 $\mu$l (20% of immunoprecipitation reaction volume) of protein A-sepharose conjugated beads (Pharmacia) preequilibrated in extraction buffer (50% slurry) and incubate for an additional 2 hr with mixing by inversion.

15. Pellet protein A-Sepharose beads by spinning in a microcentrifuge at maximum speed for 10 sec.

16. Wash protein A-Sepharose conjugated beads with 1 ml of ice-cold extraction buffer three times.

17. Elute bound proteins from the beads with SDS sample buffer, resolve eluted proteins by SDS–PAGE, and visualize components of immunoprecipitated protein complexes by immunoblotting using antibodies specific for PSD-95[3] or Kv1.4.[8] To strip and reprobe the filter, incubate the filter in stripping buffer [62 m*M* Tris (pH 6.7), 2% SDS, and 0.1% mercaptoethanol] at 60° for 30 min, block the filter with TBST containing 5% NFDM at room temperature for 30 min. Perform Western blotting as described earlier for the filter overlay assay.

[8] M. Sheng, M.-L. Tsaur, Y. N. Jan, and L. Y. Jan, *Neuron* **9,** 271 (1992).

## Colocalization of Ion Channel and Associated Protein *In Vivo*

A prerequisite for two proteins interacting *in vivo* is that they be expressed in the same cells and in the same subcellular location (at least in part). If the ion channel of interest has a specific expression pattern and a targeted subcellular localization, a similar distribution of its putative associated protein greatly supports the validity of the interaction *in vivo.* By far the best way to show colocalization is with double-labeling immunocytochemistry using specific antibodies in the two interacting proteins X and Y. Typically this is achieved using indirect immunofluorescence, with the secondary antibodies tagged by distinct fluorophores such as fluorescein (green emission) or Cy-3 (red). This requires that antibodies derived from different species (rabbit or mouse, for instance) be available for proteins X and Y; otherwise, the secondary antibodies cannot distinguish between the respective primary antibodies. At the electron microscopy (EM) level, the species-specific secondary antibodies can be labeled with gold particles of distinct size. In the available antibodies to the interacting proteins are from the same species, then the primary antibodies to X and Y can be directly tagged with different fluorophores or gold particles, but this is not an easy procedure.

Immunocolocalization can be achieved at various levels of spatial resolution. For instance, immunofluorescence confocal miroscopy can provide submicron resolution, while immunogold electron microscopy can resolve to ~10–20 nm. Obviously, the higher the resolution, the more meaningful the colocalization. For instance, two proteins can "colocalize" by light microscopy but one can be in the presynaptic membrane and the other in the postsynaptic membrane. Generally, it is impossible to distinguish pre- and postsynaptic localization at a given synapse by light microscopy alone—EM is required. Another point worth emphasizing is that colocalization does not necessarily indicate that the two proteins interact directly. Direct interaction can only be concluded from biochemical experiments *in vitro*. Nevertheless, colocalization of proteins X and Y by immunocytochemistry offers powerful support for the idea that X and Y interact *in vivo*, partly because immunostaining does not have some of the inherent biases of the biochemical approaches to showing association *in vivo* (see below). Here we provide a protocol for immunocytochemical colocalization of NMDA receptors and α-actinin-2 in rat brain.

### *Immunocolocalization of Ion Channel and Associated Protein in Rat Brain*

1. Using a Vibratome tissue sectioning system (Technical Products International, Inc., St. Louis, MO), prepare floating brain sections (50

μm thick) from anesthetized adult rats (~6 weeks age) perfused transcardiacally with 4% paraformaldehyde.

2. Permeabilize sections with phosphate-buffered saline containing either 0.1–0.3% Triton X-100 or 50% (v/v) ethanol at room temperature for 15 min.
3. Wash sections in phosphate-buffered saline at room temperature for 10 min three times.
4. Block potential nonspecific antibody binding sites by incubating the sections in antibody dilution solution (phosphate-buffered saline containing 3% horse or goat serum and 0.1% crystallized BSA) at room temperature for 1 hr. If detergent permeabilization was employed, include 0.1–0.3% Triton X-100 in the antibody dilution solution.
5. Add primary antibodies that specifically recognize NR1 receptor subunits (2 μg/ml of mouse monoclonal antibodies 54.1; PharMingen, San Diego, CA) and α-actinin-2 (1 μg/ml of rabbit polyclonal antibodies 4B2[6]) together to the tissue sections floating in the antibody dilution solution. Incubate sections at room temperature overnight.
6. Next day, wash the sections thoroughly three times in phosphate-buffered saline at room temperature for 10 min.
7. Submerge the primary-stained sections in antibody dilution solution containing species-specific fluorophore-conjugated secondary antibodies (1:1000 dilution of anti-mouse Cy3- and 1:200 dilution of anti-rabbit FITC-conjugated antibodies; Jackson ImmunoResearch, West Grove, PA). Incubate at room temperature for 2 hr.
8. Wash sections three times in phosphate-buffered saline at room temperature for 10 min.
9. To mount sections, transfer stained sections to microscope slides, add one or two drops of fluorescence mounting medium, and secure section with a glass coverslip. The fluorescently stained sections can be visualized by epifluorescence light microscopy (e.g., using a Zeiss Axioskop microscope, Thornwood, NY) or by confocal microscopy (e.g., using a confocal Bio-rad Richmond, CA, MRC 1000 microscope).

## Coimmunoprecipitation of Ion Channel and Associated Protein from Native Tissue

The most direct way to confirm that two proteins are associated *in vivo* is to purify them as an intact complex from a native tissue. Coimmunoprecipitation using antibodies specific for either protein is the most convenient way to achieving this. For ion channels and their associated proteins, this approach presents some difficulties. First, ion channels are typically not abundant proteins, thus their purification by any method is tricky. A partial purification to enrich for the ion channel may be needed prior to the immunoprecipitation to increase the yield. Second, ion channels are integral membrane proteins and they need to be solubilized by detergents before they can be purified by immunoprecipitation. Some ion channels are highly insoluble—NMDA receptors, for instance, are tightly bound to the postsynaptic density and are difficult to extract with anything less than SDS. These solublization protocols may disrupt the interaction between the ion channel and their association proteins, leading to loss of the complex.

Typically, coimmunoprecipitation is performed using antibodies to protein X to immunoprecipitate the complex, and antibodies to protein Y to detect coimmunoprecipitated protein Y by Western blotting. Ideally, this coimmunoprecipitation should be performed in both directions, i.e., antibodies to either protein should precipitate the other binding partner. It is also important to assay for the amount of cognate antigen that is precipitated (i.e., the amount of protein X brought down by anti-X antibodies) and to compare it with the amount of protein Y that is coimmunoprecipitated (i.e., the amount of protein Y brought down by anti-X antibodies). This will give you an idea of the efficiency of coimmunoprecipitation, i.e., what fraction of Y is stably associated with X, etc. For this, it will be essential to include on the immunoblots control lanes that contain known or starting amounts of proteins X and Y in the extract to be immunoprecipitated. It is rare that the quantities of protein that can be isolated by immunoprecipitation are adequate to visualize by nonspecific means such as Coomassie blue staining. In any case, the immunoglobulins used for the immunoprecipitation constitute a major portion of the precipitated proteins and their heterogeneity or impurity will complicate the interpretation of nonspecific protein staining (especially when using sensitive methods such as silver staining). In general then, the detection of the coimmunoprecipitated protein relies on its prior identification as a putative associated polypeptide and on its detection by specific antibodies in immunoblotting.

Because of this inherent "bias," and because of the high sensitivity of immunoblotting, it is critical to apply a large number of negative controls to ensure that the coimmunoprecipitation is specific. For instance, a small amount of protein X and Y may nonspecifically stick to the immunoglobulins (Igs) or to the protein A-Sepharose beads used to precipitate the Igs, and be detected by immunoblotting and interpreted as coimmunoprecipitation of these proteins. Good controls include leaving out the immunoprecipitating antibody and using unrelated immunoprecipitating antibodies. In addition, control antibodies should be used to probe the immunoblots of the precipitated proteins, since the presence of unexpected extraneous proteins in these precipitates will raise concerns about the specificity of the coimmunoprecipitation. This latter problem is especially pertinent with ion channels, which are integral membrane proteins and sometimes difficult to solubilize. Detergent treatment of membranes may release large macromolecular aggregates or even small membrane domains, both of which may include proteins not even peripherally associated with the ion channel of interest. These proteins may be falsely coimmunoprecipitated with the ion channel and interpreted to be interacting with the ion channel. To reduce this problem, it is essential to clear the detergent extracts with high-speed ultracentrifugation (100,000*g* for 30 min at 4° to pellet partly solubilized membranes and large protein aggregates), and to use negative control antibodies to probe the immunoblots of the immunoprecipitated proteins.

In conclusion, the most meaningful results are those in which the coimmunoprecipitations (1) succeed in both directions (antibodies to protein X bring down protein Y, and antibodies to protein Y bring down protein X); (2) are highly specific (immunoprecipitates contain as little as possible other than proteins X and Y); and (3) are highly efficient (the ratio of amounts of X and Y in the immunoprecipitates is close to unity). Due to the inherent difficulties of working with nonabundant membrane proteins, it is extremely challenging to obtain convincing coimmunoprecipitation results approaching the ideal ones listed above. This is especially true when trying to do immunoprecipitation studies from the highly complex mixture of proteins in the brain.

### *Methods of Solubilization*

Since solubilization is critical for any form of purification (including immunoprecipitation), it is worthwhile to optimize this step prior to immunoprecipitation. For membrane proteins such as ion channels, we generally try three types of detergent: Triton X-100 (1–2%); RIPA (1% NP-40, 0.5% deoxycholate, 0.1% SDS); and SDS (1–2%), followed by excess Triton X-100 to sequester the SDS prior to addition of antibodies. In addition, other types of detergent such as CHAPS or digitonin can also be tried. These detergents are typically used in buffers of physiologic pH (pH 7.4–8.0) containing 150 m*M* salt. Finding the right detergent is largely empirical. The SDS approach is only used when other methods fail to solubilize the ion channel, since SDS is a relatively harsh denaturing detergent.

Next, we present a protocol for showing coimmunoprecipitation of NMDA receptors and PSD-95. NMDA receptors and PSD-95 are highly insoluble except in SDS detergent.

### *Coimmunoprecipitation of Ion Channel and Associated Protein from Native Tissue*

Due to the relative insolubility of NMDA receptors and PSD-95, the coimmunoprecipitation of NMDA receptors and PSD-95 complexes from rat brain membranes requires SDS solubilization of these proteins. Surprisingly, the complex remains associated in these conditions. The following protocol is adapted from Lau *et al.*[9]

1. Prepare a rat brain membrane fraction as described earlier for the pull-down assay.
2. For each immunoprecipitation reaction add 25 $\mu$l of the membrane fraction (~200 $\mu$g protein) to an equal volume of 2× ice-cold extraction buffer [100 m*M* Tris (pH 8.0), 300 m*M* NaCl, 5 m*M* KCl, 10 m*M* EDTA, 10 m*M* EGTA, and various protease inhibitors] containing 2% SDS.
3. Incubate at 4° for 2 hr with shaking.
4. Pellet insoluble materials at 60,000 *g* at 4° for 1 hr.
5. Remove the supernatant (extracted fraction) and "neutralize" the SDS by diluting in 5 volumes (250 $\mu$l) of 1× ice-cold extraction buffer [50 m*M* Tris (pH 8.0), 150 m*M* NaCl, 2.5 m*M* KCl, 5 m*M* EDTA, 5 m*M* EGTA, and various protease inhibitors] containing 2% Triton X-100.
6. Immunoprecipitate using 5 $\mu$g/ml of immunoprecipitating antibodies specific for either PSD-95[3] or SAP97.[10] The rest of the protocol is as described earlier for coimmunoprecipitation of full-length proteins expressed in heterologous cells.

## Conclusion

The analysis of a putative ion channel associated protein involves numerous steps ranging from *in vitro* to *in vivo* studies. Proving the validity of such an interaction *in vivo* is extremely difficult, and requires genetic as well as biochemical and histological approaches. The most important point is that multiple lines of evidence are necessary to build up a case for a biologically significant direct interaction between an ion channel and its putative associated proteins.

[9] L.-F. Lau, A. Mammen, M. E. Ehlers, S. Kindler, W. J. Chung, C. C. Garner, and R. L. Huganir, *J. Biol. Chem.* **271,** 21622 (1996).

[10] E. Kim, K.-O. Cho, A. Rothschild, and M. Sheng, *Neuron* **17,** 103 (1996).

Acknowledgments

We thank Elaine Aidonidis for assistance with the manuscript, and Yi-Ping Hsueh and Jerry Lin for helpful comments. M.S. is an assistant investigator of the Howard Hughes Medical Institute.

# [21] Signal Transduction through Ion Channels Associated with Excitatory Amino Acid Receptors

*By* KIYOKAZU OGITA and YUKIO YONEDA

## Introduction

Excitatory amino acid (EAA) receptors are currently divided into two major categories according to their signal transduction processes in the mammalian central nervous system (CNS), including the ionotropic and metabotropic classes.[1] The metabotropic EAA receptor family so far comprises at least eight different subunits (mGluR1–8) and is subclassified into three groups on the basis of differences in structure and signal transduction systems. Namely, group I (mGluR1 and 5) stimulates the formation of inositol 1,4,5-trisphosphate and diacylglycerol, whereas both groups II (mGluR2 and 3) and III (mGluR4, 6, 7, and 8) induce a reduction in the amount of intracellular cyclic adenosine monophosphate (cAMP). On the other hand, ionotropic EAA receptors linked to cation channels are pharmacologically and physiologically classified into two major categories according to differential sensitivities to *N*-methyl-D-aspartic acid (NMDA). The ionotropic receptors insensitive to NMDA [non-NMDA receptors; kainic acid (KA) and DL-$\alpha$-amino-3-hydroxy-5-methylisoxazole-4-propionic acid (AMPA)] consist of nine different subunits in the rodent brain (Table I). The NMDA receptor in the rat (mouse) is composed of heteromeric assemblies between the common principal subunit, NMDAR1 ($\zeta$1), and any one of four different modulatory subunits, NMDAR2A-D ($\varepsilon$1–4). The NMDA channel is highly permeable to $Ca^{2+}$, with different ion channels coupled to non-NMDA receptors. This chapter deals with a receptor binding assay, an electrophoretic mobility shift assay (EMSA) and an immunoblotting analysis for evaluation of signal transduction from membranes to nuclei.

[1] M. Hollmann and S. Heinemann, *Ann. Rev. Neurosci.* **17,** 31 (1974).



0076-6879/99 $30.00

TABLE I
SUBUNITS OF EAA RECEPTORS

| Subunit | | Amino acids in mature protein from rat (mass in daltons)[a] | Receptor type |
|---|---|---|---|
| Mouse | Rat | | |
| $\alpha 1$ | GluR1 | 889 (99,769) | AMPA/KA |
| $\alpha 2$ | GluR2 | 862 (96,400) | AMPA/KA |
| $\alpha 3$ | GluR3 | 866 (98,000) | AMPA/KA |
| $\alpha 4$ | GluR4 | 881 (101,034) | AMPA/KA |
| $\beta 1$ | GluR5 | 875 (102,785) | KA (low affinity) |
| $\beta 2$ | GluR6 | 877 (93,891) | KA (low affinity) |
| $\beta 3$ | GluR7 | 888 (100,000) | KA (low affinity) |
| $\gamma 1$ | KA1 | 936 (105,000) | KA (high affinity) |
| $\gamma 2$ | KA2 | 965 (109,000) | KA (high affinity) |
| $\varepsilon 1$ | NMDAR2A | 1445 (163,267) | NMDA |
| $\varepsilon 2$ | NMDAR2B | 1465 (162,875) | NMDA |
| $\varepsilon 3$ | NMDAR2C | 1218 (133,513) | NMDA |
| $\varepsilon 4$ | NMDAR2D | 1329 | NMDA |
| $\zeta 1$ | NMDAR1 | 920 (103,477) | NMDA |

[a] M. Hollmann and S. Heinemann, *Ann. Rev. Neurosci.* **17,** 31 (1994).

## Receptor Binding Assay

Although receptor binding assays have been widely employed to evaluate interactions between ligands and their corresponding receptors, these studies often meet with many methodological pitfalls and artifacts.[2,3] For accurate and reproducible detection of ligand binding to each EAA receptor, therefore, the following procedures should be carefully performed.

### *Preparation of Crude Synaptic Membranes from Rat Brain*

All procedures are carried out at temperature below 4°. All solutions employed in this preparation are filtered by a nitrocellulose membrane filter with a pore size of 0.45 $\mu$m in order to avoid microbial contamination.[3]

1. Rat brains are homogenized in 15 volumes of 0.32 *M* sucrose solution (ca. 25 ml/brain) using a Teflon–glass homogenizer, followed by centrifugation at 1000*g* for 10 min.
2. Supernatants ($S_1$ fraction) are carefully collected by decantation, with attention being given to avoiding contamination with the white loose pellets,

[2] Y. Yoneda, and K. Ogita, *in* "Current Aspects of the Neuroscience" (N. N. Osborne, ed.), Vol. 3, p. 227. Macmillan Press, London, 1991.

[3] Y. Yoneda and K. Ogita, *Anal. Biochem.* **177,** 250 (1989).

so that the $P_1$ fraction including nuclei, myelin, and cell fragments is removed.

3. The $S_1$ fractions are centrifuged at 15,000*g* for 20 min followed by the removal of the supernatants. The obtained pellets (the $P_2$ fraction) including mitochondria and synaptosomes are suspended in the same volume of cold water (25 ml/brain) using an ultradisperser.

4. After the suspensions stand for more than 15 min in an ice bath, the suspensions are centrifuged at 6000*g* for 20 min. Both the supernatants and the loosely sedimentary pellets are suspended by mildly shaking the tubes.

5. The suspensions are diluted two times with cold water followed by centrifugation at 50,000*g* for 30 min. The pellets obtained are pooled as crude synaptic membrane (CSM) fractions.

6. The CSMs are washed by suspension with the same volume of 50 m*M* Tris–acetate buffer (pH 7.4) and subsequently centrifuged at 50,000*g* for 30 min. The CSMs washed once or three times are termed as "nonwashed" or "washed" membranes, respectively. The CSMs are finally suspended in 20 ml/brain of 0.32 *M* sucrose or 50 m*M* Tris–acetate buffer (pH 7.4) and can be stored at −80° until use.

7. On the day of the binding experiments, the frozen suspensions are thawed at room temperature. The thawed suspensions are washed two more times by dilution five times with 50 m*M* Tris-acetate buffer (pH 7.4) followed by centrifugation at 50,000*g* for 30 min. The pellets are resuspended in 20 ml/brain of the same buffer, and the final suspensions are employed for the binding experiments as "untreated" membranes.

*Treatment by Triton X-100 of CSMs.* CSMs should be treated with a low concentration of Triton X-100 in order to avoid the possible contribution of an association of the radioactive endogenous ligand with some membrane-bound enzymes and/or transport systems for EAA, in addition to removal of endogenous modulators.[4,5] Such a Triton treatment is carried out as follows:

1. The thawed suspensions of CSMs are diluted with 50 m*M* Tris–acetate buffer (pH 7.4) at an approximate protein concentration of 0.3 mg/ml.
2. 10% (w/v) Triton X-100 is added to the suspensions at a final concentration of 0.08–0.1% (w/v) with gentle stirring.
3. After treatment for 10 min at 2°, the suspensions are centrifuged at 50,000*g* for 30 min. The resultant pellets are once more washed with the same volume of 50 m*M* Tris-acetate buffer (pH 7.4).

[4] K. Ogita, and Y. Yoneda, *Biochem. Biophys. Res. Commun.* **153,** 51 (1988).

[5] Y. Yoneda, K. Ogita, T. Ohgaki, S. Uchida, and H. Meguri, *Biochim. Biophys. Acta* **1021,** 74 (1989).

4. The final pellets are suspended in the same buffer and employed for the binding experiments as "Triton-treated" membranes.

## *Ligands for Excitatory Amino Acid Receptors*

Table II summarizes radioactive ligands used for labeling the EAA receptors. All of these ligands could be purchased from NEN/DuPont (Boston, MA). Although the stability of the ligands differs from each other, the rate of decomposition is within 0.2–1% per month when the ligands are stored in their original solvents and at their original concentrations. The stability of the ligands is usually worsened by dilution and by storage at high temperatures. Therefore, the ligand solutions should be subdivided into small quantities for storage at −20°. On the day of the experiments, the ligands can be diluted with the original solvent to the desired concentration. The diluted ligands must be discarded after use. In particular, endogenous radioactive ligands such as [$^3$H]glutamate (Glu) and [$^3$H]glycine (Gly) are especially unstable. If these endogenous ligands are stored for a long time, they should be purified by cation-exchange chromatography using a plastic column packed with Dowex 50W-×8 resin before use.[2,3]

### *General Procedures*

REACTION MIXTURES

CSM preparations (100–200 μg of protein) in 50 m*M* Tris–acetate buffer (pH 7.4), 100 μl
50 m*M* Tris–acetate buffer (pH 7.4), 400 μl
Radiolabeled ligand, 10 μl

1. Reaction mixtures contain an aliquot (100–200 μg of protein) of CSMs and various reagents or test compounds in 0.5 ml of 50 m*M* Tris-acetate buffer (pH 7.4).

TABLE II
RADIOACTIVE LIGANDS FOR EAA RECEPTORS

| Receptor | Ligands |
|---|---|
| NMDA receptor | |
| Ion channel domain | MK-801, (+)-[3-$^3$H]- |
| NMDA recognition domain | Glu, L-[3,4-$^3$H]-; CGP39653, [*propyl*-2,3-$^3$H]- |
| | CGS19755, [*piperidinyl*-$^3$H]- |
| | CPP, (±)-[*propyl*-1,2-$^3$H]- |
| Gly recognition domain | Gly, [2-$^3$H]-; DCKA, 5,7-[3-$^3$H]- |
| AMPA receptor | AMPA, DL-α-[5-*methyl*-$^3$H]-; CNQX, [5-$^3$H]- |
| KA receptor | KA, [*vinylidene*-$^3$H]- |

2. Incubation is initiated by the addition of 10 $\mu$l of a radioactive ligand and carried out for different periods of time at different temperatures depending on the ligand used, as described later.
3. After the reaction mixtures are incubated for a constant period of time until binding reaches equilibrium, the reaction is terminated by the addition of 3 ml of ice-cold 50 m*M* Tris-acetate buffer (pH 7.4) and subsequent rapid filtration through a Whatman (Clifton, NJ) GF/B glass-fiber filter under a constant vacuum of 15 mmHg using a vacuum pump (type XX55 100 00, Millipore, Bedford, MA)
4. The filter is then rinsed with 3 ml of the same buffer four times within 10 sec. Assays are always run one by one in triplicate with variations of less than 10%.
5. Radioactivity retained on the filter is measured by a liquid scintillation spectrometer using a Triton–toluene scintillant with a counting efficiency of more than 40%.

*NMDA Receptor*

The NMDA receptor is proposed to be a receptor ionophore complex consisting of at least three distinct domains having positive modulatory properties for opening cation channels within the complex[6,7]; there are (1) an NMDA recognition domain with a high affinity for Glu, (2) a Gly recognition domain insensitive to strychnine, and (3) a polyamine domain with several indefinite profiles. These domains can be labeled by respective radioactive ligands as above under distinct individual appropriate experimental conditions described below.

*NMDA Recognition Domain.* This domain can be labeled by an endogenous agonist [3,4-$^3$H]Glu[4,5] as well as the antagonists [*propyl*-1,2-$^3$H]3-(2-carboxypiperazin-4-yl)propyl-l-phosphonic acid (CPP),[8–10] [*propyl*-2,3-$^3$H]DL-(*E*)-2-amino-4-propyl-5-phosphono-3-pentenoic acid (CGP39653),[11,12] and [*piperidinyl*-$^3$H]*cis*-4-phosphonomethyl-2-piperidine carboxylic acid (CGS19755).[13]

[6] D. T. Monaghan, R. J. Bridge, and C. W. Cotman, *Ann. Rev. Pharmacol. Toxicol.* **29,** 365 (1989).

[7] Y. Yoneda and K. Ogita, *Neurosci. Res.* **10,** 1 (1991).

[8] D. E. Murphy, J. Schneider, C. Boehm, J. Lehmann, and M. Williams, *J. Pharmacol. Exp. Ther.* **240,** 778 (1987).

[9] Y. Yoneda, K. Ogita, T. Kouda, and Y. Ogawa, *Biochem. Pharmacol.* **39,** 225 (1990).

[10] K. Ogita and Y. Yoneda, *Brain Res.* **515,** 51 (1990).

[11] M. A. Sills, G. Fagg, M. Pozza, C. Angst, D. E. Brundish, S. D. Hurt, E. J. Wilusz, and M. Williams, *Eur. J. Pharmacol.* **192,** 19 (1991).

[12] P. Zuo, K. Ogita, T. Suzuki, D. Han, and Y. Yoneda, *J. Neurochem.* **61,** 1865 (1993).

[13] D. E. Murphy, A. J. Hutchinson, S. D. Hurt, M. Williams, and M. A. Sills, *Br. J. Pharmacol.* **95,** 932 (1988).

Labeling by [$^3$H]Glu is carried out as follows. Namely, an aliquot (approximately 150 $\mu$g of protein) of "Triton-treated" membranes is incubated with 10 n*M* [$^3$H]Glu for 10 min in an ice-cold water bath at 2°. To define nonspecific binding, unlabeled NMDA at 0.1 m*M* is added to the reaction mixture. NMDA-sensitive binding is calculated by subtracting the nonspecific binding from the total binding. The NMDA-sensitive portion of [$^3$H]Glu binding occupies more than 80% of the total binding. The binding determined under these conditions is sensitive to displacement not only by NMDA but also competitive antagonists for the NMDA recognition domain such as CPP, CGP39653, and D-2-amino-5-phosphonovalaric acid (D-AP5) but not to that by AMPA and KA, which is in good agreement with the data provided by electrophysiologic studies.

Labeling by the competitive NMDA antagonists of the NMDA recognition domain is performed by incubating aliquots (approximately 150 $\mu$g of protein) of Triton-treated membranes with 2 n*M* [$^3$H]CGP39653 or 10 n*M* [$^3$H]CPP in an ice-cold water bath for 60 or 10 min, respectively. Nonspecific binding is defined by 1 m*M* nonradioactive Glu. A Scatchard analysis reveals that the dissociation constant for [$^3$H]CGP39653 binding ($K_d$ of 4–10 n*M*) is more than 10-fold lower than that for [$^3$H]CPP binding, indicating that [$^3$H]CGP39643 has the highest affinity for the NMDA recognition domain so far. However, some properties of binding of [$^3$H]CGP39653 and [$^3$H]CPP are distinct from those of [$^3$H]Glu binding.[10,12] For example, binding of [$^3$H]CGP39653 and [$^3$H]CPP is more sensitive to displacement by antagonists for the NMDA recognition domain than that by the agonists, while [$^3$H]Glu binding is potently inhibited by the agonists as compared to the antagonists. In addition, [$^3$H]CGP39653 binding, but not [$^3$H]Glu binding, is significantly inhibited by pretreatments of CSMs with sulfhydryl-reactive agents and phospholipases. Furthermore, [$^3$H]CGP39653 binding, but not [$^3$H]Glu binding, is potentiated by the addition of the selective antagonist for the polyamine domain, bis(3-aminopropyl)nonanediamine (TE393).[14] The potentiation by TE393 of [$^3$H]CGP39653 binding is selectively enhanced by pretreatment with phospholipase $A_2$, without being altered by that with phospholipases C and D as well as various glycosidases (Fig. 1). Because these data give rise to the presence of multiple forms of the NMDA recognition domain such as agonist- and antagonist-preferring forms, both [$^3$H]Glu and [$^3$H]CGP39653 should be used to characterize correctly the NMDA recognition domain by ligand binding techniques.

[14] Y. Yoneda, K. Ogita, R. Enomoto, S. Kojima, M. Shuto, A. Shirahata, and K. Samejima, *Brain Res.* **679,** 15 (1995).

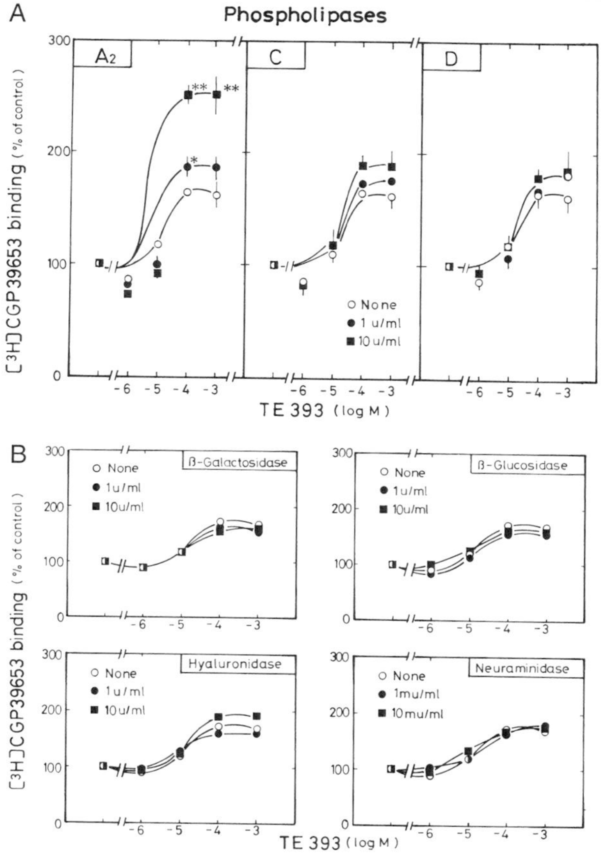


FIG. 1. Effects of pretreatment with phospholipases (A) and glycosidases (B) on the potentiation by TE393 of [$^3$H]CGP39653 binding. After "Triton-treated" membranes have been incubated with various units of each enzyme indicated at 30° for 30 min, the membranes were washed twice with 50 m*M* Tris–acetate buffer (pH 7.5) by centrifugation. An aliquot of the pretreated membranes was incubated with 2 n*M* [$^3$H]CGP39653 in buffer containing four different indicated concentrations of TE393. Values were from at least six independent experiments. *$P<0.05$, **$P<0.01$, significantly different from each value obtained with membranes pretreated in the absence of added enzymes.

*Gly Recognition Domain.* This domain is also labeled by both an endogenous agonist, [2-$^3$H]Gly,[15,16] and a competitive antagonist for the domain, [3-$^3$H]5,7-dichlorokynurenic acid (DCKA).[16] In brief, an aliquot (approximately 150 $\mu$g of protein) of Triton-treated membranes is incubated with

[15] K. Ogita, T. Suzuki, Y. Yoneda, *Neuropharmacology* **28,** 1263 (1989).
[16] Y. Yoneda, K. Ogita, and T. Suzuki, *J. Neurochem.* **55,** 237 (1990).

10 n$M$ [$^3$H]Gly or [$^3$H]DCKA for 10 min in an ice-cold water bath at 2°. Nonspecific binding of [$^3$H]Gly and [$^3$H]DCKA is defined by the addition of 1 m$M$ D-serine and 1 m$M$ Gly, respectively. The affinity for [$^3$H]DCKA binding is about 10-fold higher than that for [$^3$H]Gly binding. The possible multiplicity of the Gly recognition domain, such as agonist- and antagonist-preferring forms as seen with the NMDA recognition domain, has also been suggested by some differential pharmacologic profiles between binding of [$^3$H]Gly and [$^3$H]DCKA.[17] To characterize accurately the Gly recognition domain, therefore, it is desirable to measure both [$^3$H]Gly and [$^3$H]DCKA binding.

*Ion Channel Domain.* Ion channels opened by the activation of NMDA receptor agonists are neurochemically labeled by noncompetitive antagonists such as (+)-5-methyl-10,11-dihydro-5*H*-dibenzo[*a,d*]cyclohepten-5,10-imine (MK-801)[18–20] and *N*-[1-(2-thienyl)cyclohexyl]piperidine (TCP).[21,22] These compounds noncompetitively prevent the influx of cations permeable to the NMDA channels by associating with sites within the channels which are responsible for mediating the NMDA responses. Because [3-$^3$H]MK-801 is widely employed as a radioligand for labeling the NMDA channels, the experimental conditions are described in detail with this radioligand here.

In general, [$^3$H]MK-801 binding is performed by incubating CSM preparations (100–200 $\mu$g of protein) with 5 n$M$ [$^3$H]MK-801 for a constant period at 30° in a water bath. Nonspecific binding is defined by the addition of both 100 $\mu M$ D-AP5 (NMDA antagonist) and 100 $\mu M$ 7-chlorokynurenic acid (Gly antagonist) in order to completely inhibit the opening processes of the NMDA channels. However, other experimental conditions including the CSM preparations used, the period of incubation, and the addition of agonists should be carefully determined because [$^3$H]MK-801 binding is drastically changed by agonists including Glu, Gly, and polyamines in the reaction mixtures. For instance, [$^3$H]MK-801 binding to "nonwashed" membranes is not significantly enhanced by the addition of Glu and Gly with an incubation period of 30 min and it reaches equilibrium within 30 min, because endogenous amino acids including Glu and Gly exist in the preparations. Therefore, nonwashed membrane preparations should be incubated with 5 n$M$ [$^3$H]MK-801 for more than 30 min at 30° for the purpose of

[17] Y. Yoneda, T. Suzuki, K. Ogita, and D. Han, *J. Neurochem.* **60,** 634 (1993).

[18] E. H. F. Wong, J. A. Kemp, T. Priestley, A. R. Knight, G. N. Woodruff, and L. L. Iversen, *Proc. Natl. Acad. Sci. U.S.A.* **83,** 7107 (1986).

[19] Y. Yoneda and K. Ogita, *Brain Res.* **499,** 305 (1989).

[20] Y. Yoneda, K. Ogita, and R. Enomoto, *J. Pharmacol. Exp. Ther.* **256,** 1161 (1991).

[21] G. E. Fagg, *Neurosci. Lett.* **76,** 221 (1987).

[22] K. Ogita, T. Nabeshima, and Y. Yoneda, *J. Neurochem.* **55,** 1639 (1990).

detecting the NMDA channels fully opened by agonists for the NMDA and Gly recognition domains.

In Triton-treated membranes, by contrast, the binding of [$^3$H]MK-801 does not reach equilibrium within 24 hr in the absence of added agonists because small amounts of endogenous agonists exist in these Triton-treated membranes, which are extensively washed by the buffer following treatment with Triton X-100 at a low concentration. The binding is markedly potentiated by 10 $\mu M$ Glu alone with an acceleration of the initial association rate of the binding. The binding reaches equilibrium within 6 hr in the presence of Glu alone. Further addition of 10 $\mu M$ Gly accelerates the initial association rate of the binding in the presence of Glu alone without altering the binding at equilibrium. Moreover, the addition of 1 m$M$ spermidine markedly increases the binding at equilibrium, with acceleration of the initial association rate of the binding in the presence of both Glu and Gly. Thus, Triton-treated membrane preparations are useful for determining the effects of agonists and competitive antagonists for the domains of NMDA and Gly. In fact, [$^3$H]MK-801 binding to Triton-treated membrane preparations with an incubation period of 30 min is drastically potentiated by the addition of Glu in a concentration-dependent manner at concentrations of more than 10 n$M$, although the binding is hardly detected in the absence of added agonists. The potentiation by Glu of [$^3$H]MK-801 binding in Triton-treated membrane preparations is completely inhibited by antagonists for the NMDA recognition domains such as D-AP5, CPP, and CGP39653. Similarly, Gly antagonists, including 6,7-dichloroquinoxaline-2,3-dione and DCKA, completely abolish the potentiation by Glu and Gly. Accordingly, Triton-treated membrane preparations are useful for detecting the potentiation by the agonists or competition with the antagonists for the binding of [$^3$H]MK-801. However, the treatment with detergents may remove the proteins and lipids regulating activities of the NMDA channels as well as endogenous agonists, followed by possible changes in the functions of the NMDA channels. Therefore, "untreated" membranes, which are extensively washed by the buffer but not treated with a detergent, can also be employed for evaluating the regulations of the NMDA channels.

*Polyamine Domain.* No selective radioligands for the polyamine domain have been discovered until now. Therefore, [$^3$H]MK-801 has been used for labeling the polyamine domain, since [$^3$H]MK-801 binding in the presence of both agonists for NMDA and Gly recognition domains is further potentiated by endogenous polyamines such as spermidine and spermine with enhancement of binding at equilibrium.[20] The potentiation of binding by polyamines could be detected both after and before equilibrium, with this potentiation differing from that with Glu and Gly. Namely, potentiation by polyamines of [$^3$H]MK-801 binding is measured by incubating nonwashed membranes with 5 n*M* [$^3$H]MK-801 for 30 min at 30° without the addition of other agonists, including Glu and Gly. The addition of 1 m*M* spermidine doubles [$^3$H]MK-801 binding to nonwashed membranes. In Triton-treated membranes, polyamines have the ability to potentiate [$^3$H]MK-801 binding in the presence of added Glu and Gly at concentrations of more than 1 $\mu M$. In contrast, [$^3$H]MK-801 binding is less sensitive to polyamines in the absence of both agonists than in their presence in Triton-treated membranes.

*Non-NMDA Receptors*

Non-NMDA receptors are distinguishable by their preference to activation by agonists other than NMDA, including AMPA and KA. Molecular biological studies have demonstrated that AMPA receptors, which are also known as AMPA/KA receptors and are activated by KA as well as AMPA, consist of at least four different subunits (GluR1–4) in the rat brain, and that two types of KA binding sites with different affinities are composed of subunits of the two different groups GluR5–6 and KA1–2. These non-NMDA receptors have been labeled using [5-*methyl*-$^3$H]AMPA,[23,24] [5-$^3$H] 6-cyano-7-nitroquinoxaline-2,3-dione (CNQX),[25] and [*vinylidene*-$^3$H]KA.[24]

The binding assay of [$^3$H]AMPA is carried out by incubating an aliquot (100–150 $\mu$g of protein) of Triton-treated membranes with 10 n*M* [$^3$H]AMPA for 30 min at 2° in an ice-cold water bath in either the presence or absence of 100 m*M* KSCN. Specific binding is estimated by subtracting the nonspecific binding obtained in the presence of 1 m*M* Glu from the total binding. The specific binding of [$^3$H]AMPA is almost quadrupled by the addition of 100 m*M* KSCN. The addition of KSCN is effective in significantly decreasing the potencies to displace binding of antagonists such as CNQX and 6,7-dinitroquinoxaline-2,3-dione (DNQX) without affecting those of agonists including quisqualic acid, AMPA, Glu, and KA.[24] On the other hand, KA receptors could be labeled by the incubation of Triton-treated membranes (100–150$\mu$g of protein) with 10 n*M* [$^3$H]KA for 30 min in an ice-cold water bath. Nonspecific binding is defined by the addition of 1 m*M* Glu.

*Notes*

1. Purity of radiolabeled ligands is very important for a receptor binding assay because the data are calculated from just the radioactivity remaining on a filter. Accurate data could not easily be obtained if radiolabeled ligands

[23] D. E. Murphy, E. W. Snowhill, and M. Williams, *Neurochem. Res.* **12,** 775 (1987).

[24] K. Ogita, T. Sakamoto, D. Han, Y. Azuma, and Y. Yoneda, *Neurochem. Int.* **24,** 379 (1994).

[25] T. Honoré, J. Drejer, E. Ø. Nielsen, and M. Nielsen, *Biochemical. Pharmacol.* **38,** 3207 (1989).

that readily break down are employed in the assay. Accordingly, the authors would again recommend using radiolabeled ligands that are as fresh as possible or purified before use.

2. The binding to EAA receptors is reasonably affected by endogenous low molecular weight substances, including certain amino acids (Glu, aspartic acid, Gly, D-serine, and D-alanine, etc.), kynurenate, derivatives, and polyamines. It is therefore essential to remove these endogenous substances from membrane preparations employed by extensively washing with the buffer and/or further treatment with detergents at low concentrations.

3. In the filtration method for the receptor binding assay, it is crucial that the buffer used to finish the incubation and to rinse the filters be at a low temperature. Dissociation of a bound ligand occurs during the rinsing of the filters, which depends on the temperature of the rinsing buffer. The authors recommend that the rinsing buffer be exhaustively cooled not only by placing it on ice, but also by keeping pieces of frozen buffer. Further recommendation for an accurate and reproducible filtration assay is to filter for the rinse with a constant rhythm as rapidly as possible.

## Electrophoretic Mobility Shift Assay

EMSA has been widely employed for detecting the binding activities of DNA binding proteins, including the transcription factors for DNA with their recognition elements. EMSA is a convenient tool not only for the characterization of transcription factors, but also for studies on regulations of transcription of genes through cell signaling. In this section, EMSA procedures are explained, and studies of changes in the DNA binding activities of transcription factors through the activation of EAA receptors are described as an application model for EMSA.[26–28]

### *Preparation of Solutions Employed*

*Buffer A.* 10 m*M* HEPES–NaOH buffer (ph 7.9) containing 10 m*M* KCl, 1 m*M* ethylenediaminetetraacetic acid, trisodium salt (EDTA), 1 m*M* O,O-bis(2-aminoethyl)ethyleneglycol-N,N,N′,N′-tetraacetic acid (EGTA), 5 m*M* dithiothreitol (DTT), 10 m*M* of phosphatase inhibitors [sodium fluoride (NaF) and sodium $\beta$-glycerophosphate (GP)], and 1 $\mu$g/ml of various protease inhibitors [($\beta$-amidinophenyl) methanesulfonyl fluoride, benzamidine, leupeptin, and antipain].

Buffer A containing all ingredients except DTT, phosphatase inhibitors, and protease inhibitors is autoclaved and allowed to be stored at room

[26] K. Ogita and Y. Yoneda, *J. Neurochem.* **63,** 525 (1994).
[27] Y. Yoneda and K. Ogita, *Neurochem. Int.* **25,** 263 (1994).
[28] Y. Azuma, K. Ogita, and Y. Yoneda, *Neurochem. Int.* **29,** 289 (1996).

temperature. On the day of the experiments, DTT, phosphatase inhibitors, and protease inhibitors are added to the stock solution and then stored at 4°.

*Buffer B.* 50 m*M* Tris-HCl buffer (pH 7.5) containing 400 m*M* KCl, 1 m*M* EDTA, 1 m*M* EGTA, 5 m*M* DTT, 10% (v/v) glycerol, 10 m*M* of phosphatase inhibitors, and 1 $\mu$g/ml of protease inhibitors. Buffer B not containing DTT, phosphatase inhibitors, and protease inhibitors is autoclaved and stored at room temperature. On the day of the experiments, DTT, phosphatase inhibitors, and protease inhibitors are added to the stock solution and then stored at 4°.

*Incubation Buffer.* 50 m*M* Tris-HCl buffer (pH 7.5) containing 20 m*M* $MgCl_2$, 1 m*M* EDTA, 1 m*M* EGTA, 5 m*M* DTT, 10% (v/v) glycerol, and 1 $\mu$g/ml of protease inhibitors.

The incubation buffer is sterilized by filtration through 0.2-$\mu$m nitrocellulose membrane filter sterilized as soon as possible after preparation. $MgCl_2$ should be excluded from the incubation buffer depending on the probe used.

*Poly(dI-dC) Solution.* Poly(dI-dC) · poly(dI-dC) (Pharmacia Biotech, Uppsala, Sweden) is dissolved with the incubation buffer at a concentration of 0.1 mg/ml. The solution should be subdivided into small quantities (0.5–1 ml/tube) and stored at −20°.

*TE.* 10 m*M* Tris-HCl buffer (pH 7.5) containing 1 m*M* EDTA. TE is autoclaved and stored at room temperature.

*10 × Klenow Buffer.* 66 m*M* Tris-HCl buffer (pH 7.5) containing 0.5 *M* NaCl, 66 m*M* $MgCl_2$ and 10 m*M* DTT. This buffer is sterilized by filtration through a 0.2-$\mu$m nitrocellulose membrane filter sterilized as soon as possible after preparation. A buffer appended to the commercial Klenow fragment of DNA polymerase I could be conveniently used.

*Mixed Nucleotide Solution.* Nucleotide solutions including deoxyCTP, deoxyTTP, and deoxyGTP (Takara Biochemicals, Kyoto, Japan) are mixed and diluted with TE at the respective concentration of 1 m*M*. The solution should be subdivided into small quantities and stored at −20°.

*Acrylamide Solution (20%).* 19.3 g of acrylamide (high grade) and 0.7 g of *N,N*-methyl-bis(acrylamide) (high grade) are dissolved with pure water at a final volume of 100 ml, followed by filtration through a nitrocellulose membrane (0.45 $\mu$m) in order to remove particles.

*5 × TGE.* 30.3 g of tris(hydroxymethyl)aminomethane (Tris, electrophoresis or biochemical grade), 142.7 g of glycine, and 4.12 g of EDTA are dissolved with pure water to a final volume of 1000 ml; 5 × TGE is filtered through a nitrocellulose membrane (0.45 $\mu$m) and then autoclaved.

*Dye Solution.* 50 m*M* Tris-HCl buffer (pH 7.5) containing 0.1% bromphenol blue (BPB) and 10% (v/v) glycerol.

*Preparation of Nuclear Extracts from Mouse Brain*

Nuclear extracts from mouse brain are prepared by the method of Schreiber *et al.*[29] with some modifications.[27,28] All procedures are carried out at below 4°. All microtubes should be autoclaved before use.

1. Whole brain or each brain structure, such as the cerebral cortex, hippocampus, striatum, hypothalamus, midbrain, cerebellum, and medulla pons, is homogenized in 50 volumes of buffer A using a Dounce homogenizer (Wheaton, Millville, NJ) with a B-type pestle. The homogenates are conveniently transferred into microtubes (1.5 ml).
2. 10% (w/v) Nonidet P-40 is added to the homogenates at a final concentration of 0.6% (w/v).
3. After standing for 5 min on ice, the homogenates are centrifuged at 20,000*g* for 5 min.
4. After the supernatants have been completely removed by aspiration, the remaining pellets are suspended in 10 volumes of buffer B using a vortex mixer or a spin homogenizer for a microtube.
5. The suspensions are kept on ice to extract nuclear proteins for 30 min followed by being centrifuged at 20,000*g* for 5 min. The final supernatants are stored at −80° as nuclear extracts for EMSA, after the concentration of protein is assayed.

*Preparation of Probes*

Assays are performed with one of a double-stranded oligonucleotide with a base length of 22-mer containing each consensus sequence as a probe for the individual transcription factor. The following oligonucleotides are probes designed by the authors to detect DNA binding activities of transcription factors including activator protein-1 (AP-1), cAMP-responsive element binding protein (CREB), and c-Myc. Nucleotide sequences for these probes are almost identical to one another except for the respective consensus elements written in bold letters as follows:

| | |
|---|---|
| AP-1 | 5′-CTAGTGA**TGAGTCA**GCCGGATC-3′<br>3′-GATCACT**ACTCAGT**CGGCCTAC-5′ |
| CREB | 5′-CTAGTGA**TGACGTCA**GCCGGAT-3′<br>3′-GATCACT**ACTGCAGT**CGGCCTA-5′ |
| c-Myc | 5′-CTAGTGA**CACGTG**GCCGGATCA-3′<br>3′-GATCAC**TGTGCAC**CGGCCTAGT-5′ |

[29] E. Schreiber, P. Matthias, M. M. Müller, and W. Schaffner, *Nucleic Acids Res.* **17,** 6419 (1989).

1. The couple of synthesized oligonucleotides, except for the nucleotides underlined for each factor, are dissolved in TE. The oligonucleotides are annealed in a Klenow buffer at a final concentration of 5 $\mu M$ as follows. The oligonucleotide solutions are heated at 65–70° for 20 min followed by slowly falling temperature until the solutions reach room temperature on the heating block.

2. The annealed double-stranded oligonucleotides are labeled with [$\alpha$-$^{32}$P]deoxyATP using a Klenow fragment of DNA polymerase I. In brief, 50 $\mu$l of the reaction mixture is incubated at 25° for 30 min in Klenow buffer [10 n$M$ oligonucleotides, 40 $\mu M$ mixed nucleotides, 5–6 MBq of [$\alpha$-$^{32}$P]deoxyATP (111 TBq/mmol), and 2 units of Klenow fragment (Takara Biochemicals, Kyoto, Japan)].

3. The reactions are then stopped by the addition of 5 $\mu$l of 0.5 $M$ EDTA, and the labeled oligonucleotides are separated from the enzyme and free nucleotides by gel filtration on a Nick column (Pharmacia Biotech, Uppsala, Sweden) that has been equilibrated with TE. The labeled oligonucleotide fractions are pooled and diluted to 1 ml with TE as the probe. The probe should be subdivided into small quantities (100–200 $\mu$l) and stored at −20°.

*Preparation of Acrylamide Gel*

A method for the preparation of nondenatured acrylamide gels is described below. Composition for four pieces of 6% gels (90×80×1 mm):

Acrylamide solution (20%), 12 ml
5× TGE, 8 ml
Pure water, 20 ml
10% Ammonium persulfate (APS), 0.25 ml
*N,N,N′,N′*-Tetramethylethylenediamine (TEMED), 0.05 ml

1. Glass plates for preparation of gels are assembled according to the manufacturer's instructions.
2. Acrylamide solution, 5× TGE, and pure water are combined in a flask. The solution is then degassed by means of aspiration. The mixed solution is sufficiently cooled on ice.
3. APS and TEMED are added to the mixed solution and gently mixed, and then the solution is carefully poured to the top of assembled glass plates. Combs are inserted into the glass plates until the bottom of the teeth reach the top of the front plate. Make sure that no bubbles are trapped on the ends of the teeth. These steps should be carried out rapidly because polymerization will begin immediately after preparation of gels.

4. After the gels are polymerized, the combs are carefully removed from the glass plates. The wells are filled with 1× TGE to prevent drying of the tops of the gels and stored at 4° until use.

### *General Procedures*

EMSA is performed by incubating nuclear extracts with a radiolabeled probe in the presence of poly(dI-dC)·poly(dI-dC), which is added to protect the nonspecific binding, followed by separating the protein–probe complex from the free probe by nondenatured electrophoresis, as described below.

1. The gel plates are placed into an electrophoresis chamber, of which the reservoirs are filled by electrophoresis buffer (1× TGE: 5-fold dilution of 5× TGE with pure water). The electrophoresis chamber is placed into an ice-cold water bath, and the electrophoresis buffer is circulated using a circulating pump between the inner and outer reservoirs.
2. Ten microliters of poly(dI-dC) solution and 8 μl of nuclear extracts (3–5 μg) in buffer B are mixed in a microtube of 1.5 ml on ice.
3. Electrode plugs are attached to the proper electrodes of the electrophoresis chamber, and a prerun is started at a constant 80 V (11 V/cm high) for 30 min.
4. After the addition of 2 μl of probe, the reaction mixtures are incubated at 25° for 30 min in a water bath. Following the incubation, microtubes are put on ice for the termination.
5. After 1 μl of dye solution is added into the reaction mixtures and gently mixed, the reaction mixtures are pooled at the bottom of the microtube by centrifugation. All amounts of the reaction mixtures are loaded onto the gel after the prerun.
6. Electrophoresis is started to run at a constant voltage of 80 V (11 V/cm high) and continued until the dye front migrates to the approximately halfway position of the gel (about 90 min).
7. The gel is removed from the glass plate followed by fixation by a 10% (v/v) acetic acid–10% (v/v) methanol solution for 10–15 min until the color of the dye changes from blue to yellow-green. The fixed gel is dried on a filter paper by a gel dryer.
8. After covering the dried gel with vinylidene polymer plastic film such as Saran wrap, the radioactivities are quantitatively detected by either autoradiography using an X-ray film followed by a densitometory analysis or a bioimaging analyzer. The protein–probe complex and the free probe are detected at the upper and lower parts of the autoradiogram, respectively.

*Limited Proteolysis of Complex between Proteins and DNA Probe*

Probes for detecting transcription factors such as AP-1, CREB, and c-Myc can bind to either homodimers or heterodimers composed of different proteins. For instance, the probe for AP-1 could bind to the homodimers between Jun family proteins including c-Jun, JunB, and JunD and heterodimers between Jun family proteins and Fos family proteins (c-Fos, FosB, Fra1, Fra2, etc.). The probe for CREB also binds to homodimers between a number of family proteins such as CREB and ATFs. The probe for detecting c-Myc recognizes not only the heterodimer between c-Myc and Max, moreover, but also the Max homodimer and Max/Mad heterodimer. These findings mean that a number of different proteins could bind to a certain identical probe. Thus, proteins forming a complex with a certain probe could be different from each other depending on the nuclear extracts used. To differentiate proteins bound to a probe, electrophoresis is carried out after limited proteolysis of the protein–probe complex. Limited proteolysis of the complex results in either shifts to lower parts or disappearance of the radioactive bands. The differentiation of changes in the mobility of radioactive bands following proteolysis is supporting evidence for the distinction of proteins with affinities for the probe. The procedures for limited proteolysis are described below.

1. An aliquot (3–5 $\mu$g) of nuclear extracts is incubated with a probe in the presence of poly(dI-dC)·poly(dI-dC) at 25° for 30 min.
2. Proteases such as *Staphylococcus aureus* $V_8$ protease and trypsin are added to the reaction mixture at various different concentrations followed by an incubation at 25° for 10 min. The following procedures are performed in the same way as those described in the General Procedures section.

*Supershift Analysis*

The effects of added specific antibodies on the migration of radioactive bands in EMSA are tested in order to determine which proteins are responsible for the association with a probe. Namely, the addition of specific antibodies against proteins associating with the probe shifts upward the mobility of complex between the proteins and the probe. A supershift analysis is carried out by two kinds of procedures with distinct timings for the addition of an antibody both before and after the binding reactions as follows: (1) After nuclear extracts are incubated with an antibody in reaction media not containing a probe at 4° for overnight, an incubation with a probe is started. (2) After an aliquot of nuclear extracts has been incubated with a probe under the usual conditions, an antibody is added to the reaction mixture and incubated at 4° overnight.

### *Changes in DNA Binding Activities of Transcription Factor through in Vivo Activation of EAA Receptors*

Adult male Std-ddY mice weighing 30–35 g are housed in metallic breeding cages in a room with a light–dark cycle of 12 hr with a humidity and a temperature of 50–60% and 25°, respectively, for at least 5 days before use. Animals are intraperitoneally injected with phosphate-buffered saline (PBS), NMDA, or KA at different doses, followed by sacrifice at different periods after the administration. Nuclear extracts are prepared from different brain structures of mice, followed by being applied to EMSA.

Figure 2A (lanes S1, N1, and K1) shows that the systemic administration of NMDA and KA is effective in markedly potentiating DNA binding activities of the probe for AP-1 in the hippocampus. The selective potentiation by the injection of NMDA could be observed in the hippocampus 1–2 hr after the administration at a dose range of 50–150 mg/kg. However, no significant alteration is detected in the binding of the probe for CREB and c-Myc in any brain structures 2 hr after the administration of NMDA.[26] The administration of KA is also effective in markedly enhancing binding of a probe for AP-1 in the hippocampus 2 hr after the administration at a

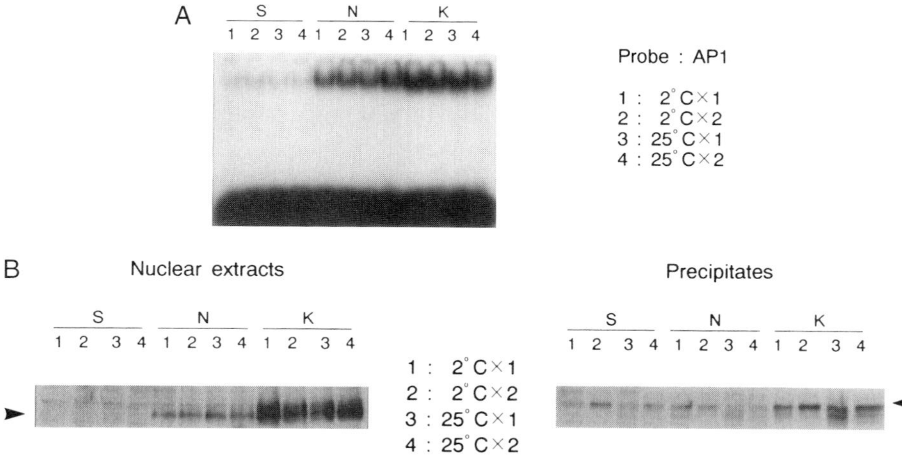


FIG. 2. Effects of different conditions for preparations of nuclear extracts on binding of a probe for AP-1 (A) and detection of proteins reactive to the anti-c-Fos antibody (B). Animals were sacrificed 2 hr after the i.p. administration of PBS (S), 100 mg/kg of NMDA (N), or 40 mg/kg of KA (K). Nuclear extracts were prepared by incubating once or twice for 30 min at 2 or 25° nuclear fractions from the hippocampus of each animal. (A) Nuclear extracts prepared under each set of conditions were applied to EMSA using a probe for the detection of AP-1. (B) The remaining precipitates as well as the nuclear extracts prepared under each set of conditions were applied to immunoblotting using the antibody against c-Fos.

dose range of 10–40 mg/kg. The potentiation by KA continues until at least 24 hr after the administration. At a dose of 40 m/kg, binding to the AP-1 probe is significantly potentiated in any brain structures except the cerebellum. Binding of a probe for CREB is significantly potentiated by the injection of KA in the hippocampus, but not in any other brain regions.[27] Thus, marked potentiation of DNA binding activities of AP-1 is induced by the *in vivo* activation of EAA receptors, including NMDA and KA receptors.

A supershift assay using an antibody reactive to c-Fos and c-Jun is carried out according to the aforementioned procedures. The addition of the anti-c-Jun antibody significantly inhibits AP-1 binding in nuclear extracts from the hippocampus of mice injected with PBS, NMDA, and KA. On the other hand, treatment by the anti-c-Fos antibody induces the appearance of a new radioactive band with lower mobility in nuclear extracts from mice injected with NMDA and KA, but not in those from mice injected with PBS. The supershift is more marked in samples of animals injected with KA than that in those with NMDA. These results suggest that facilitated expression of c-Fos is induced in the hippocampus through the activation of EAA receptors *in vivo*.

*Notes*

1. Removal of phosphatase inhibitors such as NaF and GP from all buffers used in EMSA results in a significant potentiation of binding on probes for AP-1 and CREB in a temperature-dependent fashion. In the absence of DTT, binding of probes for AP-1 and CREB is markedly decreased in nuclear extracts irrespective of the temperature (Table III). These findings mean that DNA binding activities of AP-1 and CREB might be regulated by phosphorylation and redox states within the protein molecules. Accordingly, phosphatase inhibitors and DTT should be added to all of the buffers used in order to avoid artifactual influences by endogenous phosphatases and oxidation throughout experiments.[30]

2. Extraction of proteins from the nucleus is important for the accurate detection of intracellular signal transduction by using EMSA. Therefore, different experimental procedures are employed to evaluate the efficacy of extraction of nuclear proteins as shown in Fig. 2. Nuclear proteins are extracted from the hippocampus of mice killed 2 hr after the injection of PBS, NMDA, and KA. Nuclear fractions are incubated with extraction buffer (buffer B) once or twice for 30 min at 2° or 25°. No marked changes in AP-1 binding are observed in nuclear extracts obtained under different

[30] Y. Yoneda, Y. Azuma, K. Inoue, K. Ogita, A. Mitani, L. Zhang, S. Masuda, M. Higashihara, and K. Kataoka, *Neuroscience* **79,** 1023 (1997).

TABLE III
CONDITIONS FOR HOMOGENIZATION[a]

| Condition | DNA binding (% of control) AP-1 | CREB |
|---|---|---|
| At 2° | | |
| Whole system | 100 | 100 |
| (NaF) | 107 | 116 |
| (GP) | 203 | 144 |
| (NaF and GP) | 326 | 168 |
| (DTT) | 25 | 25 |
| At 30° | | |
| Whole system | 100 | 100 |
| (NaF) | 153 | 170 |
| (GP) | 405 | 232 |
| (NaF and GP) | 813 | 300 |
| (DTT) | 22 | 38 |

[a] Whole brains of mice were homogenized in 10 m*M* HEPES–NaOH (pH 7.9) containing different ingredients indicated, followed by maintenance at 2° or 30° for 30 min. Subsequently, nuclear extracts were obtained from the individual homogenates for determination of the binding of probes of two different transcription factors.

extraction conditions (Fig. 2A). In addition, immunoreactive c-Fos protein is similarly extracted from nuclear fractions as nuclear extracts, irrespective of the extraction procedures employed, when samples from animals injected with NMDA and KA are used (Fig. 2B, left-hand side). However, immunoreactive c-Fos protein could be still detected in the precipitate after extraction, independent of the extraction processes employed, in samples from KA-injected animals (Fig. 2B, right-hand side). Accordingly, the aforementioned procedures are reasonable for the preparation of nuclear extracts.

3. It is extremely important for the quantitative detection of DNA binding activity to sufficiently cool a whole electrophoresis chamber, and to rapidly load the reaction mixtures into wells of gel.

4. Specific DNA binding should be determined by simultaneous experiments using nonradioactive oligonucleotides with point mutation at consensus sequence and with different sequences at extra-consensus regions, respectively.

5. For the purpose of quantitative analysis in EMSA, experiments should be carried out within limited ranges of protein concentrations, in which the binding is linearly increased in proportion to the amounts of the protein employed.

## Immunoblotting

Immunochemical techniques, termed immunoblotting or Western blotting, are used to detect a protein immobilized in a matrix using a polyclonal or monoclonal antibody. Immunoblotting is an extremely useful technique for identifying a particular protein or epitope out of a number of different proteins following separation by sodium dodecyl sulfate (SDS)–polyacrylamide gel electrophoresis (PAGE) or two-dimensional (2-D) PAGE. Immunoblotting is also a convenient tool when used to characterize molecules involved in cell signaling. In this section, the general procedures for immunoblotting are described, and studies on changes in the expression of particular proteins through the activation of EAA receptors are described as an application model for immunoblotting.

### *Preparation of Solutions Used*

*Homogenizing Buffer.* 20 m*M* Tris-HCl buffer (pH 7.5) containing 1 m*M* EDTA, 1 m*M* EGTA, phosphatase inhibitors (10 m*M* NaF, 10 m*M* GP, 10 m*M* sodium pyrophosphate, and 1 m*M* sodium orthovanadate), and 1 $\mu$g/ml of various protease inhibitors [($\beta$-amidinophenyl)methanesulfonyl fluoride, benzamidine, leupeptin, and antipain]. The buffer should again be adjusted to pH 7.5 using HCl after preparation and subsequently filtered through a nitrocellulose membrane filter (0.45 $\mu$m).

*Transfer Buffer.* 300 m*M* Tris containing 5% methanol (solution A); 25 m*M* Tris containing 5% methanol (solution B); and 25 m*M* Tris containing 40 m*M* 6-amino-*n*-caproic acid and 5% methanol (solution C). These buffers are employed to transfer proteins in semidry electrophoretic transfer as discontinuous buffer system. Tris–Gly buffer (25 m*M* Tris containing 192 m*M* Gly and 5% methanol) is the most commonly used buffer system for transfer. However, the discontinuous buffer system usually has higher efficiency for transfer than the Tris–Gly buffer.

*TBST.* 20 m*M* Tris-HCl (pH 7.5) containing 137 m*M* NaCl and 0.05% (w/v) Tween 20. For the purpose of decreasing nonspecific signals and backgrounds, 0.05% (w/v) Tween 20 is allowed to displace 0.1% (w/v) Tween 20. TBST should be kept at 4° until use.

*5× SDS/ME Buffer.* 50 m*M* Tris-HCl (pH 6.8) containing 50% (v/v) glycerol, 5% SDS, 10% 2-mercaptoethanol (2-ME), and 0.02% BPB.

*Lysis Buffer.* 4.8 g of urea (ultrapure grade), 0.2 g of Nonidet P-40, 0.5 ml of 2-ME, 0.4 ml of ampholyte solution, pH 5-4 (for example, Bio-Lyte from Bio-Rad), and 0.1 ml of ampholyte solution (pH 3-7) are dissolved in pure water at a final volume of 10 ml.

*Blocking Agents.* 5% Skim milk or 4% bovine serum albumin (BSA) in TBST. Skim milk and BSA free of proteases and phosphatases should be employed.

*Antibody Solution.* First antibody solution is prepared by appropriate dilution with 1% skim milk or 1% BSA dissolved in TBST. Any second antibody solutions are appropriately diluted with 1% skim milk in TBST. The first antibody is polyclonal or monoclonal antibody against the protein of interest. The second antibody is an anti-immunoglobulin G (IgG) antibody linked with horseradish peroxidase against the first antibody used.

### *Preparation of Samples from Mouse Brain*

1. Adult male Std-ddY mice weighing 30–35 g are intraperitoneally injected with PBS, NMDA, or KA at different doses, followed by sacrifice at different periods after administration.
2. The whole brain or each brain structure, such as the cerebral cortex, hippocampus, striatum, hypothalamus, midbrain, cerebellum, and medulla pons, is homogenized in 10–20 volumes of homogenizing buffer using a Dounce homogenizer with a B-type pestle. The homogenates are transferred into microtubes (1.5 ml).
3. 10% (w/v) Nonidet P-40 is added to the homogenates at a final concentration of 0.6%.
4. After standing for 5 min on ice, the homogenates are centrifuged at 20,000 *g* for 5 min.
5. The supernatants are pooled as "Nonidet-soluble" fractions. The remaining pellets are suspended in the same volume of the homogenizing buffer as the "Nonidet-insoluble" fractions. There are nuclei in the "Nonidet-insoluble" fractions.
6. Protein concentrations of each fraction obtained are assayed.
7. To prepare the samples for SDS–PAGE, each fraction is boiled for 5 min in the presence of the SDS/ME buffer. For 2-D PAGE, on the other hand, the pellets remaining from the treatment with Nonidet P-40 are dissolved in Lysis buffer, followed by centrifugation at 20,000 *g* for 10 min. The supernatants are applied to 2-D PAGE. The samples for 2-D PAGE can be prepared from "Nonidet-soluble" fractions by the addition of all reagents composing the Lysis buffer at the same concentrations.

*Protein Blotting Using Semidry Electrophoretic Transfer*

1. Both SDS–PAGE and 2-D PAGE are performed in order to separate proteins in the samples according to Laemmli[31] and O'Farrell,[32] respectively.
2. During the PAGE, the transfer can be performed as follows: (a) Soak polyvinyl difluoride (PVDF) membranes (for example, Immobilon from Millipore) of the same size as the gel in 100% methanol for 20–30 sec and then in solution B for at least 30 min. (b) Soak filter papers (for example, Whatman 3MM) of the same size as the gel in solutions for transfer (two sheets in solution A, 1 sheet/gel in solution B, and two sheets + one sheet/gel in solution C).
3. After the PAGE, the gel removed from the glass plates is soaked and gently shaken in solution B for 5–10 min to equilibrate the gel.
4. The transfer sandwich is assembled in the semidry transfer cassette as follows: Two sheets of the filter papers wetted with solution A are placed next to the anode plate. Subsequently, gel–membrane sandwiches are arranged by piling a sheet of filter paper wetted with solution B, PVDF membrane, gel, and a sheet of paper wetted with solution C in this order, and then placed on the soluton A wetting paper beside of solution B wetting paper. Further, two sheets of filter paper wetted with solution C are applied, followed by the cathode plate placed on the top. When more than two gels of the same type are transferred simultaneously, the gel–membrane sandwiches can be piled with a piece of cellophane paper of the same size as the gel inserted between the respective filter papers next to the anode and cathode plates. It is important to avoid air bubbles between gel, membrane, and filter papers.
5. The transfer is performed at a constant current of 1–2 $mA/cm^2$ for 30 min.

*Immunodetection*

1. The PVDF membrane is rinsed once by TBST in a homemade dish of Parafilm with a size larger than the membrane filter, which is placed in the wetting case. All procedures are carried out in the dish placed in the wetting case.
2. The membrane filter is blocked by gently shaking in 5% skim milk or 4% BSA in TBST in order to avoid nonspecific adsorption between the antibodies and the membrane. The blocking of the membrane

[31] U. K. Laemmli, *Nature (London)* **1227,** 680 (1970).
[32] P. H. O'Farrell, *J. Biol. Chem.* **250,** 4007 (1975).

is carried out for 1 hr or overnight at room temperature or 4°, respectively.

3. The blocking agent is poured off and the membrane is then rinsed briefly TBST with shaking.
4. After pouring off TBST, the first antibody at the appropriate dilution is added to the membrane followed by rocking gently at room temperature for at least 2 hr on a seesaw shaker. Overnight incubations are quite possible and usually increase the sensitivity for detection.
5. After the first antibody solution is poured off the membrane, the membrane is washed twice with TBST for 10 min.
6. After pouring off TBST, the second antibody is added at the appropriate dilution to the membrane followed by rocking gently at room temperature for at least 1 hr.
7. After the second antibody solution is removed from the membrane, the membrane is washed three times with TBST for 10 min.
8. After pouring off TBST, an enhanced chemiluminescence (ECL) reagent (for example, ECL from Amersham and Renaissance from DuPont/NEN) is added to the membrane and rocked for about 1 min. After wrapping the membrane by Saran wrap, detection of signals is performed by exposing the membrane to an X-ray film for an appropriate period.

*Expression of Proteins through Activation of EAA Receptors*

As mentioned above, AP-1, binding is markedly and selectively potentiated in the hippocampus through the *in vivo* activation of EAA receptors, including the NMDA and KA receptors, by an intraperitoneal administration of NMDA and KA to mice, respectively. These findings let us suppose that the facilitated expression or phosphorylation/dephosphorylation of the proteins associated with the probe for AP-1 is induced in the nucleus of the hippocampus through the activation of EAA receptors. Accordingly, proteins reactive to antibodies against some nuclear transcription factors, such as c-Fos, c-Jun, and CREB, have been detected in either nuclear extracts or "Nonidet-soluble" fractions of the hippocampus of mice injected i.p. with PBS, NMDA, or KA, using the immunoblotting analysis following SDS–PAGE and 2-D PAGE.

Although proteins reactive to the anti-c-Fos antibody are not significantly detected in the hippocampus of mice injected with PBS, an injection of MNDA induces the transient expression of proteins reactive to the anti-c-Fos antibody 1–2 hr after injection. Notably, KA induces the long-term expression of several proteins with different molecular weights that are reactive to an antibody against c-Fos (Figs. 2B and 3). Anti-c-Jun antibody-

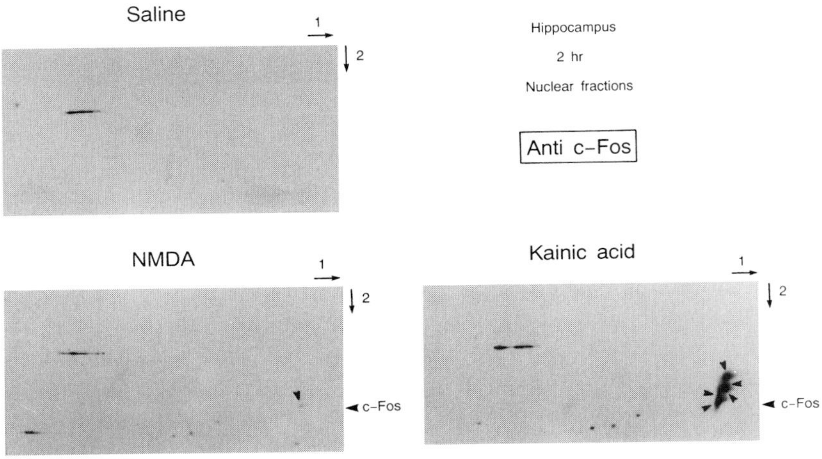


FIG. 3. Expression of proteins reactive to the antibody against c-Fos by administration of NMDA and KA. Animals were sacrificed 2 hr after the i.p. administration of PBS (saline), 100 mg/kg of NMDA, or 40 mg/kg of KA. The samples for 2-D PAGE were obtained from the Nonidet-insoluble fractions of the hippocampus of each animal. After 2-D PAGE followed by the transfer of proteins, immunodetection was performed using an antibody against c-Fos. Arrowheads indicate the signal response to the anti-c-Fos antibody.

reactive proteins are increased by NMDA as well as KA. Anti-CREB antibody-reactive proteins are also increased by KA but not by NMDA at a dose of at least 100 mg/kg. To evaluate phosphorylation and dephosphorylation through EAA signals, immunoblotting analysis has been performed using monoclonal antibodies against phosphotyrosine (PY) and phosphoserine (PS) following 2-D PAGE in hippocampal "Nonidet-insoluble" fractions of mice killed 2 hr after the i.p. administration of PBS, NMDA, and KA. The fractions obtained from mice injected with PBS have a number of proeins with different molecular weights and isoelectric points which are reactive to antibodies against PY or PS. The marked expression of new signals reactive to the anti-PY antibody is observed in fractions obtained from mice injected with NMDA and KA, while parts of the signals reactive to the anti-PS antibody disappear in the fractions obtained from mice injected with agonists for EAA receptors. These findings let us suppose that the activation of EAA receptors induces the phosphorylation of tyrosine residues and the dephosphorylation of phosphoserine residues, in particular proteins in the murine hippocampus. By using a preparative SDS–PAGE as shown in Fig. 4, purification of these proteins with responsiveness to EAA signals is attempted, although the identification of the proteins has not yet been successful.

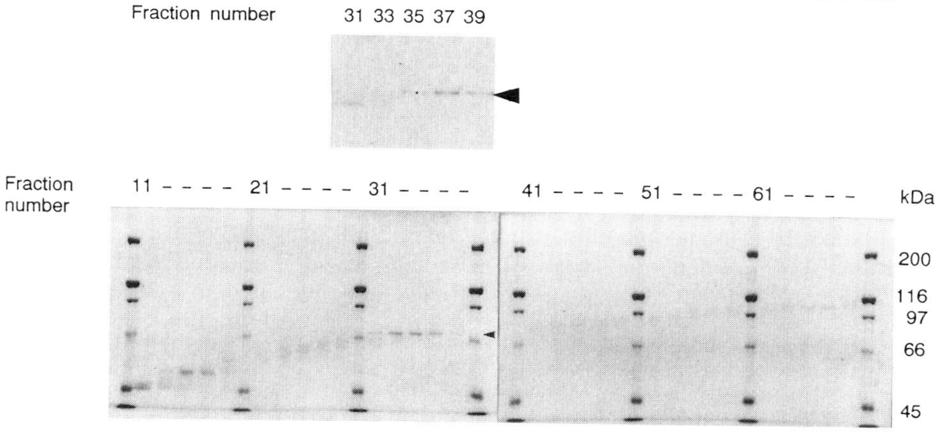


FIG. 4. Attempt at the purification of tyrosine-phosphorylated protein by preparative SDS–PAGE. *Top:* Immunodetection by the anti-PY antibody. *Bottom:* Pattern of Coomassie blue-stained protein eluted by preparative SDS–PAGE.

## *Notes*

1. Usually, the aforementioned transfer buffers work satisfactorily for semidry electrophoretic transfer. However, the conditions may have to be modified for the optimal transfer of a certain protein. For instance, the addition of methanol and SDS to the buffers greatly influences the transfer results. The addition of 20% methanol is recommended for the transfer of low molecular weight proteins, because methanol facilitates the dissociation of SDS–protein complexes and increases the hydrophobic interaction between the protein and the PVDF membrane for low molecular weight proteins. In contrast, high molecular weight proteins almost never require methanol for adequate binding to the membrane. High concentrations of methanol decrease the elution efficiency by denaturing the proteins or retarding the elution from the gel in the case of high molecular weight proteins. As for methanol, the addition of SDS to the transfer buffers has both positive and negative aspects. Namely, SDS gives the proteins a high elution efficacy from the gel due to its high charge, but decreasing the trapping of proteins to the membrane. The recommended concentrations of SDS vary between 0.01 and 0.1% when used in the cathode buffer only.[33]

[33] G. Jacobson, *in* "Protein Blotting: A Practical Approach" (B. S. Dunbar, ed.), p. 53. IRL Press, New York, 1994.

2. Because skim milk generally has more potent ability for the blocking of the PVDF membrane and is cheaper than BSA, skim milk is a good choice as an agent for blocking or the dilution of antibodies. However, skin milk cannot be employed for detection of the immunoreactivity with phosphorylated amino acids.

3. During reactions with the antibody, the antibody should be uniformly rocked on PVDF membrane without drying up. When reaction with the first antibody is carried out overnight, the membrane should be turned over.

4. If the signals are very weak, the following remedies are recommended: (a) Load more protein in the gel. (b) Incubate the first antibody overnight at 4°. (c) Use higher concentrations of the first antibody. Higher concentrations of the second antibody should not be employed because of potential increases in background and nonspecific signals.

5. If background and nonspecific signals are strong, the following remedies are recommended: (a) Use lower concentrations of antibody (especially the second antibody). (b) Increase the concentration of Tween 20 from 0.05 to 0.1% in TBST. (c) Purify the first antibody. The specificity of the first and second antibodies is the most important point for the immunodetection. Therefore, the antibodies employed should be purified at least until the IgG fraction on an affinity column packed with resins immobilized by protein A. An antibody against particular peptide should be purified using an affinity column immobilized with the peptide of the antigen.

# [22] Secondary Messenger Regulation of Ion Channels/Plant Patch Clamping

*By* SARAH M. ASSMANN and LISA ROMANO

## Introduction

Early in the history of patch clamping it was recognized that the whole-cell and inside-out configurations of the technique offered a way to manipulate signals acting from within the cell, so-called "secondary messengers," and record consequent effects on electrophysiologic responses.[1] In recent years, with the advent of sophisticated tools such as caged compounds and fluorescent ion indicators, the types of information that can be garnered by experimenters who combine electrophysiology and biochemical manipulations has grown tremendously. The purpose of this chapter is to provide an

[1] M. Cahalan and E. Neher, *Methods Enzymol.* **207,** 3 (1992).



0076-6879/99 $30.00

overview of the techniques available to manipulate and measure secondary messengers in living cells while simultaneously performing patch-clamp recording. We assume that the reader already has a basic familiarity with the patch-clamp technique. In this overview, we describe general information that will be of use regardless of the biological system under study. In addition, however, we also provide pointers on aspects of the techniques and special problems that are unique to plant systems. Many of our examples are based on the guard cell system. Guard cells have proven to be a premier system for the study of secondary messenger regulation of ion channels.[2] Guard cells are specialized cells located in the outermost tissue layer of aerial plant organs. Pairs of guard cells border microscopic pores called stomata. Through osmotic swelling and shrinking, based largely on transmembrane ion fluxes, guard cells control stomatal apertures, and thereby regulate both water vapor loss and photosynthetic $CO_2$ uptake, which occur through the stomatal pores.

## Membrane Access

To study secondary messenger regulation of ion channels via the patch-clamp technique, the first requirement is that the experimenter be able to achieve the high-resistance "gigohm" seal between the patch pipette and the cell membrane that is essential to record low-noise patch-clamp data. Below, we describe methods of membrane access that have been successfully applied to plant cells.

### *Enzymatic Access*

Each plant cell is completely encased in a cell wall. Therefore, one of the major differences between the patch clamping of animal and plant cells is the necessity in the latter case of removing the wall in order to gain membrane access. Several methods are available to achieve this. The first and still the most widespread approach is the use of cell-wall degrading enzymes[3] to digest the wall, resulting in production of what is called a plant "protoplast." This use of enzymes parallels enzymatic treatments in some animal systems, where collagenase and/or trypsin are used to break down connective tissue and expose a clean membrane surface,[4] but of course the particular enzymes involved are different. Primary plant cell walls contain

[2] S. M. Assmann, *Annu. Rev. Cell. Biol.* **9,** 345 (1993).
[3] K. Raschke and R. Hedrich, *Methods Enzymol.* **174,** 312 (1989).
[4] G. Trube, *in* "Single Channel Recording" (B. Sakmann and E. Neher, eds.). Plenum, New York and London, 1983.

carbohydrate polymers such as cellulose and pectin as major constituents, thus the enzyme mixtures used typically contain cellulases and pectinases. Secondary plant cell walls, and primary walls of some cell types in later stages of development contain, in addition to celluose and pectin, significant amounts of lignin. This phenolic polymer is very difficult to degrade. Saprophytic organisms such as fungi do produce ligninases that might be of use in protoplasting lignified plant cells. However, to our knowledge, this approach has yet to be investigated by plant electrophysiologists, with the corollary that most patch-clamp investigations to date have been performed on protoplasts from cells possessing only primary walls.

Cellulases and pectinases are commercially available (see Appendix for list of suppliers with addresses) from several sources, including Calbiochem, Worthington, Sigma, and the Japanese companies Yakult and Seishin. In our own laboratory, the enzymes from Japan are preferred. Yakult produces two cellulases, Onozuka R-10 and Onozuka RS. Anecdotally, we have found that it is easier to obtain gigohm seals on guard cell protoplasts that have been produced using Onozuka RS than on those obtained with R10. Seishin produces a very effective pectinase, Pectolyase Y-23, which we use in conjunction with cellulase RS.

The specific concentrations and ratios of cellulase and pectinase that are optimal for obtaining protoplasts differ with both species and cell type. For example, guard cell protoplasts from *Arabidopsis thaliana* are prepared using Onozuka R10 and Macerozyme R10 in a 16-hr digestion[5] while guard cells of *Vicia faba* require Onozuka RS and Pectolyase Y-23 in an approximately 1-hr digestion.[6] In contrast, when preparing mesophyll cell protoplasts from *V. faba* we use R10 cellulase and Macerozyme R10 for 1 hr.[7] Because of such variations in wall digestibility, a significant amount of time may be spent trying to find an enzyme cocktail that can break down the cell wall of cells of interest, and yet leave the resulting protoplasts in a condition suitable for patch clamping. Other factors that influence the effectiveness of the enzyme solutions include the presence or absence of ascorbic acid, digestion temperature, pH, presence and strength of agitation during tissue digestion, and extent of trituration to release protoplasts from the tissue. In addition, the removal of the cell wall leaves the protoplast vulnerable to osmotic stress. The optimum osmolality of the isolation buffers must be assessed so that the protoplasts are round, but not swollen or easily ruptured. For example, while we utilize the same enzyme mixture

[5] Z. M. Pei, K. Kuchitsu, J. M. Ward, M. Schwarz, and J. I. Schroeder, *Plant Cell* **9,** 409 (1997).

[6] H. Miedema and S. M. Assmann, *J. Membr. Biol.* **154,** 227 (1996).

[7] W. Li and S. M. Assmann, *Proc. Natl. Acad. Sci. U.S.A.* **90,** 262 (1993).

for isolation of both *Arabidopsis* and *V. faba* mesophyll cell protoplasts, we have found that it is best to utilize a 0.45 *M* sorbitol solution for *Arabidopsis* and a 0.6 *M* sorbitol solution for *V. faba* (see Table I). Unfortunately, there is little substitute for trial and error in determining the proper enzyme recipe and protocol. As a starting point, sample recipes for protoplasting solutions for several different types of plant cells are given in Table I.

Although the specific recipes for protoplast isolation may vary from tissue to tissue, a general procedure can be followed for most protoplast isolations. The first step is to prepare the tissue from which the protoplasts will be isolated. For example, to isolate guard cells or epidermal cells, the epidermes of leaves can be peeled either manually or with the aid of a blender, and these peels subjected to further treatment. This procedure may be performed using a buffer or distilled water. The second step is to place the prepared tissue in the appropriate enzyme solution. In some cases, the tissue may be subjected to two different enzyme cocktails in order to obtain a population that is enriched for one type of protoplast. For example, the guard cell isolation procedure commonly employed by our laboratory[6] uses two enzyme solutions, the first of which releases epidermal cells. The osmolality of the first enzyme solution is low ($<300$ mOsm $kg^{-1}$) so that the released epidermal protoplasts burst. The peels, which still contain the intact guard cells, are then rinsed and placed in a second enzyme solution which preferentially releases guard cells. During incubation in the enzyme solution, the protoplasts should be gently shaken and held at an appropriate temperature. In general cooler temperatures are preferred for optimal protoplast health, but this also slows the digestion process. After digestion is complete, the protoplasts are filtered through nylon mesh which allows released protoplasts to pass through freely, thus separating them from debris. The protoplasts can then be rinsed free of the enzyme solution by centrifugation in the presence of a "washing" medium. The centrifugation speed and time should be the minimum necessary to allow collection of the protoplasts. After the final centrifugation, the protoplasts should be allowed to recover for at least 1 hr on ice in the dark.

One drawback to the use of enzymes is that the commercially available enzymes are impure. Contaminating proteases may influence ion channel intactness and activity, yet such effects are difficult to assess, simply because it is impossible to patch clamp a completely walled plant cell. In the case of guard cells, double-barreled voltage-clamp recordings on intact cells have confirmed, at least qualitatively, current responses seen in protoplasts,[8] but voltage clamping is not practical for all cell types. For example, it

[8] F. Armstrong and M. R. Blatt, *Plant J.* **8,** 187 (1995).

TABLE I

General Protoplast Isolation Protocols[a]

| Species/tissue | Solutions | Enzyme 1[b] | Enzyme 2[b] | Digestion conditions | Ref. |
|---|---|---|---|---|---|
| *Vicia faba* (broad bean) guard cells (blended epidermal peels of 3, youngest, fully expanded leaves of plants 3–4 weeks old) | Blending medium: 10 m*M* MES, 5 m*M* $CaCl_2$, 0.5 m*M* ascorbic acid, 0.1% PVP40, pH 6 (KOH)<br>Basic medium: 0.45 *M* sorbitol, 0.5 mM $CaCl_2$, 0.5 m*M* $MgCl_2$, 0.5 m*M* ascorbic acid, 100 $\mu M$ $KH_2PO_4$, 5 m*M* MES, pH 5.5 (KOH) | 0.7% Cellulysin cellulase, 0.1% PVP40, 0.25% BSA, 0.5 m*M* ascorbic acid, in 55% basic medium, 45% distilled water (v/v), pH 5.5 (KOH), 10 ml total volume | 1.5% RS cellulase, 0.02% Pectolyase Y-23, 0.25% BSA, 0.5 m*M* ascorbic acid 100% basic medium, pH 3.5 for 5 min (HCl), final pH 5.5 (KOH), filter through 0.45-$\mu$m filter, 10 ml total volume | Enzyme 1: 30 min/28°; shaking: 162 excursions/min<br>Enzyme 2: 50–80 min at 17°; shaking: 108 excursions/min | *c* |
| *Arabidopsis* guard cells (blended epidermal strips of 2, 2-cm rosette leaves) | Blending medium: distilled water<br>Incubation medium: 0.5 *M* mannitol, 0.1 m*M* KCl, 0.1 m*M* $CaCl_2$, 10 m*M* ascorbic acid, 0.1% kanamycin sulfate, pH 5.5 (Tris) | 1.3% Cellulase R10, 0.7% Macerozyme R10, 0.5% BSA in incubation medium | NA | 16 hr at 22°; 70 rpm | *d* |
| *Arabidopsis* mesophyll cells (8 rosette leaves, 2 cm in length with abaxial epidermis peeled) | Basic medium (see *V. faba* guard cell protocol above) | 0.4% Macerozyme R10, 1% Cellulase R10, 0.1% PVP40, 0.2% BSA, 10 ml basic medium, pH 5.5 | Same as enzyme 1, 10 ml | Enzyme 1: 2 min vacuum infiltration, followed by 5 min shaking (160 excursions/min)<br>Enzyme 2: 45 min at 24° shaking: 108 excursions/min | *e* |

| | | | | | |
|---|---|---|---|---|---|
| *Vicia faba* mesophyll cells (1 bifolate leaf from a plant 3–4 weeks old with abaxial epidermis peeled) | 0.6 *M* Sorbitol, 1 m*M* $CaCl_2$ | Enzymes same as *Arabidopsis* mesophyll, in 0.6 *M* sorbitol, 1 m*M* $CaCl_2$ solution | Same as enzyme 1 | Enzyme 1: same as *Arabidopsis* mesophyll<br>Enzyme 2: same as *Arabidopsis* mesophyll but for 1–1.5 hr | *f* |
| *Hordeum vulgare* (barley) xylem parenchyma (minced stelar tissue from roots stripped of cortex) | 0.5 *M* Mannitol, 1 m*M* $CaCl_2$ | 2% Cellulase R10, 0.02% Pectolyase Y-23, 2% BSA, 10 m*M* sodium ascorbate, pH 5.5 ($H_2SO_4$) | NA | 2–2.5 hr at 20°; shaking: 100 excursions/min | *g* |
| *Pisum sativum* (pea) stem (epidermal strips from 7-day-old etiolated plants) and leaf (lower epidermis of young leaves in light grown plants) epidermal cells | Wash solution: 2.325% Gamborg's B5 (GIBCO), 2 m*M* $CaCl_2$, 10 m*M* MES<br>mannitol to 0.61 *M*, pH 5.5 (KOH)<br>Bath solution: 5 m*M* $CaCl_2$, 2 m*M* $MgCl_2$, 10 m*M* potassium-citrate, mannitol to 0.21 *M*, pH 5.5 | 1.7% RS cellulase, 1.7% Cellulysin cellulase, 0.026% Pectolyase Y-23, 0.2% BSA, in wash solution | NA | 5–10 min in enzyme solution, 2 × 5 min in wash solution (all steps at 30° on rotary shaker at 50 rpm)<br>Place in bath solution (in which patch-clamp experiments are to be done) | *h* |
| *Zea mays* (maize) root cortex cells (chopped cortex tissue) | 0.5 *M* sorbitol, 1 m*M* $CaCl_2$, 5 m*M* MES, pH 6 (KOH) | 0.5% PVP, 10,000 MW, 0.5% BSA, 0.8% RS cellulase, 0.08% pectolyase in 0.5 *M* sorbitol, 1 m*M* $CaCl_2$, 5 m*M* MES, pH 6.0 solution | NA | 3 hr at 28° with agitation | *i* |

*(continued)*

TABLE I (*continued*)

| Species/tissue | Solutions | Enzyme 1[b] | Enzyme 2[b] | Digestion conditions | Ref. |
|---|---|---|---|---|---|
| *Oryza sativa* (rice) root-hair protoplasts (0.5 cm of root tip of 3-day-old rice seedlings) | Basic medium (see above) with 0.40 *M* sorbital | 1.5% RS cellulase, 0.1% Pectolyase Y-23, 0.2% BSA in basic medium | NA | Place roots in a 100-$\mu$m pore nylon filter. Fix filter into 50-ml centrifuge tube which contains enzyme solution. After 10 min at room temperature, add 40 ml basic medium and centrifuge. Repeat wash step with basic medium. | *j* |

[a] All percent values are in w/v unless indicated. NA, Not applicable; BSA, bovine serum albumin; PVP40, polyvinylpyrrolidone, MW 40,000; MES, 2-(N-Morpholino)ethanesulfonic acid.

[b] Addresses of sources of enzymes are given in the Appendix: Cellulysin cellulase (Calbiochem), RS cellulase, Cellulase R10, Macerozyme R10 (Yakult Pharmaceutical Ind. Co. Ltd), Pectolyase Y-23 (Seishin Corporation), pectolyase (Sigma Chemical Co.).

[c] H. Miedema and S. M. Assmann, *J. Membr. Biol.* **154,** 227 (1996).

[d] Z. M. Pei, K. Kuchitsu, J. M. Ward, M. Schwarz, and J. I. Schroeder, *Plant Cell* **9,** 409 (1997).

[e] L. A. Romano and S. M. Assmann, unpublished observations (1997).

[f] W. Li and S. M. Assmann, *Proc. Natl. Acad. Sci. U.S.A.* **90,** 262 (1993).

[g] L. H. Wegner and K. Raschke, *Plant Physiol.* **105,** 799 (1994).

[h] J. T. M. Elzenga, C. P. Keller, and E. Van Volkenburgh, *Plant Physiol.* **97,** 1573 (1991).

[i] S. K. Roberts and M. Tester, *Plant J.* **8,** 811 (1995).

[j] C. J. Yu and W. H. Wu, unpublished observations (1997).

cannot be readily used for interior cells within plant organs (such cells can be accessed as protoplasts since the protoplasting procedure results in dissociation and digestion of the entire tissue). Another drawback to the enzymatic digestion of the cell wall is that protoplasting may remove signaling compounds such as oligosaccharins that originate from the wall itself. Finally, removal of the wall may significantly alter protoplast physiology and thus influence signal transduction chains that regulate channel activity. Most obviously, plant cells differentiate into a variety of shapes as a result of physical constraints imposed by the cell wall. When the wall is removed the resulting protoplast is inevitably spherical. Cytoskeletal rearrangements that ensue as, e.g., wall–membrane–cytoskeletal linkers are removed, may have significant effects on ion channel activity.[9,10]

### *Alternatives to Standard Enzymatic Protoplasting: Osmotic Shock and Laser-Assisted Patch Clamping*

For the above reasons, it would be helpful to have a method of membrane access that minimized or avoided enzymatic treatments. Two such approaches have recently been developed. The first involves the short-term exposure of tissue to relatively high concentrations of enzymes, followed by hypoosmotic shock.[11] The cells of the tissue are first plasmolyzed in a high-osmolality solution, and then briefly (5–10 min) exposed to a cocktail of hydrolytic enzymes. This enzyme exposure weakens the cell wall. The tissue is then placed in a low-osmolality solution, which causes the protoplasts to swell and pop out of the weakened cell walls. This technique produces viable protoplasts with a high seal success rate from several tissue types, including epidermis, mesophyll, and coleoptile cortex (see Table I, *Pisum sativum* protocol).

Depending on the system under study, the experimenter can also take advantage of inherent weaknesses in the cell wall to minimize enzyme exposure. More than a decade ago, procedures were described to obtain "subprotoplasts" from root hairs by brief enzymatic digestion.[12] Root hairs are outgrowths from root epidermal cells and their cells walls are weakest at the growing tip. Thus, enzymatic exposure results in the bulging of exposed membrane from the tip of the root hair. With gentle centrifugation,

[9] L. Thion, C. Mazars, P. Thuleau, A. Graziana, M. Rossignol, M. Moreau, and R. Ranjeva, *FEBS Lett.* **393,** 13 (1996).

[10] M. Kim, P. K. Hepler, S. O. Eun, K. S. Ha, and Y. Lee, *Plant Physiol.* **109,** 1077 (1995).

[11] J. T. M. Elzenga, C. P. Keller, and E. Van Volkenburgh, *Plant Physiol.* **97,** 1573 (1991).

[12] E. C. Cocking, *Biotechnology* **3,** 115 (1985).

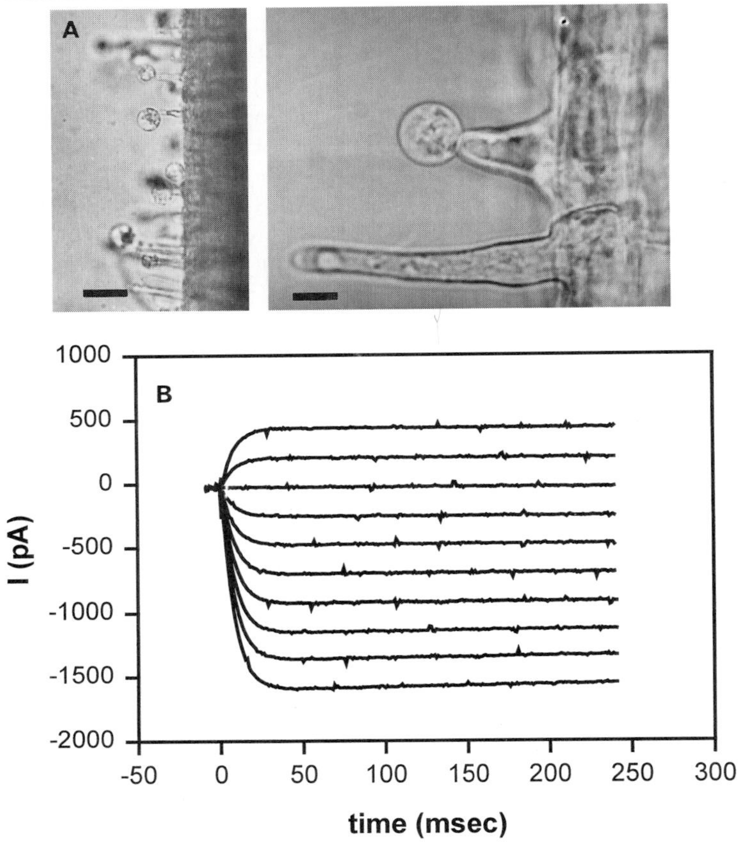


FIG. 1. (A) Emerging protoplasts from rice (*Oryza sativa*) root hairs. Bars: 70 $\mu$m (left) and 10 $\mu$m (right). (B) Whole-cell recording from rice root hair protoplast. Voltage protocol: The membrane potential was held at $-50$ mV and pulsed to potentials ranging from $-180$ to 0 mV. Pipette solution: 98 m*M* potassium glutamate, 2 m*M* KCl, 2 m*M* EGTA, 0.3 *M* sorbitol, 10 m*M* HEPES, pH 7.2 (KOH). Bath solution: 10 m*M* potassium glutamate, 2 m*M* $MgCl_2$, 5 m*M* $CaCl_2$, 0.4 *M* sorbitol, 10 m*M* HEPES, pH 6.0 (KOH). (C. J. Yu, W. H. Wu, unpublished, 1997.)

these "subprotoplasts" pinch off and can be collected. Alternatively, the exposed membrane can simply be "patched" *in situ,* thus allowing the recording of patch-clamp data from the intact root. Figure 1A shows protoplasts emerging from rice root hairs (see Table I), and Fig. 1B gives an example of the whole-cell $K^+$ currents recorded from such subprotoplasts.

Recently, our laboratory has refined a protocol that allows membrane access for patch-clamp experiments in the complete absence of enzymatic digestion of the cell wall.[13] The technique was originally applied to large algal cells,[14,15] but if adequate precision is employed, it can also be applied to the relatively small cells of higher plants. A microbeam (diameter of 0.3–1 $\mu$m) from a 337-nm UV laser is focused through the microscope onto a portion of the cell wall (Fig. 2A). The high-energy laser light ablates the wall, thus exposing the membrane. Focusing is achieved by passing the laser light through a Galilean type beam expander that has been empirically adjusted such that the laser beam is parfocal with the microscope focal plane. Once this is achieved, the beam will remain in focus as the microscope focus is changed during the course of an experiment, provided that the microscope is designed such that the carrier containing the dichroic mirror (Fig. 2A) does not change its vertical position when the microscope focus is changed. Detailed information on the equipment and optics required for the setup is given in a recent methodology.[16]

Wall ablation is enhanced by prior exposure of the tissue to Calcofluor White (Sigma), a UV-absorbing dye that binds to cell walls.[15] Damage to the protoplast from the laser beam is avoided by the precise focusing of the laser beam in the $z$ dimension, and by temporarily plasmolyzing the protoplast with high osmoticum so as to remove the protoplast from the laser light path (Fig. 2B). On deplasmolysis, the plasma membrane "blebs" through the hole in the wall and is accessible to the patch pipette (Fig. 2C). We have found that a slow rate of deplasmolysis, as provided by use of a linear gradient maker,[15] is essential to expose the membrane without rupturing the protoplast. In our experience with guard cells of *V. faba,* it is easy to obtain high-resistance seals on the exposed membrane with little or no application of suction, although this is apparently not to be true for all systems.[15] While technically more demanding than membrane exposure via protoplasting, laser-assisted patch clamping provides an alternative approach when enzymatic methods of protoplast isolation fail. Additionally, laser-assisted patch clamping allows recording of patch-clamp data *in situ.* The technique also has the potential to provide information on the spatial distribution of ion channels that cannot be obtained from spherical protoplasts.

[13] G. H. Henriksen, A. R. Taylor, C. Brownlee, and S. M. Assmann, *Plant Physiol.* **110,** 1063 (1996).

[14] A. R. Taylor and C. Brownlee, *Plant Physiol.* **99,** 1686 (1992).

[15] A. H. De Boer, B. Van Duijn, P. Giesberg, L. Wegner, G. Obermeyer, K. Köhler, and K. W. Linz, *Protoplasma* **178,** 1 (1994).

[16] G. H. Henriksen and S. M. Assmann, *Pflügers Arch.* **433,** 832 (1997).

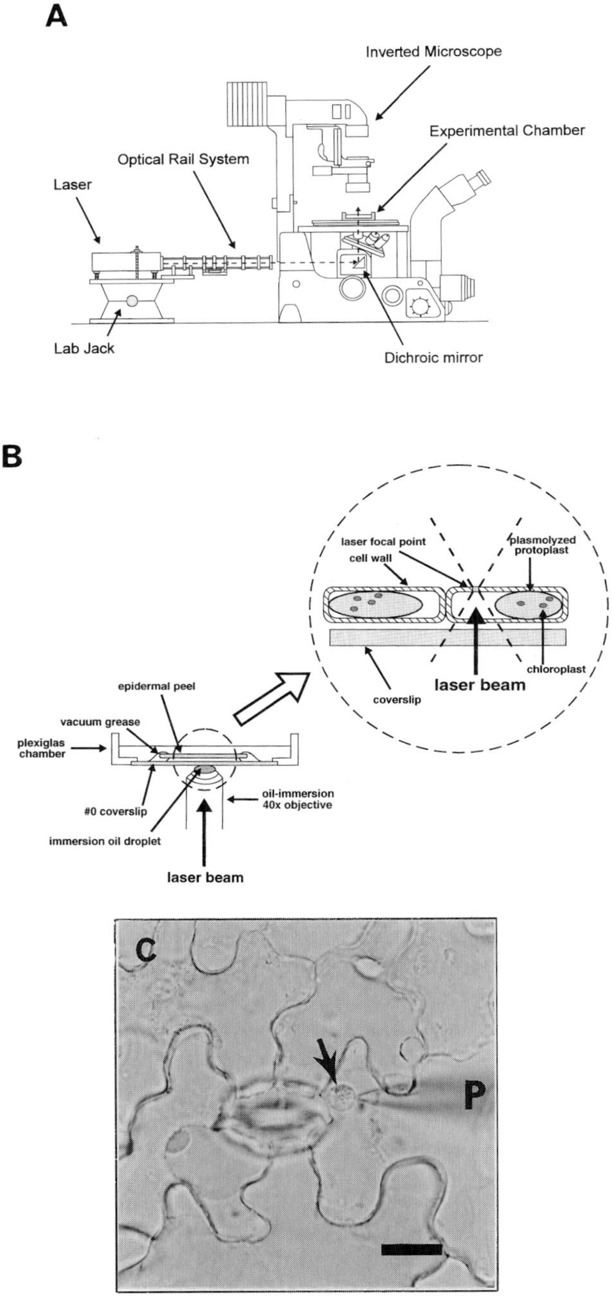
A
Inverted Microscope
Experimental Chamber
Optical Rail System
Laser
Lab Jack
Dichroic mirror
B
laser focal point
cell wall
plasmolyzed protoplast
chloroplast
laser beam
coverslip
epidermal peel
vacuum grease
plexiglas chamber
#0 coverslip
immersion oil droplet
oil-immersion 40x objective
laser beam
C
P

## Patch Clamping of Plant versus Animal Cells

The technique of patch clamping is similar for all types of cells, and the reader is referred to general reviews on this topic.[17–19] However, there are subtle differences between the patch clamping of animal cells and plant protoplasts. This section focuses on such differences.

Isolated animal cells from primary cultures can usually be maintained in nutrient media at the appropriate temperature for up to 1 week and used in patch-clamp experiments.[4] Plant protoplasts must generally be used within 12 hr of isolation. This is because protoplasts, no matter how healthy, are still missing a vital organelle, the cell wall, and are therefore fragile and more susceptible to physical, osmotic, and temperature stress. The life of protoplasts can be greatly prolonged if they are handled gently and kept on ice in the dark. In addition to the frailty of the protoplasts, protoplasts of many plant cell types are able to synthesize new cell walls after isolation, and this may eventually render the membrane too "dirty" to form tight seals.

Another difference between patch clamping of plant and animal cells is the solutions which are commonly used, especially bath solutions. In the majority of animal studies, the bath solution contains sodium, in order to mimic the extracellular environment that prevails *in vivo*. The $Na^+$ electrochemical gradient is an important source of energy for the uptake of sugars and amino acids and for pH regulation, while the resting membrane potential of animal cells is largely determined by the $K^+$ electrochemical gradient. Placing many animal cell types in a high $K^+$ bath solution could depolarize the membrane to unphysiologic levels. This could trigger the activation of voltage-gated channels, especially $Ca^{2+}$ permeable channels, and set in motion a number of $Ca^{2+}$-dependent processes, such as secretion. In contrast, even moderate $Na^+$ levels are toxic to plants, and $Na^+$ plays a more minor role in plant membrane physiology. Therefore, many plant patch-clamp experiments are performed in solutions where $K^+$ salts are the

[17] O. P. Hamill, A. Marty, E. Neher, B. Sakmann, and F. J. Sigworth, *Pflügers Arch.* **391,** 85 (1981).

[18] B. Sakmann and E. Neher, eds., "Single-Channel Recording," 2nd Ed. Plenum, New York and London, 1995.

[19] B. Rudy and L. E. Iverson, eds., *Methods Enzymol.* **207.** Academic Press, San Diego, 1992.

FIG. 2. (A) Schematic of the setup used for laser-assisted patch clamping. (B) Schematic illustrating laser-targeting of the upper wall of a plasmolyzed guard cell. (C) Laser-accessed membrane "bleb" from a guard cell of *Vicia faba*. Bar: 20 $\mu$m. [A, B, reproduced with permission from G. H. Henriksen and S. M. Assmann, *Pflügers Arch.* **433,** 832 (1997); C, reproduced with permission from G. H. Henriksen *et al., Plant Physiol.* **110,** 1063 (1996).]

predominant ions. Depending on the types of experiments, KCl, potassium glutamate, or potassium gluconate are commonly used. In our laboratory, we focus primarily on $K^+$ currents, and a typical bath solution consists of 10 or 100 m*M* KCl, 1 m*M* $CaCl_2$, 1 m*M* $MgCl_2$, 10 m*M* HEPES (or MES), pH 5.6 or 7.2, with the osmolality adjusted to 460 mOsm $kg^{-1}$ with sorbitol or mannitol. If anion channels are being studied, CsCl can be substituted for KCl. It is also important to note that in many animal patch-clamp experiments, pharmacologic blockers are used to isolate specific ion currents. This is also possible in plant systems, with the caveat that the "specificity" of the blocker must be assessed. For example, verapamil, which is used in animal systems to block $Ca^{2+}$ currents, blocks outward rectifying $K^+$ currents in many types of plant cells [personal observation (L.A.R.) (1997) and Ref. 20].

The pipette solution commonly used in our laboratory to study $K^+$ currents consists of 80 m*M* potassium glutamate, 20 m*M* KCl, 2 m*M* $MgCl_2$, 2 m*M* Mg-ATP (in Tris), 10 m*M* HEPES, pH 7.2 or 7.8. The final osmolality is adjusted to 500 mOsm $kg^{-1}$ with sorbitol or mannitol. Ethyleneglycol-*bis*($\beta$-aminoethyl)-N,N,N′,N′-tetraacetic acid (EGTA) or 1,2-*bis*(o-Aminophenoxy)ethane-N,N,N′,N′-tetraacetic acid (BAPTA) is added in various concentrations depending on the amount of $Ca^{2+}$ buffering desired for a particular experiment.

The pipettes used for patch-clamping plant protoplasts, in general, are smaller than those used for animal cells. Typically for patch clamping animal cells in the whole-cell configuration, pipette resistances of 1–5 MΩ are used. With plant protoplasts, pipette resistances of as high as 10–30 MΩ (in the solutions described above) are used for whole-cell recordings. The reason that smaller sized pipettes are required to form tight seals is unknown, but may arise from the fact that the cytoskeletal structure of the plant cell is disrupted when the cell wall is removed. This may allow a large amount of cytoplasm to be sucked into the pipette when seal formation is attempted, which tends to lower seal success rates. This problem may also be addressed by lowering the osmolality of the bath solution, which causes the protoplasts to swell.

When patch clamping was a new technique, it was widely accepted that it was extremely difficult to patch clamp plant cells. Now it is possible, at least with some types of plant cells, e.g., *V. faba* guard cell protoplasts, to obtain similar seal success rates as those enjoyed by animal patch clampers. However, after obtaining a high-resistance seal, there are a few problems a plant patch clamper may specifically encounter. The main difficulty we

[20] S. Thomine, S. Zimmermann, B. Van Duijn, H. Barbier-Brygoo, and J. Guern, *FEBS Lett.* **340,** 45 (1994).

have found is increases in series resistance. There are two main causes of such increases. The first has to do with the membrane properties of the protoplasts. Some types of protoplasts, e.g., guard cells but not mesophyll cells of *V. faba,* have membranes that can best be described as "gooey." This can be especially noticed when attempting to excise a patch for single-channel recordings. The protoplast seems to ooze along with the pipette as it is moved instead of a patch of membrane being plucked off. This membrane consistency may also promote a gradual increase in series resistance if the whole-cell configuration is maintained for prolonged periods. It is extremely important to monitor and correct for changes in series resistance during the course of an experiment, especially when a cell serves as its own control, where an increase in series resistance may be mistaken for a secondary messenger effect. The second cause of increases in series resistance is the partial or complete blocking of the pipette by a chloroplast. This problem can be extremely frustrating in cells such as mesophyll cells, which are packed with chloroplasts just the right size to plug the end of a patch pipette. The chloroplasts can be moved out of the pipette tip by suction or blowing, but this may result in loss of the seal.

Protoplasts also seem to be sensitive to ambient temperature. Most patch-clamp experiments using plant cells are conducted at room temperature. This is usually not an issue, except when room temperature becomes too high (e.g., 28°). At these high temperatures protoplast health becomes compromised.

After overcoming the technical difficulties of getting and keeping a seal, it would seem that patch clamping of plant protoplasts would be almost identical to the patch clamping of animal cells. For the most part, it is. However, it is important to note that by nature, plant cells respond to virtually every environmental stimulus. For example, while in mammals vision is restricted to particular cell types, almost every cell type in a plant contains some type of photoreceptor. Since in patch clamping the experimenter seeks to gain insight into the physiologic function of ion channels, the hope is that isolated protoplasts maintain their responsiveness to these stimuli. However, the very act of patch clamping also exposes the protoplast to a barrage of "stimuli" that the researcher must try to minimize or at best keep constant from cell to cell. For example, simply viewing cells under the microscope introduces two stimuli: light and heat. Several studies have shown effects of different wavelengths of light on various transport processes, including ion channel[21] and $H^+$-ATPase[22] activity. Such effects can be overcome by installing a green filter in the microscope light path

[21] M. H. Cho and E. P. Spalding, *Proc. Natl. Acad. Sci. U.S.A.* **93,** 8134 (1996).
[22] S. M. Assmann, L. Simoncini, and J. I. Schroeder, *Nature* **318,** 285 (1985).

(since most plant cells do not respond to this region of the spectrum) or by minimizing light exposure by turning the microscope light off after the seal is obtained. Microscope lights also generate a fair amount of heat, and at least one study reports an effect of temperature on plant ion channels.[23] Again, this problem can be overcome simply by installing heat filters and/or minimizing the amount of exposure to light. While performing simultaneous patch clamping and $Ca^{2+}$ imaging (see below) we also noticed that on days when the ambient temperature was high due to warm outdoor temperatures and large amounts of heat generated by equipment, the $Ca^{2+}$ level of patch-clamped guard cell protoplasts was abnormally high. Whether this high level of $Ca^{2+}$ was in response to the signal of high temperature or a sign of poor protoplast health is unclear. As mentioned previously, high temperature may also compromise the health of the protoplasts, even before patch clamping is initiated.

Finally, the biggest difference between patch clamping plant and animal cells is the amount of patience required of the experimenter. In many animal systems, seal formation is almost instantaneous. The plant cell patch clamper must be willing to spend sometimes up to 10 min or even longer nursing a patch along until a gigohm seal is formed. The time required varies from cell type to cell type, and also within a preparation. The key is to not abandon a cell that does not seal immediately, but to experiment to find different tricks to get the cells to seal. Sometimes doing the exact opposite of what intuition suggests works. For example, after several attempts to apply inordinate amounts of suction to get a particularly unruly batch of mesophyll protoplasts to seal, we changed our strategy to that of applying almost no suction at all, and this was successful. In our experience, if it is possible to achieve seal resistances of about 200 MΩ, then the protoplasts are "patchable" and successful seal formation should result following suitable experimentation by the investigator to find the proper pipette shape and size, suction protocol, etc. However, if seal resistances as high as 200 MΩ cannot be achieved, some modification of the enzymatic isolation protocol will probably be required.

## Secondary Messenger Regulation of Ion Channels

Once a high resistance seal and the appropriate patch-clamp configuration have been obtained, there are two basic approaches toward connecting secondary messengers with ion channel behavior. The researcher can quantify endogenous levels of a signaling compound, and correlate changes in the level of this signal with alterations in channel activity. Or, the experimenter can deliberately manipulate levels of a putative secondary messen-

[23] N. Ilan, N. Moran, and A. Schwartz, *Plant Physiol.* **108,** 1161 (1995).

ger and observe the electrophysiologic result. The following sections provide technical background on each of these topics. Such approaches have been used extensively by researchers studying animal systems, and are currently being incorporated into plant research as well. In the guard cell system, for example, electrophysiologic studies of the signal cascades underlying ion channel responses to the plant hormones abscisic acid and auxin have suggested involvement of secondary messengers also common to animal systems, including $Ca^{2+}$, pH, G proteins, inositol 1,4,5-trisphosphate ($IP_3$), and a calcineurin-like phosphatase. In addition, plant-specific signals such as a calcium-dependent protein kinase with a calmodulin-like domain have also been implicated in ion channel regulation.[24] Several recent reviews[2,25,26] on the guard cell system provide a good source of references for the reader seeking examples of how the techniques described below are actually capitalized on to address biological questions.

*Measuring Endogenous Secondary Messengers*

The availability of fluorescent indicator dyes for $Ca^{2+}$, $H^+$, and other ions, coupled with patch clamping, provides the means to measure *in vivo* changes in endogenous secondary messengers concomitantly with changes in cell electrophysiology. In our laboratory we perform simultaneous $Ca^{2+}$ measurements and whole-cell recording. Therefore, our discussion focuses on the use of $Ca^{2+}$ indicator dyes. The general principles discussed would hold equally for imaging of other secondary messengers, e.g., pH.

The ability to monitor two vitally important aspects of cellular physiology, the activity of $Ca^{2+}$-regulated ion channels via patch-clamp recording[27] and changes in intracellular $Ca^{2+}$ concentration via ratiometric $Ca^{2+}$ imaging,[28] has led to enormous insight into the impact of $Ca^{2+}$ metabolism on ion channels in animals systems.[29,30] The combination of these two powerful tools has, to our knowledge, been applied previously only once to plant protoplasts, in a study employing photometric $Ca^{2+}$ measurements with whole-cell patch clamp.[31] Grabov and Blatt[32] have also recently applied

[24] Z. M. Pei, J. M. Ward, J. F. Harper, and J. I. Schroeder, *EMBO J.* **15,** 6564 (1996).

[25] M. R. Blatt and G. Thiel, *Annu. Rev. Plant Physiol. Plant Mol. Biol.* **44,** 543 (1993).

[26] J. M. Ward, Z. M. Pei, and J. I. Schroeder, *Plant Cell* **7,** 833 (1995).

[27] B. Hille, "Ionic Channels in Excitable Membranes." Sinauer Associates, Sunderland, Massachusetts, 1992.

[28] G. Grynkiewicz, M. Poenie, and R. Y. Tsien, *J. Biol. Chem.* **260,** 3440 (1985).

[29] P.-M. Lledo, B. Somasundaram, A. J. Morton, P. C. Emson, and W. T. Mason, *Neuron* **9,** 943 (1992).

[30] A. N. van den Pol, K. Obrietan, and G. Chen, *J. Neuroscience* **16,** 4283 (1996).

[31] J. I. Schroeder and S. Hagiwara, *Proc. Natl. Acad. Sci. U.S.A.* **87,** 9305 (1990).

[32] A. Grabov and M. R. Blatt, *Planta* **201,** 84 (1997).

the techniques of intracellular $Ca^{2+}$ measurement and voltage clamp to intact guard cells still situated within an epidermal peel. The reason for the relative dearth of studies may simply be that the long time "inaccessibility" of plant protoplasts to patch-clamp studies has caused an overall lag in the types of experimental techniques that have been applied to them.

*Overview of $Ca^{2+}$ Measurement in Living Cells.* The first decision the experimenter must make when setting up for simultaneous electrophysiology and $Ca^{2+}$ quantification is what type of $Ca^{2+}$ measurement system to employ. Three options are available. From most to least expensive, these are confocal $Ca^{2+}$ imaging, standard $Ca^{2+}$ imaging, and photometry. Confocal microscopy offers extremely high resolution imaging in the $x$, $y$, and $z$ planes. It is usually the method of choice when spatial information is important,[33] unless quite fast measurements of cellular $Ca^{2+}$ changes are required, in which case photometry may be the only option. However, because of their high cost, confocal microscopes are not uniformly available, and if available they are often part of multiuser facilities, with limitations on the amount of time that the equipment is accessible to any one user.

Standard $Ca^{2+}$ imaging provides spatial information in the $x$ and $y$ dimensions. While the spatial resolution is poorer than with confocal microscopy, standard $Ca^{2+}$ imaging is also significantly less expensive. Finally, in photometry, a single fluorescent signal integrated over the entire cell or region of interest is recorded, using one or two photomultipliers (PMTs) as detectors of the emitted light. Other than the option to select a region of interest, photometry provides no spatial information, but it requires the least amount of time per measurement because essentially one "giant" pixel (i.e., the entire protoplast or selected region of interest) is instantaneously measured (as opposed to the finite scan time required for a confocal image), and because the use of highly sensitive PMTs as detectors reduces or eliminates the need for temporal signal averaging, as may be required for standard $Ca^{2+}$ imaging. Because each signal comprises many photons, photometry may also be the only workable method when signal strength is weak.

The second decision facing the researcher is which $Ca^{2+}$ indicator dye to use. The catalog from Molecular Probes provides a wealth of information on this topic.[34] One issue apparently unique to plants is that some indicator dyes partition into the large central vacuole of plant cells, and are therefore

[33] A. Hernández-Crus, F. Sala, and P. R. Adams, *Science* **247,** 858 (1990).

[34] R. P. Haugland, *in* "Handbook of Fluorescent Probes and Research Chemicals" (M. T. Z. Spence, ed.). Molecular Probes, Eugene, Oregon, 1996.

useless for measuring cytoplasmic $Ca^{2+}$ concentrations.[35,36] Dyes that are conjugated to dextran ("dextranated") avoid this problem; other dyes must be evaluated for their usefulness on a case-by-case basis.

When possible, it is always preferable to use a $Ca^{2+}$ indicator that is ratiometric, such as Indo-1 or Fura-2. A ratiometric dye is one that has a shift in either the excitation peak (Fura-2) or the emission peak (Indo-1) on binding of $Ca^{2+}$, instead of the increase in fluorescence intensity that is characteristic of nonratiometric dyes such as Calcium Green-1. The use of ratiometric dyes significantly reduces the potential artifacts of measuring intracellular $Ca^{2+}$ concentrations. Because the fluorescence intensities of two wavelengths are ratioed, the effects of photobleaching, inhomogeneous dye distribution, and nonuniform excitation intensity are canceled. In some cases it may be impossible to use Indo-1 or Fura-2, either because there is not an appropriate UV light source available, or because the experimental protocol precludes the use of UV light, e.g., if caged compounds are to be used (see below). Under these circumstances it may be possible to use pseudoratiometric dyes such as Calcium Green–Texas Red dextran (Molecular Probes). Calcium Green is a single-wavelength dye whose fluorescence intensity increases when $Ca^{2+}$ is bound. Texas Red fluorescence is totally $Ca^{2+}$ independent. These two molecules are linked via dextran molecules so that they are *de facto* equally distributed. The $Ca^{2+}$-independent fluorescence intensity of Texas Red can be used as an indicator of photobleaching and dye concentration, and the $Ca^{2+}$-dependent Calcium Green fluorescence can then be pseudoratioed against the $Ca^{2+}$-independent background of Texas Red to reduce the artifacts associated with single-wavelength $Ca^{2+}$ indicators.

*Simultaneous Patch-Clamp and Confocal $Ca^{2+}$ Imaging in Plant Protoplasts.* Recently, our laboratory has combined patch clamping and $Ca^{2+}$ quantification in the study of guard cell physiology using confocal laser scanning microscopy[37] to perform the $Ca^{2+}$ imaging. In general, the setup to perform the techniques is similar to that used in animal systems. Therefore, a relatively brief overview of the system and methods will be provided, with emphasis given to the modifications in both electrophysiology and imaging necessary for application to plant protoplasts.

The preparation of protoplasts for combined patch clamping and confocal $Ca^{2+}$ imaging is identical to the protocol for "regular" patch-clamp experiments (see above). The solutions used are also the same except that EGTA is omitted from the pipette solution, and 60 $\mu M$ Indo-1 pentapotas-

[35] D. S. Bush and R. L. Jones, *Cell Calcium* **8,** 455 (1987).
[36] D. C. Elliott and H. S. Petkoff, *Plant Science* **67,** 125 (1990).
[37] D. Schild, *Cell Calcium* **19,** 281 (1996).

sium salt (Molecular Probes) is included. A 1 mg ml$^{-1}$ stock solution of Indo-1 is almost exactly 1 m*M*, so we simply add 1 ml of pipette solution to 1 mg, store it at $-80°$ in 60-$\mu$l aliquots, and add one aliquot to 1 ml total of pipette solution. This concentration of Indo-1 allows for a good fluorescence signal, without adding excessive exogenous $Ca^{2+}$ buffering, which may alter the $Ca^{2+}$ signal that one is trying to measure. The pipette solution is filtered just before use with a 0.2-$\mu$m syringe filter and stored on ice covered with foil during the experiment.

The resistance of the pipettes used for simultaneous patch-clamp and $Ca^{2+}$ imaging is critical. Typically with guard cell protoplasts, it is easier to obtain a high-resistance seal with a pipette resistance of 15–20 M$\Omega$ (in 100 m*M* KCl bath solution/20 m*M* KCl–80 m*M* potassium glutamate pipette solution). However, this size of pipette is not suitable for $Ca^{2+}$ imaging because the resulting series resistance is about 30–40 M$\Omega$. The high series resistance leads to a low diffusion rate of dye into the protoplasts and, as a result, a poor fluorescence signal. Typically for $Ca^{2+}$ imaging, we use pipettes with resistances of 6–10 M$\Omega$. With this size of pipette it is still possible to have a high success rate of obtaining the initial seal (about 80%), and the dye loading is complete and stable within 5–10 min. There is one drawback to using large pipettes, however. The faster diffusion rate of dye into the cell also means that there is a faster rate of diffusion of soluble intracellular factors out of the cell.[38] Therefore, if the currents being studied are susceptible to rundown, the rate of rundown may also increase. With guard cells we have routinely been able to maintain cells with robust currents and $Ca^{2+}$ signals for 30 min to 1 hr.

To perform simultaneous patch-clamp measurements and confocal $Ca^{2+}$ imaging, we simply attach a patch-clamp setup to an existing Zeiss confocal laser scanning microscope. A schematic of the confocal setup is shown in Fig. 3A. For $Ca^{2+}$ imaging, Indo-1 is excited with an 8-sec scan of a 364-nm UV laser. An 80/20 beamsplitter in the laser path allows 20% of the UV laser light to reach the specimen and 80% of the fluorescence signal from the specimen to be transmitted through the light path. The fluorescence signal is then split by a 460-nm dichroic mirror, and detected simultaneously by two photomultiplier detectors that receive light of 450–505 nm and 400–435 nm through the use of appropriate emission filters. An important issue in imaging plant cells is the fluorescence of chlorophyll, which is in the red (>600-nm) range. The fluorescence of Indo-1 peaks at 400 and 480 nm. If a longpass filter (i.e., one that passes wavelengths longer than 460 nm) is placed in the position of emission filter 1, as is commonly done for Indo-1 imaging, the fluorescence from chloroplasts will contaminate the

[38] R. Horn and S. J. Korn, *Methods Enzymol.* **207,** 149 (1992).

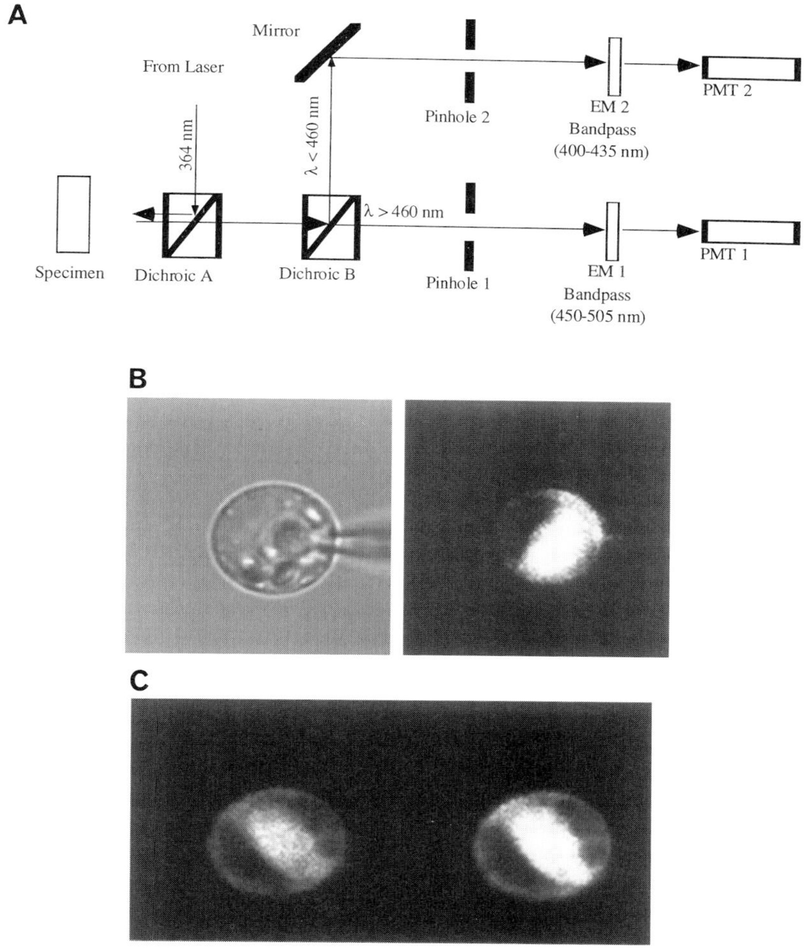


FIG. 3. (A) Diagram of confocal laser scanning microscope used for simultaneous patch clamping and $Ca^{2+}$ imaging. The same setup, with modifications, is used to release caged $Ca^{2+}$ (nitr-5). See text for details. (B) Bright-field (left) and fluorescence (right) images of a guard cell protoplast from *V. faba* loaded with Indo-1 pentapotassium salt via the patch pipette. Note that the fluorescence signal is isolated to the cytoplasmic region, indicated by the presence of chloroplasts. Pipette solution: 80 m*M* potassium glutamate, 20 m*M* KCl, 2 m*M* $MgCl_2$, 2 m*M* Mg-ATP (in Tris), 60 $\mu M$ Indo-1 pentapotassium salt, sorbitol to 500 mOsm $kg^{-1}$, 10 m*M* HEPES, pH 7.8 (KOH). Bath solution: 100 m*M* KCl, 1 m*M* $CaCl_2$, 1 m*M* $MgCl_2$, sorbitol to 460 mOsm $kg^{-1}$, 5 m*M* HEPES, 5 m*M* MES, pH 5.6 (Tris). (L. A. Romano, S. Gilroy, S. M. Assmann, unpublished.) (C) Increase in Calcium Green fluorescence showing a qualitative increase in $Ca^{2+}$ after UV photolysis of nitr-5. Pipette solution as in (B), except Indo-1 is omitted and 50 $\mu M$ Calcium Green dextran (10,000 molecular weight) and 2 m*M* nitr-5 charged with 1 mM $Ca^{2+}$ (as $CaCl_2$) are added. Bath solution same as in (B). (L. A. Romano, S. Gilroy, S. M. Assmann, unpublished).

signal received by the detector. Therefore, a 450- to 505-nm bandpass filter is installed for emission filter 1, which transmits the Indo-1 signal but not the chlorophyll fluorescence. With this modification, contamination from autofluorescence on either PMT is negligible in guard cell protoplasts.

During the course of an experiment, it is also desirable to view a transmitted light image of the cell under study, to adjust the pipette position or simply to monitor the cell's health. In our system, this is accomplished by using either a red (633-nm) or a green (543-nm) laser and a transmission detector. As mentioned earlier, isolated protoplasts may respond to light of various wavelengths, so it is important to conduct appropriate controls to make sure that whatever sources of light one uses do not produce an effect. This can be accomplished by comparing data obtained in "regular" patch-clamp experiments with data from patch-clamp/$Ca^{2+}$ imaging experiments. Also, simply subjecting patch-clamped cells to the same laser regime as used during an imaging experiment but in the absence of Indo-1 and treatment will confirm that the laser exposure is without effect.

As mentioned previously, a potential problem with measuring intracellular $Ca^{2+}$ with indicator dyes in plant protoplasts is compartmentalization of the dye into the vacuole.[35,36] Examination of transmitted light and fluorescence images of the same guard cell protoplast under study clearly shows that Indo-1 pentapotassium salt loaded via the patch pipette does not concentrate in the vacuole of this cell type (Fig. 3B). The cytoplasm is clearly distinguished in bright field from the vacuole by the presence of chloroplasts, whereas the vacuole is clear. The corresponding fluorescence image of the same cell loaded with Indo-1 shows a similar pattern.

When studying secondary messenger systems in patch clamping, the investigator tries to maintain the cell in a true physiologic state. The use of the whole-cell patch-clamp technique gives the experimenter the advantage of being able to control the contents of the cell cytosol via equilibration with the patch pipette solution. However, this also results in the loss of intracellular factors that may be crucial to signal transduction pathways. The loss of these components can be reduced or eliminated by using the perforated-patch technique, although this technique has not yet been widely applied to plant cells. To combine perforated patch recording with $Ca^{2+}$ imaging, the indicator dye must be preloaded into the cells. Investigators studying animal systems have utilized the acetoxymethyl (AM) esters of $Ca^{2+}$ indicator dyes.[29,30] However, the use of AM esters in plant systems has been difficult at best. The main problems associated with AM loading of plant protoplasts are external hydrolysis of the ester, lack of internal ester hydrolysis, and compartmentalization into the vacuole.[35,36] Some plant cell types can be loaded with Indo-1 in a pH-dependent manner. Bush and Jones[35] developed a method to load the pentapotassium salt of Indo-1 into barley aleurone protoplasts at pH 4.5. This method has also been successful in loading *Arabidopsis* roots.[39] In this method, cells are incubated for approximately 1 hr with 25 $\mu M$ Indo-1 pentapotassium salt in a buffer held at pH 4.5 with 25 m$M$ dimethylglutaric acid (DMGA). The cells are then washed twice to remove Indo-1 from the bathing media. Using this method, there is no detectable accumulation of Indo-1 into the vacuole. We have yet to combine this approach with confocal $Ca^{2+}$ imaging/electrophysiology because, in our hands, attempts to "acid load" guard cells of *V. faba* have been unsuccessful. It is possible that because of the strong pH regulation of this cell type, either the environment surrounding the cells may be of different pH than the bulk media, or the cells may be sensitive to the high concentration of DMGA needed for acid loading.

Calibration of the fluorescence signals is routinely performed *in vitro*.[39,40] This is done on a daily basis using $Ca^{2+}$ calibration standards from Molecular Probes and 6-$\mu$m polystyrene beads (Bangs Laboratories). In this procedure, 10 $\mu M$ Indo-1 (1 $\mu$l of a 1-m$M$ stock) and a small volume of bead suspension (less than 0.5 $\mu$l) is added to 100 $\mu$l of calibration solution. A small amount of this solution is placed between two coverslips and mounted on the microscope. The midplane of the beads is then focused on and imaged as described above. Focusing on the beads ensures that there are neither artifacts from imaging close to the coverslip nor edge effects from a droplet of calibration solution. *In vivo* calibrations are performed to confirm the validity of the *in vitro* approach.[29] This is accomplished by patch clamping the cell as described above, except that pipette $Ca^{2+}$ concentration is altered by adding either $CaCl_2$ (5 or 10 m$M$) or EGTA (5 to 20 m$M$). These conditions correspond to totally $Ca^{2+}$-saturated dye and totally $Ca^{2+}$-free dye, respectively.

The methods of acquisition of patch-clamp and $Ca^{2+}$ data are largely determined by the types of experiments to be performed. The two acquisition systems on our setup are not linked, i.e., they are independently triggered. The data obtained from simultaneous patch-clamp/confocal $Ca^{2+}$ imaging experiments are analyzed as follows. The electrophysiology data are acquired and analyzed using standard software such as pClamp version 6.0.3 (Axon Instruments). The Indo-1 imaging data are analyzed using IPLab Spectrum (Signal Analytics Corp.), and analysis consists of the following steps. The fluorescence intensity data from the two emission wavelengths are separated, and the background, quantified as a confocal scan on a region of the dish lacking cells, substracted. If significant autoflu-

[39] V. Legué, E. Blancaflor, C. Wymer, G. Perbal, D. Fantin, and S. Gilroy, *Plant Physiol.* **114,** 789 (1997).

[40] S. Gilroy, *Plant Cell* **8,** 2193 (1996).

orescence is present, one can obtain an average value for autofluorescence from a number of cells, and subtract this value from the data. This, however, is imprecise because each cell will have its own specific level of autofluorescence, and autofluorescence intensity will also vary from compartment to compartment (e.g., cytoplasm vs. vacuole) within the cell. Therefore, the best strategy is to load sufficient dye into the cells such that the contribution to the signal from autofluorescence is negligible. After any corrections, the fluorescence intensity values are then ratioed, and the results are presented as a map of $Ca^{2+}$ concentration across the cell. Pseudocolor is usually applied as a visual aid in distinguishing spatial distributions in $Ca^{2+}$ concentration. The $Ca^{2+}$ concentration can be derived in two ways. The first approach is to convert the ratio value using the expression:

$$[Ca^{2+}] = K_d Q[(R - R_{min})/(R_{max} - R)]$$

where $K_d$ is the dissociation constant for Indo-1, $R$ is the ratio value ($\lambda_2/\lambda_1$ where $\lambda_2$ is 400–435 nm and $\lambda_1$ is 450–505 nm), $R_{min}$ is the ratio value of a calibration solution with zero free $Ca^{2+}$, $R_{max}$ is the ratio value of a calibration solution with saturating free $Ca^{2+}$, and $Q$ is the ratio of $F_{max}/F_{min}$ for $\lambda_2$. The second approach is to construct an empirical *in vitro* $Ca^{2+}$ calibration curve using standards (Molecular Probes) that provide several free $Ca^{2+}$ concentrations, and then to plot the ratio values obtained experimentally onto this calibration curve.

When $Ca^{2+}$ imaging is performed, it is possible to obtain information about spatial changes in intracellular $Ca^{2+}$. However, it may also be beneficial to obtain overall cell averages in $Ca^{2+}$ concentration. In plant cells, a large portion of the cell volume is taken up by the vacuole, and this may present a problem. To obtain an average value of cytoplasmic $Ca^{2+}$ concentration, the largest available contiguous area of cytoplasm is measured, and black areas that correspond to the vacuole are excluded. It is important to avoid the area directly adjacent to the pipette, since it may produce artifacts due to reflection from the glass, the fact that the seal, although gigohm, is not infinite, and the fact that cellular buffering of supplied $Ca^{2+}$ diffusing from the patch pipette is not instantaneous.

*Standard $Ca^{2+}$ Imaging and Photometry.* Several companies, including Axon Instruments, Photon Technology International, and Universal Imaging, offer various configurations of imaging systems. Axon Instruments, with a history as a vendor of electrophysiology equipment, appears to be progressing most rapidly toward offering an integrated patch-clamp/standard imaging system.

The two indicators that plant biologists most commonly use at present for standard $Ca^{2+}$ imaging are Fura-2-dextran and Indo-1. Fura-2 is a dual excitation/single emission ratiometric dye, meaning that the emission signal at a single wavelength is measured following excitation of the dye by two different wavelengths. Since both excitation wavelengths for Fura-2 are in the UV region (340 and 360–380 nm) a common method to achieve the two different excitation wavelengths is to install a filter wheel with appropriate filters downstream of the standard fluorescence light source for the microscope. Excitation wavelengths can also be produced using a monochromator, if sufficient funds are available for purchase of this expensive option. Excitation light is directed to the specimen by the use of an appropriate dichroic in the microscope filter holder, and emission light (510 nm for Fura-2) is similarly directed through an appropriate filter to a detector. If spatial information is not required, the detector can be a photomultiplier. Alternatively, if somewhat lower sensitivity and temporal resolution are permissible, and spatial information is important, the detection apparatus is a camera. A bewildering variety of options is available, including cooled charge-coupled device (CCD) cameras, intensified CCD cameras, integrating video cameras, and silicon intensified (SIT) cameras. Individual camera models may differ considerably, but as a brief overview, Table II summarizes the various types of cameras in terms of cost, sensitivity, spatial resolution, speed, and data storage requirements. Not included in the table are frame transfer cameras, which are out of the price range of most

TABLE II
TYPES AND PROPERTIES OF CAMERAS USED IN IMAGING

| Camera type | Output | Price | Sensitivity | Spatial resolution | Temporal resolution | Data storage requirement per image (approximate) (Mbyte) |
|---|---|---|---|---|---|---|
| Integrating video camera | Analog (video) | Inexpensive | Low | Fair | Slow because longer integrating time required to overcome low sensitivity | 0.5 |
| Cooled CCD | Digital | Relatively inexpensive | Very high | Good (varies between cameras) | Slow because of the read-off time required (this can be shortened by choosing a smaller region of interest) | 1–2 or more |
| Intensified CCD | Analog (video) or digital | Moderate | Moderate | Good (varies between cameras) | Fast but integration may be required | 0.5 |
| SIT | Usually analog (video) | High | High to very high | Good | Fast but integration may be required | 0.5 |

investigators. These cameras are very fast because the images are electronically shuttled to a new region of the chip. Camera technology is evolving rapidly; therefore, no specific recommendations are made here. Before purchasing a system, the best strategy is simply to have a variety of cameras demonstrated, such that the most appropriate one for the application at hand can be purchased.

When performing standard $Ca^{2+}$ imaging with Fura-2, the camera (or photomultiplier) can be mounted directly on the microscope port. This is advantageous because there is no possibility that the image will shift in the camera field during the experiment or even between collection of the two successive images that comprise a ratiometric measurement as might be the case if, e.g., the detector were mounted on the vibration isolation table instead of directly on the microscope.

There are two disadvantages to the use of Fura 2-dextran. The first is that, since it is excited by UV light, this dye cannot be used to measure $Ca^{2+}$ before and after uncaging of a second messenger: the initial $Ca^{2+}$ measurement will itself uncage the dye, and, conversely, the high intensity light used for uncaging may photobleach the $Ca^{2+}$ indicator. Work-arounds to this problem have been applied to animal systems. For example, a system has been described[41] that uses low intensity light from a xenon arc lamp to measure $Ca^{2+}$ while releasing only minor amounts of caged $Ca^{2+}$ and a laser flash to release the caged compound that is sufficiently brief so as to avoid photobleaching of the $Ca^{2+}$ indicator. A specific problem with the use of Fura-2 in plant cells is that it emits toward the red region of the spectrum. For this reason, its use to date in plants has been largely restricted to achlorophyllous cells such as root cells and pollen tubes.[42,43] It may be possible to separate the Fura-2 fluorescence, with a peak at 510 nm, from the longer wavelengths of chlorophyll fluoresence by using a narrow bandpass emission filter, although we have not tested this.

The problem of chlorophyll autofluorescence can be avoided by the use instead of Indo-1, as described in the section on confocal imaging. However, Indo-1 also offers a problem for combined patch-clamp and standard imaging. Since Indo-1 is a dual emission dye, the filter wheel must be mounted directly onto the microscope on the emission side. When such a system was demonstrated in our laboratory, the amount of vibration generated by the filter wheel inevitably resulted in loss of the seal on the protoplast. We do not know whether this is a universal problem or one unique to our cell

[41] M. S. Kirby, R. W. Hadley, and W. J. Lederer, *Pflügers Arch.* **427,** 169 (1994).

[42] H. H. Felle and P. K. Hepler, *Plant Physiol.* **114,** 39 (1997).

[43] E. S. Pierson, D. D. Miller, D. A. Callaham, A. M. Shipley, B. A. Rivers, M. Cresti, and P. K. Hepler, *Plant Cell* **6,** 1815 (1994).

type (guard cells). To work around this problem, one could mount two cameras, and install a beamsplitter upstream, such that light of the appropriate wavelenths for Indo-1 emission is directed to each of the two cameras. Nikon and perhaps other microscopy companies as well do offer the necessary hardware, and it is indeed an option if photometry is to be done. However, for standard imaging, (1) it is likely to be prohibitively expensive to purchase two cameras, and (2) unless the cameras are precisely in register, down to the individual pixel, the $Ca^{2+}$ measurement will be completely erroneous. This issue does not arise in confocal microscopy of ratiometric dyes, where alignment occurs simply because both of the pixel-by-pixel images are formed simultaneously, one pixel at a time.

Given the problems described above, we are in the initial stages of investigating the applicability of Calcium Green/Texas Red dextran for $Ca^{2+}$ imaging in chlorophyllous cells. A filter wheel on the upstream side provides the two excitation wavelengths of 488 and 578 nm. A custom-made emission filter that has two windows of transmission, for the emission wavelengths of Calcium Green and Texas Red, but cuts off sharply to eliminate chlorophyll fluorescence, transmits the emitted light alternately from the two dyes, as the two dyes are alternately excited by the spinning filter wheel. We purchased this custom-made filter from Chroma, but it may also be available from other companies, such as Omega Optical.

Once decisions have been made on how to configure the imaging system and which indicator to use, simultaneous patch-clamp and standard $Ca^{2+}$ imaging experiments follow the same basic procedures for data acquisition and analysis as described earlier in the section on confocal $Ca^{2+}$ imaging.

### *Administering Exogenous Secondary Messengers*

To unravel signal transduction chains involved in ion channel modulation, channel activity may also be compared before and after administration of a regulatory molecule. Such regulators may include plant hormones, kinases/phosphatases and other regulatory proteins, antibodies against regulatory proteins, ions such as $Ca^{2+}$ and $H^+$, and synthetic regulators of endogenous signaling molecules. Examples of compounds falling in the last category are GTP$\gamma$S and GDP$\beta$S, which are nonhydrolyzable analogs of GTP and GDP. GTP$\gamma$S and GDP$\beta$S lock G proteins into active and inactive states, respectively, thereby allowing detection of involvement of a G protein in the signaling pathway under investigation.[44,45]

[44] I. McFadzean and D. A. Brown *in* "Signal Transduction: A Practical Approach" (G. Milligan, ed.). Oxford University Press, Oxford, U.K., 1992.

[45] K. Fairley-Grenot and S. M. Assmann, *Plant Cell* **3,** 1037 (1991).

If the regulatory molecule acts from the apoplastic side of the membrane, or if it is internally acting and either membrane permeant or efficacious in the inside-out patch configuration, then it is a relatively simple task to apply the regulator in the bath solution and monitor subsequent effects. However, if the regulator is effective only from within the intact cell, or if it is membrane-permeant but available in such limited amounts or at such high expense that bath application is impractical, then it is necessary to introduce the regulator via the pipette solution. There are various means of accomplishing this, and each has advantages and disadvantages.

*Comparison of Control and Experimental Cells.* One approach is simply to make up two pipette solutions, with and without the regulatory molecule in question, and then to compare recordings made from different cells using the two different solutions. This approach has the advantages of simplicity, and of enabling the researcher to compare control and treatment data obtained at precisely the same point in time during each recording, thus avoiding successive treatments on a single cell, a situation in which rundown or other temporal alterations in cell physiology may confound data interpretation. On the other hand, interpretation of results from these types of experiments can be difficult if cells within the population have variable electrophysiologic "profiles" even within the control treatment. Under such circumstances, it is most useful to record from one and the same cell before and after application of the signaling molecule, using the techniques described below. Because the recordings made will be sequential in nature, it is crucial to determine from control cells that the "baseline" response does not change throughout the duration of the recording.

*Repatching and Pipette Perfusion.* Several methods are available for successive administration of intracellular treatments. If the cells are sufficiently robust as to withstand successive sealing attempts, one can initially obtain whole-cell access and record control currents, then withdraw that patch pipette and seal onto the cell again and achieve whole-cell access using a second pipette that contains the regulatory substance. A more elegant approach is to utilize the pipette perfusion technique, in which an additional port on the pipette holder is used to introduce solution, thus forcing a replacement of the initial pipette solution with the new solution. Pipette perfusion has the practical advantage that the equipment is inexpensive, and the methodological advantage that the regulatory molecule may be subsequently removed by further perfusion, and the cell monitored for recovery of the initial response. This technique has been a standard method in the animal patch clamper's toolkit for some time.[46] A recent paper[47]

[46] J. M. Tang, J. Wang, and R. S. Eisenberg, *Methods Enzymol.* **207,** 176 (1992).

[47] F. J. M. Maathuis, A. R. Taylor, S. M. Assmann, and D. Sanders, *Plant J.* **11,** 891 (1997).

describes an inexpensive perfusion apparatus that has been used in whole-cell recordings from a variety of plant cell types. In theory, pipette perfusion is simple. The main difficulty lies in fine-tuning the perfusion so as to not disrupt the seal. The only special equipment that is required is a pipette holder that has an additional port for the perfusion capillary. Such pipette holders can be purchased from Warner Instrument Corp. in a variety of sizes to fit commonly used headstages. The perfusion capillary should be made of fused silica coated with polyimide (Polymicro Technologies). The capillary is drawn out in a flame to the appropriate size (30–50 $\mu$m in diameter[46]). The silica capillary should be filled with the initial pipette solution to remove air bubbles. The perfusion capillary is then fit into a piece of polyethylene tubing, which is threaded through the perfusion port of the pipette holder. The polyethylene tubing does not enter the pipette shaft, but stops at the end of the perfusion port. The silica capillary is then moved close to the tip of the patch pipette, taking care not to push it through the end of the pipette. The free end of the polyethylene tubing is placed in a small reservoir filled with the starting pipette solution. A seal is formed in the usual way, and after control recordings have been made, the free end of the polyethylene tubing is repositioned in a reservoir containing the pipette solution to be perfused. Suction is then applied through the usual suction port, and the perfusion can be monitored by viewing the rise of fluid in the pipette shaft. The rate of pipette perfusion will vary depending on the diameter of the perfusion capillary and the amount of suction applied. The rate of diffusion of the perfused solution will depend on both the distance of the perfusion capillary from the pipette tip and the diameter of the tip. If fluorescence microscopy is available, the inclusion of a fluorescent dye such as Lucifer Yellow in the second pipette solution can aid in determining the rate of perfusion into the cell.

*Caged Compounds.* The use of caged compounds in combination with the patch-clamp technique is a powerful probe into the secondary messenger regulation of ion channels. Caged compounds are biologically inert until they are released from their protecting group, i.e., "uncaged," by exposure to UV light, which breaks specific chemical bonds within the compound. Thus, like repatching and pipette perfusion, this technique also allows a "before" and "after" look at ion channel modulation by a chemical stimulus. A wide variety of caged compounds is available, including $Ca^{2+}$, amino acids, nucleotides, neurotransmitters, hormones, and enzymes.[34,48] Kits for the synthesis of caged compounds are also available from Molecular Probes.

As with simultaneous $Ca^{2+}$ imaging and patch clamping, the basic technique for using caged secondary messengers in combination with patch

[48] S. R. Adams and R. Y. Tsien, *Annu. Rev. Physiol.* **55,** 755 (1993).

clamping is the same as for conventional patch-clamp experiments except for the inclusion of the desired caged compound in the pipette solution. There must also be a means to UV irradiate the cell under study to release the caged probe and, if possible, to monitor the uncaging. We will describe the use of caged $Ca^{2+}$ in combination with patch clamping or guard cell protoplasts.

There are currently three types[48–50] of caged $Ca^{2+}$ available from Molecular Probes or Calbiochem. We routinely use 1-[2-Amino-5-(1-hydroxy-1-[2-nitro-4,5-methylenedioxyphenyl]methyl)phenoxy]-2-(2′-amino-5′-methylphenoxy)ethane-N,N,N′,N′ tetraacetic acid, sodium (nitr-5), because it has been used successfully in intact guard cells[51] and in barley aleurone protoplasts.[40] Another form of caged $Ca^{2+}$, o-nitrophenyl EGTA (NP-EGTA), is based on EGTA, and is also $Ca^{2+}$ selective; however, it has not been used extensively in plant systems. The other commonly used form of caged $Ca^{2+}$, 1-(2-Nitro-4,5-dimethoxyphenyl)-1,2-diaminoethane-N,N,N′,N′-tetraacetic acid (DM-nitrophen), is based on EDTA, and therefore has a high affinity for $Mg^{2+}$ as well, which limits its usefulness. The chemical structure of nitr-5 is based on BAPTA, and it is therefore highly selective for $Ca^{2+}$ over $Mg^{2+}$. Upon exposure to UV light (360 nm), the affinity of nitr-5 for $Ca^{2+}$ drops from a $K_d$ of 150 n$M$ to about 6.5 $\mu M$.[48] Figure 3 illustrates the elevation in cytosolic $Ca^{2+}$ engendered following treatment of a nitr-5-loaded guard cell protoplast with UV light.

To measure the effect of a rapid increase in intracellular $Ca^{2+}$ on ion currents in guard cell protoplasts, we use the following methods. Guard cell protoplasts are prepared as usual and the ionic compositions of the bath and pipette solutions are also the same as usual with the following exceptions. EGTA is omitted from the pipette solution and replaced with 1 or 2 m$M$ nitr-5. We prepare a 10 m$M$ stock solution of nitr-5 in pipette solution, which is stored at −80° until use. Calcium Green-1 is also added at 50 $\mu M$, from a 2-m$M$ stock solution, also prepared in pipette solution and added to the pipette solution just prior to the experiment. Calcium Green-1 is a nonratiometric dye, and is included to give a qualitative indication of the changes in intracellular $Ca^{2+}$ on photolysis of the caged $Ca^{2+}$. It is important to keep all caged probes and fluorescent indicators protected from light.

Caged $Ca^{2+}$ is actually a $Ca^{2+}$ chelator whose affinity for $Ca^{2+}$ is decreased on photolysis. Therefore, $Ca^{2+}$ must be bound to the chelator which can be released. Often, the caged $Ca^{2+}$ is charged with additional $Ca^{2+}$ before being added to the pipette solution. We have found that this some-

[49] G. C. R. Ellis-Davies and J. H. Kaplan, *Proc. Natl. Acad. Sci. U.S.A.* **91,** 187 (1994).
[50] A. M. Gurney, R. Y. Tsien, and H. A. Lester, *Proc. Natl. Acad. Sci. U.S.A.* **84,** 3496 (1987).
[51] S. Gilroy, N. D. Read, and A. J. Trewavas, *Nature* **346,** 769 (1990).

times produces abnormal whole-cell currents in guard cell protoplasts. When nitr-5 is added uncharged (i.e., with no added $Ca^{2+}$) whole-cell currents are normal and there is an increase in Calcium Green-1 fluorescence after exposure to UV light. This phenomenon most likely results from initial charging of the nitr-5 with abundant intracellular $Ca^{2+}$ as the cells attempt to regulate their $Ca^{2+}$ at normal levels (S. Gilroy, personal communication (1997) and Ref. 50). The appropriate concentrations to add of caging compound and free $Ca^{2+}$ (if any) must be determined empirically for each experimental system.

In our laboratory, the system used for releasing caged $Ca^{2+}$ and monitoring it with Calcium Green-1 is the same as that used for confocal $Ca^{2+}$ imaging with Indo-1, with the following alterations. Dichroic A is a 488 beamsplitter, which allows wavelengths 488 nm and shorter to reach the specimen and reflects wavelengths longer than 488 through the light path. A 488-nm argon laser is used to excite Calcium Green-1. Dichroic B is open, and a 515- to 565-nm bandpass emission filter is placed upstream of the channel 1 photomultiplier (Calcium Green-1 has an emission peak at 530 nm). Calcium is released by 2- to 8-sec scans of the UV laser. If a UV laser is not available, a xenon flashlamp[52] or even the standard UV source for fluorescence microscopy may be used. If a flashlamp is employed, care must be taken that the attendant electrical surge does not damage sensitive electrophysiology equipment.

There are several possible caveats to the use of nitr-5 or any caged compound. First, energy introduced into the biological system by the UV treatment may itself affect the cells, thus necessitating that control data be obtained on cells that contain no caged compound and are subjected to the identical UV treatment as the experimental cells. As for previous procedures, it is important to be certain that the indicators and lasers/light sources used in the experimental protocol do not have an ancillary, unknown effect on the cellular physiology. For example, the 488-nm light used to excite Calcium Green-1 is in the blue region of the spectrum, and plants possess several blue light receptors, which have been implicated in a variety of signaling pathways.[21,22,53] In our experience, the laser wavelengths and intensities used for confocal $Ca^{2+}$ imaging with Indo-1 or for caged $Ca^{2+}$ release have no effect on whole-cell currents of guard cells. However, Calcium Green-1 is potentially phototoxic,[54] thus the dye concentrations and excitation intensities must be carefully tested to ensure that the cell is not damaged. Second, by-products from the uncaging event may themselves

[52] G. Rapp and K. Guth, *Pflügers Arch.* **411,** 200 (1988).

[53] M. Ahmad and A. R. Cashmore, *Nature* **366,** 162 (1993).

[54] J. Eilers, R. Schneggenburger, and A. Konnerth, *in* "Single-Channel Recording" (B. Sakmann and E. Neher, eds.), 2nd Ed. Plenum, New York and London, 1995.

have a physiologic effect. For example, Blatt *et al.* hypothesized that a decrease in outward $K^+$ current that they observed on photorelease of caged $IP_3$ in guard cells actually resulted from a decrease in cytosolic pH stemming from $H^+$ release during uncaging.[2,55] Because of such possibilities, it is valuable, when feasible, to compare results obtained from the same compound chemically caged in more than one way. In the case of caged $Ca^{2+}$, the control experiment can alternatively consist of performing the uncaging in the presence of BAPTA, so that the $Ca^{2+}$ that is released is quickly sequestered. In practice, this requires high amounts of BAPTA (10–20 m*M*[56]) to abolish the rise in $Ca^{2+}$, and this may make seal formation difficult. A third issue is that, unlike the case with pipette perfusion, the signaling compound, once uncaged, cannot be subsequently removed. Finally, the caged compound approach is, of course, only available if the requisite molecule can be purchased or synthesized. Despite these potential roadblocks, the elegance and simplicity of the caged compound approach dictates that use of this technique will continue to increase.

## Concluding Remarks

After publication of the seminal paper on the patch-clamp method in 1981, it was 3 years until plant membrane biologists capitalized on this technique.[57,58] Since then, however, the gap in progress between studies of animal vs. plant ion transport systems has been steadily narrowing. It is our hope that this chapter will, in the realm of secondary messenger regulation of ion channels, contribute toward this trend.

## Appendix

Addresses of major vendors cited in this article are listed here.

Axon Instruments, Inc., 1101 Chess Drive, Foster City, CA 94404; phone: 650-571-9400, fax: 650-571-9500

Bangs Laboratories Inc., 979 Keystone Way, Carmel, IN 46032; phone: 317-844-7176

Calbiochem, P.O. Box 12087. La Jolla, CA 92039; phone: 619-450-9600, fax: 619-453-3552

Chroma Technology Corp., 72 Cotton Mill Hill, Unit A-9, Brattleboro, VT 05301; phone: 802-257-1800, fax: 802-257-9400

[55] M. R. Blatt, G. Thiel, and D. R. Trentham, *Nature* **346,** 766 (1990).

[56] S. E. Bates and A. M. Gurney, *J. Physiol.* **466,** 345 (1993).

[57] N. Moran, G. Ehrenstein, K. Iwasa, C. Bare, and C. Mischke, *Science* **226,** 835 (1984).

[58] J. I. Schroeder, R. Hedrich, and J. M. Fernandez, *Nature* **312,** 361 (1984).

GIBCO BRL, Grand Island, NY; phone: 800-828-6686, fax: 800-331-2286

Molecular Probes Inc., 4849 Pitchford Avenue, Eugene, OR 97402; phone: 541-465-8300, fax: 541-344-6504

Omega Optical, P.O. Box 573, 3 Grove Street, Brattleboro, VT 05302; phone: 802-254-2690, fax: 802-254-3937

Photon Technology International, 1 Deerpark Drive, Suite F, South Brunswick, NJ 08852; phone: 908-329-0910, fax: 908-329-9069

Polymicro Technologies Inc., 18019 North 25 Avenue, Phoenix, AZ 85023; phone: 602-375-4100, fax: 602-375-4110

Seishin Corporation, 4-13 Koamicho, Nihonbashi, Chuo-ku, Tokyo, Japan; phone: 03-3669-2876, fax: 03-3669-1684

Sigma, St. Louis, MO; phone: 800-325-3010, fax: 800-325-5052

Signal Analytics Corp., 440 Maple Avenue East, Suite 201, Vienna, VA 22180; phone: 703-281-3277, fax: 703-281-2509

Universal Imaging Corp., 502 Brandywine Parkway, West Chester, PA 19380; phone: 610-344-9410, fax: 610-344-9515

Warner Instruments Corp., 1125 Dixwell Avenue, Hamden, CT 06514; phone: 203-776-0664, fax: 203-776-1278

Yakult Pharmaceutical Ind. Co., Ltd., 1-1-19 Higashi-Shinbashi, Minato-ku, Tokyo, 105 Japan; phone: 03-3574-6766 fax: 03-3574-7254

## Acknowlegments

Research on secondary messenger regulation of ion channels in our laboratory is currently supported by grants from the U.S. Department of Agriculture (94-37304-1003), the National Science Foundation (MCB-9316319), and the NASA/NSF Network for Research on Plant Sensory Systems (MCB-9416039). We thank Dr. Simon Gilroy (Pennsylvania State University) for many valuable discussions on $Ca^{2+}$ measurement and imaging techniques, and Dr. Henk Miedema (Pennsylvania State University) for helpful comments on the manuscript. We also thank Dr. Wei-hua Wu (Beijing Agricultural University), for access to the unpublished data depicted in Fig. 1.

# Section IV

# Special Channels

# [23] ATP-Sensitive Potassium Channels

*By* M. SCHWANSTECHER, C. SCHWANSTECHER, F. CHUDZIAK, U. PANTEN, J. P. CLEMENT IV, G. GONZALEZ, L. AGUILAR-BRYAN, and J. BRYAN

## Introduction

ATP-sensitive potassium channels, or $K_{ATP}$ channels, couple changes in cellular metabolism with membrane electrical activity. In pancreatic $\beta$ cells these channels set the resting membrane potential. Increased glucose metabolism changes the ATP/ADP ratio reducing the opening of $K_{ATP}$ channels causing membrane depolarization and activation of voltage-gated $Ca^{2+}$ channels. The resulting increase in $[Ca^{2+}]_i$ stimulates insulin exocytosis. Pharmacologically distinct types of $K_{ATP}$ channels have been identified in muscle cells, where their opening would be expected to reduce electrical activity. $K_{ATP}$ channels are assembled from sulfonylurea receptors, SURs, members of the ATP-binding cassette superfamily, and members of the inwardly rectifying potassium channel family, $K_{IR}6.x$. Genetic, biochemical, and electrophysiologic data establish that the $\beta$-cell channel is assembled from SUR1 (OMIM 600509) and $K_{IR}6.2$ (OMIM 600937).[1-4] $K_{IR}6.2$ forms the pore of the channel, while SUR1 regulates channel activity and confers responsiveness to channel openers like diazoxide, pinacidil, and cromakalim, and to channel blockers, sulfonylureas, like tolbutamide and glibenclamide. SUR1 and $K_{IR}6.2$ are present in the channel complex in a 1:1 ratio in a tetrameric stoichiometry, $(SUR1/K_{IR}6.2)_4$. Loss of function mutations in SUR1 and $K_{IR}6.2$ have been shown to result in persistent hyperinsulinemic hypoglycemia of infancy (OMIM256450), a neonatal disorder characterized by unregulated insulin release despite severe hypoglycemia. The $\beta$-cell $K_{ATP}$ channel is inhibited by nanomolar concentrations of glibenclamide, and can be activated by diazoxide. The SUR1/$K_{IR}6.2$ channels have also been identified in neuronal tissue.[5-7]

SUR2A and $K_{IR}6.2$ assemble a potassium-selective, inwardly rectifying channel with the electrical and pharmacologic properties of the cardiac

[1] L. Aguilar-Bryan and J. Bryan, *Diabetes Rev.* **4,** 336 (1996).
[2] J. Bryan and Aguilar-Bryan, *Curr. Opin. Cell Biol.* **9,** 553 (1997).
[3] A. P. Babenko, L. Aguilar-Bryan, and J. Bryan, *Ann. Rev. Physiol.* **78,** 227 (1998).
[4] L. Aguilar-Bryan, J. P. Clement IV, G. Gonzalez, K. Kunjilwar, A. Babenko, and J. Bryan, *Physiol. Rev.* (in press, January issue), (1998).
[5] P. Jonas, D.-S. Koh, K. Kampe, M. Hermsteiner, and W. Vogel, *Pflügers Arch.* **418,** 68 (1991).
[6] T. Ohno-Shosaku and C. Yamamoto, *Pflügers Arch.* **422,** 260 (1992).
[7] C. Schwanstecher and D. Bassen, *Br. J. Pharmacol.* **121,** 193 (1997).



0076-6879/99 $30.00

$K_{ATP}$ channel.[8] These channels are less sensitive to glibenclamide, $IC_{50} \sim 0.3\ \mu M$, and are not responsive to diazoxide, but can be activated by pinacidil at micromolar concentrations, $IC_{50} \sim 20–100\ \mu M$. A splice variant of SUR2A, designated SUR2B, differing only in the splicing of the C-terminal 45 amino acid exons,[4] has been reported to pair with both $K_{IR}6.2$ and $K_{IR}6.1$ to form $K_{ATP}$ channels similar to those found in vascular smooth muscle.[9] Like the cardiac channels, the SUR2B channels are less sensitive to glibenclamide than the $\beta$-cell channel, but are activated by diazoxide and can be activated by lower concentrations of pinacidil than the cardiac channel. The SUR2B/$K_{IR}6.1$ channels are reported to be activated by nucleoside diphosphates, but not inhibited by ATP.[9,10]

This article describes methods used in the characterization of $K_{ATP}$ channels, including synthesis and radioiodination of two derivatives of glibenclamide that photolabel SUR1.

## Drug Synthesis

Two derivatives of glibenclamide, iodoglibenclamide[11] and azidoiodoglibenclamide,[12,13] have been critically important in the isolation and characterization of SUR1, the high-affinity sulfonylurea receptor. Although the radioiodinated species are available commercially at great cost, their synthesis, including radioiodination, can be accomplished easily by the average laboratory. We have detailed the synthesis of both derivatives here.

### *Synthesis of Iodoglibenclamide*

*Chemicals.* Chemicals (pure grade) are purchased from Aldrich Chemical Company Inc. (Wilwaukee, WI) (unless stated) and are used as received. Solvents are dried and distilled before use.

[8] N. Inagaki, T. Gonoi, J. P. Clement IV, C. Z. Wang, L. Aguilar-Bryan, J. Bryan, and S. Seino, *Neuron* **16,** 1011 (1996).

[9] S. Isomoto, C. Kondo, M. Yamada, S. Matsumoto, O. Higashiguchi, Y. Horio, Y. Matsuzawa, and Y. Kurachi, *J. Biol. Chem.* **271,** 24321 (1996).

[10] D. J. Beech, H. Zhang, K. Nakao, and T. B. Bolton, *Br. J. Pharmacol.* **110,** 573 (1993).

[11] L. Aguilar-Bryan, D. A. Nelson, Q. A. Vu, M. B. Humphrey, and A. E. Boyd III, *J. Biol. Chem.* **265,** 8218 (1990).

[12] F. Chudziak, M. Schwanstecher, H. Laatsch, and U. Panten, *J. Labelled Comp. Radiopharm.* **34,** 675 (1994).

[13] M. Schwanstecher, S. Loser, F. Chudziak, C. Bachmann, and U. Panten, *J. Neurochem.* **63,** 698 (1994).

The individual steps for making and radiolabeling iodoglibenclamide are described in detail below. Figure 1 provides an overview of the synthesis and the structures.

*Step 1: Synthesis of 4-[β-(2-Hydroxybenzenecarboxamido)ethyl]benzenesulfonamide.* Salicyclic acid 4.1 g (0.031 mol) is dissolved in 50 ml of dry acetone, 4.3 ml (0.031 mol) of triethylamine is added, and the solution is cooled to −20°. To this solution 3 ml (0.031 mol) of ethyl chloroformate is added dropwise while stirring. The resulting mixed anhydride is held at −20° for 15 min with occasional stirring.

A suspension is prepared from 6 g (0.031 mol) of *p*-(β-aminoethyl)benzenesulfonamide, 4.3 ml of triethylamine, and 50 ml of dry acetone and cooled to −20°. This suspension is added rapidly, while stirring, to the mixed anhydride prepared above. The mixture is stirred for 2 hr at 4° and then at room temperature (23°) for 4 hr. The acetone is removed by vacuum distillation and the residue acidified with 0.2 *N* HCl. The resulting precipitate is collected by filtration, washed with $H_2O$, air dried, and recrystallized from ethanol.

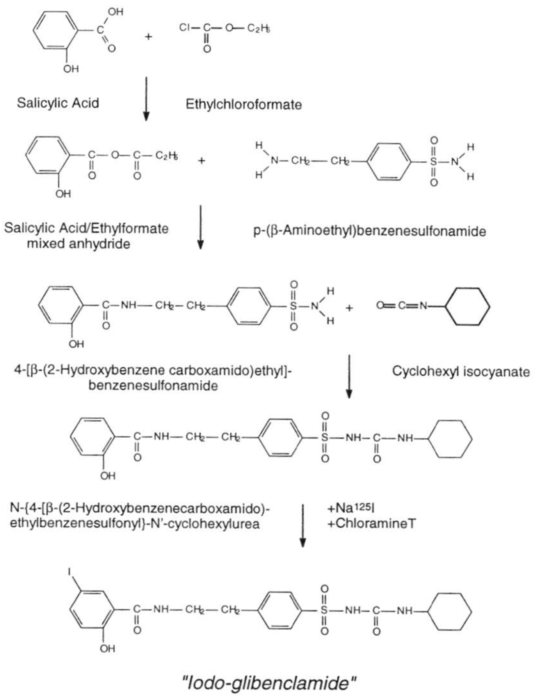


FIG. 1. Synthesis of iodoglibenclamide.

*Step 2: Synthesis of N-{4-[β-(2-Hydroxybenzenecarboxamido)ethyl]benzenesulfonyl}-N′-cyclohexylurea.* 4[β-(2-Hydroxybenzenecarboxamido) ethyl]benzenesulfonamide (12 g) is dissolved in 90 ml of acetone at 4°. NaOH (1.8 g), dissolved in 10 ml of $H_2O$, is added, with 5.75 ml of cyclohexyl isocyanate in four equal aliquots over a period of 80 min. After stirring overnight, 150 ml of $H_2O$ is added, stirred for 10 min, then filtered to remove a small amount of precipitate. Acidification by the addition of 75 ml of 2 *N* HCl produces a precipitate that is collected by filtration after 60 min, air dried, and recrystallized from ethanol.

*Step 3: Synthesis of N-{4-[β-(2-Hydroxy-5-iodobenzenecarboxamido) ethyl]benzenesulfonyl}-N′-cyclohexylurea. N*-{4-[β-(2-Hydroxybenzenecarboxamido)ethyl]benzenesulfonyl}-*N*′-cyclohexylurea, 0.5 g (1.13 mmol), and 0.26 g of NaI are dissolved in 5 ml of dimethylformamide. Chloramine-T (0.39 g), dissolved in 2 ml of dimethylformamide, is added to start the reaction. After stirring for 1 hr at 23°, the reaction is terminated by addition of 25 ml of 0.1 *N* HCl. The iodinated compound (iodoglibenclamide) is extracted into ethyl acetate and washed sequentially with equal volumes of 50 m*M* sodium bisulfite, $H_2O$, 0.5 *M* NaCl (2×), and $H_2O$ (2×). The volume of the ethyl acetate phase is reduced to approximately 5–10 ml and the iodinated product is collected and recrystallized from ethyl acetate with an approximate yield of 70%.

*Step 4: Radioiodination of N-{4[β-(2-Hydroxybenzenecarboxamido) ethyl]benzenesulfonyl}-N′-cyclohexylurea.* One microliter of a 10 m*M* solution of *N*-{4-[β-(2-hydroxybenzenecarboxamido)ethyl]benzenesulfonyl}-*N*′-cyclohexylurea dissolved in dimethylformamide is added to 2–4 μl of $Na^{125}I$ (1–4 mCi) in a solution of dilute NaOH. Iodination is initiated by the addition of 1 μl of 10 m*M* chloramine-T dissolved in dimethylformamide. The reaction is terminated after 10 min at 23° by addition of 1 μl of 14 *M* 2-mercaptoethanol and 20 μl of 50% (v/v) methanol in $H_2O$. This mixture is absorbed to a μBondapak $C_{18}$ column (Waters/Millipore, Bedford, MA) equilibrated with 50% methanol, and then separated by high-performance liquid chromatography (HPLC) using a 50–90% methanol gradient with a flow rate of 1 ml/min. A slightly concave gradient was used for increased resolution. The uniodinated and iodinated species are well resolved with retention times of 9.9 and 13.3 min, respectively. The fractions (0.5 ml) containing the $^{125}I$-labeled iodoglibenclamide are pooled and stored at −20°. The concentration of the drug, based on the specific activity of the $^{125}I$, is determined immediately after purification. The material gives a single spot on reversed-phase thin-layer chromatography when analyzed on Whatman $KC_{18}F$ plates using 80% methanol/20% 0.5 *M* NaCl as a solvent. The iodinated material cochromatographs with the unlabeled iodinated compound synthesized as described in step 3.

*Synthesis of Azidoiodoglibenclamide*

The individual steps for the synthesis and radiolabeling of azidoiodoglibenclamide are described in detail below. Figure 2 provides an overview of the synthesis and the structures.

*Step 1: Synthesis of N-Acetyl-2-phenylethylamine.* 2-Phenylethylamine (91.7 g) and 91.7 g of 2,4,6-trimethylpyridine are dissolved in 600 ml of dry dichloromethane. Acetyl chloride (59.4 g) is added dropwise with stirring at room temperature. The resulting precipitate is collected by filtration, washed with 0.1 *M* HCl, 0.1 *M* NaOH, and $H_2O$, and finally dried over anhydrous $MgSO_4$. Dichlormethane is removed by vacuum distillation.

*Step 2: Synthesis of 4-(2-Acetamidoethyl)benzenesulfonamide.* To 50 g of *N*-acetyl-2-phenylethylamine 17.5 g of chlorosulfonic acid is added dropwise with stirring at 50°. The reaction mixture is allowed to cool to room

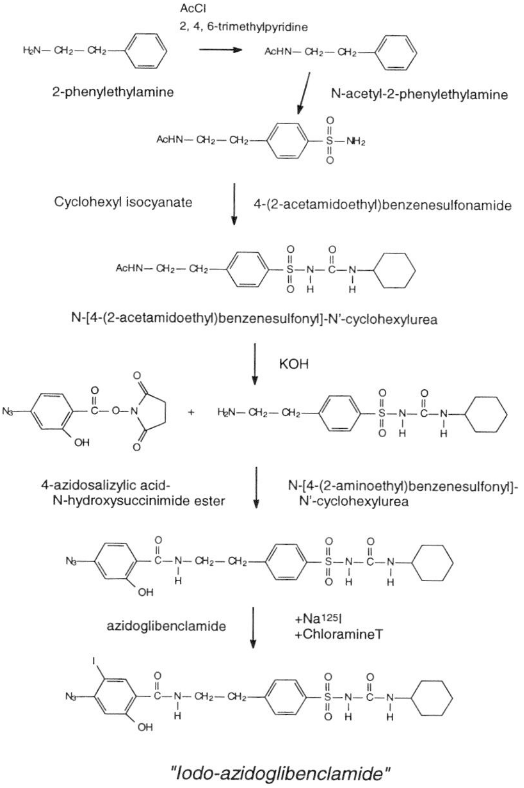


FIG. 2. Synthesis of azido[[125]I]iodoglibenclamide.

temperature, stirred for another 30 min, and then poured into ice-cold $H_2O$. The precipitated sulfochloride is collected by filtration and refluxed 15 min with a mixture of 90 ml concentrated $NH_3$ plus 135 ml $H_2O$. After heating with activated carbon and stirring overnight at 5° the resulting precipitate is collected by filtration, washed with ice-cold $H_2O$, and dried by vacuum distillation.

*Step 3: Synthesis of N-[4-(2-Acetamidoethyl)benzenesulfonyl]-N'-cyclohexylurea.* To 7.6 g of 4-(2-acetamidoethyl)benzenesulfonamide in 40 ml acetone 12.7 ml of 2 *M* NaOH is added and the solution cooled to 4°. At this temperature 5.1 g of cyclohexyl isocyanate is added dropwise while stirring. The mixture is then stirred at room temperature (23°) for 3 hr. The acetone is removed by vacuum distillation and the solid residue obtained dissolved in 0.1 *M* NaOH. The solution is adjusted to pH 9 and filtrated to remove dicyclohexylurea. The filtrate is acidified with 2 *M* HCl and the resulting precipitate collected by filtration and recrystallized from ethanol/$H_2O$.

*Step 4: Synthesis of N-[4-(2-Aminoethyl)benzenesulfonyl]-N'-cyclohexylurea.* Hydrolysis of the *N*-acetyl group is carried out by boiling 5.0 g of *N*-[4-(2-acetamidoethyl)benzenesulfonyl]-*N'*-cyclohexylurea with 7.6 g KOH for 8 hr using 34 ml of 90% (v/v) ethanol as solvent. Ethanol is removed by vacuum distillation and the residue dissolved in $H_2O$. The product is precipitated by neutralization with 2 *M* HCl, collected by filtration, and washed with ethyl acetate.

*Step 5: Synthesis of N-[4-(2-{4-Azido-2-hydroxybenzamido}ethyl)benzenesulfonyl]-N'-cyclohexylurea ($N_3$-GA).* To 6 ml dimethylformamide containing 32.5 mg *N*-[4-(2-aminoethyl)benzenesulfonyl]-*N'*-cyclohexylurea and 28 μl triethylamine 27 mg of 4-azidosalicyclic acid *N*-hydroxysuccinimide ester (Sigma, St. Louis, MO) is added at 0°. After stirring in the dark for 2 hr at 0° and then overnight at room temperature, the reaction mixture is poured into ice-cold $H_2O$, followed by acidification with 2 *M* HCl. The resulting precipitate is collected by filtration and wshed with ice-cold $H_2O$.

*Step 6: Radioiodination of $N_3$-GA.* To a NENSURE-vial containing 5 mCi of $Na^{125}I$ (2.35 nmol) in 50 μl of 10 μ*M* NaOH (NEZ-033A, NEN, Dreieich, Germany) is added $N_3$-GA (9.38 nmol, 37.5 μl of a 0.25 m*M* solution). The latter solution is prepared by dissolving 1.22 mg of $N_3$-GA in 8 ml of 10 m*M* NaOH, adding 1.95 ml of 0.5 *M* sodium phosphate buffer (pH 7.4), and adjusting to pH 7.4 with 2 *M* HCl. Iodination is started by addition of 37.5 μl of 0.25 m*M* chloramine-T dissolved in 0.1 *M* sodium phosphate buffer (pH 7.4). After 15 min at room temperature in the dark, the reaction is quenched by addition of 2 μl of 0.14 *M* 2-mercaptoethanol and mixing. Separation of the reaction mixture is carried out immediately by HPLC (ODS-Hypersil 5-$\mu$m column). The mobile phase (60% acetonitrile, 40% water, 0.07% trifluoroacetic acid) is used isocratically with a flow rate of 2 ml/min. The retention times of the radioactive peaks representing *N*-[4-(2-{4-azido-2-hydroxy-5-[$^{125}$I]iodobenzamido}ethyl)benzenesulfonyl]-*N'*-cyclohexylurea ($^{125}IN_3$-GA) and the 3,5-diiodinated derivative are 4.6 and 9.8 min, respectively. The radioactivity of the diiodinated derivative is 4% of the radioactivity of $^{125}IN_3$-GA. A minor radioactive peak is eluted just ahead of $^{125}IN_3$-GA and is excluded from the fractions containing $^{125}IN_3$-GA. The retention time of $N_3$-GA is 3.0 min, and nonradioactive UV peaks are not detected between 3.0 and 15.0 min. The fractions (333 $\mu$l) containing $^{125}IN_3$-GA are collected, counted (2.8 mCi), and stored at 4° (used for up to 2 months). Because monitoring of the eluate did not reveal chemical or radiochemical impurities in the fractions containing $^{125}IN_3$-GA, a specific activity of approximately 2100 Ci/ml can be assumed. The radiochemical yield based on $^{125}$I was 56%.

## Tissue Culture

### *Maintenance of Tissue Culture Cells*

Tissue culture cells producing SUR1 include the hamster insulin secreting cell line, HIT-T15 (passage 65-75; CRL1777 ATCC, Rockville, MD), the rat insulinoma cell line, RINm5f, and the glucagon secreting cell line, $\alpha$TC-6. Chinese hamster ovary (CHO) and COS (1, m6, and 7) cell lines do not produce SUR1 and are used for receptor and channel expression. Cells are maintained in T-175 culture flasks (Falcon) as monolayers in Dulbecco's modified Eagle's medium with high glucose (DMEM)-HG medium supplemented with 10% fetal bovine serum (FBS), 100 U/ml penicillin, and 0.1 mg/ml streptomycin. Cells are grown in 5% $CO_2$ at 37°, maintained in subconfluent cultures, fed three times a week, and subcultured as needed. To subculture, confluent cells are detached with 0.05% trypsin/EDTA, resuspended in supplemented DMEM-HG medium and replated at one-tenth the original density. For expression experiments, COS cells are plated at 50–60% confluence prior to transient transfection.

## Transfection Protocols

Cells are transfected using either a DEAE-dextran or a lipofectamine protocol.

### *DEAE-Dextran Protocol*

For $^{86}Rb^+$ efflux studies, 3-day-old cultures of COS cells are trypsinized and replated at a density of $2.0 \times 10^5$ cells per 35-mm well (six-well dish) and allowed to attach overnight. Typically, 5 $\mu$g of a SUR1 plasmid is mixed with 5 $\mu$g of a $K_{IR}$ plasmid and brought up to 7.5 $\mu$l final volume in TBS (8 g/liter NaCl; 0.38 g/liter KCl; 0.2 g/liter $Na_2HPO_4$; 3.0 g/liter Tris base; 0.15 g/liter $CaCl_2$; 0.1 g/liter $MgCl_2$, pH 7.5) before addition of DEAE-dextran (30 $\mu$l of a 5 mg/ml solution in TBS). The samples are vortexed, collected by briefly spinning in a microfuge, then incubated for 15 min at room temperature before addition of 500 $\mu$l 10% NuSerum (Collaborative Biomedical Products, Twin Oak Park, Bedford, MA) in TBS. Cells are washed twice with Hanks' balanced salt solution (HBSS), the DNA mix is added, and the cells are maintained in a 37° $CO_2$ incubator. After 4 hr the DNA mix is decanted and the cells shocked for 2 min in 1 ml HBSS + 10% dimethyl sulfoxide (DMSO), then placed in 1.5 ml of DMEM-HG + 2% FBS + 10 $\mu M$ chloroquine and kept in a 37° $CO_2$ incubator. After 4 hr, the cells are washed twice with HBSS and incubated in normal growth media until assayed (usually 36–48 hr posttransfection).

### *Lipofectamine Protocol*

COS cells are plated in six-well dishes as described above and are 70–80% confluent when transfected. Typically 1 $\mu$g of a SUR1 plasmid and 1 $\mu$g of a $K_{IR}$6.2 plasmid are mixed with 375 $\mu$l of Opti-Mem reduced serum medium (Life Technologies, Inc.) then added to 375 $\mu$l of Opti-Mem containing 9 $\mu$l of lipofectamine. After a 45-min incubation, the mixture is brought to 1 ml with Opti-Mem and added to the cells that had been washed once with 3 ml of Opti-Mem. Transfections are allowed to proceed for 5 hr, at which time the Opti-Mem is replaced with DMEM-HG medium plus 10% FBS.

Transfections are scaled based on the area of the plates used. For example, 150-mm plates used for membrane isolations are transfected with 100 $\mu$g of each plasmid using the DEAE-dextran protocol.

## Rubidium Efflux Assays

Twenty-four hours posttransfection, cells are placed in fresh media containing approximately 1 $\mu$Ci/ml $^{86}$RbCl, incubated for an additional 12–24 hr, and assayed as follows. Cells are incubated for 30 min at 25° in Krebs–Ringer solution under one of three conditions: no additions (basal), with oligomycin (2.5 $\mu$g/ml) and 2-deoxy-D-glucose (1 m$M$) (metabolically inhibited) or with oligomycin and deoxyglucose plus 1 $\mu M$ glibenclamide ($K_{ATP}$ channels inhibited). Cells are washed once in $^{86}Rb^+$-free Krebs–Ringer solution, with or without the added inhibitors, then time points are taken by removing all the medium from the cells and replacing it with fresh medium at the indicated times. Equal portions of the medium from each time point are counted, and the values summed to determine flux. Total $^{86}Rb^+$ is defined as the sum of counts from each time point plus the counts released by addition of 1% SDS to the cells at the end of the experiment. Results can be presented either as the percentage of total cellular $^{86}Rb^+$ released or as percent glibenclamide inhibitable efflux. The latter measure represents specific $K_{ATP}$ channel activity and is defined as the percent $^{86}Rb^+$ efflux from metabolically inhibited cells minus the percent $^{86}Rb^+$ efflux from metabolically inhibited plus glibenclamide inhibited cells (efflux through $K_{ATP}$ channels defined as % glibenclamide inhibitable efflux = % metabolically inhibited efflux − % metabolically and glibenclamide inhibited efflux).

## Membrane Isolation

Membranes are prepared 60–72 hr posttransfection from 10–20 150-mm dishes. Transfected cells are washed twice in phosphate-buffered saline (PBS), pH 7.4, then scraped in PBS and collected in 10-ml plastic tubes. The cells are pelleted, resuspended in 10 ml of hypotonic buffer (5 m*M* Tris-HCl, pH 7.4, 2 m*M* EDTA), and allowed to swell for 45 min on ice. Cells are then homogenized, transferred to a 15-ml glass tube, and spun at 1000 *g* for 20 min at 4° to remove nuclei and unbroken cells. The supernatant is then transferred to a polycarbonate centrifuge tube and membranes are collected by ultracentrifugation at 40,000 rpm in an 80Ti fixed-angle rotor for 2 hr. The pelleted membranes are resuspended in 3–500 $\mu$l of membrane buffer (50 m*M* Tris-HCl, pH 7.4, 5 m*M* EDTA) and stored at −80°. Typical protein concentrations are 2–5 mg/ml.

## Photolabeling Protocols

### *Photolabeling of Membranes*

[$^{125}$I]Iodoglibenclamide[11] or azido[$^{125}$]iodoglibenclamide[13] is added to samples (typically to a concentration of 5–10 n*M*) and the samples incubated at 23° for 30 min. Sample concentration was at 5–10 mg protein/ml. Samples are then transferred onto Parafilm and irradiated at 312 nm in a UV cross-linker (Model FB-UVXL-1000, Spectronics Corp., Westbury, NY) or at 356 nm by use of a hand lamp (Camag, Berlin, Germany). The photolabeling reaction is optimized in terms of energy required. In typical experiments, 1.0–1.5 J/cm$^2$ is used for iodoglibenclamide and 0.2 J/cm$^2$ for azidoiodoglibenclamide. Nonspecific labeling is evaluated by addition of unlabeled glibenclamide at a concentration of 0.1 $\mu M$.

*Photolabeling of Live Cells*

Photolabeling is carried out as described for isolated membranes.[14,15] Living cells, grown in in six-well dishes, are washed three times with PBS, then incubated in the dark with 10 n$M$ azido[$^{125}$I]iodoglibenclamide or 10 n$M$ [$^{125}$I]iodoglibenclamide in Krebs–Ringer solution supplemented with 10 m$M$ glucose. After 30 min at room temperature, the cells are irradiated in a UV cross-linker (Model FB-UVXL-1000 Spectronics Corp., Westbury, NY) at a setting of 0.9 J/cm$^2$. Excess unbound drug is removed by three 5-ml washes with PBS (1 min each). The cells are then solubilized in 250–500 $\mu$l of 2$\times$ SDS sample buffer.[16] Aliquots are separated on 8% polyacrylamide gels, stained with Coomassie blue, dried, and placed on X-ray film.[15,17]

These protocols label the core and complex glycosylated SUR1 receptors when [$^{125}$I]iodoglibenclamide is used as the probe. When azido[$^{125}$I]iodoglibenclamide is used, $K_{IR}6.2$ is cophotolabeled in addition to the core and complex glycosylated receptors.[18,19]

## Receptor Solubilization

*Preparation of Digitonin*

Digitonin (Sigma) is purchased as a powder and is not further purified. For receptor solubilization, 20% (w/v) digitonin is prepared fresh each day by addition of the detergent to deionized water, vortexing for a few seconds, and boiling in a closed test tube for 2 min. Boiling results in a relatively clear detergent solution diluted as outlined below.

*Digitonin Solubilization*

For receptor purification membranes are thawed, rehomogenized using a Teflon–glass homogenizer, and mixed with ice-cold digitonin to a final protein concentration of 5 mg/ml and 1% digitonin. Subsequent steps are performed at room temperature in the presence of a cocktail of protease inhibitors [0.1 m$M$ phenylmethylsulfonyl fluoride (PMSF), 0.1 m$M$ phenanthroline, and 0.1 m$M$ iodoacetamide]. Membranes are homogenized, then

[14] L. Aguilar-Bryan, C. G. Nichols, A. S. Rajan, C. Parker, and J. Bryan, *J. Biol. Chem.* **267,** 14934 (1992).

[15] D. A. Nelson, L. Aguilar-Bryan, and J. Bryan, *J. Biol. Chem.* **267,** 14928 (1992).

[16] U. K. Laemmli, *Nature* **227,** 680 (1970).

[17] L. Aguilar-Bryan, J. P. Clement IV, and D. A. Nelson, *Methods Enzymol. (in press),* (1998).

[18] M. Schwanstecher, S. Loser, F. Chudziak, and U. Panten, *J. Biol. Chem.* **269,** 17768 (1994).

[19] J. P. Clement IV, K. Kunjilwar, G. Gonzalez, M. Schwanstecher, U. Panten, L. Aguilar-Bryan, and J. Bryan, *Neuron* **18,** 827 (1997).

solubilized for 15 min; the solubilized receptor is separated from insoluble material by centrifugation at 100,000*g* for 1 hr at 4°.

#### *Triton X-100 or CHAPS Solubilization*

Membranes are thawed, rehomogenized using a Teflon–glass homogenizer, and mixed with ice-cold solubilization buffer to give the final concentrations: 10 mg/ml membrane protein; 0.2% (w/v) phosphatidylcholine (egg); 10% (w/v) glycerol; 115 m*M* KCl; 1 m*M* ethylenediaminetetraacetic acid (EDTA); 100 $\mu M$ PMSF; 20 m*M* *N*-2-hydroxyethylpiperazine-*N'*-2-ethanesulfonic acid (HEPES), pH 7.4. Triton X-100 or CHAPS (3-[(3-cholamidopropyl)dimethylammonio]-1-propane sulfonate) is added to 2% (w/v). The solution is gently stirred at 4° (1 hr) and centrifuged at 120,000*g* and 4° for 1 hr. The supernatant is then collected and stored at −80° until binding studies are performed.

## Partial Purification of SUR1

#### *Lectin Affinity Chromatography*

*Concanavalin A.* For purification of the 140-kDa core glycosylated species of the sulfonylurea receptor 4-ml aliquots of digitonin-solubilized receptor are cycled four times over a 1-ml concanavalin A (con A)-Sepharose (Sigma) column equilibriated with 25 m*M* Tris (pH 7.5), 0.1 *M* NaCl, 2 m*M* EDTA, 1% (w/v) digitonin. The column is washed with 8 ml of the equilibrating buffer and eluted with 4 ml of the equilibrating buffer containing 0.5 *M* methyl $\alpha$-D-mannopyranoside. The eluted protein can be stored at this stage at −80°. See also Ref. 17.

*Wheat Germ Agglutinin.* Wheat germ agglutinin (WGA; Sigma)-Sepharose was used instead of con A-Sepharose for the purification of the 150-kDa complex glycosylated receptor. The receptor was eluted with 0.3 *M* *N*-acetylglucosamine. All other manipulations are as described for the purification of the 140 kDa protein. See also Ref. 17.

#### *$Ni^{2+}$-Agarose*

*Histidine-Tagged Receptor.* $Sur1_{N\text{-}6X\text{-}HIS}$ and $SUR1_{N\text{-}6X\text{-}HIS}/K_{IR}6.x$ complexes can be partially purified by chromatography on a column of $Ni^{2+}$-agarose (Qiagen Corp., Santa Clarita, CA) equilibrated in solubilization buffer (1.0% (w/v) digitonin, 150 m*M* NaCl, 25 m*M* Tris, pH 7.4) plus 4 m*M* imidazole. For example, to demonstrate that SUR1 and $K_{IR}6.2$ form a complex, approximately 150 $\mu$g of membranes prepared from COS cells transfected with $K_{IR}6.2$ and SUR1 plasmids is photolabeled with 10 n*M* azido[$^{125}$I]iodoglibenclamide, pH 6.5. The labeled membranes are solubilized in 200 $\mu$l of solubilization buffer for 30 min on ice, then spun for 30 min at 100,000*g* at 4° (Model Tl-100, Beckman Instruments, Fullerton, CA) to remove insoluble material before passing over the $Ni^{2+}$ column four times. The column is washed with 20 ml (40 times column volume) of digitonin wash buffer (0.2% (w/v) digitonin, 150 m*M* NaCl, 25 m*M* Tris, pH 7.4, plus 4 m*M* imidazole), then eluted with wash buffer containing 100 m*M* imidazole, pH 7.4. Aliquots of each fraction are analyzed by polyacrylamide gel electrophoresis as described above.

## Additional Purification Steps

### *Reactive Green-19 Affinity Chromatography*

The eluate from con A-Sepharose is cycled twice over a 1-ml column of Reactive Green 19-agarose (Sigma) equilibrated with 50 m*M* HEPES, pH 8.5, 2 m*M* EDTA, 0.2% (w/v) digitonin. After washing with 8 ml of the equilibrating buffer, and 8 ml of the equilibrating buffer plus 0.4 *M* NaCl, the protein is eluted with 4 ml of 1.5 *M* NaCl in the equilibrating buffer.

### *Phenylboronate-10 Affinity Chromatography*

The eluate from the Reactive Green-19 purification step is diluted 1:1 with 50 m*M* HEPES, pH 8.5, 2 m*M* EDTA, 0.2% digitonin to reduce the ionic strength, then cycled twice over a 1-ml phenylboronate-10 Sepharose (Amicon, Danvers, MA) column. The phenylboronate column is washed with 8 ml of the HEPES buffer, followed by 2 ml of 0.1 *M* Tris (pH 7.5), 2 m*M* EDTA, and 0.1% (w/v) digitonin. Protein is eluted with 4 ml of 0.1 *M* Tris (pH 7.5), 2 m*M* EDTA, 0.1% (w/v) sodium dodecyl sulfate (SDS).

### *Concentration of Receptor*

Pooled samples from the various column steps are concentrated by centrifugation (3000*g* for 30 min at 4°) to 0.5 ml using Amicon 100,000 molecular weight cutoff filters. Filters are pretreated overnight at 4° with 5% (v/v) Tween 20 to prevent loss of protein.

Additional steps used in purification of SUR1 have been described.[17,20]

## Sedimentation

Sucrose gradient centrifugation is used to estimate the molecular weight of receptor and receptor/$K_{IR}$ complexes. Membranes from transfected cells

[20] L. Aguilar-Bryan, C. G. Nichols, S. W. Wechsler, J. P. Clement IV, A. E. Boyd III, G. Gonzalez, H. Herrera-Sosa, K. Nguy, J. Bryan, and D. A. Nelson, *Science* **268,** 423 (1995).

are isolated, photolabeled with azido[$^{125}$I]iodoglibenclamide, and solubilized as described above. Twelve milliliter 5–20% linear sucrose gradients (in 0.1% digitonin, 100 m*M* NaCl, 50 m*M* Tris, pH 7.4) are poured in SW41 tubes using a BioComp Gradient Master 106 (BioComp Instruments, Inc., Fredericton, NB, Canada). The solubilized proteins are loaded on top of the gradient and sedimented at 36,000 rpm in a SW-41 Ti rotor using a Beckman L8-80M ultracentrifuge for 9 hr at 4°. Fractions (0.5 ml) are collected using a Bio-Rad (Richmond, CA) model 2110 fraction collector. Seventy microliters of each fraction is combined with 15 $\mu$l of 2-mercaptoethanol and 25 $\mu$l of 5× SDS sample buffer. This mixture (100 $\mu$l) is separated on a 1.5-mm-thick 7.5% polyacrylamide gel. Gels are stained, dried, and visualized by autoradiography. The markers used are immunoglobulin M (IgM) (950 kDa), thyroglobulin (660 kDa), urease (hexamer, 545 kDa), urease (trimer, 272 kDa), catalase (240 kDa), and adolase (160 kDa).

## Binding Assays

### *Glibenclamide Binding to Particulate Receptors*

Stored membranes are thawed and rehomogenized at 4° in 50 m*M* Tris buffer, pH 7.4, either by using a glass homogenizer with Teflon pestle (10 strokes at 500 rpm) or by transferring the suspension to a 0.7-ml polypropylene tube and pressing the stoppered tube in ice-cold $H_2O$ for 40 sec against the 12-mm tip of a sonifier (Branson type B15P, pulsed mode with 40% duty cycle). Measurement of [$^3$H]glibenclamide binding is performed in incubation mixtures consisting of 0.7 ml Tris buffer (50 m*M*, pH 7.4), 100 $\mu$l [$^3$H]glibenclamide (0.1–250 n*M*), and 200 $\mu$l membrane suspension (final protein concentration 2–500 $\mu$g/ml). To reach equilibrium, incubations are carried out for 1 hr at room temperature and terminated by rapid filtration of aliquots through Whatman (Clifton, NJ) GF/B filters (25-mm diameter, soaked in ice-cold Tris buffer before use) under reduced pressure. The filters are washed three times with ice-cold Tris buffer. Filtration and washing take less than 15 sec. The $^3$H content of the filters is determined in a liquid scintillation counter after addition of 4 ml of scintillation fluid (Quickszint 402, Zinsser, Frankfurt, Germany) and equilibration for 24 hr at room temperature. Nonspecific binding is determined by incubations in the additional presence of 0.1–10 $\mu$*M* unlabeled glibenclamide. Specific binding is determined by subtracting nonspecific from total binding.

### *Glibenclamide Binding to Solubilized Receptors*

The incubation mixture consists of 0.7 ml Tris buffer (50 m*M*, pH 7.4), 100 $\mu$l [$^3$H]glibenclamide (0.1–250 nM), and 200 $\mu$l detergent extract (final protein concentration 10–500 μg/ml). To reach equilibrium, incubations are carried out for 2 hr at room temperature and terminated by addition of 1 ml of an ice-cold mixture containing 0.4% (w/v) bovine γ-globulin, 20% (w/v) polyethylene glycol, 0.4% (w/v) Triton X-100, and 50 m*M* Tris (pH 7.4). The incubation tubes are vortexed immediately after the addition of this mixture and were kept on ice for 10–20 min. Aliquots are filtered (Whatman GF/C filters, 25-mm diameter, soaked in ice-cold wash buffer before use) under vacuum and washed rapidly three times with 4 ml of a Tris buffer [50 m*M* Tris, 8% (w/v) polyethylene glycol, pH 7.4]. The [$^3$H]content of the filters and specific binding is determined as described above.

These protocols can be used to assy binding of azido[$^{125}$I]iodoglibenclamide and [$^{125}$I]iodoglibenclamide.

*[$^3$H]P1075 Binding to Particulate SUR2B*

Incubations are performed in 1-ml aliquots containing 3 n*M* [$^3$H]P1075, 100 μg/ml of resuspended COS membranes, 1 m*M* free $Mg^{2+}$, 100 μ*M* ATP, and 50 m*M* Tris buffer (pH 7.4). Incubations are carried out for 1 hr at room temperature and terminated by rapid filtration through Whatman GF/B filters as described above (see Glibenclamide binding to particulate receptors). Nonspecific binding is defined by 100 μ*M* pinacidil and amounted to 19 ± 3% ($n$ = 20) of total binding.

Acknowledgments

This work was supported by grants from the DFG (M.S.), the NIH NIDDK (J.B.), the JDFI (J.B.), the ADA (L.A-B.), and the Houston Endowment (L.A-B.).

# [24] Mechanosensitive Channels of Bacteria

*By* PAUL BLOUNT, SERGEI I. SUKHAREV, PAUL C. MOE, BORIS MARTINAC, and CHING KUNG

## Introduction

Mechanosensitive (MS) channel activities have been documented in animal, plant, and bacterial cells by patch-clamp techniques. Mechanosensitive channels are thought to be one of the principal molecular devices by which a cell detects and responds to mechanical stimuli, playing a role in the senses of touch, hearing, and balance, as well as in cardiovascular regulation. Despite their importance for much of biological life, the molecu-



0076-6879/99 $30.00

lar entities and mechanisms responsible for MS channel activities are only now being elucidated.

Perhaps the most successful means of identifying molecular players involved in mechanosensation has been the use of genetic techniques. In *Caenorhabditis elegans* several genes have been identified, and at least two are candidates for MS channels subunits.[1] However, the ability to interpret the role of proteins in mechanosensation from a strictly genetic approach is limited, and electrophysiologic evidence definitively demonstrating that these candidate genes do indeed encode MS channel subunits has yet to be reported.

As an alternative, we and others have been studying MS channels in bacteria. The MS channel activities found in *Escherichia coli* membranes fall into three categories: MscL with fast kinetics and a very large conductance of about 2.5 nS (MscL, mechanosensitive channel of large conductance); MscS with slower kinetics and a smaller conductance of approximately 0.8 nS (S for smaller); and MscM, also with slower kinetics, with a conductance of about half of MscS (M for mini).[2] These channels have now been studied by a variety of approaches. Genes responsible for MscL in *E. coli* and other bacteria have been cloned, and currently provide the only tangible MS channels for combined biophysical and molecular biological studies. MscL, a polypeptide of only 136 amino acid residues, appears to form a homohexamer with each subunit containing two transmembrance domains (Fig. 1). The channel complex apparently responds directly to stretch forces in the lipid bilayer resulting from sudden increases in osmotically induced turgor to allow the passage of solutes from the cytoplasm into the medium to readjust the turgor.[3,4]

The study of bacterial MS channels has involved several diverse approaches. MscL can now be studied using electrophysiologic, biochemical, genetic, and whole-cell physiologic techniques. Here we describe some of the diverse methodologies that have been and can be used to study the structural and functional properties of bacterial MS channels.

## Patch-Clamp Techniques Using *in Vivo* Method for Bacterial Giant Spheroplasts or Protoplasts

An *Escherichia coli* cell is only about 1 $\mu$m in diameter and 2 $\mu$m long. It also contains a strong peptidoglycan layer. Therefore, it would be a

[1] N. Tavernarakis and M. Driscoll, *Annu. Rev. Physiol.* **59,** 659 (1997).

[2] C. Berrier, M. Besnard, B. Ajouz, A. Coulombe, and A. Ghazi, *J. Membr. Biol.* **151,** 175 (1996).

[3] P. Blount, S. I. Sukharev, P. C. Moe, S. K. Nagle, and C. Kung, *Biol. Cell* **87,** 1 (1996).

[4] S. I. Sukharev, P. Blount, B. Martinac, and C. Kung, *Annu. Rev. Physiol.* **59,** 633 (1997).

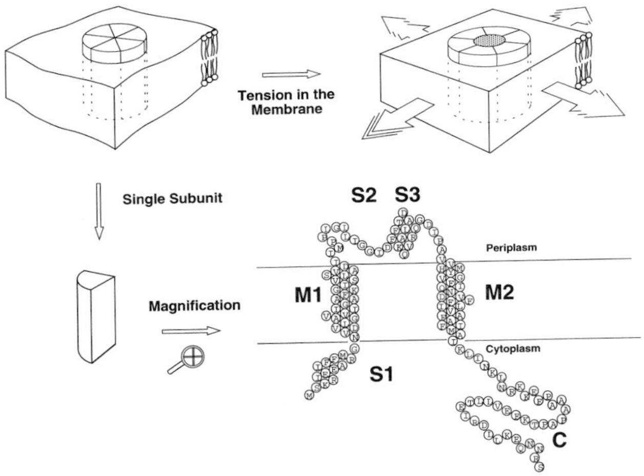


FIG. 1. A working model for MscL structure. The data suggest that MscL is a homohexameric structure, located in the bacterial cytoplasmic membrane, that gates in direct response to membrane tension. As shown in the bottom right, a single subunit contains two transmembrane domains, M1 and M2. S1 to S3 are thought to be surface structures. The amino and carboxyl (C) termini are cytoplasmic.

formidable task to patch directly. Fortunately, techniques have been developed to enlarge bacterial cells, and to allow access to the inner membrane by the pipette. The most commonly used method, a variation of the technique developed by Ruthe and Adler,[5] employs cephalexin to block septation but not growth (Fig. 2, top row; Fig. 3). The result of this treatment is the formation of long filamentous syncitia often called "snakes," which can be collapsed into patchable spheres some 5–10 $\mu$m in diameter. To achieve bacterial "snaking," a rapidly growing culture of *E. coli* at mid-log phase is diluted 1:10, treated with 60 $\mu$g/ml of cephalexin, and grown for 1.5–3 hr at 37° until nonseptate filaments of 50–150 $\mu$m are formed. Filamentous cells having a vacuolar or "salt and pepper" appearance indicate that the incubation has been extended too long. The cells are usually grown in modified Luria–Bertani medium (MLB) [0.5% yeast extract, 1% Bactotryptone (Difco Labs, Detroit, MI), 0.5% NaCl]; however, we also have had success with normal LB (then same as MLB, but 1% NaCl). In several studies, the cells were grown at 42° to ensure rapid growth; however, we have now found that temperature does not seem to be as important as once thought, and viable snakes have been created at room temperature

[5] H. J. Ruthe and J. Adler, *Biochim. Biophys. Acta* **819,** 105 (1985).

simply by increasing the incubation time. The filamentous cells can be used immediately, but are stable overnight at 4°. They are harvested, then resuspended in 0.8 *M* sucrose. Digestion at room temperature for 2–5 min with lysozyme (200 $\mu$g/ml) in the presence of DNase (50 $\mu$g/ml), 50 m*M* Tris buffer, and 6 m*M* ethylenediaminetetraacetic acid (EDTA), pH 7.2, to hydrolyze the peptidoglycan layer (cell wall) yields the giant spheroplasts. A longer treatment (of up to 10 min) appears to make more spheroplasts with a grayish rather than phase-bright appearance. These are often easier to patch, but we find that the patches are often less stable at the pressures needed to gate MS channels. Overall, this treatment apparently does not digest the peptidoglycan completely, but rather clips the polymers and weakens the cell wall, thereby allowing the swelling of *E. coli* spheroplasts. The progress of spheroplast formation is followed under a phase-contrast

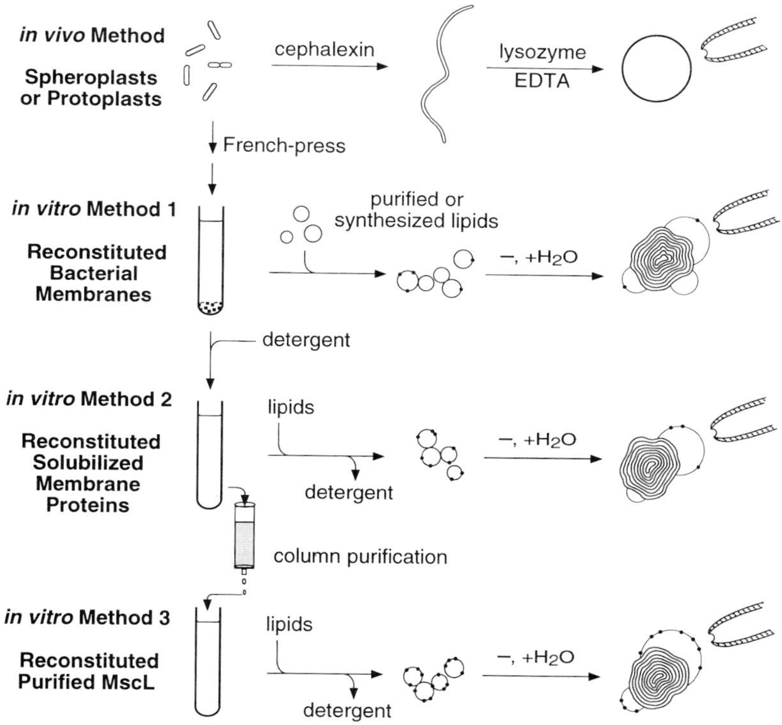


FIG. 2. Approaches for assaying and characterizing MS channel activities in bacteria. An *in vivo* method for generating giant spheroplasts (top) and three *in vitro* methods including reconstitution of bacterial membranes (method 1), solubilized bacterial membrane proteins (method 2), and purified MscL protein (method 3). See text for details.

microscope (400×) and the reaction is stopped by addition of $MgCl_2$ to a final concentration of 8 m*M* and placement of the reaction on ice. Not all snakes form spheroplasts. If desired, spheroplasts can be enriched by layering the treated cells over a 0.8 *M* sucrose cushion containing 10 m*M* each of $MgCl_2$ and Tris, pH 8.0, and spinning gently at room temperature. The amount of force required varies between preparations; a centrifugation of about 200 *g* should be tried for about 3 min, and the force gradually increased until a pellet is barely visible. The preparation, and most of the cushion is then aspirated, and the small pellet resuspended in about 1 ml of the cushion solution. The resulting preparation of spheroplasts can be patched immediately, or can be stored at −20° for several weeks.

Although this is the most commonly used method, there are alternatives such as those described by Saimi *et al.*[6] (1) *UV method: E. coli* cells irradiated with ultraviolet light form nonseptate filaments. Bacterial cultures ($OD_{590}$ 0.1–0.2, 5 mm in depth) in plastic petri dishes are shaken at 40 rpm under UV light (254 nm, 142 erg $mm^{-2}$ $sec^{-1}$) for 3 min, combined with 9 volumes of growth medium (MLB), and cultured at 42° until filaments 80–100 $\mu$m long develop (2.5–3 hr). Spheroplasts are then prepared with lysozyme as above. (2) *$Mg^{2+}$ method:* Cells are grown in LB plus 50 m*M* $MgCl_2$ at 35° to an $OD_{590}$ of 0.5, then diluted 1:10 into the same medium with the addition of 60 $\mu$g/ml cephalexin, and grown with shaking for 2 hr at 42°. This culture yields a mixture of giant round cells and long filaments, some with bulges on the sides. These bulges and giant cells can be used directly for patch-clamp studies. (3) *Mutants:* Cells of *E. coli* $lpp^-ompA^-$, lacking two major components of the outer membrane, round up in the presence of 30 m*M* $MgCl_2$. Mutant cells are cultured in LB plus 30 m*M* $MgCl_2$ at 35° to an $OD_{590}$ of 0.5, then diluted 1:10 in LB containing 60 $\mu$g/ml cephalexin plus 30 m*M* $MgCl_2$, and grown for 4 hr to form giant round cells 5–10 $\mu$m in diameter. They are usually washed twice with 0.8 *M* NaCl by centrifugation at 5000 rpm for 3 min (Sorvall SS-34 rotor, Sorvall, Newton, CT). We find an increase in seal resistance after this salt wash. Similarly, mutant *E. coli* AW693, selected for its failure to grow at high osmolarity, can be used. These cells are grown in LB plus 400 m*M* KCl to an $OD_{590}$ of about 0.6, then incubated overnight at 4° without shaking. A few of the cells are large, some are large enough (4–8 $\mu$m in diameter) to be patch-clamped directly. The number of these large cells increases to about 10% of the total by the addition of 10 m*M* $MgCl_2$ to the growth medium.

Methods have also been developed for making giant protoplasts from

[6] Y. Saimi, B. Martinac, A. H. Delcour, P. V. Minorsky, M. C. Gustin, M. R. Culbertson, J. Adler, and C. Kung, *Methods Enzymol.* **207,** 681 (1992).

the gram-positive bacteria *Streptococcus faecalis*[7,8] and *Bacillus subtilis*.[9,10] Typically, the cells are grown in CDM, a defined composition medium,[11] to an $OD_{600}$ of about 0.3. The cells are harvested and resuspended in 0.5 *M* sucrose, 10 m*M* $MgCl_2$, 50 m*M* Tris (pH 7.2), and 1.5 mg/ml lysozyme. After about 1 hr of digestion at 37°, the resulting protoplasts are diluted 1:10 into medium with 0.25 *M* sucrose and grown until protoplasts of 2–4 $\mu$m in diameter form (about 2–5 hr).

Gigohm seals form more slowly with these bacterial preparations than with most animal cells. Nonetheless, seals of several gigohms can be formed routinely. Most recordings are conducted in the on-cell or excised inside-out patch mode. Patch-clamp pipettes (Boralex, Rochester, NY) having a 1- to 2-$\mu M$ tip diameter (bubble number 4–5.5) are normally used. We often use a pipette solution of 200 m*M* KCl, 90 m*M* $MgCl_2$, 10 m*M* $CaCl_2$, and 5 m*M* HEPES adjusted to pH 5–8. We find that MscS activities appear to desensitize more readily at a lower pH, so we often use a buffer of pH 6.0 for recording MscL activity. The bath solution is usually the same as for the pipette although 0.3 *M* sucrose can be added to ensure stability of the spheroplasts. Pressure sensitivity can be assessed using a pressure transducer (Micro Switch; Omega, Stamford, CT) calibrated with a pneumatic transducer tester (Bio-Tec, Winooski, VT). The pressure required to open channels varies from patch to patch due to variability in patch geometry or membrane characteristics. However, if the pressure is normalized to MscS (applied pressure/threshold pressure required to open MscS in the same patch), the ratio of pressure required to open MscL/pressure to open MscS remains constant[12]; hence, the relative pressure sensitivities of MscL mutants or homologs expressed in a *mscL*-null *E. coli* (as discussed below) can be easily assessed.

The bacterial membranes can be directly visualized at the tip of the recording pipette. For this purpose we have used a method analogous to that originally described by Sokabe and Sachs[13] that utilizes an inverted microscope (Nikon Diaphot 200, Nikon, Melville, NY) with differential interference contrast (DIC) optics (CFN 60× Plan Apochromat objective and a high NA water-immersion DIC condenser) and a camera with contrast enhancement controls (Newicon 700 L with a 2400-07 controller, Hama-

[7] M. Zoratti and V. Petronilli, *FEBS Lett.* **240,** 105 (1988).

[8] I. Szabo, V. Petronilli, and M. Zoratti, *J. Membr. Biol.* **131,** 203 (1993).

[9] M. Zoratti, V. Petronilli, and I. Szabo, *Biochem. Biophys. Res. Comm.* **168,** 443 (1990).

[10] I. Szabo, V. Petronilli, and M. Zoratti, *Biochim. Biophys. Acta* **1112,** 29 (1992).

[11] G. S. Roth, G. D. Shockman, and L. Daneo-Moore, *J. Bacteriol.* **105,** 710 (1971).

[12] P. Blount, S. I. Sukharev, M. J. Schroeder, S. K. Nagle, and C. Kung, *Proc. Natl. Acad. Sci. U.S.A.* **93,** 11652 (1996).

[13] M. Sokabe and F. Sachs, *J. Cell Biol.* **111,** 599 (1990).

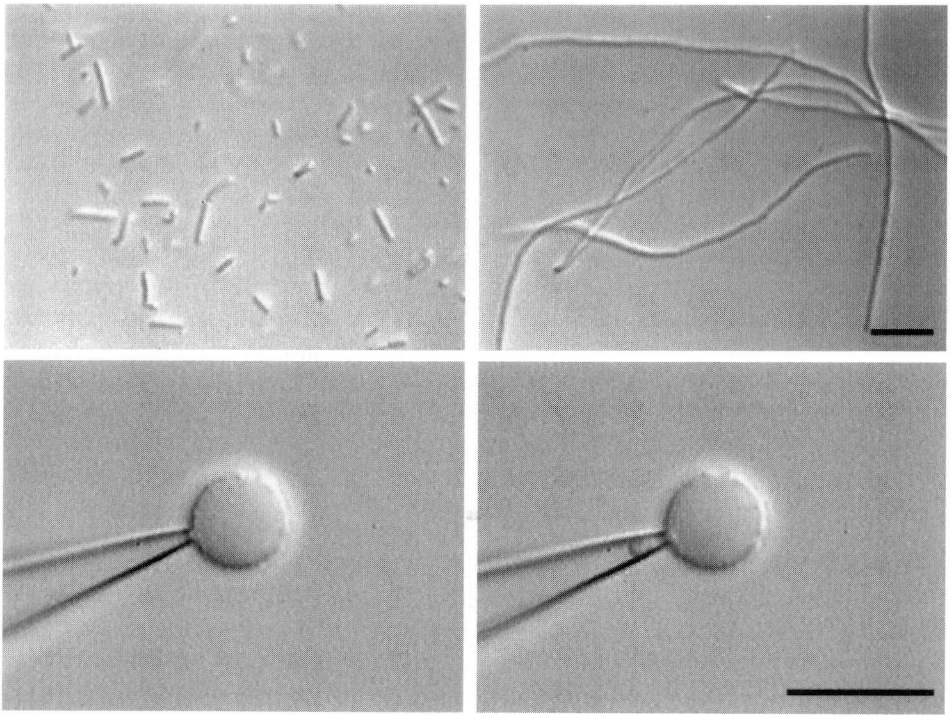

FIG. 3. Video images obtained during the process of generating and patching giant spheroplasts from *E. coli.* Rapidly growing *E. coli* cells (top left) form large filamentous cells, called snakes (top right) when treated with cephalexin for 2 hr. These are collapsed into giant spheroplasts that can be patched (bottom). On light suction applied to the pipette, the envelope of the giant spheroplast was observed to form a dome that buckles into the pipette (bottom left). On further suction, a portion of the spheroplast was drawn into the pipette, and a gigohm seal was formed (bottom right). The magnification is consistent within each horizontal set. Bars: 10 $\mu$m.

matsu Photonic Systems, Bridgewater, NJ) connected to the video port via a 4× converter lens adapter. Nonpolished pipettes of 1- to 2-$\mu$m bore diameter are used. To make the pipette axis horizontal, and thus parallel to the focal plane, the tip of the pipette (about 200 $\mu$m) is bent 15–20 degrees as described[13] to compensate for the manipulator headstage tilt. The entire seal formation process is recorded on a S-VHS tape and then representative frames are converted to a digital format using a Macintosh computer equipped with a video player system.

The following is a description of what is typically observed during gigohm seal formation (also see Fig. 3, bottom). Upon light suction inside the pipette, the envelope of an *E. coli* giant spheroplast is observed to form a dome of less than 1 $\mu$m that begins to enter the pipette. On further suction, a portion of the spheroplast is seen to be drawn into the pipette. A gigohm seal is established when the front of this portion extends, and is stabilized, some 3–7 $\mu$m inside the pipette. Once the gigohm seal is formed, the configuration appears relatively fixed except for the curvature of the patch spanning the inner wall. In contrast, membrane patches from reconstituted systems (below) have been observed to "creep" up the wall of the pipette to a much larger extent with continued pressure, thus potentially changing the size and shape of the patch. The membrane patch appears rather pliable: releasing the suction flattens it, increasing suction induces a curvature. The curvature can be measured directly from the video image and later used to assess the membrane tension acting on the channel.[14]

The images observed from *E. coli* spheroplast gigohm seal formation are consistent with the following scenario. Before seal formation, the entire bacterial envelope (the complex of the outer membrane, the peptidoglycan wall, and the inner membrane) resists suction and will buckle only slightly. Continued suction eventually ruptures the peptidoglycan layer (the "cell wall") and the outer membrane (a specialized structure, where the outer monolayer is made of lipopolysaccharides and not standard phospholipids). The inner membrane, with its conventional lipid bilayer, alone advances through the rupture into the pipette. The gigohm seal is presumably formed within, or throughout, the 3- to 7-$\mu$m length where the inner membrane plasters against the inner wall of the pipette. The MS conductances observed likely reflect MS channels in the pliable patch of inner membrane. A clearly visible ring of opaque material often accumulates at the inner rim of the pipette mouth. This may be the debris of ruptured cell wall and outer membrane.

Using the system above to observe gigohm seal formation, occasionally two membranes are observed in the electrode. The first membrane can be broken without loss of the potential gigohm seal, arguing that it is the inner membrane that is being patched. Previous investigators made a classical distinction between the two-membrane "spheroplasts" and the one-membrane "protoplasts," the latter thought to be surrounded by the plasma membrane alone. Cui *et al.*[15] digested the above spheroplasts more thoroughly and chose an optically distinct subpopulation of the lysozyme-induced giant cells for their studies; a whole-cell (or whole-protoplast) mode was achieved and the gigohm seal was undoubtedly formed with the inner membrane. Kubalski[16] similarly patched protoplasts and showed that other

[14] S. I. Sukharev, unpublished (1997).

[15] C. Cui, D. O. Smith, and J. Adler, *J. Membr. Biol.* **144,** 31 (1995).

[16] A. Kubalski, *Biochim. Biophys. Acta* **1238,** 177 (1995).

channel activities, with smaller conductances, are present in the bacterial membrane. However, the observation that the same MS channel activities, including MscL, can be observed in every patch with a good gigohm seal from either spheroplasts or protoplasts, combined with the biochemical localization of MscL to the inner membrane (discussed in the Anti-MscL Antibodies and Their Uses section later), suggests that all reported gigohm seals to *in vivo* bacterial cells are formed exclusively with the inner membrane. On the other hand, porin-like channel activities have been observed in spheroplast preparations (first identified as a "voltage-sensitive" channel[17]). These channel activities are perhaps most often observed in membranes derived from the lpp$^-$ompA$^-$ strain; however, even here they are rare and only seen in a very small subset of gigohm seals. (Note that a standard gigohm seal contains approximately the same amount of membrane as an entire bacterial cell.) The most probable explanations are that these activities are due to a very small number of porin-like channels in the inner membrane, or, perhaps more likely, a contamination of outer membrane proteins on gigohm seal formation to the inner membrane.

## Patch-Clamp Technique Using *in Vitro* Method for Reconstituted Bacterial Membranes

Membranes can be isolated from bacterial cells and fused with lipids to form patchable bilayers using a modified technique[18] after Criado and Keller[19] (see Fig. 2, second row). For this purpose, cultures are usually grown to mid- or late-log phase, washed once with a 50 m*M* potassium phosphate buffer (pH 7.2) with 5 m*M* $MgSO_4$ and 1 m*M* dithiothreitol (DTT), then French-pressed at 16,000 psi in the same buffer supplemented with 1 m*M* phenylmethylsulfonyl fluoride (PMSF). The cell homogenate can then be treated with 0.5 mg/ml lysozyme for 15 min at room temperature to digest any residual peptidoglycan associated with the membranes. This digestion appears to enhance membrane fusion as well as protein extraction (below). Unbroken cells and cell debris are pelleted at 6000 *g* at 4° for 10 min and then membranes are harvested by ultracentrifugation at 100,000 *g* at 4° for 1 hr and stored at −80°. Several modifications can be utilized: bacterial membrane vesicles can be prepared mostly inside-out from French-pressed cells or mostly outside-out from sonicated spheroplasts, or inner and outer bacterial membranes fractions can be separated

[17] M. Buechner, A. H. Delcour, B. Martinac, J. Adler, and C. Kung, *Biochim. Biophys. Acta* **1024,** 111 (1990).

[18] A. H. Delcour, B. Martinac, J. Adler, and C. Kung, *Biophys. J.* **56,** 631 (1989).

[19] M. Criado and B. U. Keller, *FEBS Lett.* **224,** 172 (1987).

via a sucrose gradient. We have found that any of these membrane fractions can be stored at $-80°$ for months without losing channel activity.

One source of lipids for this type of reconstitution is azolectin (L-$\alpha$-lectin from soybean, type II-S, Sigma, St. Louis, MO). The azolectin granules are rinsed three to four times with acetone, then dried and dissolved in chloroform. If a very clean preparation is desired, the lipids can then be purified using the chloroform–water phase separation method.[20] The lipids can also be supplemented with 10 mol% of cholesterol (Sigma) if desired. They are stored at $-20°$, under nitrogen, as a 100 mg/ml chloroform solution. Just prior to use, an aliquot containing 10–40 mg of lipid is placed in a disposable 10-ml glass tube and dried under a jet of nitrogen for 20 min. The tube is rotated so as to form a uniform, thin lipid film on the wall. The lipid film is rehydrated in 1 ml of TEE (2 m*M* EDTA, 2 m*M* ethylene glycol-bis[$\beta$-aminoethyl ether]-N,N,N′,N′-tetraacetic acid (EGTA), 10 m*M* Tris, pH 7.2) or 20 m*M* phosphate buffer (pH 7.2) with 1 m*M* DTT, for each 10 mg lipid. After hydration for approximately 10 min, the mixture is disrupted by vortexing and mild sonication (bath sonicator, Model B-12, Branson Ultrasonics, Danbury, CT) to clarity. Aliquots containing 4–10 mg of lipid are mixed with native membrane vesicles at the desired protein-to-lipid ratio (w/w; 1:600 for the outer membrane protein, 1:50 for the inner membrane protein) and centrifuged at 150,000 to 200,000 *g* (in an Airfuge or a standard ultracentrifuge, Beckman, Fullerton, CA) for 35 min. Alternatively, to avoid the high-speed centrifugation necessary for pelleting the small liposomes, the sonicated lipid suspension can be twice frozen and thawed in a dry-ice and ethanol bath, then centrifuged together with native membrane vesicles at lower acceleration (48,000 *g*) in a midrange centrifuge (Beckman J2-HS, rotor JA-18-1). The compact mixed pellet is then resuspended in a small volume (25–40 $\mu$l) of 10 m*M* MOPS (3-[N-Morpholino] propanesulfonic acid) buffer with 5% (v/v) ethylene glycol (pH 7.2). Aliquots (about 20 $\mu$l) of the suspension are placed onto a clean glass slide and subjected to a 4-hr dehydration in a desiccator at 4°. The dehydrated lipid film on the slide glass is then rehydrated at 4° overnight with a droplet of solution of 150 m*M* KCl, 0.1 m*M* EDTA, $10^{-5}$*M* $CaCl_2$, and 5 m*M* HEPES (pH 7.2). We found that a lipid concentration of at least 90 mg/ml is needed during the rehydration step to produce giant liposomes.

Proteoliposomes have been patch clamped directly for the study of channel activities.[21] However, we have found unilamellar blisters[19,22] to be more suitable for the analysis of MS channels. Blisters are formed by placing

[20] Y. Kagawa and E. Racker, *J. Biol. Chem.* **246,** 5477 (1971).
[21] D. W. Tank, C. Miller, and W. W. Webb, *Proc. Natl. Acad. Sci. U.S.A.* **79,** 7749 (1982).
[22] A. M. Correa and W. S. Agnew, *Biophys. J.* **54,** 569 (1988).

a few microliters of the rehydrated suspension in the patch-clamp recording chamber, which contains the experimental buffer plus 20–40 m*M* $MgCl_2$. The presence of $Mg^{2+}$ causes the liposomes to collapse. Within 10–60 min, faint and most likely unilamellar blisters emerge from the sides of the collapsed liposomes. Once formed, the blisters are stable for hours, as long as $Mg^{2+}$ remains in the buffer. The blisters form high-resistance seals readily and very reproducibly. Care must be taken to use very clean, sterile solutions, especially during the rehydration procedure, to avoid contamination with exogenous microbial membrane fragments. Experiments performed with liposomes from azolectin, as well as mixtures of individual native lipids, not fused with other material, yield high-resistance seals and a quiet background.

Standard recording techniques are used to examine the MS channel activities in the azolectin proteoliposomes. Patch-clamp pipettes, solutions, and gigohm seal forming procedures are similar to that described above. (Note, however, that the chamber buffer does not need the additional sucrose often used to stabilize spheroplasts.) Similarly, the patch and seal formation can be visually monitored and recorded as above.

## Patch-Clamp Technique Using *in Vitro* Method for Reconstituted Solubilized Membrane Proteins

For this technique (Fig. 2, third row), the membrane preparations (Fig. 2, third row) are isolated as above by French-press and centrifugation. To solubilize the membranes, a membrane pellet (about 0.5 g wet wt) is resuspended in 20 ml of extraction buffer [5 m*M* EDTA, 5 m*M* EGTA, 1 m*M* DTT, 50 m*M* Tris, pH 7.4, containing 2% octylglucoside (OG), 1 m*M* PMSF] and sheared in a hand-driven glass homogenizer to clarity. Insoluble particles are pelleted by ultracentrifugation (105,000 *g* 1 hr) and the protein content in the clear extract is determined using a bicinchoninic acid assay kit (Pierce Chemical, Rockford, IL). This solubilized material can then be subjected to a variety of chromatographies to enrich for MS channel activity, as long as the buffers contain 1% OG.

Lipids are purified and dried on the side of a glass tube as described above. The lipids are then solubilized in extraction buffer (more OG can be added if required to totally solubilize the lipids). Aliquots of solubilized membranes or various chromatographic fractions (10–500 $\mu$l, depending on anticipated channel content) are mixed with an equal amount of the OG-solubilized lipids (typically 4–8 mg of lipid per sample, 0.8–1.5 ml total volume) and dialyzed (Spectrapore tubing; 0.7-cm diameter, 8000 molecular weight cutoff, Spectrum, Houston, TX) against 2 liters of buffer containing 100 m*M* NaCl, 0.2 m*M* EDTA, 0.02% $NaN_3$, 5 m*M* Tris, pH 7.2, in the presence of Calbiosorb detergent-absorbing beads (Calbiochem, La Jolla,

CA) for 24 hrs, with two changes of buffer. The resulting proteoliposomes are pelleted at 160,000 *g* at 4° for 1 hr and resuspended in 30 ml of 10 m*M* MOPS buffer with 5–10% ethylene glycol. Similar to above, droplets of 20 μl of liposome suspension are placed on glass slides and subjected to dehydration (2 hr at 4° in a vacuum desiccator). The lipid films are then rehydrated by covering them with the droplets of buffer (250 m*M* KCl, 0.1 m*M* EDTA, 5 m*M* HEPES–KOH, pH 7.2) and stored overnight in petri dishes containing wet filter pads.[18] The treatment and patching of the rehydrated lipid/protein mixture is as above.

## Patch-Clamp Technique Using *in Vitro* Method for Reconstituted Purified MscL

Early studies demonstrated that the activities of MscS and MscL are both readily oberved in the total *E. coli* membrane extract on solubilization, dialysis, and patch-clamp examination. Gel filtration of the solubilized material using a Superose 6 HR 10/30 column (Pharmacia, Piscataway, NJ) readily divided the two activities into discrete and separate fractions.[23] These studies suggested that patch clamp could be used like an enzymatic assay for the biochemical purification of one of these channels (Fig. 2, fourth row). We therefore designed two independent enrichment schemes; both were highly enriched for a 17-kDa protein. Microsequencing this protein led to the cloning of the gene responsible for the MscL activity (described below).[24] Although the technique of using patch clamp as an assay is laborious, it holds promise for a limited number of channel activities that have been difficult to clone but have been found to be amenable to reconstitution into proteoliposomes.

The following are the two series of fractionations used for the biochemical enrichment of the MscL channel activity. The first series uses a DEAE-Sepharose CL-6B, followed by a hydroxyapatite column (HTP Bio-Gel, Bio-Rad, Richmond, CA). The active fractions are pooled and concentrated on a Centriprep 30 concentrator (Amicon, Bedford, MA), further fractionated with a Superose 6 HR 10/30 column (Pharmacia, Piscataway, NJ), and then chromatofocused in pH 7.2–4.0 gradient on a Pharamacia Mono P HR 5/5 column. The second series begins by cutting the total membrane extract with ammonium sulfate at 35% saturation. The supernatant is then collected by centrifugation and applied to a phenyl-Sepharose CL-6B column. The active fractions are then pooled, concentrated using a Centricon-30 concentrator, and passed through a Superose 12 HR 10/30 FPLC gel-filtration column. The final active fractions from the two series should both

[23] S. I. Sukharev, B. Martinac, V. Y. Arshavsky, and C. Kung, *Biophys. J.* **65,** 177 (1993).
[24] S. I. Sukharev, P. Blount, B. Martinac, F. R. Blattner, and C. Kung, *Nature* **368,** 265 (1994).

be enriched for the 17-kDa MscL protein, which can be visualized by SDS–PAGE and silver staining.

Since the cloning of the *mscL* gene, two methods have been used to simplify purification of the MscL protein. First, Häse *et al.*[25] expressed the MscL channel protein with a glutathione *S*-transferase (GST) tag. The cloning of the mscL gene into this GST-MscL expression system used a unique *Nru*I site, resulting in the loss of the first eight amino acid residues of MscL and the addition of two novel residues after the thrombin cleavage site. The MscL protein can now be purified as follows. Cells harboring the plasmid (pGEX1.1) are induced for 3 hr with 0.1 m*M* isopropyl β-D-thiogalactopyranoside (IPTG), harvested, resuspended in 15 m*M* NaCl, 1 m*M* EDTA, 50 m*M* Tris (pH 8.0), and lysed by addition of 0.1 mg/ml lysozyme and 1.5% OG. The mixture is then sonicated (Unisonics Pty. Ltd., Sydney, Australia), and cell debris pelleted by centrifugation (20 min at 16,000 rpm; J2-MI, Beckman). The supernatant is incubated with glutathione-Sepharose 4B beads at room temperature for 1 hr, then washed several times with phosphate-buffered saline (PBS: 137 m*M* NaCl, 2.7 m*M* KCl, 4.3 m*M* $Na_2HPO_4$, 1.4 m*M* $KH_2PO_4$, pH 7.2), and resuspended into the same buffer containing an additional 150 m*M* NaCl, 2.5 m*M* $CaCl_2$, and 50 m*M* Tris. Thrombin is then added to a final concentration of 1 unit/μg of protein and incubated for 1 hr at room temperature. OG to a final concentration of 1% is added and the beads are pelleted. The essentially purified MscL in the supernatant can then be functionally reconstituted as described above. A second method involves tagging the MscL with polyhistidine and purifying the protein with a $Ni^+$-NTA column.[26] A two-step PCR (polymerase chain reaction) amplification was used to add the chain of condons corresponding to six contiguous histidine residues on the 3′ end of the *mscL* open reading frame (ORF). The *mscL* ORF was amplified with the 5′ primer used previously to amplify the insert of p5-2-2,[24] and with the following oligonucleotide as the 3′ primer: GAT GAT GAT GAT GAT GAG AGC GGT TAT TCT GCT C; the insert of the p5-2-2 plasmid was used as the starting template. The resulting product was gel purified and used as a template for a subsequent amplification using the same 5′ primer, and the second 3′ primer with the sequence CTC GAG TTA ATG ATG ATG ATG ATG ATG AGA. This latter product was subcloned as an *Xho*I fragment into the pB10a expression vector. Now, to obtain purified protein, PB101, or PB104 bacteria containing the mscL-6His plasmid construct are cultured in LB in a 40-liter fermenter and

[25] C. C. Häse, A. C. Le Dain, and B. Martinac, *J. Biol. Chem.* **270,** 18329 (1995).

[26] P. Blount, S. I. Sukharev, P. C. Moe, M. J. Schroeder, H. R. Guy, and C. Kung, *EMBO J.* **15,** 4798 (1996).

induced with 0.8 m$M$ IPTG for 4 hr. Cells are disrupted using a French press and membrane fractions are isolated by differential centrifugation as above. The membrane pellet (5–8 g wet weight) is solubilized in 100 ml extraction buffer (50 m$M$ Na phosphate, 300 m$M$ NaCl, 35 m$M$ imidazole) containing 3% OG, 1 m$M$ PMSF at 4°. The extract is cleared by a 35-min centrifugation at 120,000 $g$ at 4° and mixed with 8 ml $Ni^+$-NTA agarose beads (Qiagen, Chatworth, CA) equilibrated with the same buffer by gentle rotation for 15 min. The beads are then packed in a 1.5-cm-diameter column and washed with 15 bed volumes of the wash buffer (extraction buffer, 1% OG), followed by elution with a 30-min linear gradient of imidazole (35–500 m$M$, 1 ml/min). The hexahistidine ($His_6$)-tagged MscL elutes in the last third of the gradient, yielding about 1 mg of pure protein from a 4- to 5-liter preparation as estimated by modified Bradford assay. The final fraction has been separated in a 12% gel by SDS–PAGE, observed by Coomassie blue staining, and estimated to contain >98% pure $His_6$-tagged MscL protein. To avoid a high concentration of imidazole in the samples, the protein is concentrated in a Centriplus-30 concentrator (Amicon) and finally transferred to 20 m$M$ phosphate buffer with 1% OG by gel filtration through a PD10 column (Pharmacia, Uppsala, Sweden). When purified, the material can be reconstituted into liposomes of azolectin or phosphatidylcholine and phosphatidylserine (1 : 1 w/w) and examined by patch clamp as above.

## Cloning *mscL* Gene and Heterologous Expression

When the 17-kDa protein was microsequenced, its sequence matched a part of an unknown ORF near minute 72 of the *E. coli* chromosome. Based on physical and molecular mapping, the corresponding λ phage clone (EC5-106) was identified and obtained from the Blattner collection.[27] The fragment that contained the putative *mscL* ORF was excised, subcloned into pBluescript $SK^-$, and sequenced, all using standard techniques. Because the *E. coli* genome sequencing project has been completed recently (www.genetics.wisc.edu), any sequence of interest to ion channel biology can now easily be located.

To show that the ORF encoding the 17-kDa protein is necessary for the MscL MS conductance, Sukharev *et al.*[24] insertionally disrupted, then reintroduced that ORF, and showed that the MS conductance was lost and then regained in both spheroplasts and on reconstitution into liposomes, as electrophysiologically determined by the methods described above. Insertional disruption is commonly performed in microbial genetics. This is

[27] D. L. Daniels and F. R. Blattner, *Nature* **325,** 831 (1987).

accomplished by first replacing a central part of the *mscL* ORF in a plasmid with a cassette containing a drug resistance gene and transcriptional stops in all three frames. In our case, we used a 3.4-kb ΩCm cassette that confers resistance to the antibiotic chloramphenicol. The ΩCm-containing linear fragment (including at least 500 base pairs of genomic DNA flanking the insertional disruption) is excised and used to transform *E. coli* selecting for chloramphenicol resistance and confirmed to be homologous recombinants by PCR. The stain we used for this type of construction (CAG732) is not ideal for patch-clamp investigation. Therefore, the disrupted *mscL* ORF was moved into AW405 by P1 transduction through chloramphenicol selection creating the *mscL*-null strain PB100 (AW405 *mscL*::CAM). To allow for better stability of plasmids, the PB100 stain was made $recA^-$ by P1 transduction from a strain containing a *Tn*10 transposition to generate PB101 (PB100-*srlA*::*Tn*10). A similar strain has recently been developed, PB104, which is similar to PB101 except that instead of being an insertional disruption of the *mscL* gene, it bears a complete deletion of *mscL* rather than an insertional disruption.[26] For expression of *mscL in trans*, PB101 or PB104 is transformed with an expression plasmid, p5-2-2, pB10a bearing the *mscL* ORF, where the gene is placed behind a *lacUV5* promoter. Note that this expression system can now be used to express not only wild-type *mscL*, but also *mscL* mutants or homologues from different bacterial species (below).

The *mscL* gene expression is not only necessary, but also sufficient for the MscL MS conductance. A functional MscL channel activity can be expressed in at least two heterologous expression systems. In both cases, near-normal MscL activities are observed on patch-clamp analysis. The first system is the heterologous expression of MscL in yeast membranes. The *mscL* ORF has been subcloned into the yeast shuttle plasmid pSJ101 ("pFusionator"[28]), which has the 2 $\mu$ origin, LEU2 selectable marker, yeast transcription and translation initiation sites, and a galactose-inducible promotor (GAL10) upstream of the multiple cloning site. The constructed shuttle plasmid can now be electroporated into competent yeast cells (in 1*M* sorbitol, at pulses of 7 kV/cm, 4-ms time constant) with subsequent selection on leucine-deficient plates. The yeast cells are cultured from a single purified transformant colony in YNB medium [yeast nitrogen base (Difco), essential amino acids] deficient of leucine, using glucose (20 gm/l) as a carbon source. Cells are collected by centrifugation and transferred into leucine-dropout YNB medium with galactose and induced for 6 hr. The cells are harvested, washed with double-distilled water, and incubated for 30 min in 10 m*M* DTT, 100 m*M* Tris, pH 9.4. Protoplasts are prepared

[28] S. Johnson, Ph.D. dissertation, University of Washington, Seattle, (1991).

by lyticase treatment in buffer consisting of 1.2 *M* sorbitol, 1% yeast extract, 2% Bacto-peptone, 1% galactose, 10 m*M* potassium phosphate, pH 7.2.[29] The protoplasts are then harvested through a sucrose cushion and French-pressed twice at 20,000 psi in 0.1 *M* sorbitol, 50 m*M* potassium acetate, 2 m*M* EDTA, 1 m*M* PMSF, 20 m*M* HEPES–KOH, pH 7.4. Cell debris is pelleted (3000 *g* at 4°, 10 min), and the membranes are then collected by ultracentrifugation (150,000 *g* at 4°, 1 hr). The membranes are resuspended in 20 $\mu$l of storage buffer (250 m*M* sucrose, 50 m*M* potassium acetate, 1 m*M* DTT, 20 m*M* HEPES–KOH, pH 7.4) and reconstituted into liposomes, as described above.

The second heterologous expression system is the cell-free transcription–translation-coupled (TnT) reticulocyte lysate system (Promega, Madison, WI). The *mscL* ORF is first subcloned into the pSP64-poly(A) vector (Promega) to be used as a template. Canine pancreatic microsomes (Promega) are added in all reactions as a necessary supplement for membrane protein translation. The specificity of the reactions for the *mscL* gene is normally assayed by protein labeling and autoradiography. A typical 25-$\mu$l reaction for protein labeling contains 12.5 $\mu$l of TnT reticulocyte lysate, 6 $\mu$l of 1 : 12 diluted TnT buffer, 0.5 $\mu$l of amino acid mix without methionine, 0.5 $\mu$l of RNasin, 0.5 $\mu$l of SP6 polymerase, 2 $\mu$l of [$^{35}$S]methionine (10 mCi/ml), 0.5 $\mu$g of vector DNA, and 2.5 $\mu$l of microsome preparation.[30] After a 2-hr incubation at 30°, microsomes are pelleted, washed with 150 m*M* NaCl, 10 m*M* MOPS–NaOH, pH 7.4, and dissolved with 1% sodium deoxycholate. Proteins are precipitated with TCA and separated in SDS–PAGE (12%), and the gel is then subjected to autoradiography to determine if the *mscL* ORF is indeed transcribed and translated. For the functional assay of MS channel activity, 50-$\mu$l reactions are performed with a full complement of unlabeled amino acids. Microsomes are washed and reconstituted into azolectin liposomes as described above.

## Cloning and Examination of MscL Homologs in Other Bacteria

Since the discovery of the gene encoding MscL in *E. coli,* homologs have been found in several diverse bacterial categories including gram-negative, gram-positive, and the cyanobacteria: gram-negative: *Erwinia carotovora, Haemophilus influenzae,* and *Pseudomonas fluorescens;* gram-positive: *Staphylococcus aureus, Bacillus subtilis,* and *Clostridium perfringens;* cyanobacterium (previously blue-green alga): *Synechocystis* sp. strain PCC6803. Most were found by homology to sequences revealed in bacterial

[29] J. A. Rothblatt and D. I. Meyer, *Cell* **44,** 619 (1986).
[30] P. Walter and G. Blobel, *Methods Enzymol.* **96,** 84 (1983).

genome sequencing projects, others by chance encounter in bacteriological research, and still others through deliberate search by Southern hybridization and by PCR. Preliminary searches in databanks available on the Internet from genome sequencing projects suggest that the following organisms also contain *mscL* homologs: *Vibrio cholerae, Deinococcus radiodurans, Streptococcus pyogenes,* and *Mycobacterium tuberculosis.* The similarity of the predicted protein sequence of these homologs in such a large variety of bacteria is striking.

To determine directly whether these structural homologs indeed encode MS channels in their native species would entail developing methods for culturing and spheroplasting these widely different organisms. To simplify matters, a system has been developed in which homologous genes can be cloned into a plasmid and expressed in a *mscL*-null *E. coli* strain, and MS channel activities examined.[31] The genes are simply amplified using standard PCR methods and specific primers, and subcloned into a pB10a-based expression plasmid; the resulting plasmid is transformed into the PB101 or PB104 strain, and spheroplasts generated and patched, all as described above.

## Anti-MscL Antibodies and their Uses

Rabbit antibodies have been raised against the *E. coli* MscL. In one case, polyclonal anti-MscL antisera were generated by injecting purified MscL obtained from the GST–MscL fusion protein discussed above.[25] In the second case, antipeptide antibodies were generated. Here, the carboxyl-terminal peptide (CEIRDLLKEQNNRS) was coupled to keyhole limpet hemocyanin (Sigma, St. Louis, MO) activated by 1-ethyl-3-(3-dimethylaminopropyl) carbodiimide hydrochloride. See protocol by Pierce for details. Rabbits were immunized according to standard procedures, and the postimmune serum was taken at 24-day intervals. Both antibody-containing sera have been screened and used in Western blots against total membrane proteins.

Anti-MscL antibodies have been used successfully in several contexts. For instance, they have been used to identify the products of MscL cross-linking experiments (below). In addition, they have been used as one of the biochemical methods for determining which of the two membranes harbors MscL protein in gram-negative bacteria. It was unclear whether the MS channels are located in the outer or the inner membrane of these cells. The arguments in this debate concerned the site of the gigohm seal in spheroplasts and the appearance of MS conductances in different mem-

[31] P. C. Moe, P. Bount, and C. Kung, *Mol. Microbiol.* **28,** 583 (1998).

brane fractions. Anti-MscL antibodies provided the most definitive molecular evidence that MscL is located in the inner (cytoplasmic), but not the outer membrane. When total membranes of *E. coli* are fractionated using a three-step sucrose gradient[32] and the fractions assayed on immunoblots using anti-MscL antibody, the 17-kDa target can be detected clearly in the inner membrane and not the outer membrane fraction.[26] This finding is corroborated by the near exclusive association of the radiolabel with the inner membrane fractions when MscL alone was labeled with [$^{35}$S]methionine. In the latter study,[33] MscL was governed by a T7 polymerase under the transcriptional control of an inducible promoter (the full procedure for labeling is given in method 3 of the cross-linking protocols below).

## Cross-Linking to Determine Number of Subunits in MscL Homomultimer

Chemically reactive compounds have been used in several systems to study stable protein complexes.[34] These compounds (cross-linkers) covalently link reactive residues that are within the reach of their two arms. They therefore can covalently link two or more proteins that are normally associated only with noncovalent bonds. We have utilized this technique to determine the number of subunits in a MscL channel. In the first method MscL is cross-linked in membranes (*in situ* technique) and then cross-linked products are visualized by Western blot. In the second, purified MscL is cross-linked and the products are visualized on SDS–PAGE by silver staining. Both methods give a ladder of bands corresponding to monomer, dimer culminating in hexamer of MscL subunits.[26] A third method[33] has also been used in which cross-linker is applied to intact cells expressing radiolabeled MscL, and the subunits and complexes are detected by autoradiography. This latter technique, however, shows mainly monomeric and dimeric MscL subunits (although the higher molecular weight forms are also seen, particularly when the cross-linker disuccinimidyl suberate, DSS, is used). The large amount of monomeric and dimeric forms of MscL observed, especially in the latter method, could either be due to unassembled subunits *in vivo*, or simply an inefficiency in cross-linking and/or an inability to visualize easily the higher molecular weight forms. A study, using independent techniques, is consistent with the hypothesis that the MscL complexes are comprised

[32] C. A. Schnaitman, *J. Bacteriol.* **104,** 890 (1970).

[33] C. C. Häse, R. F. Minchin, A. Kloda, and B. Martinac, *Biochem. Biophys. Res. Comm.* **232,** 777 (1997).

[34] S. S. Wong, "Chemistry of Protein Conjugation and Cross-Linking." CRC Press, Boca Raton, Florida 1991.

almost entirely of preformed homomultamers,[35] as depicted in Fig. 1. The three methods of cross-linking MscL subunits are detailed:

1. MscL can be cross-linked *in situ* and then examined using anti-MscL antibodies. For this purpose a 20-mg pellet of *E. coli* membrane vesicles isolated from PB104 containing p5-2-2 is resuspended in 4 ml reaction buffer (100 m*M* NaCl, 30 m*M* sodium phosphate, pH 7.5). A cross-linker, such as DSS, is added to 1 ml of this solution to the various concentrations to be tested. Tubes are rotated gently for 30 min at room temperature before quenching the reaction with Tris buffer (pH 8.0) added to 100 m*M*. Membrane vesicles are pelleted at 48,000 *g* for 30 min and resuspended in 150 $\mu$l of Laemmli sample buffer. Proteins are separated in a 12% gel by SDS–PAGE and cross-linked products visualized by Western blots using anti-MscL antibodies.
2. Purified MscL can be cross-linked and examined directly. MscL-6His proteins are purified with the above $Ni^+$-NTA column and then transferred to the reaction buffer in first protocol above with 1% OG by passing through a PD10 gel-filtration column and diluted with this buffer to 10 $\mu$g/$\mu$l. The cross linking reaction is executed as above except that DSS is added to 1.5-ml protein aliquots. The reaction products are concentrated 10-fold using Centricon-30 concentrators (Millipore, Beford, MA), and suspended in Laemmli buffer. Cross-linked products are separated in a 12% gel by SDS–PAGE and visualized by silver staining.
3. Direct application of cross-linker to bacteria in which the MscL protein has been specifically labeled. The *mscL* gene has been placed under the control of the T7 promoter. These constructs, when placed in *E. coli* strains carrying an inducible T7 RNA polymerase gene, permit the specific production and labeling of MscL with [$^{35}$S]methionine. *Escherichia coli* cells [JM109(DE3)] harboring the plasmid construct are grown to an $OD_{595}$ of 0.2–0.3, harvested and transferred to the defined medium RPMI 1640 (Difco) for 30 min, and then to RPMI 1640 medium lacking methionine for 15 min. Expression of the T7 RNA polymerase is induced for 30 min with 0.1 m*M* IPTG, then 200 $\mu$g/ml rifampicin was added for 1 hr. The nascent proteins are then labeled by adding 10 mCi of [$^{35}$S]methionine to 10 ml of cells for 10 min. The cells are then washed in MES buffer [10 m*M* 2-(morpholino)ethanesulfonic acid, 150 m*M* NaCl, pH 7.0]. The cross-linkers EDC [1-ethyl-3-(3-dimethylaminopropyl)carbodiimide hydrochloride] or DCC (*N,N'*-dicyclohexylcarbodiimide) are added to the intact cells to a final concentration of 10 m*M* in MES buffer. Alternatively, DSS, dissolved in dimethyl sulfoxide (DMSO) is added to a final concentration of 1 m*M*

[35] S. Sukharev, M. J. Schroeder, D. R. McCaslin, and C. Kung, in preparation.

in PBS, pH 8.0. Cell extracts are then separated in a 12% gel by SDS–PAGE and MscL subunits detected by autoradiography.

## Method to Study Membrane Topology

Bacterial alkaline phosphatase (PhoA) has been used in many systems as a topological reporter. The pro-PhoA protein requires modification to a mature form in the periplasm, and thus is inactive unless transported across the cytoplasmic membrane.[36] If the *phoA* gene is fused to a protein, or portion of a protein, that is transported to the periplasm, it will be active. Hence, fusion of the *phoA* gene with a gene encoding a transmembrane protein of interest, and subsequently assaying for PhoA activity from the fusion product, has provided a system that has been used extensively to study transport and membrane topology of many different proteins.[37–39] A modification of this technique, which has many advantages and could easily be adapted for almost any protein, has also been used to determine the topology of MscL subunits.[26] The results suggest a two-transmembrane model as depicted in Fig. 1.

Constructs containing the *phoA* gene, amplified by PCR, and containing terminal *Bam*HI linkers have been generated. These can now be used to insert the *phoA* gene into engineered *Bam*HI sites generated in any gene of interest. For the study of the membrane topology of MscL, we chose to insert the *phoA* gene at the position of aspartate codons in *mscL*, our target gene. These residues were chosen for two reasons: first, the codons for this amino acid provide half of the *Bam*HI consensus site, thereby facilitating the construction of *Bam*HI insertion sites for the *phoA* gene; second, as a charged amino acid, aspartate is often found in hydrophilic regions appropriate for fusion sites of this sort. We have designed two *phoA*-fusion constructs so that two kinds of in-frame fusions can be constructed for each site: a terminal fusion with 5′ *mscL* sequence followed by an intact *phoA*, and a "sandwich" fusion construct where *mscL*, or any gene of interest, is disrupted by insertion of a *phoA* ORF stripped of its stop codon. The constructions and their potential use are detailed below.

The *phoA* insertional cassettes were amplified from wild-type *E. coli* using the polymerase chain reaction (PCR).[40] Two versions were produced: first, for terminal fusions retaining the native translational stop codon of

[36] A. I. Derman and J. Beckwith, *J. Bacteriol.* **173,** 7719 (1991).
[37] C. Manoil and J. Beckwith, *Science* **233,** 1403 (1986).
[38] D. Boyd, C. Manoil, and J. Beckwith, *Proc. Natl. Acad. Sci. U.S.A.* **84,** 8525 (1987).
[39] M. Ehrmann, D. Boyd and J. Beckwith, *Proc. Natl. Acad. Sci. U.S.A.* **87,** 7574 (1990).
[40] M. A. Innis, D. H. Gelfand, J. J. Sninsky, and T. J. White, "PCR Protocols. A Guide to Methods and Applications." Academic Press, San Diego, 1990.

*phoA*. The 3′ oligonucleotide used for this construct was GGC GGG ATC CTA TTT CAG CCC CAG AGC GGC TTT. The second, for sandwich fusions lacking the stop codon, permitted in-frame translation through the distal portion and back into the target gene. The oligonucleotide used for this construct was GCG GGG ATC CTT CAG CCC CAG AGC. The same 5′ oligonucleotide, CCA GAG GAT CCT GTT CTG GAA AAC CGG GCT, was used for both versions. These primers amplified a sequence encoding the entire PhoA protein, less the signal sequence and five additional amino acids,[39] and provided in-frame *Bam*HI linkers at each end of the insert. The resulting PCR product was ligated into the *Bam*HI site of pBluescript and verified by restriction endonuclease analysis.

It is preferential to design the construct containing the target gene for mutagenesis such that it does not have a *Bam*HI site. The absence of this site obviates the need for further subcloning or partial restriction enzyme digests to insert the *phoA* gene. Therefore, for *mscL*, we subcloned the *mscL* insert into M13 using the *Kpn*I and *Hind*III restriction-endonuclease sites, thus ensuring that no *Bam*HI site remained in the construct. If *Bam*HI sites exist in the target gene, they can be removed either by subcloning portions, or cassettes, of the gene not containing this site, and replacing the cassette after the fusion is generated, or by removal of the site(s) by mutagenesis. Alternatively, additional *phoA* cassettes can be constructed, bearing terminal linker sequences not found in the gene of interest. To generate sites for *phoA* insertion within the target gene, oligonucleotides specific to each site should be designed with a 12 base overlap to introduce the unique *Bam*HI site. Mutants can be generated using standard techniques for site-directed mutagenesis; we used the Amersham Sculpture kit. Mutations are easily confirmed using *Bam*HI restriction endonuclease analysis.

Each of the *phoA* cassettes can then be ligated into the engineered *Bam*HI sites of the mutant gene of interest. Verified clones are then subcloned into the *Xho*I site of pB10a, or another expression construct. Proper orientation with respect to the promoter is verified by restriction endonuclease analysis. After transformation with this expression construct, the PhoA enzyme activity assay is performed on the DHB4 host cells induced for fusion protein expression for 2 hr by 1 m*M* IPTG. These assays and quantitation are then performed as previously described.[41]

## Reverse Genetics

The *mscL* gene can be manipulated easily by standard molecular biological means. In addition to the initial deletion and point substitution mutants

[41] E. Brickman and J. Beckwith, *J. Mol. Biol.* **96,** 307 (1975).

generated,[12,42] a MscL-MscL tandem, a MscL-MscL-MscL triple tandem, as well as functional chimeras having portions of *E. coli* MscL replaced with portions from a homolog of another bacterium have all been generated in a variety of studies.

Because *E. coli* have a native MscL, the probability of expressing functional homologs or mutants is quite high. The *mscL* gene can be placed either in the *E. coli* chromosome or in various plasmids for the purpose of subcloning or expression of functional protein. In theory, one could directly test the MS channel activities in any *E. coli* strain in which the genetic engineering is carried out. In practice, however, we have found that many "cloning" strains are more difficult to patch clamp. Therefore, as discussed above, to facilitate the *in trans* expression of these mutants, we have generated and routinely use the PB104 strain. The experimental is then easily compared with the positive control (the *mscL*-null strain transformed with p5-2-2) and the negative control (the null strain bearing an empty plasmid). By estimating the number of channels detected by patch clamp in an area of membrane equivalent to that of a single cell, we estimate that there are about 10–100 functional MscL channels in a normal bacterial cell, absolutely zero in the *mscl*-null, zero to six in a *mscl*-null containing an uninduced p5-2-2 expression plasmid, and 50–200 in a *mscl*-null containing the p5-2-2 plasmid that has been induced. In cases where it is desirable to place the engineered *mscL* in the *E. coli* chromosome, perhaps to ensure that there is only a single copy, homologous recombination (as described above for the generation of the *mscL*-null mutant) of a construct containing the mutant of interest closely linked to a drug resistance gene could be used.

## Forward Genetics

One of the strengths of microbial genetics is the ability to survey large numbers of cells, and to select for rare events. One use of genetics is to establish structure–function correlates. For a more general discussion concerning microbial forward genetics, as applied to channel research, see Saimi *et al.*[43] Sukharev *et al.*[24] showed that the loss of MscL function yields no obvious plate phenotype, at least not in the rather cushy conditions of the laboratory environment. This, however, does not exclude the use of genetic screens since experimenters can demand a "gain-of-function" phenotype. For instance, we have looked for *mscL* mutants that, when expressed, cause a growth defect observable on plates.[44] One possible reason

[42] C. C. Häse, A. C. Ledain, and B. Martinac, *J. Membr. Biol.* **157,** 17 (1997).
[43] Y. Saimi, S. H. Loukin, X-L. Zhou, B. Martinac, and C. Kung, *Methods Enzymol.* **294,** [27], 1998 (this volume).
[44] X. Ou, P. Blount, R. J. Hoffman, and C. Kung, submitted.

for this defect would be that the mutant MscL channels are more "active," gating at lower membrane tensions. Patch-clamp and whole-cell physiology have supported this hypothesis. The mutations, as divulged by sequence analysis, have revealed an interesting structure–function relationship not easily predicted *a priori*.

To perform a forward genetic study with MscL, a pB10a-based plasmid bearing the *mscL* gene is randomly mutagenized by one of a number of methods. In these studies it is important that the target, in this case the *mscl* ORF, be moved to a fresh, unmutagenized expression plasmid to rule out the possibility that any phenotype is due to a change in the expression plasmid. This can be performed either before screening or, alternatively, only on plasmids that yield an interesting phenotype. The plasmids containing the mutated gene, each bearing a different defect, are moved back into the *mscL*-null PB104. The transformant colonies on various agar plates are then examined for phenotypic changes. In the case discussed above, cells were simply plated on LB plates containing 100 $\mu M$ ampicillin, then replica plated onto the same plates, containing 1 m$M$ IPTG to induce the *mscL* gene expression. We then isolated colonies from the master plate that did not grow well on the IPTG-containing plates. Several variations of this experiment could be performed; for instance, colonies could be isolated that grow on IPTG plates containing high concentrations of an osmolite, such as NaCl, but do not grow on IPTG containing plates that do not contain the osmolite. The plasmid would then be recovered from any colony exhibiting an interesting phenotype. Bacterial cells have a tendency to mutate spontaneously at a low frequency. Therefore, to ensure that the phenotype cosegregates with the plasmid and is not due to a spontaneous mutation in the host cell, it is important to return the plasmid to a fresh host and retest the phenotype. For phenotypes that are due to the plasmid, mutation(s) are detected by sequence analysis. Multiple mutations can be difficult to interpret, so it is important that mutagenizing conditions are such that most plasmids isolated contain a single mutation in the target gene. For MscL, cells expressing the mutant can be analyzed by patch-clamp and whole-cell physiologic techniques described above and below.

## Whole-Cell Physiology: Measuring Efflux of Solutes Subsequent to Osmotic Downshock

The bacterial MS channels have been implicated in cell sensing and/or adaptation to changes in osmotic conditions. Bacteria, when challenged with a sudden decrease in osmotic environment, called osmotic downshock, jettison many solutes from the cytoplasm into the medium including potassium, adenosine triphosphate (ATP), proline, and even small proteins, while maintaining viability.[45,46] Hence, assays that measure the efflux of solutes on osmotic downshock have become important for the physiologic study of bacterial MS channels. In one study, a correlation was demonstrated between the concentration of gadolinium required to block MS channels and the concentration required to block efflux of some of the small molecules normally jettisoned from *E. coli* and *Streptococcus faecalis* in response to osmotic downshock.[47] The solutes measured were ATP, lactose, glutamate, and potassium. The search for changes in the efflux of solutes between *mscL*-null mutants and parental strains has not yet been successful, presumably because of redundancy of function (besides MscL, two additional MS activities have been identified in *E. coli*: MscS and MscM). On the other hand, we have recently found differences in the retention and efflux of potassium in cells expressing several site-directed[48] and randomly generated[44] gain-of-function MscL mutants that induce a slow-growth phenotype. For all of these efflux assays, cells are first grown in, or allowed to adapt to a high osmolarity medium or buffer (for example, medium containing an additional 300 m*M* NaCl), and the measurements are performed subsequent to dilution into, or washes with, a solution of the same osmolarity, and a solution(s) of significantly decreased osmolarity. A defined medium is usually used to control background levels of solutes. All solutions should be kept warm (37°) to avoid cold-shocking of the cells, which may exacerbate any downshock. For assays in which the solute in the medium is measured, total starting solute in the cells can be determined by taking some of the culture and measuring it after French-pressing the cells.

The measurement of solutes is straightforward. For ATP measurements, a luciferin/luciferase assay system can be used, and quantitation performed with a luminometer (for example: Turner Instruments TD-20e). If measured from intact cells subsequent to a downshock, only the ATP that is in the medium is measured. Similarly, glutamate in the medium can be measured using a commercially available assay kit (Boehringer Mannheim). Because proline and lactose are accumulated in cells grown in a high osmotic medium, accumulated levels can be measured using radiolabeled ($[^{14}C]$ or $[^{3}H]$) proline or lactose that is placed in the medium for short periods of time (about 15 min). For these latter assays, the bacteria are osmotically shocked (or not), filtered, and then by quantitating the amount of radioactivity on the filter, the solute remaining in the cell can be calculated. Finally, internal potassium can be measured by harvesting cells onto a 0.45-$\mu$m

[45] R. J. Britten and F. T. McClure, *Bacteriol. Rev.* **26,** 292 (1962).
[46] M. Schleyer, R. Schmid, and E. P. Bakker, *Arch. Microbiol.* **160,** 424 (1993).
[47] C. Berrier, A. Coulombe, I. Szabo, M. Zoratti, and A. Ghazi, *Eur. J. Biochem.* **206,** 559 (1992).
[48] P. Blount, M. J. Schroeder, and C. Kung, *J. Biol. Chem.* **272,** 32150 (1997).

nitrocellulose filter, then washing with media or buffer containing or lacking the osmolite used in the growth conditions. Hence, the downshock is actually given in the wash of the filter. In this case, potassium should be absent from the wash solution to decrease background levels of the ion. Under these conditions, we have found that 3 ml of wash which normally takes 3–5 sec, is sufficient for an osmotic shock; increasing the wash solution up to 20 ml had little effect on the values obtained, suggested that there is phase of very rapid potassium release on downshock that is essentially complete within this time. The filters are then dried at 80–95° overnight, the solutes are resuspended in 3–4 ml of double distilled water, and the solution is then assayed for potassium by flame photometry (PFP7; Buck Scientific, E. Norwalk, CT).

## Conclusions

The bacterial MS channel MscL is currently the only tangible model for studying the mechanisms by which tension in the membrane can gate a channel. This system has at its disposal several techniques for its study including electrophysiology, biochemistry, the power of microbial genetics, and whole-cell physiology. By combining these methodologies, together with possibly getting a high-resolution structural model, one day a true understanding of the interaction between functional and structural properties may integrate to yield an understanding of the molecular mechanisms of MscL channel function, which may in turn contain general principles used in other systems.

# [25] Simplified Fast Pressure-Clamp Technique for Studying Mechanically Gated Channels

*By* DON W. MCBRIDE, JR. and OWEN P. HAMILL

## Introduction

A fundamental feature of the tight seal patch-clamp technique is the existence of a suction port in the patch pipette holder. This is critical for applying suction to the membrane patch in order to form a tight seal. This feature also enables, after seal formation, the application of pressure/suction to stimulate the membrane patch mechanically. Initially, mouth-applied suction was used to obtain the seal and provide the stimulation. Although this is convenient, it lacks precision in terms of the magnitude


0076-6879/99 $30.00

and duration of the mechanical stimulus. To overcome this deficiency, we have developed a system that can apply controlled pressure waveforms to the suction port of the pipette holder. We refer to these prototypes as "pressure clamps." Although there have been previous descriptions of earlier prototypes,[1,2] we now describe a simpler but fast system and provide more details for its construction.

The basic strategy of the pressure clamp is that the desired pressure applied to the patch pipette is achieved by a balancing of pressure and suction. Central to this balancing is the use of a proportional piezoelectric valve whose opening is proportional to the applied voltage. Through feedback control of this valve, the amount of pressurized $N_2$ allowed to enter a mixing chamber can be regulated to balance the constant outflow due a continual vacuum efflux and thus achieve the desired pressure.

## Mechanical Arrangement of Simplified Pressure Clamp

Figure 1 is a schematic illustrating the mechanical arrangement of the pressure clamp. Suppliers (and addresses) are listed in the Appendix. The basic principle of the valve action has been illustrated. The system volume whose pressure is being controlled is represented by the lightly shaded region, which includes the mixing chamber and the patch pipette and holder as well as the tubing to the holder. Simple analysis of the system using the ideal gas law ($PV = nRT$, where $V$ is the system volume, and others terms have their usual meanings; see also Ref. 1) reveals that the speed with which the pressure can be changed is directly proportional to the change in the input or output flux and inversely proportional to the volume of the system [$dP/dt = (RT/V)(dn/dt)$]. Ideally, it would be preferable to control the input and output flux reciprocally, i.e., as the input flux increases the output flux decreases and vice versa. In the high-speed modified version of the pressure clamp we included two values.[2] This gave rise/fall step transition times of less than 0.5 ms. Unfortunately, during the course of several experiments, we realized that it is nearly impossible to prevent solution from entering the valve on the vacuum side and shorting it out, thus ruining it. (The piezo element in each valve has a constant 160 $V_{DC}$ across it.) Since on the pressure side dry $N_2$ is always blowing through the value into the mixing chamber, pipette solution never enters that valve. We lost several valves before we realized what was occurring. We have not found "solution-proof" piezo valves. Therefore, until we solve the

[1] D. W. McBride, Jr., and O. P. Hamill, *Pflügers Arch.* **421,** 606 (1992).

[2] D. W. McBride, Jr., and O. P. Hamill, *in* "Single Channel Recording" (B. Sakmann and E. Neher, eds), 2nd Ed., p. 329. Plenum, New York, 1995.

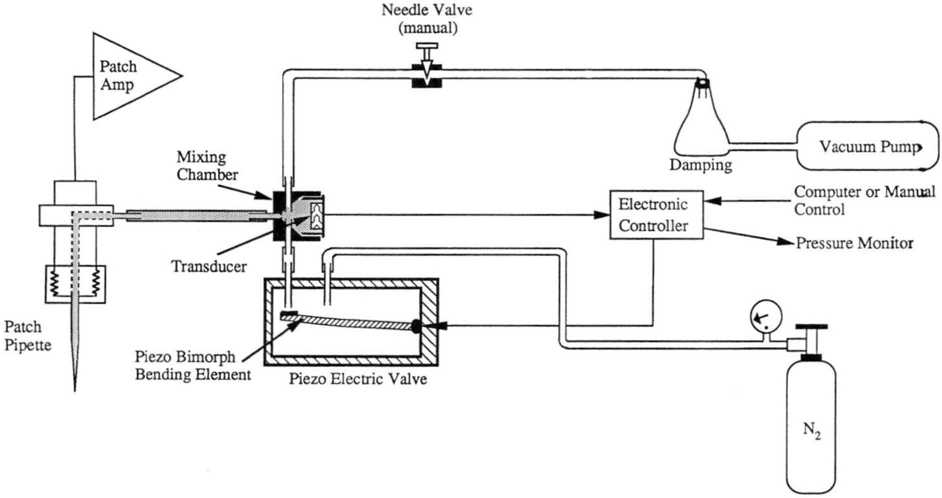


FIG. 1. Schematic diagram of the mechanical arrangement of the pressure clamp. The patch pipette holder and amplifier are shown on the left. The suction port of the pipette holder is connected by a tube to the mixing chamber. The mixing chamber contains the transducer as well as the ports leading to either a piezoelectric valve for pressure injection or to the constant vacuum. The basic mechanism of the piezoelectric valve is illustrated in which the position of the bimorph is dependent on the applied voltage. The upper port is connected to the vacuum pump through a needle valve (to adjust the suction or out flux) and damping flask (necessary to filter out cycle-to-cycle pump fluctuations). Note that the vacuum source is not critical. We have found that either house vacuum or an inexpensive commercially available vacuum pump (e.g., KNF-Neuberger) is adequate. The piezo valve is connected via tubing to the pressure regulator on the $N_2$ tank. Here again, house compressed air may be adequate if it is sufficiently clean and dry and if the pressure can be regulated appropriately (10–20 psi).

"solution-shorting" problem, we are using only a single Lee valve controlling the pressure injection into a constant vacuum output flux. The amount of output flux is adjusted by a manual needle valve or a similar type valve. We find that the speed of the pressure clamp using one piezo valve still produces a step change of ~1 ms.

Because the speed of the clamp is also inversely proportional to the volume, we have sought to minimize the system volume. In addition to the small size of the Lee valve aiding in this, we have also, as described below, incorporated the transducer into the mixing chamber (i.e., the sensor element of the transducer forms one wall of the mixing chamber). These two modifications permit closer proximity of the pressure-clamp mixing chamber to the pipette holder and thus minimize the connecting tubing.

Tubing length from the vacuum and pressure is less critical, although it can affect input/output flux.

To consolidate the transducer/mixing chamber we started with a Sen-Sym transducer, which is enclosed in a metal, transistor type TO-39 package. The actual metal can has a diameter of 0.325 in. and a height of 0.265 in. Normally the top of the can has a small-diameter hole that gives access to the sensor. By cutting off the top portion of the can (Thorlabs sells a convenient diode "can-opener"), the sensor element itself is exposed and the remaining bottom portion of the can containing the leads may be conveniently press fitted into the end of a cyclindrically shaped mixing chamber. The mixing chamber is made from a piece of 0.5-in. diameter plexiglass rod ~0.5 in. long. Into one end a hole is drilled to accept the sensor (e.g., letter O drill bit and ~0.25 in. deep). The chamber has three additional ports in it. We use #16 or #18 size hypodermic needles cut to a suitable length. Two of the ports, diametrically opposed, are perpendicular to the axis of the chamber and are drilled so as to intersect the large central transducer hole near the tip of the cone formed during its drilling. One port connects to the piezo valve and the other to the constant vacuum. The third port is along the axis of the cylinder and opposite the sensor and connects to the pipette holder. With regard to the third port care must be taken to prevent the cutoff hypodermic needle from damaging the sensor element. The total volume of the mixing chamber is about 250 $\mu$l. However, as described below, to increase the sensitivity of the pressure clamp large (either longer or bigger diameter) mixing chambers could be used.

### Electronic Control of Pressure Clamp

The basic design of the electronic controller of the pressure clamp has been retained.[1] Figure 2 shows the schematic of the electronic controller and is divided into four sections. The first section is the voltage command. There are two internal inputs and one external input (A5). The internal control includes both the manual (A6) (i.e., potentiometer) and an adjustable test pulse, which can be switched in or out. The test pulse is used during the adjustment procedure for optimizing the step response. This pulse could be supplied externally. However, because it is frequently used it has been incorporated into the controller. An LM555 timer used as an astable multivibrator (National Semiconductor, Santa Clara, CA) was used to generate a repetitive ~50-ms pulse at ~2 Hz. This pulse was then scaled by multiplying it by the output of a potentiometer using an AD633 multiplier chip (Analog Devices, available from Newark, also data sheet). The result was a test square pulse that could be varied between +5 and −5 V. The

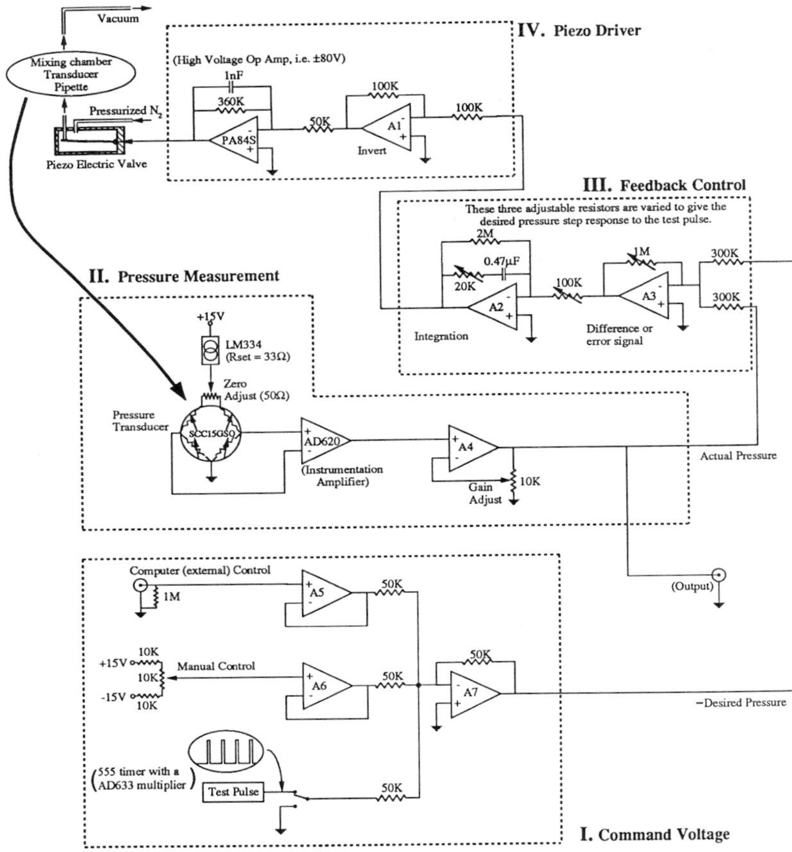


FIG. 2. Electronic control of the pressure clamp. The circuit can be divided into four blocks. (I) Command voltage. This represents the desired pressure and includes internal and external sources. The internal sources include a manual control for setting the steady-state values (A6) and a variable test pulse, which can be turned on and off by a toggle switch. The test pulse is used in the step response adjustment procedure. It could have been supplied externally but because it is frequently used, for convenience it has been incorporated into the circuitry. (II) Pressure measurement. This measures the pressure in the mixing chamber. A constant current source excites the transducer bridge whose output is measured by the instrumentation amplifier. A voltage follower (A4) is included to enable gain adjustment. (III) Feedback control. This section is composed of an initial summing amplifier (A3), which generates an error signal followed by an integrating amplifier (A2). In this section there are three adjustable resistors. While to some extent there may be some redundancy they do allow flexibility in terms of shaping the desired waveform (i.e., step response). (IV) Piezo driver. This section includes a high-voltage operational amplifier operated with a gain of about 7 and a low-pass filter. The integrated signal from section III is inverted (A1) before the high-voltage driving amp. This is necessary to provide negative feedback for the feedback loop. (All op amps used were AD712, available from Newark.)

second section of the controller (Fig. 2) is responsible for pressure measurement. The transducer used is the SCC05GSO (SynSym, Inc., Milpitas, CA). The modifications in incorporating this into the mixing chamber have been discussed above. An LM334 current regulator (National Semiconductor, available from Newark, also see data sheet) was used to excite the transducer bridge and was set at a current of ~1 mA. The AD620 (Analog Devices) instrumentation amplifier with a gain of ~500 (see data sheet for AD620) was used to measure the output of the bridge. The AD620 is an 8-pin chip which is convenient to use. The output of the AD620 is further amplified (A4) to give the desired gain of 100 mV/mm Hg. As another simplification, the Bessel filter that followed A4 has been removed. The subsequent integration was found to be sufficient. The third section is the feedback control section and is composed of a summation amplifier (A3) used to generate the error signal (between the desired and actual pressures) followed by integration or averaging (with compensation) of this error signal (A2). Three adjustable resistors have been included to adjust the response. While this may be somewhat redundant, it does allow flexibility in adjusting the desired waveform. The final stage is the piezo driver section. This includes a high-voltage op-amp (PA84S from Apex, Tucson, AZ), which is powered at ±80 V. The control signal for the pressure injection from the $N_2$ tank has been inverted (A1) so as to give overall negative feedback.

## Some Practical Tips for Construction

In the construction, we suggest that Section II (Fig. 2), pressure measurement, be assembled first. The data sheet for the transducer gives some good tips, especially with regard to adjusting the offset. The most critical part of this section is the calibration of the overall gain of 100 mV/mm Hg. To do this, a calibrated pressure transducer must be used with which to compare the pressure clamp transducer. The mixing chamber can be constructed with the transducer in place. With the sensor open to the atmosphere, adjust the zero trim potentiometer appropriately. To adjust the gain, block one of the three ports of the mixing chamber. Connect one port to a syringe and the other to the calibrated sensor. The pressure in the system can then be changed by moving the syringe plunger and monitored via the calibrated sensor. Simply adjust the gain on amplifier A4 until the two sensors agree. Two calibrated transducers (SenSym) that can be used are the 143SC series and a digital manometer. The 143SC is a simple three-pin device (power, ground, and output) which gives a ± analog signal output centered at 3.5 V. The more accurate digital manometer only has a digital readout.

A ±80 $V_{DC}$ power supply is needed to drive the piezo valve. One can be easily and inexpensively built using a dual isolation (1:1) transformer such as the Magnetek FP230-50 available from Newark Electronics. The dual transformer is necessary to obtain ±80 V. Appropriate full-wave bridges and capacitors can be used in conjunction with the TL783C, a three-pin, 125-V voltage regulator (Texas Instruments, available from Newark Electronics) to build two +80 V supplies (see the TL783C data sheets). The voltage output of the TL783C is determined by the ratio of two resistors. We use values for $R_1$ of 130 Ω and $R_2$ of 2200 Ω. Since both supplies are isolated, they can be connected in series (i.e., the positive of one connected to the ground of the other) to give one ±80 V power supply.

At the present gain of 100 mV/mm Hg the practical pressure limit of the clamp is ±100 mm Hg (i.e., ±10 V input). In studies using standard sized patch pipettes (i.e., 1–2 μm) this upper limit value is sufficient for determining stimulus–response relations of mechanically gated (MG) channels since most MG channels studied saturate at <50 mm Hg. Furthermore, this limit exceeds typical patch rupture pressures. However, there are some reports of MG channels that have saturation pressure exceeding ~100 mm Hg. To achieve higher pressures, the gain of the pressure measurement section could be reduced to, for instance, 50 mV/mm Hg [i.e., the gain adjust on A4 (Fig. 2)]. This would double the practical range to ±200 mm Hg. However, note that at high suctions the membrane can be decoupled from the underlying cytoskeleton and alter (i.e., either increase or decrease) the mechanosensitivity of a channel.[3]

When considering noise limitations and sensitivity of the pressure clamp itself (i.e., the minimal distinguishable step size), there is a reciprocal relationship between these and the speed of the clamp with regard to the volume of the system. On the one hand, larger volumes decrease the speed (i.e., for a given flux it takes longer to change the pressure in a larger volume). On the other hand, a larger volume stabilizes the pressure of the system making it less sensitive to fluctuations in input or output flux. This decreases the pressure noise of the system and also, because it requires a larger change in flux for a given change in pressure, the control can be more sensitive albeit slower. Our preference has been to optimize the time response of the clamp since oocyte and muscle MG channels are activated by moderate pressures (~10–20 mm Hg) by minimizing the mixing chamber volume. In the present configuration, the pressure rms (root mean square) noise is ~0.1 mm Hg and this presumably represents the limitation for the minimal distinguishable step size that can be applied by the clamp for the given volume of the system. This sensitivity is adequate for many MG

[3] O. P. Hamill and D. W. McBride, Jr., *Annu. Rev. Physiol.* **59,** 621 (1997).

channels, which show half saturation pressure of 10–15 mm Hg. However, some MG channels have been reported to have half-saturation pressure as low as 1–2 mm Hg. For these more sensitive MG channels the minimal pressure increments could be reduced by increasing the mixing chamber volume (i.e., making it longer or larger diameter).

Several factors influence the step response of the pressure clamp. There are three potentiometers in the feedback control section. Varying these potentiometers will affect the speed and shape of a pressure step. In addition, the setting of the needle valve on the vacuum side and the regulator pressure from the $N_2$ tank (or other pressure source) in front of the piezo valve (5–20 psi) both greatly influence the step response. Indeed, all five of these possible adjustment points determine the step response. With some experimentation, one can determine how each adjustment point affects the speed and shape of the step response so that the optimum response can be obtained.

## Appendix

We indicate here the suppliers we have used in constructing the pressure clamp.

High voltage op-amp, as well as the socket for the op-amp [Apex Microtechnology Corporation, 5980 North Shannon Road, Tucscon, AZ 85741-5230, phone: (800) 546-2739]

Piezo valve [The Lee Company, 2 Pettipaug Road, P.O. Box 424, Westbrook, CT 06498-0424, phone: (800) 533-7584]

Pressure transducer [SynSym Inc., 1804 McCarthy Boulevard, Milpitas, CA 95035, phone: (800) 392-9934]

Can opener for pressure transducer [Thorlabs, Inc., P.O. Box 366, Newton, NJ 07860-0366, phone: (201) 579-7227]

Power supply parts and electronic parts [Newark Electronics, 4801 North Ravenswood Avenue, Chicago, IL 60640-4496, phone: (773) 784-5100]

Vacuum pumps [KNF-Neuberger, 2 Black Forest Road, Trenton, NJ 08691, phone: (609) 890-8889]

## Acknowledgments

We acknowledge the NIH (RO1-AR42782), the NSF (Instrument Development for Biological Research) and the MDA for their support.

# [26] Virus Ion Channels

*By* DAVID OGDEN, I. V. CHIZHMAKOV, F. M. GERAGHTY, and A. J. HAY

## Introduction

A virus makes use of many features of the normal cellular machinery to replicate its genome, to synthesize its structural components, and to assemble them into progeny virions. Membrane-delimited systems, for example, are involved in entry of virus into cells and uncoating of the virus genome, synthesis and transport of membrane proteins, and assembly of enveloped viruses. Not surprisingly, therefore, changes may occur in the ionic permeability of cellular membranes due to nonspecific consequences of the disruption of normal cellular homeostatis by virus infection.[1] On the other hand, there are several examples of virus proteins causing specific alteration of ion flux across cellular membranes, by, e.g., (1) modification of a signal transduction pathway, as in the case of rotavirus NSP4 potentiation of chloride secretion in mouse intestinal mucosa,[2] (2) association with a cellular ion transporter, e.g., interaction of the E5 protein of human papillomavirus with the 16K subunit of the vacuolar $H^+$-ATPase proton pump inhibits acidification of endosomes in E5-expressing keratinocytes,[3] (3) formation of a novel virus-coded ion channel, e.g., the M2 and NB proteins of influenza A and B viruses, respectively, and Vpu protein of human immunodeficiency virus, HIV-1.[4–8]

## Role of M2 in Virus Infection

The M2 protein of influenza A viruses provided the first example of an animal virus membrane ion channel. Initial evidence as to its function came

[1] L. Carrasco, *Adv. Virus Res.* **45,** 61 (1995).
[2] J. B. Ball, P. Tian, C. Q.-Y. Zeng, A. P. Morris, and M. K. Estes, *Science* **272,** 101 (1996).
[3] S. W. Straight, B. Herhan, and D. J. McCance, *J. Virol.* **69,** 3185 (1995).
[4] L. H. Pinto, L. J. Holsinger, and R. A. Lamb, *Cell* **69,** 517 (1992).
[5] I. V. Chizhmakov, F. M. Geraghty, D. C. Ogden, A. Hayhurst, M. Antonou, and A. J. Hay, *J. Physiol.* **494,** 329 (1996).
[6] N. A. Sunstrom, L. S. Premkumar, A. Premkumar, G. Ewart, G. B. Cox, and P. W. Gage, *J. Membr. Biol.* **150,** 127 (1996).
[7] G. D. Ewart, T. Sutherland, P. W. Gage, and G. B. Cox, *J. Virol.* **70,** 7108 (1996).
[8] U. Schubert, A. V. Ferrer-Montiel, M. Oblatt-Montal, P. Henklein, K. Strebel, and M. Montal, *FEBS Lett.* **398,** 12 (1996).



0076-6879/99 $30.00

from biochemical studies of the inhibitory action of the anti-influenza A drug amantadine.[9–11] The M2 protein was identified as the target of amantadine action by genetic analyses of the molecular basis of drug resistance.[12] Inhibition of M2 function by amantadine was shown to affect two stages in virus infection: (1) Amantadine prevented the dissociation of the matrix–ribonucleoprotein core structure during uncoating of virus in the acidic environment of the endosomes.[13] (2) During cotransport of the virus hemagglutinin (HA) and M2 from the endoplasmic reticulum to the plasma membrane of influenza A virus-infected cells, inhibition of M2 by amantadine caused the acid-sensitive, cleaved HA to be converted to its low pH form, an effect that was counteracted by proton ionophores such as monensin. The use of HAs of differing pH sensitivities, as well as the cytochemical pH probe DAMP, permitted tritration of the acid environment of the *trans* Golgi and identified the amantadine-sensitive M2-mediated process as the dissipation of the pH gradient across the *trans* Golgi membrane. The M2 protein, a component of the virus membrane, is a homotetramer of a 97-amino-acid polypeptide. Amino acids 25 to 43 comprise the membrane spanning domain, which includes single amino acid substitutions that confer resistance to inhibition by amantadine. The implication was, therefore, that M2 forms a proton (or hydroxyl)-permeable transmembrane ionophore that permits acidification of the virus interior in the acidic endosome during endocytosis, and reduces the acidification of *trans* Golgi vesicles following infection, in particular preventing the acid-induced inactivation of newly synthesized hemagglutinin.

To study the function of M2 *in situ* presents obvious technical difficulties. The use of infected cells, in which M2 is present in the surface membrane, has the problem that cells become very leaky in whole-cell patch-clamp recording for reasons that may not be related to M2 production. The best procedure therefore is to express M2 alone in a suitable cell system and study its properties as far as possible in isolation. Alternatively, the protein can be reconstituted in vesicles and studied with flux or bilayer methods. Application of the former approach to the study of M2 is described here in detail. The broader aims are to establish a system in which the properties of previously undefined potential channels including virus proteins can be

[9] A. J. Hay, *in* "Concepts in Viral Pathogenesis III" (A. L. Notkins and M. B. A. Oldstone, eds.), p. 361. Springer Verlag, New York, 1989.

[10] R. J. Sugrue, G. Bahadur, M. C. Zambon, M. Hall-Smith, A. R. Douglas, and A. J. Hay, *EMBO J.* **9,** 3469 (1990).

[11] A. J. Hay, *Sem. Virol.* **3,** 21 (1992).

[12] A. J. Hay, A. J. Wolstenholme, J. J. Skehel, and M. H. Smith, *EMBO J.* **4,** 3021 (1985).

[13] K. Martin and A. Helenius, *Cell* **67,** 117 (1991).

assessed, and to provide a system for investigating mechanisms of action and screening antivirals targeted to ion channels.

## Permeability Characteristics of M2

The M2 protein provides a good basis to assess the suitability of expression systems for investigating virus ion channels because of the indirect evidence as to its function and the availability of a selective blocker. In these experiments, M2-mediated permeation is therefore identified by susceptibility to amantadine block. The ability of M2 to function as an $H^+$ channel has been confirmed with these methods but the conclusions concerning details of the permeability change depend on the experimental system. Fluorescent pH indicators were used to show that M2 in the plasma membrane of virus-infected cells or M2 transfected cells[14] or M2 reconstituted into lipid vesicles[15] elevated $H^+$ permeability. Experiments in *Xenopus* oocytes injected with M2 mRNA showed that expression of M2 in the surface membrane increased nonselective cation conductance,[4] later shown to produce intracellular pH changes on external acidification.[16] Expression of M2 in mammalian CV-1 cells showed a similar nonspecific cation conductance as reported for oocyte expression.[17] Experiments with the M2 protein reconstituted in lipid bilayers showed single-channel cation currents with unitary conductance of 25–90 pS or higher in 150 m*M* NaCl.[18] A study of M2 expressed in stably transfected mouse erythroleukemia (MEL) cells[5] showed a proton selective conductance increase, with negligible $Na^+$ permeability ($P_H/P_{Na} > 10^6$), and distinct proton permeation and activation processes. Also, single-channel conductance was unresolvable at high resolution, estimated at less than 0.1 pS. Thus, apparent differences in the properties of M2 with regard to selectivity and magnitude of the single-channel conductance were revealed by these different approaches. The differences cannot be easily explained and may represent variation in the properties of M2 expressed in the different systems.

In general, the desirable properties of the expression system are a high, readily inducible, stable level of expression in a cell with a low endogenous level of linear "leak" ion conductance. This permits the use of high-resolution whole-cell and patch-recording methods, and the low background will

[14] A. J. Hay, C. A. Thompson, F. M. Geraghty, A. Hayhurst, S. Grambas, and M. S. Bennett, *in* "Options for Control of Influenza II" (C. Hannoun, A. P. Kendal, H. D. Klenk, and F. L. Ruben, eds.), pp. 281–288. Excerpta Medica, Amsterdam, 1993.

[15] C. Schroeder, C. M. Ford, S. A. Wharton, and A. J. Hay, *J. Gen. Virol.* **75,** 3477 (1994).

[16] K. Shimbo, D. L. Brassard, R. A. Lamb, and L. H. Pinto, *Biophysical J.* **70,** 1335 (1996).

[17] C. Wang, R. A. Lamb, and L. H. Pinto, *Virology* **205,** 133 (1994).

[18] M. T. Tosteson, L. H. Pinto, L. J. Holsinger, and R. A. Lamb, *J. Membr. Biol.* **142,** 117 (1994).

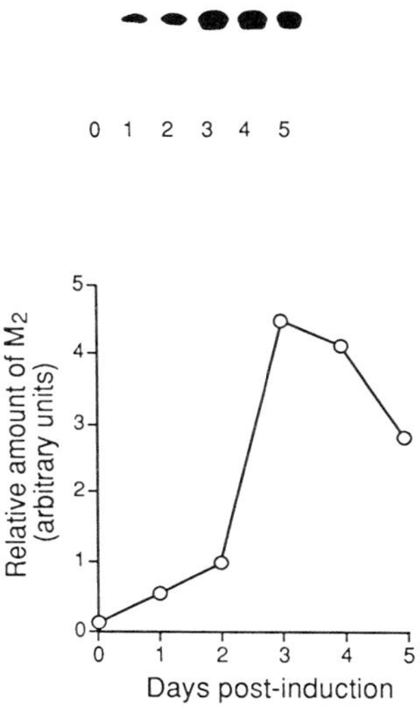


FIG. 1. Time course of the expression of M2 protein in MEL M2-39 cells. MEL cells (clone M2-39) stably transfected with M2 cDNA of influenza virus A/chicken/Germany/27 (H7N7, Weybridge strain) were incubated in medium containing 2% dimethyl sulfoxide and aliquots were taken at daily intervals (as indicated) for Western blot analysis. Top: Autoradiograph showing the accumulation of M2 with time post induction. Bottom: The relative amounts of M2, arbitrary units, were determined by microdensitometry of the autoradiograph shown above.

facilitate subtraction of leak current, important for the precise determination of reversal potentials for permeability measurements, and leak-corrected conductance.

## Expression of M2 Protein in LCR/MEL Expression System

The LCR/MEL expression system was chosen for electrophysiologic study of the M2 protein because (1) it provides a versatile mammalian system for readily obtaining cells with stable high-level expression of heterologous proteins[19] and (2) MEL cells have a low intrinsic membrane ion permeability.[20] M2 cDNA was stably integrated into the genome of MEL cells under the control of the human $\beta$-globin promoter and human globin locus control region (LCR) to allow position-independent expression induced by the induction of differentiation with dimethyl sulfoxide.

Figure 1 shows a time course of the expression of the M2 protein of

[19] M. Needham, C. Gooding, K. Hudson, M. Antoniou, F. Grosveld, and M. Hollis, *Nucleic Acids Res.* **20,** 997 (1992).

[20] P. A. Shelton, N. W. Davies, M. Antoniou, F. Grosveld, M. Needham, M. Hollis, W. J. Brammer, and W. J. Conley, *Recept. Channels* **1,** 25 (1993).

the influenza virus A/chicken/Germany/27 (Weybridge strain) by an MEL cell clone M2-39. The protein accumulated to maximum level 3–4 days after induction at which time cells were taken for patch-clamp measurements.

The structural and functional properties of M2 expressed in MEL cells were similar to those of the protein synthesized in virus-infected MDCK cells and it had a similar ability to modify cytoplasmic pH.[14]

### *Selection of Stably Transfected MEL-M2 Cells*

For reactions involving RNA, DNA, RT (reverse transcriptase), and PCR (polymerase chain reaction) use sterile HPLC water (Romil Chemicals). Manpulation of DNA involves standard methods.[21]

*Cloning of M2 cDNA.* Monolayers of Madin Darby canine kidney (MDCK) cells (approximately $10^6$ cells per 5-cm-diameter dish) are infected, at a multiplicity of 10–50 plaque-forming units (pfu) per cell with influenza virus grown in the allantoic cavity of fertile hen's eggs.[22] After incubation for 6 hr at 37° in serum-free Eagle's minimum essential medium (MEM), cells are washed with saline and lysed in 0.5% SDS, 10 m*M* sodium acetate, pH 5. The lysate is extracted once with an equal volume of water-saturated phenol at 50°, with gentle shaking to avoid shearing of DNA, cooled on ice, and the aqueous phase reextracted with cold phenol. RNA is precipitated at −20° following the addition of NaCl to 0.1 *M* and two volumes of ethanol, pelleted, washed twice with 70% (v/v) ethanol in 0.05 *M* NaCl and dissolved in water.[23] M2 cDNA is synthesized by reverse transcription using a primer, incorporating a *BglII* site and 13 nucleotides complementary to the 3′ end of M2 mRNA, in a 20-μl RT reaction containing 1 m*M* of each dNTP, 140 m*M* KCl, 10 m*M* $MgCl_2$, 20 m*M* dithiothreitol (DTT), 50 m*M* Tris-HCl, pH 8.3, 20 units ribonuclease inhibitor (Amersham) and 100 units avian myeloblastosis virus (AMV) reverse transcriptase (Boehringer Manneheim) at 42° for 60 min. M2 cDNA is amplified using the *BglII* primer and a second primer, incorporating an *Eco*RI site and 15 nucleotides (3–17) at the 5′ end of the mRNA transcript (100 ng of each) in a 50-μl PCR reaction containing 5 μl of the RT reaction, 0.25 m*M* of each dNTP, 50 m*M* KCl, 1.5 m*M* $MgCl_2$, 10 m*M* Tris-HCl, pH 8.8, 10 μg/ml bovine serum albumin and 2 units AmpliTaq DNA polymerase (Perkin-Elmer, Norwalk, CT). The PCR program consists of 35 cycles of 30 sec at 96°, 30 sec at 45°, and 90 sec at 72° followed by a final extension time of 10 min at 72°.

[21] J. Sambrook, E. F. Fritsch, and T. Maniatis, "Molecular Cloning: A Laboratory Manual," 2nd Ed. Cold Spring Harbor Laboratory Press, New York, 1989.

[22] A. J. Hay, *Virology* **60,** 398 (1974).

[23] A. J. Hay, B. Lomniczi, A. R. Bellamy, and J. J. Skehel, *Virology* **83,** 337 (1977).

The PCR product, M2 cDNA, and plasmid pEV3 DNA are digested with restriction enzymes *Eco*RI and *BglII* and electrophoresed in low melting point agarose in TAE buffer (40 m*M* Tris–acetate pH 8.0, 2 m*M* EDTA). The excised gel fragments are heated at 65° for 15 min; aliquots of the DNA are combined and ligated with T4 DNA ligase at room temperature overnight. Plasmid containing M2 cDNA, pEV3-M2, is cloned in *Escherichia coli* DH5-$\alpha$ cells, the sequence of the insert checked, and plasmid DNA prepared and purified by CsCl density gradient centrifugation.

*Transfections of MEL Cells.*[19] MEL-C88 cells[24] are maintained in $\alpha$MEM supplemented with 10% fetal calf serum (FCS). Actively dividing cells ($10^7$ per transfection) are pelleted, washed with phosphate-buffered saline (PBS), and resuspended in 0.9 ml ESB (140 m*M* NaCl, 0.7 m*M* $Na_2HPO_4$, 25 m*M* HEPES pH 7.2). Then 100 $\mu$g of pEV3-M2 plasmid DNA is linearized by *Pvu*I digestion, resuspended in 100 $\mu$l ESB, and added to the cells 10 min before electroporation with a single pulse of 250 V, 960 $\mu$F in a Bio-Rad (Richmond, CA) Gene Pulser cuvette (0.4-cm electrode gap). After 10 min, the cells are divided between two culture flasks (75 $cm^2$) containing 30 ml $\alpha$MEM, 10% FCS. Geneticin (G418; 800 $\mu$g/ml) is added after 24 hr. Cultures are incubated at 37° for 10–14 days, at which time transfected drug-resistant cells become 50% confluent (and an untransfected control culture is dead). Single MEL-M2 cell clones are obtained by limiting dilution in G418 medium in a 96-well plate and subsequently maintained in medium containing 400 $\mu$g/ml G418. The level of M2 expression by individual clones is analyzed by dot blot, and transfectants expressing high levels of M2 are selected for further study.

*Assay of M2 Expression.* Cell differentiation and expression of the M2 protein is induced by incubating cells ($10^6$/ml) in medium containing 2% dimethyl sulfoxide (DMSO). Aliquots of cells (approximately $10^6$ cells) are pelleted, washed with PBS, and lysed at 4° for 15 min in 100 $\mu$l Nonidet P-40 (NP-40) lysis buffer [1% NP-40, 150 m*M* NaCl, 2 m*M* $MgCl_2$, 1 m*M* EDTA, 20 m*M* Tris-HCl, pH 7.5, 0.2% soybean trypsin inhibitor, 10 $\mu$g/ml aprotinin, 1 m*M* phenylmethylsulfonyl fluoride (PMSF)]. Cell debris is pelleted and a 20-$\mu$l sample mixed with an equal volume of 2× sample buffer [2% sodium dodecyl sulfate (SDS), 10% glycerol, 200 m*M* DTT, 62.5 m*M* Tris-HCl, pH 6.8, 0.001% bromphenol blue], heated to 100° for 2 min, and analyzed by electrophoresis on a 12% polyacrylamide gel. The proteins are electrotransferred onto Immobilon P (Millipore, Bedford, MA). The membrane is blocked by incubating at room temperature for 30 min in PBS containing 5% newborn calf serum (PBS-CS) and then incubated

[24] A. Deisseroth, J. Barker, W. F. Anderson, and A. Nienhuis, *Proc. Natl. Acad. Sci. U.S.A.* **72,** 2682 (1975).

at room temperature for 1 hr in PBS-CS containing a 1:1000 dilution of anti-M2 rabbit serum. The antiserum is raised against a 16-amino-acid C-terminal peptide (SAVDVDDGHFVNIELE) of M2 coupled with glutaraldehyde to keyhole limpet hemocyanin (KLH). The membrane is washed three times (10 min) with PBS-0.2% Tween 20 and incubated at room temperature for 1 hr in PBS-CS containing a 1:2000 dilution of protein A–horseradish peroxidase conjugate. After washing three times with PBS the blot is developed in enhanced chemiluminescence (ECL) reagent (Amersham).

For dot blot analysis of M2 expression in large numbers of individual clones, 10 $\mu$l of NP-40 lysates is spotted onto Immobilon P membrane and the blot developed as described earlier.

## Whole-Cell Patch-Clamp Procedure

Whole-cell patch-clamping[25] of small cells such as MEL cells provides the electrical advantages of good voltage clamp, precise determination of membrane potential and good time resolution, and the ability to control the internal ionic composition in the steady state by diffusion from the patch pipette. This is important when measuring proton flux because it permits use of high intracellular concentrations of buffer to minimize pH changes resulting from $H^+$ accumulation or depletion adjacent to the membrane. $H^+$ accumulation would change the equilibrium potential for $H^+$, producing an error in the permeability estimates, and would alter the driving potential for $H^+$ current, producing an error in the conductance.

Standard whole-cell patch-clamp methods are readily applied. MEL cells generally produce stable high-resistance seals, >10 G$\Omega$, with borosilicate glass and applying minimal negative pressure. Whole-cell recordings showed capacitance in the range of 7–21 pF in transfected MEL cells. Generally, membrane currents were in the range up to 100 pA when determining current/voltage relations and relatively high series resistance, up to 20 M$\Omega$ encountered in low conductance solutions, can be tolerated.

### *Composition of External and Internal Solutions*

To optimize pH control high concentrations of pH buffer are used as impermeant ions. The patch pipette contains 90 m*M* *N*-methyl-D-glucamine (NMDG), 10 m*M* EGTA, and either 180 m*M* *N*-(2-hydroxyethyl)piperazine-*N*′-(2-ethanesulfonic acid) (HEPES), pH 7.0–8.0, or 180 m*M* 2-(*N*-

[25] O. P. Hamill, A. Marty, E. Neher, B. Sakmann, and F. J. Sigworth, *Pflugers Arch.* **391,** 85 (1981).

morpholino)ethanesulfonic acid (MES), pH 4.5–6.5. The bath contains a similar solution (280 mOsm) with 2 m$M$ $CaCl_2$ replacing EGTA. In ion selectivity experiments NaCl and other salts are substituted isosmotically for NMDG–HEPES. The presence of $CaCl_2$ facilitates seal formation; experiments performed using $CaCl_2$-free extracellular solutions show a similar pH dependence. Liquid junction potentials of up to 6 mV are present between different solutions. Potentials are corrected precisely because of the effect of small systematic errors in reversal potential on calculation of relative ion permeabilities.

*Perfusion*

Exposure of MEL cells to acid pH (<5) for longer than about 10 sec when testing M2 results in rapid deterioration manifested as an increased leak current. For this reason, and to gain kinetic information concerning activation, a fast perfusion system is used that restricts exposures to 1 sec or less. The method is essentially the U-tube method[26] in which the bath is continuously perfused slowly by a pump, and solution is applied rapidly to the cell via a 100-$\mu$m orifice in the tip of the U by suddenly stopping flow along the U-tube with a solenoid. This method produces nondisruptive solution changes in 50–100 ms and permits brief application of low pH.

A high concentration of external and internal pH buffer—HEPES or MES—is used in these experiments to optimize the rate of pH change at the cell surface and minimize internal pH changes resulting from proton flux across the plasma membrane. Figure 2 shows the time course of the increase of current in an M2 expressing MEL cell on changing from pH 7.3 to pH 4.7 with external MES concentrations of 5–50 m$M$. The speed of the pH change at the membrane is optimal at MES concentrations greater than 30 m$M$. Current through M2 is steady for several seconds after the initial change (see Fig. 4A later) indicating that the driving potential for current through M2 is not changing and therefore that there is no change of internal pH at the internal buffer concentration of 180 m$M$ used here.

*Low Background Permeability of MEL Cells*

The current/voltage ($I/V$) relation from untransfected cells and from the MEL cell clone L16 induced to express a nonchannel membrane protein, human CD4 (kindly provided by Dr. R. Daniels, National Institute of Medical Research, London, UK), is linear over the range of membrane potential from −100 mV to +80 mV, intersecting the voltage axis at 0 mV, and is unaffected by external pH ($pH_o$) tested between 9.0 and 4.3, with

[26] O. A. Krishtal and V. I. Pidoplichko, *Neuroscience* **5,** 2325 (1980).

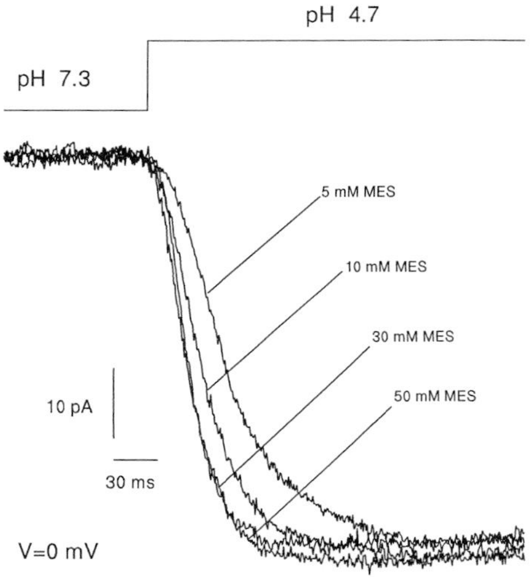


FIG. 2. Time course of current in M2 expressed in MEL cells following rapid perfusion of low pH solution of different buffering capacities. Whole-cell patch-clamp recording; $pH_o$ change from 7.3 to 4.7 buffered with MES at 5, 10, 30, or 50 m*M*. Perfusion with U-tube method.

internal pH ($pH_i$) maintained at either 7.3 or 6.0. Figure 3 shows data from a CD4 expressing cell with pH steps from pH 7.4 to pH 5.2 and an internal pH 7.4. Current/voltage data are shown for the initial current before the step ($pH_o$ 7.4, $pH_i$ = 7.4) and at the end of the step to pH 5.2. These properties are similar to those shown for rimantadine blocked cells (see Figs. 4B and 4C). This background "leak" conductance through the membrane and membrane–pipette seal has a mean of 83 pS (= 12 GΩ; range 50–200 pS) and is insensitive to the M2 inhibitor rimantadine (50 $\mu M$). The low background of endogenous channels in MEL cells has been noted in other channel expression studies.[20]

*Cell–cell Variation in M2 Expression*

The amplitude of current depends on the size of the cell and the level of M2 expression. The density of M2 current for unit membrane area was measured as pA/pF for many cells under similar conditions ($pH_i$ 7.2, $pH_o$ 4.5, −60 mV, 4 days postinduction) and showed large cell-to-cell variation from 1 to 17 pA/pF (mean 6.4 ± 5.6 SD, $n$ = 13) within the same clone. This suggests a large difference in the density of M2 expressed in the plasma membrane from cell to cell and may reflect asynchrony in induction and variation in the efficiency of the expression. Extended maintenance of cells in medium containing 400 $\mu$g/ml G418 was accompanied by a gradual

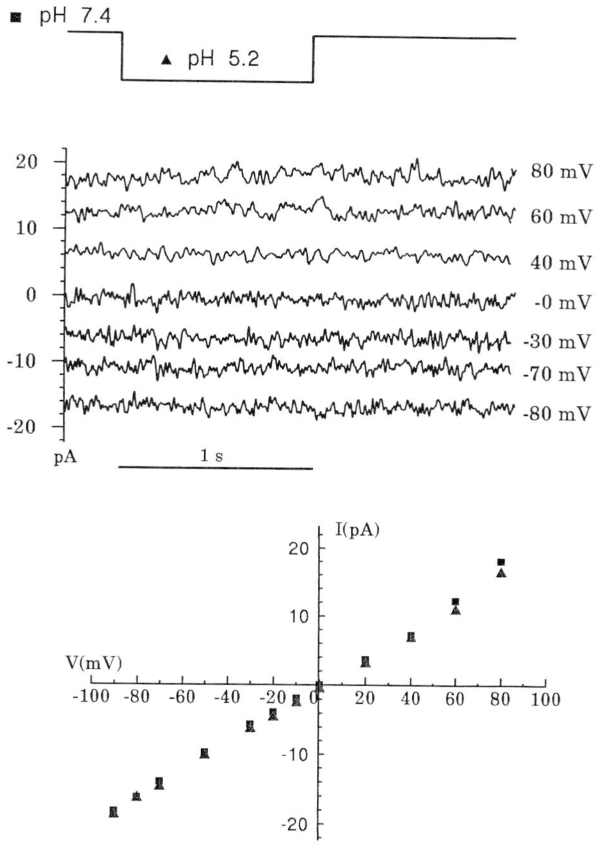


FIG. 3. pH-insensitive background leak conductance of a MEL cell expressing the nonchannel membrane protein CD4. MEL whole-cell patch clamp with internal and external solutions initially pH 7.4 with NMDG–HEPES. $pH_o$ changed to 5.2 externally by rapid perfusion for 1 sec (indicated above traces) with NMDG–MES solution repeated at different holding potentials. Data records are shown in upper panel, current/voltage curves in lower panel for initial current pH 7.4 (solid squares) and at the end of the pulse to 5.2 (solid triangles). No difference in conductance is seen and the reversal potential remained at 0 mV at the driving potential of 122 mV due to the 7.4–5.2 pH gradient.

reduction in the level and homogeneity of M2 expression within the population.

*Current/Voltage Relation for M2*

M2 current was determined by subtracting rimantadine-resistant current from total current at each potential and external pH. The residual rimanta-

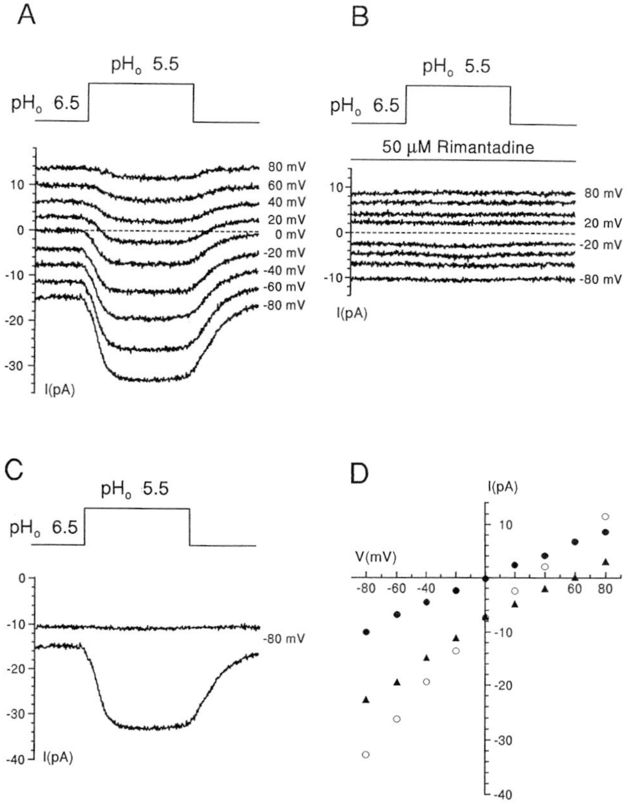


FIG. 4. Current/voltage analysis of rimantadine-sensitive M2 current. (A) Superimposed data records of membrane current evoked by external pH change from 6.5 to 5.5 at membrane potentials from −80 mV to +80 mV. (B) Current traces with pH change from 6.5 to 5.5 in the same cell in the period 5–10 min after applying 50 $\mu M$ rimantadine. (C) Current traces with pH change 6.5 to 5.5 at −80 mV just prior to (lower trace) and 5 min after applying 50 $\mu M$ rimantadine. (D) *I/V* plots: open circles, total membrane current at pH 5.5 measured 0.5 sec after solution change, $E_{rev} = +32$ mV; filled circles, rimantadine-resistance current, $E_{rev} = 0$ mV; filled triangles, difference current due to M2, $E_{rev} = +60$ mV. $E_H = +58$ mV for internal pH 6.5. [Data from I.V. Chizhmakov *et al.*, *J. Physiol.* **494,** 329 (1996) with permission.]

dine-resistant current gives a measure of the background leak and was used to define the current due to M2 by subtraction from the total current, as illustrated by Fig. 4 for changes from pH 6.5 to pH 5.5, with internal pH 6.5. Figures 4A and 4B show current traces at different membrane potentials before and after inhibition of M2 by rimantadine, respectively. The rimantadine-resistant current, like the current in untransfected cells, was insensitive to changes in $pH_o$ (Fig. 4B) and was linear with respect to membrane potential (Fig. 4D). Figure 4C shows a direct comparison of the current, at −80 mV, at pH 6.5 before and 5 min after addition of 50 $\mu M$ rimantadine; the rimantadine-sensitive current under these conditions accounted for approximately 32% of the initial current of −15pA. Figure 4D shows $I/V$ plots for total current (open circles), rimantadine-resistant leak (closed circles), and the net rimantadine-sensitive current through M2 (triangles) obtained by subtraction at $pH_i$ 6.5, $pH_o$ 5.5. The specificity of inhibition of M2 by rimantadine was confirmed by similar experiments in cells expressing a rimantadine-resistant mutant M2 protein with glutamic acid in place of glycine-34,[12] which showed no inhibition with 50 $\mu M$ rimantadine.

## Ion Selectivity: Estimation of Relative Permeabilities from Reversal Potentials

The pH-evoked currents in conditions where $H^+$ ions (or $OH^-$) are the only small permeant ions, described above, showed that $H^+$ ions are permeant in M2 but do not indicate the permeability relative to other physiologic ions. This can be estimated from the reversal (zero current) potential for bi-ionic conditions by assuming independent permeation and using the Goldman–Hodgkin–Katz equation,[27] e.g., for $H^+$ and $Na^+$:

$$E_{rev} = (RT/zF)\ln(P[H^+]_o + [Na^+]_o)/(P[H^+]_i + [Na^+]_i)$$

where $RT/zF$ has its usual meaning and is 25 mV at 20°, and $P$ is the permeability ratio of $H^+$ to $Na^+$ defined by this relation. Reversal potentials ($E_{rev}$) were estimated by linear interpolation of least squares fits to two or three data points of the $I/V$ relation on each side of zero current. The zero current intercept and its statistical error were calculated from the best-fit parameters to get some measure of the precision of the estimate. Systematic errors such as changes in liquid junction potential were carefully corrected. Zero current potentials, $E_{rev}$, determined from $I/V$ plots for M2 current were compared with the $H^+$ equilibrium potentials, $E_H$, calculated for the pH gradient with the Nernst equation. These are indicated in the $I/V$ plots for M2 current shown in Figs. 4D and 5A for different pH gradients in NMDG/HEPES/MES solutions. In each case $E_{rev}$ is close to $E_H$. Data are summarized in Fig. 5C, which shows that the reversal potentials (solid circles) determined in NMDG–HEPES/MES solutions were close to the equilibrium potentials, indicated by the continuous line, calculated for transmembrane $H^+$ concentration gradients (bottom abscissa) in the range 25-

[27] A. L. Hodgkin and B. Katz, *J. Physiol.* **108,** (1949).

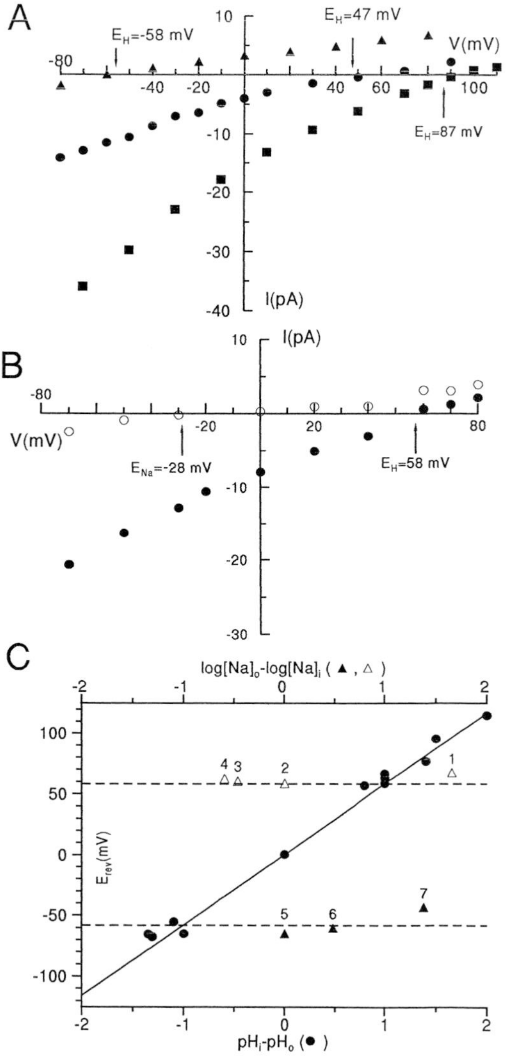


FIG. 5. Reversal potentials of rimantadine-sensitive M2 current. (A) *I/V* plots of current evoked by $pH_o$ change from 7.3 (=$pH_i$) to 6.5 (circles, $E_H$ = 47 mV, $E_{rev}$ = 56 mV) or 5.8 (squares, $E_H$ = 87 mV, $E_{rev}$ = 93 mV) in the same cell; and rimantadine-sensitive current on going from pH 6.0 to 7.0 in a different cell (triangles, $E_H$ = −58 mV, $E_{rev}$ = −55 mV). (B) *I/V* plots of rimantadine-sensitive current on going from pH 7.0 internal and external (open circles, $E_H$ = 0 mV) to pH 6.0 external (filled circles, $E_H$ = 58 mV). External NaCl 10 m*M*, internal NaCl 30 m*M*, $E_{Na}$ = −28 mV. (C) Reversal potentials plotted against transmembrane pH gradient (filled circles, lower abscissa) and log $Na^+$ concentration gradient (triangles,

fold outward to 100-fold inward. Linear regression of $E_{rev}$ on pH gradient gave a slope of 56.7 ± 1.7 mV (SD) compared to the theoretical of 58.2 mV for $E_H$ at 20°. The standard deviation (SD) of the data points around the best-fit line, ±6.8 mV, indicates the overall error in estimating $E_{rev}$. The equilibrium potentials calculated for components of the solutions other than $H^+$ (or equivalently $OH^-$)—NMDG, HEPES, MES, $Ca^{2+}$, and $Cl^-$—could not account for the $E_{rev}$ measured.

To test the ion selectivity of M2, other inorganic ions were included in internal and external solutions. Figure 5B shows that when $pH_o$ was changed from $pH_o$ 7.0 to $pH_o$ 6.0 with $pH_i$ of 7.0 in the presence of an outward $Na^+$ gradient ($[Na^+]_i = 30$ m*M*, $[Na^+]_o = 10$ m*M*, $E_{Na} = -28$ mV), $E_{rev}$ changed from 0 mV to +61 mV, close to $E_H$ of +58 mV, in the direction away from $E_{Na} = -28$ mV. This result is contrary to the conclusion reported elsewhere[4] that $Na^+$ permeability in M2 increases on lowering external pH. Figure 5C shows the zero current potentials determined at internal and external concentrations of NaCl up to 120 m*M* but with $E_H$ set at +58 mV (open triangles) or −58 mV (filled triangles), plotted against the logarithm of the $Na^+$ concentration gradient (top abscissa). The values did not change with the equilibrium potentials for either the $Na^+$ gradient (indicated by the continuous line as for pH) or the $Cl^-$ gradient. However, they did correspond with $E_H$ for the pH gradient, either +58 mV (open triangles) or −58 mV (filled triangles) indicated by the dotted lines. Linear regression of $E_{rev}$ on pH gradient in the presence of NaCl gave a slope of 59.2 ± 2.9 mV (SD), which compares with the theoretical value of 58.2 mV for $E_H$ at 20°, and is not significantly different from the value of 56.7 ± 1.7 mV obtained in the absence of NaCl. There is no significant deviation from the slope expected for a $H^+$ (or $OH^-$) selective membrane. Assuming the deviation is due to $Na^+$ permeability the Goldman–Hodgkin–Katz equation was used to estimate the permeability to $Na^+$ relative to $H^+$ from the maximum difference observed between $E_{rev}$ and $E_H$, 14.4 mV with 5 m*M* NaCl, pH 6.0, inside and 120 m*M* NaCl, pH 7.0, outside the cell, giving a permeability ratio of $Na^+$ to $H^+$ of $6 \times 10^{-7}$ in the worst case.

---

upper abscissa). Filled triangles, data at $E_H = -58$ mV; open triangles, data at $E_H = +58$ mV. Internal/external $[Na^+]$ (m*M*) given by numbers adjacent to points were (1) 2/90, (2,5) 5/5, (3) 30/10, (4) 40/10, (6) 10/30, and (7) 5/120. Solid line has Nernst slope 58.2 mV/decade; dotted lines indicate +58 mV and −58 mV. Least squares regression of $E_{rev}$ on $pH_i - pH_o$ has slope of 56.7 ± 1.7 mV for Na-free and 59.2 ± 2.9 mV for Na-containing solutions. Overall error SD for the estimation of $E_{rev}$ was ± 6.8 mV. [From I.V. Chizhmakov, *et al. J. Physiol.* **494,** 329 (1996) with permission.]

*Permeability Estimates without Leak Subtraction*

The precise estimation of M2 reversal potentials is facilitated by the ability to subtract the rimantadine-resistant leak. The error that might arise in data from an MEL expression system if this were not possible, e.g., for ionic conductances without a blocker, can be estimated from the data shown in Figs. 4D and 5C before leak subtraction. Figure 4D shows that the deviation from the proton equilibrium potential for $pH_i$ 6.5, $pH_o$ 5.5 was approximately −30 mV before leak subtraction in this example. Figure 6 shows the deviation due to the leak for the data of Fig. 5C before correction.

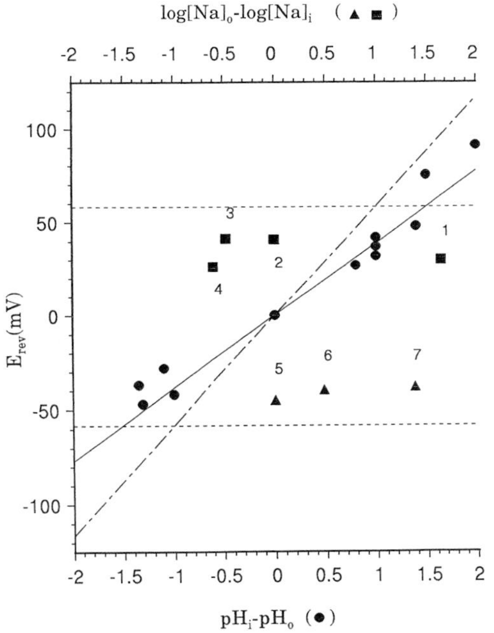


FIG. 6. Reversal potentials in M2, determined without leak subtraction, plotted against $H^+$ and $Na^+$ gradients. Same experiments as Fig. 4, reversal potentials measured without subtracting rimantadine-resistant leak current plotted against transmembrane pH gradient (filled circles, lower abscissa) and log $Na^+$ concentration gradient (triangles and squares, upper abscissa). Filled triangles, data at $E_H = -58$ mV; filled squares, data at $E_H = +58$ mV. Internal/external $[Na^+]$ (m*M*) given by numbers adjacent to points were (1) 2/90, (2) 5/5, (3) 30/10, (4) 40/10, (5) 5/5, (6) 10/30, and (7) 5/120. Dash-dot line, Nernst slope-58.2 mV/decade; dotted lines, +58 mV and −58 mV. Least squares regression of $E_{rev}$ on $pH_i - pH_o$ (solid line) has slope 35 mV/decade.

The filled circles are the reversal potentials before subtraction plotted against the pH gradient, giving a slope of 35 mV compared with the theoretical slope for a $H^+$ selective membrane of 58 mV, shown by the dot–dash line. More importantly, the plot of reversal potentials against the log $Na^+$ gradient shows no dependence of reversal potential on $Na^+$ gradient, the potentials lying close to the proton equilibrium potential (indicated by the lines at $\pm$58 mV) rather than $Na^+$ potential (shown by the dot–dash line). Without leak subtraction the ratio $P_H$ *to* $P_{Na}$ was in the worst case $1.25 \times 10^4$. Thus, an estimate of relative permeabilities can be obtained without a specific inhibitor, facilitated by the pH insensitivity and linear potential dependence of the leak, shown for MEL cells to be unaffected by a step to low $pH_o$ and to be linear with membrane potential (Figs. 3 and 4B).

### *Estimation of Single-Channel Parameters*

The resolutions of whole-cell and membrane patch are similar to those in other small cells, approximately 2 and 0.5 pA, respectively, at 1 kHz. Experiments with M2 showed no single-channel transitions at this resolution, indicating a very small single-channel current, or possibly an open probability close to 1, variation in cell current amplitude occurring as a change in open current amplitude with no resolved gating transitions. Similar results were found when the variances of whole-cell currents were analyzed and compared to those with rimantadine block. No variance increase was usually seen, indicating a unitary current of less than 10 fA in whole-cell conditions. With mean currents >100 pA under these conditions this indicates a large number of channels contributing to the current. These results differ from those obtained in bilayer recordings, which showed a high single-channel conductance nonselective among monovalent cations. Variance and spectral analysis can be done with methods described, e.g., in Refs. 28 and 29.

### *Combined Microfluorimetry and Whole-Cell Patch Clamp*

The relative permeabilities of two or more permeant ions that interact, e.g., by blocking or activating channels, cannot be calculated from reversal potentials with the GHK equation because the condition of independence

[28] P. T. A. Gray, *in* "Microelectrode Techniques for Cell Physiology" (D. Ogden, ed.), Chap. 8, 2nd Ed. Company of Biologists, Cambridge, United Kingdom, 1994.

[29] J. Dempster, "Computer Analysis of Electrophysiological Signals." Academic Press, London, 1993.

does not hold. M2 transfected MEL cells have been used in fluorimetric measurements of $H^+$ permeability with pH-sensitive indicators loaded in the cell via the AM-ester method.[14] This approach can be extended in single-cell experiments to investigate permeation of $H^+$ and a coion by using the change of fluorescence of a pH indicator. $H^+$ flux is estimated from the rate of change of the concentration of protonated/nonprotonated fluorophores, and the whole-cell current recorded to give the total ionic flux. Calculation predicts a 1 m*M* change of protonated fluorophore concentration in 1 sec with proton flux corresponding to 100 pA commonly seen with M2 activation in MEL cells. This method permits separation of the proton flux from other ion fluxes with procedures applied previously to $Ca^{2+}$ and $Na^+$ permeation in glutamate-gated ion channels.[30]

## Concluding Remarks

The LCR/MEL mammalian cell expression system has proved successful for the expression of functional M2 protein at levels comparable to those in virus-infected cells and are suitable for the detailed study of ion conductance. Cell toxicity due to the activity of certain mutant M2 proteins has been circumvented by selection of transfected clones in the presence of rimantadine. Without the availability of a suitable reversible inhibitor, toxicity due to the ion conductance may limit the usefulness of stable transfection systems such as that used here.

The low intrinsic conductance of MEL cells and the ability to identify unambiguously amantadine-sensitive current allowed discrimination of the small M2 proton conductance. Precise control of the ion composition of both external and internal environments is crucial for the precise definition of ion selectivity and differential effects of ions applied outside and inside the cell. Experiments in *Xenopus* oocytes which do not afford intracellular accessibility proved less conclusive in defining the selectivity of M2 or in demonstrating intrinsic ion conductance of the NB protein of influenza B virus.[31] Similar experiments in MEL cells expressing NB have shown clear proton and $Cl^-$ selective conductances even in the absence of a specific inhibitor.[32]

[30] R. Schneggenberger, Z. Zhou, A. Konnerth, and E. Neher, *Neuron* **11,** 133 (1993).

[31] K. Shimbo, D. L. Brassard, R. A. Lamb, and L. H. Pinto, *Biophys. J.* **69,** 1819 (1995).

[32] I. Chizhmakov, D. C. Ogden, T. Betakova, and A. J. Hay, *Biophys. J.* **74,** A319 (1998).

# [27] Ion Channels in Microbes

*By* YOSHIRO SAIMI, STEPHEN H. LOUKIN, XIN-LIANG ZHOU, BORIS MARTINAC, and CHING KUNG

## Introduction

Small unicells dominate our planet in biochemical variety, ecological breadth, certainly in number, and arguably even in mass. Multicellular animals and plants that are visible to the naked eyes form only a small part of the biodiversity[1] (Fig. 1). Classification and general biology of microorganisms are vast subjects beyond the scope of this review, but can be found in comprehensive texts such as Brock *et al.*[2] and Margulis *et al.*[3]

Although ion channel studies began with and continue to center on animals, a comprehensive study of ion channels must include microbes. Gated conductances, channel genes, proteins, and even crystal structures have been amply documented in microbes. Traditionally, only a few large microbes such as *Paramecium* and *Chara* were amenable to direct electrophysiologic examination. However, the advent of patch-clamp techniques[4] opened the field of ion channel studies in smaller cells including small animal, plant, or microbial cells. The major part of this chapter reviews electrophysiologic methods as applied to representative microbes. A classical and well-established technique is to use planar lipid bilayers (bilayer lipid membranes, BLMs) to study reconstituted microbial materials. Although not thoroughly reviewed, some references to the application of this technique are also provided here. We also list some references to channels found in microbial organelles such as mitochondria and vacuoles. The last two sections deal with unique opportunities for the study of structure–function relationships of ion channels in microbes.

Several foreign channel genes have been expressed and analyzed in such microbial workhorses as *Escherichia coli* and the budding yeast *Saccharomyces cerevisiae*. Furthermore, these microbes allow mutant selection, based on agar-plate phenotypes after nondirected, random mutagenesis of channel genes which can be expressed in them (Fig. 2). The strategies and

[1] N. R. Pace, *Science* **276,** 734 (1997).

[2] T. D. Brock, M. T. Madigan, J. M. Martinko, and J. Parker, "Biology of Microorganisms," 7th Ed. Prentice Hall, Englewood Cliffs, New Jersey, 1994.

[3] L. Margulis, J. O. Corliss, M. Melkonian, D. J. Chapman, "Handbook of Protoctista." Jones and Barlett Publishers, Boston, 1990.

[4] O. P. Hamill, A. Marty, E. Neher, B. Sakmann, and F. J. Sigworth, *Pflugers Arch.* **391,** 85 (1981).



0076-6879/99 $30.00

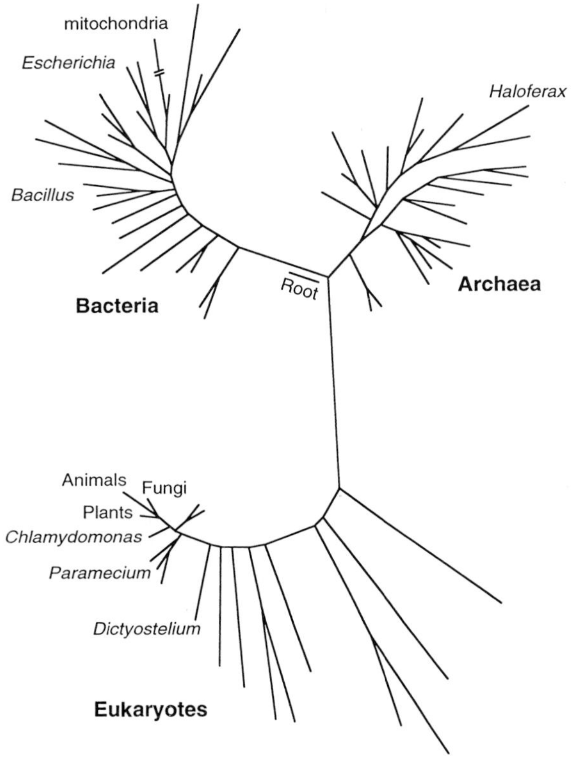


FIG. 1. Universal phylogenetic tree of known organisms based on small subunit rRNA sequences. Note the division into three domains: bacteria, archaea, and eukaryotes and the limited diversity of plants and animals in this scale. See Ref. 1 for details. [Modified from N. R. Pace, *Science,* **276,** 734 (1997) with permission.]

methods of such studies are also reviewed here. A full treatment of molecular biological methods on *E. coli* can be found in Gerhardt *et al.*[5] and one on yeast in Guthrie and Fink.[6]

## Methods for Ciliates: *Paramecium*

Ciliates are large unicells. Various species of *Paramecium* range from 100 to 300 $\mu$m in length and were therefore the first microbes studied using classical recording methods. Classical and modern methods are given here

[5] P. Gerhardt, R. G. E. Murray, W. A. Wood, and N. R. Krieg, "Methods for General and Molecular Bacteriology." American Society for Microbiology, Washington, D.C., 1994.

[6] C. Guthrie and G. R. Fink, "Guide to Yeast Genetics and Molecular Biology." Harcourt Brace Jovanovich, San Diego, California, 1991.

for *Paramecium* as a representative of ciliated protists. Paramecia can easily be cultured in dilute vegetable infusions where they feed on undefined bacterial flora. Monoxenic media with single food species, e.g., *Aerobacter aerogenes,* are available, as are axenic media with chemical nutrients but no food organisms.[7] For electrical recordings, a single paramecium is first transferred into an "adaptation solution," then penetrated with electrodes and examined while bathed in an "experimental solution." Commonly used are freshwater species (*P. caudatum, P. tetraurelia*), so the above solutions are of low ionic strength, and can both be as simple as 4 m*M* KCl, 1 m*M* $CaCl_2$, 1 m*M* Tris (or HEPES), pH 7.2. Millimolar NaCl, $MgCl_2$, etc., can be added to the experimental solution for various investigations.

A paramecium cell is completely covered with some 5000 cilia and is nearly in constant motion. Capturing, immobilizing, and recording are all performed on a microscope stage. A common method of capturing entails withdrawing solution in a drop until the paramecium in the drop is held still by surface tension. The drop hangs from a coverslip stationed above a perfusible bath. The tip of a penetrating microelectrode or an empty needle tethers the cell before solution is added back and the drop fuses with the bath.[8] Additional electrodes can then be placed in the cell. As an alternative to the hanging drop method under an upright microscope, one can use an inverted microscope where the drop with the cell is facing upward.[9] Though the basic protocol is the same as above, reduced submerged surface area of the electrodes in the latter method makes voltage clamp more efficient. The electrodes can be coated with polystyrene film to further reduce capacitance. Resting, receptor, or action potentials are usually recorded with one or two KCl electrodes of high resistance. Voltage-clamp experiments require electrodes of lower resistance, about 10 MΩ.[9,10] CsCl-filled electrodes have also been used to inhibit most of the $K^+$ current.[9] Variations on these methods have been successfully applied to other ciliates such as a marine *Paramecium,*[11] *Didinium,*[12] *Stylonychia,*[13,14] and *Euplotes.*[15,16]

The *Paramecium* cell is too large and its surface too complex for direct

[7] H.-D. Gortz, "Paramecium." Springer-Verlag, New York, 1988.

[8] Y. Naitoh and R. Eckert, *in* "Experiments in Physiology and Biochemistry" (G. A. Kierkut, ed.), Vol. 5, p. 17. Academic Press, London, 1972.

[9] Y. Saimi and C. Kung, *Science* **218,** 153 (1982).

[10] D. Oertel, S. J. Schein, and C. Kung, *Nature* **268,** 120 (1977).

[11] J. Bernal and B. E. Ehrlich, *J. Exptl. Biol.* **176,** 117 (1993).

[12] J. Pernberg and H. Machemer, *J. Exptl. Biol.* **198,** 2537 (1995).

[13] J. W. Deitmer, *J. Physiol.* **355,** 137 (1984).

[14] J. W. Deitmer, *J. Physiol.* **380,** 551 (1986).

[15] T. Kruppel, *J. Membr. Biol.* **133,** 263 (1993).

[16] T. Kruppel and F. Wissing, *Cell Calcium* **19,** 229 (1996).

patch-clamp experimentation. However, blisters can be induced and the small vesicles detached from the cell for patch-clamp purposes. Cells in late logarithmic to early stationary phase of growth are preferred. Several such cells can be transferred to a chamber where blisters are induced in 100–150 m$M$ sodium glutamate or NaCl, $10^{-5}$ $M$ $Ca^{2+}$, 5 m$M$ HEPES, pH 7.0. The 5-min blistering process can be monitored under a phase-contrast microscope. Blisters adhering to the cell surface can be dislodged by passing the *Paramecium* cell in and out of a glass suction pipette of a narrow bore. Detached vesicles, 5–20 $\mu$m in diameter, are then picked up with a suction pipette and laid down on the bottom of a separate chamber containing the seal-forming solution (100 m$M$ $K^+$ or $Na^+$, 10 m$M$ $MgCl_2$, 5 m$M$ HEPES, pH 7.0, with $10^{-8}$ to $10^{-4}$ $M$ $Ca^{2+}$ depending on the experiments). The blisters are double layered. Depending on the osmolarity difference between the blistering and the gigaohm seal-forming solution, the blisters may stay round or start to peel to give rise to a vesicle surrounded only by the inner, probably the alveolar, membrane capped with a hemispherical outer, plasma membrane. Gigaohm seals can be formed on either membrane. The inner membrane appears electrically silent. The outer membrane contains a variety of channels.[17,18] After gigaohm seal formation, the membrane patches can be excised by air exposure or shaking the vesicle off by tapping the pipette holder. The excised membrane can be closed into small vesicles if the free $Ca^{2+}$ level in the seal forming solution is too high. Fire-polished, coated pipettes of Boralex glass are favored. In eucaryotes, the membrane that covers the cilia are continuous with the body membrane. Forcing live paramecia through a narrow-bore pipette can shear off cilia. Each detached cilium retains its microtubular axoneme inside, and is osmotically active indicating that its membrane apparently reseals. These ciliary vesicles can be transferred to the recording chamber, form gigaseal with recording pipettes, and show activities of several channel types.[17,18]

Ciliary membrane vesicles from *Paramecium,* freshly prepared or previously frozen, have also been incorporated into planar lipid bilayer showing two types of channel activities.[19] Similarly, ciliary membrane vesicles of *Tetrahymena* have also been successfully incorporated into planar lipid bilayer.[20–22] Several conductances have also been observed, after reconsti-

[17] Y. Saimi and B. Martinac, *J. Membr. Biol.* **112,** 79 (1989).

[18] Y. Saimi, B. Martinac, A. H. Delcour, P. V. Minorsky, M. C. Gustin, M. R. Culbertson, J. Adler, and C. Kung, *Methods Enzymol.* **207,** 681 (1992).

[19] B. E. Ehrlich, A. Finkelstein, M. Forte, and C. Kung, *Science* **225,** 427 (1984).

[20] Y. Oosawa and M. Sokabe, *Am. J. Physiol.* **249,** C177 (1985).

[21] Y. Oosawa, M. Sokabe, and M. Kasai, *Cell Struct. Funct.* **13,** 51 (1988).

[22] Fujiwara-Hirashima, K. Anzai, M. Takahashi, and Y. Kirino, *Biochim. Biophys. Acta* **1280,** 207 (1996).

tuting *Paramecium* cortex protein with azolectin and examining the liposomes with patch clamp.[23] For this work, cells harvested from culture are washed with buffer and are then packed in 250 m*M* sucrose, 3 m*M* ethylenediaminetetraacetic acid (EDTA), 20 m*M* Tris-HCl, pH 7.8. Packed cells are then resuspended in the same solution containing 1 m*M* phenylmethylsulfonyl fluoride (PMSF) and 10 $\mu$g/ml of leupeptin to inhibit endogenous protease. Cells in this suspension are homogenized with a prechilled Potter homogenizer. The homogenate is then subjected to a series of centrifugations to collect the cell cortices.[23,24] This material can be stored at $-80^\circ$ and then thawed for the reconstitution without losing activities. To reconstitute, 75 $\mu$l stock azolectin was dried in a stream of nitrogen gas, and then dissolved in 10 m*M* Tris, 0.1 m*M* EDTA, pH 7.2, with either 2% (w/v) 3-[(3-cholamidopropyl)dimethylammonio]-1-propane sulfonate (CHAPS) or 2% (w/v) octylglucoside (OG). Thawed cortical preparation is added at a desired protein-to-lipid ratio and mixed thoroughly. The mixture is then dialyzed against 10 m*M* Tris, 100 m*M* NaCl, 0.2 m*M* EDTA, 0.02% $NaN_3$, pH 7.2, for 24 hr with one change of buffer. After dialysis, the mixture is spun at 50,000 rpm, 1 hr at 4° and the precipitate is resuspended in 10 m*M* MOPS, 5% ethylene glycol, pH 7.4. This material is then added on a microscope slide and subjected to dehydration overnight in a desiccator at 4°, followed by rehydration in 150 m*M* KCl, 0.1 m*M* EDTA, $10^{-5}$ *M* $Ca^{2+}$, 5 m*M* HEPES, pH 7.2, at 4° for 4 hr. The rehydrated material is then added to 100 m*M* KCl, 40 m*M* $MgCl_2$, $10^{-5}$ *M* $CaCl_2$, 5 m*M* HEPES, pH 7.2, for seal formation. The patch-clamp pipettes (Boralex, Rochester Sci., Rochester, NY) are filled with 100 m*M* KCl, $10^{-1}$ *M* $CaCl_2$, 5 m*M* HEPES, pH 7.2.

The voltage-dependent anion channel of mitochondria was first discovered on reconstituting *Paramecium* material into planar lipid bilayers. For this discovery and its methods see Schein *et al.*[25]

## Methods for Fungi

### *Budding Yeast*

The unicellular budding yeast *Saccharomyces cerevisiae* has been used extensively in genetics and molecular biological research. At least two types of conductances have been routinely observed in the plasma membrane of this yeast, an outwardly rectifying $K^+$ conductance[26] and a mechanosensitive

[23] X.-L. Zhou, C. W. M. Chan, Y. Saimi, and C. Kung, *J. Membr. Biol.* **144,** 199 (1995).
[24] N. Stelly, J. P. Mauger, M. Claret, and A. Adoutte, *J. Cell Biol.* **113,** 103 (1991).
[25] S. J. Schein, M. Columbini, and A. V. Finkelstein, *J. Membr. Biol.* **30,** 99 (1976).
[26] M. C. Gustin, B. Martinac, Y. Saimi, M. R. Culbertson, and C. Kung, *Science* **233,** 1195 (1986).

conductance.[27] The gene encoding the $K^+$ channel has been recognized from genome sequence,[28–31] expressed heterologously,[28,30–32] and genetically dissected.[32] Yeasts are mononucleated small unicells. A yeast cell is about 5 $\mu$m in diameter and has an external cell wall. Here a general method is described based on the experience of two laboratories.[29,33] For preparation of spheroplasts, haploid, diploid, or tetraploid yeast is first cultured in a liquid medium at 30° overnight. Either rich growth medium (e.g., YEPD)[6] or defined minimal medium (e.g., SD and SG)[6] has been used successfully depending on the experiments. It is imperative that the yeasts to be examined be proliferating in the medium. Only growing cells produce easily workable spheroplasts. It remains very difficult to generate useful spheroplasts from mutant strains that grow poorly even in rich medium. The culture does not have to be large. Because the wall-digesting enzymes (below) are costly, small cultures, ca. 1 ml, are recommended. A booster culture of 1 or 2 hr the next morning by a 1 to 20 dilution into fresh medium can be used, but is not necessary. Yeast cells are washed once with 50 m*M* $KH_2PO_4$, 50 m*M* 2-mercaptoethanol, adjusted to pH 7.2 with KOH. Washing is accomplished by centrifugation (2000 rpm, for 5 min in a table-top centrifuge). The pellet is resuspended and incubated in 1 ml of the same solution at 30° for 30 min with occasional shaking. To generate protoplasts, a 100-$\mu$l aliquot of cells is mixed with 0.9 ml of 1.6 *M* sorbitol, 50 m*M* $KH_2PO_4$, 50 m*M* 2-mercaptoethanol, pH 7.2. Sorbitol is added here to provide an osmotic support for the emerging spheroplasts, no longer protected by the cell wall. Additionally, this solution also contains the wall-digesting enzymes: 2 mg/ml zymolyase (100T, ICN, Costa Mesa, CA, or Kogyo Co., Seikagaku, Japan), 2 mg/ml glucuronidase (Sigma, St. Louis, MO) and 30 mg/ml bovine serum albumin (BSA). After about 45 min of incubation at 30°, the preparation is pelleted and resuspended in 250 m*M* KCl, 5 m*M* $MgCl_2$, 5 m*M* MES, adjusted to pH 7.2 with Tris base, in preparation for patch-clamp experimentation. The solution may also contain 10 m*M* $CaCl_2$ and 1% glucose.[33]

[27] M. C. Gustin, X.-L. Zhou, B. Martinac, and C. Kung, *Science* **242,** 762 (1988).

[28] K. A. Ketchum, W. J. Joiner, A. J. Sellers, L. K, Kaczmarek, and S. A. Goldstein, *Nature* **376,** 690 (1995).

[29] X.-L. Zhou, B. Vaillant, S. H. Loukin, C. Kung, and Y. Saimi, *FEBS Lett.* **373,** 170 (1995).

[30] F. Lesage, E. Guillemare, M. Fink, F. Duprat, M. Lazdunski, G. Romey, and J. Barhanin, *J. Biol. Chem.* **271,** 4183 (1996).

[31] J. D. Reid, W. Lukas, R. Shafaatian, A. Bertl, C. Scheurmann-Kettner, H. R. Guy, and R. A. North, *Recept. Channels* **4,** 51 (1996).

[32] S. H. Loukin, B. Vaillant, X.-L. Zhou, E. P. Spalding, C. Kung, and Y. Saimi, *EMBO J.* **16,** 4817 (1997).

[33] A. Bertl, J. D. Reid, H. Sentenac, and C. Slayman, *J. Expt. Botany* **48,** 405 (1997).

This period of enzyme digestion as well as the osmolarity of the digestion buffer should be adjusted depending on the growth conditions of the various strains in various cultures. Slow-growing cells require longer periods of digestion and lower osmolarity to encourage the emergence of spheroplasts. Such spheroplasts tend to be fragile, however. Spheroplasts are made fresh for each day's use. Frozen and thawed material from yeast has not been rendered useful in patch-clamp study.

Various recording configurations (on-cell, excised inside-out and outside-out, and whole-cell modes) can be and have been used. Seals are usually obtained on free-floating spheroplasts; a spheroplast is moved off the bottom by physical perturbations and then captured at the mouth of the pipette by suction. Seals higher than 10 gigohm in resistance can be formed but take longer time and appear to require more suction than with animal cells. A 3-ml syringe with a two-way valve connected by tubing to the port on the pipette holder is used to apply the strong and prolonged suction necessary.

A pressure transducer (PX143, Omega Engineering, Stamford, CT) in line is used for monitoring the applied pressure. At least 5–20 cm Hg appears to be necessary for seal formation. During maintained application of suction the resistance will sometimes abruptly decrease (probably owing to the pinching off of surface membrane blebs), only to start slowly increasing again on an application of a slightly stronger suction. Usually, gigohm seals will still develop with such spheroplasts. Recording at seal resistances of less than 10 $G\Omega$ is not recommended. Patch pipettes are made from Boralex glass, pulled to a "bubble number"[34] around 3.2 (8–10 $M\Omega$). The pipette solution varies with experiments. For whole-cell recording, this solution enters the cytoplasm and should be kept at a neutral pH, e.g., 180 m*M* KCl, 0.5 m*M* EGTA, 5 m*M* HEPES, adjust to pH 7.2 with KOH.[29] An additonal 5 m*M* $MgCl_2$ helps to form gigohm seals. Some investigators include 4 m*M* adenosine triphosphate (ATP) in the pipette solution.[35] For on-cell or excised inside-out patch mode, the pipette solution should have a low pH, e.g., 50 m*M* KCl, 0.1 m*M* $CaCl_2$, 250 m*M* sorbitol, pH 5.5–5.7, unbuffered.[33]

The membranes of organelles inside yeast cells have also been studied. The mitochondrial VDAC has been studied in great detail by mutations and by reconstitution into BLMs. For a review see Colombini *et al.*[36] and

[34] D. P. Corey and C. F. Stevens, *in* "Single-Channel Recording" (B. Sakmann and E. Neher, eds.), p. 53. Plenum Press, New York, 1983.

[35] A. Bertl, J. A. Anderson, C. L. Slayman, and R. F. Gaber, *Proc. Natl. Acad. Sci. U.S.A.* **92,** 2701 (1995).

[36] M. Colombini, E. Blachly-Dyson, and M. Forte, *in* "Ion Channels" (T. Narahashi, ed.), Vol. 4, p. 169. Plenum Press, New York, 1996.

references therein. Methods have also been developed to incorporate yeast vacuolar membranes into BLMs and channel activities observed by Wada *et al.*[37,38] Methods to study channel activities by patch clamping the mitochondrion and the vacuole extruded from each individual yeast cells can be found in Saimi *et al.*[18]

### *Fission Yeast*

The fission yeast *Schizosaccharomyces pombe* is vastly different from the budding yeast, but has also been developed as a model organism to understand eukaryotic processes.[39] Methods to generate surfaces suitable for patch-clamp experiments have also been developed for it. The procedure is not a global cell–wall digestion but to weaken the wall at one end, thereby allowing the protoplast to emerge as a protuberance at that end.[40] Fission yeast cells are first cultured overnight at 30° in 5 ml of YEPD medium (2% glucose, 2% peptone, 1% yeast extract). A second culture is started the following morning in 2 ml of YEPD inoculated with 0.3 ml of the overnight culture. After 4 hr of incubation at 30° and the culture reaching an $OD_{600}$ of ca. 0.3, cells are washed twice and then resuspended in double-distilled water to $OD_{600}$ of ca. 0.8. This suspension (375 $\mu$l) is then mixed with 275 $\mu$l of 0.8 *M* sorbitol and 100 $\mu$l of zymolyase stock (2 mg/ml in 0.8 *M* sorbitol of zymolyase, 60 U/mg) and then incubated at 30° for 30 min with occasional gentle shaking. The zymolyase-treated cells are then added to 3 ml of the following solution: 120 m*M* KCl, 50 m*M* $MgCl_2$, 0.1 m*M* EGTA, 5 m*M* HEPES, adjusted to pH 7.2 with KOH. They are then collected by centrifugation and resuspended in 0.5 ml of the same solution. After these treatments, blebs can be found at one pole of a small number of cells. These blebs appear to contain cytoplasm and still be physically adhered to the cells, in most instances. Gigaohm seal can be formed instantaneously by small suction with this membrane, unlike that of budding yeast. This may reflect the difference in how the plasma membranes of the two species are revealed during the spheroplasting procedures.

### *Bean Rust*

The rust fungus *Uromyces appendiculatus* infects bean plant by growing a germ tube from a spore and entering the interior of a leaf through a

[37] Y. Wada, Y. Ohsumi, M. Tanifugi, M. Kasai, and Y. Anraku, *J. Biol. Chem.* **25,** 17260 (1987).

[38] M. Tanifuji, M. Sato, Y. Wada, Y. Anraku, and M. Kasai, *J. Membr. Biol.* **106,** 47 (1988).

[39] A. Nasim, P. Young, and B. F. Johnson, "Molecular Biology of the Fission Yeast." Academic Press, New York, 1989.

[40] X.-L. Zhou and C. Kung, *EMBO J.* **11,** 2869 (1992).

stoma. The germ tube recognizes the minute surface geometry of the lips that surround the stoma and not by chemotaxis.[41] Patch-clamp experiments showed that there are mechanosensitive ion channels in the germ tube.[42] For these experiments, urediospores are germinated in water for some 2 hr at 16°–19° until the germ tubes are 40–70 $\mu$m long. The emergence of the germ tube from the spore is monitored under a microscope. Germ tubes are used before the entire cytoplasm moves out of the spore.

To generate protoplasts, germlings are digested for 25 to 40 min with 5 mg/ml Novozyme (Novo BioLabs, Bagsvaerd, Denmark) in 0.5 *M* sorbitol, 17 m*M* MES, pH 6.0. All protoplasts form from the growing ends of the germling, presumably the loci most susceptible to wall breakage. The protoplasts are filtered through a 20-$\mu$m mesh Nytex cloth (Tetko, Inc., Elmsford, NY), collected by centrifugation at 2000*g* for 4 min, and then resuspended in sorbitol without Novozyme. Protoplasts that appear dark and smooth when viewed with phase-contrast optics are best for patch-clamp work. Standard techniques are used having the pipettes filled with 290 m*M* KCl, $10^{-5}$ *M* $CaCl_2$, 5 m*M* HEPES, adjusted to pH 7.2, with KOH, and the bath with 220 m*M* KCl, 50 m*M* $MgCl_2$, 5 m*M* HEPES, pH 7.2, with KOH. Seals form more easily in this preparation than with protoplasts of budding yeast. The first configuration is invariably the on-cell patch mode. Ten to 30 mm Hg continued suction over minutes can convert the on-cell to the whole-cell mode.

### *Neurospora*

The ordinary cells of the mycelial bread mold *Neurospora* can be some 100 $\mu$m long but only 3–5 $\mu$m in diameter, and thus not suitable for electrophysiologic examination. Instead, spherical cells, 15–25 $\mu$m in diameter, are used. This fungus yields spherical cells when conidia are inoculated, germinated, and cultured in 18% (v/v) ethylene glycol together with 1% (w/v) glucose and Vogel's minimal salts.[21] These spherical cells can be harvested and washed over Millipore filters (3 $\mu$m, type SM, Millipore Corp, Bedford, MA) and resuspended in 20 m*M* dimethylglutaric acid and 1 m*M* $CaCl_2$, adjusted to pH 5.8 with KOH ($[K^+]$ = 25 m*M*) with ethylene glycol. After a progressive dilution of the ethylene glycol, cells are ready for electrode penetration.[43,44] Individual cells are selected and held for impalement with the aid of a suction pipette, which is withdrawn after penetration with KCl-filled electrodes. To keep the cytoplasmic composi-

[41] H. C. Hoch, R. C. Staples, B. Whitehead, J. Comeau, and E. D. Wolf, *Science* **235,** 1659 (1987).
[42] X.-L. Zhou, M. A. Stumpf, H. C. Hoch, and C. Kung, *Science* **253,** 1415 (1991).
[43] M. R. Blatt, A. Rodriguez-Navarro, and C. L. Slayman, *J. Membr. Biol.* **72,** 223 (1983).
[44] M. R. Blatt, A. Rodriguez-Navarro, and C. L. Slayman, *J. Membr. Biol.* **98,** 169 (1987).

tion near physiologic, 50 or 100 m*M* potassium acetate can be used to fill the pipettes.[45]

Patch-clamp recording has been accomplished on protoplasts from the growing hyphal tube of *Neurospora.*[46] To prepare protoplasts, apical fragments of hyphae attached to a dialysis strip are first rinsed with a solution containing 100 mOsm/kg sorbitol, 1 m*M* $MgCl_2$, 0.1 m*M* $CaCl_2$, 100 m*M* KCl, 2% sucrose, 10 m*M* 1,4-piperazinediethanesulfonic acid (PIPES) buffer, adjusted to pH 5.8 with KOH, and then transferred and incubated in the same solution containing 2 mg/ml each of Novozyme 234 (InterSpex Products, Inc., Foster City, CA) and $\beta$-glucuronidase (Sigma). After the emergence of protoplasts at the tips of the hyphal tubes, the osmolarity of the bath is decreased to extrude the protoplasts through pores along the partially digested cell walls. Micropipettes of "bubble number"[34] 4 are used without fire polish. They are filled with 150 m*M* KCl, 1 m*M* $CaCl_2$, 1 m*M* $MgCl_2$, 10 m*M* PIPES, pH 5.8. High suction (ca. 100 mm Hg) for tens of seconds is used to form seals. Usually, seals can be more easily formed and are more stable with protoplasts released from the hyphal tips within 2–3 min of incubation with the digestive enzymes.

## Methods for *Dictyostelium*

The slime mold *Dictyostelium* has been used extensively in the study of morphogenesis and differentiation and in ameboid motility. During its reproductive phase, nutrient deprivation causes the unicellular amebas to aggregate to form a multicellular structure. Patch-clamp experiments have been reported. In these studies, after induction of differentiation by nutrient removal, the amebic cells are examined in the on-cell mode.[47,48] Cells are induced by washing and shaking in Sorensen phosphate buffer, pH 6.0, at a cell density of $10^7$/ml at room temperature. In preparation for patch clamping, 20 $\mu$l of cells is diluted in 1 ml of 10 m*M* calcium chloride, acetate, or cyclamate, and 1 m*M* HEPES, pH 7.0. Pipettes drawn from hematocrit capillaries are used and are filled with the same solution. Activity of two channels passing outward $K^+$ currents and one passing inward $Ca^{2+}$ current have been observed.[47,48]

[45] A. Rodriguez-Navarro, M. R. Blatt, and C. L. Slayman, *J. Gen. Physiol.* **87,** 649 (1986).
[46] N. N. Levina, R. R. Lew, G. J. Hyde, and I. B. Heath, *J. Cell Sci.* **108,** 3405 (1995).
[47] U. Mueller, D. Malchow, and K. Hartung, *Biochim. Biophys. Acta.* **857,** 287 (1986).
[48] U. Mueller and K. Hartung, *Biochim. Biophys. Acta.* **1026,** 204 (1990).

## Methods for Algae and Flagellates

### *Characeae*

The internodal cells of *Chara, Nitellopis,* etc., can be some 1 mm in diameter and 10 cm long. As botanists' answer to giant squid axons, these cells have been studied extensively with the conventional methods employing microelectrodes as well as those using metal wires or petroleum jelly gaps.[49–52]

Patch-clamp studies have been made on-cell or with "cytoplasmic droplets." For on-cell study, internodal cells are dissected out of plants grown in artificial pond water and bathed in 1 m*M* $CaCl_2$, 1 or 3 m*M* KCl, 5 HEPES, pH 7.5, with 180 m*M* sorbitol. Plasmolysis is induced by increasing sorbitol to 360 m*M*. A small incision is made in a part of the cell wall from which the protoplast has withdrawn. A patch-clamp pipette is then advanced through the slit and forms a seal with the plasma membrane. The patch-clamp recording can be done while the *Chara* cell is separated by a petroleum jelly gap into two electrically isolated parts in two baths. Extracellular electrodes connecting the two baths can be used to monitor space-averaged membrane voltage or to send pulses to trigger action potentials. The activities of two types of $Cl^-$ conductance and a $K^+$ conductance have been observed during a triggered action potential.[53] Cytoplasmic droplets can be obtained by cutting an internodal cell and releasing its cytoplasm into 250 m*M* KCl and 5 m*M* $CaCl_2$. Membrane enclosed vesicles are formed naturally from this extruded cytoplasm.[54,55] The outer membrane of such vesicles is apparently of tonoplast origin and contains $K^+$ channels.[56,57]

### *Chlamydomonas*

The small green flagellate, *Chlamydomonas reinhardtii,* has been developed into a "green yeast" and is now rather amenable to various genetic and molecular biological manipulations.[58] Recently, in an effort to understand

[49] A. B. Hope and G. P. Findlay, *Plant Cell Physiol.* **5,** 377 (1964).
[50] M. J. Beilby and M. R. Blatt, *Plant Physiol.* **82,** 417 (1986).
[51] A. Bertl, *J. Membr. Biol.* **109,** 9 (1989).
[52] K. A. Fairley and N. A. Walker, *J. Membr. Biol.* **98,** 191 (1987).
[53] U. Homann and G. Thiel, *J. Membr. Biol.* **141,** 297 (1994).
[54] S. Draber, R. Schultze, and U.-P. Hansen, *J. Membr. Biol.* **123,** 183 (1991).
[55] S. Draber, R. Schultze, and U.-P. Hansen, *Biophys. J.* **65,** 1553 (1993).
[56] A. Bertl, C. L. Slayman, and D. Gradmann, *J. Membr. Biol.* **132,** 183 (1993).
[57] D. R. Laver, C. A. Cherry, and N. A. Walker, *J. Membr. Biol.* **155,** 263 (1997).
[58] U. W. Goodenough, *Cell* **70,** 533 (1992).

the electric events in response to light, local currents have been examined when different parts of the cell body are captured by suction electrodes.[59] Using a cell-wall deficient mutant, *cw2,* 100–200 MΩ seals can be formed between the suction pipette and the cell. By this means, membrane currents produced by the cell on pipette suctions have been detected.[60]

## Methods for Bacteria

### *Escherichia coli Outer Membrane*

*Escherichia coli* is the best studied and most understood organism. It is also most commonly used as a tool in genetic engineering. Commonly used methods in general and molecular bacteriology can be found in Gerhardt *et al.*[5]

The first type of channels studied in bacteria is the porins. These 16-stranded antiparallel $\beta$-sheet barrel pores of large conductances are found in the outer membrane of many gram-negative bacteria. To date, bacterial porins remain the only channel proteins whose structures are known in atomic dimensions, deduced from three-dimensional (3-D) X-ray crystallography.[61,62] Much work has been done using the planar bilayer technique (see Benz[63] for a review on the findings using this and other methods). Some work uses patch-clamp methods to study rapid kinetic events as well as voltage gating,[64,65] pressure gating,[66] and modulation.[67] In these studies outer-membrane vesicles of *E. coli* are reconstituted into liposomes[68–70] as modified from the method of Criado and Keller.[71] Detail of these methods can be found in Blount *et al.*[71a] in this volume.

[59] C. Nonnengasser, E.-M. Holland, H. Harz, and P. Hegemann, *Biophys. J.* **70,** 932 (1996).
[60] K. Yoshimura, *J. Exptl. Biol.* **199,** 295 (1996).
[61] M. S. Weiss, U. Abele, J. Weckesser, W. Welte, E. Schiltz, and G. E. Schultz, *Science* **254,** 1627 (1991).
[62] S. W. Cowan, T. Schirmer, G. Rummel, M. Steiert, R. Ghosh, R. A. Pauptit, J. N. Jansonius, and J. P. Rosenbusch, *Nature* **358,** 727 (1992).
[63] R. Benz, *in* "Bacterial Cell Wall" (J.-M. Ghuysen and R. Hakenbeck, eds.), pp. 397–423. Elsevier Science Publishers, Amsterdam, 1994.
[64] A. H. Delcour, J. Adler, and C. Kung, *J. Membr. Biol.* **119,** 267 (1991).
[65] C. Berrier, M. Besnard, and A. Ghazi, *J. Membr. Biol.* **156,** 105 (1997).
[66] A. C. Le Dain, C. C. Hase, J. Tommassen, and B. Martinac, *EMBO J.* **15,** 3524 (1996).
[67] A. H. Delcour, J. Adler, C. Kung, and B. Martinac, *FEBS Lett.* **304,** 216 (1991).
[68] C. Berrier, A. Coulombe, C. Houssin, and A. Ghazi, *FEBS Lett.* **259,** 27 (1989).
[69] A. H. Delcour, B. Martinac, J. Adler, and C. Kung, *J. Membr. Biol.* **112,** 267 (1989).
[70] A. H. Delcour, B. Martinac, J. Adler, and C. Kung, *Biophys. J.* **56,** 631 (1989).
[71] M. Criado and B. U. Keller, *FEBS Lett.* **224,** 272 (1987).
[71a] P. Blount, S. I. Sukharev, P. C. Moe, B. Martinac, and C. Kung, *Methods Enzymol.* **294,** [24], 1998, (this volume).

*Escherichia Coli Inner Membrane*

Although individual *E. coli* cells are too small for direct patch-clamp experimentation, several methods have been worked out to generate giant *E. coli* cells for this purpose. In addition, functional reconstitution of membrane, solubilized membrane fractions, and a purified mechanosensitive channel protein are practicable. Details for the methods of rendering *E. coli* amenable to patch-clamp investigation can be found in the chapter by Blount *et al.*[71a] in this volume.

*Bacillus subtilis*

*Bacillus subtilis* is the most studied gram-positive bacterium. The cephalexin–lysozyme procedure developed to study *E. coli* has been modified for the study of *B. subtilis.*[72,73] *Bacillus subtilis* is grown overnight in 10 ml of 2% Bacto-peptone (Difco, Detroit, MI) in the presence (crucial) of 0.5% (w/v) NaCl and 25 m*M* KCl. A 1 : 100 dilution of the culture is made into 80 ml of the same fresh medium as above, and the bacteria are grown at 35° to an $A_{590}$ of 0.8–0.9. Cells are harvested by centrifugation in a Sorvall SS-34 rotor at 5000 rpm for 15 min. The cell pellet is resuspended in 70 ml of 0.5 *M* sucrose, 16 m*M* $MgSO_4$, and 0.5 *M* potassium phosphate buffer (pH 7.0) in a 1-liter flask. The cell suspension is incubated at 35° for 1 hr in the presence of 2 mg/ml lysozyme. Protoplasts are harvested by centrifugation in a Sorvall SS-34 rotor at 5000 rpm for 15 min. The protoplasts are resuspended in 1.6 ml of the same solution as above. The protoplasts must be grown further to larger sizes in order to be patch clamped. A 1:25 dilution (4–10 ml total) of the protoplast suspension is made with the succinate medium (0.5 *M* sodium succinate, 0.1 *M* KCl, 1 m*M* $MgSO_4$, 10 mg/ml casamino acid, 0.6 mg/ml fructose bisphosphate) and then added (per ml) with 40 $\mu$l of 5% casamino acid and 10 $\mu$l of 10% yeast extract. The culture is shaken slowly (50–75 rpm) at 33° for 24 hr before the giant protoplasts are harvested by centrifugation in a Sorvall SS-34 rotor at 5000 rpm for 10 min. The protoplast pellet is resuspended gently into 10 ml of the activating solution (1 *M* sucrose, 6 m*M* KCl, 20 m*M* $NaHCO_3$, and 10 m*M* $MgSO_4$, pH 7.5) and incubated at 35° for 90 min. This treatment activates the autolysin. RNase and DNase (0.8 mg/ml each) can be added to the suspension during this time to reduce viscosity, but this is optional. Protoplasts are diluted directly into the recording chamber, where

[72] M. Zoratti, V. Petronilli, and I. Szabo, *Biochem. Biophys. Res. Commun.* **168,** 443 (1990).
[73] B. Martinac, A. H. Delcour, M. Buechner, J. Adler, and C. Kung, *in* "Comparative Aspects of Mechanoreceptor Systems" (F. Ito, ed.), p. 3. Springer-Verlag, 1992.

gigaohm seals are formed on them with patch pipettes. Similar studies have been made on another gram-positive bacterium, *Streptococcus faecalis*.[74]

## Methods for Archaea

Archaea are small unicells that form a third domain of organisms (Fig. 1). They have molecular features distinct from those of bacteria and eukaryotes.[1] Many archaea thrive in extreme conditions such as hot springs and brines. Therefore, a challenge to the study of these organisms is often their cultivation. One archeon, *Haloferax volcanii* cells, can be grown in a medium containing 3.35 *M* NaCl, 170 m*M* $MgCl_2$, 200 m*M* $CaCl_2$, 26 m*M* KCl, 6.6 m*M* $NaHCO_3$, 5.4 m*M* NaBr, and 5 g/liter yeast extract for 3–4 days at 37° to $OD_{650}$ of 1.0 to 1.5.[75] One liter of culture yields approximately 500 mg dry weight of cells. For the isolation of cell envelope, cells are pelleted at 7500 rpm for 20 min at room temperature and resuspended in several milliters of double distilled water. Once the cells are resuspended, $CaCl_2$ is added to a final concentration of 10 m*M*. Thereafter, 20 $\mu$g/ml DNase is added and after 54 min $MgCl_2$ is also added to a final concentration of 5 m*M*. Cells are then French pressed twice at 8500 psi and centrifuged at 7500 rpm for 15 min. The supernatant is layered on top of a discontinuous sucrose gradient of 60 and 20% sucrose and ultracentrifuged at 49,000 rpm for 3 hr. The gradient yields a black pellet and pink membranes at the 60% boundary. For further experiments, the two fractions can be pelleted separately or together at 90,000 rpm for 30 min. The pellet is resuspended in 10 m*M* HEPES plus 5% (v/v) ethylene glycol. The resuspended membrane vesicles are aliquoted and stored at −20° or −70°. For patch-clamp experiments cell envelope membranes are mixed with azolectin liposomes at the desired protein to lipid ration (w/w, usually 50) and fused into giant proteoliposomes by dehydration and rehydration as described previously.[70] Rehydration is performed either in 100 or 200 m*M* KCl, 10 m*M* HEPES, pH 7.2. Using this method, one type of voltage-dependent porin-like conductance[75] and two types of mechanically gated conductances have been observed.[76]

## Heterologous Channels Expressed in Microbes

Inwardly rectifying $K^+$ channels from plants (*KAT1* or *AKT1*) or from animals (*gpIRK*) have been expressed in yeast. *KAT1* has further been

[74] M. Zoratti and V. Petronilli, *FEBS Lett.* **240,** 105 (1988).

[74a] S. I. Sukharev, B. Martinac, P. Blount, and C. Kung, "Methods: A Companion to *Methods Enzymol.* **6,** 51 (1994).

[75] M. Besnard, B. Martinac, and A. Ghazi, *J. Biol. Chem.* **272,** 992 (1997).

[76] A. C. Le Dain, A. Kloda, A. Ghazi, and B. Martinac, *Biophys. J.* **72,** A267 (1997).

examined through comprehensive mutations of specified regions. See Nakamura and Gaber[76a] in this volume for details.

When cRNAs for the $\alpha$, $\beta$, $\gamma$, and $\delta$ subunits of rat muscle acetylcholine receptor are injected into isolated internodal cells of *Chara corallina* and cells incubated for 24 hr, they apparently express functional channel. When cytoplasmic droplets of these cells are released from cut cells (see above) and examined in the excised patch mode, unit conductances typical of acetylcholine receptors are observed.[77]

The genes that encode the rod cyclic nucleotide gated (CNG) channel of bovine and one from *Drosophila* have been expressed in *E. coli*. This was done to determine the membrane topology of these channel proteins using the enzyme domain-fusion method. (See chapter by Blount *et al.*[71a] in this volume for details of this method.) The results of the fusion study confirm a membrane topology generally assumed.[78] Whether such protein forms functional channels in *E. coli* has not been tested. Similarly, an inward rectifying ATP-sensitive $K^+$ channel (ROMK1) fused with glutathione *S*-transferase has been expressed in *E. coli* and examined using an atomic force microscope.[79]

## Plate Phenotypes and Forward Genetics

A chief advantage in working with microbes over the worm, fly, mustard, and mouse is the ability to screen very large numbers of individuals and therefore to identify, by growth phenotype, rare mutants, recombinants, transductants, or transformants. The chief disadvantage in using microbes, especially with respect to ion channel research, is phenotypic limitation. In these microbes, there are no or only poor equivalents of leg shaking, fainting, blindness, touch insensitivity, or learning deficiency. Most useful microbial phenotypes are growth differences in various culture conditions. Thus to make use of the power of microbial genetics in any aspect of biology, the function, dysfunction, or malfunction of any gene product needs to be translated into observable plate phenotypes, usually in the ability or inability to form colonies on certain plates. One such ingenious use of plate phenotype is the rescue of $K^+$-uptake mutant yeast by the expression of foreign inward rectifiers (see ref. 76a).

Site-directed mutageneses or chimera formations are common practices

[76a] R. L. Nakamura and R. F. Gaber, *Methods Enzymol.* **293,** [6], (1998).
[77] H. Luhring and V. Witzemann, *FEBS Lett.* **361,** 65 (1995).
[78] D. K. Henn, A. Baumann, and U. B. Kaupp, *Proc. Natl. Acad. Sci. U.S.A.* **92,** 7425 (1995).
[79] R. M. Henderson, S. Schneider, Q. Li, D. Hornby, S. J. White, and H. Oberleithner, *Proc. Natl. Acad. Sci. U.S.A.* **93,** 8756 (1996).

in channel research. Such reverse genetic works begin with a cloned and expressed gene. Changes are made in sites or areas hypothesized to be of certain functional significance, usually based on comparisons of homologous channel genes from different organisms or tissues. Forward genetics, on the other hand, is a prospective approach, not limited to testing specific hypotheses. Classic examples are the collection of behavioral mutants such as *Shaker* and *ether-a-go-go* in *Drosophila,* in which the defects were traced to ion channels through analyses of the behavioral and electrophysiologic phenotypes. This strategy for gene hunting in the genome can also be applied to site hunting within a gene, i.e., the same phenotype-to-genotype approach can also be used to prospect for sites or areas within an individual channel protein that are of functional importance. Ideally, the entire channel gene is randomly mutagenized and mutant organisms are selected that yield *in vivo* phenotypes suggesting possible channel malfunctions. The mutant genes are then subcloned and sequenced while the mutant channels are expressed and electrophysiologically examined. Amino acid substitutions can then be correlated with possible changes in unit conductance, ion selectivity, voltage sensitivity, kinetics, etc. The structure–function correlations discovered through this route will not be based on any hypothesis *a priori.* However, this method of discovering important functional sites cannot yet be applied to animals or plants, not even model systems such as the worm or the fly, because of their low rate of transformation and the small numbers of specimens that can be studied at a time.

Microbes are commonly used in forward genetics. Figure 2 diagrams one kind of experiment showing how microbes can be used in ion channel studies. A cloned channel gene is placed behind an inducible promoter in a plasmid with a selectable marker. This plasmid, not the microbe genome, is randomly mutagenized. The mutagenized plasmid pool is then used to transform the microbe; successful transformants are selected by the presence of the marker. Transformants are then plated on one nutrient plate without the inducer and replica-plated onto one with the inducer. Pairs of such plates are compared later. Colonies that appear on plates without but do not appear on plates with the inducer are retrieved and retested. The plasmid from the selected transformant is then isolated and its channel gene sequenced to locate the mutation(s). This channel gene in the original plasmid may be expressed briefly before the microbe is examined by patch clamp. Alternatively this channel gene can be subcloned into a different plasmid for the expression in a heterologous system, e.g., oocytes, where no growth is required. Parallel to the electrophysiologic analysis, microbial physiology can also be studied by comparing the wild type with the mutants *in vivo,* e.g., by comparing the fluxes of ions or other solutes measured as changes in cell contents before and after a treatment.

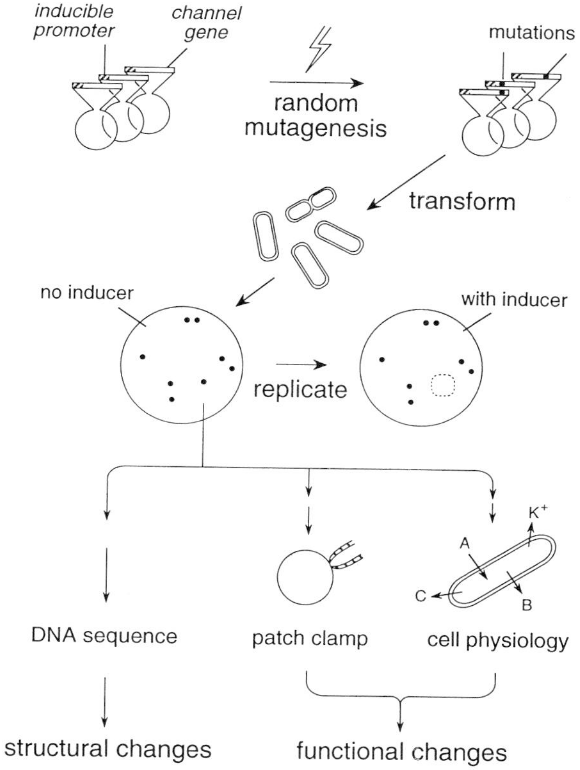


FIG. 2. A forward-genetic strategy to study molecular structure–function relationships of ion channels. A plasmid bearing the channel gene is randomly mutagenized and then used to transform a microbe (a bacterium or yeast). The transformants are plated on agar without the inducer and replica-plated on agar with the inducer, which turns on the promoter upstream of the channel gene. Colonies arising from the former but not the latter are retrieved. The mutated channel genes therein are sequenced while the mutant channels are examined for biophysical alterations and possibly for cell-physiologic defects. This scheme select mutants whose overly active channels hamper growth. It has so far been applied to channels native to yeast[32] or *E. coli*[81] but should be applicable to heterologous channels functionally expressed in microbes. See text and Refs. 32 and 81 for details.

One such study in yeast has been reported.[32] Here the native yeast outwardly rectifying $K^+$ channel, Ykc1, is investigated. The plasmid contains a *URA3* gene, which confers uracil prototrophy. *YKC1* was inserted behind the promoter *GAL1,* which is repressed by glucose but induced by galactose. In this experiment, the *in vitro* mutagenesis was achieved by incubating the plasmid DNA with hydroxylamine.[80] Other methods of mutagenesis, such as the use of error-prone PCR[81] or growing the plasmids in XL1-Red, a mutator strain of *E. coli* (Strategene, La Jolla, CA), have also been tried successfully. The extent of mutagenesis was monitored by transforming, with the treated plasmids, a strain of *E. coli* (MH1066) which relies on the plasmid-borne *URA3* for growth. The rate at which the uracil prototrophy is lost is used to measure the mutation rate.[4] A Ura$^-$ yeast strain whose native *YKC1* had been disrupted was used as recipient. Yeast cells were electroporated in the presence of the mutagenized plasmid.[82] Transformants were then allowed to form colonies on uracil-deficient nutrient plates with glucose, which represses expression of the plasmid-borne *YKC1*. Such colonies were replica plated onto the same medium with galactose, which induces *YKC1* expression.

Colonies which failed to proliferate on galactose were isolated from the glucose plates and rescreened for cosegregation of galactose sensitivity with the plasmid-borne Ura$^+$ phenotype.[6] Plasmids from each of the selected transformant are recovered and their *YKC1* gene sequenced to locate the mutations. The *YKC1* open reading frame (ORF) was excised and spliced into an oocyte expression plasmid, which was then injected into oocyte where the expressed Ykc1 currents were scrutinized (Ref. 83). The results of this study were unexpected. All mutant *YKC1* gives that block growth when expressed on galactose induction are located near the cytoplasmic ends of the transmembrane segments following either of the duplicated "pore" loop, called the "Post-P" or "PP" region. Further, all such mutant channels showed a profound loss of all closed states that have conventional kinetics. The results indicate that PP is likely a determinant of the closed state conformations.[32] A parallel forward-genetic study on a mechanosensitive channel of *E. coli*, MscL, also yielded valuable insights into the mechanosensitive mechanism[81] (see chapter by Blount *et al.*[71a] in this volume).

[80] H. C. Lee, Y. P. Toung, Y. S. Tu, and C. P. Tu, *J. Biol. Chem.* **270,** 99 (1995).
[81] X.-R. Ou, P. Blount, R. Hoffman, and C. Kung, submitted.
[82] M. Grey and M. Brendel, *Curr. Genetics* **22,** 335 (1992).
[83] T. M. Shih, R. D. Smith, L. Toro, and A. L. Goldin, *Methods Enzymol.* **293,** [29], (1998).

# [28] Design and Characterization of Gramicidin Channels

*By* DENISE V. GREATHOUSE, ROGER E. KOEPPE II, LYNDON L. PROVIDENCE, S. SHOBANA, and OLAF S. ANDERSEN

## Introduction

Gramicidin channels are miniproteins that are comprised of two tryptophan-rich transmembrane subunits. Gramicidin channels are useful models for understanding the principles that govern the structure and function of proteins (and lipids) in biological membranes. Through modifications of the parent gramicidin A (gA)[1] sequence, formyl-L-Val$^{1}$-Gly$^{2}$-L-Ala$^{3}$-D-Leu$^{4}$-L-Ala$^{5}$-D-Val$^{6}$-L-Val$^{7}$-D-Val$^{8}$-L-Trp$^{9}$-D-Leu$^{10}$-L-Trp$^{11}$-D-Leu$^{12}$-L-Trp$^{13}$-D-Leu$^{14}$-L-Trp$^{15}$-ethanolamine,[2] gramicidins have been designed for specific purposes. For example: (1) The native sequence has been labeled with stable isotopes at specific positions for solid-state nuclear magnetic resonance (NMR) determinations of backbone and side-chain conformations.[3–7] (2) The native sequence has been altered to enable investigations of the sequence dependence of backbone folding, conformation, and helix sense.[8–11] (3) Func-

[1] CD, circular dichroism; DCC, dicyclohexylcarbodiimide; DCM, dichloromethane; DIEA, diisopropylethylamine; DMF, dimethylformamide; Eq, equivalent (molar); DMPC, dimyristoylphosphatidylcholine; DS, double-stranded; Fmoc, 9-fluorenylmethoxycarbonyl; gA, gramicidin A; gLW, formyl-L-Val-Gly-L-Ala-D-Leu-L-Ala-D-Val-L-Val-D-Val-L-Leu-D-Trp-L-Leu-D-Trp-L-Leu-D-Trp-L-Leu-ethanolamine; HBTU, 2-(1*H*-benzotriazol-1-yl)-1,1,3,3-tetramethyluronium hexafluorophosphate; HMP, *p*-hydroxymethylphenoxymethyl; HOBt, 1-hydroxybenzotriazole; LH, left-handed; $LN_2$, liquid nitrogen; MD, molecular dynamics; NMP, *N*-methylpyrrolidone; *p*-NPF, *p*-nitrophenylformate; PPY, 4-pyrrolidinopyridine; RH, right-handed; SE-HPLC, size-exclusion, high-performance liquid chromatography; SS, single-stranded; THF, tetrahydrofuran; TLC, thin-layer chromatography.

[2] R. Sarges and B. Witkop, *J. Am. Chem. Soc.* **87,** 2011 (1965).

[3] R. S. Prosser, J. H. Davis, F. W. Dahlquist, and M. A. Lindorfer, *Biochemistry* **30,** 4687 (1991).

[4] R. R. Ketchem, W. Hu, and T. A. Cross, *Science* **261,** 1457 (1993).

[5] K. C. Lee, S. Huo, and T. A. Cross, *Biochemistry* **34,** 857 (1995).

[6] W. Hu, N. D. Lazo, and T. A. Cross, *Biochemistry* **34,** 14138 (1995).

[7] R. E. Koeppe II, J. A. Killian, T. C. B. Vogt, B. De Kruijff, M. J. Taylor, G. L. Mattice, and D. V. Greathouse, *Biochemistry* **34,** 9299 (1995).

[8] R. E. Koeppe II, L. L. Providence, D. V. Greathouse, F. Heitz, Y. Trudelle, N. Purdie, and O. S. Andersen, *Proteins* **12,** 49 (1992).

[9] R. E. Koeppe II, D. V. Greathouse, A. Jude, G. Saberwal, L. L. Providence, and O. S. Andersen, *J. Biol. Chem.* **269,** 12567 (1994).

[10] R. E. Koeppe II and O. S. Andersen, *Annu. Rev. Biophys. Biomol. Struct.* **25,** 231 (1996).

[11] O. S. Andersen, G. Saberwal, D. V. Greathouse, and R. E. Koeppe II, *Ind. J. Biochem. Biophys.* **33,** 331 (1996).




0076-6879/99 $30.00

tional side-chain dipoles have been introduced or removed for studies of the mechanism of ion transport.[12–15] (4) Sequence changes have allowed the engineering of voltage-dependent gramicidin channels for studies of the mechanism of channel gating.[16–18] (5) Fatty acid derivatives of gA[19] are useful models to investigate covalent lipid–peptide interactions.[20] (6) Gramicidins have been developed into useful molecular force transducers to monitor lipid bilayer–membrane protein interactions.[21,22]

In this article, we describe methods for the solid-phase synthesis of gramicidins and acylgramicidins, for the purification of the resulting peptides and acylpeptides, and for the characterization of the channels that are formed by these molecules in phospholipid membranes, using spectroscopic, chromatographic, computational, and electrophysiologic approaches.

## Synthesis Methods

### *Preparation of N-Formylamino Acid*

Some formylamino acids can be purchased from a chemical supply company; when this option is not available, the N-terminal amino acid of synthetic gramicidin can be formylated in one of two ways. By the method of Weiss and Koeppe,[23] fresh formic-acetic anhydride is prepared by adding 25 ml (0.66 mol) formic acid (95–97%) dropwise, while stirring in an ice bath, to 50 ml (0.53 mol) acetic anhydride and then warming the reaction to 50° for 15 min. This solution can be stored for up to 1 week at 4°.

Add 70 ml of the formic-acetic anhydride to 1 g of dry amino acid in a 250-ml round-bottom flask, flush with $N_2$, and stir at 4° for 18–24 hr. Re-

[12] J. L. Mazet, O. S. Andersen, and R. E. Koeppe II, *Biophys. J.* **45,** 263 (1984).

[13] E. W. B. Russell, L. B. Weiss, F. I. Navetta, R. E. Koeppe II, and O. S. Andersen, *Biophys. J.* **49,** 673 (1986).

[14] J. T. Durkin, R. E. Koeppe II, and O. S. Andersen, *J. Mol. Biol.* **211,** 221 (1990).

[15] M. D. Becker, D. V. Greathouse, R. E. Koeppe II, and O. S. Andersen, *Biochemistry* **30,** 8830 (1991).

[16] J. T. Durkin, L. L. Providence, R. E. Koeppe II, and O. S. Andersen, *J. Mol. Biol.* **231,** 1102 (1993).

[17] S. Oiki, R. E. Koeppe II, and O. S. Andersen, *Biophys. J.* **66,** 1823 (1994).

[18] S. Oiki, R. E. Koeppe II, and O. S. Andersen, *Proc. Natl. Acad. Sci. U.S.A.* **92,** 2121 (1995).

[19] R. E. Koeppe II, J. A. Paczkowski, and W. L. Whaley, *Biochemistry* **24,** 2822 (1985).

[20] R. E. Koeppe II, T. C. B. Vogt, D. V. Greathouse, J. A. Killian, and B. De Kruijff, *Biochemistry* **35,** 3641 (1996).

[21] J. A. Lundbæk, P. Birn, J. Girshman, A. J. Hansen, and O. S. Andersen, *Biochemistry* **35,** 3825 (1996).

[22] O. S. Andersen, C. Nielsen, A. M. Maier, J. A. Lundback, M. Goulian, and R. E. Koeppe II, *Methods Enzymol.* **294,** [10], (1998) (*this volume*).

[23] L. B. Weiss and R. E. Koeppe II, *Int. J. Pept. Protein Res.* **26,** 305 (1985).

move the solvent by rotary evaporation under high vacuum (0.005 mm Hg), with dry ice and liquid $N_2$ ($LN_2$) traps, and dry the sample further under high vacuum for 24 hr. The formyl amino acid can be recrystallized from either water (formyl-Tyr, -Phe, -Leu, -Ile), ethyl acetate (formyl-Val), acetone (formyl-Ala), or methanol (formyl-Gly).

An alternate method for the formation of amino acids is modified from Kanska.[24] Add 4 ml (0.04 mol) of acetic anhydride, dropwise, to 1 g of amino acid dissolved in 12 ml (0.32 mol) formic acid (95–97%), while stirring in an ice bath, flush with $N_2$, and stir at 4° for 18–24 hr. The solvent is removed and the formylated amino acid recrystallized as above.

The quality of the product should be tested by thin-layer chromatography (TLC) with silica gel plates using *n*-butanol : acetic acid : distilled $H_2O$ (6.5 : 1.5 : 2.0, v/v) as the solvent. After drying, the plates are sprayed with ninhydrin solution (0.2 g in 100 ml butanol) and heated to 110° for 15 min to develop color. The nonformylated amino acid will give a blue/purple color, whereas the formylated amino acid will be colorless (or may appear faintly pink). The plates can then be developed in an $I_2$ saturated chamber, where some formylated amino acids can be detected by the formation of a brown spot.

### *Synthesis of N-Fmoc-Amino Acids*

In most instances *N*-Fmoc-protected amino acids can be purchased from either Advanced ChemTech (Louisville, KY) or Bachem (King of Prussia, PA). When a modified amino acid, or an isotopically labeled amino acid, is required for a synthesis, the *N*-Fmoc-protected amino acid can be prepared using a method adapted from Fields *et al.*[25]: 2 mmol of amino acid are dissolved in 8 ml 10% $Na_2CO_3$; this solution is pipetted dropwise into 3 mmol *N*-Fmoc-oxysuccinimide in 10 ml dimethoxyethane at 0° (a precipitate forms immediately). After stirring for 1 hr, 5–10 ml acetone is added, and the reaction is allowed to proceed overnight at 20–25°. The formation of product is confirmed by TLC using chloroform : methanol : acetic acid (95 : 10 : 3). After drying, the plates are reacted with ninhydrin as above. The unreacted amino acid will react with ninhydrin, whereas the *N*-Fmoc-protected amino acid, which runs faster, will react with $I_2$. The precipitate is removed by filtration (Büchner funnel and Whatman, Clifton, NJ, #44 filter), the pH of the yellow filtrate is adjusted to 7.0 by dropwise addition of concentrated HCl, and the volume is reduced by rotary evaporation to ~5 ml. Ethyl acetate (30 ml) is added to the concentrated solution and the

[24] M. Kanska, *J. Radioanalyt. Nucl. Chem.* **87,** 95 (1984).

[25] C. G. Fields, G. B. Fields, R. L. Noble, and T. A. Cross, *Int. J. Pept. Protein Res.* **33,** 298 (1989).

mixture is stirred at 20–25° for 4 hr or overnight. The ethyl acetate solution is extracted two times with 30 ml 0.1 *N* HCl, one time with saturated NaCl; dried over $MgSO_4$, filtered, and dried under reduced pressure to a yellow oil. The oil is rinsed twice by dissolving in ether and redrying under reduced pressure, and finally left under high vacuum (0.005 mm Hg, with $LN_2$ trap) to dry completely for 24 hr. The *N*-Fmoc-amino acids can be recrystallized from ether or ether/hexane with yields of 40–60%.

### *Choice of Resin for Synthesis of Peptide*

In most cases, 1% polystyrene-divinylbenzene *p*-alkoxybenzyl alcohol resin,[26] with the first *N*-Fmoc-amino acid covalently attached, is the resin of choice for synthesis of gramicidin. These resins can be purchased from either Bachem or Advanced ChemTech. Occasionally it is necessary to attach the first *N*-Fmoc-amino acid to the resin, for instance, when a modified or specifically labeled amino acid is required at the C terminus of the peptide. In this case, 1% polystyrene-divinylbenzene-HMP resin from Perkin Elmer/Applied Biosystems Division (Foster City, CA) can be used to receive the first amino acid (see below).

### *Fmoc Solid-Phase Peptide Synthesis*

Gramicidin analogs (0.005, 0.010, 0.020, 0.10, or 0.25 mmol scale) can be synthesized using Fastmoc chemistry on an automated synthesizer, such as the Applied Biosystems Model 431 peptide synthesizer. When using HMP resin, the first *N*-Fmoc-protected amino acid (4- to 5-fold molar excess) is attached to the resin using DCC (1 Eq), with 4-dimethylaminopyridine (0.1 Eq) added as a catalyst. Unreacted hydroxyl groups remaining on the resin should be capped with excess benzoic anhydride (at least 3 mmol) to prevent their reacting with HBTU-activated *N*-Fmoc-amino acids in subsequent steps.[27] Deprotection of the *N*-Fmoc protecting group at each cycle is done with 20% piperidine in NMP. At each step the next *N*-Fmoc-protected amino acid is dissolved in NMP with 0.9 Eq of 0.45 *M* HBTU/HOBt in DMF. In a typical reaction, 2.1 ml NMP and 2 g 0.45 *M* HBTU/HOBt in DMF are used per millimole of *N*-Fmoc amino acid, and activation is initiated by the addition of DIEA (2 Eq). The activated *N*-Fmoc-amino acid is transferred to the resin with coupling times of between 10 min and 1 hr, depending on the synthesis scale and length of the peptide. For difficult couplings, or in cases where it may be necessary to use less

[26] S. S. Wang, *J. Am. Chem. Soc.* **95,** 1328 (1973).

[27] C. G. Fields, D. H. Lloyd, R. L. MacDonald, K. M. Otteson, and R. L. Noble, *Peptide Res.* **4,** 95 (1991).

than the normal 4- to 5-fold molar excess of *N*-Fmoc-protected amino acid (see below), unreacted amines remaining on the resin can be capped with a solution of 0.5 *M* acetic anhydride, 0.125 *M* DIEA, and 0.015 *M* HOBt in NMP (5 ml per 0.1 mmol resin) in NMP.

When synthesizing a gramicidin analog with an isotopically labeled or nontypical amino acid at a specific site, a slightly modified strategy is employed to conserve the expensive and valuable *N*-Fmoc-amino acid. For example, during the incorporation of a deuterated amino acid, a 2-fold molar excess (rather than a 4- to 5-fold excess) is used, and the efficiency of the coupling is monitored using a quantitative ninhydrin test.[28] When the coupling is at least 90% complete, a second coupling is done with nondeuterated amino acid (4- to 5-fold excess); finally the resin is capped as described above with acetic anhydride. A similar strategy may be followed for atypical synthetic amino acids; when at least 90% coupling efficiency has been achieved, the resin is capped as above with acetic anhydride. In either case, the remaining amino acids after the unusual one in the synthesis are incorporated using the normal 4- to 5-fold molar excess of *N*-Fmoc-amino acid.

The synthesis of most gramicidin analogs is completed by the addition of an *N*-formylamino acid, previously prepared and recrystallized as described above. When the *N*-terminal amino acid is formyl-L-Ala or formyl-D-Ala, some racemization may occur during the coupling step, resulting in the formation of two products, one with formyl-L- and the other with formyl-D-Ala. To prevent racemization, the peptide can instead be formylated using *p*-NPF. For this modified procedure, the N-terminal amino acid is added to the resin as *N*-Fmoc-alanine. When the coupling is complete, the *N*-Fmoc group is removed with 20% piperidine, and the resin is washed extensively with NMP and finally DCM. After drying overnight under high vacuum, the resin is transferred to a glass scintillaton vial (with a Teflon-lined cap) and *p*-NPF (5-fold molar excess) is added, which has been previously prepared and recrystallized according to the following procedure[29]: 25 mmol DCC in 15 ml THF are added dropwise to 25 mmol *p*-NPF in 1.2 ml 97% formic acid (30 mmol) and 30 ml THF while stirring at 0°. The mixture is left at 20–25° for 1 hr, and then the dicyclohexylurea precipitate is removed by filtration through a medium frit glass funnel. Most of the yellow filtrate is eliminated by rotary evaporation, and the remaining 3–5 ml, in which yellow crystals have formed, is removed under high vacuum, with two traps. The *p*-NPF is recrystallized from THF. To initiate the formylation reaction, a solution containing 10 ml DMF and 40 $\mu$l *N*-methyl-

[28] V. K. Sarin, S. B. H. Kent, J. P. Tam, and R. B. Merrifield, *Anal. Biochem.* **117,** 147 (1981).
[29] K. Okawa and S. Hase, *Bull. Chem. Soc. Japan* **36,** 754 (1963).

morpholine at 0° is added to the dried peptide/resin/*p*-NPF, and this is stirred in the dark at 4° for 18–24 hr. The volume of the solution varies with the amount of resin being formylated; as a general rule, the volume should be sufficient to completely cover the resin plus an additional 2–3 mm to provide adequate solvation. The resulting yellow liquid is removed by filtration through a medium frit glass funnel, and the resin is washed extensively with methanol and DCM to remove as much of the yellow color as possible.

*Cleavage of Peptide from Resin*

Once the synthesis is complete, the resin containing the peptide is dried under high vacuum for at least 18–24 hr and is transferred to a glass scintillation vial (with a Teflon-lined cap). In most instances, synthetic gramicidin analogs are cleaved from the resin by adding 10% distilled ethanolamine in DMF to the vial, flushing with $N_2$, and stirring the resin in a 55° water bath for 4 hr. An exception to this is deformylgramicidin, where a 50% solution of ethanolamine in DMF and 18–24 hr is required to provide adequate cleavage. As above for the formylation reaction, the volume of the cleavage solution should be sufficient to completely cover the resin plus an additional 2–3 mm extra to provide adequate solvation.

Once the cleavage is complete, a small aliquot of the cleavage solution (containing peptide) is removed for analysis by HPLC (10 $\mu$l cleavage solution into 490 $\mu$l methanol). The resin is filtered through a medium frit glass funnel and rinsed multiple times with methanol and DCM to release the peptide into the solvent. For very hydrophobic sequences, it may be necessary to add several rinses with trifluoroethanol or hexafluoro-2-propanol prior to the final DCM rinses. The filtrate containing the peptide is dried by rotary evaporation to a minimal volume (approximately the volume of the original cleavage solution prior to rinsing the resin) and transferred to a 50-ml polypropylene centrifuge tube. The gramicidin is then precipitated with six to seven volumes of distilled $H_2O$. After incubation on ice for 30 min or at 4° overnight, the precipitated peptide is collected by centrifuging for 2 hr at 14,000 rpm and 4°. The supernatant is discarded, the white pellet resuspended in a minimal volume of methanol, reprecipitated with distilled $H_2O$, and finally dried under high vacuum for 24 hr to a white powder.

The dried gramicidin will generally dissolve in several milliliters of methanol. The yield can be quantified by measuring the absorbance at 280 nm using an extinction coefficient[30] of approximately $n(5{,}200\ M^{-1}\ \text{cm}^{-1})$,

[30] G. L. Turner, J. F. Hinton, R. E. Koeppe II, J. A. Parli, and F. S. Millett, *Biochim. Biophys. Acta* **756,** 133 (1983).

where $n$ is the number of Trp's in the sequence. If the powder does not dissolve immediately, heating to 45–50° for 30 min will often help. For extremely hydrophobic sequences, a 50/50 mixture of methanol/chloroform sometimes is required.

### *Acylation of Gramicidin*

The ethanolamine hydroxyl group of gramicidin can be palmitoylated by the method of Vogt *et al.*[31] To a 10-ml Wheaton reaction vial, add 50 mg dried gramicidin (~25 μmol) and 50 mg DCC (250 μmol), suspend in 2 ml dry benzene (stored over BaO), seal with a Teflon-lined cap, and mix well by vortexing. To this mixture add 65 mg palmitic acid (250 μmol) and 20 mg lyophilized PPY (125 μmol), flush with $N_2$, and stir in the dark for 18–24 hr at 20–25°. The reaction should be tested for completeness by TLC using chloroform : methanol : water (100 : 10 : 1); the spots can be detected by heating the silica plates after spraying with 10% $H_2SO_4$. The $R_f$ values for gramicidin and palmitoylgramicidin are 0.2 and 0.5, respectively.[31]

On completion the reaction suspension is dissolved by bringing the volume to 5 ml with methanol, and the acylated gramicidin is separated from the reactants by Sephadex LH-20 (Pharmacia-LKB, obtained from Sigma Chemical Co., St. Louis, MO) gel filtration liquid chromatography. The 5-ml solution is applied to two 96- × 1.5-cm glass columns, connected in series, and eluted with 100% methanol. The column is pumped at 0.45 ml/min (~2–6 psi, 20–25°), and the materials elute in the following volumes: acylgramicidin, 140 ml; gramicidin, 145 ml; palmitic acid, 200 ml; PPY and other excess starting reagents, 260 ml.[20] The fractions containing the acylated gramicidin are analyzed by reversed-phase HPLC (as described above), pooled, dried by rotary evaporation, dissolved in a minimal volume of methanol, and quantitated by measuring the absorbance at 280 nm (as described above).

## Purification Methods

Modern automated solid-phase peptide synthesizers (e.g., Model 431A or 433A synthesizer from the Applied Biosystems Division of Perkin-Elmer) often yield gramicidins that are sufficiently pure (>98%) for spectroscopic measurements. For example, many solid-state NMR experiments have been performed using specifically $^{2}$H- or $^{15}$N-labeled gramicidins that were not chromatographically purified.[6,32] The synthetic peptide must, nev-

[31] T. C. B. Vogt, J. A. Killian, R. A. Demel, and B. De Kruijff, *Biochim. Biophys. Acta* **1069,** 157 (1991).

[32] R. E. Koeppe II, J. A. Killian, and D. V. Greathouse, *Biophys. J.* **66,** 14 (1994).

ertheless, be analyzed for purity by analytical reversed-phase HPLC[23] and for the correct molecular weight by mass spectrometry.

Reversed-phase chromatography is usually sufficient to purify the hydrophobic synthetic gramicidins. Octyl($C_8$)- or phenyl-silica columns are preferred[19,33] because the gramicidins can be easily eluted with methanol/water gradients. Octadecyl($C_{18}$)-silica columns bind the gramicidins very tightly at 20–25° and demand either higher temperatures or more expensive solvent systems for the elution process.

Small-molecule impurities, or reagents that are left over from the synthesis, are generally removed by the precipitation and wash steps discussed earlier. If necessary, the final peptide can be additionally cleaned by LH-20 chromatography, as described above for the acylgramicidins. Deformyl impurities, if present, can be removed by either reversed-phase chromatography on octyl-silica[23] or ion-exchange chromatography on Dowex AGMP50.[23] Sequence variants or residue-deleted impurities, if present, can be removed by preparative reversed-phase chromatography.[19,33]

Prior to single-channel measurements, the gramicidins must always be purified by reversed-phase chromatography.[34] These experiments require small amounts of very highly purified gramicidins, particularly for experiments with hybrid channels[14] or in any case where multiple channel types will be observed in the same membrane.[10] The most effective procedure is to use an analytical, high-resolution octyl-silica column, collect only the top of the peak of interest by hand,[23] and then rechromatograph on a second clean octyl-silica column.[23] This procedure of "double" HPLC purification yields gramicidins that exhibit "clean," i.e., unimodal, single-channel current transition histograms (see below).[35,36]

## Biophysical Characterization

### *Spectroscopic and Chromatographic Methods*

*Circular Dichroism (CD) Spectroscopy.* Gramicidin A adopts different types of conformations in different solvent environments, and the predominant conformation in a membrane may vary when the amino acid sequence or the membrane environment is changed. In phospholipids with acyl chains

[33] R. E. Koeppe II and L. B. Weiss, *J. Chromatogr.* **208,** 414 (1981).

[34] D. D. Busath, O. S. Andersen, and R. E. Koeppe II, *Biophys. J.* **51,** 79 (1987).

[35] G. L. Mattice, R. E. Koeppe II, L. L. Providence, and O. S. Andersen, *Biochemistry* **34,** 6827 (1995).

[36] L. L. Providence, O. S. Andersen, D. V. Greathouse, R. E. Koeppe II, and R. Bittman, *Biochemistry* **34,** 16404 (1995).

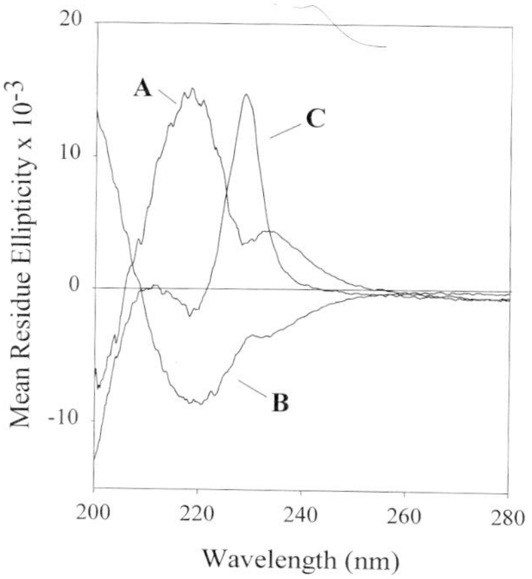


FIG. 1. The use of CD spectra to diagnose the backbone conformations of gramicidins in aqueous DMPC (0.1 m*M* gramicidin, 2.8 m*M* lipid). (A) The spectrum of gA represents a predominant RH SS $\beta^{6.3}$-helical conformation, that of the ion-conducting channel. (B) The spectrum of gLW suggests a change in the helix sense of the predominant conformation. (C) The spectrum of [Val$^5$]gLW suggests a return to a RH helix sense (for the major conformation), but with a different backbone motif (that is, in fact, DS instead of SS; see Fig. 2). Parameters: 1-mm path, 1-sec response time, 50 nm/min scan speed, 0.2-nm step resolution, 0.1-nm bandwidth. Units of mean residue ellipticity: (deg) (cm$^2$) (decimol$^{-1}$). Each spectrum is the average of six scans. For details, see Greathouse *et al.*[37]

longer than eight carbons,[37] gA folds into its characteristic RH, SS, two-subunit, $\beta^{6.3}$-helical membrane-spanning channel conformation[38] whose CD spectrum is characterized by peaks of positive ellipticity at 218 nm and 235 nm.[39–41] When solubilized in micelles formed by shorter chain phospholipids,[37] or in some organic solvents, gA adopts several DS, predominantly LH, $\beta$-helical conformations[42,43] and gives a CD spectrum with negative ellipticity from about 205 to 250 nm.[41] CD spectra between 190 and 260 nm therefore provide "fingerprints" of the strand interactions and helix sense of backbone folding for gramicidins of different sequence in membranes or other environments.

As an example, we illustrate in Fig. 1 CD spectra that are observed in aqueous DMPC for three gramicidins: gA, formyl-L-Val$^1$-Gly$^2$-L-Ala$^3$-D-

[37] D. V. Greathouse, J. F. Hinton, K. S. Kim, and R. E. Koeppe II. *Biochemistry* **33,** 4291 (1994).
[38] D. W. Urry, *Proc. Natl. Acad. Sci. U.S.A.* **68,** 672 (1971).
[39] D. W. Urry, A. Spisni, and M. A. Khaled, *Biochem. Biophys. Res. Commun.* **88,** 940 (1979).
[40] L. Masotti, A. Spisni, and D. W. Urry, *Cell Biophys.* **2,** 241 (1980).
[41] B. A. Wallace, W. R. Veatch, and E. R. Blout, *Biochemistry* **20,** 5754 (1981).
[42] W. R. Veatch, E. T. Fossel, and E. R. Blout, *Biochemistry* **13,** 5249 (1974).
[43] V. F. Bystrov and A. S. Aresen'ev, *Tetrahedron* **44,** 925 (1988).

Leu$^4$-L-Ala$^5$-D-Val$^6$-L-Val$^7$-D-Val$^8$-L-Leu$^9$-D-Trp$^{10}$-L-Leu$^{11}$-D-Trp$^{12}$-L-Leu$^{13}$-D-Trp$^{14}$-L-Leu$^{15}$-ethanolamine ("gLW"), and the position-5 variant [Val$^5$]gLW, in which Ala$^5$ is replaced by Val$^5$. The spectrum of gA is characteristic of the RH $\beta$-helical channel,[37] and the approximately mirrored spectrum of gLW suggests a LH $\beta$-helical channel that has been verified by two-dimensional NMR.[44] A single additional sequence change at position 5 leads to a completely different CD spectrum for [Val$^5$]gLW, suggesting that the predominant conformation has changed from LH, SS to RH, DS.[45] This type of result shows the power of CD as a tool for monitoring the backbone folding of gramicidins. CD spectra are also useful for characterizing the folding and orientation of membrane-incorporated $\alpha$ helices.[46] The CD spectra become even more informative when they are done on peptides (gramicidins or $\alpha$ helices) in oriented multilayers, because one in this case can monitor whether the bilayer-associated peptides are inserted into the bilayer (perpendicular to the plane of the membrane) or absorbed at the membrane/solution interface (parallel to the plane of the membrane).[46,47]

*Size-Exclusion HPLC.* SE-HPLC is a useful technique to assess the presence of SS and DS conformations. Microliter volumes of a gramicidin in a particular organic solvent or aqueous lipid (or detergent) suspension are injected into THF, which is pumped at ca. 1.0 ml/min as the elution solvent for the chromatography. THF is useful because it acts to dilute the original solvent and disrupt phospholipid membranes (or detergent micelles), while at the same time preserving the folding of individual gramicidin molecules.[48] Disruption of the phospholipid (or detergent) assembly results in the dissociation of the six intermolecular hydrogen bonds that stabilize the SS channel conformation of gramicidin, so that this conformation will elute as monomers. DS conformations, by contrast, remain intact and elute as the dimers that are stabilized by about 28 intermolecular hydrogen bonds.[48]

As an example, we show in Fig. 2 the SE-HPLC elution profiles for aqueous DMPC dispersions of gA, gLW, and [Val$^5$]gLW (the same molecules whose CD spectra are shown in Fig. 1). The chromatograms show that gA and gLW exist in DMPC membranes as primarily SS structures,

[44] D. Greathouse, N. Le, R. E. Koeppe II, J. Hinton, G. Saberwal, L. Providence, and O. Andersen, *Biophys. J.* **68,** A151 (1995).

[45] N. H. Le, D. V. Greathouse, R. E. Koeppe II, G. Saberwal, L. L. Providence, and O. S. Andersen, *Biophys. J.* **70,** A79 (1996).

[46] J. A. Killian, I. Salemink, M. De Planque, G. Lindblom, R. E. Koeppe II, and D. V. Greathouse, *Biochemistry* **35,** 1037 (1996).

[47] H. H. J. De Jongh, E. Goormaghtigh, and J. A. Killian, *Biochemistry* **33,** 14521 (1994).

[48] M. C. Bañó, L. Braco, and C. Abad, *J. Chromatogr.* **458,** 105 (1988).

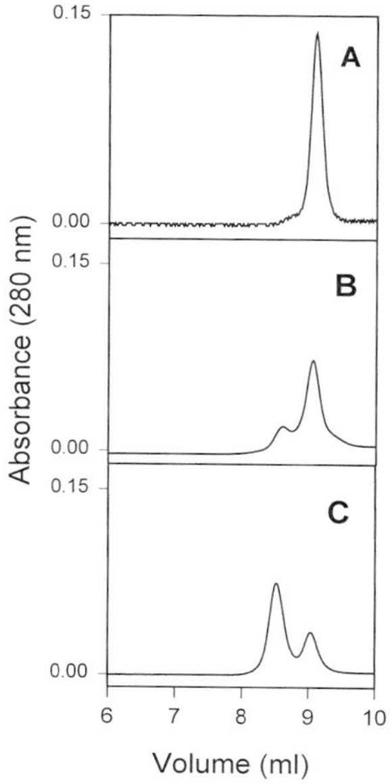


FIG. 2. The use of size-exclusion chromatography to determine the molecularity of gramicidins in aqueous DMPC. DS dimers elute at approximately 8.5 min, whereas SS conformations elute at about 9.0 min. (A) gA is >95% SS. (B) gLW is predominantly SS, but with ~20% DS conformers. (C) [Val[5]]gLW is predominantly DS, with ~25% SS conformers. Conditions: Ultrastyrogel (7 $\mu$m, $10^3$ Å) column (Waters Corp., Milford, MA), 7.8 × 300 mm; THF elution solvent at 1.0 ml/min; 5 $\mu$l of 0.5 m*M* gramicidin, 14 m*M* lipid injected at 22–23°.

whereas [Val[5]]gLW is predominantly DS in DMPC. Figure 2 shows that gA, gLW, and [Val[5]]gLW are <5, 30, and 74% DS in DMPC vesicle dispersions, respectively, in agreement with the inferences from the CD spectra in Fig. 1. Qualitatively, the two methods lead to similar conclusions: CD provides information about the backbone folding and helix sense without revealing the molecularity; SE-HPLC provides direct information about the molecularity without revealing anything about the backbone folding (or the helix sense).

*Solution NMR.* The membrane-spanning channel structure of gA has not been solved by X-ray crystallography. The several crystal structures of gA[49–52] are all DS conformers that differ from the SS channel. In phospho-

[49] B. A. Wallace and K. Ravikumar, *Science* **241,** 182 (1988).

[50] D. A. Langs, *Science* **241,** 188 (1988).

[51] D. A. Langs, G. D. Smith, C. Courseille, G. Precigoux, and M. Hospital, *Proc. Natl. Acad. Sci. U.S.A.* **88,** 5345 (1991).

[52] D. A. Doyle and B. A. Wallace, *J. Mol. Biol.* **266,** 963 (1997).

lipid membranes, gramicidin/lipid assemblies have not yielded diffraction quality crystals, nor do they tumble with sufficient speed to be tractable by solution NMR. Gramicidin/detergent assemblies (using either SDS or dodecylphosphocholine) tumble rapidly enough to give well-resolved NMR spectra, and these have yielded a high-resolution structure.[53,54] Importantly, the CD spectrum of detergent-incorporated gA is very similar to that of bilayer-incorporated gA.[55] This means that the structure of gA in detergents is very similar to the structure in bilayers.

When used for two-dimensional NMR spectroscopy, SDS-$d_{25}$ (Cambridge Isotope Laboratories, Andover, MA) should be recrystallized from 95% ethanol.[7] Samples containing 2–6 m*M* synthetic gramicidin of the desired sequence and 200–600 m*M* SDS-$d_{25}$ in 80% $H_2O$, 10% $D_2O$, and 10% trifluoroethanol-$d_3$ are prepared at pH $\leq$ 6.7, as described by Mattice *et al.*[35] Several types of spectra can be recorded, including $^1H$–$^1H$ correlated (COSY), nuclear Overhauser enhancement (NOESY), and homonuclear Hartman–Hahn (HOHAHA or TOCSY). The spectra are obtained typically at 55°,[35] using standard methods.[56–59] The basic strategy is to obtain the sequential assignment of backbone and side-chain $^1H$ resonances[56] and identify as many through-space nuclear Overhauser (NOE) connectivities as possible.[57,58] These results can be used for two different purposes. First, long-range NOEs indicate the basic secondary structure or helical fold of the gramicidin, including the helix sense.[7,9,35,55] This method of characterizing the backbone serves to validate and extend conclusions based on CD spectroscopy and SE-HPLC (above). Second, distance constraints based on the NOEs can be used to calculate full three-dimensional structures of the sequence-altered gramicidins, as has been accomplished for gA.[53,54,60,61]

*Solid-State NMR.* Solid-state NMR takes advantage of the restricted molecular motions of membrane-bound proteins[62,63] and has been effectively applied to gramicidins. When applying this method, spectral resolution and sensitivity are improved by isotopic labeling and by the macro-

[53] A. S. Arsen'ev, A. L. Lomize, I. L. Barsukov, and V. F. Bystrov, *Biol. Membr.* **3,** 1077 (1986).

[54] A. L. Lomize, V. Y. Orekhov, and A. S. Arsen'ev, *Bioorg. Khimiya* **18,** 182 (1992) (in Russian).

[55] A. S. Arsen'ev, I. L. Barsukov, V. F. Bystrov, A. L. Lomize, and Yu. A. Ovchinnikov, *FEBS Lett.* **186,** 168 (1985).

[56] K. Wüthrich, "NMR of Proteins and Nucleic Acids," pp. 292. Wiley, New York, 1986.

[57] J. L. Markley, *Methods Enzymol.* **176,** 12 (1989).

[58] A. Bax, *Annu. Rev. Biochem.* **58,** 223 (1989).

[59] G. M. Clore and A. M. Gronenborn, *Annu. Rev. Biophys. Chem.* **20,** 29 (1991).

[60] Z. Zhang, S. M. Pascal, and T. A. Cross, *Biochemistry* **31,** 8822 (1992).

[61] Y. Chen, A. Tucker, and B. A. Wallace, *Biophys. J.* **70,** A79 (1996).

[62] S. J. Opella, *Biol. Magn. Reson.* **9,** 177 (1990).

[63] S. O. Smith and O. B. Peersen, *Annu. Rev. Biophys. Biomol. Struct.* **21,** 25 (1992).

scopic alignment of samples relative to an external magnetic field.[63] Gramicidins can be labeled either uniformly by biosynthesis[64] or by amide- and indole-hydrogen exchange,[65] or specifically (at particular sites) by chemical synthesis, as described above. Sample alignment is achieved by coating glass slides with gramicidin-containing phospholipid multilayers[66] or by using strong magnetic fields.[67]

Orientational constraints based on $^{15}N$ chemical shifts, $^{15}N-^{1}H$ and $^{15}N-^{13}C$ dipolar interactions, and/or $^{2}H$ quadrupolar interactions have been used to define the entire gA structure,[4] the backbone folding and helix sense,[68–70] backbone dynamics,[71,72] side-chain orientations and dynamics,[5,6,32,73] and side-chain perturbation due to sequence changes or covalent fatty acylation.[7,20,74]

In typical experiments with oriented gramicidins in DMPC, as few as 3 $\mu$mol[32] or as many as 15 $\mu$mol of gramicidin[73] are codissolved in an organic solvent with DMPC at a lipid:peptide ratio between 15:1[68] and 8:1.[73] Early methods used mixtures of benzene/ethanol or methanol/water as the cosolvent; we prefer to resuspend a dried mixture of gramicidin/lipid in pure chloroform.[20] Gramicidin A itself is insoluble in chloroform, but it dissolves readily in the presence of the lipid and readily folds into the transmembrane channel conformation after the chloroform is removed and the sample rehydrated. The use of chloroform minimizes the waiting time that is required for converting folding intermediates, which may be temporarily trapped because of a "solvent history" dependence,[75–77] into the channel conformation.

As a standard procedure,[20] the use of 4.8- × 23.0-mm glass plates, thickness no. 00 (0.06–0.08 nm) from Marienfeld Laboratory Glassware

[64] L. K. Nicholson, F. Moll, T. E. Mixon, P. V. LoGrasso, J. C. Lay, and T. A. Cross, *Biochemistry* **26,** 6621 (1987).

[65] R. S. Prosser, S. I. Daleman, and J. H. Davis, *Biophys J.* **66,** 1415 (1994).

[66] F. I. Moll and T. A. Cross, *Biophys. J.* **57,** 351 (1990).

[67] J. H. Davis, *Biochemistry* **27,** 428 (1988).

[68] R. Smith, D. E. Thomas, F. Separovic, A. R. Atkins, and B. A. Cornell, *Biophys. J.* **56,** 307 (1989).

[69] L. K. Nicholson and T. A. Cross, *Biochemistry* **28,** 9379 (1989).

[70] A. W. Hing, S. P. Adams, D. F. Silbert, and R. E. Norberg, *Biochemistry* **29,** 4144 (1990).

[71] R. S. Prosser and J. H. Davis, *Biophys J.* **66,** 1429 (1994).

[72] K. C. Lee, W. Hu, and T. A. Cross, *Biophys. J.* **65,** 1162 (1993).

[73] W. Hu, K. C. Lee, and T. A. Cross, *Biochemistry* **32,** 7035 (1993).

[74] D. V. Greathouse, J. Hatchett, A. R. Jude, R. E. Koeppe II, L. L. Providence, and O. S. Andersen, *Biophys. J.* **72,** A396 (1997).

[75] P. V. LoGrasso, F. I. Moll, and T. A. Cross, *Biophys. J.* **54,** 259 (1988).

[76] J. A. Killian, K. U. Prasad, D. Hains, and D. W. Urry, *Biochemistry* **27,** 4848 (1988).

[77] S. Arumugam, S. Pascal, C. L. North, W. Hu, K. C. Lee, M. Cotten, R. R. Ketchem, M. Xu, M. Brenneman, F. Kovacs, F. Tian, A. Wang, S. Huo, and T. A. Cross, *Proc. Natl. Acad. Sci. U.S.A.* **93,** 5872 (1996).

(Bad Mergentheim, Germany) and DMPC from Avanti Polar Lipids (Alabaster, AL) is recommended. A dried film consisting of 6 $\mu$mol gramicidin and 60 $\mu$mol DMPC is prepared, resuspended in 0.8 ml of chloroform and applied evenly to 50 glass plates. After extensive drying, the plates are stacked in a glass cuvette and hydrated with about 38 $\mu$l of $^2$H-depleted water. The cuvette should be tightly sealed by using epoxy to attach a small glass end plate, and incubated a minimum of 72 hr at 40°. The extent of phospholipid alignment is monitored by $^{31}$P NMR.[66,78] When using $^2$H labels, deuterium NMR spectra are recorded using the quadrupole echo sequence, with full-phase cycling.[79] Cross-polarized $^{13}$C and $^{15}$N spectra are obtained using standard methods.[73,80,81]

Sharp resonance lines can be obtained when samples are macroscopically aligned in the magnetic field. The positions of the resonance lines are determined by the average orientations of the labeled sites with respect to the magnetic field. For deuterium-labeled compounds which undergo fast motional averaging about a molecular long axis, the observed quadrupolar splitting is determined by:

$$\Delta\nu_q = (3/2)\,(e^2qQ/h)\,(1/2[3\cos^2\Theta - 1])(1/2[3\cos^2\phi - 1])\,(1/2[3\cos^2\beta - 1]) \tag{1}$$

in which $e^2qQ/h$ is the C–D quadrupolar coupling constant, $\Theta$ the angle between a C–D bond or C–$CD_3$ bond and the axis of motional averaging, $\phi$ is 0° for C–D bonds or ~109.5° (the tetrahedral angle for rapidly reorienting $CD_3$ groups), and $\beta$ is the angle between the axis of motional averaging and the magnetic field, $H_o$.[78] When there is rapid reorientation about a molecular long axis, Eq. (1) predicts that $\Delta\nu_q$ values are 2-fold larger for $\beta = 0°$ than for $\beta = 90°$.[78] The $^2$H quadrupolar splittings, along with dipolar interactions and $^{15}$N chemical shifts, are used as constraints for molecular modeling.[4,6,20] Powder-pattern spectra from nonaligned samples give essential information about global and local dynamics.[6,72,82] The dynamic constraints influence fundamentally the determination of average local side-chain conformations.[82]

For $^{13}$C- and $^{15}$N-labeled gramicidins, computer simulations of powder-pattern spectra together with resonance frequencies that are measured in

[78] J. A. Killian, M. J. Taylor, and R. E. Koeppe II, *Biochemistry* **31,** 11283 (1992).

[79] J. H. Davis, K. R. Jeffrey, M. Bloom, M. I. Valio, and T. P. Higgs, *Chem. Phys. Lett.* **42,** 390 (1976).

[80] A. Pines, M. C. Gibby, and J. S. Waugh, *J. Chem. Phys.* **59,** 569 (1973).

[81] F. Separovic, J. Gehrmann, T. Milne, B. A. Cornell, S. Y. Lin, and R. Smith, *Biophys. J.* **67,** 1495 (1994).

[82] K. C. Lee and T. A. Cross, *Biophys. J.* **66,** 1380 (1994).

oriented-sample spectra are used to determine the magnitudes and orientations of the principle values of the nuclear spin interaction tensors.[73,83] These tensors in turn serve as constraints in the development and testing of molecular models of the membrane-incorporated gramicidin channels.[6,81] These methods are general and can be applied well to other membrane-incorporated peptides.[46]

### *Molecular Modeling*

Molecular dynamics (MD) simulations are used to gain insights into the molecular basis for macroscopic properties of biomolecules, which can be estimated based on computational evaluation of the underlying microscopic interactions. MD simulations thus facilitate the understanding and interpretation of experimental results and allow for semiquantitative predictions. The power of the method is the capability to interpolate and extrapolate experimental data into regions that are difficult to access in the laboratory.[84]

MD simulations require a set of initial atomic positions and velocities. The initial configuration of a molecule can be obtained from X-ray or NMR structures, or from model building; and the initial velocities are obtained using a Maxwellian distribution for a very low temperature. The initial phase of the simulation, termed the equilibration period, depends on the relaxation time of the property of interest. Once the system is equilibrated, the production phase begins. During the production phase, the potential and kinetic energies are monitored to evaluate the stability of the simulation. The trajectories are stored, which allows one to calculate statistical equilibrium averages for the quantities of interest. Thermodynamic parameters such as the difference in free energy and entropy between any two well-defined states thus can be obtained. The connection between MD simulations and equilibrium statistical thermodynamics is commonly denoted as "free-energy simulations."[85] The approach has been described in detail.[86–89]

[83] F. Separovic, K. Hayamizu, R. Smith, and B. A. Cornell, *Chem. Phys. Lett.* **181,** 157 (1991).

[84] W. J. van Gunsteren and H. J. C. Berendsen, *Angew. Chem. Int. Ed. Engl.* **29,** 992 (1990).

[85] D. A. Case, *Prog. Biophys. Molec. Biol.* **52,** 39 (1988).

[86] D. L. Beveridge and F. M. DiCapua, *in* "Computer Simulation in Biomolecular Systems. Theoretical and Experimental Applications" (W. F. van Gunsteren and P. K. Weiner, eds.), pp. 1–26. ESCOM Science Publishers B. V., Amsterdam, 1989.

[87] M. Mezei and D. L. Beveridge, *Ann. N.Y. Acad. Sci.* **482,** 1 (1986).

[88] W. F. van Gunsteren, *in* "Computer Simulation in Biomolecular Systems. Theoretical and Experimental Applications" (W. F. van Gunsteren and P. K. Weiner, eds.), pp. 27–59. ESCOM Science Publishers B. V., Amsterdam, 1989.

[89] J. A. McGammon and S. C. Harvey, *in* "Dynamics of Proteins and Nucleic Acids," p. 234. Cambridge University Press, Cambridge, United Kingdom, 1987.

In typical applications, one computes the difference in Helmholtz free energy between two well-defined systems, e.g., two different mutants that can be transformed into one another by a well-defined algorithm. The algorithm works as follows. Consider a molecule of mutant I (with an Ala at a given position) and a molecule of mutant II (with a Val at that position). One transforms mutant I into mutant II using a process of molecular "alchemy," by constructing a series of mixed potential functions for the hybrid residue (Ala changing toward Val). The mixed potential function can be described by a coupling parameter $\lambda$, which varies between 0 and 1: $E_\lambda = \lambda E_{\mathrm{II}} + (1 - \lambda)E_{\mathrm{I}}$, where $E_{\mathrm{I}}$ and $E_{\mathrm{II}}$ denote the potential functions for mutant I and mutant II, respectively. The free energy difference then can be expressed as a function of $\lambda$, $\Delta A = A(\lambda = 1) - A(\lambda = 0)$. To calculate the energy difference, the simulation is started at state I and proceeds to state II, generating a set of intermediate configurations at different values of $\lambda$. The energy difference between two intermediate states, $A(\lambda)$ and $A(\lambda + \delta\lambda)$, is determined as:

$$A(\lambda + \delta\lambda) - A(\lambda) = -kT \ln \left\langle \exp\left[\frac{-[V(\lambda + \delta\lambda) - V(\lambda)]}{kT}\right]\right\rangle$$

where $k$ is Boltzmann's constant and $T$ is the temperature in degrees Kelvin. The free-energy change between states I and II is simply the sum of the intermediate free-energy differences.

An important issue in free-energy simulations is how one approaches the two end states: $\lambda \rightarrow 0$ and $\lambda \rightarrow 1$. Problems may arise because the initial and final states put the corresponding groups at different positions (e.g., H or $CH_3$; as Ala "swells" into Val or Val "shrinks" to Ala), which may give rise to numerical difficulties. To resolve this problem it is necessary to chose a nonlinear progression of $\lambda$ (Fig. 3) and to keep the $\lambda$ increments small; this should ensure a smooth, well-equilibrated simulation.[87,90,91]

Most computational studies on gramicidins have been done to understand the energetics of ion permeation,[92–97] the structure and dynamics of the proton wire in the channel,[98] or the behavior of gA in lipid bilayers.[99]

[90] A. J. Cross, *Chem. Phys. Lett.* **128,** 198 (1986).
[91] S. Shobana, B. Roux, R. E. Koeppe II, and O. S. Andersen, *Biophys. J.* **72,** A395 (1997).
[92] J. Åqvist and A. Warshel, *Biophys. J.* **56,** 171 (1989).
[93] S.-W. Chiu, S. Subramanian, E. Jakobsson, and J. A. McCammon, *Biophys. J.* **56,** 253 (1989).
[94] B. Roux and M. Karplus, *Biophys. J.* **53,** 297 (1988).
[95] B. Roux and M. Karplus, *J. Am. Chem. Soc.* **115,** 3250 (1993).
[96] R. Elber, D. P. Chen, D. Rojewska, and R. Eisenberg, *Biophys. J.* **68,** 906 (1995).
[97] B. Roux, B. Prod'hom, and M. Karplus, *Biophys. J.* **68,** 876 (1995).
[98] P. Regis and B. Roux, *Biophys. J.* **71,** 19 (1996).
[99] T. B. Woolf and B. Roux, *Proc. Natl. Acad. Sci. U.S.A.* **91,** 11631 (1994).

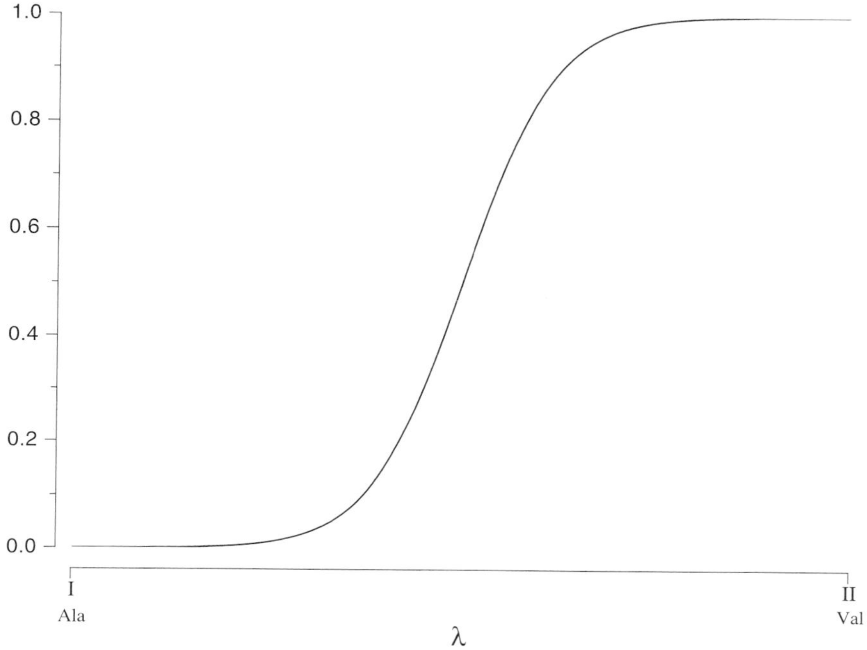


FIG. 3. The nonlinear progression of $\lambda$ from 0 (state I) to 1 (state II) in a molecular dynamics simulation using mixed potential functions to simulate a hybrid residue in order to represent intermediate states in the theoretical transformation of mutant I to mutant II (see text).

A particularly attractive use of MD simulations on the gramicidins is the use of free-energy simulations to determine the conformational preferences of polymorphic gramicidin analogs, e.g., gLW.[10] The structural preferences of gLW can be switched by single amino acid substitutions, which allows one to examine the molecular basis for the macroscopic behavior by means of MD simulations coupled with free-energy perturbation. To do so, a SS, RH gLW is constructed based on a refined $\beta^{6.3}$ gA structure. The SS, LH dimer, for which high-resolution data are not available, can be created from RH gA by applying Venketachalam–Urry transformations[100]: $\phi_L^R \rightarrow -\phi_D^S$; $\phi_D^R \rightarrow -\phi_L^S$; $\psi_L^R \rightarrow -\psi_D^S$; $\psi_D^R \rightarrow -\psi_L^S$, where the subscripts denote the amino acid chirality and the superscripts denote helix sense. Further, the RH and LH gA monomers are minimized using backbone constraints so that the gA backbone conformation is preserved. The RH and LH gLW

[100] C. M. Venkatachalam and D. W. Urry, *J. Comput. Chem.* **4,** 461 (1983).

monomers and dimers are obtained from the RH and LH gA monomers by doing appropriate amino acid substitutions. The Helmholtz free-energy difference between the SS, RH and SS, LH channels then can be computed by free-energy perturbation simulations using the thermodynamic cycle approach.[89] Assuming that $\Delta A$ is approximately equal to the difference in Gibbs free energy ($\Delta G$), one can estimate the conformational preference of the mutant channel as

$$\Delta\Delta G_{\rm II}^{0'} = \Delta\Delta G_{\rm I,expt}^{0'} + (\Delta A_{\rm I\to II}^{\rm L} - \Delta A_{\rm I\to II}^{\rm R})$$

where $\Delta\Delta G_{\rm II}^{0'}$ denotes the predicted LH/RH energy difference of the mutant channel, $\Delta\Delta G_{\rm I,expt}^{0}$ is the experimental value found for the reference channel, $\Delta A_{\rm I\to II}^{\rm L}$ the calculated free-energy difference associated with I $\rightarrow$ II substitution for the LH channel, and $\Delta A_{\rm I\to II}^{\rm R}$ the corresponding difference for the RH channel. This method allows direct comparison between theory and experiment, as well as extensions to "fractional mutants" and regions of conformational space that are not accessible experimentally.

*Electrophysiologic Methods: Single-Channel Measurements*

Electrophysiologic methods—usually single-channel measurements—are used not only to characterize channel function, but also to determine the structural identity of bilayer-incorporated, membrane-spanning gramicidin channels.[8,9,11,14,16,101] The bilayer is formed across a hole in a partition that separates two chambers (e.g., White[102]), which contain the desired electrolyte solutions. Electrical contact to the electrolyte solutions is usually through Ag/AgCl electrodes, which may be directly inserted into the chamber or connected via pipettes (or agar bridges).

To achieve the signal/noise performance that is necessary for single-channel measurements, it is important to keep the membrane capacitance[103] as low as possible. This is conveniently done by using a small hole in the partition that separates the two chambers,[104] or by using the bilayer punch[105] to isolate a small membrane from a larger bilayer. In the bilayer-punch method, planar bilayers are formed from *n*-decane solutions (2.5% w/v) of the desired lipid across a hole (~1.6 mm in diameter) in a Teflon partition

[101] J. T. Durkin, L. L. Providence, R. E. Koeppe II, and O. S. Andersen, *Biophys. J.* **62,** 145 (1992).

[102] S. H. White *in* "Ion Channel Reconstitution" (C. Miller, ed.), pp. 3–35. Plenum, New York, 1986.

[103] R. D. Hotchkiss, *Anv. Enzymol. Relat. Areas Mol. Biol.* **4,** 153, (1944).

[104] E. Neher, J. Sandblom, and G. Eisenman, *J. Membr. Biol.* **40,** 97 (1978).

[105] O. S. Andersen, *Biophys. J.* **41,** 119 (1983).

that separates the two electrolyte solutions. Small membranes are isolated at the tip of glass pipettes, with a tip diameter of ~30 $\mu$m, that have been fire-polished and coated with trioctylsilane (at ~100° for ~10 sec). The ensuing "small" membranes have an area of ~$10^3$ $\mu m^2$ and a resistance ~$10^{12}$ $\Omega$, which allows for single-channel measurements[105,106] using conventional patch-clamp amplifiers.

A convenient lipid for single-channel measurements is diphytanoylphosphatidylcholine (DPhPC) from Avanti Polar Lipds, Inc. Usually the lipid is suspended in *n*-decane, e.g., 99.9% pure from Wiley Organics (Columbus, OH). Alkali metal chloride salts should be AR grade or better; before use they should be roasted at 500° for 24 hr (and stored over $CaSO_4$ in an evacuated desiccator). The water can be high-quality deionized water, e.g., Milli-Q water (Millipore Corp., Bedford, MA). Usually the electrolyte solutions are unbuffered (pH ~6); but they must be made up the day of the experiment in glassware cleaned in chromic sulfuric acid (cf. Busath *et al.*[34]).

When added to the aqueous solutions bathing a bilayer, during stirring, the very hydrophobic gramicidin adsorbs to the bilayer. Even though the gramicidins are hydrophobic, they cross lipid bilayers poorly—because the four Trp residues, with their indole NH moieties, anchor the ethanolamide–CO-terminal end of the molecule at the membrane–solution interface.[107] It is therefore important to add the gramicidins symmetrically to both aqueous solutions. With symmetric addition, channel activity is seen within a few minutes after addition. The amount of added gramicidin is adjusted (from a few p*M* to n*M*) to give a single-channel appearance rate of about one transition per second, which optimizes the ability to distinguish among different channel types (if present). The current signal is digitized and sampled on a personal computer. Single-channel current transitions can be detected off-line or on-line using an absolute level-crossing algorithm[108] or a differential level-crossing algorithm.[105] In either case, the current signal is converted into a series of idealized channel events, from which the current–transition amplitudes and channel lifetimes are determined, and used to construct the corresponding current–transition amplitude or channel lifetime histograms.

Figure 4 shows channel activity observed with gA that has been purified on a preparative HPLC column; several channel types can be seen in Fig. 4A. Figure 5 shows channel activity seen with gA that has been twice purified on an analytical HPLC column (Fig. 6). In Fig. 5A, only one

[106] R. U. Muller and O. S. Andersen, *J. Gen. Physiol.* **80,** 427 (1982).
[107] A. M. O'Connell, R. E. Koeppe II, and O. S. Andersen, *Science* **250,** 1256 (1990).
[108] D. Colquhoun and B. Sakmann, *J. Physiol.* **369,** 501 (1985).

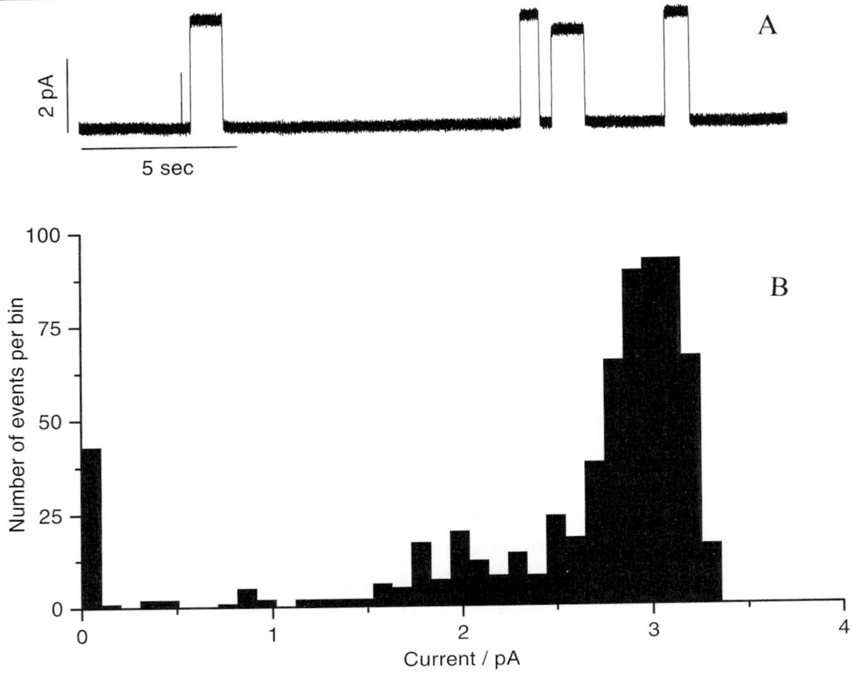


FIG. 4. Channel formation with gA purified on a preparative HPLC column. (A) Single-channel current trace observed after the addition of gA to both sides of the bilayer. Several channel types can be seen, with the predominant type being gA channels. The first (brief) and fourth events are not gA channels. (B) Current transition amplitude histogram. There are 928 events in the histogram with 684, or 74%, in the rather wide peak at 2.9 pA ($\pm$0.2 pA); 1.0 *M* NaCl; 200 mV.

channel type is apparent. The visual impression from the current traces (Figs. 4A and 5A) is substantiated in current–transition amplitude histograms (Figs 4B and 5B). The histogram obtained with the preparative sample (Fig. 4B) has a broad distribution of current–transition amplitudes, whereas the histogram obtained with the analytical sample has a narrow distribution (Fig. 5B). Such well-defined, narrow histograms are essential if electrophysiologic methods are to be used as tools for structure identification (see below). When the analog is clean, i.e., when the current–transition amplitude histogram looks like Fig. 5B, the lifetime distribution can be described by a single exponential distribution (Fig. 7). This latter result provides further evidence for a single kinetically homogenous channel population.

In addition to characterizing the catalytic ability (ion flux through the

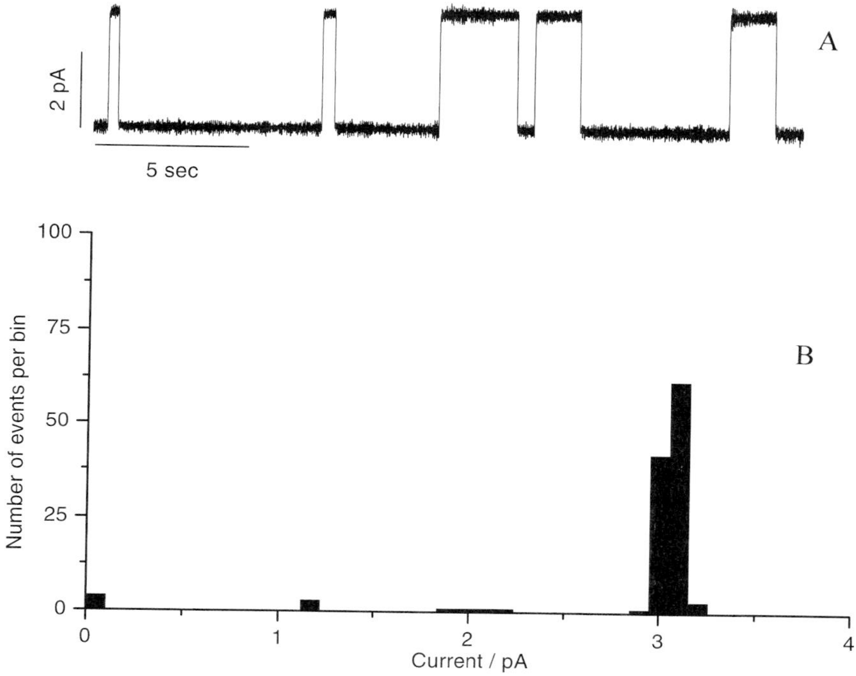


FIG. 5. Channel formation with gA twice purified on an analytical HPLC column. (A) Single-channel current trace observed after the addition of gA to both sides of the bilayer. Only one channel type can be seen. (B) Current transition amplitude histogram. There are 119 events in the histogram with 108, or 91%, in the gA peak at 3.0 pA (±0.1 pA); 1.0 *M* NaCl; 200 mV.

channel) of a channel, electrophysiologic methods can be used to identify the structure of channels formed by an amino acid-substituted gramicidin analog. This is possible because the dimeric channel structure allows one to use single-channel methods to determine whether a mutant gramicidin forms channels that have the same structure as those formed by some reference gramicidin.[14] This functional approach to structure identification is based on the following considerations (cf. Andersen *et al.*[11]):

1. The reference gramicidin forms only a single type of membrane-spanning (conducting) channel.
2. Standard (i.e., SS) gramicidin channels form by the transmembrane dimerization of two SS $\beta^{6.3}$-helical monomers.[107]
3. Membrane-spanning dimers cannot form between monomers of opposite helix sense.[8,36]

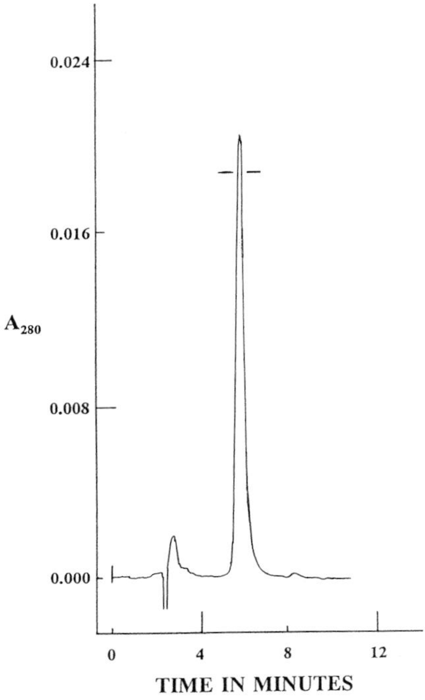


FIG. 6. HPLC purification of a synthetic gramicidin for single-channel analysis. A peptide that has been previously purified by reversed-phase HPLC is reinjected onto a freshly washed reversed-phase column and eluted as shown. Only the top portion of the peak above the indicated marks is collected for use in single-channel experiments. Conditions: Zorbax 5 $\mu$m octyl-silica ($C_8$) column (4.6 × 250 mm) eluted with 85% methanol, 15% water at 1.5 ml/min; 1800 psi; 22–23°.

4. DS gramicidin channels assemble at one side of the bilayer and insert into the membrane to form the long-lived membrane-spanning channels.[101,109]

If an unknown gramicidin analog forms one single channel type, the structural identity of the channels can be established through symmetric/asymmetric addition of the analog and through heterodimer formation experiments. If channels form only when the analog is added to the electrolyte solution on both sides of the bilayer, the channels are SS dimers. If the analog forms channels when added to only one side of a bilayer–at a rate that is comparable to that observed when the analog

[109] Single-stranded gA channels form at a very low rate after asymmetric addition of gA,[107] but that does not affect the analysis.

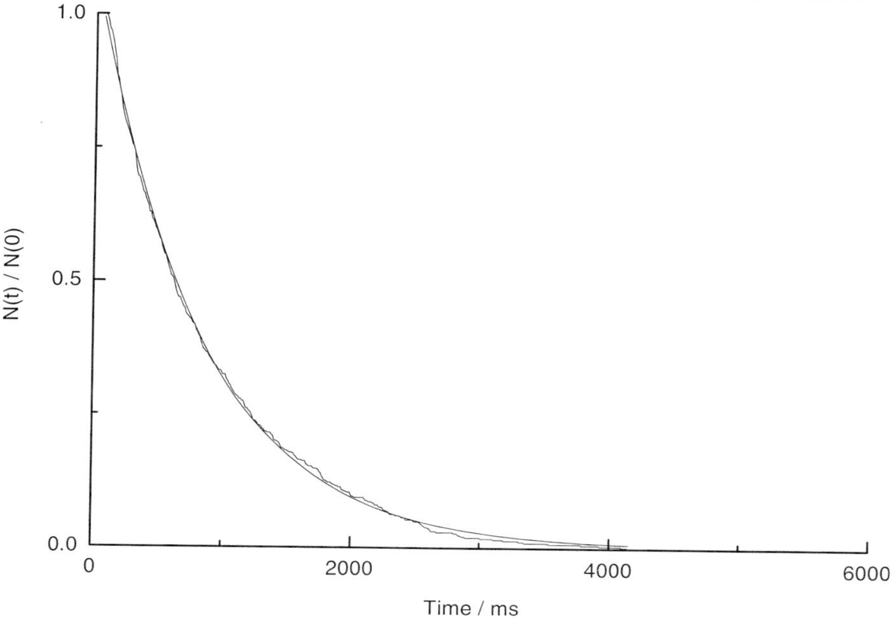


FIG. 7. Lifetime duration histogram for gA channels, based on experiments with gA that was twice purified on an analytical HPLC column. The results are fitted by a single exponential distribution: $[N(t)/N(0)] = \exp(-t/\tau)$, where $N(0)$ denotes the number of channels at time zero, and $N(t)$ denotes the number of channels with lifetimes longer than time $t$. $\tau = 660$ ms; 1.0 $M$ NaCl; 200 mV.

is present on both sides of the membranes—the channels are DS dimers. For analogs that form SS channels, the helix handedness can be established by heterodimer formation: if an analog forms hybrid channels with gA (but not with an enantiomeric reference compound), the analog forms SS, RH $\beta^{6.3}$-helical channels. Figure 8 shows a current–transition amplitude histogram for a [Ala$^1$, D-Ala$^2$]gA/gA heterodimer formation (hybrid channel) experiment. The histogram shows three peaks, corresponding to the symmetric [Ala$^1$, D-Ala$^2$]gA channels (at 3.5 pA), the symmetric gA channels (at 2.8 pA), and an intermediate channel population that consists of [Ala$^1$, D-Ala$^2$]gA/gA heterodimers (at 3.2 pA). The corresponding lifetime distributions are shown in Fig. 8; again the hybrid channels have characteristics that are intermediate to those of the two symmetric channel types.

For SS $\beta^{6.3}$-helical channels, the structural identity of the analog channel type can be further defined by comparing the hybrid

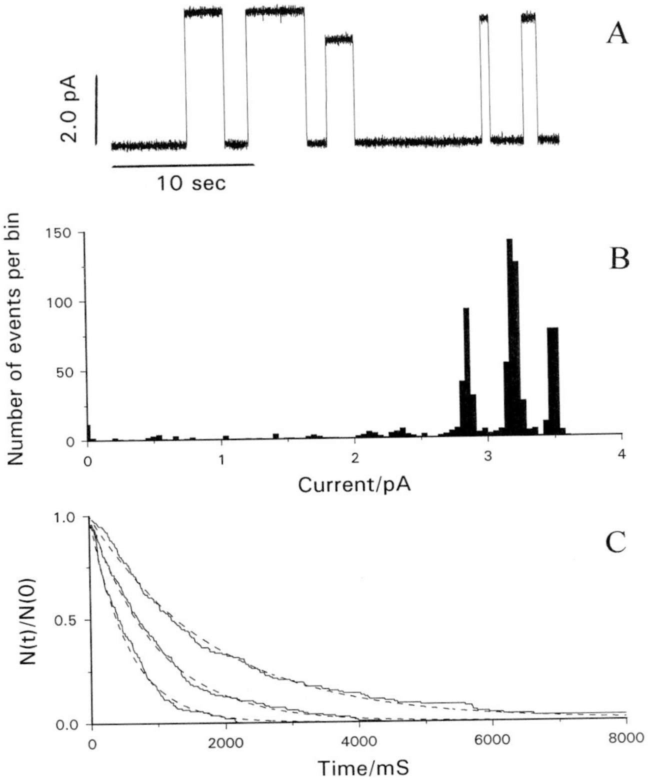


FIG. 8. Hybrid channel formation between [Ala[1], D-Ala[2]]gA and gA. (A) Single-channel current trace observed after addition of both gramicidins to both sides of the bilayer. There are three channel types: symmetric [Ala[1], D-Ala[2]]gA (the first two events); symmetric gA channels (the third event); and heterodimeric [Ala[1], D-Ala[2]]gA/gA channels (the last two events). (B) Current transition amplitude histogram. There are 829 events in the histogram. The three peaks represent (from left to right) symmetric gA channels, $i$ = 2.8 pA, 187 events (23%); heterodimeric [Ala[1], D-Ala[2]]gA/gA channels, $i$ = 3.2 pA, 363 events (44%); symmetric [Ala[1], D-Ala[2]]gA channels, $i$ = 3.5 pA, 172 events (21%). (C) Duration distributions for the three channel types. The solid curves denote normalized survivor plots; the interrupted curves denote maximum-likelihood fits of single-exponential distributions to the results. gA channels have the shortest duration ($\tau$ = 560 ms); the [Ala[1], D-Ala[2]]gA channels have the longest duration ($\tau$ = 1800 ms); [Ala[1], D-Ala[2]]gA/gA hybrid channels have an intermediate duration ($\tau$ = 980 ms). [Modified from G. L. Mattice *et al., Biochemistry* **34,** 6827 (1995) with permission. Copyright 1995 American Chemical Society.]

channel appearance rates and average lifetimes to those predicted from the appearance rates and average lifetimes of the two symmetric channel types.[14,16] The analysis is based on the following considerations (cf. Andersen *et al.*[11]):

1. The formation of heterodimers between two analogs, A and B (Fig. 8), is described by the formal reaction:

$$A_2 + B_2 \rightleftharpoons AB + BA$$

where the two heterodimer orientations can be distinguished if the energy profile for ion movement through the heterodimeric channels is asymmetric—such that the current in the A $\rightarrow$ B direction differs from the current in the B $\rightarrow$ A direction.[12,13,110] This usually is the case.

2. Single-channel experiments allow one to measure the relative appearance rates for the two symmetrical channels ($f_{AA}$ and $f_{BB}$) and for the heterodimeric channels ($f_{AB}$ and $f_{BA}$) simply by measuring the number of events in each peak in the current–transition amplitude histograms (e.g., Fig. 8). The corresponding dissociation rate constants can be determined from the lifetime distributions, as the reciprocal lifetimes ($\tau_{AA}$, $\tau_{BB}$, $\tau_{AB}$, and $\tau_{BA}$).

3. The difference in activation energy for heterodimer formation relative to homodimer formation ($\Delta\Delta G_f^\ddagger$) can be estimated from the channel appearance rates[16,111]:

$$\Delta\Delta G_f^\ddagger = \frac{-RT}{2} \ln \left\{ \frac{f_{AB}\, f_{BA}}{f_{AA}\, f_{BB}} \right\} \tag{2}$$

Similarly, the difference in activation energy for channel dissociation ($\Delta\Delta G_d^\ddagger$) can be determined from the average channel lifetimes as

$$\Delta\Delta G_d^\ddagger = \frac{RT}{2} \ln \left\{ \frac{\tau_{AB}\, \tau_{BA}}{\tau_{AA}\, \tau_{BB}} \right\} \tag{3}$$

The standard free energy for the heterodimeric channels relative to the symmetric channels ($\Delta\Delta G^0$) is given by[16]:

$$\Delta\Delta G^0 = \Delta\Delta G_f^\ddagger - \Delta\Delta G_d^\ddagger = \frac{-RT}{2} \ln \left\{ \frac{f_{AB}\, f_{BA}}{f_{AA}\, f_{BB}} \frac{\tau_{AB}\, \tau_{BA}}{\tau_{AA}\, \tau_{BB}} \right\} \tag{4}$$

[110] V. Fonseca, P. Daumas, L. Ranjalahy-Rasoloarijao, F. Heitz, R. Lazaro, Y. Trudelle, and O. S. Andersen, *Biochemistry* **31,** 5340 (1992).

[111] For Eq. (2) to be valid, it is important for the analogs to be added symmetrically, to both sides of the bilayer. Asymmetric distribution of the analogs between the two sides of the bilayer will increase the probability of observing heterodimer channels, as compared to the two symmetric channel types.

4. If there are no dimer-specific interactions between the monomers, i.e., if the monomer interactions in AB can be predicted simply from the monomer interactions in $A_2$ and $B_2$, then $\Delta\Delta G_f^\ddagger$, and $\Delta\Delta G_d^\ddagger$, and $\Delta\Delta G^0$ will all be zero, or

$$f_{AB} f_{BA} = f_{AA} f_{BB} \tag{5}$$

and

$$\tau_{AB}\ \tau_{BA} = \tau_{AA}\ \tau_{BB} \tag{6}$$

Dimer-specific interactions will be reflected primarily in the dissociation rate constants (cf. Durkin *et al.*[16]).

The validity of the heterodimer method for establishing the structural identity of channels formed by mutant gramicidins has been verified in experiments where the helix sense of mutant channels was investigated by CD spectroscopy, 2-D NMR spectroscopy, and electrophysiology.[9,35] The results obtained using these three independent methods are in full agreement.

## Summary

This article summarizes methods for the chemical synthesis and biophysical characterization of gramicidins with varying sequences and labels. The family of gramicidin channels has developed into a powerful model system for understanding fundamental properties, interactions, and dynamics of proteins and lipids generally, and ion channels specifically, in biological membranes.

# [29] Functional Analyses of Aquaporin Water Channel Proteins

*By* Peter Agre, John C. Mathai, Barbara L. Smith, and Gregory M. Preston

## Introduction

Water is the most abundant component of all living organisms, so the entry and release of water from cells must be considered a fundamental process of life. Red blood cells and multiple "leaky" epithelia such as renal tubules, salivary glands, and choroid plexus have long been thought to



0076-6879/99 $30.00

contain water-selective channels or pores,[1] which have been studied by several biophysical techniques.[2] Although the last 20 years have featured major advances in the recognition and molecular cloning of numerous salt, ion, and sugar transport proteins, the molecular identity of the membrane water transporters has eluded investigators until recently. This in part has resulted from the relatively high diffusional permeability of water through simple lipid bilayers, causing a significant background water permeability that foiled attempts to clone water channels by expression. Efforts to isolate candidate water transport molecules from leaky epithelia were also without success, and the ubiquity of water and its simple molecular structure has precluded chemical modifications for affinity labeling.

Identification of the first recognized water transport molecule resulted from the serendipitous observation made possible by the identification, purification, and cDNA cloning of a previously unknown 28-kDa membrane protein from red cells and renal tubules.[3–5] The distribution, abundance, and physical nature of this protein suggested that it might be the sought-after water channel, and this was demonstrated by expression of the cRNA in *Xenopus laevis* oocytes, which then became highly permeable to water.[6] Now designated AQP1, this protein is the archetypal member of the aquaporins,[7,8] a large family of water transport molecules found in vertebrates, invertebrates, plants, and microorganisms.[9] A large body of aquaporin literature is emerging, but most experimental methods are derived from those established for AQP1, which are outlined here.

## II. Purification of Red Cell AQP1 Protein

AQP1 is unique among membrane channels, since the native protein may be purified in milligram amounts from human red cells. Purification is made simple by the limited solubility of this protein in *N*-lauroylsarcosine.[3,4] Thus, while AQP1 makes up ~2.4% of the total red cell membrane protein, it is >50% pure after extraction of cytoskeleton-stripped membrane vesicles

[1] A. Finkelstein, "Water Movement through Lipid Bilayers, Pores, and Plasma Membranes: Theory and Reality." Wiley and Sons, New York, 1987.

[2] A. K. Solomon, *Methods Enzymol.* **173,** 192 (1989).

[3] B. M. Denker, B. L. Smith, F. P. Kuhajda, and P. Agre, *J. Biol. Chem.* **263,** 15634 (1988).

[4] B. L. Smith and P. Agre, *J. Biol. Chem.* **266,** 6407 (1991).

[5] G. M. Preston and P. Agre, *Proc. Natl. Acad. Sci. U.S.A.* **88,** 11110 (1991).

[6] G. M. Preston, T. P. Carroll, W. B. Guggino, and P. Agre, *Science* **256,** 385 (1992).

[7] P. Agre, S. Sasaki, and M. J. Chrispeels, *Am. J. Physiol. Renal* **265,** F461 (1993a).

[8] P. Agre, G. M. Preston, B. L. Smith, J. S. Jung, S. Raina, C. Moon, W. B. Guggino, and S. Nielsen, *Am. J. Physiol. Renal* **265,** F463 (1993b).

[9] M. J. Chrispeels and P. Agre, *Trends Biochem. Sci.* **19,** 421 (1994).

in *N*-lauroylsarcosine. Our current preparation technique is modified from published methods.[3,4,10–12]

*Materials*

- Two units of anticoagulated human red blood cells (American Red Cross, local affiliate)
- BPF4 high-efficiency leukocyte removal filters (Pall Biomedical, East Hills, NY)
- Wash buffer: 150 m*M* NaCl, 7.5 m*M* sodium phosphate (pH 7.4), 1 m*M* NaEDTA
- Lysis buffer: 7.5 m*M* sodium phosphate (pH 7.4), 1 m*M* NaEDTA, 10 $\mu M$ leupeptin, 0.5 m*M* diisopropyl fluorophosphate (DFP)
- 1 *M* potassium iodide
- Sarcosine buffer: 1% (w/v) *N*-lauroylsarcosine, 1 m*M* $NH_4HCO_3$, 1 m*M* $NaN_3$, 1 m*M* dithiothreitol (DTT) (pH 7.8)
- Triton X-100 (except where noted, reagents were from Sigma, St. Louis, MO)
- Octylglucoside (*n*-octyl-$\beta$-D-glucopyranoside, Calbiochem, San Diego, CA)
- Chromatography buffer: 20 m*M* Tris-HCl (pH 7.8), 1 m*M* $NaN_3$, 1 m*M* DTT
- 0.22-$\mu$m Millex GV membranes (Millipore, Bedford, MA)
- 10- × 100-mm POROS Q/F column and 10- × 100-mm POROS HQ/F column (PerSeptive Biosystems, Framingham, MA)

*Procedures*

*Preparation of Red Cell Membrane Vesicles.* Two units of anticoagulated human red cells (~200 ml packed red cells/unit) obtained within 5 days after blood donation are passed through leukocyte removal filters, and washed with wash buffer. Membranes are prepared by lysing the washed red cells in hypotonic lysis buffer, and membrane skeleton proteins are extracted with 1 *M* potassium iodide.[13] The extracted membrane vesicles from the 400 ml of red cells are further extracted by shaking for 1 hr at 22° in 800 ml of sarcosine buffer and pelleted at 30,000*g* for 4 hr at 4° in a Beckman JA-14 rotor. The pellet is washed once in 7.5 m*M* sodium phosphate (pH 7.4) and solubilized by shaking for 1 hr at 22° in 1.2 liters of

[10] S. Nielsen, B. L. Smith, E. I. Christensen, M. Knepper, and P. Agre, *J. Cell Biol.* **120,** 371 (1993).

[11] M. L. Zeidel, S. V. Ambudkar, B. L. Smith, and P. Agre, *Biochemistry* **31,** 7436 (1992).

[12] M. L. Zeidel, S. Nielsen, B. L. Smith, S. V. Ambudkar, A. B. Maunsbach, and P. Agre, *Biochemistry* **33,** 1606 (1994).

[13] V. Bennett, *Methods Enzymol.* **96,** 313 (1983).

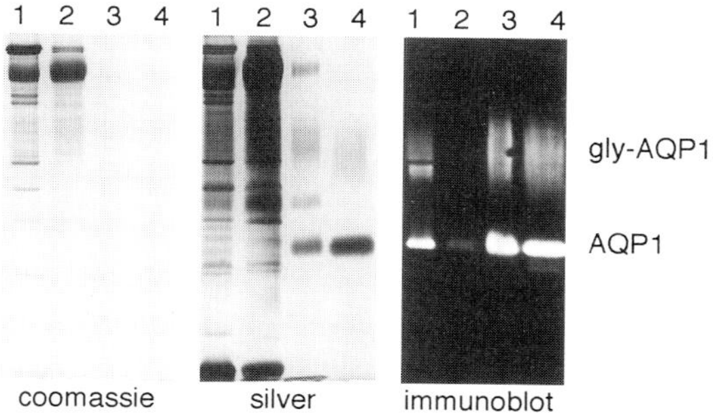


FIG. 1. Purification of AQP1 protein from human red blood cells. Panels correspond to membranes analyzed by SDS–PAGE with Coomassie staining (left), silver staining (middle), or after transfer to an immunoblot incubated with anti-AQP1 IgG and visualized by ECL (Amersham). Individual lanes: 1, whole red cell membranes (6 $\mu$g); 2, *N*-lauroylsarcosine-soluble proteins (6 $\mu$g); 3, *N*-lauroylsarcosine-insoluble proteins (0.4 $\mu$g); 4, pure AQP1 (0.4 $\mu$g). Note that AQP1 is a homotetramer comprised of three nonglycosylated subunits (AQP1) and one subunit with a polylactosaminoglycan attached (Gly-AQP1).[3,4]

chromatography buffer containing 4% Triton X-100 (v/v). After a 4-hr centrifugation at 4° as above, the supernatant is filtered through 0.22-$\mu$m Millex GV membranes. AQP1 is the major protein present at this stage when analyzed by SDS–PAGE with silver staining (Fig. 1).

*Chromatographic Procedures.* A 600-ml aliquot of the filtrate is loaded onto a 10- × 100-mm POROS Q/F column equilibrated with chromatography buffer containing 0.1% Triton X-100 driven at 3 ml/min by an FPLC apparatus (Pharmacia, Piscataway, NJ). The UV monitor is used to measure $A_{280}$, but the baseline must be adjusted to compensate for absorption by 0.1% Triton X-100. The column is washed with 40 ml of chromatography buffer and eluted with a 120 ml 0.2–0.6 *M* NaCl gradient at 4 ml/min while $A_{280}$ is recorded (elution no. 1). Peaks eluted at ~0.3 *M* NaCl from two runs are combined, diluted 7- to 10-fold with chromatography buffer containing 1.2% octylglucoside and loaded onto a 10- × 100-mm POROS HQ/F column equilibrated with the same buffer while running at 2 ml/min. The column is washed until the $A_{280}$ baseline is recovered, and then eluted at 1 ml/min with a 40-ml gradient of 0–0.6 *M* NaCl in the same buffer (elution no. 2). The peak fractions eluting at ~0.3 *M* NaCl are combined and diluted 15- to 20-fold in chromatography buffer containing 1.2% octylglucoside and reloaded onto the same POROS HQ/F column at 2 ml/min, washed with 40 ml, and concentrated by elution at 1 ml/min with a 0–0.6 *M* NaCl gradient of 15 ml (elution no. 3). Peak fractions collected in 250-$\mu$l volumes elute at ~0.3 *M* NaCl and range in concentration from 1 to 3 mg/ml (BCA protein method, Pierce, Rockford, IL). The fractions are snap frozen in dry ice and stored at $-80°$ until reconstitution. SDS–PAGE analysis with silver staining reveals the final product to be ~99% pure (Fig. 1), and a two-unit purification yields approximately 5 mg of AQP1.

*Comments*

If the highest purity is not essential or final elution into octylglucoside is not required, substantially larger yields of less concentrated AQP1 may be obtained by eliminating anion-exchange elution no. 2 or no. 3. Because the critical step in the purification comes at the *N*-lauroylsarcosine extraction, the subsequent anion-exchange steps may be varied. We have found that the POROS columns have excellent flow properties but may deteriorate after repeated exposure to detergents and often cannot be regenerated. Also, somewhat different elution profiles have been noted for columns with different lot numbers. Thus, chromatography fractions are routinely analyzed by SDS–PAGE with visualization by silver staining, since the protein is poorly visible after Coomassie blue staining (Fig. 1). Other prepacked, high-performance anion-exchange columns may also be used (e.g., Mono Q HR, Pharmacia). Alternatively, less expensive anion-exchange beads may be purchased in bulk, packed into columns, and periodically replaced with fresh material (e.g., Source 15Q or 30Q, Q-Sepharose HP or Fast Flow, Pharmacia).

## III. Expression of AQP1 in Yeast

Temperature-sensitive *sec6-4* mutant strains of yeast are a powerful system for producing membrane proteins for functional and structural studies. Although AQP1 can be purified from red cells (Section II), other aquaporins are less abundant and are expressed in complex tissues, making their purifications impractical. The yeast expression system also offers some advantages to the oocyte system (Section VII), since failure of AQP1 mutants to increase the water permeability when expressed in oocytes often reflects defective targeting to the plasma membrane.[14] Quantities of up to 1 mg of native AQP1 and mutant forms of AQP1 can be generated with these methods, although for most studies smaller scale preparations are convenient and provide approximately 100 $\mu$g yields. The same approach is used for preparation of AQP2 protein.[15]

[14] J. S. Jung, R. V. Bhat, G. M. Preston, W. B. Guggino, J. M. Baraban, and P. Agre, *Proc. Natl. Acad. Sci. U.S.A.* **91,** 13052 (1994).

[15] L. A. Coury, J. C. Mathai, B. V. R. Prasad, J. L. Brodsky, P. Agre, and M. L. Zeidel, *Am. J. Physiol. Renal* **274,** F34 (1998).

*Materials*

Yeast strains: temperature-sensitive SY1 strain of *Saccharomyces cerevisiae* (*MATα, ura3-52, leu2-3,112, his4-619, sec6-4, GAL*[16]), (Source: Dr. C. W. Slayman, Department of Genetics, Yale University School of Medicine, New Haven, CT)

Defined minimal media: yeast nitrogen base, 6.7 g/liter (Bio 101, Vista, CA) 20 mg/liter histidine, 30 mg/liter leucine, 2% raffinose

YEP–galactose medium: 0.5% yeast extract; 1% Bacto-peptone (Difco Labs, Detroit MI); 2% galactose (w/v)

Spheroplasting medium: 1.4 *M* sorbitol, 50 m*M* $K_2HPO_4$ (pH 7.5), 10 m*M* $NaN_3$, 40 m*M* 2-mercaptoethanol

Lysis buffer: 0.8 *M* sorbitol, 10 m*M* triethanolamine (pH 7.2), 1 m*M* NaEDTA

Lyticase and concanavalin A (Con A) (Sigma Chemical Co., St. Louis MO)

100 m*M* $MnCl_2$ and 100 m*M* $CaCl_2$

Dounce homogenizer with pestle A (Wheaton, Millville, NJ)

*Procedures*

*pYES2-AQP1 Plasmid Construction and Yeast Cultures.* Aquaporin cDNAs are cloned into the pYES2 plasmid (Invitrogen, Carlsbad, CA) using standard molecular biological methods, with orientation confirmed by restriction digestions (978-bp DNA fragment containing human *AQP1* is released with *Bcl*I and *Xba*I) and dideoxynucleotide sequencing. The SY1 strain is transformed with pYES2-AQP1 by the lithium acetate method[16a] and is maintained in uracil-deficient defined media. One liter of culture (200 ml media in five 1000-ml culture flasks) is grown at room temperature to mid-log phase ($OD_{600} \sim 1$) and cells are harvested by centrifugation at 2000*g* for 10 min at room temperature. Production and accumulation of secretory vesicles is induced by transfer to 1 liter of rich YEP–galactose medium during overnight incubation at 37°. Sodium azide is added to a final concentration of 10 m*M* 10 min prior to centrifugation at 2000*g* for 10 min at 4°.

*Spheroplast Preparation and Secretory Vesicle Isolation.* The yeast cells are suspended to a density of 50 $OD_{600}$ units/ml in spheroplasting medium containing lyticase (0.2 mg/ml) and incubated at 37° for 45 min. Spheroplasts are pelleted by low-speed centrifugation (1000*g*, 15 min, 4°) and resuspended in the same volume of spheroplasting media containing 1 m*M* $MnCl_2$ and 1 m*M* $CaCl_2$ and incubated with Con A (25 mg/1600 $OD_{600}$ units) for 15 min at 0°. Spheroplasts are again pelleted by centrifugation, resuspended

[16] R. K. Nakamoto, R. Rao, and C. W. Slayman, *J. Biol. Chem.* **266,** 7940 (1991).

[16a] H. Ito, Y. Fukuda, K. Murata, and A. Kimura, *J. Bacteriol.* **153,** 163 (1983).

at 80 $OD_{600}$ units/ml in cold lysis buffer, and lysed at 4° with 25 strokes of the pestle A in a Dounce homogenizer. The lysate is spun at 10,000*g* for 10 min at 4° to remove unbroken cells, nuclei, mitochondria, and cell debris. Secretory vesicles are collected by centrifugation of supernatant at 100,000*g* for 1 hr in a Beckman Type 50 Ti rotor at 4°.[16] Protein expression is confirmed by immunoblotting with anti-AQP1.[4] For many studies, the isolated membranes may be used directly; however, AQP1 may be partially purified from isolated secretory membrane vesicles by extraction with 0.6% *N*-lauroylsarcosine (w/v) yielding aquaporin proteins that are ~50% pure at up to 1 mg/liter of culture. The protein may be solubilized in 4% Triton X-100 (v/v) or 1.2% octylglucoside (w/v) using conditions outlined in Section II.

## IV. Reconstitution of AQP1 into Proteoliposomes

Investigators studying bacterial transport proteins have routinely characterized the functions of purified membrane proteins after reconstitution into lipid bilayers. Although it is expensive, the detergent octylglucoside is particularly useful, since it has a relatively high micellar concentration. By rapidly lowering the octylglucoside concentration, solubilized phospholipids will form liposomes with integral membrane proteins imbedded in the lipid bilayer.[17,18] We have adapted this method to reconstitute purified human red cell AQP1 into proteoliposomes.[11,12]

### *Reconstitution of Purified Red Cell AQP1*

*Materials*

Octylglucoside (octyl-β-D-glucopyranoside, Calbiochem, San Diego, CA), 15% (w/v) in water, stored at −80°

*Escherichia coli* lipids (bulk phospholipid, Avanti Polar Lipids, Birmingham AL) depleted of neutral lipid by acetone/ether wash,[17] 50 mg/ml in 2 m*M* 2-mercaptoethanol; stored at −70° under nitrogen

100 μg AQP1 protein in 1.2% octylglucoside buffer (Section II)

Carboxyfluorescein (Molecular Probes, Eugene, OR)

Millipore filter: Pore size 1.2 μm, filter type RA (Millipore Co., Bedford, MA)

Individual stocks solutions: 500 m*M* Tris-HCl (pH 7.5), 10 m*M* $NaN_3$, 100 m*M* DTT, and 500 m*M* MOPS [3-(*N*-morpholino)propanesulfonic acid, pH 7.5]

[17] S. V. Ambudkar and P. C. Maloney, *Methods Enzymol.* **125,** 558 (1986a).

[18] S. V. Ambudkar and P. C. Maloney, *J. Biol. Chem.* **261,** 10079 (1986b).

Reconstitution buffer: 50 m*M* MOPS (pH 7.5), 150 m*M* *N*-methyl-D-glucamine chloride (neutralized with HCl to pH 7.0), 10–15 m*M* carboxyfluorescein, 1 m*M* DTT, and 0.5 m*M* phenylmethylsulfonyl fluoride (PMSF)

*Procedure.* The reconstitutions are performed in a final volume of 1.0 ml of 50 m*M* Tris-HCl (pH 7.5), 1 m*M* $NaN_3$, 1 m*M* DTT, 1.25% (w/v) octylglucoside, 100 $\mu$g pure AQP1 protein (Section II), and 9 mg of bath-sonicated, purified *E. coli* phospholipid. The AQP1 and phospholipid mixture is vortex mixed and incubated for 20 min at 0°. Proteoliposomes form at room temperature by rapidly injecting the mixture through a 23-gauge needle into 25 ml of reconstitution buffer. Liposomes with simple lipid bilayer membranes are prepared identically, except AQP1 protein is not included. Proteoliposomes or liposomes are then collected by centrifugation for 1 hr at 123,000*g* in a Beckman Type 42.1 rotor at 4°. The pellets are resuspended in 8 ml of reconstitution buffer and then centrifuged at 10,000*g* for 15 min in a Beckman Type 50 Ti rotor to remove any lipid aggregates. The supernatant is then centrifuged for 1 hr at 152,000*g* in the same rotor. Proteoliposomes or liposomes are resuspended in 0.3 ml of reconstitution buffer for up to 24 hr at 4°.

Proteoliposome and liposome sizes are determined by negative staining electron microscopy or by dynamic light scatter in a DynaPro-801 (Protein Solutions, Charlottesville, VA). Most preparations yield vesicles 0.1–0.15 $\mu$m in diameter. Proteoliposomes contain multiple intramembranous particles (each representing an AQP1 tetramer) which may be examined by freeze-fracture electron microscopy (Fig. 2). Protein was measured by a modification of the Schaffner and Weissman method.[19] The final concentration of sodium dodecyl sulfate (SDS) in the assay is increased by 1%, and the protein is precipitated with 12% trichloroacetic acid and vacuum filtered onto a 48-mm-diameter Millipore filter disk.[18] Phospholipid is estimated by assuming 70% recovery in the proteoliposomes but may be accurately determined by the method of Hallen.[20]

### *Comments*

In a typical reconstitution, about one-half of the purified red cell AQP1 protein and two-thirds of the pure phospholipid are incorporated into proteoliposomes. The protein : lipid ratios may be varied from 1 : 125 to 1 : 25 (or even higher) and yield correspondingly increased water permeabilities.[12] The purified phospholipid is comprised of phosphatidylethanolamine

[19] W. Schaffner and C. Weissman, *Anal. Biochem.* **56,** 502 (1973).

[20] R. M. Hallen, *J. Biochem. Biophys. Methods* **2,** 251 (1980).

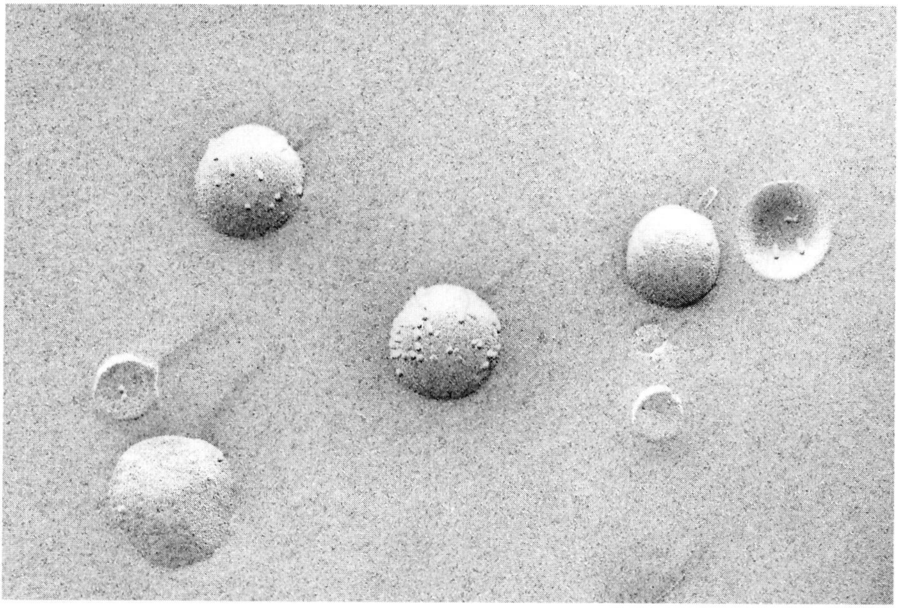

FIG. 2. Freeze-fracture electron micrographs of proteoliposomes reconstituted with AQP1 at a protein : lipid ratio of 1 : 50. Intramembranous particles represent AQP1 tetramers. Magnification: ×90,000. [Modified and reprinted from M. L. Zeidel *et al.*, *Biochemistry* **33,** 1606 (1994) with permission. Micrograph provided by Søren Nielsen and Arvid Maunsbach, Department of Cell Biology, Institute of Anatomy, University of Aarhus, Denmark.]

(70%), phosphatidylglycerol (15%), and cardiolipin (15%)[21]; however, experiments with other phospholipids yield proteoliposomes that are functionally similar.[12] Internal volumes of 1 μl/mg phospholipid are obtained for proteoliposomes.[18]

## *Reconstitution of AQP1 Expressed in Yeast*

### *Materials*

Secretory vesicles or membranes (8–12 mg/ml) containing AQP1 (Section II)

Solubilization buffer: 1.5% (w/v) octylglucoside, 50 m*M* MOPS (pH 7.5), 20% (v/v) glycerol, 1 m*M* DTT, 0.4% *E. coli* phospholipids, 0.5 m*M* PMSF

Reconstitution buffer: (see Section IV)

[21] C.-C. Chen and T. H. Wilson, *J. Biol. Chem.* **259,** 10150 (1984).

*Procedures*

SOLUBILIZATION OF SECRETORY VESICLES. Solubilization buffer should be mixed well before addition of octylglucoside. One milliliter of secretory vesicles (8–12 mg/ml) is mixed with 3 ml of solubilization buffer, incubated for 20 min on ice, and centrifuged at 100,000*g* for 1 hr at 4°. The supernatant is used for reconstitution.

RECONSTITUTION OF LIPOSOMES. Reconstitution is performed in a final volume of 1.14 ml containing 1.4% octylglucoside, 6 mg of bath-sonicated *E. coli* phospholipid, and 1.0 ml of the above supernatant. This mixture is incubated on ice for 20 min and proteoliposomes are formed by rapid injection of the mixture into 25 ml of reconstitution buffer at room temperature. Proteoliposomes are collected and washed by centrifugation as described in Section IV.

*Comments*

Although the methods for reconstitution of heterologously expressed AQP1 and other aquaporins are still being established, the technique promises an approach for physical analysis of these proteins. Note that as a cost-effective alternative, carboxyfluorescein may also be loaded into proteoliposomes by overnight incubation of proteoliposomes at 0° in 1 ml of reconstitution buffer containing 15 m*M* of carboxyfluorescein. The carboxyfluorescein-loaded vesicles are then washed just prior to use by centrifugation as described in Section IV.

## V. Water Permeability of AQP1 Proteoliposomes

The water permeability of AQP1 has been determined by comparing wild-type red cells and AQP1-deficient red cells.[22] The possibility of measuring water permeability of purified, reconstituted AQP1 should be of greater general usefulness and is described here. By directly measuring the coefficients of osmotic water permeability of proteoliposomes containing known concentrations of AQP1, the unit permeability may be calculated for the AQP1 tetramer, which is comprised of four independently functional subunits. Moreover, by substituting other solutes for sucrose, the permeability of a variety of osmolytes may be measured.

[22] J. C. Mathai, S. Mori, B. L. Smith, G. M. Preston, N. Mohandas, M. Collins, P. C. M. van Zijl, M. L. Zeidel, and P. Agre, *J. Biol. Chem.* **271,** 1309 (1996).

*Materials*

Anti-fluorescein antibody (Molecular Probes).

Isotonic solution: reconstitution buffer without carboxyfluorescein (Section IV); measured osmolality 356 mOsm by using an osmometer such as the Vescor 5100C vapor pressure osmometer (Vescor Inc., Logan, UT).

Hypertonic solution: add pure sucrose to isotonic solution to increase total osmolality of the solution 3-fold. For example, add 19.0 g of sucrose to 100 ml of reconstitution buffer (736 mOsm + 356 mOsm = 1092 mOsm) and mix 1 : 1 with isotonic solution by stopped flow. The final osmolality will be approximately twice the original osmolality (724 mOsm). Note that osmolalities of sucrose solutions are not directly related to calculated molarities but are listed in *The Handbook of Chemistry and Physics,* CRC Press, Boca Raton, FL.

AQP1 inhibitors: 0.5 m*M* $HgCl_2$ and 1 m*M* *p*-chloromercuribenenesulfonate.

*Procedure*

Measurements of the coefficient osmotic water permeability ($P_f$) are performed on AQP1 proteoliposomes and control liposomes (Section IV). Extravesicular carboxyfluorescein is removed by centrifugation (Section IV) just before the water permeability measurements, if carboxyfluorescein is loaded into proteoliposomes by overnight incubation. $P_f$ is measured as described[11] by abruptly exposing proteoliposomes to an increase in extravesicular osmolality using a stopped-flow fluorimeter (SF.17MV, Applied Photophysics, Leatherhead, UK) with a measured dead time of 0.7 ms. Excitation wavelength is 490 ± 1.5 nm, using a monochrometer, and emission wavelength is >515 nm using a cut-on filter (Oriel Corp., Stratford, CT). Extravesicular carboxyfluorescein fluorescence is completely quenched by using anti-fluorescein antibody of 1 : 1000 dilution. Vesicles act as perfect osmometers, and relative volume (absolute volume divided by initial volume) is linearly related to relative fluorescence (absolute fluorescence divided by initial fluorescence). Averaged data from multiple determinations are fitted to single exponental curves using software provided by Applied Photophysics.[11] Fitting parameters are then used to determine $P_f$ exactly as described[11,23] using the equation:

$$dV(t)/dt = (P_f)(SAV)(MVW)\{[C_{in}/V(t)] - C_{out}\}$$

[23] N. Priver, E. C. Rabon, and M. L. Zeidel, *Biochemistry* **21,** 2459 (1993).

where $V(t)$ is the relative volume as a function of time, $SAV$ is the surface area to volume ratio, $MVW$ is the molar volume of water (18 $cm^3$/mol), and $C_{in}$ and $C_{out}$ are the initial concentrations of total solute inside and outside the vesicle.

*Comments*

At 37°, typical measurements yield estimates of $P_f \sim 500 \times 10^{-4}$ cm/sec (for AQP1 proteoliposomes) versus $\sim 100 \times 10^{-4}$ cm/sec (for control liposomes). Note that the background permeability rapidly declines at lower temperatures, yielding a relatively larger value of $P_f$ to background. By measuring the $P_f$ at a range of temperatures (8–22°), the Arrhenius activation energies may be computed from the slope of a plot of $P_f$ and absolute temperature (ln $P_f$ vs. $1/T$, K). The slope multiplied by gas constant $R$ (1.986 cal/K/mol) will give the activation energy. AQP1 proteoliposomes are routinely found to have $E_a < 5$ kcal/mol, which is equivalent to the diffusion of water in bulk solution. In contrast, control liposomes yield values of >10 kcal/mol, which represents a significant barrier to water flow. When incubated for 30 min at 37° in submillimolar concentrations of the known inhibitors $HgCl_2$ or *p*-chloro-mercuribenzene sulfonate, the $P_f$ of AQP1 proteoliposomes is reduced to ~20% of the original level, indicating that this method may be very useful for screening for new potential inhibitors.

The number of AQP1 tetramers per milliliter of suspension is calculated from the amount of protein/ml in each preparation and the molecular weight of the AQP1 tetramer (4 × 28.5 kDa). The measured total entrapped volume/mg phospholipid, the calculated volume of each proteoliposome, and the total number of proteoliposomes/ml of suspension are calculated from the total phospholipid content. The number of AQP1 tetramers/ml divided by the number of proteoliposomes/ml yields the numbers of AQP1 tetramers/proteoliposome. Thus the subunit permeability may be calculated by multiplying the $P_f$ (cm/sec) by the surface area of each proteoliposome ($6.2 \times 10^{-10}$ $cm^2$) and dividing by four times the number of AQP1 tetramers reconstituted per proteoliposome. This typically yields values of $\sim 5 \times 10^{-10}$ cm/sec/AQP1 subunit (equivalent to $2 \times 10^9$ water molecules per subunit per second).

Heterologous expression of AQP1 in yeast will permit the assessment of osmotic water permeability of wild-type or site-directed AQP1 mutants. This may be performed directly on the secretory vesicles isolated from yeast or reconstituted vesicles containing expressed AQP1. Osmotic water permeability is measured and calculated as describe above by abrupt exposure of vesicles to increased external osmolality using the stopped-flow fluorimeter (Fig. 3).

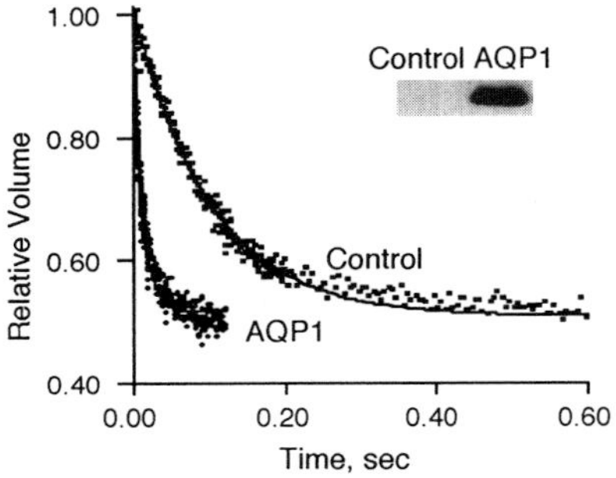


FIG. 3. Measurement of osmotic water permeability ($P_f$) of reconstituted proteoliposomes. Tracings are from reconstituted membranes analyzed by stopped-flow transfer to hyperosmolar solution (100% increase in osmolality). AQP1 proteoliposomes reconstituted with vesicle protein from AQP1 transformed *sec*⁻ yeast ($P_f = 70$ μm/sec) are compared to control proteoliposomes reconstituted with vesicle protein from untransformed *sec*⁻ yeast ($P_f = 6.2$ μm/sec). *Inset:* Immunoblot (2 μg protein per lane) of control proteoliposomes and AQP1 proteoliposomes.

## VI. Homology Cloning of Aquaporins by Degenerate Oligonucleotide PCR

All members of the aquaporin family of water channels are encoded by DNA sequences that represent the duplication of an ancestral gene. Each tandem repeat contains the highly conserved NPA motif (asparagine-proline-alanine) and other conserved amino acids.[5] Since the cloning and expression of AQP1, multiple aquaporin homologs have been cloned using degenerate oligonucleotide primers corresponding to the most highly conserved regions of the AQP1 molecule for amplification by polymerase chain reaction (PCR) of the intervening sequence (Fig. 4). The amplified cDNAs are then used to probe the appropriate libraries for full-length cDNAs. Examples where this approach was used include the cloning of *AQP4* from brain,[14] *AQP5* from salivary gland,[24] and *aqpZ* from *E. coli*.[25]

### *Materials*

10× PCR reaction buffer: 100 m*M* Tris-HCl (pH 8.3 at 25°), 500 m*M* KCl, 15 m*M* $MgCl_2$, 0.1% w/v gelatin. Incubate at 50° to melt the gelatin, filter sterilize, and store at −20° in aliquots.

[24] S. Raina, G. M. Preston, W. B. Guggino, and P. Agre, *J. Biol. Chem.* **270,** 1508 (1995).
[25] G. Calamita, W. R. Bishai, G. M. Preston, W. B. Guggino, and P. Agre, *J. Biol. Chem.* **270,** 29063 (1995).

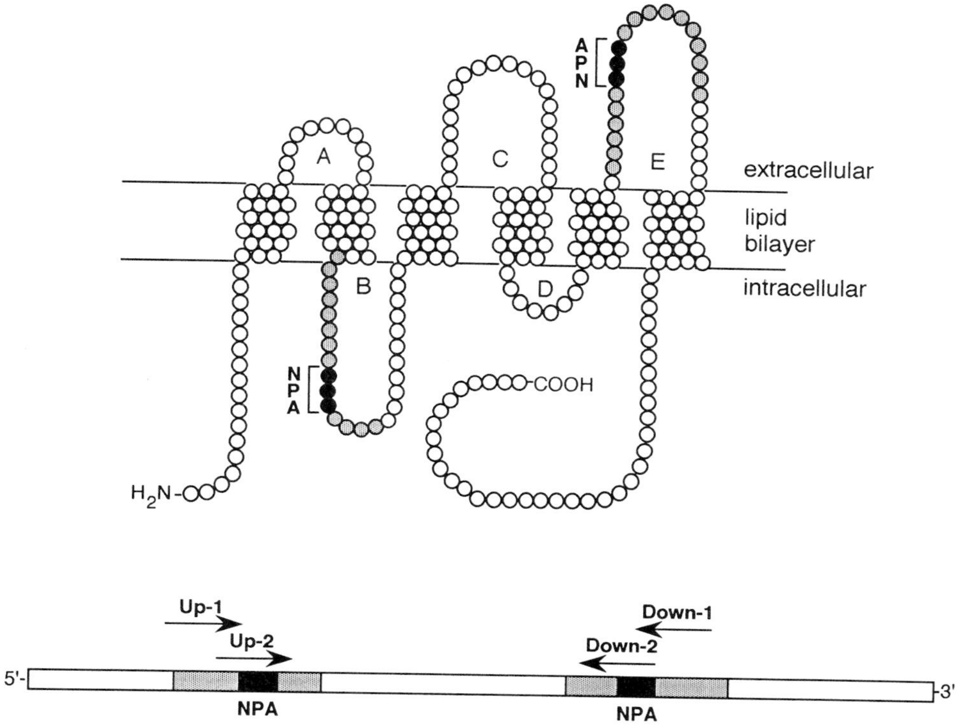


FIG. 4. Highly conserved domains with aquaporins are used for homology cloning by PCR. *Top:* Membrane topology model of AQP1 showing areas of high conservation among different members of the aquaporin family (loop B and loop E). Residues shared among most aquaporins (shaded circles) and NPA motifs absolutely conserved among all aquaporins (black circles). *Bottom:* Model of aquaporin cDNA showing sites for hybridization of forward (Up-1 and Up-2) and reverse (Down-1 and Down-2) degenerate oligonucleotide primers (see text).

dNTP stock solution (1.25 m*M* dATP, dGTP, dCTP, dTTP) made by diluting commercially available deoxynucleotides with sterile water.

Thermostable DNA polymerase (such as AmpliTaq DNA polymerase, Perkin-Elmer-Cetus Norwalk, CT) supplied at 5 units/$\mu$l.

Mineral oil.

A programmable thermal cycler machine (available from Perkin-Elmer-Cetus, MJ Research Inc., Stratagene, and other manufacturers).

The following degenerate oligonucleotide primers should be purified by reversed-phase HPLC or by elution from acrylamide gels. The primers should be resuspended at 20 pmol/$\mu$l in sterile water, and stored at $-20°$ in aliquots.

*Forward primers:*

Up-1 5′-STB GGN CAY RTB AGY GGN GCN CA-3′

Up-2 5′-G GGA TCC GCH CAY NTN AAY CCH GYN GTN GTN AC-3′

*Reverse primers:*

Down-1 5′-GC DGR NSC VAR DGA NCG NGC NGG-3′

Down-2 5′-CGG AAT TCG DGC DGG RTT NAT NSH NSM NCC-3′

Degenerate codes: B = G, C, or T (not A); D = A, G, or T (not C); H = A, C, or T (not G); K = G or T (keto); M = A or C (amino); N = any (A, G, C, or T); R = A or G (purine); S = G or C (strong); V = A, C, or G (not T); W = A or T (weak); Y = C or T (pyrimidine). The underlined sequences correspond to *Bam*HI and *Eco*RI restriction sites. A related series of degenerate oligonucleotide primers with broader reactivity for nonmammalian aquaporins has also been published.[26]

The DNA template can be almost any DNA sample, including a single-stranded cDNA from a reverse transcription reaction, DNA from a phage library, or genomic DNA. The DNA is heat denatured at 99° for 10 min and stored at $-20°$.

Chloroform.

Tris-saturated phenol: prepared using ultrapure redistilled crystalline phenol as recommended by the supplier (GIBCO-BRL, Gaithersburg, MD). Use polypropylene or glass tubes for preparation and storage.

PC9: mix equal volumes of buffer-saturated phenol (pH > 7.2) and chloroform, extract twice with an equal volume of 100 m*M* Tris (pH 9.0), separate phases by centrifugation at room temperature for 5 min at 2000*g*, and store at 4° to $-20°$ for up to 1 month.

7.5 *M* Ammonium acetate is preferred to sodium acetate for precipitation of DNA because nucleotides and primers generally do not precipitate. Dissolve ammonium acetate in water, filter through 0.2-$\mu$m membrane, and store at room temperature.

For the elution of specific PCR-amplified DNA products from agarose gels, we have observed consistent results with the QIAEX gel extrac-

[26] G. M. Preston, *in* "Methods in Molecular Biology" (B. A. White, ed.), Vol. 67, pp. 433–449. Human Press Inc., Totowa, New Jersey, 1997.

tion kit (Qiagen Inc., Santa Clarita, CA) for products from 50 bp to 5 kbp, however several other kits are available.

*Procedures*

*PCR Amplification Reaction.* The DNA template should be PCR amplified with all four possible pair-combinations of forward and reverse primers (Up-1 and Down-1, Up-1 and Down-2, Up-2 and Down-1, Up-2 and Down-2), as well as the individual degenerate primers to determine if any of the bands amplified are derived from one of the degenerate primer pools. A DNA-free control is required to assess if there is any contaminating DNA in any of the other reagents. Pipette into 0.5-ml microcentrifuge tubes in the following order: 58.5 $\mu$l sterile water (0.2 $\mu$m filtered and autoclaved); 10 $\mu$l 10× PCR reaction buffer; 16 $\mu$l 1.25 m$M$ dNTP stock solution; 5.0 $\mu$l primer Up-1 or Up-2; 5.0 $\mu$l primer Down-1 or Down-2; 5.0 $\mu$l heat denatured DNA (1–100 ng). If several reactions are being set up concurrently, a master reaction mix can be made up of the reagents used in all of the reactions. Samples are briefly vortex mixed and spun for 10 sec in a microfuge. Each sample is overlaid with two to three drops of mineral oil. Amplify by hot-start PCR by pausing the thermocycler at step 4–cycle 1 and add 0.5 $\mu$l AmpliTaq DNA polymerase to each tube. Cycle in the following protocol: step 1, 95° for 3 min; step 2, 95° for 10 sec (denaturation); step 3, 50° for 30 sec (annealing); step 4, 72° for 30 sec (extension); step 5, cycle 29 or 34 times to step 2; step 6, 72° for 4 min; step 7, 10° hold.

*DNA Isolation and Gel Electrophoresis Analysis.* The reaction tubes are removed from the thermal cycler, 200 $\mu$l chloroform is added, and the tubes are spun for 10 sec in a microfuge to separate oil–chloroform layer from the aqueous layer. The aqueous layers are transferred to clean microfuge tubes, and the AmpliTaq DNA polymerase is removed by extracting the aqueous phase twice with 100 $\mu$l PC9. The tubes are spun for 2 min to separate the lower organic layer from the upper aqueous layer, and the aqueous layers are transferred to clean microfuge tubes. Ammonium acetate–ethanol precipitation of cDNA in 100-$\mu$l samples is performed by adding 50 $\mu$l 7.5 $M$ ammonium acetate plus 350 $\mu$l 100% ethanol. The samples are vortex mixed for 15 sec, placed on ice for 15 min, and the DNA pelleted at 12,000$g$ for 15 min at 4° in a microfuge. The aqueous waste is decanted, replaced with 250 $\mu$l 70% ethanol, and briefly vortex-mixed and spun 5 min at 4°. The ethanol is decanted and the pellets are allowed to dry inverted at room temperature before resuspending in 20 $\mu$l sterile water. Aliquots of the PCR fragments (2–10 $\mu$l) may be resolved by gel electrophoresis through 2–3% NuSieve agarose gels (FMC BioProducts, Rockland,

ME). The DNA is stained by soaking the gel for 5–30 min in about 10 volumes of water containing 1 $\mu$g/ml ethidium bromide and photographed under UV light. Since the NPA motifs are separated by 110–140 amino acids within all known members of the aquaporin gene family, PCR amplification of the known aquaporin cDNAs using these internal degenerate primers will yield products of approximately 400 bp (Fig. 4).

*Secondary PCR Amplifications and DNA Purification.* Based on the results from gel electrophoresis of the PCR amplified DNA products, multiple subsequent approaches may be made: (1) the initial DNA sample may be PCR amplified under different conditions (altered $MgCl_2$ concentration, annealing temperature, or primers); (2) a different DNA template may be used for PCR amplification; (3) one or more bands from the agarose gel may be eluted for subcloning or reamplification by PCR; (4) the product may be reamplified by PCR with an internal pair of degenerate primers. Options 1 and 2 are self-explanatory. Depending on whether no band is visible on the agarose gel or if multiple product bands are detected, the amount of template DNA, the $MgCl_2$ concentration, or the primer annealing temperatures may be altered. The goal is to obtain cDNA products that are detectable as bands on the agarose gels which are in the 340- to 420-bp range. If some products are detected in this size range, they may be purified with the QIAEX gel extraction kit (Qiagen Inc.) and either reamplified with the same primer set, an internal primer set (to be designed), or the cDNAs may be directly subcloned. When attempting to identify an aquaporin cDNA homolog from a tissue that is known to express recognized aquaporins, the final PCR sample may be enriched for novel homologs. Because the degenerate oligonucleotide primers are designed from the sequences of known members of the aquaporin family, these primers will likely be biased for those homologs. For example, AQP1 is abundant in the capillaries in soft tissues throughout the body, but AQP1 is absent within the salivary gland. To identify a salivary homology of the aquaporin gene family, a rat salivary gland cDNA library was used which also contained AQP1 cDNAs presumably from the surrounding capillaries.[24] The rat cDNA library was first amplified with the external set of degenerate primers, Up-1 and Down-1. The PCR amplified products were then digested with the restriction enzyme *Pst*I, which cuts between the NPA motifs of rat AQP1. This was followed by reamplification with the internal pair of primers (Up-2 and Down-2). The PCR fragments of approximately 400 bp were cloned and sequenced. Obviously this strategy would not work if the resulting cDNA (*AQP5*) also contained a *Pst*I site. By trying different restriction enzymes that cut DNA infrequently (6- to 8-bp recognition sites), multiple novel homologs may be identified.

After identifying a novel aquaporin cDNA by cloning and sequencing the PCR products, the resulting PCR clone should be used to screen a cDNA library for a recombinant with a full-length insert. Alternatively the resulting DNA sequence can be used to design primers for cloning the 5′ and 3′ ends of the cDNA by anchor PCR.[26] A disadvantage of anchor PCR is that it yields fragmented cDNA sequences, which need to be spliced together for expression studies.

*Comments*

All PCR reactions should be set up in either a sterile laminar flow hood or a dead air box using pipette tips containing filters (aerosol-resistant tips) or positive displacement pipettors to prevent the contamination of samples, primers, nucleotides, and reaction buffers by DNA. If the PCR reaction is going to be reamplified by PCR, all possible intervening steps should also be performed in a sterile hood with the same precautions to prevent DNA contamination. These precautions should also be extended to all extractions and reactions on the nucleic acid (RNA or DNA) through the last PCR reaction. Likewise, all primers, nucleotides, and reaction buffers for PCR should be made up and aliquoted using similar precautions. All buffers for PCR should be made with great care using sterile disposable plastic or baked glass, and restricted for use with aerosol-resistant pipette tips. The surfaces and pipettors should be cleaned with 10% bleach (made fresh daily) and rinsed with distilled deionized water between amplifications to destroy contaminating DNA, which could be a source of amplification artifact. Standard PCR reaction buffers contain 15 m*M* $MgCl_2$ (1.5 m*M* final concentration). In many cases, changing the $MgCl_2$ concentration will have significant consequences on the amplification of specific bands, and various concentrations between 0.5 and 5.0 m*M* $MgCl_2$ have been successful. Note that organic solvents (phenol and chloroform) and ethidium bromide are hazardous materials and should always be handled with caution while wearing gloves and eye protection. The institutional hazardous waste department should be contacted regarding the proper disposal procedures.

## VII. Water Permeability of AQP1 Expressed in Oocytes

Female *Xenopus laevis* lay their eggs in freshwater ponds, so the water permeability of the oocytes is extremely low. This observation prompted Professor Erich Windhager (Cornell University Medical School, New York, NY) to propose this system for functional analysis of cDNAs encoding putative water channels. The suggestion was followed during the first functional expression of AQP1[6] and all other aquaporins.

*Materials*

An oocyte expression construct, such as pSP64T (ref. 27)
Various restriction endonucleases
T4 DNA polymerase and ligase
Calf intestinal alkaline phosphatase
DH5 $\alpha$-competent *E. coli*
DEPC–water (ribonucleases are inactivated in diethylpyrocarbonate-treated water)
T7 RNA polymerase (50 U/$\mu$l) and 5× T3/T7 buffer (GIBCO-BRL)
ACUG mix: 10 m*M* ATP, 10 m*M* CTP, 10 m*M* UTP, and 2 m*M* GTP
50 m*M* DTT, made up in DEPC–water, stored in aliquots at $-20°$
10 m*M* G(5′)ppp(5′)G cap (Pharmacia) in DEPC–water, stored in aliquots at $-20°$
RNasin (40 U/$\mu$l) and RQ1 DNase (1 U/$\mu$l) (Promega, Madison, WI)
Female *Xenopus laevis* and storage facilities
Modified Barth's solution: 88 m*M* NaCl, 1.0 m*M* KCl, 2.4 m*M* $NaHCO_3$; 15 m*M* Tris-HCl (pH 7.6); 0.3 m*M* $Ca(NO_3)_2$, 0.4 m*M* $CaCl_2$, 0.8 m*M* $MgSO_4$, sodium penicillin 0.1 mg/ml, streptomycin sulfate, 0.1 mg/ml, 0.5 m*M* theophylline; make up to 1000 ml with water

*Procedures*

*Preparation of cDNA Expression Constructs.* The human *AQP1* cDNA is available in a *Xenopus* oocyte expression construct from the American Type Culture Collection (Rockville, MD) and plasmids containing rat *AQP4,* rat *AQP5,* and *E. coli aqpZ* are also available (Table I). The entire coding regions for other aquaporin cDNAs must be spliced into the *Bgl*II

TABLE I
AQUAPORIN cDNAs[a]

| Aquaporin | Species | Catalog No. |
|---|---|---|
| AQP1 | Human | 95674, 95675, 99538[b] |
| AQP4 | Rat | 87184 |
| AQP5 | Rat | 87185 |
| AqpZ | *E. coli* | 87386 |

[a] Available through American Type Tissue Collection (Tel: 800-638-6597; Fax: 301-816-4361; e-mail: sales@atcc.org; Internet: http://www.atcc.org).
[b] In *X. laevis* expression vector.

site of pSP64T or another suitable *Xenopus* oocyte expression construct.[27] For blunt-end ligation reactions, the *Bgl*II restriction site overhangs are filled in with T4 DNA polymerase, followed by calf intestinal alkaline phosphatase to remove the 5′-phosphates from the DNA. For cohesive ligation reactions, the plasmid is treated with the phosphatase following *Bgl*II digestion. Control ligations are set up with the vector alone to assess efficiency of phosphatase treatment. The ligation reactions are transformed into DH5 $\alpha$-competent bacteria. Plasmid DNA minpreparations are prepared from several of the resulting colonies, and a clone is identified with the correct cDNA insert orientation relative to the 5′ and 3′ UTRs of the expression construct by digesting the plasmid DNA with different restriction enzymes followed by agarose gel electrophoresis and DNA sequencing. A clone with the insert in the reverse orientation may be used as a negative control in the expression experiments.

*Preparation of cRNAs for Oocyte Expression.* Approximately 50 $\mu$g of each expression cDNA is digested with a restriction enzyme that cuts on the 3′ end of the construct, producing either a blunt end or a 3′ overhang. This is extracted twice with PC9, once with chloroform, and is precipitated by ammonium acetate/ethanol. The precipitate is resuspended at 1 mg/ml in DEPC–water. The following cRNA synthesis reaction is set up at room temperature in a 0.5-ml sterile microfuge tube with the following reagents added in order: 44.5 $\mu$l DEPC–water, 20 $\mu$l 5× T3/T7 buffer, 10 $\mu$l ACUG mix, 10 $\mu$l 50 m*M* DTT, 6 $\mu$l 10 m*M* GppG cap, 5 $\mu$l linearized DNA, 2.5 $\mu$l RNasin, 2 $\mu$l T7 RNA polymerase. These are mixed gently and incubated at 37° for 90 min. A 2.5-$\mu$l aliquot of RNasin and 4 $\mu$l of RQ1 DNase I are added, mixed, and incubated at 37° for 30 min. The DNA digestion reactions are stopped with 1 $\mu$l of 500 m*M* NaEDTA and are extracted twice with 100 $\mu$l PC9 and once with chloroform. The aqueous supernatants are transferred to a clean sterile tube, and precipitated on ice with ammonium acetate/ethanol. The pellet is washed with 500 $\mu$l 70% ethanol, spun 5 min at 4°, decanted, and residual fluid is removed with a dry paper tissue. The cRNA is resuspended in 20 $\mu$l DEPC–water with 2 $\mu$l RNasin. (*Note:* Vacuum drying RNA is faster, but overly dried RNA pellets are often difficult to resuspend. All cRNAs must be kept on ice.) The absorbancy is read at 260 and 280 nm on a spectrophotometer with 0.5–2 $\mu$l RNA. Concentrations and purity are calculated (1 U $OD_{260}$ = 40 $\mu$g/ml ssRNA; $OD_{260}/OD_{280} \sim 2.0$ for pure ssRNA). The material is diluted with DEPC–water to 1 mg/ml and stored in 2-$\mu$l aliquots at −80°. A small aliquot of the cRNA may be visualized by an RNA gel with known amounts of other

[27] P. A. Krieg and D. A. Melton, *Nucleic Acids Res.* **12,** 7057 (1984).

*in vitro* synthesized RNA molecules to verify concentration, purity, and lack of degradation.

*Preparation and Microinjection of Xenopus Oocytes.* The amphibia are anesthetized on ice, and stage V and VI oocytes are isolated[28] and stored overnight at 18° in modified Barth's buffer (MBS). Oocytes are injected with 50 nl of water (control injected oocytes) or 50 nl of water containing 0.1–50 ng cRNA. Injected oocytes are incubated at 18° in MBS for 1–5 days with daily changes of the buffer to promote good viability.

*Measurement of Osmotic Water Permeability.* A computer-interfaced videomicroscope is needed for determination of oocyte swelling (available from Nikon and other manufacturers (see Fig. 5). The hypotonic buffer (70 mOsm) is prepared by diluting one volume of MBS with two parts distilled water. Single oocytes are transferred from isotonic MBS (200 mOsm) to hypotonic MBS (at 22°), and digitized images are collected by computer (Universal Imaging Corporation, West Chester, PA) for up to 2 min or whenever the oocyte ruptures. The surface area of sequential images is calculated (Image-1 software, version 4.01B, Universal Imaging) assuming the oocytes are spheres without microvilli:

$$V = (4/3) \times (\text{area}) \times (\text{area}/\pi)^{1/2}$$

The relative changes in oocyte volume with time for up to 2 min, $d(V/V_o)/dt$, are fit by computer to a quadratic polynomial, and the initial rate of swelling is calculated. The osmotic water permeability ($P_f$, cm/sec $\times 10^{-4}$) is determined between 10 and 20 sec, using the average initial oocyte volume ($V_o = 9 \times 10^{-4}$ cm$^3$), the average initial oocyte surface area ($S = 0.045$ cm$^2$), the molar ratio of water ($V_w = 18$ cm$^3$/mol), and the following formula[29]:

$$P_f = V_o[d(V/V_o)/dt]/[SV_w(\text{osm}_{in} - \text{osm}_{out})]$$

In addition to water permeability measurements, other small uncharged molecules such as glycerol have also been shown to permeate AQP3.[30] This may be measured by incubating the oocytes in buffer containing isotopically labeled glycerol in the absence of an osmotic gradient followed by solubilization of the washed oocytes in SDS and scintillation counting. Aquaporins are not believed to conduct ions or protons,[31] so electrophysiologic techniques are not described here.

*Oocyte Membrane Immunoblot Analysis.* It is often useful to confirm

[28] L. Lu, C. Montrose-Rafizadeh, T.-C. Hwang, and W. B. Guggino, *Biophys. J.* **57,** 117 (1990).

[29] R. Zhang, K. A. Logee, and A. S. Verkman, *J. Biol. Chem.* **265,** 15375 (1990).

[30] K. Ishibashi, S. Sasaki, K. Fushimi, S. Uchida, M. Kuwahara, H. Saito, T. Furukawa, K. Nakajima, Y. Yamaguchi, T. Gojobori, and F. Marumo, *Proc. Natl. Acad. Sci. U.S.A.* **91,** 6269 (1994).

[31] P. Agre, M. D. Lee, S. Devidas, and W. B. Guggino, *Science* **275,** 1490 (1997).

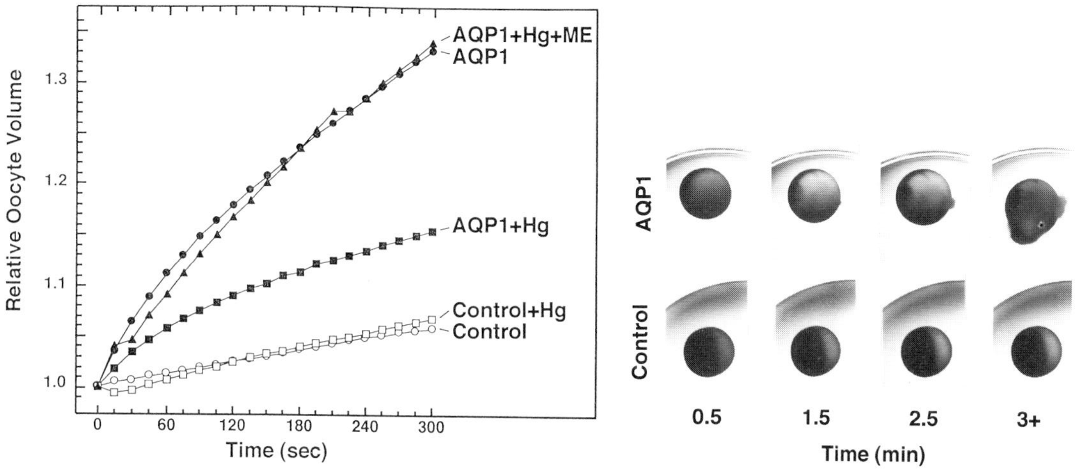


FIG. 5. Time course and $Hg^{2+}$ inhibition of osmotic swelling of oocytes expressing AQP1. *Left:* Oocytes were injected with 50 nl of water (control) or 50 nl of water containing 1 ng of cRNA (AQP1). After 72 hr, the oocytes were transferred from 200 to 70 mOsm modified Barth's buffer, and changes in size were measured by video microscopy. This permitted determination of $P_f$ (AQP1, $\sim 200 \times 10^{-4}$ cm/sec; control, $\sim 20 \times 10^{-4}$ cm/sec). Where indicated (+Hg), the oocytes were incubated for 5 min in buffer containing 0.3 m*M* $HgCl_2$ followed by osmotic swelling. After incubation for 5 min in 0.3 m*M* $HgCl_2$, other oocytes (AQP1 + Hg + ME) were then incubated for 15 min in buffer containing 5 m*M* 2-mercaptoethanol followed by osmotic swelling. *Right:* Series of still photographs of oocytes during osmotic swelling from video micrographs taken at indicated intervals. [Modified and reprinted from G. M. Preston *et al., Science* **256,** 385 (1992) and G. M. Preston *et al., J. Biol. Chem.* **268,** 17 (1993) with permission.]

that the oocytes expressed the aquaporin of interest, particularly when testing site-directed mutant forms.[32] Groups of 5–10 oocytes are transferred with modified Barth's buffer into 1.5-ml microcentrifuge tubes on ice. After chilling for 5 min, the buffer is removed and the oocytes are lysed in 0.5–1 ml ice-cold hypotonic lysis buffer (7.5 m*M* $Na_2HPO_4$, pH 7.4, 1 m*M* NaEDTA) in the presence of protease inhibitors (20 $\mu$g/ml PMSF, 1 $\mu$g/ml pepstatin A, 1 $\mu$g/ml leupeptin, 1 : 2000 diisopropyl fluorophosphate) by repeatedly vortex agitating and pipetting. The yolk and cellular debris are pelletted at 750*g* for 5 min at 4°, and membranes are then pelleted from the supernant at 16,000*g* for 30 min at 4°. The floating yolk is removed from the top of the tubes with a cotton applicator, and the supernatant is removed. The membrane pellets are gently washed once with an equal volume of ice-cold hypotonic lysis buffer and resuspended in 10 $\mu$l of 1.25% (w/v) SDS per oocyte and electrophoresed into a 12% SDS–polyacrylamide gel, transferred to nitrocellulose, incubated with a 1 : 1000 dilution of affinity-purified anti-AQP1[4] or another anti-aquaporin immunoglobulin G (IgG), and visualized by ECL (enhanced chemiluminescence) Western blotting detection system (Amersham Corp., Arlington Heights, IL).

## Acknowledgments

This work was supported in part by research grants from the National Institutes of Health to P.A. and a postdoctoral fellowship from the American Heart Association, Maryland Affiliate, to J.C.M. We thank our colleagues Suresh V. Ambudkar, Mark L. Zeidel, Larry A. Coury, and William B. Guggino for their efforts in developing these methods.

[32] G. M. Preston, J. S. Jung, W. B. Guggino, and P. Agre, *J. Biol. Chem.* **268,** 17 (1993).

# Section V

# Toxins and Other Membrane Active Compounds

# [30] Pore-Blocking Toxins as Probes of Voltage-Dependent Channels

*By* ROBERT J. FRENCH and SAMUEL C. DUDLEY, JR.

## Introduction

Our goal in this article is to outline some areas in which understanding of voltage-dependent $Na^+$ channels has been, and may further be, illuminated by judicious experiments with pore-blocking toxins. We describe briefly both experimental methods and analytical approaches that are providing new glimpses of the molecular basis of $Na^+$ channel function. These studies have the potential to yield not only new details about the molecular geometry of the ion-conducting pore, but also to provide clues about the packing of channel domains, about the interactions between charged ligands and the pore, about interactions between pairs of blocking ions within the pore, and about the proximity and interactions of the voltage-sensitive gating machinery with the charged entities within the pore.

Our discussion centers around studies using derivatives of the highly selective $Na^+$ channel-blocking peptide, $\mu$-conotoxin GIIIA (to be abbreviated throughout as $\mu$CTX), but results from these studies will be discussed in the context of work on $K^+$ channels using the charybdotoxin (ChTX) family of peptides, and $Na^+$ channel studies with the heterocyclic guanidinium toxins which include tetrodotoxin (TTX) and saxitoxin (STX).

## Sodium Channels: Molecular Context of Functional Studies

### *A. Transmembrane Topology and Terminology*

An enormous amount has been learned about the structure of voltage-dependent $Na^+$ and $K^+$ channels.[1,2] A growing list of primary sequences has been reported for the $\alpha$ subunits of $Na^+$ channels,[3–12] including a TTX-

[1] M. Noda, *Ann. N.Y. Acad. Sci.* **707,** 20 (1993).

[2] R. G. Kallen, S. A. Cohen, and R. L. Barchi, *Mol. Neurobio.* **7,** 383 (1993).

[3] M. Noda, H. Takahashi, T. Tanabe, M. Toyosato, Y. Furutani, T. Hirose, M. Asai, S. Inayama, T. Miyata, and S. Numa, *Nature* **312,** 121 (1984).

[4] M. Noda, T. Ikeda, T. Kayano, H. Suzuki, H. Takeshima, M. Kurasaki, H. Takahashi, and S. Numa, *Nature* **320,** 188 (1986).

[5] M. Noda, T. Ikeda, H. Suzuki, H. Takahashi, T. Takahashi, M. Kuno, and S. Numa, *Nature* **322,** 826 (1986).

[6] L. Salkoff, A. Butler, A. Wei, N. Scavarda, K. Giffen, C. Ifune, R. H. Goodman, and G. Mandel, *Science* **237,** 744 (1987).



0076-6879/99 $30.00

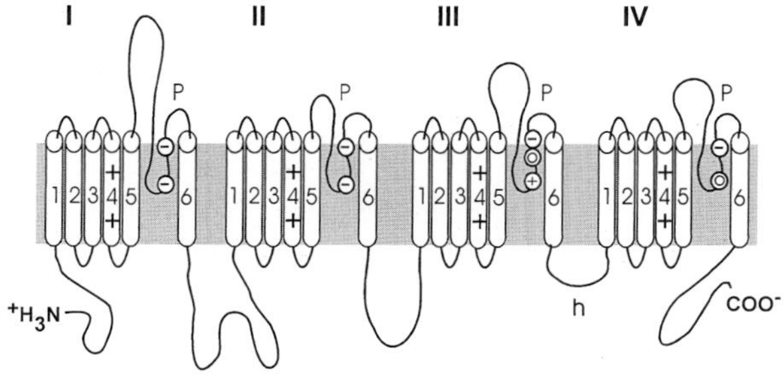


FIG. 1. Proposed transmembrane topology of the sodium channel α subunit, supported by the results of various authors and redrawn after Catterall.[17,157] The N- and C-terminal ends of the peptide are cytoplasmic. The putative pore-forming (P) loops, with the positions of residues on the inner and outer charged rings, which are important in binding of guanidinium toxins (TTX, STX, and μCTX) and in determining channel selectivity, are indicated. Also shown are the positively charged S4 segments—major components of the voltage sensor controlling activation gating—and the cytoplasmic loop (h) connecting domains III and IV, the putative "hinged lid" of the fast inactivation gate.

sensitive, μCTX-sensitive isoform from rat skeletal muscle, rSkM1.[13] The sequences, of approximately 2000 amino acids, show four repeating homologous domains (I–IV), each having six probable transmembrane helices, S1–S6[3], and two additional short segments (SS1 and SS2, which together comprise the P region), within the "extracellular" S5–6 loop, which probably fold back into the membrane to form at least part of the ion-conducting pore (Fig. 1).[14,15] Expression of the gene for only the α subunit of one of these channels yields conductances with most of the identifying properties

[7] M. E. Gellens, A. L. George, Jr., L. Chen, M. Chahine, R. Horn, R. L. Barchi, and R. G. Kallen, *Proc. Natl. Acad. Sci. U.S.A.* **89,** 554 (1992).

[8] A. L. George, Jr., T. J. Knittle, and M. M. Tamkun, *Proc. Natl. Acad. Sci. U.S.A.* **89,** 4893 (1992).

[9] S. Gautron, G. Dos Santos, D. Pinto-Henrique, A. Koulakoff, F. Cros, and Y. Berwald-Netter, *Proc. Natl. Acad. Sci. U.S.A.* **89,** 7272 (1992).

[10] R. G. Kallen, Z. Sheng, J. Yang, L. Chen, R. B. Rogart, and R. L. Barchi, *Neuron* **4,** 233 (1990).

[11] R. B. Rogart, L. L. Cribbs, L. K. Muglia, D. D. Kephart, and M. W. Kaiser, *Proc. Natl. Acad. Sci. U.S.A.* **86,** 8170 (1989).

[12] P. A. V. Anderson, M. A. Holman, and R. M. Greenberg, *Proc. Natl. Acad. Sci. U.S.A.* **90,** 7419 (1993).

[13] J. S. Trimmer, S. S. Cooperman, S. Tomiko, J. Zhou, S. M. Crean, M. B. Boyle, R. G. Kallen, Z. Sheng, R. L. Barchi, F. J. Sigworth, R. H. Goodman, W. S. Agnew, and G. Mandel, *Neuron* **3,** 33 (1989).

[14] H. R. Guy and F. Conti, *TINS* **13,** 201 (1990).

[15] R. MacKinnon, *Neuron* **14,** 889 (1995).

of the respective channel type, including ion selectivity, toxin sensitivity, and voltage-dependent activation, although inactivation speeds up with coexpression of a $\beta$ subunit.[16] Within the voltage-dependent channel superfamily of $K^+$, $Na^+$, and $Ca^{2+}$ channels, there is extensive homology among $\alpha$ subunits.[17] Among different $Na^+$ channels, two functionally important regions show a high degree of conservation and are very similar among the four homologous domains: the positively charged, arginine- and lysine-rich S4 segments, and the P (SS1–SS2) regions in the S5–6 loops. Each of these chemically disparate regions appears to have a significant interaction with $\mu$CTX, the former probably being a long-range electrostatic repulsion, while the latter appears to be an intimate binding.

### *B. Voltage Sensor*

The highly conserved S4 segments possess positively charged lysine or arginine as approximately every third residue, as well as numerous hydrophobic residues, and they adjoin highly hydrophobic segments that show a leucine zipper motif.[18] At least some of the positive residues of S4 (about 25 charges distributed among domains I–IV) are essential for the normal voltage sensitivity of the channel.[19] Observations of the state dependence of accessibility to chemical modification of cysteine-substituted S4 residues have provided direct evidence for translocation of S4 residues across the membrane during voltage-dependent opening of $Na^+$ channels.[20,21] Similar conclusions were reached for *Shaker* $K^+$ channels from studies using fluorescent labeling within S4.[22,23] Shifts in $Na^+$ channel half-activation voltage, associated with the binding of a $\mu$CTX derivative, indicate an inhibition of channel opening by the peptide, and are consistent with a substantial outward movement of S4 charges associated with channel opening[24] (see Section V,A later).

[16] D. E. Patton, L. L. Isom, W. A. Catterall, and A. L. Goldin, *J. Biol. Chem.* **269(26),** 17649 (1994).

[17] W. A. Catterall, *Science* **242,** 50 (1988).

[18] K. McCormack, J. T. Campanelli, M. Ramaswami, M. K. Mathew, M. A. Tanouye, L. E. Iverson, and B. Rudy, *Nature* **340,** 103 (1989).

[19] W. Stühmer, F. Conti, H. Suzuki, X. Wang, M. Noda, N. Yahagi, H. Kubo, and S. Numa, *Nature* **339,** 597 (1989).

[20] N. Yang and R. Horn, *Neuron* **15,** 213 (1995).

[21] N. Yang, A. L. George, Jr., and R. Horn, *Neuron* **16,** 113 (1996).

[22] L. M. Mannuzzu, M. M. Moronne, and E. Y. Isacoff, *Science* **271,** 213 (1996).

[23] H. P. Larsson, O. S. Baker, D. S. Dhillon, and E. Y. Isacoff, *Neuron* **16,** 387 (1996).

[24] R. J. French, E. Prusak-Sochaczewski, G. W. Zamponi, S. Becker, A. S. Kularatna, and R. Horn, *Neuron* **16,** 407 (1996).

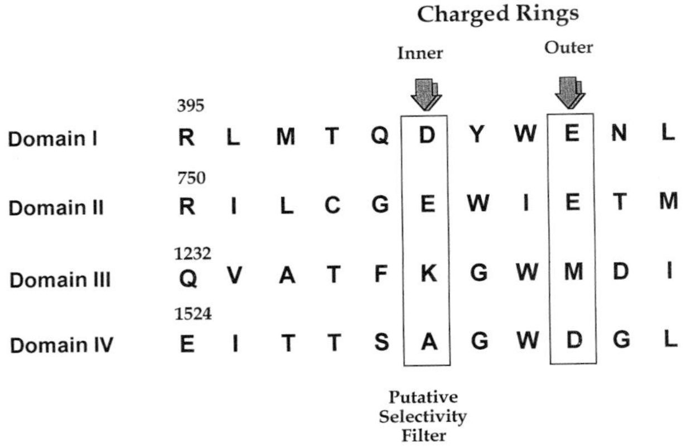

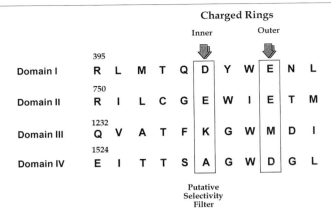

FIG. 2. Primary sequences of parts of the extracellular loops linking transmembrane segments S5 and S6 in each repeat showing the P-loop residues (from the rSkM1 Na channel $\alpha$ subunit). These segments include residues that have been the focus of the most detailed analysis of $\mu$CTX–Na interactions to date, but more distant residues also appear to make significant contributions to the interaction.

## *C. Pore Region*

Mutagenesis of the $Na^+$ channel $\alpha$ subunit has revealed, in the P region (Figs. 1 and 2), two groups of negative residues (glutamate or aspartate), spread over the four domains, which are critical for TTX/STX sensitivity (e.g., Refs. 25–37). The charge-changing mutations which reduce sensitivity to the guanidinium toxins, in general, reduce conductance and sensitivity

[25] M. Noda, H. Suzuki, S. Numa, and W. Stühmer, *FEBS Lett.* **259,** 213 (1989).

[26] H. Terlau, S. H. Heinemann, W. Stühmer, M. Pusch, F. Conti, K. Imoto, and S. Numa, *FEBS Lett.* **293,** 93 (1991).

[27] S. H. Heinemann, H. Terlau, and K. Imoto, *Pflügers Arch.* **422,** 90 (1992).

[28] K. J. Kontis and A. L. Goldin, *Molec. Pharm.* **43,** 635 (1993).

[29] M. M. Stephan, J. F. Potts, and W. S. Agnew, *J. Membr. Biol* **137,** 1 (1994).

[30] I. Favre, E. Moczydlowski, and L. Schild, *J. Gen. Physiol.* **106,** 203 (1995).

[31] I. Favre, E. Moczydlowski, and L. Schild, *Biophys. J.* **71,** 3110 (1996).

[32] P. H. Backx, D. T. Yue, J. H. Lawrence, E. Marban, and G. F. Tomaselli, *Science* **257,** 248 (1992).

[33] R. G. Tsushima, R. A. Li, and P. H. Backx, *J. Gen. Physiol.* **109,** 463 (1997).

[34] R. G. Tsushima, R. A. Li, and P. H. Backx, *J. Gen. Physiol.* **110,** 59 (1997).

[35] J. Satin, J. W. Kyle, M. Chen, P. Bell, L. L. Cribbs, H. A. Fozzard, and R. B. Rogart, *Science* **256,** 1202 (1992).

[36] S. C. Dudley, H. Todt, G. Lipkind, and H. A. Fozzard, *Biophys. J.* **69,** 1657 (1995).

[37] M. T. Péres-García, N. Chiamvimonvat, R. Ranjan, J. R. Balser, G. F. Tomaselli, and E. Marban, *Biophys. J.* **72,** 989 (1997).

to block by $Ca^{2+}$ or $Mg^{2+}$.[38] Also in this region (IS5–6) is a location at which substitution of Cys for Tyr (Y401C in rSkM1) changes a channel from the high TTX/STX sensitivity ($K_d$ values in the nanomolar range) characteristic of skeletal muscle to the low TTX/STX sensitivity ($K_d$ values in the micromolar range), and high sensitivity to the divalents, $Zn^{2+}$ and $Cd^{2+}$, typical of heart.[32,35] Even more dramatic are single or paired mutations (K1422E, with or without A1714E in the rat brain II channel) in domains III and IV that confer $Ca^{2+}$ channel-like conduction properties on a $Na^+$ channel.[39] Elegant studies using cysteine-scanning mutagenesis[40] have begun to place detailed constraints on the conformation of the pore-forming loops.[33,34,37,41–45] Overall, there is a convincing case that the P region contains the major determinants of ion conduction, selectivity, and sensitivity to blocking drugs and toxins (e.g., see reviews in Refs. 46 and 47).

### *D. Block by μ-Conotoxins*

An unsolved puzzle is μCTX's specific and potent block of one TTX-sensitive type of $Na^+$ channel (adult skeletal muscle, and muscle-derived eel electroplax) but its lack of effect on others (brain and heart). Despite a number of studies, no single residue has been found that has an overriding influence on either the potency or specificity of μCTX. By contrast, substitution of a single glutamate can abolish TTX and STX sensitivity in the rat brain 2 channel (the mutation E387Q of SS2 within the S5–6 loop; see Ref. 25). The equivalent mutation in rSkM1, E403Q, also eliminates TTX/STX block, but reduces μCTX sensitivity by only about 4-fold.[29] The Y401C mutation, which reduces the TTX sensitivity of rSkM1 to near that of heart (or rSkM2) $Na^+$ channels, diminishes the effect of μCTX somewhat, but

[38] M. Pusch, M. Noda, W. Stühmer, S. Numa, and F. Conti, *Eur. Biophys. J.* **20(3),** 127 (1991).

[39] S. H. Heinemann, H. Terlau, W. Stühmer, K. Imoto, and S. Numa, *Nature* **356,** 441 (1992).

[40] M. H. Akabas, D. A. Stauffer, M. Xu, and A. Karlin, *Science* **258,** 307 (1992).

[41] N. Chiamvimonvat, B. O'Rourke, T. J. Kamp, R. G. Kallen, F. Hofmann, V. Flockerzi, and E. Marban, *Circ. Res.* **76,** 325 (1995).

[42] M. T. Pérez-García, N. Chiamvimonvat, E. Marban, and G. F. Tomaselli, *Proc. Natl. Acad. Sci. U.S.A.* **93,** 300 (1996).

[43] N. Chiamvimonvat, M. T. Pérez-García, R. Ranjan, E. Marban, and G. F. Tomaselli, *Neuron* **16,** 1037 (1996).

[44] T. Yamagishi, M. Janecki, E. Marban, and G. F. Tomaselli, *Biophys. J.* **73,** 195 (1997).

[45] J.-P. Bénitah, R. Ranjan, T. Yamagishi, M. Janecki, G. F. Tomaselli, and E. Marban, *Biophys. J.* **73,** 603 (1997).

[46] K. Imoto, *Ann. N.Y. Acad. Sci.* **707,** 38 (1993).

[47] G. F. Tomaselli, P. H. Backx, and E. Marban, *Circ. Res.* **72(3),** 491 (1993).

considerable sensitivity is retained.[48] The latter authors tested several chimeric constructs and 22 amino acid block mutations that altered the P region, with the result that any change in the rSkM1 sequence reduced μCTX potency. The greatest effect was a 24-fold reduction when domain I from rSkM1 was linked to domains II, III, and IV of rSkM2. Replacing only domain I in the rSkM1 gave a 9-fold reduction. The block of domain I and IV mutants pointed to the C-terminal end of the P region (SS2) being more important, with a maximal change of 9-fold when ISS2 was replaced. These data are consistent with a μCTX site formed by the SS2 segments as a group, but reveal no dominant contacts. Rather, these authors suggested that μCTX's docking site is considerably more extensive than that for TTX, consistent with conclusions from a study of a family of μCTX derivatives.[49]

## III. Pore-Blocking Toxins as Probes of Ion Channels

### *A. Sodium Channel Pore Blockers: Structural Features and Phenomenology*

Two groups of natural, hydrophilic toxins block current through voltage-dependent sodium channels when applied from the extracellular side: the heterocyclic guanidinium toxins[50]—tetrodotoxin, saxitoxin (molecular weights ~300) and their derivatives—and the peptide μ-conotoxins[51–55] (molecular weight ~2600 for the GIII toxins from *Conus geographus,* and about 50% larger for conotoxin GS, which appears related to the $Ca^{2+}$ channel-blocking ω-conotoxins). Each of these toxins binds sodium channels with 1 : 1 stoichiometry, blocking single-channel current in an all-or-

[48] L.-Q. Chen, M. Chahine, R. G. Kallen, R. L. Barchi, and R. Horn, *FEBS Lett.* **309(3),** 253 (1992).

[49] S. Becker, E. Prusak-Sochaczewski, G. Zamponi, A. G. Beck-Sickinger, R. D. Gordon, and R. J. French, *Biochemistry* **31(35),** 8229 (1992).

[50] C. Y. Kao and S. R. Levinson, eds., *Ann. N.Y. Acad. Sci.* **479** (1986).

[51] L. J. Cruz, W. R. Gray, B. M. Olivera, R. D. Zeikus, L. Kerr, D. Yoshikami, and E. Moczydlowski, *J. Biol. Chem.* **260(16),** 9280 (1985).

[52] Y. Ohizumi, H. Nakamura, J. Kobayashi, and W. A. Catterall, *J. Biol. Chem.* **261,** 6149 (1986).

[53] Y. Yanagawa, T. Abe, M. Satake, S. Odani, J. Suzuki, and K. Ishikawa, *Biochemistry* **27,** 6256 (1988).

[54] J.-M. Lancelin, D. Kohda, S.-I. Tate, Y. Yanagawa, T. Abe, M. Satake, and F. Inagaki, *Biochemistry* **30,** 6908 (1991).

[55] J. M. Hill, P. F. Alewood, and D. J. Craik, *Biochemistry* **35,** 8824 (1996).

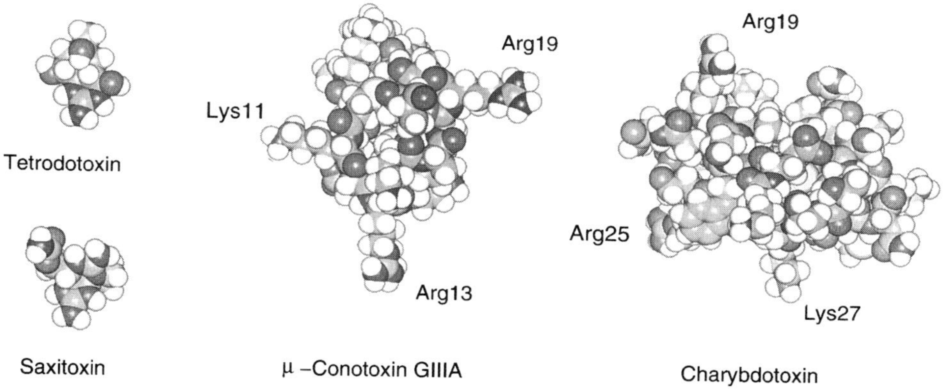


FIG. 3. Space-filling models of the structures of the $Na^+$ channel-blocking guanidinium toxins, tetrodotoxin, saxitoxin, μ-conotoxin GIIIA, and the $K^+$ channel blocker, charybdotoxin. For comparison, models are oriented with the putative pore-blocking cationic groups pointing downward. This critical positive charge is provided by a guanidinium group for TTX, STX, and μCTX (side chain of R13), and a lysine side chain for ChTX (Lys-27). The binding (lower) surface of ChTX is rotated forward slightly for better visibility.

none manner with mean blocked times ranging from seconds[56–59] for the guanidinium toxins to minutes for the GIII conotoxins.[49,51] Each of the toxins possesses a guanidinium group that is thought to enter the $Na^+$ channel pore and block flow of current (see Section V,D for more details). Their specific and potent action makes them useful probes to explore the details of pore structure.

### *B. Probes for Molecular Dentistry: Charybdotoxin and μ-Conotoxin*

Miller coined the term "molecular dentistry" for the use of the rigid peptide blocker, charybdotoxin[60–62] (Fig. 3), as a template to define the

[56] B. K. Krueger, J. F. Worley III, and R. J. French, *Nature* **303,** 172 (1983).

[57] R. J. French, J. F. Worley III, and B. K. Krueger, *Biophys. J.* **45,** 301 (1984).

[58] E. Moczydlowski, S. S. Garber, and C. Miller, *J. Gen. Physiol.* **84,** 665 (1984).

[59] E. Moczydlowski, S. Hall, S. S. Garber, G. R. Strichartz, and C. Miller, *J. Gen. Physiol.* **84,** 687 (1984).

[60] F. Bontems, C. Roumestand, B. Gilquin, A. Ménez, and F. Toma, *Science* **254,** 1521 (1991).

[61] F. Bontems, B. Gilquin, C. Roumestand, A. Ménez, and F. Toma, *Biochemistry* **31,** 7756 (1992).

[62] F. Bontems, C. Roumestand, P. Boyot, B. Gilquin, Y. Doljansky, A. Ménez, and F. Toma, *Eur. J. Biochem.* **196,** 19 (1991).

extracellular mouth of the $K^+$ channels to which it binds.[63–68] This has provided a picture of the outer vestibule of two ChTX-sensitive potassium channels, the large calcium-activated (BK) and the voltage-dependent *Shaker* channel. For *Shaker,* the analysis has progressed rapidly using mutated forms of the channel to identify residues in a protein with which the peptide interacts. Reconstruction of the interacting surface of the channel protein is based on the assertion that ChTX blocks $K^+$ channels by physical occlusion of the pore[69,70] and on the following axioms: (1) that the surface of the protein must be closely complementary with that of the peptide toxin in order to produce binding of high affinity and selectivity, (2) that the backbone structure of the toxin, determined by two-dimensional nuclear magnetic resonance (NMR) in aqueous solution, is unperturbed when the toxin is bound to the channel, (3) that the conducting pore is occluded by a known residue, lysine-27, of ChTX, and (4) that BK channels show a general homology with the tetrametric structure of *Shaker*. These studies provide a departure point for developing strategies to explore different parts of the structures of other channels for which high-affinity, specific peptide ligands are available.

Based on comparison of chemically conservative and chemically radical mutants of ChTX, it is clear that both nonionic interactions—based on intimate fit of the peptide to the protein surface—and long-range electrostatic interactions are important to the high-affinity binding of the peptide.[63,64,71–73] Although different charybdotoxin-related peptides show specificity for different potassium channel subtypes, and there are well-defined differences of detail between the deduced vestibule structures for the BK and *Shaker* channels,[64,65] the general structure of the outer mouths of these channels is very similar. Evidence for a fit of intimate detail between the peptide and the channel protein comes from the dramatic changes in toxin affinity produced by chemically conservative mutations in either toxin or channel.[64,65,74]

[63] C.-S. Park and C. Miller, *Biochemistry* **31,** 7749 (1992).
[64] P. Stampe, L. Kolmakova-Partensky, and C. Miller, *Biochemistry* **33(2),** 443 (1994).
[65] S. A. N. Goldstein, D. J. Pheasant, and C. Miller, *Neuron* **12,** 1377 (1994).
[66] R. MacKinnon and C. Miller, *Science* **245,** 1382 (1989).
[67] R. MacKinnon, L. Heginbotham, and T. Abramson, *Neuron* **5,** 767 (1990).
[68] J. Vázquez, P. Feigenbaum, V. F. King, G. J. Kaczorowski, and M. L. Garcia, *J. Biol. Chem.* **265(26),** 15564 (1990).
[69] R. MacKinnon and C. Miller, *J. Gen. Physiol.* **91,** 335 (1988).
[70] C.-S. Park and C. Miller, *Neuron* **9,** 307 (1992).
[71] M. Stocker and C. Miller, *Proc. Natl. Acad. Sci. U.S.A.* **91,** 9509 (1994).
[72] P. Stampe, L. Kolmakova-Partensky, and C. Miller, *Biophys. J.* **62,** 8 (1992).
[73] D. Naranjo and C. Miller, *Neuron* **16,** 123 (1996).
[74] S. A. N. Goldstein and C. Miller, *Biophys. J.* **62,** 5 (1992).

TABLE I
KINETIC PARAMETERS FOR CHANNEL BLOCK OF SKELETAL MUSCLE $Na^+$ CHANNELS IN PLANAR LIPID BILAYERS BY $\mu$CTX DERIVATIVES[a]

| Residue number | Substitution | Charge | $K_d$ (n$M$) | $k_{off}$ ($10^{-4}$/sec) | $k_{on}$ ($10^{-9}$/sec $M$) |
|---|---|---|---|---|---|
| | WT | 6 | 32 | 88 | 2761 |
| 1 | R1Q | 5 | 276 | 517 | 1874 |
| 2 | D2N | 7 | 83 | 83 | 996 |
| 6,7 | Hyp6,7P | 6 | 57 | 164 | 2960 |
| | Hyp6,7,17P | 6 | ~2000 | | |
| 8 | K8Q | 5 | 726 | 386 | 531 |
| | K8,9Q | 4 | ~700 | | |
| | K8,9,11Q | 3 | ~54000 | | |
| 9 | K9Q | 5 | 226 | 91 | 403 |
| 11 | K11Q | 5 | 337 | 58 | 172 |
| 12 | D12N | 7 | 51.9 | 149 | 2390 |
| 13 | R13Q | 5 | 3249 | 1189 | 366 |
| | R13K | 6 | ~2600 | | |
| 14 | Q14E | 5 | 3147 | | |
| 16 | K16Q | 5 | 862 | 431 | 500 |
| | Hyp17P | 6 | 837 | 770 | 920 |
| | HypQR17,18,19 DRQ | 5 | ~46000 | | |
| 19 | R19Q | 5 | 1394 | 191 | 137 |

[a] Previously untabulated data from the study of Becker *et al.* [S. Becker, E. Prusak-Sochaczewski, G. Zamponi, A. G. Beck-Sickinger, R. D. Gordon, and R. J. French, *Biochemistry* **31(35),** 8229 (1992)], plus unpublished experiments.

For $\mu$CTX[54,75,76] (Fig. 3), no conservative or neutralizing single residue mutation has been identified that leads to a change in affinity of much greater than the mutations R13Q and R13A (~100-fold; see Table I, and Refs. 49 and 75). These mutations point to R13 as the single most important residue in determining the high-affinity binding by $\mu$CTX. It is the working hypothesis of our groups and others, that the side chain of arginine-13 protrudes far enough into the pore to occlude current flow when $\mu$CTX binds. However, this hypothesis deserves more rigorous experimental testing. Generally consistent with the idea is that R13Q only partially blocks single-channel current.[24,49] However, whereas replacing the pore blocking K27 of ChTX with a neutral residue abolishes the interaction with internal potassium,[70] the R13Q mutant of $\mu$CTX *does* interact with diethylammo-

[75] K. Sato, Y. Ishida, K. Wakamatsu, R. Kato, H. Honda, H. Nakamura, M. Ohya, D. Kohda, F. Inagaki, J. Lancelin, and Y. Ohizumi, *J. Biol. Chem.* **266(26),** 16989 (1991).

[76] K. Wakamatsu, D. Kohda, H. Hatanaka, J.-M. Lancelin, Y. Ishida, M. Oya, H. Nakamura, F. Inagaki, and K. Sato, *Biochemistry* **31,** 12577 (1992).

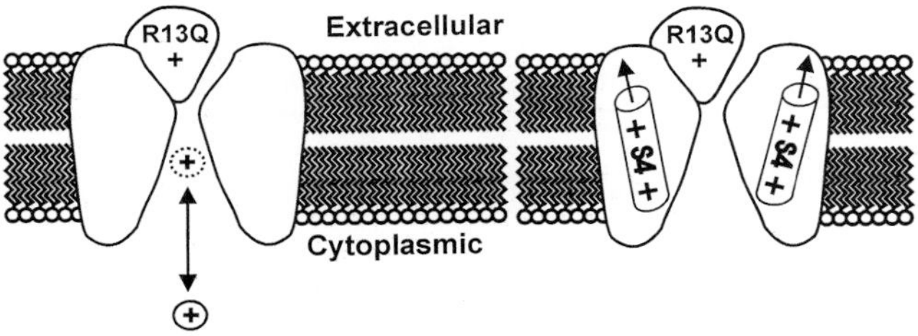


FIG. 4. Diagram suggesting inhibitory electrostatic interactions between the bound, partially blocking μCTX derivative, R13Q, and either a pore-blocking cation (cytoplasmically applied diethylammonium, DEA), or the voltage-sensing S4 segments. Both DEA block and channel opening (activation) involve movement of positive charge toward the bound, polycationic R13Q (nominal net charge, +5), which would be expected to inhibit each process. Inhibition is, in fact, observed as a shift in the curve depicting voltage dependence of channel block, or activation, to more depolarized voltages. See Refs. 24 and 145 for data and details of the analysis.

nium (DEA; see Fig. 4), which enters and blocks the sodium channel pore from the cytoplasmic side. Further mutagenesis of μCTX may reveal single substitutions that cause more dramatic changes in affinity, but it seems, at present, that the high-affinity binding of μCTX results from the summed effects of numerous relatively weak interactions.

Some general features of the toxins, and the toxin–channel interactions for ChTX and μCTX, are compared and contrasted in Table II. The two toxins share features that make them well suited for use as molecular "dental" probes. They have known, well-defined structures, with backbone conformations firmly constrained by three disulfide bonds, and bind with high affinity and specificity to the outer mouths of their respective channel targets.

One mechanistic difference is that the voltage dependence of ChTX block of BK channels is clearly dependent on binding of the permeant ion, whereas no such interaction has been reported for μCTX (but see Section V,B). The strong interaction between $K^+$ and ChTX may depend on the presence of high-affinity $K^+$ binding sites in the pore,[77,78] whereas $Na^+$ binds with relatively low affinity to sodium channels[79] and the interaction with DEA is relatively weak. A striking feature is that for charybdotoxin, the critical residues for binding are concentrated on a rather flat interactive surface. In contrast, for μCTX, contributions to high-affinity binding appear to be distributed more evenly among several residues. This, coupled with

[77] J. Neyton and C. Miller, *J. Gen. Physiol.* **92,** 569 (1988).
[78] J. Neyton and C. Miller, *J. Gen. Physiol.* **92,** 549 (1988).
[79] A. Ravindran, H. Kwiecinski, O. Alvarez, G. Eisenman, and E. Moczydlowski, *Biophys. J.* **61,** 494 (1992).

TABLE II

FEATURES OF TWO TOXIN–CHANNEL PAIRS: ChTX AND THE *Shaker* K CHANNEL, AND μCTX AND THE SKELETAL MUSCLE $Na^+$ CHANNEL

| Parameter | ChTX/*Shaker* or BK | μCTX GIIIA/rSkM1 |
|---|---|---|
| Toxin | | |
| Number of | | |
| amino acids | 37 | 22 |
| –S–S– bonds | 3 | 3 |
| Interaction surface | Localized | Distributed |
| E dependence | Via internal K[a,b] | Intrinsic? |
| Pore | | |
| Symmetry | Tetrameric (*Shaker*) | Pseudosymmetric |
| Occupancy $K^+/Na^+$ | High, >3[c–f] | Low, ≥1[g–l] |
| Affinity, permeant ion | High[m,n] | Low[i,o,p] |
| Toxin "contact" points—individual contributions to binding | Strong[q] | Relatively weak |

[a] C.-S. Park and C. Miller, *Neuron* **9,** 307 (1992).
[b] R. MacKinnon and C. Miller, *J. Gen. Physiol.* **91,** 335 (1988).
[c] A. L. Hodgkin and R. D. Keynes, *J. Physiol.* (*Lond.*) **128,** 61 (1955).
[d] T. B. Begenisich and P. DeWeer, *J. Gen. Physiol.* **76,** 83 (1980).
[e] T. B. Begenisich and C. Smith, *in* "The Squid Axon" (P. F. Baker, ed.), p. 353. Academic Press, New York, 1984.
[f] P. Stampe and T. Begenisich, *Biophys. J.* **68,** A129, (abstract), (1995).
[g] T. B. Begenisich and D. Busath, *J. Gen. Physiol.* **77,** 489 (1981).
[h] D. Busath and T. B. Begenisich, *Biophys. J.* **40,** 41 (1982).
[i] A. Ravindran, H. Kwiecinski, O. Alvarez, G. Eisenman, and E. Moczydlowski, *Biophys. J.* **61,** 494 (1992).
[j] R. J. French, J. F. Worley III, W. F. Wonderlin, A. S. Kularatna, and B. K. Krueger, *J. Gen. Physiol.* **103,** 447 (1994).
[k] D. Naranjo and R. Latorre, *Biophys. J.* **64,** 1038 (1993).
[l] E. Moczydlowski, *Biophys. J.* **64,** 1051 (1993).
[m] J. Neyton and C. Miller, *J. Gen. Physiol.* **92,** 549 (1988).
[n] J. Neyton and C. Miller, *J. Gen. Physiol.* **92,** 569 (1988).
[o] E. Moczydlowski, S. S. Garber, and C. Miller, *J. Gen. Physiol.* **84,** 665 (1984).
[p] R. J. French, J. F. Worley III, and B. K. Krueger, *in* "Ionic Channels in Cells and Model Systems" (R. Latorre, ed.), p. 273. Plenum Press, New York, 1986.
[q] S. A. N. Goldstein, D. J. Pheasant, and C. Miller, *Neuron* **12,** 1377 (1994).

the different shapes and sizes of the two toxins, suggests quite different shapes for the vestibules of the target $K^+$ and $Na^+$ channels.

The fact that the target proteins for ChTX are probably homotetrameric, whereas μCTX binds to a pseudosymmetric, four-domain single polypeptide, has a number of implications. Presumably, there are four permissive orientations for binding of ChTX, while it is likely that there is a single unique binding orientation for μCTX. This may partly account, on purely statistical grounds, for the faster association rate constant for ChTX binding

TABLE III
KINETIC PARAMETERS FOR CHANNEL BLOCK BY TETRODOTOXIN AND $\mu$-CONOTOXIN OF RAT SKELETAL MUSCLE Na CHANNELS, AND ChTX BLOCK OF LARGE Ca-ACTIVATED K CHANNELS

| Toxin–channel | TTX/rSkM1[a] | ChTX/BK[b] | $\mu$CTX/rSkM1[c] |
|---|---|---|---|
| $K_d$ (n$M$) | 29 | 9 | 32 |
| $k_{off}$ ($sec^{-1}$) | 0.12 | 0.062 | 0.009 |
| $k_{on}$ ($10^6$ $M^{-1}$ $sec^{-1}$) | 4.0 | 7.0 | 0.28 |

[a] X. Guo, A. Uehara, A. Ravindran, S. H. Bryant, S. Hall, and E. Moczydlowski, *Biochemistry* **26,** 7546 (1987).
[b] P. Stampe, L. Kolmakova-Partensky, and C. Miller, *Biochemistry* **33(2),** 443 (1994).
[c] S. Becker, E. Prusak-Sochaczewski, G. Zamponi, A. G. Beck-Sickinger, R. D. Gordon, and R. J. French, *Biochemistry* **31(35),** 8229 (1992).

than for $\mu$CTX (see next section and Table III). Obviously, the task of searching for interacting residues on the channel is reduced in magnitude by about 4-fold for ChTX by the channel symmetry. A reward for the extra effort required to identify the binding site residues for $\mu$CTX, however, is that identification of these points of contact should enable the spatial packing of the repeat domains of the Na channel $\alpha$ subunit to be unambiguously determined (see also Section VI).

### *C. Kinetic Bases of High-Affinity Binding*

Kinetic parameters for the toxins TTX, ChTX, and $\mu$CTX are summarized in Table III. For each of these particular toxin–channel combinations, equilibrium dissociation constants fall within a 4-fold range. Charybdotoxin appears as slightly more potent at equilibrium because of both a lower dissociation (off) rate constant and higher association (on) rate constant than for TTX. The striking entry in the table, however, is for $\mu$CTX. Although it has about the same equilibrium dissociation constant as TTX, the kinetic basis for that value is quite different. Both the off rate constant and the on rate constant are about an order of magnitude lower than for TTX. Thus, $\mu$CTX binds to rSkM1 >10-fold longer than TTX, and about as long as ChTX binds to its highest known affinity receptor, a mutated form of the *Shaker* channel.[74] Despite the slow on rate, $\mu$CTX association, like that for ChTX,[80] may be a diffusion-limited process with a low probability of successful collisions (but see Ref. 81). The fact that $\mu$CTX affinity

[80] C. Miller, *Biochemistry* **29,** 5320 (1990).
[81] G. E. Kirsch, M. Alam, and H. A. Hartmann, *Biophys. J.* **67,** 2305 (1994).

seems to result from the summed effect of several relatively weak interactions, together with the slow association rate, seems consistent with the formation of a relatively unstable encounter complex (see Section 8.2.3 in Creighton[82]) as a first step in binding. For ChTX, almost all mutations speed dissociation, but have little effect on the association rate. The especially slow association rate of μCTX (much less than the theoretical diffusion-limited collision rate) means that its equilibrium binding affinity can potentially be substantially increased or decreased by changes in either on or off rates. This makes available another dimension of analysis of the interaction between the toxin and its receptor.

## IV. μ-Conotoxins and Sodium Channels: Experimental Approaches

### A. *Mutating the Toxin*

Direct chemical synthesis has proved to be the method of choice for production of μCTX derivatives. Although costs of this approach are significant, they are not prohibitive, given the relatively small size of the toxin (22 residues). For μCTX derivatives, direct synthesis has the following advantages over the alternate possibility of recombinant production in bacteria, which has been used successfully to produce charybdotoxin derivatives[64,65,83] and other peptide toxins[84]:

1. The hydroxyproline residues, which are present in the native sequence, and at least one of which is important for μCTX binding, can be inserted directly. *Escherichia coli* expression systems do not allow this.
2. With an appropriate choice of the synthesis resin, the peptide is obtained with the C-terminal residue in the native, amidated form. Although an enzyme system is available to catalyze terminal amidation posttranslationally,[85,86] this procedure would involve additional labor and expense, and, to date, *in vitro* protocols have not been developed for the conotoxin peptides.
3. Multimilligram quantities of purified peptide can be routinely obtained with relatively small-scale syntheses, whereas a bacterial ex-

[82] T. E. Creighton, "Proteins: Structures and Molecular Properties." W. H. Freeman and Company, San Francisco, 1993.

[83] C.-S. Park, S. F. Hausdorff, and C. Miller, *Proc. Natl. Acad. Sci. U.S.A.* **88,** 2046 (1991).

[84] M. J. Gallagher and K. M. Blumenthal, *J. Biol. Chem.* **267(20),** 13958 (1992).

[85] A. F. Bradbury and D. G. Smyth, *TIBS* **112,** 115 (1991).

[86] B. A. Eipper, S. L. Milgram, E. Jean Husten, H.-Y. Yun, and R. E. Mains, *Prot. Sci.* **2,** 489 (1993).

pression system would require multiliter cultures for this level of production.

4. A final complication with bacterial expression is that proteolytic enzymes (including enterokinase 2 and factor Xa), routinely used to cleave the linear ChTX peptide from the expressed gene 9 fusion protein, did not cleanly cleave the conotoxin peptide from the fusion protein in preliminary attempts to develop this method.

*1. Peptide Synthesis.* Synthesis of conotoxin derivatives generally follows the methods of S. Becker.[49] The linear peptides are synthesized by solid-phase synthesis using 9-fluorenylmethoxycarbonyl (Fmoc) chemistry.[87,88] The following side-chain protection groups are used: Cys, trityl; Gln, trityl; Asn, trityl; Arg, Pmc or Pbf; Lys, Boc; Asp, tBu; Hyp, tBu; Thr, tBu; His, trityl; and Glu, tBu. The protected amino acids are used with free carboxylic acid groups. The syntheses are performed on a polystyrene-based Rink amide resin (approximately 0.5 mmol/g), which delivers a peptide amide on cleavage of the linear peptide from the resin. Coupling of the single Fmoc amino acids was performed using the 2-(1H-benzotriazole-1-yl)-1,1,3,3-tetramethyluronium hexafluorophosphate (HBTU)/N-hydroxybenzotriazole (HOBT)/N,N′-diisopropylethylamine (DIPEA) method. Single couplings were used for all except the first and the final five amino acids, for which double couplings were used. Syntheses are performed on an Applied Biosystems 431A synthesizer (PE Biosystems, Foster City, CA).

The raw peptides are air oxidized and purified as previously described,[89,90] with the following modifications. The crude linear peptides are initially desalted on Sephadex G-10/20% (v/v) acetic acid, then purified by preparative high-performance liquid chromatography (HPLC) to ~90% homogeneity as determined by analytical HPLC. Air oxidation to form the disulfide bonds and to fold the peptide into the active conformation is performed as described[89] except that the volatile buffer ammonium acetate (50 m*M*), containing 2-mercaptoethanol (2-ME; 20 $\mu$l in 300 ml of solution), is used (peptide concentration, ~1 mg/ml), and after cyclization, the mixture is acidified with glacial acetic acid before lyophilizing. Cyclization is monitored by analytical HPLC, and is usually complete after 2–3 days at 4°.

Following the folding of the peptide by air oxidation, toxin derivatives are purified to near homogeneity by HPLC (≥95%). Purified peptides are characterized by quantitative amino acid analysis, and in some cases by

[87] E. Atherton, H. Fox, D. Harkiss, C. J. Logan, R. C. Sheppard, and B. J. Williams, *J. Chem. Soc. D Chem. Commun.* **13,** 537 (1978).

[88] J. Meienhofer, M. Waki, E. P. Heimer, T. J. Lambros, R. C. Makofske, and C.-D. Chang, *Int. J. Peptide Prot. Res.* **13,** 35 (1979).

[89] S. Becker, E. A. Atherton, and R. D. Gordon, *Eur. J. Biochem.* **185,** 79 (1989).

[90] S. Becker, R. Liebe, and R. D. Gordon, *FEBS. Lett.* **272,** 152 (1990).

mass spectroscopic (electrospray) molecular weight determination. Amino acid analyses are generally in good agreement with the theoretical predictions except for hydroxyproline (Hyp) and cysteine which give somewhat lower apparent values (Hyp ~80–90% and Cys ~65–80% of predicted amounts).

As a preliminary check that folded structures do not deviate qualitatively from that of the native toxin, one-dimensional $^1$H NMR spectra are recorded at 15° in aqueous solution containing 5% $D_2O$ at 500 MHz. The proton chemical shifts of the R13X derivatives were generally similar, with the exception of the shifts of the Asp-12 and Glu-14, as would be expected in response to a substitution at the adjacent position 13. Qualitative nuclear Overhauser effect (NOE) data indicate that the basic secondary structure remains the same in all cases (D. McIntyre, P. Hwang, I. Sierralta, and R. French, 1998, unpublished). The channel-blocking activity of derivatives is examined using recording from single sodium channels incorporated into planar bilayers (as in Refs. 49 and 51).

### *B. Mutating the Channel*

Early in the study of μCTX binding to the $Na^+$ channel, it was realized that even though the site 1 toxins showed competitive inhibition for a single binding site,[52,91–97] at the molecular level, their specific interactions with the channel are not identical. For example, although E403Q of domain I (adult rat skeletal muscle numbering) dramatically reduced STX and TTX affinity, the mutation had a modest 3.8-fold effect on μCTX blocking efficacy.[29] On the other hand, E758Q caused a 375-fold reduction in TTX block and the largest reduction in μCTX affinity of any single mutation to date (48-fold).[36] Therefore, site-directed mutagenesis appears to be more specific than competition experiments for delineating the exact binding site and gives a closer look at the specific interactions of a bound toxin at equilibrium. The added detail provided by mutagenesis will be necessary to construct valid molecular models.

In contrast to binding by STX and TTX, a unique aspect of μCTX binding is that no single mutation of the channel seems to be critically important for affinity. Mutations of any one of a number of critical carboxyl residues in the outer vestibule virtually eliminate TTX and STX bind-

[91] F. V. Barnola, R. Villegas, and G. Camejo, *Biochim. Biophys. Acta* **298,** 84 (1973).
[92] D. Colquhoun, R. Henderson, and J. M. Ritchie, *J. Physiol.* **227,** 95 (1972).
[93] R. Henderson, J. M. Ritchie, and G. R. Strichartz, *J. Physiol.* **235,** 783 (1973).
[94] E. Moczydlowski, A. Uehara, X. Guo, and J. Heiny, *Ann. N.Y. Acad. Sci.* **479,** 269 (1986).
[95] E. Moczydlowski, B. M. Olivera, W. R. Gray, and G. R. Strichartz, *Proc. Natl. Acad. Sci. U.S.A.* **83,** 5321 (1986).
[96] W. Ulbricht, H. H. Wagner, and J. Schmidtmayer, *Ann. N.Y. Acad. Sci.* **479,** 68 (1986).
[97] H. H. Wagner and W. Ulbricht, *Pflügers Arch.* **359,** 297 (1975).

ing[25,26,28,29,42] but neutralization of some of these same residues has only modest effects on μCTX binding.[29,36,48,98] This suggests that the μCTX binding energy is distributed more equally among a number of contact points and that these points probably coordinate less among themselves to engender high-affinity μCTX binding. Demonstrating the wide distribution of μCTX contact points, Y1220C—more than 15 amino acids from the pore-forming region of domain III—improves μCTX binding (S. C. Dudley, G. Lipkind, and H. A. Fozzard, 1995, unpublished observation) implying that these segments must fold back toward the central lumen of the pore.

This wide contact surface has advantages and disadvantages for molecular dentistry. Probably because of the wide contact surface of μCTX with the channel, mutagenesis has more limited effects on affinity. As discussed later, mutagenesis experiments that change, but do not eliminate, binding allow for mutant cycle analysis, which can address the character and strength of an individual channel/toxin interaction. Also, the large contact surface allows for the possibility to probe more of the channel structure. Disadvantages include the need for more precise measurements to recognize the effect of any mutation and the need to make many more mutations in order to fully characterize the mechanism of binding. With multiple, low-affinity contact points, the common strategy of comparing the primary sequences of channel isoforms with differing toxin affinities to highlight residues involved in the binding is less likely to be successful.[48,98]

*1. Strategies: Chimera Studies and Block Mutations versus Point Mutagenesis.* Of course, the strategy for mutagenesis depends on the goals. Initially, finding a binding site usually requires looking through a long primary sequence of unknown structure. This problem is most often approached with expression cloning, block deletions, or chimera construction (if isoforms exist with different affinities for the ligand). In the case of μCTX, the latter two approaches have been used. These approaches have ruled out a large section of the S5–6 linker of domain I as the source of the isoform affinity differences[99] and chimeras between the high-affinity skeletal muscle and the low-affinity denervated skeletal muscle channels (equivalent to the heart isoform) have pointed out that μCTX interacts with all of the domains and is without a dominant interaction with any one of them.[48] The large channel contact area of μCTX increases the likelihood that isoform differences are determined by steric hindrances as well as

[98] M. Chahine, L.-Q. Chen, N. Fotouhi, R. Walsky, D. Fry, V. Santarelli, R. Horn, and R. G. Kallen, *Recept. Channels* **3,** 161 (1995).

[99] E. Bennett, M. S. Urcan, S. S. Tinkle, A. G. Koszowski, and S. R. Levinson, *J. Gen. Physiol.* **109,** 327 (1997).

specific attractions. Experimental approaches to address this possibility are still rudimentary in the absence of direct structural data.

Point mutations have been the method of choice for dissecting the complex interactions at a ligand–receptor complex. Considerations include which residues to mutate, what to introduce, and how to avoid or detect unanticipated mutations. The choice of amino acids to mutate depends on the type of interactions that are likely to be important. For example, since the net charge of $\mu$CTX is +6 at neutral pH, carboxyl groups on the channel are likely to be important for binding. Also, aromatic residues can interact with positively charged channel ligands[27,30,32,35,40,42,48,100–106] and are logical targets for mutagenesis. Groups participating in hydrophobic interactions are much harder to predict, and alanine scanning mutagenesis, in which an entire length of amino acids is individually mutated to alanines, is often tried.[107–110]

The group introduced is a compromise between the desire to see the largest effect and to preserve the channel structure. Conservative mutations, alanine scanning, and cysteine scanning[37,40,42,43] have been used with success. Conservative mutations have the advantage of being the least disruptive but can fail to cause a detectable alteration in an existing interaction. Alanine substitutions appear to be well tolerated especially in hydrophobic regions and generally cause sufficient change in the interaction that detection is possible. Also, cysteine scanning is well tolerated, but the introduced cysteine chemistry may not be ideal when using toxins as probes. For example, Pérez-García *et al.*[42] noted an increase in TTX affinity with M1240C of domain III, which is the opposite effect on affinity of introducing either Gln or Lys into the analogous position of the rat brain II Na channel.[26]

*2. Tactics: Choice of Mutagenic Methods.* The design of oligonucleotides and the details of oligonucleotide-directed mutagenesis are beyond the

[100] L. Heginbotham and R. MacKinnon, *Neuron* **8(3),** 483 (1992).
[101] R. A. Kumpf and D. A. Dougherty, *Science* **261,** 1708 (1993).
[102] Q. Lü and C. Miller, *Science* **268,** 304 (1995).
[103] R. MacKinnon and G. Yellen, *Science* **250,** 276 (1990).
[104] J. Satin, J. W. Kyle, Z. Fan, R. B. Rogart, H. A. Fozzard, and J. C. Makielski, *Biophys. J.* **66,** 1353 (1994).
[105] D. A. Dougherty, and D. A. Stauffer, *Science* **250,** 1558 (1990).
[106] D. A. Dougherty, *Science* **271,** 163 (1996).
[107] B. C. Cunningham and J. A. Wells, *Science* **244,** 1081 (1989).
[108] D. S. Ragsdale, J. C. McPhee, T. Scheuer, and W. A. Catterall, *Proc. Natl. Acad. Sci. U.S.A.* **93,** 9270 (1996).
[109] D. S. Ragsdale, J. C. McPhee, T. Scheuer, and W. A. Catterall, *Science* **265,** 1724 (1994).
[110] J. C. McPhee, D. S. Ragsdale, T. Scheuer, and W. A. Catterall, *J. Biol. Chem.* **270(20),** 12025 (1995).

scope of this article (for an introduction see Trower[111]), but a general discussion of common approaches and pitfalls is included below. Most mutagenesis requires DNA polymerization after annealing an oligonucleotide with the desired, new codon to a single-stranded template containing the original channel coding sequence. Alternatively, if closely approximated restriction sites exist or can be engineered, a synthetic, double-stranded DNA cassette can be used to introduce insertions or mutations.[112] In usual cases, DNA polymerization is necessary and is carried out over a limited range of the template using polymerase chain reaction (PCR) or may be conducted over the entire vector containing the template as is done in most commercially obtained mutagenesis kits. One advantage of PCR is that polymerization can be carried out over a limited range so that the risk of spurious mutations is easier to assess. A disadvantage is that the resultant DNA segment often must be isolated and cloned into larger templates. Oligonucleotide-initiated DNA polymerization using the entire vector containing the template channel obviates the need for subsequent ligations. Other advantages of this technique over a PCR-based method include the ease of introducing multiple, distant mutations simultaneously (with some reduction in efficiency). Nevertheless, the risk of spurious mutations increases with the length of DNA polymerized, and larger vectors such as those containing the $Na^+$ channel often require more than one oligonucleotide initiation point for efficient replication.

DNA segments containing the desired mutation can be made by PCR using either two, three,[113] or four primer techniques.[114] For easy ligation into a template or vector, the amplified DNA segment usually spans two unique restriction sites. If the mutation desired is within about 50 base pairs of one of the restriction sites, then two-primer PCR is most desirable. In this case, a single-step PCR amplification can be undertaken that is directed by a mutant primer incorporating one unique restriction site and one primer designed so that the product contains another unique restriction site at the opposite end. Most of the time, a convenient, unique restriction site is unavailable near the desired mutation, and three- or four-primer PCR must be used in lieu of introducing such sites. As before, outside

[111] M. K. Trower, ed., "Methods in Molecular Biology." Humana Press, Totowa, New Jersey, 1996.

[112] A. F. Worrall, *in* "Methods in Molecular Biology. DNA-Protein Interactions: Principles and Protocols" (G. G. Kneale, ed.), Vol. 30, p. 199. Humana Press, Totowa, New Jersey, 1994.

[113] S. Bowman, J. A. Tischfield, and P. J. Stambrook, *Technique* **2,** 254 (1990).

[114] R. Higuchi, "PCR Protocols: A Guide to Methods and Applications," p. 177. Academic Press, San Diego, 1990.

primers are designed that define the extent of the desired PCR fragment containing unique restriction sites on either end. Before the final PCR segment can be amplified, an intermediate amplification must take place to introduce the mutation. The products of this step are used in the final PCR amplification. In three-primer mutagenesis, the intermediate step is directed by one primer containing the mutation and the appropriate outside primer partner. The second step either amplifies a minor product of the first PCR reaction that spans the entire desired length[113] or uses the product of the first reaction, which incorporates the desired mutation, as a megaprimer to amplify the original template.[115] In four-primer mutagenesis, two inside oligonucleotides, at least one of which directs the desired mutation, are used with their appropriate outside primer partners to create two PCR fragments that span the entire length from the two outside restriction sites. These two pieces are isolated, annealed at an overlapping segment, and used to form the template for the final amplification step.

While requiring one less oligonucleotide, the disadvantage of three-primer-directed mutagenesis is that since the native DNA is used as the template the second amplification step yields a mixture of native and mutated DNA segments. The four-primer technique ensures that any final product contains the mutation, but difficulty may be encountered when annealing the two isolated pieces to form the template for the final reaction.

Oligonucleotide-initiated DNA polymerization of an entire single-stranded template requires only one annealing step and potentially no ligations, but, as usually performed, this technique creates only a small number of mutant strands that must be separated from those containing the native channel by some selection process favoring the mutant. In one early scheme, selection was accomplished by using a template containing uracil in place of thymine. Polymerization was carried out incorporating thymine, and the resultant heteroduplex DNA was transformed into bacteria that favored the thymine-containing strand.[116,117] Another method fixes the native strands to a solid-phase support, so that after polymerization, denaturing liberates only the mutant strand.[118,119] Commonly used commercial kits employ a second, simultaneous selection mutation that either reconstitutes an antibiotic resistance that only allows growth of bacteria transformed with the mutant vector (Altered Sites System, Promega, Madison,

[115] G. Sankar and S. S. Sommer, *BioTechniques* **8,** 404 (1990).
[116] T. A. Kunkel, *Proc. Natl. Acad. Sci. U.S.A.* **82,** 488 (1985).
[117] T. A. Kunkel, J. D. Roberts, and R. A. Zakour, *Methods Enzymol.* **154,** 367 (1987).
[118] C. C. Wang, L. P. Fernando, and P. S. Low, *Nucleic Acids Res.* **24,** 1378 (1996).
[119] M. P. Weiner, K. A. Felts, T. G. Simcox, and J. C. Braman, *Gene* **126,** 35 (1993).

WI) or that eliminates a unique restriction site preventing digestion of the mutant vector (Unique Site Elimination Mutagenesis Kit, Pharmacia Biotech, Piscataway, NJ).

When ensuring that only the intended mutation has been introduced, detection of unanticipated missense mutations is problematic, and a single missense mutation can have profound effects on the parameter of interest. For example, a single mutation of an uncharged residue in the domain II, S4 segment of rat brain IIa clone is responsible for a 20-mV shift in the voltage dependence of activation.[120] This problem can be addressed by minimizing the length of DNA subject polymerization, by reducing the amount of amplification, and by using DNA polymerases with higher fidelity. Sequencing the entire polymerized region is reasonable for small segments, but many techniques polymerize over the entire template making sequencing for large channels like the $Na^+$ channel somewhat impractical. In general, two separate mutant clones with the same behavior have been taken as evidence that no spurious missense mutations have occurred.

### *C. Complementary Mutations of Toxin and Channel*

Effects of toxin or channel mutations on affinity are not generally the result of loss of one specific, individual interaction between a pair of residues on the channel and toxin, respectively. For example, a single carboxyl group in the outer vestibule might be expected to attract several positive charges on $\mu$CTX, and neutralization of this group would have a larger effect on the affinity than the elimination of any single toxin/channel interacting pair. The contribution to the binding energy of one interacting pair of residues can be estimated by mutant cycle analysis.[73,121–129]

*1. Mutant Cycle Analysis.* In mutant cycle analysis, the separate effects of mutations on the channel and on the toxin are assessed, and a prediction

[120] V. J. Auld, A. L. Goldin, D. S. Krafte, W. A. Catterall, H. A. Lester, N. Davidson, and R. J. Dunn, *Proc. Natl. Acad. Sci. U.S.A.* **87,** 323 (1990).

[121] J. Aiyar, J. M. Withka, J. P. Rizzi, D. H. Singleton, G. C. Andrews, W. Lin, J. Boyd, D. C. Hanson, M. Simon, B. Dethlefs, C. Lee, J. E. Hall, G. A. Gutman, and K. G. Chandy, *Neuron* **15,** 1169 (1995).

[122] R. Ranganathan, J. H. Lewis, and R. MacKinnon, *Neuron* **16,** 131 (1996).

[123] P. Hidalgo and R. MacKinnon, *Science* **268,** 307 (1995).

[124] A. R. Fersht, A. Matouschek, and L. Serrano, *J. Molec. Biol.* **224,** 771 (1992).

[125] A. Horovitz and A. R. Fersht, *J. Molec. Biol.* **214,** 613 (1990).

[126] A. Horovitz, L. Serrano, B. Avron, M. Bycroft, and A. R. Fersht, *J. Molec. Biol.* **216,** 1031 (1990).

[127] G. Schreiber and A. R. Fersht, *J. Molec. Biol.* **248,** 478 (1995).

[128] L. Serrano, A. Horovitz, B. Avron, M. Bycroft, and A. R. Fersht, *Biochemistry* **29,** 9343 (1990).

[129] J. Aiyar, J. P. Rizz, G. A. Gutman, and K. G. Chandy, *J. Biol. Chem.* **271(49),** 31013 (1996).

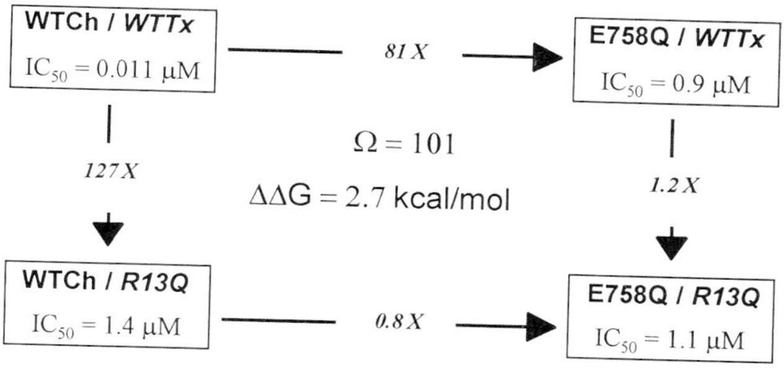


FIG. 5. Thermodynamic mutant cycle for the most strongly coupled pair of residues so far identified in the binding of μCTX to the rSkM1 $Na^+$ channel—R13 of μCTX and E758 of the outer charged ring in domain II of the channel. An even larger coupling coefficient results from analysis of the pair of charge-reversing mutations, R13D and E758K.

is made for the resultant affinity if the two mutations are combined assuming their effects are independent and additive. The actual result is compared to the independence expectation. Under some conditions, the ratio of the expected and actual results, Ω, is a quantitative measure of the interaction between the two residues mutated. By convention when Ω is $<1$, the inverse value is reported; Ω values of 1 are usually taken to suggest no interaction between the pair, but an interaction is not rigorously excluded. On the other hand, a larger Ω indicates a larger energy of interaction. In the case of the barnase–barstar interaction where structural data are available, values greater than 2.4 are associated with close approximation of the residue pair and are greater than twice the variance of the measure.[127]

With a pair of channel and toxin mutants known to affect affinity, calculating an Ω requires accurate determination of the blocking efficacy of the four possible combinations of the wild-type (w.t.) channel, wild-type toxin, mutated channel, and mutated toxin (Fig. 5). Coupling coefficients for each toxin–channel pair can be calculated as follows:

$$\Omega = X_1/X_2 = Y_1/Y_2$$

where $X_1$ is the (affinity of mutant toxin : w.t. channel)/(affinity of w.t. toxin : w.t. channel); $X_2$ is (affinity of mutant toxin : mutant channel)/(affinity of w.t. toxin : mutant channel); $Y_1$ is (affinity of w.t. toxin : mutant channel)/(affinity of w.t. toxin : w.t. channel); and $Y_2$ is (affinity of mutant toxin : mutant channel)/(affinity of mutant toxin : w.t. channel).

Given that ratios are used in this calculation, the measure of "affinity" can be the equilibrium dissociation or association constant, or, if rigorous measurements of these parameters are impractical, $IC_{50}$.

Under certain circumstances, the logarithm of the coupling coefficient is proportional to the energy of interaction between the pair ($\Delta\Delta G = RT \ln \Omega$). These circumstances include no movement of either residue on mutation of the partner and no local structural changes that are different when one or both of the partners are present.[126] If these two conditions are met and all of the interaction between the pair is eliminated by the mutations, then $\Delta\Delta G$ equals the energy of interaction relative to the energy of the fully solvated individual partners.

The value of $\Delta\Delta G$ should be less than or equal to the change in energy of mutating any single residue on the channel or toxin. In practice, this is not always the case, however, and there are several reasons why $\Delta\Delta G$ might not reflect accurately the interaction energy. Calculating an interaction energy assumes that the pair of mutations eliminates any interaction between them so that $\Omega$ represents the total energy of the interaction. Some mutation pairs may not eliminate all of this interaction energy or may result in new forms of interaction that change the value of $\Omega$. This can be recognized experimentally if the $\Omega$ values are dependent on the groups substituted. If this occurs, $\Delta\Delta G$ no longer represents the energy of interaction,[130] but the dependence of $\Omega$ on the groups substituted can be used to probe the nature of the interactions between the two sites.[131] Quantitative errors in $\Delta\Delta G$ can result from failing to take into account energies of hydration, changes in counterion concentrations, movement of residues, or strain in the proteins brought about by the mutations. Significant structural rearrangement can usually be excluded if mutant pairs can be found that improve binding in comparison to the effects of either mutation alone. These combinations are rare (e.g., Naranjo and Miller[73]) suggesting that it is much easier to disrupt existing interactions than to establish new ones. Also, an energetically significant mutant cycle in the background of multiple, nearby noninteracting pairs suggests that no significant structural rearrangements have occurred.

*2. Nontraditional Applications of Mutant Cycles*. Some less conventional applications of mutant cycle analysis have proven useful. First, mutant cycle analysis can be used to assess the interaction of a residue pair during formation of the ligand–receptor activated state.[124,132] This requires rate constant data for the ligand before and after mutations of the ligand and the channel. Second, untraditional "mutations" such as changes in ion concentration can be used to delineate relationships between the ion and the mutated site.[122] For example, the effect of *Shaker* $K^+$ channel Y445F

[130] G. A. Faiman and A. Horovitz, *Prot. Eng.* **9,** 315 (1996).

[131] A. R. Fersht, A. J. Wilkinson, P. Carter, and G. Winter, *Biochemistry* **23,** 5858 (1985).

[132] A. Matouschek, J. T. Kellis, Jr., L. Serrano, and A. R. Fersht, *Nature* **340,** 122 (1989).

mutation on Agitoxin2 binding was dependent on the external $K^+$ concentration. Third, ligands need not be peptides to be "mutated," and STX analogs have been used to map interactions with the $Na^+$ channel outer vestibule (unpublished observation). Finally, the interaction energy of a pair may depend on the presence of a third amino acid as is the case for the Asp-12 : Arg-110 salt bridge in barnase that is influenced by Asp-8.[126] These energetic relationships can be sorted out by multidimensional mutant cycles.[125]

*3. Electrostatic Compliance and Measurement of Interresidue Distances.* Analysis of electrostatic compliance is an ingenious approach developed by Stocker and Miller.[71] It complements mutant cycle analysis of toxin–channel interactions. This technique isolates the charge-dependent interaction between two specified residues, even though each residue in the bound complex may participate in an arbitrary number of interactions with additional residues. In contrast to mutant cycle analysis, it is desirable that the pair not have any close-range nonelectrostatic interactions such those found in a salt bridge (Ref. 82, section 4.1.2). Electrostatic compliance analysis allows quantification of purely electrostatic interactions while mutant cycle analysis allows evaluation of the strength of any type of close interactions. Electrostatic compliance between two groups $i$ and $j$, $\sigma(i, j)$, is defined as

$$\sigma(i, j) = \frac{\partial^2(\ln K_d)}{\partial z_j \partial q_i}$$

where $q_i$ and $z_j$ refer to the charges of the two groups. A larger value of $\sigma$ implies a stronger interaction.

To measure electrostatic compliance, it is necessary to express neutral, positive, and negative mutations at both the channel and the toxin interacting sites, and the affinities of all nine combinations of mutant toxins and channels must be determined. The ln $K_d$ values for each toxin mutation are plotted as a function of the charge at the interacting channel site, preferably for more than one substitution of each charge. Then, the slopes of these three lines are plotted as a function of the charge at the toxin site. The electrostatic compliance, $\sigma$, is the slope of this relationship. If the relationship is nonlinear, then the two groups have indirect or nonelectrostatic (e.g., conformational) interactions that preclude a simple interpretation of electrostatic compliance. On the other hand, if these plots are linear, $\sigma$ can provide a measure of the strength of the electrostatic interaction between the two residues. If the Debye–Hückel relationship provides a reasonable description of the electric potential between the residues, the distance between them may be calculated directly from the electrostatic compliance (see also Section IV,D for alternate force field assumptions).

Electrostatic compliance has been used successfully to evaluate interactions between pairs of residues on *Shaker*[71] and Kv1.3 channels[121] and various peptide toxin blockers similar to $\mu$CTX. The technique has not yet been applied to $\mu$CTX–$Na^+$ channel interactions, but R19 of $\mu$CTX would appear to be a good candidate for study. R19A causes a significant reduction in affinity, but R19K retains wild-type affinity, suggesting that the R19–$Na^+$ channel interaction is exclusively electrostatic.[75]

### *D. Using the Data: Modeling of Outer Vestibule*

While many generally applicable concepts about ligand–receptor binding can be enumerated using the techniques above, one ultimate goal is to employ the information gained as constraints to refine molecular models of the outer vestibule and selectivity filter of the $Na^+$ channel. The techniques above provide evidence of interacting pairs and estimate the energy of this interaction. Using these energies requires that the structure of the toxin be known and that the distance between toxin–channel interacting pairs can be estimated.

The interaction energy is a function of the distance separating the pair, but in order to calculate the distance, an assumption about the distribution of the electric field between the pair must be made. The true representation of the electric potential ($\varphi$) in the vicinity of a toxin–channel interface as a function of distance is unknown. Common approaches are to assume a Coulombic relationship ($\varphi \propto 1/r$; where $r$ is the radius of separation) or a Debye–Hückel relationship that accounts for counterion shielding [$\varphi \propto (1/r) \exp - (r/r_D)$, where $r_D$ is the Debye length]. If the charged group(s) in question are only partially exposed at an interface, additional complications arise (see Section V,A). Even more complexity can be added by assuming dielectric discontinuities or distance-dependent dielectric constants to address changes in the distribution of $\varphi$ as a result of protein–solvent interfaces. Numerical solutions of the Poisson–Boltzmann equation address these considerations and allow for calculations of the electric potential around objects of complex shapes. These calculations closely approximate experimental results (see Honig and Nicholls[133] for a review).

These calculations often deal with the solvent effects through the use of a single, effective dielectric constant ($\varepsilon_{eff}$). The physical interpretation or appropriate mathematical description of a dielectric constant on the scale of the interactions explored in toxin–channel interfaces is unclear. These interactions take place between two polarizable proteins in a partially solvated interface. Therefore, $\varepsilon_{eff}$ might be assumed to be somewhere be-

[133] B. Honig and A. Nicholls, *Science* **268,** 1144 (1995).

tween the value for a protein and that for water. Changes in $\mu$CTX affinity with point mutations of the channel have been used to estimate $\varepsilon_{\mathrm{eff}}$ at the interface with the outer vestibule. In this case, the resulting value for $\varepsilon_{\mathrm{eff}}$ was 31, a reasonable value falling between the protein and water limits.[36] A value in this range would seem to be a reasonable first approximation for calculations.

The distance calculations constrain the tertiary structure without directly addressing the secondary structure in the binding domain. Rules such as those promulgated by Chou and Fasman[134] are used to make secondary structural predictions from the primary sequence. These predictions are combined with the interaction constraints to arrive at a molecular model. This is the approach taken by Lipkind and Fozzard[135] in developing the first model of the outer vestibule and selectivity filter of the $Na^+$ channel based on the complementary binding surface for STX and TTX. Subsequently, this model has been adequately reconciled with experimental observations concerning $\mu$CTX's interaction with the channel.[36]

By necessity, the structure generated using this approach is a static picture representing the conformation of the channel–toxin complex at equilibrium. Two unresolved concerns about this structure are interrelated. How well does this static structure represent the possibly more flexible channel in the absence of toxin,[34,35] and how much does toxin binding deform the native structure of the channel? There is likely to be a fundamental limit on the flexibility of the $Na^+$ channel in the outer vestibule region since wide structural fluctuations would be inconsistent with maintaining selectivity or specificity of high-affinity toxin binding. On the other hand, some motion is necessary for conduction.[136] The amount of energy available for protein deformation should be a function of the toxin affinity. Therefore, there is always a compromise between the complementarity of the binding site to the toxin, which is assumed to be higher with higher affinities, and the increased affinity resulting from a closer fit induced by bending the channel out of its usual conformation. Because toxin binding is reversible and the product of multiple, dispersed interactions, probing with $\mu$CTX might be expected to deform the channel structure less than the binding of group IIB metals to introduced cysteines ($K_d$ values $\approx 10^{-50}$ $M$) or than the use of chemical modifiers such as sulfhydryl reagents.

With these relatively crude techniques, multiple rounds of modeling, prediction, and experimental testing are necessary to exclude alternative

[134] P. Y. Chou and G. D. Fasman, *Annu. Rev. Biochem.* **47,** 251 (1978).

[135] G. M. Lipkind and H. A. Fozzard, *Biophys. J.* **66,** 1 (1994).

[136] C. Singh, R. Sankararamakrishnan, S. Subramaniam, and E. Jakobsson, *Biophys. J.* **71,** 2276 (1996).

structures and refine the most likely model. Apart from these repeated cycles, molecular dynamics simulations have been employed to assess model validity and behavior. Even with a relatively early model,[137] interesting observations have been made about the role of water in the pore, the distance over which selectivity occurs, and the forces at work to bring about selectivity.[136,138]

## V. Partially Blocking Toxins as Tools to Study Gating and Permeation

A remarkable result emerged from an initial single-channel study of the blocking action $\mu$CTX derivatives. A neutralizing substitution of arginine 13 yielded a peptide (R13Q), which induced discrete, but *incomplete,* block of single-channel currents to a level about 30% of the full, unblocked single-channel single current.[49] Individual, partial blocking events are, on the average, seconds in duration, despite the substantial decrease in binding affinity associated with the substitution of R13. Thus, superposed on this residual current, it is possible to observe the channel's intrinsic gating fluctuations, or block of the residual currents by more rapidly acting, lower affinity blockers such as quaternary ammonium local anesthetic analogs.[24] In effect, the binding and dissociation of the peptide functions as a transient and reversible structural mutation of the channel, in which the highly charged ligand, R13Q (nominal net charge, +5), attaches and detaches from the channel protein. Functions of the bound and unbound channel can be directly monitored in adjacent segments of one single-channel record, allowing evaluation of electrostatic, or other, effects of the ligand on conduction, gating, or block by other agents.

Subsequently, it has been found that other charge-changing substitutions (R13A, R13D) in the peptide result in incomplete blocking action (K. Hui, I. Sierralta, and R. French, 1996, unpublished and Ref. 139), opening the possibility of mapping the effects of charge changes at different points on the peptide structure. If the effects of peptide binding are wholly or partly electrostatic, these derivatives can function as a precise set of electrical probes to perturb channel function and reveal responses of the channel to changes in the local electrostatic potential at a precisely defined site.

[137] H. R. Guy and S. R. Durell, *in* "Ion Channels and Genetic Diseases" (D. C. Dawson *et al.,* eds.), p. 1. The Rockfeller University Press, New York, 1995.

[138] R. Sankararamakrishnan, S. Subramaniam, and E. Jakobsson, *Biophys. J.* **72,** A360, (abstract), (1997).

[139] N. S. Chang, R. J. French, G. M. Lipkind, H. A. Fozzard, and S. Dudley, Jr., *Biochemistry* **37,** 4407 (1998).

### A. *Voltage Sensor Movement and Location*

Phenomenologically, the effect of the binding of the peptide R13Q on voltage-dependent channel activation resembles that of increasing extracellular divalent ion concentration—the activation curve shifts toward more depolarized potentials reflecting the greater electrical work required to open the channels. Mechanistically, there is an important difference. Whereas the gating shift induced by $Ca^{2+}$ (e.g., Refs. 140–142) results from a generalized neutralization of extracellular surface charges on both the channel protein and the membrane lipid[143] (but see Ref. 144) by some combination of binding or screening,[141] the effect of the peptide results from its binding to a specific site on the channel, rather than from a generalized surface charge effect.[145] Nonetheless, the effects of both $Ca^{2+}$ and the peptide are consistent with an electrostatic inhibition of the outward movement of voltage-sensing gating charge, as depicted by the diagram in Fig. 4. This does not, of course, exclude the possibility of nonelectrostatic mechanisms also playing a role. However, if one takes as a working hypothesis that the interaction is entirely electrostatic, it is possible to obtain an empirical estimate of the extra electrical work required, in the presence of the bound peptide, to open a channel. Then, if the charge distribution on the bound peptide and on the moving voltage sensor are known, possible trajectories of physical movement of the voltage-sensing charge can, in principle, be calculated.

Some calculations of this type have been done.[24,145] One key consideration is to account for the partial screening, by counterions in the solution,[146,147] of charges on the bound peptide when calculating the contribution of the peptide charges to electrostatic potential. A detailed critique of the additional assumptions and approximations involved is beyond the scope of this article, but this initial analysis raises the hope that in the future, more precise calculations may allow rigorous experimental testing of detailed

[140] B. Frankenhaeuser and A. L. Hodgkin, *J. Physiol.* (*Lond.*) **137,** 218 (1957).

[141] B. Hille, A. M. Woodhull, and B. I. Shapiro, *Phil. Trans. R. Soc. Lond. B.* **270,** 301 (1975).

[142] R. Hahin and D. T. Campbell, *J. Gen. Physiol.* **82,** 785 (1983).

[143] S. Cukierman, W. C. Zinkand, R. J. French, and B. K. Krueger, *J. Gen. Physiol.* **92,** 431 (1988).

[144] C. M. Armstrong and G. Cota, *Proc. Natl. Acad. Sci. U.S.A.* **88,** 6528 (1991).

[145] R. J. French and R. Horn, *in* "From Ion Channels to Cell-to-Cell Conversations" (R. Latorre *et al.*, eds.), p. 67. Plenum Press, New York, 1997.

[146] R. T. Mathias, G. J. Baldo, K. Manivannan, and S. McLaughlin, *in* "Electrified Interfaces in Physics, Chemistry and Biology" (R. Guidelli, ed.), p. 473. Kluwer Academic Publishers, The Netherlands, 1992.

[147] F. H. J. Stillinger, *J. Chem. Phys.* **35(5),** 1584 (1961).

structural models of the channel, and evaluation of charge redistribution that occurs during the conformational changes involved in channel opening.

### *B. Locating Intrapore Binding Sites*

A simpler computational scenario results from analysis of the weakening, by R13Q, of the action of cationic blockers such as diethylammonium (DEA) that enter the channel through its cytoplasmic mouth. Because the DEA approaches from the bulk solution—electrostatically an infinite distance from the bound peptide—it is possible to estimate directly the physical distance between the bound DEA and the peptide from the shift in blocking equilibrium determined from a plot of fractional block of single-channel current versus voltage. By contrast, the starting position for gating charge movement, for calculations outlined in the previous section, is not known, and the distance moved by the gating charge and its final position relative to the peptide charge can be obtained explicitly only when a particular starting position is assumed. Given that the side chain of the neutral Q13 is the one expected to lie closest to the bound DEA, it seems clear that the remaining peptide charges must contribute to the repulsive interaction. Thus, pending more detailed studies with a variety of $\mu$CTX derivatives, the positions of bound blocking ions like DEA can be calculated only with respect to an operationally defined center of charge of the peptide.

There is a striking difference between interaction of R13Q and DEA, and that of ChTX derivatives with ions in the large conductance potassium (BK) channel. In the latter case, the voltage dependence of ChTX block appears to arise exclusively from the repulsion between a $K^+$ ion bound in the pore and the K27 side chain—the voltage dependence is abolished when K27 is replaced by a neutralizing mutation. This suggests that charges on ChTX, with the exception of K27, are completely screened from ions in the BK channel pore. Unfortunately, to date, it has not been possible to explore the electrostatic interactions of residue 27 in ChTX with charge-reversing substitutions. The distance between K27 and the adjacent $K^+$ in the BK pore has been estimated,[70] although the authors suggested that the exact calculated value (~5 Å) be treated "merely as a visual aid to idle speculation about the underlying structural features." Such skepticism seems healthy, but if the "idle speculations" lead to testable hypotheses about interactions within the channel, the calculations will have made a useful contribution to our understanding.

### *C. Factors Controlling Selectivity and Permeation*

It seems clear that $\mu$CTX binds in the outer vestibule of the $Na^+$ channel's conducting pore, and that the most intimate association is between

R13 of the toxin and Glu-758 of the pore's outer charged ring, rather than with the residues of the inner ring (Fig. 2) that are thought to comprise the selectivity filter. The interaction with Glu-758 is significantly weakened by any substitution for R13. Thus, although not vastly distant, the bound R13X derivatives may be sufficiently removed from the selectivity filter that they do not greatly perturb the channel's intrinsic ability to discriminate among different ion species. On the other hand, the single-channel conductance is strongly influenced by the nature of the residue at position 13. It remains to be determined whether the partially blocking peptides made by a charge-changing substitution for R13 merely alter conductance by nonselective means, such as reducing the capture volume for ion access to the pore or the local monovalent cation concentration, or whether the channel's ability to discriminate among different ions is drastically altered when such peptides are bound. Intuitively, it seems rudely intrusive to place a peptide with a net charge of +4 or +5 strategically in the mouth of a cation-conducting pore—perhaps most surprising is that there is any residual conductance at all! Preliminary data suggest that changes in selectivity in the partially blocked channels are not dramatic (K. Hui and R. French, 1997, unpublished observations). If that is confirmed by ongoing studies, these derivatives may teach us more about the way in which ion entry into a channel can be tuned to modulate conductance than about the channel's mechanism of selection among different ion species.

### *D. Mechanism of Pore Block*

Based on experiments showing that guanidinium ions pass through the $Na^+$ channel selectivity filter but the guanidinium-containing toxins, STX and TTX, block conduction, Hille[148] proposed that these toxins blocked the channel by interacting with the selectivity filter within the outer vestibule. A wealth of mutagenesis data places the binding site of these toxins at the outer vestibule.[26–28,30,32,35,36,42,48] Mutant cycle analysis has been used to show that the N1 hydroxyl group of neosaxitoxin interacts with domain I selectivity filter residue (unpublished observation, 1997, Penzotti and Dudley). Therefore, it appears likely that Hille's original hypothesis, that STX and TTX sterically occlude the narrowest part of pore and thus prevent current flow, is correct.

Several lines of evidence suggest that the guanidinium group of R13 of μCTX plays a similarly important role in the block of current when μCTX is bound to $Na^+$ channels. Mutant cycle analysis suggests that R13 of bound μCTX is located in the outer vestibule.[139] R13 substitutions result in residual

[148] B. Hille, *Biophys. J.* **15,** 615 (1975).

current when μCTX is bound to the channel,[24,49,139,145] and the residual current increases progressively when R13 is substituted by lysine, glutamine, alanine, and aspartate. Despite this, it is not clear whether any of the residues sterically prevent $Na^+$ passage through the pore, rather than make it progressively less likely by electrostatic repulsion as residue is made more positive.

R13 need not block current in the same way as the guanidinium groups of STX and TTX, however, and consistent with this idea, there are no reports of residual currents with STX or TTX analogs. Several scenarios exist whereby the presence of R13 of μCTX might block the channel but different substitutions at the R13 position might modulate a residual current. Possibilities include (1) variable strength of repulsive interactions between $Na^+$ and R13X derivatives positioned in the outer vestibule, (2) variable coordination, by the toxin derivatives, of channel residues necessary for conduction, (3) incremental conformational changes that alter pore size depending on the residue at the R13 site, or (4) progressive modulation of how tightly the body of μCTX is nestled into the pore as a function of the strength of the residue 13–channel interaction. The occurrence of residual currents with toxin derivatives having nominal net charges of +5 (R13Q, R13A) and +4 (R13D) argues against a generalized electrostatic repulsion of $Na^+$ by charges on the body of the toxin being responsible for block. Thus, a repulsive exclusion mechanism (as opposed to steric exclusion) would have to be highly local and centered around residue 13. Consistent with this possibility are observations that leave open the possibility that R13K and R13Orn may conduct some residual current (Ref. 145, unpublished observation, 1997, Hui, Sierralta and French). Even native μCTX may allow a small residual current,[99] but this is difficult to confirm using macroscopic current measurements in heterologous expression systems that may express a low density of toxin-insensitive channels.[149] The various possibilities remain to be explored in detail.

## VI. Questions for the Future

Many questions remain for the future. As more of the points of contact between the μCTX and the $Na^+$ channel are elucidated, some of these issues should be settled. For example, the order in which the domains are packed is unknown. In fact, recent molecular models of the outer vestibule (viewed from the extracellular surface) have arranged the pore-forming loops contributed by each domain in either counterclockwise[37,43,137] or clockwise directions.[42,45,135] The multiple points of contact between μCTX

[149] C. Ukomadu, J. Zhou, F. J. Sigworth, and W. S. Agnew, *Neuron* **8,** 663 (1992).

and the channel hold the promise of resolving this important issue. On a broader front, other toxins offer the possibility to explore different parts of the surface of various channels,[150–152] including sodium channels[153] and potassium channels,[154] and the potential of peptide toxins to reveal details of calcium channel structure remains largely untapped.[150–152,155,156]

## Acknowledgments

We thank Dr. Gregory Lipkind, of the University of Chicago Cardiology Molecular Modeling Core, and Chris Bladen for providing figures, and to Dr. Denis McMaster for providing details of the peptide synthesis protocols. Our research is supported by the Medical Research Council of Canada and the National Institutes of Health, USA, P01-HL20592. We are grateful to Dr. Harry Fozzard for numerous discussions, and ongoing support and encouragement. R.J.F. is a Medical Scientist of the Alberta Heritage Foundation for Medical Research and a Medical Research Council Distinguished Scientist.

[150] B. M. Olivera, J. Rivier, C. Clark, C. A. Ramilo, G. P. Corpuz, F. C. Abogadie, E. E. Mena, S. R. Woodward, D. R. Hillyard, and L. J. Cruz, *Science* **249,** 257 (1990).
[151] B. M. Olivera, J. Rivier, J. K. Scott, D. R. Hillyard, and L. J. Cruz, *J. Biol. Chem.* **266(33),** 22067 (1991).
[152] B. M. Olivera, G. P. Miljanich, J. Ramachandran, and M. E. Adams, *Annu. Rev. Biochem.* **63,** 823 (1994).
[153] J. M. McIntosh, A. Hasson, M. E. Spira, W. R. Gray, W. Li, M. Marsh, D. R. Hillyard, and B. M. Olivera, *J. Biol. Chem.* **270(28),** 16796 (1995).
[154] K. J. Swartz and R. MacKinnon, *Neuron* **15,** 941 (1995).
[155] I. M. Mintz, *J. Neurosci.* **14(5),** 2844 (1994).
[156] I. M. Mintz, V. J. Venema, M. E. Adams, and B. P. Bean, *Proc. Natl. Acad. Sci. U.S.A.* **88,** 6628 (1991).
[157] W. A. Catterall, *Curr. Opin. Cell Biol.* **6,** 607 (1994).

# [31] *Conus* Peptides as Probes for Ion Channels

*By* J. MICHAEL MCINTOSH, BALDOMERO M. OLIVERA, and LOURDES J. CRUZ

## Introduction

*Conus* peptides are increasingly used as tools for investigating ion channels. The 500 species of predatory cone snails each produces a complex venom that has a large number of biologically active peptides. The majority of *Conus* peptides characterized to date appear to be targeted to different types of ion channels. It is estimated that the venom of each *Conus* species has between 50 and 200 peptides. Because of the remarkable divergence that occurs when cone snails speciate, the complement of venom peptides



0076-6879/99 $30.00

in any one *Conus* species is distinct from that of any other. Thus, many thousands of peptides that affect ion channel function are present in *Conus* venoms but only a miniscule fraction of these have been characterized biochemically. An even smaller number have been used as tools in neurobiology. However, there is little doubt that as more of these peptides become available to the neurobiological community, an increasing number will be used as ligands for characterizing ion channel structure and function. Because of their relatively small size, most of these peptides can be chemically synthesized, and thus be made widely available.

## Biochemical Overview of *Conus* Peptides

The *Conus* venom peptides can be divided into two general groups: (1) multiply disulfide-bonded peptides from 12 to 50 amino acids in length (most under 30 residues). Generically, these are called conotoxins, and (2) other peptidic venom components that are not disulfide-rich; these either completely lack disulfide bonds or have a single disulfide linkage. The latter are a heterogeneous group of peptides with several distinct families.

In the following sections, we focus first on *Conus* peptides that are targeted to ligand-gated ion channels, followed by peptides that are targeted to voltage-gated ion channels. The last section discusses practical considerations for using *Conus* peptides. It should be noted parenthetically that in much of the literature of the late 1980s and early 1990s, the term *conotoxin* was routinely used to refer to one specific molecule out of the many tens of thousands of *Conus* peptides—this was ω-conotoxin GVIA, the first natural toxin known to inhibit voltage-gated calcium channels. Given the very large number of *Conus* peptides, it is no longer appropriate to use the term *conotoxin* for this one peptide. In this review, *conotoxin* will be used generically for all multiply disulfide-bonded *Conus* peptides.

For neurobiologists, the major interest in *Conus* peptides is that they are highly subtype-specific ligands. For several ion channel targets, *Conus* peptides are the most specific ligands known. For example, among ligands that target voltage-gated sodium channels, μ-conotoxin GIIIA has unprecedented specificity for the skeletal muscle subtype. This isoform is among the set of sodium channels that are tetrodotoxin and saxitoxin sensitive. However, μ-conotoxin GIIIA is much more specific than either of the guanidinium toxins; it has a preference for the skeletal muscle isoform by at least three orders-of-magnitude over other tetrodotoxin-sensitive subtypes.[1,2] This high subtype selectivity is proving to be a general feature of

[1] L. J. Cruz, W. R. Gray, B. M. Olivera, R. D. Zeikus, L. Kerr, D. Yoshikami, and E. Moczydlowski, *J. Biol. Chem.* **260,** 9280 (1985).

[2] T. Gonoi, Y. Ohizumi, H. Nakamura, J. Kobayashi, and W. A. Catterall, *J. Neurosci.* **7,** 1728 (1987).

TABLE I
CLASSES OF *Conus* PEPTIDES AND THEIR MACROMOLECULAR TARGETS

| Peptide class | Characteristic structural features (number of amino acids) | Mode of action |
|---|---|---|
| α-Conotoxins | CC—C—C (12–19) | Competitive inhibitor of nicotinic ACh receptor |
| αA-Conotoxins | CC—C—C—C—C (25–30) | Competitive inhibitor of nicotinic ACh receptor |
| ψ-Conotoxins | CC—C—C—CC (24) | Noncompetitive inhibitor of nicotinic ACh receptor |
| Conantokins | γ-carboxylate residues, Cys residues not necessary (17–27) | Noncompetitive inhibitor of NMDA receptor |
| μ-Conotoxins | CC—C—C—CC (22) | Sodium channel blocker; competes with saxitoxin and tetrodotoxin for site I |
| μO-Conotoxins | C—C—CC—C—C (31) | Sodium channel blocker; does not compete with saxitoxin for site I binding |
| μ-Conotoxins | CC—C—C—C—C (17) | Blocks molluscan sodium channels |
| δ-Conotoxins | C—C—CC—C—C (27–31) | Delays sodium channel inactivation; binds to site VI of the channel |
| κ-Conotoxins | C—C—CC—C—C (27) | Potassium channel blocker |
| ω-Conotoxins | C—C—CC—C—C (24–29) | Calcium channel blocker |

*Conus* peptides. As a consequence, with more isoforms of ion channel families being cloned and characterized, and the need for subtype-specific ligands increasing, *Conus* peptides will undoubtedly be increasingly used to discriminate functionally between closely related molecular forms of ion channels. In many ways, having a very highly subtype-specific *Conus* peptide ligand provides a complementary approach to having a gene knockout of one particular ion channel isoform.

An overview of the *Conus* peptides known to affect ion channel function is given in Table I.

## *Conus* Peptides Targeting Ligand-Gated Ion Channels

Four families of *Conus* peptides are known to target ligand-gated ion channels; three of these target nicotinic acetylcholine receptors (nAChRs). These include the α-conotoxins, the αA-conotoxins, and the ψ-conotoxins. The first two families are believed to be competitive antagonists of the nicotinic receptor, while the ψ-conotoxins have recently been shown to be noncompetitive antagonists. To date, peptides in all three families have been found that target the skeletal muscle subtype of nicotinic receptors. However, all *Conus* peptides characterized so far that preferentially inhibit *neuronal* nicotinic receptors belong to the $\alpha$-conotoxin family.

The other group of peptides that target ligand-gated ion channels is the conantokins; these are unusual *Conus* peptides that have been shown to antagonize the NMDA (*N*-methyl-D-aspartate) subclass of glutamate receptors.

Preliminary evidence for *Conus* peptides that target other ligand-gated ion channels such as the 5HT3 receptor has been obtained, but a complete biochemical characterization of these peptides is not yet published.[3]

## *Conus* Peptides Targeting Skeletal Muscle Subtype of Nicotinic Acetylcholine Receptors

### $\alpha$-*Conotoxins*

One group of $\alpha$-conotoxins is known to target the skeletal muscle subtype of nicotinic receptors (the "$\alpha$3/5 subfamily"). Characteristically, these have three amino acids between the second and third cysteine residues, and five amino acids between the third and fourth cysteine residues of the peptide. The sequences of all $\alpha$-conotoxins of this subfamily are shown in Table II. Among these is the very first *Conus* peptide that was biochemically characterized, $\alpha$-conotoxin GI. Certain members of a second subfamily of $\alpha$-conotoxins, the "$\alpha$4/7 subfamily" also target the muscle receptor. One example is $\alpha$-conotoxin EI.[4]

The $\alpha$3/5 subfamily of $\alpha$-conotoxins is the best characterized with respect to high targeting specificity for the muscle receptor. $\alpha$-Conotoxin MI has been shown to discriminate between the $\alpha/\delta$ and the $\alpha/\gamma$ interface of the mammalian nicotinic acetylcholine receptor by approximately $10^4$. When the nicotinic receptor from *Torpedo* is used, $\alpha$-conotoxin SIA has been shown to discriminate totally between the two ligand-binding sites (in this case targeting to the $\alpha/\gamma$ interface of the *Torpedo* receptor). $\alpha$-Conotoxins MI and GI have been shown to be inactive at neuronal nAChRs including $\alpha_2\beta_2$, $\alpha_2\beta_4$, $\alpha_3\beta_2$, $\alpha_3\beta_4$, $\alpha_4\beta_2$, and $\alpha_4\beta_4$ subtypes. Additionally, they do not block $\alpha_7$ and $\alpha_9$ homomers in contrast to the long $\alpha$-neurotoxins from elapiid snakes, such as $\alpha$-bungarotoxin. Thus, compared to $\alpha$-bungarotoxin, peptides such as $\alpha$-contoxin MI appear to be much more highly specific.

[3] L. J. England, J. Imperial, R. Jacobsen, A. G. Craig, J. Gulyas, J. Rivier, D. Julius, and B. M. Olivera, Seratonin Symposium, San Francisco (1997).

[4] J. S. Martinez, B. M. Olivera, W. R. Gray, A. G. Craig, D. R. Groebe, S. N. Abramson, and J. M. McIntosh, *Biochemistry* **34,** 14519 (1995).

TABLE II
STRUCTURE AND SPECIFICITY OF $\alpha$-CONOTOXINS

Disulfide bond arrangement:

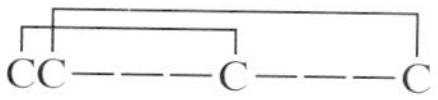

| $\alpha$-Conotoxin | Source | Primary structure | Site preference | Ref. |
|---|---|---|---|---|
| Targeted to skeletal muscle nAChR | | | | |
| | $\alpha$3/5 Subfamily | | | |
| GI | *Conus geographus* | E**CC**NPA**C**GRHYS**C**[a] | Mouse: $\alpha/\delta$ subunit interface | 5–7,42,43 |
| GIA | *Conus geographus* | E**CC**NPA**C**GRHYS**C**GK[a] | | 42 |
| GII[c] | *Conus geographus* | E**CC**HPA**C**GKHFS**C**[a] | | 42 |
| MI | *Conus magus* | GR**CC**HPA**C**GKNYS**C**[a] | Mouse: $\alpha/\delta$ subunit interface | 5–7,43,44 |
| SI | *Conus striatus* | I**CC**NPA**C**GPKYS**C**[a] | | 45 |
| SIA[c] | *Conus striatus* | Y**CC**HPA**C**GKNFD**C**[a] | *Torpedo:* $\alpha/\gamma$ subunit interface | 46,47 |
| SII[c] | *Conus striatus* | G**CCC**NPA**C**GPNYG**C**GTS**CS**[b] | | 48 |
| | $\alpha$4/7 Subfamily | | | |
| EI | *Conus ermineus* | RDO**CC**YHPT**C**NMSNPQI**C**[a] | *Torpedo:* $\alpha/\delta$ subunit interface | 4 |
| Targeted to neuronal nAChRs | | | | |
| | $\alpha$4/7 subfamily | | | |
| MII | *Conus magus* | G**CC**SNPV**C**HLEHSNL**C**[a] | Rat: $\alpha_3\beta_2$ subunit interface | 49 |
| PnIA | *Conus pennaceus* | G**CC**SLPP**C**AANNPDY**C**[a] | *Aplysia:* neuronal nAChR | 12 |
| PnIB | *Conus pennaceus* | G**CC**SLPP**C**ALSNPDY**C**[a] | *Aplysia:* neuronal nAChR | 12 |
| AuIA/B/C | *C. aulicus* | Unpublished | Rat: $\alpha_3\beta_4$ subunit interface | 50 |
| | Other | | | |
| ImI | *Conus imperialis* | G**CC**SDPR**C**AWR**C**[a] | Rat: $\alpha_7$ nAChR; *Aplysia:* neuronal nAChR | 13,51–53 |

[a] C-terminal $\alpha$-carboxyl group is amidated.
[b] C-terminal $\alpha$-carboxyl group is the free acid.
[c] Disulfide bond arrangement has not been determined for GII, SIA, or SII, but very likely is conserved.

It is noteworthy that a small subset of the $\alpha$3/5 family shows a much greater differential affinity for teleost nicotinic receptors versus mammalian nicotinic receptors. The majority of the peptides in this subfamily ($\alpha$-conotoxins GI, MI, and SIA) have high affinity for all skeletal muscle nicotinic receptors; in contrast, peptides such as $\alpha$-conotoxin SI have a dramatically lower affinity for the mammalian skeletal muscle nicotinic receptors.[5]

In contrast to the $\alpha$3/5 conotoxins which have high affinity for the mammalian $\alpha/\delta$ but not the $\alpha/\gamma$ interface in mammalian muscles, but not

[5] D. R. Groebe, J. M. Dumm, E. S. Levitan, and S. N. Abramson, *Molec. Pharmacol.* **48,** 105 (1995).

TABLE III
STRUCTURE OF $\alpha$A-CONOTOXINS, $\psi$-CONOTOXINS, and CONANTOKINS

| Conotoxin | Source | Primary structure[a] | Ref. |
|---|---|---|---|
| Competitive muscle nAChR antagonists | | | |
| Disulfide bond arrangement: | | 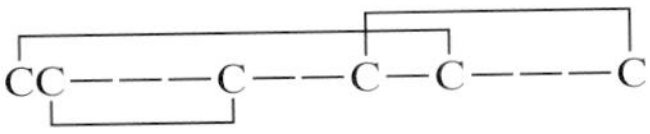 | |
| $\alpha$A-EIVA | *Conus ermineus* | G**CC**GPYONAA**C**HO**C**G**C**KVGROOY**C**DROSGG[b] | 54 |
| $\alpha$A-EIVB | *Conus ermineus* | G**CC**GKYONAA**C**HO**C**G**C**TVGROOY**C**DROSGG[b] | 54 |
| $\alpha$A-PIVA | *Conus purpurascens* | G**CC**GSYONAA**C**HO**C**S**C**KDROSY**C**GQ[b] | 55 |
| Noncompetitive muscle nAChR antagonists | | | |
| Disulfide bond arrangement: | | 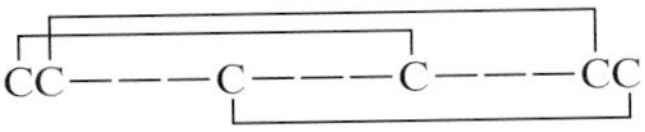 | |
| $\psi$-PIIIE | *Conus purpurascens* | HOO**CC**LYGK**C**RRYOG**C**SSAS**CC**QR[b] | 56 |
| Noncompetitive NMDA receptor antagonists | | | |
| Conantokin-G | *Conus geographus* | GE$\gamma\gamma$LQ$\gamma$NQ$\gamma$LIR$\gamma$KSN[b] | 14,15 |
| Conantokin-T | *Conus tulipa* | GE$\gamma\gamma$YQKML$\gamma$NLR$\gamma$AEVKKNA[b] | 17 |

[a] $\gamma$, $\gamma$-Carboxyglutamate; O, *trans*-4-hydroxyproline.
[b] C-terminal $\alpha$-carboxyl group is amidated.

the $\alpha/\gamma$ interface in *Torpedo*,[6,7] the $\alpha$4/7 conotoxin EI shows high affinity for the $\alpha/\delta$ interface in both systems and can be used as a selective probe for the $\alpha/\delta$ site in *Torpedo*.[4] The structures of several $\alpha$-conotoxins have been solved both by nuclear magnetic resonance (NMR) techniques and, more recently, by X-ray crystallography.

### *$\alpha$A-Conotoxins*

Like $\alpha$-conotoxins of the $\alpha$3/5 subfamily, $\alpha$A-conotoxins are believed to be competitive antagonists of skeletal muscle nicotinic receptors (Table III). It has been demonstrated that in contrast to $\alpha$-conotoxin MI, $\alpha$A-conotoxin EIVA from the fish-hunting species *Conus ermineus* has almost equal affinity for the two ligand-binding sites of the nicotinic receptor. Indeed, $\alpha$A-conotoxin EIVA exhibited a higher affinity than any other *Conus* peptide for the $\alpha/\gamma$ ligand-binding site of the mouse skeletal muscle nicotinic receptor. Thus, $\alpha$-conotoxins and $\alpha$A-conotoxins that target the skeletal muscle nicotinic receptor subtype have different specificity for the two ligand-binding sites of mammalian receptors. Clearly, the different

[6] R. M. Hann, O. R. Pagán, and V. A. Eterovic, *Biochemistry* **33,** 14058 (1994).
[7] Y. N. Utkin, F. H. Kobayashi, and V. I. Tsetlin, *Toxicon* **32,** 1153 (1994).

structures reflect different "microsite" interactions[8] even though both groups of peptides are competitive antagonists. The structures of two $\alpha$A-conotoxins have been solved by NMR.

### *ψ-Conotoxins*

A novel noncompetitive nicotinic receptor antagonist has been described, ψ-conotoxin PIIIE from *Conus purpurascens.* At least two other peptides belonging to this family have been discovered (R. Jacobsen and B. Olivera, unpublished results). ψ-Conotoxin PIIIE has been shown to inhibit the skeletal muscle subtype of nicotinic receptors expressed in oocytes, although it has a significantly higher affinity for the *Torpedo* receptor compared to the homologous mouse receptor. The structure of ψ-conotoxin PIIIE has been determined by multidimensional NMR.

### *Conus Peptides Targeted to Neuronal Subtypes of Nicotinic Receptors*

All *Conus* ligands for neuronal subtypes of nicotinic receptors in mammalian systems belong to the $\alpha$-conotoxin family. The most specific of such peptides described to date is $\alpha$-conotoxin MII. This peptide has a very high affinity and target specificity for the $\alpha_3\beta_2$ subtype of neuronal nicotinic receptors. This peptide was used to demonstrate that at least two presynaptic subtypes of neuronal nicotinic receptor are involved in striatal dopamine release, one of which contains an $\alpha_3\beta_2$ interface.[9] Additionally, MII has been used to pharmacologically dissect nicotinically mediated synaptic transmission in chick parasympathetic ciliary ganglion. At this ganglion, MII selectively inhibits the slowly decaying versus rapidly decaying current.[10] A combination of MII and IMI has been used to distinguish subpopulations of nAChRs in frog sympathetic ganglion.[11] The NMR structure of $\alpha$-conotoxin MII has recently been solved. A variety of data suggest that $\alpha$-conotoxin MII is a Janus ligand, with two interacting interfaces. One interface is proposed to specifically cause rapid association with the $\beta_2$ subunit, and the other to cause functional block and very slow dissociation from the $\alpha_3$ subunit.

A variety of *Conus* peptides have also been shown to target the $\alpha_7$ subtype of nicotinic receptors. The first one of these characterized was $\alpha$-conotoxin IMI from *Conus imperialis* venom. In addition to its specificity

[8] B. M. Olivera, J. Rivier, C. Clark, C. A. Ramilo, G. P. Corpuz, F. C. Abogadie, E. E. Mena, S. R. Woodward, D. R. Hillyard, and L. J. Cruz, *Science* **249,** 257 (1990).

[9] J. M. Kulak, T. A. Nguyen, B. M. Olivera, and J. M. McIntosh, *J. Neurosci.* **17,** 5263 (1997).

[10] E. M. Ullian, J. M. McIntosh, and P. B. Sargent, *J. Neurosci.* **17,** 7210 (1997).

[11] S. F. Tavazoie, M. F. Tavazoie, J. M. McIntosh, B. M. Olivera, and D. Yoshikami, *Br. J. Pharmacol.* **120,** 995 (1996).

for $\alpha_7$ in mammalian systems, this peptide has been used to discriminate between different types of nicotinic receptors in molluscan systems. Other $\alpha$-conotoxins have recently been discovered that target the $\alpha_7$ subtype with significantly higher affinity than $\alpha$-conotoxin ImI (J. M. McIntosh, unpublished results).

A number of peptides from *Conus aulicus* venom ($\alpha$-conotoxins AUIA, AuIB, and AuIC), which prefer the $\alpha_3\beta_4$ subtype of neuronal nicotinic receptor, have been characterized. However, the sequences of these peptides have not yet been published.

Some of the $\alpha$-conotoxins have been shown to act potently at molluscan nAChRs. The first reported peptides were $\alpha$-conotoxins PnIA and PnIB from *C. pennaceus.*[12] The peptides block the nAChR of cultured *Aplysia* neurons. More recently, $\alpha$-conotoxin ImI was shown to be a selective antagonist of subpopulations of *Apylsia* nAChRs.[13]

## *Conus Peptides Targeting NMDA Receptors*

The conantokins, which are perhaps the most novel family of *Conus* peptides have been shown to be NMDA receptor antagonists.[14] In contrast to the conotoxins, conantokins are not multiply disulfide-bonded but instead have a very unusual post-translational modification, the $\gamma$-carboxylation of glutamate residues to $\gamma$-carboxyglutamate (Gla). The discovery of the first member of this family, conantokin-G, established that this unusual post-translational modification could occur outside mammalian systems.[15]

Three conantokins have been characterized so far, conantokin-G from *C. geographus,*[16] conantokin-T from *C. tulipa,*[17] and conantokin-R from *C. radiatus.*[18] These peptides were purified from venom by following an unusual *in vivo* activity in mammals: the ability to induce a sleep-like state in young mice (under 2 weeks of age). Thus, in the earlier papers describing these peptides (before they were found to be NMDA receptor antagonists) they are referred to as "sleeper peptides."

[12] M. Fainzilber, A. Hasson, R. Oren, A. L. Burlingame, D. Gordon, M. E. Spira, and E. Zlotkin, *Biochemistry* **33,** 9523 (1994).

[13] J. Kehoe, M. Spira, and J. M. McIntosh, *Soc. Neurosci.* **22,** 267 (1996).

[14] E. E. Mena, M. F. Gullak, M. J. Pagnozzi, K. E. Richter, J. Rivier, L. J. Cruz, and B. M. Olivera, *Neurosci. Lett.* **118,** 241 (1990).

[15] J. M. McIntosh, B. M. Olivera, L. J. Cruz, and W. R. Gray, *J. Biol. Chem.* **259,** 14343 (1984).

[16] B. M. Olivera, J. M. McIntosh, L. J. Cruz, F. A. Luque, and W. R. Gray, *Biochemistry* **23,** 5087 (1984).

[17] J. A. Haack, J. Rivier, T. N. Parks, E. E. Mena, L. J. Cruz, and B. M. Olivera, *J. Biol. Chem.* **265,** 6025 (1990).

[18] H. S. White, R. T. McCabe, F. Abogadie, J. Torres, J. E. Rivier, I. Paarmann, M. Hollmann, B. M. Olivera, and L. J. Cruz, *J. Neurosci. Abst.* **23,** 2164 (1997).

The conantokins are the only natural peptides known to inhibit NMDA receptors. So far, all natural conantokins tested cause inhibition of a variety of NMDA receptor isoforms, albeit with very different affinities. No other subclass of glutamate receptors that have been examined are inhibited by the conantokin peptides. A report has demonstrated that conantokins have potential as anticonvulsant compounds, exhibiting great potency in an audiogenic seizure mouse model, with a very high protective index compared to commercial anticonvulsant compounds.[18]

Several structural investigations have been carried out on the conantokins using circular dichroism and NMR techniques.[19–21] These studies are in general agreement that conantokins are highly structured peptides, with $\alpha$-helical structure as well as a distorted $3_{10}$ helix. For conantokin-G at least, the peptide becomes more structured in the presence of divalent cations. Like the Gla-containing peptides of the blood clotting cascade, conantokin-G binds acidic membranes in the presence of $Ca^{2+}$ ions.[19]

It has recently been shown that the conantokins are initially translated as a large prepropeptide precursor; the mature peptide is found in the C-terminal end in a single copy. In the excised region, which is N terminal to the mature conantokin-encoding C-terminal region, a recognition signal sequence is present that facilitates vitamin K-dependent carboxylation of selected glutamate residues in the mature peptide region.[22] Thus, in contrast to the conotoxins where structure is largely stabilized by the multiple disulfide cross-links, in the conantokin family of peptides the structure is stabilized by the presence of mutiple $\gamma$-carboxyglutamate (Gla) residues, appropriately spaced for a helical configuration to be assumed. Sequences in the prepropeptide precursor that do not appear in the mature peptide play an important role in the post-translational conversion of Glu to Gla.

## *Conus* Peptides That Target Voltage-Gated Ion Channels

### *Overview*

The most widely used *Conus* peptides in neurobiology are those that target voltage-gated calcium channels; these all belong to the $\omega$-conotoxin

[19] R. A. Myers, J. River, and B. M. Olivera, *J. Neurosci.* **16,** 958 (1990).

[20] N. Skjaebaek, K. J. Nielsen, R. J. Lewis, P. Alewood, and D. J. Craik, *J. Biol. Chem.* **272,** 2291 (1997).

[21] A. C. Rigby, J. D. Baleja, B. C. Furie, and B. Furie, *Biochemistry* **36,** 6906 (1997).

[22] P. K. Bandyopadhyay, C. J. Colledge, C. S. Walker, L.-M. Zhou, D. R. Hillyard, and B. M. Olivera, *J. Biol. Chem.* submitted (1997).

family (see Table IV). Several different *Conus* peptide families target voltage-gated sodium channels; the first of these discovered were the $\mu$-conotoxins, which are $Na^+$ channel blockers.[1] The $\delta$-conotoxins are a family of *Conus* peptides that inhibits sodium channel inactivation.[23] Finally, the $\mu$O-conotoxins also are sodium channel antagonists,[24] but do not appear to act on the same site as the $\mu$-conotoxins and have a different structural motif (see Table V). The first *Conus* peptide that targets a voltage-gated potassium channel, $\kappa$-conotoxin, has been characterized.[25]

## *Conus Peptides That Target Voltage-Gated Calcium Channels*

The literature on the $\omega$-conotoxins that target voltage-gated calcium channels is very extensive, but in this article, only a very brief overview is presented. For a more comprehensive review, the reader is referred to Olivera *et al.*[26] and Dunlap *et al.*[27]

The first $\omega$-conotoxin that was biochemically characterized was $\omega$-conotoxin GVIA from *C. geographus* venom, followed by $\omega$-conotoxin MVIIA from *C. magnus* venom. These were the first natural peptide toxins that inhibited voltage-gated calcium channels. In mammalian systems, these two peptides are very highly subtype-specific, targeting voltage-gated calcium channel complexes that contain an $\alpha_{1B}$ subunit (which correspond to what is known as the "N-type" Ca current).

Note that these peptides may have broader selectivity in lower vertebrates (see a discussion in Olivera *et al.*[26]). In the literature, there has been a tendency to assume that any voltage-gated calcium channel that is sensitive to $\omega$-conotoxin GVIA or MVIIA must be $\alpha_{1B}$ containing (i.e., an N-type calcium channel), while any voltage-gated calcium channel resistant to these peptides must be of a different subtype. Although there are no known exceptions so far to this generalization in mammalian systems, there is reason to suspect that the correlation will not hold in lower vertebrates, and almost certainly does *not* apply to invertebrates.

The structures of both $\omega$-conotoxins GVIA and MVIIA have been reported by several laboratories, using multidimensional NMR techniques. Some structure–function studies have been carried out. Both peptides have

[23] K.-J. Shon, A. Hasson, M. E. Spira, L. J. Cruz, W. R. Gray, and B. M. Olivera, *Biochemistry* **33,** 11420 (1994).

[24] M. Fainzilber, O. Kofman, E. Zlotkin, and D. Gordon, *J. Biol. Chem.* **269,** 2574 (1994).

[25] K. Shon, M. Stocker, H. Terlau, W. Stühmer, R. Jacobsen, C. Walker, M. Grilley, M. Watkins, D. R. Hillyard, W. R. Gray, and B. M. Olivera, *J. Biol. Chem.* in press (1997).

[26] B. M. Olivera, G. Miljanich, J. Ramachandran, and M. E. Adams, *Ann. Rev. Biochem.* **63,** 823 (1994).

[27] K. Dunlap, J. I. Luebke, and T. J. Turner, *Trends Neurosci.* **18,** 89 (1995).

TABLE IV

STRUCTURE AND SPECIFICITY OF THE CALCIUM CHANNEL BLOCKERS, ω-CONOTOXINS

| ω-Conotoxin | Source | Primary structure | Specificity | Ref.[b] |
|---|---|---|---|---|
| | | Disulfide linkages: 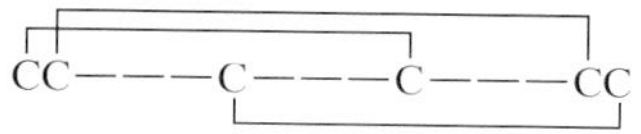 | | 60,61 |
| GVIA | *Conus geographus* | **C**KSOGSS**C**SOTSYN**CC**RS**C**NOYTKR**C**Y[a] | N-type calcium channels ($\alpha_{1B}$ subunit) | 16 |
| GVIIA | *Conus geographus* | **C**KSOGTO**C**SRGMRD**CC**TS**C**LLYSNK**C**RRY[a] | | 57 |
| MVIIA | *Conus magus* | **C**KGKGAK**C**SRLMYD**CC**TGS**C**RSGK**C**[a] | N-type calcium channels ($\alpha_{1B}$ subunit) | 58 |
| MVIIB | *Conus magus* | **C**KGKGAS**C**HRTSYD**CC**TGS**C**NRGK**C**[a] | | 58 |
| MVIIC | *Conus magus* | **C**KGKGAP**C**RKTMYD**CC**SGS**C**GRRGK**C**[a] | P/Q- and N-type calcium channels ($\alpha_{1B}$ and $\alpha_{1A}$) | 29 |
| MVIID | *Conus magus* | **C**QGRGAS**C**RKTMYN**CC**SGS**C**NRGR**C**[a] | P/Q- and N-type calcium channels ($\alpha_{1B}$ and $\alpha_{1A}$) | 59 |
| SVIA | *Conus striatus* | **C**RSSGSP**C**GVTSI**CC**GR**C**YRGK**C**T[a] | | 48 |
| SVIB | *Conus striatus* | **C**KLKGQS**C**RKTSYD**CC**SGS**C**GRSGK**C**[a] | N- and P/Q-type calcium channels ($\alpha_{1A}$ and $\alpha_{1B}$) | 48 |
| TxVIIA | *Conus textile* | **C**KQADEP**C**DVFSLD**CC**TGI**C**LGV**C**MV[c] | Dihydropyridine-sensitive currents in *Aplysia* | 30 |

[a] C-terminal α-carboxyl group is amidated.
[b] See also reviews for primary references.[26,27,62–67]
[c] C-terminal amide is the free acid.

TABLE V

STRUCTURE AND SPECIFICITY OF SODIUM AND POTASSIUM CHANNEL LIGANDS FROM *Conus*

| Conotoxin | Source | Primary structure | Specificity | Ref. |
|---|---|---|---|---|
| | | Disulfide linkages:<br>$\mu$-conotoxins (III-family)<br>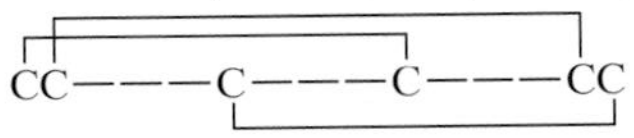<br>$\mu$O-, $\delta$- and $\kappa$-conotoxins<br>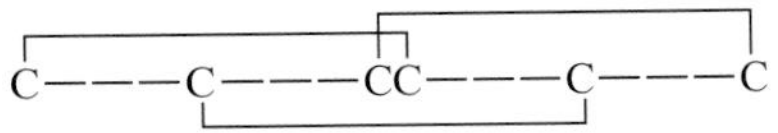 | | 23,25,38,72 |
| Blockers of voltage-sensitive sodium channels | | | | |
| $\mu$-GIIIA | *Conus geographus* | RD**CC**TOOKK**C**KDRQ**C**KOQR**CC**A[a] | Skeletal muscle Na channel ($\mu$l subtype); binds to site I | 1,68–70 |
| $\mu$-GIIIB | *Conus geographus* | RD**CC**TPPRK**C**KDRR**C**KPMK**CC**AGR[a] | Same | 1,2,68,70 |
| $\mu$-GIIIC | *Conus geographus* | RD**CC**TPPRK**C**KDRR**C**KPMK**CC**AGR[a] | Same | 1 |
| $\mu$O-MrVIA | *Conus marmoreus* | A**C**RKKWEY**C**IVPIIGFIY**CC**PGLI**C**GPFV**C**V[b] | Molluscan neurons (~100 n$M$); type II $Na^+$ channels and $Na^+$ channels in cultured rat hippocampal cells; block of rapidly inactivating $Ca^{2+}$ current at higher concentrations (>1 $\mu M$) | 36–38 |

| | | | | |
|---|---|---|---|---|
| μO-MrVIB | *Conus marmoreus* | A**C**SKKWEY**C**IVPILGFVY**CC**PGLI**C**GPFV**C**V[a] | Same | Same |
| μ-PnIVA | *Conus pennaceus* | **CC**KYGWT**C**LLG**C**SP**C**G**C**[b] | Tetrodotoxin-insensitive molluscan $Na^+$ channels | 39 |
| μ-PnIVB | *Conus pennaceus* | **CC**KYGWT**C**WLG**C**SP**C**G**C**[b] | Same | 39 |
| Ligands that delay inactivation of voltage-sensitive sodium channels | | | | |
| δ-GmVIA | *Conus gloriamaris* | VKP**C**RKEGQL**C**DPIFQN**CC**RGWN**C**VLF**C**V[b] | Molluscan neurons; shifts voltage-dependent activation curve to more negative potentials and inactivation curve to more positive potentials | 23 |
| δ-TxVIA | *Conus textile* | W**C**KQSGEM**C**NLLDQN**CC**DGY**C**IVLV**C**T[b] | Molluscan neurons; binding to mammalian $Na^+$ channels with no apparent physiologic effects and acts to protect against toxic effects of other toxins binding to the same site | 24,32,71 |
| δ-PVIA | *Conus purpurascens* | EA**C**YAOGTF**C**GIKOGL**CC**SEF**C**LPGV**C**FG[b] | Rat brain type II $Na^+$ channel; rat hippocampal neurons; vertebrate neuromuscular junction | 33,34 |
| NgVIA | *Conus nigropunctatus* | SK**C**FSOGTF**C**GIKOGL**CC**SVR**C**FSLF**C**ISFE[b] | Molluscan and vertebrate $Na^+$ channels; δ-TxVIA is a partial antagonist of NgVIA | 35 |
| Potassium channel blocker | | | | |
| κ-PVIIA | *Conus purpurascens* | **C**RIONQK**C**FQHLDD**CC**SRK**C**NRFNK**C**V[b] | *Shaker* $K^+$channel | 25,34 |

[a] C-terminal α-carboxyl group is amidated.
[b] C-terminal α-carboxyl group is the free acid.

been radiolabeled, and used productively in binding experiments, and in autoradiographic studies (for example, see Filloux *et al.*[28]).

In electrophysiologic experiments, $\omega$-conotoxin GVIA is used to inhibit $\alpha_{1B}$-containing complexes irreversibly, while $\omega$-conotoxin MVIIA is the ligand of choice when a high-affinity but reversible block is desired. Several other homologs of these peptides have been described in the literature (see Table IV).

A second group of $\omega$-conotoxins inhibits both $\alpha_{1B}$- and $\alpha_{1A}$-containing calcium channel complexes. These have broader specificity than the $\alpha_{1B}$-specific $\omega$-conotoxins described above. The most widely used of these peptides is $\omega$-conotoxin MVIIC, which has been used to discriminate between different subclasses of voltage-gated calcium channels. Both $\omega$-conotoxins MVIIC and MVIID clearly inhibit the so-called "P/Q subclasses" of voltage-gated calcium channels, which are widely believed to contain an $\alpha_{1A}$ subunit, although the precise correspondence of P-type and Q-type calcium currents as described by electrophysiologic investigations to $\alpha_{1A}$-containing calcium channel complexes is still uncertain.

The structure of $\omega$-conotoxin MVIIC has been reported.[29] This peptide has been radiolabeled and used for binding studies. Richard Tsien and coworkers have proposed that $\omega$-conotoxin MVIIC can serve as a key reagent in discriminating between P- and Q-type calcium currents, but this view has not been universally accepted.[27]

Additional $\omega$-conotoxins which inhibit voltage-gated $Ca^{2+}$ channels in invertebrate systems, particularly in mollusks, have been reported.[30] However, although these peptides have been biochemically characterized, their specificity for particular calcium channel subtypes has not yet been established. In certain cases, peptides that were originally isolated as being voltage-gated calcium channel antagonists have proved to be more potent as sodium channel inhibitors.

## *Conus* Peptides That Target Voltage-Gated Sodium Channels

### *$\mu$-Conotoxins*

The $\mu$-conotoxins were the first polypeptide toxins to compete for the same site on $Na^+$ channels as the well-established guanidinium toxins which

[28] F. Filloux, A. Schapper, S. R. Naisbitt, B. M. Olivera, and J. M. McIntosh, *Develop. Brain Res.* **78,** 131 (1994).

[29] D. R. Hillyard, V. D. Monje, I. M. Mintz, B. P. Bean, L. Nadasdi, J. Ramachandran, G. Miljanich, A. Azimi-Zoonooz, J. M. McIntosh, L. J. Cruz, J. S. Imperial, and B. M. Olivera, *Neuron* **9,** 69 (1992).

[30] M. Fainzilber, J. C. Lodder, R. C. van der Schors, K. W. Li, Z. Yu, A. L. Burlingame, W. P. Geraerts, and K. S. Kits, *Biochemistry* **35,** 8748 (1996).

target sodium channels, tetrodotoxin and saxitoxin. In the nomenclature of Catterall,[31] all of these toxins bind to site I, which is believed to be the outer vestibule of theion channel pore. The $\mu$-conotoxins were originally characterized from *C. geographus* venom, but more recently another $\mu$-conotoxin was isolated and characterized from *C. purpurascens.* As noted earlier, the $\mu$-conotoxins have narrower subtype specificity than the guanidinium toxins. Like the critical guanidinium moiety in saxitoxin and tetrodotoxin, there is believed to be a key arginine in all $\mu$-conotoxins that have been characterized. However, it has been suggested that the guanidinium group of arginine does not in fact interact with the same residues on the voltage-gated ion channel as does the guanidinium group on tetrodotoxin. The structure of several $\mu$-conotoxins, including some analogs, has been described by several groups using NMR techniques.

### *δ-Conotoxins*

The first $\delta$-conotoxin was originally called a "King-Kong peptide" from *C. textile* venom, because it elicited a peculiar symptomatology when injected into lobsters. It was subsequently shown using electrophysiological methods that the peptide delayed inactivation of voltage-gated sodium channels in *Aplysia* ganglion cells.[24,32] Another $\delta$-conotoxin from a snail-hunting *Conus,* $\delta$-conotoxin GmVIA, has also been characterized.[23]

A $\delta$-conotoxin from a fish-hunting cone snail, $\delta$-conotoxin PVIA from *C. purpurascens* venom, has been isolated and chemically synthesized. This peptide has been shown to be important for the very rapid stunning effect of *C. purpurascens* venom on prey.[33,34] This peptide is believed to play a key role in the prey capture strategy of this fish-hunting cone snail.[34] Like the $\delta$-conotoxins from snail-hunting *Conus* venoms, $\delta$-conotoxin PVIA also causes a delay in inactivation.

A conotoxin, NgVIA, that delays inactivation of molluscan and vertebrate sodium channels has been isolated[35] and appears to act on a receptor site distinct from that of $\delta$-TXVIA.

It is notable that although the $\delta$-conotoxins have the same disulfide bonding pattern as the $\omega$-conotoxins, they differ strikingly in the type of amino acids found in the loop regions between disulfide linkages. While $\omega$-conotoxins largely have hydrophilic and positively charged amino acids,

[31] W. A. Catterall, *Physiol. Rev.* **72,** S15 (1992).

[32] S. R. Woodward, L. J. Cruz, B. M. Olivera, and D. R. Hillyard, *EMBO J.* **1,** 1015 (1990).

[33] K. Shon, M. M. Grilley, M. Marsh, D. Yoshikami, A. R. Hall, B. Kurz, W. R. Gray, J. S. Imperial, D. R. Hillyard, and B. M. Olivera, *Biochemistry* **34,** 4913 (1995).

[34] H. Terlau, K. Shon, M. Grilley, M. Stocker, W. Stühmer, and B. M. Olivera, *Nature* **381,** 148 (1996).

[35] M. Fainzilber, J. C. Lodder, K. S. Kits, O. Kofman, I. Vinnitsky, J. Van Rietschoten, E. Zlotkin, and D. Gordon, *J. Biol. Chem.* **270,** 1123 (1995).

in all δ-conotoxins there is a preponderance of hydrophobic residues. It was proposed that the δ-conotoxins bind to a unique site on voltage-gated sodium channels, which has been called site VI. Given the very hydrophobic nature of these peptides, this site may be at least partially in the lipid bilayer.[24]

Because fast inactivation of voltage-gated sodium channels is generally believed to be mediated by a cytoplasmic "ball" region of the ion channel complex, the δ-conotoxins present an intriguing mechanistic puzzle in that they cause an inhibition of fast inactivation from the extracellular side of the membrane.

### *μO-Conotoxins*

Two peptides from the snail-hunting species *C. marmoreus,* μO-conotoxins MrVIA and MrVIB, were shown to block voltage-gated sodium channels.[36,37] They differ from the μ-conotoxins in being more closely related to the δ-conotoxins than to the μ-conotoxins, and also in being the first polypeptide inhibitors that inhibit conductance through $Na^+$ channels that do not compete for binding with tetrodotoxin/saxitoxin, and clearly target a different site.[38] Furthermore, in contrast to the μ-conotoxins, these peptides act more broadly on different voltage-gated sodium channel subtypes, and a wide variety of different voltage-gated sodium channels are inhibited.

Two conotoxins from *C. pennaceus,* μ-PnIVA and μ-PnIVB, were found by Fainzilber *et al.*[39] to block the tetrodotoxin-insensitive molluscan sodium channels. These peptides are structurally distinct from the originally described μ-conotoxins (e.g., μ-conotoxin GIIIA) and are named with a Roman numeral IV to indicate this difference.

## *Conus* Peptides That Target Voltage-Gated Potassium Channels

So far, only one *Conus* peptide has been shown to inhibit a voltage-gated potassium channel, κ-conotoxin PVIIA from *C. purpurascens* venom. This peptide has a disulfide bonding pattern generally similar to the ω-conotoxins, but instead of inhibiting voltage-gated calcium channels it targets potassium channels. Although the peptide is active both in lower

[36] M. Fainzilber, R. van der Schors, J. C. Lodder, K. W. Li, W. P. Geraerts, and K. S. Kits, *Biochemistry* **34,** 5364 (1995).

[37] J. M. McIntosh, A. Hasson, M. E. Spira, W. Li, M. Marsh, D. R. Hillyard, and B. M. Olivera, *J. Biol. Chem.* **270,** 16796 (1995).

[38] H. Terlau, M. Stocker, K. Shon, J. M. McIntosh, and B. M. Olivera, *J. Neurosci.* (1996).

[39] M. Fainzilber, T. Nakamura, A. Gaathon, J. C. Lodder, K. S. Kits, A. L. Burlingame, and E. Zlotkin, *Biochemistry* **34,** 8649 (1995).

vertebrate systems (where together with δ-conotoxin PVIA, it appears to be responsible for the very fast stunning effect of venom injection on the prey), and shows activity in mammalian systems as well, no vertebrate potassium channel subtype has yet been identified as being targeted by κ-conotoxin PVIIA. However, the well-characterized *Drosophila Shaker* channel is a κ-conotoxin PVIIA target.[25]

There is preliminary evidence for a number of peptides unrelated in structure to κ-conotoxin PVIIA which also inhibit voltage-gated potassium channels. However, the biochemical characterization of these peptides is still in progress, and has not been published. It will be interesting to compare the subtype specificity of these peptides with κ-conotoxin PVIIA. Given the vast diversity of potassium channels, it seems likely that the *Conus* venom system will provide many novel peptides that target potassium channels in the future.

## Some Practical Considerations in Handling *Conus* Peptides

### *Solubility*

*Conus* peptides are soluble in aqueous solutions. In general, a stock concentration of 500 $\mu M$ may be prepared without difficulty. Some peptides are soluble at higher concentration. Care should be taken, however, to ensure that peptide is actually in solution. Adding buffer to lyophilized peptide often gives the appearance of dissolving the peptide, when, in fact, a suspension has been created. This usually can be detected by holding the mixture up to a light and inspecting for particulates or cloudiness. Examining the solution under a dissecting microscope is often helpful. Certain peptides such as the μO- and δ-conotoxins are much less soluble and require the addition of organic solvents such as dimethyl sulfoxide (DMSO) or acetonitrile to achieve higher micromolar stock concentrations.

### *Storage*

*Conus* peptides are most stable in lyophilized form. For transport over a few days, they can be safely shipped at room temperature. For longer periods they should be stored at −20° or −80°. Static charge can cause the lyophilized peptide powder to "fly" out of the test tube. If static is encountered, use of an antistatic gun eliminates the problem. Particularly after transport of peptide, it is wise to centrifuge the container to ensure that peptide will not exit the tube on opening. As a side note, peptides lyophilize in a somewhat unpredictable fashion. Small quantities of peptide lyophilized side by side in a rotary evaporator often appear as either a very visible white powder, or a nearly invisible crystalline substance. The latter can easily be mistaken for "no peptide in the tube" without close inspection.

Peptides solutions can also be stored. For immediate use, solutions are generally kept at room temperature or on ice. For longer storage, solutions are frozen at $-20°$ to $-80°$. With some peptides we have noted decreased activity after repeated freeze–thaw cycles. High-performance liquid chromatography (HPLC) of these peptide solutions suggests that loss of peptide in solution, rather than peptide breakdown, is occurring. To avoid this, we routinely make aliquots of solutions such that a given aliquot will not need to be thawed more than two or three times prior to consumption.

We often store peptides in HPLC elution buffer consisting of 0.1% trifluoroacetic acid (TFA) and acetonitrile/$H_2O$. We have found that with long-term storage, however, some peptides (e.g., $\alpha$-conotoxin EI) undergo degradation, which is consistent with deamination as measured by mass spectrometry. We presume that this is secondary to the acidic pH, and therefore avoid long-term storage under these conditions.

*Nonspecific Adsorption*

Many *Conus* peptides are hydrophobic in nature and have a tendency to "stick" to glassware and plasticware. At nanomolar concentrations and below, this can lead to significant changes in solution concentration of peptide. To avoid this, we often add 0.1 mg/ml lysozyme or 0.1–1.0 mg/ml bovine serum albumin (BSA) to the solution.

Lyophilization of small quantities of peptide (less than 1 nmol) can lead to significant loss of peptide to container walls. We have found that the addition of carrier protein (e.g., 10–50 $\mu$g of lysozyme) to the solution prior to lyophilization largely circumvents this problem. Conodipine-M[40] (a phospholipase $A_2$ from *C. magus*) is a particularly striking example. The apparent $IC_{50}$ shifts by two orders of magnitude to the right without the utilization of carrier protein.

The use of carrier protein is not always sufficient to prevent nonspecific adsorption, particularly at low peptide concentrations. We have found, for example, that static bath application of $\alpha$-conotoxins to *Xenopus* oocyte recording chambers leads to an apparent 10-fold decrease in potency compared to preparations where the solution is applied as a continuous flow.[41]

[40] J. M. McIntosh, F. Ghomashchi, M. H. Gelb, D. J. Dooley, S. J. Stoehr, A. B. Giordani, S. R. Naisbitt, and B. M. Olivera, *J. Biol. Chem.* **270,** 3518 (1995).

[41] S. C. Harvey, J. M. McIntosh, G. E. Cartier, F. N. Maddox, and C. W. Luetje, *Mol. Pharmacol.* **51,** 336 (1997).

[42] W. R. Gray, A. Luque, B. M. Olivera, J. Barrett, and L. J. Cruz, *J. Biol. Chem.* **256,** 4734 (1981).

Radioiodinated *Conus* peptides may be particularly sticky. We routinely siliconize (Sigmacote, Sigma, St. Louis, MO) pipette tips and test tubes (including the caps) when using iodinated peptides and assess radioactivity after solution transfer (e.g., pipette tips) using a gamma counter. We also gamma count final reaction tubes as a measure of true radioactivity concen-

[43] H.-J. Kreienkamp, S. M. Sine, R. K. Maeda, and P. Taylor, *J. Biol. Chem.* **269,** 8108 (1994).

[44] J. M. McIntosh, L. J. Cruz, M. W. Hunkapiller, W. R. Gray, and B. M. Olivera, *Arch. Biochem. Biophys.* **218,** 329 (1982).

[45] C. G. Zafaralla, C. Ramilo, W. R. Gray, R. Karlstrom, B. M. Olivera, and L. J. Cruz, *Biochemistry* **27,** 7102 (1988).

[46] R. A. Myers, G. C. Zafaralla, W. R. Gray, J. Abbott, L. J. Cruz, and B. M. Olivera, *Biochemistry* **30,** 9370 (1991).

[47] R. M. Hann, O. R. Pagán, L M. Gregory, T. Jácome, and V. A. Eterovic, *Biochemistry* **36,** 9051 (1997).

[48] C. A. Ramilo, G. C. Zafaralla, L. Nadasdi, L. G. Hammerland, D. Yoshikami, W. R. Gray, R. Kristipati, J. Ramachandran, G. Miljanich, B. M. Olivera, and L. J. Cruz, *Biochemistry* **31,** 9919 (1992).

[49] G. E. Cartier, D. Yoshikami, W. R. Gray, S. Luo, B. M. Olivera, and J. M. McIntosh, *J. Biol. Chem.* **271,** 7522 (1996).

[50] S. Luo, D. Yoshikami, G. E. Cartier, R. Jacobsen, B. M. Olivera, and J. M. McIntosh, *J. Neurosci. Abst.* **23,** 384 (1997).

[51] J. M. McIntosh, D. Yoshikami, E. Mahe, D. B. Nielsen, J. E. Rivier, W. R. Gray, and B. M. Olivera, *J. Biol. Chem.* **269,** 16733 (1994).

[52] D. S. Johnson, J. Martinez, A. B. Elgoyhen, S. S. Heinemann, and J. M. McIntosh, *Mol. Pharmacol.* **48,** 194 (1995).

[53] E. F. R. Pereira, M. Alkondon, J. M. McIntosh, and E. X. Albuquerque, *J. Pharmacol. Exp. Ther.* **278,** 1472 (1996).

[54] R. Jacobsen, D. Yoshikami, M. Ellison, J. Martinez, W. R. Gray, G. E. Cartier, K. Shon, D. R. Groebe, S. N. Abramson, B. M. Olivera, and J. M. McIntosh, *J. Biol. Chem.* **36,** 22531 (1997).

[55] C. Hopkins, M. Grilley, C. Miller, K.-J. Shon, L. J. Cruz, W. R. Gray, J. Dykert, J. Rivier, D. Yoshikami, and B. M. Olivera, *J. Biiol. Chem.* **270,** 22361 (1995).

[56] K. Shon, M. Grilley, R. Jacobsen, G. E. Cartier, C. Hopkins, W. R. Gray, M. Watkins, D. R. Hillyard, J. Rivier, J. Torres, D. Yoshikami, and B. M. Olivera, *Biochemistry* in press (1997).

[57] B. M. Olivera, W. R. Gray, R. Zeikus, J. M. McIntosh, J. Varga, J. Rivier, V. de Santos, and L. J. Cruz, *Science* **230,** 1338 (1985).

[58] B. M. Olivera, L. J. Cruz, V. de Santos, G. LeCheminant, D. Griffin, R. Zeikus, J. M. McIntosh, R. Galyean, J. Varga, W. R. Gray, and J. Rivier, *Biochemistry* **26,** 2086 (1987).

[59] V. D. Monje, J. Haack, S. Naisbitt, G. Miljanich, J. Ramachandran, L. Nasdasdi, B. M. Olivera, D. R. Hillyard, and W. R. Gray, *Neuropharmacology* **32,** 1141 (1993).

[60] Y. Nishiuchi and S. Sakakibara, *FEBS Lett.* **148,** 260 (1982).

[61] Y. Nishiuchi, K. Kumagaye, Y. Noda, T. X. Watanabe, and S. Sakakibara, *Biopolymers* **25,** 561 (1986).

[62] R. J. Miller, *Science* **235,** 46 (1987).

[63] R. W. Tsien, D. Lipscombe, D. V. Madison, K. R. Bley, and A. P. Fox, *Trends Neurosci.* **11,** 431 (1988).

[64] B. P. Bean, *Annu. Rev. Physiol.* **51,** 367 (1989).

tration. Iodinated peptides may also stick to dust particles introduced into solution, for example, by pipette tips. This can lead to scatter of signal in receptor binding assays. To avoid this, stock solutions of radiolabeled peptide are centrifuged (e.g., in an Eppendorf microfuge) to pellet such particles prior to solution use.

[65] P. Hess, *Annu. Rev. Neurosci.* **13,** 337 (1990).
[66] R. W. Tsien, P. T. Ellinor, and W. A. Horne, *Trends Pharmacol. Sci.* **12,** 349 (1991).
[67] T. P. Snutch, *Curr. Biol.* **2,** 247 (1992).
[68] S. Sato, H. Nakamura, Y. Ohizumi, J. Kobayashi, and Y. Hirata, *FEBS Lett.* **155,** 277 (1983).
[69] E. Moczydlowski, B. M. Olivera, W. R. Gray, and G. R. Strichartz, *Proc. Natl. Acad. Sci. U.S.A.* **83,** 5321 (1986).
[70] M. M. Stephan, J. F. Potts, and W. S. Agnew, *J. Membr. Biol.* **137,** 1 (1994).
[71] M. Fainzilber, D. Gordon, A. Hasson, M. E. Spira, and E. Zlotkin, *Eur. J. Biochem.* **202,** 589 (1991).
[72] Y. Hidaka, K. Sato, H. Nakamura, J. Kobayashi, Y. Ohizumi, and Y. Shimonishi, *FEBS Lett.* **1,** 29 (1990).

# [32] Scorpion Toxins as Tools for Studying Potassium Channels

*By* MARIA L. GARCIA, MARKUS HANNER, HANS-GÜNTHER KNAUS, ROBERT SLAUGHTER, and GREGORY J. KACZOROWSKI

## Introduction

Ion channels play a fundamental role in control of cell excitability. Thus, their activity is largely involved in modulation of contractility of muscle cells, and in release of hormones and neurotransmitters from endocrine and neuronal cells. Out of all the families of ion channels, $K^+$ channels represent the largest and most diverse group of proteins. Gating of these proteins occurs through conformational changes that are controlled by voltage and/or ligand binding. Therefore, $K^+$ channels can be broadly divided into two groups: voltage-dependent and ligand-activated channels. A number of techniques have become available during the last few years for studying ion channel structure and function. Electrophysiology affords determination of biophysical parameters that are inherent to each individual ion channel. With the use of molecular biology, a large amount of information regarding the structure and existence of subfamilies of $K^+$ channels has become available due to molecular cloning of cDNAs encoding these



0076-6879/99 $30.00

proteins.[1] However, two major questions are still the subject of further investigation. These concern the molecular composition of given channels as expressed *in vivo*, as well as the physiologic role that channels play in cell function. To explore these questions, high-affinity, selective modulators for a given channel must be found. Progress in this area has occurred due to discovery of high-affinity peptidyl inhibitors of $K^+$ channels in venom of different organisms, such as scorpions, snakes, bees, spiders, and sea anemone.[2] These peptidyl inhibitors have been useful in development of the pharmacology of $K^+$ channels, and have also been employed in binding reactions as a marker during the purification of channels from native tissues. This has allowed determination of a channel's subunit composition, as well as identification of specific auxiliary subunits of $K^+$ channels. These auxiliary proteins are very important for channel function because they cause profound effects on both the biophysical and pharmacologic properties of the pore-forming subunit. In addition, due to their well-understood mechanism of action, peptidyl inhibitors of $K^+$ channels have allowed the identification and molecular characterization of the pore-forming region of these proteins, and determination of subunit stoichiometry.

Scorpion venoms constitute a rich source of peptidyl inhibitors of $K^+$ channels.[2] These peptides typically consist of 37–39 amino acids and display significant sequence homology among themselves. They usually contain six Cys residues that form three disulfide bridges, thus providing the peptide with a very rigid structure. Production of significant quantities of these peptides can be accomplished by either solid-phases synthesis or recombinant techniques. This has allowed determination of the three-dimensional solution structure of some of these peptides by nuclear magnetic resonance (NMR) techniques. The overall structures so far obtained are very conserved and possess an $\alpha$-helical region that is linked by disulfide bonds to a three-strand antiparallel $\beta$ sheet. Knowledge of the three-dimensional structure of these peptides, combined with site-directed mutagenesis, has allowed identification of those residues critical for binding to the channel pore, and have provided a picture of the interaction surface with $K^+$ channels. Importantly, all residues that are crucial for interaction with the channel are located on one face of the $\beta$ sheet, while the $\alpha$-helical region does not make contact with the channel directly. Complementary mutagenesis targeting residues in the channel pore has then allowed derivation of models describing shape and dimensions of the receptor in the vestibule of the channel.

[1] A. Wei, T. Jegla, and L. Salkoff, *Neuropharmacology* **35,** 805 (1996).

[2] M. L. Garcia, M. Hanner, H.-G. Knaus, R. Koch, W. Schmalhofer, R. S. Slaughter, and G. J. Kaczorowski, *Adv. Pharmacol.* **39,** 425 (1997).

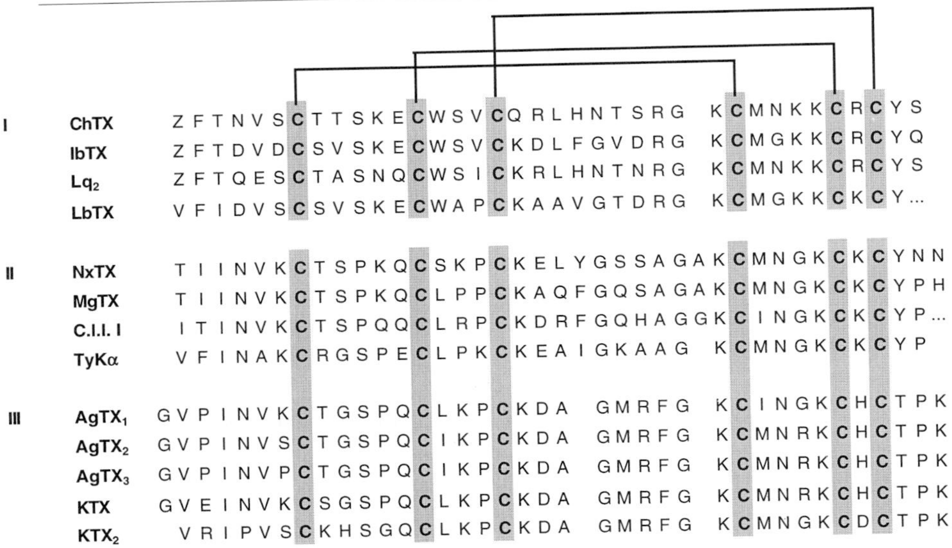


FIG. 1. Comparison of the amino acid sequences of charybdotoxin (ChTX), iberiotoxin (IbTX), *Leiurus Centruvoides quinquestriatus* toxin 2 (Lq2), limbatustoxin (LbTX), noxiustoxin (NxTX), margatoxin (MgTX), *C. limpidus limpidus* toxin I (C.l.l. 1), tityustoxin-K$\alpha$ (TyK$\alpha$), agitoxin 1 ($AgTX_1$), agitoxin 2 ($AgTX_2$), agitoxin 3 ($AgTX_3$), kaliotoxin (KTX), and kaliotoxin 2 ($KTX_2$). The sequences have been aligned with respect to the six cysteine residues which are in boldface type. The position of the disulfide bonds is indicated.

Peptidyl inhibitors from scorpion venoms can be subdivided into different groups based on sequence conservation and specificity[2] (Fig. 1). Members of the first group inhibit $Ca^{2+}$-activated $K^+$ channels, although only iberiotoxin (IbTX) and limbatustoxin (LbTX) are specific for the high-conductance $Ca^{2+}$-activated $K^+$ channel. All other groups inhibit voltage-gated $K^+$ channels. However, only members of the $K_V1$ family are sensitive to inhibition; $K_V2$, $K_V3$, and $K_V4$ type channels are not blocked by these inhibitors. Within the $K_V1$ family, only $K_V1.1$, 1.2, 1.3, 1.6, and 1.7 are targets of these peptides; $K_V1.4$ and 1.5 are refractory to inhibition.

In this review, we discuss methods for (1) purifying peptides from venom sources, (2) synthesis of the peptides by recombinant techniques, (3) radiolabeling of peptides in biologically active form, and (4) use of radiolabeled peptides to characterize high-affinity receptors in native tissue.

## Purification of Peptidyl Inhibitors of $K^+$ Channels from Scorpion Venoms

Scorpion venoms contain a vast number of components with activities directed against ion channels. For example, they constitute a rich source of peptidyl $Na^+$ channel modulators.[3] A distinguishing feature of $K^+$ channel inhibitors is the existence of a large number of positively charged residues present in these peptides. Thus, it is possible to take advantage of this property to achieve a simple, efficient, and highly reproducible purification procedure. For all peptides that have been purified in our laboratory [charybdotoxin (ChTX), IbTX, LbTX margatoxin (MgTX), agitoxin 1–3 ($AgTX_{1-3}$)], we have employed two consecutive chromatographic steps: cation-exchange chromatography on a Mono S column (Pharmacia, Piscataway, NJ), and reversed-phase chromatography on a $C_8$ or $C_{18}$ column.[4–8] This has always afforded the production of pure material as judged by amino acid sequence, amino acid composition, and mass spectroscopic analyses.

The first step in peptide purification involves separation of venom components on a Mono S HR5/5 or HR 10/10 column, depending on the amount of material being processed, using a high-performance liquid chromatography (HPLC) system. Lyophilized venom is initially resuspended in 20 m*M* sodium borate, pH 9.0, at a final protein concentration of ca. 5 mg/ml using vortex agitation. The suspension is then subjected to centrifugation for 15 min at 27,000*g* at 4° to remove insoluble material. Before applying the soluble material onto the Mono S column, the sample must be filtered through a Millex-GV 0.2-μm pore size filter (Millipore, Bedford, MA) to remove particulate material that could cause obstruction in the HPLC system, thereby increasing back pressure of the column. We do not recommend subjecting the pellet obtained after initial centrifugation to a second extraction. Even after processing the soluble material as described earlier, we frequently observe an increase in the pressure of the system, up to the limit tolerated by the column. If this occurs, the direction of the column must be reversed with continued pumping of buffer until the pressure decreases to normal values.

Before injecting sample, the ion-exchange column should be equilibrated with 20 m*M* sodium borate, pH 9.0, at a flow rate of 0.5 ml/min

[3] W. A. Catterall, *Ann. Rev. Pharmacol. Toxicol.* **20,** 15 (1980).

[4] G. Gimenez-Gallego, M. A. Navia, J. P. Reuben, G. M. Katz, G. J. Kaczorowski, and M. L. Garcia, *Proc. Natl. Acad. Sci. U.S.A.* **85,** 3329 (1988).

[5] A. Galvez, G. Gimenez-Gallego, J. P. Reuben, L. Roy-Contancin, P. Feigenbaum, G. J. Kaczorowski, and M. L. Garcia, *J. Biol. Chem.* **265,** 11083 (1990).

[6] M. Garcia-Calvo, R. J. Leonard, J. Novick, S. P. Stevens, W. Schmalhofer, G. J. Kaczorowski, and M. L. Garcia, *J. Biol. Chem.* **268,** 18866 (1993).

[7] M. L. Garcia, M. Garcia-Calvo, P. Hidalgo, A. Lee, and R. MacKinnon, *Biochemistry* **33,** 6834 (1994).

[8] J. Novick, R. J. Leonard, V. F. King, W. Schmalhofer, G. J. Kaczorowski, and M. L. Garcia, *Biophys. J.* **59,** 78a (1991).

(HR 5/5 column) or 2 ml/min (HR 10/10 column) depending on the Mono S column employed. The sample is then applied and absorbance monitored at 280 nm. Because of the basic pH conditions, a large amount of material absorbing at 280 nm is not retained by the column and appears in the void volume. We have never observed any biological activity in this material against the $K^+$ channels tested and, therefore, it can be discarded. Once the absorbance returns to baseline values, the retained material is eluted with a linear gradient of NaCl in 20 m*M* sodium borate, pH 9.0; 0.75 *M*/hr for HR 5/5 or 0.5 *M*/hr for HR 10/10 columns. Individual peaks are collected manually and used for determination of their biological activity. These fractions can be stored at −70° until further processing is accomplished.

The second step of purification is achieved by using a $C_{18}$ reversed-phase HPLC column (25- × 0.46-cm, 5-μm particle size, or 25- × 1-cm, 5-μm particle size, depending on the amount of material to be processed; the Separations Group). In some cases, we have also employed a $C_8$ reversed-phase HPLC column and have obtained nearly identical results. The column is equilibrated with 10 m*M* trifluoroacetic acid (TFA) at a flow rate of either 0.5 ml/min (25- × 0.46-cm column) or 3 ml/min (25- × 1-cm column), depending on the size of the column/amount of material to be loaded. Because most of the peptides are highly charged and do not contain many hydrophobic surfaces in their native conformation, it is very important to have the column well equilibrated in starting buffer. Failure to do this may lead to lack of retention of peptides by the column. Fractions of interest from the Mono S column are then directly applied, without further processing, to the reversed-phase column. Elution is achieved with a linear gradient of organic solvent. We have successfully employed a combination of 2-propanol/acetonitrile (2:1) in 4 m*M* TFA, although acetonitrile by itself is also an appropriate solvent for elution. The gradient can be applied from 0 to 40% over either a 30-min period (25- × 0.46-cm column) or a 60-min period (25- × 1-cm column) depending on column size. For best results, it is better to monitor absorbance of eluting material, at least, at two different wavelengths (e.g., 280 and 235 nm). Some peptides have little or no aromatic amino acid content and, therefore, give very small or sometimes undetectable absorbance at 280 nm. Components are separated manually and, in most cases, well-defined peaks with good baseline separation can be obtained.

For testing of biological activity, it is suggested that small amounts of material be subjected to lyophilization, and then reconstituted in any buffer containing high ionic strength (e.g., 100 m*M* NaCl). We also suggest including 0.1% (w/v) bovine serum albumin (BSA) in the resuspension buffer to prevent loss of peptide due to binding to glass or plastic surfaces. Remaining material can be stored at −70°. Once a fraction of interest is identified, and because only small amounts of material are needed for most experiments, it is desirable to lyophilize samples in small aliquots of ca. 20 $\mu$g and store them at −70°. Material stored in this way is usually stable for very long periods of time. When an aliquot is needed, the peptide is resuspended as indicated above and can be stored at 4° for several months without loss of biological activity. We do not routinely subject toxin solutions to repetitive freeze–thaw cycles as we have not investigated such procedures with respect to toxin stability. It is important to note that $K^+$ channel peptides are typically highly positively charged molecules that will stick to glass surfaces unless a high ionic strength buffer is used to prevent such an interaction from taking place. This is particularly important when drying the peptides in glass tubes. If water is used as the sole resuspension agent, it is likely that most of the peptides will be lost by absorption onto the surface of the tube.

Although some material may appear to be chromatographically pure, this does not necessarily mean that it is homogeneous. For example, two different peptide entities may elute together, or a major absorbance peak may contain a component that does not have significant absorbance at the wavelength monitored. To resolve this later situation, it is helpful to monitor absorbance of eluting material at two different wavelengths. If more than one component has eluted together, it is often possible to ascertain that this situation exists by applying the material again to the reversed-phase column, and selecting more shallow gradient elution conditions. Despite all of these precautions, it is still necessary to characterize the material of interest in terms of its amino acid composition, amino acid sequence, and mass spectroscopic properties. When performing automated Edman degradation using most commercially available instruments, it is important to be aware that the last residue of the peptide is usually washed off the filter support, and this amino acid must, therefore, be determined by an independent means. Thus, it is important that the amino acid composition be determined after acid hydrolysis of the peptide, and that it matches the composition obtained by sequence. Furthermore, both of these parameters must correlate well with results obtained by mass spectroscopy.

Finally, it is imperative that a peptide of interest be synthesized to confirm that the amino acid sequence determined corresponds to the biological activity of interest. This can be accomplished by either of two independent methods: solid-phase synthesis or biosynthesis by recombinant techniques. This latter approach is discussed in detail in the next section. The solid-phase synthesis of some $K^+$ channel inhibitory peptides has been accomplished, and fully reduced peptides have been oxidized to yield material that is indistinguishable from samples purified from crude scorpion venom.[9–13] This has allowed confirmation of the identities of some peptides, and has further provided a means by which to obtain different variants of these agents for structure–activity relationship studies.

## Synthesis of $K^+$ Channel Inhibitory Peptides by Recombinant Techniques

The production of $K^+$ channel inhibitory peptides by recombinant techniques has become a very popular approach, not only for obtaining large quantities of material at relatively low cost, but also for producing peptide variants with which to carry out mechanistic studies in order to identify those residues important for channel inhibition.[13–16] In general terms, a gene encoding the peptide of interest is inserted into an *Escherichia coli* expression vector. The peptide is produced as part of a fusion protein, folded, cleaved, and then purified to homogeneity by conventional methods.

The first step of this process consists of constructing an artificial gene for the peptide of interest, and inserting it into an appropriate expression vector. We have successfully employed two types of vectors. In pCSP105, the resulting construct encodes a fusion protein of the viral T7 gene 9 product with the $K^+$ channel inhibitory peptide, where the two proteins are separated by either a Factor $X_a$ cleavage site, or an enteropeptidase site. In pG9, six histidine residues are inserted between the fusion protein and the Factor $X_a$ cleavage site, at the beginning of the $K^+$ channel inhibitory peptide sequence (Fig. 2). This latter construct may facilitate purification of the fusion protein through application of $Ni^{2+}$-affinity chromatographic techniques. However, in our experience, this type of chromatographic step appears to work only with some constructs, and does not represent a very significant advantage in isolation of the fusion protein.

An appropriate strain of *E. coli*, such as BL21(DE3), is then transformed

[9] E. E. Sugg, M. L. Garcia, J. P. Reuben, A. A. Patchett, and G. J. Kaczorowski, *J. Biol. Chem.* **265,** 18745 (1990).

[10] B. A. Johnson and E. E. Sugg, *Biochemistry* **31,** 8151 (1992).

[11] M. A. Bednarek, R. M. Bugianesi, R. J. Leonard, and J. P. Felix, *Biochem. Biophys. Res. Commun.* **198,** 619 (1994).

[12] E. Drakopoulou, J. Cotton, H. Virelizier, E. Bernardi, A. R. Schoofs, M. Partiseti, D. Choquet, G. Gurrola, L. D. Possani, and C. Vita, *Biochem. Biophys. Res. Commun.* **213,** 901 (1995).

[13] J. Aiyar, J. M. Withka, J. P. Rizzi, D. H. Singleton, G. C. Andrews, W. Lin, J. Boyd, D. C. Hanson, M. Simon, B. Dethlefs, C.-L. Lee, J. E. Hall, G. A. Gutman, and K. G. Chandy, *Neuron* **15,** 1169 (1995).

[14] R. Ranganathan, J. H. Lewis, and R. MacKinnon, *Neuron* **16,** 131 (1996).

[15] P. Stampe, L. Kolmakova-Partensky, and C. Miller, *Biochemistry* **33,** 443 (1994).

[16] S. A. N. Goldstein, D. J. Pheasant, and C. Miller, *Neuron* **12,** 1377 (1994).

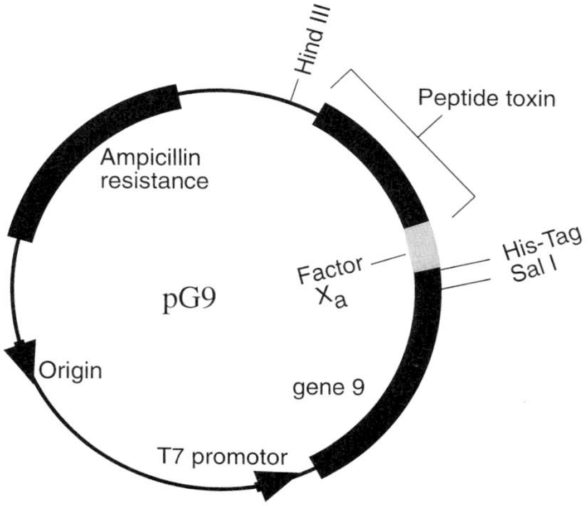


FIG. 2. Design of the synthetic peptide gene. The plasmid map of $K^+$ channel peptides shows the location of the synthetic peptide gene, the Factor $X_a$ cleavage site, the His tag, and T7 gene 9 fusion protein.

with the corresponding plasmid and a single colony is selected to inoculate the culture medium. The cells are grown at 30–37° with shaking, and when the optical density of the culture suspension at 650 nm has reached 0.6–0.7, the culture is induced with 0.5 m*M* isopropylthiogalactoside (IPTG), followed by further incubation for 3 hr at 37°. Induction of the fusion protein can be assayed by employing 12% SDS–PAGE gels and subsequently staining these with Coomassie blue. After destaining, a major product at ca. 57,000 should be observed on testing the induced culture. If no induction of fusion protein is seen, one should check for proper orientation of the toxin gene, as well as be sure that the cells do not grow above 0.7 optical density units, before addition of IPTG. If induction is successful, cells are pelleted by centifugation and washed once with 50 m*M* NaCl, 10 m*M* Tris-HCl, pH 8.0, 1 m*M* ethylenediaminetetraacetic acid (EDTA), 1 m*M* dithiothreitol (DTT). Cells are then resuspended in 10 ml/liter of culture using the buffer described earlier. Cells are frozen in liquid $N_2$ and can be stored at −70° until further processing. In our experience, one person can easily manage the purification of up to four different peptides at once. For this reason, we store cells at −70° until sufficient material is ready for purification. Cells are thawed on ice and the volume is adjusted to 50 ml/liter of culture with 50 m*M* NaCl, 10 m*M* Tris-HCl, pH 8.0, 1 m*M* EDTA, 1 m*M* DTT, 100 $\mu M$ PMSF (phenylmethylsulfonyl fluoride), 0.5 mg/ml lysozyme. After incubation on ice for 1 hr, the sample is subjected to sonication four times for 1-min intervals, and then subjected to centrifugation at 27,000*g* for 15 min. The supernatant is collected and the nucleic acids are removed by precipitation with streptomycin sulfate. It is recommended that at each step, samples be subjected to 12% SDS–PAGE analysis to monitor purification progress.

The next step involves purification of the fusion protein by ion-exchange chromatography. For this purpose, the supernatant is loaded onto a DEAE-Sepharose fast flow column equilibrated with 50 m*M* NaCl, 10 m*M* Tris-HCl, pH 8.0, 1 m*M* EDTA, 1 m*M* DTT, 100 μ*M* PMSF. After extensive washing with equilibration buffer, the fusion protein is batch eluted with 350 m*M* NaCl, 10 m*M* Tris-HCl, pH 8.0, 1 m*M* EDTA, 1 m*M* DTT, 100 μ*M* PMSF, followed by dialysis overnight at 4° against 100 m*M* NaCl, 20 m*M* Tris-HCl, pH 8.0, 0.5 m*M* 2-mercaptoethanol. During the dialysis step, folding of the peptide (i.e., formation of disulfide linkages) takes place.

Purified fusion protein must undergo enzymatic cleavage in order to release the $K^+$ channel inhibitory peptide. To accomplish this, either of two procedures can be employed. In one of these, Factor $X_a$ is used to release the peptide from the fusion protein. For this purpose, $CaCl_2$ is added to a final concentration of 3 m*M*, followed by addition of 1 μg of Factor $X_a$ per milligram of fusion protein, and the digestion mixture is incubated overnight at room temperature. The time-course of digestion can be followed by SDS–PAGE, but in most cases an overnight incubation is sufficient. We have found this procedure to give reproducible results for given constructs. However, with some peptides, Factor $X_a$ causes a cleavage upstream from its recognition site. We interpret this occurrence as the result of either of two different phenomena: (1) inability of Factor $X_a$ to reach its primary site of action (most likely due to steric factors), and/or (2) lack of specificity of Factor $X_a$. Indeed, if the released peptide containing a part of the fusion protein is purified and subsequently subjected to further incubation with Factor $X_a$, no further digestion is observed, suggesting that the conformation of the peptide prevents enzyme from reaching its binding site.

A different and less expensive way of achieving release of the $K^+$ channel inhibitory peptide from its fusion protein is through the use of trypsin. For this purpose, $CaCl_2$ is added to a final concentration of 5 m*M*, and sample is incubated with 5 μg trypsin per milligram of fusion protein for 30 min at room temperature. In this case, time of incubation with enzyme is critical since longer incubations could lead to cleavage within the peptide. This procedure, however, works well with all constructs tested so far, and it is less expensive than those protocols involving Factor $X_a$. Even in those situations where Factor $X_a$ produced wrong cleavage of the fusion protein, it is possible to treat the resulting sample with trypsin to release the appropriate peptide.

The $K^+$ channel inhibitory peptides must next be purified from the rest of the digestion mixture. This can be accomplished, as described in the previous section, by loading the sample onto a Mono S HR 10/10 column equilibrated with 20 m*M* sodium borate, pH 9.0. The fusion protein and various proteolytic fragments are not retained by the column under these conditions. Highly positively charged peptides, such as ChTX, MgTX, noxiustoxin (NxTX), and $AgTX_{1-3}$, can be loaded directly onto the column, whereas less positively charged peptides, such as IbTX, must first be dialyzed against 20 m*M* sodium borate, pH 9.0, before loading. Bound peptides are eluted in the presence of a linear gradient of NaCl in 20 m*M* sodium borate, pH 9.0 (0.5 *M*/hr) at a flow rate of 2 ml/min. A major absorbance peak representing the peptide of interest is usually observed. Certain peptides, such as ChTX and IbTX, have their amino-terminal group cyclized in the form of pyroglutamine. Because cyclization is important for the biological activity of these peptides, this reaction must be accomplished before the last purification step. For this purpose, HPLC-grade acetic acid is added to the Mono S sample at a final concentration of 5%, and the mixture is incubated at either 65° (ChTX) or 45° (IbTX) until full cyclization of the N-terminal amino group is achieved. Cyclization is followed by N-terminal sequencing, since the peptide species with the blocked N-terminal is resistant to Edman degradation. For ChTX, cyclization takes place during an overnight incubation with acetic acid, whereas for IbTX, this process may take up to 2 days, perhaps due to the lower incubation temperature used. We have found that at higher temperatures, IbTX is not stable in the presence of acetic acid, as indicated by the appearance of different chromatographic species with time.

The final stage of peptide purification is achieved on reversed-phase chromatography, using a semipreparative $C_{18}$ reversed-phase HPLC column, employing the same conditions as those described in the previous section. The composition of the purified peptide should be checked by N-terminal sequencing, amino acid hydrolysis, and/or mass spectroscopic analysis. Production of peptides through recombinant techniques is an easy way of obtaining large quantities of material at low cost. The yields of purified peptide per liter of *E. coli* culture vary depending on the construct, but it is highly reproducible for any given peptide. For instance, 10–15 mg of MgTX per liter of culture can routinely be obtained, whereas for ChTX, the yields are lower: 1–5 mg of peptide per liter of culture. We believe that these differences in yield are related to the efficiency of the cleavage step since amounts of purified fusion protein produced are similar in all cases.

Radiolabeling of $K^+$ Channel Inhibitory Peptides

The goal for radiolabeling $K^+$ channel inhibitory peptides is to obtain a biologically active derivative that can be used to identify the target protein in tissues of interest, and further to develop the molecular pharmacology of given $K^+$ channels. Radiolabeled peptides at high specific activity can be produced after reaction of peptidyl Tyr or His residues with $Na^{125}I$. Alternatively, Lys residues can be covalently modified by reaction with $^{125}I$-labeled Bolton–Hunter reagent. Because certain Lys residues in $K^+$ channel inhibitory peptides are crucial for their activity,[16,17] this latter method has not been successfully employed to date. Labeling of Tyr residues in ChTX and MgTX has been successful and, thus, these radiolabeled peptides have played a crucial role in characterizing $K^+$ channels.[18,19] Other peptides such as $AgTX_{1-3}$ do not contain Tyr residues, or, as in the case of IbTX, iodination of the Tyr residue leads to complete loss of biological activity. In these situations, it is possible to engineer amino acid residues into the sequence so that they can be subsequently modified by incorporation of a radiolabeled tag at a position that is not critical for biological activity.[20–23]

Iodination of peptides such as ChTX, MgTX, and IbTX-D19Y/Y36F has been achieved using either of two methods: the Iodo-Gen (Pierce, Rockford, IL) or glucose oxidase/lactoperoxidase protocols. The latter utilizes beads to which the enzymes have been immobilized. Use of this system usually yields chromatograms displaying fewer reaction side products, due to the milder oxidizing condition of the procedure. Unfortunately, the manufacturer of these beads, Bio-Rad (Richmond, CA), has discontinued their production. In the reaction with Iodo-Gen, the water-insoluble reagent is immobilized on the surface of a glass vial. A solution of peptide and $Na^{125}I$ is added, and after a certain period of time, the reaction mixture is removed and injected onto a reversed-phase HPLC column for purification of the radiolabeled peptide.

For a typical iodination, a solution of Iodo-Gen reagent is prepared in acetone and an aliquot corresponding to 0.5–1.0 $\mu$g of reagent is placed at the bottom of a vial and dried under argon or nitrogen. Once dried, the

[17] C.-S. Park, and C. Miller, *Neuron* **9,** 307 (1992).

[18] J. Vazquez, P. Feigenbaum, G. Katz, V. F. King, J. P. Reuben, L. Roy-Contancin, R. S. Slaughter, G. J. Kaczorowski, and M. L. Garcia, *J. Biol. Chem.* **264,** 20902 (1989).

[19] H.-G. Knaus, R. O. A. Koch, A. Eberhart, G. J. Kaczorowski, M. L. Garcia, and R. S. Slaughter, *Biochemistry* **34,** 13627 (1995).

[20] S. K. Aggarwal and R. MacKinnon, *Neuron* **16,** 1169 (1996).

[21] H.-G. Knaus, C. Schwarzer, R. O. A. Koch, A. Eberhart, G. J. Kaczorowski, H. Glossmann, F. Wunder, O. Pongs, M. L. Garcia, and G. Sperk, *J. Neurosci.* **16,** 955 (1996).

[22] E. Shimony, T. Sun, L. Kolmakova-Partensky, and C. Miller, *Prot. Eng.* **7,** 503 (1994).

[23] A. Koschak, R. O. Koch, J. Liu, G. J. Kaczorowski, P. H. Reinhart, M. L. Garcia, and H.-G. Knaus, *Biochemistry* **36,** 1943 (1997).

vial can be rinsed several times with buffer to remove any aqueous soluble contaminants. Then, a solution containing 10–20 $\mu$g of peptide in 100 m*M* sodium phosphate, pH 7.3–8.0, is added, followed by addition of 2–5 $\mu$Ci of $Na^{125}I$ (2200 Ci/mmol). After mixing, the vial is capped and the reaction is allowed to proceed at room temperature for 10–15 min. One minute before the end of the reaction, the vial is open to the air, the iodination mixture is removed with a syringe, and it is injected onto a $C_{18}$ reversed-phase HPLC column that has been equilibrated with 10 m*M* TFA. All of these procedures should be carried out inside a hood with appropriate ventilation; special care should be taken when loading the iodination mixture onto the HPLC column because the low pH buffer employed could lead to the generation of radioactive iodine gas, which will permeate skin and accumulate in the thyroid gland. For this reason, the loading and washing volumes from the HPLC column are always collected in 1 *N* NaOH to block formation of iodine gas. HPLC column elution conditions will depend on the peptide under investigation, and it is recommended that the iodination reaction be performed with unlabeled NaI first, in order to optimize the separation conditions. With ChTX, for instance, a linear gradient of 2-propanol/acetonitrile 2:1 (5–14%, 40 min) at a flow rate of 0.5 ml/min leads to a well-defined and well-separated peak corresponding to monoiodotyrosine ChTX, based on the specific activity of the radiolabeled peptide and on sequence analysis of the modified toxin. Because the amounts of material used in these procedures is low, changes in absorbance should be monitored at 210 nm. To store radiolabeled peptides for extended periods of time, the fraction of interest is made 0.1% (w/v) in BSA, lyophilized, and reconstituted with 100 m*M* NaCl, 20 m*M* Tris-HCl, pH 7.4. Aliquots of this material containing ca. 5$\mu$Ci are frozen in liquid $N_2$ and stored at $-70°$. When needed, fractions are thawed and then stored at 4°. These handling conditions have been found to be optimal in that no loss of biological activity of the labeled peptide has been observed, even after several months of storage. In most cases, in addition to the monoiodotyrosine derivative, a second minor peak corresponding to the diiodotyrosine-peptide is obtained. We usually do not retain this material since in many cases its biological activity is undefined.

An alternative method for introducing a radiolabel into these peptides makes use of an analog in which a Cys residue has been placed at a position in the peptide that is not critical for bioactivity. The free sulfydryl is then reacted with [$^3$H]N-ethyl-maleimide (NEM) to produce material with full biological activity, but at lower specific activity than that obtained with $Na^{125}I$ labeling (e.g., ca. 60 Ci/mmol). This procedure has been successfully employed to radiolabel ChTX and IbTX making use of position 19,[22,23] and with $AgTX_1$ and $AgTX_2$ at position 20.[20] Note that this type of reaction is not restricted solely to radiolabeling of the peptide; other fluorescent and

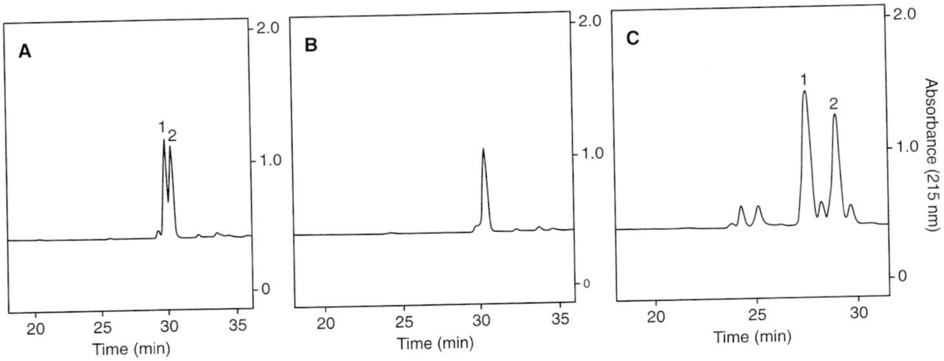


FIG. 3. Synthesis of [$^3$H]IbTX-D19C-NEM. An aliquot of IbTX-D19C in 50 m*M* sodium phosphate, pH 7.0, was treated with 2 m*M* (A) or 10 m*M* (B) DTT for 1 hr at room temperature. Peaks 1 and 2 in (A) with retention times of 29.8 and 30.2 min, respectively, represent the unreduced and reduced forms at position 19 of IbTX-D19C. In (B) the retention time of the reduced peptide is 30.3 min. (C) Reduced IbTX-D19C prepared as in (B) was reacted with 1 mCi [$^3$H]NEM (56 Ci/mmol) for 1 hr at 37°. Samples were loaded onto a $C_{18}$ reversed-phase column equilibrated with 10 m*M* TFA. Elution was achieved in the presence of a linear gradient (7–25%, 30 min) of 2-propanol/acetonitrile (2 : 1) in 4 m*M* TFA. Peaks 1 and 2 with retention times of 27.5 and 28.9 min, respectively, have identical specific activity and their biological activity is also indistinguishable.

biotinylated derivatives of NEM can also be reacted to yield interesting peptidyl ion channel probes. However, there is one consideration to note with this approach which concerns the fact that a newly introduced Cys residue is not initially reactive on purification of the peptide, since it does not exist as a free sulfhydryl group. Instead, it exists in a disulfide bridge with either 2-mercaptoethanol or another toxin molecule. It is necessary, therefore, to selectively reduce this disulfide bond without disrupting the integrity of the natural disulfide bridges of the peptide that are crucial for biological activity before labeling with the sulfhydryl reagent. To accomplish this, peptides are incubated in 50 m*M* sodium phosphate, pH 7.0, with 1–10 m*M* DTT at room temperature for 1 hr. The reaction mixture is then applied to a $C_{18}$ reversed-phase HPLC column. Elution conditions must be optimized for each particular peptide preparation. The peptide containing the free-sulfhydryl group should be easily identified by a change in retention time when compared with unreduced material (Fig. 3A). The change in retention time may not be very large and, therefore, appropriate elution conditions should be used to achieve good separation. In the event that the separation is suboptimal, this should not preclude continuation of the procedure since unreduced material will not react with NEM. In our experience, optimal recovery of $C_{19}/C_{20}$ reduced peptide approximates 50%. Although higher concentrations of either DTT or increases in pH can lead to full disappearance of starting material (Fig. 3B), they also cause appearance of other forms of the peptide with much longer retention times in which additional disulfide bonds have been reduced. This material displays no biological activity and can be discarded. Reduced peptide can be concentrated to about 20 $\mu$l using a Speed-Vac, and then 100 $\mu$l of 50 m$M$ sodium phosphate is added. It is important to use reduced peptide for reaction with [$^3$H]NEM as soon as possible after its production. We have observed that, in some cases, reduced peptide can dimerize via formation of a disulfide linkage, even when stored at $-70°$.

For the alkylating reaction, one vial containing 1 mCi of [$^3$H]NEM in pentane is opened and the content transferred to a glass tube. The solvent is evaporated under $N_2$ to about 300 $\mu$l, and then 200 $\mu$l of 50 m$M$ sodium phosphate, pH 7.0, is added. After mixing by vortex, the remaining pentane is evaporated, the [$^3$H]NEM solution is added to the peptide solution, and the vial is rinsed with an additional 200 $\mu$l of 50 m$M$ sodium phosphate, pH 7.0, which is then also transferred to the peptide solution. The reaction mixture is allowed to react for 1 hr at 37°, and then injected onto a $C_{18}$ reversed-phase HPLC column equilibrated with 10 m$M$ TFA. Elution conditions should be optimized for each individual peptide, and alkylated material can be easily identified by the change in retention time. In addition, radioactivity should be associated with this material. If good separation is achieved, the specific activity of the alkylated peptide should correspond to that of the [$^3$H]NEM employed for the reaction, ca. 60 Ci/mmol. It is worth noting that in the case of reaction of IbTX-$C_{19}$ with [$^3$H]NEM, two well-separated radiolabeled peaks are obtained in equivalent amounts, and both display identical biological activity (Fig. 3C). The exact nature of these two species is unknown, but this may correspond to the production of two isomers of IbTX-$C_{19}$-NEM.

## Receptor Binding Studies

The goal of radiolabeling a given $K^+$ channel inhibitory peptide is so that it can be used to characterize its receptor in native tissues. This includes characterizing the kinetics of ligand association and dissociation, the molecular pharmacology of the channel, and the tissue distribution of the channel.[18,19,23–26] Additionally, radiolabeled peptides have been successfully em-

[24] J. Vazquez, P. Feigenbaum, V. F. King, G. J. Kaczorowski, and M. L. Garcia, *J. Biol. Chem.* **265,** 15564 (1990).

[25] H.-G. Knaus, O. B. McManus, S. H. Lee, W. A. Schmalhofer, M. Garcia-Calvo, L. M. H. Helms, M. Sanchez, K. Giangiacomo, J. P. Reuben, A. B. Smith III, G. J. Kaczorowski, and M. L. Garcia, *Biochemistry* **33,** 5819 (1994).

[26] O. B. McManus, G. H. Harris, K. M. Giangiacomo, P. Feigenbaum, J. P. Reuben, M. E. Addy, J. F. Burka, G. J. Kaczorowski, and M. L. Garcia, *Biochemistry* **32,** 6128 (1993).

ployed in the purification of native ion channels.[27–30] As a matter of fact, all auxiliary ($\beta$) subunits of ion channels have been identified and subsequently characterized through the purification of the corresponding ion channel complex using binding of channel probes as a monitor of channel purification. There are different ways in which to monitor peptide–$K^+$ channel interactions, such as autoradiographic techniques or binding of ligand to intact cells or tissues, but perhaps the most common studies are those in which the peptide interaction is measured in highly enriched membrane preparations.

Three major considerations should be taken into account when considering the development of a binding reaction involving $K^+$ channel inhibitory peptides discussed in this chapter. The first one concerns the potencies of the peptides as inhibitors of $K^+$ channels; under physiologic conditions, some of these peptides display nanomolar affinities, but their potency can be greatly enhanced by selecting appropriate binding reaction conditions (i.e., by lowering ionic strength of the incubation medium, which will enhance electrostatic interaction between positively charged residues on the peptide and negatively charged residues in the channel's vestibule.[31–33] Second, problems arising from the handling of these peptides should be considered. The easiest way to separate bound from free ligand is by using filtration techniques and, therefore, the filters used must have specific characteristics, or be treated accordingly to minimize binding of the positively charged peptides. We have found that GF/C glass fiber filters (Whatman, England) presoaked in 0.5–1.0% (w/v) polyethyleneimine (Sigma, St. Louis, MO) provide a low background binding support if high ionic strength buffer (e.g., 100 m*M* NaCl, 20 m*M* Tris-HCl, pH 7.4) is used to quench the binding reaction. This high ionic strength quench buffer is also important to induce immediate dissociation of toxin bound to low-affinity sites. We should also consider the physical properties of the test tubes in which the binding reaction is carried out. If low ionic strength media are used, then glass tubes should be avoided because peptides may adhere to these surfaces; under such conditions, use of polystyrene tubes is recommended. In addition, inclusion of 0.1% (w/v) BSA in the incubation medium would also

[27] M. Garcia-Calvo, H.-G. Knaus, O. B. McManus, K. M. Giangiacomo, G. J. Kaczorowski, and M. L. Garcia, *J. Biol. Chem.* **269,** 676 (1994).

[28] K. M. Giangiacomo, M. Garcia-Calvo, H.-G. Knaus, T. J. Mullmann, M. L. Garcia, and O. McManus, *Biochemistry* **34,** 15849 (1995).

[29] D. N. Parcej and J. O. Dolly, *Biochem. J.* **257,** 899 (1989).

[30] H. Rehm, and M. Lazdunski, *Proc. Natl. Acad. Sci. U.S.A.* **85,** 4919 (1988).

[31] K. M. Giangiacomo, M. L. Garcia, and O. B. McManus, *Biochemistry* **31,** 6719 (1992).

[32] R. MacKinnon and C. Miller, *J. Gen. Physiol.* **91,** 335 (1988).

[33] S. Candia, M. L. Garcia, and R. Latorre, *Biophys. J.* **63,** 583 (1992).

prevent peptide from being absorbed onto the surface of tubes. Finally, the last consideration is the source of receptor. For instance, ChTX blocks two different types of $K^+$ channels: high-conductance $Ca^{2+}$-activated $K^+$ channels, and voltage-dependent $K^+$ channels. If one considers using this peptide as a ligand, it is important to select a tissue in which either, but not both, of these channels is present. Moreover, use of highly enriched plasma membrane preparation should give a higher ratio of specific to nonspecific binding and should be employed if at all feasible.

Other parameters which are optimal for each particular ligand–receptor interaction should be determined according to each individual situation. For example, the incubation time necessary to reach equilibrium for any particular ligand can vary considerably from about 1 hr for [$^{125}$I]MgTX or [$^{125}$I]ChTX,[19,24] to close to 60–72 hr for [$^{125}$I]IbTX.[23]

## Summary

The search for peptidyl inhibitors of $K^+$ channels is a very active area of investigation. In addition to scorpion venoms, other venom sources have been investigated; all of these sources have yielded novel peptides with interesting properties. For instance, spider venoms have provided peptides that block other families of $K^+$ channels (e.g., $K_V2$ and $K_V4$) that act via mechanisms which modify the gating properties of these channels.[34–36] Such inhibitors bind to a receptor on the channel that is different from the pore region in which the peptides discussed in this chapter bind.[37] In fact, it is possible to have a channel occupied simultaneously by both inhibitor types. It is expected that many of the methodologies concerning peptidyl inhibitors from scorpion venom, which have been developed in the past and outlined above, will be extended to the new families of $K^+$ channel blockers currently under development.

[34] M. C. Sanguinetti, J. H. Johnson, L. G. Hammerland, P. R. Kelbaugh, R. A. Volkmann, N. A. Saccomano, and A. L. Mueller, *Molec. Pharmacol.* **51,** 491 (1997).
[35] K. J. Swartz and R. MacKinnon, *Neuron* **15,** 941 (1995).
[36] K. J. Swartz and R. MacKinnon, *Neuron* **18,** 665 (1997).
[37] K. J. Swartz and R. MacKinnon, *Neuron* **18,** 675 (1997).

# [33] Potassium Ion Channel Inactivation Peptides

*By* RUTH D. MURRELL-LAGNADO

## Introduction

Voltage-gated $K^+$ channels activate with depolarization of the membrane potential and subsequently inactivate. *Shaker* $K^+$ channels inactivate very rapidly with a time constant of 2–3 ms. Aldrich and co-workers showed that this rapid inactivation arises from an amino(N)-terminal inactivation domain acting as a tethered particle that blocks the open channel.[1,2] This mechanism is analogous to the "ball-and-chain" model proposed by Bezanilla and Armstrong[3] for inactivation of $Na^+$ channels. Removing the cytoplasmic N-terminal domain of *Shaker* $K^+$ channels disrupts rapid inactivation, and internal application of a synthetic peptide corresponding to this domain (residues 1–20) restores inactivation.[1,2] The rapidly inactivating mammalian channels, Raw3 (Kv3.4) and RCK4 (Kv1.4), also undergo N-type inactivation[4,5] and more recently it has been shown that one of the soluble $\beta$ subunits of voltage-gated $K^+$ channels (Kv$\beta$1) can dramatically speed inactivation by a similar mechanism involving the N-terminal inactivation domain of the $\beta$ subunit itself.[6]

Several pieces of evidence indicate that block of *Shaker* channels by the inactivation domain and synthetic (ShB) peptide involves a single blocking particle physically occluding the pore. First, the Kv channels have a tetrameric arrangement of $\alpha$ subunits contributing to a central conductance pore, but only a single amino terminal inactivation domain is required for N-type inactivation.[7] Second, the kinetics of block by the ShB peptide is consistent with a simple bimolecular reaction; the blocking rate depends linearly on peptide concentration and the unblocking rate is independent of concentration.[8] Third, N-type inactivation and peptide block are coupled

[1] T. Hoshi, W. N. Zagotta, and R. W. Aldrich, *Science* **250,** 533 (1990).
[2] W. N. Zagotta, T. Hoshi, and R. W. Aldrich, *Science* **250,** 568 (1990).
[3] F. Bezanilla and C. M. Armstrong, *J. Gen. Physiol.* **70,** 549 (1977).
[4] J. P. Ruppersberg, M. Stocker, O. Pongs, S. H. Heinemann, R. Frank, and M. Koenen, *Nature* **352,** 711.
[5] J. P. Ruppersberg, R. Frank, O. Pongs, and M. Stocker, *Nature* **353,** 657 (1991).
[6] J. Rettig, S. H. Heinemann, F. Wunder, C. Lorra, D. N. Parcej, J. O. Dolly, and O. Pongs, *Nature* **369,** 289 (1994).
[7] R. MacKinnon, R. W. Aldrich, and A. W. Lee, *Science* **262,** 757 (1993).
[8] R. D. Murrell-Lagnado and R. W. Aldrich, *J. Gen. Physiol.* **102,** 949 (1993).


METHODS IN ENZYMOLOGY, VOL. 294
0076-6879/99 $30.00

TABLE I
SEQUENCES OF INACTIVATION PEPTIDES

| Nomenclature | Sequence |
|---|---|
| ShB | MAAVAGLYGLGEDRQHRKKQ |
| ShC | MQMILVAGGSLPKSSQ |
| ShD | MAHITTTHGSLSGATR |
| Raw3 | MISSVCVSSYRGKKSGNKPPSKTCLKEEMA |
| RCK4 | MEVAMVSAESSGCNSHMPYGYAAQARARERERLAHSR |

to channel activation,[9] suggesting that they can only gain access to the binding site when the channel is in the open state. Fourth, they are both slowed by internal tetraethylammonium (TEA),[8,10] which is known to block the channel by occluding the pore from the inside, suggesting that binding of TEA competes with binding of the inactivation peptide. Finally, recovery from inactivation and peptide block are speeded by raising the concentration of $K^+$ on the opposite (external) side of the channel.[11,12] Although the inactivation domain of the *Shaker* channel has a net positive charge, N-type inactivation shows no apparent voltage sensitivity over the range of $-30$ to $+50$ mV.[9] Peptide binding does show some voltage dependence, but increasing the net charge on the peptide does not increase this voltage dependence.[12,13] Thus, the majority of the blocking particle does not appear to enter into the membrane electric field, but binds instead at the mouth of the pore.

The ShB inactivation peptide (Table I) has been shown to block not only the *Shaker* channel but also a variety of other channels, including $Ca^{2+}$- activated $K^+$ channels,[14-17] a $K^+$ channel from epithelial cells,[18] and the cyclic nucleotide-gated channel.[19] By modifying the peptide and measuring changes in the binding energy a coherent picture has emerged of the features

[9] W. N. Zagotta and R. W. Aldrich, *J. Gen. Physiol.* **95,** 29 (1990).
[10] K. L. Choi, R. W. Aldrich, and G. Yellen, *Proc. Natl. Acad. Sci. U.S.A.* **88,** 5092 (1991).
[11] S. D. Demo and G. Yellen, *Neuron* **7,** 743 (1991).
[12] R. D. Murrell-Lagnado and R. W. Aldrich, *J. Gen. Physiol.* **102,** 977 (1993).
[13] L. Toro, M. Ottolia, E. Stefani, and R. Latorre, *Biochemistry* **33,** 7220 (1994).
[14] C. D. Foster, S. Chung, W. N. Zagotta, R. W. Aldrich, and I. B. Levitan, *Neuron* **9,** 229 (1992).
[15] L. Toro, E. Stefani, and R. Latorre, *Neuron* **9,** 237 (1992).
[16] C. R. Solaro and C. J. Lingle, *Science* **257,** 1694 (1992).
[17] P. S. L. Beirao, N. W. Davies, and P. R. Stanfield, *J. Physiol.* **474** 269 (1994).
[18] W. P. Dubinsky, W. O. Mayorga, and S. G. Schultz, *Proc. Natl. Acad. Sci. U.S.A.* **89,** 1770 (1992).
[19] R. H. Kramer, E. Goulding, and S. A. Siegelbaum, *Neuron* **12,** 655 (1994).

TABLE II

BLOCKADE OF *SHAKER* $K^+$ CHANNELS (ShBΔ6-46), $Ca^{2+}$-ACTIVATED $K^+$ CHANNELS (MAXI $K_{CA}$), AND mKv1.1 CHANNELS BY INACTIVATION PEPTIDES[a]

| | | Shaker (ShBΔ6-46)[b] | | | Maxi $K_{Ca}$[c] | | |
|---|---|---|---|---|---|---|---|
| Peptide | Charge | $K_d$ ($\mu M$) | $k_{on}$ ($\times 10^6$ $M^{-1}$ $sec^{-1}$) | $k_{off}$ ($sec^{-1}$) | $K_d$ ($\mu M$) | $k_{on}$ ($\times 10^6$ $M^{-1}$ $sec^{-1}$) | $k_{off}$§ ($sec^{-1}$) |
| **ShB 1-20** | +2 | 2.9 ± 0.03 | 4.7 ± 0.38 | 14 ± 0.87 | 160 ± 9 | 1.4 ± 0.3 | 224 |
| acetyl ShB[d] | +2 | 3.6 ± 1.3 | 1.6 ± 0.2 | 5.7 ± 3.6 | | | |
| Δ A3A5 | +2 | 4.1 ± 0.67 | 3.7 ± 0.37 | 15 ± 2.0 | | | |
| ins A6, 7, 8[e] | +2 | 5.8 ± 1.9 | 1.9 ± 0.09 | 11 ± 3.5 | | | |
| A2V, A3V, A5V | +2 | | | | 5.5 ± 1.1 | 1.6 ± 0.18 | 8.8 |
| G6V, G9V | +2 | | | | 58.6 ± 8.7 | 0.93 ± 0.25 | 54 |
| G6V, G9V, G11V | +2 | | Low solubility | | | | |
| L7E | +1 | >300 | | | | | |
| L7Q | +2 | 22 ± 3.1 | 3.8 ± 0.45 | 84 ± 6.0 | | | |
| L7A | +2 | 8.8 ± 1.4 | 5.0 ± 0.7 | 43 ± 3.4 | | | |
| L7I | +2 | 3.7 ± 0.48 | 4.7 ± 0.47 | 17 ± 1.5 | | | |
| L7F | +2 | 2.1 ± 0.32 | 5.0 ± 0.62 | 11 ± 0.84 | | | |
| Y8F | +2 | 2.6 ± 0.90 | 2.7 ± 0.47 | 6.9 ± 2.1 | | | |
| Y8K | +3 | 34.7 ± 8.5 | 4.9 ± 1.2 | 6.9 ± 2.1 | | | |
| L10Q | +2 | 8.0 ± 1.4 | 2.9 ± 0.27 | 23 ± 3.3 | | | |
| L10K | +3 | 9.6 ± 1.2 | 4.7 ± 0.57 | 44 ± 1.2 | | | |
| L10P | +2 | 11 | 2.8 | 32 | | | |
| ins P10 | +2 | 4.9 ± 0.84 | 4.1 ± 0.52 | 20 ± 2.3 | | | |
| ins P6 | +2 | 13 ± 1.9 | 2.9 ± 0.27 | 39 ± 4.0 | | | |
| G11P | +2 | 6 | 3 | 18 | | | |
| E12Q | +3 | 1.5 ± 0.29 | 11 ± 1.3 | 16 ± 2.6 | | | |
| E12Q, D13Q | +4 | 0.53 ± 0.10 | 27 ± 4.5 | 14 ± 1.3 | | | |
| E12N, D13N | +4 | | | | 10.7 ± 0.7 | 7.3 ± 2.4 | 78 |
| E12K, D13K | +6 | 0.17 ± 0.02 | 140 ± 16 | 24 ± 1.5 | 3.6 ± 0.3 | 15 ± 0.7 | 54 |
| E12Q, D13Q, R14Q, R17Q | +2 | 10 ± 1.1 | 2.7 ± 0.09 | 2.8 ± 2.8 | | | |
| E12K, D13K, K18E, K19D | +2 | 7.4 | 3.9 | 29 | | | |
| R14Q | +1 | | | | 309 ± 59 | 0.69 ± 0.03 | 213 |
| R17Q | +1 | | | | 474 ± 31 | 0.32 ± 0.06 | 152 |
| H16Q, R17Q, K18Q, K19Q | −1 | 140 ± 39 | 0.29 ± 0.06 | 43 ± 6.3 | | | |
| R17Q, K18Q, K19Q | −1 | 140 ± 20 | 0.22 ± 0.02 | 30 ± 2.6 | | | |
| K18Q, K19Q | 0 | | | | 259 ± 92 | 0.36 ± 0.07 | 93 |
| K19Q | +1 | | | | 168 ± 5.5 | 1.1 ± 0.13 | 185 |
| **ShC 1-16** | +1 | 13 ± 1.7 | 0.61 ± 0.06 | 7.7 ± 0.75 | | | |
| ShC 1-16 M3M(O)[f] | +1 | >300 | | | | | |
| ShC 1-20 | −2 | >300 | | | | | |
| **ShD d1-22** | +1 | 14 ± 1.2 | 5.5 ± 0.35 | 7.6 ± 4.4 | | | |
| **Raw3 1-28** | +4 | 1.8 ± 0.024 | 5.8 ± 0.47 | 11 ± 1.1 | | | |

TABLE II (*continued*)

| Peptide | Charge | Shaker (ShBΔ6-46)[b] $K_d$ ($\mu M$) | $k_{on}$ ($\times 10^6\ M^{-1}$ sec$^{-1}$) | $k_{off}$ (sec$^{-1}$) | Maxi $K_{Ca}$[c] $K_d$ ($\mu M$) | $k_{on}$ ($\times 10^6\ M^{-1}$ sec$^{-1}$) | $k_{off}$§ (sec$^{-1}$) |
|---|---|---|---|---|---|---|---|
| | | mKv1.1[g] | | | | | |
| **Raw3 1-28** | +4 | 8.4 ± 1.5 | 0.31 ± 0.05 | 2.4 ± 0.3 | | | |
| S-AAM Raw3[h] | +4 | 175 ± 38 | 0.15 ± 0.03 | 19 ± 1.7 | | | |
| Δ1-5 | +4 | 33 ± 12 | 0.05 ± 0.01 | 0.01 | | | |
| Δ25-28 | +5 | 5.7 ± 1.9 | 0.78 ± 0.16 | 3.4 ± 0.9 | | | |
| Δ16-28 | +3 | 36 ± 6.9 | 0.09 ± 0.04 | 2.5 ± 0.7 | | | |
| Δ1-15 | +1 | >300 | | | | | |
| Δ1-5, Δ25-28 | +5 | >300 | | | | | |
| C6S | +4 | 12 ± 5.3 | 0.37 ± 0.09 | 3.1 ± 0.7 | | | |
| C24S | +4 | 22 ± 7.5 | 0.20 ± 0.04 | 3.2 ± 0.6 | | | |
| C6S, C24S | +4 | 44 ± 9.0 | 0.11 ± 0.02 | 4.2 ± 1.0 | | | |
| C6A, C24A | +4 | 52 ± 18 | 0.13 ± 0.06 | 4.3 ± 0.7 | | | |
| K26A | +3 | 24 ± 9 | 0.19 ± 0.09 | 2.6 ± 0.8 | | | |
| K26A, E27A | +4 | 7.9 ± .31 | 0.32 ± 0.09 | 1.7 ± 0.3 | | | |
| K26A, E27A, E28A | +5 | 9.1 ± 2.7 | 0.36 ± 0.09 | 2.6 ± 0.5 | | | |

[a] Peptide block of ShBΔ6-46 expressed in *Xenopus* oocytes was measured by application of peptides to the internal surface of macropatches, at +50 mV. Block of maxi $K_{Ca}$ channels incorporated into lipid bilayers was measured at 0 mV. $K_d$ values were calculated from Hill plots and $k_{off}$§ values were calculated from ($K_d k_{on}$). Block of mKv1.1 expressed in Chinese hamster ovary cells was measured by application of peptides via the recording electrode in the whole-cell patch-clamp mode, at +70 mV. These peptides were N-terminal acetylated. Charge represents the net charge of the amino acid side groups within the peptide but does not include the N-terminal positive charge.

[b] Values from R. D. Murrell-Lagnado and R. W. Aldrich,[8] except *, which is from Ciorba *et al.*[29]

[c] Values from L. Toro *et al.*[13]

[d] Acetylating N terminus of ShB did not significantly affect block of ShBΔ6-46.

[e] "ins" indicates amino acids inserted into the peptide.

[f] Substitution of methionine 3 for oxidized methionine.[29]

[g] Values from G. J. Stephens *et al.*[26]

[h] Attachment of acetamidomethyl side chains to the sulfurs on both cysteine residues.[26]

required for blocking these channels (Table II).[8,12,13] The nonpolar nature of the first 10 amino acids is important for stabilizing the binding of the ShB peptide. Substituting fewer hydrophobic residues at position 7 increases the dissociation rate constant, whereas making the peptide more hydrophobic induces tighter binding. Positive charges within the next 10 amino acids are important for increasing the rate of diffusion of the peptide toward the binding site via long-range electrostatic interactions. Changing the net charge on the peptide primarily affects the association rate constant ($k_{on}$).

Substitutions that switch the position of charged residues or which neutralize equal numbers of positive and negative charges have relatively little effect on the binding energy. Thus, the precise nature of the side groups or specific localization of the charges does not appear to be important, indicating that these residues are not involved in specific, short-range interactions. These features of part hydrophobic and part positively charged are shared by the Raw3 (Kv3.4) and RCK4 (Kv1.4) inactivation peptides (Table I). It suggests that there is a hydrophobic pocket at the mouth of the channel pore and a negative potential in the vicinity of this pocket.

## Using Inactivation Peptides as Probes of Inner Mouth of Channel Pores

The peptide toxins have been extremely useful as probes of the *outer* mouth of channel pores.[20] They interact in a highly specific manner with particular ion channels and inhibit in the 1-p*M* to 10-n*M* concentration range. In some cases their solution structure has been solved by multidimensional nuclear magnetic resonance (NMR) methods and they have been shown to be rigid, compact molecules. In contrast, the inactivation peptides generally inhibit in the 100-n*M* to 100-$\mu M$ concentration range and the interactions involved appear to be less specific in nature. A relatively low-affinity interaction is suitable for an inactivation gate but less suitable for a probe of channel structure. Although the inactivation peptides are unlikely to provide information about the precise spatial arrangement of channel amino acids, they can be used to identify regions of the channel that are in the vicinity of the inner vestibule of the pore, and residues within these regions whose side chains are accessible to the aqueous media.

By making mutations within the S4–S5 linker region of the *Shaker* channel and demonstrating alterations in the rate of N-type inactivation, Isacoff *et al.*[21] identified this region as forming at least part of the receptor for the amino-terminal inactivation domain and, thus, it is likely to contribute to forming the pore. Holmgren *et al.*[22] used the approach of cysteine substitution mutagenesis, and subsequent chemical modification to show that attachment of a positively charged thioalkyl group to a cysteine at position 391, within the S4–S5 linker of the *Shaker* channels, produced a strong reduction in the ShB peptide-binding affinity. This reduction in affinity was mainly by a decrease in $k_{on}$, whereas attachment of a negatively charged group to C391 increased $k_{on}$. These results indicate that the side group of

[20] C. Miller, *Neuron* **15,** 5 (1995).

[21] E. Y. Isacoff, Y. N. Jan, and L. Y. Jan, *Nature* **353,** 86 (1991).

[22] M. Holmgren, M. E. Jurman, and G. Yellen, *J. Gen. Physiol.* **108,** 195 (1996).

amino acid 391 is at an internally accessible location, in or near the receptor site. An advantage of this approach is that by performing modification experiments during single-channel recording they were able to show that, in the absence of peptide, the chemical modifications did not alter the gating of the channel. Binding of the inactivation peptide is dependent on the activation state of the channel and mutations that affect the activation process could indirectly alter the apparent affinity.

Ideally a probe of channel structure would itself adopt a stable structure. Circular dichroism and NMR experiments suggest that the ShB peptide is unstructured in aqueous solution.[23,24] However, Antz *et al.*[25] have recently reported that the 30 amino acid Raw3 inactivation peptide does adopt a stable structure in aqueous solution and it is compact with two exposed phosphorylation sites and two spatially close cysteine residues. The 37-amino-acid RCK4 inactivation peptide appears to be more flexible but with a well-defined $\alpha$ helix and $\beta$ turn. Although the two molecules adopt rather different structures, the charged and hydrophobic surface domains have a similar spatial distribution. Modifications to the Raw3 peptide have suggested that both charged and hydrophobic features of the peptide are important for block of the Kv1.1 channel[26] (see Table II). The exposed nature of the serines and cysteine residues within the Raw3 peptide is consistent with previous experiments that have demonstrated the regulation of N-type inactivation and peptide block of Kv3.4, by phosphorylation of serines 15 and 21[27] and by redox potential.[4] Phosphorylation of S15 and S21 reduces the net positive charge of the inactivation domain and disrupts long-range electrostatic interactions. Oxidation of cysteine-6 is thought to inhibit inactivation as a result of disulfide bonds forming between the amino terminal domains.

In view of the compact, stable structure adopted by the Raw3 inactivation peptide this may prove to be an attractive choice for probing channel structure at the inner mouth of the pore, although at present we know less about how this peptide blocks channels than we do about ShB peptide blockade. Because the binding interactions involved are not highly specific the inactivation peptides could be useful for investigating a variety of cation

[23] C. W. B. Lee, R. W. Aldrich, and L. M. Gierasch. *Biophys. J.* **61,** A379 (1992).

[24] G. Fernandez-Ballester, F. Gavilanes, J. P. Albar, M. Criado, J. A. Ferragut, and J. M. Gonzalez-Ros, *Biophys. J.* **68,** 858 (1995).

[25] C. Antz, M. Geyer, B. Fakler, M. K. Schott, H. R. Guy, R. Frank, J. P. Ruppersberg, and H. R. Kalbitzer, *Nature* **385,** 272 (1997).

[26] G. J. Stephens, D. G. Owen, A. Opalko, M. R. Pisano, W. H. MacGregor, and B. Robertson, *J. Physiol.* **496,** 145 (1996).

[27] M. Covarrubias, A. Wei, L. Salkoff, and T. B. Vyas, *Neuron* **13,** 1403 (1994).

selective ion channels that are likely to have hydrophobic and negatively charged residues in the vicinity of the pore.

## Purification of Peptides

Peptides can be synthesized by an in-house facility or are conveniently ordered from a number of different companies. They should be amidated at their carboxyl-terminal end. Pure peptide should be a white powder with no obvious odor. Companies/facilities should provide mass spectrometry and analytic HPLC data on the peptide to indicate its purity. The impurities that can occur include side-chain protection groups that are employed during synthesis and are not completely removed during the deprotection phase, peptides with an internal deletion, and N- and C-terminal truncated peptides. If the peptide has an unpleasant odor it indicates the presence of thiols.

Purification is generally carried out by reversed-phase HPLC using a $C_{18}$ column. To purify 150–200 mg of peptide, a preparative column of at least 25 mm in width and 200 mm in length is required, with flow rates of 10–20 ml/min. The pore size should be 10–30 nm and the particle size 15 $\mu$m. The analytical column should have the same medium and pore size, but a narrow bore and flow rate of 1 ml/min. The mobile phase frequently used includes water, acetonitrile, and 0.05% trifluoroacetic acid (TFA), and this needs to be deaerated by stirring under vacuum for about 30 min.

The procedure for purifying crude peptide is as follows:

1. Run a sample (<1 mg in water) through an analytical column to determine the percent acetonitrile at which the peptide and any impurities elute. A gradient of between 10 and 40% acetonitrile during step 3 (see Table III) is generally suitable for eluting the inactivation peptides. The ShB peptide elutes at ~25% using a Bio-Rad (Richmond, CA) Hi-Pore RP-318 column (250 × 4.6 mm). The peptides are detected in the eluant by monitoring the absorbance at 220 nm.
2. The gradient for the preparative run is generally set to be 12% below and 5% above the percent acetonitrile at which the peptide elutes during the analytical run ($x$%). Peptides tend to elute slightly earlier during the preparative run than during the analytical run. The different fractions are collected during step 3 and these can subsequently be analyzed and pooled.

After purification the peptide should be dried. A centrifuge evaporator eliminates any danger of loss of peptide by spattering.

TABLE III
PROTOCOL FOR PREPARATIVE HPLC RUN

Solution A: Water + 0.05% TFA
Solution B: Acetonitrile + 0.05% TFA
Flow rate: 10 ml $min^{-1}$

| Step | Time (min) | Gradient |
|---|---|---|
| 1 | 5 | 0% B/100% A |
| 2 | 1 | 0 to $(x - 12)$% B |
| 3 | 20 | $(x - 12)$% to $(x + 5)$% B |
| 4 | 5 | $(x + 5)$% B to 100% B |
| 5 | 10 | 100% B |
| 6 | 5 | 100% to 0% B |
| 7 | 10 | 0% B/100% A |

## Storage

Most peptides are very stable when stored lyophilized in a desiccator at −20°. Stock solutions can also be aliquoted and stored at −20°.

## Determining Concentration

Spectroscopic methods can be used to determine the concentration of peptides with tyrosine or tryptophan residues.[28] For peptides containing tyrosines but no tryptophans, the absorption should be measured at 275.5 nm and the molar extinction coefficient calculated according to Eq. (1)

$$\varepsilon_{275.5} = M_{\mathrm{Tyr}}1500 + N_{\mathrm{cys}}155 \tag{1}$$

where $M$ and $N$ are the number of moles of tyrosine and cysteine per mole of peptide. For peptides containing tryptophans the absorption should be measured at either 288 or 280 nm and the molar extinction coefficients calculated according to Eqs. (2) and (3):

$$\varepsilon_{288} = L_{\mathrm{Trp}}4815 + M_{\mathrm{tyr}}385 + N_{\mathrm{cys}}75 \tag{2}$$
$$\varepsilon_{280} = L_{\mathrm{Trp}}5690 + M_{\mathrm{tyr}}1280 + N_{\mathrm{cys}}120, \tag{3}$$

where $L$ is the number of moles of tryptophan per mole of peptide. Thus, for the ShB peptide $\varepsilon_{275.5}$ is 1500 and for the Raw3 peptide $\varepsilon_{275.5}$ is 1790. The concentration is calculated by dividing the absorbance by the molar

[28] H. Edelhoch, *Biochemistry* **6,** 1948 (1967).
[29] M. A. Ciorba, S. H. Heinemann, H. Weissbach, N. Brot, and T. Hoshi, *Proc. Natl. Acad. Sci. U.S.A.* **94(18),** 9932 (1997).

extinction coefficient. When determining the absorbance, a sample of the peptide stock solution should be diluted (1 : 100) in 6.0 *M* guanidine hydrochloride (pH 6.5)–0.02 *M* phosphate buffer, to ensure that the peptide is highly unfolded.

## Application of Peptides

The inactivation peptides are usually applied to the internal surface of isolated membrane patches or to channels incorporated into lipid bilayers. In some cases they have also been applied via the recording electrode in the whole-cell patch-clamp mode. The block of voltage-gated $K^+$ channels by the inactivation peptides is strongly coupled to channel opening. Therefore, for those channels that activate rapidly with depolarizing voltage steps the rate of block of macroscopic currents can be determined without the use of rapid perfusion techniques by applying the peptide at voltages at which the channel is fully deactivated. If activation is not considerably faster than peptide binding, then the decay phase will reflect both channel opening and peptide binding kinetics, in which case it is necessary to use fast perfusion methods or single-channel analysis to determine the association and dissociation rate constants.

## Analyzing Kinetics of Channel Block

Block of *Shaker* channels by the inactivation peptide involves a single molecule physically occluding the pore, when the channel is in the open state (O). Therefore, the following simple bimolecular reaction model can be used to analyze the kinetics of blockade:

$$\mathrm{C} \underset{\beta}{\overset{\alpha}{\rightleftharpoons}} \mathrm{O} \underset{k_{\mathrm{off}}}{\overset{\text{peptide} \quad k_{\mathrm{on}}}{\rightleftharpoons}} \mathrm{B}$$

When using the inactivation peptides to probe the structure of other channels it is important to demonstrate that a similar pore blocking mechanism is involved. According to this scheme, on application of peptide (or following channel activation by a depolarizing voltage step), the macroscopic current will relax exponentially to a new level with a time constant, ($\tau$) equal to $1/([\text{peptide}]k_{\text{on}} + k_{\text{off}})$. The fraction of current remaining, ($I_{\text{ub}}$) is equal to $k_{\text{off}}/([\text{peptide}]k_{\text{on}} + k_{\text{off}})$. Therefore,

$$k_{\text{on}} = (1 - I_{\text{ub}})/\tau[\text{peptide}] \tag{4}$$

$$k_{\text{off}} = I_{\text{ub}}/\tau \tag{5}$$

For the *Shaker* channel there is an additional slower component of inactivation that is independent of peptide block, and this has to be corrected for when calculating $I_{ub}$. The relaxation phase is fitted with the sum of two exponentials and

$$I_{ub} = a_{slow}/(a_{slow} + a_{fast}) \tag{6}$$

where $a_{slow} + a_{fast}$ represent the amplitudes of the fast and slow components, respectively.

When analyzing single-channel records, the $K_d$ can be calculated by fitting a Hill equation to a plot of normalized channel open probability versus peptide concentration.[13] The association rate constant can be obtained from the slope of a plot of $1/\tau_{open}$ versus peptide concentration, where $\tau_{open}$ is the time constant of the exponential fit to the open time distribution. The dissociation rate constant is equal to $1/\tau_{blocked}$, where $\tau_{blocked}$ is the time constant of the exponential fit to the blocked time distribution.

## Acknowledgments

I thank R. W. Aldrich and E. Perozo for helpful comments on the chapter.

# [34] Interactions of Snake Dendrotoxins with Potassium Channels

*By* William F. Hopkins, Margaret Allen, and Bruce L. Tempel

## Introduction

Peptide toxins isolated from the venom of a number of Elapidae snake species have proven useful as high-affinity probes for the characterization of ligand- and voltage-gated ion channels. A good example of the utility of snake toxins has been the use of the dendrotoxins in the isolation and characterization of voltage-gated potassium channels.[1,2] The dendrotoxins are seven peptide toxins isolated from the venom of three species of African mamba snakes. The published amino acid sequences of five dendrotoxins

[1] H. Rehm and M. Lazdunski, *Proc. Natl. Acad. Sci. U.S.A.* **85,** 4919 (1988).

[2] V. E. Scott, D. N. Parcej, J. N. Keen, J. B. C. Findlay, and J. O. Dolly, *J. Biol. Chem.* **265,** 20094 (1990).





0076-6879/99 $30.00

TABLE I
DENDROTOXIN SUBFAMILIES[a]

| Toxin | Sequence |
| --- | --- |
| | 1 10 20 30 40 50 |
| δ-DTX | AAKYCKLP VR YGPCKKKI PS FYYKWKAK QC LPFDYSGC GG NANRFKTI EE CRRTCVG |
| DTX-K | -------- L- I----R-- -- -------- -- -------- -- -------- -- ------- |
| DTX-DV14 | -------- -- -------- -- -------- -- Y------- -- -------- -- ------- |
| | 1 10 20 30 40 50 |
| α-DTX | ZPRRKLCILH RN PGRCYDKI PA FYYNQKKK QC ERFDWSGC GG NSNRFKTI EE CRRTCIG |
| DTX-I | --L------- -- ------Q- -- -------- -- -G-T---- -- -------- -- ------RK |

[a] Complete β- and γ-DTX sequences have not been published. Dashes indicate amino acid identity with topmost member of the subfamily. The single-letter amino acid code is A, alanine; R, arginine; N, asparagine; D, aspartic acid; C, cysteine; Q, glutamine; E, glutamic acid; G, glycine; H, histidine; I, isoleucine; L, leucine; K, lysine; M, methionine; F, phenylalanine; P, proline; S, serine; T, threonine; W, tryptophan; Y, tyrosine; V, valine.

are shown in Table I. These dendrotoxins can be grouped into two subfamilies on the basis of amino acid sequence homology. This article describes methods used to investigate the interaction of four of these dendrotoxins, α-DTX, γ-DTX, and δ-DTX from *Dendroaspis angusticeps,* and DTX-I from *Dendroaspis polylepis,* with mKv1.1, a mouse member of the *Shaker* subfamily of voltage-gated potassium channel α subunits.[3] As will become apparent, each of the dendrotoxins can display unique properties as high-affinity ligands for this potassium channel subunit, even though two of them have a high degree of amino acid sequence homology (α-DTX and DTX-I).

## Dendrotoxins: General Methodologic Considerations

An obvious first consideration for studies using dendrotoxins is their source. Most of the dendrotoxins are now commercially available with a high degree of purity from a number of reputable vendors. In our early experiments with DTX-I, we obtained crude *D. polylepis* venom from Sigma Chemical Co. (St. Louis, MO) and purified the toxin using a modification of the method of Benishin *et al.*[4] Later we obtained DTX-I and the other dendrotoxins described here from Alomone Labs (Jerusalem, Israel). The results we obtained for DTX-I from the two sources were substantially identical. All of the dendrotoxins have high aqueous solubility and are

[3] B. L Tempel, Y. N. Jan, and L. Y. Jan, *Nature* **332,** 837 (1988).
[4] C. G. Benishin, R. G. Sorensen, W. E. Brown, B. K. Krueger, and M. P. Blaustein, *Mol. Pharmacol.* **34,** 152 (1988).

stable (maintain high potency) for months if stock solutions are kept frozen at −20°. Dendrotoxins in solution are stable for many hours at room temperature, but quite often lose potency if left out overnight. We make a concentrated stock solution using the same solution as would bathe the experimental preparation (frog Ringer's solution for the experiments described here), and add microliter aliquots to experimental solutions shortly before an experiment. With the exception of δ-DTX, the dendrotoxins discussed here are safe and easy to work with. At concentrations greater than 1 n*M*, δ-DTX does not readily wash out of the tubing and chamber that constitute the perfusion system. Fortunately, there is little reason to use concentrations this high, since it is very potent against mKv1.1 in the *Xenopus* oocyte expression system.

Another increasingly popular method for obtaining dendrotoxins is to express them as fusion proteins in bacterial expression systems.[5,6] The advantages of this approach are that in some cases milligram quantities of toxin can be obtained at relatively low cost, and panels of mutant toxins can be obtained using carefully selected oligodeoxynucleotides to construct the synthetic dendrotoxin gene. This approach would be indispensible for those wishing to perform mutagenesis experiments to determine amino acid residues on the toxin critically involved in high-affinity binding to potassium channels or reciprocal mutagenesis experiments involving both the toxin and the channel.[5–8] The disadvantage is that it can take a considerable amount of time and effort to be reasonably certain that the synthetic dendrotoxin has the potency and properties of the native toxin.

Another important consideration is the potassium channel expression system (*Xenopus* oocyte, stably transfected mammalian cells, etc.) in which to study dendrotoxin interactions. The various types of expression systems have been discussed elsewhere.[9,10] A major factor contributing to the discrepancies in the toxin–channel literature is the system in which the channels are expressed. To the extent possible, the expression system should be chosen with the specific questions one wishes to answer in mind.

[5] J. M. Danse, E. G. Rowan, S. Gasparini, F. Ducancel, H. Vatanpour, L. C. Young, G. Poorheidari, E. Lajeunesse, P. Drevet, R. Menez, S. Pinkasfeld, J.-C. Boulain, A. L. Harvey, and A. Menez, *FEBS Lett.* **356,** 153 (1994).

[6] P. Reid, F. Wang, M. Olson, J. Schmidt, D. Parcej, O. Dolly, and L. Smith, *Toxicon* **34,** 289 (1996).

[7] P. Hidalgo and R. MacKinnon, *Science* **268,** 307 (1995).

[8] J. Aiyar, J. M. Withka, J. P. Rizzi, D. H. Singleton, G. C. Andrews, W. Lin, J. Boyd, D. C. Hanson, M. Simon, B. Dethlefs, C.-L. Lee, J. E. Hall, G. A. Gutman, and K. G. Chandy, *Neuron* **15,** 1169 (1995).

[9] E. Sigel, *J. Membr. Biol.* **117,** 201 (1990).

[10] W. Stuhmer and A. B. Parekh, *in* "Single Channel Recording" (B. Sakmann and E. Neher, eds.), 2nd Ed., p. 341. Plenum Press, New York and London, 1995.

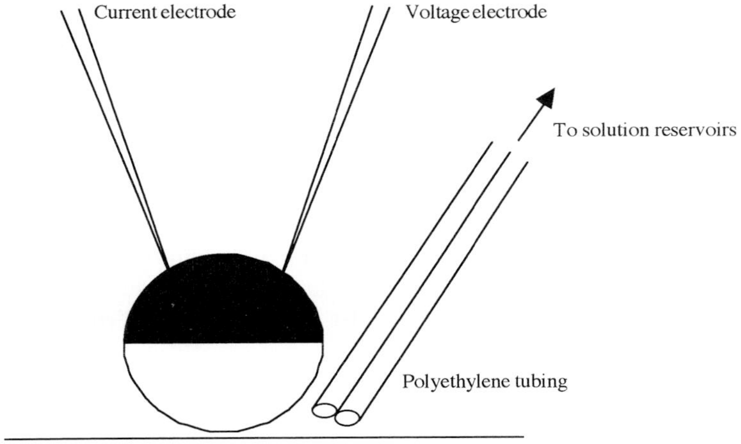


FIG. 1. Schematic diagram showing the whole oocyte rapid perfusion setup. The PE tubing is positioned adjacent to the oocyte near the bottom of the recording chamber. The PE tubing is connected to high-volume reservoirs via larger gauge plastic tubing. The oocyte is concurrently voltage clamped with two microelectrodes.

## Kinetics of Dendrotoxin Block of Potassium Channels

The "on" and "off" kinetics of dendrotoxins on mKv1.1 expressed in *Xenopus* oocytes can be simply and accurately investigated using the experimental arrangement shown schematically in Fig. 1. Two sections of polyethylene (PE) tubing (i.d., 0.86 mm; o.d., 1.27 mm; PE90; Clay Adams, Parsippany, NJ) are glued together and positioned with a micromanipulator adjacent to the underside of an oocyte, which is voltage clamped using two microelectrodes. The PE tubing is connected to two 500-ml reservoirs via larger plastic ("intravenous drip") tubing. The reservoirs are positioned about 1 m above the chamber holding the oocyte, and the flow through each tube is controlled with manual valves, although a solenoid-controlled system could be used as well. This gravity-driven flow system can completely change the composition of the solution bathing the oocyte within 5 sec, as measured by the size and polarity of potassium tail currents in response to changing the potassium ion concentration from 2.5 to 90 m$M$ (Fig. 2). Some trial-and-error adjustments of the positioning of the tubing may be required to achieve this. Although a great deal of fluid turbulence is observed with this technique, it is very rarely sufficient to disrupt the recording. Figure 2 shows that complete solution switching can be achieved repeated from the same oocyte. Although 5-second switching is not rapid by most standards, it is rapid in the context of changing the bathing solution of a

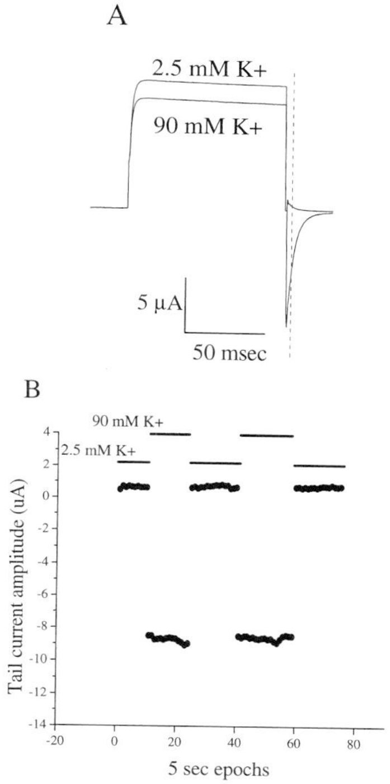


FIG. 2. (A) Potassium currents (mKv1.1 in *Xenopus* oocytes) elicited by voltage pulses to 50 mV from a holding potential of −70 mV in normal Ringer's solution with 2.5 m*M* $K^+$ (top trace with outward tail current) and in high $K^+$-Ringer's solution with 90 m*M* $K^+$ (bottom trace with inward tail current). The dashed line indicates where the tail current measurement was made. (B) Tail currents measured as in (A) at 5-sec intervals during solution switching between 2.5 m*M* (top data points) and 90 m*M* $K^+$ (bottom data points) showing complete switching and equilibration within 5 sec.

whole *Xenopus* oocyte and, more importantly, it is rapid enough to capture the on and off kinetics of dendrotoxin block.

Figure 3 shows the raw data from such an experiment with $\alpha$-DTX. Whole-cell potassium currents (peak current measured at 50 mV from a holding potential of −70 mV) were elicited at 5-sec intervals and $\alpha$-DTX was applied until a steady-state block was achieved. The toxin was then washed out until control amplitude currents were observed. The time course of blocking and unblocking was always described well by single exponential functions from which $\tau_{on}$, the time constant for approach to equilibrium block, and $\tau_{off}$, the time constant for recovery from block, could be obtained. The single exponential on and off kinetics are consistent with a simple

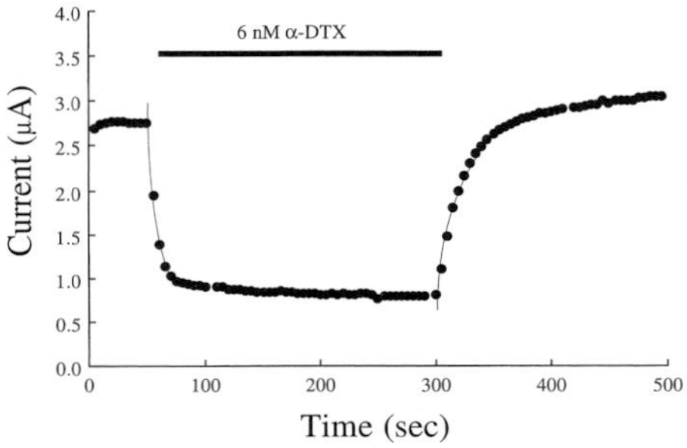


FIG. 3. Peak potassium current amplitude sampled at 5-sec intervals showing the onset and recovery from block by 6 n$M$ $\alpha$-DTX. Exponential curve fits to the onset and recovery from block are also shown from which $\tau_{on}$ and $\tau_{off}$ were derived. The $\tau_{on}$ was 9 sec and the $\tau_{off}$ was 28 sec. Same voltage protocol and procedure as in Fig. 2.

bimolecular reaction scheme between the channel and the toxin. The second-order association rate constant, $k_{on}$, and the first-order dissociation rate constant, $k_{off}$, are related to the time constants by the following equations:

$$\tau_{on} = \frac{1}{k_{on}[DTX] + k_{off}} \tag{1}$$

$$\tau_{off} = \frac{1}{k_{off}} \tag{2}$$

where [DTX] is the concentration of toxin. As a further check, these rate constants are related to the fraction of unblocked channels, $F_u$, by the following equation:

$$F_u = \frac{k_{off}}{k_{on}[DTX] + k_{off}} \tag{3}$$

The rate constants are therefore straightforwardly determined from the kinetic data from these "rapid" perfusion experiments. The equilibrium dissociation constant, $K_d$, is then easily obtained from the rate constants:

$$K_d = \frac{k_{off}}{k_{on}} \tag{4}$$

To determine amino acid residues that comprise the high affinity binding site on mKv1.1 for $\alpha$-DTX and DTX-I, we mutated singly and in combination three residues that are in the putative pore-forming region of mKv1.1 into the corresponding residues in $\alpha$-DTX- and DTX-I-insensitive mKv1.3,

TABLE II
KINETIC DATA FOR DTX-I AND $\alpha$-DTX ON mKv1.1, TWO SINGLE AND ONE DOUBLE AMINO ACID MUTATION IN mKv1.1[a]

| Toxin | $k_{on}$ ($M^{-1}$ s$^{-1}$) | $k_{off}$ (s$^{-1}$) | $K_d$ (n$M$) | $IC_{50}$ (n$M$) |
|---|---|---|---|---|
| DTX-I | | | | |
| mKv1.1 | $1.04 \times 10^7$ | 0.029 | 3 | 3 |
| 1.1-A352P | $1.45 \times 10^7$ | 0.015 | 1 | 2 |
| 1.1-E353S | $1.50 \times 10^6$ | 0.005 | 3 | 5 |
| 1.1-A352P + E353S | $1.80 \times 10^6$ | 0.006 | 3 | 6 |
| $\alpha$-DTX | | | | |
| mKv1.1 | $1.25 \times 10^7$ | 0.025 | 2 | 2 |
| 1.1-A352P | $1.22 \times 10^7$ | 0.027 | 2 | 2 |
| 1.1-E353S | $2.50 \times 10^5$ | 0.012 | 48 | 50 |
| 1.1-A352P + E353S | $2.43 \times 10^5$ | 0.013 | 53 | 55 |

[a] $k_{on}$ and $k_{off}$ calculated according to Eqs. (1–3). $K_d$ calculated according to Eq. (4). $IC_{50}$ calculated from best fitting drug binding isotherm to concentration–inhibition data under "steady-state" conditions.

another mouse *Shaker*-like potassium channel $\alpha$ subunit.[11,12] A similar strategy was followed by Hurst *et al.*[13] Three of these mutant subunits expressed well enough in *Xenopus* oocytes to obtain acceptable kinetic data for both $\alpha$-DTX and DTX-I. The results are given in Table II. The $IC_{50}$ values given in Table II were obtained independently by sampling potassium currents (peak current at 50 mV) in control Ringer's solution and in the presence of increasing concentrations of the toxin to obtain a concentration–inhibition curve. A drug-binding isotherm was then fit to these data to obtain an $IC_{50}$. Several features of these data are worthy of note. First, the correspondence between the $K_d$ values obtained from the kinetic data and the $IC_{50}$ values is quite good. Second, the pattern of the data is different for the two toxins. For both toxins, the A352P mutation (alanine to proline) had little effect on the toxin affinity or kinetics. However, the results were strikingly different for the E353S mutation. For DTX-I, both the on and off rate constants were somewhat smaller than the wild-type channel, but the $K_d$ was unchanged. For $\alpha$-DTX, there was more than 20-fold increase in the $K_d$ with this mutation that was accounted for by an almost 100-fold decrease in the association rate constant. For both $\alpha$-DTX and DTX-I, the effect of the double mutation (A352P + E353S) was very similar to the effect of the

[11] K. G. Chandy, C. B. Williams, R. H. Spencer, B. A. Aguilar, S. Ghanshani, B. L Tempel, and G. A. Gutman, *Science* **247,** 973 (1990).

[12] W. F. Hopkins, M. L. Allen, K. M. Houamed, and B. L Tempel, *Pflugers Arch.* **328,** 382 (1994).

[13] R. S. Hurst, A. E. Busch, M. P. Kavanaugh, P. B. Osborne, R. A. North, and J. P. Adelman, *Mol. Pharmacol.* **40,** 572 (1991).

TABLE III
EFFECT OF E353S MUTATION ON BLOCKING AFFINITY OF FOUR DENDROTOXINS FOR mKv1.1

| Toxin | Test concentration (n*M*) | Unblocked fraction of current | | Increase in $IC_{50}$ |
|---|---|---|---|---|
| | | mKv1.1 | 1.1-E353S | |
| γ-DTX | 1 | 0.31 ± 0.06 | 0.99 ± 0.0 | Over 100-fold |
| α-DTX | 10 | 0.21 ± 0.04 | 0.92 ± 0.03 | Over 10-fold |
| DTX-I | 10 | 0.34 ± 0.02 | 0.31 ± 0.05 | None |
| δ-DTX | 1 | 0.18 ± 0.08 | 0.20 ± 0.04 | None |

E353S mutation alone. It is clear that the glutamate to serine mutation had a large effect on the affinity of α-DTX for mKv1.1, while for DTX-I this mutation had little or no effect. Therefore, these mutations account for none of the high-affinity binding of DTX-I for mKv1.1 whereas they account for some but not all of the high-affinity binding of α-DTX to mKv1.1. The tyrosine residue at position 379 is another candidate residue, but the Y379H mutant alone did not express functional channels. Surprisingly, a triple mutant (A352P + E353S + Y379H) did express to a small degree, and these channels had low affinities for α-DTX and DTX-I.[12,13] Unfortunately, we were unable to obtain acceptable kinetic data from these channels.

The data in the previous paragraph suggest that the glutamate residue at position 353 of mKv1.1 is important for high-affinity binding of α-DTX, but not DTX-I. We partially extended this approach to two other dendrotoxins, γ-DTX and δ-DTX, and the results are shown in Table III. The most striking effect of this mutation was on the blocking affinity of γ-DTX, where the channels have been rendered virtually insensitive to the toxin. At the other extreme is δ-DTX, whose blocking affinity was not affected by this mutation. It is therefore clear that the dendrotoxins studied here are not a homogeneous family of toxins with respect to their high-affinity binding interactions with mKv1.1.

## Use of Low Ionic Strength Solutions to Investigate Role of Electrostatic Interactions in Dendrotoxin Blocking Affinity

The results in the previous section, in which the E353S mutation affected the blocking affinity of α-DTX and γ-DTX but not DTX-I or δ-DTX, suggest the hypothesis that electrostatic interactions between some of the polycationic dendrotoxins and negatively charged residues on the external surface of the channel could contribute to high-affinity binding. This hypothesis could be tested explicitly by converting the glutamate (negative charge) residue at position 353 instead to glutamine (neutral charge) or perhaps lysine (positive charge). This approach could be extended to other negatively charged residues in the pore-forming region, but in the case of mKv1.1, the position 353 glutamate is the most obvious candidate based on available data. Before doing mutagenesis experiments, a simple way to further justify doing them is to investigate the effect of ionic strength on toxin blocking affinity.[14] The rationale for this is that if through-space electrostatic interactions contribute to the binding affinity of a ligand for a receptor, the binding affinity should be sensitive to the charge-screening ability of cations and anions in the aqueous solution in which the binding takes place. The binding free energy of the interaction should increase if the ionic strength of the solution is reduced because there are fewer ions in the solution to screen through-space interactions based on charge.

These experiments can be quite easily done using two-electrode voltage clamp in *Xenopus* oocytes. The standard Ringer's solution that we used consisted of (in m*M*): NaCl 115, KCl 2.5, $CaCl_2$ 1.8, and HEPES 10 (pH 7.2). In the low ionic strength experiments, NaCl was reduce to 40 m*M* with the remainder replaced by 75 m*M* sucrose to maintain osmolarity, with all other constituents remaining the same. If one wishes to switch between these two solutions while recording from the same oocyte, care must be taken to avoid, or at least be aware of changes in junction potentials that occur whenever there are changes in the chloride ion concentration.[15] We prefer to do the low ionic strength experiments separately from the experiments in normal Ringer's solution. Figure 4 shows typical results from such experiments for $\alpha$-DTX and $\delta$-DTX. In normal Ringer's solution, 2 n*M* $\alpha$-DTX blocked about half of the potassium current elicited by a voltage pulse to 50 mV. In contrast, the same concentration of $\alpha$-DTX blocked nearly all of the current in the low ionic strength solution using the same voltage clamp protocol, and this result was observed in two other oocytes. The results with $\delta$-DTX were strikingly different. In normal Ringer's solution, 10 p*M* $\delta$-DTX blocked about 20% of the current and this degree of block was also observed in the low ionic strength solution in a different oocyte. Similar results were observed in five other oocytes in low ionic strength saline. For $\alpha$-DTX, the $IC_{50}$ in normal saline was found to be 2 n*M*, while in low ionic strength saline, it was estimated to be 200 p*M*. In contrast, the $IC_{50}$ for $\delta$-DTX in normal Ringer's solution was 29 p*M* and was 42 p*M* in low ionic strength saline. We can conclude from these experiments that electrostatic interactions play little or no role

[14] R. MacKinnon and C. Miller, *Science* **245,** 1382 (1989).

[15] E. Neher, *Methods Enzymol.* **207,** 123 (1992).

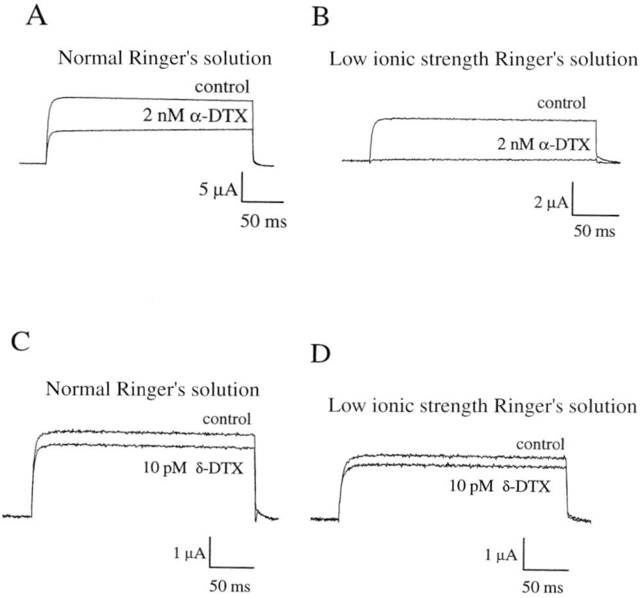


FIG. 4. Effects of low ionic strength solutions on degree of block by $\alpha$-DTX and $\delta$-DTX. *Top*: Traces shows block by 2 n*M* $\alpha$-DTX in normal Ringer's solution (A) and in low ionic strength solution (B). *Bottom*: Traces displays block by 10 p*M* $\delta$-DTX in normal Ringer's solution (C) and in low ionic strength solution (D). Data are representative of at least three oocytes for each experimental condition.

in the blocking affinity of $\delta$-DTX for mKv1.1 but play a considerable role for $\alpha$-DTX.[16] These results are consistent with the data presented in Tables II and III and lend credence to the hypothesis that the binding affinity of $\alpha$-DTX for mKv1.1 mediated by the glutamate residue at position 353 is electrostatic in nature. Furthermore, the data suggest that this residue is not importantly involved in the binding of $\delta$-DTX to mKv1.1 nor is any other charged residue. These hypotheses could be more conclusively tested by a combination of mutagenesis and changing ionic strength of the bath solution.[14,17] However, it is again clear that the dendrotoxins are not a homogeneous family of toxins with respect to the biophysical nature of their interactions with the mKv1.1 potassium channel subunit.

[16] J. Tytgat, T. Debont, E. Carmeliet, and P. Daenens, *J. Biol. Chem.* **270,** 24776 (1995).
[17] M. Stocker and C. Miller, *Proc. Natl. Acad. Sci. U.S.A.* **91,** 9509 (1994).

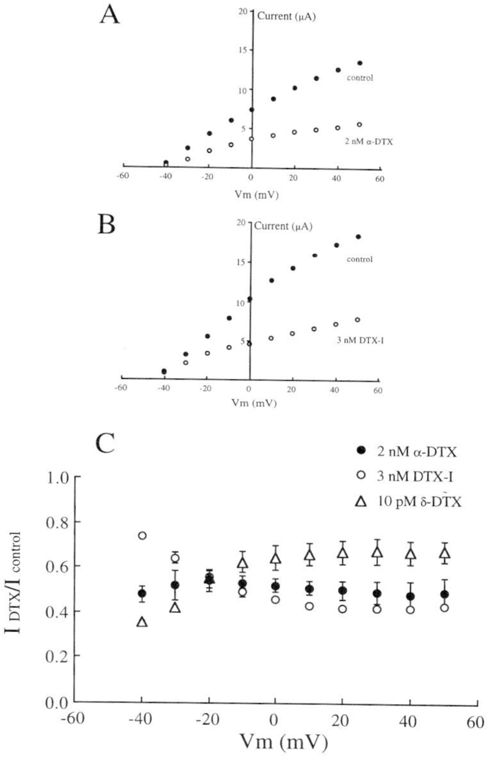


FIG. 5. (A) Potassium current–voltage relations sampled in normal saline and then in the presence of 2 n*M* α-DTX (B) Same for another oocyte with 3 n*M* DTX-I. (C) Fraction of unblocked potassium current as a function of voltage for 2 n*M* α-DTX, 3 n*M* DTX-I, and 10 p*M* δ-DTX. Data points represent the means for three oocytes for each denodrotoxin. Error bars are standard errors of the mean and are smaller than the symbols where they do not appear.

## Voltage Dependence of Dendrotoxin Block

Several studies have concluded that dendrotoxin inhibition of potassium channels is not voltage dependent.[16,18] We have investigated this issue for three of the dendrotoxins and found that for only one of them, α-DTX, is this conclusion valid. Figure 5A and 5B show current–voltage relations for mKv1.1 potassium currents in control Ringer's solution and then in the presence of 2 n*M* α-DTX and 3 n*M* DTX-I, respectively. Figure 5C displays the fraction of unblocked current in the presence of α-DTX, DTX-I, and

[18] D. G. Owen, A. Hall, G. Stephens, J. Stow, and B. Robertson, *Br. J. Pharmacol.* **120,** 1029 (1997).

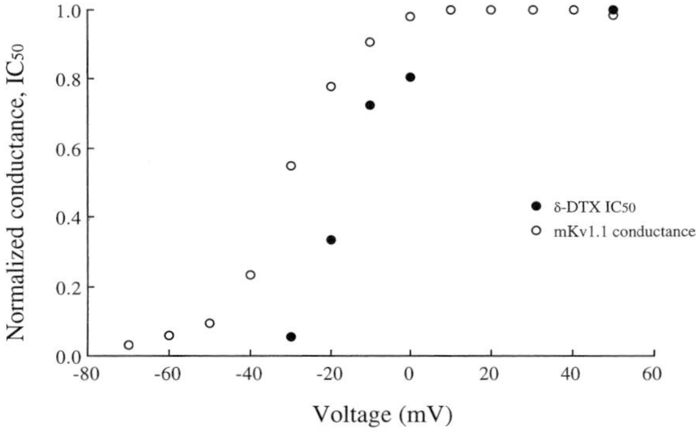


FIG. 6. Normalized mKv1.1 potassium conductance and normalized δ-DTX $IC_{50}$ values plotted as a function of voltage. The conductance–voltage data represent the mean of nine oocytes and the error bars are smaller than the symbols in each case. The $IC_{50}$ values were derived by fitting drug-binding isotherms to concentration–inhibition data for δ-DTX at five different voltages. Ten oocytes were used in this analysis.

δ-DTX as a function of voltage. Each of the dendrotoxins displayed a unique profile, with only α-DTX showing little or no voltage dependence of block. The most striking voltage dependences for the other two dendrotoxins were observed at membrane potentials of less than 0 mV, with DTX-I showing greater block at more positive potentials and δ-DTX showing less. Figure 6 is a normalized plot of mKv1.1 conductance data and the normalized δ-DTX $IC_{50}$ both as functions of voltage. There is a striking similarity in the slopes of the two curves, with both appearing to saturate at voltages positive to 0 mV. These data are consistent with the hypothesis that potassium ion flux through the channel destabilizes the binding interaction of δ-DTX with the channel and that the apparent voltage dependence of δ-DTX block is actually due to the voltage dependence of potassium channel conductance. The DTX-I data cannot be so explained, since the block increased as voltage was made more positive. It is clear, however, that both DTX-I and δ-DTX displayed at least apparent voltage-dependent block of mKv1.1. Failure to observe voltage-dependent block for dendrotoxins other than α-DTX in previous studies may have been due to restricting the voltage range over which the analysis was conducted.

## Conclusions and Future Prospects

Compared to scorpion toxins, the dendrotoxins have received relatively little attention and use as probes for the study of potassium channels. This is probably because of their larger size compared to the scorpion toxins, and also because the most highly studied potassium channel subunit, *Shaker* B, is insensitive to dendrotoxins. This situation seems to be changing as more groups are turning to dendrotoxins to potentially gain greater insights into the structure and function of an apparently small group of mammalian potassium channel $\alpha$ subunits.[19,20] This endeavor will be greatly facilitated by high-resolution structural information that has become available for the dendrotoxins,[6,21] and the ability to synthesize recombinant dendrotoxins that will enable mutagenesis to be performed on the toxins.[5,6]

The present work has endeavored to present some relatively simple methods for the study of dendrotoxins and their interactions with potassium channels. These techniques become more powerful when they are combined with other approaches. We have alluded to some of these, but there are numerous others that can be combined with a modicum of creativity. A theme that recurs repeatedly throughout this work is, by any measure, the marked heterogeneity of the dendrotoxins in spite of their high degree of amino acid sequence homology in some cases. "Dendrotoxin" is commonly discussed in the literature as if it were a single entity. We hope that the data presented here dispel that misconception and, more importantly, stimulate a great deal more work using dendrotoxins as probes for potassium channel structure and function.

[19] W. F. Hopkins, J. L. Miller, and G. P. Miljanich, *Curr. Pharmaceut. Des.* **2,** 389 (1996).
[20] G. E. Kirsch, J. A. Drewe, S. Verma, A. M. Brown, and R. H. Joho, *FEBS Lett.* **278,** 55 (1991).
[21] T. Skarzynski, *J. Mol. Biol.* **224,** 671 (1992).

# [35] Use of Planar Lipid Bilayer Membranes for Rapid Screening of Membrane Active Compounds

*By* TAJIB A. MIRZABEKOV, ANATOLY Y. SILBERSTEIN, and BRUCE L. KAGAN

## Introduction

Planar lipid bilayer membranes (BLMs) have been in use for more than three decades for the study of membrane active compounds.[1] BLMs provide a unique environment that allows the assay of the functional activity of

[1] P. Mueller, D. O. Rudin, H. T. Tien, and W. C. Wescott, *Nature (Lond.)* **194,** 979 (1962).

carriers and channels translocating ions across membrane.[2,3] Although BLMs are *in vitro* systems, which have been used extensively for the measurement of biophysical properties of carriers and channels, their relevance to *in vivo* phenomena has been demonstrated repeatedly through the more recent technology of patch clamping[4] which has shown that channel properties measured in BLMs correspond qualitatively and often quantitatively to those observed in whole cells. In this review we describe a new BLM perfusion chamber that allows rapid, reversible, and quantitative assay of compounds on the bilayers.

One of the major advantages of the *in vitro* bilayer system is its remarkable sensitivity in that the ionic current flowing through a single channel can be readily observed. Furthermore, the opening and closing of this single ionic channel can be easily detected and modification of the channel properties by voltage, pH, ionic composition, blockers, mutation, and chemical reagents can also be quantified. This sensitivity also confers specificity on the assay, because identifiable channel properties, such as single-channel conductance, channel lifetime, voltage dependence, and ionic selectivity, can serve as a "fingerprint" to identify uniquely a given channel.

A second major advantage of BLMs is their freedom from many potential sources of confounding error. The system has a well-defined set of components including water, salts, buffers, lipids, and organic solvent, making it feasible to observe the effects of the introduction of new proteins and peptides into the BLM. In the absence of added membrane active material, the system is quite predictably dull, showing little permeability to ions, no voltage dependence, and no interesting transport behavior. However, the introduction of peptide or protein channel-forming agents that spontaneously incorporate into the bilayer dramatically alters the situation. The system is readily manipulated to study the transport properties of these agents. Furthermore, a wide variety of ionic compositions, lipids, temperatures, and pH conditions can be employed to study the biophysical properties of the channels in question over a range not usually possible in physiological situations. This allows important biophysical questions to be addressed in addition to the more traditional physiologic questions.

The stability of BLMs is also an important asset in their utility. Early on, BLMs were shown to be capable of lasting as long as 120 days, the typical lifetime of a red blood cell, and it is common for BLMs to have long lifetimes even in the presence of added channels. However, it is clear

[2] C. Miller, "Ion Channel Reconstitution." Plenum Press, New York, 1986.
[3] B. L. Kagan and Yu. Sokolov, *Methods Enzymol.* **235,** 699 (1994).
[4] B. Sakmann and E. Neher, "Single Channel Recording." Plenum Press, New York, 1983.

that some membrane active compounds do weaken BLM stability and can eventually lead to BLM destruction. Indeed, it is likely that for many membrane active toxic compounds their toxic mechanism involves the disruption of membrane integrity.[5-7]

Another attractive feature of BLMs is their simplicity. The number of defined parameters is small. The electrical properties are predictable and well defined, and interpretation of data is often straightforward.

BLMs do have some well-known shortcomings. The assay for channel activity in BLMs is often poorly reproducible due to a variety of factors including geometric shape of the chamber, conformation of the BLM, accessibility of the BLM, presence of a variable unstirred layer directly in front of the BLM, variability of BLM surface area, and the presence of "microlenses" of organic solvent. These factors combine to make it difficult sometimes to observe reproducibly the same quantitative channel activity for a given concentration of protein or peptide in the solution. Even with the most efficient channel formers, only a small fraction of the added protein incorporates into the membrane (approximately 1 in $10^5$ molecules).[8,9] This makes BLMs difficult to use as an assay for ion channel activity and has hindered their use in the purification of channels and of putative channel activities from various sources.

The poor stability of BLMS in response to hydrostatic pressure gradients is also a limiting factor. This can limit the ability to add or subtract volume from the solution, and can also hinder the ability to change the solution bathing the membrane without breaking the membrane. Furthermore, it can slow the speed at which solutions can be exchanged in the aqueous compartments. The slowness of solution changes often required can limit the ability to look at rapid transitions in the functional status of channels.

Noise is frequently a problem in BLM recording. The issues of electrical, vibrational, mechanical, and other noise have been discussed in detail elsewhere.[1,10] Noise reduction is frequently an important goal of chamber and membrane design, so that smaller currents can be measured on faster time scales.

The requirement for organic solvent in the BLM may alter membrane and channel properties from the *in vivo* situation. "Solvent-free" BLMs

[5] B. L. Kagan, M. E. Selsted, T. Ganz, and R. I. Lehrer, *Proc. Natl. Acad. Sci. U.S.A.* **87,** 210 (1990).

[6] T. A. Mirzabekov, M-C. Lin, and B. L. Kagan, *J. Biol. Chem.* **271,** 1988 (1996).

[7] M-C. Lin, T. A. Mirzabekov, and B. L. Kagan, *J. Biol. Chem.* **272,** 44 (1997).

[8] S. J. Schein, B. L. Kagan, and A. Finkelstein, *Nature* (*Lond.*) **276,** 159 (1978).

[9] B. L. Kagan, M. Colombini, and A. Finkelstein, *Proc. Natl. Acad. Sci. U.S.A.* **78,** 4950 (1981).

[10] O. Alvarez, *in* "Ion Channel Reconstitution" (C. Miller, ed.), pp. 115–130. Plenum Press, New York, 1986.

dramatically reduce the amount of solvent present, but are significantly less stable.[11]

## A New Bilayer Membrane System

In this article we present a new BLM system included chamber, electrodes, and membrane supports, for rapid screening of new compounds for membrane activity. The system has several advantages over older designs: (1) Solutions can be exchanged rapidly and completely. (2) Membranes are stable to these solution changes. (3) The volumes employed are quite small, which allows sample size to be in the microliter range. Thus, repeated measurements can be made on material that is in very short supply. (4) The use of solution changes allows the reversibility of sample effects to be tested handily. (5) Because the system allows complete volume exchange, multiple samples can be tested for activity on a single membrane. Of course, once activity is obtained, a new membrane must be formed, but inactive samples can be simply washed out and the next sample tested on the same membrane. This system also allows positive controls with known channel formers to be used to calibrate and quantify activity. (6) Active compounds may be applied to the membrane in high concentrations (without dilution).

The chamber is also extremely easy to use. It comes apart quite rapidly, and is made from inexpensive materials. It is easily and efficiently cleaned without the use of volatile organic solvents. The system is also remarkably quiet due to the fact that the membrane itself has very little solution surface area, and thus is largely insulated from a great deal of electrical and vibrational noise that normally interferes with recording. The electrodes are made from common materials and easily customized to the requirements of the chamber. Virtually any type of standard electronic recording devices such as voltage clamps, current clamps, patch clamp amplifiers, etc., can be used with this system. Examples of studies in which this new system can be used include (1) screening fractions of protein preparations for channel activity, (2) screening for channel blocks, (3) screening peptides for channel activity, and (4) screening pharmaceuticals.

## Planar Lipid Membrane Setups and Chambers

### *Perfusion Setup for Painted Membranes*

The formation of painted, or solvent-containing, bilayer membranes in a rapid perfusion chamber has been described in limited detail.[12,13] The

[11] M. Montal and P. Mueller, *Proc. Natl. Acad. Sci. U.S.A.* **69,** 3561 (1972).

[12] T. A. Mirzabekov, C. Ballarin, P. Zatta, M. Nicolini, and C. M. Sorgato, *J. Membr. Biol.* **33,** 129 (1993).

[13] A. Ya. Silberstein, *Biol. Membr.* **6,** 1317 (1989).

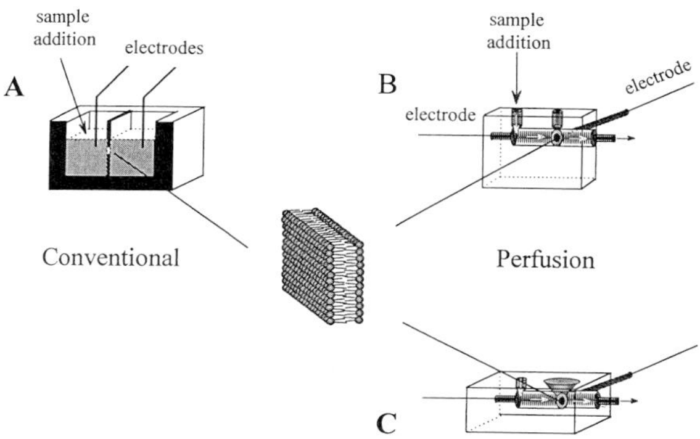


FIG. 1. Chambers for the conventional and perfusion planar membrane setups. (A) Traditional planar membrane chamber. Membrane here is made on an aperture in the Teflon film separating two 0.3- to 2-ml volumes, *cis* and *trans* sides of the chamber. Samples of compounds are added as stocks and have to be concentrated because of the increase in volume of the *cis* or *trans* side of chamber where the compound is added. Increases in the volume due to addition of a large volume of diluted compound can result in a hydrostatic pressure gradient on the membrane leading to membrane rupture. Magnetic stirrers are needed on both sides of the chamber for mixing of added compounds. Because of unstirred layers bordering the membrane, added compounds reach the membrane only after a few (3–5) minutes. Each new compound must be added to a cleaned chamber with a newly formed membrane. (B) Perfusion chamber for "painted" membranes. Membranes were made from heptane solutions of lipids by the use of a pipette inserted into the right vertical hole to the level of the hole in the Teflon tubing (black circle). Samples were premixed with a salt solution at the necessary concentration and dropped in the left vertical hole. Solution dropped out was collected in a reservoir. Membranes that had a slight concave shape were immediately accessible for the additives, and if the added solution contained membrane active compounds they reacted with the membrane immediately (in 1–3 sec). New compounds were added until a membrane active compound was found. Then compound-free solution was added, the membrane was reformed, and screening was continued. (C) Perfusion planar membrane system for the solvent-free membrane or membranes made from native membrane vesicles. The lipid dissolved in hexane is gently spread to the surface of the right side conical hole. Membrane is made 20–30 min later after evaporation of hexane. In the case where membranes were made from vesicles or liposomes, 20–30 $\mu$l of them were added to the same hole and incubated for 30 min for formation of lipid (or lipid–protein) monolayer on the surface of the solution. The membrane was made by gentle lowering and raising of the salt solution surface level across the Teflon tubing hole.

BLM setup used was specifically designed to allow perfusion of one chamber in a few seconds. The *cis* chamber of 30- to 50-$\mu$l total volume was a 2-mm-diameter tube bored longitudinally into a rectangular Lucite block (10 $\times$ 25 $\times$ 30 mm). The *trans* chamber was a 2-mm-diameter cylinder bored perpendicularly to the *cis* chamber (Fig. 1). The *trans* chamber was fitted with a 15-mm-long Teflon tube (with an inner diameter of 0.25, 0.5, or 1 mm; outer diameter, 2 mm) on which the membrane was formed. A flexible plastic (Tygon) tube was fitted over the Teflon tube to connect the electrode. A plastic threaded sleeve fitted to the Teflon tube and the hole allowed easy fixation of the Teflon tube to the chamber. On insertion of the *trans* electrode, the *trans* side became a closed volume, decreasing noise and increasing stability to hydrostatic pressure because the water solution is practically noncompressible. The perfusing medium was dropped into the *cis* chamber at one end and flowed out by gravity from the other end. The *cis* chamber was also accessed vertically by three "wells," the center one directly above the *trans* chamber. These wells allow variable points for insertion of sample and escape for air bubbles in the system. The bilayer lipid membrane was formed on an air bubble of a 13–15 mg/ml solution of lipids in *n*-heptane (or *n*-decane) at the end of Teflon tubing with a 0.25-, 0.5-, or 1-mm diameter using 10-$\mu$l small pipette tips, the end of which was cut at an angle of 45°. The tip was dipped in a lipid solution, shaken, and then dipped into the solution of the *cis* chamber. The hemisphere of a bubble was pressed out of the end of the tip facing the Teflon tubing. The bubble hemisphere was gently contacted to the Teflon tubing and simultaneously the bubble was gently sucked back by the release of finger pressure on a tube connected to the pipette tip.

Use of *n*-decane as a solvent resulted in formation of membranes that have been suggested to have a smaller percentage of residual solvent in the membrane and less membrane thickness, closer to the thickness of a native membrane. Heptane offers the advantage of more rapid thinning of the membrane to a bilayer. The construction of the chamber allowed substitution of the solution in one (*cis*) compartment within several seconds.[13] After formation, the brightly colored membrane becomes optically black within 20–30 sec. Turning "black" is evidence that the BLM's thickness has become much smaller than the wavelength of visible light. For visual observation of the color change of the membrane we used a stereo microscope in which one eyepiece was replaced by the light source. The light reflected from the membrane was visually detected with the other eyepiece of the microscope. The solution in the *cis* side was replaced with other solutions usually of the same salt composition but containing compounds to be screened for activity. After the initial incorporation of membrane active ingredients, the newly added solution could be washed out and substituted by the original one. Membranes were typically stable for periods >60 min and had conductances of less than 10 pS up to voltages of ±150 mV. Stability to voltage and membrane longevity depended to some extent on the specific lipids and solutions used.

### *Perfusion Chamber for Solvent-Free Membranes and Membranes Made of Native Membrane Vesicles*

We also developed a perfusion planar membrane system for the formation of "solvent-free" membranes,[11] and membranes made from liposomes or native membrane vesicles.[14] The construction of this chamber is similar to the perfusion chamber for the painted (solvent-containing) planar lipid membranes. Like the solvent-containing membranes, the membrane was made on the end of Teflon tubing. The *trans* chamber became a closed volume after formation of the membrane. This kept the membrane stable and prevented it from breaking under the fast changes of hydrostatic pressure (up to 10–15 mm) on the membrane during salt solution perfusion on the *cis* side. (Note that "solvent-free" membranes formed from monolayers are not usually very sensitive to hydrostatic pressure, whereas solvent-containing BLMs are typically very sensitive.)

The chamber itself was made of Teflon and had a size of $20 \times 20 \times 25$ mm. The 2-mm hole was made through the long axis at 15 mm from the bottom and 12 mm from the front surface (Fig. 1C). Two holes with diameters of 2 and 4 mm were made from the surface down, perpendicular to the horizontal 2-mm-diameter hole (*cis* chamber). These two holes were made at a distance of 7 mm from the right and left edges of the Teflon chamber top and served to allow addition of samples to the chamber and for the formation of membranes, respectively. The planar lipid membrane was formed from the lipid monolayer covering the 4-mm-diameter hole. For formation of the monolayer, 10 $\mu$l of 10 mg/ml lipid in hexane was applied to the surface of the salt solution in the 4-mm-hole, which was kept at approximately 2 mm above the end of the hexadecane-coated end of the Teflon tubing. Instead of hexadecane, squalene can also be used. This reduces membrane stability unless divalent cations are added at millimolar concentrations. For precoating with hexadecane, the end of the Teflon tubing was first dipped in a 20 mg/ml solution of hexadecane in pentane and air dried for 10 min to allow full evaporation of pentane. The Teflon tubing was exposed to the 4-mm hole. For membrane formation the sale solution surface near the Teflon tubing was lowered below the end of the Teflon tubing (for formation of one lipid monolayer) and again raised above it (second monolayer). After formation of the membrane, the salt solution level could be varied up to 5–10 mm above the membrane during salt solution perfusion, and the membrane remained stable. During manipulation with the salt solution levels, the hole in the Teflon tubing where the membrane was made was kept under visual observation. A low-magnification (10–12×) stereoscopic microscope with a long focus distance (about 100 mm) containing the light source in place of one eyepiece was used as before.

A rectangular oscillating voltage (1 Hz, 10 or 20 mV) was applied between the electrodes and membrane formation was monitored using capacitance measurements. Membrane capitance was estimated using the

[14] H. Schindler, *FEBS Lett.* **122(1),** 77 (1980).

calibration on standard 100- to 1000-pF capacitors. The amplitudes of charging–recharging currents on capacitors were measured at a standard oscillating voltage.

It is clear that when solvent-free membranes were made, the diameter of the membrane was practically equal to the inner diameter of the Teflon tubing. Two kinds of tubing were used in experiments, and they had an inner diameter of 100 and 250 $\mu$m. Tubing was purchased from Pharmacia-Upjohn, which sells Teflon tubing as kits for HPLC and FPLC chromatography.

## Materials

### *Lipids*

Lipids are purchased from Avanti Polar Lipids, Inc. (Birmingham, AL). The purity of lipids used for planar lipid membrane experiments is essential for the formation of stable membranes with low baseline conductance, <10 pS. All lipids are stored at $-20°$ under $N_2$. Most lipids are stored as 10–20 mg/ml solutions in chloroform. A few [asolectin, diphythanoyl-phosphatidylcholine (DPPC)] are stored as lyophilized powders. For membrane experiments, 1–2 mg of lipid or lipid mixture is dried under a stream of $N_2$ and dissolved in *n*-heptane at a final concentration of 13–15 mg/ml. In the case of membranes made from lipid mixtures, lipids dissolved in chloroform are mixed first and then dried and dissolved in heptane. It has been previously shown that the lipid composition of actual BLMs corresponds well to the composition of the lipid mixture used for formation of the bilayer.[15] Pure DPPC membranes are very stable and usually are not very sensitive to added channel formers. We use DPPC membranes for the study of channel-forming compounds with high membrane activity, such as porins or gramicidin. Membranes made from asolectin (soybean phosphatide extract, granulated, 45% (w/w) phosphocholine content), a phospholipid mixture containing a high percentage of unsaturated fatty acid hydrocarbon chains, are very sensitive to added compounds, and are used often in the screening experiments. The percentage of lipids with a net negative charge in the membrane as well as the percentage of the sterol (cholesterol) can be varied between 0 and 50%.

### *Electrodes*

Ag | AgCl electrodes with agar bridges are used for current measurements on BLMs. The total volumes of both chambers with salt solution

[15] R. U. Muller and A. Finkelstein, *J. Gen. Physiol.* **60,** 285 (1972).

bathing the membrane are very small, around 30–50 μl for the *cis* chamber and 10–20 μl for the *trans* chamber. Therefore, it is necessary to use electrodes that have (1) small volumes which are comparable with the volumes of the chambers and (2) agar prepared in salt solutions with the same ionic strength as the solution used in the membrane experiments.

After membrane formation the *trans* side chamber is closed. Therefore, if electrodes have a volume much bigger than the *trans* side, it could lead to two unfavorable results: any small temperature increase could break the membrane due to the thermal expansion of the *trans* side chamber salt solution volume with subsequent "pressing out" of the membrane. Under these conditions it would be difficult to form stable, long-lasting membranes. On the other hand a small temperature decrease could lead to compression of the *trans* side volume with resulting "sucking in" of the bilayer membrane from the end of Teflon tubing inside of it. As a result the bilayer membrane could be less accessible to added compounds.

The electrodes we use are simple to prepare and do not suffer from the shortcoming mentioned above. Electrodes are made from the lower part of 200-μl micropipette tips filled by 2% agar prepared in the salt solutions used in experiments (usually 100 m*M* or 1 *M* KCl). Silver wire (1.5 cm in length) having a 1-mm diameter with one-half electrolytically covered by silver chloride is used to make an electrode. The chloride end is covered by 5–10 thin layers of Parafilm leaving 3–4 mm uncovered at the end of the electrode. The wider end of a 200-μl plastic pipette tip is cut to 15–20 mm of length, filled with hot agar, and the silver | silver chloride electrode pressed into it. The end of the tip on the side of the wire is covered by a couple of Parafilm layers, which prevent the drying of the agar. Electrodes are stored in the same salt solutions as those used for the experiments. One electrode, connected to the amplifier headstage, is inserted into a 10-mm soft silicon tube sleeve over the membrane supporting Teflon tube. A second electrode is inserted into the left end of the *cis* chamber. Electrode asymmetry is always less than 1 mV.

### *Chamber Cleaning*

The perfusion chamber is rinsed in a stream of distilled water for 10–15 sec, then the Teflon tubing is disconnected, shaken a few times in a chloroform : methanol (2 : 1 v/v) solution, and dried. For experiments we filter salt solutions through detergent-free antibacterial filters (e.g., Millipore, Bedford, MA, 0.22 mm) although we found this step not to be essential.

### *Recording Equipment*

Virtually any standard voltage-clamp recording equipment will work with this chamber. Membrane formation is assessed by monitoring of membrane capacitance and resistance. Data are digitized and stored on VHS tape and played back for later analysis. An Axopatch 1C amplifier (Axon Instruments, Sunnyvale, CA) with headstage CV-3B is used for measuring membrane current. For data acquisition, a digital tape recorder and video cassette recorder allow recording of large amounts of data. A storage oscilloscope is used for monitoring membrane capacitance and single-channel recordings.

## Applications for Perfusion Planar Lipid Membrane Technique

### *Single-Channel Reconstitution*

BLMs have been intensively used in the identification and reconstitution of channel-forming proteins.[2] The perfusion planar membrane system has several advantages over the traditional membrane system. It is approximately 20–100 times faster. For example, with the new chamber it takes *Borrelia* porin about 5–20 sec to reach a steady-state conductance (hundreds of channels). At the same concentration with a standard chamber, we have to wait for 10–30 min to reach a steady-state conductance. The perfusion planar membrane setup requires sample volumes 10–100 times smaller than typical setups. A solution with a membrane active compound immediately (within 1–3 sec) reaches the planar membrane. In Fig. 2, the *cis* solution was substituted by solution containing 10 mg/ml of porin isolated from *Borrelia burgdorferi.* The resulting channel formation stops immediately after washout out of porin-containing solution. This procedure could be repeated many times. This property of the setup allows us to easily reconstitute and study the properties of a single channel in the membrane. Long-lasting single channels of the mitochondrial outer membrane channel, VDAC (voltage-dependent anion channel, also called mitochondrial porin), were easily reconstituted in BLMs made in the perfusion chamber and the transport properties of the channel were studied.[12] The new chamber also allows screening of compounds (such as blockers or other modifiers) on reconstituted ion channels as well as studying the reversibility of the interaction of these compounds. By addition of salt solutions at different temperatures or pH, these parameters can also be changed very rapidly. In our experiments we found that planar lipid membranes can be formed and remain stable at temperatures between 4° and 50° and in a pH range between pH 3 and 10.

### *Interactions of Amyloidogenic Peptides with Membranes*

Deposition of proteins called amyloid proteins is observed in a wide variety of animal and human diseases. Amyloid proteins share certain

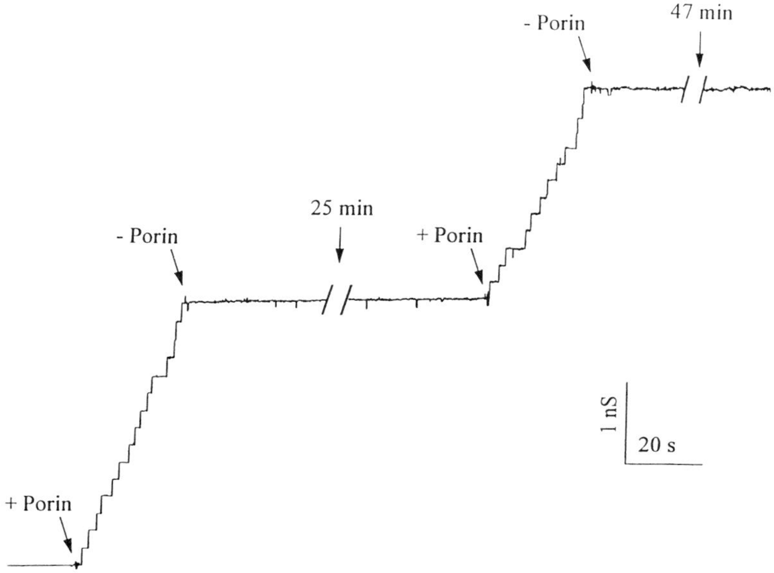


FIG. 2. Rapid initiation and termination of *Borrelia* porin channel activity with the new perfusion chamber. The record shows current as a function of time while the membrane was held at constant voltage (+10 mV). At the first arrow (+porin), solution containing porin was introduced into the chamber. The conductance begins to increase instantaneously and insertion of single porin channels can be seen. At the second arrow (−porin), porin-free solution was introduced and the insertion events stop at once. As the rest of the record demonstrates, the insertion of porin channels can be turned on and off rapidly and repeatedly. This immediacy and reversibility, coupled with the small sample volumes needed, make this chamber ideally suited for testing the activity of multiple samples in succession rapidly.

microscopic and biochemical properties and have recently been implicated in the cell death and tissue pathology underlying these illnesses. Our recent work using the perfusion membrane system has shown that at least three of these amyloid-forming peptides can form ion channels at cytotoxic concentrations, and we have proposed that channel formation is a mechanism of toxicity.

Alzheimer's disease (AD) pathology is characterized by plaques, tangles, and neuronal cell loss. The main constituent of plaques is $\beta$-amyloid peptide (A$\beta$), a 39–42 residue peptide that has been linked to disruption of calcium homeostasis and neurotoxicity *in vitro*. We have demonstrated that a neurotoxic fragment of A$\beta$, A$\beta$ (25–35) spontaneously inserted in planar lipid membranes forms weakly selective, voltage-dependent, ion-permeable channels. We suggest that channel formation may be involved in the pathogenesis of AD and that A$\beta$ (25–35) may be the active channel-forming segment.[16]

Amylin is a 37-amino-acid cytotoxic constituent of amyloid deposits found in the islets of Langerhans of patients with type II diabetes. Extracellular accumulation of this peptide results in damage to insulin-producing $\beta$-cell membranes and cell death. We have shown that at cytotoxic concentrations, amylin forms voltage-dependent, relatively nonselective, ion-permeable channels in planar phospholipid bilayer membranes.[6]

Prions cause neurodegenerative disease in animals and humans. It has been shown that a 21-residue fragment of the prion protein (106–126) could be toxic to cultured neurons. We found that this peptide forms ion-permeable channels in planar lipid bilayer membranes. These channels are freely permeable to common physiologic ions, and their formation is significantly enhanced by "aging" and/or low pH. We suggest that channel formation is the cytotoxic mechanism of action of amyloidogenic peptides found in prion-related disease.[7]

*Search for New Membrane Active Compounds*

A fast and highly sensitive planar membrane setup allows a new approach to the screening and identification of membrane active compounds in complex organic mixtures of nature and synthetic origin.

It has been shown that the mechanism of microbicidal activity by many antibiotics is based on the formation of ion-conducting channels in the host cell membranes.[5,7,8] Among the classes of membrane active antibiotics are the polyenes, e.g., amphotericin B[13]; the host defense peptides, e.g., defensins,[5] magainins[17]; and the ion channel forming exotoxins, yeast killer toxin[18] and colicins.[8] All these compounds, when added to the membrane bathing aqueous solution, interacted with the BLM and induced ion currents at very low concentrations ranging from $10^{-6}$ to $10^{-12}$ g/ml. On the other hand, compounds that are not membrane related do not induce any changes in membrane currents at concentrations up to $10^{-3}$ g/ml. Therefore, the planar lipid membrane technique can identify membrane active compounds present in a complex mixture even at concentrations as low as $10^{-3}$–$10^{-9}$ part of the total. Use of the traditional planar membrane system for the screening of new membrane active compounds was practically difficult and time consuming. The speed and low sample volume of the perfusion planar

[16] T. A. Mirzabekov, M-C. Lin, W. Yuan, P. Marshall, M. Carman, K. Tomaselli, I. Lieberburg, and B. L. Kagan, *Biochem. Biophys. Res. Commun.* **202,** 1142 (1994).

[17] H. Duclohier, *Toxicology* **87,** 175 (1994).

[18] B. L. Kagan, *Nature (Lond.)* **302,** 709 (1983).

membrane technique can be applied to the screening of such mixtures as bacterial metabolites, plant extracts, combinatorial peptide libraries, mutations of known membrane active antimicrobials, and peptides synthesized on the basis of their predicted membrane active structure.

Use of the perfusion planar lipid membrane setup in such screening could potentially result in the discovery of new bioactive compounds of bacterial, plant, animal, or synthetic origin (such as antibacterials, antivirals, toxins) in short time periods and at low cost. Automation of this system is also feasible since the same membrane can be used over and over again until an "active" sample is encountered.

### *Identification, Purification, and Characterization of Bacterial Outer Membrane Transport Proteins: Porins*

The use of the perfusion planar lipid membrane technique allowed fast identification, isolation, and characterization of transport proteins (porins) in the outer membrane of bacteria. In the Lyme disease-associated spirochete *B. burgdorferi,* we identified, isolated, and characterized three different porin channels.[19–21] The method was based on solubilization of bacterial membranes, separation of solubilized membrane proteins into fractions (100–150) eluted from an HPLC column followed by immediate screening on the BLM for channel-forming activity (Fig. 2) allowing precise localization of the porin-containing protein peak. Since Triton X-100 solubilized *Borrelia* porin channels lose channel-forming activity within 20–40 hr after isolation, this rapidity is essential. This methodology can be easily applied to the isolation and study of membrane proteins of other disease-related bacteria, whose porins or pathogenic toxins might be quite labile.

### *Microemulsions*

Microemulsions are emulsified drug delivery systems used for improving dissolution and delivery (oral) of hydrophilic drugs. Surfactants, one of three to five basic compounds of microemulsions, are amphiphilic molecules necessary to stabilize the water-in-oil or oil-in-water conformation of the emulsion. Membrane-active amphiphilic surfactants are often cytotoxic. This toxicity strongly correlates with the ability of the surfactants to permea-

[19] J. Skare, E. Shang, D. Foley, D. R. Blanco, C. I. Champion, T. A. Mirzabekov, Y. Sokolov, B. L. Kagan, R. Miller, and M. Lovett, *J. Clin. Invest.* **96,** 2380 (1995).

[20] J. T. Skare, T. A. Mirzabekov, E. Shang, D. R. Blanco, H. Erjument-Bromage, P. Tempst, B. L. Kagan, J. N. Miller, and M. A. Lovett, *J. Bacteriol.* **178,** 4909 (1996).

[21] J. T. Skare, T. A. Mirzabekov, E. Shang, D. R. Blanco, B. L. Kagan, J. N. Miller, and M. A. Lovett, *J. Bacteriol.* in press (1997).

bilize and/or destroy the cellular membrane. Therefore, for the development of nontoxic microemulsions, it is important to know how basic compounds of microemulsions interact with membranes.

The planar lipid membrane setup is a good instrument in the search for less toxic surfactants. It can help predict quickly and quantitatively the nontoxic concentrations of surfactants. Questions that can be directed to planar membrane studies include these; How do the microemulsions interact with membranes? How do they change lipid membrane permeability and stability? What is the role of membrane lipid composition, solution pH, temperature, and ionic strength in microemulsion interactions with membranes?

### *Enveloped Viruses*

The perfusion planar membrane system is a useful instrument in the study of viral fusion. Using this system we screened synthetic peptides corresponding to Moloney murine leukemia viral and human immunodeficiency viral envelope protein fusion peptides. A number of point mutated peptides were screened for the ability to induce ion leakage and break membranes.[22] Results of these studies were used for directed mutational *in vivo* analysis of viral fusion–replication.

## Acknowledgments

This work was supported by grants from the Alzheimer's Association, the NIMH (MH 01174), the University of California AIDS Research Program, and the UCLA Alzheimer's Disease Center.

[22] Y. Y. Rosenberg, Y-L. Zhang, H. Aguilar-Carreno, T. A. Mirzabekov, M. M. Januszeski, W. Hubble, B. L. Kagan, and W. F. Anderson, *Virology* (in review).

# Section VI

# Reagent and Information Sources

# [36] Antibodies to Ion Channels

*By* ANGELA VINCENT, IAN HART, ASHWIN PINTO, and F. ANNE STEPHENSON

## Introduction

Antibodies to ion channels have been of considerable use in research on the distribution of different channels and the association of ion channel subunits with other subunits or with associated proteins. Tables I–XVI describe both polyclonal and monoclonal antibodies and give an indication of the uses to which they have been put. Note that many antibodies will not necessarily have been tested for all purposes, nor will their cross-reactivity with different species be defined.

Polyclonal antibodies have been raised against purified proteins, recombinant polypeptides, or synthetic peptides, and in the latter case the antibodies can easily be affinity purified on peptide conjugated to Sepharose using conventional techniques. Most of the immunolocalization work, for instance, uses affinity-purified antibodies.

Many of the antibodies listed were originally tested by Western blotting and, therefore, were chosen more for their reactivity with denatured protein than with the intact molecule. Nevertheless, many can also be used for immunoprecipitation of solubilized proteins, or as immunohistochemical or immunoelectron microscopic reagents. Where there is sufficient information on reactivity with native or denatured protein, this has been added to the tables.

Antibodies raised against purified proteins often bind to extracellular determinants, whereas antibodies raised against synthetic peptides or recombinant polypeptides often do not bind well to the extracellular surface of native proteins. Thus for identification of ion channels in nonpermeabilized cell lines, antisera raised against the native protein are best. A novel alternative approach is to use a specific neurotoxin to label cell surface receptors or ion channels. Recently, a monoclonal antibody against $\omega$-agatoxin IVA has been used in conjunction with the toxin, that binds to P-type voltage-gated calcium channels, to immunolocalized P-type channels on cerebellar Purkinje cells (see Table XII).

Antibodies are not the only way of identifying the expression of a particular ion channel. In particular, the use of neurotoxins and specific pharmacologic ligands can be very useful in affinity purification of proteins, and in localizing channels *in situ*. *In situ* hybridization using specific probes is an obvious alternative method to immunohistochemical or autoradiographic studies.

# TABLE I

## SPONTANEOUS HUMAN ANTIBODIES AGAINST ION CHANNELS[a]

| Ion channels | Source of sera | Subunit specificity | Region | Detection of antibody by: | Sources for further information | Refs. |
|---|---|---|---|---|---|---|
| Muscle AChR | Patients with myasthenia gravis | Mainly $\alpha_1$ but very variable; $\alpha_{67-76}$ represents main immunogenic region but many antibodies bind to other subunits or partly overlapping sites | Extracell | Immunoprecipitation of $^{125}$I-$\alpha$-BuTx-labeled AChR | Vincent (Oxford)<br>Tzartoz (Athens) | *a*<br>*b* |
| | Fab fragments cloned from myasthenia gravis combinatorial libraries | $\alpha_1$, $\gamma_1$ | Extracell | Immunoprecipitation of $^{125}$I-$\alpha$-BuTx-labeled AChR | Vincent (Oxford) | *c* |
| Fetal form of muscle AChR | Mothers of babies with antibody-mediated fetal arthrogryposis | Principally $\gamma_1$ | Extracell | Inhibition of fetal AChR function but not adult | Vincent (Oxford) | *d* |
| GluR1 | Patients with paraneoplastic disorders | | | IH, Agonist-like activity | Rogers (Salt Lake City) | *e* |
| GluR3 | Children with Rasussen's encephalitis | | Extracell p372–395 | WB, IP, IH | Rogers (Salt Lake City) | *f* |
| GluR4 | Patients with paraneoplastic disorders | | | IH | Rogers (Salt Lake City) | *e* |
| GluR5 | Patients with paraneoplastic disorders | | | WB | Rogers (Salt Lake City) | *e* |
| GluR6 | Patients with paraneoplastic disorders | | | IH | Rogers (Salt Lake City) | *e* |
| VCCC | Patients with Lambert Eaton myasthenic syndrome | Mainly $\alpha_1$ | Extracell | Immunoprecipitation of $^{125}$I-$\omega$-conotoxin-labeled VGCC | Lang (Oxford) | *g* |
| VGKC KCNA6 | Patients with acquired neuromyotonia | | Extracell | Immunoprecipitation of $^{125}$I-dendrotoxin-labeled VGKCs or KCNA6 subunits expressed in oocytes | Hart (Liverpool)<br>Vincent (Oxford) | *h*<br>*i* |
| VGKC KCNA1a,2 | Patients with acquired neuromyotonia | | Probably extracell | Immunohistochemistry of oocytes expressing KCNA subunit | Hart (Liverpool) | *i* |

*Key:* The following abbreviations are used in Tables I–XVI: T, *Torpedo*; rt, rat; m, mouse; h, human; WB, Western blotting; IH, immunohistochemistry; IP, immunoprecipitation; IEP, immunoelectron microscopy; AChR, acetylcholine receptor; VGCC, voltage-gated calcium channel; VGKC, voltage-gated potassium channel; and GluR, glutamate receptor (usually AMPA). Extracell, extracellular epitope; Cyt, cytoplasmic epitope.

*Key to references:*
[a] S. Tzartos, M. T. Cung, P. Demange, *et al., Mol. Neurobiol.* **5,** 1 (1991); [b] A. Vincent, P. J. Whiting, M. Schluep, *et al., Ann. N.Y. Acad. Sci.* **505,** 106 (1987); [c] J. Farrar, S. Portolano, N. Willcox, *et al., Int. Immunol.* **9,** 1311 (1997); [d] S. Riemersma, A. Vincent, D. Beeson, *et al., J. Clin Invest.* **98,** 2358 (1996); [e] L. C. Gahring, R. E. Twyman, J. E. Greenlee, and S. W. Rogers, *Molecul. Med.* **1,** 245 (1995); [f] S. W. Rogers, P. I. Andrews, L. C. Gahring, *et al., Science* **265,** 648 (1994); [g] B. Lang and J. Newsom-Davis, *Springer Semin. Immunopathol.* **17,** 3 (1995); [h] P. Shillito, P. C. Molenaar, A. Vincent, *et al., Ann. Neurol.* **38,** 714 (1995); [i] I. Hart, C. Water, A. Vincent, *et al., Ann. Neurol.* **41,** 238 (1997).

## TABLE II
### Rat Monoclonal Antibodies Raised Against *Torpedo* or Mammalian Muscle AChR

| Name or number isotype (if known) | Species and source of antigen | Subunit | Region and peptide (if known) | Binding to native (N) or denatured (D) epitope | Species specificity (if known) | Uses | Commercial or academic source | Refs. |
|---|---|---|---|---|---|---|---|---|
| Rat | | | | | | | | |
| 1,2,4 | *Torpedo* AChR | $\alpha_1$ | MIR extracell | N | T | IP, IEM | Tzartos S (Athens) | *a, b* |
| 6 IgG$_1$ | *Torpedo* AChR | $\alpha_1 + \alpha_3$ | MIR $\alpha_{67-76}$ | N > D | T, rt, m, h | WB, IP, IH, IEM | Tzartos S (Athens) | *a, c* |
| 35, 42 IgG$_1$ | Eel AChR | $\alpha_1 + \alpha_3$ | MIR $\alpha_{67-76}$ | N >>> D | T, rt, m, h | IP, IH, IEM | Tzartos S (Athens) | *c, d, e* |
| 198 | Human AChR | $\alpha_1 + \alpha_3$ | MIR $\alpha_{67-76}$ | N > D | T, rt, m, h | WB, IP, IH, IEM | Tzartos S (Athens) | *c, f, g* |
| 202, 145 | Human AChR | $\alpha_1$ | MIR | N >>> D | rt, m, h | IP | Tzartos S (Athens) | *b, e* |
| 190, 192 | | | | N | | | | |
| 3, 5 | *Torpedo* $\alpha$ | $\alpha_1$ | Cyt $\alpha_{351-360}$ | N = D | T | WB, IP, IEM | Tzartos S (Athens) | *a, h* |
| 8 | *Torpedo* $\alpha$ | $\alpha_1$ | Cyt $\alpha_{370-378}$ | N = D | T | WB, IP, IEM | Tzartos S (Athens) | *a, h* |
| 152, 153, 155, 154 | *Torpedo* $\alpha$ | $\alpha_1$ | Cyt $\alpha_{373-380}$ | N = D | T, rt, m, h | WB, IP, IH, IEM | Tzartos S (Athens) | *c, h, i* |
| 12 | *Torpedo* AChR | $\alpha_1$ | Near-MIR | N | T | IP, IEM | Tzartos S (Athens) | *a, b* |
| 14 | *Torpedo* AChR | non-$\alpha_1$ | Near-MIR | N | T | IP | Tzartos S (Athens) | *a, b* |
| 64 | Calf AChR | $\alpha_1$ | Extracell | N >> D | rt, m, h | IP | Tzartos S (Athens) | *b, i* |
| 73 | Calf AChR | $\beta_1$ | Extracell | N >>> D | rt, m, h | IP | Tzartos S (Athens) | *i* |

(*continued*)

TABLE II (*continued*)

| Name or number isotype (if known) | Species and source of antigen | Subunit | Region and peptide (if known) | Binding to native (N) or denatured (D) epitope | Species specificity (if known) | Uses | Commercial or academic source | Refs. |
|---|---|---|---|---|---|---|---|---|
| 111, 124, 148, 151 | *Torpedo* $\beta$ | $\beta_1$ | Cyt $\beta_{354-359}$ | N = D | T, rt, m, h | WB, IP, IH, IEM prefers non-phosphorylated | Tzartos S (Athens) | *i, j* |
| 125 | *Torpedo* $\beta$ | $\beta_1$ | Cyt $\beta_{408-414}$ | N = D | T | nk | Tzartos S (Athens) | *i, j* |
| 117 | *Torpedo* $\beta$ | $\beta_1$ | Cyt $\beta_{343-352}$ | N = D | T | nk | Tzartos S (Athens) | *i, j* |
| 66, 67 | Calf AChR | $\gamma_1$ | Extracell | N >>> D | h | IP | Tzartos S (Athens) | *i* |
| 154, 168 $IgG_1$ | *Torpedo* $\gamma$ | $\gamma_1$-Torp; $\varepsilon_1$-mammal. | Cyt $\gamma_{364-370}$ | N = D | T, rt, m, h | WB, IP | Tzartos S (Athens) | *i, k* |
| 165 | *Torpedo* $\gamma$ | $\gamma_1$, $\varepsilon_1$ | Cyt $\gamma_{364-370}$ | N = D | T | WB, IP | Tzartos S (Athens) | *i, k* |
| 7 | *Torpedo* $\gamma$ | $\delta_1$, $\varepsilon_1$ | Cyt $\delta_{389-395}$ $\gamma_{382-387}$ | N = D | T, m, h | WB, IP | Tzartos S (Athens) | *i, l* |
| 137 $IgG_{2a}$ | *Torpedo* $\gamma$ | $\delta_1$ | Cyt $\delta_{374-391}$ | N = D | T, rt, m, h | WB, IP | Tzartos S (Athens) | *i, l* |
| 141 $IgG_{2a}$ | *Torpedo* $\gamma$ | $\delta_1$ | Cyt $\delta_{385-392}$ | N = D | T, rt, m, h | WB, IP | Tzartos S (Athens) | *b, h* |
| 166 $IgG_{2a}$ | *Torpedo* AChR | $\delta_1$ | Cyt $\delta_{300-410}$ | N + D | T, Rana, Xenopus | IH (amphibians) | Tzartos S (Athens) | |

[a] S. J. Tzartos and J. L. Lindstrom, *Proc. Natl. Acad. Sci. U.S.A.* **77,** 755 (1980).
[b] H. Loutrari, A. Kokla, N. Trakas, and S. J. Tzartos, *Clin. Exp. Immunol.* **109,** 538 (1997).
[c] S. J. Tzartos, A. Kokla, S. Walgrave, and B. Conti-Tronconi. *Proc. Natl. Acad. Sci. U.S.A.* **85,** 2899 (1988).
[d] S. J. Tzartos, D. E. Rand, B. E. Einarson, and J. M. Lindstrom, *J. Biol. Chem.* **256,** 8635 (1981).
[e] I. Papadouli, C. Sakarellos, and S. T. Tzartos, *Eur. J. Biochem.* **211,** 227 (1993).
[f] S. Tzartos, L. Langeberg, S. Hochschwender, and J. Lindstrom, *FEBS Lett.* **158,** 116 (1983).
[g] A. Mamalaki, N. Trakas, S. J. Tzartos, *Eur. J. Immunol.* **23,** 1839 (1993).
[h] S. J. Tzartos and M. S. Remoundos, *Eur. J. Biochem.* **207,** 915 (1992).
[i] S. Tzartos, L. Langeberg, S. Hockschwender, L. Swanson, and J. Lindstrom, *J. Neuroimmunol.* **10,** 235 (1986).
[j] S. J. Tzartos, C. Valcana, R. Kouvatsou, and A. Kokla, *EMBO J.* **12,** 5141 (1993).
[k] S. J. Tzartos, E. Tzartos, and J. S. Tzartos, *FEBS Letts.* **363,** 195 (1995).
[l] S. J. Tzartos, R. Kouvatsou, and E. Tzartos, *Eur. J. Biochem.* **228,** 463 (1995).

TABLE III

MOUSE MONOCLONAL ANTIBODIES AGAINST MUSCLE AChR

| Name or number isotype (if known) | Species and source of antigen | Subunit specificity | Region and peptide (if known) | Binding to native (N) or denatured (D) epitope | Species specificity (if known) | Uses | Commercial or academic source | Ref |
|---|---|---|---|---|---|---|---|---|
| D6$\alpha$ $IgG_{2b}$ | Human muscle AChR | $\alpha_1$ | MIR | N >>> D | h, m, rt, calf | WB, IP, IH | Vincent (Oxford) Serotec (Kidlington) | *a*, *b*, *c* |
| G10$\alpha$ C3a $IgG_1$ | Human muscle AChR | $\alpha_1$ | MIR | N >>> D | h, m, calf h, m, calf | WB, IP, IH | Vincent (Oxford) Serotec (Kidlington) | *a*, *b*, *c* |
| B3$\beta$ $IgG_1$ | Human muscle AChR | $\beta_1$ | Extracell | N >>> D | h, calf | WB, IP, IH | Vincent (Oxford) Serotec (Kidlington) | *a*, *b*, *c* |
| C7$\delta$, $IgG_{2a}$ G3$\delta$, $IgG_{2b}$ | Human muscle AChR | $\delta_1$ | Extracell | N >>> D | h, m, calf h | WB, IP, IH | Vincent (Oxford) Serotec (Kidlington) | *a*, *b*, *c* |
| B8$\gamma$, C2$\gamma$, C9$\gamma$, F8$\gamma$ $IgG_1$ | Human muscle AChR | $\gamma_1$ | Extracell | N >> D | h, calf fetal-type only | WB, IP, IH | Vincent (Oxford) Serotec (Kidlington) | *a*, *b*, *c* |
| WF6 | Electric organ | $\alpha_1$ | Extracell | N | Agonist binding site of *Torpedo* and chick AChR | Acts as agonist and then blocks | Maelicke (Munich) | *d* |

[a] P. J. Whiting, A. Vincent, M. Schluep, and J. Newsom-Davis, *J. Neuroimmunol.* **11,** 223 (1986).
[b] F. Heidenreich, A. Vincent, A. Roberts, and J. Newsom-Davis, *Autoimmunity* **1,** 285 (1998).
[c] L. Jacobson, D. Beeson, and A. Vincent, in preparation.
[d] J. Bufler, S. Kahlert, S. Tzartos, K. V. Toyka, A. Maelicke, and C. Franke, *J. Physiol. Lond.* **492,** 107 (1996).

TABLE IV
RABBIT POLYCLONAL ANTIBODIES TO HUMAN MUSCLE AChR

| Name or number, isotype (if known) | Species and source of antigen | Subunit | Region | Native or denatured | Species specificity | Uses | Source for further information | Refs. |
|---|---|---|---|---|---|---|---|---|
| Anti-$\alpha$ | Two overlapping $\alpha$ peptides | $\alpha_1$ | Cyt $\alpha_{309-368}$ | N + D | h, m | WB IP IF IH | Vincent (Oxford) | *a*, *b* |
| Anti-$\varepsilon$ | Two overlapping $\varepsilon$ peptides | $\varepsilon_1$ | Cyt $\varepsilon_{341-413}$ | N + D | h | WB IP IF IH | Vincent (Oxford) | *a*, *b* |
| Anti-$\gamma$ | Two overlapping $\gamma$ peptides | $\gamma_1$ | Cyt $\gamma_{337-410}$ | N + D | h | WB IP IF IH | Vincent (Oxford) | *a* |

[a] D. Beeson, M. Omar, I. Bermudez, A. Vincent, and J. Newson-Davis, *Neurosci. Lett.* **207,** 57 (1996).
[b] C. R. Slater, C. Young, S. J. Wood, *et al., Brain* **120,** 1513 (1997).

TABLE V

MONOCLONAL ANTIBODIES RAISED AGAINST NEURONAL ACHR

| Name or number isotype (if known) | Species and source of antigen | Subunit | Region and peptide (if known) | Native (N) or denatured (D) | Species specificity | Uses | Soure for further information | Ref. |
|---|---|---|---|---|---|---|---|---|
| 321–326 | Chick recombinant | $\alpha_2$ | Cyt | | Ck | nk | Lindstrom (Philadelphia) | See *a* for a review and much further information |
| 313–315 $IgG_{2a}$ | Chick recombinant | $\alpha_3$ | Cyt $\alpha3_{315-441}$ | N + D | Ck | 1H, 1P | Lindstrom (Philadelphia) | |
| 286 IgM | Chick brain AChR | $\alpha_4$ | | N + D | Ck, rt, h | WB | Lindstrom (Philadelphia) | |
| 289 IgM | Chick brain AChR | $\alpha_4$ | $\alpha4_{330-511}$ | N + D | Ck | WB | Lindstrom (Philadelphia) | |
| 292 $IgG_1$ | Rat brain AChR | $\alpha_4$ | | N + D | Rt | WB | Lindstrom (Philadelphia) | |
| 293 $IgG_{2a}$ | Rat brain AChR | $\alpha_4$ | | N + D | Ck, rt, h | WB | Lindstrom (Philadelphia) | |
| 299 $IgG_1$ | Rat brain AChR | $\alpha_4$ | Extracell | N + D | Ck, rt, h | WB, IH | Lindstrom (Philadelphia) | |
| 268 $IgG_{1/2a}$ | Chick brain AChR | $\alpha_5$ | Extracell $\alpha5_{91-100}$ | D | Ch, h | WB | Lindstrom (Philadelphia) | |
| 306, 307 $IgG_1$ | a-BuTx-binding rat brain AChR | $\alpha_7$ | Cyt $\alpha7_{380-400}$ | N (ck) + D | Ck, rt, h | IH (but not mammalian) | Lindstrom (Philadelphia) | |
| 320 IgG | Recombinant a7 cyt | $\alpha_7$ | Cyt $\alpha7_{380-400}$ | N + D | Ck | | Lindstrom (Philadelphia) | |
| 308 $IgG_{2b}$ | Recombinant a8 cy5 | $\alpha_8$ | Cyt $\alpha8_{323-342}$ | N + D | Ck | IH, AP | Lindstrom (Philadelphia) | |
| 270 $IgG_{2a}$ | Purified ck brain AChR | $\beta_2$ | Extracell | N.D | Ck, m, rt | IH, AP | Lindstrom (Philadelphia) | |
| 287 IgM | Purified ck brain AChR | $\beta_2$ | Cyt | D | Ck | | Lindstrom (Philadelphia) | |
| 290 $IgG_1$ | Purified rt brain AChR | $\beta_2$ | Extracell | N | Ck, rt, b, h | IH, AP | Lindstrom (Philadelphia) | |
| 295, 297, 298 $IgG_{2a}$ | Purified rt brain AChR | $\beta_2$ | Extracell | N | Ck, rt, b, h | | Lindstrom (Philadelphia) | |

[a] J. Lindstrom, *Ion Channels* **4,** 377 (1996).

Key: Ck, chick.

TABLE VI
RABBIT POLYCLONAL ANTIBODIES RAISED AGAINST SYNTHETIC PEPTIDES OF EXTRACELLULAR DOMAIN OF NEURONAL AChRs

| Subunit | Region and peptide | Native (N) or denatured (D) | Species specificity | Uses | Source for further information | Ref. |
|---|---|---|---|---|---|---|
| Rat | | | | | | |
| a2 | Extracell 68-81 | D > N? | Not tested | Tested on WB<br>Some binding to transfected cells | Patrick (Houston) | *a* |
| α3 | Extracell 68-81 | D > N? | Not tested | Tested on WB<br>Some binding to transfected cells | Patrick (Houston) | *a* |
| α4 | Extracell 68-81 | D > N? | Not tested | Tested on WB<br>Some binding to transfected cells | Patrick (Houston) | *a* |
| α5 | Extracell 68-81 | D > N? | Not tested | Tested on WB<br>Some binding to transfected cells | Patrick (Houston) | *a* |
| α6 | Extracell 68-81 | D > N? | Not tested | Tested on WB<br>Some binding to transfected cells | Patrick (Houston) | *a* |
| β2 | Extracell 68-81 | D > N? | Not tested | Tested on WB<br>Some binding to transfected cells | Patrick (Houston) | *a* |
| β3 | Extracell 68-81 | D > N? | Not tested | Tested on WB<br>Some binding to transfected cells | Patrick (Houston) | *a* |
| β4 | Extracell 68-81 | D > N? | Not tested | Tested on WB<br>Some binding to transfected cells | Patrick (Houston) | *a* |

[a] S. Neff, K. Dineley-Miller, D. Char, M. Quik, and J. Patrick, *J. Neurochem.* **64,** 332 (1995).

## TABLE VII

MONOCLONAL AND POLYCLONAL ANTIBODIES TO GLYCINE AND $GABA_A$ RECEPTORS

| Name or number (if relevant) | Antigen: peptide (p-) or recombinant antigen (r-) | Subunit | Region | Uses | Source for further information | Refs. |
|---|---|---|---|---|---|---|
| 4a IgG1 | Purified glycine receptor | $\alpha$ and $\beta$ | | IP WB | Betx (Frankfurt) | *a, b, c* |
| 7a IgG1 | Purified glycine receptor | $\alpha$ | | WB | Betz (Frankfurt) | *a, b, c* |
| bd17 monoclonal | Native $GABA_A$ receptor | $\beta2$, $\beta3$ | | | Mohler (Zurich) | *d, e* |
| bd24 monoclonal | Native $GABA_A$ receptor | Human and bovine a1 N-terminus | | | Mohler (Zurich) | *e* |
| Rabbit polyclonal sera to $GABA_A$ | p324–341 | $\alpha1$ | Cyt | IP, WB | Stephenson (London) | *f, g* |
| | p414–424 | $\alpha2$ | Cyt | IP, WB | Stephenson (London) | *f, g* |
| | p454–467 | $\alpha3$ | Cyt | IP, WB | Stephenson (London) | *f, g* |
| | p1–9 | $\alpha1$ | Extracell | WB, IH | Sieghart (Vienna) | *h, i* |
| | p1–14 (rat) | $\alpha1$ | Extracell | WB | Stephenson (London) | Pollard and Stephenson (unpublished) |
| | p328–382 | $\alpha1$ | Cyt | WB, IP | Sieghart (Vienna) | *j* |
| | p324–338 (rat) | $\alpha1$ | | WB IH | Seighart (Vienna) | *h, k* |
| | p416–424 | $\alpha2$ | Cyt | WB | Sieghert (Vienna) | *h, k* |
| | n459–467 | $\alpha3$ | Cyt | WB | Sieghart (Vienna) | *h, k* |
| | p517–523 | $\alpha4$ | Extracell | WB | Sieghart (Vienna) | *h, k* |
| | p427–433 | $\alpha5$ | Cyt | WB | Sieghart (Vienna) | *h, k* |
| | p1–15 (rat) | $\alpha5$ | Extracell | WB IH | Stephenson (London) | *l, m* |
| | p1–16 (bovine) | $\alpha$6-N | Extracell | IP, WB, IH | Stephenson (London) | *i, l* |
| | p429–434 (rat) | $\alpha6$ | Extracell | IP, IH | Sieghart (Vienna) | *i, n* |
| | p423–434 | $\alpha6$ | Extracell | IP, WB | Stephenson (London) | Pollard and Stephenson (unpublished) |
| | Recombinant cytoplasmic loops | $\alpha1$ | Cyt | WB IP | Whiting (UK) | *o* |
| | Recombinant cytoplasmic loops | $\alpha2$ | Cyt | WB IP | Whiting (UK) | *o* |
| | Recombinant cytoplasmic loops | $\alpha3$ | Cyt | WB IP | Whiting (UK) | *o* |
| | Recombinant cytoplasmic loops | $\alpha5$ | Cyt | WB IP | Whiting (UK) | *o* |
| | p382–393 | $\beta1$ | | WB IP | Stephenson (London) | *p, q* |
| | p380–391 | $\beta3$ | | WB IP | Stephenson (London) | *p, q* |
| | p381–395 | $\beta2$ | | WB IP | Stephenson (London) | *p, q* |
| | p379–394 | $\beta3$ | | WB IP | Stephenson (London) | *p, q* |

*(continued)*

TABLE VII (*continued*)

| Name or number (if relevant) | Antigen: peptide (p-) or recombinant antigen (r-) | Subunit | Region | Uses | Source for further information | Refs. |
|---|---|---|---|---|---|---|
| | p345–408 (rat) | $\beta3$ | | IP IH WB IEM | Sieghart (Vienna) | *r* |
| | p1–15 | $\gamma2$ | Extracell | WB IP IH | Mohler (Zurich) | *s* |
| | p336–350 | $\gamma2$ | Cyt | WB IP IH | Mohler (Zurich) | *t* |
| | p1–15 | $\gamma2$ | Extracell | WB IP IH | Stephenson (London) | *u, v* |
| | p316–352 | $\gamma2$ | Cyt | WB IP | Sieghart (Vienna) | *n* |
| | p339–353 (rat) | $\gamma2$ | | WB IH | Siegel (Cleveland) | *w* |
| | p1–9 | $\gamma3$ | Extracell | WB IP | Siegel (Cleveland) | *w* |
| | p322–372 | $\gamma3$ | Cyt | WB IP | Siegel (Cleveland) | *w* |
| | p1–44 (rat) | $\delta$ | | | | *j* |
| | p324–338 (rat) | $\alpha1$ | nk | WB IP IH IF | Pharmingen | *w* |
| | p (rat) | $\alpha3$ | nk | IH | Pharmingen | *w* |
| | p (rat) | $\alpha4$ | nk | IH | Pharmingen | *w* |
| | p (rat) | $\alpha6$ | nk | IH | Pharmingen | *w* |

[a] F. Pfeiffer, R. Simler, G. Grenningloh, and H. Betz, *Proc. Natl. Acad. Sci. U.S.A.* **81,** 7224 (1984).
[b] G. Meyer, J. Kirsch, H. Betz, and D. Langosch, *Neuron* **15,** 563 (1995).
[c] J. Kirsch, I. Wolters, A. Triller, and H. Betz, *Nature* **366,** 745 (1993).
[d] P. Schoch, J. G. Richards, P. Haring, *et al., Nature* **314,** 168 (1985).
[e] M. Ewert, B. D. Shivers, H. Luddens, *et al., J. Cell Biol.* **110,** 2043 (1990).
[f] M. J. Duggan and F. A. Stephenson, *J. Biol. Chem.* **265,** 3831 (1990).
[g] M. J. Duggan and F. A. Stephenson, *J. Neurochem.* **53,** 132 (1989).
[h] J. Zezula and W. Sieghart, *FEBS Lett.* **284,** 15 (1991).
[i] Z. Nusser, W. Sieghart, F. A. Stephenson, and P. Somogyi, *J. Neurosci.* **16,** 103 (1996).
[j] B. Mossier, M. Togel, K. Fuchs, and W. Sieghart, *J. Biol. Chem.* **269,** 25777 (1994).
[k] W. Kern and W. Sieghart, *J. Neurochem.* **62,** 764 (1994).
[l] S. Pollard, M. J. Duggan, and F. A. Stephenson, *J. Biol. Chem.* **268,** 3753 (1993).
[m] A. Jones, E. R. Korpi, R. M. McKernan, *et al., J. Neurosci.* **17,** 1350 (1997).
[n] M. Togel, B. Mossier, K. Fuchs, and W. Sieghart, *J. Biol. Chem.* **269,** 12993 (1994).
[o] McKernan *et al. Neuron* **7,** 667 (1991).
[p] S. Pollard and F. A. Stephenson, *Biochem. Soc. Trans.* **25,** 5475 (19p7).
[q] S. Pollard, M. J. Duggan, and F. A. Stephenson, *FEBS Lett.* **295,** 81 (1991).
[r] D. Benke, J. M. Fritschy, A. Trzeciak, *et al., J. Biol. Chem.* **269,** 27100 (1994).
[s] D. Benke, S. Mertens, A. Trzeciak, D. Gillessen, and H. Mohler, *J. Biol. Chem.* **266,** 4478 (1991).
[t] Z. U. Khan, A. Guitterez, and A. L. De Blas, *J. Neurochem.* **63,** 371 (1994).
[u] F. A. Stephenson, M. J. Duggan, and S. Pollard, *J. Biol. Chem.* **265,** 21160 (1990).
[v] U. Greferath, U. Grunert, J. M. Fritschy, A. Stephenson, H. Mohler, and H. Wassle, *J. Comp. Neurol.* **353,** 553 (1995).
[w] L. S. Nadler, E. R. Guirguis, and R. E. Siegel, *J. Neurobiol.* **25,** 1533 (1994).

For reviews see:

R. M. McKernan and P. J. Whiting, *Trends Neurosci.* **19,** 139 (1996).

F. A. Stephenson, *Biochem. J.* **310,** 1 (1995).

## TABLE VIII
### Monoclonal and Polyclonal Antibodies to NMDA Receptors

| Name | Subunit | Region | Peptide (p-) or recombinant antigen (r-) | Uses | Source for further information | Refs. |
|---|---|---|---|---|---|---|
| Monoclonal 54.4 | Rat NMDAR1 | Transmembrane and cytoplasmic | rNMDAR1 660-811 | WB IH | Heinemann (San Diego) | *a* |
| Monoclonal 54.1 | Rat NMDAR1 | Transmembrane and cytoplasmic | rNMDAR1 660-811 | WB IH IEM | Heinemann (San Diego) | *b, c, d* |
| Monoclonal 54.2 | Rat NMDAR1 | Transmembrane and cytoplasmic | rNMDAR1 660-811 | WB IH IEM | Heinemann (San Diego) | *b, c, d* |
| Monoclonal NR1-TM3/4 | Rat NMDAR1 | Transmembrane and cytoplasmic | rNMDAR1 660-811 | WB IH | Heinemann (San Diego) | *c* |
| Polyclonal antisera: | | | | | | |
| NR1 827 AP | Rat NMDAR1 | Transmembrane and cytoplasmic | rNMDAR1 660-811 | WB IH | Heinemann (San Diego) | *e* |
| NR1-NH2 | Rat NMDAR1 | N terminus | p19-38 | WB | Huganir (Baltimore) | *f* |
| NR1-COOH | Rat NMDAR1 | C terminus | r-919-938 | WB | Huganir (Baltimore) | *g* |
| NR2A | Rat NMDAR1 | C terminus | el-1247-1464 (fusion protein) | WB | Huganir (Baltimore) | *f* |
| NR2B | Rat NMDAR1 | C terminus | p1463-I482 | WB IF | Huganir (Baltimore) | *f* |
| NR1A antiphospho-peptide | Rat NMDAR1 | Cyt | r891-902-phosphorylated p884-895-phosphorylated | WB IF WB of tryptic digests | Huganir (Baltimore) | *h, i* |
| | Rat NMDAR1 | C terminus | | WB IH | Wenthold (Bethesda) | *j* |

[a] N. J. Sucher, N. Brose, D. L. Deitcher, M. Awobuluyi, G. P. Gasic, H. Bading, C. L. Cepko, M. E. Greenberg, R. Jahn, S. F. Heinemann, *et al., J. Biol. Chem.* **268,** 22299 (1993).
[b] N. Brose, G. W. Huntley, Y. Stern-Bach, G. Sharma, J. H. Morrison, and S. F. Heinemann, *J. Biol. Chem.* **269,** 16780 (1994).
[c] S. J. Siegel, N. Brose, W. G. Janssen, G. P. Gasic, R. Jahn, and S. F. Heinemann, *Proc. Natl. Acad. Sci. U.S.A.* **91,** 564 (1994).
[d] G. W. Huntley, J. C. Vickers, W. Janssen, N. Brose, S. F. Heinemann, and J. H. Morrison, *J. Neurosci.* **14,** 3603 (1994).
[e] N. Brose, G. W. Huntley, Y. Stern-Bach, G. Sharma, J. H. Morrison, and S. F. Heinemann, *J. Biol. Chem.* **269,** 16780 (1994).
[f] L. F. Lau and R. L. Huganir, *J. Biol. Chem.* **270,** 20036 (1995).
[g] W. G. Tingley, K. W. Roche, A. K. Thompson, and R. L. Huganir, *Nature* **364,** 70 (1993).
[h] W. G. Tingley, M. D. Ehlers, K. Kameyama, C. Doherty, J. B. Ptak, C. T. Riley, and R. L. Huganir, *J. Biol. Chem.* **272,** 5157 (1997).
[i] L. Mei, C. A. Doherty, and R. L. Huganir, *J. Biol. Chem.* **269,** 12254 (1994).
[j] R. S. Petralia, N. Yokotani, and R. J. Wenthold, *J. Neurosci.* **14,** 667 (1994).

TABLE IX

MONOCLONAL AND POLYCLONAL ANTIBODIES TO AMPA RECEPTORS[a]

| Name or number isotype (if known) | Peptide (p-) or recombinant antigen (r-) | Subunit or subtype | Region | Uses | Source for further information | Refs.[b] |
|---|---|---|---|---|---|---|
| Monoclonal antibodies | | | | | | |
| 3a11 | 17–430 | GluR2 + 4 | N terminus | WB IH IP IEM | Huganir | *b, c, d* |
| Mouse $IgG_{2a}$ | | | | | | |
| 4F5 | N-terminal extracellular domain fusion protein | GluR5/6/7 | Extra N terminus | IH IF IEM | Pharmingen | *e* |
| Mouse IgM | | | | | | |
| 1F1 | C-terminal 13 aas | | C terminus | IH IEM | Somogyi (Oxford) | *f* |
| Rabbit polyclonal antisera | | | | | | |
| 67 | 877–889 | GluR1 | Extracell | IP WB IAP | Wenthold (Bethesda) | *g* |
| 23 | 369–381 | GluR 2 | Extracell | IP WB | Wenthold (Bethesda) | *g* |
| 33 | 834–844 | GluR 2 | Extracell | IP WB | Wenthold (Bethesda) | *g* |
| 24 | 372–383 | GluR 3 | Extracell | IP | Wenthold (Bethesda) | *g* |
| 34 | 838–848 | GluR 3 | Extracell | | Wenthold (Bethesda) | *g* |
| 35 | 838–848 | GluR 3 | Extracell | IP WB | Wenthold (Bethesda) | *g* |
| 22 | 868–881 | GluR 4 | Extracell | IP WB | Wenthold (Bethesda) | *g* |
| 25 | 850–862 | GluR 2 + 3 | Extracell | IP WB IAP | Wenthold (Bethesda) | *g* |
| | 894–907 | GluR 1 | C terminus | WB IH IF IP (2,4,4c) | Huganir (Baltimore) | *h* |
| | | GluR 2/3/4c | | WB IH IF | Chemicon Int Incr | *i* |

| | | | | | | |
|---|---|---|---|---|---|---|
| | | GluR 4 | | WB IH IF | Chemicon Int Incr (London) | *i* |
| | | GluR 6/7 | C-terminus | WB IH | Chemicon Int Incr (London) | *i* |
| GluR1 | p876–889 | | | WB IH | Huganir (Baltimore) | *j, k* |
| GluR1 | p872–889 | | | WB IH ?IP | Wenthold (Bethesda) | *j, k* |
| GluR2/3/4c | p843–862 | | | WB IH | Huganir (Baltimore) | *j, k* |
| GluR2/3/4c | p850–862 | | | WB (2,3) | Wenthold (Bethesda) | *j, k* |
| GluR4 | p862–881 | | | WB IH | Huganir (Baltimore) | *j, k* |
| GluR4 | p868–881 | | | WB IH | Wenthold (Bethesda) | *j, k* |
| GluR6/7 | p863–877 | | | WB IH IP | Huganir (Baltimore) | *j, k* |
| GluR6/7 | p864–877 | | | WB IH IP | Wenthold (Bethesda) | *j, k* |
| GluR3 | r245–457 | | p245–274 p372–395 | WB IH agonist like activity | Rogers (Salt Lake City) | *l* |

[a] Most of the antibodies raised against sequences of rat AMPA receptors. Other antibodies against synthetic peptides of glutamate receptors are available from Pharmingen.
[b] C. D. Blackstone, S. J. Moss, L. J. Martin, A. I. Levey, D. L. Price, and R. L. Huganir, *J. Neurochem.* **58,** 1118 (1992).
[c] C. D. Blackstone, A. I. Levey, L. J. Martin, D. L. Price, and R. L. Huganir, *Ann. Neurol.* **31,** 680 (1992).
[d] L. A. Raymond, C. D. Blackstone, and R. L. Huganir, *Nature* **361,** 637 (1993).
[e] G. W. Huntley, S. W. Rogers, T. Moran, W. Janssen, N. Archin, J. C. Vickers, K. Cauley, S. F. Heinemann, and J. H. Morrison, *J. Neurosci.* **13,** 2965 (1993).
[f] Z. Nusser, W. Sieghart, F. A. Stephenson, and P. Somogyi, *J. Neurosci.* **16,** 103 (1996).
[g] R. J. Wenthold, N. Yokotani, K. Doi, and K. Wada, *J. Biol. Chem.* **267,** 501 (1992).
[h] A. M. Craig, C. D. Blackstone, R. L. Huganir, and G. Banker, *Neuron* **10,** 1055 (1993).
[i] L. J. Martin, C. D. Blackstone, R. L. Huganir, and D. L. Price, *Neuron* **9,** 259 (1992).
[j] R. B. Puchalski, J.-C. Louis, N. Brose, S. F. Traynelis, J. Egebjerg, V. Kukekov, R. J. Wenthold, S. W. Rogers, F. Lin, T. Moran, J. H. Morrison, and S. F. Heinemann, *Neuron* **13,** 131 (1994).
[k] M. E. Rubio, and R. J. Wenthold, *Neuron* **19,** 939 (1997).
[l] R. E. Twyman, L. C. Gahring, J. Spiess, and S. W. Rogers, *Neuron* **14,** 755 (1995).

TABLE X
RABBIT POLYCLONAL ANTIBODIES TO VOLTAGE-GATED CALCIUM CHANNELS

| Name or number | Species and source of antigen | Reactivity with native (N) or denatured (D) | Subunit | Region | Species specificity | Uses | Source for further information | Ref. |
|---|---|---|---|---|---|---|---|---|
| Anti-$\alpha_{1A}$ | Human, GST fusion protein 1041–1202 | N + D | $\alpha_{1A}$ | Cyt | Human, rat | WB IH | Beattie (Surrey, UK) | *a* |
| Anti-$\alpha_{1B}$ | Human, GST fusion protein 983–1106 | N + D | $\alpha_{1B}$ | Cyt | Human, rat | WB IH | Beattie (Surrey, UK) | *a* |
| Anti-$\alpha_{1E}$ | Human, GST fusion protein 984–1099 | N + D | $\alpha_{1E}$ | Cyt | Human, rat | WB IH | Beattie (Surrey, UK) | *a* |
| Anti-$\beta_{1b}$ | Human, GST fusion protein 431–598 | N + D | $\beta_{1b}$ | Cyt | Human, rat | WB IH | Beattie (Surrey, UK) | *a* |
| Anti-$\beta_2$ | Human, GST fusion protein 554–660 | N + D | $\beta_2$ | Cyt | Human, rat | WB IH | Beattie (Surrey, UK) | *a* |
| Anti-$\beta_3$ | Human, GST fusion protein 418–483 | N + D | $\beta_3$ | Cyt | Human, rat | WB IH | Beattie (Surrey, UK) | *a* |
| Anti-$\beta_4$ | Human, GST fusion protein | N + D | $\beta_4$ | a.a. 410–520 cyt | Human, rat | WB IH | Beattie (Surrey, UK) | *a* |

[a] R. E. Beattie *et al.*, *Brain Res. Prot.* **1,** 307 (1997).

TABLE XI

RABBIT POLYCLONAL ANTIBODIES TO VGCC

| Name or number | Species and source of antigen | Reactivity with native (N) or denatured (D) | Subunit specificity | Region | Species specificity | Uses | Source for further information | Refs. |
|---|---|---|---|---|---|---|---|---|
| Anti-CNA1 | Rat, Peptide 865–881 | N + D | $\alpha_{1A}$ | II-III Cyt | Rat, mouse | IH WB IP | Catterall (Seattle) Alomone Labs | *a* |
| Anti-CNA3 | Rat, Peptide 882–896 | N + D | $\alpha_{1A}$ | II-III Cyt | Rat | IH WB IP | Catterall (Seattle) Alomone Labs | *a* |
| Anti-CNA5 | Rat, GST fusion protein 842–981 | N + D | $\alpha_{1A}$ | II-III Cyt | Rat | IH WB IP | Catterall (Seattle) Alomone Labs | *b* |
| Anti-CNA6 | Human, GST fusion protein 569–712 | N + D | $\alpha_{1A}$ | II-III Cyt | Rat | IH WB | Catterall (Seattle) Alomone Labs | *b* |
| Anti-NBI-1 | Rabbit, peptide 845–861 | N + D | $\alpha_{1A}$ | II-III Cyt | Rat | IH WB | Catterall (Seattle) Alomone Labs | *b* |
| Anti-NBI-2 | Rabbit, peptide 904–918 | N + D | $\alpha_{1A}$ | II-III Cyt | Rat | IH WB | Catterall (Seattle) Alomone Labs | *b* |
| Anti-CNB1 | Rat, peptide 851–867 | N + D | $\alpha_{1B}$ | II-III Cyt | Rat, mouse | IH WB | Catterall (Seattle) Alomone Labs | *c* |
| Anti-CNC1 | Rat, peptide 815, 835 | N + D | $\alpha_{1C}$ | II-III Cyt | Rat, mouse | IH WB | Catterall (Seattle) Alomone Labs | *d* |
| Anti-CND1 | Rat, peptide 809–825 | N + D | $\alpha_{1D}$ | II-III Cyt | Rat | IH WB | Catterall (Seattle) Alomone Labs | *d* |
| Anti-pan $\alpha_1$ | ? species, peptide 1382–1400 | N + D | $\alpha_{1S}$ | | Rat, mouse | IH WB | Alomone Labs | *e* |

(*continued*)

TABLE XI (*continued*)

| Name or number | Species and source of antigen | Reactivity with native (N) or denatured (D) | Subunit specificity | Region | Species specificity | Uses | Source for further information | Refs. |
|---|---|---|---|---|---|---|---|---|
| Anti-$\beta_{1b}$ | Rat, GST fusion protein 428–597 | N + D | $\beta_{1b}$ | Cyt | Rabbit | WB<br>IP | Campbell (Iowa) | *f* |
| Anti-$\beta_{2a}$ | Rat, GST fusion protein 462–578 | N + D | $\beta_2$ | Cyt | Rabbit | WB<br>IP | Campbell (Iowa) | *f* |
| Anti-$\beta_3$ | Rat, GST fusion protein 369–484 | N + D | $\beta_3$ | Cyt | Rabbit | WB<br>IP | Campbell (Iowa) | *f* |
| Anti-$\beta_4$ | Rat, GST fusion protein 419–519 | N + D | $\beta_4$ | Cyt | Rabbit | WB<br>IP | Campbell (Iowa) | *f* |
| Anti-$\alpha_2$ | Rabbit, peptide 839–856 | | $\alpha_2$ | | | | Campbell (Iowa) | *g* |
| Anti-$\alpha_{1A}$ | a.a. 965–983 | N + D | $\alpha_{1A}$ | Cyt | Rabbit | WB<br>IH | Froehner (North Carolina) | *h* |
| Anti-$\alpha_{1A}$ | IVS5-S6 | N + D | $\alpha_{1A}$ | Extracell | Rabbit | WB<br>IH | Froehner (North Carolina) | *i* |

[a] T. Sakurai, *et al., J. Biol. Chem.* **270,** 21234 (1995).
[b] T. Sakurai, *et al., J. Cell Biol.* **134,** 511 (1996).
[c] R. E. Westenbroek, *et al., Neuron* **9,** 1099 (1992).
[d] J. W. Hell, *et al., J. Cell Biol.* **123,** 949 (1993).
[e] J. Striessnig, H. Glossmann, and W. A. Catterall, *Proc. Natl. Acad. Sci. U.S.A.* **87,** 9108 (1990).
[f] V. E. Scott, *et al., J. Biol. Chem.* **271,** 3207 (1996).
[g] C. A. Gurnett, *et al., J. Biol. Chem.* **270,** 9035 (1995).
[h] A. H. Ousley and S. C. Froehner, *Proc. Natl. Acad. Sci. U.S.A.* **91,** 12263 (1994).
[i] E. L. Barry, *et al., J. Neurosci.* **15,** 274 (1995).

TABLE XII

MONOCLONAL ANTIBODIES TO VGCC-SPECIFIC NEUROTOXINS FOR IMMUNOLOCALIZATION STUDIES

| Name or number | Immunogen | Epitope | Uses | Senior author | Ref. |
|---|---|---|---|---|---|
| Anti-agatoxin IVA | Purified peptide | $\omega$-Aga-IVA binding sites ($\alpha_{1A}$) | IH using $\omega$-Aga-IVA as first layer | Beattie (Surrey) | *a* |
| Anti-conotoxin GVIA | Purified peptide | $\omega$-CTx-GVIA binding sites ($\alpha_{1B}$) | IH using $\omega$-CTx-GVIA as first layer | Beattie (Surrey) | *a* |

[a] S. E. Gillard, *et al., Neuropharmacology* **36**(3), 405 (1997).

TABLE XIII

MOUSE MONOCLONAL ANTIBODIES TO VOLTAGE-GATED POTASSIUM CHANNELS

| Name | Channel subunit | Region | Epitope if known | Immunogen | Uses | Senior author or commercial company | Refs. |
|---|---|---|---|---|---|---|---|
| K1C3 (rat) | Kv1.1 | cyt | 141 aas of C terminus | Kv1.1 (RCK1) | WB IH | Pongs (Hamburg) | *a* |
| K20/78 | Kv1.1 | | C terminus | Peptide | | Upstate Biotechnology | |
| Anti-Kv1.2 | Kv1.2 | cyt | C terminus | Peptide | IH IEM | Dolly (London) | *b* |
| K14/16 | Kv1.2 | | C terminus | Peptide | | Trimmer (Stony Brook, NY) | *c, d* |
| $IgG_{2a}$ | | | | | | Upstate Biotechnology | |
| K13/31 | Kv1.4 | | N terminus | Peptide | | Trimmer (Stony Brook, NY) | *c, d* |
| $IgG_1$ | | | | | | Upstate Biotechnology | |
| K7/45 | Kv1.5 | | C terminus | Peptide | | Trimmer (Stony Brook, NY) | *c, d* |
| $IgG_1$ | | | | | | Upstate Biotechnology | |
| K19/36 | Kv1.6 | | C terminus | Peptide | | Trimmer (Stony Brook, NY) | *c, d* |
| $IgG_3$ | | | | | | Upstate Biotechnology | |
| D4/11 | Kv2.1 | | C terminus | Peptide | | Trimmer (Stony Brook, NY) | *c, d* |
| $IgG_1$ | | | | | | Upstate Biotechnology | |
| K37/89 | Kv2.2 | | N terminus | Peptide | | Trimmer (Stony Brook, NY) | *c, d* |
| $IgG_{2a}$ | | | | | | Upstate Biotechnology | |
| K9/40 | Kvb1 | | N terminus | Peptide | | Trimmer (Stony Brook, NY) | *c, d* |
| $IgG_{2b}$ | | | | | | | |
| K17/70 | Kvb2 | | N terminus | Peptide | | Trimmer (Stony Brook, NY) | *c, d* |
| $IgG_1$ | | | | | | | |

[a] S. Reinhardt-Maelicke *et al., J. Rec. Research* **13,** 513 (1993).
[b] N. M. McNamara *et al., Eur. J. Neurosci.* **8,** 688 (1996).
[c] Z. Bekele-Arcuri, M. F. Matos, L. Manganas, B. W. Strassle, B. W. Monaghan, K. J. Rhodes, and J. S. Trimmer, *Neuropharmacology* **35,** 851 (1996).
[d] G. Shi, K. Nakahira, S. Hammond, L. E. Schechter, K. J. Rhodes, and J. S. Trimmer, *Neuron* **16,** 843 (1996).

TABLE XIV

Rabbit Polyclonal Antibodies Raised against Synthetic Peptides of Voltage-Gated Potassium Channels

| Name | Channel subunit | Native (N) or denatured (D) | Region | Epitope if known | Uses | Senior author or commercial company | Refs. |
|---|---|---|---|---|---|---|---|
| Anti-Kv1.1 (mouse) | Kv1.1 | N = D | Cyt | C terminus | WB IP IH | Tempel (Seattle) | *a* |
| Anti-Kv1.1 (rat) | Kv1.1 | D | Cyt | C terminus | WB bovine | Pongs (Hamburg) | *b* |
| Anti-Kv1.1 (rat) | Kv1.1 | N = D | Cyt | C terminus aa 354–495 | WB IH | Pongs (Hamburg) | *c* |
| Anti-Kv1.1 | Kv1.1 | | Cyt | aa 458–476 | | Trimmer (Stony Brook, NY) | *d* |
| KCN1A (human) | KCN1A | N = D | Extracell | S3-S4 | IH IP | Hart (Liverpool, UK) | *e* |
| Anti-Kv1.2 | Kv1.2 | N = D | Cyt | C terminus | WB IP IH | Tempel (Seattle) | *a* |
| Anti-Kv1.2 | Kv1.2 | D | Cyt | C terminus | WB (bovine) | Pongs (Hamburg) | *b* |
| Anti-Kv1.2 | Kv1.2 | N = D | Cyt | p468–486 | WB IH | Jan (San Francisco) | *f* |
| Anti-Kv1.2 | Kv1.2 | N = D | Cyt | p422–498 | WB IH | Pongs (Hamburg) | *c* |
| Anti-Kv1.3 | Kv1.3 | D | Extracell | S1–S2 and S3–S4 | WB (h, m, Drosophila, yeast) | Gutman (Irvine, CA) | *g* |
| Anti-Kv1.3 | Kv1.3 | N = D | Cyt | aa 409–525 | WB IH | Pongs (Hamburg) | *c* |
| Anti-Kv1.3 | Kv1.3 | D | Cyt | aa 456–474 | WB | Slaughter (Merck Research Labs) | *h* |
| Anti-Kv 1.4N | Kv1.4 | N = D | Cyt | aa 13–37 | WB IH | Jan (San Francisco) | *i* |
| Anti-Kv1.4 | Kv1.4 | D | Cyt | C terminus | WB (bovine) | Pongs (Hamburg) | *b* |
| Anti-Kv1.4 | Kv1.4 | N = D | Cyt | aa 578–655 | WB IH | Pongs (Hamburg) | *c* |
| Anti-Kv1.5 | Kv1.5 | D | Extracell | aa 272–312, S1–S2 | WB rat heart | Levitan (Pittsburgh) | *j* |
| Anti-Kv1.5 | Kv1.5 | N = D | Cyt | aa 542–602, C terminus | WB IH rat heart | Nerbonne (St. Louis) Trimmer (Stony Brook, NY) Upstate Biotechnology | *k, l* |
| Anti-Kv1.5N | Kv1.5 | N = D | Cyt | N terminus | WB IH human heart | Mays (Nashville) | *m* |
| Anti-Kv1.5S1-2 | Kv1.5 | N = D | Extracell | S1–S2 extracellular | WB IH human heart | Mays (Nashville) | *m* |
| Anti-Kv1.5 | Kv1.5 | | Cyt | aa 586–602 | WB | Trimmer (Stony Brook, NY) | *p* |
| Anti-Kv1.5 | Kv1.5 | N | Cyt | | WB rat brain | Roy (Conneticut) | *n* |
| Anti-Kv1.6 | Kv1.6 | D | Cyt | C terminus | WB bovine brain | Pongs (Hamburg) | *b* |
| Anti-Kv1.6 | Kv1.6 | N = D | Cyt | aa 438–530, C terminus | IH WB rat brain | Pongs (Hamburg) | *c* |
| Anti-Kv1.6 | Kv1.6 | | | aa 506–524 | | Trimmer (Stony Brook, NY) | *d* |
| drk1 | DRK1 (Kv2.1) | N = D | | aa 506–533 | WB IH Confocal | Trimmer (Stony Brook, NY) | *o, p* |
| KC | DRK1 (Kv2.1) | N = D | Cyt | aa 837–853 | WB IH Confocal | Trimmer (Stony Brook, NY) | *o, p* |

*(continued)*

TABLE XIV (*continued*)

| Name | Channel subunit | Native (N) or denatured (D) | Region | Epitope if known | Uses | Senior author or commercial company | Refs. |
|---|---|---|---|---|---|---|---|
| Anti-Kv2.1 | DRK1 (Kv2.1) | N = D | Cyt | aa 853–857, C terminus | WB IH rat heart | Nerbonne (St. Louis) | *k* |
| Anti-Kv3.1b | Kv3.1b | N = D | Cyt | 19 aa, C terminus | WB IH | Rudy (New York); Alomone Labs | *q* |
| Anti-Kv3.2 | Kv3.2 | N = D | Cyt | aa 184–204, C terminus | IH IP | Rudy (New York); Alomone Labs | *r* |
| Anti-Kv3.4 | Kv3.4 | N = D | Cyt | C terminus | IH WB | Pongs (Hamburg) | *c* |
| Anti-Kv4.2N | Kv4.2 | N = D | Cyt | aa23-42, N terminus | WB IH | Jan (San Francisco) | *i* |
| Anti-Kv4.2C | Kv4.2 | N = D | Cyt | aa484-502, C terminus | NB IH | Jan (San Francisco) | *d* |
| Anti-SHB N22 *Drosophila* | Shaker B | N = D | Cyt | N terminus | WB IP IEM | Schultz (Houston) | *s, t* |
| Anti-AKT1 pore plant | AKT1 | D | Pore | aa 250–258 | WB cross reacts with maxi-K, DRK1, KAT1 | Berkowitz (New Brunswick) | *u* |
| Anti-minK | minK (Kvs1) slowly activating | N | | full sequence | IH guinea pig myocytes | Kass (New York) | *v* |
| Anti-ERG (human) | IKr (delayed rectifier) | N = D | | aa 321–335<br>aa 820–834 | IH ferrit heart | Morales (Durham) | *w* |
| ROMK1-GST | ROMK1 Kir1.1a (inward rectifier) | N | | Not stated | IH rat kidney | White (Scheffield, UK) | *x* |
| Anti-IRK | IRK1 (Kir2.1) (inward) rectifier) | N = D | Cyt | N terminus | WB IH | Schwartz (Stanford) | *y* |
| C-1 | IRK1 (Kir2.1) (inward rectifier) | N | Cyt | aa 376–403, C terminus | IH | Miyashita (Tokyo) | *z* |
| Anti-GIRK1C1 | GIRK1 (kir3.1) (inward rectifier) | D > N | Cyt | aa 488–501, C terminus | WB IP IH IEM | Kurachi (Osaka, Japan) | *aa, bb* |
| C-2 | GIRK1 (kir3.1) (inward rectifier) | N | Cyt | aa 346–375 | IH | Miyashita (Tokyo) | *z* |
| Anti-CIR | CIR (kir 3.4) (inward rectifier) | N = D | Cyt | N terminus | WB IH IP | Iizuka (Nippoa Boehringer Ingelheim, Japan) | *cc* |
| Anti-KAB-2 | kir4.1 KAB inward rectifier | N = D | Cyt | C terminus aa, 366–379 | WB IH IEM rat kidney | Ito (Yamagata) | *dd* |
| ROMK1 | ROMK1 (inward rectifier) | N = D | | | WB IH | Alomone Labs | |
| GIRK1 | GIRK1 (inward rectifier) | N = D | | | WB IH | Alomone Labs | |
| GIRK2 | GIRK2 (inward rectifier) | N = D | | | WB IH | Alomone Labs | |

| | | | | | | | |
|---|---|---|---|---|---|---|---|
| Anti-RACTK1 | RACTK1 | N = D | Extracell + Cyt | aa 268–283 cross-linked to aa 205–220 | WB IH rabbit kidney | Suzuki (Minamikawachi, Japan) | *ee* |
| Anti-Slo | Slo (voltage and calcium gated) | N = D | | aa 913–926 | WB IP IH | Knaus (Innsbruck) | *ff* |

[a] H. Wang *et al., Nature* **365,** 75 (1993).
[b] V. E. Scott *et al., Biochem.* **33,** 1617 (1994).
[c] R. W. Veh *et al., Eur. J. Neurosci.* **7,** 2189 (1995).
[d] K. Nakahira, G. Shi, K. K. Rhodes, and J. S. Trimmer, *J. Biol. Chem.* **271,** 7084 (1996).
[e] I. K. Hart *et al., Ann. Neurol.* **41,** 238 (1997).
[f] M. Sheng, M. L. Tsaur, Y. N. Jan, and L. Y. Jan, *J. Neurosci.* **14,** 2408 (1994).
[g] R. H. Spencer, K. G. Chandy, and G. A. Gutman, *Biochem. Biophys. Res. Commun.* **191,** 201 (1993).
[h] L. M. Helms, *et al., Biochem.* **36,** 3737 (1997).
[i] M. Sheng, M. L. Tsaur, Y. N. Jan, and L. Y. Jan, *Neuron* **9,** 271 (1992).
[j] K. Takimoto and E. S. Levitan, *Circ. Res.* **75,** 1006 (1994).
[k] D. M. Barry, J. S. Trimmer, J. P. Merlie, and J. M. Nerbonne, *Circ. Res.* **77,** 361 (1995).
[l] J. S. Trimmer, *Proc. Natl. Acad. Sci. U.S.A.* **88,** 10764 (1991).
[m] D. J. Mays, J. M. Foose, L. H. Philipson, and M. M. Tamkun, *J. Clin. Invest.* **96,** 282 (1995).
[n] M. L. Roy *et al., Glia.* **18,** 177 (1996).
[o] N. Sharma *et al., J. Cell Biol.* **123,** 1835 (1993).
[p] J. S. Trimmer. *FEBS Lett.* **324,** 205 (1993).
[q] M. Weiser *et al., J. Neurosci.* **15,** 4298 (1995).
[r] H. Moreno *et al., J. Neurosci.* **15,** 5486 (1995).
[s] W. P. Dubinsky, O. Mayorga-Wark, L. T. Garretson, and S. G. Schultz, *Am. J. Physiol.* **265,** C548 (1993).
[t] M. Li, Y. N. Jang, and L. Y. Jan, *Science* **257,** 1225 (1992).
[u] F. Mi and G. A. Berkowitz, *PNAS* **92,** 3386 (1995).
[v] L. C. Freeman and R. S. Kass, *Circ. Res.* **73,** 968 (1993).
[w] M. V. Brahmajothi, M. J. Morales, K. A. Reimer, and H. C. Strauss, *Circ. Res.* **81,** 128 (1997).
[x] Q. Li, G. Cope, D. Hornby, and S. White, *J. Physiol.* **489,** 93P (1995).
[y] H. Mi *et al., J. Neurosci.* **16,** 2421 (1996).
[z] T. Miyashita and Y. Kubo, *Brain Res.* **750,** 251 (1997).
[aa] A. Inanobe *et al., Biochem. Biophys. Res. Commun.* **217,** 1238 (1995).
[bb] K. I. Morishige *et al., Biochem. Biophys. Res. Commun.* **220,** 300 (1996).
[cc] M. Iizuka *et al., Neuroscience* **77,** 1 (1997).
[dd] M. Ito *et al., FEBS Lett.* **338,** 11 (1996).
[ee] M. Suzuki *et al., Am. J. Physiol.* **269,** C496 (1995).
[ff] H. G. Knaus *et al., J. Neurosci.* **16,** 955 (1996).

*Additional references:*

K. J. Rhodes, S. A. Keilbaugh, N. X. Barrezueta, K. L. Lopez, and J. S. Trimmer, *J. Neurosci.* **15,** 5360 (1995).
K. J. Rhodes, N. X. Barrezueta, M. M. Monaghan, Z. Bekele-Arcuri, K. Nakahira, S. Nawoschlik, L. E. Schechter, and J. S. Trimmer, *J. Neurosci.* **16,** 4846 (1996).

TABLE XV
MONOCLONAL ANTIBODIES TO CFTR

| Name | Immunized species, nature of antibody | Region | Peptide (p-) or recombinant antigen (r-) | Uses | Senior Author or Company | Refs. |
|---|---|---|---|---|---|---|
| CF1 | Mouse mAbs | Cyt | Peptides, see 1 | WB IH IF | Banting (Bristol) | *a* |
| CF2 | | Cyt | | | Banting (Bristol) | *a* |
| CF3 | | Extracell | | | Banting (Bristol) | *a* |
| CF4 | | Extracell | | | Banting (Bristol) | *a* |
| CF5 | | Cyt | | | Banting (Bristol) | *a* |
| CF6 | | Cyt | | | Banting (Bristol) | *a* |
| CF7 | | Cyt | | | Banting (Bristol) | *a* |
| CF8 | | Cyt | | | Banting (Bristol) | *a* |
| L11E8 | Mouse $IgG_1$ | NBF1 | | IH IP | Riordan (Toronto) | *b* |
| M3A7 | Mouse $IgG_1$ | NBF2 | | WB IH IP | Riordan (Toronto) | *b* |
| L12b4 | Mouse $IgG_{2a}$ | R domain | | WB IH IP | Riordan (Toronto) | *b* |
| 14H10 | | Mid R domain | | IH IP | Riordan (Toronto) | *b* |
| M13-1 | $IgG_{1k}$ | not stated | p729–736 | WB IP IF | Genzyme | |
| M24-1 | $IgG_{2ak}$ | not stated | p1377–1480 | WB IP | Genzyme | |
| | Polyclonal | pre NBF | p415–427 | WB | Genzyme | *c* |
| | | R domain | p724–746 | WB | Genzyme | *c* |
| MATG 1016 | $IgG_{2q}$ | | p101–117 | WB IP | Transgene | *d* |
| MATG 1016 | $IgG_1$ | | p101–117 | WB IP | Transgene | *d* |
| MATG 1061 | $IgG_{2a}$ | NBF | r503–507/509–515 | IP | Transgene | *d* |
| MATG 1101 | $IgG_1$ | R domain | p722–734 | WB | Transgene | *d* |
| MATG 1102 | $IgG_{2a}$ | R domain | p722–734 | IP | Transgene | *d* |
| MATG 1103 | $IgG_1$ | R domain | p722–734 | WB | Transgene | *d* |
| MATG 1104 | $IgG_1$ | R domain | p722–734 | WB IP | Transgene | *d* |
| MATG 1105 | $IgG_1$ | R domain | p722–734 | WB IP | Transgene | *d* |
| MATG 1106 | $IgG_1$ | R domain | p722–734 | IP | Transgene | *d* |
| MATG 1107 | $IgG_1$ | R domain | p722–734 | WB IP | Transgene | *d* |

[a] J. Walker, J. Watson, C. Homes, A. Edelman, and G. Banting, *J. Cell Sci.* **108,** 2433 (1995).
[b] N. Kartner, O. Augustinas, T. J. Jensen, A. L. Naismith, and J. R. Riordan, *Nature Genet.* **1,** 321 (1992).
[c] P. L. Zeitlin, I. Crawford, L. Lu, *et al., Proc. Natl. Acad. Sci. U.S.A.* **89,** 344 (1992).
[d] See Transgene for information. See also N. Kortner and J. R. Riordan, *Method Enzymol.* **292,** 629 (1998).

TABLE XVI

Monoclonal and Polyclonal Antibodies to Photoreceptor cGMP-Gated Ion Channels

| Name | Subunit | Region | Peptide (p-) or recombinant antigen (r-) | Immunised species, nature of antibody | Uses | Source | Refs. |
|---|---|---|---|---|---|---|---|
| 1D1 | α | | Purified bovine rod outer segment protein | Mouse | IEM | Molday (British Columbia) | *a, b* |
| 2G11 | α | | | Mouse | | Molday (British Columbia) | *c* |
| 6E7 | α | 93–115 | Bovine ROS channel | Mouse | WB IP IH IF | Moday (British Columbia) | *d* |
| 1F6 | α | | Rod and cone channels | Mouse | WB | Molday (British Columbia) | *e* |
| 5E11 | β GARP part | | Purified 240-kDa protein | Mouse | WB IH IEM | Molday (British Columbia) | *d, f* |
| 4B2 | β GARP part | | ROS 240-kDa protein | Mouse | WB | Molday (British Columbia) | *f* |
| 3C9 | β part | | ROS 240-kDa protein | Mouse | | Molday (British Columbia) | *g* |
| PPc6N | α | 93–115 | ROS channels | | WB | Molday (British Columbia) | *a* |
| Ab331 | α | 93–102 | ROS channels | | WB | Molday (British Columbia) | *a* |
| PPc32K | β | 1292–1334 | Peptide 1292–1334 | | | Molday (British Columbia) | *e* |
| PPcCC1 | | | Recombinant channel | Rabbit AP | | Molday (British Columbia) | *e* |
| 63-4 | | | Recombinant channel | Rabbit AP | | Molday (British Columbia) | *e* |

[a] R. S. Molday, L. L. Molday, A. Dose, I. Clark-Lewis, M. Illing, N. J. Cook, E. Eismann, and U. B. Kaupp, *J. Biol. Chem.* **266,** 21917 (1991).
[b] C. Colville and R. S. Molday, *J. Biol. Chem.* **271,** 32968 (1996).
[c] T.-Y. Chen, M. Illing, L. L. Molday, Y.-T. Hsu, K.-W. Yau, and R. S. Molday, *Proc. Natl. Acad. Sci. U.S.A.* **91,** 11757 (1994).
[d] Y.-T. Hsu and R. S. Molday, *Nature* **361,** 76 (1993).
[e] W. Bonigk, W. Altenhofen, F. Muller, A. Dose, M. Illing, R. S. Molday, and U. B. Kaupp, *Neuron* **10,** 865 (1993).
[f] L. L. Molday, N. J. Cook, U. B. Kaupp, and R. S. Molday, *J. Biol. Chem.* **265,** 18690 (1990).
[g] H. G. Korschen, M. Illing, R. Seifert, F. Sesti, A. Williams, S. Gotzes, C. Colville, F. Muller, A. Dose, M. Godde, L. Molday, U. B. Kaupp, and R. S. Molday, *Neuron* **15,** 627 (1995).

The references given in Tables I–XVI refer to the main papers in which the antibodies were described. The name of the senior author or the commercial company is also given. Many of the companies distribute information via the Internet.

## Spontaneous Antibodies in Human Disease

One of the intriguing aspects of ion channel research is the extent to which they are involved in human disorders. Thus not only are they targets for genetic disorders[1] but also for autoimmune diseases in which spontaneous antibodies lead to loss of ion channel and neurologic dysfunction (Table I). Many of these spontaneous antisera have well-defined functional effects which have generally not been achieved with experimentally produced antibodies.

## Ligand-Gated Receptors

### *Acetylcholine Receptors*

The acetylcholine receptor was the first ion channel protein to be purified, cloned, and sequenced. Polyclonal antisera and monoclonal antibodies were raised against the purified detergent-solubilized, alpha-neurotoxin-purified protein before it was fully cloned and characterized. Polyclonal antibodies to individual purified subunits were also raised and polyclonal sera to defined sequences became availabile as the genes were cloned. As a result of the earlier work of Jon Lindstrom and Socrates Tzartos, there are many available monoclonal antibodies against *Torpedo* and human acetycholine (AChR), many of which cross-react with other species. Some of these bind to the main immunogenic region (MIR), a determinant on the two $\alpha$ subunits which is distinct from the neurotoxin-binding site (Table II). The studies performed on characterizing epitopes on AChR have provided paradigms for subsequent investigations on other ion channels.

Several antibodies raised against human AChR were also obtained by Paul Whiting, of which four are specific for the fetal form ($\alpha_2\beta\delta\varepsilon$) and do not bind to the adult form ($\alpha_2\beta\gamma\delta$); these monoclonals have been shown to bind specifically to the $\gamma$ subunit on Western blots of recombinant proteins (Table III). In general, though, anti-AChR antibodies bind much better to the native AChR than to subunits on Western blots, or to peptides. This phenomenon emphasizes the extent to which antibodies raised against the intact proteins do not recognize linear sequences. The exception, interestingly, is antibodies directed toward cytoplasmic epitopes. These antibodies, both monoclonal and polyclonal, often bind strongly to both denatured sequences and the intact (solubilized protein or permeabilized cells) molecule

[1] L. Ptacek, *Neuromusc. Dis.* **7,** 250 (1977).

(e.g., Table IV). They can sometimes be induced by immunization against the whole recombinant peptide. Thus, it has become clear that the easiest way to induce a specific antibody to an individual AChR subunit, for instance, is to immunize against a peptide sequence of the cytoplasmic domain. These observations on the immunogenicity of AChR have taught us useful lessons regarding the best ways to raise antibodies to other ion channels.

Animals immunized with purified AChR frequently develop signs and symptoms of experimental myasthenia gravis. Patients with myasthenia gravis have spontaneous antibodies directed toward their muscle AChRs that cross-react poorly with AChRs of other species. These antibodies are very variable in specificity between patients, however, and some MG sera show functional effects on AChR function. A particularly notable example is the complete inhibition of fetal AChR function by sera from women whose babies suffer from fetal paralysis leading to joint contractures and other abnormalities. These antibodies do not affect adult AChR function, and often the women themselves are clinically normal (see Table I).

Monoclonal and polyclonal antibodies have also been raised, the former mainly by Jon Lindstrom and colleagues, against neuronal forms of AChR (Tables V and VI).

### *Glycine and GABA$_a$ Receptors*

Glycine and GABA$_A$ receptors were first purified in the 1980s. The purified receptors were first used for the production of monoclonal antibodies, and later polyclonal antibodies to peptide sequences or to fusion proteins were raised (Table VII). Since the GABA$_A$ and glycine receptor subunits are highly homologous, antibodies specific for particular isoforms required the use of sequences from the N- or C-terminals, or cytoplasmic loops. Most of the antibodies work in blotting and immunoprecipitation assays across different species, but single amino acid changes can result in the loss of reactivity. For immunocytochemical studies, particularly subcellular localization by electron microscopy, antibodies directed at extracellular determinants have been the most successful. None of the antibodies have been demonstrated to affect function.

### *NMDA and AMPA Receptors*

Several groups have made monoclonal antibodies or polyclonal antisera to different peptide sequences of the various NMDA receptors (Table VIII) or AMPA receptors (Table IX). It is not clear whether there are advantages to the different preparations, and indeed a close comparison suggests that there are few differences. These antisera have been very productive in terms of immunohistochemical studies; they are not always as reliable when it comes to immunoprecipitation, and specificity for the subunit should not be inferred without direct evidence (see Refs. i and j in Table VIII).

Antibodies to some of the GluR isoforms have been found in a childhood form of epilepsy and in a few patients with paraneoplastic (cancer-associated) neurologic disorders (see Table I). Their significance is not yet clear.

## Voltage-Gated Ion Channels

### *Voltage-Gated Calcium Channels*

Attempts to raise antibodies to native voltage-gated calcium channels (VGCCs) have been complicated by the difficulty in obtained sufficient quantities of purified VGCC complex for immunization. Antibodies have been successfully raised against cytoplasmic epitopes derived either from recombinant fusion proteins or peptides. These antibodies are largely generated against the highly variable II–III cytoplasmic domain and generally show marked specificity for the appropriate $\alpha_1$ subunit (Tables X and XI). There appear to be few successful attempts to raise antibodies to the extracellular domains of VGCCs. An interesting technique to overcome these problems has been described by Gillard *et al.* (see Table XII). Monoclonal antibodies to the peptide neurotoxins $\omega$-Aga-IVA, specific for the P/Q-type VGCC ($\alpha_{1A}$ subunit), and $\omega$-CTx–GVIA, specific for the N-type VGCC ($\alpha_{1B}$ subunit), have been raised in mice. These antibodies can then be used to detect $\omega$-Aga-IVA or $\omega$-CTx–GVIA binding sites in paraformaldehyde-fixed tissue by incubation with the appropriate peptide neurotoxin prior to application of antibody. Specific antibodies to the different $\beta$ subunits have been generated by immunization with recombinant fusion proteins derived from the hypervariable carboxy terminus of the $\beta$ subunit gene.

Patients with the Lambert Eaton myasthenic syndrome have antibodies directed against $\omega$-conotoxin-MVIIC binding VGCCs. These antibodies are highly specific for the disease state, and bind poorly to other forms of VGCC. It is not yet clear whether some of these antibodies directly inhibit function, but they lead to down-regulation of VGCC in cell lines (Table I).

### *Voltage-Gated Potassium Channels*

Voltage-gated potassium channels (VGKCs) are very heterogenous, oligomeric proteins. Most of the antibodies have been raised against synthetic peptides, fusion proteins, or recombinant subunits. Monoclonal antibodies for C-terminal domains exist for many subunits (Table XIII). Polyclonal antisera can be used to demonstrate the distribution of VGKCs by immunohistochemistry and some immunoprecipitate solubilized VGKCs. The large number of groups involved, and the interest of pharmaceutical companies, means that many antibodies are available (Table XIV).

Antibodies to VGKCs are present in some patients with acquired neuromyotonia, an autoimmune disease that results in muscle twitching and cramps as a result of neuronal hyperexcitability. It is not yet clear whether these antibodies are mainly directed against a single subtype of VGKC; early evidence suggests that they may be widely cross-reactive or that there are several different antibody specificities in some sera (Table I).

*Voltage-Gated Sodium Channels*

There are few available antibodies to voltage-gated sodium channels. Polyclonal antibodies raised in rabbit against brain type 1 $Na^+$ channel p 465–481 recognize the $\alpha$ subunit in an intracellular loop between domains I and II on WB, and may be useful for immunohistochemistry and immunoprecipitation.[2,3] It is marketed by Alomone Laboratories. A polyclonal antibody to the III–IV loop region is obtainable from Upstate Biotechnology.

*cGMP-Gated Channels and CFTR*

A number of monoclonal antibodies have been generated against CGMP-gated ion channels of the rod outer segment and CFTR (Tables XV and XVI).

Commercial Products

Several of the antibodies or similar ones are available through commercial companies as listed below. Only those which are described in the commercial literature are listed in Tables I–XVI. It is worth checking with the companies for new products, and looking for information on the Internet.

Voltage-gated ion channel antibodies:

Alomone Labs
Headquarters
Shatner Center 3 PO Box 4287
Jerusalem 91042
Israel
Tel 972-2-652-8002
Fax 97202065205233
E-mail: alomone@netvision.net.il
http://www.alomone.com

CFTR antibodies:

Transgene S.A.
11 rue de Molsheim
67082 Strasbourg Cedex
France

[2] G. Gordon, H. Moskowitz, M. Eitan, C. Warner, W. A. Catterall, and E. Zlotkin, *Proc. Natl. Acad. Sci. U.S.A.* **27,** 8682 (1987).

[3] R. E. Westenbroek, J. L. Noebels, and W. A. Catterall, *J. Neurosci,* **12,** 2259 (1992).

Tel 33-388-279100
Fax 33-388-279111

CFTR antibodies:
Genzyme
http://www.genzyme.com

Calcium, sodium, potassium channel and glutamate receptor antibodies:
Upstate Biotechnology
http://www.upstate biotech.com or www.biosignals.com

Glutamate and GABA receptor antibodies:
Boehringer Ingelheim Bioproducts
http://www.bi-bioproducts.de

Human acetylcholine receptor antibodies:
Serotec (Immunotec)
22 Bankside, Station Approach
Kidlington, Oxford OX5 1JE
England
Tel 44-1865-852700
Fax 44-1865-373899

Potassium channels antibodies:
Merck Research Labs
Rahway, New Jersey
07065 USA

## Acknowledgments

We thank S. Tzartos, J. Lindstrom, H. Betz, R. Huganir, R. S. Molday, M. Li, and J. S. Trimmer for sending information.

# [37] Internet Information on Ion Channels: Issues of Access and Organization

*By* Edward C. Conley

## Methods for Making "Biological Sense" of Genome Information

The use of the Internet (and in particular the World Wide Web) as a device for systematic interrelation of detailed scientific information is now well established. The medium has shown remarkable development within



0076-6879/99 $30.00

the last few years, to the point that web-based client–server technologies are dominating the design of new operating systems. These forms of distributed network computing will have major applications in biology, particularly where the computational power can lie at a remote site. Even today, several public domain and commercial databases can perform computation-intensive processes from a simple web browser irrespective of the host computer used. This is possible because the client application generally focuses on a single datatype and searches can be specified by an on-screen interactive form (e.g., querying genomic sequence, tissue-selective cDNA sequence, amino acid sequence, genome map position, structural coordinates, bibliographic data). Beyond these examples, however, few applications help make biological sense of the burgeoning sequence data collections. In consequence, while much information can be retrieved about *individual* proteins that contribute to the electrical properties of cells (the excitability proteins[1] including ion channels), tools for understanding their *interactive* properties (e.g., as multiprotein machines) in defined cell-type lineages are rare. Meanwhile, the expansive output of experimental data pertaining to native and cloned ion channel expression, sequence analysis, structure–function, electrophysiology, and pharmacology makes it predictable that these fields will become increasingly information-driven. To enable scientists to cope with this burden of new information (particularly outside their immediate discipline), the Internet should be able to play a key role in distributing networked tools for data organization, retrieval, visualization, and mining. In serving this role for understanding the electrical functions of cells, it is thus timely to consider issues such as access, organization, and federation of different types of ion channel data using the Internet as a central coordinating tool. Despite current optimism shown by proponents of "bioinformatics," it should be emphasized that the process of interlinking (federation) of more than a few different datasets will be a complex and labor-intensive process (and one that will not happen without a great deal of coordination and specific investment). Present techniques cannot easily make accurate predictions about the biological function of ion channels from raw sequence data. Despite this, databases can efficiently correlate similarities in sequence with respective known functions; this aspect is explored in detail later. Internet databases actually have very limited applications as storage devices—it is their application in answering scientific questions that makes them interesting and important for the future.

[1] H. A. Lester, *Science* **241,** 1057 (1988). See also H. O. Hsu, E. Huang, X. C. Yang, A. Karschin, C. Labarca, A. Figl, B. Ho, N. Davidson, and H. A. Lester, *Biophys. J.* **65,** 1196 (1993), which combined heterologous coexpression of channel types and numerical simulations to analyze mechanisms of action potential generation.

The dynamic content and direction of Internet applications are so changeable that it is a mistake to pay too much attention to its present state of development. Significant, common problems arising from slow data transmission speeds, excess traffic, and incompatible software will motivate investment in new technologies, expanding the uses of the web beyond those originally conceived. This article can therefore only be offered as a perspective on the magnitude and difficulty of the problems to be solved in creating a federated collection of ion channel databases. Perhaps the greatest of these problems is not technical but societal, connected with the "use culture" of Internet resources. For instance, how do we bring a sense of ownership to Internet-sourced information so that it gains credibility and value among respective specialist groups? Peer review will probably *never* work with large sets of on-line information, and this will require fundamental shifts in the ways that accuracy of data is verified. As suggested below, new types of modeling, simulation, anatomic atlas, and sequence analytical software which retrieve and filter data from Internet repositories will help break this new ground. In this novel mode of information gathering, Internet searches will be performed for real data that are needed for integration with a user's particular experimental system: In essence, to make biological sense of the highly complex sets of relationships between specific channel genes/proteins and multiple types of functional information, the structure and meaning that the above classes of software can provide are required. Overall, there appears to be a need to combine retrieval of "curated" information on ion channels with systematic access to collections of web-based analytical programs that perform a specified range of tasks. This underlines an essential aim of data organization via the Internet: complete flexibility for use of retrieved data. New data accession methods will be invented by reference to how data are used; for the ion channel biologists access will most likely take place from an interface that specializes in types of questions he/she needs answered. While there are a number of excellent web sites offering small applications (e.g., Java applets) and other resources for molecular signaling[2] these are more suited for teaching than research. Dedicated effort is still required to provide a truly interdisciplinary web-based data accession model for the ion channel community. A prototype mechanism for interdisciplinary accessions is given by the NCSA's Biology

[2] Some hyperlinks to these sites have been placed on the *Ion Channel Network* web site (www.le.ac.uk/csn/index.html) as an on-line companion to this article. The site supports the development of systematic nomenclatures for all ion channel gene families as an intuitive method for data organization and accession on the internet, thus illustrating the database federation strategies described herein.

Workbench project,[3] which offers a strategy for federation of multiple ion channel database collections (Fig. 1), at least for cloned ion channels. The Workbench protocols already form an advanced, working model for multiple ion channel sequence and structure database integration across the web. In addition to specially written routines that work within any web browser, the Workbench approach uses the Internet to deliver datasets for use with plug-ins or helper applications installed on local machines that permit display of data within the web page. While structural biology is well served by these tools,[4] use of this approach with other key information types in molecular physiology and pharmacology offers a new strategy for interoperating presently disparate datasets (Table I). These information-handling technologies will clearly need to span a wide range of disciplines, including web-accessible single-cell electrical modeling/simulator tools, annotated gene expression atlases, and sequence analysis packages. This is an exceedingly demanding specification.

## "Compacting" Channel-Specific Information in "Federated" Cell Electrical Simulation Databases

As introduced above, the next few years could see the development of many more types of web-based tools (or web-linked helper applications) able to analyze the origin and control of electrical signaling in cells, tissues, and even whole organisms. At the cell-type level, several of these approaches will build on the original Hodgkin–Huxley (H-H) model for electrical excitability, as the classic model for quantitative computational simulation. In its embodiment within the neural simulation/database programs GENESIS[5] and NEURON,[6] the H-H model provides a link between the

[3] The Biology Workbench is being developed in The Department of Molecular and Integrative Physiology, Center for Biophysics and Computational Biology, National Center for Supercomputing Applications (NCSA, Urbana, Illinois); see bioweb.ncsa.uiuc.edu, which includes a link to the generic Workbench interface. The specific application of the generic toolbox to ion channels under the title *Ion Channel Workbench* has been proposed and is under review at the time of writing.

[4] Software tools for structural biology are covered in detail within other articles of this volume and are only briefly described here.

[5] An aim of the GEneral NEural SImulation System (GENESIS) simulator (see also footnote 24) is to federate itself with a growing number of web databases containing neuroscientific data. Of pertinence to this article, this site discusses inherent limitations in conventional database designs including accuracy and relevance of the data entered, conflicting data, data compression, promoting participation and connection between the data and their functional significance. See also J. M. Bower, *Trends Neurosci.* **15,** 411 (1992).

[6] NEURONDB, part of Senselab (senselab.med.yale.edu/neurondb/) provides an advanced set of web-based tools to support research on neuron properties. These properties include

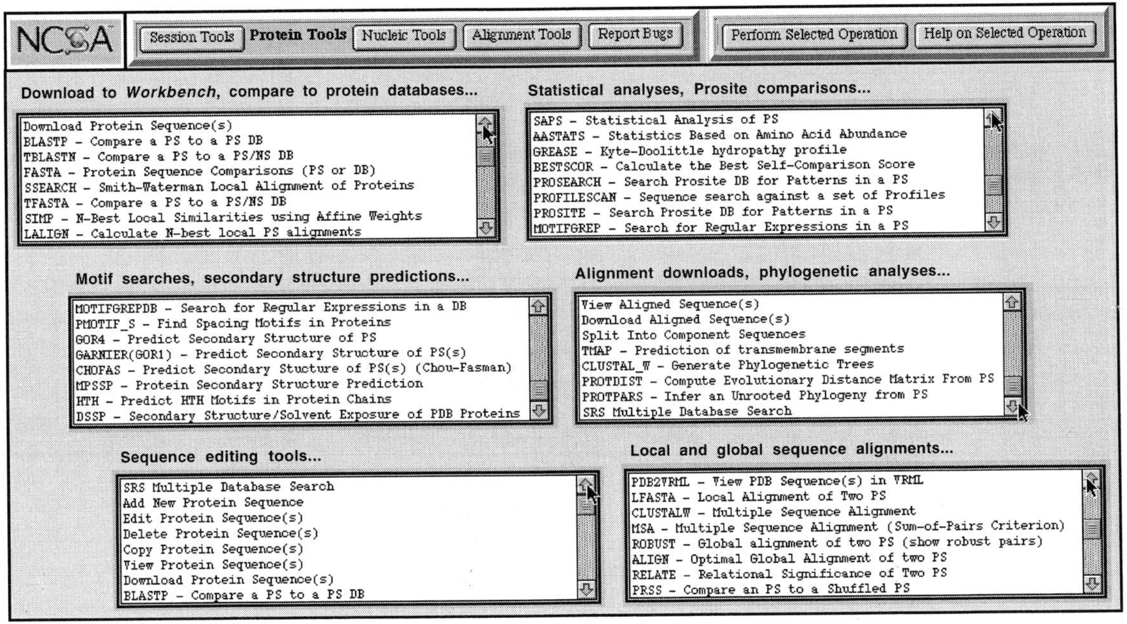


FIG. 1. Sample functions of analytical programs in the Biology Workbench model being developed at the National Center for Supercomputing Applications (NCSA). The Biology Workbench (biology.ncsa.uiuc.edu), created by the Computational Biology Group of the NCSA provides a working model for database federation applicable to ion channel information distributed on the web. A proposal for an interoperating system also incorporating physiologic, pharmacologic, and gene expression data of the types described in the text (the Ion Channel Workbench) is under review. Presently, the Workbench approach provides seamless integration to numerous molecular biology databases and analysis programs for such purposes as homology searches, sequence alignment, and creation of phylogenetic trees from a web browser interface. The Workbench technology is being extended to structure analysis, refinement, and prediction together with genomics and molecular simulations. The Workbench analysis environment thus achieves interrelation of multiple databases (between which information can be shared) and a set of tools to manipulate the data. Uniquely, this environment is accessible by all researchers free of charge, reducing dependence on local workstations by providing network access to the NCSA's formidable computational power. The Workbench is both sequence and object centric: every operation performed by the user is targeted at performing some task with one or more sequences (the objects can be databases, tools, sequences, alignments or structures; as long as the databases have commonly describable objects, they can become federated).

TABLE I
KEY AREAS FOR INTEROPERATION OF ION CHANNEL DATA TYPES VIA THE INTERNET

| Area/topic | Need/criteria | Potential solutions |
|---|---|---|
| Internet-retrievable parameter data for single channel types | Compact encapsulation of electrophysiologic data for cloned ion channels (see text) and their mutants | Simulator elements for tools like the Modeler's Workspace (see text) |
| Internet-retrievable parameter data for specified cell-type models | Functionally connected databases of cell-type electrical activities with high fidelity to experimental data | Simulator elements for tools like GENESIS and NEURON (see text) |
| Internet-retrievable, tissue-selective *in situ* ion channel expression atlases | Visual, searchable representations of channel mRNA and protein distributions, mapped to anatomic atlases | E.g., for brain, anatomic template-based databases (following neuroanatomic atlases; Fig. 4) |
| Internet-retrievable 3-D developmental gene expression datasets | Resolution of expression lineages/developmental fates (late embryonic to adult stages correspond to database of tissue distributions) | Jackson Labs/Edinburgh/MRC GXD (Gene Expression Database, see Fig. 3) |
| Analysis of *cis*-acting, genomic expression elements and *trans*-acting factors | Analyzing expression determinants in genomic primary sequence against Internet-accessible database of elements and factors | TRANSFAC database or similar (see text) |
| Controlled vocabularies for Internet searches on defined channel variants | Single-access point for resolving naming conflicts using declarative queries from web browsers | Web sites supporting systematic nomenclatures that can be formulated intuitively by researchers (see text and Fig. 5) |
| Comparative sequence analyses (nucleic acid and protein, individual and multiple) | Web-accessible genome and cell-type selective multiple sequence retrieval and comparative analyses | Biology Workbench-type toolbox programs (Fig. 1, see text) |
| Bibliographic object linking | Unique web-addressible hyperlinks connecting to citation-plus-abstract detail. Option for linking to full-text retrievals subject to access privileges | Simple, intuitive, linking from any other database on the web without knowledge of accession numbers (e.g., PubMed citation matcher, see text) |

voltage-gated conductances, neurotransmitter receptors, and neurotransmitter substances. The senselab project (funded by the Human Brain Project) has pioneered the federation of different databases (focused on the olfactory pathway) as a means to collect, analyze, and disseminate multidisciplinary neuroscience information.

properties of individual ion channel proteins and the complex behavior of networks of anatomically and physiologically "realistic" nerve cells (see footnote 24 later). An emerging concept is that many electrical modeling parameters (originally conceived as mathematical abstractions) have correlates in amino acid sequences of proteins contributing to cell-type electrical activity. Thus, many complex electrophysiologic behaviors (action potential shapes, firing patterns, modulation properties, etc.) can be quantitatively described within a cell-type model by interaction of known protein expression/distributions with their unique functional properties (gating, modulation, ion selectivity, etc.) As discussed on the GENESIS website (footnote 24) simulations proabably represent the most compact form of data possible. In principle, a simulation capable of replicating all features of a biologic system it mimics can reconstruct whatever dataset is of interest from first principles.[6a] For example, a correctly parameterized Hodgkin–Huxley model can contain all the information necessary to reconstruct voltage and current clamp records for a single population of channels. Simulator approaches that incorporate real morphology (i.e., 3-D photomicrograph images of cells) and channel parameter data in their optimization strategies have the capacity to form functionally connected databases for both cell-types *and* the ion channels expressed in them.

This functional connectness has major advantages for representing/reproducing extremely large datasets about ion channels and for making them accessible over the web. For example, quantitative modeling permits incorporation of time series data from ion imaging or patch-clamp electrophysiology, and can compare how well simulations reproduce real cell-type behavior. Moreover, evolutionary algorithm (EA) computational methods[7] are capable of recombining real channel parameter datasets to select high-fitness solutions so that (in effect) they can be used as databases of cell-type electrical activities by experimentalists (Fig. 2). For the first time, the EA approach directly associates electrical modeling and ion channel expression, while simulators form part of an integrated computing environment spanning data from genes/gene products to patterned electrical activity in whole organisms. Other (prototype) simulation environments have been created enabling selection of cellular morphologies, cellular compart-

[6a] This function is analogous to the more familiar example of flight simulators (as used for training pilots), which also compute the behavior of a complex system with high concordance to real-world measurements. Simulations optimized using evolutionary algorithm techniques (Fig. 2) can achieve ultra-high fitness in reproducing the behavior of biological systems, and should reasonably offset the skepticism which some experimentalists have for modeling approaches.

[7] R. M. Eichler-West, E. De-Schutter, and G. L. Wilcox, "Using Evolutionary Algorithms to Search for Control Parameters in a Nonlinear Partial Differential Equation in Evolutionary Algorithms and High Performance Computing." Springer-Verlag, New York. 1997.

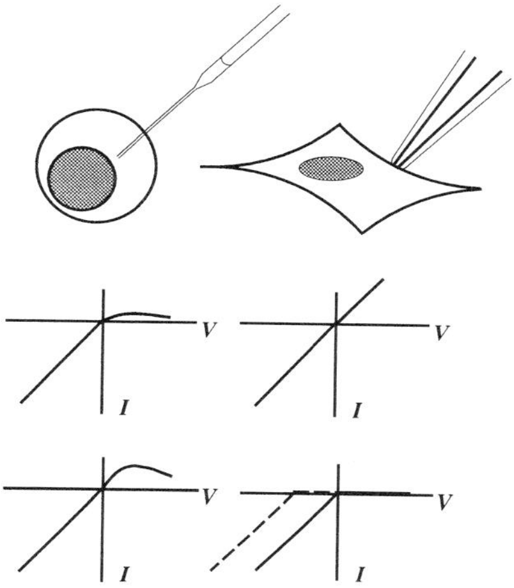


FIG. 2. (A) **Associating cell-electrical activity and ion channel functional expression data through Internet-distributable models.** Compacting large datasets on single subtypes of ion channels expressed in heterologous cells for Internet distribution. In principle, large amounts of ion channel-specific functional data can be distributed over the Internet in a compact form by adopting cell-electrical simulator approaches (see text). Mathematical representations of specified heterologous cell/channel combinations could be accessed from distributed sites on the Internet by constructing hyperlinks using restricted vocabularies capable of distinguishing all ion channel variants (Fig. 5). This approach would also render much physiologic and pharmacologic data on thousands of cloned channels and their mutants interoperable with the sequence, structure, and gene expression databases described in this article. By analogy to sequence and structure retrieval methods, a local simulator application could download and open correctly parameterized models (see below); for example, these models might contain all the information necessary to reconstruct voltage and current clamp records for a single population of channels. The approach would also be applicable to efficient description of interactive properties of channel components with defined receptor/second messenger systems or pharmacologic modulators. In these cases, raw data or signal transduction rules affecting suppression or potentiation of the channel activity could be incorporated by users (e.g., when coupled to specific receptor subtypes or in the presence of modulators). In contrast to arbitrary listings of unlinked facts about a defined channel subtype, simulators retain functional connectness and permit diverse functional properties to be combined *in silico* (see below). Application of these techniques would harness the Internet to make a greater diversity of datatypes accessible to molecular biologists, physiologists and pharmacologists. See Fig. 2(B).

ments and channel-type simulations from a database of such objects.[8] These objects can be combined into a model that can be simulated by GENESIS or NEURON. Conceivably, this interface could be applied to recombinant channel expression options within standard heterologous cell-type models

[8] For example, a prototype for Caltech's database and model construction interface known as the Modeler's Workspace maintained by Jenny Forss can be accessed at http://smaug-gw.caltech.edu:8081.

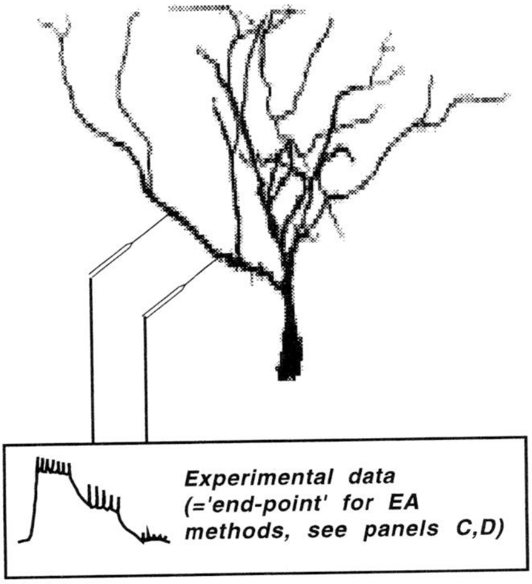


FIG. 2. (B) Parameter optimization as the rate-limiting step for establishment of databases for complex cell types. Most simulation work to date has focused on much more complex systems than the single-channel/heterologous cell datasets described above. Much effort has been directed toward the reproduction of native neuronal electrical activities by derivation of appropriate channel parameter sets for incorporation into models. In constructing these, trial-and-error approaches to parameter optimization generally require unworkable amounts of computing time, even for supercomputers. Recently, however, the use of evolutionary algorithms (EA, panel D) has demonstrated an efficient and robust method for optimized parameter fitting in complex systems where the end result has been (or can be) experimentally determined. This method, pioneered for CNS simulations by Rogene Eichler West and George Wilcox, is capable of producing high-fitness models that can potentially be used as databases by experimentalists. Although still not applied in the context of a simulator like NEURON or GENESIS (see text), EA methods may represent an essential approach for integrating large quantities of ion channel information in models prior to Internet distribution. Conventional methods of information handling would not be able to cope with the morphologic variability of native cells (e.g., neurons), the overwhelming diversity of channel subtypes, their interactions with other molecules, their nonlinear subcellar distributions and modulatory properties. See Fig. 2(C).

[e.g., oocyte, COS cell, Chinese hamster ovary (CHO), cell, etc.] efficiently representing the behavior of thousands of cloned channels and their mutants: Internet distribution of parameter sets on a channel subtype or variant (specified by a systematic name, see Fig. 5 later) that loads into the simulator (as a helper application or web-based interactive session) is much more efficient than conventional databasing of large quantities of raw (descriptive) data. Importantly, this could make electrical modeling techniques more accessible to molecular biologists, physiologists, or pharmacologists who might not otherwise be inclined to interact with computer models. Simulator approaches could also help minimize errors when attempting to match currents produced by native channels to those formed by recombi-

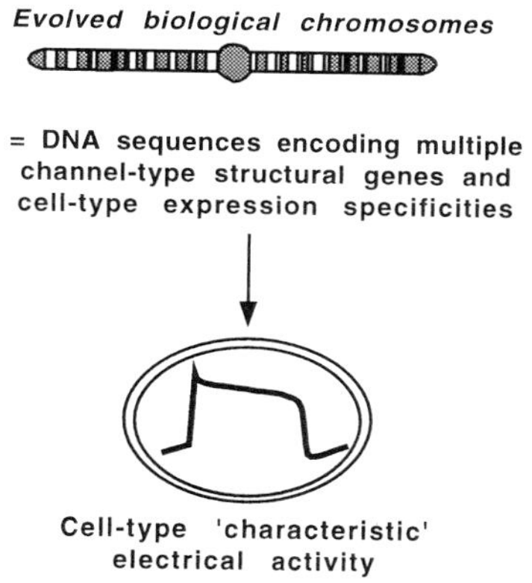


FIG 2. (C) Evolution of cell-type selective channel expression, subcellular distribution, and assembly patterns. Ancient cell-based life forms (surviving in ion-rich aqueous environments on the primeval earth) would likely have acquired ionic partitioning mechanisms at an very early stage. Molecular elements capable of selective ion transfer across cellular and intracellular compartments would also likely to have been important for driving primordial physiologic processes. Over the course of genomic, cellular, and organismal evolution, the genes encoding these basic components have been continuously reshuffled on evolving chromosomes by processes of gene duplication, segregation, recombination, and mutational divergence. These processes have produced immense variations in cell-excitation control (subject to selection for biological fitness within the individual cell-types of the organism). In contemporary cells, therefore, the origin and control of cell-electrical activities are primarily determined by the selective expression of ion channels and other excitability proteins. This expression-control is remarkably robust, and can account for the high stability of electrical signaling *in vivo* (for example, firing patterns characteristic of single identified neurons or action potential shapes in different parts of the cardiac conduction system). The centrality of ionic flux in the development of cell signaling (and the ability to characterize it at a single-cell level, see text) make a case for creating information systems that can analyze or reconstruct complex signaling patterns on the basis of expressed component parts. See Fig. 2(D).

nant channel subunits: without the rigorous testing for a wide range of functional parameters related by a cell-type model, it is too easy to conclude that a given channel subunit may contribute to a native current. Historically, superficial comparisons of native versus cloned channel currents (e.g., based on similarities of inactivation behavior or resemblance of pharmacologic sensitivity profiles) have led to major errors in phenotype interpretation. Cell-type modeling is sophisticated enough to generate tests for multiple alternative hypotheses for the origin of a measured current under defined experimental conditions. Thus, although ionic currents in a given cell are not biological objects like channel proteins, electrical models (incorporating protein distribution data; see below) may permit interoperation of native and cloned channel data. Construction of simulation databases with real channel data might also permit modelers to develop sets of predictive rules

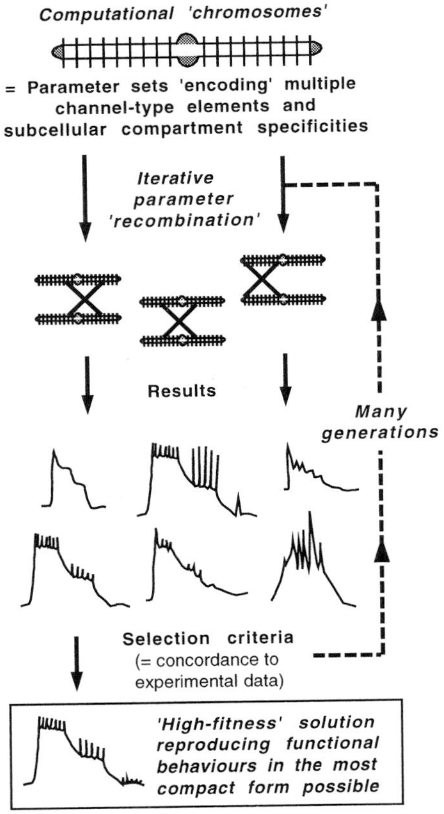


FIG. 2. (D) Recombination of channel type and subcellular distribution parameters evolving high-fitness models. By analogy to the processes of natural selection (above), computational methods based on EAs can generate solutions for cellular models that converge on factual (experimental) data for a cell type. The principal advantage of EA methods is their efficiency, finding solutions that would otherwise take hundreds of centuries of computing time by trial-and-error methods. Complex simulations that invoke computational chromosomes as multicompartment models (see figure), run independent simulations for many channel conductances within each compartment. Parameter sets (the chromosomes) thus link the electrical patterns a cell supports and the component genes it expresses. Under selection, populations of chromosomes generated from (initially) randomized parameter settings evolve toward high-fitness solutions (by criteria based on concordance to experimental data). Following this principle, those population members with low fitness become extinct, while those with high fitness are recombined by various crossover operators to produce the next generation of chromosomes (until a population with the desired fitness is achieved). EA methods are only limited by the criteria for fitness, which therefore must be defined with high precision. When high-fitness solutions are input into a simulator, the resulting model is capable of revealing the sensitivity of the system to various modulators of the component conductances.

to help fill-in missing data. In practice, simulators may network access (data mine) the parameter sets it needs to build its cell-type electrical models, and this will be aided by web-accessible data being organized in compatible repositories (see discussion of CORBA later). In addition to presenting compact interactive summaries of knowledge on the electrical signaling patterns of a cell type, sophisticated modeling approaches such as these may find application in analysis of cellular plasticity associated with multiple mechanisms of channel modulation.

## Significance of Developmental Gene Expression Data

Although neurophysiologic applications will continue to dominate developments in database models, the scope of the nervous system alone may actually be too limiting for many ion channel biologists. Certainly there are many ion channels that are not detectable in the nervous system (although such conclusions need to be tempered by the fact that a channel may still have a functional role while being of very low abundance). A more compelling reason for extending database efforts beyond the nervous system, however, is that signaling protein complexes incorporating the same channel proteins occur in multiple cell-type lineages (and that there is value in comparing their respective functions). Multiple cell-type expression reflects a dynamic pattern of ion channel gene activation (or silencing) that is characteristic of the ion channel gene in specified developmental compartments and time windows (compare Fig. 2). It is in these combined patterns of gene expression that underly corecruitment of appropriate channels, receptors, pumps, transporters, etc., to a cell-type lineage or a developmental compartment. Web-accessible gene expression databases will undoubtedly be needed to comprehend and query these complex developmental processes that ultimately control excitability in every tissue. Furthermore, the expression properties of intercellular channels (i.e., connexin proteins that form gap junctions) are of special significance to understanding mechanisms of developmental compartment formation. In developing multicellular organisms, intercellular channels establish both compartment formation per se, and resultant morphogen diffusion gradients that play an central role in deciding embryonic cell fate (and hence set the rules for the next round of gene activation/silencing). Major advances are currently being made in terms of organizing web-based access to this type of expression data, mainly through projects like the Jackson/MRC/Edinburg mouse gene expression database (GXD, Fig. 3, which uses an extensive atlas of mouse development[9] for its templates) and various brain anatomy databases

[9] M. H. Kaufman, "The Atlas of Mouse Development." Academic Press, San Diego, 1992.

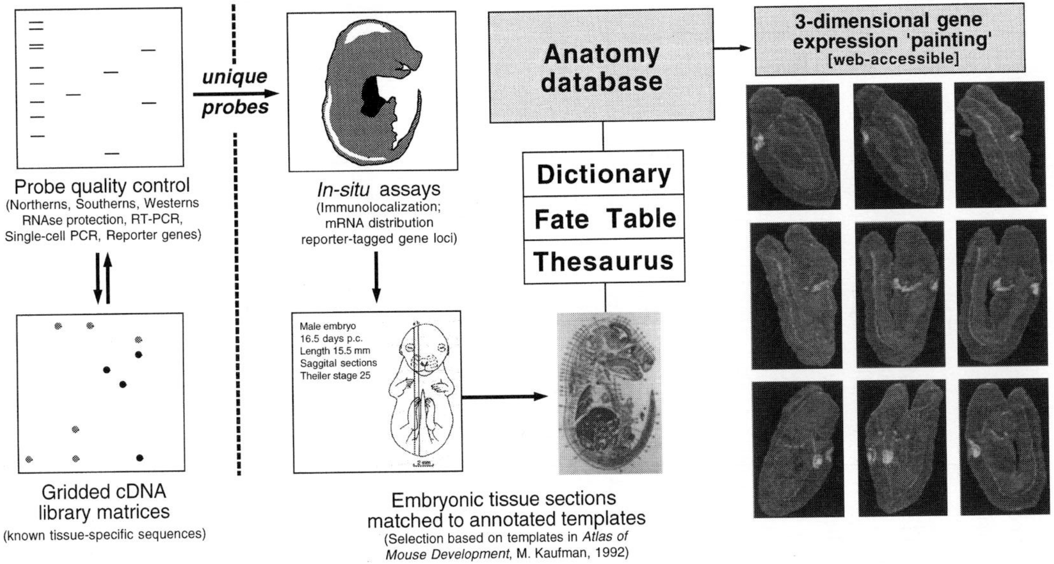


FIG. 3. Mapping mouse developmental gene expression lineages with GXD. The Gene Expression Database (GXD, http://genex.hgu.mrc.ac.uk/) aims to integrate multiple types of expression data (e.g., immunolocalization, mRNA localization, and lower resolution assays based on Northern and Western blots, RT–PCR, RNAase protection) and place it in developmental context. In its *in situ* applications, the GXD has much to offer the ion channel community, particularly for Internet accession of developmental gene expression information mapped to named anatomic structures (see text). Stage-specific cDNA libraries are capable of analysing broad changes in gene expression patterns as development proceeds, but *in situ* assay are capable of much higher resolution, albeit they are difficult to standardize. As emphasized in the figure, any comparative application demands rigorous quality control for molecular probes or gene reporters for the *in situ* applications (demonstrating at minimum, unique sequence, low background signals). For *in situ* expression assays, the textual annotations in GXD are complemented by 2-D images of original expression data that are indexed via the terms from the dictionary. The "time and space" description of gene expression is described by a controlled Dictionary of Anatomical Terms that is part of an Anatomy Database. The latter also provides a standard nomenclature for queries relating gene expression to developmental anatomy (compare Fig. 5). The dictionary names the tissues and structures for each developmental stage, while fate tables list the progenitor(s) and derivative(s) for each structure, enabling analysis of differentiation pathways. In the future, a 3-D atlas component will provide high-resolution digital representations of mouse anatomy (reconstructed from serial sections of single embryos at each representative developmental stage) enabling "painting" of gene expression patterns on to a 3-D graphical display (as shown). Images of serial histologic sections (generally from the same embryos illustrated in the landmark work "The Atlas of Mouse Development" by Matthew Kaufman) are being used to construct full gray-level, voxel images of mouse embryos at successive stages of development (approximately daily intervals) in which histologic detail is visible in any plane of section. This information and the standardized nomenclature will be held in an object-oriented database which will be accessible via the Internet.

(Fig. 4). We should strongly emphasize that these projects are very much in their infancy, and that expression data entry demands rigorous quality control if different sets of results are to be compared *in silico*. Furthermore, while *in situ* hybridization and immunolocalization techniques have the highest positional resolution, they are not trivial to standardize or perform.

Multiple gene expression patterns (mapped to different regions within a brain slice or whole embryo) will ultimately need correlation with the various sequence elements upstream of multiple ion channel structural genes. A web-accessible database known as TRANSFAC maintained at the GBF[10] on eukaryotic *cis*-acting genome regulatory elements ("sites") and *trans*-acting proteins (transcription factors that bind to and act through these sites) is of potential application to this problem. Interoperation of these data in annotated gene expression atlases will form a bridge between cell types, the excitability proteins they express, and "emergent" electrophysiologic/pharmacologic properties demonstrated by experiment. When these tools are in place, one could begin to envisage Internet-accessible ion channel expression data federated with *in situ* mapped functional data. In the case of the web-accessible brain databases, neurophysiologists, neuropharmacologists, and molecular biologists could conceivably share digitized anatomic atlases for interaccession purposes. This is not feasible at the moment.

## Prediction of Signal-Transducing Molecular Assemblies Containing Channel Effectors

It is difficult to comprehend the vast amount of new primary sequence information that is being deposited in the web-accessible databases daily. This will not slow down for the forseeable future—rates of data throughput will increase when new sequencing technologies and additional centers for human genome sequencing come on-line. Various databasing approaches are being evolved to keep pace with these. In particular, knowledge from completely sequenced genomes (presently several bacteria and yeast, but eventually including a collection of model organisms like *Caenorhabditis elegans,* zebrafish, *Fugu,* and ultimately, human) will permit further insights into receptor and ion channel evolution. These "solved" genomes are now offering scope for modeling complete signal transduction pathways that are not yet feasible for higher eukaryotes. In the case of higher organisms, however, it may be possible to use databases to study the relationships between cell-type-specific gene expression patterns and the electrical phe-

[10] The Gesellschaft für Biotechnologische Forschung mbH (GBF). TRANSFAC (http://transfac.gbf.de/TRANSFAC/index.html) is being piloted for CORBA compliance at the European Bioinformatics Institute.

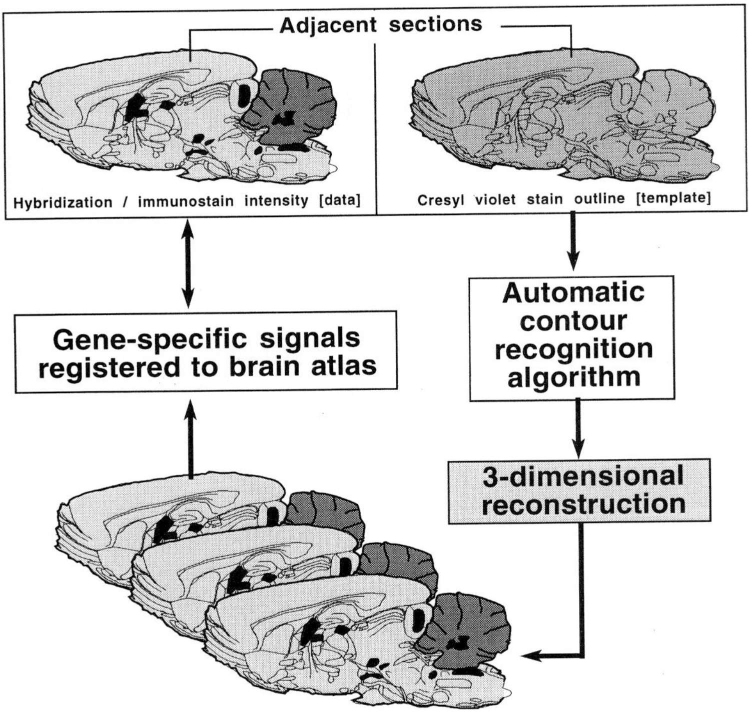


FIG. 4. Mapping ion channel, receptor, and other protein/mRNA distributions using automatic structure identification algorithms to attribute data to given neuroanatomic structures and/or cell groups. Neuroanatomic databases that can store, analyze, and compare multiple gene expression datasets present a great technical challenge. Software is being developed that is capable of superposing (registering, attributing) rat brain *in situ* hybridization/immunolocalization data for a unique molecular probe or reporter (see Fig. 3) on to a standard annotated neuroanatomic atlas. This approach [under development; see A. Danckaert *et al., Soc. Neurosci. Abstr.* 246.1, p. 616 (1996)] uses a 3-D reconstruction of adult Wistar rat brain (obtained by averaging many thousands of images collected from several animals) in both coronal and saggital cutting planes. Like the GXD (Fig. 3) the Wistar brain 3-D reconstruction relates to a standardized nomenclature provided by an established anatomic atlas (in this case, that of Swanson, 1992). In use, adjacent brain sections would be (1) stained with cresyl violet and (2) subjected to the hybridization/immunolocalization procedure. The Nissl (cresyl violet) stained section is then used as a template to drive the automatic attribution of the adjacent hybridization/immunolocalization signals to anatomic structures. For Internet-based ion channel molecular, physiologic, pharmacologic, and computational approaches, the data interoperation capacities of neuroanatomic databases may constitute a "Rosetta stone."

notypes of individual cells. For example, a by-product of interoperating transcriptional motif data (e.g., as in TRANSFAC) with cell-type-selective gene expression data [e.g., as in a GXD embryo template or a brain expression atlas, (Figs. 3 and 4)] may be the formulation of finite rules that can explain and predict combinatorial gene expression patterns in a given cell lineage. This introduces the concept that an ion channel (or other) gene expression pattern can be "solved" in the sense that the dataset will be essentially complete (i.e., like a protein structure or completely sequenced genome). Retrospective analysis of coexpression patterns for groups of structurally unrelated, but functionally interactive signaling proteins (deduced from querying anatomically mapped expression databases) could unveil much evidence for preferred subunit coassemblies. Evidently, these multiprotein signaling machines (comprising specified combinations of receptors/transducers and channel effector protein subtypes) constitute a basic operational unit resulting from processes of natural selection. The realization that multiprotein machines form critical functional units is apparent in many branches of biology. In following the functional interactions predicted from expression data, ion channel biologists have the important advantage that electrical signaling interactions can largely be verified at the level of the single cell type (and the results can be entered in the same anatomic database framework that was used to make the prediction). Other clues to the definition of these complexes may also be entered into the framework following diverse experimental approaches (protein structural interactions, phosphorylation–dephosphorylation sensitivities, modulator sensitivities, discovery of new protein assembly motifs, etc.). By integrating these types of observation, some signaling machines incorporating ion channels are known in great detail (e.g., several receptor–G protein–channel interactions) and the next few years will see characterization of many more. In deducing signal transduction pathways, it may be acceptable to omit transducer elements (e.g., G proteins) because the participation of these can be inferred by the type of electrical signal induced by the receptor agonist. In dealing with large amounts of such data, many people would find it useful to query hypothetical receptor–effector models in a "what-if" mode, e.g., what electrical responses occur in the (specified) channel effector if a given set of stimuli (voltage, ligands, modulators, etc.) is applied to the receptor–effector complex.

## Interoperability of Ion Channel Datasets through Systematic Gene Family Nomenclatures

Beyond the on-line sequence or bibliographic databases (and a few nonsystematic specialist data sites), there is presently little federated ion channel information on the web. Even if a set of core databases on ion channels could be originated, several unanswered questions would remain, such as the best methods for providing database infrastructure, prioritization for datatype inclusion/exclusion, data entry/interlinking/curation, etc. Experience gained with the pilot Ion Channel Network site (ICN, www.le.ac.uk/csn/) has, however, enabled some perspective to be gained on these questions. Certainly, the concept of a systematic framework (library) providing a "logical place to look" for information when the need arises is popular among users from many disciplines. As presently implemented, however, there has had to be a strong bias toward cloned ion channels forming any prototype for this framework. This limitation arises because of ambiguities arising when defining channels in native cells (the complex issues relating to "native versus cloned" channel identities are discussed on the ICN site). As a first step, therefore, much consideration has been given to systematic naming or specification for indexing information associated with specific ion channel genes/proteins. This process has also identified the need for a unified front end to a query-driven database framework specific to named ion channels. This front end would need to support declarative[11] queries on multiple databases (like genome/cDNA/protein structure/cell-type databases, functional data, and abstracted journal publications) maintained at (perhaps hundreds of) external web sites. For linking many primary data sources like this, formulation of unambiguous queries appears critical for valid data retrieval about a specified ion channel gene or protein.[12] To permit an analytical toolbox such as the proposed Ion Channel Workbench (NCSA, see footnote 3 and Fig. 1) to perform comprehensive correlations of sequence–function relationships for specified channels, naming must also be able to encompass many thousands of site-directed mutations created in the laboratory. It is certainly now feasible to apply systematic naming schemes for known ion channel genes/proteins (analogous to those first proposed for Kv channels and exemplified for database indexing purposes in Fig. 5).[13] There remains a schism between those classifications of channels based entirely on gene/protein primary sequence relatedness (as in the systematic nomenclatures used by the ICN) versus those that use functional attributes based on pharmacologic or physiologic properties. For the specific application of indexing (organizing) and retrieving (accessing) data, sequence-based classifications are far superior,

[11] A declarative query specifies what is being searched for (see example in text) rather than a specific navigation path, e.g., 'Data/Channels/Kir/Kir2/Kir2.3.'

[12] For accession outside of the Ion Channel Network/Workbench search engines, channels, and project divisions must be given distinctive names (i.e., would not tend to return false hits from random character strings when typed into a general web search engine).

[13] G. A. Gutman and K. G. Chandy, *Semi. Neurosci.* **5,** 101 (1993).

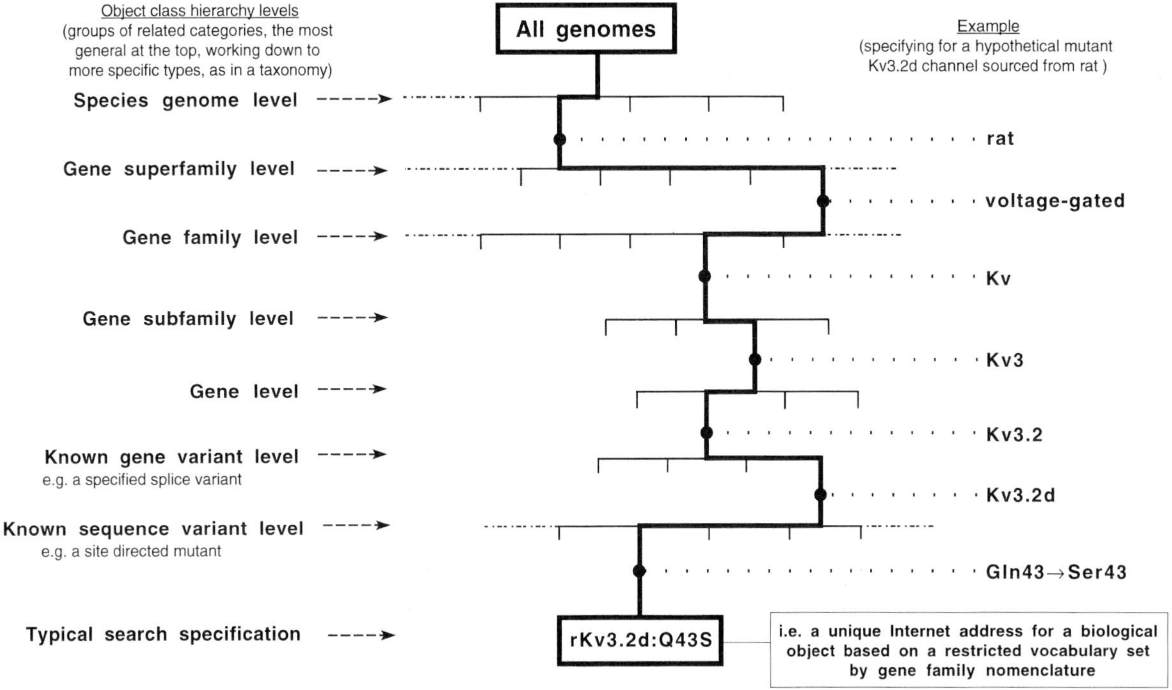


FIG. 5. Specifying unique Internet addresses for ion channel objects by adoption of systematic gene family nomenclatures as restricted search vocabularies. At least in relation to cloned channel genes and their products, systematic nomenclatures (see text) effectively establish defined search criteria for indexing/retrieving information on unique biological objects on the Internet. For those not accustomed to the derivation of systematic nomenclatures, the figure illustrates how a unique name can be used for retrieval purposes (in this case, a hypothetical site directed mutant of an alternative splice variant derived from a specific ion channel gene). Systematic nomenclatures could also be used to indicate the generality of data (e.g., whether the data applies to the same channel gene in different species or between channel genes in different subfamilies but the same family). Significantly, naming is based on sequence relatedness and not on functional properties, so the alphanumeric shorthand name indicates the natural gene family relationships. If this type of indexing scheme was adopted for excitability protein data across the Internet, simple alphabetical gene family prefixes (preferably with a mnemonic function like Kv, Kir, RYR, egc.) would need to be proposed and accepted for each gene family (or orphan gene). Gene family trees based on the simple criterion of amino acid identity (see text) have already been published for most channel and receptor families; in these cases application of numerical designations to the tree (by analogous means to the potassium channel families) is reasonably straightforward.

because they provide stable descriptors of biological objects (whereas functional criteria generally can not). The degree of amino acid identity between channels provides a straightforward and objective approach to defining subfamily nomenclatures. Using these simple principles, there should be no impediment to producing an internally consistent nomenclature for all ion channel gene families[14] (with which everyone in the field agrees and supports). For retaining specificity of data access in web searches (and in general literature/keyword searches), the acceptance of systematic nomenclatures as controlled vocabularies for use in index/search terms would be a significant step forward. In practice, this would mean specification of the channel variant by controlled vocabulary at the point of entering the searchable literature (e.g., in keywords assigned by the journal or abstracting service).

At this point, it is acknowledged that systematic nomenclatures cannot (and do not attempt to) solve the major problems inherent in native channel classifications. There is no simple equivalence between genes, native channels, and currents, so it is a major error to consider using the simple gene nomenclature (above) to index any property that cannot be directly related to the channel gene or gene product. Native channel currents will probably always be named on the basis of biophysical, pharmacologic, or physiologic criteria, and this multiplicity of approaches has led to some confusion in the literature. The continuous difficulties arising from native channel/current classifications raise a serious question about whether they can ever be applied with rigor and without controversy or uncertainty for search purposes. The dilemmas and limitations arising when classifying one class of ligand-gated channels by functional criteria have been exemplified within a draft document submitted by members of the IUPHAR committee on $GABA_A$ receptor nomenclature.[15] In this report, preliminary classifications were proposed on the basis that it was very difficult to equate subtypes recognized from recombinant coexpression with *in vivo* subtypes. Difficulties also arose from the complexity and inaccessibility of the CNS (and from the co-occurrence of multiple subtypes of GABA receptors in small regions or even a single neuron). Moreover, it was presumed that for multisubunit families of channels, many more receptor subtypes can be created *in vitro* than were likely to occur *in vivo*. These difficulties are illustrative of the incompatibilities between the molecular-centric and function-centric ap-

[14] To date, only the alphanumeric nomenclatures of the Kv and Kir families have entered use; the principle of adopting similar alphanumeric "tags" for other ions channel families is discussed under www.le.ac.uk/csn/e00/00-38-nomenclists.html.

[15] IUPHAR Committee Report on GABA Receptor Nomenclature, Sub-types of $\gamma$-Aminobutyric $Acid_A$ Receptors: Classification on the Basis of Subunit Structure and Function, *Pharmacol. Rev.*, submitted (1998).

proaches to classification, which if used together would create significant incongruities in web-indexed documents. Superimposition of gene-based information elements within the types of cell model-organized databases and annotated anatomic atlases described earlier may help heal the schism between these classifications. A further useful step would be the development of interoperating tools that permit receptor/channel protein distribution data to be correlated with functional data (e.g., patch-clamp data or selective radioligand binding to receptor subtypes) pertaining to real tissues and/or heterologous expression systems. For example, although many neuronal connectivity databases have been created to map neurotransmitter usage in identified brain regions, none of these is able to interoperate receptor or channel protein distribution data at any level. If this was done, it might help deduce rules for integration of many different types of receptor signals that inhibit or potentiate firing by opening or closing specific effector channels active in specified regions. Despite present limitations, therefore, the incorporation of native channel data into a molecular or anatomic framework should not be seen as an insoluble problem, but one for which the interoperating (interdatabase accession) tools are not yet available. Providing these is still a formidable task for software developers, who need to articulate these complex biological problems in great detail (i.e., understand the structure of the data) before they can commit to writing code.

### Data Structures able to Support Intuitive Access and Declarative Queries

As indicated above, continuing and extensive dialogue between ion channel biologists and software developers is crucial for realizing the concept of a federated set of Internet-accessible ion channel databases. Programmers often ask biologists to be more concerned with the abstract issues of data-type relatedness (e.g., by data modeling) than about the fine details of software structuring. Their message usually is that the requirements for various user-accessible naming systems or coding should be designed after one has solidified a data model. In the special case of naming ion channel genes/proteins for web access, ambiguity should be minimized, and the system should support intuitive formulation of search tags. Search tags are character strings that provide reliable database entry points by articulating a unique biological object in a complex system containing many thousands of such objects. Though often misunderstood, this is needed by biologists and has little to do with data structure. For example, a systematic gene nomenclature tag (like "rKv3.2d:Q43S" exemplified in Fig. 5) could be derived intuitively by users to retrieve information of relevance to the stated mutation.[16] Analogous names could be used to specify any other known channel variant from any species. Beyond this simple device for channel naming, a well-designed database will need to represent a natural set of the relationships between all data types about the specified channels (as in modeling approaches described earlier). In doing this, it should *not* be necessary to impose a hierarchy or any other relationship on the data (although it is important to articulate, in precise terms, what categorizations of information are important to retrieve). This need to broadly categorize ion channel information led to the proposal[17] of an "entry-fieldname" structure to break down the kinds of information people commonly need to retrieve (Fig. 6). Although this did not constitute a database design, it did at least provide a pilot structure for a web-based library of ion channel-related information. The limited structure shown in Fig. 6 was therefore proposed in response to broad, typical questions that biologists were imagined to ask for each channel subtype: What compounds block it? What compounds open it? What receptors are coupled to it? When does it appear during development? What diseases are associated with mutations in the gene? What is the activation threshold of the channel? In what cell types is it expressed? What is it coexpressed with? What happens when you knock out the gene? What does its current–voltage relation look like? What has it been called in the literature? What happens when you coexpress it with another subunit? Does it localize in some part of the cell, and if so, what does this mean in functional terms? etc. This "inquisitorial" model for database construction has a serious limitation that obstructs its usefulness—one can never anticipate all the questions (contrast with the simulator-type databases, above, which can replicate all features of the system). The fieldname approach (Fig. 6) or its future derivatives, will therefore only serve as an index to information, and *not* the sole hierarchical organizing system for it. Any new database design that emerges will need to be searchable by declarative queries[18] and this will require some reorganization of the entry-field model to support those queries based on the relationships asserted in the data model. Any data restructuring should also permit

[16] There may actually be any number of entry point "synonyms"—accession numbers, ID numbers, or something similar—to represent the same biological object in a computer model, but only systematic nomenclatures can be formulated accurately without prior knowledge of the name.

[17] Introduction and layout of entries, Entry 02 in E. C. Conley, and W. J. Brammar, "The Ion Channels FactsBook," Vols. I to IV. Academic Press, London, 1995, 1996, 1998, 1999.

[18] The terms entered in a declarative query [square brackets in example shown] may be a numerical range, experimental conditions, or possibilities chosen from a list; named ligands, compounds, toxins, or other modulators may be specified from within an external database.

**NOMENCLATURES [NOM]**
f01: Abstract/general description
f02: Category (sortcode)
f03: Channel designation
f04: Current designation
f05: Gene family
f06: Subtype classifications
f07: Trivial names

**EXPRESSION [EXP]**
f08: Cell-type expression index
f09: Channel density
f10: Cloning resource
f11: Developmental regulation
f12: Isolation probe
f13: mRNA distribution
f14: Phenotypic expression

**SEQUENCE ANALYSES [SEQ]**
f15: Protein distribution
f16: Subcellular locations
f17: Transcript size
f18: Chromosomal location
f19: Encoding
f20: Gene organization
f21: Homologous isoforms

**STRUCTURE & FUNCTIONS [S&F]**
f22: Protein MW - purified
f23: Protein MW - calc.
f24: Sequence motifs
f25: Southerns
f26: Amino acid composition
f27: Domain arrangement
f28: Domain conservation
f29: Domain functions
f30: Protein topography
f31: Protein interactions
f32: Protein phosphorylation

**ELECTROPHYSIOLOGY [ELE]**
f33: Activation
f34: Current type
f35: Current-voltage relation
f36: Dose-response
f37: Inactivation
f38: Kinetic model
f39: Rundown
f40: Selectivity
f41: Single-channel data
f42: Voltage sensitivity

**PHARMACOLOGY [PHA]**
f43: Blockers
f44: Channel modulation
f45: Eq. dissociation constant
f46: Hill coefficient
f47: Ligands
f48: Openers
f49: Receptor-transd. ints
f50: Receptor agonists
f51: Receptor antagonists
f52: Receptor inverse agonists

**INFORMATION RETRIEVAL [INF]** f53: Database listings f54: Gene mapping locus designation
f55: Miscellaneous information f56: Related sources & reviews f57: Feedback **REFERENCES [REF]**

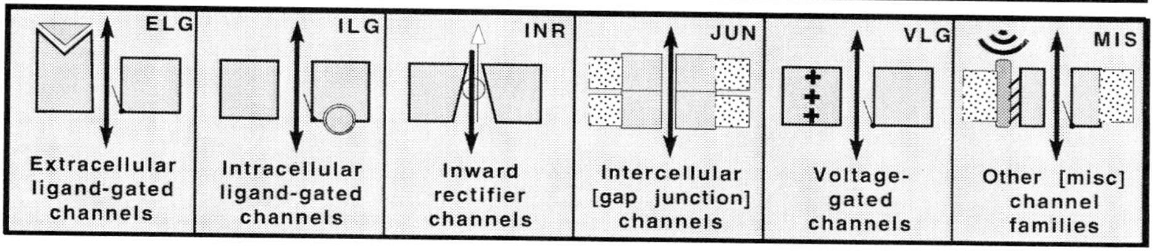

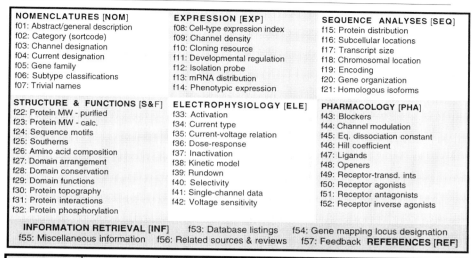

FIG. 6. A rudimentary organizational hierarchy for information about ion channel molecules. As further described on the Ion Channel Network site (www.le.ac.uk/csn/) information on named channel types can be grouped for accession in entries under common headings repeating in a fixed order. In the upper figure, broad sections (capitalized) each have related fieldnames delineating certain predefined topic areas (f01 to f57 as listed in the table). This breakdown of information into sections and fields when broadly applied across all ion channel subtypes (lower figure) is straightforward, but has clear limitations for accession of multiple datatypes on the Internet (see text). Despite these limitations, the entry-field framework still provides a useful model for indexing channel information, though *not* the sole hierarchical system for organizing it.

independence from (solely) keyword or full-text searching methods. A hypothetical example of a declarative query follows:

```
retrieve channel [systematic names] AND [PubMed links] AND [H-H pa-
rameters] that:
selects [potassium ions] AND
interacts with [any species] [PDZ 1+2 domain segment] AND
possesses consensus [≥1][protein kinase A consensus site] AND
possesses consensus [≥2][tyrosine kinase consensus sites] AND
immunodistributed to [Rat][Purkinje neurons] AND
```

```
immunolocalized to [any species][exclusively][postsynaptic densi-
ties] AND
potentiated when heterologously coupled to [β1-adrenoreceptors] in
[oocytes] AND
inhibited when heterologously coupled to [M3 receptors] in [oo-
cytes] AND
gene maps to [human] chromosome [14q]
where underlined terms are standard query types in the toolbox
while [bracketed] terms are choices or ranges (perhaps selected
from a pop-up menu of possibilities); limits to qualitative terms
such as 'selects', 'potentiated' or 'inhibited' would be defined by
specific criteria.
```

While this hypothetical example makes little biological sense, it does illustrate how a specialist user (such as a modeler looking for numerical parameters and data sources) could specifically retrieve that information from the network by means of a declarative query. Importantly, this approach does not set the database structure in stone (i.e., it is not as rigid as a conventional relational-type database) and more complex queries can be handled by elementary Boolean logic (e.g., by pop-up menus interposing AND, OR, NOT between query terms). All of these features are important to consider if the eventual database is to be easy to use, update, curate, and federate. Optimizing the design could incorporate the so-called "object-oriented" database paradigm (as used in GENESIS), which takes further the principles of identifying natural data relationships described above.

## Optimizing Accession by Object Specification and Indexing

In a proposed ion channel database referring to many data collections distributed around the web, objects could be broadly defined as anything stored that might be rationally addressed or queried by users. The specification of objects is thus crucial in the design of query and updating interfaces (which in general, should avoid showing implementation detail). Random examples[19] of objects drawn from the range of disciplines covered by a web-accessible framework might include (1) genetic maps [and the elements that go into building them, such as genes, yeast artificial chromosomes (YAC), primer binding sites, restriction sites, etc.], (2) protein structures (and any elements, such as structural motifs, residue types, ligand binding sites, etc.), (3) electrophysiologic records (within GENESIS/NEURON models and elements, like H-H parameters, see above), and/or (4) pharma-

[19] It should be emphasized that the random examples of objects and their elements given have no special significance in the database organization, but are merely illustrative of the kinds of information that might be included. The listings in Fig. 6 suggest many other potential objects.

cologic data (toxin structures, compound structures, potencies, selectivities, perhaps linking to a commercial database such as Ion Channel Modulators[20]). Applying an indexing scheme to named objects should make searching the database more straightforward for users unfamiliar with the data structure (but who, it is assumed, know how to use web-based forms to formulate Boolean queries). There is probably no limit to the number of objects one can define—objects can always be added or deleted if searches are ambiguous in practice. To coordinate efforts between multiple sites on the web is perhaps the most difficult task of all—and it is a problem common to any projects requiring interoperation of multiple databases. As a potential solution to this, a well-established object orientation model for linking biological databases is being piloted at the European Bioinformatics Institute (EBI, Cambridge, UK) using the Common Object Request Broker Architecture (CORBA). In simple terms, CORBA is an international standard specifying a coherent framework in which distributed applications can interoperate. CORBA actually consists of a number of software elements that include an interface definition language (used to specify the data as objects, see above), an object request broker (which manages requests between client applications and object implementations), an interface repository (which maintains details of existing interface definitions), and other elements that specify generic services (CORBAservices) and dynamic interface construction (DII). The CORBA standard is capable of delivering independence from operating system, programming language, file location, or data file format. As such, it will be crucial for the linking multiple biological datasets about ion channels, their distribution, or their functional activities (like those listed in Table I). CORBA's robust infrastructure will enable distributed applications and data sources to work together (although many new client applications will need to be written that make use of the standard to provide integrated views of the distributed data). Logically, some of these could become part of the Ion Channel Workbench toolbox (see footnote 3; Fig. 1).

## Toward a Universally Addressable Bibliographic Data Collection

In the coming years, web-accessible electronic journal archives containing primary data will form increasingly usable resources for all scientists.

[20] The Ion Channel Modulators (ICM) subset of The Investigational Drugs Database (IDdb) focuses on ion channels as therapeutic drug targets and monitors drug and biotechnology patents, chemical structures, pharmaceutical actions, and clinical reports. Since the majority of drug development data is indexed either by compound name, company, or broad mechanism of action, search engines may be developed that "link-out" to specialist subscription-based sites like these (providing appropriate access privileges exist).

It is perhaps useful to think of the electronic journal article forming component parts of a database repository of information that needs to be indexed to be accessed. Robust addressing systems have already been devised to retrieve dispersed bibliographic information about topics specified in web-based forms (e.g., the NCBI's citation matcher[21]). These systems are of great practical value, because web targeting to cited articles can be intuitive (by construction of a "formula" hyperlink), thereby freeing the user from needing arbitrary document ID numbers to achieve specific bibliographic retrievals. As web accession of full-text electronic journal articles becomes commonplace over the next few years, hyperlinks built from citation matching techniques will no doubt encourage their use as primary data repositories without the need to recompile the information in a centralized database. In principle, citation matching permits either a web author or reader to cite and/or access basic bibliographic data by input of minimal identifier elements for the article (links to the full-text article from publishers web sites may also exist if access privileges permit). Since a citation matcher is a bibliographic database front end, input of any unique combination of search terms (via a Boolean AND request in a web-based form) will be sufficient for correct retrieval. In practice, this is as simple as inputting (for example) "first author surname" AND "volume number" AND "first page number"—the search engine then returns documents matching all the criteria (see example, below). Citation linking can be applied to any existing reference list; commercially available reference manager programs can easily reformat a personal reference collection into hyperlinks so that the article abstracts are automatically retrievable on the web.

```
Example of a citation retrieval via PubMed citation matcher search
string using limited knowledge of volume number (=16), first page
number (=321) and author surname (=Slesinger) - [typed into URL box
on one line with no spaces].

http://www4.ncbi.nlm.nih.gov/htbin-post/PubMed/wgetcit?
display=abstract&format=html&volume=16&page=321&author=Slesinger
```

This exemplifies the power for database accession where object specification is derived by combining simple search terms to make a hyperlink.[22] This mode of search/retrieval can, of course, be applied to many types of channel objects, including ones named by systematic nomenclatures (Fig. 5). Automatic construction of hyperlinks (i.e., by software) from search term input

[21] An introduction to the PubMed system (which supports the citation matching principle) is given in http://www4.ncbi.nlm.nih.gov/PubMed/.

[22] In practice, citation matching approaches are far more flexible than exemplified here (e.g., using other search term types like journal name, year, title, etc.).

is thus likely to be an important technique for web database interaccession in the future.

## Internet Information on Ion Channels: Present Realities and Future Developments

Presently, a scan of the major Internet search engines (AltaVista, Yahoo, Lycos, etc.) with a text string specifying "ion channel" typically retrieves a heterogenous collection of web sites, researcher home pages, and a long list of citations within various bibliographic archives. Refinement of these queries can generate a few links of direct interest, but it is still more usual to be forced to "trawl" through large numbers of electronic documents (obviating the advantage of an electronic search system). In addition to their nonspecificity, these generic web retrievals may also suffer instability, so that bookmarks created at one point in time return "not found" errors when activated at a later date. Most significantly, generalized web searches (1) have no control or guarantee on the reliability or quality of information retrieved, (2) have no mechanism to permit direct comparison of similar types of information between members of a specified group of ion channels, and (3) lack a systematic framework for data comparison leading to uneven, inconsistent (or no) coverage on a topic. Evidently, these problems present formidable obstacles to accessibility of Internet distributed information on ion channels. While some individuals have invested great effort in providing sets of information on the web that overlap their special interest, there is still no "meta-system" that oversees the linkage between these potentially useful contributions (in practice, enabling interested readers to point their browsers toward the resource with reasonable expectation of finding the information of interest). Given the diversity of molecular entities and functions which constitute electrical signaling complexes, this is not a straightforward task, and before getting lost in possibilities, a critical, common sense appraisal of factors that make data-based[23] information on ion channel signaling gain added value over simple bibliographic citation is always valuable. Internet databases will be of maximum utility when users of modeling programs can extract the data they need quickly and transparently from across the web. The best organizational scheme should be linked to the functional relationships between the data (as exemplified in the GENESIS environment) representing the most natu-

[23] Throughout this article, the term *database* has been used in a limited sense of providing organized access to data; databases may actually take many different forms, e.g., a simulation program, a parameter model, an indexed list. The optimal form of data storage/retrieval will vary with the type of data.

ral form of data organization, precisely because it is intuitive. Using CORBA, it may be possible to develop generalized retrieval tools to help access data elsewhere on the web irrespective of an individual database's front end. In questioning whether the end results of a database are actually worth the protracted development efforts involved, an acid test might be whether a databasing, modeling, or simulation process enables researches to resolve questions which would otherwise be difficult or impossible to ask (as with the sequence databases). What data types could conceivably benefit from "organization" into collections of web-accessible databases? What added value will follow when data are in a searchable form? The invention of appropriate solutions to information retrieval demands clear communication between biologists who can provide domain expert knowledge and those who know how to plan and engineer user-friendly software (e.g., systems analyst/programmers). Such dialogue is also essential to capture the natural relationships between different types of ion channel data and therefore aid the design of analytical/search engines for attractive database front ends. As suggested by the developers of GENESIS,[24] interoperability and integration of data resources into digital libraries (data repositories) will motivate the development of information retrieval and other tools that provide (1) multiple forms of access to research data as well as (2) overviews describing the origins and contexts of the research, and (3) metadata (i.e., data about the data) indicating the appropriate uses of the data and its availability or distribution. In particular, metadata standards (e.g., CORBA, this article) will provide specialist users with information about such things as the content, format, availability, and appropriate uses of data, whereas data transfer standards will help facilitate the exchange of data over the Internet. With continued creative effort, these standards may enable federation of diverse datasets, helping to realize the huge potential of the Internet to solve problems in ion channel biology. Like a large jigsaw puzzle, however, many disconnected pieces must interlock before a bigger picture can emerge.

## Acknowledgments

This article has been largely distilled from discussions with others interested in web-accessible information framework centered on ion channel signaling. In particular, I wish to thank the following colleagues for their continued input: Bill Brammar of the Department

[24] See the GENESIS hypertext primer (www.bbb.caltech.edu:80/hbp/GOOD/GOOD.html), based on a chapter to appear in "Progress in Neuroinformatics" (S. Koslow and M. Huerta, eds.), Vol. 1. Lawrence Erlbaum Associates. See also J. M. Bower and D. Beeman, "The Book of GENESIS," 2nd Ed. Springer-Verlag, Berlin, 1997, which includes a CD-ROM with the GENESIS simulator TELOS.

of Biochemistry, University of Leicester; Eric Jakobsson and Shankar Subramanian at the Dept. of Integrative Physiology, National Center for Supercomputing Applications (NCSA), Urbana, Illinois; Gordon Shepherd, Jason Mirsky, Prakash Nadkarni, and Emmanouil Skoufos at Yale Medical School; George Wilcox and Ihab Awad at the Center for Neuroscientific Databases, Minnesota Supercomputer Institute, Minneapolis; Rogene Eichler West and Jim Bower at Caltech, Pasadena (and Dave Beeman for content quoted from the GENESIS web site); and Tom Flores at the European Bioinformatics Institute (EBI), Hinxton Park, Cambridge. I also wish to thank the Medical Research Council, The Department of Pathology, University of Leicester, and Zeneca Pharmaceuticals PLC for generous support.

# Author Index

Numbers in parentheses are footnote reference numbers and indicate that an author's work is referred to although the name is not cited in the text.

## A


Abad, C., 534
Abbott, J., 609(46), 623
Abe, T., 580, 583(54)
Abele, U., 518
Abogadie, F. C., 605, 611, 612, 613(18)
Abramow, M., 304, 306(2)
Abramson, S. N., 608, 609, 609(4), 610(4, 54), 623
Abramson, T., 582
Ackerley, C., 227, 230(3), 231(3), 241(3)
Ackers, G. K., 133, 135(33)
Adair, B., 152(100), 179
Adams, M. E., 27, 92, 93, 94, 96(14), 100, 101(3, 10, 11, 14), 108(11), 110(10, 11, 14), 111(9), 114(10), 605
Adams, P. R., 426
Adams, S. P., 537
Adams, S. R., 437
Addy, M. E., 637
Adelman, J. P., 299, 300(64), 655, 656(13)
Adjadj, E., 94
Adler, J., 460, 462, 465, 466, 510, 514(18), 518, 519
Adoutte, A., 511
Adriaenssens, P., 100
Adrian, M., 137
Aebi, U., 135, 144(17, 18), 147(17, 18)
Affolter, H., 296(49), 297
Agard, D. A., 165
Aggarwal, S. K., 634, 635(20)
Aggeli, A., 66, 83(30), 91(30)
Agnew, W. S., 467, 576, 578, 579(29), 589(29), 590(29), 604, 616(70), 624
Agre, P., 148, 179, 550, 551, 552, 553(3, 4), 554, 556(11, 12), 558(12), 559, 560(11), 562, 562(5, 14), 567(6), 570, 572
Aguilar, B. A., 655
Aguilar-Bryan, L., 445, 446, 454, 455(17), 456, 456(17)
Aguilar-Carreno, 674
Ahamed, B., 247, 249, 249(7), 258(7), 260(12)
Ahlers, M., 151
Ahmad, M., 439
Aiba, A., 3
Aiyar, J., 594, 598(121), 630, 651
Ajouz, B., 459
Akabas, M. H., 319, 579, 591(40)
Akey, C. W., 143
Akke, M., 98, 113(32)
Akong, M., 26, 46(18)
Ala, P., 150
Alam, M., 586
Albar, J. P., 645
Albuquerque, E. X., 187, 609(53), 623
Aldrich, R. W., 46, 640, 641, 641(8), 643(8, 12), 645
Aleksandrov, A., 299(65), 300
Alewood, P. F., 580, 613
Alkan, S. S., 108
Alkondon, M., 609(53), 623
Allard, P., 98
Allbritton, N. L., 191, 197(7), 199(7), 206(7), 207
Allen, A. C., 187
Allen, M., 649, 655, 656(12)
Allen, T. G. J., 191
Alon, N., 227
Als-Nielsen, J., 135(10), 136
Altenbach, C., 60, 73, 73(4)
Altenhofen, W., 699
Alvarez, O., 296(46), 297, 584, 585, 663
Amador, M., 45
Ambudkar, S. V., 552, 556, 556(11, 12), 557(18), 558(12, 18), 560(11)
Amos, L. A., 136, 143, 165(12)
Amudi, N. M., 200, 202(26)
Andersen, O. S., 208, 211, 212, 212(24), 214, 214(24), 217, 218, 219(42), 220, 221(27, 37, 38), 222, 223, 223(24), 224, 289, 290, 525, 526, 532, 532(10, 14), 534, 536(2, 35),

537, 540, 542, 542(8, 9, 11, 14), 543, 543(10, 34), 545(8, 11, 14, 36, 107), 546(101, 107), 549, 549(10–14), 550 (9, 35)
Anderson, D. J., 36
Anderson, J. A., 513
Anderson, M., 227, 241(2)
Anderson, P. A. V., 575(12), 576
Anderson, W. F., 495, 674
Andreasen, G. L., 126
Andree, J., 232
Andrews, G. C., 594, 598(121), 630, 651
Andrews, P. I., 678
Angal, S., 103
Angelides, K. J., 48
Angst, C., 389
Angus, J. A., 100
Anholt, R., 292
Anraku, Y., 514
Antoniou, M., 490, 492(5), 493, 495(19), 498(20)
Antz, C., 645
Anzai, K., 510
Apell, H. J., 77
Åqvist, J., 540
Aracava, Y., 187
Arata, Y., 94, 97(13), 101(13), 110(13)
Archin, N., 689
Aresen'ev, A. S., 533
Arias, H. R., 64
Armstrong, C. M., 601, 640
Armstrong, F., 413
Arnau, M. J., 116
Arneric, S. P., 36
Arrondo, J. L. R., 80, 82
Arsen'ev, A. S., 94, 536
Arshavsky, V. Y., 469
Arumugam, S., 537
Arvan, P., 319
Arvidsson, K., 98
Asai, M., 575, 576(3)
Assmann, S. M., 410, 411, 412, 413(6), 416, 419, 421, 423, 425(2), 429, 435, 436, 439(22), 440(2)
Asturias, J. F. J., 153
Atherton, E. A., 588, 589(89)
Atkins, A. R., 537
Auer, M., 142
Augustinas, O., 698
Augustine, G. J., 30
Auld, V. J., 594
Avila-Sakar, A. J., 150
Avron, B., 594, 596(126), 597(126)
Awobuluyi, M., 687
Axelsen, P. H., 86
Azimi-Zoonooz, A., 106, 615(29), 618
Azuma, Y., 394, 395, 397(28), 402

B

Babenko, A. P., 445
Bachmann, C., 446, 453(13)
Backx, P. H., 578, 579, 579(32, 33, 34), 591(32), 599(34), 603(32)
Baggetto, L. G., 232, 238(18)
Bahadur, G., 491
Baker, O. S., 577
Baker, T. S., 143, 174
Bakker, E. P., 481
Baldo, G. J., 601
Baldwin, J. M., 136, 165, 178(22), 179(85)
Baldwin, P. A., 208
Baleja, J. D., 613
Ball, J. B., 490
Ballarin, C., 664, 670(12)
Balser, J. R., 578, 579(37), 591(37), 604(37)
Bandekar, J., 60, 80
Bandyopadhyay, P. K., 613
Banker, G., 689
Bañó, M. C., 534
Banting, G., 698
Baraban, J. M., 354, 368(4), 554, 562(14)
Barak, Y., 342
Barbier-Brygoo, H., 422
Barchi, R. L., 575, 575(7, 10, 11), 576, 580, 590(48), 591(48), 603(48)
Bare, C., 440
Barhanin, J., 512
Barish, M. E., 300
Barker, J. L., 495, 672
Barklis, E., 151
Barnola, F. V., 589
Barrantes, F. J., 64, 208
Barrett, J., 609(42), 622
Barrezueta, N. X., 697
Barry, D. M., 697
Barry, E. L., 692
Barsukov, I. L., 536

Bartucci, R., 73, 74(51)
Bassen, D., 445
Basus, V. J., 92, 94(6), 117
Bates, S. E., 440
Baukrowitz, T., 241
Baumann, A., 521
Baumeister, W., 136, 144(17, 18), 147(17, 18)
Bax, A., 112, 536
Baxter, G. T., 190
Bayley, H., 137, 232, 238(12)
Bean, B. P., 94, 101(10), 106, 110(10), 114(10), 605, 615(29, 64), 618, 623
Bear, C. E., 227, 228, 228(7), 230(3, 7), 231(3), 232(11), 235(11), 240(7), 241(3, 7, 9, 10), 242(9), 299(78), 303, 319, 324
Beattie, R. E., 690, 693
Beaudet, A. L., 227
Beauwens, R., 304, 306(2), 309, 313(9), 315(9)
Beavis, R. C., 109
Beawens, R., 308, 314(7)
Becker, C. M., 261, 262, 262(3, 5)
Becker, M. D., 526
Becker, S., 577, 580, 581(49), 583, 583(23, 49), 586, 588, 588(49), 589(89, 90), 595(24), 600(24, 49), 601(24), 604(24, 49)
Beckman, R. A., 120, 125(8)
Beckmann, E., 136, 178(22)
Beck-Sickinger, A. G., 580, 581(49), 583, 583(49), 586, 588(49), 600(49), 604(49)
Beckwith, J., 477, 478
Bednarek, M. A., 630
Beech, D. J., 446
Beeman, D., 730
Beeson, D., 678, 681, 682
Beeston, B. E. P., 165
Begenisich, T. B., 585
Beggs, A. H., 375, 377(6), 381(6)
Beilby, M. J., 517
Beirao, P. S., 641
Bekele-Arcuri, Z., 694, 697
Bell, P., 578, 579(35), 591(35), 599(35), 603(35)
Bellamy, A. R., 143, 174, 494
Bellare, J. R., 155(103), 179
Ben-Azis, R., 265
Bencherif, M., 45
Benishin, C. G., 650
Bénitah, J.-P., 579, 604(45)
Benke, D., 686
Bennett, E., 590, 604(99)
Bennett, M. K., 52, 342, 343(5)
Bennett, M. S., 492, 494(14), 506(14)
Bennett, V., 552
Benos, D. J., 296(59, 60), 297, 312
Benz, R., 518
Berckhan, K. J., 20, 36(5)
Berdiev, B. K., 296(60), 297
Berendsen, H. J. C., 539
Berg, S., 296(54), 297
Berkowitz, G. A., 697
Berman, H. A., 126
Bermudez, I., 682
Bernal, J., 509
Bernard, M., 459
Bernardi, E., 630
Bernassau, J.-M., 94
Bernhardt, J., 356, 357(6), 360(6), 366(6), 367(6), 368(6)
Berrier, C., 459, 518
Berriman, J. A., 137, 155(105), 174, 179
Bertl, A., 512, 513, 513(33), 517
Bertrand, S., 117
Berwald-Netter, Y., 575(9), 576
Besnard, M., 518, 520
Betz, H., 260, 261, 262, 262(2–5), 263, 263(1, 2), 264, 265(13), 269, 270(16), 273, 686, 697
Beveridge, D. L., 539, 540(87)
Beyreuther, K., 261
Bezanilla, F., 640
Bezprozvanny, I., 296(55), 297
Bhat, R. V., 554, 562(14)
Biamonti, C., 116
Biel, M., 248
Bienvenüe, A., 208
Birn, P., 211, 212(24), 214(24), 223(24), 526
Bishai, W. R., 562
Bitner, S., 92
Bittman, R., 218, 532, 545(36)
Bittner, M. A., 52
Blachly-Dyson, E., 298, 513
Blackstone, C. D., 353, 354, 355, 366, 368(4), 689
Blanc, E., 94
Blancaflor, E., 431
Blanco, D. R., 672, 673, 673(20)
Blatt, M. R., 413, 425, 439, 440, 515, 516, 517
Blattner, F. R., 469, 471, 471(24), 479(24)
Blaustein, M. P., 650
Blazing, M. A., 66

Blenkenburg, R., 151
Bley, K. R., 615(63), 623
Blobel, G., 473
Bloom, M., 209, 538
Blount, P., 458, 459, 463, 469, 470, 471(24), 472(26), 474, 475(26), 477(26), 479, 479(12, 24), 481, 518, 519(71a), 521(71a), 523(81), 524, 524(71a)
Blout, E. R., 80, 533
Blumenthal, K. M., 587
Blundell, T., 92
Boat, T. F., 227
Bochkarev, A., 150
Boden, N., 66, 83(30), 91(30)
Bodenhausen, G., 117
Boehm, C., 389
Boekeman, E. J., 136
Boheim, G., 290, 291(18)
Bokman, A. M., 92
Bolard, J., 330
Bolton, T. B., 446
Bonigk, W., 699
Bonilla-Marin, M., 320
Bontems, F., 94, 581
Bormann, J., 187, 264, 270(16)
Born, M., 300
Bothner-By, A. A., 117
Boulain, J.-C., 100, 651, 661(5)
Bower, J. M., 707, 730
Bowersox, S., 92
Bowman, S., 592, 593(113)
Box, A., 113, 114(70)
Boyd, A. E. III, 446, 456
Boyd, D., 477
Boyd, J., 594, 598(121), 630, 651
Boyle, M. B., 576
Boyle, W. J., 358
Boyot, P., 94, 581
Braco, L., 534
Bradbury, A. F., 587
Bradbury, E. M., 80
Brahmajothi, M. V., 697
Braman, J. C., 593
Brammar, W. J., 724
Brammer, W. J., 493, 498(20)
Brassard, D. L., 492
Brauner, J. W., 83
Braunschweiler, L., 117
Brayn, J., 454
Breer, H., 292
Brendel, M., 524
Brenner, R., 20
Breton, J., 91
Bretschneider, F., 180, 182, 187(3)
Bretzel, G. J., 300
Breuer, W., 308
Brickey, D. A., 366
Brickman, E., 478
Bridges, R. J., 227, 228(7), 230(7), 240(7), 241(7), 299(78), 303, 324, 389
Brillantes, A. M., 299(79), 303
Brink, J., 159(107, 108), 179
Brisson, A., 137, 150, 155(102), 179
Britten, R. J., 481
Brock, T. D., 507, 520(1)
Brodsky, J. L., 554
Bronner, C., 344
Brophy, P. J., 75
Brose, N., 687, 689
Brot, N., 643(29), 647
Brotherus, J. R., 64, 66(20)
Brotherus, M. O., 64, 66(20)
Brown, D. A., 435
Brown, L., 80
Brown, M. F., 208
Brown, P. O., 151
Brown, R. D., 126
Brown, W. E., 650
Brownlee, C., 419
Bruan, D. G., 108
Brümger, A. T., 95
Brundish, D. E., 389
Brunschwig, J.-P., 293
Brüschweiler, R., 95
Brust, P. F., 20, 31(4)
Bryan, J., 445, 446, 454, 456
Bryant, S. H., 288, 293(7), 296(7), 302(7), 586
Buck, L. B., 249
Buechner, M., 466, 519
Bufler, J., 681
Bugianesi, R. M., 630
Bullough, P. A., 136
Burg, M., 296(59), 297
Burgi, D. S., 194
Burka, J. F., 637
Burke, N. V., 57
Burlingame, A. L., 609(12), 612, 615(30), 616(39), 618, 620
Burnashev, N., 27, 32(22), 36(22), 41
Burns, J. A., 105

Busath, D. D., 532, 543(34), 585
Busch, A. E., 655, 656(13)
Bush, D. S., 427, 430(35)
Butcher, S. J., 155(105), 179
Butler, A., 575
Butler, J. C., 105
Bycroft, M., 594, 596(126), 597(126)
Byler, D. M., 60, 81(7)
Bystrov, V. F., 94, 533, 536
Byus, C., 198

## C

Cabiaux, V., 79
Cacan, R., 342
Caffrey, M., 208
Cahalan, M. D., 45, 410
Calamita, G., 562
Callahan, D. A., 434
Camejo, G., 589
Cameron, D. G., 78
Campanelli, J. T., 577
Campbell, D. T., 601
Campbell, K. P., 296(52), 297
Candia, S., 638
Cantor, C. R., 128
Carman, M., 672
Carmeliet, E., 658, 659(16)
Carney, M., 26, 46(18)
Carrasco, L., 490
Carrasco, N., 304
Carrascosa, J. L., 165
Carroll, T. P., 551, 567(6)
Cartaud, J., 70
Carter, P. J., 133, 596
Cartier, G. E., 609(49, 50), 610(54, 56), 622, 623
Cascio, M., 273
Case, D. A., 539
Cashmore, A. R., 439
Cass, A., 320
Castresana, J., 80, 82
Caswell, A. H., 293
Catterall, W. A., 27, 576(157), 577, 580, 589(52), 591, 594, 605, 606, 616(2), 619, 627, 692, 703
Cauley, K., 689
Cecchi, X., 296(46), 297
Celda, B., 116
Cepko, C. L., 687
Cerrier, C., 481
Ceska, T. A., 136, 178(22)
Chahine, M., 575(7), 576, 580, 590, 590(48), 591(48), 603(48)
Chait, B. T., 109
Chak, A., 119
Champion, C. I., 672, 673(20)
Chan, C. W. M., 511
Chan, K. W., 51
Chandy, K. G., 594, 598(121), 630, 651, 655, 697, 720
Chang, C.-D., 588
Chang, J.-J., 137
Chang, J. Y., 108
Chang, N. S., 600, 604(139)
Chang, X.-B., 227, 241(4)
Changeux, J.-P., 70, 117, 119, 126, 212
Chanturiya, A., 292
Chapman, D. J., 82, 507, 520(1)
Char, D., 684
Chavez-Noriega, L. E., 20, 36(5)
Chelay, S., 137
Chen, C., 270, 302(82), 303
Chen, C.-C., 558
Chen, D. P., 540
Chen, G., 425
Chen, L., 366, 575(7, 10), 576
Chen, L.-Q., 580, 590, 590(48), 591(48), 603(48)
Chen, M., 578, 579(35), 591(35), 599(35), 603(35)
Chen, T.-Y., 247, 249, 249(7), 258(7), 260(12), 699
Chen, Y., 536
Cheng, A., 165, 179
Cheng, L., 49
Cheng, Y.-L., 66, 83(30), 91(30)
Chernomordik, L., 292
Cherry, C. A., 517
Chiamvimonvat, N., 578, 579, 579(37), 590(42), 591(37, 42, 43), 603(42), 604(37, 42, 43)
Chien, T.-F., 238
Chirgadze, Y. N., 80
Chirgwin, J., 198
Chiu, D. T., 190
Chiu, S.-W., 540
Chiu, W., 136, 137, 150, 159(107, 108), 179

Chizhmakov, I. V., 490, 492(5), 500, 503
Cho, K.-O., 384
Cho, M. H., 423, 439(21)
Choi, K. L., 641
Choquet, D., 630
Chou, P. Y., 599
Chrispeels, M. J., 551
Christensen, E. I., 552
Chudziak, F., 445, 446, 453(13), 454
Chung, D., 100
Chung, S., 296(45), 297, 641
Chung, W. J., 384
Ciaranello, R., 92
Ciorba, M. A., 643(29), 647
Citra, M. J., 86
Clapham, D. E., 299(65), 300
Claret, M., 511
Clark, C., 605, 611
Clark, N. A., 60, 77, 82(9)
Clark-Lewis, I., 253, 254(20), 255(20), 699
Clement, J. P. IV, 445, 446, 454, 455(17), 456, 456(17)
Clift-O'Grady, L., 342
Clore, G. M., 95, 97, 536
Cocking, E. C., 417
Cohen, F. S., 319, 320, 321, 322, 326, 328(2), 331(6)
Cohen, J. B., 117, 118, 119, 120, 120(3), 125, 125(8), 126, 126(3)
Cohen, L. B., 45
Cohen, S. A., 575
Colin, F., 304
Collaborative Computational Project, No. 4, 174
Colledge, C. J., 613
Collins, F., 227
Collins, M., 559
Colman, D. R., 135(10), 136
Colombini, M., 298, 513, 663
Colquhoun, D., 204, 543, 589
Columbini, M., 511
Colville, C., 247, 249(8), 255(8), 257(8), 259(8), 699
Comeau, J., 515
Conley, E. C., 704, 724
Conley, W. J., 493, 498(20)
Conti, F., 576, 577, 578, 579, 590(26), 591(26), 603(26)
Conti-Tronconi, B., 680
Cook, N. J., 247, 249, 252, 252(13), 253, 254, 254(13, 19, 20), 255, 255(20), 256(13, 16, 19), 257, 257(13), 258(13), 259(13, 21), 699
Cooperman, S. S., 576
Cope, G., 697
Copello, J. A., 296(54), 297
Corey, D. P., 371, 513, 516(34)
Corliss, J. O., 507, 520(1)
Cormack, B. P., 49
Cornell, B. A., 537, 538, 539, 539(81)
Coronado, R. J., 290, 291(15), 296(42, 49, 52, 56), 297
Corpuz, G. P., 605, 611
Correa, A. M., 467
Correll, L. A., 125
Costa, A.C.S., 187
Cota, G., 601
Cotman, C. W., 389
Cotton, J., 94, 630
Coulombe, A., 459, 481, 518
Courseille, C., 535
Courtney, M. J., 366
Coury, L. A., 554
Covarrubias, M., 645
Cowan, S. W., 518
Cox, G. B., 490
Craescu, C. T., 94
Craig, A. G., 608, 609(4), 610(4)
Craig, A. M., 375, 377(6), 381(6), 689
Craik, D. J., 92, 580, 613
Crean, S. M., 576
Creighton, T. E., 587
Criado, M., 208, 466, 467(19), 518, 645
Cribbs, L. L., 575(11), 576, 578, 579(35), 591(35), 599(35), 603(35)
Crkvenjakov, R., 198
Crona, J. H., 20
Cros, F., 575(9), 576
Cross, A. J., 540
Cross, T. A., 525, 527, 531(6), 536, 537, 537(4–6), 538, 538(4, 6, 66, 72, 73), 539(6, 73)
Crowther, R. A., 165
Cruciani, R. A., 672
Cruz, L. J., 106, 580, 581(51), 605, 606, 609(42, 44–46, 48), 610(14, 15, 17, 55), 611, 612, 613(18), 614(1), 615(16, 29, 48, 57, 58), 616(1), 617(32), 618, 619, 622, 623
Cubitt, A. B., 49
Cui, C., 465
Cukierman, S., 601

Culbertson, M. R., 462, 510, 511, 514(18)
Cullis, P. R., 211, 239
Cung, M. T., 678
Cunningham, B. C., 591
Curz, L. J., 614, 616(23), 617(23), 619(23)
Czajkowsky, D. M., 137

## D

Daenens, P., 658, 659(16)
Daggett, L. P., 20, 26, 34(8), 46(18), 46(19)
Dahlquist, F. W., 525
Dai, G., 304
Daleman, S. I., 537
Dalton, L. R., 68
Damen, J., 239
Danckaert, A., 717
Daneo-Moore, L., 463
Dani, J. A., 3, 17, 18(29), 19(29), 27, 45
Daniels, D. L., 471
Danse, J. M., 651, 661(5)
Darbon, H., 94
Darst, S. A., 150, 151
Daumas, P., 549
Dauplais, M., 94
Davidson, N., 594, 705
Davies, N. W., 493, 498(20), 641
Davis, H. T., 155(103), 179
Davis, J. H., 525, 537, 538
Davis, P. B., 299(74, 75), 301, 303(74, 75)
Dawson, C. R., 232
Dayie, K. T., 98, 113, 113(33)
Deal, C. R., 20, 31(4), 34(8)
De Blas, A. L., 686
De Boer, A. H., 419
Debont, T., 658, 659(16)
Decker, G. L., 300
de Curtis, I., 342, 348(10)
Deisseroth, A., 495
Deitcher, D. L., 687
Deitmer, J. W., 509
de Jager, P. A., 67
De Jongh, H. H., 534
De Kruijff, B., 211, 239, 525, 526, 531, 531(20), 536(7), 537(20), 538(20)
Delaglio, F., 113
Delaney, K. R., 4
Delbruck, M., 58
Delcour, A. H., 462, 466, 469(18), 510, 514(18), 518, 519
Demange, P., 678
de Medeiros, C. L. C., 94
Demel, R. A., 531
Demo, S. D., 641
Dempster, J., 505
Denker, B. M., 551, 553(3)
De Panque, M., 534, 539(46)
DePue, P. L., 97, 112(27)
Derkach, V. A., 366
Derman, A. I., 477
DeRosier, D., 164(110), 179
de Santos, V., 615(57, 58), 623
De-Schutter, E., 710
Dethlefs, B., 594, 598(121), 630, 651
Devaux, P. F., 70, 208
Devidas, S., 570
Devillers-Thiery, A., 117
De Vos, A. M., 135
DeWeer, P., 585
Dhallan, R. S., 247, 249(7), 258(7)
Dhillon, D. S., 577
DiCapua, F. M., 539
Diliberto, E. J., 189
Dineley-Miller, K., 684
Dix, J., 45
Dixon, J. E., 108
Dodt, H. U., 5, 11(19), 13(19)
Doel, M. T., 103
Doherty, A. M., 97, 112(27)
Doherty, C., 364, 687
Doi, K., 689
Doige, C. A., 238
Doljansky, Y., 94, 581
Dolly, J. O., 638, 640, 649
Dolly, O., 651, 661(6)
Dong, M., 232, 238(18)
Donnelly-Roberts, D., 36
Dooley, D. J., 622
Dorset, D. L., 136
Dose, A., 247, 249(8), 253, 254(20), 255(8, 20), 257(8), 259(8), 699
Dos Santos, G., 575(9), 576
Dougherty, D. A., 591
Douglas, A. R., 491
Downie, A. R., 80
Downing, K. H., 136, 143, 161, 165, 178(22), 179(85)
Doyle, D. A., 535

Draber, S., 517
Drake, A. F., 232
Drakopoulou, E., 630
Dreessen, J., 26, 46(18)
Drejer, J., 394
Drevet, P., 100, 651, 661(5)
Dreyer, E. B., 119, 125
Driscoll, M., 459
Drumm, M. L., 227, 299(74, 75), 301, 303(74, 75)
Dryselius, S., 182, 187(5)
Dubinsky, W. P., 641, 697
Dubochet, J., 137, 145(99), 179
Ducancel, F., 100, 651, 661(5)
Dudel, J., 182
Dudley, S. C., 578, 589(36), 590(36), 599(36), 600, 603(36), 604(139)
Dudley, S. C., Jr., 575
Duggan, M. J., 686
Dumm, J. M., 609
Dumont, J. E., 304, 306(2)
Dumont, J. N., 266
Dunbar, J. B., Jr., 97, 112(27)
Dunlap, K., 614, 615(27), 618(27)
Dunn, R. J., 594
Duprat, F., 512
Durell, S. R., 600, 604(137), 672
Durkin, J. R., 526, 542(16), 549(16), 550 (16)
Durkin, J. T., 223, 526, 532(14), 542, 542(14), 545(14), 546(101), 549(14)
Duzgunes, N., 238
Dykert, J., 610(55), 623


## E


East, J. E., 298, 299(62)
Ebbinghaus, U., 45
Eberhart, A., 280, 634, 637(19)
Eckert, R., 509
Edelhoch, H., 647
Edelstein, S. J., 143
Edidin, M., 56
Edwards, A. M., 150
Egebjerg, J., 689
Ehlers, M. D., 364, 687
Ehlers, M. E., 384
Ehrenberg, A., 98
Ehrenstein, G., 440
Ehring, G. R., 45
Ehrlich, B. E., 288, 296(55), 297, 299(79), 303, 509, 510
Ehrmann, M., 477
Eibl, H., 208
Eichler-West, R. M., 710
Eilers, J., 439
Einarson, B. E., 680
Eipper, B. A., 587
Eisele, J. L., 117
Eisenberg, M., 218
Eisenberg, R. S., 436, 437(46), 540
Eisenman, G., 290, 542, 584, 585
Eismann, E., 253, 254(20), 255(20), 699
Eitan, M., 703
Elber, R., 540
Elgoyhen, A. B., 609(52), 623
Ellena, J. F., 66
Ellinor, P. T., 300, 615(66), 624
Elliot, A., 80
Elliott, D. C., 427, 430(36)
Elliott, K. J., 20, 36(5)
Ellis, S. B., 20, 26, 31(4), 36(5), 46(18)
Ellis-Davies, G. C. R., 438
Ellison, M., 610(54), 623
Elmer, L. W., 48
Elson, E., 48
Elzenga, J. T. M., 416, 417
Emson, P. C., 425, 431(29)
Engel, A., 135, 144(17, 18), 147(17, 18), 148, 179
England, L. J., 608
Ennis, P. D., 155(101), 179, 210, 215(19)
Enomoto, R., 390
Entrikin, D. W., 139
Epand, R. M., 208
Erikson, H. P., 296(51), 297
Erjument-Bromage, H., 673
Ernst, R. R., 117
Esmann, M., 66, 70
Estbs, M. K., 490
Eterovic, V. A., 609(6, 47), 610, 623
Eun, S. O., 417
Evans, E. A., 209
Evans, R. J., 187
Ewart, G., 490

Ewert, M., 686
Ewing, A. G., 194

F

Fagg, G. E., 389, 392
Faiman, G. A., 596
Fainzilber, M., 609(12), 612, 614, 615(30), 616(36, 39), 617(24, 35, 71), 618, 619, 619(24), 620, 620(24), 624
Fairley, K. A., 517
Fairley-Grenot, K., 435
Fajer, P., 62, 67, 67(14)
Fakler, B., 645
Falkow, S., 49
Fan, Z., 591
Fannon, A. M., 135(10), 136
Fantin, D., 431
Farrar, J., 678
Farrow, N. A., 98, 113
Fasman, G. D., 80, 81, 599
Favre, I., 287, 578, 591(31), 603(31)
Feigenbaum, P., 275, 582, 627, 634, 637, 637(18)
Feigenson, G. W., 208
Feldman, D. H., 20
Felix, J. P., 276, 630
Felle, H. H., 434
Feller, M. B., 4
Felts, K. A., 593
Fenwick, E. M., 182, 186(2), 187(2)
Fernandez, J. M., 440
Fernandez-Ballester, G., 645
Fernando, L. P., 593
Ferragut, J. A., 645
Ferrer-Montiel, A. V., 490
Fersht, A. R., 133, 594, 595(127), 596, 596(124, 126), 597(125, 126)
Fields, C. G., 527, 528
Figl, A., 705
Fill, M., 296(52), 297
Filloux, F., 618
Finbow, M. E., 82
Findlay, G. P., 517
Findlay, J. B. C., 59, 60(1), 65, 66, 66(22, 23), 67(22), 76(1), 82, 83(22, 30), 91(30), 649
Fine, R. E., 342
Fink, G. R., 508, 524(6)
Fink, M., 512
Finkel, A., 290, 326
Finkelstein, A. V., 319, 320, 322, 328(2), 510, 511, 551, 663, 668, 672(8)
Finn, J. T., 247, 249(2)
Fisher, H. W., 158
Fishman, H. A., 191, 197(7), 199(7), 200, 202(26), 203, 203(8), 206(7), 207
Fleischer, S., 296(54), 297
Flinn, J. P., 100
Flockerzi, V., 579
Flor, P., 26, 46(19)
Fluhler, E. N., 45
Folander, K., 197, 286
Foley, D., 672, 673(20)
Fomina, A. F., 697
Fonseca, V., 549
Fonteneau, P., 344
Foose, J. M., 697
Ford, C. M., 492
Ford, R. C., 136, 144(17), 147(17)
Foret, F., 193
Forman-Kay, J. D., 113
Forscher, P., 368
Forte, M., 298, 510, 513
Fossel, E. T., 533
Foster, C. D., 641
Fotouhi, N., 590
Fox, A. P., 615(63), 623
Fox, H., 588
Fox, R. O., 273
Fozzard, H. A., 578, 579(35), 589(36), 590(36), 591, 591(35), 599, 599(35, 36), 600, 603(35, 36), 604(135, 139)
Frank, R., 640, 645(4)
Franke, C., 182, 681
Frankenhaeuser, B., 601
Franklin, T., 82
Fraser, R. D. B., 80
Freed, J. H., 67
Freeman, L. C., 697
Fremont, V., 94
French, R. J., 289, 290, 292(8), 296(8), 302(8, 9), 326, 340, 341(1), 575, 577, 580, 581, 581(49), 583, 583(23, 49), 585, 586, 588(49), 589, 595(24, 145), 600, 600(24, 49), 601, 601(24), 604(24, 49, 139, 145)
Freund, J., 111

Friend, D. S., 139
Fries, E., 309
Fringeli, M., 77
Fringeli, U. P., 77
Frings, S., 41
Fritsch, E. F., 494
Fritschy, J. M., 686
Froehner, S. C., 692
Froncisz, W., 61
Fry, D., 590
Fuchs, K., 686
Fuerst, W., 163
Fujiwara-Hirashima, 510
Fujiyoshi, Y., 136, 178(23, 24), 179, 179(23, 24)
Fuller, N., 238
Fuller, S. D., 151, 155(105), 179
Furie, B. C., 613
Furman, T. C., 104
Furth, A. J., 238
Furukawa, T., 570
Furutani, Y., 575, 576(3)
Fushimi, K., 570

## G

Gaathon, A., 616(39), 620
Gaber, R. F., 513, 521
Gadaleta, S. J., 83
Gadbois, T., 92
Gadsby, D., 241
Gage, P. W., 490
Gaggar, A., 190
Gahring, L. C., 678, 689
Gallagher, M. J., 587
Galley, K., 227, 228, 232(11), 235(11), 241(10), 242(10), 319
Galvez, A., 627
Galyean, R., 615(58), 623
Galzi, J. L., 117
Gammon, C. M., 187
Ganetzky, B., 287
Ganz, T., 663, 672(5)
Garami, E., 227, 319
Garavito, R. M., 91, 135
Garber, S. S., 581, 585
Garcia, M. L., 274, 275, 276, 277, 277(6), 280, 281(6, 9, 10), 283(9), 285, 286, 287, 297, 582, 624, 625, 626(2), 627, 630, 634, 635(23), 637, 637(18, 19, 23), 638, 639(23)
Garcia-Anoveros, J., 371
Garcia-Calvo, M., 276, 277, 277(6), 281(6, 9, 10), 283(9), 285, 627, 637, 638
Garigan, D., 191, 197(7), 199(7), 206(7), 207
Garlick, R. L., 135
Garner, C. C., 384
Garretson, L. T., 697
Garrett, T. P. J., 135
Garty, H., 244, 307
Garzia-López, R. A., 190
Gasic, G. P., 687
Gasparini, S., 651, 661(5)
Gautron, S., 575(9), 576
Gavilanes, F., 645
Gealy, R., 697
Geer, J., 109
Geer, P. V. D., 358
Gehrmann, J., 538, 539(81)
Gelb, M. H., 622
Gelfand, D. H., 477
Gellens, M. E., 575(7), 576
George, A. L., Jr., 575(7, 8), 576, 577
Geraerts, W. P., 615(30), 616(36), 618, 620
Geraghty, F. M., 490, 492, 492(5), 494(14), 506(14)
Gerami, E., 228, 232(11), 235(11), 241(10), 242(10)
Gerhardt, P., 508, 518(5)
Gericke, A., 83
Gerstein, M., 155(104), 179
Gesmar, H., 114
Geuze, H. J., 342, 348(12)
Geyer, M., 645
Ghashani, S., 655
Ghazi, A., 459, 481, 518, 520
Ghomashchi, F., 622
Ghosh, R., 518
Giangiacomo, K. M., 277, 281(9, 10), 283(9), 285, 637, 638
Gibby, M. C., 538
Gierasch, L. M., 645
Gies, R. D., 270
Giesberg, P., 419
Giffen, K., 575
Gill, S. J., 128
Gillessen, D., 686
Gilquin, B., 94, 581
Gilroy, S., 429, 431, 438, 438(40)
Gilula, N. B., 139, 144, 144(36), 175(49), 177, 178(36)

Gimenez-Gallego, G., 627
Giordani, A. B., 622
Girshman, J., 211, 212(24), 214(24), 218, 223(24), 526
Gish, G., 113
Glaeser, R. M., 137, 144
Glisin, V., 198
Glossmann, H., 634, 692
Gobind Khorana, H., 73
Goda, Y., 3
Godde, M., 41, 247, 249(8), 255(8), 257(8), 259(8), 699
Gogol, E., 144
Gojobori, T., 570
Goldberg, A. F., 244, 303
Goldberg, M. S., 95
Goldin, A. L., 199, 264, 524, 577, 578, 590(28), 594, 603(28)
Goldin, S. M., 238
Goldstein, S. A., 100, 512, 582, 585, 587(65), 630, 634(16)
Golstein, P. E., 304, 306(2), 308, 309, 313(9), 314(7), 315(9)
Goñi, F. M., 80, 82
Gonoi, T., 446, 606, 616(2)
Gonzalez, G., 445, 454, 456
González, J. E., 45
Gonzalez-Beltran, C., 320
Gonzalez-Ros, J. M., 645
Goodenough, U. W., 517
Gooding, C., 493, 495(19)
Goodman, H. M., 198
Goodman, R. H., 575, 576
Goormaghtigh, E., 79, 534
Gopalakrishnan, M., 36
Gordon, D., 100, 614, 617(24, 35, 71), 619, 619(24), 620(24), 624
Gordon, G., 703
Gordon, R. D., 580, 581(49), 583, 583(49), 586, 588, 588(49), 589(89, 90), 600(49), 604(49)
Gormann, C. M., 270
Görne-Tschelnokow, U., 83, 88(69), 92(69)
Gortz, H.-D., 509
Gotzes, S., 247, 249(8), 255(8), 257(8), 259(8), 699
Gouauz, J. E., 137
Gouda, H., 94, 97(13), 101(13), 110(13)
Goulding, E., 641
Goulian, M., 208, 214, 221(27), 289, 526
Gowen, B. E., 155(105), 179
Grabov, A., 425
Gradmann, D., 517
Graham, D., 261, 262(2, 4), 263(2)
Grambas, S., 492, 494(14), 506(14)
Grant, S. G., 3
Gray, P. T. A., 505
Gray, R., 3, 17, 18(29), 19(29)
Gray, W. R., 94, 101(11), 108(11), 110(11), 580, 581(51), 589, 605, 606, 608, 609(4, 42, 44–46, 48, 49, 51), 610(4, 15, 54–56), 612, 614, 614(1), 615(16, 57–59), 616(1, 23, 25, 69), 617(23, 25, 33), 619, 619(23), 621(25), 622, 623, 624
Graziana, A., 417
Greathouse, D. V., 217, 218, 223, 289, 525, 526, 531, 531(20), 532, 533, 534, 534(37), 536(2, 7), 537, 537(20, 32), 538(20), 539(46), 542(8, 9, 11), 545(8, 11, 36), 549(11), 550(9)
Green, F., 292
Green, M. R., 197
Greenberg, M. E., 687
Greenberg, R. M., 575(12), 576
Greengard, P., 366
Greenhalgh, D. A., 73
Greeningloh, G., 262, 264, 686, 697
Greenlee, J. E., 678
Greferath, U., 686
Gregory, L. M., 609(47), 623
Gregory, R., 227, 241(2)
Grey, M., 524
Griffin, D., 615(58), 623
Griffith, O. H., 64, 66(20)
Grilley, M., 610(55, 56), 614, 616(25), 617(25, 33, 34), 619, 621(25), 623
Grimes, M., 342
Grisshammar, R., 139
Grodzicki, R. L., 273
Groebe, D. R., 608, 609, 609(4), 610(4, 54), 623
Gronenborn, A. M., 95, 97, 536
Groom, M., 145(99), 179
Gross, L. A., 49
Grosveld, F., 493, 495(19), 498(20)
Grote, E., 342
Grübel, G., 135(10), 136
Gruner, S. M., 216
Grünhagen, H.-H., 126
Grunnert, U., 686

Grunwald, M. E., 247, 249(2)
Grynkiewicz, G., 4, 22, 25(15), 425
Grzelczak, Z., 227
Guan, X., 139
Guern, J., 422
Guggino, W. B., 551, 554, 562, 562(14), 567(6), 570, 572
Guillemare, E., 512
Guirguis, E. R., 686
Gullak, M. F., 610(14), 612
Gulyas, J., 608
Gunderson, K. L., 241
Guo, X., 288, 293(7), 296(7), 302(7), 586, 589
Gurdon, J. B., 197
Gurevitz, M., 100
Gurnett, C. A., 692
Gurney, A. M., 438, 440
Gurrola, G., 630
Guruprasad, K., 92
Gustin, M. C., 462, 510, 511, 512, 514(18)
Guth, K., 439
Guthrie, C., 508, 524(6)
Gutknecht, J., 209
Gutman, G. A., 594, 598(121), 630, 651, 655, 697, 720
Guy, H. R., 470, 472(26), 475(26), 477(26), 512, 576, 600, 604(137), 645, 672


## H


Ha, K. S., 417
Haack, J. A., 610(17), 612, 615(59), 623
Haase, W., 247
Hadley, R. W., 434
Hagiwara, S., 425
Hahin, R., 601
Haims, D., 537
Hake, D. J., 108
Hall, A. R., 617(33), 619, 659
Hall, J. E., 319, 322, 328(11), 594, 598(121), 630, 651
Hall, S. H., 288, 293(7), 296(7), 302(7), 581, 586
Hallen, R. M., 557
Hall-Smith, M., 491
Hamill, O. P., 267, 482, 483, 485(1), 488, 496, 507, 524(4)
Hammerland, L. G., 609(48), 615(48), 623, 639
Hammond, S., 694
Hampson, D. R., 366, 368(15)
Han, D., 389, 390(12), 392, 394
Han, W., 57
Handby, W. E., 80
Hanill, O. P., 421
Hanke, W., 232, 249, 252(13), 254(13), 255, 256(13), 257, 257(13), 258(13), 259(13), 288, 290, 291(18, 19), 292, 337
Hankovszky, H. O., 64, 66
Hann, R. M., 609(6, 47), 610, 623
Hanner, M., 274, 624, 625, 626(2)
Hanrahan, J. W., 227, 230(3), 231(3), 241(3, 4)
Hans, M., 20, 31(4)
Hansen, A. J., 211, 212(24), 214(24), 223(24), 526
Hansen, U.-P., 517
Hanson, D. C., 594, 598(121), 630, 651
Hanstein, W. G., 144
Hareland, M. L., 125, 129(14)
Haring, P., 686
Harkiss, D., 588
Harootunian, A. T., 21, 25(13)
Harper, J. F., 425
Harpold, M. M., 20, 31(4), 36(5)
Harrick, N. J., 86
Harris, G. H., 637
Harrison, S. C., 135
Hart, I. K., 677, 678, 697
Hartmann, H. A., 586
Hartung, K., 515
Harvey, A. L., 94, 651, 661(5)
Harvey, S. C., 539, 542(89), 622
Harz, H., 518
Hase, C. C., 470, 474(25), 475, 479, 518
Hase, S., 529
Hasson, A., 605, 609(12), 612, 614, 616(23, 37), 617(23, 71), 619(23), 620, 624
Hatanaka, H., 583
Hatchett, J., 537
Hatefi, Y., 144
Hatt, H., 182
Haugland, R. P., 9, 21, 22(12), 25(12), 45(12), 126, 426, 437(34)
Hausdorf, S. F., 100, 108(43)
Hausdorff, S. F., 587
Havel, T. F., 95
Havelka, W. A., 165
Hawkes, A., 204
Hawrot, E., 92, 94(6)

Hay, A. J., 490, 491, 492, 492(5), 494, 494(14), 506(14)
Hayamizu, K., 539
Hayhurst, A., 490, 492, 492(5), 494(14), 506(14)
He, K., 217
Heath, I. B., 515
Hebert, H., 137, 179
Hebert, S. C., 193
Hediger, M. A., 193
Hedrich, R., 411, 440
Hefti, A., 136, 144(17, 18), 147(17, 18)
Hegemann, P., 518
Heginbotham, L., 582, 591
Heidenreich, F., 681
Heidmann, T., 126
Heim, R., 49, 50, 51(10), 52
Heimburg, T., 65, 66(22), 67(22), 82, 83, 83(22), 84(71)
Heimer, E. P., 588
Heinemann, S., 385, 386
Heinemann, S. F., 32, 687, 689
Heinemann, S. H., 300, 578, 579, 590(26), 591(26, 27), 603(26, 27), 640, 643(29), 645(4), 647
Heinemann, S. S., 609(52), 623
Heiny, J., 589
Heitz, F., 525, 542(8), 545(8), 549
Helenius, H., 309
Helfrich, P., 210, 214(22)
Helfrich, W., 209
Hell, J. W., 692
Helliwell, J., 232
Hellman, B., 182, 187(5)
Helms, L. M. H., 287, 637, 697
Helpler, P. K., 417
Hemming, S. A., 150
Hemminga, M. A., 67, 71
Henderson, R., 136, 144, 155(104), 165, 165(12), 178(22), 179, 179(85), 521, 589
Hendrickson, W. A., 135(10), 136
Henklein, P., 490
Henn, C., 136, 144(17), 147(17)
Henn, D. K., 521
Henriksen, G. H., 419, 421
Hepler, P. K., 434
Herhan, B., 490
Hermanns-Borgmeyer, I., 261, 262(3)
Hermsteiner, M., 445
Hernández-Crus, A., 426
Herrera-Sosa, H., 456
Hershenson, F. M., 262
Hervé, M., 100
Herz, J. M., 126
Hess, G., 12
Hess, P., 290, 296(50), 297, 615(65), 624
Hess, S. D., 20, 26, 34(8), 46(18)
Hewryk, M., 228, 232(11), 235(11), 241(10), 242(10)
Heymann, J. B., 179
Heymann, J. A. W., 165
Hidaka, Y., 616(72), 624
Hidalgo, C., 292, 293, 296(32)
Hidalgo, P., 135, 594, 627, 651
Hideg, K., 64, 66, 70
Hider, R. C., 232
Higashiguchi, O., 446
Higashihara, M., 402
Higashijima, T., 112
Higgs, T. P., 538
Higuchi, R., 592
Hill, A. V., 129
Hill, J. M., 580
Hille, B., 205, 318, 337, 425, 601, 603
Hillenkamp, F., 109
Hillyard, D. R., 106, 605, 610(56), 611, 613, 614, 615(29, 59), 616(25, 37), 617(25, 32, 33), 618, 619, 620, 621(25), 623
Hilschmann, N., 108
Hing, A. W., 537
Hinkle, P. C., 331, 332(21)
Hinkle, P. M., 52
Hinton, J. F., 530, 533, 534, 534(37)
Hirai, T., 136, 178(23), 179, 179(23)
Hirata, Y., 616(68), 624
Hirose, T., 575, 576(3)
Hirth, C., 100
Hiu, S.-W., 208
Hjerten, S., 232, 238(16)
Ho, B., 705
Ho, R. J., 239
Ho, S., 4
Hobaugh, M. R., 137
Hoch, H. C., 515
Hoch, W., 261, 262(5)
Hochmuth, R. M., 209
Hochschwender, S., 680
Hodges, R. S., 82
Hodgkin, A. L., 501, 585, 601
Hoenger, A., 136, 144(17), 147(17)

Hoffman, B. B., 92
Hoffman, R., 523(81), 524
Hoffman, R. J., 479
Hofmann, F., 248, 579
Hokin, L. E., 64, 66(20)
Holcombe, V., 259
Holland, E.-M., 518
Hollinshead, M., 342
Hollis, M., 493, 495(19), 498(20)
Hollmann, M., 32, 385, 386, 612, 613(18)
Hollosi, M., 80, 81
Holman, M. A., 575(12), 576
Holmgren, M., 644
Holsinger, L. J., 490, 492, 492(4)
Holub, K. E., 93, 94, 101(11), 108(11), 110(11), 111(9)
Holz, R. W., 52
Holzenburg, A., 82, 235
Homann, U., 517
Homes Aedelman, C., 698
Homo, J.-C., 137
Honda, H., 583, 598(75)
Honig, B., 598
Honoré, T., 394
Hoogstraten, C. G., 117
Hope, A. B., 517
Hopfer, U., 307
Hopkins, C., 610(55, 56), 623
Hopkins, W. F., 649, 655, 656(12)
Horio, Y., 446
Horn, R., 428, 575(7), 576, 577, 580, 583(23), 590, 590(48), 591(48), 595(24, 145), 600(24), 601, 601(24), 603(48), 604(24, 145)
Hornby, D., 521, 697
Horne, R. W., 165
Horne, W. A., 615(66), 624
Horovitz, A., 594, 596, 596(126), 597(125, 126)
Horváth, L. I., 59, 64, 65, 66, 66(22, 23), 67(22), 68, 68(2), 69(2), 70, 71, 72(44), 73, 74(51), 75, 83(22, 30), 91(30)
Hoshi, T., 640, 643(29), 647
Hospital, M., 535
Hospod, F. E., 12
Hotchkiss, R. D., 542
Houamed, K. M., 655, 656(12)
Houssin, C., 518
Howe, J. R., 271, 299(77), 301, 303(77)
Hoyt, K., 48, 51(3), 57(3)
Hsiao, A. H., 190
Hsu, G., 143
Hsu, H. O., 705
Hsu, Y.-T., 249, 252(14), 253(14), 254(14), 256(11, 14), 257, 257(11, 14), 258(11, 14, 22), 699
Hu, L. R., 322
Hu, W., 525, 531(6), 537, 537(4, 6), 538(4, 6, 72, 73), 539(6, 73)
Huan, L.-J., 228, 232(11), 235(11)
Huang, E., 705
Huang, H. W., 210, 214(21), 217
Huang, K. S., 232, 238(12)
Huang, L. Y., 239, 366
Huang, X. D., 194, 366, 368(15)
Hubbell, W. L., 60, 73, 73(4), 208
Hubble, W., 674
Hucho, F., 83, 88(69), 92(69)
Hudson, K., 493, 495(19)
Huerta, M., 730
Huganir, R. L., 353, 354, 355, 356, 357(6), 360(6), 364, 366, 366(6), 367(6), 368(4, 6), 384, 687, 688, 689
Hughes, T. E., 271
Hui, K., 600
Humbert, P., 100
Humblet, C. C., 92, 97, 112(28)
Humphrey, M. B., 446
Hunkapiller, M. W., 609(44), 623
Hunter, T., 358
Huntley, G. W., 687, 689
Huo, S., 525, 537(5)
Hurst, R. S., 655, 656(13)
Hurt, S. D., 389
Hurwitz, R., 259
Husain, Y., 135
Hutchinson, A. J., 389
Hwang, P., 589
Hwang, T.-C., 241, 570
Hyberts, S. G., 95
Hyde, G. J., 515
Hyde, J. S., 61, 68

## I

Iannuzzi, M., 227
Ifune, C., 575
Iizuka, M., 697

Ikeda, T., 575
Ikemoto, N., 293
Ilan, N., 424
Illing, M., 247, 249(8), 253, 254(20), 255(8, 20), 257(8), 259(8), 699
Imoto, K., 300, 578, 579, 590(26), 591(26, 27), 603(26, 27)
Imperial, J. S., 106, 608, 615(29), 617(33), 618, 619
Inagaki, F., 580, 583, 583(54), 598(75)
Inagaki, N., 446
Inanobe, A., 697
Inayama, S., 575, 576(3)
Ingolia, T. D., 104
Innis, M. A., 477
Inoue, K., 402
Ireland, C. M., 94
Isacoff, E. Y., 577, 644
Isai, H., 94, 97(13), 101(13), 110(13)
Isenberg, G., 182
Ishibashi, K., 570
Ishida, Y., 583, 598(75)
Ishikawa, K., 580
Ismailov, I. I., 296(60), 297
Isom, L. L., 577
Isomoto, S., 446
Israelachvili, J. N., 209
Isselbacher, K. J., 307
Ito, M., 697
Iversen, L. L., 392
Iverson, L. E., 421, 577
Iwasa, K., 440

## J

Jachec, C., 6, 20, 34(8), 46(18, 19)
Jackson, M., 80, 81, 82, 82(58, 59)
Jacobsen, R., 608, 609(50), 610(54, 56), 614, 616(25), 617(25), 621(25), 623
Jacobson, G., 409
Jacobson, I., 191, 194(9), 196(9), 203, 203(8, 9)
Jacobson, L., 681
Jácome, T., 609(47), 623
Jaffe, D. B., 9
Jahn, R., 687
Jähnig, F., 68, 89, 91(76)
Jakobsson, E., 210, 214(22), 540, 599, 600, 600(136)
Jamin, N., 94
Jan, L. Y., 379, 644, 650, 697
Jan, Y. N., 371, 379(3), 384(3), 644, 650, 697
Janecki, M., 579, 604(45)
Janeczek, J., 300
Jang, Y. N., 697
Jankowski, J. A., 189
Jansen, K. A. J., 71
Jansonius, J. N., 518
Janssen, W. G., 687, 689
Januszeski, M. M., 674
Jap, B. K., 136, 144(18), 147(18), 148, 179
Jardemark, K., 191, 194(9), 196(9), 203, 203(8, 9)
Jarvet, J., 98
Jean Husten, E., 587
Jeanloz, R. W., 117
Jeffrey, K. R., 538
Jegla, T., 273, 274, 286(1), 625
Jen, J., 366
Jensen, J. W., 208
Jensen, T., 227, 228(7), 230(3, 7), 231(3), 240(7), 241(3, 7), 299(78), 303, 324, 698
Jia-ling, C., 227
Jobayashi, Y., 112
Johannsson, A., 208
Johnson, B. A., 630
Johnson, B. F., 514
Johnson, D. A., 126
Johnson, D. S., 609(52), 623
Johnson, E. C., 20, 26, 31(4), 34(8), 36(5), 46(18)
Johnson, J. H., 639
Johnson, R. J., 342
Johnson, S., 472
Johnston, D., 9, 12
Joiner, W. J., 512
Jonas, P., 445
Jones, A., 686
Jones, P. C., 82
Jones, R., 151
Jones, R. L., 427, 430(35)
Jones, T. A., 174
Jorgenson, J. W., 191, 192(10)
Jost, P. C., 64, 66(20)
Jovin, T. M., 56
Jubb, J. S., 144
Jude, A. R., 525, 536(2), 537, 542(9), 550(9)
Julius, D., 608
Jung, J. S., 551, 554, 562(14), 572

Jurman, M. E., 644
Justice, A., 92

## K

Kaczmarek, L. K., 512
Kaczorowski, G. J., 274, 275, 276, 277, 277(6), 280, 281(6, 9), 283(9), 286, 287, 297, 582, 624, 625, 626(2), 627, 630, 634, 635(23), 637, 637(18, 19, 23), 638, 639(23)
Kaduk, C., 83, 88(69), 92(69)
Kagan, B. L., 661, 662, 663, 672, 672(5, 7, 8), 673, 673(20), 674
Kagawa, Y., 467
Kahle, C., 232
Kahlert, S., 681
Kain, S. R., 49
Kaiser, M. W., 575(11), 576
Kalbitzer, H. R., 111, 645
Kallen, R. G., 575, 575(7, 10), 576, 579, 580, 590, 590(48), 591(48), 603(48)
Kallio, K., 49
Kamboj, S., 353
Kameyama, K., 364, 687
Kamp, T. J., 579
Kampe, K., 445
Kandrach, A., 238
Kanska, M., 527
Kao, C. Y., 580
Kao, J. P. Y., 21, 22, 25(13)
Kaplan, J. H., 438
Kapur, J., 187
Karanjia, S., 151
Karas, M., 109
Karger, B. L., 193
Karlin, A., 119, 579, 591(40)
Karlish, S. J., 244, 307
Karlstrom, R., 609(45), 623
Karplus, M., 95, 540
Karschin, A., 705
Kartner, N., 227, 228(7), 230(3, 7), 231(3), 240(7), 241(3, 7), 299(78), 303, 324, 698
Kasahara, M., 331, 332(21)
Kasai, M., 212, 510, 514, 515(21)
Kass, R. S., 697
Kataoka, K., 402
Katayama, K., 109
Kato, R., 583, 598(75)
Katz, B., 501
Katz, G. M., 627, 634, 637(18)
Kauakami, Y., 109
Kaufman, M. H., 715, 716
Kaupp, U. B., 41, 247, 249, 249(1, 8), 252, 252(13), 253, 254(13, 19, 20), 255, 255(8, 20), 256(13, 16), 257, 257(8, 13), 258, 258(13), 259(8, 13, 19), 521, 699
Kauppinen, J. K., 78
Kavanaugh, M. P., 655, 656(13)
Kavéus, U., 137
Kawagoe, K. T., 189
Kawai, T., 109
Kawano, S., 297
Kay, C. M., 113
Kay, L. E., 98, 112(30), 113, 113(30, 34, 35), 114(70)
Kayano, T., 575
Keen, J. N., 649
Kehoe, J., 609(13), 612
Keilbaugh, S. A., 697
Kelbaugh, P. R., 639
Keller, B. U., 466, 467(19), 518
Keller, C. P., 416, 417
Kellis, J. T., Jr., 596
Kelly, M. L., 331, 333, 335(22)
Kelly, R. B., 342
Kemp, B. E., 362
Kemp, J. A., 392
Kennedy, C., 182, 187, 187(4)
Kennedy, M. B., 371
Kennedy, R. T., 189
Kent, S. B. H., 529
Kephart, D. D., 575(11), 576
Kerem, B.-S., 227
Kern, W., 686
Kerner, J. A., 26, 46(19)
Kerr, J. A., 126
Kerr, L., 580, 581(51), 606, 614(1), 616(1)
Ketchem, R. R., 525, 537(4), 538(4)
Ketchum, K. A., 512
Kettenmann, H., 262
Keynes, R. D., 585
Khaled, M. A., 533
Khorana, H. G., 232, 238(12)
Kideg, K., 65, 66(22), 67(22), 83(22)
Kidera, A., 136, 178(23), 179(23)
Killian, J. A., 217, 525, 526, 531, 531(20), 534, 536(7), 537, 537(20, 32), 538, 538(20), 539(46)

Kim, E., 371, 373(5), 377, 379(3), 384, 384(3)
Kim, J. I., 94, 97(13), 101(13), 110(13)
Kim, K. S., 533, 534(37)
Kim, M., 417
Kimhi, Y., 342
Kimura, Y., 136, 178(23), 179(23)
Kindler, S., 384
King, C., 510, 514(18)
King, G. I., 238
King, V. F., 275, 582, 627, 634, 637, 637(18)
Kingston, R. E., 26
Kinne, K. H., 306
Kinne, R., 306
Kinsey, W. H., 300
Kirby, M. S., 434
Kirino, Y., 510
Kirsch, A., 261, 264
Kirsch, G. E., 586
Kirsch, J., 260, 263(1), 686
Kistler, J., 136, 144(17), 147(17)
Kits, K. S., 615(30), 616(36, 39), 617(35), 618, 619, 620
Kjeldgaard, M., 174
Klapperstück, M., 182, 187(3)
Kloda, A., 475, 520
Klotz, I. M., 128, 130(28)
Knaus, H.-G., 274, 277, 280, 281(9, 10), 283(9), 286, 624, 625, 626(2), 634, 635(23), 637, 637(19, 23), 638, 639(23), 697
Knaus, P., 261
Knepper, M., 552
Knight, A. R., 392
Knittle, T. J., 575(8), 576
Knopfel, T., 26, 46(18)
Knowles, P. F., 62, 65, 66, 66(23), 83(30), 91(30)
Knudson, M., 296(52), 297
Kobayashi, F. H., 609(7), 610
Kobayashi, J., 580, 589(52), 606, 616(2, 68, 72), 624
Koch, K.-W., 252, 256(16), 258
Koch, R. O., 274, 280, 625, 626(2), 634, 635(23), 637(19, 23), 639(23)
Kochman, R. L., 262
Koczorowski, G. J., 285
Koenen, M., 640, 645(4)
Koeppe, R. E. II, 208, 212, 217, 218, 223, 289, 525, 526, 530, 531, 531(20), 532, 532(10, 14, 19, 23), 533, 534, 534(37), 536(2, 7, 35), 537, 537(20, 32), 538, 538(20), 539(46), 540, 541(10), 542, 542(8, 9, 11, 14, 16), 543, 543(34), 545(8, 11, 14, 36, 107), 546(101, 107), 549(12–14, 16), 550(9, 16, 35)
Kofman, O., 614, 617(24, 35), 619, 619(24), 620(24)
Koga, S., 238
Koh, D.-S., 445
Kohda, D., 580, 583, 583(54), 598(75)
Köhler, K., 419
Kohno, T., 94, 97(13), 101(13), 110(13)
Kojima, S., 390
Kokla, A., 680
Kolmakova-Partensky, L., 92, 582, 586, 587(64), 630, 634, 635(22)
Kondakov, V. I., 94
Kondo, C., 446
Konishi, S., 94, 97(13), 101(13), 110(13)
Konnerth, A., 3, 439, 506
Kontis, K. J., 578, 590(28), 603(28)
Kooser, R. G., 67
Kopito, R. R., 241
Korn, S. J., 428
Kornau, H.-C., 371
Kornberg, R. D., 150, 151, 153
Korpi, E. R., 686
Korschen, H. G., 247, 249(8), 255(8), 257(8), 259(8), 699
Koschak, A., 634, 635(23), 637(23), 639(23)
Koslow, S., 730
Kossiakoff, A. A., 135
Koszowski, A. G., 590, 604(99)
Koulakoff, A., 575(9), 576
Kouvatsou, R., 680
Kovacs, R. J., 296(57), 297
Kovatchev, P., 65, 66, 66(22, 23), 67(22), 83(22, 30), 91(30)
Kraffe, D. S., 199
Krafte, D. S., 594
Kramer, G., 45
Kramer, R. H., 641
Kreienkamp, H.-J., 609(43), 623
Kreig, P. A., 197
Krespi, V., 320
Krieg, N. R., 508, 518(5)
Krieg, P. A., 569
Krimm, S., 60, 80
Krishtal, O. A., 180, 181(1), 182(1), 187(1), 497
Kristipati, R., 609(48), 615(48), 623

Krodel, E. K., 120, 125(8)
Krueger, B. K., 289, 292(8), 296(8), 302(8, 9), 581, 585, 601, 650
Kruppel, T., 509
Kubalek, E. W., 151
Kubalski, A., 465
Kubo, H., 577
Kubo, Y., 697
Kuchitsu, K., 412, 416
Kuhajda, F. P., 551, 553(3)
Kühlbrandt, W., 135, 136, 142, 144(16), 147(16), 178(24), 179(24)
Kuhn, R., 26, 46(19)
Kuhnt, U., 12
Kuhry, J., 344
Kuhse, J., 260, 263, 263(1), 264, 265(13)
Kuisk, I. R., 125
Kukekov, V., 689
Kulak, J. M., 611
Kularatna, A. S., 577, 583(23), 585, 595(24), 600(24), 601(24), 604(24)
Kumagaye, K., 93, 109, 615(61), 623
Kumar, A., 117
Kumar, N. M., 139, 144(36), 177, 178(36)
Kumpf, R. A., 591
Kung, C., 458, 459, 462, 463, 466, 469, 469(18), 470, 471(24), 472(26), 474, 475(26), 476, 477(26), 479, 479(12, 24), 481, 507, 509, 510, 511, 512, 513(29), 514, 515, 518, 519, 519(71a), 521(71a), 523(32, 81), 524, 524(32, 71a)
Kunicki, T., 152(100), 179
Kunjilwar, K., 445, 454
Kunkel, T. A., 593
Kuno, M., 575
Kurachi, Y., 446
Kurasaki, M., 575
Kurz, B., 617(33), 619
Kuwada, M., 109
Kuwahara, M., 570
Kwiecinski, H., 584, 585
Kwong, P. D., 135(10), 136
Kyle, J. W., 578, 579(35), 591, 591(35), 599(35), 603(35)

L

Laatsch, H., 446
Labarca, C., 705
Labarca, P., 288, 292, 296(42), 297, 337
Laemmli, U. K., 454
Lagrutta, A., 299, 300(64)
Lai, F. A., 296(51), 297
Lajeunesse, E., 651, 661(5)
Lakey, J. H., 91
Lamb, R. A., 490, 492, 492(4)
Lambros, T. J., 588
Lancelin, J.-M., 580, 583, 583(54), 598(75)
Landau, E. M., 135, 178(5)
Landry, Y., 344
Landwehrmeyer, G. B., 26, 46(18)
Lane, C. D., 197
Lang, B., 678
Langeberg, L., 680
Langosch, D., 261, 264, 270(16), 273, 686
Langs, D. A., 535
Larsson, H. P., 577
Lasic, D. D., 323
Lasser-Ross, N., 9
Latorre, R., 287, 288, 290, 291(15), 292, 293(6), 296(32, 59), 297, 337, 585, 638, 641, 643(13), 649(13)
Lau, L.-F., 384, 687
Lau, Y. H., 293
Laube, B., 260, 263, 264, 265(13), 270(16)
Läuger, P., 77
Laver, D. R., 517
Lawrence, J. H., 578, 579(32), 591(32), 603(32)
Lay, J. C., 537
Lazaro, R., 549
Lazdunski, M., 512, 638
Lazo, N. D., 525, 531(6), 537(6), 538(6), 539(6)
Le, N. H., 534
Leammli, 406
LeCheminant, G., 615(58), 623
Lecoq, A., 94
Led, J. J., 114
Ledain, A. C., 470, 474(25), 479, 518, 520
Lederer, W. J., 434
Lee, A. G., 209
Lee, A. W., 627, 640
Lee, C., 594, 598(121)
Lee, C. J., 297
Lee, C.-L., 630, 651
Lee, C. W. B., 645
Lee, D., 228, 241(10), 242(10)
Lee, H. C., 524

Lee, J., 117
Lee, K. C., 525, 537, 537(5), 538, 538(72, 73), 539(73)
Lee, M. D., 570
Lee, S. H., 637
Lee, T. T., 200, 202(26)
Lee, Y., 417
Leeman, S. E., 342
Lefevre, J.-F., 98, 113(33)
Legrana, J.-F., 135(10), 136
Le Grice, S. F., 151
Legué, V., 431
Lehmann, J., 389
Lehmann, M. S., 135(10), 136
Lehrer, R. I., 663, 672(5)
Lennarz, W. J., 300
Leonard, R. J., 274, 285, 287, 627, 630
Lepault, J., 137, 165, 179(85)
Lepers, A., 342
Leppik, R., 117
Lesage, F., 512
Lester, H. A., 199, 438, 594, 705
Leszczyzyn, D. J., 189
Levey, A. I., 355, 689
Levina, N. N., 515
Levinson, S. R., 580, 590, 604(99)
Levis, R. A., 202
Levitan, E. S., 48, 51(3), 57, 57(3), 609, 697
Levitan, I. B., 296(45), 297, 299(76), 301, 303(76), 641
Levy, O., 304
Lew, M. J., 100
Lew, R. R., 515
Lewis, J. H., 594, 596(122), 630
Lewis, R. J., 613
Lewis, R. N. A. H., 82
Li, C., 227, 228, 228(7), 230(7), 232(11), 235(11), 240(7), 241(7, 9, 10), 242(9), 319
Li, C. H., 299(78), 303, 324
Li, D., 48, 51(3), 57, 57(3)
Li, H., 148, 179
Li, J., 249, 260(12)
Li, K. W., 615(30), 616(36), 618, 620
Li, M., 686, 697
Li, Q., 521, 697
Li, R. A., 578, 579(33, 34), 599(34)
Li, S. F. Y., 192
Li, W., 412, 416, 605, 616(37), 620
Li, Y.-C., 113
Liao, J., 232, 238(16)
Liao, M. J., 232, 238(12)
Liaw, C., 26, 46(18)
Liebe, R., 588, 589(90)
Lieberburg, I., 672
Lillard, S. J., 189
Liman, E. R., 249
Limenez-Gallego, G., 627
Lin, F., 689
Lin, F. F., 26, 46(19)
Lin, J., 375, 377(6), 381(6)
Lin, M.-C., 663, 672, 672(7)
Lin, S. Y., 538, 539(81)
Lin, W., 594, 598(121), 630, 651
Lindblom, G., 534, 539(46)
Lindorfer, M. A., 525
Lindsay, A. R. G., 296(53), 297
Lindstrom, J., 36, 290, 291(16), 680, 683
Lingle, C. J., 641
Linstedt, A. D., 342
Linz, K. W., 419
Lipkind, G. M., 578, 589(36), 590(36), 599, 599(36), 600, 603(36), 604(135, 139)
Lipscombe, D., 615(63), 623
Lipton, S. A., 33, 36
Lis, L. J., 238
Lisman, J. E., 9
Littauer, U. Z., 342
Liu, J., 135, 634, 635(23), 637(23), 639(23)
Liu, M., 249, 260(12)
Liu, N., 342
Liu, Q.-Y., 296(51), 297
Livshits, V. A., 72
Lledo, P.-M., 425, 431(29)
Lloyd, D. H., 528
Lodder, J. C., 615(30), 616(36, 39), 617(35), 618, 619, 620
Loew, L. M., 45
Loftus, D., 48
Logan, C. J., 588
Logee, K. A., 570
Logotheris, D. E., 51
LoGrasso, P. V., 537
Löhn, M., 182, 187(3)
Lok, S., 227
Loll, P. J., 135
Lomize, A. L., 536
Lomniczi, B., 494
London, E., 232, 238(12)
Lopez, K. L., 697
Lorra, C., 640

Loser, S., 446, 453(13), 454
Louis, J.-C., 689
Loukin, H., 507
Loukin, S. H., 479, 512, 513(29), 523(32), 524(32)
Loutrari, H., 680
Love, Z., 151
Lovett, M. A., 672, 673, 673(20)
Low, P. S., 593
Lowe, A. W., 342
Lowe, P. A., 103
Lu, C.-C., 20
Lu, L., 570
Lü, Q., 591
Luddens, H., 686
Ludtke, S. J., 217
Ludwig, A., 248
Ludwig, D. S., 151
Luebke, J. I., 614, 615(27), 618(27)
Luetje, C. W., 45, 622
Luhring, H., 521
Lukacs, K. D., 191, 192(10)
Lukas, R. J., 45
Lukas, W., 512
Lund, P. E., 182, 187(5)
Lundbæk, J. A., 208, 211, 212(24), 214(24), 220, 221(37, 38), 223(24), 289, 526
Lunney, E. A., 97, 112(28)
Luo, S., 609(49, 50), 623
Luque, A., 609(42), 622
Luque, F. A., 612, 615(16)
Lurtz, M. M., 117, 125, 127, 129(14)
Lynn, P. A., 163

M

Ma, J., 296(52), 297, 299(74, 75), 301, 303(74, 75)
Maathuis, F. J. M., 436
MacDonald, J. F., 366, 368(15)
MacDonald, R. C., 322
MacDonald, R. I., 322
MacDonald, R. L., 187, 528
MacGregor, W. H., 643(26), 645
Machemer, H., 509
Machen, T. E., 22
Mackerer, C. R., 262
MacKinnon, R., 135, 247, 576, 582, 585, 591, 594, 596(122), 605, 627, 630, 634, 635(20), 638, 639, 640, 651, 657, 658(14)
MacLeish, P., 258
Macura, S., 117
Madan, D., 299(76), 301, 303(76)
Maddox, F. N., 622
Madigan, M. T., 507, 520(1)
Madison, D. V., 3, 615(63), 623
Maeda, R. K., 609(43), 623
Maelicke, A., 45, 126, 681
Maer, A. M., 208, 220, 221(38), 289
Magalei, D., 263, 265(13)
Magee, J. C., 12
Mahe, E., 609(51), 623
Maier, A. M., 526
Mains, R. E., 587
Maiorov, V. N., 94
Makielski, J. C., 591
Makino, S., 238
Makofske, R. C., 588
Malchow, D., 515
Malenka, R. C., 3, 353
Malinow, R., 3
Maloney, P. C., 556, 557(18), 558(18)
Mamalaki, A., 680
Mammen, A., 384
Mammen, A. L., 353, 356, 357(6), 360(6), 366(6), 367(6), 368(6)
Mandel, G., 213, 575, 576
Manganas, L., 694
Maniatis, T., 197, 494
Manivannan, K., 601
Mannella, C. A., 144
Mannuzzu, L. M., 577
Manoil, C., 477
Mansbach, A. B., 552, 556(12), 558(12)
Mantsch, H. H., 78, 80, 81, 82, 82(58, 59)
Marbaix, G., 197
Marban, E., 578, 579, 579(32, 37), 590(42), 591(32, 37, 42, 43), 603(32, 42), 604(37, 42, 43, 45)
Marell, P., 187
Margulis, L., 507, 520(1)
Marie, J. S., 208
Markham, R., 165
Markley, J. L., 117, 536
Markram, H., 366
Marks, A. R., 299(79), 303
Markwardt, F., 180, 182, 187(3)
Maroufi, A., 20

Marsh, D., 59, 60, 60(1), 62, 63, 63(16), 64, 64(14, 17), 65, 65(3), 66, 66(22, 23, 25), 67, 67(22), 68, 68(2), 69(2), 70, 71, 72, 72(44), 73, 73(48), 74(51), 75, 76(1), 82, 83, 83(12, 22, 30), 84(71), 85, 87(10), 89, 89(10), 90(10), 91(10, 30, 76)
Marsh, M., 605, 616(37), 617(33), 619, 620
Marshall, J., 271, 281, 299(77), 301, 303(77)
Marshall, P., 672
Marston, F. A. O., 103
Marti, T., 73
Martin, L. J., 355, 689
Martinac, B., 458, 459, 462, 466, 469, 469(18), 470, 471(24), 474(25), 475, 479, 479(24), 507, 510, 511, 512, 514(18), 518, 519, 519(71a), 520, 521(71a), 524(71a)
Martinez, G., 82
Martinez, J., 608, 609(4, 52), 610(4, 54), 623
Martinko, J. M., 507, 520(1)
Marty, A., 182, 186(2), 187(2), 200, 267, 421, 496, 507, 524(4)
Marubio, L. M., 20, 31(4)
Marumo, F., 570
Mason, W. T., 425, 431(29)
Masotti, L., 533
Masuda, S., 402
Mathai, J. C., 550, 554, 559
Mathew, M. K., 577
Mathias, R. T., 601
Mathies, R., 218
Matos, M. F., 694
Matouschek, A., 594, 596, 596(124)
Matsumoto, S., 446
Matsushima, M., 136, 178(23), 179(23)
Matsuzawa, Y., 446
Matthias, P., 397
Mattice, G. L., 525, 532, 536(7, 35), 548, 550(35)
Mauger, J. P., 511
Maulet, Y., 261
Mayer, M. L., 46
Mayoraga-Wark, O., 697
Mayorga, W. O., 641
Mays, D. J., 697
Mazars, C., 417
Mazet, J. L., 526, 549(12)
McAlister, M., 238
McBride, D. W., Jr., 482, 483, 485(1), 488
McCabe, R. T., 612, 613(18)
McCallum, C. D., 208
McCammon, J. A., 540
McCance, D. J., 490
McCaromack, K., 577
McCarthy, K. D., 187
McCaslin, D. R., 309, 476
McCloskey, M. A., 189
McClure, F. T., 481
McConnell, H. M., 190
McConnell, P., 92
McCray, G., 270
McCue, A. F., 20
McDermott, J., 151
McDowall, A. W., 137
McElhaney, R. N., 82
McFadzean, I., 435
McGammon, J. A., 539, 542(89)
McGehee, D. S., 36
McIntosh, J. M., 106, 605, 608, 609(4, 13, 44, 49–53), 610(4, 15, 54), 611, 612, 615(16, 29, 57, 58), 616(37, 38), 618, 620, 622, 623
McIntyre, D., 589
McKernan, R. M., 686
McLaughlin, S., 224, 601
McManus, O. B., 274, 277, 281(9, 10), 283(9), 284, 285, 287, 637, 638
McNamara, N. M., 694
McNamee, M. G., 66
McPhee, J. C., 591
McPherson, A., 142
Medynski, D. C., 125
Meera, P., 281, 287, 297
Meguri, H., 387, 389(5)
Mei, L., 687
Meienhofer, J., 588
Meissner, G., 296(51), 297
Melkonian, M., 507, 520(1)
Mellema, J. E., 163
Meller, P. H., 151
Melton, D. A., 197, 569
Mena, E. E., 605, 610(14, 17), 611, 612
Menco, B. P., 322
Mendelsohn, R., 83
Menez, A., 94, 100, 581, 651, 661(5)
Menez, R., 651, 661(5)
Mercer, K. L., 151
Merlie, J. P., 697
Merrifield, R. B., 529
Mertens, S., 686
Metcalfe, J. C., 208
Methfessel, C., 45, 290, 291(18)

Meunier, S., 94
Meyer, D. I., 473
Meyer, G., 686
Mezai, M., 539, 540(87)
Mi, F., 697
Mi, H., 697
Michel, H., 135
Middleton, R. E., 296(44), 297, 299(80), 303, 303(44)
Miedema, H., 412, 413(6), 416
Miick, S. M., 82
Miledi, R. E., 263
Milgram, S. L., 587
Miljanich, G., 27, 92, 101(3), 106, 117, 605, 609(48), 615(29, 48, 59), 618, 623
Miller, C., 92, 100, 108(43), 244, 245, 246(40), 288, 290, 292, 296(28, 42–44), 297, 299(73, 80), 301, 302(73), 303, 303(30, 44, 73), 319, 320, 321(7), 337, 467, 581, 582, 583(70), 584, 585, 586, 587, 587(64, 65), 591, 594(73), 597(71), 598(71), 602(70), 610(55), 623, 630, 634, 634(16), 635(22), 638, 644, 657, 658, 658(14), 662
Miller, D. D., 434
Miller, D. L., 190
Miller, J. N., 673
Miller, R. J., 20, 31(4), 615(62), 623, 672, 673(20)
Millett, F. S., 530
Millhauser, G., 82
Milligan, R. A., 137
Milne, T., 538, 539(81)
Minchin, R. F., 475
Minorsky, P. V., 462, 510, 514(18)
Minta, A., 22
Mintz, I. M., 94, 101(10), 106, 110(10), 114(10), 605, 615(29), 618
Mirzabekov, T. A., 661, 663, 664, 670(12), 672, 672(7), 673, 673(20), 674
Mischke, C., 440
Misell, D. L., 163(77), 165, 176(77)
Mitani, A., 402
Mitchell, S. S., 94
Mitra, A. K., 135, 148, 149, 174, 175(57), 179
Mitra, R. D., 52
Mitsuoka, K., 136, 178(23), 179, 179(23)
Mixon, T. E., 537
Miyakawa, H., 9
Miyashita, T., 697
Miyata, T., 575, 576(3)
Miyazawa, A., 136, 178(23), 179(23)
Miyazawa, T., 80, 112
Mobashery, N., 222
Moczydlowski, E., 281, 287, 288, 293(6, 7), 296(7), 297, 299(77), 301, 302(7), 303(77), 578, 580, 581, 581(51), 584, 585, 586, 589, 591(31), 603(31), 606, 614(1), 616(1, 69), 624
Moe, P. C., 458, 459, 470, 472(26), 474, 475(26), 477(26), 518, 519(71a), 521(71a), 524(71a)
Moffatt, D. J., 78
Mohandas, N., 559
Mohler, H., 686
Molday, L. L., 246, 247, 249(8), 250, 252(15), 253, 254(19, 20), 255(8, 20), 257(8), 259(8, 19), 699
Molday, R. S., 246, 247, 249, 249(3, 8), 250, 252(14, 15), 253, 253(14), 254(14, 19, 20), 255(8, 20), 256(11, 14), 257, 257(8, 11, 14), 258(11, 14, 22), 259(8, 19), 699
Molenaar, P. C., 678
Molinari, E. J., 36
Moll, F. I., 537, 538(66)
Molloy, R., 271
Monaghan, B. W., 694
Monaghan, D. T., 389
Monaghan, M. M., 697
Monje, V. D., 106, 615(29, 59), 618, 623
Montal, M., 290, 291, 291(16), 292, 490, 664
Montana, V., 45
Monteggia, L. M., 36
Montelione, G. T., 113, 116
Montgomery, R. A. P., 290
Montrose-Rafizadeh, C., 570
Monyer, H., 41
Moody, M., 165
Moon, C., 551
Morabito, M., 299(77), 301, 303(77)
Morales, M. J., 697
Moran, J., 105
Moran, N., 424, 440
Moran, T., 20, 34(8), 689
Moreau, M., 417
Moreno, H., 697
Moreno-Bello, M., 320
Morgan, D. W., 276
Morgenstern, R., 179
Mori, S., 559
Morin, P. J., 342

Morishige, K. I., 697
Moriyama, R., 238
Moronne, M. M., 577
Morr, J., 273
Morris, A. P., 490
Morrison, J. H., 20, 34(8), 687, 689
Morton, A. J., 425, 431(29)
Moscho, A., 191, 194(9), 196(9), 203, 203(8, 9)
Moskowitz, H., 703
Moss, G. W. J., 271, 281, 299(77), 301, 303(77)
Moss, L. G., 49
Moss, S. J., 354, 355, 366, 368(4), 689
Mossier, B., 686
Mouritsen, O. G., 209
Mueller, A. L., 639
Mueller, P., 291, 661, 663(1), 664, 670(1)
Mueller, U., 515
Mueller-Neuteboom, S., 145(99), 179
Muga, A., 80, 82
Muglia, L. K., 575(11), 576
Muhandiram, D. R., 98, 113(35)
Muhandiram, R., 98, 112(30), 113, 113(30)
Muller, F., 247, 249(8), 255(8), 257(8), 259(8), 699
Müller, M. M., 299(76), 301, 303(76), 397
Muller, R. U., 543
Muller, R. V., 668
Mulligan, R., 227, 241(2)
Mullmann, T. J., 277, 281(10), 638
Multhaup, G., 261
Munujos, P., 274, 280
Murata, K., 136, 178(23), 179, 179(23)
Murer, H., 316
Murphy, D. E., 389, 394
Murphy, R., 100
Murphy, T. H., 354, 368(4)
Murray, R. G. E., 508, 518(5)
Murrell-Lagnado, R. D., 640, 641, 641(8), 643(8, 12)
Myers, R. A., 609(46), 613, 623

N

Nabedryk, E., 91
Nabeshima, T., 392, 394(22)
Nadasdi, L., 46, 609(48), 615(29, 48), 618, 623
Nadasdki, L., 100
Nadaski, L., 106
Nadler, L. S., 686
Nagai, U., 112
Nagle, S. K., 459, 463, 479(12)
Naini, A. A., 299(73), 301, 302(73), 303(73)
Nairn, A. C., 241, 366
Naisbitt, S. R., 615(59), 618, 622, 623
Naismith, A. L., 698
Naismith, L., 227, 230(3), 231(3), 241(3)
Naitoh, Y., 509
Nakahira, K., 694, 697
Nakajima, K., 109, 570
Nakamoto, R. K., 555
Nakamura, H., 580, 583, 589(52), 598(75), 606, 616(2, 68, 72), 624
Nakamura, R. L., 521
Nakamura, T., 616(39), 620
Nakanishi, S., 65
Nakao, K., 446
Nakashima, H., 238
Naranjo, D., 582, 585, 594(73)
Narasimhan, L., 92
Nasdasdi, L., 615(59), 623
Nasim, A., 514
Naudat, V., 94
Naumann, D., 83, 88(69), 92(69)
Navarro, J., 208
Navetta, F. I., 526, 549(13)
Navia, M. A., 627
Nawoschlik, S., 697
Near, J. A., 189
Needham, M., 493, 495(19), 498(20)
Neff, S., 684
Negulescu, P. A., 22
Neher, E., 3, 24, 27, 30, 32(22), 36(22), 182, 186(2), 187(2), 199, 200, 202(24), 204(24), 205(23, 24), 267, 300, 410, 421, 496, 506, 507, 524(4), 542, 657, 662
Nelson, D. A., 446, 454, 455(17), 456, 456(17)
Nelson, K., 307
Nelson, M. T., 296(57), 297
Nerbonne, J. M., 697
Neubig, R. R., 117, 120(3), 126(3)
Nevskaja, N. A., 80
Nevskaka, N. A., 80
Newcomb, R., 92
Newman, G. C., 12
Newsom-Davis, J., 678, 681, 682
Neyton, J., 584, 585
Nguy, K., 456
Nguyen, T. A., 611

Nicholls, A., 598
Nicholls, D. G., 366
Nichols, C. G., 454, 456
Nicholson, L. K., 113, 537
Nicolini, M., 664, 670(12)
Nicoll, R. A., 3, 353
Nicolson, G. L., 208, 209(12), 224(12)
Nielsen, C., 208, 214, 221(27), 222, 289, 526
Nielsen, D. B., 609(51), 623
Nielsen, E. Ø., 394
Nielsen, K. J., 92, 613
Nielsen, M., 394
Nielsen, S., 551, 552, 556(12), 558(12)
Nienhuis, A., 495
Niethammer, M., 371, 373(5), 379(3), 384(3)
Nigh, E., 375, 377(6), 381(6)
Niidome, T., 109
Niles, W. D., 319, 320, 321, 322, 331(6)
Nirmala, N. R., 98
Nishiuchi, Y., 93, 615(60, 61), 623
Nishizawa, Y., 109
Noble, R. L., 527, 528
Noda, M., 575, 576(3), 577, 578, 579, 590(25)
Noda, T., 93
Noda, Y., 615(61), 623
Noebels, J. L., 703
Nogales, E., 143
Nonnengasser, C., 518
Norberg, R. E., 537
Norris, T. M., 94, 100, 101(11), 108(11), 110(11)
North, R. A., 512, 655, 656(13)
Norton, R. S., 92, 100
Novick, J., 627
Nozaki, Y., 323
Numa, S., 575, 576(3), 577, 578, 579, 590(25, 26), 591(26), 603(26)
Nussberger, S., 193
Nusser, Z., 689

## O

Oberleithner, H., 521
Obermeyer, G., 419
Oblatt-Montal, M., 490
O'Brien, R. J., 356, 357(6), 360(6), 366(6), 367(6), 368(6)
Obrietan, K., 425
O'Connell, A. M., 217, 543, 545(107), 546(107)
Odani, S., 580
O'Donovan, M. J., 4
Oertel, D., 509
Oesterhelt, D., 155(104), 165, 179
O'Farrell, 406
Offord, J., 92
Ogden, D. C., 490, 492(5)
Ogita, K., 385, 386, 387, 388(2, 3), 389, 389(4, 5), 390, 390(10, 12), 391, 392, 393(20), 394, 394(22), 395, 397(27, 28), 401(26), 402, 402(27)
Ogura, A., 12
Ohgaki, T., 387, 389(5)
Ohizumi, Y., 580, 583, 589(52), 598(75), 606, 616(2, 68, 72), 624
Ohkubo, H., 65
Ohno-Shosaku, T., 445
Ohsumi, Y., 514
Ohya, M., 583, 598(75)
Oiki, S., 526
Okawa, K., 529
Okayama, H., 270, 302(82), 303
Olans, L., 296(59), 297
Olivera, B. M., 27, 92, 94, 101(3), 106, 580, 581(51), 605, 606, 608, 609(4, 42, 44–46, 48–51), 610(4, 14, 15, 17, 54–56), 611, 612, 613, 613(18), 614, 614(1), 615(16, 26, 29, 48, 57–59), 616(1, 23, 25, 37, 38, 69), 617(23, 25, 32–34), 618, 619, 619(23), 620, 621(25), 622, 623, 624
Olofsson, A., 137, 150
Olson, M., 651, 661(6)
Omar, M., 682
Omecinsky, D. O., 93, 94, 97, 111(9), 112 (27)
Ondrias, K., 299(79), 303
Onoue, H., 296(54), 297
Oorschot, V., 342, 348(12)
Oosawa, Y., 510, 515(21)
Opalko, A., 643(26), 645
Opella, S. J., 136, 536
Orekhov, V. Y., 536
Oren, R., 609(12), 612
Ormo, M., 49
O'Rourke, B., 579
Orwar, O., 190, 191, 194(9), 196(9), 203, 203(8, 9)
Osborne, P. B., 655, 656(13)

Osipchuk, Y. V., 45
Ostermeier, C., 135
Otteson, K. M., 528
Ottolia, M., 287, 296(58), 297, 641, 643(13), 649(13)
Ou, X., 479, 523(81), 524
Ousley, A. H., 692
Ovchinnikov, Yu. A., 536
Owen, D. G., 643(26), 645, 659
Owicki, J. C., 190
Oxford, G. S., 187, 368
Oya, M., 583

## P

Paarmann, I., 612, 613(18)
Pace, N. R., 507, 508, 520(1)
Paczkowski, J. A., 526, 532(19)
Pagán, O. R., 609(6, 47), 610, 623
Pagnozzi, M. J., 610(14), 612
Palfrey, I., 342
Páli, T., 71, 72, 72(44), 73, 74(51), 82
Pallaghy, P. K., 92, 100
Pallanck, L., 287
Palmer, A. G. III, 98, 99, 113(32, 36)
Panten, U., 445, 446, 453(13), 454
Papadouli, I., 680
Papahadjopoulos, D., 238, 239, 239(30)
Papineni, R. V. L., 117, 131
Parce, J. W., 190
Parcej, D. N., 638, 640, 649, 651, 661(6)
Parekh, A. B., 300, 651
Parent, L., 296(56), 297
Park, C.-S., 100, 108(43), 582, 583(70), 585, 587, 602(70), 634
Parker, C., 454
Parker, J., 507, 520(1)
Parks, T. N., 610(17), 612
Parli, J. A., 530
Parsegian, V. A., 238, 335
Partin, K. M., 46
Partiseti, M., 630
Pascal, S. M., 113, 536
Patchett, A. A., 630
Patel, H. J., 12
Patneau, D. K., 46
Patrick, J., 45, 684
Patthi, S., 20
Patton, D. E., 577
Pattus, F., 91
Paul, S., 227, 241(2)
Pauptit, R. A., 518
Pawson, T., 113
Pearson, R. B., 362
Pebay-Peyroula, E., 135, 178(5)
Pedersen, J. A., 66
Pedersen, S. E., 117, 118, 119, 120, 125, 125(8), 127, 129(14), 131
Peersen, O. B., 536, 537(63)
Pei, Z. M., 412, 416, 425
Pelhate, M., 100
Peng, J. W., 98, 113(31)
Peng, S., 298
Peng, Y.-W., 247, 249(7), 258(7)
Penin, F., 232, 238(18)
Pentoney, S. L., 194
Perbal, G., 431
Perczel, A., 80, 81
Pereira, E. F. R., 609(53), 623
Péres-García, M. T., 578, 579(37), 591(37), 604(37)
Pérez, G., 299, 300(64)
Pérez-García, M. T., 579, 590(42), 591(42, 43), 603(42), 604(42, 43)
Pernberg, J., 509
Perrotto, J., 307
Perschini, A., 52
Petkoff, H. S., 427, 430(36)
Petralia, R. S., 687
Petronilli, V., 463, 519, 520
Pfeiffer, F., 261, 262(2, 4), 263(2), 686, 697
Pheasant, D. J., 100, 296(44), 297, 299(80), 303, 303(44), 582, 585, 587(65), 630, 634(16)
Philipson, L. H., 20, 31(4), 697
Phillips, D., 94, 101(12), 110(12)
Phillips, G. N., 49
Phillips, M., 297
Piattoni-Kaplan, M., 36
Pick, U., 322
Picot, D., 135
Pictet, R., 198
Pidgeon, C., 104
Pidoplichko, V. I., 180, 181(1), 182(1), 187(1), 497
Pierson, E. S., 434
Pillet, L., 100

Russ, J. C., 165, 171(79)
Russell, E. W. B., 526, 549(13)
Ruthe, H. J., 460
Rutter, W. J., 198
Ruysschaert, J.-M., 79

## S

Sabatini, B. L., 4
Saberwal, G., 525, 534, 536(2), 542(9, 11), 545(11), 549(11), 550(9)
Sacaan, A. I., 26, 46(18), 46(19)
Saccomano, N. A., 94, 101(12), 110(12), 639
Sachs, F., 463, 464(13)
Saffman, P. G., 58
Saggau, P., 3, 4, 12, 17(28)
Saimi, Y., 462, 479, 507, 509, 510, 511, 512, 513(29), 514(18), 523(32), 524(32)
Saito, H., 570
Sakakibara, S., 93, 615(60, 61), 623
Sakamoto, T., 394
Sakarellos, C., 680
Sakmann, B., 5, 11(19), 13(19), 27, 32(22), 36(22), 41, 199, 202(24), 204(24), 205(23, 24), 267, 300, 421, 496, 507, 524(4), 543, 662
Sakurai, T., 692
Sala, F., 426
Salama, G., 45
Salemink, I., 534, 539(46)
Salkoff, L., 273, 274, 286(1), 575, 625, 645
Salmon, E. D., 164(110), 179
Sambrook, J., 494
Samejima, K., 390
Sanchez, M., 637
Sandblom, J., 542
Sanguinetti, M. C., 639
Sankar, G., 593
Sankararamakrishnan, R., 599, 600, 600(136)
Santarelli, V., 590
Sar, P. C., 66
Sargent, P. B., 611
Sarges, R., 525
Sariban-Sohraby, S., 296(59), 297
Sarin, V. K., 529
Sasaki, S., 551, 570
Sassaroli, M., 51
Satake, M., 580, 583(54)
Sather, W. A., 300
Satin, J., 578, 579(35), 591, 591(35), 599(35), 603(35)
Sato, K., 94, 97(13), 101(13), 110(13), 583, 598(75), 616(72), 624
Sato, M., 514
Sato, S., 616(68), 624
Sattsangi, S., 339
Saudek, V., 117
Sautiere, P., 94
Sautter, A., 248
Sawada, K., 109
Sawyer, D. B., 212
Sayre, D., 163
Scarborough, G., 142
Scatchard, G., 129
Scavarda, N., 575
Schabtach, E., 151
Schaffner, W., 397, 557
Schapper, A., 618
Schechter, L. E., 694, 697
Schein, S. J., 509, 511, 663, 672(8)
Scheller, R. H., 191, 197(7), 199(7), 200, 202(26), 203, 203(8), 206(7), 207, 342
Schenker, L. T., 371
Scherler, G. F., 144
Scherphof, G., 239
Scheuer, T., 591
Scheurmann-Kettner, C., 512
Scheybani, T., 136, 144(17, 18), 147(17, 18)
Schild, D., 427
Schild, L., 578, 591(31), 603(31)
Schiltz, E., 518
Schimmel, P. R., 128
Schindler, H., 91, 290, 291(17), 667
Schirmer, T., 518
Schleyer, M., 481
Schlief, T., 300
Schlue, W.-R., 288, 337
Schluep, M., 678, 681
Schmalhofer, W. A., 274, 280, 625, 626(2), 627, 637
Schmid, M. F., 136, 150, 151
Schmid, R., 481
Schmidt, J., 123, 651, 661(6)
Schmidt-Krey, I., 179
Schmidtmayer, J., 589
Schmieden, V., 263, 264, 270(16)
Schmitt, B., 261, 262(3), 273
Schmutz, M., 150

Schnaitman, C. A., 475
Schneggenberger, R., 506
Schneggenburger, R., 3, 439
Schneider, J., 389
Schneider, S., 521
Schoch, P., 686
Schoemaker, J. M., 103
Schoepfer, R., 36
Schofield, P. R., 262, 264
Schonerr, R., 300
Schoofs, A. R., 630
Schoolnik, G. K., 151
Schoonmaker, S., 20
Schoppa, N. E., 273
Schott, M. K., 645
Schrattenholz, A., 45
Schreiber, E., 397
Schreiber, G., 594, 595(127)
Schreiber, S. L., 94, 101(12), 110(12)
Schroeder, C., 492
Schroeder, J. I., 412, 416, 423, 425, 439(22), 440
Schroeder, M. J., 463, 470, 472(26), 475(26), 476, 477(26), 479(12), 481
Schroeder, T. J., 189
Schubert, V., 490
Schuerholz, T., 290, 291(17)
Schultz, G. E., 518
Schultz, P., 137
Schultz, S. G., 641, 697
Schultze, R., 517
Schutzbach, J. S., 208
Schwanstecher, C., 445
Schwanstecher, M., 445, 446, 453(13), 454
Schwartz, A., 424
Schwarz, M., 412, 416
Schwarzer, C., 634
Scott, A., 299(79), 303
Scott, J. K., 605
Scott, V. E., 649, 692, 697
Scriven, L. E., 155(103), 179
Seeburg, P. H., 41, 262, 371
Segal, M., 366
Seifert, R., 41, 247, 249(8), 255(8), 257(8), 259(8), 699
Seigneuret, M., 208
Seino, S., 446
Sellers, A. J., 512
Selsted, M. E., 663, 672(5)
Sen, A., 208
Sener, A., 304, 308, 309, 313(9), 314(7), 315(9)
Sentenac, H., 512, 513(33)
Separovic, F., 537, 538, 539, 539(81)
Serrano, L., 594, 595(127), 596, 596(124, 126), 597(126)
Sesti, F., 247, 249(8), 255(8), 257(8), 259(8)
Sesti, R., 699
Shafaatian, R., 512
Shang, E., 672, 673, 673(20)
Shao, Z., 137
Shapiro, B. I., 601
Shapiro, L., 135(10), 136
Sharma, G., 687
Sharma, N., 697
Sharom, F. J., 238
Shaw, K. P., 187
Shear, J. B., 191, 197(7), 199(7), 206(7), 207
Sheehan, B., 174
Shelton, P. A., 493, 498(20)
Shen, T. F., 262
Sheng, M., 371, 373(5), 375, 377, 377(6), 379, 379(3), 381(6), 384, 384(3), 697
Sheng, Z., 575(10), 576
Shepherd, F. H., 235
Sheppard, R. C., 588
Sheridan, R. D., 12
Sherman, M. B., 159(108), 179
Sherman-Gold, R., 290
Shevell, J. L., 276
Shi, G., 694, 697
Shih, T. M., 524
Shillito, P., 678
Shimada, I., 94, 97(13), 101(13), 110(13)
Shimbo, K., 492
Shimomura, O., 49
Shimonishi, Y., 616(72), 624
Shimony, E., 634, 635(22)
Shine, J., 198
Shirahata, A., 390
Shivers, B. D., 686
Shobana, S., 223, 289, 525, 540
Shockman, G. D., 463
Shoelson, S. E., 113
Sholomenko, G., 4
Shon, K., 94, 610(54–56), 614, 616(23, 25), 617(23, 25, 33, 34), 619, 619(23), 621(25), 623
Shustak, C., 137
Shuto, M., 390
Siegel, R. E., 686

Siegel, S. J., 687
Siegelbaum, S. A., 247, 641
Sieghart, W., 686, 689
Sierralta, I., 589, 600
Sigel, E., 651
Sigworth, F. J., 7, 267, 273, 289, 421, 496, 507, 524(4), 576, 604
Sikerwar, S. S., 144
Silberstein, A. Ya., 661, 664, 666(13), 672(13)
Silbert, D. F., 537
Sills, M. A., 389
Silva, A. J., 3
Silva, C. M., 52
Silvius, J. R., 64, 66(20)
Simcox, T. G., 593
Simerson, S., 20
Simler, R., 261, 262(4), 686, 697
Simon, M., 594, 598(121), 630, 651
Simoncini, L., 423, 439(22)
Simons, K., 342, 343(5), 348(10)
Sine, S. M., 124, 609(43), 623
Singelton, D. H., 594, 598(121)
Singer, A. U., 113
Singer, S. J., 208, 209(12), 224(12)
Singh, C., 599, 600(136)
Singh, J., 92
Singh, T., 92
Singleton, D. H., 630, 651
Sinha, S. R., 4
Sitsapesan, R., 290
Sizun, P., 94
Skare, J. T., 672, 673, 673(20)
Skehel, J. J., 491, 494
Skjaebaek, N., 613
Sklenar, V., 117
Skvoretz, R., 20, 34(8)
Slater, C. R., 682
Slaughter, R. S., 274, 276, 624, 625, 626(2), 634, 637(18, 19)
Slayman, C. L., 512, 513, 513(33), 515, 516, 517
Slayman, C. W., 555
Sleytre, U. B., 165
Smart, T. G., 366
Smilowitz, H., 290, 296(50), 297
Smith, A., 227, 241(2)
Smith, A. B. III, 637
Smith, B. L., 148, 179, 550, 551, 552, 553(3, 4), 556(11, 12), 558(12), 559, 560(11)
Smith, C., 585
Smith, D. O., 465
Smith, F. R., 133, 135(33)
Smith, G. A., 208
Smith, G. D., 535
Smith, J. M., 165
Smith, L., 651, 661(6)
Smith, M., 276, 277(6), 281(6), 286
Smith, M. C., 104
Smith, M. H., 491
Smith, P. E., 116
Smith, R., 537, 538, 539, 539(81)
Smith, R. D., 524
Smith, S. O., 536, 537(63)
Smyth, D. G., 587
Snel, M. M. E., 72, 73(48), 75
Sninsky, J. J., 477
Snowhill, E. W., 394
Snutch, T. P., 624
Sobel, A., 119
Soderling, T. R., 366
Soerensen, O. W., 117
Sokabe, M., 463, 464(13), 510, 515(21)
Sokolov, Y., 662, 672, 673(20)
Solaro, C. R., 641
Solomon, A. K., 551
Somasundaram, B., 425, 431(29)
Sommer, S. S., 593
Somoguyi, P., 686, 689
Song, G., 92, 94(6)
Song, J., 94
Song, L., 137
Sonstrom, N. A., 490
Sontheimer, H., 262
Sorensen, R. G., 650
Soreq, H., 265
Sorgato, C. M., 664, 670(12)
Souza, D., 227, 241(2)
Spalding, E. P., 423, 439(21), 512, 523(32), 524(32)
Sparrman, M., 232, 238(16)
Spector, I., 342
Spencer, R. H., 655, 697
Sperk, G., 634
Spira, M. E., 605, 609(12, 13), 612, 614, 616(23, 37), 617(23, 71), 619(23), 620, 624
Spisni, A., 533
Springer, E., 239
Stabinsky, Y., 103
Stambrook, P. J., 592, 593(113)

Stampe, P., 92, 582, 585, 586, 587(64), 630
Standaert, D., 26, 46(19)
Stanfield, P. R., 641
Stanley, E. F., 672
Staples, R. C., 515
Stauffer, D. A., 579, 591(40)
Steele, J. C., Jr., 232
Stefani, E., 287, 296(47, 48), 297, 641, 643(13), 649(13)
Steiert, M., 518
Stelly, N., 511
Stephan, M. M., 578, 579(29), 589(29), 590(29), 616(70), 624
Stephanson, F. A., 677, 689
Stephens, G. J., 643(26), 645, 659
Stephens, R. L., 117
Stephenson, F. A., 686
Stern, J. H., 258
Stern-Bach, Y., 687
Stetzer, E., 45
Stevens, C. F., 3, 366, 513, 516(34)
Stevens, S. P., 627
Stewart, M., 136, 165(13)
Stieger, B., 316
Stillinger, F. H. J., 601
Stocker, J. W., 46
Stocker, M., 582, 597(71), 598(71), 614, 616(25, 38), 617(25, 34), 619, 620, 621(25), 640, 645(4), 658
Stoehr, S. J., 622
Stoorvogel, W., 342, 348(12)
Stopar, D., 71
Storch, A., 45
Stow, J., 659
Stoylova, S., 150
Straight, S. W., 490
Strassle, B. W., 694
Strauss, H. C., 697
Strecker, A., 83, 88(69), 92(69)
Strichartz, G. R., 581, 589, 616(69), 624
Striessnig, J., 692
Strnad, N. P., 125
Strotmann, J., 232
Stryer, L., 218
Stuart, G. J., 5, 11(19), 13(19)
Stühmer, W., 205, 300, 577, 578, 579, 590(25, 26), 591(26), 603(26), 614, 616(25), 617(25, 34), 619, 621(25), 651
Stumpf, M. A., 515
Suarez-Isla, B. A., 290, 291(16)
Subbarao, N. K., 322
Subramaniam, S., 155(104), 179, 599, 600, 600(136)
Subramanian, S., 540
Sucher, N. J., 687
Südhof, T. C., 342
Sugg, E. E., 285, 630
Sugrue, R. J., 491
Sukharev, S. I., 458, 459, 463, 465, 469, 470, 471(24), 472(26), 475(26), 476, 477(26), 479(12, 24), 518, 519(71a), 521(71a), 524(71a)
Sullivan, J. P., 36
Sumikawa, K., 263, 264
Sun, S., 227, 230(3), 231(3), 241(3)
Sun, T., 299(73), 301, 302(73), 303(73), 634, 635(22)
Sun, Y.-M., 287
Sundler, R., 238
Surewicz, W. K., 82
Surrey, T., 89, 91(76)
Susi, H., 60, 81(7)
Sutherland, T., 490
Suzuki, H., 575, 577, 578, 590(25)
Suzuki, J., 580
Suzuki, M., 697
Suzuki, T., 389, 390(12), 391, 392
Swanson, L., 680
Swanson, R., 197, 286, 287
Swartz, K. J., 605, 639
Sykes, B. D., 98, 113(35)
Szabo, I., 463, 481, 519
Szalay, M., 143
Szilagyi, L., 109
Szyperski, T., 116

## T

Tabcharani, J. A., 227, 241(4)
Takahashi, H., 575, 576(3)
Takahashi, M., 510
Takahashi, T., 300, 575
Takeshima, H., 575
Takeshita, K, 322
Takimoto, K., 48, 51(3), 57, 57(3), 697
Takumi, T., 65

Taleb, O., 269
Talmon, Y., 155(103), 179
Tam, J. P., 529
Tamazaki, T., 98, 112(30), 113(30)
Tamkun, M. M., 575(8), 576, 697
Tamm, L. K., 60, 82(11), 89, 91(11, 76)
Tanabe, T., 575, 576(3)
Tanford, C., 309, 323
Tang, J. M., 436, 437(46)
Tanifugi, M., 514
Tanifuji, M., 514
Tank, D. W., 4, 12(10), 17(10), 18(10), 467
Tanouye, M. A., 577
Taraschi, T. F., 239
Tarczy-Hornoch, K., 100
Tarr, G. E., 135
Tate, C. G., 139
Tate, S.-I., 580, 583(54)
Tatulian, S. A., 60, 82(11), 89, 91(11, 76)
Tavazoie, M. F., 611
Tavazoie, S. F., 611
Taverna, F. A., 366, 368(15)
Tavernarakis, N., 459
Taylor, A. R., 419, 436
Taylor, K. A., 137
Taylor, M. J., 525, 536(7), 538
Taylor, P., 124, 126, 609(43), 623
Tejero, R., 116
Telford, J. N., 319
Tempel, B. L., 649, 650, 655, 656(12)
Tempst, P., 673
Teramoto, T., 109
Terlau, H., 578, 579, 590(26), 591(26, 27), 603(26, 27), 614, 616(25, 38), 617(25, 34), 619, 620, 621(25)
Testa, C. M., 26, 46(18)
Tester, M., 416
Thanabal, V., 94, 96(14), 97, 101(10, 14), 110(10, 14), 112(27, 28), 114(10)
Thelestam, 137
Thevand, A., 94
Thiel, G., 425, 440, 517
Thion, L., 417
Thomas, D. D., 68
Thomas, D. E., 537
Thomine, S., 422
Thompson, A., 135(10), 136
Thompson, A. K., 687
Thompson, C. A., 492, 494(14), 506(14)
Thompson, D., 151
Thompson, S., 227, 241(2)
Thon, 163(109), 179
Thuleau, P., 417
Tian, P., 490
Tiedemann, H., 300
Tiedemann, J. H., 300
Tien, H. T., 291, 661, 663(1), 670(1)
Tiley, C. T., 364
Tingley, W. G., 353, 364, 687
Tinkle, S. S., 590, 604(99)
Tischfield, J. A., 592, 593(113)
Todd, A. P., 82
Todt, H., 578, 589(36), 590(36), 599(36), 603(36)
Togel, M., 686
Toivio-Kinnucan, M., 208
Toma, F., 94, 581
Tomaselli, G. F., 578, 579, 579(32, 37), 590(42), 591(32, 37, 42, 43), 603(32, 42), 604(37, 42, 43, 45)
Tomaselli, K., 672
Tomiko, S., 576
Tommassen, J., 518
Tooze, J., 342
Torchia, D. A., 98, 113, 114(70)
Toro, L., 281, 287, 296(47, 48, 58), 297, 299, 300(64), 524, 641, 643(13), 649(13)
Torres, J., 610(56), 612, 613(18), 623
Tosteson, M. T., 492
Toung, Y. P., 524
Toyka, K. V., 681
Toyosato, M., 575, 576(3)
Trakas, N., 680
Traynelis, S. F., 689
Trémeau, O., 100
Trentham, D. R., 440
Trewavas, A. J., 438
Triller, A., 686
Trimmer, J. S., 576, 694, 697
Trinick, J., 155(106), 179
Trotter, M., 194
Trower, M. K., 592
Trube, G., 411, 421(4)
Trudelle, Y., 525, 542(8), 545(8), 549
Trzeciak, A., 686
Tsaur, M.-L., 379, 697
Tsetlin, V. I., 609(7), 610
Tsien, R. W., 46, 290, 296(50), 297, 300, 615(63, 66), 623, 624

Tsien, R. Y., 4, 5, 21, 22, 25(13, 15), 45, 49, 50, 51(10), 52, 425, 437, 438
Tsui, L.-C., 227, 230(3), 231(3), 241(3)
Tsushima, R. G., 578, 579(33, 34), 599(34)
Tu, C. P., 524
Tu, Y. S., 524
Tucker, A., 536
Turnbull, P. J. H., 66, 83(30), 91(30)
Turner, G. L., 530
Turner, T. J., 614, 615(27), 618(27)
Twyman, R. E., 678, 689
Tygat, J., 658, 659(16)
Typke, D., 148
Tzartos, E., 680
Tzartos, J. S., 680
Tzartos, S. J., 678, 680

U

Uchida, S., 387, 389(5), 570
Uehara, A., 288, 293(7), 296(7), 302(7), 586, 589
Ukomadu, C., 604
Ulbricht, W., 589
Ullian, E. M., 611
Ullrich, A., 198
Ultsch, M., 135
Unger, V. M., 135, 139, 144(36), 165, 177, 178(36)
Unwin, N., 66, 82(29), 88(29), 137, 144, 155(34), 210, 215(19)
Unwin, P. N. T., 136, 137, 143, 144, 155(101, 102), 165(12), 179, 210, 215(19)
Urcan, M. S., 20, 34(8), 590, 604(99)
Urrutia, A., 20, 26, 36(5), 46(18)
Urry, D. W., 533, 537, 541
Utkin, Y. N., 609(7), 610

V

Vaca, L., 296(48), 297
Vaillant, B., 512, 513(29), 523(32), 524(32)
Valcana, C., 680
Valdivia, C., 297
Valdivia, H. H., 297
Valdivia, R., 49
Valentino, K., 92
Valenzuela, C. F., 126
Valio, M. I., 538
Valpuesta, J. M., 165
van den Pol, A. N., 425
van der Schors, R. C., 615(30), 616(36), 618, 620
van der Steen, A. T., 239
Van Duijn, B., 419, 422
van Gunsteren, W. F., 116, 539
van Gunsteren, W. J., 539
van Hoek, A. N., 148, 149, 174, 175(57), 179
Van Rietschoten, J., 94, 617(35), 619
van Schaik, R. C., 116
Van Volkenburgh, E., 416, 417
van Zijl, P. C. M., 559
Varga, J., 615(57, 58), 623
Varney, M. A., 20, 26, 34(8), 46(19)
Vasquez, J., 634, 637(18)
Vassylyev, D. G., 136, 178(23), 179(23)
Vatanpour, H., 651, 661(5)
Vaughan, P., 194
Vaz, W. L. C., 56
Vazquez, J., 275, 276, 277(6), 281(6), 582, 637
Veatch, W. R., 218, 533
Veh, R. W., 697
Veliçelebi, G., 20, 26, 34(8), 36(5), 46(18, 19)
Velimirovic, B., 299(65), 300
Venatachalam, C. M., 541
Venema, V. J., 605
Veraga, C., 292, 296(32)
Verbert, A., 342
Vergara, C., 293
Verkman, A. S., 148, 149, 174, 175(57), 179, 570
Vernino, S., 3, 45
Veyna-Burke, N., 48, 51(3), 57(3)
Vickers, J. C., 687, 689
Villa, C., 144
Villegras, R., 589
Vincent, A., 677, 678, 681, 682
Vinnitsky, I., 617(35), 619
Virelizier, H., 630
Vissavajjhala, P., 366
Vita, C., 94, 630
Viveros, O. H., 189
Vogel, W., 445
Vogt, T. C. B., 525, 526, 531, 531(20), 536(7), 537(20), 538(20)
Volkmann, R. A., 94, 101(12), 110(12), 639
Volland, W. V., 67

Vu, Q. A., 446
Vyas, T. B., 645

## W

Wada, H. G., 190
Wada, K., 689
Wada, Y., 514
Wagner, G., 95, 97, 98, 112(26, 28), 113, 113(31, 33), 117
Wagner, H. H., 589
Wakamatsu, K., 583, 598(75)
Waki, M., 588
Walgrave, S., 680
Walker, C. S., 613, 614, 616(25), 617(25), 621(25)
Walker, J., 698
Walker, M., 155(106), 179
Walker, N. A., 517
Wallace, B. A., 533, 535, 536
Wallingford, R. A., 194
Wallner, M., 281, 287, 297
Walter, A., 209
Walter, P., 473
Walz, T., 148, 179
Wan, K., 290, 291(16)
Wandinger-Ness, A., 342, 343(5)
Wang, A. C., 112
Wang, C., 492
Wang, C. C., 593
Wang, C. Z., 446
Wang, D. N., 136, 178(24), 179(24)
Wang, F., 651, 661(6)
Wang, H., 697
Wang, J., 135, 436, 437(46)
Wang, L.-Y., 366, 368(15)
Wang, S. S., 528
Wang, W., 228, 241(10), 242(10)
Wang, X., 577
Wang, Y., 227, 228, 232(11), 235(11), 319
Ward, J. M., 412, 416, 425
Warmke, J. W., 280
Warner, C., 703
Warren, C. D., 117
Warshel, A., 540
Washburn, M. S., 20
Wassle, H., 686
Watanabe, T. X., 93, 109, 615(61), 623
Water, C., 678
Watkins, J. C., 238, 239(30)
Watkins, M., 610(56), 614, 616(25), 617(25), 621(25), 623
Watkins, S. C., 48, 51(3), 57, 57(3)
Watson, J., 698
Watt, A., 163
Watt, M., 163
Watts, A., 63, 66
Waugh, J. S., 538
Webb, W. W., 467
Weber, M., 119
Wechsler, S. W., 456
Weckesser, J., 518
Wegner, L. H., 416, 419
Wei, A., 274, 286(1), 575, 625, 645
Weiner, S., 152(100), 179
Weiser, M., 697
Weiss, L. B., 526, 532, 532(23), 549(13)
Weiss, M. S., 518
Weissbach, H., 643(29), 647
Weissman, C., 557
Welsh, M., 227, 241(2)
Welte, W., 518
Welton, A. F., 276
Wenthold, R. J., 687, 688, 689
Werner, P., 261
Wescott, W. C., 291, 661, 663(1), 670(1)
Westenbroek, R. E., 692, 703
Westler, W. M., 117
Whaley, W. L., 526, 532(19)
Wharton, S. A., 492
Whisenant, T., 26, 46(19)
White, H., 155(106), 179
White, H. S., 612, 613(18)
White, M. M., 296(43), 297
White, S., 103, 697
White, S. H., 238, 290, 542
White, S. J., 521
White, T. J., 477
Whitehead, B., 515
Whitesides, G. M., 105
Whiting, P. J., 36, 678, 681, 686
Whorlow, S. L., 100
Wiener, M. C., 148, 149, 174, 175(57)
Wightman, R. M., 189
Wijesuriya, D. C., 190
Wilcox, G. L., 710
Wilcox, N., 678
Wilkens, S., 151

Wilkinson, A. J., 133, 596
Williams, A. J., 247, 249(8), 255(8), 257(8), 259(8), 288, 290, 296(53), 297, 699
Williams, B. J., 588
Williams, C. B., 655
Williams, M. E., 20, 31(4), 389, 394
Williams, R. C., 158
Wilmsen, U., 290, 291(18)
Wilson, E., 152(100), 179
Wilson, T. H., 558
Wilusz, E. J., 389
Winter, G., 133, 596
Wissing, F., 509
Withka, J. M., 594, 598(121), 630, 651
Witkop, B., 525
Witzemann, V., 521
Wohlfart, P., 247
Wolf, E. D., 515
Wolf, S. G., 143
Wolff, D., 296(46), 297
Wolstenholme, A. J., 491
Wolters, I., 686
Wonderlin, W. F., 290, 326, 339, 340, 341(1), 585
Wong, E. H. F., 392
Wong, S. S., 475
Wood, S. J., 682
Wood, W. A., 508, 518(5)
Woodbury, D. J., 242, 245, 246(40), 292, 303(30, 31), 319, 320, 321(7), 322, 328(11), 331, 333, 335(22)
Woodhull, A. M., 601
Woodland, H. R., 197
Woodruff, G. N., 392
Woodward, S. R., 605, 611, 617(32), 619
Woolf, T. B., 540
Worley, J. F. III, 289, 292(8), 296(8), 302(8, 9), 581, 585
Worrall, A. F., 592
Wouters, D., 94
Wright, C. E., 100
Wright, P. E., 98, 99, 113(36)
Wu, C. S., 232
Wu, J.-Y., 45
Wu, L. G., 4, 12, 17(28)
Wu, W. H., 416, 418
Wu, Y., 217
Wunder, F., 634, 640
Wüthrich, K., 94, 95, 111, 116, 117, 536
Wyman, J., 128
Wymer, C., 431
Wyszynski, M., 371, 373(5), 375, 377(6), 381(6)

## X

Xie, J., 299(74, 75), 301, 303(74, 75)
Xu, M., 579, 591(40)

## Y

Yahagi, N., 577
Yakehiro, M., 17, 18(29), 19(29)
Yakel, J. L., 366
Yamada, M., 446
Yamagishi, T., 579, 604(45)
Yamaguchi, Y., 570
Yamamoto, C., 445
Yamashiro, D., 92, 100
Yamazaki, T., 98, 113(35)
Yan, Y., 135
Yanagawa, Y., 580, 583(54)
Yang, D. S., 150
Yang, F., 49
Yang, J., 300, 575(10), 576
Yang, J. T., 232
Yang, N., 577
Yang, T., 49
Yang, X. C., 705
Yau, K.-W., 247, 249, 249(2, 7), 258(7), 260(12), 699
Yeager, M., 135, 138, 139, 143, 144, 144(36), 148, 149, 152(100), 165, 174, 175(47, 49, 57), 177, 178(36), 179
Yee, W., 4
Yehezkely, I., 12
Yellen, G., 591, 641, 644
Yeung, E. S., 189
Yokotani, N., 687, 689
Yoneda, Y., 385, 386, 387, 388(2, 3), 389, 389(4, 5), 390, 390(10, 12), 391, 392, 393(20), 394, 394(22), 395, 397(27, 28), 401(26), 402, 402(27)
Yoshida, M., 51
Yoshikami, D., 580, 581(51), 606, 609(48–51), 610(54–56), 611, 614(1), 615(48), 616(1), 617(33), 619, 623

Yoshimura, K., 518
Young, A. B., 26, 46(18, 19)
Young, C., 682
Young, L. C., 651, 661(5)
Young, N. M., 82
Young, P., 514
Youvan, D. C., 52
Yu, C. J., 416, 418
Yu, H., 94, 101(12), 110(12)
Yu, X., 238
Yu, Z., 615(30), 618
Yuan, W., 672
Yue, D. T., 578, 579(32), 591(32), 603(32)
Yun, H.-Y., 587

## Z

Zafaralla, C. G., 609(45), 623
Zafaralla, G. C., 609(46, 48), 615(48), 623
Zagotta, W. N., 247, 640, 641
Zakour, R. A., 593
Zambon, M. C., 491
Zampighi, G., 144
Zamponi, G. W., 577, 580, 581(49), 583, 583(23, 49), 586, 588(23, 49), 595(24), 600(24, 49), 601(24), 604(24, 49)
Zare, R. N., 190, 191, 194, 194(9), 196(9), 197(7), 199(7), 200, 202(26), 203, 203(8, 9), 206(7), 207
Zasloff, M., 672
Zatta, P., 664, 670(12)
Zeidel, M. L., 552, 554, 556(11, 12), 558, 558(12), 559, 560, 560(11)
Zeikus, R. D., 580, 581(51), 606, 614(1), 615(57, 58), 616(1), 623
Zeilinger, C., 252, 256(16)
Zemlin, F., 136, 165, 178(22), 179(85)
Zeng, C. Q.-Y., 490
Zezula, J., 686
Zhang, H., 446
Zhang, J. P., 300
Zhang, L., 402
Zhang, O., 98
Zhang, R., 570
Zhang, Y.-L., 674
Zhang, Y.-P., 82
Zhang, Z., 536
Zhao, J., 299(74), 301, 303(74)
Zhao, X., 151
Zhou, J., 576, 604
Zhou, L.-M., 613
Zhou, X.-L., 479, 507, 511, 512, 513(29), 514, 515, 523(32), 524(32)
Zhou, Z., 3, 27, 32(22), 36(22), 506
Zielenski, J., 227
Zilberberg, N., 100
Zimmerberg, J., 292, 319, 328(2)
Zimmermann, S., 422
Zinkand, W. C., 601
Zinn, K., 197
Zinn-Justin, S., 100
Zlotkin, E., 100, 609(12), 612, 614, 616(39), 617(24, 35, 71), 619, 619(24), 620, 620(24), 624, 703
Zong, X., 248
Zoratti, M., 463, 481, 519, 520
Zucker, R. S., 4
Zulauf, M., 136, 144(17, 18), 147(17, 18)
Zuo, P., 389, 390(12)

# Subject Index

## A


ω-Agatoxin-IVA, antibody, 677, 693
ω-Agatoxin-IVB
biological assays, 108–109
expression and purification in *Escherichia coli*
expression vectors, 101
gene cloning, 101
inclusion body isolation, 103
large-scale expression, 103–104
leader peptide cleavage, 108
matrix-assisted laser desorption mass spectrometry, 109
nickel affinity chromatography, 104–105
recombinant protein design, 100–101
refolding conditions, 106–107
reversed-phase high-performance liquid chromatography, 105–106
small-scale expression, 103
structure determination by nuclear magnetic resonance
buffers, 111
confirmation of properly-refolded proteins, 109–110
dynamic analysis with relaxation studies, 112–116
heteronuclear-based determinations, 97–98, 111–112
homonuclear proton-based determination, 96–97, 109–110
isotope labeling, 99–100, 116–117
relaxation times, senstivity to correlation time, 98
rigidity in solution, 93–95
suitability for nuclear magnetic resonance, 92–93
Alzheimer's disease, β-amyloid insertion into planar lipid membranes, 670–672
γ-Aminobutyric acid-A receptor
antibodies, 685–686, 701
nomenclature, 722
α-Amino-3-hydroxy-5-methylisoxazole-4-propioninc acid receptor, *see also* Excitatory amino acid receptors
antibodies, 688–689, 701–702
domain labeling with radioligands, 394
drug screening with fluorescence assay of calcium
activity of various drugs, 36
advantages of assay, 46–47
limitations of assay, 45–46
pharmacologic validation, 26–27
radioactive filter assays
ligand selection, 388
reaction mixtures, 388–389
synaptic membrane preparation from rat brain, 386–387
Triton X-100 treatment of crude fractions, 387–388
stable cell line expression of recombinant channels, 25–26
AMPA receptor, *see* α-Amino-3-hydroxy-5-methylisoxazole-4-propioninc acid receptor
Amylin, insertion into planar lipid membranes, 672
Antibody, ion channels
α-amino-3-hydroxy-5-methylisoxazole-4-propioninc acid receptors, 688–689, 701–702
γ-aminobutyric acid-A receptors, 685–686, 701
autoimmune antibodies in humans, 678, 700–703
calcium channels, voltage-gated
autoimmunity, 702
channel antibodies, 690–692, 702
neurotoxin antibodies, 693, 702
commercial availability, 703–704
cyclic nucleotide-gated channels, 699, 703
cystic fibrosis transmembrane conductance regulator, 698, 703
GluR1 phosphorylation, site-specific antibodies, 364–366
glycine receptor, 685, 701

*N*-methyl-D-aspartate receptors, 687, 701–702
nicotinic acetylcholine receptors
autoimmunity, 701
muscle receptors
mouse monoclonal antibodies, 681, 700–701
rabbit polyclonal antibodies, 682, 701
rat monoclonal antibodies, 679–680, 700
neuronal receptors
monoclonal antibodies, 683
rabbit polyclonal antibodies raised against extracellular domain peptides, 684
potassium channels, voltage-gated
autoimmunity, 703
mouse monoclonal channel antibodies, 694, 702
rabbit polyclonal antibodies against synthetic peptides, 695–697, 702
sodium channels, voltage-gated, 703
Aquaporin I
discovery, 551
expression in *Xenopus* oocytes
complementary DNA expression construct preparation, 568–569
complementary RNA
injection, 570
preparation, 569–570
materials, 568
osmotic water permeability measurement, 570
Western blot analysis, 570, 572
expression in yeast
materials, 555
plasmid construction, 555
reconstitution into proteoliposomes, 558–559
spheroplast preparation and secretory vesicle isolation, 555–556
homology cloning by degenerate oligonucleotide polymerase chain reaction
amplification reaction, 565, 567
gel electrophoresis analysis, 565–566
materials, 562–565
overview, 562
secondary amplification and DNA purification, 566–567
purification from red cells
anion-exchange chromatography, 553–554
materials, 552
membrane vesicle preparation, 552–553
reconstitution into proteoliposomes, 556–558
two-dimensional crystallization by *in vitro* reconstitution
dialysis, 148–150
electron cryomicroscopy, image processing, 165–167, 169, 171, 173–175
lipids for reconstitution, 147–148
water permeability assay
activation energy, 561
calculations, 559–561
carboxyfluorescein fluorescence quenching, 560
materials, 560
stopped-flow analysis, 561
tetramer density determination, 561
ATP-sensitive potassium channel
components in assembly, 445–443
expression in *Escherichia coli*, 521
expression in tissue culture cells
maintenance of cells, 451
membrane isolation, 453
transfection, 451–452
rubidium efflux assay, 452–453
solubilization
3-[(3-cholamidopropyl)dimethylammonio]-1-propane sulfonate, 455
digitonin, 454–455
Triton X-100, 455
sucrose gradient centrifugation, 456–457
SUR1
binding assays
gibenclamide, 457–458
P1075, 458
concentration of receptor, 456
dye affinity chromatography, 456
lectin affinity chromatography, 455
nickel affinity chromatography, 455–456
phenylboronate-10 affinity chromatography, 456
photolabeling

live cells, 454
membranes, 453
radioligand synthesis
azidoiodoglibenclamide, 449–451
iodoglibenclamide, 446–448
types, 442–443
Azidoiodoglibenclamide, radioligand synthesis, 449–451

## B

Baculovirus–insect cell expression system
cystic fibrosis transmembrane conductance regulator expression and purification
gel filtration chromatography, 233–235
hydroxyapatatite chromatography, 233, 235
nickel affinity chromatography, 236–237
overview, 228–230
solubilization in detergents
sodium dodecyl sulfate, 232–233
sodium pentadecafluorooctanoic acid, 235–236
transfer vector construction, 230–231
glycine receptor, 273
Binding isotherm, analysis
competitive inhibition data analysis, 130–131
curve fitting, 129–130
subtraction of nonspecific binding, 128–129
Biology Workbench project, *see* Internet resources, ion channels
Biosensor, *see* Capillary electrophoresis; Gramicidin; Sniffer-patch detector
Black lipid membrane, *see* Planar lipid bilayer
BLM, *see* Black lipid membrane

## C

Calcium
calmodulin fluorescence assay, 52
fluorescence assay with Fluo-3
calibration, 24–25
detection wavelength, 21
dye loading, 23–24
microtiter plate reading, 22–23
fluorescence assay with ratio fluorescence probes
calibration of fluorescence signal, 7–9, 24–25
cell loading, 4–5, 23–24
hippocampal brain slice measurement, 11–12
indicator selection criteria, 10–11
instrumentation
detectors, 7
filters, 6
light source, 6
microscope, 5
microtiter plate reader, 21, 23
presynapyic terminal measurements
calibration of fluorescence signal, 14–15
data acquisition, 13–14
dye loading, 12–13
single terminal measurements in mossy fibers, 15–19
principle, 4, 21–22
simultaneous patch-clamp analysis of plant cells
caged calcium studies, 437–440
calibration of fluorescence signal, 431–432
cameras for imaging, 433–434
compartmentalization of dye, 430
confocal laser scanning microscopy, 427–428, 430–432
dye selection, 426–427, 434–435
Indo-1 loading, 430–431
instrumentation, 426, 428, 430, 433–434
light and heat responses of protoplasts, 423–424
photometry and imaging of ratio fluorescence dyes, 432–435
pipette resistance, 428
pipettes, 422
protoplast membrane exposure, 411–419
seal formation, 422–424
ion channel activity, drug screening with fluorescence assays
advantages of assay, 46–47

$\alpha$-amino-3-hydroxy-5-methylisoxazole-4-propioninc acid/kainate receptor, 36
cyclic nucleotide-gated channel, 41
limitations of assay, 45–46
*N*-methyl-D-aspartate receptor, 32, 34–35
nicotinic acetylcholine receptor, 36–37, 39, 41–45
pharmacologic validation, 26–27
voltage-gated calcium channel, 27, 30–32
Calcium channel, *see* Voltage-gated calcium channel
Capillary electrophoresis
buffers, 194
fractured electrophoresis capillary fabrication, 194–196
patch-clamp detection
buffer vial maintenance, 200
cell preparation, 196–197
current-to-voltage relationships, 205
electrophoretic migration rate calculation, 204
instrumentation, 201–202
neurotransmitter detection, principle, 189–191, 193
outside-out patch recording, 205
patch clamping, 199
pipette solutions, 200
regeneration of capillary, 200
resolution in neurotransmitter separations, 203
sample volume calculation, 202
spectral analysis of currents, 204
principles
electro-osmotic flow, 192
electrophoretic migration velocity, 192
electrophoretic mobility, 192
net migration velocity, 192–193
theoretical plate number, 192
two-electrode voltage-clamp detection
cell preparation, 197–199
instrumentation, 205
neurotransmitter detection, principle, 189–191, 193
stability of detection system, 206–207
CD, *see* Circular dichroism spectroscopy
CFTR, *see* Cystic fibrosis transmembrane conductance regulator
Channel density, determination in isolated vesicles, 333–337
Charybdotoxin receptor, *see also* Scorpion venom peptides
calcium sensitivity, 275, 287
comparison with $\mu$-conotoxin GIIIA blocking of sodium channels, 584–586
fusion of reconstituted receptors into lipid bilayers
calcium regulation, 284
osmotic gradient enhancement, 283
patch-clamp analysis, 283–284
polarity, 283
tetraethylammonium blocking, 284–285
kinetic basis of high-affinity binding, 585–587
mechanism of blocking, 582, 584–586
molecular dentistry concept of binding, 581–582
purification from bovine tracheal smooth muscle
anion-exchange chromatography, 278–279
cation-exchange chromatography, 279–280
charybdotoxin binding assay, 275, 277, 281–282
hydroxylapatite chromatography, 279
membrane preparation, 276
solubilization, 275–277
sucrose density gradient centrifugation, 279–282
wheat germ agglutinin affinity chromatography, 278
reconstitution into liposomes, 282–283
reconstruction in planar bilayers, 299–300
stoichiometry of binding, 274
subunits
amino acid analysis, 285–286
isolation, 285
quaternary structure, 281
T-tubule membrane content, 292–293
Chloride channel, *see* Cystic fibrosis transmembrane conductance regulator
Circular dichroism spectroscopy, gramicidin A secondary structure, 532–534
Common Object Request Broker Architecture, ion channel database accession optimization, 727, 730

Conantokin
N-methyl-D-aspartate receptor binding, 612–613
posttranslational processing, 613
structure, 610, 612–613
types, 612
Concentration jump, *see* U-tube
Confocal scanning laser microscopy
fluorescence recovery after photobleaching of potassium channels, 54–56
simultaneous patch-clamp analysis of plant cells, 427–428, 430–432
α-Conotoxin
IMI, 611–612
MII, 611
nicotinic acetylcholine receptor binding, 608–612
sequences and binding specificities, 608–610
αA-Conotoxin
nicotinic acetylcholine receptor binding, 610–611
structures, 610
δ-Conotoxin
inhibition of sodium channel inactivation, 614, 617, 619–620
solubility, 621
types and structures, 619–620
ψ-Conotoxin
nicotinic acetylcholine receptor binding, 611
structure, 610
κ-Conotoxin PVIIA, voltage-gated potassium channel binding, 620–621
μ-Conotoxin, *see also* μ-Conotoxin GIIIA
arginine interactions in sodium channel binding, 618–619
structures and sodium channel binding specificity, 616
μ-Conotoxin GIIIA, sodium channel blocking
binding site, 580, 584, 590, 602, 604–605
comparison with charybdotoxin blocking of potassium channels, 584–586
electrostatic compliance analysis, interresidue distance measurement, 597–598
kinetics of binding, 585–586
mutagenesis effects on binding sensitivity
channel mutation
alanine scanning mutagenesis, 591
chimera studies and block mutations, 590–591
conservative mutations, 591
cysteine scanning mutagenesis, 591
multiple primer techniques, 592–593
oligonucleotide-initiated polymerization of entire vector, 592–594
point mutations, 579, 589–591
ligand mutation, 583–584
mutant cycle analysis
applications, 596–597
coupling coefficient calculation, 595
interaction energy, 596
principle, 594–595
partial blocking toxins
intrapore binding site determination, 602
mechanism of pore block, 603–604
mutant types, 600
selectivity and permeation factor analysis, 602–603
voltage sensor movement and location analysis, 601–602
peptide synthesis of analogs, 587–589
specificity for skeletal muscle isoform, 606
μO-Conotoxin
sodium channel binding, 620
solubility, 621
ω-Conotoxin
antibodies, 693
types and structures, 615, 618
voltage-gated calcium channel binding, 614, 618
*Conus* venom peptides, *see also specific conotoxins*
complexity of mixtures, 605–606
classification and nomenclature, 606–608
solubility, 621
storage, 621–622
nonspecific adsorption, 622–624
CORBA, *see* Common Object Request Broker Architecture
Crystallization, *see* Two-dimensional crystallization, membrane proteins
Cyclic nucleotide-gated channel
antibodies, 699, 703
calmodulin regulation, 249
drug screening with fluorescence assay of calcium

activity of various drugs, 41
advantages of assay, 46–47
limitations of assay, 45–46
pharmacologic validation, 26–27
expression in *Escherichia coli*, 521
functions, 246–247
purification of bovine rod channel
anion-exchange chromatography, 254, 259
8-bromo-cyclic GMP affinity chromatography, 259
calmodulin affinity chromatography, 252–253, 255–256, 258
dye affinity chromatography, 254, 259
gel electrophoresis in purity analysis, 254–255, 257
immunoaffinity chromatography, 253–254, 258–260
rod outer segment membranes
preparation, 250–252
solubilization, 252
solution preparation, 249–250
reconstitution
channel activity measurement, 256–258
detergent dialysis, 256
L-*cis*-diltiazem inhibition, 257–258
subunits of sensory channels, 247–249
Cystic fibrosis transmembrane conductance regulator
antibodies, 698, 703
ATPase activity, 228, 241, 243–244
chloride channel
assays, 244–246
function, 227, 240–241
expression and purification in baculovirus–insect cell system
gel filtration chromatography, 233–235
hydroxyapatatite chromatography, 233, 235
nickel affinity chromatography, 236–237
overview, 228–230
solubilization in detergents
sodium dodecyl sulfate, 232–233
sodium pentadecafluorooctanoic acid, 235–236
transfer vector construction, 230–231
phosphorylation, 241, 243
proteoliposome fusion assay with nystatin, 241–242, 245–246
reconstitution
detergent removal, 237–238
LiDS reconstitution, 239
lipid selection, 238–239
liposome preparation, 240
sodium pentadecafluorooctanoic acid reconstitution, 239–240
reconstruction in planar bilayers, 301, 303

## D

Dendrotoxin
handling, 650–651
potassium channel blocking
electrostatic interactions, 656–658
kinetic analysis, 653–655
mutation analysis of channel residue interactions, 654–656
voltage dependence, 659–670
*Xenopus* oocyte characterization, 652–653, 657
recombinant protein expression systems for study, 651
types and sequences, 649–650, 661
DTX, *see* Dendrotoxin

## E

EAA receptors, *see* Excitatory amino acid receptors
Electron cryomicroscopy, structural analysis of membrane proteins
advantages over other structure analysis techniques, 135–137, 179
bacteriorhodopsin, 179
cold stage insertion into microscope, 158
cryotransfer of grid to cold stage, 157–158
crystal quality and resolution, 138
crystal transfer to grids, 153
frozen specimen preparation, 153–157
image appraisal, 162–163, 165
image processing, 165–167, 169, 171, 173–175
low-dose cryomicroscopy
exposure mode, 160–161
focus mode, 159–160

scanning and recording, 161–162
search mode, 158–159
map interpretation
3.5 Angström resolution or better, 178
5 to 10 Angström resolution, 177–178
10 to 15 Angström resolution, 177
15 to 30 Angström resolution, 175
microscopes, 138
overview, 138, 140
sample quantity requirements, 137
Electron spin resonance, *see* Spin-label electron spin resonance
Electrophoretic mobility shift assay, excitatory amino acid receptor transcription factors
acrylamide gel preparation, 398–399
electrophoresis, 399, 403
limited proteolysis of complexes in validation, 400
nuclear extract preparation from mouse brain, 397, 402–403
probe preparation, 397–398
quantitative analysis, 404
receptor activation studies *in vivo*, 401–402
solution preparation, 395–396
supershift assay, 400, 402
Electrostatic compliance analysis, interresidue distance measurement, 597–598
EMSA, *see* Electrophoretic mobility shift assay
ESR, *see* Electron spin resonance
Excitatory amino acid receptors, *see also specific receptors*
electrophoretic mobility shift assay of transcription factors
acrylamide gel preparation, 398–399
electrophoresis, 399, 403
limited proteolysis of complexes in validation, 400
nuclear extract preparation from mouse brain, 397, 402–403
probe preparation, 397–398
quantitative analysis, 404
receptor activation studies *in vivo*, 401–402
solution preparation, 395–396
supershift assay, 400, 402
radioactive filter assays
ligand selection, 388
reaction mixtures, 388–389
synaptic membrane preparation from rat brain, 386–387
Triton X-100 treatment of crude fractions, 387–388
Western blot analysis
blotting with semidry electrophoretic transfer, 406, 409
immunodetection, 406–407, 410
receptor activation studies *in vivo*, 407–408
sample preparation from mouse brain, 405
solution preparation, 404–405
troubleshooting, 410

## F

Filter overlay assay, ion channel associated proteins, 373–375
Fluid mosaic membrane model, limitations, 208–209
Fluo-3, calcium measurements
calibration, 24–25
detection wavelength, 21
dye loading, 23–24
microtiter plate reading, 22–23
Fluorescence microscopy, *see* Calcium; Confocal scanning laser microscopy; Immunofluorescence microscopy
Fluorescence recovery after photobleaching
applications, 49
potassium channel–green fluorescent protein constructs
artifacts, corrections, and calibrations, 56–58
channel mobility measurement, 53–54, 58
confocal scanning laser microscopy measurements, 54–56
Fluorescence resonance energy transfer
green fluorescent protein with blue-shifted protein
calmodulin linker in calcium fluorescence assay, 52
protease assays, 52
protein–protein interactions in potassium channels, 52–53
principle, 52

Forward genetics
MscL studies, 479–480, 524
overview of microbial ion channel studies, 522–524
plate phenotypes of microbes, 521–522
Fourier transform infrared spectroscopy
amide band assignments, 79–83
amide hydrogen–deuterium exchange, 83–84
angular orientation of secondary structural elements to membrane, 87–91
attenuated total reflection, 77, 90
Fourier deconvolution, band narrowing, and band fitting, 78–79
infrared dichroism, 84–87
instrumentation, 76–77
ion channel applications, 59–60, 75–76, 91
sample cells, 77
sensitivity, 59
FRAP, *see* Fluorescence recovery after photobleaching
FRET, *see* Fluorescence resonance energy transfer
FTIR, *see* Fourier transform infrared spectroscopy
Fura-2, calcium measurements
calibration, 7–9, 24–25
cell loading, 4–5, 21, 23–24
instrumentation
detectors, 7
filters, 6, 22–23
light source, 6
microscope, 5
microtiter plate reader, 21, 23
presynaptic terminals
population measurements
calibration of fluorescence signal, 14–15
data acquisition, 13–14
dye loading, 12–13
single terminal measurements in mossy fibers
advantages, 15–16
calibration of fluorescence signal, 19
data acquisition, 17–18
dye loading, 17
principle, 4, 21–22
simultaneous patch-clamp analysis of plant cells, 433–435

## G

$GABA_A$ receptor, *see* $\gamma$-Aminobutyric acid-A receptor
Gap junction, two-dimensional crystallization
electron cryomicroscopy, image processing, 165–167, 169, 171, 173–175
membrane isolation from enriched recombinant preparations, 139–142
*in situ* crystallization, 144, 147
GENESIS simulation program, 707, 709–710, 713, 726–727, 730
GFP, *see* Green fluorescent protein
Glibenclamide
radioligand synthesis
azidoiodoglibenclamide, 449–451
iodoglibenclamide, 446–448
SUR1 binding assays
particulate receptors, 457
solubilized receptors, 457–458
Glutamate receptor, *see also* Excitatory amino acid receptors
GluR1 phosphorylation
extraction of labeled receptors, 355–356
functional analysis by patch-clamp whole-cell recording
basal phosphorylation, 368, 370
intracellular patch perfusion solutions, 367–368
site-directed mutants, 370
stimulation of phosphorylation, 366–368
fusion proteins
phosphorylation *in vitro*, 362–363
rationale for study, 361–362
immunoprecipitation of labeled receptors, 354, 356–357
investigation in native tissues, 353–354
phosphoaminoacid analysis, 360–361
phosphopetide mapping
overview, 357–358
proteolytic digestion, 358
two-dimensional thin-layer chromatography, 359–360
radiolabeling in cell culture, 354–355
site-directed mutagenesis of phosphorylation sites, 363–364, 370

site-specific antibodies, generation and application, 364–366
radioactive filter assays
ligand selection, 388
reaction mixtures, 388–389
synaptic membrane preparation from rat brain, 386–387
Triton X-100 treatment of crude fractions, 387–388
types, 385
Glycine receptor
antibodies, 685, 701
baculovirus–insect cell expression system, 273
purification from mammalian spinal cord
aminostrychnine affinity chromatography, 262–263
membrane preparation, 261–262
solubilization, 262
strychnine binding, 260
subunits in spinal cord, 260
transient expression systems
human embryonic kidney cells
calcium phosphate transfection, 270, 272–273
cell culture and harvesting, 271–272
plasmid preparation, 270–271
vectors, 270
*Xenopus* oocytes
applications of expression system, 263–264
injection capillary preparation, 268
injection of cytoplasma, 269
injection of nucleus, 269–270
messenger RNA isolation from neural tissue, 264
preparation of oocytes, 266–267
transcription of complementary DNA *in vitro*, 264–266
vector linearization, 265
Gramicidin
applications in research, 525–526
conformational analysis of analogs
circular dichroism spectroscopy, 532–534
molecular dynamics simulations, 539–542
nuclear magnetic resonance
solid-state studies, 536–539
solution studies, 535–536
single-channel measurements
electrodes, 542
functional approach to structure identification by dimerization, 545–547, 549–550
membrane capacitance minimization, 542–543
purity requirements, 543–544
reconstitution, 543
size-exclusion high-performance liquid chromatography, 534–535
molecular force transducer in membranes
changes in membrane deformation energy, 220–222
channel lifetime measurement, 222–223
heterodimer formation approach, 223
membrane conductance measurements, 220–222
monomer–dimer equilibrium, 217–218, 220
phenomological spring constant determination, 221, 223
peptide synthesis of analogs
acylation, 531
cleavage of peptide from resin, 530–531
*N*-Fmoc-amino acid synthesis, 527–528
N-formylamino acid preparation, 526–527
purification of product, 531–532
resin selection, 528
solid-phase synthesis, 528–530
sequence of gramicidin A, 525
Green fluorescent protein
commercial vectors for fusion protein construction, 49–50
fluorescence resonance energy transfer with blue-shifted proteins
calmodulin linker in calcium fluorescence assay, 52
protease assays, 52
protein–protein interactions in potassium channels, 52–53
potassium channel fusion protein expression, 51–52
spectral variants, 49–51

## H


High-performance liquid chromatography
ω-agatoxin-IVB, 105–106
conformational analysis of gramicidin analogs with size-exclusion chromatography, 534–535
scorpion venom peptide inhibitors 628
Hippocampal brain slice, calcium measurement with fluorecent probes, 11–12
HPLC, *see* High-performance liquid chromatography


## I


Immunoblotting, *see* Western blot analysis
Immunofluorescence microscopy, colocalization of ion channel associated proteins, 372, 380–381
Immunoprecipitation, coimmunoprecipitation of ion channel associated proteins
full-length proteins expressed in heterologous cells, 377–380
native tissue analysis, 381–384
principle, 372
Indo-1, *see* Calcium
Influenza A virus, *see* M2
Infrared spectroscopy, *see* Fourier transform infrared spectroscopy
Internet resources, ion channels
accession optimization by object specification and indexing, 726–728, 730–731
Biology Workbench project, federation of databases, 706–707
Common Object Request Broker Architecture, 727, 730
compacting channel-specific information in federated cell electrical stimulation databases, 705, 707, 709–713, 715
developmental gene expression data, 715, 718
GENESIS simulation program, 707, 709–710, 713, 726–727, 730
genome information strategies, 704–707
intuitive access and declarative queries, supporting data structures, 723–726
Ion Channel Network site, 720
peer review, 706
prediction of signal-transducing molecular assemblies containing channel effectors, 718–719
search engines, 729
systematic gene family nomenclature and interoperability of data sets, 719–720, 722–723, 730
TRANSFAC database, 718
universally addressable bibliographic data collection, 728–729
Iodide channel
iodine uptake assays
inward potassium gradient assay, 306–307
outward anionic gradient assay of selectivity, 314–315
outward iodide gradient assay
membrane preparations, 307–308
reconstituted proteoliposomes, 312–313
reconstitution into proteoliposomes
iodide uptake
activation energy, 318
imposed iodide gradient dependence, 314
kinetic properties, 316–318
lipid dependency, 313
protein dependency, 313
selectivity and specificity, 314–316
liposome preparation, 310–311
plasma membrane vesicle preparation from thyroid, 305–306
proteoliposome reconstitution, 311–312
solubilization, 308–310
thyroid function, 304
Iodoglibenclamide, radioligand synthesis, 446–448
Ion channel associated proteins
coimmunoprecipitation
full-length proteins expressed in heterologous cells, 377–380
native tissue analysis, 381–384
principle, 372
comparison to ion channel subunits, 371
filter overlay assay, 373–375
genetic approaches in identification, 372
immunofluorescence colocalization, 372, 380–381
pull-down assay, 373, 375–377

yeast two-hybrid screening in identification, 371, 373

## K

Kainate receptor, *see also* Excitatory amino acid receptors
domain labeling with radioligands, 394
drug screening with fluorescence assay of calcium
activity of various drugs, 36
advantages of assay, 46–47
limitations of assay, 45–46
pharmacologic validation, 26–27
radioactive filter assays
ligand selection, 388
reaction mixtures, 388–389
synaptic membrane preparation from rat brain, 386–387
Triton X-100 treatment of crude fractions, 387–388
stable cell line expression of recombinant channels, 25–26
subunits, 386

## L

Lectin affinity chromatography, SUR1, 455

## M

M2
amantadine inhibition, 491
expression in mouse erythroleukemia cells
gene cloning and amplification, 494
plasmid preparation, 495
rationale, 493
time course of expression, 493–494
transfection, 495
Western blot analysis, 495–496
whole cell patch-clamp analysis
background permeability, 497–498
cell-to-cell variation, 498–499
combined measurements with fluorescent pH indicators, 505–506
current/voltage relation, 499–501
ion selectivity, estimation from reverse potentials, 501, 503–505
overview, 496
perfusion, 497
single-channel parameter estimation, 505
solution composition, 496–497
function in influenza A virus infection, 490–491
permeability characteristics, 492–493, 501, 503–505
structure, 491
Maxi-potassium channel, *see* Charybdotoxin receptor
Mechanosensitive channels
*Escherichia coli*, *see* MscL
gene identification in *Caenorhabditis elegans*, 459
overview, 458–459
pressure clamp
construction tips, 487–489
electronic control, 485, 487
mechanical arrangement, 483–485
principle, 482–483
system volume minimization, 484–485
transducer/mixer chamber, 485
valves, 483–484
*Saccharomyces cerevisiae* channels, patch-clamp analysis, 511–514
Membrane bilayer, *see also* Planar lipid bilayer
carriers versus channels as conformation reporter proteins, 216–217
deformation energy, 214–216
fluid mosaic membrane model, limitations, 208–209
gramicidin as molecular force transducer
changes in membrane deformation energy, 220–222
channel lifetime measurement, 222–223
heterodimer formation approach, 223
membrane conductance measurements, 220–222
monomer–dimer equilibrium, 217–218, 220
phenomological spring constant determination, 221, 223
perturbation and channel function, 211–212, 214

protein conformational changes and bilayer perturbations, 210–211, 223–224
*N*-Methyl-D-aspartate receptor, *see also* Excitatory amino acid receptors
α-actinin-2 association, pull-down assay, 375–377
antibodies, 687, 701–702
conantokin binding, 612–613
domain labeling with radioligands
glycine recognition domain, 391–392
ion channel domain, 392–393
*N*-methyl-D-aspartate recognition domain, 389–390
polyamine domain, 393–394
drug screening with fluorescence assay of calcium
activity of various drugs, 32, 34–35
advantages of assay, 46–47
limitations of assay, 45–46
pharmacologic validation, 26–27
PSD-95 association
filter overlay assay, 373–375
immunofluorescence colocalization, 380–381
radioactive filter assays
ligand selection, 388
reaction mixtures, 388–389
synaptic membrane preparation from rat brain, 386–387
Triton X-100 treatment of crude fractions, 387–388
stable cell line expression of recombinant channels, 25–26
subunits, 385
Molecular dynamics simulation, gramicidin A conformational analysis, 539–542
MscL
antibody generation and applications, 474–475
cross-linking of subunits, 475–477
enrichment of channel activity, 469–470
forward genetics studies, 479–480, 524
gene cloning and heterologous expression in yeast or cell-free reticulocyte lysate systems, 471–473
homolog cloning from other bacteria, 473–474
membrane topology analysis using alkaline phosphatase as reporter gene, 477–478
patch clamping of bacteria
filament generation from *Escherichia coli*
magnesium formation, 462
mutants, 462
snaking approach, 460–461
ultraviolet formation, 462
protoplast generation, 462–463
reconstituted bacterial membranes, 466–468
reconstituted solubilized membrane proteins, 468–469
seal formation, 463–466
spheroplast formation, 461–462
purification of recombinant channels, 470–471
reverse genetics studies, 479
solute efflux measurement following osmotic downshock, 480–482
structure, 459
Mutant cycle analysis
applications, 596–597
coupling coefficient calculation, 595
interaction energy, 596
principle, 594–595


## N


NAChR, *see* Nicotinic acetylcholine receptor
Nicotinic acetylcholine receptor
antibodies
autoimmunity, 701
muscle receptors
mouse monoclonal antibodies, 681, 700–701
rabbit polyclonal antibodies, 682, 701
rat monoclonal antibodies, 679–680, 700
neuronal receptors
monoclonal antibodies, 683
rabbit polyclonal antibodies raised against extracellular domain peptides, 684
binding isotherms, analysis

competitive inhibition data analysis, 130–131
curve fitting, 129–130
subtraction of nonspecific binding, 128–129
binding sites and heterogeneity, 117–118
conotoxin binding
$\alpha$-conotoxin, 608–612
$\alpha$A-conotoxin, 610–611
$\psi$-conotoxin, 611
desensitization, 117–118
drug screening with fluorescence assay of calcium
activity of various drugs, 36–37, 39, 41–45
advantages of assay, 46–47
limitations of assay, 45–46
pharmacologic validation, 26–27
fluorescence binding assays
advantages and disadvantages, 118–119
ligand selection
acetylcholine antagonists, 126
noncompetitive site ligands, 119, 126–127
membrane preparation, 119
noncompetitive binding assay, 127
heterologous expression in *Chara corallina*, 521
radioligand binding assays
acetylcholine assays for allosteric interaction analysis, 121
$\alpha$-bungarotoxin assays
cells, 124–125
membranes, 122–124
ligand selection, 119
membrane preparation, 119
microcentrifuge assay, 120–121
noncompetitive agonists, effects on binding to agonist site, 125
noncompetitive binding analysis with phencyclidine or ethidium, 121–122
sensitivity, 118
stable cell line expression of recombinant channels, 25–26
thermodynamic cycle analysis
allosteric effects on second ligand binding, 132–133
double-mutant analysis, 133–135
principle, 132
NMDA receptor, *see* *N*-Methyl-D-aspartate receptor
NMR, *see* Nuclear magnetic resonance
Nuclear magnetic resonance
gramicidin A conformational analysis
solid-state studies, 536–539
solution studies, 535–536
peptide ion channel ligands, structure determination
buffers, 111
dynamic analysis with relaxation studies, 112–116
heteronuclear-based determinations, 97–98, 111–112
homonuclear proton-based determinations, 95–97, 109–110
isotope labeling, 99–100, 116–117
relaxation times, senstivity to correlation time, 98
rigidity in solution, 93–95
suitability for nuclear magnetic resonance, 92–93
Nystatin
cystic fibrosis transmembrane conductance regulator assay by proteoliposome fusion, 241–242, 245–246
nystatin/ergosterol-induced vesicle fusion with planar bilayers
channel density calculation, 333–337
enhancing vesicle delivery to bilayers, 329–330
limitations, 337–338
lipid-to-protein ratio, 332–333
native fusigenic vesicle preparation, 330–334
osmotic gradient, 328, 338
overview, 320–321
stirring, 329, 338
vesicle lipids, 321–322
vesicle size, 322–325

# P

Patch-clamp, *see also* Pressure clamp
archaea, 520

*Bacillus subtilis*, 519–520
biosensor, *see* Capillary electrophoresis
*Characeae*, 517
charybdotoxin receptor reconstituted into bilayers, 283–284
*Chlamydomas reinhardtii*, 517–518
*Dictyostelium*, 516
*Escherichia coli*
filament generation
magnesium formation, 462
mutants, 462
snaking approach, 460–461
ultraviolet formation, 462
outer membrane, 518
protoplast generation, 462–463
reconstituted bacterial membranes, 466–468
reconstituted solubilized membrane proteins, 468–469
seal formation, 463–466
spheroplast formation, 461–462
GluR1 phosphorylation, whole-cell recording in functional analysis
basal phosphorylation, 368, 370
intracellular patch perfusion solutions, 367–368
site-directed mutants, 370
stimulation of phosphorylation, 366–368
M2 expressed in mouse erythroleukemia cells, whole cell patch-clamp analysis
background permeability, 497–498
cell-to-cell variation, 498–499
combined measurements with fluorescent pH indicators, 505–506
current/voltage relation, 499–501
ion selectivity, estimation from reverse potentials, 501, 503–505
overview, 496
perfusion, 497
single-channel parameter estimation, 505
solution composition, 496–497
*Neurospora*, 515–516
*Paramecium*, 508–511
planar bilayers, reconstituted channels, 289–290, 337
plant cells
comparison to animal cell patch clamping, 421–424
exogenous second messenger administration
caged calcium studies, 437–440
comparison of control and experimental cells, 436
repatching and pipette perfusion, 436–437
light and heat responses of protoplasts, 423–424
pipettes, 422
protoplast membrane exposure
enzymatic access, 411–417
laser-assisted patch clamping, 419
osmotic shock, 417–418
seal formation, 422–424
second messengers in ion channel regulation, 424–425, 435
simultaneous calcium dye imaging
calibration of fluorescence signal, 431–432
cameras for imaging, 433–434
compartmentalization of dye, 430
confocal laser scanning microscopy, 427–428, 430–432
dye selection, 426–427, 434–435
Indo-1 loading, 430–431
instrumentation, 426, 428, 430, 433–434
photometry and imaging of ratio fluorescence dyes, 432–435
pipette resistance, 428
solution compositions, 421–422
*Saccharomyces cerevisiae*, 511–514
*Schizosaccharomyces pombe*, 514
*Uromyces appendiculatus*, 514–515
PCR, *see* Polymerase chain reaction
Peptide mapping, phosphopeptides of GluR1
overview, 357–358
proteolytic digestion, 358
two-dimensional thin-layer chromatography, 359–360
Peptide synthesis
$\mu$-conotoxin GIIIA analogs, 587–589
gramicidin analogs
acylation, 531
cleavage of peptide from resin, 530–531
*N*-Fmoc-amino acid synthesis, 527–528

N-formylamino acid preparation, 526–527
purification of product, 531–532
resin selection, 528
solid-phase synthesis, 528–530
scorpion venom peptide inhibitors, 629–630
Phosphorylation, ion channels
cystic fibrosis transmembrane conductance regulator, 241, 243
glutamate receptors
extraction of labeled receptors, 355–356
functional analysis by patch-clamp whole-cell recording
basal phosphorylation, 368, 370
intracellular patch perfusion solutions, 367–368
site-directed mutants, 370
stimulation of phosphorylation, 366–368
fusion proteins
phosphorylation *in vitro*, 362–363
rationale for study, 361–362
immunoprecipitation of labeled receptors, 354, 356–357
investigation in native tissues, 353–354
phosphoaminoacid analysis, 360–361
phosphopetide mapping
overview, 357–358
proteolytic digestion, 358
two-dimensional thin-layer chromatography, 359–360
radiolabeling in cell culture, 354–355
site-directed mutagenesis of phosphorylation sites, 363–364, 370
site-specific antibodies, generation and application, 364–366
Photoaffinity labeling, SUR1
live cells, 454
membranes, 453
Planar lipid bilayer
bilayer formation
folding, 291
painting, 291, 327–328
polystyrene cups with holes, preparation, 326–327
capacitance measurements, 327–328
channel insertion
cross validation, 296–297
cystic fibrosis transmembrane conductance regulator, 301, 303
examples of membrane preparations and reconstituted ion channels, 295–296
high-conductance calcium-activated potassium channels, 299–300
IRK1, 300
mammalian cell lines as starting materials, 300–301, 303
nystatin/ergosterol-induced fusion
channel density calculation, 333–337
enhancing vesicle delivery to bilayers, 329–330
limitations, 337–338
lipid-to-protein ratio, 332–333
native fusigenic vesicle preparation, 330–334
osmotic gradient, 328, 338
overview, 320–321
stirring, 329, 338
vesicle lipids, 321–322
vesicle size, 322–325
rationale, 287–288, 319
recombinant channels, 297–301, 303
*Shaker*, 298, 301
techniques, overview, 291–292, 319
*Xenopus* oocyte membranes, 299–300
microbial membrane vesicle incorporation
*Paramecium*, 510–511
*Saccharomyces cerevisiae* organelles, 513–514
patch-clamp recording, 289–290, 337, 662–663
screening of membrane active compounds
advantages, 662–663
amyloidogenic peptides, interactions with membranes, 670–672
antibiotic screening, 672–673
automation, 673
electrodes, 669
limitations, 663–664
lipids, 668
measuring system
chamber cleaning, 669
overview of features, 664
perfusion chamber for solvent-free

and native membrane vesicles, 666–668
perfusion setup for painted membranes, 664–666
porin channels, 673
recording equipment, 669–670
sensitivity, 662
single-channel reconstitution studies, 670
surfactant toxicity screening for microemulsions, 673–674
viral fusion analysis, 674
stability, 662–663
T-tubule membranes
high-conductance calcium-activated potassium channels, 292–293
preparation of planar bilayers from rat muscle, 293–295
voltage-sensitive sodium channels, 292–293
Polymerase chain reaction
aquaporin homology cloning by degenerate oligonucleotide polymerase chain reaction
amplification reaction, 565, 567
gel electrophoresis analysis, 565–566
materials, 562–565
overview, 562
secondary amplification and DNA purification, 566–567
multiple primer mutagenesis, 592–593
Potassium channel, *see also* ATP-sensitive potassium channel; *Shaker*-type potassium channel; Ykc1
antibodies, voltage-gated channels
autoimmunity, 703
mouse monoclonal channel antibodies, 694, 702
rabbit polyclonal antibodies against synthetic peptides, 695–697, 702
classification, 274, 286, 624
κ-conotoxin PVIIA, voltage-gated channel binding, 620–621
dendrotoxin blocking
electrostatic interactions, 656–658
kinetic analysis, 653–655
mutation analysis of channel residue interactions, 654–656
voltage dependence, 659–670
*Xenopus* oocyte characterization, 652–653, 657
fluorescence recovery after photobleaching
applications, 49
artifacts, corrections, and calibrations, 56–58
channel mobility measurement, 53–54, 58
confocal scanning laser microscopy measurements, 54–56
fluorescence resonance energy transfer in protein–protein interactions, 52–53
green fluorescent protein fusion protein construction, 49–50
high-conductance calcium-activated channel, *see* Charybdotoxin receptor
inwardly-rectifying channel IRK1, reconstruction in planar bilayers, 300
potassium channel fusion protein expression, 51–52
scorpion venom peptide inhibitors, *see also* Charybdotoxin receptor
applications, 625
binding assays, 637–639
classification, 626
expression of recombinant peptides in *Escherichia coli*
extraction, 631–632
induction, 631
purification of histidine-tagged proteins, 632–633
transformation, 630–631
vectors, 630
peptide synthesis, 629–630
purification
assessment of purity, 6290
cation-exchange chromatography, 627–628
reversed-phase high-performance liquid chromatography, 628
storage, 628–629
radiolabeling
iodination, 634–635
tritiation by cysteine alkylation, 635–637
sequence homology, 626
specificity, 626
structural overview, 625

tetramerization, 47–48
transport vesicle isolation
neuroblastoma cells, 342–350
squid giant axon, 340–341
velocity sedimentation on Ficoll gradients, 342–348
Pressure clamp
construction tips, 487–489
electronic control, 485, 487
mechanical arrangement, 483–485
principle, 482–483
system volume minimization, 484–485
transducer/mixer chamber, 485
valves, 483–484
Presynaptic terminal
calcium assay with ratio fluorescence probes
population measurements
calibration of fluorescence signal, 14–15
data acquisition, 13–14
dye loading, 12–13
single terminal measurements in mossy fibers
advantages, 15–16
calibration of fluorescence signal, 19
data acquisition, 17–18
dye loading, 17
diameter, 3
Prion protein, insertion into planar lipid membranes, 672
Pull-down assay, ion channel associated proteins, 373, 375–377

## R

Raw3, *see Shaker*-type potassium channel
RCK4, *see Shaker*-type potassium channel
Rubidium efflux assay, ATP-sensitive potassium channels, 452–453

## S

Scorpion venom peptides, *see also* Charybdotoxin receptor
applications, 625
binding assays, 637–639
classification, 626
expression of recombinant peptides in *Escherichia coli*
extraction, 631–632
induction, 631
purification of histidine-tagged proteins, 632–633
transformation, 630–631
vectors, 630
peptide synthesis, 629–630
potassium channel inhibition specificity, 626
purification
assessment of purity, 629
cation-exchange chromatography, 627–628
reversed-phase high-performance liquid chromatography, 628
storage, 628–629
radiolabeling
iodination, 634–635
tritiation by cysteine alkylation, 635–637
sequence homology, 626
structural overview, 625
*Shaker*-type potassium channel
charybdotoxin binding, 582, 584–586
inactivation
mechanism, 640
peptides
concentration determination, 647–648
delivery of peptides, 648
kinetics of blocking, 640, 648–649
net charge effects, 641
pore inner mouth probing, 644–646
purification, 646
Raw3, 640–645
RCK4, 644
ShB, 640–645
storage, 647
tetraethylammonium slowing of inactivation, 641
PSD-95 association, coimmunoprecipitation assay, 377–380
reconstruction in planar bilayers, 298, 301
ShB
concentration determination, 647–648

purification, 646
*Shaker*-type potassium channel blocking
delivery of peptides, 648
kinetics of blocking, 640, 648–649
net charge effects, 641
pore inner mouth probing, 644–646
tetraethylammonium slowing of inactivation, 641
storage, 647
Single-channel measurements, gramicidin channels
electrodes, 542
functional approach to structure identification by dimerization, 545–547, 549–550
membrane capacitance minimization, 542–543
purity requirements, 543–544
reconstitution, 543
Sniffer-patch detector
applications, 191
spatial resolution, 191
Sodium channel
antibodies, voltage-gated channels, 703
blocking peptides, *see also* μ-Conotoxin
conotoxin binding
arginine interactions in sodium channel binding, 618–619
δ-conotoxin inhibition of sodium channel inactivation, 614, 617, 619–620
μO-conotoxin, 620
electrostatic compliance analysis, inter-residue distance measurement, 597–598
mechanism, 581, 603–604
mutant cycle analysis
applications, 596–597
coupling coefficient calculation, 595
interaction energy, 596
principle, 594–595
stoichiometry of binding, 580–581
types, 580
ligands delaying inactivation, 614, 617
outer vestibule modeling
chemical modification probing, 599
effective dielectric constant determination, 598–599
secondary structure prediction, 599
pore region, 578–580, 584, 590, 602, 604–605
selectivity and permeation factor analysis, 602–603
transmembrane topology, 575–577
transport vesicle isolation
neuroblastoma cells, 342–350
squid giant axon, 340–341
velocity sedimentation on Ficoll gradients, 342–348
voltage-sensitive channels in T-tubule membranes, 292–293
voltage sensor, 577, 601–602
Spin-label electron spin resonance
instrumentation, 61–62
ion channel applications, 59–60, 65–66, 70
labeling of lipids and proteins, 60–63
lipid–protein interactions
difference spectroscopy, 63–64
number of lipid asciation sites on protein, 64
stoichiometry determination, 64–66
sample cells, 62
saturation and relaxation enhancements, 70–75
saturation transfer measurements, 66–70
sensitivity, 59
Sulfonylurea receptor, *see* ATP-sensitive potassium channel
SUR1 *see* ATP-sensitive potassium channel


## T


Thermodynamic cycle analysis
allosteric effects on second ligand binding, 132–133
double-mutant analysis, 133–135
principle, 132
Thin-layer chromatography
phosphoaminoacid analysis, 360–361
phosphopeptide mapping, 359–360
TLC, *see* Thin-layer chromatography
Topology, *see* Transmembrane topology
Transmembrane topology
MscL analysis using alkaline phosphatase as reporter gene, 477–478
sodium channels, 575–577

Two-dimensional crystallization, membrane proteins
aquaporin I, *in vitro* reconstitution
dialysis, 148–150
lipids for reconstitution, 147–148
gap junctions
membrane isolation from enriched recombinant preparations, 139–142
*in situ* crystallization, 144, 147
lipid monolayer crystallization, 150–153
monitoring by negative stain electron microscopy, 143
structural analysis, *see* Electron cryomicroscopy, structural analysis of membrane proteins
variables, 142–143

## U

U-tube
advantages as concentration jump system, 180, 187, 189
equipment, 183, 185
exchange solutions
flow rate, 180
time required for complete exchange, 181–182
volume, 180
mounting, 183
operation, 186–187
preparation, 185
principle of function, 181–183
testing, 185–186
troubleshooting, 187

## V

Vesicle sizing, 322–325
VGCC, *see* Voltage-gated calcium channel
Voltage clamp biosensor, *see* Capillary electrophoresis
Voltage-gated calcium channel
antibodies
autoimmunity, 702
channel antibodies, 690–692, 702
neurotoxin antibodies, 693, 702
ω-conotoxin binding, 614, 618
drug screening with fluorescence assay of calcium
activity of various drugs, 27, 30–32
advantages of assay, 46–47
limitations of assay, 45–46
pharmacologic validation, 26–27
stable cell line expression of recombinant channels, 25–26

## W

Western blot analysis
aquaporin I expressed in *Xenopus* oocytes, 570, 572
excitatory amino acid receptors
blotting with semidry electrophoretic transfer, 406, 409
immunodetection, 406–407, 410
receptor activation studies *in vivo*, 407–408
sample preparation from mouse brain, 405
solution preparation, 404–405
troubleshooting, 410
M2 expression in mouse erythroleukemia cells, 495–496

## X

*Xenopus laevis* oocyte
aquaporin I expression
complementary DNA expression construct preparation, 568–569
complementary RNA
injection, 570
preparation, 569–570
materials, 568
osmotic water permeability measurement, 570
Western blot analysis, 570, 572
glycine receptor, transient expression
applications of expression system, 263–264
injection capillary preparation, 268
injection of cytoplasma, 269
injection of nucleus, 269–270

messenger RNA isolation from neural tissue, 264
preparation of oocytes, 266–267
transcription of complementary DNA *in vitro*, 264–266
vector linearization, 265
membrane insertion into planar bilayers, 299–300
two-electrode voltage-clamp detection for capillary electrophoresis, 197–199

## Y

Yeast two-hybrid screening, ion channel associated protein identification, 371, 373
Ykc1, forward genetics analysis, 523–524

ISBN 0-12-182195-1

90038

9 780121 821951